Distance Formula

Length of segment PQ =

$$\sqrt{(x_1 - x_2)^2 + (y_1 - y_2)^2}$$

Slope

If $x_1 \neq x_2$, slope of line PQ =

$$\frac{y_2 - y_1}{x_2 - x_1}$$

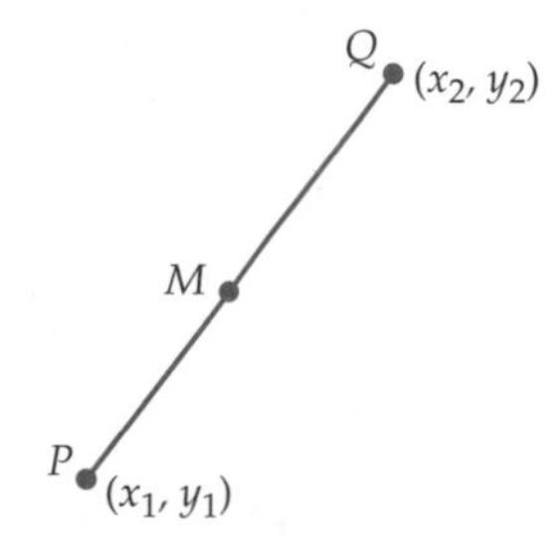

Midpoint Formula

Midpoint M of segment PQ =

$$\left(\frac{x_1 + x_2}{2}, \frac{y_1 + y_2}{2}\right)$$

The equation of the straight line through (x_1, y_1) with slope m is $y - y_1 = m(x - x_1)$.
The equation of line with slope m and y-intercept b is $y = mx + b$.

Rectangular and Parametric Equations for Conic Sections

Circles
Center (h, k), radius r

$$(x - h)^2 + (y - k)^2 = r^2$$

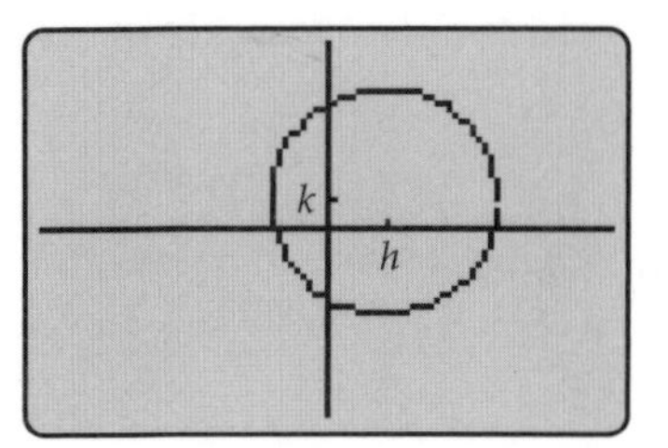

$$x = r\cos t + h \quad y = r\sin t + k \qquad (0 \le t \le 2\pi)$$

Ellipse
Center (h, k)

$$\frac{(x - h)^2}{a^2} + \frac{(y - k)^2}{b^2} = 1$$

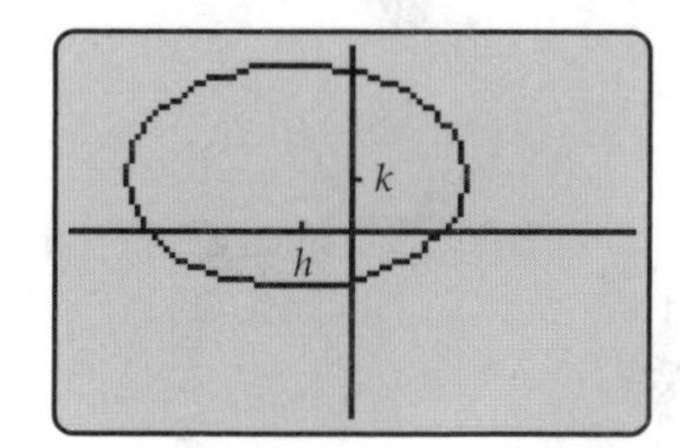

$$x = a\cos t + h \quad y = b\sin t + k \qquad (0 \le t \le 2\pi)$$

Parabola
Vertex (h, k)

$$(x - h)^2 = 4p(y - k)$$

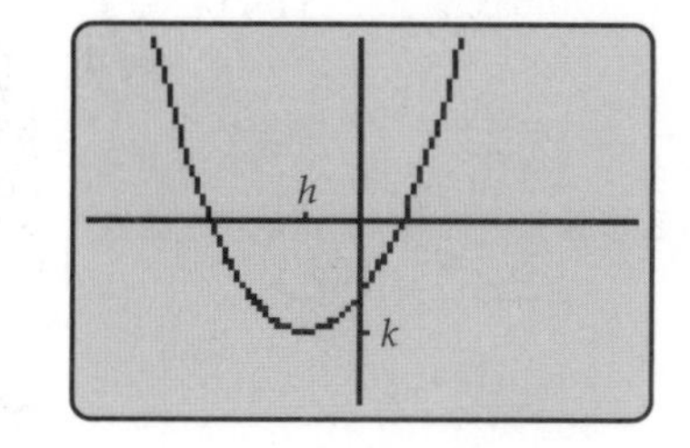

$$x = t \quad y = \frac{(t - h)^2}{4p} + k \qquad (t \text{ any real})$$

Parabola
Vertex (h, k)

$$(y - k)^2 = 4p(x - h)$$

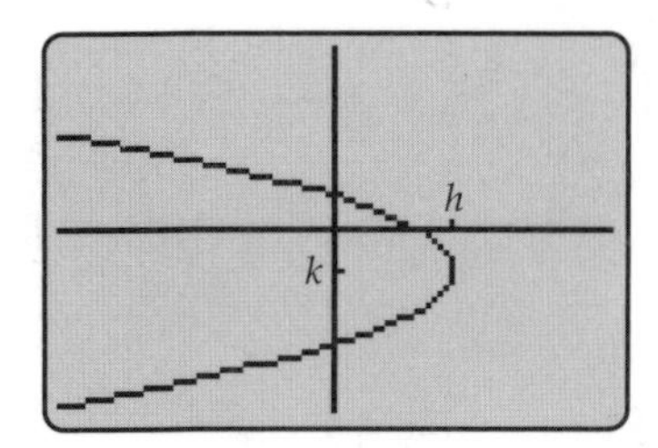

$$x = \frac{(t - k)^2}{4p} + h \quad y = t \qquad (t \text{ any real})$$

Hyperbola
Center (h, k)

$$\frac{(x - h)^2}{a^2} - \frac{(y - k)^2}{b^2} = 1$$

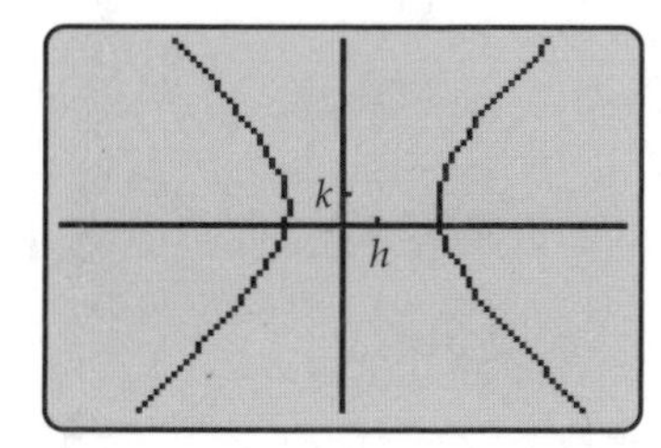

$$x = \frac{a}{\cos t} + h \quad y = b\tan t + k \qquad (0 \le t \le 2\pi)$$

Hyperbola
Center (h, k)

$$\frac{(y - k)^2}{a^2} - \frac{(x - h)^2}{b^2} = 1$$

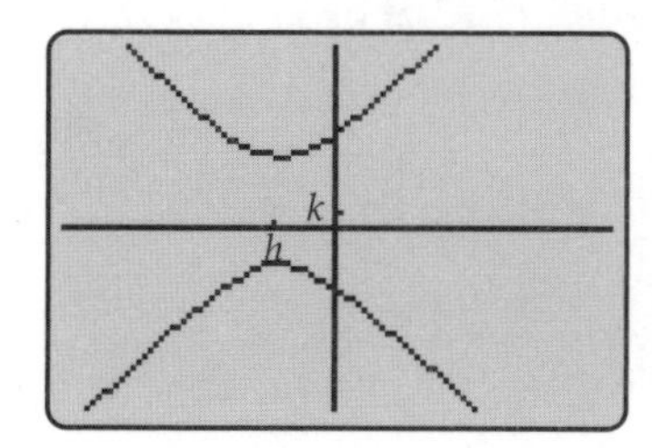

$$x = b\tan t + h \quad y = \frac{a}{\cos t} + k \qquad (0 \le t \le 2\pi)$$

Contemporary Precalculus

A Graphing Approach

Third Edition

Contemporary Precalculus

A Graphing Approach

Thomas W. Hungerford
Cleveland State University

Australia • Canada • Mexico • Singapore • Spain
United Kingdom • United States

THOMSON
BROOKS/COLE

Publisher: Emily Barosse
Developmental Editor: James D. LaPointe
Acquisitions Editor: Angus McDonald, Liz Covello
Marketing Strategist: Julia Downs
Production Manager: Alicia Jackson
Project Editor: Ellen Sklar
Text Designer/ Art Director: Cara Castiglio

Copy Editor: Linda Davoli
Illustrator: Academy Art
Cover Printer: Phoenix Color
Compositor: TechBooks
Printer: Von Hoffmann Press
Cover Image: Amana Images/Photonica

COPYRIGHT © 2000 Brooks/Cole, a division of Thomson Learning, Inc. Thomson Learning™ is a trademark used herein under license.

ALL RIGHTS RESERVED. No part of this work covered by the copyright hereon may be reproduced or used in any form or by any means—graphic, electronic, or mechanical, including but not limited to photocopying, recording, taping, Web distribution, information networks, or information storage and retrieval systems—without the written permission of the publisher.

Printed in the United States of America
4 5 6 7 8 9 10 06 05 04 03 02

For more information about our products, contact us at:
Thomson Learning Academic Resource Center
1-800-423-0563

For permission to use material from this text, contact us by:
Phone: 1-800-730-2214
Fax: 1-800-730-2215
Web: http://www.thomsonrights.com

CONTEMPORARY PRECALCULUS: A GRAPHING APPROACH, Third Edition

Library of Congress Control Number: 99-66124

ISBN 0-03-025989-4

Brooks/Cole—Thomson Learning
511 Forest Lodge Road
Pacific Grove, CA 93950
USA

Asia
Thomson Learning
5 Shenton Way #01-01
UIC Building
Singapore 068808

Australia
Nelson Thomson Learning
102 Dodds Street
South Melbourne, Victoria 3205
Australia

Canada
Nelson Thomson Learning
1120 Birchmount Road
Toronto, Ontario M1K 5G4
Canada

Europe/Middle East/Africa
Thomson Learning
High Holborn House
50/51 Bedford Row
London WC1R 4LR
United Kingdom

Latin America
Thomson Learning
Seneca, 53
Colonia Polanco
11560 Mexico D.F.
Mexico

Spain
Paraninfo Thomson Learning
Calle/Magallanes, 25
28015 Madrid, Spain

Dedicated to the Parks sisters,

whose presence in my life has greatly enriched it:

To my aunt,

Irene Parks Mills

And to the memory of my mother,

Grace Parks Hungerford

and my aunt,

Florence M. Parks

Preface

This book is intended to provide the mathematical background needed in calculus by students with two or three years of high school mathematics. It integrates graphing technology into the course without losing sight of the fact that the underlying mathematics is the crucial issue. Mathematics is presented in an informal manner that stresses meaningful motivation, careful explanations, and numerous examples, with an ongoing focus on real-world problem solving.

The concepts that play a central role in calculus are explored from algebraic, graphical, and numerical perspectives. Students are expected to participate actively in the development of these concepts by using graphing calculators (or computers with suitable software), as directed in the *Graphing Explorations,* either to complete a particular discussion or to explore appropriate examples.

Changes in the Third Edition

A number of changes have been made in response to user requests.

Data Analysis and Mathematical Models Optional sections have been added, covering linear, polynomial, exponential and logarithmic models that can be constructed from data by using the regression capabilities of a calculator (Section 2.5, Excursion 4.4.A, and Section 5.5).

Trigonometry Chapter 6 has been significantly reordered, so as to have a more coherent and consistent presentation. As before, trigonometric functions are introduced as functions of a real number, rather than functions of an angle. This viewpoint is used consistently throughout the chapter (with all triangle trigonometry now in Chapter 7). Instructors now have considerably more flexibility than they did in the old arrangement. In particular, those who want to introduce all six trig functions simultaneously or who want to cover triangle trigonometry early can do so.

Rearrangement of Chapters 1–4 Material that is review for most students (including lines) is now in Chapter 1, with graphing technology being introduced in Chapter 2. Functions (including rates of change) are treated in Chapter 3, while Chapter 4 covers polynomial and rational functions (including quadratic functions).

Updated Technology Both the Technology Tips and the presentation of the texts have been revised to reflect the fact that most students have a computer or calculator at least at the level of the TI-83, Sharp 9600, Casio 9850, or HP-38, and many have next-generation calculators, such as the TI-89. Although these newer machines do not affect the underlying mathematical concepts, the wider range of tools that they make available requires some changes in how certain topics are approached.

Several parts of the book have been rewritten to improve clarity, including Sections 3.7 (Inverse Functions) and 4.3 (Real Roots of Polynomials). Other improvements to make the book more convenient to use are as follows.

Real-Data Applications A variety of new examples and exercises based on real-world data are now included in the text.

Chapter Openers Each chapter now begins with a brief example of an application of the mathematics treated in that chapter, together with a reference to an appropriate exercise. The opener also lists the titles of the sections in the chapter and, when necessary, provides a diagram showing their interdependence.

Discovery Projects Each chapter now ends with an investigative problem (suitable for small-group work) that allows students to apply some of the mathematics from the chapter in solving a real-world problem.

Functional Use of Color The effective use of a multi-color format makes the book more attractive and readable for both instructor and students.

An Electronic Companion to Precalculus CD-ROM The concepts in the text will be strengthened through the use of a CD-ROM that will be packaged with it. This dynamic and interactive CD-ROM covers the key concepts through multiple representations.

Mathematical and Pedagogical Features

The mathematical approaches to important topics in the second edition have been retained.

Functional notation and its uses are thoroughly treated.

The natural exponential and logarithmic functions are emphasized because of their central role in calculus.

Trigonometric functions of real numbers—the ones most widely used in calculus—are introduced first, with traditional triangle trigonometry treated later.

Parametric graphing is introduced early and used thereafter to illustrate such concepts as inverse functions, the definition of trigonometric functions, and the graphs of conic sections.

Average rates of change—a crucial concept for calculus—are fully treated and the calculator is used to explore the intuitive connections between average and instantaneous rates of change.

In addition to the *Graphing Explorations* mentioned in the second paragraph of this preface, all the other pedagogical features of the second edition are included here.

Cautions Students are alerted to common errors and misconceptions (both mathematical and technological) by clearly marked Caution boxes (formerly called Warnings).

Exercises Exercise sets proceed from routine calculations and drill to exercises requiring some thought, including graph interpretation and word problems. Some sets include problems labeled *Thinkers,* most of which are not difficult, but simply in a different form from what students may have seen before; a few of the *Thinkers* are quite challenging. Answers for odd-numbered problems are given in the back of the book, and solutions for these problems are in the Students Solutions Manual.

Chapter Reviews Each Chapter concludes with a list of important concepts (referenced by section and page number), a summary of important facts and formulas, and a set of review questions.

Algebra Review Basic algebra is reviewed in an appendix, which can be omitted by well-prepared students or covered as an introductory chapter if necessary.

Geometry Review Frequently used facts from plane geometry are summarized, with examples, in an appendix.

Finally, several technology assistance features have been retained and updated.

Technology Tips Although the discussion of technology in the text proper is as generic as possible, the Technology Tips in the margin provide information and assistance with carrying out various procedures on specific calculators.

Calculator Investigations Exercise sets in the early sections of the book are often preceded by Calculator Investigations that encourage students to become familiar with the capabilities and limitations of calculators (since many of them often don't realize what a calculator can—or can't—do).

Program Appendix This appendix provides a small number of programs that are useful either for updating older calculators (such as a table maker program for the TI-85) or for carrying out certain procedures discussed in the text (such as synthetic division).

Acknowledgments

I am particularly grateful to

Ann Steen, Santa Fe Community College

who served as a reviewer of the previous edition and supplied more than one hundred new exercises for this edition, to

Edward Miller, Lewis–Clark State College

who designed the Discovery Projects at the end of each chapter, and to our accuracy reviewer

Sudhir Goel, Valdosta State University

who examined (and corrected where necessary) the examples and exercises. Their work has greatly improved the final product.

My sincere thanks go to the following reviewers who provided many helpful suggestions for improving this edition of the text:

David Blankenbaker, University of New Mexico
Margaret Donlan, University of Delaware
Patricia Dueck, Arizona State University
Betsy Farber, Bucks County Community College
Alex Feldman, Boise State University
Frances Gulick, University of Maryland
John Hamm, University of New Mexico
Lonnie Hass, North Dakota State University
Ann Lawrance, Wake Technical Community College
Matthew Liu, University of Wisconsin—Stevens Point
Sergey Lvin, University of Maine
George Matthews, Onondaga Community College
Nancy Matthews, University of Oklahoma
William Miller, Central Michigan University
Jack Porter, University of Kansas
Robert Rogers, University of New Mexico
Barbara Sausen, Fresno City College
Stuart Thomas, University of Oregon
Jan Vandever, South Dakota State University
Judith Wolbert, Kettering University

Thanks also go to those reviewers who provided feedback for the previous edition of this text:

Deborah Adams, Jacksonville University
Kelly Bach, University of Kansas
Bettyann Daley, University of Delaware
Betty Givan, Eastern Kentucky University
William Grimes, Central Missouri State University
Charles Laws, Cleveland State Community College
Martha Lisle, Prince George's Community College
Ruth Meyering, Grand Valley State University
Philip Montgomery, University of Kansas

Roger Nelsen, Lewis and Clark College
Hugo Sun, California State University at Fresno
Bettie Truitt, Black Hawk College

Finally, I want to thank the people who have prepared the various supplements that are available to instructors and students who use this book:

Instructor's Resource Manual: Matt Foss, North Hennepin Community College

Student Resource Manual: Matt Foss, North Hennepin Community College

Test Bank: Bruce Hoelter, Raritan Valley Community College; Elizabeth Hoelter, Moravian Academy

Graphing Calculator Manual: Joan McCarter, Arizona State University

Projects for Precalculus: Janet Anderson & Todd Swanson, Hope College; Robert Keeley, Calvin College

Digital Video Applications CD-ROM: Lori Palmer & Carolyn Hamilton, Utah Valley State College

It is a pleasure to acknowledge the invaluable assistance of the Saunders staff, particularly,

James LaPointe, Developmental Editor
Ellen Sklar, Project Editor
Cara Castiglio, Art Director

Their fine work has made my job much easier.

The last word goes to the love of my life, my wife Mary Alice, who has provided understanding and support when it was most needed.

Thomas W. Hungerford
Cleveland, Ohio
September 1999

Table of Contents

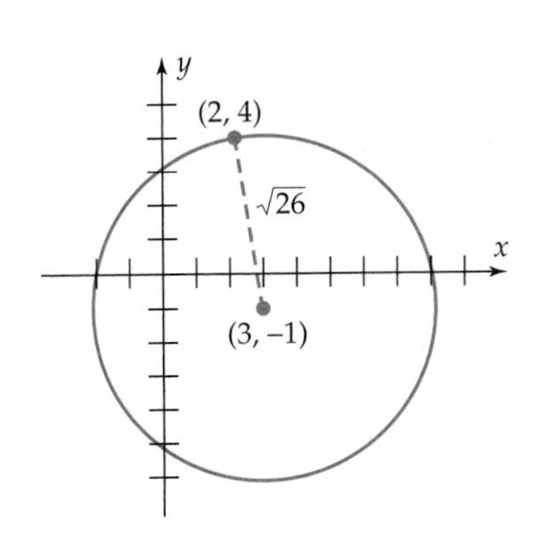

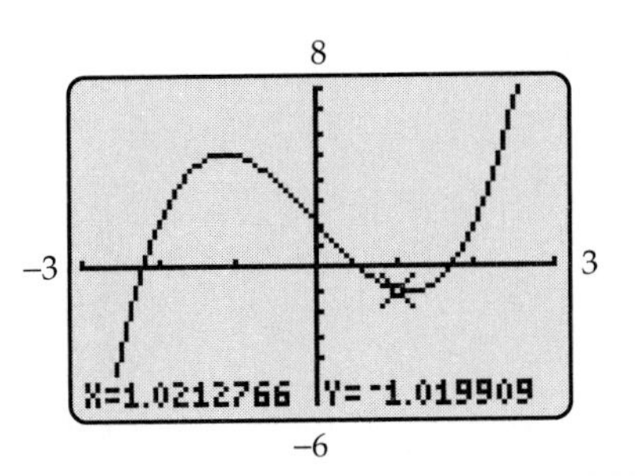

Chapter 3 Functions and Graphs 117

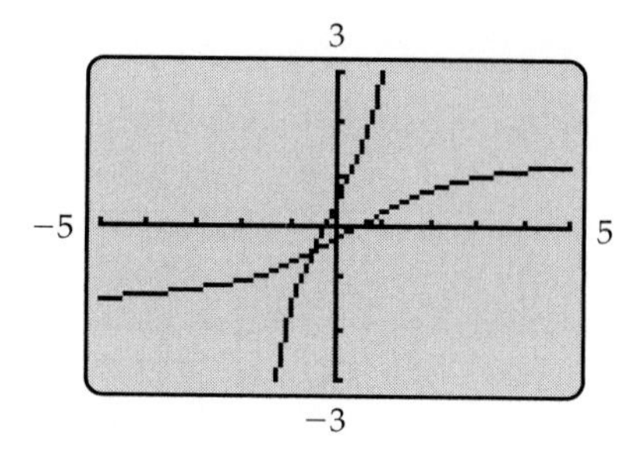

Chapter 4 Polynomial and Rational Functions 205

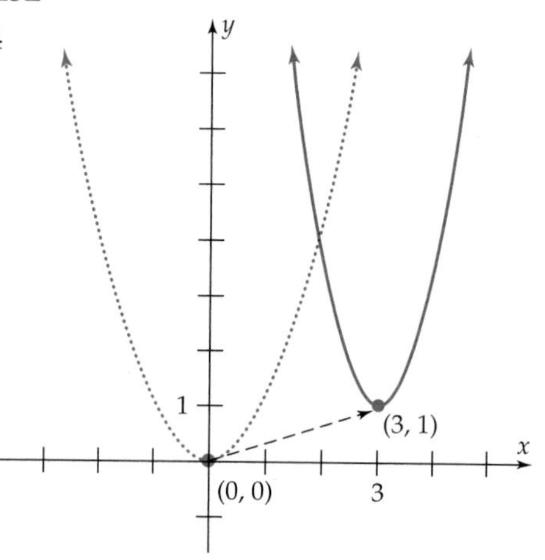

Chapter 5 Exponential and Logarithmic Functions 305

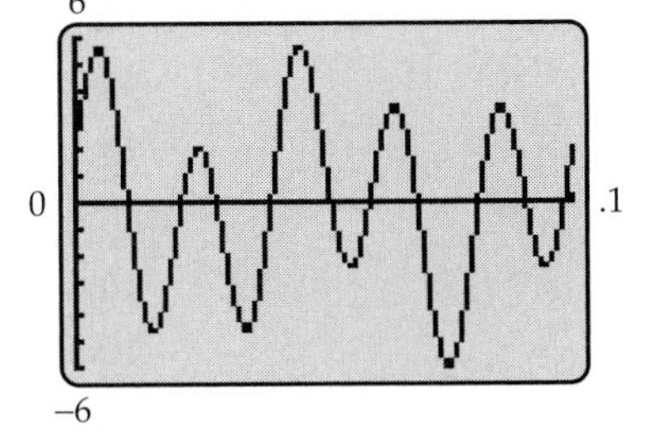
6
0
.1
−6

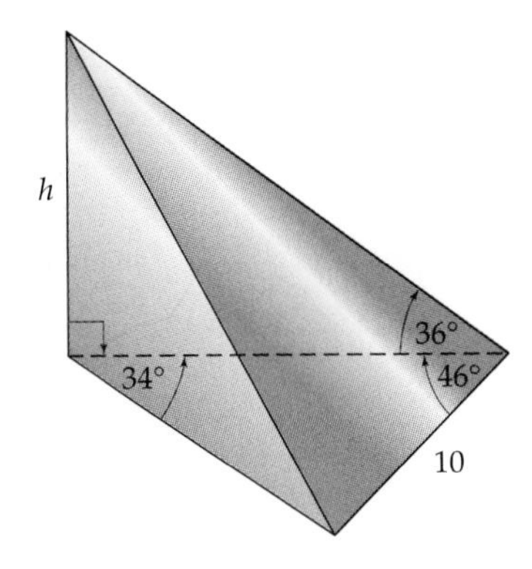
h
36°
34°
46°
10

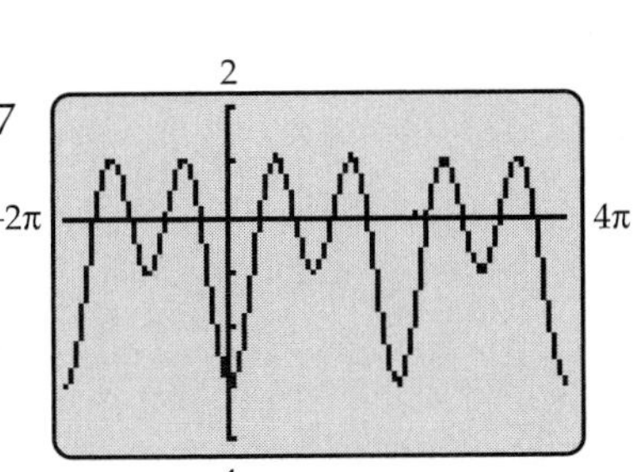
2
−2π
4π
−4

Chapter 9 Applications of Trigonometry 563

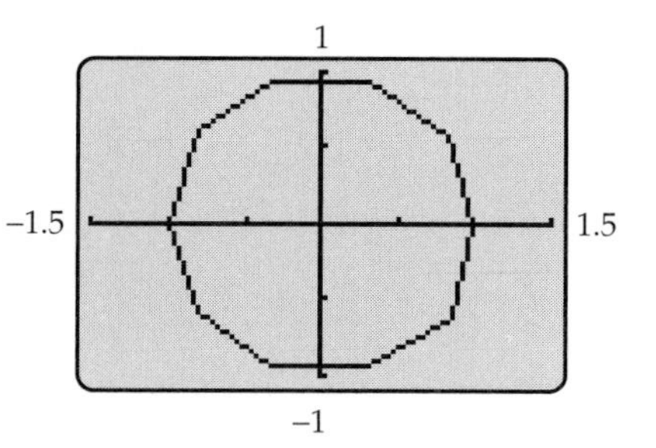

Chapter 10 Analytic Geometry 609

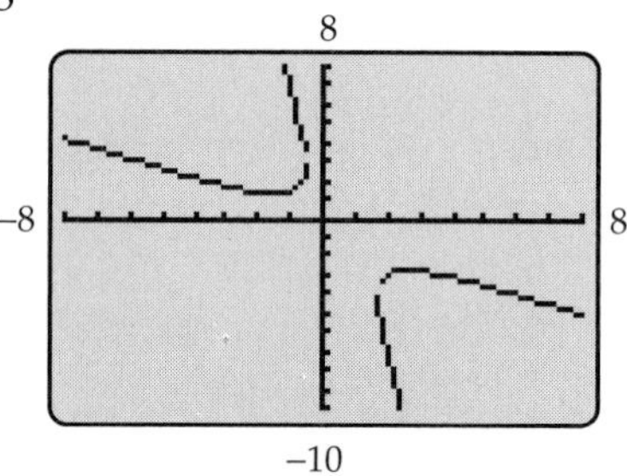

Chapter 11 Systems of Equations 681

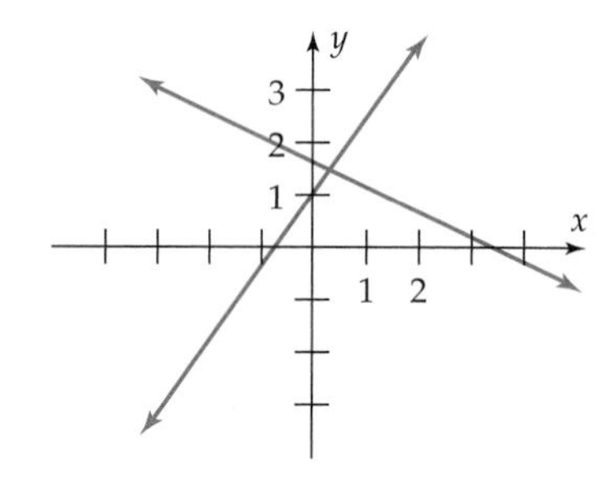

Chapter 12 Discrete Algebra 731

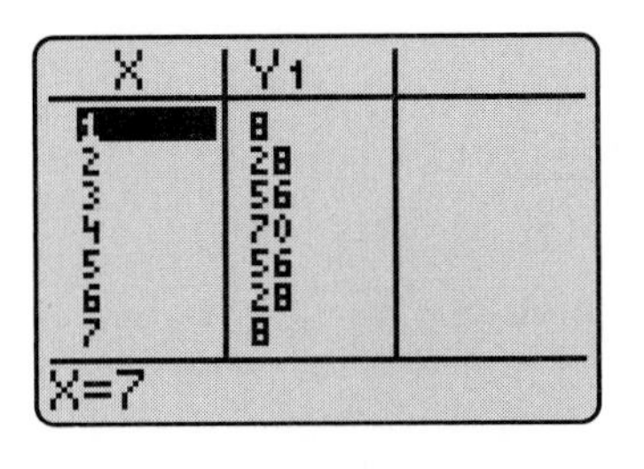

To the Instructor

The following information should assist you in planning a workable syllabus for your course.

Prerequisites The Algebra Appendix is a prerequisite for the entire book and may be covered as Chapter 0, if necessary. Chapter 1 is also review material that may be omitted by well-prepared classes.

Interdependence of Chapters and Sections The chart on the facing page shows the interdependence of chapters. A similar chart appears at the beginning of each chapter, showing the interdependence of sections within the chapter. Note, in particular, that Chapters 4–6 (polynomial, exponential/logarithmic, and trigonometric functions) may be covered in any order.

Excursions Sections labeled Excursion are usually related to the preceding section and are never prerequisites for other sections of the text. The "Excursion" label is designed solely to make syllabus planning easier and is not intended as a kind of value judgement on the topic in question.

Limits and Continuity An optional chapter on these topics (that may be covered after Chapter 6) is published separately and is available at nominal cost to schools that adopt this text. Please contact your local sales representative for details.

In this text "calculator" means "graphing calculator." You and your students should be aware of the following facts about calculators.

Minimal Technology Requirements It is assumed that each student has either a computer with appropriate software or a calculator at least at the level of a TI-82. Among current calculator models that meet or exceed this minimal requirement are TI-82 through TI-92, Sharp 9600, HP-38, HP-48, and Casio 9850 and 9970. Although students with less powerful calculators may be able to handle much of the material, they will be at a disadvantage at a number of points.

(text continues on p. xxii)

Interdependence of Chapters

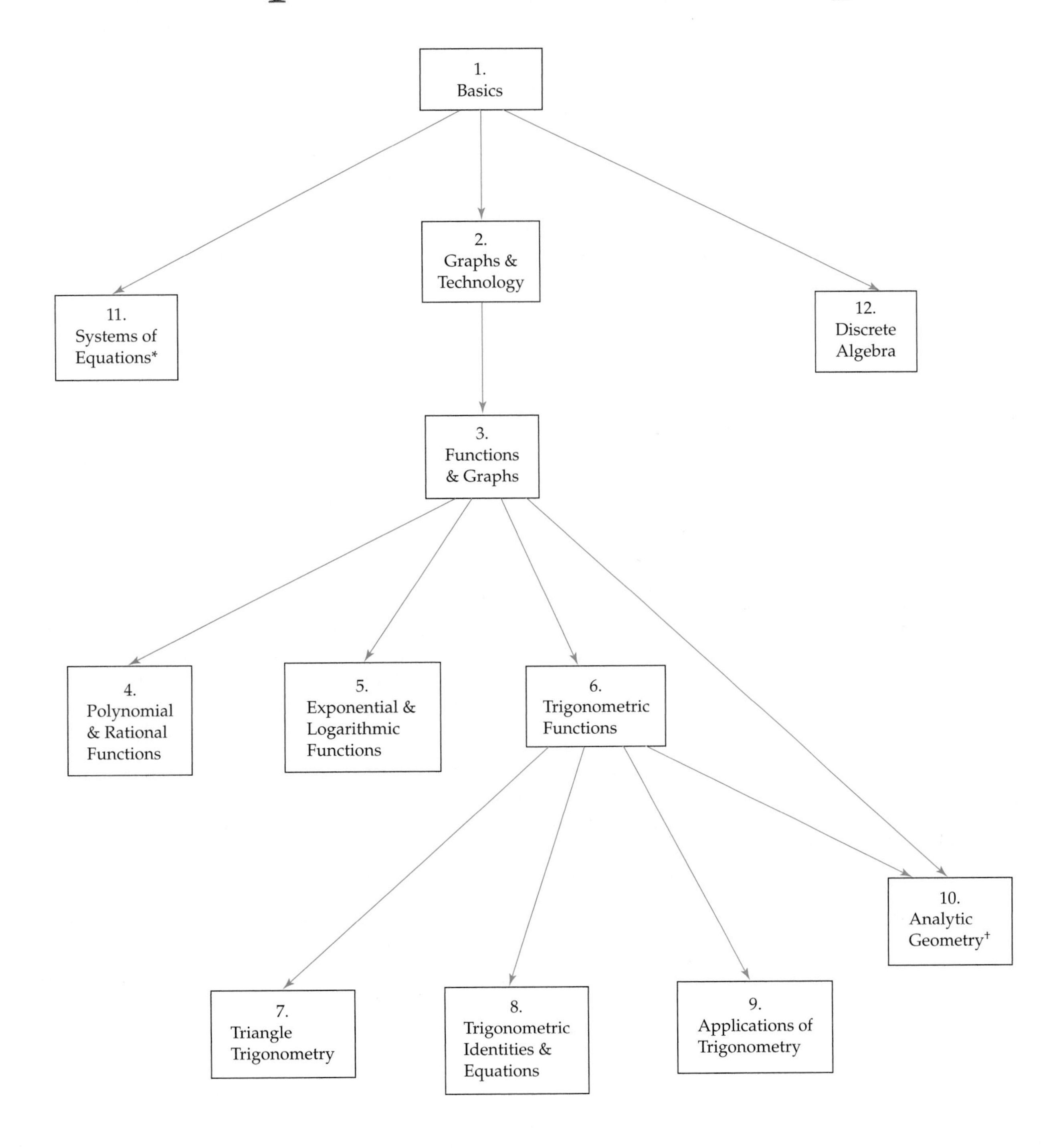

*Section 2.1 (Graphs and Graphing Calculators) is a prerequisite for Section 11.4 (Systems of Nonlinear Equations).

†Standard equations for conics (Sections 10.2 and 10.3) may be covered without using trigonometry.

Technology Tips To avoid much clutter, only a limited number of calculators are specifically mentioned in the Technology Tips. However, unless noted otherwise,

Technology Tips for the TI-83 also apply to the TI-82;

Technology Tips for the TI-86 also apply to the TI-85;

Technology Tips for the TI-89 also apply to the TI-92.

Technology Tips for the Casio 9850 also apply to the Casio 9970.

There are no Tips specifically for HP-48 calculators since they use entirely different operating systems than other calculators.

Supplements

Instructors who adopt this text may receive, free of charge, the following items

Electronic Companion to Precalculus CD-ROM This dynamic and interactive CD-ROM, which is packaged with the text, covers key concepts using multiple representations.

Instructor's Resource Manual Written by Matt Foss of North Hennepin Community College, this manual contains detailed solutions to all the exercises and end-of-chapter Review Questions to assist the instructor in the classroom and in grading assignments. Solutions to the Discovery Projects at the end of each chapter are also included.

Digital Video Applications CD-ROM This innovative ancillary packaged with the Instructor's Resource Manual is designed to show students how and where precalculus concepts arise in real-life. Over 20 engaging interviews conducted by Lori Palmer of Utah Valley State College with professionals from such fields as aviation, food services, banking, and environmental science motivate the key concepts from the text. Each vignette is accompanied by two problems written by Carolyn Hamilton of Utah Valley State College to test students' understanding of the underlying mathematical ideas and skills. Answers are provided on disk; detailed solutions are in the Instructor's Resource Manual.

Test Bank Written by Bruce Hoelter of Raritan Valley Community College, this manual provides over 2000 multiple-choice and open-ended questions arranged in 5 forms per chapter, each form containing about 30 questions. Master answer sheets and a complete answer section are included. Additionally, the Test Bank also includes a set of 20 practice problems for each chapter of the text with complete solutions.

Computerized Test Bank The computerized test bank contains all the test bank questions and allows instructors to prepare quizzes and examinations quickly and easily. Instructors may also add questions or modify existing ones. The Computerized Test Bank has gradebook capabilities for recording and tracking students' grades. Instructors have the opportunity to post and administer a test over a network or on the Web. Additionally, user-friendly printing capability accommodates all printing platforms.

Graphing Calculator Manual Written by Joan McCarter of Arizona State University, this manual covers several of the latest calculator models in more detail than do the Technology Tips in the text.

Projects for Precalculus Written by Hope College's Janet Andersen and Todd Swanson and Robert Keeley of Calvin College, this supplement stems from a popular and successful NSF-sponsored program in reform precalculus. The authors have conducted numerous workshops and compiled the feedback of over 100 class testers to create excellent precalculus applications. This manual contains carefully prepared and tested activities that promote conceptual understanding and active learning.

Video Series Free to adopters, the videotape package consists of 12 VHS videotapes, one for each chapter in the book. Each tape is an hour long and further develops the concepts of the chapter. On-location footage is utilized to introduce an extended application at the beginning of each tape. This application is explained fully at the end of the tape.

Website The Website (www.brookscole.com/math_d) offers additional resources to both instructors and students in conjunction with the adoption of the text. An on-line glossary allows students to confirm their understanding of important terms and theorems in the text. Web-based projects created by Gene Fiorini of Shippensburg University help develop the concepts in the text through researching Websites that cover a variety of topics.

WebCT Course This Web-based teaching and learning tool is free to instructors and students in conjunction with the adoption of the text. It allows instructors—with or without technical expertise—to create highly effective Web-based learning, communication, and collaboration environments. A full array of educational tools including testing, student tracking, access control, and much more are available.

Brooks/Cole, a division of Thomson Learning, may provide complimentary instructional aids and supplements or supplement packages to those adopters qualified under our adoption policy. Please contact your sales representative for more information. If as an adopter or potential user you receive supplements you do not need, please return them to your sales representative or send them to

Attn: Returns Department
Troy Warehouse
465 South Lincoln Drive
Troy, MO 63379

To the Student

This text assumes the use of technology, so you should be aware of the following facts.

Terminology In this text "calculator" means "graphing calculator." All discussions of calculators, with obvious modifications, apply to graphing software for computers.

Minimal Technology Requirements In order to use this text effectively, you must have either a computer with appropriate software or a calculator at least at the level of a TI-82. Among current models that meet or exceed this minimal requirement are the TI-82 through TI-92, the Sharp 9600, HP-38, HP-48, and Casio 9850 and 9970. Although students with less powerful calculators may be able to handle much of the material, they will be at a disadvantage at a number of points.

The following features of the text will enable you to get the most out of your calculator.

Technology Tips Some of the Technology Tips in the margin tell you the proper menus or keys to be used on specific calculators to carry out procedures mentioned in the text. Other Tips offer general information or helpful advice for performing a particular task on a calculator.

As a general rule, the only calculators mentioned in the Technology Tips are the TI-83, TI-86, TI-89, Sharp 9600, HP-38, Casio 9850*. However, unless noted otherwise,

Technology Tips for the TI-83 also apply to the TI-82;
Technology Tips for the TI-86 also apply to the TI-85;
Technology Tips for the TI-89 also apply to the TI-92;
Technology Tips for the Casio 9850 also apply to the Casio 9970.

*There are no Tips in the text specifically for HP-48 calculators, which use an entirely different operating system than other calculators.

Calculator Investigations You may not be aware of the full capabilities of your calculator (or some of its limitations). The Calculator Investigations (which appear just before the exercise sets in some of the earlier sections of the book) will help you to become familiar with your calculator and to maximize the mathematical power it provides. Even if your instructor does not assign these investigations, you may want to look through them to be sure you are getting the most you can from your calculator.

With all this talk about calculators, don't lose sight of this crucial fact:

Technology is only a *tool* for doing mathematics.

You can't build a house if you only use a hammer. A hammer is great for pounding nails, but useless for sawing boards. Similarly, a calculator is great for computations and graphing, but it is not the right tool for every mathematical task. To succeed in this course, you must develop and use your algebraic and geometric skills, your reasoning power and common sense, and you must be willing to work.

The key to success is to use all of the resources at your disposal: your instructor, your fellow students, your calculator (and its instruction manual), and this book. Here are some tips for making the most of these resources.

Ask Questions. Remember the words of Hillel:

The bashful do not learn.

There is no such thing as a "dumb question" (assuming, of course, that you have attended class, taken notes, and read the text). Your instructor will welcome questions that arise from a serious effort on your part.

Read the Book. Not just the homework exercises, but the rest of the text as well. There is no way your instructor can possibly cover the essential topics, clarify ambiguities, explain the fine points, and answer all your questions during class time. You simply will not develop the level of understanding you need to succeed in this course and in calculus unless you read the text fully and carefully.

Be an Interactive Reader. You can't read a math book the way you read a novel or history book. You need pencil, paper, and your calculator at hand to work out the statements you don't understand and to make notes of things to ask your fellow students and/or your instructor.

Do the Graphing Explorations. When you come to a box labeled "Graphing Exploration," use your calculator as directed to complete the discussion. Typically, this will involve graphing one or more equations and answering some questions about the graphs. Doing these explorations as they arise will improve your understanding and clarify issues that might otherwise cause difficulties.

Do Your Homework. Remember that

Mathematics is not a spectator sport.

You can't expect to learn mathematics without doing mathematics, any more than you could learn to swim without getting wet. Like swimming or dancing or reading or any other skill, mathematics takes practice. Homework assignments are where you get the practice that is essential for passing this course and succeeding in calculus.

Supplements

The following items are available at no cost to students.

Electronic Companion to Precalculus CD-ROM This dynamic and interactive CD-ROM, which is packaged with the text, covers key concepts from a variety of different perspectives.

Website The Website (www.brookscole.com/math_d) includes an online glossary to assist you to confirm your understanding of important concepts and theorems in the text. Projects created by Gene Fiorini of Shippensburg University help develop the concepts in the text by researching Websites that cover a variety of topics.

WebCT Course If your instructor has made arrangements, this feature will provide an array of educational tools to assist you in learning the material in the course.

Students using Contemporary Precalculus may purchase the following supplements.

Student Resource Manual This manual, also written by Matt Foss, is comprised of two distinct parts. The first part consists of detailed solutions to all the odd-numbered Exercises and end-of-chapter Review Questions. Specific instructions for solving graphing calculator problems are included, as are accurate representations of graphing calculator screens. The Student Resource Manual also contains the **Math in Practice: An Applied Video Companion CD-ROM** and accompanying problems. This innovative ancillary is designed to show students how and where precalculus concepts arise in real life. Over 20 engaging interviews conducted by Lori Palmer of Utah Valley State College with professionals from such fields as aviation, food services, banking, and environmental science motivate the key concepts from the text. Each vignette is accompanied by two problems written by Carolyn Hamilton of Utah Valley State College to test students' understanding of the underlying mathematical ideas and skills. The problems and answers are given on the CD-ROM; they are also reprinted in this manual with sufficient space to show student work.

Graphing Calculator Manual Written by Joan McCarter of Arizona State University, this manual covers several of the latest calculator models in more detail than do the Technology Tips in the text.

Projects for Precalculus Written by Hope College's Janet Andersen and Todd Swanson and Robert Keeley of Calvin College, this manual contains activities that promote a better conceptual understanding of the major topics covered in precalculus.

ESATUTOR 2000 This computer software package contains hundreds of problems and answers that correspond with every section of the text. Students can complete a pretest to evaluate their level of understanding of the concepts in each chapter. Additionally, students can complete post-tests to ensure they have grasped the primary learning objectives. The software comes with a built-in graphing calculator. Students interested in purchasing this software package should refer to the marketing material inside the back cover of this text.

Core Concept Video This single videotape contains the most important topics covered in the full video series that is given to each school that uses *Contemporary Precalculus.* This take-home tutorial can be used as a preview of what is to be covered in class, as an aid to completing homework assignments, or as a tool to review for a test.

Basics

On a clear day, can you see forever?

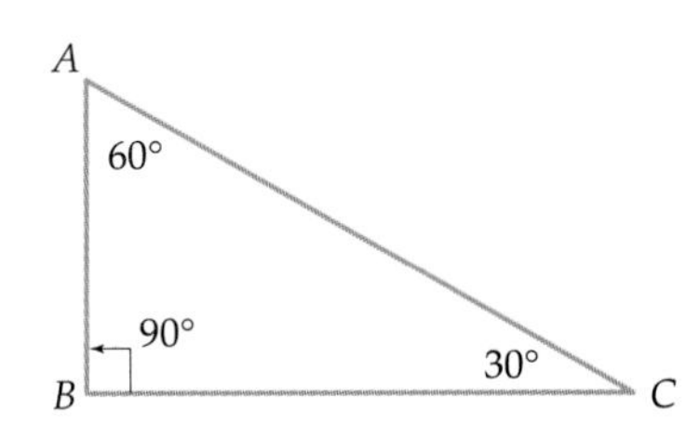

If you are at the top of the Sears Tower in Chicago, how far can you see? In earlier centuries, the lookout on a sailing ship was posted on the highest mast because he could see farther from there than from the deck. How much farther? These questions, and similar ones, can be answered (at least approximately) by using basic algebra and geometry. See Exercises 71 and 72 on page 13.

Interdependence of Sections

1.1 → 1.2
1.1 → 1.3 → 1.4

Chapter Outline

This chapter reviews the essential facts about real numbers, equations, the coordinate plane, and lines that are needed in this course and in calculus. The Algebra Review Appendix at the end of the book is a prerequisite for this material.

1.1 The Real Number System

You have been using **real numbers** most of your life. They include the **natural numbers** (or **positive integers**): 1, 2, 3, 4, ... and the **integers:**

$$\ldots, -5, -4, -3, -2, -1, 0, 1, 2, 3, 4, 5, \ldots.$$

A real number is said to be a **rational number** if it can be expressed as a fraction $\frac{r}{s}$, with r and s integers and $s \neq 0$; for instance

$$\frac{1}{2}, \qquad -.983 = -\frac{983}{1000}, \qquad 47 = \frac{47}{1}, \qquad 8\tfrac{3}{5} = \frac{43}{5}.$$

Alternatively, rational numbers may be described as numbers that can be expressed as terminating decimals, such as $.25 = \frac{1}{4}$, or as nonterminating repeating decimals in which a single digit or a block of digits repeats for ever, such as

$$\frac{5}{3} = 1.66666\cdots \qquad \text{or} \qquad \frac{58}{333} = .174174174\cdots.$$

A real number that cannot be expressed as a fraction with integer numerator and denominator is called an **irrational number.** Alternatively, an irrational number is one that can be expressed as a nonterminating, nonrepeating decimal (no block of digits repeats forever). For example, the number π, which is used to calculate the area of a circle, is irrational.*

*This fact is difficult to prove. In the past you may have used 22/7 as π; a calculator might display π as 3.141592654. However, these numbers are just *approximations* of π (close, but not quite *equal* to π).

More information about decimal expansions of real numbers is given in Excursion 1.1.A.

The real numbers are often represented geometrically as points on a **number line,** as in Figure 1–1. We shall assume that there is exactly one point on the line for every real number (and vice versa) and use phrases such as "the point 3.6" or "a number on the line." This mental identification of real numbers and points on the line is often extremely helpful.

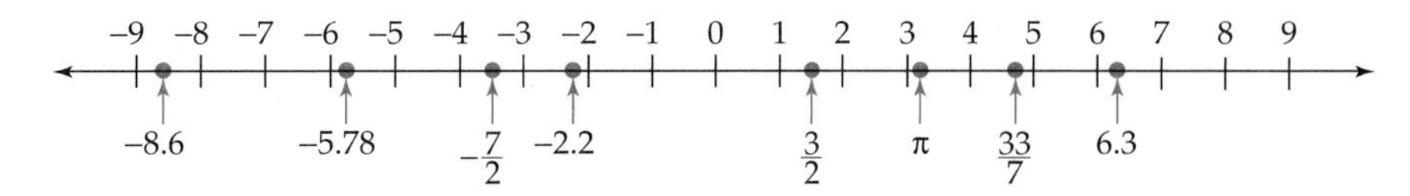

Figure 1–1

Arithmetic and Scientific Notation

We assume that you are familiar with the basic properties of real number arithmetic, particularly this fact:

Distributive Law

For all real numbers *a, b, c*

$$a(b + c) = ab + ac \quad \textbf{and} \quad (b + c)a = ba + ca.$$

The distributive law doesn't usually play a direct role in easy computations, such as $4(3 + 5)$. Most people don't say $4\cdot 3 + 4\cdot 5 = 12 + 20 = 32$. Instead, they mentally add the numbers in parentheses and say 4 times 8 is 32. But when symbols are involved, you can't do that, and the distributive law is essential. For example, $4(3 + x) = 4\cdot 3 + 4x = 12 + 4x$.

Recall that any positive real number may be written as the product of a power of 10 and a number between 1 and 10. For example,

$$356 = 3.56 \times 100 = 3.56 \times 10^2$$

$$1{,}563{,}427 = 1.563427 \times 1{,}000{,}000 = 1.563427 \times 10^6$$

$$.072 = 7.2 \times 1/100 = 7.2 \times 10^{-2*}$$

$$.000862 = 8.62 \times 1/10{,}000 = 8.62 \times 10^{-4}.$$

A number written in this form is said to be in **scientific notation.** Scientific notation is useful for computations with very large or very small numbers.

*Negative exponents are explained in the first section of the Algebra Review Appendix.

Example 1 $(.00000002)(4{,}300{,}000{,}000) = (2 \times 10^{-8})(4.3 \times 10^{9})$

$$= 2(4.3)10^{-8+9} = (8.6)10^{1} = 86. \quad \blacksquare$$

When calculators display numbers in scientific notation, they omit the 10 and use an E to indicate the exponent. For instance,

$$7.235 \times 10^{-12} \qquad \text{is displayed as} \qquad 7.235 \text{ E } -12$$

To enter a number in scientific notation, for example 5.6×10^{73}, type in 5.6, then press the EE key (labeled EEX or EXP or 10^x on some calculators), and type in 73. Calculators automatically switch to scientific notation whenever a number is too large or too small to be displayed in the standard way. If you try to enter a number with more digits than the calculator can handle, such as 45,000,000,333,222,111, a typical calculator will approximate it using scientific notation as 4.500000033 E 16, that is, as 45,000,000,330,000,000.

CAUTION

Your calculator may not always obey the laws of arithmetic when dealing with very large or very small numbers. For instance, we know that

$$(1 + 10^{19}) - 10^{19} = 1 + (10^{19} - 10^{19}) = 1 + 0 = 1,$$

but a calculator may round off $1 + 10^{19}$ as 10^{19} (instead of the correct number 10,000,000,000,000,000,001). So the calculator computes $(1 + 10^{19}) - 10^{19}$ as $10^{19} - 10^{19} = 0$ (try it!).

Order

The statement $c < d$, which is read "**c is less than d**," and the statement $d > c$ (read "**d is greater than c**") mean exactly the same thing:

c lies to the *left* of d on the number line.

For example, Figure 1–1 shows that $-5.78 < -2.2$ and $4 > \pi$.

The statement $c \le d$, which is read "**c is less than or equal to d**," means

Either c is less than d or c is equal to d.

Only one part of an "either . . . or" statement needs to be true for the entire statement to be true. So the statement $5 \le 10$ is true because $5 < 10$, and the statement $5 \le 5$ is true because $5 = 5$. The statement $d \ge c$ (read "**d is greater than or equal to c**") means exactly the same thing as $c \le d$.

The statement $b < c < d$ means

$$b < c \qquad \text{and simultaneously} \qquad c < d.$$

For example, $3 < x < 7$ means that x is a number that is strictly between 3 and 7 on the number line (greater than 3 and less than 7). Similarly, $b \le c < d$ means

$$b \le c \qquad \text{and simultaneously} \qquad c < d,$$

and so on.

Certain sets of numbers, defined in terms of the order relation, appear frequently enough to merit special notation. Let c and d be real numbers with $c < d$. Then

Interval Notation

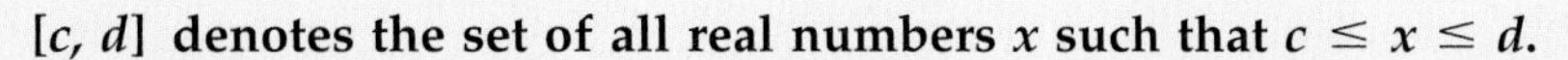

$[c, d]$ denotes the set of all real numbers x such that $c \le x \le d$.

(c, d) denotes the set of all real numbers x such that $c < x < d$.

$[c, d)$ denotes the set of all real numbers x such that $c \le x < d$.

$(c, d]$ denotes the set of all real numbers x such that $c < x \le d$.

All four of these sets are called **intervals** from c to d. The numbers c and d are the **endpoints** of the interval. $[c, d]$ is called the **closed interval** from c to d (both endpoints included and *square* brackets) and (c, d) is called the **open interval** from c to d (neither endpoint included and *round* brackets). For example,*

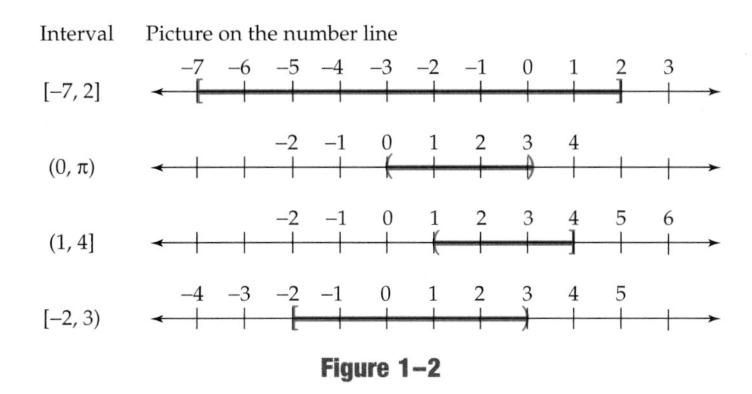

Figure 1–2

If b is a real number, then the half-line extending to the right or left of b is also called an interval. Depending on whether or not b is included, there are four possibilities:

Interval Notation

$[b, \infty)$ denotes the set of all real numbers x such that $x \ge b$.

(b, ∞) denotes the set of all real numbers x such that $x > b$.

$(-\infty, b]$ denotes the set of all real numbers x such that $x \le b$.

$(-\infty, b)$ denotes the set of all real numbers x such that $x < b$.

*In Figures 1–2 and 1–3 a round bracket such as) or (indicates that the endpoint is *not* included, whereas a square bracket such as] or [indicates that the endpoint is included.

For example,

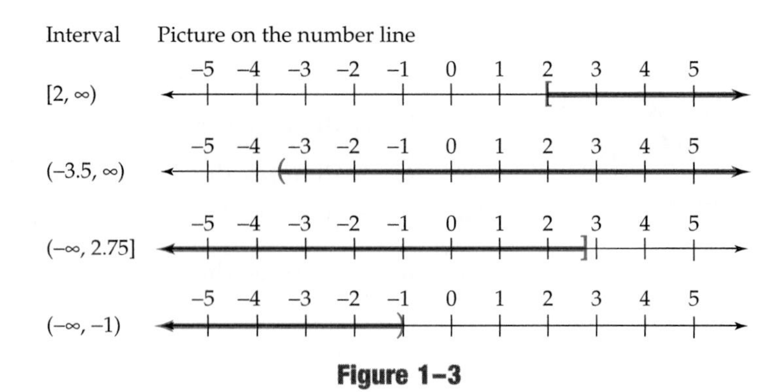

Figure 1–3

In a similar vein, **$(-\infty, \infty)$ denotes the set of all real numbers.**

NOTE The symbol ∞ is read "infinity," and we call the set $[b, \infty)$ "the interval from b to infinity." The symbol ∞ does *not* denote a real number; it is simply part of the notation used to label the first two sets of numbers defined in the previous box. Analogous remarks apply to the symbol $-\infty$, which is read "negative infinity."

Technology Tip

To enter a negative number, such as -5, on most calculators, you must use the negation key: (–) 5. If you use the subtraction key on such calculators and enter – 5, the display will read

ANS – 5

which tells the calculator to subtract 5 from the previous answer.

Negative Numbers and Negatives of Numbers

The **positive numbers** are those to the right of 0 on the number line, that is,

$$\text{All numbers } c \text{ with } c > 0.$$

The **negative numbers** are those to the left of 0, that is,

$$\text{All numbers } c \text{ with } c < 0.$$

The **nonnegative** numbers are the numbers c with $c \geq 0$.

The word "negative" has a second meaning in mathematics. The **negative *of a* number** c is the number $-c$. For example, the negative of 5 is -5, and the negative of -3 is $-(-3) = 3$. Thus the negative of a negative number is a positive number. Zero is its own negative since $-0 = 0$. In summary,

Negatives

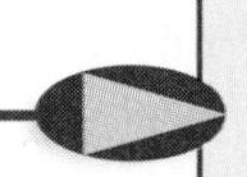

The negative of the number c is $-c$.

If c is a positive number, then $-c$ is a negative number.

If c is a negative number, then $-c$ is a positive number.

Square Roots

The **square root** of a nonnegative real number d is defined as the nonnegative number whose square is d and is denoted $\sqrt{d}$. For instance,

$$\sqrt{25} = 5 \qquad \text{because} \qquad 5^2 = 25.$$

In the past you may have said that $\sqrt{25} = \pm 5$ since $(-5)^2$ is also 25. It is preferable, however, to have a single unambiguous meaning for the symbol $\sqrt{25}$. So in the real number system, the term "square root" and the radical symbol $\sqrt{\ }$ always denote a *nonnegative* number. To express -5 in terms of radicals, we write $-5 = -\sqrt{25}$.

Although $-\sqrt{25}$ is a real number, the expression $\sqrt{-25}$ is *not defined* in the real numbers because there is no real number whose square is -25. In fact since the square of every real number is nonnegative,

No negative number has a square root in the real numbers.

Some square roots can be found (or verified) by hand, such as

$$\sqrt{225} = 15 \qquad \text{and} \qquad \sqrt{1.21} = 1.1.$$

Usually, however, a calculator is needed to obtain rational *approximations* of roots. For instance, we know that $\sqrt{87}$ is between 9 and 10 because $9^2 = 81$ and $10^2 = 100$. A calculator shows that $\sqrt{87} \approx 9.327379$.*

You can easily convince yourself that $\sqrt{cd} = \sqrt{c}\sqrt{d}$ for any nonnegative real numbers c and d. Sums, however, are different:

CAUTION

If c and d are positive real numbers, then

$$\sqrt{c + d} \neq \sqrt{c} + \sqrt{d}.$$

For example,

$$\sqrt{9 + 16} = \sqrt{25} = 5, \qquad \text{but} \qquad \sqrt{9} + \sqrt{16} = 3 + 4 = 7.$$

Technology Tip

To compute an expression such as $\sqrt{7^2 + 51}$ on a calculator you must use parentheses:

$$\sqrt{\ }(7^2 + 51).$$

Without the parentheses the calculator will compute

$$\sqrt{7^2} + 51 = 7 + 51 = 58$$

instead of the correct answer

$$\sqrt{7^2 + 51} = \sqrt{49 + 51} = \sqrt{100} = 10.$$

Absolute Value

On an informal level most students think of absolute value like this:

The absolute value of a nonnegative number is the number itself.

The absolute value of a negative number is found by "erasing the minus sign."

If $|c|$ denotes the absolute value of c, then, for example, $|5| = 5$ and $|-4| = 4$.

This informal approach is inadequate, however, for finding the absolute value of a number such as $\pi - 6$. It doesn't make sense to "erase the minus sign" here. So we must develop a more precise definition. The statement $|5| = 5$ suggests that the absolute value of a positive number ought to be the number itself. For negative numbers, such as -4, note that $|-4| = 4 = -(-4)$, that is, the absolute value of the negative

* $\approx$ means "approximately equal."

number -4 is the *negative* of -4. These facts are the basis of the formal definition:

Absolute Value

The *absolute value* of a real number c is denoted $|c|$ and is defined as follows:

$$\text{If } c \geq 0, \text{ then } |c| = c.$$

$$\text{If } c < 0, \text{ then } |c| = -c.$$

Example 2

(a) $|3.5| = 3.5$ and $|-7/2| = -(-7/2) = 7/2$.

(b) To find $|\pi - 6|$ note that $\pi \approx 3.14$, so that $\pi - 6 < 0$. Hence, $|\pi - 6|$ is defined to be the *negative* of $\pi - 6$, that is,

$$|\pi - 6| = -(\pi - 6) = -\pi + 6.$$

(c) $|5 - \sqrt{2}| = 5 - \sqrt{2}$ because $5 - \sqrt{2} \geq 0$. ■

Here are the important facts about absolute value:

Properties of Absolute Value

1. $|c| \geq 0$ **and** $|c| > 0$ **when** $c \neq 0$.
2. $|c| = |-c|$
3. $|cd| = |c| \cdot |d|$
4. $\left|\dfrac{c}{d}\right| = \dfrac{|c|}{|d|}$ $\quad (d \neq 0)$

Technology Tip

To compute absolute values on a calculator, use the ABS key and parentheses. To find $|9 - 3\pi|$, for example, key in

ABS(9 − 3π).

The ABS key is on the keyboard of HP-38 and in the NUM submenu of the MATH menu of TI and SHARP 9600. It is in the NUM submenu of the Casio 9850 OPTN menu.

Example 3 Here are examples of some of the properties listed in the box.

2. $|3| = 3$ and $|-3| = 3$, so that $|3| = |-3|$.
3. If $c = 6$ and $d = -2$, then

$$|cd| = |6(-2)| = |-12| = 12$$

and

$$|c| \cdot |d| = |6| \cdot |-2| = 6 \cdot 2 = 12$$

so that $|cd| = |c| \cdot |d|$.

4. If $c = -5$ and $d = 4$, then

$$\left|\frac{c}{d}\right| = \left|\frac{-5}{4}\right| = \left|-\frac{5}{4}\right| = \frac{5}{4} \quad \text{and} \quad \frac{|c|}{|d|} = \frac{|-5|}{|4|} = \frac{5}{4},$$

so that $\left|\dfrac{c}{d}\right| = \dfrac{|c|}{|d|}$. ■

When c is a positive number, then $\sqrt{c^2} = c$, but when c is negative, this is *false*. For example, if $c = -3$, then

$$\sqrt{c^2} = \sqrt{(-3)^2} = \sqrt{9} = 3 \qquad (\textit{not } -3),$$

so that $\sqrt{c^2} \neq c$. In this case, however, $|c| = |-3| = 3$, so that $\sqrt{c^2} = |c|$. The same thing is true for any negative number c. It is also true for positive numbers (since $|c| = c$ when c is positive). In other words,

Square Roots of Squares

For every real number c,

$$\sqrt{c^2} = |c|.$$

When dealing with long expressions inside absolute value bars, do the computations inside first, and then take the absolute value.

Example 4

(a) $|5(2 - 4) + 7| = |5(-2) + 7| = |-10 + 7| = |-3| = 3.$

(b) $4 - |3 - 9| = 4 - |-6| = 4 - 6 = -2.$ ■

CAUTION

When c and d have opposite signs, $|c + d|$ is *not equal* to $|c| + |d|$. For example, when $c = -3$ and $d = 5$, then

$$|c + d| = |-3 + 5| = 2,$$

but $$|c| + |d| = |-3| + |5| = 3 + 5 = 8.$$

The caution shows that $|c + d| < |c| + |d|$ when $c = -3$ and $d = 5$. In the general case, we have the following fact.

The Triangle Inequality

For any real numbers c and d,

$$|c + d| \leq |c| + |d|.$$

Distance on the Number Line

Observe that the distance from -5 to 3 on the number line is 8 units:

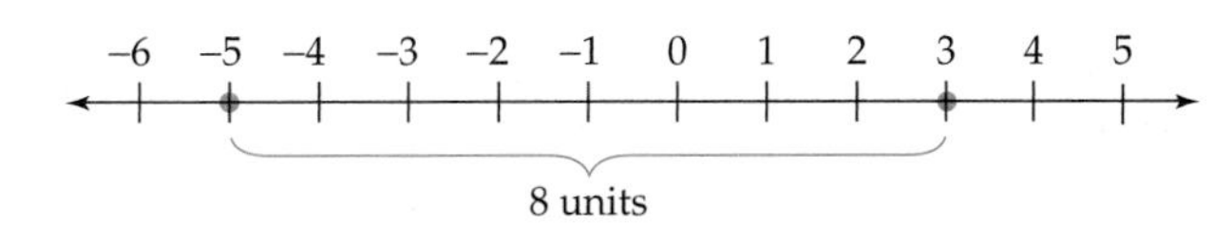

Figure 1–4

This distance can be expressed in terms of absolute value by noting that $|(-5) - 3| = 8$. That is, the distance is the *absolute value of the difference* of the two numbers. Furthermore, the order in which you take the difference doesn't matter; $|3 - (-5)|$ is also 8. This reflects the geometric fact that the distance from -5 to 3 is the same as the distance from 3 to -5. The same thing is true in the general case:

Distance on the Number Line

The distance between c and d on the number line is the number

$$|c - d| = |d - c|.$$

Example 5 The distance from 4.2 to 9 is $|4.2 - 9| = |-4.8| = 4.8$, and the distance from 6 to $\sqrt{2}$ is $|6 - \sqrt{2}|$. ■

In the special case when $d = 0$, the distance formula shows that $|c - 0| = |c|$. Hence,

Distance to Zero

$|c|$ is the distance between c and 0 on the number line.

Algebraic problems can sometimes be solved by translating them into equivalent geometric problems. The key is to interpret statements involving absolute value as statements about distance on the number line.

Example 6 Solve the equation $|x + 5| = 3$ geometrically.

Solution We rewrite it as $|x - (-5)| = 3$. In this form it states that

*The distance between x and -5 is 3 units.**

Figure 1–5 shows that -8 and -2 are the only two numbers whose distance to -5 is 3 units:

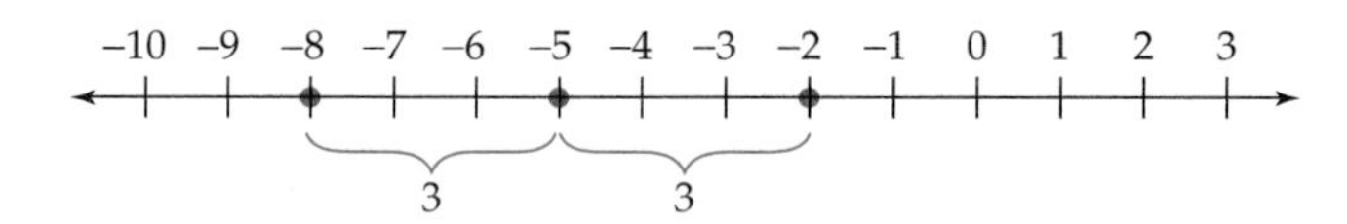

Figure 1–5

Thus $x = -8$ and $x = -2$ are the solutions of $|x + 5| = 3$. ■

*It's necessary to rewrite the equation first because the distance formula involves the *difference* of two numbers, not their sum.

Example 7 The solutions of $|x - 1| \geq 2$ are all numbers x such that

The distance between x and 1 is greater than or equal to 2.

Figure 1–6 shows that the numbers 2 or more units away from 1 are the numbers x such that

$$x \leq -1 \qquad \text{or} \qquad x \geq 3.$$

So these numbers are the solutions of the inequality. ■

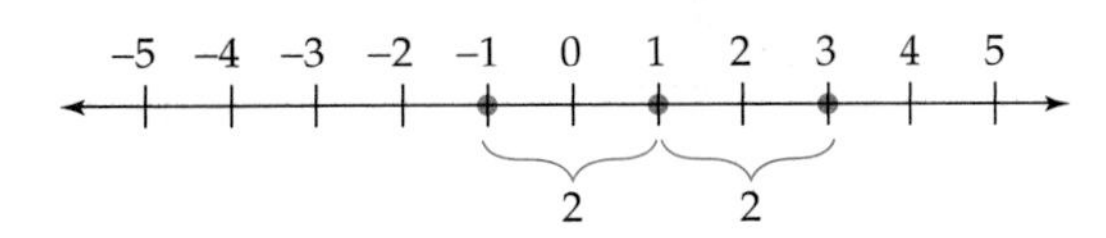

Figure 1–6

Example 8 The solutions of $|x - 7| < 2.5$ are all numbers x such that

The distance between x and 7 is less than 2.5.

Figure 1–7 shows that the solutions of the inequality, that is, the numbers within 2.5 units of 7, are the numbers x such that $4.5 < x < 9.5$, that is, the interval (4.5, 9.5). ■

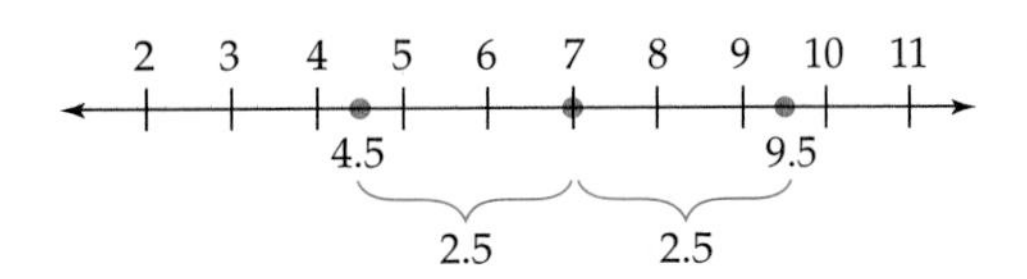

Figure 1–7

Calculator Investigations 1.1

1. **Edit and Replay** Consider the equation $y = x^3 + 6x^2 - 5$.
   (a) Find the value of y when $x = -7$ by keying in

   (*) $$(-7)^3 + 6(-7)^2 - 5$$

   and pressing ENTER.†

   (b) Find the value of y when $x = 9$ by editing your previous calculation as follows. On TI, press SECOND ENTER; on Sharp 9600, press SECOND ENTRY; on Casio 9850 touch the left or right arrow key; on HP-38, use the up arrow key until the previous computation is shaded, then press COPY. This activates the **replay** feature: Your screen now returns to the previous calculation (*). Use the left/right arrow keys and the DEL(ete) key to move through the

†Here and throughout the book, Casio users should read "EXE" in place of "ENTER."

equation and replace −7 by 9. Then press ENTER, and the result of the new computation is displayed.

(c) Use the replay feature again to find the value of y when $x = 108$. Use the INS(sert) key to avoid unnecessary retyping.

(d) [This part does not work on Casio.] Press the replay keys repeatedly; each time you do, one of the preceding calculations will appear. Go back to the first one (*), and compute it again.

2. Mathematical Operations

(a) Key in each of the following, and explain why your answers are different. [See the Technology Tip on page 8.]

ABS (−) 9 + 2 ENTER and ABS ((−) 9 + 2) ENTER

(b) Find INT in the NUM submenu of the MATH or OPTN menu [on HP-38, use FLOOR in the REAL submenu of the MATH menu]. Find out what this command does when you follow it by a number and ENTER. [Remember that a number may be positive or negative, an integer or a fraction or a decimal.] Casio users will sometimes get different answers from users of other brands.

3. Symbolic Calculations

(a) In order to store the number 2 in memory A of a calculator, type

2 STO ▶ A ENTER (TI-85/86); or

2 STO ▶ ALPHA A ENTER (TI-82/83/89; Sharp 9600; HP-38); or

2 → ALPHA A EXE (Casio 9850).

If you now key in ALPHA A ENTER, what does the calculator display?

(b) In a similar fashion store the number 5 in memory B and −10 in memory C. Then, using the ALPHA keys, display this expression on the screen: B + C/A. If you press ENTER, what happens? Explain what the calculator is doing.

(c) Experiment with other expressions, such as $B^2 - 4AC$.

4. Inequalities

Find out what happens when you key in each of these statements and press ENTER:

$$8 < 9, \quad 8 < 5, \quad 9 > 2, \quad 9 > 10.$$

[Inequality symbols are in the TEST menu on the TI keyboard or the TESTS submenu of the HP-38 MATH menu or the INEQ submenu of the Sharp 9600 MATH menu.]

Exercises 1.1

1. Draw a number line, and mark the location of each of these numbers: 0, −7, 8/3, 10, −1, −4.75, 1/2, −5, and 2.25.

In Exercises 2–7, express the given numbers (based on 1998 estimates) in scientific notation.

2. Population of the world: 5,926,467,000
3. Population of the United States: 270,312,000
4. U.S. civilian labor force: 137,523,000
5. Number of unemployed U.S. workers: 6,529,000
6. Radius of a hydrogen atom: .00000000001 meters

7. Width of a DNA double helix: .000000002 meters

In Exercises 8–11, express the given number in normal decimal notation.

8. Speed of light in a vacuum: 2.9979×10^8 m/sec

9. Average distance from earth to sun: 1.50×10^{11} m

10. Electron charge: 1.602×10^{-27} C

11. Proton mass: 1.6726×10^{-19} kg

12. One light-year is the distance light travels in a 365-day year. The speed of light is about 186,282.4 miles per second.
- **(a)** How long is one light-year (in miles)? Express your answer in scientific notation.
- **(b)** Light from the North Star takes 680 years to reach earth. How many miles is the North Star from earth?

In Exercises 13–22, express the given statement in symbols.

13. -4 is greater than -8.

14. -17 is less than 14.

15. π is less than 100.

16. x is nonnegative.

17. y is less than or equal to 7.5.

18. z is greater than or equal to -4.

19. t is positive.

20. d is not greater than 2.

21. c is at most 3.

22. z is at least -17.

In Exercises 23–27, fill the blank with $<$, $=$, or $>$ so that the resulting statement is true.

23. -6 ____ -2 **24.** 5 ____ -3

25. $3/4$ ____ $.75$ **26.** 3.1 ____ π

27. $1/3$ ____ $.33$

In Exercises 28–34, fill the blank so as to produce two equivalent statements. For example, the arithmetic statement "a is negative" is equivalent to the geometric statement "the point a lies to the left of the point 0."

Arithmetic Statement	Geometric Statement
28. $a \geq b$	____________
29. ____________	a lies c units to the right of b
30. ____________	a lies between b and c
31. $a - b > 0$	____________
32. a is positive	____________
33. ____________	a lies to the left of b
34. $a + b < c \quad (b > 0)$	____________

In Exercises 35–44, simplify, and write the given number without using absolute values.

35. $|3 - 14|$ **36.** $|(-2)3|$ **37.** $3 - |2 - 5|$

38. $-2 - |-2|$ **39.** $|(-13)^2|$ **40.** $-|-5|^2$

41. $|\pi - \sqrt{2}|$ **42.** $|\sqrt{2} - 2|$ **43.** $|3 - \pi| + 3$

44. $|4 - \sqrt{2}| - 5$

In Exercises 45–50, fill the blank with $<$, $=$, or $>$ so that the resulting statement is true.

45. $|-2|$ ____ $|-5|$ **46.** 5 ____ $|-2|$

47. $|3|$ ____ $-|4|$ **48.** $|-3|$ ____ 0

49. -7 ____ $|-1|$ **50.** $-|-4|$ ____ 0

In Exercises 51–56, draw a picture on the number line of the given interval.

51. $(0, 8]$ **52.** $(0, \infty)$ **53.** $[-2, 1]$

54. $(-1, 1)$ **55.** $(-\infty, 0]$ **56.** $[-2, 7)$

In Exercises 57–62, use interval notation to denote the set of all real numbers x that satisfy the given inequality.

57. $5 \leq x \leq 8$ **58.** $-2 \leq x \leq 7$

59. $-3 < x < 14$ **60.** $7 < x < 135$

61. $x \geq -8$ **62.** $x \geq 12$

In Exercises 63–70, find the distance between the given numbers.

63. -3 and 4 **64.** 7 and 107

65. -7 and $15/2$ **66.** $-3/4$ and -10

67. π and 3 **68.** π and -3

69. $\sqrt{2}$ and $\sqrt{3}$ **70.** π and $\sqrt{2}$

71. If you are h feet above the ground, then because of the curvature of the earth, the maximum distance you can see is approximately d miles, where

$$d = \sqrt{1.5h + (3.587 \times 10^{-8})h^2}.$$

- **(a)** How far can you see from the SkyDeck viewing area of the Sears Tower in Chicago (1353 ft high)?
- **(b)** If you were allowed to stand on the top of the Sears Tower (1454 ft high), how much farther could you see?

72. Suppose you are k miles (not feet) above the ground. The radius of the earth is approximately

3960 miles and the point where your line of sight meets the earth is perpendicular to the radius of the earth at that point, as shown in the figure.

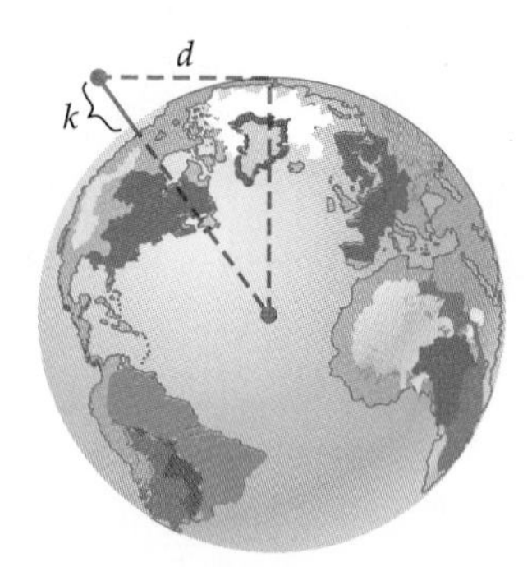

(a) Use the Pythagorean Theorem (see the Geometry Review Appendix) to show that

$$d = \sqrt{(3960 + k)^2 - 3960^2}.$$

(b) Show that the equation in part (a) simplifies to $d = \sqrt{7920k + k^2}$.

(c) If you are h feet above the ground, then you are $h/5280$ miles high (why?). Use this fact and the equation in part (b) to obtain the formula used in Exercise 71.

73. A broker predicts that over the next six months the price, p, of a particular stock will not vary from its current price of \$25.75 by more than \$4. Use absolute value to express this prediction as an inequality.

74. At Statewide Insurance, each department's expenses are reviewed monthly. A department can fail to pass the "budget variance test" in a category if either (i) the absolute value of the difference between actual expenses and the budget is more than \$500; or (ii) the absolute value of the difference between the actual expenses and the budget is more than 5% of the budgeted amount. Which of the following items fails the budget variance test? Explain your answers.

Item	Budgeted Expense (\$)	Actual Expense (\$)
Wages	220,750	221,239
Overtime	10,500	11,018
Shipping and Postage	530	589

In Exercises 75–82, write the given expression without using absolute values.

75. $|t^2|$

76. $|-2 - y^2|$

77. $|b - 3|$ if $b \geq 3$

78. $|a - 5|$ if $a < 5$

79. $|c - d|$ if $c < d$

80. $|c - d|$ if $c \geq d$

81. $|u - v| - |v - u|$

82. $\dfrac{|u - v|}{|v - u|}$ if $u \neq v$, $u \neq 0$, $v \neq 0$

In Exercises 83 and 84, explain why the given statement is true for any numbers c and d. [Hint: Look at the properties of absolute value on pages 8 and 9.]

83. $|(c - d)^2| = c^2 - 2cd + d^2$

84. $\sqrt{9c^2 - 18cd + 9d^2} = 3|c - d|$

In Exercises 85–90, express the given geometric statement about numbers on the number line algebraically, using absolute values.

85. The distance from x to 5 is less than 4.

86. x is more than 6 units from c.

87. x is at most 17 units from -4.

88. x is within 3 units of 7.

89. c is closer to 0 than b is.

90. x is closer to 1 than to 4.

In Exercises 91–94, translate the given algebraic statement into a geometric statement about numbers on the number line.

91. $|x - 3| < 2$

92. $|x - c| > 6$

93. $|x + 7| \leq 3$

94. $|u + v| \geq 2$

95. Match each of the following graphs with an appropriate absolute value equation or inequality.

(a) [10, 24]

(b) 10 24

(c)

(d) 10 24

(e)

i. $|x - 17| = 7$

ii. $|x - 17| = -7$

iii. $|x - 17| \leq 7$

iv. $|x - 17| \geq -7$

v. $|x - 17| > 7$

96. Explain geometrically why this statement is always false:

$|c - 1| < 2$ and simultaneously $|c - 12| < 3$.

In Exercises 97–108, use the geometric approach explained in the text to solve the given equation or inequality.

97. $|x| = 1$ **98.** $|x| = 3/2$

99. $|x - 2| = 1$ **100.** $|x + 3| = 2$

101. $|x + \pi| = 4$ **102.** $|x - \frac{3}{2}| = 5$

103. $|x| < 7$ **104.** $|x| \geq 5$

105. $|x - 5| < 2$ **106.** $|x - 6| > 2$

107. $|x + 2| \geq 3$ **108.** $|x + 4| \leq 2$

Thinkers

109. Explain why the statement $|a| + |b| + |c| > 0$ is algebraic shorthand for "at least one of the numbers a, b, c, is different from zero."

110. Find an algebraic shorthand version of the statement "none of the numbers a, b, c, is zero."

1.1.A EXCURSION Decimal Representation of Real Numbers

Every rational number can be expressed as a terminating or repeating decimal. For instance $3/4 = 0.75$. To express $15/11$ as a decimal, divide the numerator by the denominator:

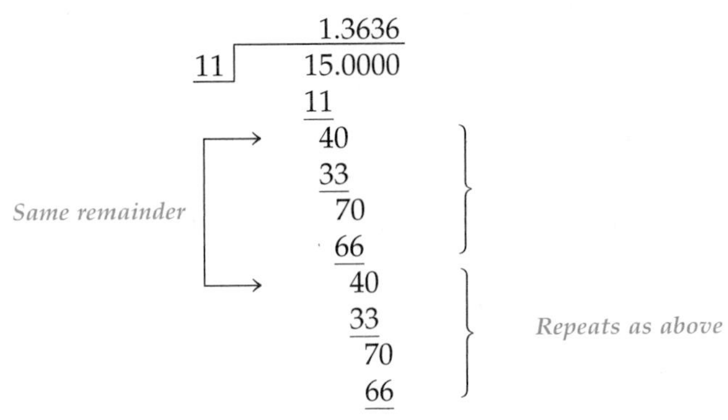

Since the remainder at the first step (namely 4) occurs again at the third step, it is clear that the division process goes on forever with the two-digit block "36" repeating over and over in the quotient $15/11 = 1.3636363636\cdots$.

The method used in the preceding example can be used to express any rational number as a decimal. During the division process some remainder *necessarily repeats.** If the remainder at which this repetition starts is 0, the result is a repeating decimal ending in zeros—that is, a terminating decimal (for instance, $.75000\cdots = .75$). If the remainder at which the repetition starts is nonzero, then the result is a nonterminating repeating decimal, as in the example above.

A typical calculator displays only the first ten digits of a number in decimal form (although it uses several additional digits in its internal computations). For example, a calculator might display the decimal expansion of $1/17$ as .0588235294, although the actual expansion has a repeating block of 16 digits:

*For instance, if you divide a number by 11 as in the example above, the only possible remainders at each step are the 11 numbers 0, 1, 2, 3, 4, 5, 6, 7, 8, 9, and 10. Hence after *at most* 12 steps, some remainder must occur for a second time (it happened at the third step in the excursion). At this point (if there are no new digits in the dividend) the division process and hence the quotient begin to repeat.

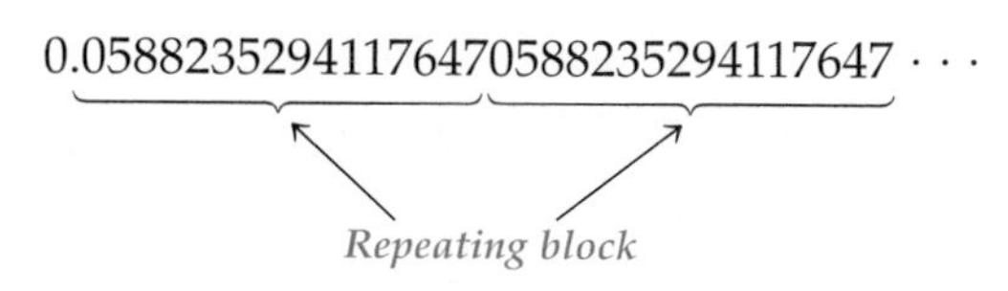

Thus, a calculator can contain the *exact* value of a rational number only if its decimal expansion terminates after approximately ten decimal places. Techniques for obtaining the full decimal expansion of a rational number from a calculator are discussed in Exercise 25.

Conversely, there is a simple method for converting any repeating decimal into a rational number.

Example 1 Write $d = .272727\cdots$ as a rational number.

Technology Tip

The FRAC key on most TI calculators automatically converts decimals to fractions (subject to some limitations). The same thing can be done on HP-38 by choosing FRACTION number format in the MODE menu and then entering the decimal. Conversion programs for Sharp and Casio are in the Program Appendix.

Solution Assuming that the usual rules of arithmetic hold, we see that

$$100d = 27.272727\cdots \qquad \text{and} \qquad d = .272727\cdots$$

Now subtract d from $100d$:

$$\begin{aligned} 100d &= 27.272727\cdots \\ -d &= -.272727\cdots \\ \hline 99d &= 27 \end{aligned}$$

Dividing both sides of this last equation by 99 shows that $d = 27/99 = 3/11$. ■

Irrational Numbers

Many nonterminating decimals are *nonrepeating* (that is, no block of digits repeats forever), such as $.202002000200002\cdots$ (where after each 2 there is one more zero than before). Although the proof is too long to give here, it is in fact true that every nonterminating and nonrepeating decimal represents an *irrational* real number.

Conversely every irrational number can be expressed as a nonterminating and nonrepeating decimal (no proof to be given here).

Since a calculator can deal with only the first 10–12 digits of a number in decimal form, *a calculator cannot contain the exact value of any irrational number.* Furthermore, there are no easy calculator techniques for carrying out the decimal expansion of an arbitrary irrational number to a specified number of decimal places, as there are for rational numbers.

Since every real number is either a rational number or an irrational one, the preceding discussion can be summarized as follows.

Decimal Representation

1. **Every real number can be expressed as a decimal.**
2. **Every decimal represents a real number.**
3. **The terminating decimals and the nonterminating repeating decimals are the rational numbers.**
4. **The nonterminating, nonrepeating decimals are the irrational numbers.**

Calculator Investigation 1.1.A

1. FRAC key If your calculator has a FRAC key or program (see the Program Appendix), test its limitations by entering each of the following numbers and then pressing the FRAC key.

   (a) .058823529411 **(b)** .0588235294117
   (c) .058823529411724 **(d)** .0588235294117985

   Which of your answers are correct? [*Hint:* Look at the decimal expansion of 1/17 on page 16.]

Exercises 1.1.A

In Exercises 1–6, express the given rational number as a repeating decimal.

1. 7/9 **2.** 2/13 **3.** 23/14
4. 19/88 **5.** 1/19 (long) **6.** 9/11

In Exercises 7–14, state whether a calculator can express the given number exactly.

7. 2/3 **8.** 7/16 **9.** 1/64 **10.** 1/22
11. $3\pi/2$ **12.** $\pi - 3$ **13.** 1/.625 **14.** 1/.16

In Exercises 15–21, express the given repeating decimal as a fraction.

15. .373737 ⋯ **16.** .929292 ⋯

17. 76.63424242 ⋯ [*Hint:* Consider $10{,}000d - 100d$, where $d = 76.63424242\cdots$.]

18. 13.513513 ⋯ [*Hint:* Consider $1000d - d$, where $d = 13.513513\cdots$.]

19. .135135135 ⋯ [*Hint:* See Exercise 18.]

20. .33030303 ⋯ **21.** 52.31272727 ⋯

22. If two real numbers have the same decimal expansion through three decimal places, how far apart can they be on the number line?

Thinkers

23. Use the methods in Exercises 15–21 to show that both .74999 ⋯ and .75000 ⋯ are decimal expansions of 3/4. [Every terminating decimal can also be expressed as a decimal ending in repeated 9's. It can be proved that these are the only real numbers with more than one decimal expansion.]

Finding remainders with a calculator

24. If you use long division to divide 369 by 7, you obtain:

$$\begin{array}{rrl} & 52 & \leftarrow \textit{Quotient} \\ \textit{Divisor} \rightarrow & 7\overline{)369} & \leftarrow \textit{Dividend} \\ & \underline{35} & \\ & 19 & \\ & \underline{14} & \\ & 5 & \leftarrow \textit{Remainder} \end{array}$$

If you use a calculator to find $369 \div 7$, the answer is displayed as 52.71428571. Observe that the integer part of this calculator answer, 52, is the quotient when you do the problem by long division. The usual "checking procedure" for long division shows that

$7 \cdot 52 + 5 = 369$ or equivalently, $369 - 7 \cdot 52 = 5$.

Thus, the remainder is

$$\text{Dividend} - (\text{divisor})\left(\begin{array}{c}\text{integer part of} \\ \text{calculator answer}\end{array}\right).$$

Use this method to find the quotient and remainder in these problems:

(a) $5683 \div 9$ **(b)** $1{,}000{,}000 \div 19$
(c) $53{,}000{,}000 \div 37$

In Exercises 25–30, find the decimal expansion of the given rational number. All these expansions are too long to fit in a calculator, but can be readily found by using the hint in Exercise 25.

25. 6/17 [*Hint:* The first part of dividing 6 by 17 involves working this division problem: $6{,}000{,}000 \div 17$. The method of Exercise 24 shows that the quotient is 352,941 and the remainder is 3. Thus the decimal expansion of 6/17 begins

.352941, and the next block of digits in the expansion will be the quotient in the problem 3,000,000 ÷ 17. The remainder when 3,000,000 is divided by 17 is 10, so the next block of digits in the expansion of 6/17 is the quotient in the problem 10,000,000 ÷ 17. Continue in this way until the decimal expansion repeats.]

26. 3/19 **27.** 1/29 **28.** 3/43 **29.** 283/47

30. 768/59

31. (a) Show that there are at least as many irrational numbers (nonrepeating decimals) as there are terminating decimals. [*Hint:* With each terminating decimal associate a nonrepeating decimal.]

(b) Show that there are at least as many irrational numbers as there are repeating decimals. [*Hint:* With each repeating decimal, associate a nonrepeating decimal by inserting longer and longer strings of zeros: for instance, with .11111111 ⋯ associate the number .101001000100001 ⋯ .]

1.2 Solving Equations Algebraically

This section deals with equations such as

$$3x - 6 = 7x + 4, \qquad x^2 - 5x + 6 = 0, \qquad 2x^4 - 13x^2 = 3.$$

A **solution** of an equation is a number that, when substituted for the variable x, produces a true statement.* For example, 5 is a solution of $3x + 2 = 17$ because $3 \cdot 5 + 2 = 17$ is a true statement. To **solve** an equation means to find all its solutions. Throughout this chapter we shall deal only with **real solutions,** that is, solutions that are real numbers.

Two equations are said to be **equivalent** if they have the same solutions. For example, $3x + 2 = 17$ and $x - 2 = 3$ are equivalent because 5 is the only solution of each one.

Basic Principles for Solving Equations

Performing any of the following operations on an equation produces an equivalent equation:

1. **Add or subtract the same quantity from both sides of the equation.**
2. **Multiply or divide both sides of the equation by the same *nonzero* quantity.**

The usual strategy in equation solving is to use these Basic Principles to transform a given equation into an equivalent one whose solutions are known.

A **first-degree,** or **linear, equation** is one that can be written in the form

$$ax + b = 0$$

for some constants a, b, with $a \neq 0$. Every first-degree equation has exactly one solution, which is easily found.

*Any letter may be used for the variable.

Example 1 To solve $3x - 6 = 7x + 4$, we use the Basic Principles to transform this equation into an equivalent one whose solution is obvious:

$$3x - 6 = 7x + 4$$

Add 6 to both sides: $$3x = 7x + 10$$

Subtract 7x from both sides: $$-4x = 10$$

Divide both sides by −4: $$x = \frac{10}{-4} = -\frac{5}{2}$$

Since $-5/2$ is the only solution of this last equation, $-5/2$ is the only solution of the original equation, $3x - 6 = 7x + 4$. ■

Quadratic Equations

A **second-degree,** or **quadratic, equation** is an equation that can be written in the form

$$ax^2 + bx + c = 0$$

for some constants a, b, c, with $a \neq 0$. There are several techniques for solving such equations. We begin with the **factoring method,** which makes use of this property of the real numbers:

Zero Products

If a product of real numbers is zero, then at least one of the factors is zero; in other words,

$$\textbf{If } cd = 0, \textbf{ then } c = 0 \textbf{ or } d = 0 \textbf{ (or both).}$$

Example 2 To solve $3x^2 - x = 10$, we first rearrange the terms to make one side 0 and then factor:

Subtract 10 from each side: $$3x^2 - x - 10 = 0$$

Factor left side: $$(3x + 5)(x - 2) = 0$$

If a product of real numbers is 0, then at least one of the factors must be 0. So this equation is equivalent to:

$$3x + 5 = 0 \quad \text{or} \quad x - 2 = 0$$
$$3x = -5 \qquad\qquad x = 2$$
$$x = -5/3$$

Therefore the solutions are $-5/3$ and 2. ■

CAUTION
To guard against mistakes, check your solutions by substituting each one in the *original* equation to make sure it really *is* a solution.

The solutions of $x^2 = 7$ are the numbers whose square is 7. Although 7 has just *one* square root, there are *two* numbers whose square is 7, namely,

$\sqrt{7}$ and its negative, $-\sqrt{7}$. So the solutions of $x^2 = 7$ are $\sqrt{7}$ and $-\sqrt{7}$, or in abbreviated form, $\pm\sqrt{7}$. The same argument works for any positive real number d:

The solutions of $x^2 = d$ are $\sqrt{d}$ and $-\sqrt{d}$.

We now use a slight variation of this idea to develop a method for solving quadratic equations that don't readily factor. The method depends on this fact: A polynomial of the form $x^2 + bx$ can be changed into a perfect square by adding a suitable constant.* For example, $x^2 + 6x$ can be changed into a perfect square by adding 9:

$$x^2 + 6x + 9 = (x + 3)^2.$$

This process is called **completing the square.**

To complete the square in $x^2 + 6x$, we added $9 = 3^2$. Note that 3 is one-half the coefficient of x in the original polynomial $x^2 + 6x$. The same idea works in the general case. The multiplication pattern for perfect squares shows that for any real number b:

$$\left(x + \frac{b}{2}\right)^2 = x^2 + 2\left(\frac{b}{2}\right)x + \left(\frac{b}{2}\right)^2 = x^2 + bx + \left(\frac{b}{2}\right)^2.$$

Therefore,

Completing the Square

To complete the square in $x^2 + bx$, add the square of one-half the coefficient of x, namely, $(b/2)^2$. This produces a perfect square:

$$x^2 + bx + \left(\frac{b}{2}\right)^2 = \left(x + \frac{b}{2}\right)^2$$

The following example shows how completing the square can be used to solve quadratic equations.

Example 3 To solve $x^2 + 6x + 1 = 0$, we first rewrite the equation as $x^2 + 6x = -1$. Next we complete the square on the left side by adding the square of half the coefficient of x, namely, $(6/2)^2 = 9$. In order to have an equivalent equation, we must add 9 to *both* sides:

$$x^2 + 6x + 9 = -1 + 9$$

Factor left side: $$(x + 3)^2 = 8$$

Thus $x + 3$ is a number whose square is 8. The only numbers whose squares equal 8 are $\sqrt{8}$ and $-\sqrt{8}$. So we must have:

*A quadratic polynomial $x^2 + bx + c$ is a **perfect square** if it factors as $(x + d)^2$ for some constant d.

$$x + 3 = \sqrt{8} \quad \text{or} \quad x + 3 = -\sqrt{8}$$

$$x = \sqrt{8} - 3 \quad \text{or} \quad x = -\sqrt{8} - 3.$$

Therefore the solutions of the original equation are $\sqrt{8} - 3$ and $-\sqrt{8} - 3$, or in more compact notation, $\pm\sqrt{8} - 3$. ■

CAUTION

Completing the square only works when the coefficient of x^2 is 1. In an equation such as

$$5x^2 - x + 2 = 0$$

you must first divide every term on both sides by 5, and *then* complete the square.

We can use the completing-the-square method to solve *any* quadratic equation:*

$$ax^2 + bx + c = 0$$

Divide both sides by a:

$$x^2 + \frac{b}{a}x + \frac{c}{a} = 0$$

Subtract $\frac{c}{a}$ from both sides:

$$x^2 + \frac{b}{a}x = -\frac{c}{a}$$

Add $\left(\frac{b}{2a}\right)^2$ to both sides:†

$$x^2 + \frac{b}{a}x + \left(\frac{b}{2a}\right)^2 = \left(\frac{b}{2a}\right)^2 - \frac{c}{a}$$

Factor left side:

$$\left(x + \frac{b}{2a}\right)^2 = \left(\frac{b}{2a}\right)^2 - \frac{c}{a}$$

Find common denominator for right side:

$$\left(x + \frac{b}{2a}\right)^2 = \frac{b^2}{4a^2} - \frac{c}{a} = \frac{b^2 - 4ac}{4a^2}$$

Since the square of $x + \frac{b}{2a}$ equals $\frac{b^2 - 4ac}{4a^2}$, we must have:

$$x + \frac{b}{2a} = \pm\sqrt{\frac{b^2 - 4ac}{4a^2}} = \pm\frac{\sqrt{b^2 - 4ac}}{2a}$$

Subtract $\frac{b}{2a}$ from both sides:

$$x = \frac{-b}{2a} \pm \frac{\sqrt{b^2 - 4ac}}{2a} = \frac{-b \pm \sqrt{b^2 - 4ac}}{2a}$$

We have proved:

The Quadratic Formula

The solutions of the quadratic equation $ax^2 + bx + c = 0$ are:

$$x = \frac{-b \pm \sqrt{b^2 - 4ac}}{2a}$$

You should memorize the quadratic formula.

*If you have trouble following any step here, do it for a numerical example, such as the case when $a = 3$, $b = 11$, $c = 5$.

†This is the square of half the coefficient of x.

Example 4 Solve $x^2 + 3 = -8x$.

Solution Rewrite the equation as $x^2 + 8x + 3 = 0$, and apply the quadratic formula with $a = 1$, $b = 8$, and $c = 3$:

$$x = \frac{-b \pm \sqrt{b^2 - 4ac}}{2a} = \frac{-8 \pm \sqrt{8^2 - 4 \cdot 1 \cdot 3}}{2 \cdot 1}$$

$$= \frac{-8 \pm \sqrt{52}}{2} = \frac{-8 \pm \sqrt{4 \cdot 13}}{2}$$

$$= \frac{-8 \pm 2\sqrt{13}}{2} = -4 \pm \sqrt{13}.$$

Therefore the equation has two distinct real solutions, $-4 + \sqrt{13}$ and $-4 - \sqrt{13}$. ■

Example 5 Solve $x^2 - 194x + 9409 = 0$.

Solution Use a calculator and the quadratic formula with $a = 1$, $b = -194$, and $c = 9409$:

$$x = \frac{-b \pm \sqrt{b^2 - 4ac}}{2a} = \frac{-(-194) \pm \sqrt{(-194)^2 - 4 \cdot 1 \cdot 9409}}{2 \cdot 1}$$

$$= \frac{194 \pm \sqrt{37636 - 37636}}{2} = \frac{194 \pm 0}{2} = 97.$$

Thus 97 is the only solution of the equation. ■

Example 6 Solve $2x^2 + x + 3 = 0$.

Solution Using the quadratic formula with $a = 2$, $b = 1$, and $c = 3$:

$$x = \frac{-b \pm \sqrt{b^2 - 4ac}}{2a} = \frac{-1 \pm \sqrt{1^2 - 4 \cdot 2 \cdot 3}}{2 \cdot 2} = \frac{-1 \pm \sqrt{1 - 24}}{4}$$

$$= \frac{-1 \pm \sqrt{-23}}{4}.$$

Since $\sqrt{-23}$ is not a real number, this equation has *no real solutions* (that is, no solutions in the real number system). ■

The expression $b^2 - 4ac$ in the quadratic formula is called the **discriminant.** As the last three examples illustrate, the discriminant determines the *number* of real solutions of the equation $ax^2 + bx + c = 0$.

Real Solutions of a Quadratic Equation

Discriminant $b^2 - 4ac$	Number of Real Solutions of $ax^2 + bx + c = 0$	Example
> 0	Two distinct real solutions	$x^2 + 8x + 3 = 0$
$= 0$	One real solution	$x^2 - 194x + 9409 = 0$
< 0	No real solutions	$2x^2 + x + 3 = 0$

The quadratic formula and a calculator can be used to solve any quadratic equation with nonnegative discriminant. Experiment with your calculator to find the most efficient sequence of keystrokes for doing this.

Technology Tip

Most calculators have built-in polynomial equation solvers that will solve quadratic and other polynomial equations. See Calculator Investigation 1 at the end of this section. A quadratic formula program for other calculators is in the Program Appendix.

Example 7 Use a calculator to solve $3.2x^2 + 15.93x - 7.1 = 0$.

First compute $\sqrt{b^2 - 4ac} = \sqrt{15.93^2 - 4(3.2)(-7.1)}$, and store the result in memory D. Then the solutions of the equation are given by

$$\frac{-15.93 + D}{2(3.2)} \approx .411658467 \quad \text{and} \quad \frac{-15.93 - D}{2(3.2)} \approx -5.389783347.$$

Remember that these answers are *approximations,* so they may not check exactly when substituted in the original equation. ■

Higher Degree Equations

A **polynomial equation of degree n** is one that can be written in the form

$$a_nx^n + \cdots + a_3x^3 + a_2x^2 + a_1x + a_0 = 0,$$

where n is a positive integer, each a_i is a constant, and $a_n \neq 0$. For instance,

$$4x^6 - 3x^5 + x^4 + 7x^3 - 8x^2 + 4x + 9 = 0$$

is a polynomial equation of degree 6. As a general rule, polynomial equations of degree 3 and above are best solved by the numerical or graphical methods presented in Section 2.2. However, some such equations can be solved algebraically by making a suitable substitution, as we now see.

Example 8 To solve $4x^4 - 13x^2 + 3 = 0$, substitute u for x^2, and solve the resulting quadratic equation:

$$4x^4 - 13x^2 + 3 = 0$$
$$4(x^2)^2 - 13x^2 + 3 = 0$$
$$4u^2 - 13u + 3 = 0$$
$$(u - 3)(4u - 1) = 0$$
$$u - 3 = 0 \quad \text{or} \quad 4u - 1 = 0$$
$$u = 3 \qquad\qquad 4u = 1$$
$$u = \frac{1}{4}$$

Since $u = x^2$, we see that

$$x^2 = 3 \qquad \text{or} \qquad x^2 = \frac{1}{4}$$

$$x = \pm\sqrt{3} \qquad\qquad x = \pm\frac{1}{2}$$

Hence, the original equation has four solutions: $-\sqrt{3}, \sqrt{3}, -1/2, 1/2$. ■

Example 9 To solve $x^4 - 4x^2 + 1 = 0$, let $u = x^2$:

$$x^4 - 4x^2 + 1 = 0$$

$$u^2 - 4u + 1 = 0$$

The quadratic formula shows that

$$u = \frac{-(-4) \pm \sqrt{(-4)^2 - 4\cdot 1\cdot 1}}{2\cdot 1} = \frac{4 \pm \sqrt{12}}{2}$$

$$= \frac{4 \pm \sqrt{4\cdot 3}}{2} = \frac{4 \pm 2\sqrt{3}}{2} = 2 \pm \sqrt{3}$$

Since $u = x^2$, we have the equivalent statements:

$$x^2 = 2 + \sqrt{3} \qquad \text{or} \qquad x^2 = 2 - \sqrt{3}$$

$$x = \pm\sqrt{2 + \sqrt{3}} \qquad\qquad x = \pm\sqrt{2 - \sqrt{3}}$$

Therefore the original equation has four solutions. ■

Calculator Investigation 1.2

1. **Polynomial Equation Solvers** If you have one of the calculators mentioned below, use its polynomial equation solver to solve the equation $3.2x^2 + 15.93x - 7.1 = 0$, as follows. Bring up the solver by choosing POLY on the TI-86 keyboard, or POLY in the Sharp 9600 TOOL menu, or EQUATION in the Casio 9850 main menu (and selecting POLYNOMIAL). Enter 2 as the degree of the polynomial (called its "order" on TI), enter the coefficients of the polynomial, and press SOLVE (or EXE on Sharp 9600). On HP-38, use the POLYNOM submenu of the MATH menu, and key in POLYROOT([3.2, 15.93, −7.1]), including both parentheses and brackets; then press ENTER. On TI-89, use the ALGEBRA menu, and key in: SOLVE($3.2x^2 + 15.93x - 7.1 = 0, x$); then press ENTER. How do the answers you obtained compare with those in Example 7? Use the same technique to solve the following equations.

   (a) $x^2 + 2x - 4 = 0$ **(b)** $x^2 + 5x + 2 = 0$.

Exercises 1.2

In Exercises 1–12, solve the equation by factoring.

1. $x^2 - 8x + 15 = 0$
2. $x^2 + 5x + 6 = 0$
3. $x^2 - 5x = 14$
4. $x^2 + x = 20$
5. $2y^2 + 5y - 3 = 0$
6. $3t^2 - t - 2 = 0$
7. $4t^2 + 9t + 2 = 0$
8. $9t^2 + 2 = 11t$
9. $3u^2 + u = 4$
10. $5x^2 + 26x = -5$
11. $12x^2 + 13x = 4$
12. $18x^2 = 23x + 6$

In Exercises 13–16, solve the equation by completing the square.

13. $x^2 - 2x = 12$
14. $x^2 - 4x - 30 = 0$
15. $x^2 - x - 1 = 0$
16. $x^2 + 3x - 2 = 0$

In Exercises 17–28, use the quadratic formula to solve the equation.

17. $x^2 - 4x + 1 = 0$
18. $x^2 - 2x - 1 = 0$
19. $x^2 + 6x + 7 = 0$
20. $x^2 + 4x - 3 = 0$
21. $x^2 + 6 = 2x$
22. $x^2 + 11 = 6x$
23. $4x^2 - 4x = 7$
24. $4x^2 - 4x = 11$
25. $4x^2 - 8x + 1 = 0$
26. $2t^2 + 4t + 1 = 0$
27. $5u^2 + 8u = -2$
28. $4x^2 = 3x + 5$

In Exercises 29–34, find the number *of real solutions of the equation by computing the discriminant.*

29. $x^2 + 4x + 1 = 0$
30. $4x^2 - 4x - 3 = 0$
31. $9x^2 = 12x + 1$
32. $9t^2 + 15 = 30t$
33. $25t^2 + 49 = 70t$
34. $49t^2 + 5 = 42t$

In a simple model of the economy (by J. M. Keynes), equilibrium between national output and national expenditures holds when $Y = C + I$, where Y is the national income, C is consumption (which depends on the national income), and I is the amount of investment. In Exercises 35–38, solve the equilibrium equation for Y under the given conditions:

35. $C = 70 + .8Y$ and $I = 90$
36. $C = 300 + .75Y$ and $I = 450$
37. $C = 220 + .6Y$ and $I = 300$
38. $C = 640 + .7Y$ and $I = 350$

In a more complete model of the economy, the equilibrium equation is $Y = C + I + G + (X - M)$, where Y, C, and I are as in Exercises 35–38, G is government spending, X is exports, and M is imports. In Exercises 39–42, solve the equilibrium equation for Y under the given conditions.

39. $C = 120 + .9Y$, $M = 20 + .2Y$, $I = 140$, $G = 150$, and $X = 60$
40. $C = 60 + .85Y$, $M = 35 + .2Y$, $I = 95$, $G = 145$, and $X = 50$
41. $C = 300 + .875Y$, $M = 80 + .25Y$, $I = 390$, $G = 420$, and $X = 110$
42. $C = 499 + .8Y$, $M = 120 + .25Y$, $I = 695$, $G = 895$, and $X = 110$

In Exercises 43–46, solve the equation and check your answers. [Hint: First, eliminate fractions by multiplying both sides by a common denominator.]

43. $1 - \frac{3}{x} = \frac{40}{x^2}$
44. $\frac{4x^2 + 5}{3x^2 + 5x - 2} = \frac{4}{3x - 1} - \frac{3}{x + 2}$
45. $\frac{2}{x^2} - \frac{5}{x} = 4$
46. $\frac{x}{x - 1} + \frac{x + 2}{x} = 3$

In Exercises 47–56, solve the equation by any method.

47. $x^2 + 9x + 18 = 0$
48. $3t^2 - 11t - 20 = 0$
49. $4x(x + 1) = 1$
50. $25y^2 = 20y + 1$
51. $2x^2 = 7x + 15$
52. $2x^2 = 6x + 3$
53. $t^2 + 4t + 13 = 0$
54. $5x^2 + 2x = -2$
55. $\frac{7x^2}{3} = \frac{2x}{3} - 1$
56. $25x + \frac{4}{x} = 20$

In Exercises 57–60, use a calculator to find approximate solutions of the equation.

57. $4.42x^2 - 10.14x + 3.79 = 0$
58. $8.06x^2 + 25.8726x - 25.047256 = 0$
59. $3x^2 - 82.74x + 570.4923 = 0$
60. $7.63x^2 + 2.79x = 5.32$

In Exercises 61–68, find all real solutions of the equation exactly.

61. $y^4 - 7y^2 + 6 = 0$
62. $x^4 - 2x^2 + 1 = 0$
63. $x^4 - 2x^2 - 35 = 0$
64. $x^4 - 2x^2 - 24 = 0$
65. $2y^4 - 9y^2 + 4 = 0$
66. $6z^4 - 7z^2 + 2 = 0$
67. $10x^4 + 3x^2 = 1$
68. $6x^4 - 7x^2 = 3$

In Exercises 69–72, find a number k such that the given equation has exactly one real solution.

69. $x^2 + kx + 25 = 0$
70. $x^2 - kx + 49 = 0$
71. $kx^2 + 8x + 1 = 0$
72. $kx^2 + 24x + 16 = 0$

In Exercises 73–75, the discriminant of the equation $ax^2 + bx + c = 0$ (with a, b, c integers) is given. Use it to determine whether or not the solutions of the equation are rational numbers.

73. $b^2 - 4ac = 25$

74. $b^2 - 4ac = 0$

75. $b^2 - 4ac = 72$

76. Find the error in the following "proof" that $6 = 3$.

	$x = 3$
Multiply both sides by x:	$x^2 = 3x$
Subtract 9 from both sides:	$x^2 - 9 = 3x - 9$
Factor each side:	$(x - 3)(x + 3) = 3(x - 3)$
Divide both sides by $x - 3$:	$x + 3 = 3$
Since $x = 3$:	$3 + 3 = 3$
	$6 = 3$

77. Find a number k such that 4 and 1 are the solutions of $x^2 - 5x + k = 0$.

78. Suppose a, b, c are fixed real numbers such that $b^2 - 4ac \geq 0$. Let r and s be the solutions of $ax^2 + bx + c = 0$.

(a) Use the quadratic formula to show that $r + s = -b/a$ and $rs = c/a$.

(b) Use part (a) to verify that $ax^2 + bx + c = a(x - r)(x - s)$.

(c) Use part (b) to factor $x^2 - 2x - 1$ and $5x^2 + 8x + 2$.

1.2.A EXCURSION Absolute Value Equations

If c is a real number, then by the definition of absolute value $|c|$ is either c or $-c$ (whichever one is positive). This fact can be used to solve absolute value equations algebraically.

Example 1 To solve $|3x - 4| = 8$, apply the fact stated above with $c = 3x - 4$. Then $|3x - 4|$ is either $3x - 4$ or $-(3x - 4)$, so that

$$\begin{aligned} 3x - 4 &= 8 & \text{or} \qquad -(3x - 4) &= 8 \\ 3x &= 12 & -3x + 4 &= 8 \\ x &= 4 & -3x &= 4 \\ & & x &= -4/3. \end{aligned}$$

So there are two possible solutions of the original equation $|3x - 4| = 8$. You can readily verify that both 4 and $-4/3$ actually are solutions. ■

Example 2 Solve $|x + 4| = 5x - 2$.

Solution The left side of the equation is either $x + 4$ or $-(x + 4)$ (why?). Hence,

$$\begin{aligned} x + 4 &= 5x - 2 & \text{or} \qquad -(x + 4) &= 5x - 2 \\ -4x + 4 &= -2 & -x - 4 &= 5x - 2 \\ -4x &= -6 & -6x &= 2 \\ x &= \frac{-6}{-4} = \frac{3}{2} & x &= \frac{2}{-6} = -\frac{1}{3}. \end{aligned}$$

We must check each of these possible solutions in the original equation

$$|x + 4| = 5x - 2.$$

We see that $x = 3/2$ is a solution because

$$\left|\frac{3}{2} + 4\right| = \frac{11}{2} \quad \text{and} \quad 5\left(\frac{3}{2}\right) - 2 = \frac{11}{2}.$$

However, $x = -1/3$ is not a solution since

$$\left|-\frac{1}{3} + 4\right| = \frac{11}{3} \quad \text{but} \quad 5\left(-\frac{1}{3}\right) - 2 = -\frac{11}{3}. \quad \blacksquare$$

Example 3 Solve the equation $|x^2 + 4x - 3| = 2$.

Solution The equation is equivalent to

$$\begin{aligned} x^2 + 4x - 3 &= 2 \qquad \text{or} \qquad & -(x^2 + 4x - 3) &= 2 \\ x^2 + 4x - 5 &= 0 & -x^2 - 4x + 3 &= 2 \\ & & -x^2 - 4x + 1 &= 0 \\ & & x^2 + 4x - 1 &= 0 \end{aligned}$$

The first of these equations can be solved by factoring and the second by the quadratic formula:

$$(x + 5)(x - 1) = 0 \quad \text{or} \quad x = \frac{-4 \pm \sqrt{4^2 - 4\cdot 1\cdot(-1)}}{2\cdot 1}$$

$$x = -5 \quad \text{or} \quad x = 1 \qquad\qquad x = \frac{-4 \pm \sqrt{20}}{2} = \frac{-4 \pm 2\sqrt{5}}{2}$$

$$x = -2 + \sqrt{5} \quad \text{or} \quad x = -2 - \sqrt{5}$$

Verify that all four of these numbers are solutions of the original equation. ■

Exercises 1.2.A

In Exercises 1–12, find all real solutions of each equation.

1. $|2x + 3| = 9$
2. $|3x - 5| = 7$
3. $|6x - 9| = 0$
4. $|4x - 5| = -9$
5. $|2x + 3| = 4x - 1$
6. $|3x - 2| = 5x + 4$
7. $|x - 3| = x$
8. $|2x - 1| = 2x + 1$
9. $|x^2 + 4x - 1| = 4$
10. $|x^2 + 2x - 9| = 6$
11. $|x^2 - 5x + 1| = 3$
12. $|12x^2 + 5x - 7| = 4$
13. In statistical quality control, one needs to find the proportion of the product that is not acceptable. The upper and lower control limits are found by solving the following equation (in which $\overline{p}$ is the mean percent defective, and n is the sample size) for CL.

$$|CL - \overline{p}| = 3\sqrt{\frac{\overline{p}(1 - \overline{p})}{n}}$$

Find the control limits when $\overline{p} = .02$ and $n = 200$.

1.3 The Coordinate Plane

Just as real numbers are identified with points on the number line, ordered *pairs* of real numbers can be identified with points in the plane. To do this, draw two number lines in the plane, one vertical and one horizontal, as in Figure 1–8. The horizontal line is usually called the ***x*-axis,** and the vertical line the ***y*-axis,** but other letters may be used if desired. The point where the axes intersect is the **origin.** The axes divide the plane into four regions, called **quadrants,** that are numbered as in Figure 1–8.

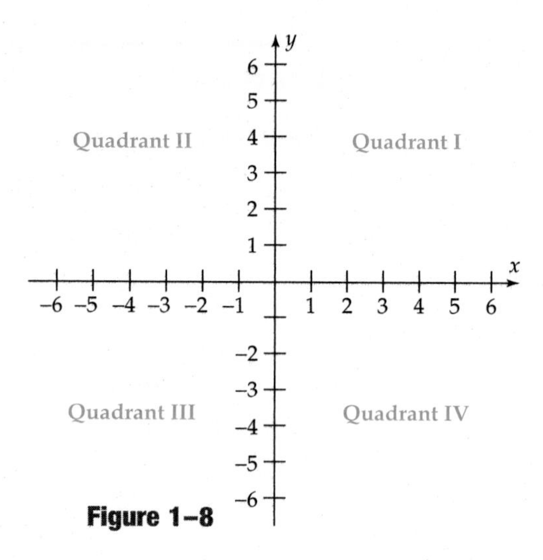

Figure 1–8

If P is a point in the plane, draw vertical and horizontal lines through P to the coordinate axes, as shown in Figure 1–9. These lines intersect the x-axis at some number c and the y-axis at d. We say that P has **coordinates** (c, d). The number c is the ***x*-coordinate** of P, and d is the ***y*-coordinate** of P. The plane is said to have a **rectangular** (or **Cartesian**) **coordinate system.**

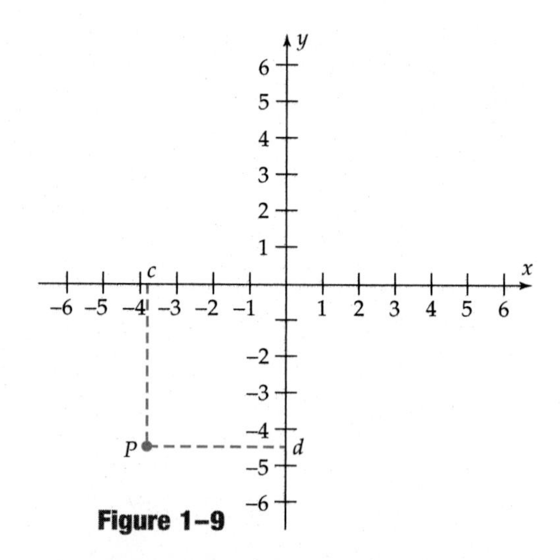

Figure 1–9

You can think of the coordinates of a point as directions for locating it. For instance, to find $(4, -3)$, start at the origin and move 4 units to the right along the x-axis, then move 3 units downward, as shown in Figure 1–10, which also shows other points and their coordinates.

> **CAUTION**
>
> The coordinates of a point are an *ordered* pair. Figure 1–10 shows that the point P with coordinates $(-5, 2)$ is quite different from the point Q with coordinates $(2, -5)$. The same numbers (2 and -5) occur in both cases, but in *different order.*

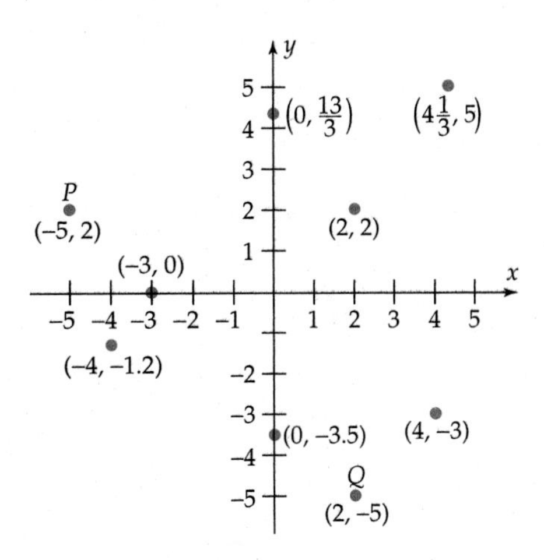

Figure 1–10

Example 1 The following table, from the Federal Election Commission, shows the total amount of money (in millions of dollars) contributed to all congressional candidates in selected years.

Year	1988	1990	1992	1994	1996
Amount	276	284	392	418	500

One way to represent this data graphically is to represent each year's total by a point; for instance, (1988, 276) and (1990, 284). Alternatively, to avoid using very large numbers, we can let x be the number of years since 1988, so that $x = 0$ is 1988, $x = 2$ is 1990 and so on. In this case, we plot the points (0, 276), (2, 284), and so on to obtain the **scatter plot** in Figure 1–11. Connecting these data points with line segments produces the **line graph** in Figure 1–12. ■

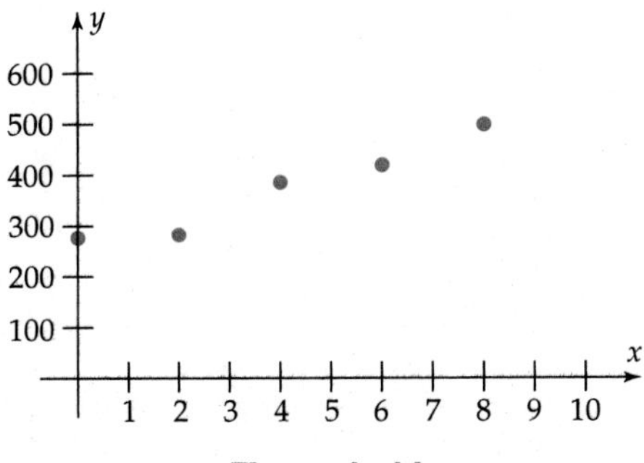

Figure 1–11

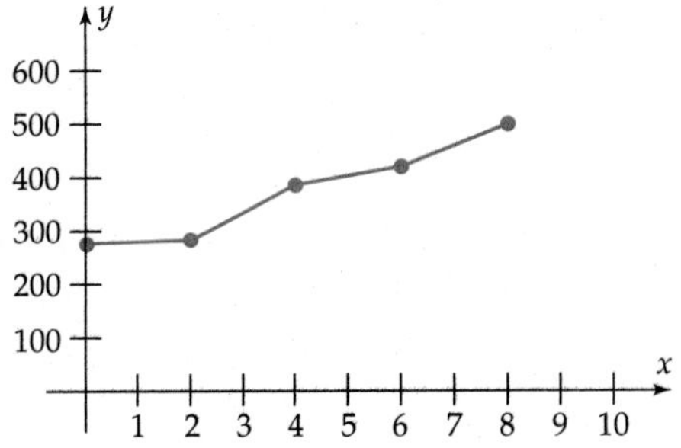

Figure 1–12

The Distance Formula

We shall often identify a point with its coordinates and refer, for example, to the point (2, 3). When dealing with several points simultaneously, it is customary to label the coordinates of the first point (x_1, y_1), the second point (x_2, y_2), the third point (x_3, y_3), and so on.* Once the plane is coordinatized, it's easy to compute the distance between any two points:

The Distance Formula

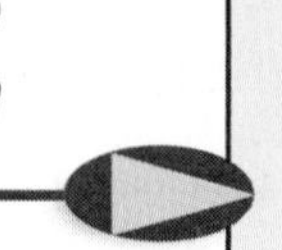

The distance between points (x_1, y_1) and (x_2, y_2) is

$$\sqrt{(x_1 - x_2)^2 + (y_1 - y_2)^2}.$$

Before proving the distance formula, we shall see how it is used.

Example 2 To find the distance between the points $(-1, -3)$ and $(2, -4)$ in Figure 1–13, substitute $(-1, -3)$ for (x_1, y_1) and $(2, -4)$ for (x_2, y_2) in the distance formula:

Distance formula:	$\text{distance} = \sqrt{(x_1 - x_2)^2 + (y_1 - y_2)^2}$
Substitute:	$= \sqrt{(-1 - 2)^2 + (-3 - (-4))^2}$
Simplify:	$= \sqrt{(-3)^2 + (-3 + 4)^2}$
	$= \sqrt{9 + 1} = \sqrt{10}.$

Figure 1–13

The order in which the points are used in the distance formula doesn't make a difference. If we substitute $(2, -4)$ for (x_1, y_1) and $(-1, -3)$ for (x_2, y_2), we get the same answer:

$$\sqrt{[2 - (-1)]^2 + [-4 - (-3)]^2} = \sqrt{3^2 + (-1)^2} = \sqrt{10}. \quad \blacksquare$$

Example 3 To find the distance from (a, b) to $(2a, -b)$, where a and b are fixed real numbers, substitute a for x_1, b for y_1, $2a$ for x_2, and $-b$ for y_2 in the distance formula:

$$\sqrt{(x_1 - x_2)^2 + (y_1 - y_2)^2} = \sqrt{(a - 2a)^2 + (b - (-b))^2}$$
$$= \sqrt{(-a)^2 + (b + b)^2} = \sqrt{a^2 + (2b)^2}$$
$$= \sqrt{a^2 + 4b^2} \quad \blacksquare$$

CAUTION
$\sqrt{a^2 + 4b^2}$ cannot be simplified. In particular, it is *not* equal to $a + 2b$.

Proof of the Distance Formula Figure 1–14 shows typical points P and Q in the plane. We must find length d of line segment PQ.

* "x_1" is read "x-one" or "x-sub-one"; it is a *single symbol* denoting the first coordinate of the first point, just as c denotes the first coordinate of (c, d). Analogous remarks apply to y_1, x_2, and so on.

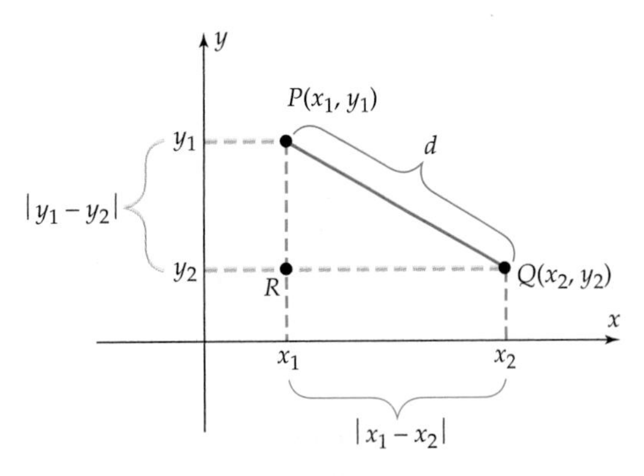

Figure 1–14

As shown in Figure 1–14, the length of RQ is the same as the distance from x_1 to x_2 on the x-axis (number line), namely, $|x_1 - x_2|$. Similarly, the length of PR is the same as the distance from y_1 to y_2 on the y-axis, namely, $|y_1 - y_2|$. According to the Pythagorean Theorem* the length d of PQ is given by:

$$(\text{Length } PQ)^2 = (\text{length } RQ)^2 + (\text{length } PR)^2$$

$$d^2 = |x_1 - x_2|^2 + |y_1 - y_2|^2$$

Since $|c|^2 = |c| \cdot |c| = |c^2| = c^2$ (because $c^2 \geq 0$), this equation becomes:

$$d^2 = (x_1 - x_2)^2 + (y_1 - y_2)^2$$

Since the length d is nonnegative, we must have

$$d = \sqrt{(x_1 - x_2)^2 + (y_1 - y_2)^2} \quad \blacksquare$$

The distance formula can be used to prove the following useful fact (see Exercise 78).

The Midpoint Formula

The midpoint of the line segment from (x_1, y_1) to (x_2, y_2) is

$$\left(\frac{x_1 + x_2}{2}, \frac{y_1 + y_2}{2}\right)$$

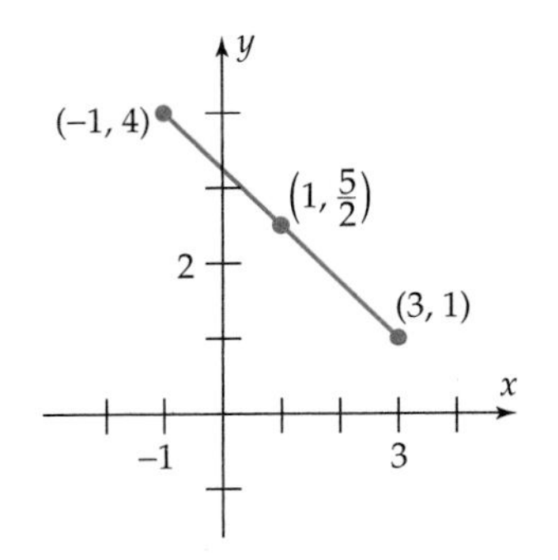

Figure 1–15

Example 4 To find the midpoint of the segment joining $(-1, 4)$ and $(3, 1)$, use the formula in the box with $x_1 = -1$, $y_1 = 4$, $x_2 = 3$, and $y_2 = 1$. The midpoint is

$$\left(\frac{x_1 + x_2}{2}, \frac{y_1 + y_2}{2}\right) = \left(\frac{-1 + 3}{2}, \frac{4 + 1}{2}\right) = \left(1, \frac{5}{2}\right)$$

as shown in Figure 1–15. ■

*See the Geometry Review Appendix.

Graphs

A **graph** is a set of points in the plane. Some graphs are based on data points, such as Figures 1–11 and 1–12. Other graphs arise from equations, as follows. A **solution** of an equation in variables x and y is a pair of numbers such that the substitution of the first number for x and the second for y produces a true statement. For instance, $(3, -2)$ is a solution of $5x + 7y = 1$ because

$$5 \cdot 3 + 7(-2) = 1$$

and $(-2, 3)$ is *not* a solution because $5(-2) + 7 \cdot 3 \neq 1$. The **graph of an equation** in two variables is the set of points in the plane whose coordinates are solutions of the equation. Thus the graph is a *geometric picture of the solutions.*

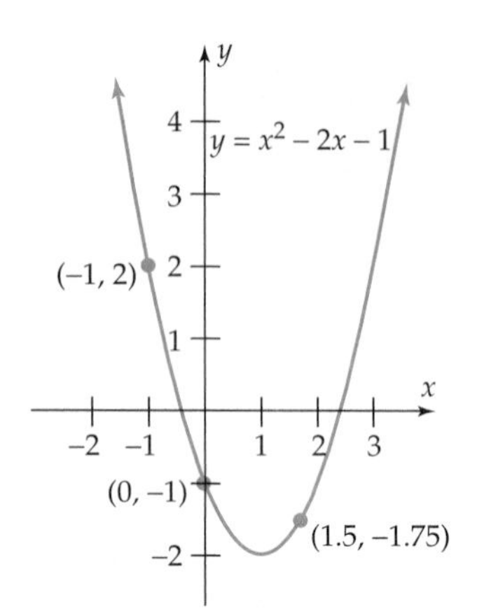

Figure 1–16

Example 5 The graph of $y = x^2 - 2x - 1$ is shown in Figure 1–16. You can readily verify that each of the points whose coordinates are labeled is a solution of the equation. For instance, $(0, -1)$ is a solution because $-1 = 0^2 - 2(0) - 1$. ■

Circles

If (c, d) is a point in the plane and r a positive number, then the **circle with center (c, d) and radius r** consists of all points (x, y) that lie r units from (c, d), as shown in Figure 1–17. According to the distance formula, the statement that "the distance from (x, y) to (c, d) is r units" is equivalent to:

$$\sqrt{(x - c)^2 + (y - d)^2} = r$$

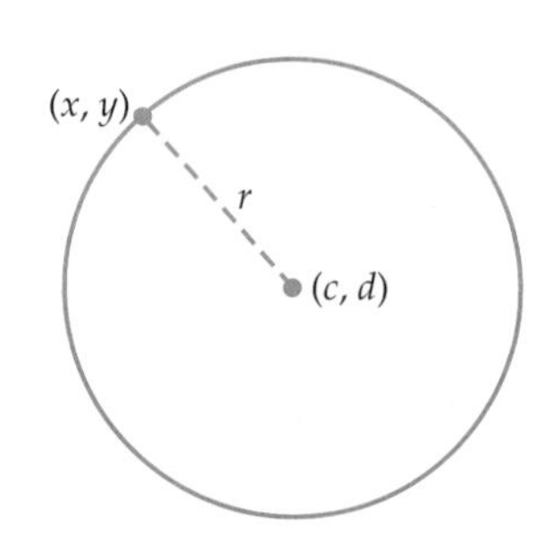

Figure 1–17

Squaring both sides shows that (x, y) satisfies this equation:

$$(x - c)^2 + (y - d)^2 = r^2$$

Reversing the procedure shows that any solution (x, y) of this equation is a point on the circle. Therefore,

Circle Equation

The circle with center (c, d) and radius r is the graph of

$$(x - c)^2 + (y - d)^2 = r^2.$$

We say that $(x - c)^2 + (y - d)^2 = r^2$ is the **equation of the circle** with center (c, d) and radius r. If the center is at the origin, then $(c, d) = (0, 0)$ and the equation has a simpler form:

Circle at the Origin

The circle with center $(0, 0)$ and radius r is the graph of

$$x^2 + y^2 = r^2.$$

Example 6

(a) Letting $r = 1$ shows that the graph of $x^2 + y^2 = 1$ is the circle of radius 1 centered at the origin, as shown in Figure 1–18. This circle is called the **unit circle.**

(b) The circle with center $(-3, 2)$ and radius 2, shown in Figure 1–19, is the graph of the equation

$$((x - (-3))^2 + (y - 2)^2 = 2^2$$

or equivalently,

$$(x + 3)^2 + (y - 2)^2 = 4. \quad \blacksquare$$

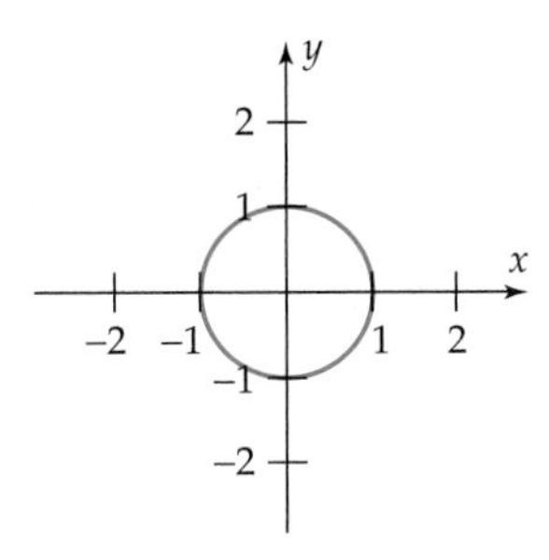

Figure 1–18

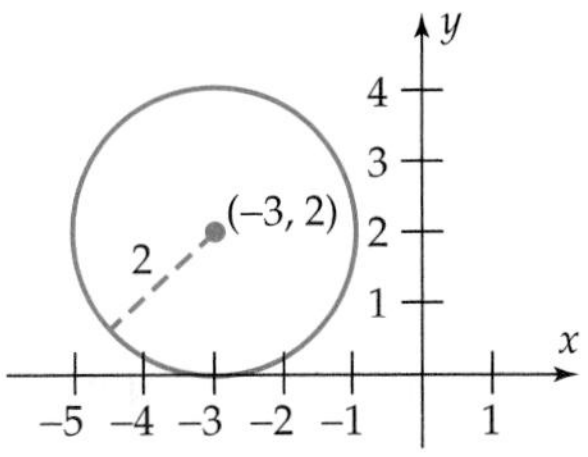

Figure 1–19

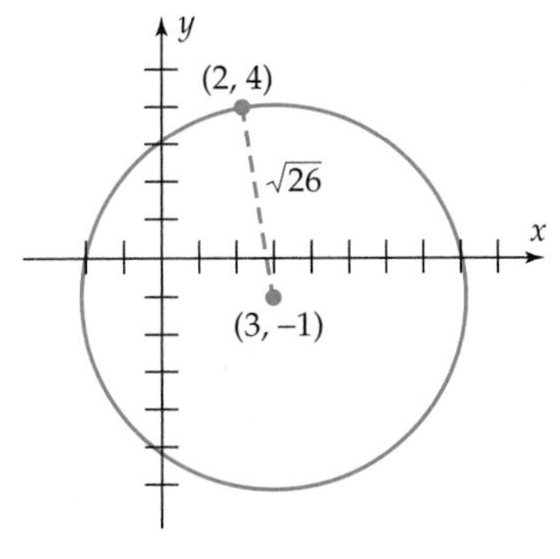

Figure 1–20

Example 7 Find the equation of the circle with center $(3, -1)$ that passes through $(2, 4)$.

Solution We must first find the radius. Since $(2, 4)$ is on the circle, the radius is the distance from $(2, 4)$ to $(3, -1)$ as shown in Figure 1–20, namely,

$$\sqrt{(2 - 3)^2 + (4 - (-1))^2} = \sqrt{1 + 25} = \sqrt{26}$$

The equation of the circle with center at $(3, -1)$ and radius $\sqrt{26}$ is

$$(x - 3)^2 + (y - (-1))^2 = (\sqrt{26})^2$$

$$(x - 3)^2 + (y + 1)^2 = 26$$

$$x^2 - 6x + 9 + y^2 + 2y + 1 = 26$$

$$x^2 + y^2 - 6x + 2y - 16 = 0. \quad \blacksquare$$

The equation of any circle can always be written in the form

$$x^2 + y^2 + Bx + Cy + D = 0$$

for some constants B, C, D, as in Example 7 (where $B = -6$, $C = 2$, $D = -16$). Conversely, the graph of such an equation can always be determined.

Example 8 To find the graph of $3x^2 + 3y^2 - 12x - 30y + 45 = 0$, we divide both sides by 3 and rewrite the equation as

$$(x^2 - 4x) + (y^2 - 10y) = -15.$$

Next we complete the square in both expressions in parentheses (see page 20). To complete the square in $x^2 - 4x$, we add 4 (the square of half the coefficient of x) and to complete the square in $y^2 - 10y$ we add 25 (why?). In order to have an equivalent equation we must add these numbers to *both* sides:

$$(x^2 - 4x + 4) + (y^2 - 10y + 25) = -15 + 4 + 25$$

$$(x - 2)^2 + (y - 5)^2 = 14$$

Since $14 = (\sqrt{14})^2$, this is the equation of the circle with center (2, 5) and radius $\sqrt{14}$. ■

Exercises 1.3

1. Find the coordinates of points *A*–*I*.

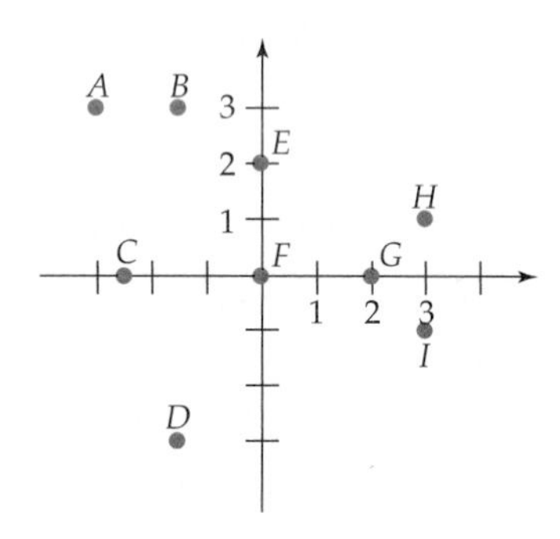

In Exercises 2–5, find the coordinates of the point P.

2. P lies 4 units to the left of the y-axis and 5 units below the x-axis.
3. P lies 3 units above the x-axis and on the same vertical line as $(-6, 7)$.
4. P lies 2 units below the x-axis and its x-coordinate is three times its y-coordinate.
5. P lies 4 units to the right of the y-axis and its y-coordinate is half its x-coordinate.

In Exercises 6–8, sketch a scatter plot and a line graph of the given data. In each case, let the x-axis run from 0 to 10, with x = 0 corresponding to 1990, x = 1 to 1991, and so on.

6. The table shows the number of climbers in recent years at Mt. McKinley (*Source:* National Park Service). Mark the y-axis in units of 100.

Year	1992	1993	1994	1995	1996	1997
Climbers	935	1108	1277	1220	1148	1109

7. The table shows the number of missions by National Park Service rangers to rescue climbers on Mt. McKinley each year (*Source:* National Park Service).

Year	1992	1993	1994	1995	1996	1997
Missions	22	14	20	12	13	10

8. The tuition and fees at public four-year colleges in the fall of each year are shown in the table [*Source:* National Center for Education Statistics (1990–1994) and the College Board (1995–1997)].

Year	Tuition & Fees
1990	$2159
1991	$2410
1992	$2349
1993	$2537

Year	Tuition & Fees
1994	$2681
1995	$2811
1996	$2975
1997	$3111

9. The graph, which is based on data from the U.S. Department of Energy, shows approximate average gasoline prices (in cents per gallon) between 1985 and 1996, with $x = 0$ corresponding to 1985.

(a) Estimate the average price in 1987 and in 1995.

(b) What was the approximate percentage increase in the average price from 1987 to 1995?

(c) In what years was the average price at least \$1.10 per gallon?

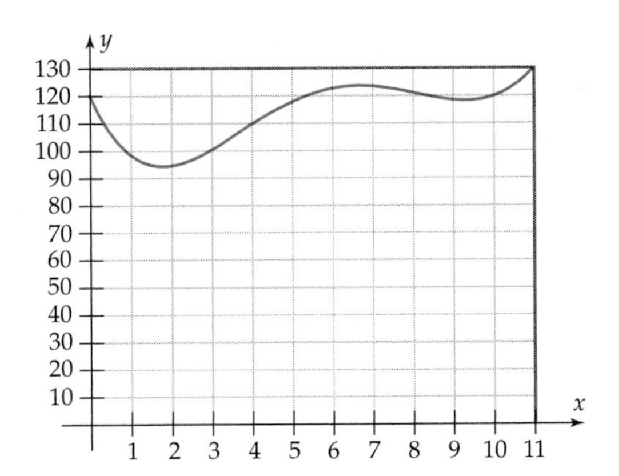

10. The graph, which is based on data from the U.S. Department of Commerce, shows the approximate amount of personal savings as a percent of disposable income between 1960 and 1995, with $x = 0$ corresponding to 1960.

(a) In what years during this period were personal savings largest and smallest (as a percent of disposable income)?

(b) In what years were personal savings at least 7% of disposable income?

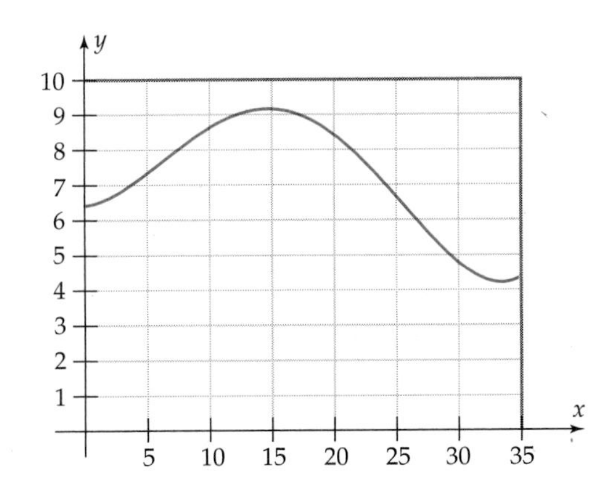

11. (a) If the first coordinate of a point is greater than 3 and its second coordinate is negative, in what quadrant does it lie?

(b) What is the answer in part (a) if the first coordinate is less than 3?

12. In what quadrant(s) does a point lie if the product of its coordinates is

(a) positive? **(b)** negative?

13. (a) Plot the points (3, 2), (4, −1), (−2, 3), and (−5, −4).

(b) Change the sign of the y-coordinate in each of the points in part (a), and plot these new points.

(c) Explain how the points (a, b) and $(a, -b)$ are related graphically. [*Hint:* What are their relative positions with respect to the x-axis?]

14. (a) Plot the points (5, 3), (4, −2), (−1, 4), and (−3, −5).

(b) Change the sign of the x-coordinate in each of the points in part (a), and plot these new points.

(c) Explain how the points (a, b) and $(-a, b)$ are related graphically. [*Hint:* What are their relative positions with respect to the y-axis?]

In Exercises 15–22, find the distance between the two points and the midpoint of the segment joining them.

15. $(-3, 5), (2, -7)$

16. $(2, 4), (1, 5)$

17. $(1, -5), (2, -1)$

18. $(-2, 3), (-3, 2)$

19. $(\sqrt{2}, 1), (\sqrt{3}, 2)$

20. $(-1, \sqrt{5}), (\sqrt{2}, -\sqrt{3})$

21. $(a, b), (b, a)$

22. $(s, t), (0, 0)$

23. According to the Information Technology Industry Council, there were about 12 million personal computers sold in the United States in 1992 and about 36 million in 1998.

(a) Represent the data graphically by two points.

(b) Find the midpoint of the line segment joining these points.

(c) How might this midpoint be interpreted? What assumptions, if any, are needed to make this interpretation?

24. A standard baseball diamond (which is actually a square) is shown in the figure on the next page. Suppose it is placed on a coordinate plane with home plate at the origin, first base on the positive x-axis, and third base on the positive y-axis. The unit of measurement is feet.

(a) Find the coordinates of first, second, and third base.

(b) If the left fielder is at the point (50, 325), how far is he from first base?

(c) How far is the left fielder in part (b) from the right fielder, who is at the point (280, 20)?

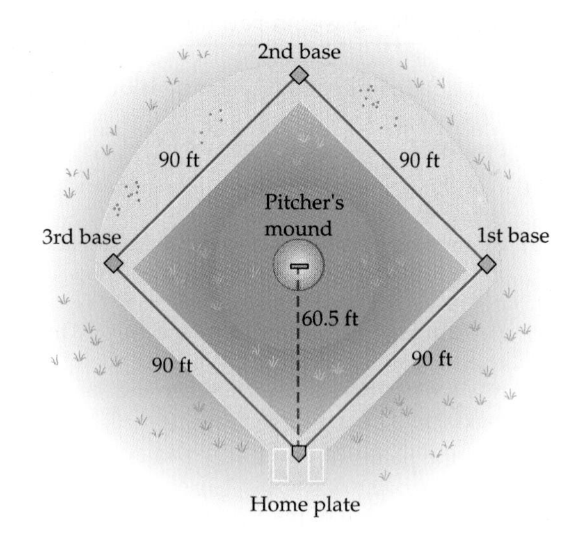

In Exercises 25–30, determine whether the point is on the graph of the given equation.

25. $(1, -2)$; $3x - y - 5 = 0$

26. $(2, -1)$; $x^2 + y^2 - 6x + 8y = -15$

27. $(6, 2)$; $3y + x = 12$

28. $(1, -2)$; $3x + y = 12$

29. $(3, 4)$; $(x - 2)^2 + (y + 5)^2 = 4$

30. $(1, -1)$; $\frac{x^2}{2} + \frac{y^2}{3} = 1$

In Exercises 31–34, find the equation of the circle with given center and radius r.

31. $(-3, 4)$; $r = 2$

32. $(-2, -1)$; $r = 3$

33. $(0, 0)$; $r = \sqrt{2}$

34. $(5, -2)$; $r = 1$

In Exercises 35–38, sketch the graph of the equation.

35. $(x - 2)^2 + (y - 4)^2 = 1$

36. $(x + 1)^2 + (y - 3)^2 = 9$

37. $(x - 5)^2 + (y + 2)^2 = 5$

38. $(x + 6)^2 + y^2 = 4$

In Exercises 39–44, find the center and radius of the circle whose equation is given.

39. $x^2 + y^2 + 8x - 6y - 15 = 0$

40. $15x^2 + 15y^2 = 10$

41. $x^2 + y^2 + 6x - 4y - 15 = 0$

42. $x^2 + y^2 + 10x - 75 = 0$

43. $x^2 + y^2 + 25x + 10y = -12$

44. $3x^2 + 3y^2 + 12x + 12 = 18y$

In Exercises 45–47, show that the three points are the vertices of a right triangle, and state the length of the hypotenuse. [You may assume that a triangle with sides of lengths a, b, c is a right triangle with hypotenuse c provided that $a^2 + b^2 = c^2$.]

45. $(0, 0), (1, 1), (2, -2)$

46. $\left(\frac{\sqrt{2}}{2}, 0\right), \left(\frac{\sqrt{2}}{2}, \frac{\sqrt{2}}{2}\right), (0, 0)$

47. $(3, -2), (0, 4), (-2, 3)$

48. What is the perimeter of the triangle with vertices $(1, 1)$, $(5, 4)$, and $(-2, 5)$?

In Exercises 49–52, determine which of graphs A, B, C best describes the given situation.

49. You have a job that pays a fixed salary for the week. The graph shows your salary.

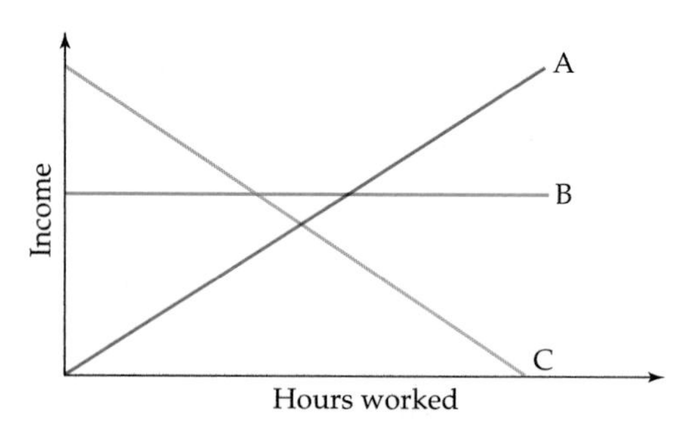

50. You have a job that pays an hourly wage. The graph shows your salary.

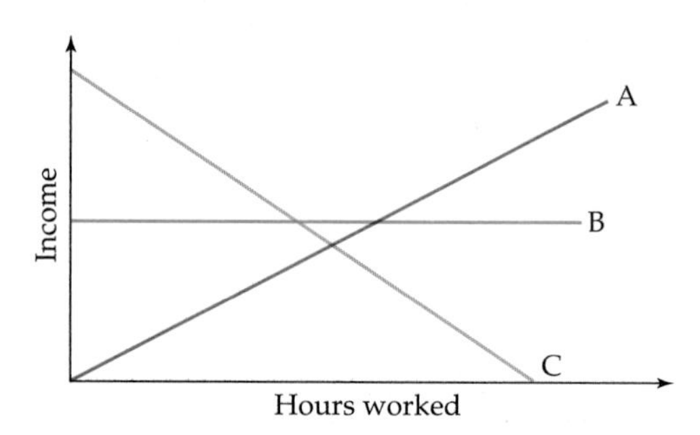

51. You inflate a birthday balloon. The graph shows the amount of air in the balloon.

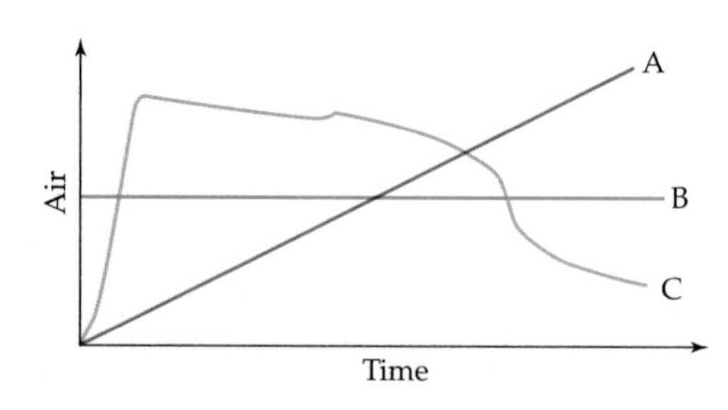

52. Alison's wading pool is filled with a hose by her big sister Emily and she plays in the pool. When they are finished, Emily empties the pool. The graph shows the water level of the pool.

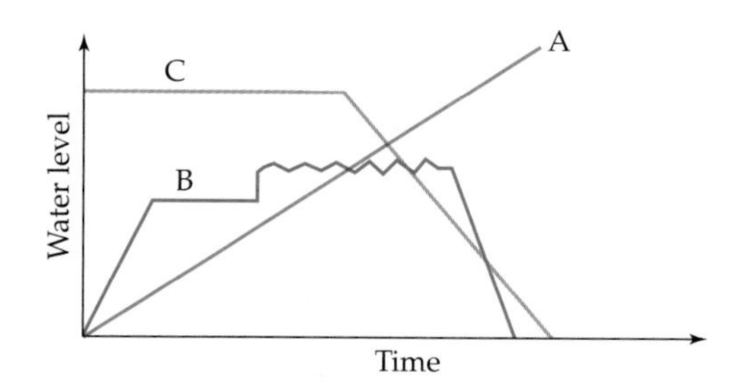

In Exercises 53–60, find the equation of the circle.

53. Center (2, 2); passes through the origin.

54. Center $(-1, -3)$; passes through $(-4, -2)$.

55. Center (1, 2); intersects x-axis at -1 and 3.

56. Center (3, 1); diameter 2.

57. Center $(-5, 4)$; tangent (touching at one point) to the x-axis.

58. Center $(2, -6)$; tangent to the y-axis.

59. Endpoints of diameter are (3, 3) and $(1, -1)$.

60. Endpoints of diameter are $(-3, 5)$ and $(7, -5)$.

61. One diagonal of a square has endpoints $(-3, 1)$ and $(2, -4)$. Find the endpoints of the other diagonal.

62. Find the vertices of all possible squares with this property: Two of the vertices are (2, 1) and (2, 5). [*Hint:* There are three such squares.]

63. Do Exercise 62 with (c, d) and (c, k) in place of (2, 1) and (2, 5).

64. Find the three points that divide the line segment from $(-4, 7)$ to $(10, -9)$ into four parts of equal length.

65. Find all points P on the x-axis that are 5 units from (3, 4). [*Hint: P* must have coordinates $(x, 0)$ for some x and the distance from P to (3, 4) is 5.]

66. Find all points on the y-axis that are 8 units from $(-2, 4)$.

67. Find all points with first coordinate 3 that are 6 units from $(-2, -5)$.

68. Find all points with second coordinate -1 that are 4 units from (2, 3).

69. Find a number x such that (0, 0), (3, 2), and $(x, 0)$ are the vertices of an isosceles triangle, neither of whose two equal sides lie on the x-axis.

70. Do Exercise 69 if one of the two equal sides lies on the positive x-axis.

71. Show that the midpoint M of the hypotenuse of a right triangle is equidistant from the vertices of the triangle. [*Hint:* Place the triangle in the first quadrant of the plane, with right angle at the origin so that the situation looks like the figure.]

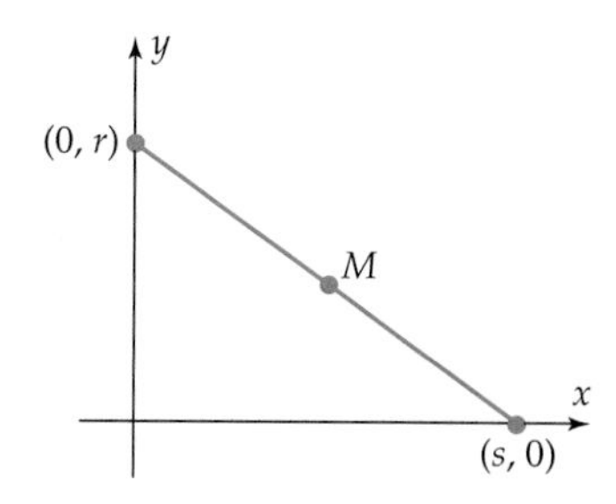

72. Show that the diagonals of a parallelogram bisect each other. [*Hint:* Place the parallelogram in the first quadrant with a vertex at the origin and one side along the x-axis, so that the situation looks like the figure.]

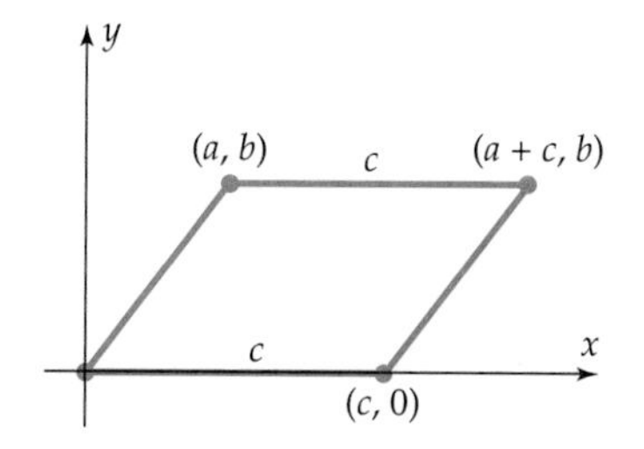

73. Show that the diagonals of a rectangle have the same length. [*Hint:* Place the rectangle in the first quadrant of the plane and label its vertices appropriately, as in Exercises 71–72.]

74. If the diagonals of a parallelogram have the same length, show that the parallelogram is actually a rectangle. [*Hint:* See Exercise 72.]

Thinkers

75. For each nonzero real number k, the graph of $(x - k)^2 + y^2 = k^2$ is a circle. Describe all possible such circles.

76. Suppose every point in the cordinate plane is moved 5 units straight up.

(a) To what point does each of these points go: $(0, -5)$, (2, 2), (5, 0), (5, 5), (4, 1)?

(b) Which points go to each of the points in part (a)?

(c) To what point does (a, b) go?

(d) To what point does $(a, b - 5)$ go?

(e) What point goes to $(-4a, b)$?

(f) What points go to themselves?

77. Let (c, d) be any point in the plane with $c \neq 0$. Prove that (c, d) and $(-c, -d)$ lie on the same straight line through the origin, on opposite sides of the origin, the same distance from the origin. [*Hint:* Find the midpoint of the line segment joining (c, d) and $(-c, -d)$.]

78. *Proof of the Midpoint Formula* Let P and Q be the points (x_1, y_1) and (x_2, y_2) respectively and let M be the point with coordinates

$$\left(\frac{x_1 + x_2}{2}, \frac{y_1 + y_2}{2}\right).$$

Use the distance formula to compute the following:

(a) The distance d from P to Q;
(b) The distance d_1 from M to P;
(c) The distance d_2 from M to Q.
(d) Verify that $d_1 = d_2$.
(e) Show that $d_1 + d_2 = d$. [*Hint:* Verify that $d_1 = \frac{1}{2}d$ and $d_2 = \frac{1}{2}d$.]
(f) Explain why parts (d) and (e) show that M is the midpoint of PQ.

1.4 Lines

When you move from a point P to a point Q on a line,* two numbers are involved, as illustrated in Figure 1–21:

(i) The vertical distance you move (the **change in** y);
(ii) The horizontal distance you move (the **change in** x).

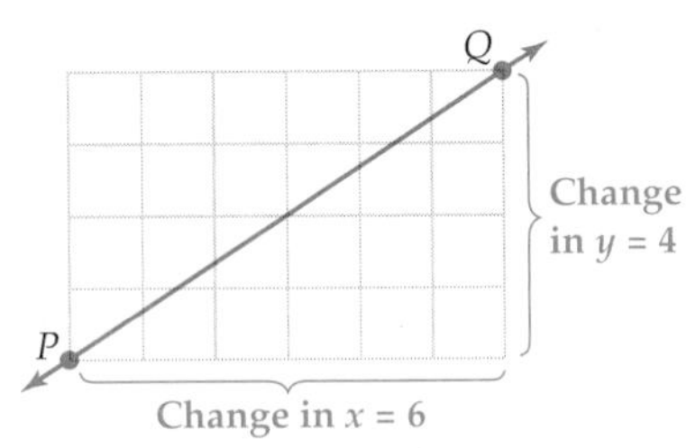

(a)

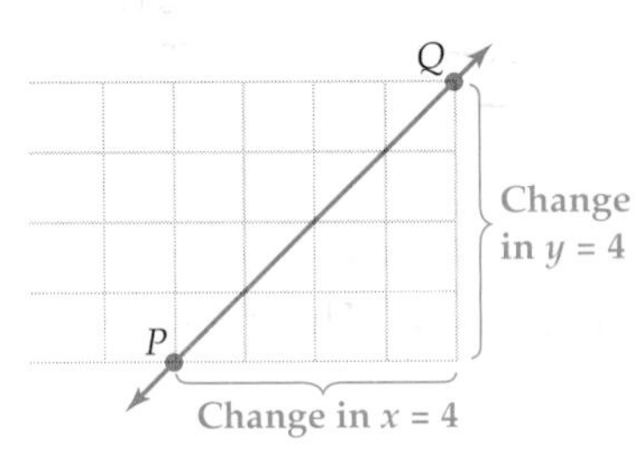

(b)

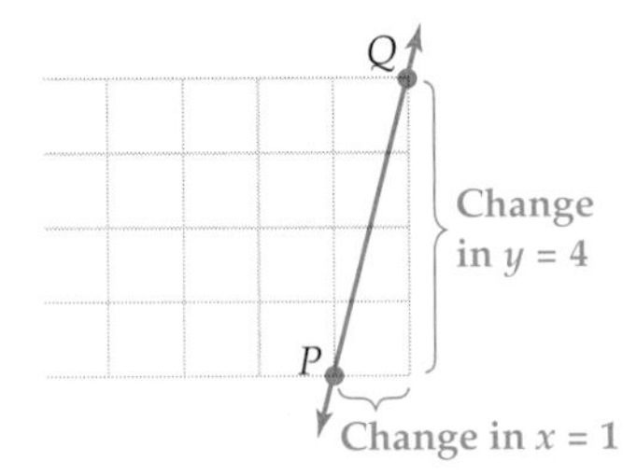

(c)

Figure 1–21

The number $\dfrac{\text{change in } y}{\text{change in } x}$ measures the steepness of the line: the steeper the line, the larger the number. In Figure 1–21, the grid allowed us to measure the change in y and the change in x. When the coordinates of P and Q are given, then

The change in y is the difference of the y-coordinates of P and Q;

The change in x is the difference of the x-coordinates of P and Q;

as shown in Figure 1–22. Consequently, we have the following definition.

Figure 1–22

Slope of a Line

If (x_1, y_1) and (x_2, y_2) are points with $x_1 \neq x_2$, then the *slope* of the line through these points is the number

$$\frac{\textbf{change in } y}{\textbf{change in } x} = \frac{y_2 - y_1}{x_2 - x_1}.$$

*In this section, "line" means "straight line" and movement is from left to right.

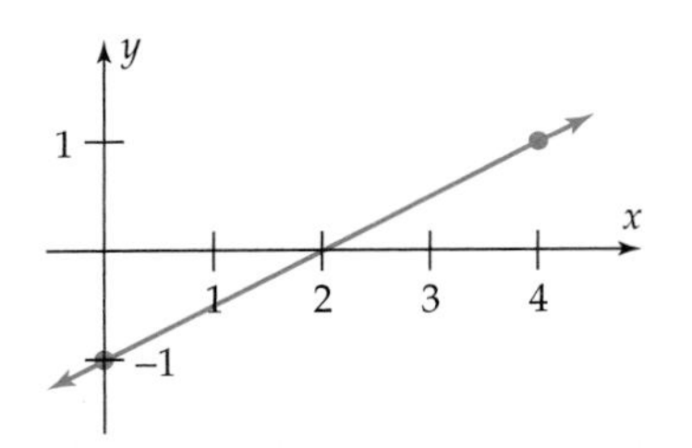

Figure 1–23

Example 1 To find the slope of the line through (0, −1) and (4, 1) (see Figure 1–23), we apply the formula in the previous box with $x_1 = 0$, $y_1 = -1$ and $x_2 = 4$, $y_2 = 1$:

$$\text{Slope} = \frac{y_2 - y_1}{x_2 - x_1} = \frac{1 - (-1)}{4 - 0} = \frac{2}{4} = \frac{1}{2}.$$

The order of the points makes no difference; if you use (4, 1) for (x_1, y_1) and (0, −1) for (x_2, y_2) you obtain the same number:

$$\text{Slope} = \frac{y_2 - y_1}{x_2 - x_1} = \frac{-1 - 1}{0 - 4} = \frac{-2}{-4} = \frac{1}{2}.$$ ■

CAUTION

When finding slopes, you must subtract the *y*-coordinates and *x*-coordiantes in the same order. With the points (3, 4) and (1, 8), for instance, if you use 8 − 4 in the numerator, you must use 1 − 3 in the denominator (*not* 3 − 1).

Example 2 The lines shown in Figure 1–24 are determined by these points:

L_1: (−1, −1) and (0, 2) L_2: (0, 2) and (2, 4) L_3: (−6, 2) and (3, 2)

L_4: (−3, 5) and (3, −1) L_5: (1, 0) and (2, −2).

Their slopes are as follows.

$$L_1: \frac{2 - (-1)}{0 - (-1)} = \frac{3}{1} = 3$$

$$L_2: \frac{4 - 2}{2 - 0} = \frac{2}{2} = 1$$

$$L_3: \frac{2 - 2}{3 - (-6)} = \frac{0}{9} = 0$$

$$L_4: \frac{-1 - 5}{3 - (-3)} = \frac{-6}{6} = -1$$

$$L_5: \frac{-2 - 0}{2 - 1} = \frac{-2}{1} = -2$$

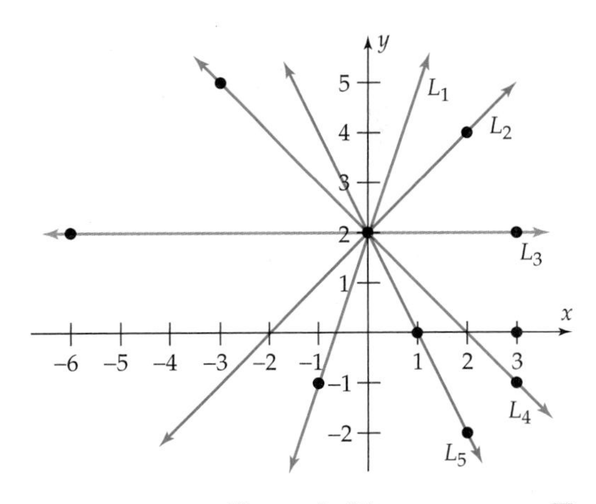

Figure 1–24 ■

Example 2 illustrates how the slope measures the steepness of the line, as summarized below:

Properties of Slope

The slope of a nonvertical line is a number m that measures how steeply the line rises or falls.

If $m > 0$, the line rises from left to right; the larger m is, the more steeply the line rises. *[Lines L_1 and L_2 in Example 2]*

If $m = 0$, the line is horizontal. *[Line L_3]*

If $m < 0$, the line falls from left to right; the larger $|m|$ is, the more steeply the line falls. *[Lines L_4 and L_5]*

Slope-Intercept Form

A nonvertical line intersects the y-axis at a point with coordinates $(0, b)$ for some number b (because every point on the y-axis has first coordinate 0). The number b is called the **y-intercept** of the line. For example, the line in Figure 1–23 has y-intercept -1 because it crosses the y-axis at $(0, -1)$.

Let L be a nonvertical line with slope m and y-intercept b. Then $(0, b)$ is a point on L. Let (x, y) be any other point on L. Using the points $(0, b)$ and (x, y) to compute the slope of L (see Figure 1–25), we have:

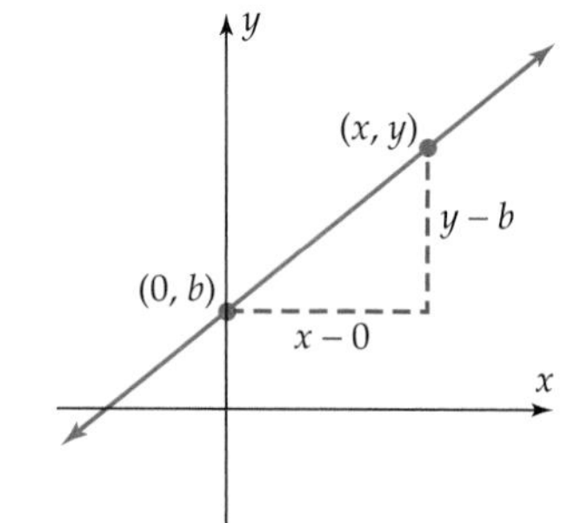

Figure 1–25

$$\text{Slope of } L = \frac{y - b}{x - 0}.$$

Since the slope of L is m, this equation becomes

$$m = \frac{y - b}{x}$$

$$mx = y - b$$

$$y = mx + b$$

Thus, the coordinates of any point on L satisfy the equation $y = mx + b$. So we have this fact:

Slope-Intercept Form

The line with slope m and y-intercept b is the graph of the equation

$$\boldsymbol{y = mx + b.}$$

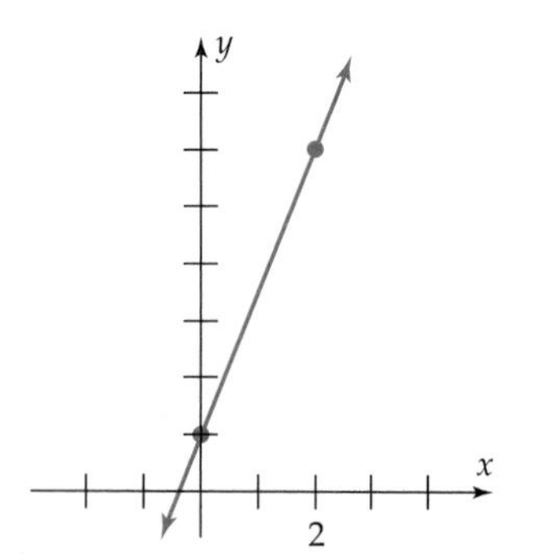

Figure 1–26

Example 3 Graph the equation $2y - 5x = 2$.

Solution We begin by solving the equation for y:

$$2y = 5x + 2$$

$$y = 2.5x + 1.$$

Therefore, its graph is the line with slope 2.5 (coefficient of x) and y-intercept 1 (constant term), which means that $(0, 1)$ is on the line. When $x = 2$, then $y = 2.5(2) + 1 = 6$, so $(2, 6)$ is also on the line. Plotting the line through these points produces Figure 1–26. ■

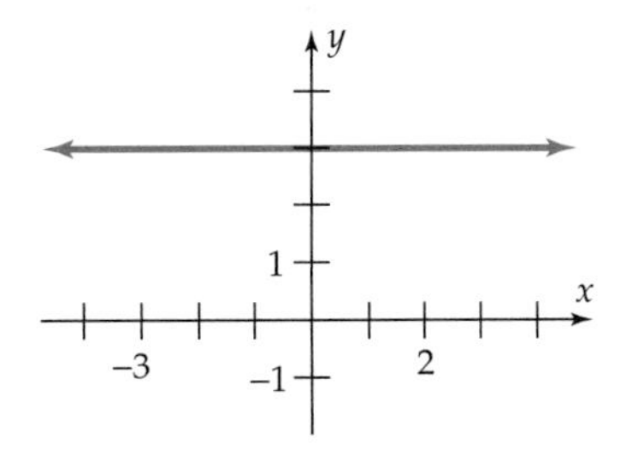

Figure 1–27

Example 4 Describe and sketch the graph of the equation $y = 3$.

Solution We can write $y = 3$ as $y = 0x + 3$. So its graph is a line with slope 0, which means the line is horizontal, and y-intercept 3, which means the line crosses the y-axis at 3. This is sufficient information to obtain the graph in Figure 1–27. ■

Example 5 An office buys a new computer system for $7000. Five years later its value is $800. Assuming linear depreciation, what was its value two years after it was purchased?

Solution Linear depreciation means that the equation which gives the value y of the system in year x is linear, and hence of the form $y = mx + b$ for some constants m and b. Since the system is worth $7000 new ($x = 0$), we have

$$y = mx + b$$
$$7000 = m \cdot 0 + b$$
$$b = 7000$$

so that the equation is $y = mx + 7000$. Since the system is worth $800 after 5 years (i.e., $y = 800$ when $x = 5$), we have

$$y = mx + 7000$$
$$800 = m \cdot 5 + 7000$$
$$-6200 = 5m$$
$$m = \frac{-6200}{5} = -1240.$$

The depreciation equation is therefore $y = -1240x + 7000$. The value of the system after two years ($x = 2$) is

$$y = -1240(2) + 7000 = \$4520. \quad ■$$

Example 6 A factory that makes can openers has fixed costs (for building, fixtures, machinery, etc.) of $26,000. The variable cost (materials and labor) for making one can opener is $2.75.

(a) What is the total cost of making 1000 can openers? 20,000? 40,000?
(b) What is the average cost per can opener in each case?

Solution

(a) Since each can opener costs $2.75, the variable costs for making x can openers is $2.75x$. The total cost y of making x can openers is

$$y = \text{variable costs} + \text{fixed costs} = 2.75x + 26{,}000.$$

The cost of making 1000 can openers is

$$y = 2.75x + 26{,}000 = 2.75(1000) + 26{,}000 = \$28{,}750.$$

Similarly, the cost of making 20,000 can openers is

$$y = 2.75(20{,}000) + 26{,}000 = \$81{,}000$$

and the cost of 40,000 is

$$y = 2.75(40{,}000) + 26{,}000 = \$136{,}000.$$

(b) The average cost per can opener in each case is the total cost divided by the number of can openers. So the average cost per can opener is:

For 1000:	\$28,750/1000 = \$28.75 per can opener;
For 20,000:	\$81,000/20,000 = \$4.05 per can opener;
For 40,000:	\$136,000/40,000 = \$3.40 per can opener. ■

We have seen that the geometrical interpretation of slope is that it measures the "steepness" of the line. Examples 5 and 6 show that slope can also be interpreted as a *rate of change.* In Example 5, the computer depreciates \$1240 per year, meaning that its value changes at a rate of -1245 per year and -1245 is the slope of the depreciation equation

$$y = -1245x + 7000.$$

In Example 6, the total cost of making can openers increases at a rate of \$2.75 per can opener, and 2.75 is the slope of the cost equation $y = 2.75x + 26{,}000$. Rates of change will be considered further in Section 3.6.

The Point-Slope Form

Suppose the line L passes through the point (x_1, y_1) and has slope m. Let (x, y) be any other point on L. Using the points (x_1, y_1) and (x, y) to compute the slope m of L (see Figure 1–28), we have

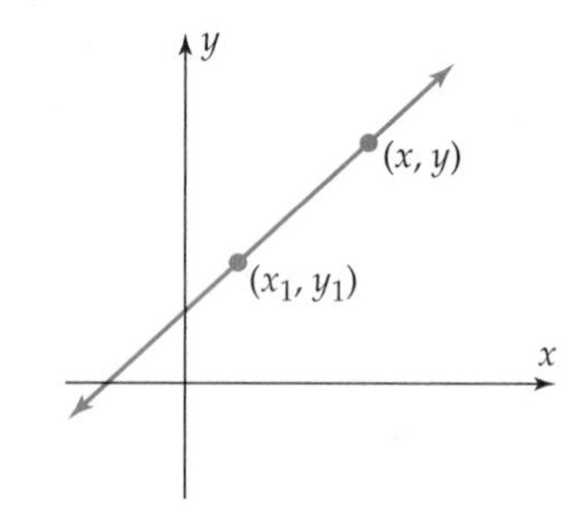

Figure 1–28

$$\frac{y - y_1}{x - x_1} = \text{slope of } L$$

$$\frac{y - y_1}{x - x_1} = m$$

$$y - y_1 = m(x - x_1)$$

Thus, the coordinates of every point on L satisfy the equation $y - y_1 = m(x - x_1)$ and we have this fact:

Point-Slope Form

The line with slope m through the point (x_1, y_1) is the graph of the equation

$$y - y_1 = m(x - x_1).$$

Example 7 Find the equation of the line with slope 2 through the point (1, −6).

Solution Substitute 2 for m and (1, −6) for (x_1, y_1) in the point-slope equation:

$$
\begin{aligned}
y - y_1 &= m(x - x_1) \\
y - (-6) &= 2(x - 1) && \textit{[Point-Slope Form]} \\
y + 6 &= 2x - 2 \\
y &= 2x - 8. && \textit{[Slope-Intercept Form]} \quad \blacksquare
\end{aligned}
$$

Example 8 According to the National Center for Health Statistics, the U.S. infant mortality rate declined at an approximately linear rate from 10 per 1000 live births in 1988 to 8 per 1000 in 1994.

(a) Find an equation that gives the mortality rate y in year x.

(b) Use this equation to estimate the mortality rate in 1991 and 1997.

Solution

(a) Let $x = 0$ correspond to 1988, so that $x = 6$ corresponds to 1994. Then the given information can be represented by the points (0, 10) and (6, 8). We must find the equation of the line through these points. Its slope is

$$\frac{8 - 10}{6 - 0} = \frac{-2}{6} = -\frac{1}{3}.$$

Now we use the slope $-1/3$ and one of the points (0, 10) or (6, 8) to find the equation of the line. It doesn't matter which point, since both lead to the same equation.

$$
\begin{aligned}
y - y_1 &= m(x - x_1) & \qquad y - y_1 &= m(x - x_1) \\
y - 10 &= -\frac{1}{3}(x - 0) & \qquad y - 8 &= -\frac{1}{3}(x - 6) \\
y &= -\frac{1}{3}x + 10 & \qquad y - 8 &= -\frac{1}{3}x + 2 \\
& & \qquad y &= -\frac{1}{3}x + 10
\end{aligned}
$$

(b) Since 1991 corresponds to $x = 3$, the approximate mortality rate in 1993 was

$$y = -\frac{1}{3}x + 10 = -\frac{1}{3} \cdot 3 + 10 = -1 + 10 = 9 \text{ per 1000 births.}$$

and the approximate mortality rate in 1997 ($x = 10$) was

$$y = -\frac{1}{3}x + 10 = -\frac{1}{3} \cdot 10 + 10 = -\frac{10}{3} + 10 = \frac{20}{3} \approx 6.67 \text{ per 1000 births.} \quad \blacksquare$$

Parallel and Perpendicular Lines

The slope of a line measures how steeply it rises or falls. Since parallel lines rise or fall equally steeply, the following fact should be plausible (see Exercises 66–67 for a proof).

Parallel Lines

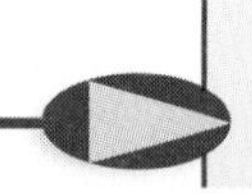

Two nonvertical lines are parallel exactly when they have the same slope.

Example 9 Find the equation of the line L through $(2, -1)$ that is parallel to the line M whose equation is $3x - 2y + 6 = 0$.

Solution First find the slope of M by rewriting its equation in slope-intercept form:

$$\begin{aligned} 3x - 2y + 6 &= 0 \\ -2y &= -3x - 6 \\ y &= \frac{3}{2}x + 3. \end{aligned}$$

Therefore, M has slope $3/2$. The parallel line L must have the same slope, $3/2$. Since $(2, -1)$ is on L, we can use the point-slope form to find its equation:

$$\begin{aligned} y - y_1 &= m(x - x_1) \\ y - (-1) &= \frac{3}{2}(x - 2) && \text{[Point-Slope Form]} \\ y + 1 &= \frac{3}{2}x - 3 \\ y &= \frac{3}{2}x - 4 && \text{[Slope-Intercept Form]} \end{aligned}$$ ■

Two lines that meet in a right angle (90° angle) are said to be **perpendicular.** As you might suspect, there is a close relationship between the slopes of two perpendicular lines.

Perpendicular Lines

Two nonvertical lines are perpendicular exactly when the product of their slopes is -1.

A proof of this fact is outlined in Exercise 68.

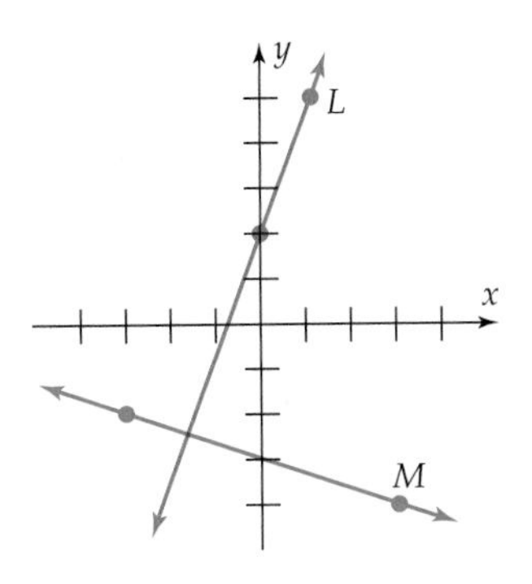

Figure 1–29

Example 10 In Figure 1–29 the line L through (0, 2) and (1, 5) appears to be perpendicular to the line M through (−3, −2) and (3, −4). We can confirm this fact by computing the slopes of these lines:

$$\text{Slope } L = \frac{5-2}{1-0} = 3 \qquad \text{and} \qquad \text{slope } M = \frac{-4-(-2)}{3-(-3)} = \frac{-2}{6} = -\frac{1}{3}.$$

Since $3(-1/3) = -1$, the lines L and M are perpendicular. ■

Example 11 Find the equation of the perpendicular bisector of the line segment with endpoints (−5, −4) and (7, 2).

Solution The perpendicular bisector M goes through the midpoint of the line segment from (−5, −4) and (7, 2). The Midpoint Formula (page 31) shows that this midpoint is

$$\left(\frac{x_1 + x_2}{2}, \frac{y_1 + y_2}{2}\right) = \left(\frac{-5+7}{2}, \frac{-4+2}{2}\right) = (1, -1).$$

The line L through (−5, −4) and (7, 2) has slope

$$\frac{y_2 - y_1}{x_2 - x_1} = \frac{2-(-4)}{7-(-5)} = \frac{6}{12} = \frac{1}{2}.$$

Since M is perpendicular to L, we have (slope M)(slope L) = −1, so that

$$\text{Slope } M = \frac{-1}{\text{slope } L} = \frac{-1}{1/2} = -2.$$

Thus, M is the line through (1, −1) with slope −2, and its equation is:

$$y - (-1) = -2(x - 1) \qquad \textit{[Point-Slope Form]}$$

$$y = -2x + 1. \qquad \textit{[Slope-Intercept Form]}$$ ■

Vertical Lines

The preceding discussion does not apply to vertical lines, whose equations have a different form from those examined earlier.

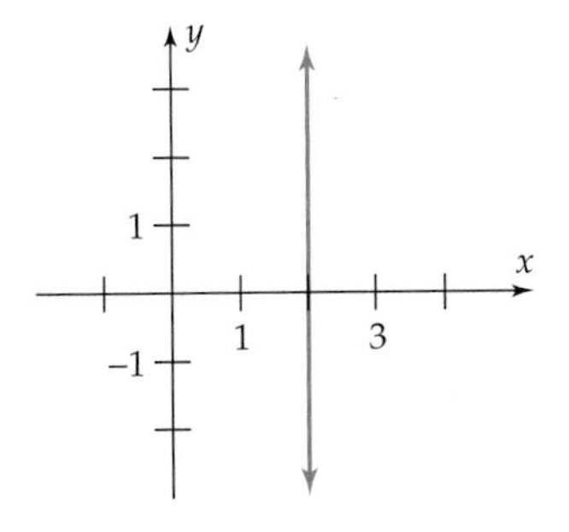

Figure 1–30

Example 12 Every point on the vertical line in Figure 1–30 has first coordinate 2. Thus, every point on the line satisfies $x + 0y = 2$. Thus, the line is the graph of the equation $x = 2$. If you try to compute the slope of this line, say using (2, 1) and (2, 4), you obtain $\frac{4-1}{2-2} = \frac{4}{0}$, which is not defined. ■

Examples 3–12 illustrate the following facts (where A, B, C, b, and c are constants, with at least one of A or B nonzero).

Equations and Lines

The graph of the equation $Ax + By = C$ is a straight line.

A horizontal line has slope 0 and an equation of the form $y = b$.

A vertical line has undefined slope and an equation of the form $x = c$.

Exercises 1.4

1. For which of the line segments in the figure is the slope
   (a) largest? (b) smallest?
   (c) largest in absolute value? (d) closest to zero?

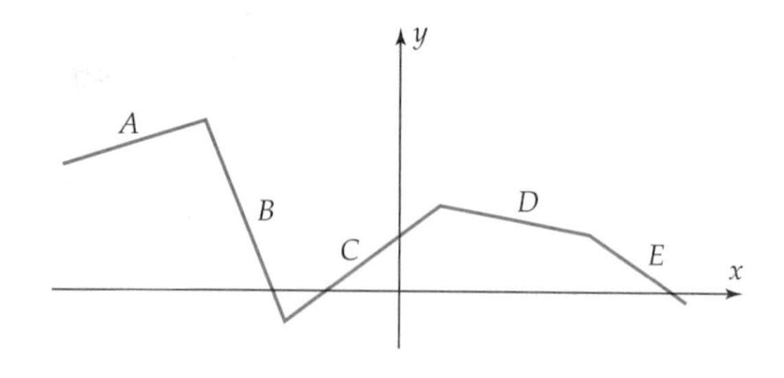

2. The doorsill of a campus building is 5 ft above ground level. To allow wheelchair access, the steps in front of the door are to be replaced by a straight ramp with constant slope 1/12, as shown in the figure. How long must the ramp be? [The answer is *not* 60 ft.]

In Exercises 3–6, find the slope and y-intercept of the line whose equation is given.

3. $2x - y + 5 = 0$
4. $3x + 4y = 7$
5. $3(x - 2) + y = 7 - 6(y + 4)$
6. $2(y - 3) + (x - 6) = 4(x + 1) - 2$

In Exercises 7–10, find the slope of the line through the given points.

7. $(1, 2); (3, 7)$
8. $(-1, -2); (2, -1)$
9. $(1/4, 0); (3/4, 2)$
10. $(\sqrt{2}, -1); (2, -9)$

In Exercises 11–14, find a number t such that the line passing through the two given points has slope −2.

11. $(0, t); (9, 4)$
12. $(1, t); (-3, 5)$
13. $(t + 1, 5); (6, -3t + 7)$
14. $(t, t); (5, 9)$

15. Let L be a nonvertical straight line through the origin. L intersects the vertical line through $(1, 0)$ at a point P. Show that the second coordinate of P is the slope of L.
16. On one graph, sketch five line segments, not all meeting at a single point, whose slopes are five different positive numbers. Do this in such a way that the left-hand line has the largest slope, the second line from the left the next largest slope, and so on.

In Exercises 17–20, find the equation of the line with slope m that passes through the given point.

17. $m = 1; (3, 5)$
18. $m = 2; (-2, 1)$
19. $m = -1; (6, 2)$
20. $m = 0; (-4, -5)$

In Exercises 21–24, find the equation of the line through the given points.

21. $(0, -5)$ and $(-3, -2)$
22. $(4, 3)$ and $(2, -1)$
23. $(4/3, 2/3)$ and $(1/3, 3)$
24. $(6, 7)$ and $(6, 15)$

In Exercises 25–28, determine whether the line through P and Q is parallel or perpendicular to the line through R and S, or neither.

25. $P = (2, 5)$, $Q = (-1, -1)$ and $R = (4, 2)$, $S = (6, 1)$

26. $P = (0, 3/2)$, $Q = (1, 1)$ and $R = (2, 7)$, $S = (3, 9)$

27. $P = (-3, 1/3)$, $Q = (1, -1)$ and $R = (2, 0)$, $S = (4, -2/3)$

28. $P = (3, 3)$, $Q = (-3, -1)$ and $R = (2, -2)$, $S = (4, -5)$

In Exercises 29–31, determine whether the lines whose equations are given are parallel, perpendicular, or neither.

29. $2x + y - 2 = 0$ and $4x + 2y + 18 = 0$

30. $3x + y - 3 = 0$ and $6x + 2y + 17 = 0$

31. $y = 2x + 4$ and $.5x + y = -3$

32. Use slopes to show that the points $(-4, 6)$, $(-1, 12)$, and $(-7, 0)$ all lie on the same straight line.

33. Use slopes to determine if $(9, 6)$, $(-1, 2)$, and $(1, -3)$ are the vertices of a right triangle.

34. Use slopes to show that the points $(-5, -2)$ $(-3, 1)$, $(3, 0)$, and $(5, 3)$ are the vertices of a parallelogram.

In Exercises 35–42, find an equation for the line satisfying the given conditions.

35. Through $(-2, 1)$ with slope 3.

36. y-intercept -7 and slope 1.

37. Through $(2, 3)$ and parallel to $3x - 2y = 5$.

38. Through $(1, -2)$ and perpendicular to $y = 2x - 3$.

39. x-intercept 5 and y-intercept -5.

40. Through $(-5, 2)$ and parallel to the line through $(1, 2)$ and $(4, 3)$.

41. Through $(-1, 3)$ and perpendicular to the line through $(0, 1)$ and $(2, 3)$.

42. y-intercept 3 and perpendicular to $2x - y + 6 = 0$.

43. Find a real number k such that $(3, -2)$ is on the line $kx - 2y + 7 = 0$.

44. Find a real number k such that the line $3x - ky + 2 = 0$ has y-intercept -3.

If P is a point on a circle with center C, then the tangent line to the circle at P is the straight line through P that is perpendicular to the radius CP. In Exercises 45–48, find the equation of the tangent line to the circle at the given point.

45. $x^2 + y^2 = 25$ at $(3, 4)$ [*Hint:* Here C is $(0, 0)$ and P is $(3, 4)$; what is the slope of radius CP?]

46. $x^2 + y^2 = 169$ at $(-5, 12)$

47. $(x - 1)^2 + (y - 3)^2 = 5$ at $(2, 5)$

48. $x^2 + y^2 + 6x - 8y + 15 = 0$ at $(-2, 1)$

49. Let A, B, C, D be nonzero real numbers. Show that the lines $Ax + By + C = 0$ and $Ax + By + D = 0$ are parallel.

50. Let L be a line that is neither vertical nor horizontal and which does not pass through the origin. Show that L is the graph of $\frac{x}{a} + \frac{y}{b} = 1$, where a is the x-intercept and b is the y-intercept of L.

51. Sales of a software company increased linearly from \$120,000 in 1996 to \$180,000 in 1999.
- **(a)** Find an equation that expresses the sales y in year x (where $x = 0$ corresponds to 1996).
- **(b)** Estimate the sales in 2001.

52. Carbon dioxide (CO_2) concentration is measured regularly at the Mauna Loa observatory in Hawaii. The mean annual concentration in parts per million in various years is given in the table.

Year	Concentration (ppm)
1965	319.9
1970	325.3
1980	338.5
1988	351.3

- **(a)** Do the data points lie on a single line? How do you know?
- **(b)** Write a linear equation that uses the data from 1965 and 1988 to model CO_2 concentration over time. What does this model say the concentration should have been in 1970? Does the model overestimate or underestimate the concentration?
- **(c)** Write a linear equation that uses the data from 1980 and 1988 to model CO_2 concentration over time. What does this model say the concentration should have been in 1965? Does the model overestimate or underestimate the concentration?
- **(d)** What do the two models say about the concentration in the year 1995? Which model do you think is the most accurate? Why?

53. According to the Bureau of Debt of the U.S. Department of the Treasury, the national debt was about 2125 billion dollars in 1986 and about 5225 billion dollars in 1996.

(a) Find a linear equation that approximates the national debt y (in billions of dollars) in year x (with $x = 0$ corresponding to 1986).

(b) Use the equation of part (a) to estimate the national debt in 1991 and 2000. [For comparison purposes, the actual national debt in 1991 was 3665.3 billion dollars.]

54. At sea level, water boils at 212°F. At a height of 1100 ft, water boils at 210°F. The relationship between boiling point and height is linear.

(a) Find an equation that gives the boiling point y of water at a height of x feet.

Find the boiling point of water in each of the following cities (whose altitudes are given).

(b) Cincinnati, OH (550 ft)

(c) Springfield, MO (1300 ft)

(d) Billings, MT (3120 ft)

(e) Flagstaff, AZ (6900 ft)

55. A small plane costs \$600,000 new. Ten years later, it is valued at \$150,000. Assuming linear depreciation, find the value of the plane when it is 5 years old and when it is 12 years old.

56. In 1950, the death rate from heart disease was about 511 per 100,000 people. In 1996, the rate had decreased to 359 per 100,000.

(a) Assuming the rate decreased linearly, find an equation that gives the number y of deaths per 100,000 from heart disease in year x, with $x = 0$ corresponding to 1950. Round the slope of the line to one decimal place.

(b) Use the equation in part (a) to estimate the death rate in 1980 and 2000.

57. According to the National Center for Health Statistics of the U.S. Department of Health and Human Services, total health care expenditures were approximately 698 billion dollars in 1990 and 989 billion dollars in 1995. The growth in health care costs was approximately linear.

(a) Find an equation that gives the approximate health care costs y in year x (with $x = 0$ corresponding to 1990).

(b) Use the equation in part (a) to estimate the health care costs in 1993 and 1999.

58. The profit p (in thousands of dollars) on x thousand units of a specialty item is $p = .6x - 14.5$. The cost c of manufacturing x items is given by $c = .8x + 14.5$.

(a) Find an equation that gives the revenue r from selling x items.

(b) How many items must be sold for the company to break even (i.e., for revenue to equal cost)?

59. A publisher has fixed costs of \$180,000 for a mathematics text. The variable costs are \$25 per book. The book sells for \$40. Find equations that give

(a) The cost c of making x books

(b) The revenue r from selling x books

(c) The profit p from selling x books

(d) What is the publisher's break-even point (see Exercise 58(b))?

60. If the fixed costs of a manufacturer are \$1000 and it costs \$2000 to produce 40 items, find a linear equation that gives the total cost of making x items.

61. The Whismo Hat Company has fixed costs of \$50,000 and variable costs of \$8.50 per hat.

(a) Find an equation that gives the total cost y of producing x hats.

(b) What is the average cost per hat when 20,000 are made? 50,000? 100,000?

Use the graph and the following information for Exercises 62–64. Rocky is an "independent" ticket dealer who markets choice tickets for Los Angeles Lakers home games (California currently has no laws against scalping). Each graph shows how many tickets will be demanded by buyers at a particular price. For instance, when the Lakers play the Chicago Bulls, the graph shows that at a price of \$160, no tickets are demanded. As the price (y-coordinate) gets lower, the number of tickets demanded (x-coordinate) increases.

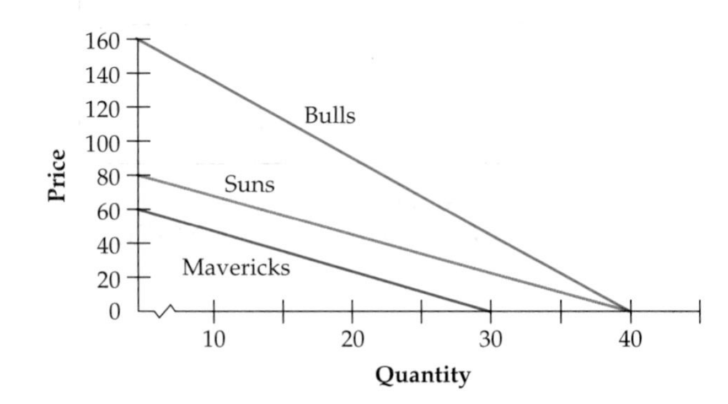

62. Write a linear equation that relates the quantity x of tickets demanded at price y when the Lakers play the

(a) Dallas Mavericks

(b) Phoenix Suns

(c) Chicago Bulls

[*Hint:* In each case, use the points where the graph crosses the two axes to determine its slope.]

63. Use the equations from Exercise 62 to find the number of tickets Rocky would sell at a price of \$40 for a game against the
(a) Mavericks **(b)** Bulls

64. Suppose Rocky has 20 tickets to sell. At what price could he sell them all when the Lakers play the
(a) Mavericks **(b)** Suns

65. A dietician has a "rule of thumb" for determining the target weights for her clients, based on their heights. For women, the target weight is 100 lb plus 5 lb for each inch over 5 ft in height. For men it is 106 lb plus 6 lb for each inch over 5 ft in height.
(a) Fill in the following tables.

Woman's Target Weight	
x inches over 5 ft	**Weight (lb)**
0	100
1	
2	
3	
6	
12	
15	

Man's Target Weight	
x inches over 5 ft	**Weight (lb)**
0	106
1	
2	
3	
6	
12	
15	

(b) Write an equation that gives a woman's target weight y in terms of the number x of inches that her height exceeds 5 ft.
(c) What is the target weight for a woman who is 5 ft 9 in tall?
(d) Do parts (b) and (c) for men.

66. Prove that nonvertical parallel lines L and M have the same slope, as follows. Suppose M lies above L and choose two points (x_1, y_1) and (x_2, y_2) on L.
(a) Let P be the point on M with first coordinate x_1. Let b denote the vertical distance from P to (x_1, y_1). Show that the second coordinate of P is $y_1 + b$.
(b) Let Q be the point on M with first coordinate x_2. Use the fact that L and M are parallel to show that the second coordinate of Q is $y_2 + b$.
(c) Compute the slope of L using (x_1, y_1) and (x_2, y_2). Compute the slope of M using the points P and Q. Verify that the two slopes are the same.

67. Show that two nonvertical lines with the same slope are parallel. [*Hint:* The equations of distinct lines with the same slope must be of the form $y = mx + b$ and $y = mx + c$ with $b \neq c$ (why?). If (x_1, y_1) were a point on both lines, its coordinates would satisfy both equations. Show that this leads to a contradiction and conclude that the lines have no point in common.]

68. This exercise provides a proof of the statement about slopes of perpendicular lines in the box on page 44. First, assume that L and M are nonvertical perpendicular lines that both pass through the origin. L and M intersect the vertical line $x = 1$ at the points $(1, k)$ and $(1, m)$ respectively, as shown in the figure.

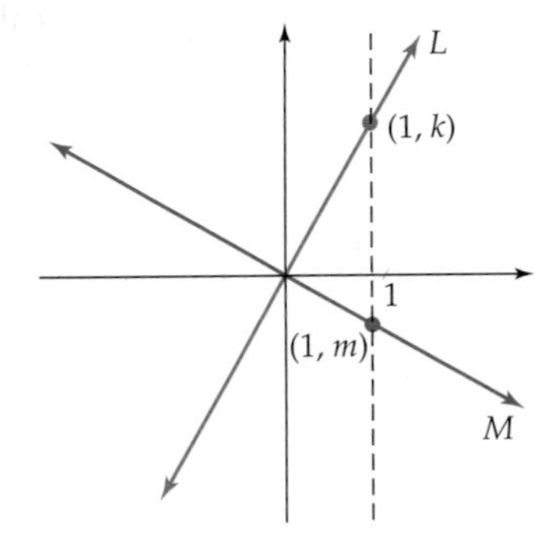

(a) Use $(0, 0)$ and $(1, k)$ to show that L has slope k. Use $(0, 0)$ and $(1, m)$ to show that M has slope m.

(b) Use the distance formula to compute the length of each side of the right triangle with vertices (0, 0), (1, k), and (1, m).

(c) Use part (b) and the Pythagorean Theorem to find an equation involving k, m, and various constants. Show that this equation simplifies to $km = -1$. This proves half of the statement.

(d) To prove the other half, assume that $km = -1$ and show that L and M are perpendicular as follows. You may assume that a triangle whose sides a, b, c satisfy $a^2 + b^2 = c^2$ is a right triangle with hypotenuse c. Use this fact, and do the computation in part (b) in reverse (starting with $km = -1$) to show that the triangle with vertices (0, 0), (1, k), and (1, m) is a right triangle, so that L and M are perpendicular.

(e) Finally, to prove the general case when L and M do not intersect at the origin, let L_1 be a line through the origin that is parallel to L, and M_1 a line through the origin that is parallel to M. Then L and L_1 have the same slope, and M and M_1 have the same slope (why?). Use this fact and parts (a)–(d) to prove that L is perpendicular to M exactly when $km = -1$.

69. Show that the diagonals of a square are perpendicular. [*Hint:* Place the square in the first quadrant of the plane, with one vertex at the origin and sides on the positive axes. Label the coordinates of the vertices appropriately.]

Chapter 1 Review

Important Concepts

Section 1.1	Real numbers, integers, rationals, irrationals	2
	Number line	3
	Distributive law	3
	Scientific notation	3
	Order ($<, \leq, >, \geq$)	4
	Intervals, open intervals, closed intervals	5
	Negatives	6
	Square roots	7
	Absolute value	8
	Distance on the number line	10
Excursion 1.1.A	Repeating and nonrepeating decimals	15
Section 1.2	Basic principles for solving equations	18
	First-degree equations	18
	Quadratic equations	19
	Factoring	19
	Completing the square	20
	Quadratic formula	21
	Discriminant	22–23
	Higher degree equations	23
Excursion 1.2.A	Absolute value equations	26
Section 1.3	Coordinate plane, x-axis, y-axis, quadrants	28
	Scatter plots and line graphs	29
	Distance formula	30
	Midpoint formula	31
	Graph of an equation	32
	Circle, center, radius	33
	Equation of the circle	33
	Unit circle	34
Section 1.4	Change in x, change in y	38
	Slope	38
	Properties of slope	40
	y-intercept	40
	Slope-intercept form of the equation of a line	40
	Fixed and variable costs	41
	Point-slope form of the equation of a line	42
	Slopes of parallel lines	44
	Slopes of perpendicular lines	44
	Equations of horizontal and vertical lines	45–46

Important Facts and Formulas

- $|c - d|$ = distance from c to d on the number line
- *Quadratic Formula:* If $a \neq 0$, then the solutions of $ax^2 + bx + c = 0$ are
$$x = \frac{-b \pm \sqrt{b^2 - 4ac}}{2a}.$$
- If $a \neq 0$, then the number of real solutions of $ax^2 + bx + c = 0$ is 0, 1, or 2, depending on whether the discriminant $b^2 - 4ac$ is negative, zero, or positive.
- *Distance Formula:* The distance from (x_1, y_1) to (x_2, y_2) is
$$\sqrt{(x_1 - x_2)^2 + (y_1 - y_2)^2}.$$

- *Midpoint Formula:* The midpoint of the line segment from (x_1, y_1) to (x_2, y_2) is
$$\left(\frac{x_1 + x_2}{2}, \frac{y_1 + y_2}{2}\right).$$
- Equation of the circle with center (c, d) and radius r:
$$(x - c)^2 + (y - d)^2 = r^2$$
- The slope of the line through (x_1, y_1) and (x_2, y_2) (where $x_1 \neq x_2$) is
$$\frac{y_2 - y_1}{x_2 - x_1}.$$
- Nonvertical parallel lines have the same slope.
- Two lines (neither vertical) are perpendicular exactly when the product of their slopes is -1.
- The equation of the line with slope m and y-intercept b is
$$y = mx + b.$$
- The equation of the line through (x_1, y_1) with slope m is
$$y - y_1 = m(x - x_1).$$

Review Questions

1. Fill the blanks with one of the symbols $<$, $=$, or $>$ so that the resulting statement is true.
(a) 142 ____ $|-51|$ **(b)** $\sqrt{2}$ ____ $|-2|$
(c) -1000 ____ $\frac{1}{10}$ **(d)** $|-2|$ ____ $-|6|$
(e) $|u - v|$ ____ $|v - u|$, where u and v are fixed real numbers.

2. List two real numbers that are *not* rational numbers.

3. Express in scientific notation:
(a) 12,320,000,000,000,000 **(b)** .0000000000789

4. Express in decimal notation:
(a) 4.78×10^8 **(b)** 6.53×10^{-9}

5. Express in symbols:
(a) y is negative, but greater than -10;
(b) x is nonnegative and not greater than 10.

6. Express in symbols:
(a) $c - 7$ is nonnegative; **(b)** .6 is greater than $|5x - 2|$.

7. Express in symbols:
(a) x is less than 3 units from -7 on the number line;
(b) y is farther from 0 than x is from 3 on the number line.

8. Simplify: $|b^2 - 2b + 1|$

9. Solve: $|x - 5| = 3$

10. Solve: $|x + 2| = 4$

11. Solve: $|x + 3| = \frac{5}{2}$

12. Solve: $|x - 5| \leq 2$

13. Solve $|x + 2| \leq 2$

14. Solve: $|x - 1| > 4$

15. **(a)** $|\pi - 7| =$ ____ **(b)** $|\sqrt{23} - \sqrt{3}| =$ ____

16. If c and d are real numbers with $c \neq d$ what are the possible values of $\frac{c - d}{|c - d|}$?

17. Express in interval notation:
(a) The set of all real numbers that are strictly greater than -8;
(b) The set of all real numbers that are less than or equal to 5.

18. Express in interval notation:
(a) The set of all real numbers that are strictly between -6 and 9;
(b) The set of all real numbers that are greater than or equal to 5, but strictly less than 14.

19. Express $.282828\cdots$ as a fraction.

20. Express $.362362362\cdots$ as a fraction.

21. Solve for x:

$$2\left(\frac{x}{5}+7\right)-3x=\frac{x+2}{5}-4$$

22. Solve for x in terms of y: $xy+3=x-2y$

23. Solve for x: $3x^2-2x+5=0$

24. Solve for y: $3y^2-2y=5$

25. Solve for z: $5z^2+6z=7$

26. Solve for x: $325x^2+17x-127=0$

27. Find the *number* of real solutions of the equation $20x^2+12=31x$.

28. For what value of k does the equation $kt^2+5t+2=0$ have exactly one real solution for t?

In Questions 29–32, find all real solutions of the equation. Do not approximate.

29. $x^4-11x^2+18=0$

30. $x^6-4x^3+4=0$

31. $|3x-1|=4$

32. $|2x-1|=x+4$

33. Find the distance from $(1,-2)$ to $(4,5)$.

34. Find the distance from $(3/2, 4)$ to $(3, 5/2)$.

35. Find the distance from (c, d) to $(c-d, c+d)$.

36. Find the midpoint of the line segment from $(-4, 7)$ to $(9, 5)$.

37. Find the midpoint of the line segment from (c, d) to $(2d-c, c+d)$.

38. Find the equation of the circle with center $(-3, 4)$ that passes through the origin.

39. **(a)** If $(1, 1)$ is on a circle with center $(2, -3)$, what is the radius of the circle?
(b) Find the equation of the circle in part (a).

40. Sketch the graph of $3x^2+3y^2=12$.

41. Sketch the graph of $(x-5)^2+y^2-9=0$.

42. Find the center and radius of the circle whose equation is

$$x^2+y^2-2x+6y+1=0.$$

43. Which of statements (a)–(d) are descriptions of the circle with center $(0,-2)$ and radius 5?
(a) The set of all points (x, y) that satisfy $|x|+|y+2|=5$.
(b) The set of all points whose distance from $(0, -2)$ is 5.
(c) The set of all points (x, y) such that $x^2+(y+2)^2=5$.
(d) The set of all points (x, y) such that $\sqrt{x^2+(y+2)^5}=5$.

44. If the equation of a circle is $3x^2+3(y-2)^2=12$, which of the following statements is true?
(a) The circle has diameter 3.
(b) The center of the circle is $(2, 0)$.
(c) The point $(0, 0)$ is on the circle.
(d) The circle has radius $\sqrt{12}$.
(e) The point $(1, 1)$ is on the circle.

45. The graph of one of the equations below is *not* a circle. Which one?

(a) $x^2 + (y + 5)^2 = \pi$ **(b)** $7x^2 + 4y^2 - 14x + 3y^2 - 2 = 0$

(c) $3x^2 + 6x + 3 = 3y^2 + 15$ **(d)** $2(x - 1)^2 - 8 = -2(y + 3)^2$

(e) $\frac{x^2}{4} + \frac{y^2}{4} = 1$

46. The point $(7, -2)$ is on the circle whose center is on the midpoint of the segment joining $(3, 5)$ and $(-5, -1)$. Find the equation of this circle.

47. The national unemployment rates for 1990–1996 were as follows. (*Source:* U.S. Department of Labor, Bureau of Labor Statistics)

Year	1990	1991	1992	1993	1994	1995	1996
Rate (%)	5.6	6.8	7.5	6.9	6.1	5.6	5.4

Sketch a scatter plot and a line graph for these data, letting $x = 0$ correspond to 1990.

48. The table shows the average speed (mph) of the winning car in the Indianapolis 500 race in selected years.

Year	1980	1982	1984	1986	1988	1990	1992	1994	1996
Speed (mph)	143	162	164	171	145	186	134	161	148

Sketch a scatter plot and a line graph for these data, letting $x = 0$ correspond to 1980.

49. (a) What is the y-intercept of the graph of the line

$$y = x - \frac{x - 2}{5} + \frac{3}{5}?$$

(b) What is the slope of the line?

50. Find the equation of the line passing through $(1, 3)$ and $(2, 5)$.

51. Find the equation of the line passing through $(2, -1)$ with slope 3.

52. Find a point on the graph of $y = 3x$ whose distance to the origin is 2.

53. Find the equation of the line that crosses the y-axis at $y = 1$ and is perpendicular to the line $2y - x = 5$.

54. (a) Find the y-intercept of the line $2x + 3y - 4 = 0$.

(b) Find the equation of the line through $(1, 3)$ that has the same y-intercept as the line in part (a).

55. Find the equation of the line through $(-4, 5)$ that is parallel to the line through $(1, 3)$ and $(-4, 2)$.

56. Sketch the graph of the line $3x + y - 1 = 0$.

57. As a balloon is launched from the ground, the wind is blowing it due east. The conditions are such that the balloon is ascending along a straight line with slope $1/5$. After 1 hour the balloon is 5000 ft vertically above the ground. How far east has the balloon blown?

58. The point (u, v) lies on the line $y = 5x - 10$. What is the slope of the line passing through (u, v) and the point $(0, -10)$?

In Questions 59–65, determine whether the statement is true or false.

59. The graph of $x = 5y + 6$ has y-intercept 6.

60. The graph of $2y - 8 = 3x$ has y-intercept 4.
61. The lines $3x + 4y = 12$ and $4x + 3y = 12$ are perpendicular.
62. Slope is not defined for horizontal lines.
63. The line in the figure has positive slope.
64. The line in the figure does not pass through the third quadrant.
65. The y-intercept of the line in the figure is negative.

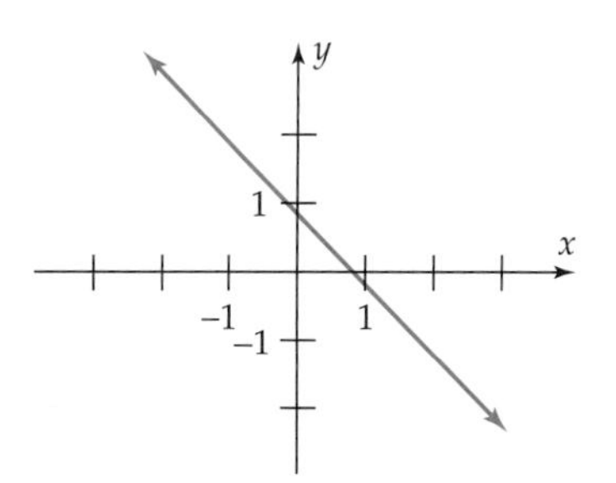

66. Consider the *slopes* of the lines shown in the figure below. Which line has the slope with the largest *absolute value*?

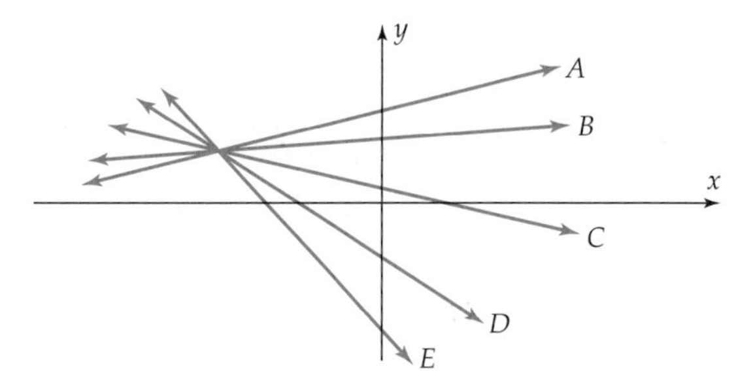

67. Which of the following lines rises most steeply from left to right?
    (a) $y = -4x - 10$ **(b)** $y = 3x + 4$
    (c) $20x + 2y - 20 = 0$ **(d)** $4x = y - 1$
    (e) $4x = 1 - y$
68. Which of the following lines is *not* perpendicular to the line $y = x + 5$?
    (a) $y = 4 - x$ **(b)** $y + x = -5$
    (c) $4 - 2x - 2y = 0$ **(d)** $x = 1 - y$
    (e) $y - x = \frac{1}{5}$
69. Which of the following lines does *not* pass through the third quadrant?
    (a) $y = x$ **(b)** $y = 4x - 7$
    (c) $y = -2x - 5$ **(d)** $y = 4x + 7$
    (e) $y = -2x + 5$
70. Let a, b be fixed real numbers. Where do the lines $x = a$ and $y = b$ intersect?
    (a) Only at (b, a).
    (b) Only at (a, b).
    (c) These lines are parallel, so they don't intersect.
    (d) If $a = b$, then these are the same line, so they have infinitely many points of intersection.
    (e) Since these equations are not of the form $y = mx + b$, the graphs are not lines.
71. What is the y-intercept of the line $2x - 3y + 5 = 0$?
72. For what values of k will the graphs of $2y + x + 3 = 0$ and $3y + kx + 2 = 0$ be perpendicular lines?
73. The average life expectancy increased linearly from 62.9 years for a person born in 1940 to 75.4 years for a person born in 1990.
    (a) Find an equation that gives the average life expectancy y of a person born in year x, with $x = 0$ corresponding to 1940.
    (b) Use the equation in part (a) to estimate the average life expectancy of a person born in 1980.

74. The population of San Diego grew in an approximately linear fashion from 334,413 in 1950 to 1,151,977 in 1994.

(a) Find an equation that gives the population y of San Diego in year x (with $x = 0$ corresponding to 1950).

(b) Use the equation in part (a) to estimate the population of San Diego in 1975 and 2000.

In Exercises 75–78, match the given information with the graph, and determine the slope of the graph.

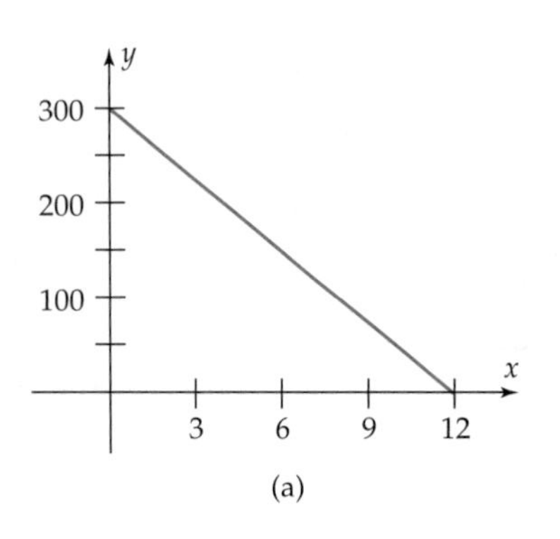

(a)

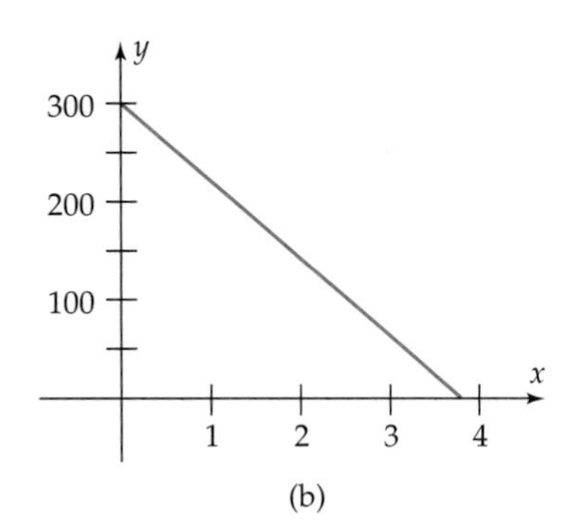

(b)

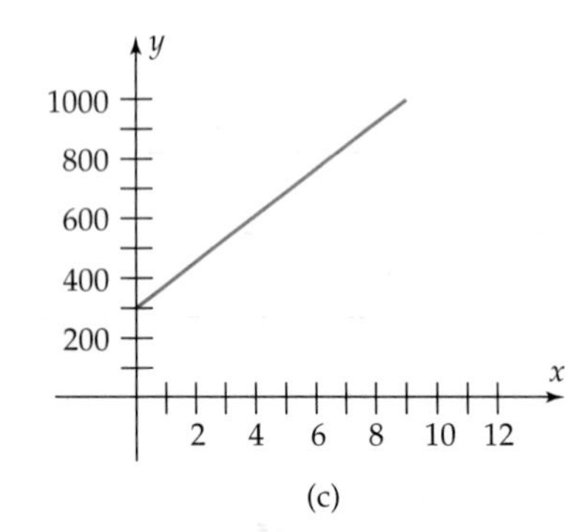

(c)

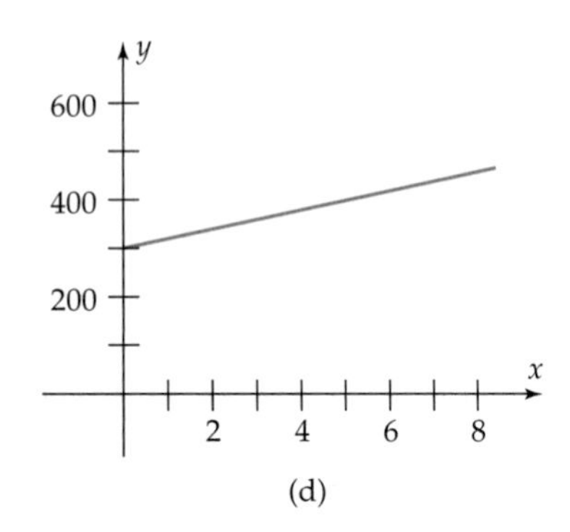

(d)

75. A salesman is paid \$300 per week plus \$75 for each unit sold.

76. A person is paying \$25 per week to repay a \$300 loan.

77. A gold coin that was purchased for \$300 appreciates \$20 per year.

78. A CD player that was purchased for \$300 depreciates \$80 per year.

Discovery Project 1

Modeling the Real World with Lines

Mathematical models are often used to represent the physical world. These models may take many forms, including equations, formulas, and graphs. When the model accurately reflects the real world, it may enable us to predict certain outcomes without actually using physical experimentation. Other models, such as the one presented here, result from different ways of measuring a particular quantity. They make it easy to compare the different measurement systems.

There are two common units which we (the average citizens of the world) use to measure temperature. One is the Fahrenheit scale, the usual one seen on U.S. television; the other is the Celsius, or centigrade, scale, the usual one seen on Canadian television. The two scales are linearly related and calibrated using the freezing and boiling points of water at sea level.

Temperature Scale	Fahrenheit Scale	Celsius Scale
Water Freezes	32°	0°
Water Boils	212°	100°

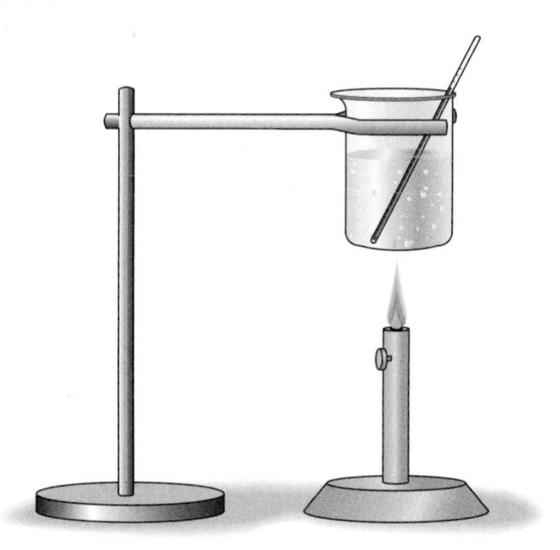

1. Write a formula that relates Fahrenheit temperature F to Celsius temperature C. Your answer should be in the form of $F = mC + b$.
2. Write a similar formula that relates Celsius to Fahrenheit.

3. If the temperature changes 1° Fahrenheit, how many degrees Celsius does the temperature change? What happens to the Fahrenheit temperature when the Celsius temperature changes 1°? What do the answers to these questions have to do with the models from questions 1 and 2?
4. Common wisdom (which is often neither) has it that "normal" human body temperature is 98.6° Fahrenheit. What is this temperature in degrees Celsius?
5. Where do the two scales agree? That is, when is the temperature in degrees Fahrenheit equal to the temperature in degrees Celsius?
6. In the Kelvin scale, water freezes at 273°. One Kelvin is the same size as 1° Celsius. At what Kelvin temperature does water boil?

Chapter

2 Graphs and Technology

Hamburger and fries—will there be anything else?

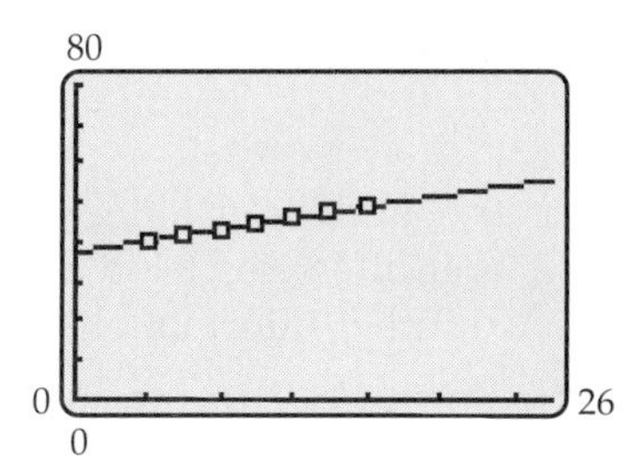

The average number of take-out meals purchased (per person) each year increased by almost 50% from 1984 to 1996. Suppose this trend continues and you have the data for selected years during 1984–1996. Then you can use your calculator to construct a linear model to predict the number of take-out meals (per person) in future years. See Exercise 21 in Section 2.5.

Interdependence of Sections

2.1 → 2.2 → 2.3 → 2.4
2.1 ↘ 2.5

Chapter Outline

Graphing technology is introduced and explored in this chapter. A variety of techniques are presented that enable you to solve complicated equations and to deal with real-world problems. As you will see, the effective use of this technology requires a sound knowledge of both algebra and geometry.

2.1 Graphs and Graphing Calculators

The traditional method of graphing an equation "by hand" is as follows: Construct a table of values with a reasonable number of entries, plot the corresponding points, and use whatever algebraic or other information is available to make an "educated guess" about the rest.

Example 1 The graph of $y = x^2$ consists of all points (x, x^2), where x is a real number. You can easily construct a table of values and plot the corresponding points, as in Figure 2–1. These points suggest that the graph looks like the one in Figure 2–2, which is obtained by connecting the plotted points and extending the graph upward. ■

x	$y = x^2$
−2.5	6.25
−2	4
−1.5	2.25
−1	1
−.5	.25
0	0
.5	.25
1	1
1.5	2.25
2	4
2.5	6.25

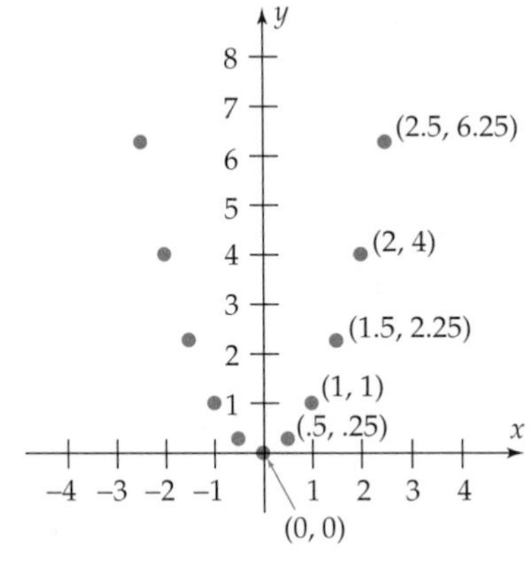

Figure 2–1

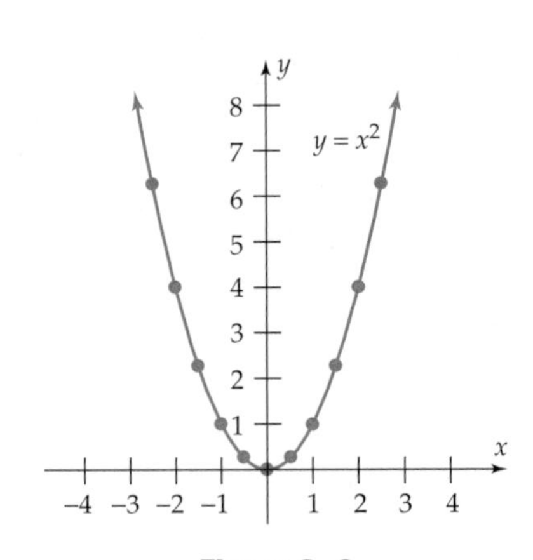

Figure 2–2

Graphing technology generally improves speed, convenience, and accuracy in graphing, provided you have sufficient mathematical and technical knowledge to get the most out of your calculator or computer. The rest of this section deals with the latter issue.

Viewing Windows

The first step in graphing an equation with a calculator is to choose a **viewing window** (or **viewing rectangle**), that is, the portion of the coordinate plane that will appear on the screen.

Example 2 The viewing window for the graph in Figure 2–3 is the rectangular region indicated by the dashed blue lines. It includes all points (x, y) whose coordinates satisfy $-4 \le x \le 5$ and $-3 \le y \le 6$. To display this viewing window on a calculator, press the WINDOW (or RANGE or V-WINDOW or PLOT SETUP) key,* and enter the appropriate numbers, as shown in Figure 2–4 for a TI-83 (other calculators are similar). The settings Xscl = 1 and Yscl = 1 put the tick marks 1 unit apart on each axis.† This is usually the best setting for small viewing windows, but not for large ones. ■

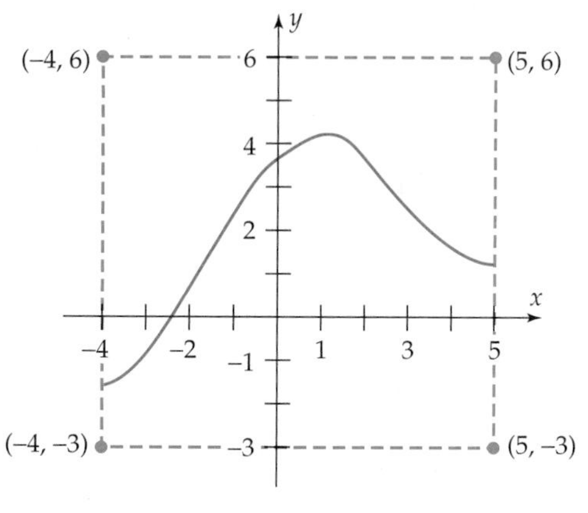

Figure 2–3

```
WINDOW
 Xmin=-4
 Xmax=5
 Xscl=1
 Ymin=-3
 Ymax=6
 Yscl=1
 Xres=1
```

Figure 2–4

NOTE The heading GRAPHING EXPLORATION indicates that you are to use your graphing calculator or computer software as directed to complete the discussion.

GRAPHING EXPLORATION

Set up a viewing window with $-200 \le x \le 200$ and $-30 \le y \le 30$ and Xscl = 1, Yscl = 1. Then press GRAPH (or PLOT) to display this window on the screen. (If a graph also appears, ignore it; just look at the axes). Can you distinguish the tick marks on each axis? Now press WINDOW or RANGE or PLOT SETUP), and change the settings to Xscl = 20 and Yscl = 5, so that adjacent tick marks will be 20 units apart on the x-axis and 5 units apart on the y-axis, and press GRAPH (or PLOT) again. Can you distinguish the tick marks now?

*On TI-85/86, press GRAPH first, then WINDOW or RANGE.

†On HP-38 "Xscl" and "Yscl" are called "Xtick" and "Ytick". Some calculators do not have an Xres setting. On those that do, it should normally be set at 1 (or at "detail" on HP-38).

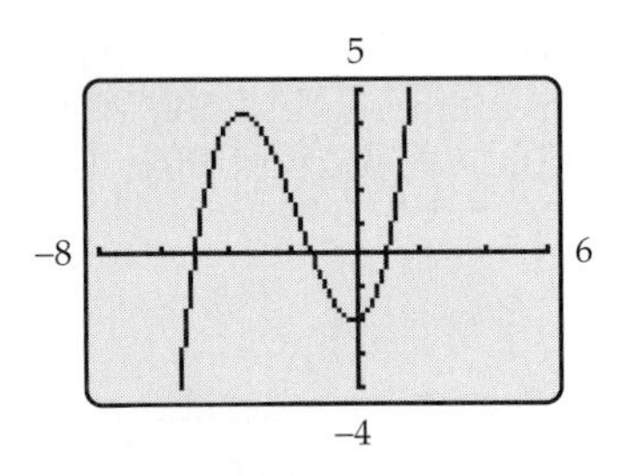

Figure 2–5

In the calculator screens in this book, the viewing window is indicated by the numbers at the ends of the axes. For instance, the window in Figure 2–5 has $-8 \le x \le 6$ and $-4 \le y \le 5$. Since a 14-unit-long segment of the x-axis is shown, the tick marks on the axis are two units apart (Xscl = 2). Similarly, the tick marks on the 9-unit-long segment of the y-axis are one unit apart (Yscl = 1).

Graphing with a Calculator

A graphing calculator or computer graphing program graphs in the same way you would graph by hand, but uses many more points. It plots 95 or more points and simultaneously connects them with line segments. A step-by-step procedure for calculator graphing is presented in the next example.

Example 3 Use a calculator to graph the equation

$$2x^3 - 8x - 2y + 4 = 0.$$

Solution

Step 1 *Choose a (preliminary) viewing window.* Since we don't know yet where the graph lies in the plane, we'll try the viewing window with $-10 \le x \le 10$ and $-10 \le y \le 10$. If that doesn't work, we'll try another window.

Step 2 *Enter the equation in the calculator.* Since calculators can only graph equations in the form y = expression in x, we first solve the given equation for y:

$$2x^3 - 8x - 2y + 4 = 0$$

Rearrange terms: $\quad -2y = -2x^3 + 8x - 4$

Divide by −2: $\quad y = x^3 - 4x + 2.$

Now call up the **equation memory** (press Y= on TI and Sharp 9600, SYMB on HP-38, or GRAPH (Main Menu) on Casio 9850). Next, use the Technology Tip in the margin to enter the equation, as shown in Figure 2–6 for TI-83 (other calculators are similar).

Technology Tip

When entering an equation for graphing, use the "variable" key rather than the ALPHA X key. It has a label like X, T, θ or X, T, θ, n or X, θ, T or x-VAR or X/θ/T/n. On TI-89, however, use the x key on the keyboard.

Step 3 *Graph the equation.* Press GRAPH (or PLOT or DRAW) to obtain Figure 2–7. Because of the limited resolution of a calculator screen, the graph appears to consist of short adjacent line segments rather than a smooth unbroken curve.

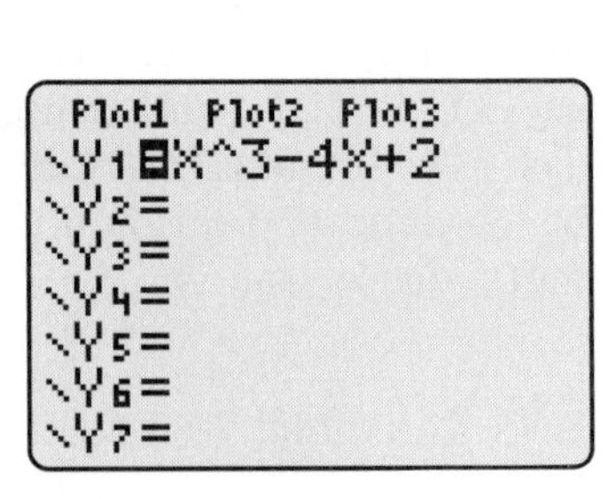

Figure 2–6

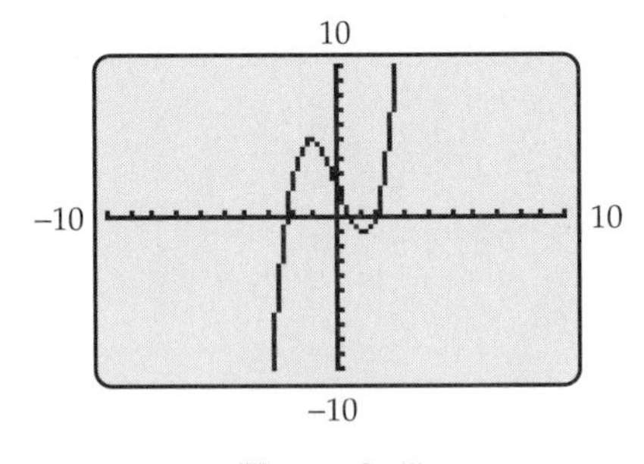

Figure 2–7

Step 4 *If necessary, adjust the viewing window for a better view.* The point where the graph crosses the y-axis isn't clear in Figure 2–7, so we change the viewing window (Figure 2–8) and press GRAPH (or PLOT or DRW) to obtain Figure 2–9, which shows clearly that the graph crosses the y-axis at 2. ■

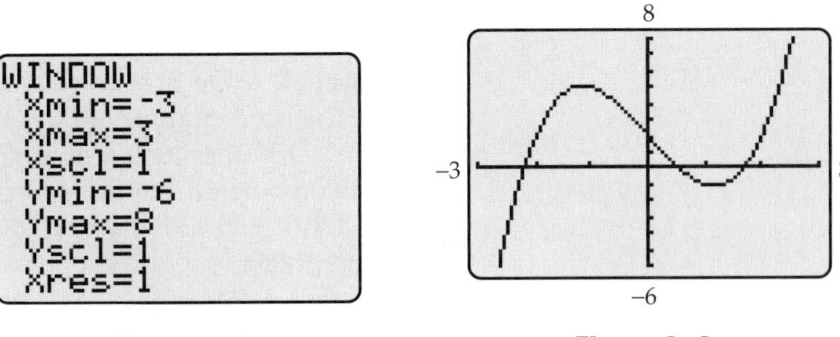

Figure 2–8 **Figure 2–9**

To see the points that the calculator actually plotted in Figure 2–9, press TRACE and a flashing cursor appears on the graph.* Use the *left* and *right* arrow keys to move the cursor along the graph. As you do, the coordinates of the point the cursor is on appear at the bottom of the screen (Figure 2–10).

Technology Tip

On most calculators an equation is "on" if its equal sign is shaded and "off" if its equal sign is clear. On TI-89 and HP-38 an equation that is "on" has a check mark ✓ next to it.

Only the equations that are "on" will be graphed when you press GRAPH (or PLOT). To turn an equation "on" or "off," place the cursor on its equal sign and press ENTER (TI-83, Sharp 9600), or move the cursor to the equation and press SELECT (TI-86, Casio 9850) or CHECK (HP-38) or F4 (TI-89).

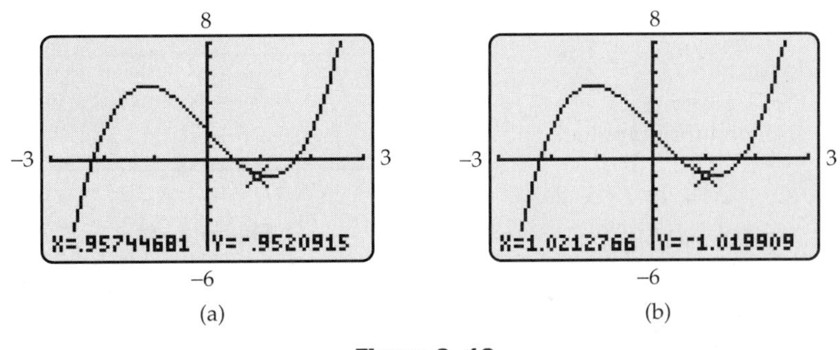

Figure 2–10

The trace cursor displays only points that the calculator plotted. In Figure 2–10, it does not land on $(1, -1)$, which is on the graph of $y = x^3 - 4x + 2$ (why?), but was not plotted. Instead it lands on two nearby points that were plotted.

An equation stays in the equation memory until you delete it. If there are several equations in the memory and you want to graph only one of them, you must turn that equation "on" and the others "off," as explained in the Tip in the margin.

Now that basic graphing is clear, we'll examine some useful calculator graphing tools.

*TRACE is automatically on after graphing on HP-38.

Example 4 The Cortopassi Computer Company can produce a maximum of 100,000 computers a year. Their annual profit is given by

$$y = -.003x^4 + .3x^3 - x^2 + 5x - 4000,$$

where y is the profit (in thousands of dollars) from selling x thousand computers. Use graphical methods to estimate how many computers should be sold in order to make the largest possible profit.

Solution We first choose a viewing window. The number x of computers is nonnegative, and no more than 100,000 can be produced, so that $0 \leq x \leq 100$ (because x is measured in thousands). The profit y may be positive or negative (the company could lose money). So we try a window with $-25{,}000 \leq y \leq 25{,}000$ and obtain Figure 2–11. For each point on the graph,

> The x-coordinate is the number of thousands of computers produced;
>
> The y-coordinate is the profit (in thousands) on that number of computers.

The largest possible profit occurs at the point with the largest y-coordinate, that is, the highest point in the window.

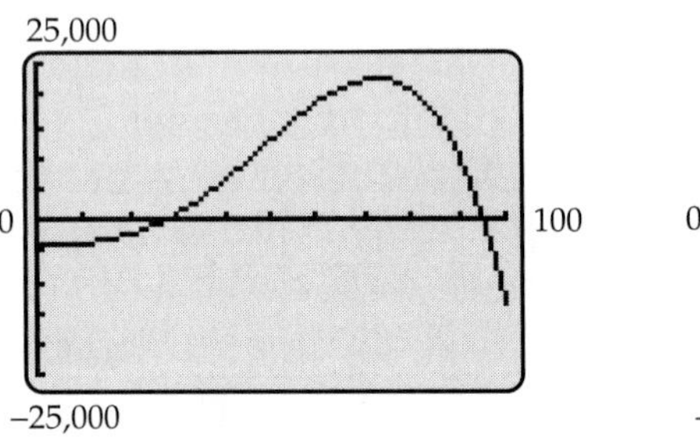

Figure 2–11

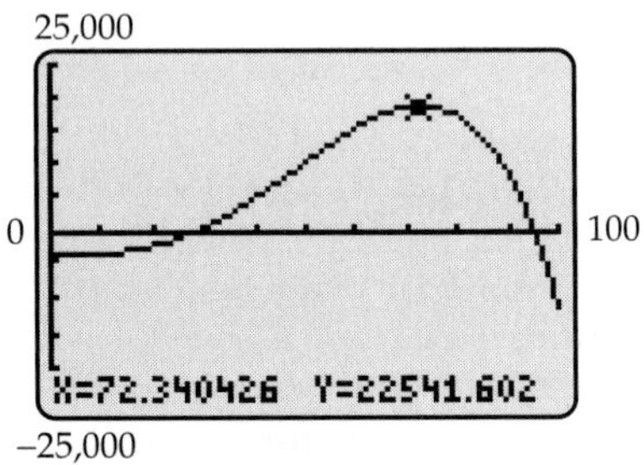

Figure 2–12

Technology Tip

The ZOOM menu is on the keyboard of TI-83, Sharp 9600, and Casio 9850. It is a submenu of the TI-86/89 GRAPH menu and of the HP-38 PLOT menu.

To set the factors, look for FACT or ZFACT or (SET) FACTORS in the ZOOM menu (or in its MEMORY submenu on TI-83).

There are three ways to approximate the highest point.

Use TRACE: Move the cursor along the graph, watching until the y-coordinate takes its largest value (Figure 2–12). Because different calculators plot different points, the trace feature on your calculator may produce a slightly different result. Because x and y are measured in thousands, the point (72.340426, 22541.602) indicates that producing about 72,340 computers gives a profit of approximately \$22,541,602. Since the trace cursor displays only those points that the calculator actually plotted when it drew the graph, this point is not necessarily the highest point in the window.

Use ZOOM: Select ZOOM-IN on the ZOOM menu, move the cursor to approximately the highest point, and hit ENTER. Figure 2–13 shows the result of zooming in by a factor of 4 and using TRACE again: the highest point appears to be (72.87234, 22547.542), meaning that producing about 72,872 computers gives a profit of about

$22,547,542. Using larger zoom factors or zooming-in more than once would produce a more accurate answer.

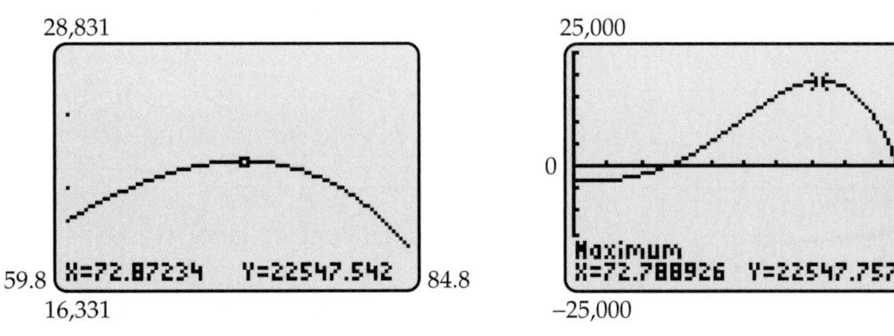

Figure 2–13 **Figure 2–14**

Use the MAXIMUM FINDER: This feature automatically finds the highest point with the greatest degree of accuracy that the calculator is capable of. The maximum finder on a TI-83 produced Figure 2–14, which shows that making about 72,789 computers results in a profit of $22,547,757. Calculus can be used to confirm that this is accurate. ■

Technology Tip

The graphical maximum finder is in the CALC menu of TI-83 and Sharp 9600, and in the G-SOLVE menu of Casio 9850. It is in the MATH submenu of the TI-89 and TI-86 GRAPH menu (labeled FMAX in the latter). It is labeled EXTREMUM in the HP-38 PLOT FCN menu.

On Some TI calculators, you must first select a left (or lower) bound, meaning an x-value to the left of the highest point, and a right (or upper) bound, meaning an x-value to its right, and make an initial guess. On other calculators, you may have to move the cursor near the point you are seeking.

GRAPHING EXPLORATION

Graph $y = .3x^3 + .8x^2 - 2x - 1$ in the window with $-5 \leq x \leq 5$ and $-5 \leq y \leq 5$. Use your maximum finder and the Tip in the margin to approximate the coordinates of the highest point to the left of the y-axis. Then use your minimum finder (in the same menu) to approximate the coordinates of the *lowest* point to the right of the y-axis. How do these answers compare with the ones you get by using the trace feature (but not zoom-in)?

Special Viewing Windows

Throughout this book and in the ZOOM menu of most calculators, the viewing window with $-10 \leq x \leq 10$ and $-10 \leq y \leq 10$ is called the **standard viewing window** (or the **default window** on Sharp 9600). Although it's often a good place to start when you don't have any other information, it may not always be the best choice.

Example 5 Graph the circle $x^2 + y^2 = 4$ on a calculator.

Solution Solving the equation for y shows that

$$y = \sqrt{4 - x^2} \qquad \text{or} \qquad y = -\sqrt{4 - x^2}.$$

Graphing both of these equations on the same screen will produce the graph of the circle. We choose the standard viewing window and enter both equations in the equation memory (Figure 2–15); note that the equal sign is shaded in both equations, indicating that both will be graphed. Pressing GRAPH (or PLOT or DRAW) produces Figure 2–16. The circle doesn't look round, so we change the viewing window and obtain Figure 2–17, which does look like a circle. ■

Figure 2–15

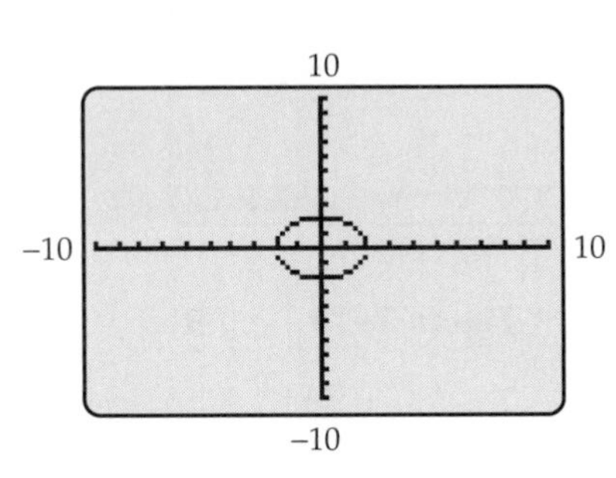

Figure 2–16

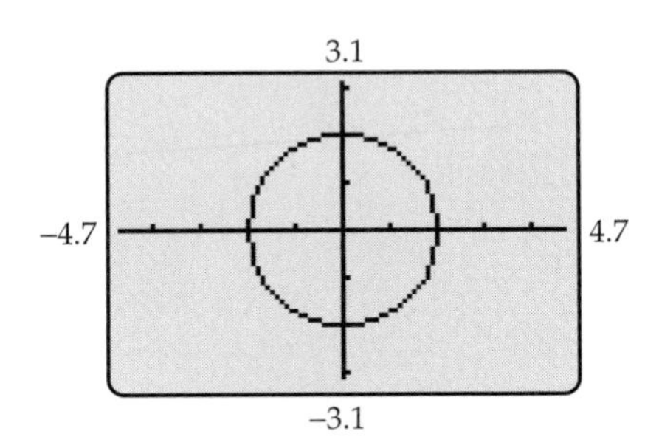

Figure 2–17

The viewing window in Figure 2–17, in which the circle looks round, is an example of a **square viewing window,** one in which a one-unit segment on the x-axis is the same length on the screen as a one-unit segment on the y-axis. Because calculator screens are wider than they are high, the y-axis in a square window must be shorter than the x-axis (about two-thirds as long on standard-width screens, such as the TI-83 screen in Figure 2–17).* In particular, the standard viewing window ($-10 \le x \le 10$ and $-10 \le y \le 10$) is *not* square because the 20 one-unit segments on the y-axis occupy less space on the screen than the 20 one-unit segments on the x-axis. You should use square windows when you want circles to look round and perpendicular lines to look perpendicular, but use any convenient window for other graphs.

Technology Tip

To change the current viewing window to a square one, use SQUARE (or ZSQUARE or ZOOMSQR or SQR) in the ZOOM menu.

GRAPHING EXPLORATION

The lines $y = .5x$ and $y = -2x + 2$ are perpendicular (why?). Graph them in the standard viewing window. Do they look perpendicular? Now graph them in a square window (see the Tip in the margin). Do they look perpendicular?

The method used to graph the circle $x^2 + y^2 = 4$ in Example 5 can be used to graph any equation that can be solved for y.

*It should be about three-fifths as long on TI-86 and about one-half as long on other wide-screen models, such as Casio 9850, HP-38, Sharp 9600, and TI-89.

Example 6 To graph $12x^2 - 4y^2 + 16x + 12 = 0$, solve the equation for y:

$$4y^2 = 12x^2 + 16x + 12$$
$$y^2 = 3x^2 + 4x + 3$$
$$y = \pm\sqrt{3x^2 + 4x + 3}.$$

Every point on the graph of the equation is therefore on the graph of either

$$y = \sqrt{3x^2 + 4x + 3} \qquad \text{or} \qquad y = -\sqrt{3x^2 + 4x + 3}.$$

GRAPHING EXPLORATION

Graph the previous two equations on the same screen. The result will be the graph of the original equation. ■

Technology Tip

The screen widths (in pixels) of commonly used calculators are:

TI-83 and Casio 9800: 95

TI-86 and Sharp 9600: 127

Casio 9850: 127

HP-38: 131

TI-89: 159

TI-92: 239

When using the trace feature, you probably noticed that your calculator typically graphs points with first coordinates like 2.340425, rather than 2.3 or 2.34. You can sometimes rectify this situation by using the following facts. A calculator screen consists of tiny rectangles, called pixels, which are darkened to indicate a graphed point. Suppose the screen is 127 pixels wide (see the Tip in the margin). In a viewing window with $0 \le x \le 126$, the 127 pixels on the x-axis represent the 127 numbers 0, 1, 2, ..., 126, which divide the axis into 126 pieces of length 1. Since the trace cursor moves one horizontal pixel at a time, its x-coordinate will change by one unit each time you press the arrow key.

Similarly, if the screen is 95 pixels wide, then the 95 pixels on the x-axis divide it into 94 equal pieces. Thus, if the x-axis has a length of 9.4 (for instance, $-4.7 \le x \le 4.7$), then the distance between adjacent pixels (the distance the trace cursor moves each time) is $9.4/94 = .1$. A screen in which the horizontal distance between adjacent pixels is .1 is sometimes called a **decimal window.**

Choosing a viewing window carefully can make the trace feature much more convenient, as the following Exploration demonstrates.

Technology Tip

For a decimal window that is also square, use ZDECIMAL or ZOOMDEC (in the TI-83/89 ZOOM menu), DECIMAL (in the Sharp 9600 ZOOM menu and the HP-38 VIEWS menu), and INIT (in the Casio 9850 V-WINDOW menu). In the TI-86 ZOOM menu, ZDECIMAL gives a decimal window that is not square.

GRAPHING EXPLORATION

Graph $y = (5/9)(x - 32)$, which relates the temperature x in degrees Fahrenheit and the temperature y in degrees Celsius, in a viewing window with $-40 \le y \le 40$ and $0 \le x \le k$, where k is chosen so that adjacent pixels are 1 unit apart (see the first Technology Tip). Use the trace feature to determine the Celsius temperatures corresponding to 20°F and 77°F.

Now use a window with $32 - k/20 \le x \le 32 + k/20$ (where k is as above). This is a decimal window with (32, 0) at its center. Graph the equation again. Use the trace to determine the Celsius equivalent of 33.8°F.

Complete Graphs

A viewing window is said to display a **complete graph** if it shows all the important features of the graph (peaks, valleys, points where it touches an axis, etc.) and suggests the general shape of the portions of the graph that aren't in the window. Many different windows may show a complete graph. It's usually best to use a window that is small enough to show as much detail as possible.

In later chapters we shall develop algebraic facts that will enable us to know when certain graphs are complete. For the present, however, the best you may be able to do is try several different windows to see which, if any, appear to display a complete graph.

Example 7 Sketch a complete graph of

$$y = .007x^5 - .2x^4 + 1.332x^3 - .004x^2 + 10.$$

Technology Tip

Most calculators have an "auto scaling" feature. Once the range of x values has been set, the calculator selects a viewing window that includes all the points on the graph whose x-coordinates are in the chosen range. It's labeled ZOOMFIT or ZFIT in the TI ZOOM menu, AUTO in the Casio 9850 and Sharp 9600 ZOOM menus, and AUTO-SCALE in the HP-38 VIEWS menu. This feature can eliminate some guesswork, but may produce a window so large that it hides some of the features of the graph.

Solution Four different viewing windows for this graph are shown in Figure 2–18. Graph (a) (the standard window) is certainly not complete since it shows no points to the right of the y-axis. Graph (b) is not complete since it indicates that parts of the graph lie outside the window.

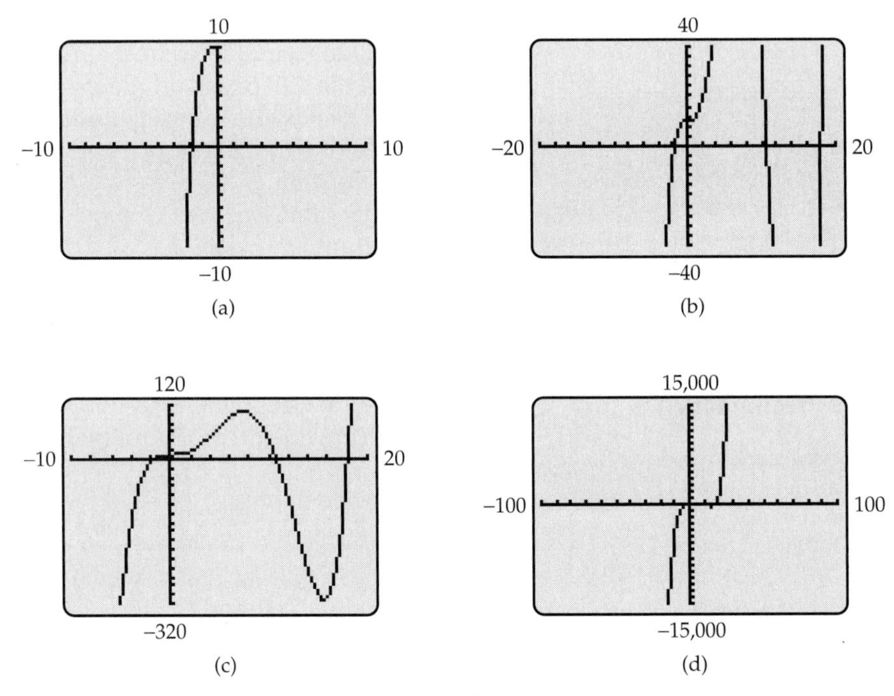

Figure 2–18

Graph (d) tends to confirm what graph (c) suggests, that the graph keeps climbing sharply forever as you move to the right and that it keeps falling sharply forever as you move to the left. Because of its large scale, however, graph (d) doesn't show the features of the graph near the origin. So we conclude that graph (c) is probably a complete graph since it shows important features (twists and turns) near the origin, as well as suggesting the shape of the graph farther out. ■

Technology Tip

When you get a blank screen, press TRACE and use the left/right arrow keys. The coordinates of points on the graph will be displayed at the bottom of the screen, even though they aren't in the current viewing window. Use these coordinates as a guide for selecting a viewing window in which the graph does appear.

Example 8 If you graph $y = -2x^3 + 26x^2 + 18x + 50$ in the standard viewing window, you get a blank screen (try it!). In such cases you can usually find at least one point on the graph by setting $x = 0$ and determining the corresponding value of y (the y-intercept of the graph). If $x = 0$ here, then $y = 50$, so the point (0, 50) is on the graph. Consequently, the y-axis of our viewing window should extend well beyond 50. An alternative method of finding some points on the graph is in the Tip in the margin.

GRAPHING EXPLORATION

Find a complete graph of this equation. [*Hint:* The graph crosses the x-axis once and has one "peak" and one "valley."] ■

As a general rule, you should follow the directions in the following box when graphing equations.

Graphing Conventions

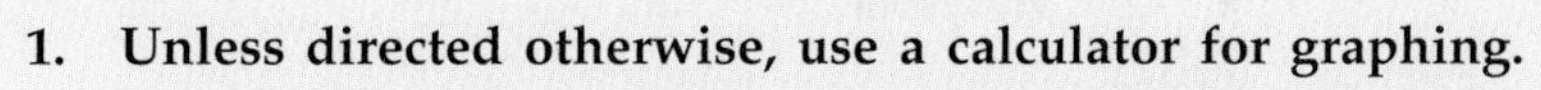

1. **Unless directed otherwise, use a calculator for graphing.**
2. **Complete graphs are required unless a viewing window is specified or the context of a problem indicates that a partial graph is acceptable.**
3. **If the directions say "obtain the graph," "find the graph," or "graph the equation," you need not actually draw the graph on paper. For review purposes, however, it may be helpful to record the viewing window used.**
4. **The directions "sketch the graph" mean "draw the graph on paper, indicating the scale on each axis." This may involve simply copying the display on the calculator screen, or it may require work if the calculator display is misleading.**

Calculator Investigations

1. Tick Marks

(a) Set Xscl = 1 so that adjacent tick marks on the x-axis are one unit apart. Find the largest range of x values such that the tick marks on the x-axis are clearly distinguishable and appear to be equally spaced.

(b) Do part (a) with y in place of x.

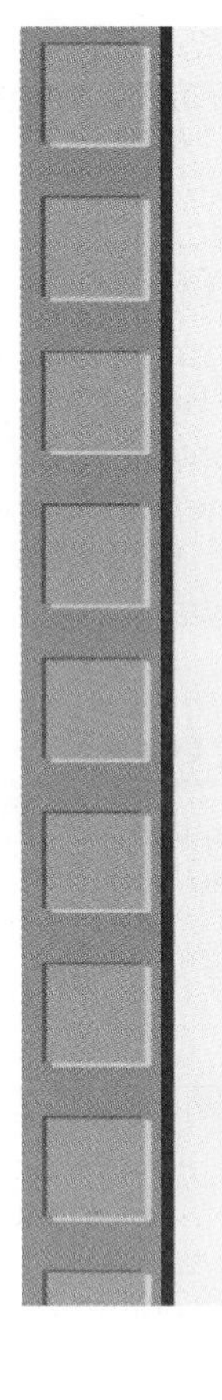

2. **Viewing Windows** Look in the ZOOM menu (or the VIEWS menu on HP-38) to find out how many built-in viewing windows your calculator has. Take a look at each one.
3. **Maximum/Minimum Finders** Use your minimum finder to approximate the x-coordinates of the lowest point on the graph of $y = x^3 - 2x + 5$ in the window with $0 \leq x \leq 5$ and $-3 \leq y \leq 8$. The correct answer is

$$x = \sqrt{\frac{2}{3}} \approx .816496580928.$$

How good is your approximation?

4. **Square Windows** Find a square viewing window on your calculator that has $-10 \leq x \leq 10$.
5. **Dot Graphing Mode** A calculator graphs by plotting points and connecting them. To see which points it plots, without the connecting segments, change your graphing mode by selecting Dot or DrawDot in the MODE menu of TI-83 or the FORMT submenu of the TI-86 GRAPH menu or the STYLE submenu of the TI-89 Y= menu or the STYLE 1 submenu of the Sharp 9600 FORMAT menu. In the Casio 9850 SETUP menu, set the DRAWTYPE to PLOT and in the second page of the HP-38 PLOT SETUP menu, uncheck CONNECT. After changing the graphing mode, graph $y = .5x^3 - 2x^2 + 1$ in the standard window. Try some other equations as well.

Exercises 2.1

Exercises 1–4 are representations of calculator screens. State the approximate coordinates of the points P and Q.

1.

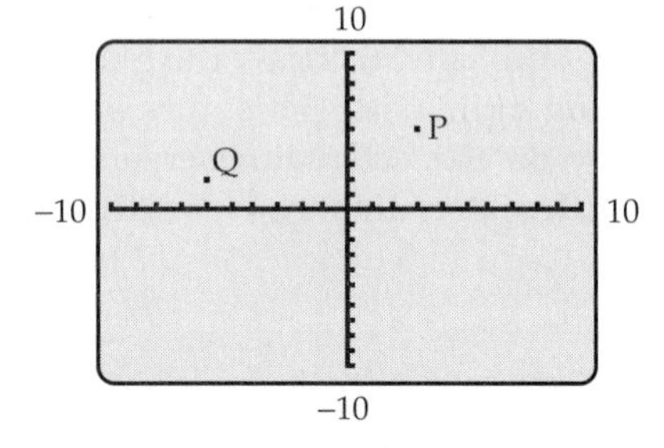

2.

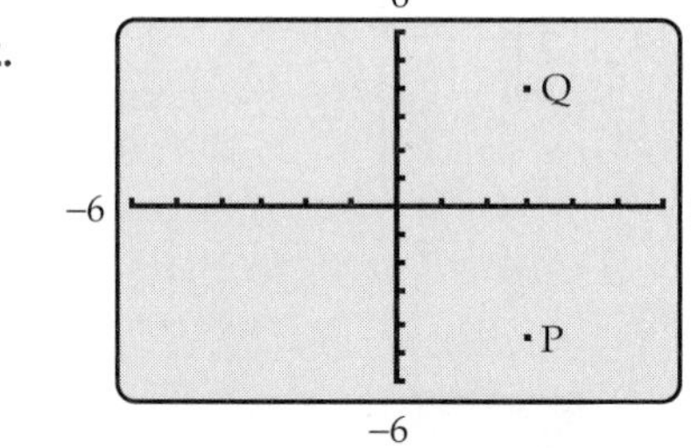

3.

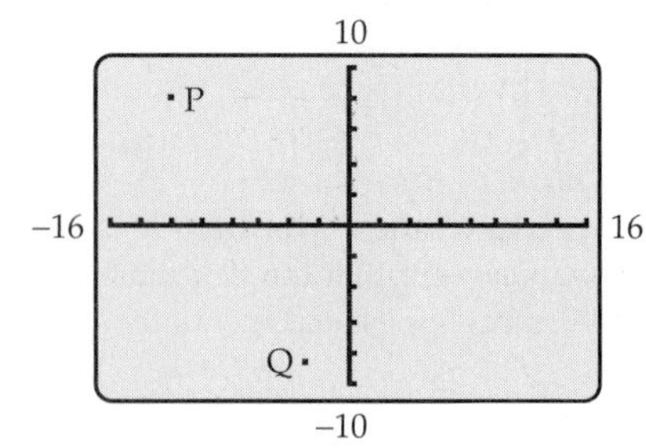

4.

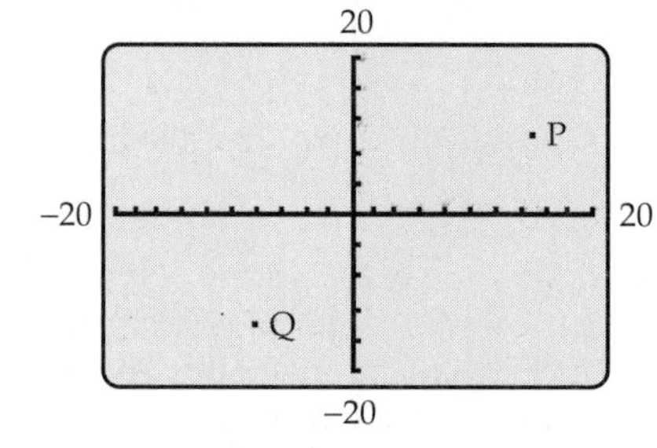

In Exercises 5–10, graph the equation by hand by plotting no more than six points and filling in the rest of the graph as best you can. Then use the calculator to graph the equation and compare the results.

5. $y = |x - 2|$ **6.** $y = \sqrt{x + 5}$

7. $y = x^2 - x$ **8.** $y = x^2 + x + 1$

9. $y = x^3 + 1$ **10.** $y = \dfrac{1}{x}$

In Exercises 11–16, find the graph of the equation in the standard window.

11. $3 + y = .5x$ **12.** $y - 2x = 4$

13. $y = x^2 - 5x + 2$ **14.** $y = .3x^2 + x - 4$

15. $y = .2x^3 + .1x^2 - 4x + 1$

16. $y = .2x^4 - .2x^3 - 2x^2 - 2x + 5$

In Exercises 17–22, determine which of the following viewing windows gives the best view of the graph of the given equation.

a. $-10 \le x \le 10; \quad -10 \le y \le 10$
b. $-5 \le x \le 25; \quad 0 \le y \le 20$
c. $-10 \le x \le 10; \quad -100 \le y \le 100$
d. $-20 \le x \le 15; \quad -60 \le y \le 250$
e. None of a, b, c, d gives a complete graph.

17. $y = 18x - 3x^2$ **18.** $y = 4x^2 + 80x + 350$

19. $y = \dfrac{1}{3}x^3 - 25x + 100$ **20.** $y = x^4 + x - 5$

21. $y = x^2 + 50x + 625$ **22.** $y = .01(x - 15)^4$

23. A toy rocket is shot straight up from ground level and then falls back to earth; wind resistance is negligible. Use your calculator to determine which of the following equations has a graph whose portion above the x-axis provides the most plausible model of the path of the rocket.
(a) $y = .1(x - 3)^3 - .1x^2 + 5$
(b) $y = -x^4 + 16x^3 - 88x^2 + 192x$
(c) $y = -16x^2 + 117x$
(d) $y = .16x^2 - 3.2x + 16$
(e) $y = -(.1x - 3)^6 + 600$

24. Monthly profits at DayGlo Tee Shirt Company appear to be given by the equation

$$y = -.00027(x - 15{,}000)^2 + 60{,}000,$$

where x is the number of shirts sold that month, and y is the profit. DayGlo's maximum production capacity is 15,000 shirts per month.
(a) If you plan to graph the profit equation, what range of x values should you use? [*Hint:* You can't make a negative number of shirts.]
(b) The president of DayGlo wants to motivate the sales force (who are all in the profit-sharing plan), so he asks you to prepare a graph that shows DayGlo's profits increasing *dramatically* as sales increase. Using the profit equation and the x range from part (a), what viewing window would be suitable?
(c) The City Council is talking about imposing more taxes. The president asks you to prepare a graph showing that DayGlo's profits are essentially flat. Using the profit equation and the x range from part (a), what viewing window would be suitable?

In each of the applied situations in Exercises 25–28, find an appropriate viewing window for the equation (that is, a window that includes all the points relevant to the problem, but does not include large regions that are not relevant to the problem, and has easily readable tick marks on the axes). Explain why you chose this window. See the Hint in Exercise 24(a).

25. A cardiac test measures the concentration y of a dye x seconds after a known amount is injected into a vein near the heart. In a normal heart $y = -.006x^4 + .14x^3 - .053x^2 + 179x$.

26. Beginning in 1905 the deer population in a region of Arizona rapidly increased because of a lack of natural predators. Eventually food resources were depleted to such a degree that the deer population completely died out. In the equation $y = -.125x^5 + 3.125x^4 + 4000$, y is the number of deer in year x, where $x = 0$ corresponds to 1905.

27. A winery can produce x barrels of red wine and y barrels of white wine, where

$$y = \frac{200{,}000 - 50x}{2000 + x}.$$

28. The concentration of a certain medication in the bloodstream at time x hours is approximated by the equation

$$y = \frac{375x}{.1x^3 + 50},$$

where y is measured in milligrams per liter. After two days the medication has no effect.

29. (a) Graph $y = \dfrac{1}{x^2 + 1}$ in the standard window.
(b) Does the graph appear to stop abruptly partway along the x-axis? Use the trace feature to explain why this happens. [*Hint:* In this viewing window, each pixel represents a rectangle that is approximately .32 units high.]
(c) Find a viewing window with $-10 \le x \le 10$, which shows a complete graph that does not fade into the x-axis.

30. (a) Graph $y = x^3 - 2x^2 + x - 2$ in the standard window.

(b) Use the trace feature to show that the portion of the graph with $0 \le x \le 1.5$ is not actually horizontal. [*Hint:* All the points on a horizontal segment must have the same y-coordinate (why?).]

(c) Find a viewing window that clearly shows that the graph is not horizontal when $0 \le x \le 1.5$.

In Exercises 31–32, use zoom-in or a maximum/minimum finder to determine the highest and lowest point on the graph in the given window.

31. $y = .4x^4 - 3x^2 + 4x + 3$ $(0 \le x \le 5$ and $-5 \le y \le 5)$

32. $y = .07x^5 - .3x^3 + 1.5x^2 - 2$ $(-3 \le x \le 2$ and $-6 \le y \le 6)$

33. The population y of New Orleans (in thousands) in year x of the twentieth century is approximated by

$$y = .000046685x^4 - .0108x^3 + .7194x^2 - 9.2426x + 305 \quad (0 \le x \le 100),$$

where $x = 0$ corresponds to 1900. According to this model, in what year was the population largest?

34. The number y of LP records sold (in millions) between 1990 and 1996 is approximated by

$$y = \frac{x^4}{120} - \frac{121x^3}{540} + \frac{149x^2}{72} - \frac{799x}{108} + \frac{311}{30},$$

where $0 \le x \le 6$ and $x = 0$ corresponds to 1990. In what year were the fewest records sold?

In Exercises 35–40, use your algebraic knowledge to state whether or not the two equations have the same graph. Confirm your answer by graphing the equations in the standard window.

35. $y = |x + 3|$ and $y = |x| + 3$

36. $y = |x| - 4$ and $y = |x - 4|$

37. $y = \sqrt{x^2}$ and $y = |x|$

38. $y = \sqrt{x^2 + 6x + 9}$ and $y = |x + 3|$

39. $y = \sqrt{x^2 + 9}$ and $y = x + 3$

40. $y = \dfrac{1}{x^2 + 2}$ and $y = \dfrac{1}{x^2} + \dfrac{1}{2}$

41. (a) Confirm the accuracy of the factorization $x^2 - 5x + 6 = (x - 2)(x - 3)$ graphically. [*Hint:* Graph $y = x^2 - 5x + 6$ and $y = (x - 2)(x - 3)$ on the same screen. If the factorization is correct, the graphs will be identical (which means that you will see only a single graph on the screen).]

(b) Show graphically that $(x + 5)^2 \ne x^2 + 5^2$. [*Hint:* Graph $y = (x + 5)^2$ and $y = x^2 + 5^2$ on the same screen. If the graphs are different, then the two expressions cannot be equal.]

True or False *In Exercises 42–44, use the technique of Exercise 41 to determine graphically whether the given statement is possibly true or definitely false. (We say "possibly true" because two graphs that appear identical on a calculator screen may actually differ by small amounts or at places not shown in the window.)*

42. $x^3 - 7x - 6 = (x + 1)(x + 2)(x - 3)$

43. $(1 - x)^6 = 1 - 6x + 15x^2 - 20x^3 + 15x^4 - 6x^5 + x^6$

44. $x^5 - 8x^4 + 16x^3 - 5x^2 + 4x - 20 = (x - 2)^2(x - 5)(x^2 + x + 1)$

In Exercises 45–54, use the techniques of Examples 5 and 6 to graph the equation in a suitable square viewing window.

45. $x^2 + y^2 = 9$

46. $y^2 = x + 2$

47. $3x^2 + 2y^2 = 48$

48. $25(x - 5)^2 + 36(y + 4)^2 = 900$

49. $(x - 4)^2 + (y + 2)^2 = 25$

50. $9x^2 + 4y^2 = 36$

51. $4x^2 - 9y^2 = 36$

52. $9y^2 - x^2 = 9$

53. $9x^2 + 5y^2 = 45$

54. $x = y^2 - 2$

In Exercises 55–60, obtain a complete graph of the equation by trying various viewing windows. List a viewing window that produces this complete graph. (Many correct answers are possible; consider your answer as correct if your window shows all the features in the window given in the answer section.)

55. $y = 7x^3 + 35x + 10$

56. $y = x^3 - 5x^2 + 5x - 6$

57. $y = \sqrt{x^2 - x}$

58. $y = 1/x^2$

59. $y = -.1x^4 + x^3 + x^2 + x + 50$

60. $y = .002x^5 + .06x^4 - .001x^3 + .04x^2 - .2x + 15$

In Exercises 61–64, graph all four equations on the same screen, using a sufficiently large square viewing window, and answer this question: What is the geometric relationship of graphs (b), (c), and (d) to graph (a)?

61. (a) $y = x^2$ **(b)** $y = x^2 + 5$ **(c)** $y = x^2 - 5$ **(d)** $y = x^2 - 2$

62. (a) $y = \sqrt{x}$ **(b)** $y = \sqrt{x - 3}$ **(c)** $y = \sqrt{x + 3}$ **(d)** $y = \sqrt{x - 6}$

63. (a) $y = \sqrt{x}$ **(b)** $y = 2\sqrt{x}$ **(c)** $y = 3\sqrt{x}$ **(d)** $y = \frac{1}{2}\sqrt{x}$

64. (a) $y = x^2$ **(b)** $y = -x^2$ **(c)** $y = -\frac{1}{2}x^2$ **(d)** $y = -2x^2$

In Exercises 65–67, graph the two given equations and the equation $y = x$ on the same screen, using a sufficiently large square viewing window, and answer this question: What is the geometric relationship between graphs (a) and (b)?

65. **(a)** $y = x^3$ **(b)** $y = \sqrt[3]{x}$

66. **(a)** $y = \frac{1}{2}x^3 - 4$ **(b)** $y = \sqrt[3]{2x + 8}$

67. **(a)** $y = 5x - 15$ **(b)** $y = .2x + 3$

68. Put your calculator in *radian* mode, and use the viewing window given by $0 \le x \le 6.28$ and $-2 \le y \le 2$.*

(a) Graph $y = \sin x$*
(b) Graph $y = \sin(2x)$
(c) Graph $y = \sin(3x)$
(d) Based on parts (a)–(c), what do you think the graphs of $y = \sin(4x), y = \sin(5x)$, $y = \sin(6x)$, and so on, will look like? Use the calculator to verify your answer.
(e) Based on part (d), what do you think the graphs of $y = \sin(50x)$ and $y = \sin(100x)$ will look like? What does a calculator display instead? What might explain the graphs of the calculator?

2.2 Solving Equations Graphically and Numerically

Algebraic techniques for solving linear and quadratic equations were considered in Section 1.2, where we saw that the quadratic formula provides exact solutions for every second-degree polynomial equation and for a few higher degree equations. For most other equations, however, there are no formulas. For such equations, graphical and numerical approximation methods are the only practical alternatives. These methods depend on the connection between equations and graphs, so we begin with that.

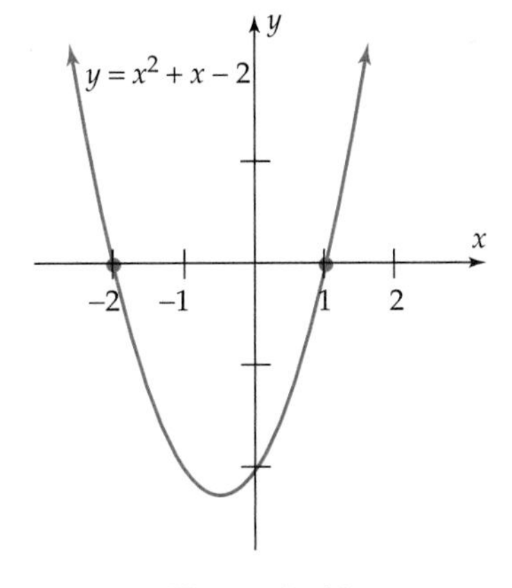

Figure 2–19

In the coordinate plane all points on the x-axis have second coordinate 0. Consequently, when a graph intersects the x-axis, the intersection point has coordinates of the form $(a, 0)$ for some real number a. The number a is called an ***x*-intercept** of the graph. For example, the graph of $y = x^2 + x - 2$ in Figure 2–19 has x-intercepts at -2 and 1 because it intersects the x-axis at $(-2, 0)$ and $(1, 0)$. To say that $(1, 0)$ is on the graph of $y = x^2 + x - 2$ means that $x = 1$ and $y = 0$ satisfy the equation:

$$y = x^2 + x - 2$$
$$0 = 1^2 + 1 - 2.$$

In other words, the x-intercept 1 is a *solution* of the equation $x^2 + x - 2 = 0$. Similarly, the other x-intercept, -2, is also a solution of $x^2 + x - 2 = 0$ because $(-2)^2 + (-2) - 2 = 0$. The same argument works in the general case.

Solutions and Intercepts

The real solutions of a one-variable equation of the form

$$\textbf{Expression in } x = 0$$

are the x-intercepts of the graph of the two-variable equation

$$y = \textbf{expression in } x.$$

*You don't need to know what radian mode is or what "sin" means. Just use the calculator key with this label.

Therefore, to solve an equation, you need only find the x-intercepts of its graph, as illustrated in the next example.

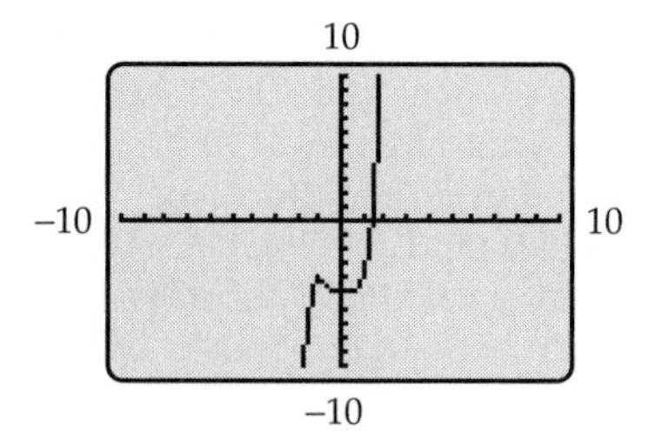

Figure 2–20

Example 1 Solve the equation $x^5 + x^2 = x^3 + 5$ graphically.

Solution We first rewrite the equation so that one side is zero:

$$x^5 - x^3 + x^2 - 5 = 0.$$

Now we graph $y = x^5 - x^3 + x^2 - 5$ in the standard viewing window (Figure 2–20). There is one x-intercept (solution of the equation) between 1 and 2. Assuming that this is the only x-intercept of the graph, this means that the original equation has exactly one real number solution.* It may be approximated graphically in three ways.

Manual Zoom-in. Repeatedly change the viewing window to show the part of the graph near the x-intercept in greater and greater detail, as shown in Figure 2–21. At each step the displayed portion of the x-axis is decreased by a factor of 1/10, and the Xscl adjusted accordingly.

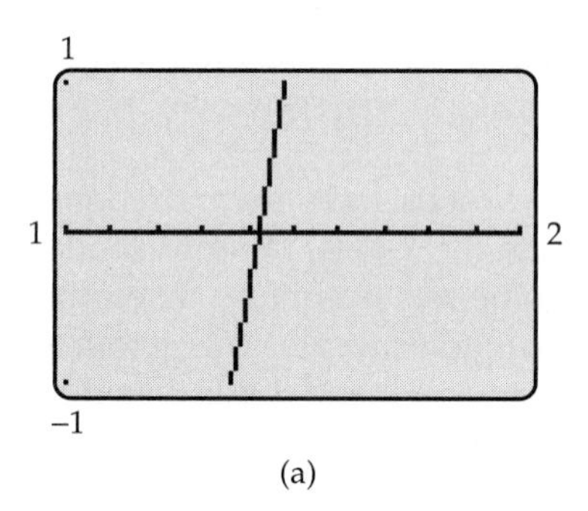

(a)

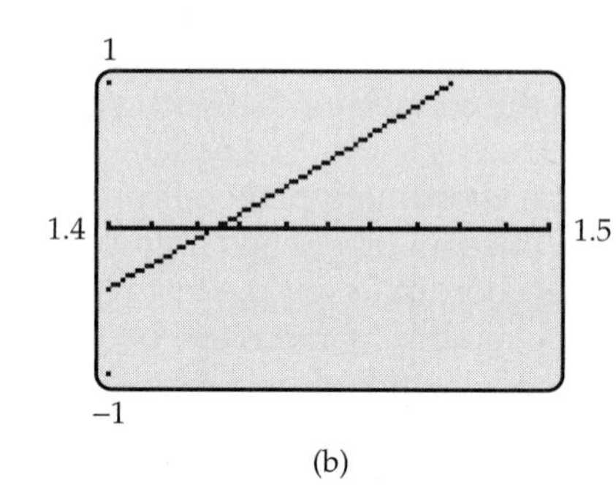

(b)

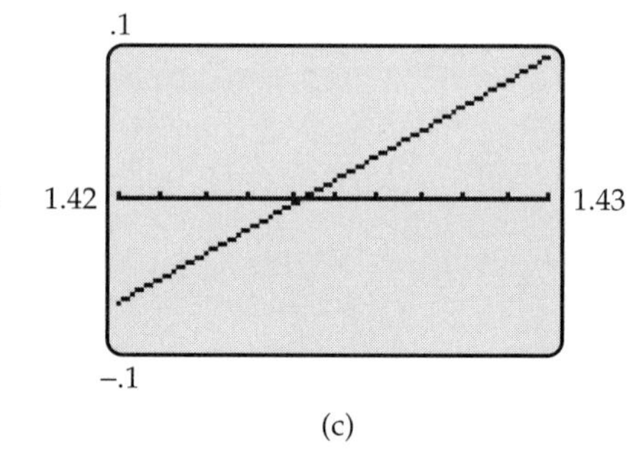

(c)

Figure 2–21

In graph (a), the intercept (solution) is between 1.4 and 1.5. Graph (b) shows that it is between 1.42 and 1.43. Finally, Graph (c) shows that the solution is between 1.424 and 1.425, at approximately 1.4242. Since adjacent tick marks in Figure 2–21(c) are .001 units apart, the maximum possible error in this approximation is .001.

Automatic Zoom-in. Finding a very small window, such as Figure 2–21(c), can be done in just a few keystrokes by using either ZOOM-IN or BOX on the ZOOM menu. The procedures vary on different calculators, so check your instruction manual. After zooming in, the tick marks probably won't show, so you will have to reset the Xscl setting in order to read the approximate solution and determine its accuracy. When using ZOOM-IN, it's convenient to set the zoom factors at 10 (see the Technology Tip on page 64).

*Unless stated otherwise in the examples and exercises of this section, you may assume that the standard viewing window includes all x-intercepts of a graph.

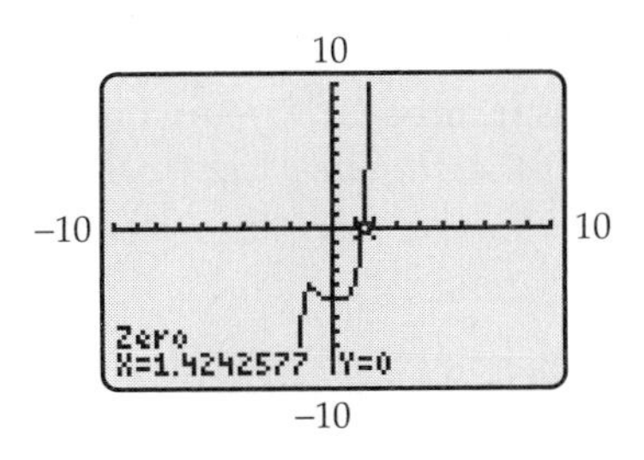

Figure 2–22

Graphical Root Finder. The easiest way to find the solution (x-intercept) with a high degree of accuracy is to use the graphical root finder on your calculator. On some TI calculators, you must select a left (or lower) bound, meaning an x-value to the left of the intercept, and a right (or upper) bound, meaning an x-value to the right of the intercept, and to make an initial guess. On other calculators, you may have to move the cursor near the x-intercept you are seeking. Check your instruction manual for details. A typical root finder produced Figure 2–22, which shows that the solution (x-intercept) is $x \approx 1.4242577$. ■

Technology Tip

The graphical root finder is labeled ROOT (or ZERO or X-INCPT) in the TI-83 CALC menu, the Sharp 9600 CALC menu, the Casio 9850 G-SOLVE menu, the MATH submenu of the TI-86/89 GRAPH menu, and the FCN submenu of the HP-38 PLOT menu.

Example 2 To approximate the solution of

$$x^4 - 4x^3 + 3x^2 - x - 2 = 0,$$

we graph $y = x^4 - 4x^3 + 3x^2 - x - 2$ (Figure 2–23). The graph shows two solutions (x-intercepts): one between -1 and 0, and the other between 3 and 4. A graphical root finder (Figure 2–24) shows that the negative solution is $x \approx -.52417$. ■

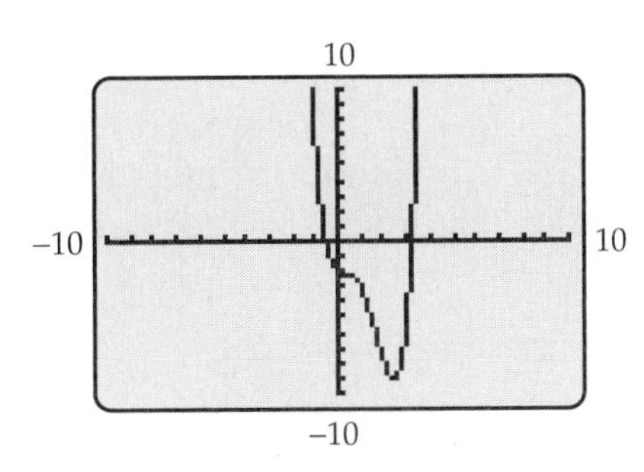

Figure 2–23

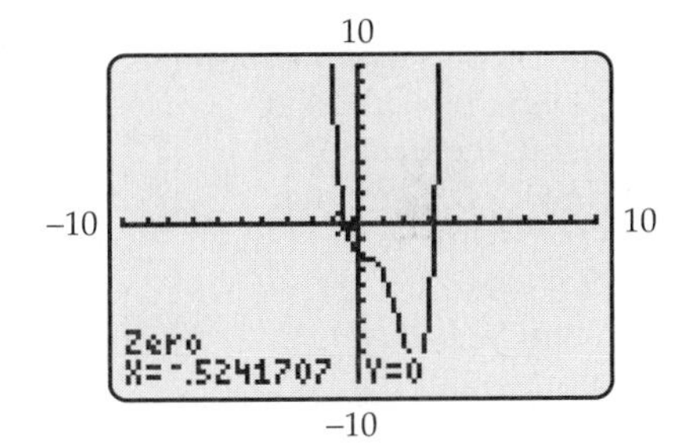

Figure 2–24

GRAPHING EXPLORATION

Use zoom-in or a graphical root finder to approximate the positive solution of the equation in Example 2.

The Intersection Method

The next two examples illustrate an alternative graphical method of approximating solutions of equations.

Example 3 Solve $|x^2 - 4x - 3| = x^3 + x - 6$.

Solution Let $y_1 = |x^2 - 4x - 3|$ and $y_2 = x^3 + x - 6$, and graph both equations on the same screen (Figure 2–25 on the next page). Consider the point where the two graphs intersect. Since it is on the graph of y_1, its second coordinate is $|x^2 - 4x - 3|$, and since it is also on the graph of y_2, its second coordinate is $x^3 + x - 6$. So for this number x, we must have $|x^2 - 4x - 3| = x^3 + x - 6$. In other words, *the x-coordinate of the intersection point is the solution of the equation.*

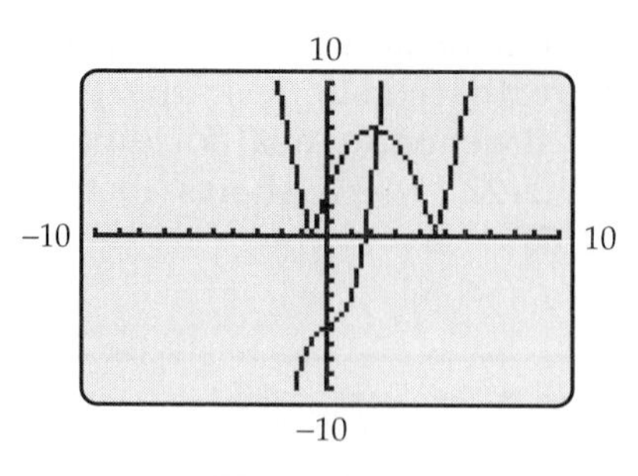

Figure 2–25

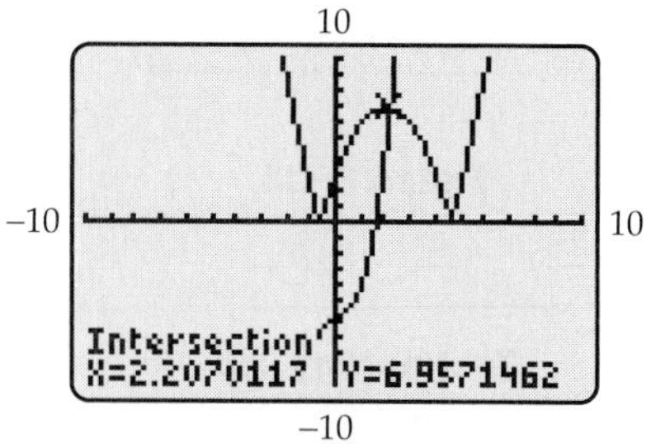

Figure 2–26

This coordinate can be approximated by zooming in or by using a graphical intersection finder (see the Tip in the margin), as shown in Figure 2–26. Therefore, the solution of the original equation is $x \approx 2.207$. ■

Technology Tip

The graphical intersection finder is labeled INTERSECT or ISECT or ISCT or INTSCT in the TI-83 CALC menu, the Sharp 9600 CALC menu, the Casio 9850 G-SOLVE menu, the MATH submenu of the TI-86/89 GRAPH menu, and the FCN submenu of the HP-38 PLOT menu.

Example 4 To solve $x^2 - 2x - 6 = \sqrt{2x + 7}$, we graph $y_1 = x^2 - 2x - 6$ and $y_2 = \sqrt{2x + 7}$ on the same screen. Figure 2–27 shows that there are two solutions (intersection points). According to an intersection finder (Figure 2–28), the positive solution is $x \approx 4.3094$. ■

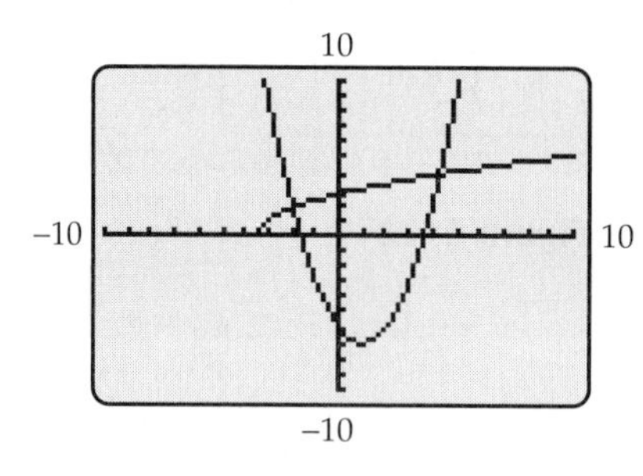

Figure 2–27

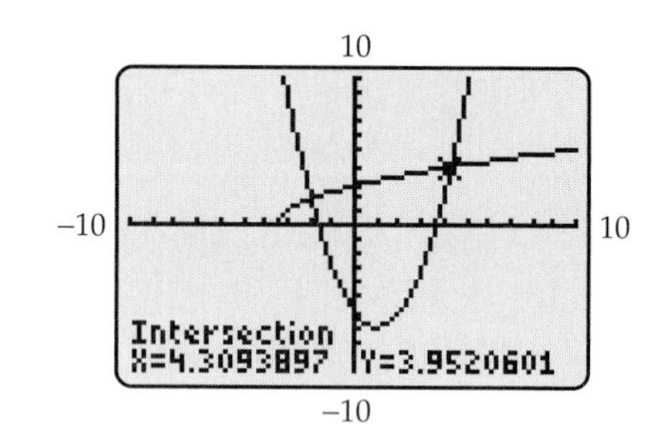

Figure 2–28

GRAPHING EXPLORATION

Use a graphical intersection finder to approximate the negative solution of the equation in Example 4.

Numerical Methods

In addition to graphical root finders, most calculators have an **equation solver** that can approximate solutions of any equation (one at a time, except on TI-89), typically by using Newton's method (which involves calculus). Usually you must enter the equation and an initial estimate of the solution, and possibly an interval in which to search.

Several calculators also have **polynomial solvers** that will find all the solutions of a polynomial equation at once. However, those on Sharp 9600 and Casio 9850 are limited to equations of degree 2 or 3. See Calculator Investigation 1 on page 24 for details.

Technology Tip

The equation solver is labeled SOLVE or SOLVER. It is on the TI-86 and Sharp 9600 keyboards, in the TI-83 MATH menu, in the TI-89 ALGEBRA menu, in the HP-38 LIB menu, and in the EQUA sub-menu of the Casio 9850 Main Menu. The syntax varies, depending on the calculator, so check your instruction manual.

Example 5 When asked to search the interval $-10 \leq x \leq 10$ with an initial guess of 3, the equation solver on a TI-83 found one solution of

$$4x^5 - 12x^3 + 8x - 1 = 0,$$

namely, $x \approx .128138691376$ (Figure 2–29). The POLY solver on a TI-86 produced that solution and four others (Figure 2–30). ■

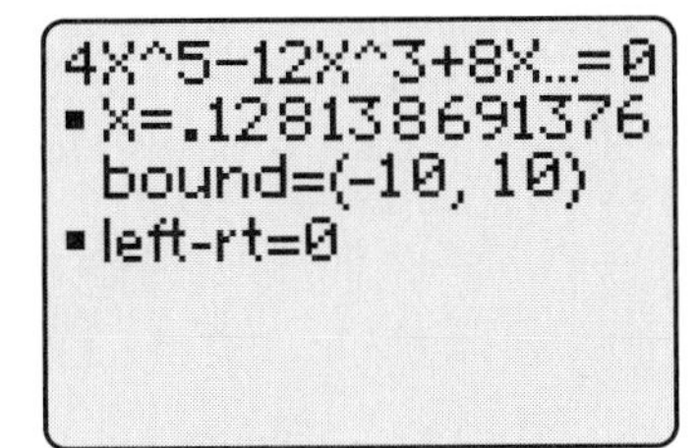

Figure 2–29

Figure 2–30

Dealing with Technological Quirks

Graphical root finders and equation solvers may fail to find some solutions of an equation, particularly when the graph of the equation touches, but does not cross, the x-axis. So if your calculator doesn't show any x-intercepts on a graph or if its root finder gives an error message, you should try an alternative approach, if possible.

Example 6 If you attempt to solve $\sqrt{x^4 + x^2 - 2x - 1} = 0$ by graphing the equation $y = \sqrt{x^4 + x^2 - 2x - 1}$, as in Figure 2–31, and using a root finder, you may get an error message, as we did on four out of six models. If you use an equation solver instead, the results may also be disappointing (three solvers produced an error message). The difficulty can be eliminated by using the fact that *the only number whose square root is zero is zero itself.* Thus, the only time that $\sqrt{x^4 + x^2 - 2x - 1} = 0$ is when $x^4 + x^2 - 2x - 1 = 0$. So we need only solve $x^4 + x^2 - 2x - 1 = 0$, which is easily done on a calculator.

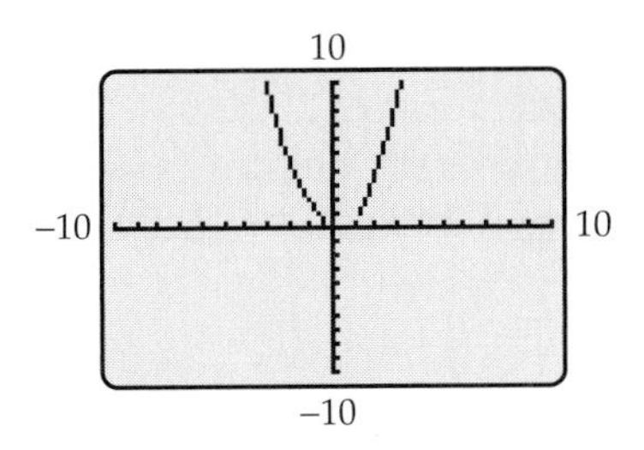

Figure 2–31

GRAPHING EXPLORATION

One solution of $x^4 + x^2 - 2x - 1 = 0$, and hence of the original equation, is $x \approx -.4046978$. Use a graphical root finder, an equation solver, or a polynomial solver to find the other solution. ■

Even though your calculator might work directly in Example 6, there may be similar cases where it doesn't. So the technique of Example 6 should normally be used with all equations of the form "square root of a quantity = 0" (but not on other equations involving square roots): Set the expression under the radical equal to 0, and solve.

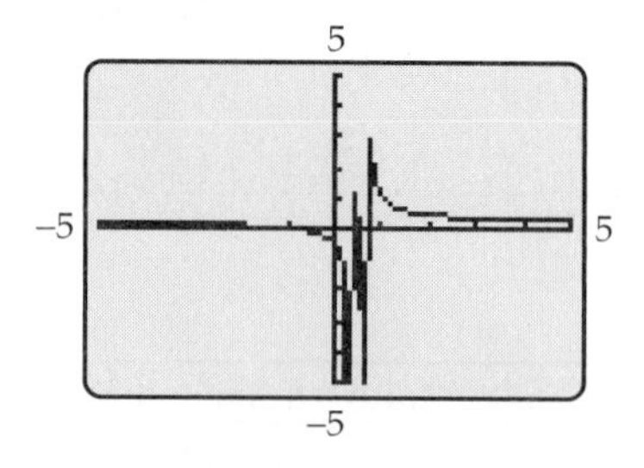

Figure 2–32

Example 7 If you try to solve $\dfrac{2x^2 + x - 1}{9x^2 - 9x + 2} = 0$ by graphing $y = \dfrac{2x^2 + x - 1}{9x^2 - 9x + 2}$ you may get "garbage," as in Figure 2–32. Now you could zoom in for a clearer picture, but it's easier to use the fact that *a fraction is zero only when its numerator is zero and its denominator is nonzero.* Finding the numbers that make the numerator 0, means solving $2x^2 + x - 1 = 0$, which is easily done.

$$2x^2 + x - 1 = 0$$

$$(2x - 1)(x + 1) = 0$$

$$x = 1/2 \qquad \text{or} \qquad x = -1$$

You can readily verify that neither of these numbers make the denominator 0. Hence the solutions of the original equation are 1/2 and −1. ■

The technique of Example 7 is recommended for equations of the form "fraction = 0": set the numerator equal to 0, and solve. Those solutions that do not make the denominator 0 are the solutions of the original equation. A number that makes *both* numerator and denominator 0 is *not* a solution because 0/0 is not defined.

Applications

Graphical and numerical solution methods can be very helpful in dealing with applied problems, since approximate solutions are perfectly adequate for many real-life situations.

Example 8 According to data from the U.S. Bureau of the Census, the approximate population y (in millions) of Chicago and Los Angeles between 1950 and 2000 are given by:

Chicago $\quad y = .00003046x^3 - .0023x^2 + .02024x + 3.62,$

Los Angeles $\quad y = .0000113x^3 - .000922x^2 + .0538x + 1.97,$

where $x = 0$ corresponds to 1950. In what year did the two cities have the same population?

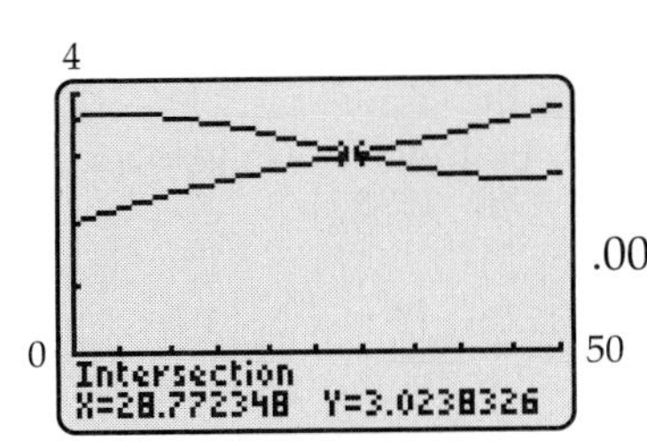

Figure 2–33

Solution We are asked to find the value for x for which

$$\text{Population of Chicago} = \text{Population of Los Angeles}$$

$$.00003046x^3 - .0023x^2 + .02024x + 3.62 = .0000113x^3 - .000922x^2 + .0538x + 1.97$$

We can solve this equation by graphing the left and right sides and finding their intersection point, as in Figure 2–33. Therefore, the populations are the same when $x \approx 28.77$, that is, late in the year 1978. ■

Exercises 2.2

In Exercises 1–6, determine graphically the number *of solutions of the equation, but don't solve the equation. You may need a viewing window other than the standard one to find all the* x*-intercepts.*

1. $x^5 + 5 = 3x^4 + x$
2. $x^3 + 5 = 3x^2 + 24x$
3. $x^7 - 10x^5 + 15x + 10 = 0$
4. $x^5 + 36x + 25 = 13x^3$
5. $x^4 + 500x^2 - 8000x = 16x^3 - 32{,}000$
6. $6x^5 + 80x^3 + 45x^2 + 30 = 45x^4 + 86x$

In Exercises 7–20, use graphical approximation (zoom-in or a root finder or an intersection finder) to find a solution of the equation in the given interval.

7. $x^3 + 4x^2 + 10x + 15 = 0; \quad (-3, -2)$
8. $x^3 + 9 = 3x^2 + 6x; \quad (1, 2)$
9. $x^4 + x - 3 = 0; \quad (0, \infty)$
10. $x^5 + 5 = 3x^4 + x; \quad (-\infty, 1)$
11. $\sqrt{x^4 + x^3 - x - 3} = 0; \quad (0, \infty)$
12. $\sqrt{8x^4 - 14x^3 - 9x^2 + 11x - 1} = 0; \quad (-\infty, 0)$
13. $\sqrt{\frac{2}{5}x^5 + x^2 - 2x} = 0; \quad (0, \infty)$
14. $\sqrt{x^4 + x^2 - 3x + 1} = 0; \quad (0, 1)$
15. $x^2 = \sqrt{x + 5}; \quad (-2, -1)$
16. $\sqrt{x^2 - 1} - \sqrt{x + 9} = 0; \quad (3, 4)$
17. $\dfrac{2x^5 - 10x + 5}{x^3 + x^2 - 12x} = 0; \quad (-\infty, 0)$
18. $\dfrac{3x^5 - 15x + 5}{x^7 - 8x^5 + 2x^2 - 5} = 0; \quad (1, \infty)$
19. $\dfrac{x^3 - 4x + 1}{x^2 + x - 6} = 0; \quad (1, \infty)$
20. $\dfrac{4}{x + 2} - \dfrac{3}{x + 1} = 0; \quad (0, \infty)$ [*Hint:* Write the left side as a single fraction.]

In Exercises 21–34, use algebraic, graphical, or numerical methods to find all real solutions of the equation, approximating when necessary.

21. $2x^3 - 4x^2 + x - 3 = 0$
22. $6x^3 - 5x^2 + 3x - 2 = 0$
23. $x^5 - 6x + 6 = 0$
24. $x^3 - 3x^2 + x - 1 = 0$
25. $10x^5 - 3x^2 + x - 6 = 0$
26. $\frac{1}{4}x^4 - x - 4 = 0$
27. $2x - \frac{1}{2}x^2 - \frac{1}{12}x^4 = 0$
28. $\frac{1}{4}x^4 + \frac{1}{3}x^2 + 3x - 1 = 0$
29. $\dfrac{5x}{x^2 + 1} - 2x + 3 = 0$
30. $\dfrac{2x}{x + 5} = 1$
31. $|x^2 - 4| = 3x^2 - 2x + 1$
32. $|x^3 + 2| = 5 + x - x^2$
33. $\sqrt{x^2 + 3} = \sqrt{x - 2} + 5$
34. $\sqrt{x^3 + 2} = \sqrt{x + 5} + 4$

In Exercises 35–40, find an exact solution of the equation in the given interval. (For example, if the graphical approximation of a solution begins .3333, check to see if 1/3 is the exact solution. Similarly, $\sqrt{2} \approx 1.414$; so if your approximation begins 1.414, check to see if $\sqrt{2}$ is a solution.)

35. $3x^3 - 2x^2 + 3x - 2 = 0; \quad (0, 1)$
36. $4x^3 - 3x^2 - 3x - 7 = 0; \quad (1, 2)$
37. $12x^4 - x^3 - 12x^2 + 25x - 2 = 0; \quad (0, 1)$
38. $8x^5 + 7x^4 - x^3 + 16x - 2 = 0; \quad (0, 1)$
39. $4x^4 - 13x^2 + 3 = 0; \quad (1, 2)$
40. $x^3 + x^2 - 2x - 2 = 0; \quad (1, 2)$

Exercises 41–44 deal with exponential, logarithmic, and trigonometric equations, which will be dealt with in later chapters. If you are familiar with these concepts, solve each equation graphically.

41. $10^x - \frac{1}{4}x = 28$

42. $x + \sin\left(\frac{1}{2}x\right) = 4$

43. $\ln x - x^2 + 3 = 0$

44. $e^x - 6x = 5$

45. According to data from the National Center for Education Statistics and the College Board, the average cost y of tuition and fees (in thousands of dollars) at public four-year institutions in year x is approximated by the equation

$$y = .00044x^3 - .00039x^2 + .114x + 2.195,$$

where $x = 0$ corresponds to 1990. If this model continues to be accurate, in what year will tuition and fees reach $4000?

46. Use the equation in Example 8 to determine the year in which the population of Los Angeles reached 2.6 million.

47. According to data from the U.S. Department of Health and Human Services, the cumulative number y of AIDS cases (in thousands) diagnosed in the United States during 1982–1993 is approximated by

$$y = 3.223x^2 - 14.81x + 17.75 \quad (2 \le x \le 13),$$

where $x = 0$ corresponds to 1980. In what year did the cumulative number of cases reach 250,000?

48. **(a)** How many real solutions does the equation

$$.2x^5 - 2x^3 + 1.8x + k = 0$$

have when $k = 0$?

(b) How many real solutions does it have when $k = 1$?

(c) Is there a value of k for which the equation has just one real solution?

(d) Is there a value of k for which the equation has no real solutions?

2.3 Applications of Equations

Actual problem situations are usually described verbally. In order to solve such problems you must interpret this verbal information and express it as an equivalent mathematical problem. The following guidelines may be helpful.

Setting up Applied Problems

1. ***Read*** **the problem carefully, and determine what is asked for.**
2. ***Label*** **the unknown quantities by letters (variables), and, if appropriate, draw a picture of the situation.**
3. ***Translate*** **the verbal statements in the problem and the relationships between the known and unknown quantities into mathematical language.**
4. ***Consolidate*** **the mathematical information into an equation in one variable that can be solved or an equation in two variables that can be graphed in order to determine at least one of the unknown quantities.**

Here are some examples of how these guidelines are applied.

Example 1 Set up the following problem: The average of two real numbers is 41.125, and their product is 1683. What are the numbers?

Solution *Read:* We are asked for two numbers. *Label:* Call the numbers x and y. *Translate:*

English Language	Mathematical Language
Two numbers	x and y
Their average is 41.125.	$\frac{x+y}{2} = 41.125$
Their product is 1683.	$xy = 1683$

Consolidate: One technique to use when you have two unknowns is to express one in terms of the other and use this to obtain an equation in one variable. In this case we can do that by solving the second equation for y:

$$xy = 1683$$
$$y = \frac{1683}{x}$$

and substituting the result in the first equation:

$$\frac{x+y}{2} = 41.125$$
$$\frac{x + \frac{1683}{x}}{2} = 41.125$$

Multiply both sides by 2:
$$x + \frac{1683}{x} = 82.25$$

The solution of this equation is one of the numbers, and $1683/x$ is the other. ■

Example 2 Set up the following problem: A rectangle is twice as long as it is wide. If it has an area of 24.5 square inches, what are its dimensions?

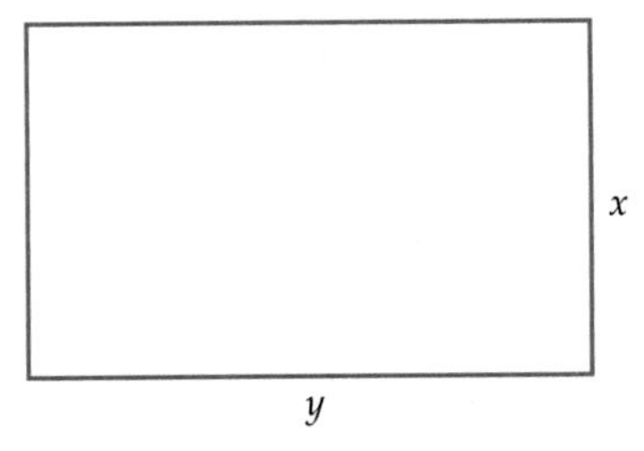

Figure 2–34

Solution *Read:* We are asked to find the length and width. *Label:* Let x denote the width and y the length, and draw a picture of the situation, as in Figure 2–34. *Translate:* Use the fact that the area of a rectangle is length × width.

English Language	Mathematical Language
The width and length of the rectangle	x and y
The length is twice the width.	$y = 2x$
The area is 24.5 square inches.	$xy = 24.5$

Consolidate: Substitute $y = 2x$ in the area equation:

$$xy = 24.5$$
$$x(2x) = 24.5$$

So the equation to be solved is $2x^2 = 24.5$. ■

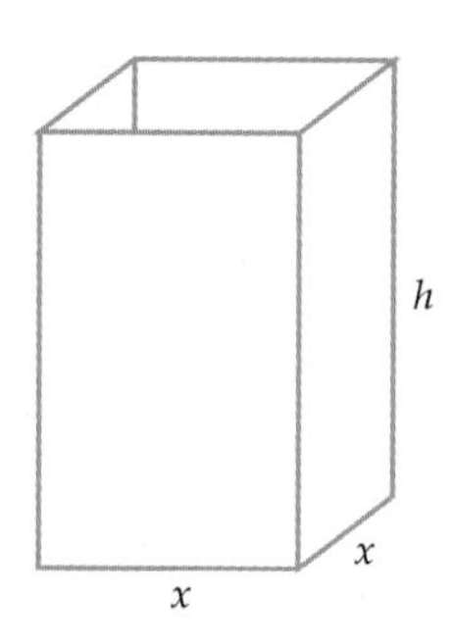

Figure 2–35

Example 3 Set up this problem: A rectangular box with a square base and no top is to have a volume of 30,000 cubic centimeters. If the surface area of the box is 6000 square centimeters, what are its dimensions?

Solution *Read:* We must find the length, width, and height of the box. *Label:* Let x denote the length. Since the base is square, the length and width are the same. Let h denote the height, as in Figure 2–35. *Translate:* Recall that the volume of a box is given by the product length × width × height and that the surface area is the sum of the area of the base and the area of the four sides of the box. Then we have these translations:

English Language	Mathematical Language
The length, width, and height	x, x, and h
The volume is 30,000 cm^3.	$x^2h = 30{,}000$
The surface area is 6000 cm^2.	$x^2 + 4xh = 6000$

Consolidate: We have two equations in two variables, so we solve the first equation for h

$$h = \frac{30{,}000}{x^2}$$

and substitute this result in the second equation:

$$x^2 + 4x\left(\frac{30{,}000}{x^2}\right) = 6000$$

$$x^2 + \frac{120{,}000}{x} = 6000$$

The solution of this last equation will provide the solution of the problem. ■

Once you are comfortable with the process for setting up problems, you can often do much of it mentally. In the rest of this section the setup process will be simplified or shortened whenever doing so won't cause any confusion.

Setting up a problem is only half the job. You must then solve the equation you have obtained. Whenever possible, linear and quadratic equations should be solved algebraically, giving exact answers. Other equations may require graphical or numerical methods to find approximate solutions. When you get an answer, by whatever means, it is extremely important that you

Interpret your answers in terms of the original problem. Do they make sense? Do they satisfy the required conditions?

In particular, an equation may have several solutions, some of which may not make sense in the context of the problem. For instance, distance can't be negative, the number of people in a room cannot be a proper fraction, etc.

Applications

We begin with some problems involving interest. Recall that 8% means .08 and that "8% *of* 227" means ".08 *times* 227," that is, $.08(227) = 18.16$. The basic rule of annual simple interest is:

Interest = rate × amount.

Example 4 A high-risk stock pays dividends at a rate of 12% per year, and a savings account pays 6% interest per year. How much of a \$9000 investment should be put in the stock and how much in the savings account in order to obtain a return of 8% per year on the total investment?

Solution *Label:* Let x be the amount invested in stock. Then the rest of the \$9000, namely, $(9000 - x)$ dollars, goes in the savings account. *Translate:* We want the total return on \$9000 to be 8%, so we have:

$$\left(\begin{array}{l}\text{Return on x dollars}\\ \text{of stock at 12\%}\end{array}\right) + \left[\begin{array}{l}\text{Return on } (9000 - x)\\ \text{dollars of savings at 6\%}\end{array}\right] = 8\% \text{ of } \$9000$$

$$(12\% \text{ of } x \text{ dollars}) + [6\% \text{ of } (9000 - x) \text{ dollars}] = 8\% \text{ of } \$9000$$

$$\begin{aligned} .12x + .06(9000 - x) &= .08(9000)\\ .12x + .06(9000) - .06x &= .08(9000)\\ .12x + 540 - .06x &= 720\\ .12x - .06x &= 720 - 540\\ .06x &= 180\\ x &= \frac{180}{.06} = 3000. \end{aligned}$$

Therefore, \$3000 should be invested in stock and $(9000 - 3000) = \$6000$ in the savings account. If this is done, the total return will be 12% of \$3000 (\$360) plus 6% of \$6000 (\$360), a total of \$720, which is precisely 8% of \$9000. ■

Example 5 A car radiator contains 12 quarts of fluid, 20% of which is antifreeze. How much fluid should be drained and replaced with pure antifreeze in order that the resulting mixture be 50% antifreeze?

Solution Let x be the number of quarts of fluid to be replaced by pure antifreeze.* When x quarts are drained, there are $12 - x$ quarts of fluid left in the radiator, 20% of which is antifeeze. So we have:

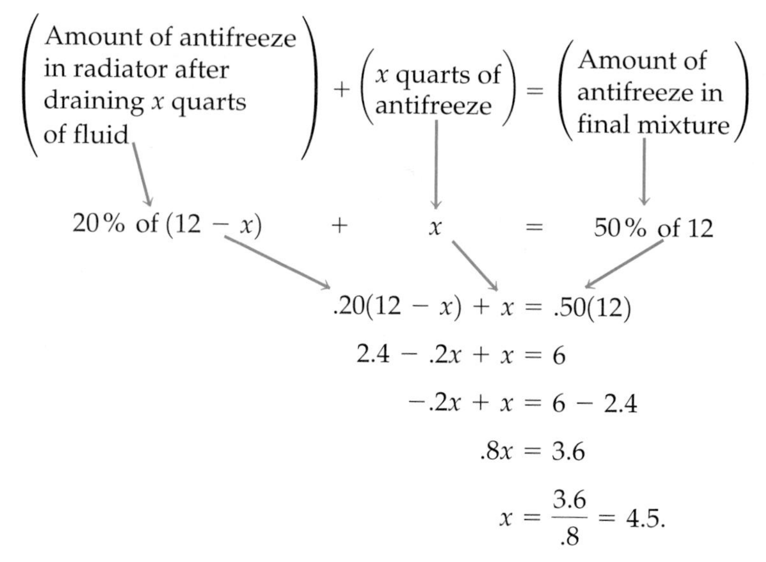

$$
\begin{aligned}
.20(12 - x) + x &= .50(12)\\
2.4 - .2x + x &= 6\\
-.2x + x &= 6 - 2.4\\
.8x &= 3.6\\
x &= \frac{3.6}{.8} = 4.5.
\end{aligned}
$$

Therefore, 4.5 quarts should be drained and replaced with pure antifreeze. ■

The basic formula for problems involving distance and a uniform rate of speed is:

Distance = rate × time.

For instance, if you drive at a rate of 55 mph for 2 hours, you travel a distance of $55 \cdot 2 = 110$ miles.

Example 6 A pilot wants to make an 840-mile round trip from Cleveland to Peoria and back in 5 hours flying time. There will be a headwind of 30 mph going to Peoria, and it is estimated that there will be a 40-mph tailwind returning to Cleveland. At what constant air speed should the plane be flown?

Solution Let x be the engine speed of the plane. On the trip to Peoria, a distance of 420 miles, the actual speed will be $x - 30$ because of the headwind. Since Rate·time = distance, the time to Peoria will be $\frac{\text{distance}}{\text{rate}} = \frac{420}{x - 30}$. On the return trip the actual speed will be $x + 40$ because of the tailwind and the time will be

$$\frac{\text{Distance}}{\text{Rate}} = \frac{420}{x + 40}.$$

*Hereafter we omit the headings Label, Translate, and so on.

Therefore:

$$5 = \begin{pmatrix}\text{Time from Cleveland}\\ \text{to Peoria}\end{pmatrix} + \begin{pmatrix}\text{Time from Peoria}\\ \text{to Cleveland}\end{pmatrix}$$

$$= \frac{420}{x-30} + \frac{420}{x+40}$$

Multiplying both sides by the common denominator $(x - 30)(x + 40)$ and simplifying, we have

$$5(x-30)(x+40) = \frac{420}{x-30}\cdot(x-30)(x+40) + \frac{420}{x+40}\cdot(x-30)(x+40)$$

$$5(x-30)(x+40) = 420(x+40) + 420(x-30)$$

$$(x-30)(x+40) = 84(x+40) + 84(x-30)$$

$$x^2 + 10x - 1200 = 84x + 3360 + 84x - 2520$$

$$x^2 - 158x - 2040 = 0$$

$$(x-170)(x+12) = 0$$

$$x - 170 = 0 \quad \text{or} \quad x + 12 = 0$$

$$x = 170 \qquad\qquad x = -12.$$

Obviously, the negative doesn't apply. Since we multiplied both sides by a quantity involving the variable, we must check that 170 actually is a solution of the original equation. It is; so the plane should be flown at a speed of 170 mph. ■

The preceding examples used only algebraic models (equations in one variable). Sometimes a diagram (geometrical model) is helpful in visualizing the situation and setting up an appropriate equation.

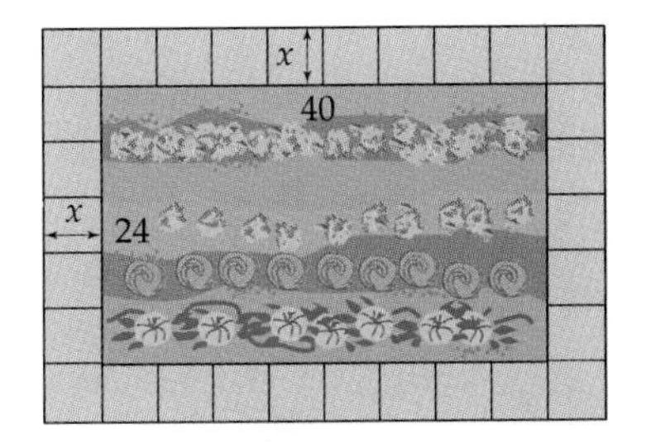

Figure 2–36

Example 7 A landscaper wants to put a cement walk of uniform width around a rectangular garden that measures 24 by 40 feet. She has enough cement to cover 660 square feet. How wide should the walk be in order to use up all the cement?

Solution Let x denote the width of the walk (in feet) and draw a picture of the situation (Figure 2–36).

The length of the outer rectangle is $40 + 2x$ (the garden length plus walks on each end) and its width is $24 + 2x$.

$$\begin{pmatrix}\text{Area of outer}\\ \text{rectangle}\end{pmatrix} - \begin{pmatrix}\text{Area of}\\ \text{garden}\end{pmatrix} = \text{Area of walk}$$

$$\text{Length}\cdot\text{Width} - \text{Length}\cdot\text{Width} = 660$$

$$(40 + 2x)(24 + 2x) - 40\cdot 24 = 660$$

$$960 + 128x + 4x^2 - 960 = 660$$

$$4x^2 + 128x - 660 = 0$$

Dividing both sides by 4 and applying the quadratic formula yields

$$x^2 + 32x - 165 = 0$$

$$x = \frac{-32 \pm \sqrt{(32)^2 - 4\cdot 1\cdot(-165)}}{2\cdot 1}$$

$$x = \frac{-32 \pm \sqrt{1684}}{2} \approx \begin{cases} 4.5183 \\ \text{or} \\ -36.5183 \end{cases}$$

Only the positive solution makes sense in the context of this problem. The walk should be approximately 4.5 feet wide. ■

Example 8 A rectangular box with a square base and no top is to have a volume of 30,000 cubic centimeters. If the surface area of the box is 6000 square centimeters and the box is required to be higher than it is wide, what are its dimensions?

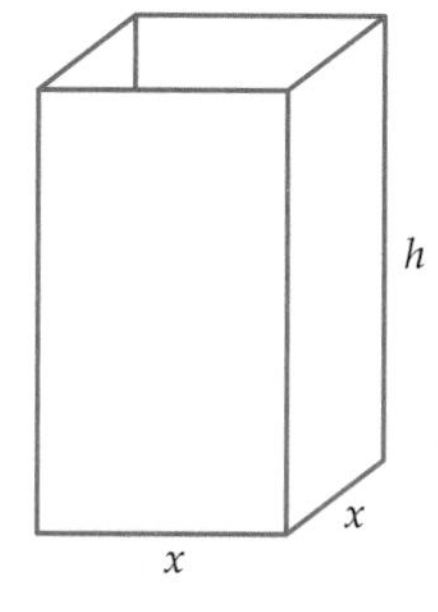

Figure 2–37

Solution This is essentially Example 3 with an extra condition (the box must be higher than it is wide). If the length, width, and height of the box are x, x, and h as shown in Figure 2–37, then as we saw in Example 3,

$$x^2 + \frac{120{,}000}{x} = 6000 \qquad \text{and} \qquad h = \frac{30{,}000}{x^2}.$$

Multiplying both sides of the first equation by x (which is nonzero because it is a length), we obtain

$$x^3 + 120{,}000 = 6000x$$

$$x^3 - 6000x + 120{,}000 = 0.$$

This equation can be solved graphically. The graph of $y = x^3 - 6000x + 120{,}000$ in Figure 2–38 has one negative x-intercept (solution) and two positive ones.

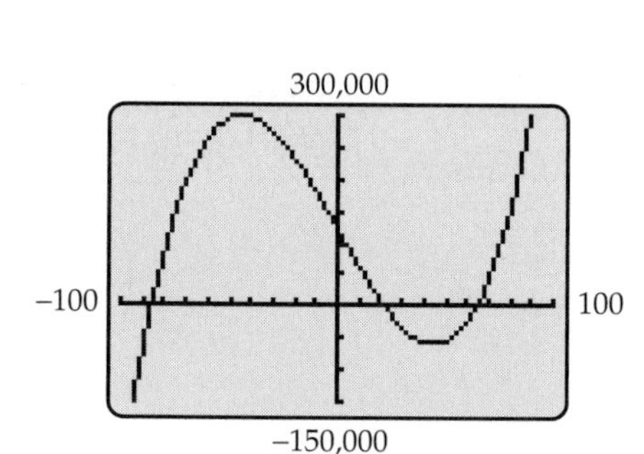

Figure 2–38

The negative solution does not apply here. The tick marks on the x-axis in Figure 2–38 are ten units apart, so one positive solution is near $x = 20$ and the other between $x = 60$ and $x = 70$. If $60 < x < 70$, however, then the height $h = 30{,}000/x^2$ will be smaller than the width x (for instance, if $x = 64$, then $h = 30{,}000/64^2 \approx 7.3$). So the solution near $x = 20$ is the only one that applies to this situation.

GRAPHING EXPLORATION

Use zoom-in or a graphical root finder to obtain an approximation of this solution. Then find the corresponding value of h. ■

Example 9 A box (with no top) of volume 1000 cubic inches is to be made from a 22 × 30 inch sheet of cardboard by cutting squares of equal size from each corner and folding up the flaps, as shown in Figure 2–39. If the box must be at least 4 inches high, what size square should be cut from each corner?

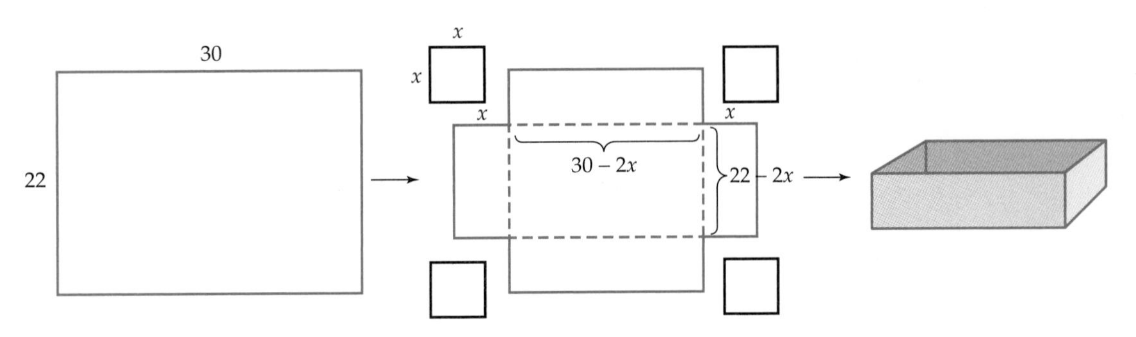

Figure 2–39

Solution Let x denote the length of the side of the square to be cut from each corner. The dashed rectangle in Figure 2–39 is the bottom of the box. Its length is $30 - 2x$ as shown in the figure. Similarly, the width of the box will be $22 - 2x$ and its height will be x inches. Therefore,

$$\text{Length} \times \text{Width} \times \text{Height} = \text{Volume of box}$$

$$(30 - 2x)\cdot(22 - 2x)\cdot x = 1000$$

$$(600 - 104x + 4x^2)x = 1000$$

$$4x^3 - 104x^2 + 660x - 1000 = 0$$

Since the cardboard is 22 inches wide, x must be less than 11 (otherwise you can't cut out two squares of length x). Since x is a length, it is positive. So we need only find solutions of the equation between 0 and 11. We graph $y = 4x^3 - 104x^2 + 660x - 1000$ in a window with $0 \le x \le 11$. A complete graph isn't needed here, only the x-intercepts. Figure 2–40 shows that there are two x-intercepts (solutions). The one between 2 and 3 is not relevant here because x is the height of the box, which must be at least 4 inches. A root finder shows that the other solution is $x \approx 6.47$ (Figure 2–41).

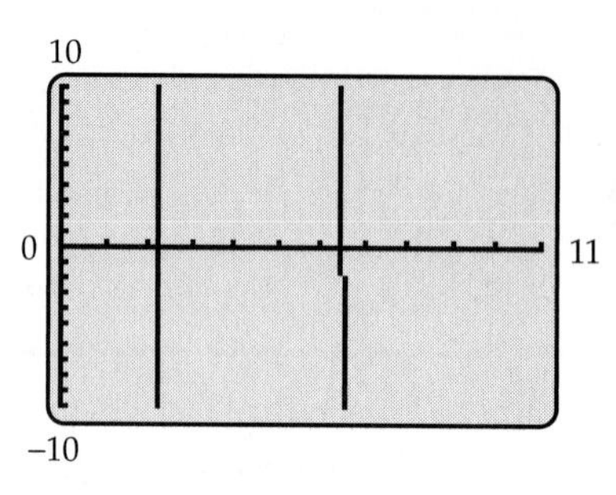

Figure 2–40

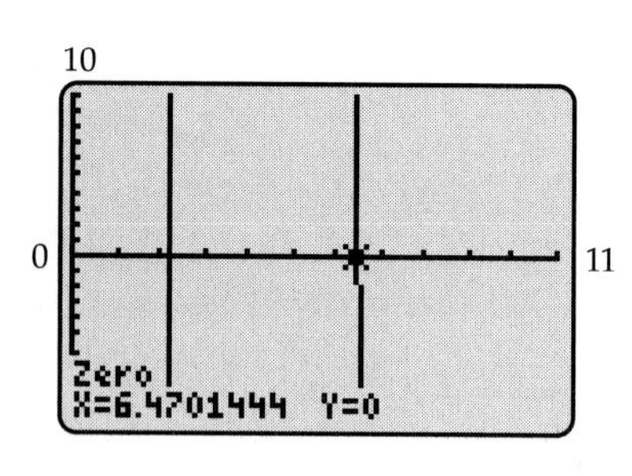

Figure 2–41

Therefore, a 6.47×6.47 inch square should be cut from each corner, resulting in a box whose dimensions are:

Length: $30 - 2x \approx 30 - 2(6.47) = 17.06$ inches

Width: $22 - 2x \approx 20 - 2(6.47) = 9.06$ inches

Height: $x \approx 6.47$ inches ■

Exercises 2.3

In Exercises 1–4, a problem situation is given.

(a) Decide what is being asked for, and label the unknown quantities.

(b) Translate the verbal statements in the problem and the relationships between the known and unknown quantities into mathematical language, using a table as in Examples 1–3. The table is provided in Exercises 1 and 2. You need not find an equation to be solved.

1. The sum of two numbers is 15 and the difference of their squares is 5. What are the numbers?

English Language	Mathematical Language
The two numbers	
Their sum is 15.	
The difference of their squares is 5.	

2. The sum of the squares of two consecutive integers is 4513. What are the integers?

English Language	Mathematical Language
The two integers	
The integers are consecutive.	
The sum of their squares is 4513.	

3. A rectangle has perimeter of 45 centimeters and an area of 112.5 square centimeters. What are its dimensions?

4. A triangle has area 96 square inches and its height is two-thirds of its base. What are the base and height of the triangle?

In Exercises 5–8, set up the problem by labeling the unknowns, translating the given information into mathematical language, and finding an equation that will produce the solution to the problem. You need not solve this equation.

5. A worker gets an 8% pay raise and now makes \$1600 per month. What was the worker's old salary?

6. A merchant has 5 pounds of mixed nuts that cost \$30. He wants to add peanuts that cost \$1.50 per pound and cashews that cost \$4.50 per pound to obtain 50 pounds of a mixture that costs \$2.90 per pound. How many pounds of peanuts are needed?

7. The diameter of a circle is 16 cm. By what amount must the radius be decreased in order to decrease the area by 48π square centimeters?

8. A corner lot has dimensions 25 by 40 yards. The city plans to take a strip of uniform width along the two sides bordering the streets in order to widen these roads. How wide should the strip be if the remainder of the lot is to have an area of 844 square yards?

In the remaining exercises, solve the applied problems.

9. You have already invested \$550 in a stock with an annual return of 11%. How much of an additional \$1100 should be invested at 12% and how much at 6% so that the total return on the entire \$1650 is 9%?

10. If you borrow \$500 from a credit union at 12% annual interest and \$250 from a bank at 18% annual interest, what is the *effective annual interest rate* (that is, what single rate of interest on \$750 would result in the same total amount of interest)?

11. A radiator contains 8 quarts of fluid, 40% of which is antifreeze. How much fluid should be drained and replaced with pure antifreeze so that the new mixture is 60% antifreeze?

12. A radiator contains 10 quarts of fluid, 30% of which is antifreeze. How much fluid should be drained and replaced with pure antifreeze so that the new mixture is 40% antifreeze?

13. Two cars leave a gas station at the same time, one traveling north and the other south. The northbound car travels at 50 mph. After 3 hours the cars are 345 miles apart. How fast is the southbound car traveling?

14. An airplane flew with the wind for 2.5 hours and returned the same distance against the wind in 3.5 hours. If the cruising speed of the plane was a constant 360 mph in air, how fast was the wind blowing? [*Hint:* If the wind speed is r miles per hour, then the plane travels at $(360 + r)$ mph with the wind and at $(360 - r)$ mph against the wind.]

15. The average of two real numbers is 41.125 and their product is 1683. What are the numbers? [*Hint:* See Example 1.]

16. A rectangle is twice as long as it is wide. If it has an area of 24.5 inches, what are its dimensions? [*Hint:* See Example 2.]

17. A 13-foot-long ladder leans on a wall. The bottom of the ladder is 5 feet from the wall. If the bottom is pulled out 3 feet farther from the wall, how far does the top of the ladder move down the wall? [*Hint:* The ladder, ground, and wall form a right triangle. Draw pictures of this triangle before and after the ladder is moved. Use the Pythagorean Theorem to set up an equation.]

18. In Example 6 of Section 1.4, how many can openers must be produced in order to have an average cost per can opener of \$3?

19. Red Riding Hood drives the 432 miles to Grandmother's house in 1 hour less than it takes the Wolf to drive the same route. Her average speed is 6 mph faster than the Wolf's average speed. How fast does each drive?

20. To get to work Sam jogs 3 kilometers to the train, then rides the remaining 5 kilometers. If the train goes 40 kilometers per hour faster than Sam's constant rate of jogging and the entire trip takes 30 minutes, how fast does Sam jog?

Background for Exercises 21 – 24 *If an object is thrown upward, dropped, or thrown downward and travels in a vertical line subject only to gravity (with wind resistance ignored), then the height h of the object above the ground (in feet) after t seconds is given by:*

$$h = -16t^2 + v_0t + h_0$$

where h_0 is the initial height of the object at starting time $t = 0$, and v_0 is the initial velocity (speed) of the object at time $t = 0$. The value of v_0 is taken as positive if the object starts moving upward at time $t = 0$ and negative if the object starts moving downward at $t = 0$. An object that is dropped (rather than thrown downward) has initial velocity $v_0 = 0$.

21. How long does it take an object to reach the ground if
(a) it is dropped from the top of a 640-foot-high building?
(b) it is thrown downward from the top of the same building, with an initial velocity of 52 feet per second?

22. You are standing on a cliff 200 feet high. How long will it take a rock to reach the ground if
(a) you drop it?
(b) you throw it downward at an initial velocity of 40 feet per second?
(c) How far does the rock fall in 2 seconds if you throw it downward with an initial velocity of 40 feet per second?

23. A rocket is fired straight up from ground level with an initial velocity of 800 feet per second.
(a) How long does it take the rocket to rise 3200 feet?
(b) When will the rocket hit the ground?

24. A rocket loaded with fireworks is to be shot vertically upward from ground level with an initial velocity of 200 feet per second. When the rocket reaches a height of 400 feet on its upward trip the fireworks will be detonated. How many seconds after lift-off will this take place?

25. The dimensions of a rectangular box are consecutive integers. If the box has volume of 13,800 cubic centimeters, what are its dimensions?

26. Find a real number that exceeds its cube by 2.

27. The surface area S of the right circular cone at the left in the figure below is given by $S = \pi r\sqrt{r^2 + h^2}$. What radius should be used to produce a cone of height 5 inches and surface area 100 square inches?

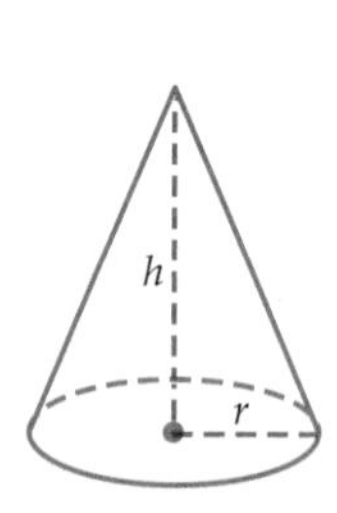

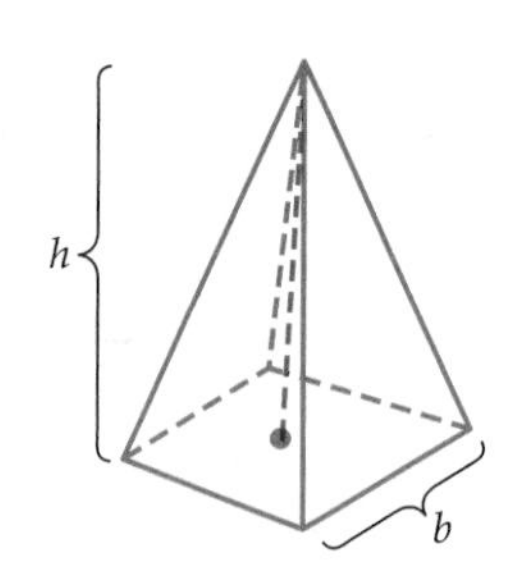

28. The surface area of the right square pyramid at the right in the figure above is given by $S = b\sqrt{b^2 + 4h^2}$. If the pyramid has height 10 feet and surface area 100 square feet, what is the length of a side b of its base?

29. Suppose that the open-top box being made from a sheet of cardboard in Example 9 is required to have at least one of its dimensions *greater* than 18 inches. What size square should be cut from each corner?

30. A homemade loaf of bread turns out to be a perfect cube. Five slices of bread, each .6 inch thick, are cut from one end of the loaf. The remainder of the loaf now has a volume of 235 cubic inches. What were the dimensions of the original loaf?

31. A rectangular bin with an open top and volume of 38.72 cubic feet is to be built. The length of its base must be twice the width and the bin must be at least 3 feet high. Material for the base of the bin costs \$12 per square foot and material for the sides costs \$8 per square foot. If it costs \$538.56 to build the bin, what are its dimensions?

Thinker

32. One corner of an 8.5×11 inch piece of paper is folded over to the opposite side, as shown in the figure. The area of the darkly shaded triangle at the lower left is 6 square inches and we want to find the length x.
(a) Take a piece of paper this size and experiment. Approximately, what is the largest value x could have (and still have the paper look like the figure)? With this value of x, what is the approximate area of the triangle? Try some other possibilities.
(b) Now find an exact answer by constructing and solving a suitable equation. Explain why one of the solutions to the equation is not an answer to this problem.

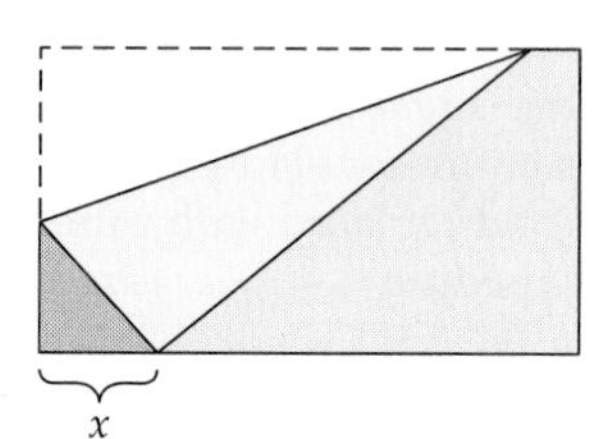

2.4 Optimization Applications*

Many real-life situations require you to find the largest or smallest quantity satisfying certain conditions. For instance, automotive engineers want to design engines with maximum fuel efficiency. Similarly, a cereal manufacturer who needs a box of volume 300 cubic inches might want to know the dimensions of the box that requires the least amount of cardboard (and hence is cheapest). The exact solutions of such minimum/maximum problems require calculus. However, graphing technology can provide very accurate approximate solutions as we shall see. Before reading the following examples, you should review the guidelines for setting up applied problems (page 80).

Example 1 Find two negative numbers whose product is 50 and whose sum is as large as possible.

Solution Let x and z be the two negative numbers, and let y be their sum. Then $xz = 50$ and $y = x + z$. Solving $xz = 50$ for z we have $z = 50/x$ so that

$$y = x + z = x + 50/x.$$

We must find the value of x that makes y as large as possible. Since x must be negative, we graph $y = x + 50/x$ in a window with $-20 \leq x \leq 0$ (Figure 2–42).

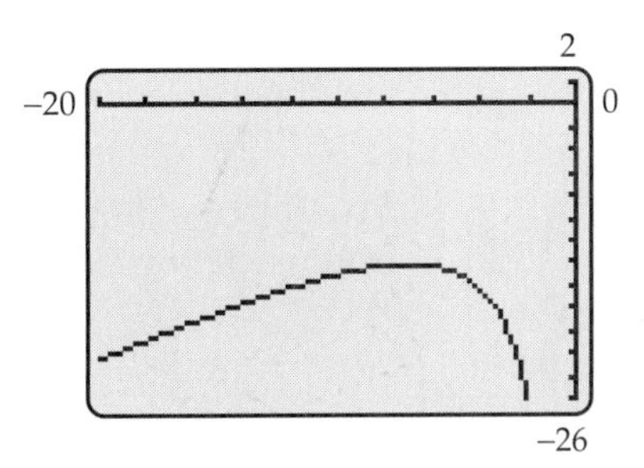

Figure 2–42

Each point (x, y) on this graph represents one of the possibilities:

x represents one of the two negative numbers ($50/x$ is the other);

y is the sum of the two negative numbers.

Since we want y to be as large as possible, we must find the point on the graph with largest y-coordinate, that is, the highest point on this part of the graph.

*This section may be omitted or postponed until Chapter 4 (Polynomial and Rational Functions). It will be used only in occasional examples and in clearly identifiable exercises.

GRAPHING EXPLORATION

Use zoom-in or a maximum finder to approximate the highest point. Its first coordinate x is one of the two numbers, and $50/x$ is the other. ■

Example 2 A box with no top is to be made from a 22×30 inch sheet of cardboard by cutting squares of equal size from each corner and bending up the flaps, as shown in Figure 2–43. To the nearest hundredth of an inch, what size square should be cut from each corner in order to obtain a box with the largest possible volume, and what is the volume of this box?

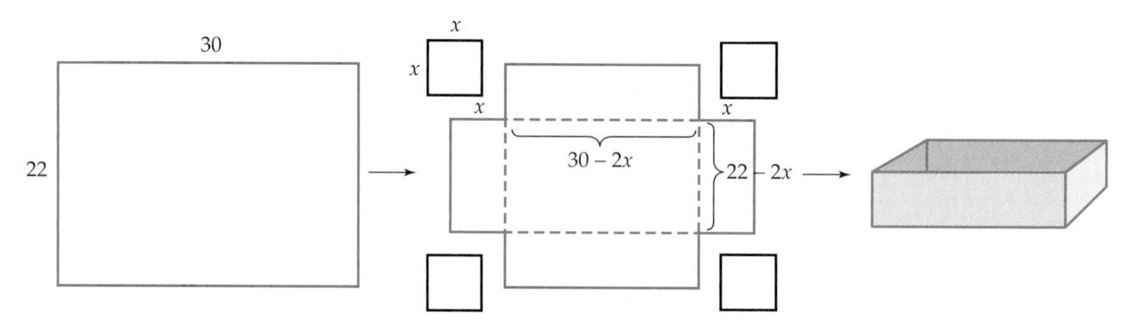

Figure 2–43

Solution Let x denote the length of the side of the square to be cut from each corner. Then,

$$\begin{aligned}\text{Volume of box} &= \text{Length} \times \text{Width} \times \text{Height} \\ &= (30 - 2x)\cdot(22 - 2x)\cdot x \\ &= 4x^3 - 104x^2 + 660x\end{aligned}$$

Thus the equation $y = 4x^3 - 104x^2 + 660x$ gives the volume y of the box that results from cutting a square of side x from each corner. Since the shortest side of the cardboard is 22 inches, the length x of the side of the cut-out square must be less than 11 (why?).

Each point on the graph of $y = 4x^3 - 104x^2 + 660x$ $(0 < x < 11)$ in Figure 2–44 represents one of the possibilities:

The x-coordinate is the size of the square to be cut from each corner;

The y-coordinate is the volume of the resulting box.

Figure 2–44

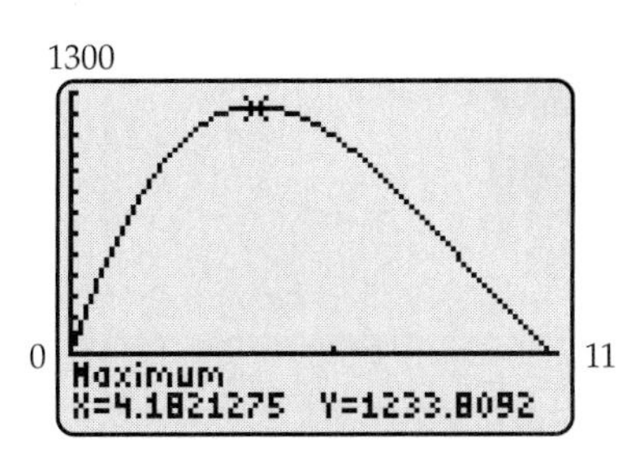

Figure 2–45

The box with the largest volume corresponds to the point with the largest y-coordinate, that is, the highest point in the viewing window. A maximum finder (Figure 2–45) shows that this point is approximately (4.182, 1233.809). Therefore, a square measuring approximately 4.18×4.18 inches should be cut from each corner, producing a box of volume approximately 1233.81 cubic inches. ■

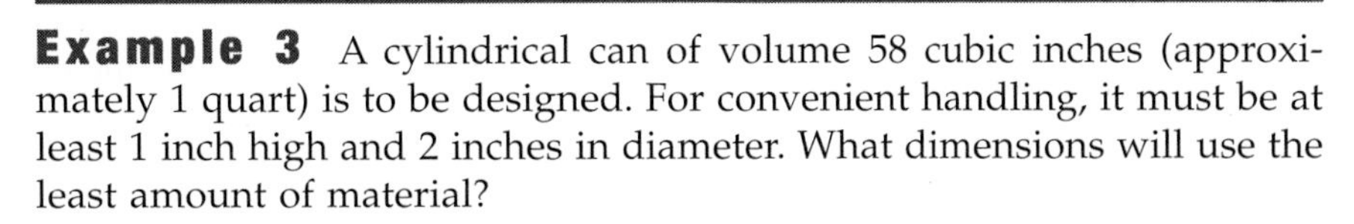

Example 3 A cylindrical can of volume 58 cubic inches (approximately 1 quart) is to be designed. For convenient handling, it must be at least 1 inch high and 2 inches in diameter. What dimensions will use the least amount of material?

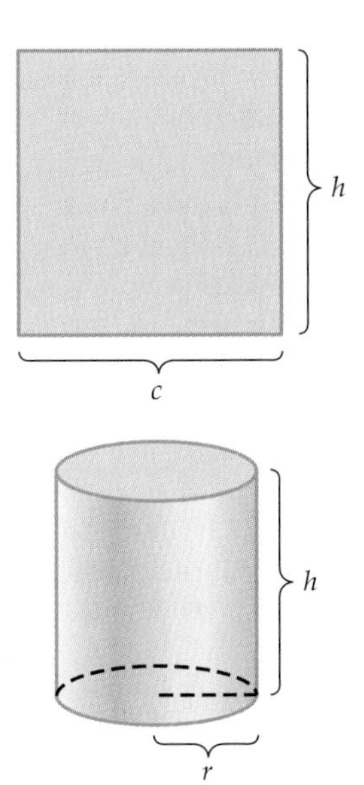

Figure 2–46

Solution We can construct a can by rolling a rectangular sheet of metal into a tube and then attaching the top and bottom, as shown in Figure 2–46. The surface area of the can (which determines the amount of material) is

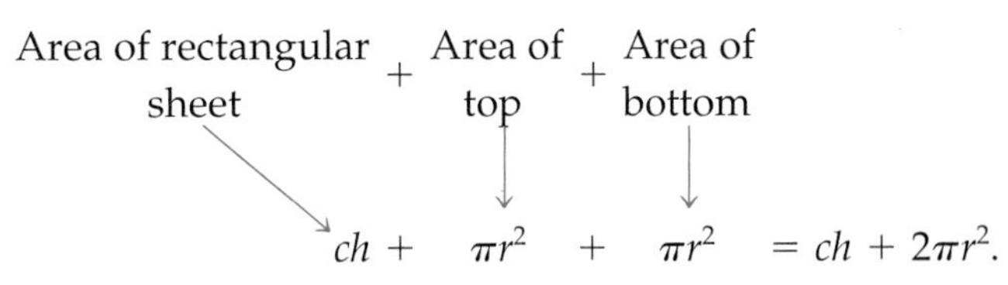

$$ch + \pi r^2 + \pi r^2 = ch + 2\pi r^2.$$

When the sheet is rolled into a tube, the width c of the sheet is the circumference of the end of the can, so that $c = 2\pi r$ and, hence,

$$\text{Surface area} = ch + 2\pi r^2 = 2\pi rh + 2\pi r^2.$$

The volume of a cylinder of radius r and height h is $\pi r^2 h$. Since the can is to have volume 58 cubic inches, we have

$$\pi r^2 h = 58, \qquad \text{or equivalently,} \qquad h = \frac{58}{\pi r^2}.$$

Therefore,

$$\text{Surface area} = 2\pi rh + 2\pi r^2 = 2\pi r\left(\frac{58}{\pi r^2}\right) + 2\pi r^2 = \frac{116}{r} + 2\pi r^2.$$

Note that r must be greater than 1 (since the diameter $2r$ must be at least 2). Furthermore, r cannot be more than 5 (if $r > 5$ and $h \geq 1$, then the volume $\pi r^2 h$ would be at least $\pi \cdot 25 \cdot 1$, which is greater than 58).

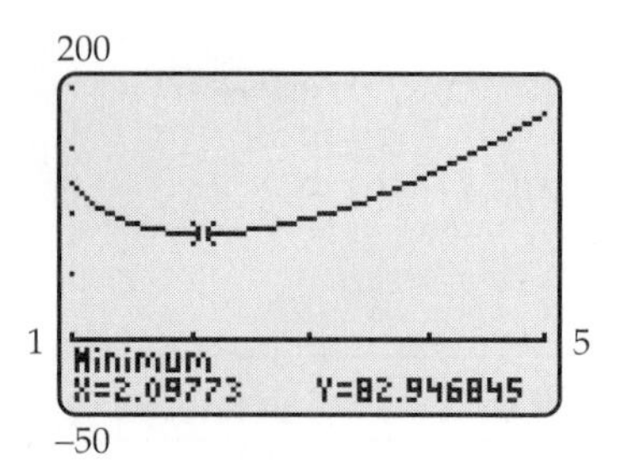

Figure 2–47

The situation can be represented by the graph of the equation $y = 116/x + 2\pi x^2$, as in Figure 2–47. The x-coordinate of each point represents a possible radius and the y-coordinate the surface area of the corresponding can. We must find the point with the smallest y-coordinate, that is, the lowest point on the graph. A graphical minimum finder (Figure 2–47) shows that the coordinates of this point are approximately (2.09773, 82.946845). If the radius is 2.09773, then the height is $58/(\pi 2.09773^2) \approx 4.1955$. As a practical matter, it would probably be best to round to one decimal place and construct a can of radius 2.1 and height 4.2 inches. ■

Exercises 2.4

In Exercises 1–6, find the coordinates of the highest or lowest point on the part of the graph of the equation in the given viewing window. Only the range of x-coordinates for the window are given; you must choose an appropriate range of y-coordinates.

1. $y = 2x^3 - 3x^2 - 12x + 1$; highest point when $-3 \le x \le 3$

2. $y = 2x^6 + 3x^5 + 3x^3 - 2x^2$; lowest point when $-3 \le x \le 3$

3. $y = \dfrac{4}{x^2} - \dfrac{7}{x} + 1$; lowest point when $-10 \le x \le 10$

4. $y = \dfrac{1}{x^2 + 2x + 2}$; highest point when $-5 \le x \le 5$

5. $y = \dfrac{x^2(x+1)^3}{(x-2)^2(x-4)^2}$; highest point when $-1 \le x \le 0$ [*Hint:* Think small.]

6. $y = \dfrac{x^2(x+1)^3}{(x-2)(x-4)^2}$; lowest point when $-10 \le x \le -1$

7. Find the highest point on the part of the graph of $y = x^3 - 3x + 2$ that is shown in the given window. The answers are not all the same.
(a) $-2 \le x \le 0$ **(b)** $-2 \le x \le 2$
(c) $-2 \le x \le 3$

8. Find the lowest point on the part of the graph of $y = x^3 - 3x + 2$ that is shown in the given window.
(a) $0 \le x \le 2$ **(b)** $-2 \le x \le 2$
(c) $-3 \le x \le 2$

9. Find the dimensions of the rectangular box with a square base and no top that has volume 30,000 cubic centimeters and the smallest possible surface area. [*Hint:* See Example 3 in Section 2.3.]

10. An open-top box with a square base is to be constructed from 120 square centimeters of material. What dimensions will produce a box
(a) of volume 100 cubic centimeters?
(b) with largest possible volume?

11. A 20-inch square piece of metal is to be used to make an open-top box by cutting equal-sized squares from each corner and folding up the sides (as in Example 2). The length, width, and height of the box are each to be less than 12 inches. What size squares should be cut out to produce a box with
(a) volume 550 cubic inches?
(b) largest possible volume?

12. A cylindrical waste container with no top, a diameter of at least 2 feet, and a volume of 25 cubic feet is to be constructed. What should its radius be if
(a) 65 square feet of material are to be used to construct it?
(b) the smallest possible amount of material is to be used to construct it? In this case, how much material is needed?

13. If $c(x)$ is the cost of producing x units, then $c(x)/x$ is the *average cost* per unit.* Suppose the cost of producing x units is given by $c(x) = .13x^3 - 70x^2 + 10{,}000x$ and that no more than 300 units can be produced per week.
(a) If the average cost is \$1100 per unit, how many units are being produced?
(b) What production level should be used in order to minimize the average cost per unit? What is the minimum average cost?

*Depending on the situation, a unit of production might consist of a single item or several thousand items. Similarly, the cost of x units might be measured in thousands of dollars.

14. A manufacturer's revenue (in cents) from selling x items per week is given by $200x - .02x^2$. It costs $60x + 30{,}000$ cents to make x items.
   (a) Approximately how many items should be made each week in order to make a profit of \$1100? (Don't confuse cents and dollars.)
   (b) How many items should be made each week in order to have the largest possible profit? What is that profit?

15. (a) A company makes novelty bookmarks that sell for \$142 per hundred. The cost (in dollars) of making x hundred bookmarks is $x^3 - 8x^2 + 20x + 40$. Because of other projects, a maximum of 600 bookmarks per day can be manufactured. Assuming that the company can sell all the bookmarks it makes, how many should it make each day to maximize profits?
   (b) Due to a change in other orders, as many as 1600 bookmarks can now be manufactured each day. How many should be made in order to maximize profits?

16. If the cost of material to make the can in Example 3 is 5 cents per square inch for the top and bottom and 3 cents per square inch for the sides, what dimensions should be used to minimize the cost of making the can? [The answer is not the same as in Example 3.]

17. A certain type of fencing comes in rigid 10-foot-long segments. Four uncut segments are used to fence in a garden on the side of a building, as shown in the figure. What value of x will result in a garden of the largest possible area and what is that area?

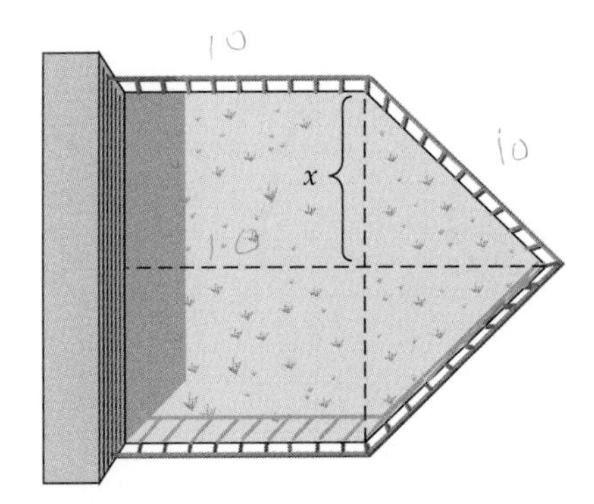

18. A rectangle is to be inscribed in a semicircle of radius 2, as shown in the figure. What is the largest possible area of such a rectangle? [*Hint:* The width of the rectangle is the second coordinate of the point P (why?) and P is on the top half of the circle $x^2 + y^2 = 4$.

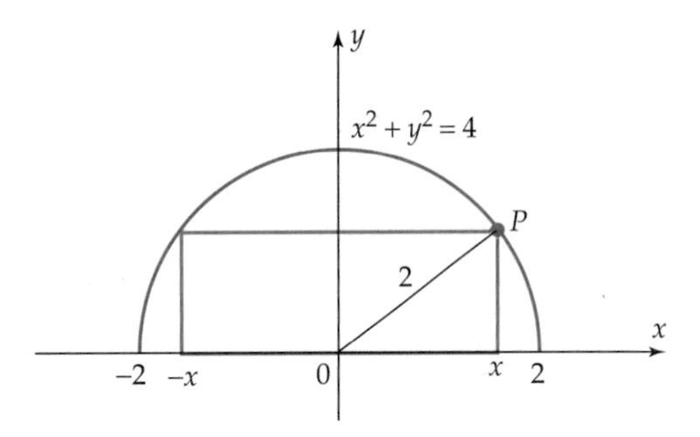

19. Find the point on the graph of $y = 5 - x^2$ that is closest to the point (0, 1) and has positive coordinates. [*Hint:* The distance from the point (x, y) on the graph to (0, 1) is $\sqrt{(x-0)^2 + (y-1)^2}$; express y in terms of x.]

20. A manufacturer's cost (in thousands of dollars) of producing x thousand units is $x^3 - 6x^2 + 15x$ dollars and the revenue (in thousands) from x thousand units is $9x$ dollars. What production level(s) will result in the largest possible profit?

21. A hardware store sells ladders throughout the year. It costs \$20 every time an order for ladders is placed and \$10 to store a ladder until it is sold. When ladders are ordered x times per year, then an average of $300/x$ ladders are in storage at any given time. How often should the company order ladders each year in order to minimize its total ordering and storage costs? [*Be careful:* The answer must be an integer.]

22. A mathematics book has 36 square inches of print per page. Each page has a left side margin of 1.5 inches and top, bottom, and right side margins of .5 inch. If a page cannot be wider than 7.5 inches, what should its length and width be in order to use the least amount of paper?

2.5 Linear Models*

People working in business, medicine, agriculture, and other fields frequently want to know the relationship between two quantities. For instance,

How does money spent on advertising affect sales?

What effect does a fertilizer have on crop yield?

Do large doses of certain vitamins lengthen life expectancy?

In many such situations there is sufficient data available to construct a **mathematical model,** such as an equation or graph, which provides the desired answers or predicts the likely outcome in cases not included in the data. In this section we consider applications in which the data can be modeled by a linear equation. More complicated models will be considered in later sections.

Example 1 The following table, from the U.S. Department of Commerce, shows the poverty level for a family of four in selected years (families whose income is below this level are considered to be in poverty).

Year	1990	1991	1992	1993	1994	1995	1996
Income	\$13,359	13,942	14,335	14,763	15,141	15,569	16,036

We let $x = 0$ correspond to 1990 and write the incomes in thousands (for instance, 13.359 in place of 13,359). Plotting the data points (0, 13.359), (1, 13.942), and so on, we obtain the scatter plot in Figure 2–48. The fact that these points are almost in a straight line suggests that a linear equation should provide a suitable model. One such equation is

$$y = .43x + 13.43,$$

whose graph is shown in Figure 2–49.

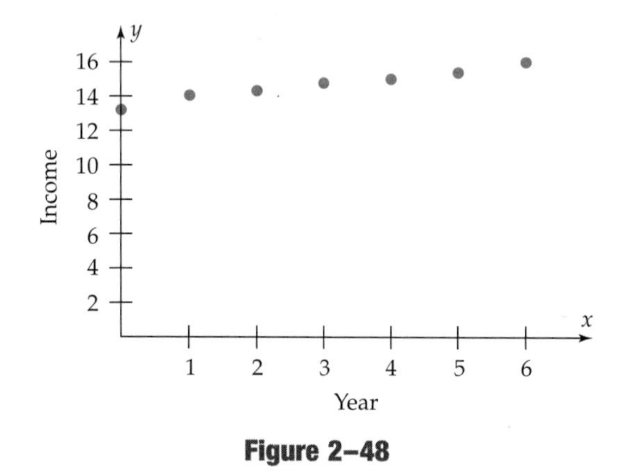

Figure 2–48

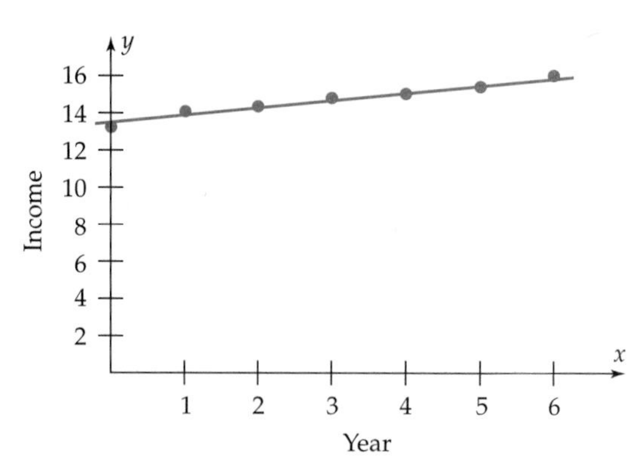

Figure 2–49

*This section is optional. It will regularly be used in clearly identifiable exercises, but not elsewhere in the text.

The graph appears to fit the data quite well. In fact, the poverty level given by the equation differs from the actual level each year by no more than \$82, as the following table demonstrates.

Year x	0	1	2	3	4	5	6
Actual	13,359	13,942	14,335	14,763	15,141	15,569	16,036
Model	13,430	13,860	14,290	14,720	15,150	15,580	16,010

Assuming that this model remains valid after 1996, the poverty level in 2000 ($x = 10$) should be approximately

$$y = .43x + 13.43 = .43(10) + 13.43 = \$17{,}730. \quad \blacksquare$$

Example 1 shows how a typical model approximates the actual data and makes it possible to predict the outcome in other cases. However, it does not deal with questions such as

How do you find an equation that models the data?

How do you determine which equation is the best model for the data?

These questions are best answered with a simplified example.

Example 2 The weekly amount spent on advertising and the weekly sales revenue of a small store over a five-week period are shown in the table.

Advertising Expenditure x (in hundreds of dollars)	1	2	3	4	5
Sales Revenue y (in thousands of dollars)	2	2	3	3	5

The data points are (1, 2), (2, 2), (3, 3), (4, 3), and (5, 5). Figure 2–50 shows three linear models. Model (a) was obtained by taking the data points (1, 2) and (3, 3) and finding the equation of the line they determine (as in Section 1.4). Model (b) was obtained similarly, using the data points (2, 2) and (5, 5). Model (c) was obtained by a method discussed below.

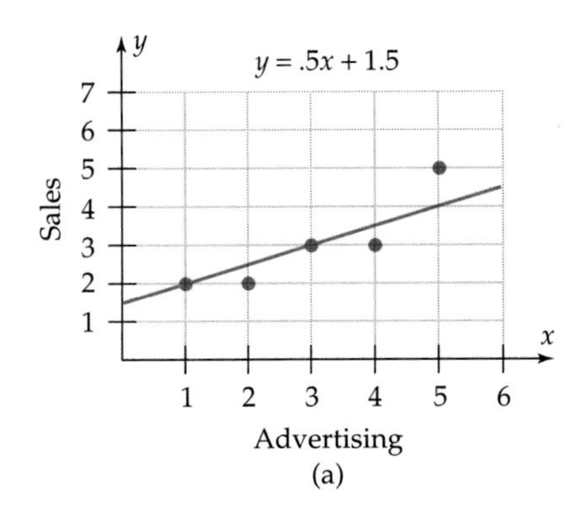

(a)

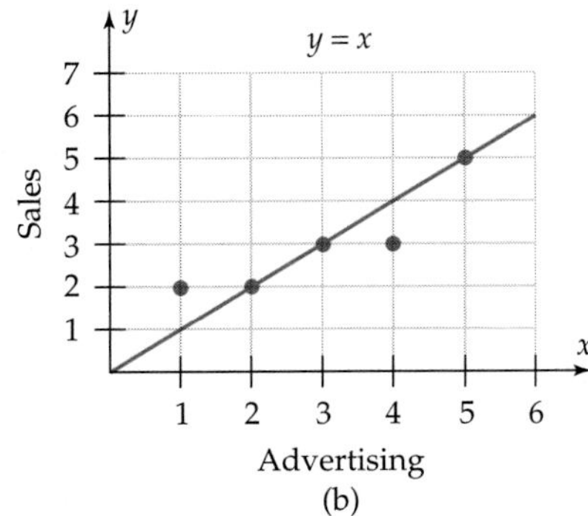

(b)

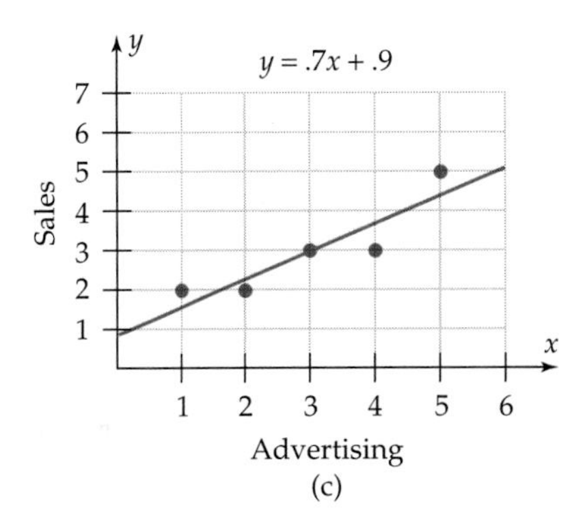

(c)

Figure 2–50

To determine the best model, we compute for each one the difference between the actual revenue r and the revenue y given by the model. If the data point is (x, r) and the corresponding point on the model is (x, y), then the difference $r - y$ measures the error in the model for that particular value of x. Graphically, this is the (positive or negative) vertical distance between the two points. Here are the errors for all data points in each model.

$y = .5x + 1.5$

Data Point (x, r)	**Point on Model** (x, y)	**Error** $r - y$
(1, 2)	(1, 2)	0
(2, 2)	(2, 2.5)	−.5
(3, 3)	(3, 3)	0
(4, 3)	(4, 3.5)	−.5
(5, 5)	(5, 4)	1
		sum 0

(a)

$y = x$

Data Point (x, r)	**Point on Model** (x, y)	**Error** $r - y$
(1, 2)	(1, 1)	1
(2, 2)	(2, 2)	0
(3, 3)	(3, 3)	0
(4, 3)	(4, 4)	−1
(5, 5)	(5, 5)	0
		sum 0

(b)

$y = .7x + .9$

Data Point (x, r)	**Point on Model** (x, y)	**Error** $r - y$
(1, 2)	(1, 1.6)	.4
(2, 2)	(2, 2.3)	−.3
(3, 3)	(3, 3)	0
(4, 3)	(4, 3.7)	−.7
(5, 5)	(5, 4.4)	.6
		sum 0

(c)

As the tables show, the sum of the errors is 0 in each case because positive and negative errors cancel out. Although this may suggest that the models are reasonably good ones, the sum of the errors cannot be used to decide whether one model is better than another. However, if we use the sum of the *squares* of the errors, there is no canceling and we obtain a different sum for each model:

$y = .5x + 1.5$

Data Point (x, r)	Model Point (x, y)	Error $r - y$	Squared Error $(r - y)^2$
(1, 2)	(1, 2)	0	0
(2, 2)	(2, 2.5)	−.5	.25
(3, 3)	(3, 3)	0	0
(4, 3)	(4, 3.5)	−.5	.25
(5, 5)	(5, 4)	1	1
			sum 1.5

(a)

$y = x$

Data Point (x, r)	Model Point (x, y)	Error $r - y$	Squared Error $(r - y)^2$
(1, 2)	(1, 1)	1	1
(2, 2)	(2, 2)	0	0
(3, 3)	(3, 3)	0	0
(4, 3)	(4, 4)	−1	1
(5, 5)	(5, 5)	0	0
			sum 2

(b)

$y = .7x + .9$

Data Point (x, r)	Model Point (x, y)	Error $r - y$	Squared Error $(r - y)^2$
(1, 2)	(1, 1.6)	.4	.16
(2, 2)	(2, 2.3)	−.3	.09
(3, 3)	(3, 3)	0	0
(4, 3)	(4, 3.7)	−.7	.49
(5, 5)	(5, 4.4)	.6	.36
			sum 1.1

(c)

Using the sum of the squares of the errors as a measure of accuracy has the effect of emphasizing large errors (those with absolute value greater than 1) because the square is greater than the error and minimizing small errors (those with absolute value less than 1) because the square is less than the error. By this measure, the best of the three models is $y = .7x + .9$ because the sum of the squares of its errors is smallest. ■

It can be proved that for any set of data points there is one and only one line for which the sum of the squares of the errors is as small as possible. This line is called the **least squares regression line,** and the computational process for finding its equation (which is built into most calculators) is called **linear regression.** Linear regression was used to obtain model (c) in Example 2, as well as the linear equation in Example 1.

Example 3 The total number of farm workers (in millions) in selected years is shown in the following table. (*Source:* Economic Research Service of the U.S. Department of Agriculture)

Year	Workers	Year	Workers	Year	Workers
1900	29.030	1950	59.230	1985	106.210
1920	42.206	1960	67.990	1990	117.490
1930	48.686	1970	79.802	1994	120.380
1940	51.742	1980	105.06		

Use linear regression to find an equation that models this data. Use the equation to estimate the number of farm workers in 1975 and 2000.

Solution Let $x = 0$ correspond to 1900, so that the data points are (0, 29.030), (20, 42.206), … ,(94, 120.380). We display the calculator's statistics editor and enter the data points: x-coordinates are the first list and y-coordinates (in the same order) are the second list (Figure 2–51).* Using the statistical plotting feature of the calculator to plot the data points, we obtain Figure 2–52, which shows that the data is approximately linear.

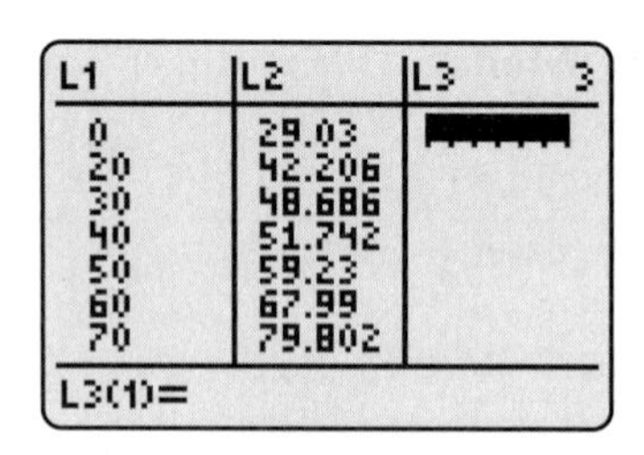

Figure 2–51

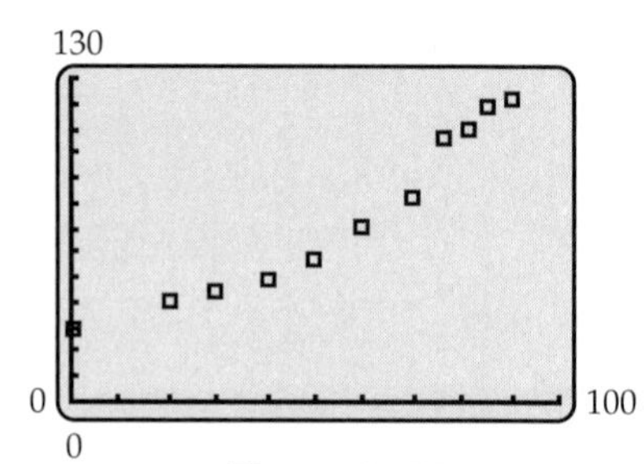

Figure 2–52

Now we use the linear regression feature to obtain Figure 2–53, which shows that the equation of the least squares regression line is

$$y = 1.011599433x + 18.33145006.$$

When you do this on most calculators, you can simultaneously store the regression equation in the equation memory, so that it can be graphed along with the data points (Figure 2–54).

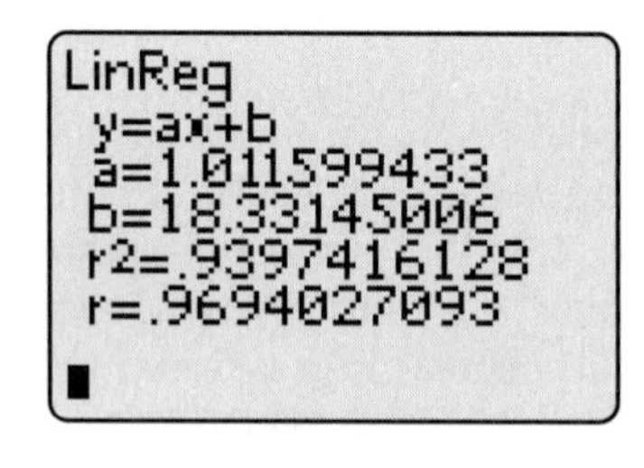

Figure 2–53

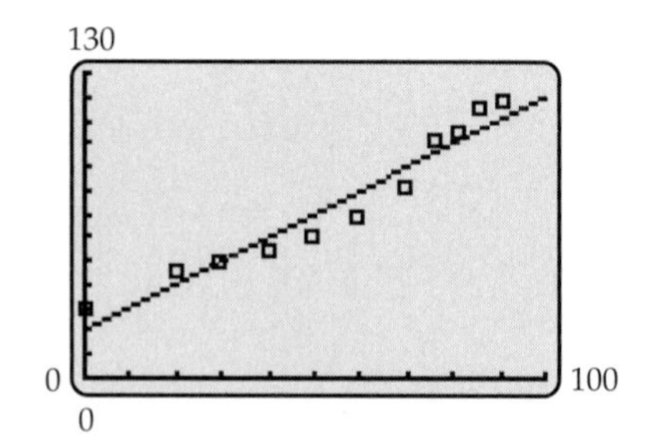

Figure 2–54

*Consult the Technology Tip after the example for details on how to carry out this and subsequent steps in the process.

Figure 2–54 suggests that the regression line provides a reasonable model for approximating the number of farm workers in a given year. If $x = 75$, then

$$y = 1.011599433(75) + 18.33145006 \approx 94.201$$

and if $x = 100$,

$$y = 1.011599433(100) + 18.33145006 \approx 119.491$$

Therefore, there were approximately 94,201,000 farm workers in 1975 and 119,491,000 in 2000. ■

Technology Tip

Most calculators allow you to store three or more statistics graphs (identified by number); the following directions assume that the first one is used.

To call up statistics editor, use these commands:

TI-83 and Sharp 9600: STAT EDIT (lists are $L_1, L_2, \ldots$);

TI-85/86: STAT EDIT (built-in lists are x-stat, y-stat; it is usually better to create your own lists; we use L1 and L2 as list names here);*

TI-89: APPS DATA/MATRIX EDITOR NEW; then choose DATA as the TYPE, enter a VARIABLE name (we use L here), and key in ENTER (lists are $C_1, C_2, \ldots$);

Casio 9850: STAT (lists are List 1, List 2, ...);

HP-38: LIB STATISTICS (lists are $C_1, C_2, \ldots$).

To graph the data points that have been entered as lists (x-coordinates in the first list, corresponding y-coordinates in the second), choose an appropriate viewing window (TI and Sharp only). On TI-85, use STAT DRAW SCAT to plot the points determined by the lists currently chosen in the STAT EDIT screen. On other calculators, use these commands to enter the setup screen:

TI-83 and Sharp 9600: STAT-PLOT (on keyboard); choose PLOT 1;

TI-86: STAT; choose PLOT and PLOT 1;

TI-89: from the Data Editor, choose PLOT-SETUP (F_2) and DEFINE (F_1);

Casio: STAT GRPH SET;

HP-38: LIB STATISTICS PLOT-SETUP and SYMB.

Key in the appropriate information [on/off (TI-83/86 and Sharp), graph number (Casio and HP), lists to be used, graph type (scatter plot), the mark to be used for data points, and on HP-38, the viewing window]. Then press GRAPH (on TI, Sharp) or GPH 1 (on Casio) or PLOT (on HP-38).

(Tip continues)

*When statistical computations are run on TI-85/86, the lists used are automatically copied into the x-stat and y-stat lists, replacing whatever was there before. So use the x-stat and y-stat lists only if you don't want to save them. See your instruction manual to find out how to create new lists in the statistics editor.

To produce the least squares regression line, store its equation as y_1 in the equation memory, and graph it, use these commands:

TI-83: STAT CALC LinReg $(ax + b)$ L_1, L_2, Y_1;

TI-85: STAT CALC L1, L2 LINR; then press STREG and enter y_1 as the name;

TI-86: STAT CALC Lin R L1, L2, y_1;

TI-89: From the Data Editor, choose CALC (F_5); enter LIN REG as CALCULATION TYPE, C_1 as x, C_2 as y; and choose $y_1(x)$ in STORE REGEQ;

Sharp 9600: STAT REG Rg_ax + b (L_1, L_2, Y_1) [after exiting the statistics editor and returning to the home screen].

Then press GRAPH or DRAW to obtain the graph of the equation and the data points (assuming PLOT 1 has not been turned off). On HP-38, after plotting the data points, use MENU FIT to graph the regression line and SYMB to see its equation. On Casio 9850, after plotting the data points, press X for the equation of the regression line and DRAW for its graph.

You may have noticed that Figure 2–53 contains a number r (and its square), in addition to the coefficients of the regression line equation. The number r, which is called the **correlation coefficient,** is a statistical measure of how well the least squares regression line fits the data points. It is always between -1 and 1. The closer the absolute value of r is to 1, the better the fit. For instance, the regression line in Example 3 is a good fit since $r \approx .97$. When $|r| = 1$, the fit is perfect: all the data points are on the regression line. Conversely, a regression coefficient near 0 indicates a poor fit.

Example 4 The numbers of unemployed people in the labor force (in millions) for 1984–1995 are as follows. (*Source:* U.S. Department of Labor, Bureau of Labor Statistics)

Year	Unemployed	Year	Unemployed	Year	Unemployed
1984	8.539	1988	6.701	1992	9.613
1985	8.312	1989	6.528	1993	8.940
1986	8.237	1990	7.047	1994	7.996
1987	7.425	1991	8.628	1995	7.404

Is a linear equation a good model for this data?

Solution After entering the data as two lists in the statistics editor (with $x = 0$ corresponding to 1980), you can test it graphically or analytically.

Graphical: Plotting the data points (Figure 2–55) shows that they do not form a linear pattern (unemployment tends to rise and fall).

Analytical: Linear regression (Figure 2–56) produces an equation whose correlation coefficient is $r \approx .092$, a number very close to 0, which indicates that the regression line is a very poor fit for the data.

Therefore, a linear equation is not a good model for this data. ■

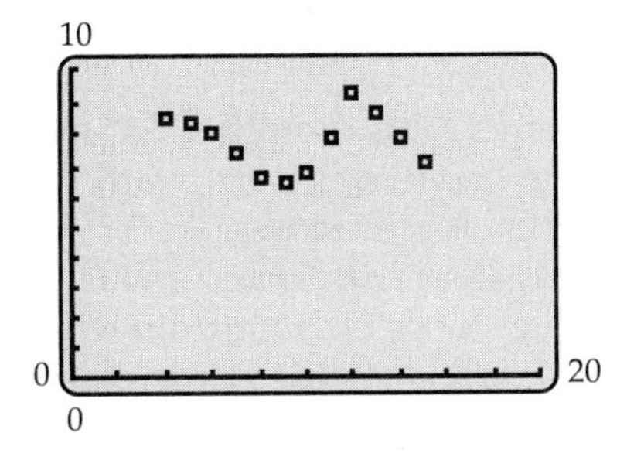

Figure 2–55

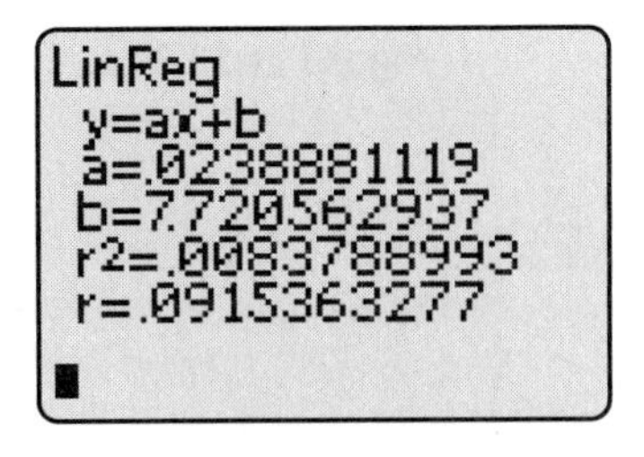

Figure 2–56

GRAPHING EXPLORATION

Enter the data from Example 4 in the statistics editor of your calculator. Graph the data points. Graph the least squares regression line on the same screen to see how poorly it fits the data.

Example 5 Forty people were randomly selected for a survey that asked their annual income and how many hours they watched TV each day. The results were as follows:

Income	TV	Income	TV	Income	TV	Income	TV
\$12,000	8	\$56,000	3	\$48,000	3	\$51,000	5
\$18,000	6	\$12,000	7	\$20,000	4	\$22,000	5
\$26,000	5	\$24,000	4	\$14,000	4	\$22,000	6
\$21,000	6	\$28,000	7	\$96,000	0	\$15,000	5
\$16,000	7	\$31,000	5	\$33,000	4	\$92,000	2
\$35,000	5	\$53,000	4	\$29,000	6	\$75,000	3
\$85,000	3	\$39,000	3	\$16,000	7	\$42,000	4
\$68,000	2	\$80,000	2	\$64,000	0	\$17,000	6
\$17,000	7	\$88,000	1	\$77,000	5	\$53,000	4
\$17,000	7	\$31,000	3	\$45,000	3	\$73,000	4

If possible, find a linear model for this data.

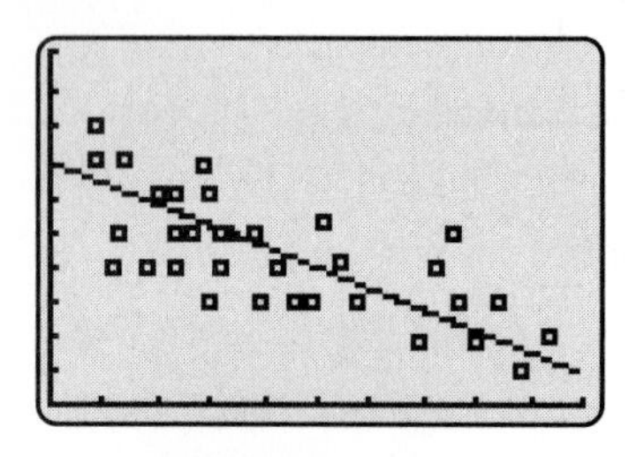

Figure 2–57

Solution We express incomes in thousands so that the data points are (12, 8), (18, 6), etc. We enter the data points in the statistics editor of a calculator and find that the equation of the least squares regression line (rounded) is $y = -.06x + 6.85$, with correlation coefficient $r \approx -.78$. Since $|r|$ is relatively close to 1, the regression line is a fairly good model for the data, as shown in Figure 2–57. ■

The correlation coefficient r always has the same sign as the slope of the least squares regression line. So when r is negative, as in Example 5, the regression line moves downward from left to right (see Figure 2–57). In other words, as x increases, y decreases. In such cases, we say that the data has a **negative correlation.** When r is positive, as in Example 3, the regression line slopes upward from left to right (see Figure 2–54), and we say that the data has a **positive correlation:** as x increases, y also increases. When r is close to 0 (regardless of sign), we say that there is **no correlation,** as in Example 4.

Exercises 2.5

1. **(a)** In Example 2, find the equation of the line through the data points (1, 2) and (5, 5).
   (b) Compute the sum of the squares of the errors for this line. Is it a better model than any of the models in the example? Why?
2. The linear model in Example 1 is the least squares regression line with coefficients rounded. Find the correlation coefficient for this model.
3. **(a)** In Example 3, find the slope of the line through the data points for 1920 and 1994.
   (b) Find the equation of the line through these two data points.
   (c) Which model predicts the higher number of farm workers in 2010: the line in part (b) or the regression line found in Example 3?
4. If you consider only the data for 1992–1994 in Example 4, is there a positive or negative or no correlation?

In Exercises 5–8, determine whether the given scatter plot of the data indicates that there is a positive correlation, negative correlation, or very little correlation.

5.

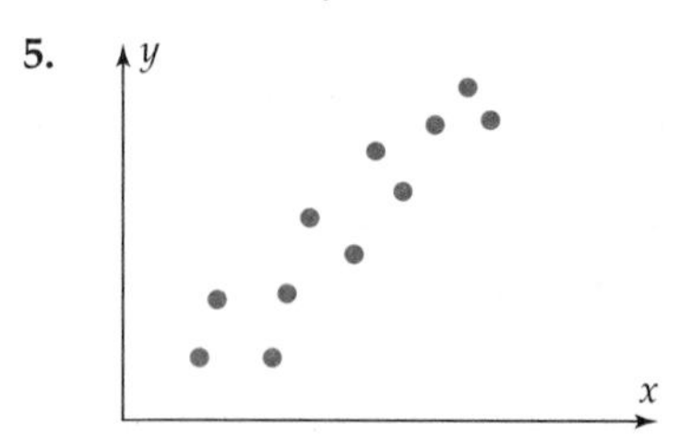

6.

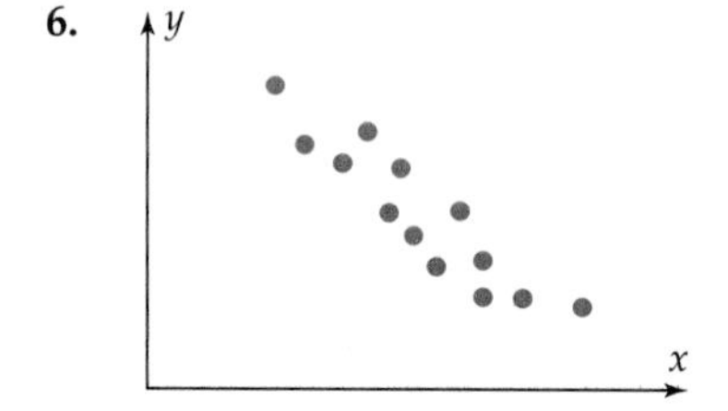

7.

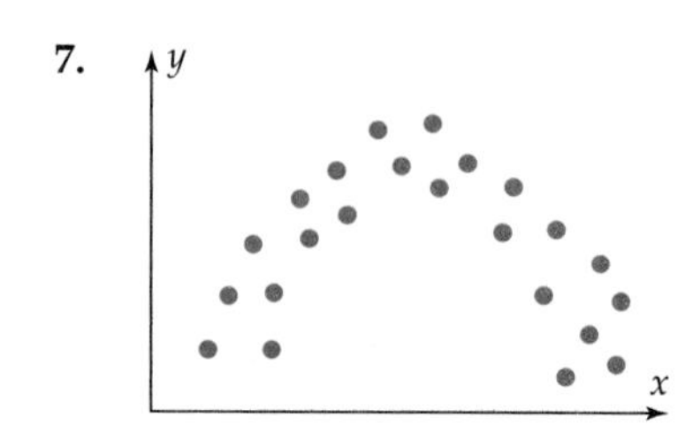

8.

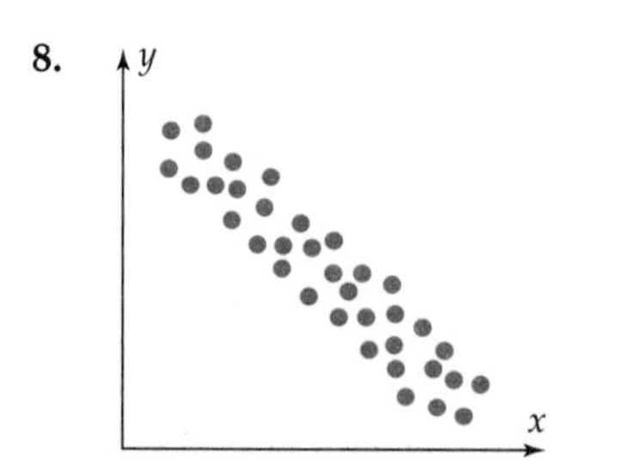

In Exercises 9–14, construct a scatter diagram for the data and answer these questions: (a) Does the data appear to be linear? (b) If so, is there a positive or negative correlation?

9. The U.S. gross domestic product (GDP) is the total value of all goods and services produced in the United States. The table shows the per capita GDP in billions of constant 1992 dollars (adjusted for inflation). Let $x = 0$ correspond to 1990.

Year	Per Capita Gross Domestic Product
1990	$24,743
1991	24,058
1992	24,311
1993	24,638
1994	25,150
1995	25,599
1996	25,829
1997	26,622

10. The table shows the monthly premium (in dollars) for a term life insurance policy for female nonsmokers. Let x represent age and y premiums.

Age	Premium
25	$11.57
30	11.66
35	11.83
40	13.05
45	16.18
50	21.32
55	29.58

11. The table shows the percent of persons in the United States below the U.S. poverty level in selected years. Let $x = 0$ correspond to 1960.

Year	Percent below poverty level
1960	22.2
1965	17.3
1970	12.6
1975	12.3
1980	13.0
1985	14.0
1990	13.5
1991	14.2
1992	14.8
1993	15.1
1994	14.5
1995	13.8
1996	13.7

12. The vapor pressure y of water depends on the temperature x, as given in the table.

Temperature (°C)	Pressure (mm Hg)
0	4.6
10	9.2
20	17.5
30	31.8
40	55.3
50	92.5
60	149.4
70	233.7
80	355.1
90	525.8
100	760

13. The table shows the U.S. Bureau of Census population data for St. Louis, Missouri. Let $x = 0$ correspond to 1950.

Year	Population
1950	856,796
1970	622,236
1980	452,801
1990	396,685
1994	368,215

14. The table shows the U.S. disposable income (personal income less personal taxes) in billions of dollars. (*Source:* Bureau of Economics Analysis, U.S. Dept. of Commerce). Let $x = 0$ correspond to 1990.

Year	Disposal Personal Income
1990	4179.4
1991	4356.8
1992	4626.7
1993	4829.2
1994	5052.7
1995	5355.7
1996	5608.3

15. The table gives the annual U.S. consumption (in million of pounds) of beef and poultry. (*Source:* U.S. Dept. of Agriculture)

Year	Beef	Poultry
1990	24,031	22,151
1991	24,113	23,270
1992	24,261	24,394
1993	24,006	25,099
1994	25,125	25,754
1995	25,533	25,940
1996	25,875	26,614

(a) Make scatter plots for both beef and poultry consumption, using the actual years (1990, 1991, etc.) as x in each case.

(b) Without graphing, use your knowledge of slopes to determine which of the following equations models beef consumption and which one models poultry consumption. Confirm your answer by graphing.

$y_1 = 717.44x - 1{,}405{,}160$ $\quad y_2 = 329.86x - 632{,}699$

16. The table gives the median weekly earnings of full-time workers 25 years and older by the amount of education. (*Source:* U.S. Bureau of Labor Statistics).

Year	Less than 4 years of high school	High School 4 years	College 1–3 years	College 4 years or more
1980	$222	$266	$304	$376
1982	248	302	351	438
1984	263	323	382	486
1986	278	344	409	525
1988	288	368	430	585
1990	304	386	476	639

(a) Make four scatter plots, one for each educational group, using $x = 0$ to correspond to 1980.

(b) Four linear models are given below. Match each model with the appropriate data set.

$$y_1 = 16.06x + 312 \qquad y_2 = 7.79x + 228$$
$$y_3 = 25.64x + 379 \qquad y_4 = 11.7x + 273$$

In Exercises 17–23, use the linear regression feature of your calculator to find the required model.

17. The table shows the number of deaths per 100,000 people from heart disease.

Year	1950	1960	1970	1980	1990	1996
Deaths	510.8	521.8	496.0	436.4	368.3	358.6

(a) Find a linear model for this data ($x = 0$ is 1950).

(b) In the unlikely event that the linear model in part (a) remains valid far into the future, will there be a time when death from heart disease has been completely eliminated? If so, when would this occur?

18. The table shows the share of total U.S. household income received by the poorest 20% of households and the share received by the wealthiest 5% of households. (*Source:* U.S. Census Bureau)

Year	**Lowest 20%**	**Top 5%**
1985	4	17.0
1990	3.9	18.6
1995	3.7	21.0
1996	3.7	21.4

(a) Find a linear model for the income share of the poorest 20% of households ($x = 0$ is 1985).

(b) Find a linear model for the income share of the wealthiest 5% of households.

(c) What do the slopes of the two models suggest for each model?

(d) Assuming these models remain accurate, will the income gap between the wealthy and the poor grow, stay about the same, or decline in the year 2000?

19. The table shows the percent of federal aid given to college students in the form of loans in selected years at a particular college.

Year (in which school year begins)	**Loans (%)**
1975	18
1978	30
1984	54
1987	66
1990	78

(a) Find a linear model for this data, with $x = 0$ corresponding to 1975.

(b) Interpret the meaning of the slope and the y-intercept.

(c) If the model remains accurate in the future, what percentage of federal student aid are loans in 2000?

20. The table shows the percent of federal aid given to college students in the form of grants or work-study in selected years at the college of Exercise 19.

Year (in which school year begins)	**Grants and Work-Study (%)**
1975	82
1978	70
1984	46
1987	34
1990	22

(a) Find a linear model for this data, with $x = 0$ corresponding to 1975.

(b) Graph the model from part (a) and the model from Exercise 19 on the same axes. What appears to be the trend in the federal share of financial aid to college students?

(c) In what year is the percent of federal aid the same for loans as for grants and work-study?

21. The table on the next page gives the average number of take-out meals purchased at restaurants per person in selected years. (*Source:* NPD Group's Crest Service)

Year	Average number of take-out meals per person, annually
1984	43
1986	48
1988	53
1990	55
1992	57
1994	61
1996	65

(a) Make a scatter plot of the data, with $x = 0$ corresponding to 1980.
(b) Find a linear model for the data.
(c) According to the model, what was the average number of take-out meals purchased in 1993? in 2000?

22. The table shows the median time (in months) after the application has been made for the Food and Drug Administration to approve a new drug. (*Source:* U.S. Food and Drug Administration)

Year	Median time to approval
1986	32.9
1987	29.9
1988	27.2
1989	29.3
1990	24.3
1991	22.1
1992	22.6
1993	23.0
1994	17.5
1995	15.9
1996	14.3

(a) Make a scatter plot of the data, with $x = 0$ corresponding to 1980.
(b) Find a linear model for the data.
(c) What are the limitations of this model? [*Hint:* What does it say about approval in the year 2005?]

23. The data below give production (x) and consumption (y) of primary energy* in quadrillion Btus for a sample of countries in 1995.

Australia (7.29, 4.43)	Japan (3.98, 21.42)
Brazil (4.55, 6.76)	Mexico (8.15, 5.59)
Canada (16.81, 11.72)	Poland (3.74, 3.75)
China (35.49, 35.67)	Russia (39.1, 26.75)
France (4.92, 9.43)	Saudi Arabia (20.34, 3.72)
Germany (5.42, 13.71)	South Africa (6.08, 5.51)
India (8.33, 10.50)	United States (69.1, 88.28)
Indonesia (6.65, 3.06)	United Kingdom (10.57, 9.85)
Iran (9.35, 3.90)	Venezuela (8.22, 2.53)

(a) Make a scatter plot of the data.
(b) Find a linear model for the data. Graph the model with the scatter diagram.
(c) In 1995, what three countries were the world's leading producers and consumers of energy?
(d) As a general trend what does it mean if a country is "above" the linear model?
(e) As a general trend what does it mean if a country is "below" the linear model?
(f) Identify any countries that appear to differ dramatically from most of the others.

*Production and consumption include petroleum, natural gas, coal, net hydroelectric, nuclear, geothermal, solar, wind, electric power, and biofuels.

Chapter 2 Review

Important Concepts

Review Questions

In Questions 1–6,

(a) Determine which of the viewing windows a–e shows a complete graph of the equation.

(b) For each viewing window that does not show a complete graph, explain why.

(c) Find a viewing window that gives a "better" complete graph than windows a–e (meaning that the window is small enough to show as much detail as possible, yet large enough to show a complete graph).

a. Standard viewing window

b. $-10 \le x \le 10, -200 \le y \le 200$

c. $-20 \le x \le 20, -500 \le y \le 500$

d. $-50 \le x \le 50, -50 \le y \le 50$

e. $-1000 \le x \le 1000, -1000 \le y \le 1000$

1. $y = .2x^3 - .8x^2 - 2.2x + 6$

2. $y = x^3 - 11x^2 - 25x + 275$

3. $y = x^4 - 7x^3 - 48x^2 + 180x + 200$

4. $y = x^3 - 6x^2 - 4x + 24$

5. $y = .03x^5 - 3x^3 + 69.12x$

6. $y = .00000002x^6 - .0000014x^5 - .00017x^4 + .0107x^3 + .2568x^2 - 12.096x$

In Questions 7–10, sketch a complete graph of the equation, and give reasons why it is complete.

7. $y = x^2 - 10$

8. $y = x^3 + x + 4$

9. $y = \sqrt{x - 5}$

10. $y = x^4 + x^2 - 6$

In Questions 11–14, sketch a complete graph of the equation.

11. $y = x^2 - 13x + 43$

12. $y = |x|$

13. $y = |x + 5|$

14. $y = 1/x$

In Questions 15–22, solve the equation graphically. You need only find solutions in the given interval.

15. $x^3 + 2x^2 = 11x + 6; \quad [0, \infty)$

16. $x^3 + 2x^2 = 11x + 6; \quad (-\infty, 0)$

17. $x^4 + x^3 - 10x^2 = 8x + 16; \quad [0, \infty)$

18. $2x^4 + x^3 - 2x^2 + 6x + 2 = 0; \quad (-\infty, -1)$

19. $\dfrac{x^3 + 2x^2 - 3x + 4}{x^2 + 2x - 15} = 0; \quad (-10, \infty)$

20. $\dfrac{3x^4 + x^3 - 6x^2 - 2x}{x^5 + x^3 + 2} = 0; \quad [0, \infty)$

21. $\sqrt{x^3 + 2x^2 - 3x - 5} = 0; \quad [0, \infty)$

22. $\sqrt{1 + 2x - 3x^2 + 4x^3 - x^4} = 0; \quad (-5, 5)$

23. A jeweler wants to make a 1-ounce ring consisting of gold and silver, using \$200 worth of metal. If gold costs \$600 per ounce and silver \$50 per ounce, how much of each metal should she use?

24. A calculator is on sale for 15% less than the list price. The sale price, plus a 5% shipping charge, totals \$210. What is the list price?

25. Karen can do a job in 5 hours and Claire can do the same job in 4 hours. How long will it take them to do the job together?

26. A car leaves the city traveling at 54 mph. One-half hour later, a second car leaves from the same place and travels at 63 mph along the same road. How long will it take for the second car to catch up with the first?

27. A 12-foot-long rectangular board is cut in two pieces so that one piece is four times as long as the other. How long is the bigger piece?

28. George owns 200 shares of stock, 40% of which are in the computer industry. How many more shares must he buy in order to have 50% of his total shares in computers?

29. A square region is changed into a rectangular one by making it 2 feet longer and twice as wide. If the area of the rectangular region is three times larger than the area of the original square region, what was the length of a side of the square before it was changed?

30. The radius of a circle is 10 inches. By how many inches should the radius be increased so that the area increases by 5π square inches?

31. The cost of manufacturing x caseloads of ballpoint pens is $\dfrac{600x^2 + 600x}{x^2 + 1}$ dollars. How many caseloads should be manufactured in order to have an *average cost* of \$25? [Average cost was defined in Exercise 13 of Section 2.4.]

32. An open-top box with a rectangular base is to be constructed. The box is to be at least 2 inches wide, twice as long as it is wide, and have a volume of 150 cubic inches. What should the dimensions of the box be if the surface area is to be

(a) 90 square inches? **(b)** as small as possible?

33. A farmer has 120 yards of fencing and wants to construct a rectangular pen, divided in two parts by an interior fence, as shown in the figure. What should the dimensions of the pen be in order to enclose the maximum possible area?

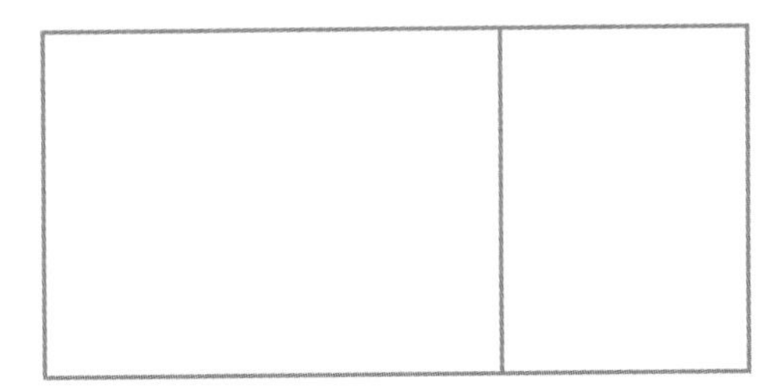

34. The top and bottom margins of a rectangular poster are each 5 inches and each side margin is 3 inches. The printed material on the poster occupies an area of 400 square inches. Find the dimensions that will use the least possible amount of posterboard.

35. A rectangle has one side on the x-axis and its other two corners sit on the graph of $y = 9 - x^2$, as shown in the figure. What value of x gives a rectangle of maximum area?

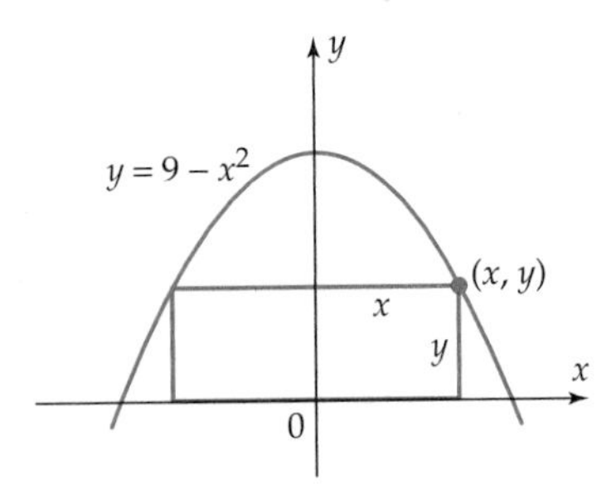

36. The window in the figure has a rectangular bottom, with a semicircle of radius r lying on top of it, and a perimeter of 40 feet. In order that the window have the maximum possible area, what should r and h be?

37. The table shows the monthly premium (in dollars) for a term life insurance policy for women who smoke.

Age (years)	25	30	35	40	45	50	55	60
Premium	19.58	20.10	20.79	25.23	34.89	48.55	69.17	98.92

(a) Make a scatter plot of the data, using x for age and y for premiums.
(b) Does the data appear to be linear?

38. For which of the following scatter plots would a linear model be reasonable? Which sets of data show positive correlation, and which show negative correlation?

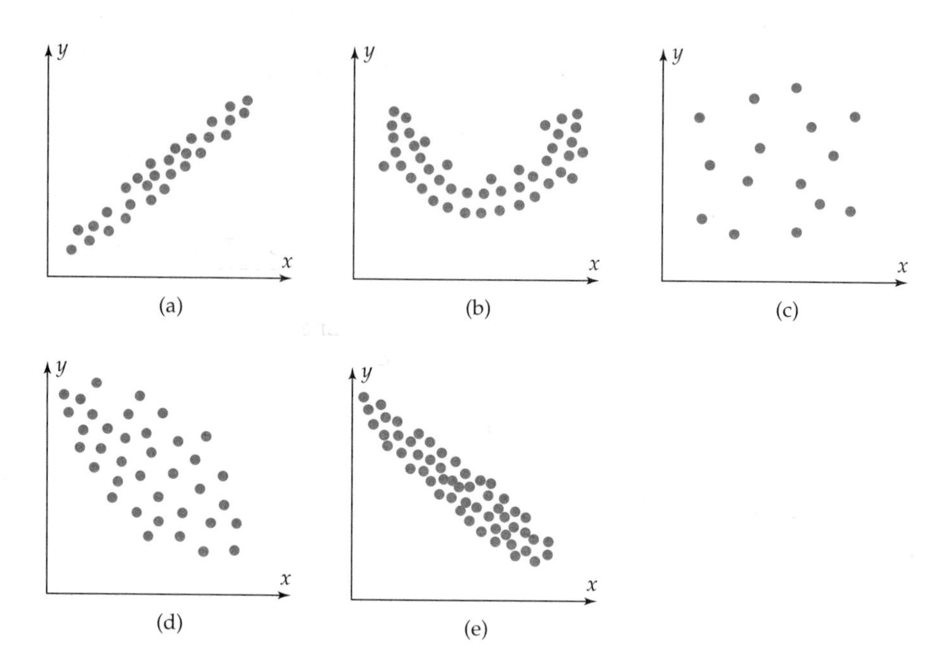

Exercises 39–40 refer to the following table, which shows the percentage of jobs that are classified as managerial and the percentage of male and female employees who are managers.

Year (since 1990)	Managerial Jobs (%)	Female Managers (%)	Male Managers (%)
−8	12.32	6.28	16.81
−5	12.31	6.85	16.67
−2	12.00	7.21	16.09
0	11.83	7.45	15.64
1	11.79	7.53	15.52
3	11.43	7.65	14.79
5	11.09	7.73	14.10

39. **(a)** Make scatter plots of each data set (managerial jobs, female managers, male managers).

(b) Match the following linear models with the correct data set. Explain your choices.

$$y_1 = .11x + 7.34 \qquad y_2 = -.09x + 11.74 \qquad y_3 = -.21x + 15.48$$

40. **(a)** According to the models in Exercise 39, is the percentage of female or male managers increasing at the greater rate?

(b) Use the models to predict the percentage of female managers and the percentage of male managers in the year 2000.

(c) In what year do the models predict that the percentage of female managers will surpass the percentage of male managers?

41. The table shows the average hourly earnings of production workers. (*Source:* U.S. Bureau of Labor Statistics)

Year	1980	1982	1984	1986	1988	1990	1992	1994	1996
Hourly Earnings ($)	6.66	7.68	8.32	8.76	9.28	10.01	10.57	11.12	11.81

(a) Find a linear model for this data, with $x = 0$ corresponding to 1980.

(b) Use the model to estimate the average hourly wage in 1991 and in 2001. The actual average in 1991 was $10.32. How far off is the model?

42. The table shows the winning times (in minutes) for men's 1500-meter freestyle swimming at the Olympics in selected years.

Year	1912	1924	1936	1948	1960	1972	1984	1996
Time	22.00	20.11	19.23	19.31	17.33	15.88	15.09	14.94

(a) Find a linear model for this data, with $x = 0$ corresponding to 1900.

(b) The Olympic record of 14.72 minutes was set in 1992 by Kieren Perkins of Australia. How accurately did your model estimate his time?

(c) How long is this model likely to remain accurate? Why?

43. The table shows the total amount of charitable giving (in billions of dollars) in the United States during recent years.

Year	Total Charitable Giving
1986	83.79
1987	89.99
1988	98.13
1989	108.73
1990	111.48
1991	117.22
1992	121.09
1993	126.46
1994	129.84
1995	143.84
1996	150.70

(a) Find a linear model for this data, with $x = 0$ corresponding to 1980.
(b) Use your model to estimate the approximate total giving in 1995 and 2002.

44. The table shows, for selected states, the percent of high school students in the class of 1997 who took the SAT and the average SAT math score.

State	Students Who Took SAT (%)	Average Math Score
Connecticut	79	507
Delaware	65	498
Georgia	60	481
Idaho	15	539
Indiana	87	497
Iowa	6	601
Montana	22	548
Nevada	32	509
New Jersey	69	508
New Mexico	12	545
North Dakota	5	595
Ohio	25	536
Pennsylvania	72	495
South Carolina	56	474
Washington	46	523

(a) Make a scatter plot of average SAT math score y and percent x of students who took the SAT.
(b) Find a linear model for the data.
(c) What is the slope of your linear model? What does this mean in the context of the problem?
(d) Here are the data on four additional states. How well does the model match the actual figures for these states?

State	Maryland	Arizona	Alaska	Hawaii
Students Taking SAT (%)	9	29	48	50
Average Math Score	566	522	517	512

Discovery Project 2

Breaking Even at the Espresso Cart

A local resident owns an espresso cart and has asked you to provide an analysis based on last summer's data. To simplify things, only data for Mondays is provided. The data includes the amount the workers were paid each day, the number of cups sold, the cost of materials (frothy milk and such), and the total revenue for the day. The owner also must spend $40 each operating day for rent for her location and payment on a business loan. Sales taxes have been cleaned out of the data, so you need not consider them. Amounts have been rounded to the nearest dollar.

Date	Pay ($)	Cups Sold	Materials Cost ($)	Total Revenue ($)
June 02	68	112	55	202
June 09	60	88	42	119
June 16	66	81	33	125
June 23	63	112	49	188
June 30	63	87	38	147
July 07	59	105	45	159
July 14	57	116	49	165
July 21	61	122	52	178
July 28	64	100	48	193
August 04	58	80	36	112
August 11	65	96	42	158
August 18	57	108	52	162
August 25	64	93	47	166

The owner is, of course, interested in making a profit. She wishes to know (on the average) how many cups of espresso must be sold each day to break even without raising prices. She would also like to know how much she needs to raise prices in order to break even most days. After you tell her this information, she will decide what the best course of action is.

1. Find a linear regression model for the daily cost as a function of the number of cups sold. Be sure to include the pay for the workers, the fixed daily cost, and the cost of materials.
2. Find a linear regression model for the daily revenue as a function of the number of cups sold.
3. Use the two models you have created to locate the break-even point. That is, find the minimum number of cups for the value of the revenue model to equal or exceed the value of the cost model. Remember that the number of cups must be an integer.
4. The slope portion of the revenue model represents the average selling price of a cup of espresso. What slope in the revenue model would cause the break-even point to be less than 80 cups?
5. How much would the owner of the espresso cart need to raise her prices to break even every day, assuming that at least 80 cups will be sold each day?

Functions and Graphs

Looking for a house?

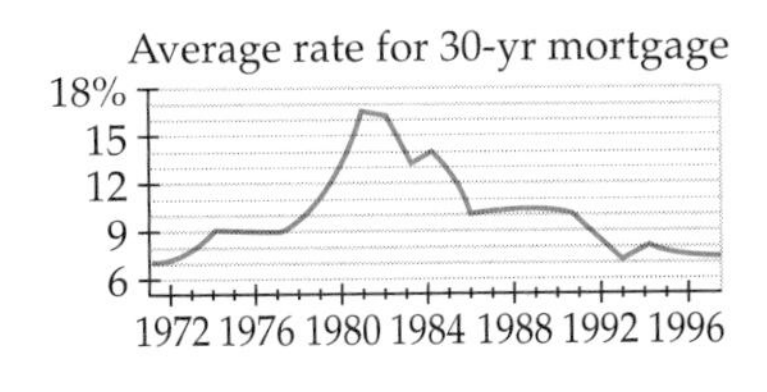

If you buy a house, you'll probably need a mortgage. If you can get a low interest rate, your monthly payments are lower (or alternatively, you can afford a more expensive house). The timing of your purchase can make a difference because mortgage interest rates constantly fluctuate. In mathematical terms, rates are a function of time. The graph of this function provides a picture of how interest rates change. See Exercise 82 on page 149.

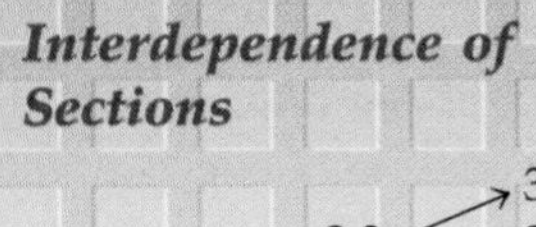

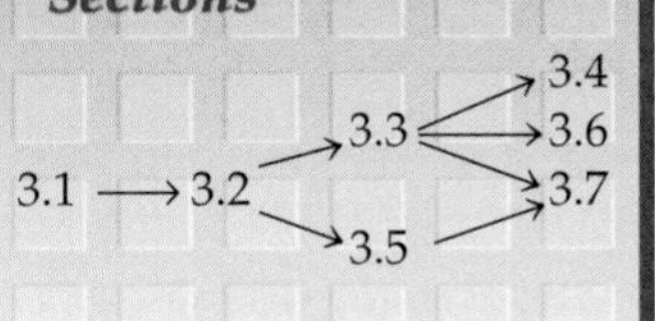

Chapter Outline

The concept of a function and functional notation are central to modern mathematics and its applications. In this chapter you will be introduced to functions and operations on functions, learn how to use functional notation, and develop skill in constructing and interpreting graphs of functions.

3.1 Functions

To understand the origin of the concept of function it may help to consider some "real-life" situations in which one numerical quantity depends on, corresponds to, or determines another.

Example 1 The amount of income tax you pay depends on the amount of your income. The way in which the income determines the tax is given by the tax law. ■

Example 2 The weather bureau records the temperature over a 24-hour period in the form of a graph (Figure 3–1). The graph shows the temperature that corresponds to each given time. ■

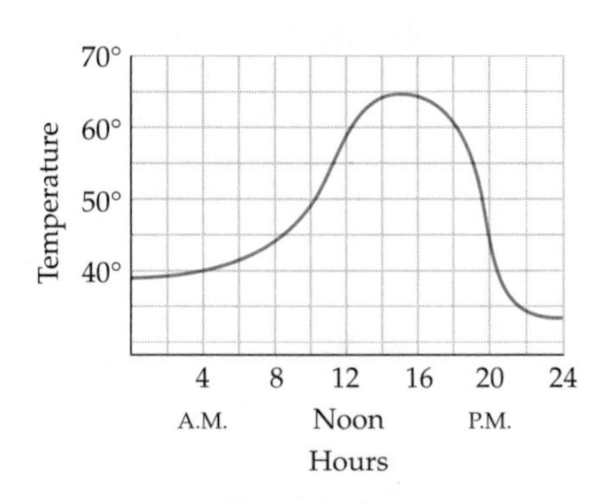

Figure 3–1

Example 3 Suppose a rock is dropped straight down from a high place. Physics tells us that the distance traveled by the rock in t seconds is $16t^2$ feet. So the distance depends on the time. ■

These examples share several common features. Each involves two sets of numbers, which we can think of as inputs and outputs. In each case there is a rule by which each input determines an output, as summarized here.

	Set of Inputs	Set of Outputs	Rule
Example 1	All incomes	All tax amounts	The tax laws
Example 2	Hours since midnight	Temperatures during the day	Time/temperature graph
Example 3	Seconds elapsed after dropping the rock	Distance rock travels	Distance $= 16t^2$

Each of these examples may be mentally represented by an idealized calculator that has a single operation key: a number is entered [*input*], the rule key is pushed [*rule*], and an answer is displayed [*output*]. The formal definition of function incorporates these common features (input/rule/output), with a slight change in terminology.

Functions

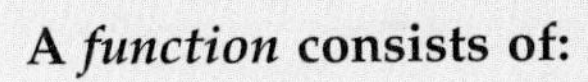

A *function* consists of:

A set of inputs (called the *domain*);

A *rule* by which each input determines one and only one output;

A set of outputs (called the *range*).

The phrase "one and only one" in the definition of the rule of a function may need some clarification. In Example 2, for each time of day (input), there is one and only one temperature (output). But it is quite possible to have the same temperature (output) at different times (inputs). In general,

For each input (number in the domain), the rule of a function determines exactly one output (number in the range). But different inputs may produce the same output.

Although real-world situations, such as Examples 1–3, are the motivation for functions, much of the emphasis in mathematics courses is on the functions themselves, independent of possible interpretations in specific situations, as illustrated in the following examples.

Technology Tip

The greatest integer function is denoted INT or FLOOR in the NUM submenu of the MATH menu of TI and Sharp 9600. It is denoted FLOOR in the REAL submenu of the HP-38 MATH menu, and INTG in the NUM submenu of the Casio 9850 OPTN menu.

Example 4 For each real number s that is not an integer, let $[s]$ denote the *integer* that is closest to s on the *left* side of s on the number line; if s is itself an integer, we define $[s] = s$. Here are some examples:

$$[-4.7] = -5, \quad [-3] = -3, \quad [-1.5] = -2,$$

$$[0] = 0, \quad \left[\frac{5}{3}\right] = 1, \quad [\pi] = 3.$$

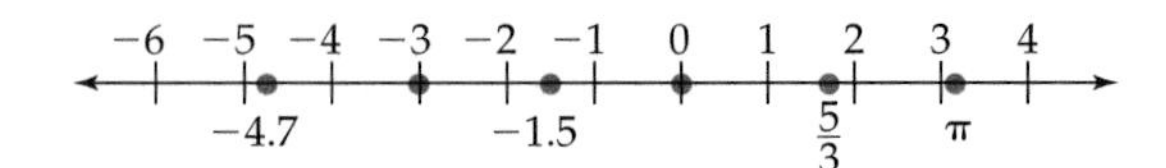

Figure 3–2

The **greatest integer function** is the function whose domain is the set of all real numbers, whose range is the set of integers, and whose rule is

For each input (real number) x, the output is the integer $[x]$. ■

Functions Defined by Equations and Graphs

Equations in two variables are *not* the same things as functions. However, many equations can be used to define functions.

Example 5 The equation $4x - 2y^3 + 5 = 0$ can be solved uniquely for y:

$$2y^3 = 4x + 5$$

$$y^3 = 2x + \frac{5}{2}$$

$$y = \sqrt[3]{2x + \frac{5}{2}}.$$

If a number is substituted for x in this equation, then exactly one value of y is produced. So we can define a function whose domain is the set of all real numbers and whose rule is

The input x produces the output $\sqrt[3]{2x + 5/2}$.

In this situation we say that the equation defines **y as a function of x.**

The original equation can also be solved for x:

$$4x = 2y^3 - 5$$

$$x = \frac{2y^3 - 5}{4}.$$

Now if a number is substituted for y, exactly one value of x is produced. So we can think of y as the input and the corresponding x as the output and say that the equation defines **x as a function of y.** ■

Example 6 If you solve the equation

$$y^2 - x + 1 = 0$$

for y, you obtain

$$y^2 = x - 1$$
$$y = \pm\sqrt{x - 1}.$$

This equation does *not* define y as a function of x because, for example, the input $x = 5$ produces two outputs: $y = \pm 2$. ■

Example 7 A group of students drives from Cleveland to Seattle, a distance of 2350 miles, at an average speed of 52 mph.

(a) Express their distance from Cleveland as a function of time.

(b) Express their distance from Seattle as a function of time.

Solution

(a) Let t denote the time traveled in hours after leaving Cleveland and D the distance from Cleveland at time t. Then the equation that expresses D as a function of t is

$$D = \text{Distance traveled in } t \text{ hours at 52 mph} = 52t.$$

(b) At time t the car has traveled $52t$ miles of the 2350-mile journey, so the distance K remaining to Seattle is given by $K = 2350 - 52t$. This equation expresses K as a function of t. ■

Technology Tip

The table setup screen is labeled TBLSET on the TI-83/89 and Sharp 9600 keyboards and NUM SETUP on the HP-38 keyboard. It is labeled TBLSET in the TI-86 TABLE menu and RANG in the Casio 9850 TABLE menu.

The increment is labeled ΔTBL on TI, TBLSTEP on Sharp 9600, NUMSTEP on HP-38, and PITCH on Casio 9850.

The table type is labeled INDPNT on TI, INPUT on Sharp 9600, and NUMTYPE on HP-38.

Graphing calculators are designed to deal with equations that define y as a function of x. The table feature of a calculator is a convenient way to evaluate such functions (that is, to produce the outputs from various inputs).*

Example 8 The equation $y = x^3 - 2x + 3$ defines y as a function of x. Use the table feature to find the outputs for each of the following inputs:

(a) $-3, -2, -1, 0, 1, 2, 3, 4, 5$ (b) $-5, -11, 8, 7.2, -.44$

Solution

(a) To use the table feature, we first enter $y = x^3 - 2x + 3$ in the equation memory, say as y_1. Then we call up the setup screen (see the Tip in the margin and Figure 3–3 on the next page) and enter the *starting number* (-3), the *increment* (the amount the input changes for each subsequent entry, which is 1 here), and the *table type* (AUTO, which

*TI-85 does not have a built-in table feature, but a program to provide one is in the Program Appendix.

means the calculator will compute all the outputs at once).* Then press TABLE to obtain the table in Figure 3–4. To find values that don't appear on the screen in Figure 3–4, use the up and down arrow keys to scroll through the table.

(b) With an apparently random list of inputs, as here, we change the table type to ASK (or USER or BUILD YOUR OWN).† Then key in each value of x, and hit ENTER. This produces the table one line at a time, as in Figure 3–5. ■

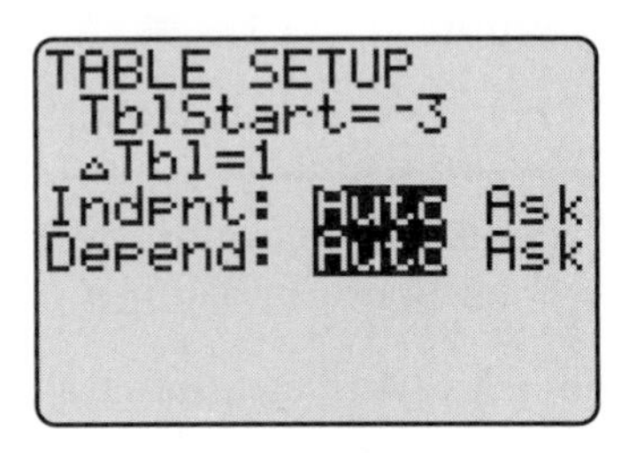

Figure 3–3

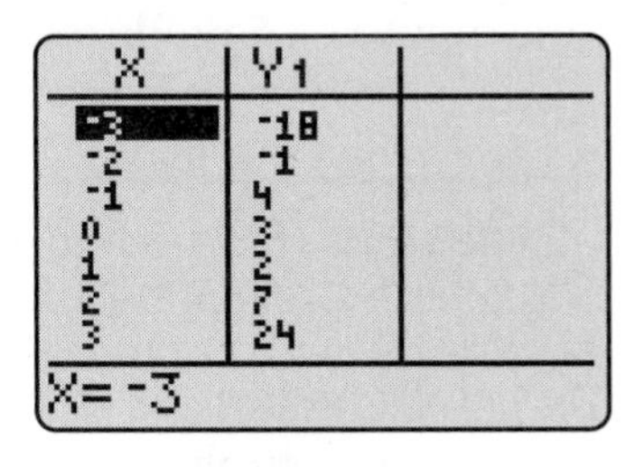

Figure 3–4

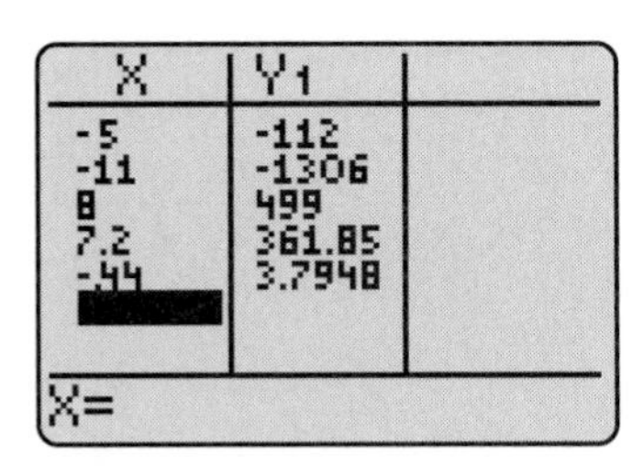

Figure 3–5

CALCULATOR EXPLORATION

Construct a table of values for the function in Example 8 that shows the outputs for these inputs: 2, 2.4, 2.8, 3.2, 3.6, and 4. What is the increment here?

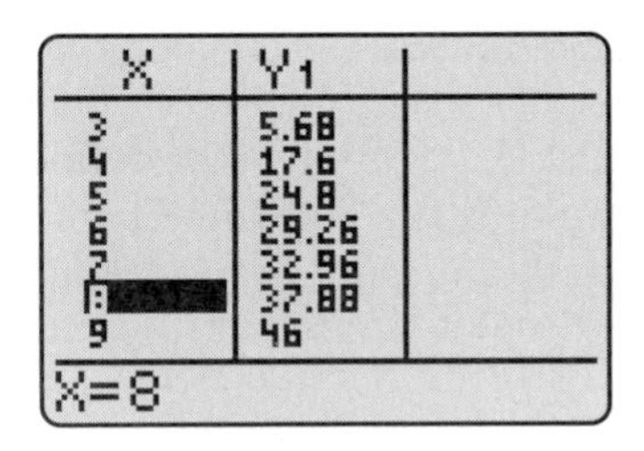

Figure 3–6

Example 9 The number of public schools (K–12) that have computer networks can be approximated by the equation

$$y = .33x^3 - 6.32x^2 + 43.95x - 78.2 \quad (4 \le x \le 10),$$

where $x = 4$ corresponds to 1994 and y is in thousands. In this case, the number y of computer networks is a function of the year x. What is the first year in which the number of networks exceeds 40,000?

Solution We make a table of values for the function (Figure 3–6). Since y is in thousands, it shows that the number of networks is approximately 37,880 in 1998 ($x = 8$) and 46,000 in 1999 ($x = 9$). ■

Example 10 The graph in Figure 3–7 defines a function whose rule is:

For input x, the output is the unique number y such that (x, y) is on the graph.

*On Casio calculators, there is no table type selection, but you must enter a maximum value of x.

†This type of table is not available on TI-85 or Casio calculators. However, TI-85 users have an effective alternative (see Calculator 1).

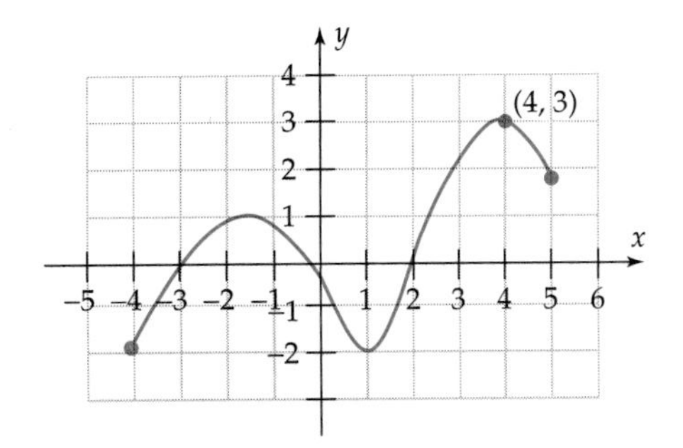

Figure 3–7

Input 4, for example, produces output 3 because (4, 3) is on the graph. Similarly, $(-3, 0)$ is on the graph, which means that input -3 produces output 0. Since the first coordinates of all points on the graph (the inputs) lie between -4 and 5, the domain of this function is the interval $[-4, 5]$. The range is the interval $[-2, 3]$ because all the second coordinates of points on the graph (the outputs) lie between -2 and 3. ■

Calculator Investigation

1. **Function Evaluation on TI-85/86** The equation $y = x^2 - 3.7x + 4.5$ defines y as a function of x.
   - **(a)** Enter this equation as y_1 in the equation memory, and return to the home screen by pressing 2nd QUIT. To determine the value of y when $x = 29$, key in EVAL—29—ENTER (you will find EVAL in the MISC submenu of the MATH menu). Use this method to find the value of y when $x = .45$ and 611.
   - **(b)** Enter two more equations in the equation memory: $y_2 = x^3 - 6$ and $y_3 = 5x + 2$. Now when you key in EVAL 29 ENTER, the value of all three functions at $x = 29$ will be displayed as a list, which you can scroll through by using the arrow keys.
   - **(c)** In the equation memory, turn off y_2, and return to the home screen. If you use EVAL now, what happens?
2. **Custom Menu on TI-85/86** If you press CUSTOM, the calculator displays a menu that may be blank. You may fill in any entries you want, so that frequently used operations (such as EVAL and FRAC) may be accessed quickly from the CUSTOM menu, rather than searching for them in various menus and submenus. Check your instruction manual for details.

Exercises 3.1

In Exercises 1–4, determine whether or not the given table could possibly be a table of values of a function. Give reasons for your answer.

1.

Input	-2	0	3	1	-5
Output	2	3	-2.5	2	14

2.

Input	-5	3	0	-3	5
Output	7	3	0	5	-3

3.

Input	-5	1	3	-5	7
Output	0	2	4	6	8

4.

Input	1	-1	2	-2	3
Output	1	-2	± 5	-6	8

Exercises 5–10 deal with the greatest integer function of Example 4, which is given by the equation $y = [x]$. Compute the following values of the function:

5. $[6.75]$ **6.** $[.75]$ **7.** $[-4/3]$ **8.** $[5/3]$ **9.** $[-16.0001]$

10. Does the equation $y = [x]$ define x as a function of y? Give reasons for your answer.

In Exercises 11 – 18, determine whether the equation defines y as a function of x or defines x as a function of y.

11. $y = 3x^2 - 12$ **12.** $y = 2x^4 + 3x^2 - 2$

13. $y^2 = 4x + 1$ **14.** $5x - 4y^4 + 64 = 0$

15. $3x + 2y = 12$ **16.** $y - 4x^3 - 14 = 0$

17. $x^2 + y^2 = 9$ **18.** $y^2 - 3x^4 + 8 = 0$

Exercises 19 – 22, each equation defines y as a function of x. Create a table that shows the values of the function for the given values of x.

19. $y = x^2 + x - 4$; $x = -2, -1.5, -1, \ldots, 3, 3.5, 4.$

20. $y = x^3 - x^2 + 4x + 1$; $x = 3, 3.1, 3.2, \ldots, 3.9, 4$

21. $y = \sqrt{4 - x^2}$; $x = -2, -1.2, -.04, .04, 1.2, 2$

22. $y = |x^2 - 5|$; $x = -8, -6, \ldots, 8, 10, 12$

Exercises 23 – 26 refer to Example 1. Assume that the state income tax law reads as follows.

Annual Income	Amount of Tax
Less than \$2000	0
\$2000–\$6000	2% of income over \$2000
More than \$6000	\$80 plus 5% of income over \$6000

23. Find the output (tax amount) that is produced by each of the following inputs (incomes):

\$500 \$1509 \$3754
\$6783 \$12,500 \$55,342

24. Find four different numbers in the domain of this function that produce the same output (number in the range).

25. Explain why your answer in Exercise 24 does *not* contradict the definition of a function (in the box on page 119).

26. Is it possible to do Exercise 24 if all four numbers in the domain are required to be greater than 2000? Why or why not?

27. The amount of postage required to mail a first-class letter is determined by its weight. In this situation, is weight a function of postage? Or vice versa? Or both?

28. Could the following statement ever be the rule of a function?

For input x, the output is the number whose square is x.

Why or why not? If there is a function with this rule, what is its domain and range?

29. Find an equation that expresses the area A of a circle as a function of its

(a) radius r **(b)** diameter d

30. Find an equation that expresses the area of a square as a function of its

(a) side x **(b)** diagonal d

31. A box with a square base of side x is four times higher than it is wide. Express the volume V of the box as a function of x.

32. The surface area of a cylindrical can of radius r and height h is $2\pi r^2 + 2\pi rh$. If the can is twice as high as the diameter of its top, express its surface area S as a function of r.

33. Suppose you drop a rock from the top of a 400-ft-high building. Express the distance D from the rock to the ground as a function of time t. What is the range of this function? [*Hint:* See Example 3.]

34. A bicycle factory has weekly fixed costs of $26,000. In addition, the material and labor costs for each bicycle are $125. Express the total weekly cost C as a function of the number x of bicycles that are made.

35. The table at right (from Metropolitan Life Insurance Company) relates a large-framed woman's height to the weight at which the woman should live longest. In this situation, is weight a function of height? Is height a function of weight? Justify your answer.

Height (in shoes)	Weight (in pounds, in indoor clothing)
4′10″	118–131
4′11″	120–134
5′0″	122–137
5′1″	125–140
5′2″	128–143
5′3″	131–147
5′4″	134–151
5′5″	137–155
5′6″	140–159
5′7″	143–163
5′8″	146–167
5′9″	149–170
5′10″	152–173
5′11″	155–176
6′0″	158–179

36. The table below (from U.S. Census Bureau data) shows the percentage of white households that were single-parent households in various years.

Year	Percent
1970	11.1
1980	14.3
1990	17
1993	17.8
1994	18

(a) Make a table of values for the equation $y = -4.539 + \sqrt{\dfrac{x - 1947.73}{.091}}$, which defines y as a function of x, with the x-values including the years shown in the Census Bureau table.

(b) How do the values of y in your table compare with the percents in the Census Bureau table? Does this equation seem to provide a reasonable model of the Census Bureau data?

(c) How might you estimate the percent of single-parent white households in 1985 and 1992? What are your estimates?

(d) Assuming this model remains reasonably accurate, in what year will 20% of white households be single-parent households?

Use the figure at the top of the next page for Exercises 37–43. Each of the graphs in the figure defines a function as in Example 10.

37. State the domain and range of the function defined by graph (a).

38. State the output (number in the range) that the function of Exercise 37 produces from the following inputs (numbers in the domain): -2, -1, 0, 1.

39. Do Exercise 38 for these numbers in the domain: $1/2, 5/2, -5/2$.

40. State the domain and range of the function defined by graph (b).

41. State the output (number in the range) that the function of Exercise 40 produces from the following inputs (numbers in the domain): -2, 0, 1, 2.5, -1.5.

42. State the domain and range of the function defined by graph (c).

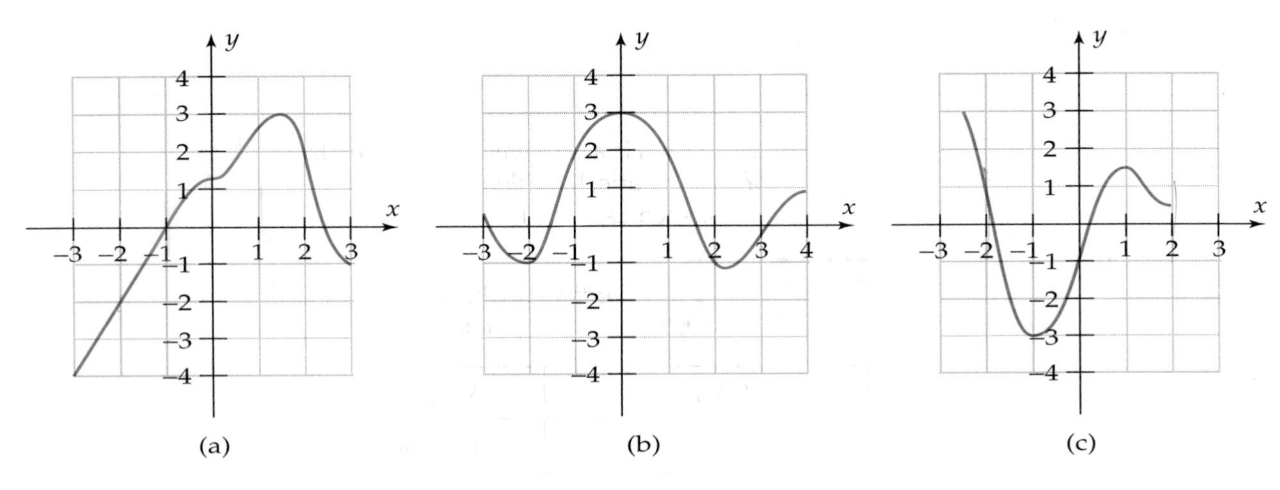

43. State the output (number in the range) that the function of Exercise 42 produces from the following inputs (numbers in the domain): -2, -1, 0, $1/2$, 1.

44. Explain why none of the graphs in the figure below defines a function according to the procedure in Example 10. What goes wrong?

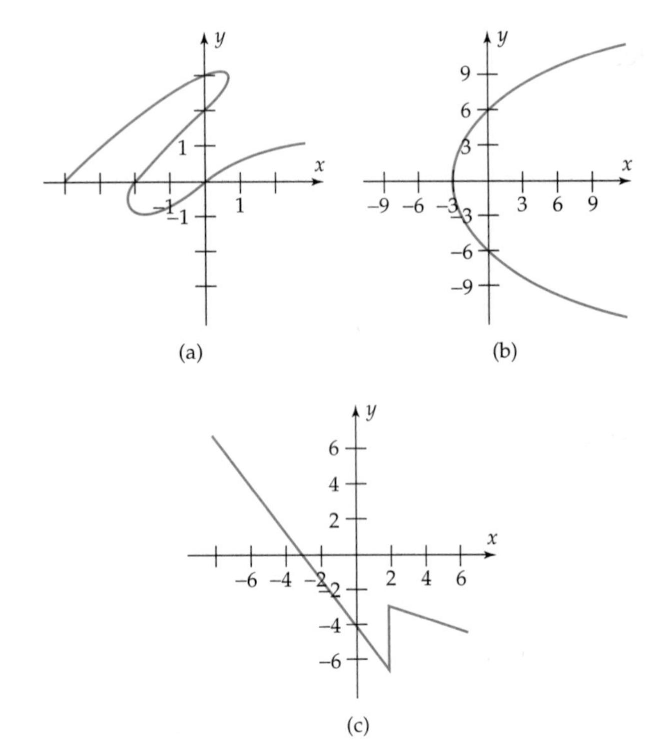

Thinkers

45. Consider the function whose rule uses a calculator as follows: "press COS, and then press LN; then enter a number in the domain, and press ENTER."* Experiment with this function, then answer the following questions. You may not be able to prove your answers—just make the best estimate you can based on the evidence from your experiments.

- **(a)** What is the largest set of real numbers that could be used for the domain of this function? [If applying the rule to a number produces an error message or a complex number, that number cannot be in the domain.]
- **(b)** Using the domain in part (a), what is the range of this function?

46. Do Exercise 45 for the function whose rule is "press 10^x, and then press TAN; then enter a number in the domain, and press ENTER."

47. The *integer part* function has the set of all real numbers (written as decimals) as its domain. The rule is "for each input number, the output is the part of the number to the left of the decimal point." For instance, the input 37.986 produces the output 37, and the input -1.5 produces the output -1. On most calculators, the integer part function is denoted "iPart." On calculators that use "Intg" or "Floor" for the greatest integer function, the integer part function is denoted by "INT."

- **(a)** For each nonnegative real number input, explain why both the integer part function and the greatest integer function [Example 4] produce the same output.
- **(b)** For which negative numbers do the two functions produce the same output?
- **(c)** For which negative numbers do the two functions produce different outputs?

*You don't need to know what these keys mean in order to do this exercise.

3.2 Functional Notation

Functional notation is a convenient shorthand language that facilitates the analysis of mathematical problems involving functions. It arises from real-life situations, such as the following.

Example 1 According to the Connecticut Department of Revenue Services, the 1998 state income tax rates were as follows.

Taxable Income	Amount of Tax
$15,000 or less	3% of income
More than $15,000	$450 plus 4.5% of income over $15,000

Let I denote income, and write $T(I)$ (read "T of I") to denote the amount of tax on income I. In this shorthand language, $T(7500)$ denotes "the tax on an income of $7500." The sentence "The tax on an income of $7500 is $225" is abbreviated as $T(7500) = 225$. Similarly, $T(25{,}000) = 900$ says that the tax on an income of $25,000 is $900. There is nothing that forces us to use the letters T and I here:

> **Any choice of letters will do, provided we make clear what is meant by these letters.** ■

Example 2 Recall that a falling rock travels $16t^2$ feet after t seconds. Let $d(t)$ stand for the phrase "the distance the rock has traveled after t seconds." Then the sentence "The distance the rock has traveled after t seconds is $16t^2$ feet" can be abbreviated as $d(t) = 16t^2$. For instance,

$$d(1) = 16 \cdot 1^2 = 16$$

means "the distance the rock has traveled after 1 second is 16 feet" and

$$d(4) = 16 \cdot 4^2 = 256$$

means "the distance the rock has traveled after 4 seconds is 256 feet." ■

CAUTION

The parentheses in $d(t)$ do *not* denote multiplication as in the algebraic equation $3(a + b) = 3a + 3b$. The entire symbol $d(t)$ is part of a *shorthand language*. In particular,

$$d(1 + 4) \text{ is } \textit{not} \text{ equal to } d(1) + d(4).$$

For we saw above that $d(1) = 16$ and $d(4) = 256$, so that $d(1) + d(4) = 16 + 256 = 272$. But $d(1 + 4)$ is "the distance traveled after $1 + 4$ seconds," that is, the distance after 5 seconds, namely, $16 \cdot 5^2 = 400$. In general,

Functional notation is a convenient shorthand for phrases and sentences in the English language. It is *not* the same as ordinary algebraic notation.

Functional notation is easily adapted to mathematical settings, in which the particulars of time, distance, etc., are not mentioned. Suppose a function is given. Denote the function by f and let x denote a number in the domain. Then

$f(x)$ denotes the output produced by input x.

For example, $f(6)$ is the output produced by the input 6. The sentence

"y is the output produced by input x according to the rule of the function f"

is abbreviated

$$y = f(x),$$

which is read "y equals f of x." The output $f(x)$ is sometimes called the **value** of the function f at x.

In actual practice, functions are seldom presented in the style of domain, rule, range, as they have been here. Usually, you will be given a phrase such as "the function $f(x) = \sqrt{x^2 + 1}$." This should be understood as a set of directions:

Technology Tip

Functional notation can be used directly on TI-82/83/86/89, Sharp 9600, and HP-38. For example, if the function is entered in the function memory as y_1, then keying in $y_1(5)$ ENTER evaluates the function at $x = 5$.

You will find y_1 in the FUNCTION submenu of the TI-82/83 Y-VARS menu or in the EQVARS submenu of the Sharp 9600 VARS menu. On TI-86/89 and HP-38, type in y_1 on the keyboard.

Functional notation cannot be used this way on TI-85 or Casio 9850 (you'll get an answer, but it usually will be wrong).

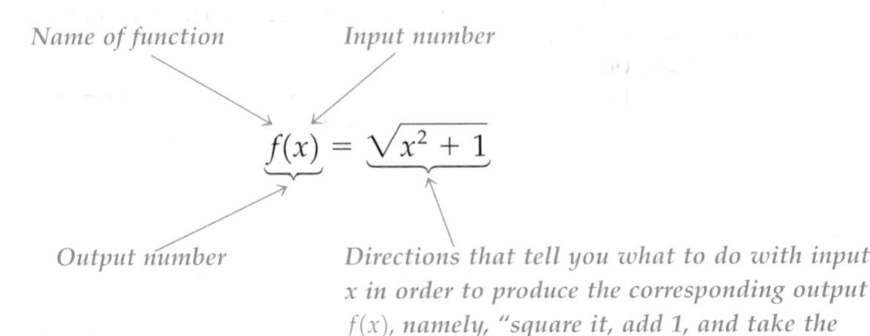

For example, to find $f(3)$, the output of the function f for input 3, simply replace x by 3 in the formula:

$$f(x) = \sqrt{x^2 + 1}$$
$$f(3) = \sqrt{3^2 + 1} = \sqrt{10}.$$

Similarly, replacing x by -5 and 0 shows that

$$f(-5) = \sqrt{(-5)^2 + 1} = \sqrt{26} \quad \text{and} \quad f(0) = \sqrt{0^2 + 1} = 1.$$

Example 3 The expression $h(x) = \dfrac{x^2 + 5}{x - 1}$ defines the function h whose rule is:

For input x, the output is the number $\dfrac{x^2 + 5}{x - 1}$.

Find each of the following.

$$h(\sqrt{3}), \quad h(-2), \quad h(-a), \quad h(r^2 + 3), \quad h(\sqrt{c} + 2).$$

Solution To find $h(\sqrt{3})$ and $h(-2)$, replace x by $\sqrt{3}$ and -2, respectively, in the rule of h:

$$h(\sqrt{3}) = \frac{(\sqrt{3})^2 + 5}{\sqrt{3} - 1} = \frac{8}{\sqrt{3} - 1} \quad \text{and} \quad h(-2) = \frac{(-2)^2 + 5}{-2 - 1} = -3.$$

The value of the function h at any quantity, such as $-a$, $r^2 + 3$, etc., can be found by using the same procedure: *replace x in the formula for $h(x)$ by that quantity:*

$$h(-a) = \frac{(-a)^2 + 5}{-a - 1} = \frac{a^2 + 5}{-a - 1}$$

$$h(r^2 + 3) = \frac{(r^2 + 3)^2 + 5}{(r^2 + 3) - 1} = \frac{r^4 + 6r^2 + 9 + 5}{r^2 + 2} = \frac{r^4 + 6r^2 + 14}{r^2 + 2}$$

$$h(\sqrt{c + 2}) = \frac{(\sqrt{c + 2})^2 + 5}{\sqrt{c + 2} - 1} = \frac{c + 2 + 5}{\sqrt{c + 2} - 1} = \frac{c + 7}{\sqrt{c + 2} - 1}. \quad \blacksquare$$

Technology Tip

One way to evaluate a function $f(x)$ is to enter its rule as an equation $y = f(x)$ in the equation memory and use TABLE or (on TI-85/86) EVAL; see Example 8 or Calculator Investigation 1 in Section 3.1.

When functional notation is used in expressions such as $f(-x)$ or $f(x + h)$, the same basic rule applies: Replace x in the formula by the *entire* expression in parentheses.

Example 4 If $f(x) = x^2 + x - 2$, then

$$f(-x) = (-x)^2 + (-x) - 2 = x^2 - x - 2$$

Note that in this case $f(-x)$ is *not* the same as $-f(x)$, because $-f(x)$ is the negative of the number $f(x)$, that is,

$$-f(x) = -(x^2 + x - 2) = -x^2 - x + 2. \quad \blacksquare$$

Example 5 If $f(x) = x^2 - x + 2$ and $h \neq 0$, find

(a) $f(x + h)$ (b) $f(x + h) - f(x)$ (c) $\dfrac{f(x + h) - f(x)}{h}$.

Solution

(a) Replace x by $x + h$ in the rule of the function:

$$f(x) = x^2 - x + 2$$
$$f(x + h) = (x + h)^2 - (x + h) + 2 = x^2 + 2xh + h^2 - x - h + 2.$$

(b) By part (a),

$$\begin{aligned} f(x + h) - f(x) &= [(x + h)^2 - (x + h) + 2] - [x^2 - x + 2] \\ &= [x^2 + 2xh + h^2 - x - h + 2] - [x^2 - x + 2] \\ &= x^2 + 2xh + h^2 - x - h + 2 - x^2 + x - 2 \\ &= 2xh + h^2 - h \end{aligned}$$

(c) By part (b), we have

$$\frac{f(x + h) - f(x)}{h} = \frac{2xh + h^2 - h}{h} = \frac{h(2x + h - 1)}{h} = 2x + h - 1. \quad \blacksquare$$

If f is a function, then the quantity $\dfrac{f(x+h)-f(x)}{h}$, as in Example 5(c), is called the **difference quotient** of f. Difference quotients, whose significance is explained in Section 3.6, play an important role in calculus.

As the preceding examples illustrate, functional notation is a specialized shorthand language. Treating it as ordinary algebraic notation may lead to mistakes.

CAUTION **Common Mistakes with Functional Notation**

Each of the following statements may be *false:*

1. $f(a+b) = f(a) + f(b)$
2. $f(a-b) = f(a) - f(b)$
3. $f(ab) = f(a)f(b)$
4. $f(ab) = af(b)$
5. $f(ab) = f(a)b$

Example 6 Here are examples of three of the errors listed in the caution box.

1. If $f(x) = x^2$, then
$$f(3+2) = f(5) = 5^2 = 25.$$
But
$$f(3) + f(2) = 3^2 + 2^2 = 9 + 4 = 13.$$
So $f(3+2) \neq f(3) + f(2)$.

3. If $f(x) = x + 7$, then
$$f(3\cdot 4) = f(12) = 12 + 7 = 19.$$
But
$$f(3)f(4) = (3+7)(4+7) = 10\cdot 11 = 110.$$
So $f(3\cdot 4) \neq f(3)f(4)$.

5. If $f(x) = x^2 + 1$, then
$$f(2\cdot 3) = (2\cdot 3)^2 + 1 = 36 + 1 = 37.$$
But
$$f(2)\cdot 3 = (2^2 + 1)3 = 5\cdot 3 = 15.$$
So $f(2\cdot 3) \neq f(2)\cdot 3$. ■

Domains

When the rule of a function is given by a formula, as in Examples 3–5, its domain (set of inputs) is determined by the following convention.

Domain Convention

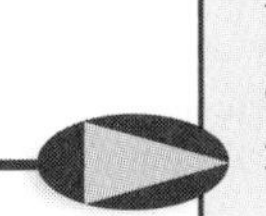

Unless specific information to the contrary is given, the domain of a function f includes every real number (input) for which the rule of the function produces a real number as output.

Thus, the domain of a polynomial function such as $f(x) = x^3 - 4x + 1$ is the set of all real numbers, since $f(x)$ is defined for every value of x. In cases where applying the rule of a function leads to division by zero or to the square root of a negative number, however, the domain may not consist of all real numbers.

Example 7 Find the domain of the function given by

(a) $k(x) = \dfrac{x^2 - 6x}{x - 1}$ (b) $f(u) = \sqrt{u + 2}$

Solution

(a) When $x = 1$, the denominator of $\dfrac{x^2 - 6x}{x - 1}$ is 0 and the fraction is not defined. When $x \neq 1$, however, the denominator is nonzero and the fraction *is* defined. Therefore, the domain of the function k consists of all real numbers *except* 1.

(b) Since negative numbers do not have real square roots, $\sqrt{u + 2}$ is a real number only when $u + 2 \geq 0$, that is, when $u \geq -2$. Therefore, the domain of f consists of all real numbers greater than or equal to -2, that is, the interval $[-2, \infty)$. ■

Example 8 A **piecewise-defined** function is one whose rule includes several formulas, such as

$$f(x) = \begin{cases} 2x + 3 & \text{if } x < 4 \\ x^2 - 1 & \text{if } 4 \leq x \leq 10. \end{cases}$$

Find each of the following.

(a) $f(-5)$ (b) $f(8)$ (c) $f(k)$

(d) The domain of f.

Solution

(a) Since $-5 < 4$, the first part of the rule applies:

$$f(-5) = 2(-5) + 3 = -7.$$

(b) Since 8 is between 4 and 10, the second part of the rule applies:

$$f(8) = 8^2 - 1 = 63.$$

(c) We cannot find $f(k)$ unless we know whether $k < 4$ or $4 \leq k \leq 10$.

(d) The rule of f gives no directions when $x > 10$, so the domain of f consists of all real numbers x with $x \leq 10$, that is, $(-\infty, 10]$. ■

Applications

The domain convention does not always apply when dealing with applications. Consider, for example, the distance function for falling objects, $d(t) = 16t^2$ (see Example 2). Since t represents time, only nonnegative values of t make sense here, even though the rule of the function is defined for all values of t. Analogous comments apply to other applications.

A real-life situation may lead to a function whose domain does not include all the numbers for which the rule of the function is defined.

Example 9 A glassware factory has fixed expenses (mortgage, taxes, machinery, etc.) of $12,000 per week. It costs 80 cents to make one cup (labor, materials, shipping). A cup sells for $1.95. At most 18,000 cups can be manufactured each week.

(a) Express the weekly revenue as a function of the number x of cups made.

(b) Express the weekly costs as a function of x.

(c) Find the domain and the rule of the weekly profit function.

Solution

(a) If $R(x)$ is the weekly revenue from selling x cups, then

$$R(x) = (\text{price per cup}) \times (\text{number sold})$$
$$R(x) = 1.95x.$$

(b) If $C(x)$ is the weekly cost of manufacturing x cups, then

$$C(x) = (\text{cost per cup}) \times (\text{number sold}) + (\text{fixed expenses})$$
$$C(x) = .80x + 12{,}000.$$

(c) If $P(x)$ is the weekly profit from selling x cups, then

$$P(x) = \text{Revenue} - \text{Cost}$$
$$P(x) = R(x) - C(x)$$
$$P(x) = 1.95x - (.80x + 12{,}000) = 1.95x - .80x - 12{,}000$$
$$P(x) = 1.15x - 12{,}000$$

Although this rule is defined for all real numbers x, the domain of the function P consists of the possible number of cups that can be made each week. Since you can only make whole cups and the maximum production is 18,000, the domain of P consists of all integers from 0 to 18,000. ■

Example 10 Let P be the profit function in Example 9.

(a) What is the profit from selling 5000 cups? From 14,000 cups?

(b) What is the break-even point?

Solution

(a) We evaluate the function $P(x) = 1.15x - 12{,}000$ at the required values of x:

$$P(5000) = 1.15(5000) - 12{,}000 = -\$6250$$
$$P(14{,}000) = 1.15(14{,}000) - 12{,}000 = \$4100.$$

Thus, sales of 5000 cups produce a loss of \$6250, while sales of 14,000 produce a profit of \$4100.

(b) The break-even point occurs when revenue equals costs (that is, when profit is 0). So we set $P(x) = 0$ and solve for x:

$$1.15x - 12{,}000 = 0$$
$$1.15x = 12{,}000$$
$$x = \frac{12{,}000}{1.15} \approx 10{,}434.78$$

Thus, the break-even point occurs between 10,434 and 10,435 cups. There is a slight loss from selling 10,434 cups and a slight profit from selling 10,435. ■

Exercises 3.2

In Exercises 1 and 2, find the indicated values of the function by hand and *by using the table feature of a calculator (or the EVAL key on TI-85/86). If your answers do not agree with each other or with those at the back of the book, you are either making algebraic mistakes or incorrectly entering the function in the equation memory.*

1. $f(x) = \dfrac{x-3}{x^2+4}$

(a) $f(-1)$ **(b)** $f(0)$ **(c)** $f(1)$ **(d)** $f(2)$ **(e)** $f(3)$

2. $g(x) = \sqrt{x+4} - 2$

(a) $g(-2)$ **(b)** $g(0)$ **(c)** $g(4)$ **(d)** $g(5)$ **(e)** $g(12)$

Exercises 3–24 refer to these three functions:

$$f(x) = \sqrt{x+3} - x + 1 \qquad g(t) = t^2 - 1$$
$$h(x) = x^2 + \frac{1}{x} + 2$$

In each case find the indicated value of the function.

3. $f(0)$ **4.** $f(1)$

5. $f(\sqrt{2})$ **6.** $f(\sqrt{2} - 1)$

7. $f(-2)$ **8.** $f(-3/2)$

9. $h(3)$ **10.** $h(-4)$

11. $h(3/2)$ **12.** $h(\pi + 1)$

13. $h(a + k)$ **14.** $h(-x)$

15. $h(2 - x)$ **16.** $h(x - 3)$

17. $g(3)$ **18.** $g(-2)$

19. $g(0)$ **20.** $g(x)$

21. $g(s + 1)$ **22.** $g(1 - r)$

23. $g(-t)$ **24.** $g(t + h)$

In Exercises 25–32, compute:

(a) $f(r)$ **(b)** $f(r) - f(x)$ **(c)** $\dfrac{f(r) - f(x)}{r - x}$

In part (c) assume $r \neq x$ and simplify your answer.

Example: *If $f(x) = x^2$, then*

$$\frac{f(r) - f(x)}{r - x} = \frac{r^2 - x^2}{r - x} = \frac{(r + x)(r - x)}{r - x} = r + x.$$

25. $f(x) = x$

26. $f(x) = -10x$

27. $f(x) = 3x + 7$

28. $f(x) = x^3$

29. $f(x) = x - x^2$

30. $f(x) = x^2 + 1$

31. $f(x) = \sqrt{x}$

32. $f(x) = 1/x$

In Exercises 33–40, assume $h \neq 0$. Compute and simplify the difference quotient

$$\frac{f(x+h) - f(x)}{h}$$

33. $f(x) = x + 1$

34. $f(x) = -10x$

35. $f(x) = 3x + 7$

36. $f(x) = x^2$

37. $f(x) = x - x^2$

38. $f(x) = x^3$

39. $f(x) = \sqrt{x}$

40. $f(x) = 1/x$

41. The rule of the function f is given by the graph, as in Example 10 of Section 3.1. Find

(a) The domain of f

(b) The range of f

(c) $f(-3)$

(d) $f(-1)$

(e) $f(1)$

(f) $f(2)$

42. The rule of the function g is given by the graph, as in Example 10 of Section 3.1. Find

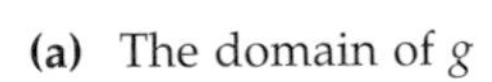
(a) The domain of g

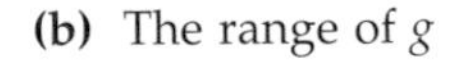
(b) The range of g

(c) $g(-3)$

(d) $g(-1)$

(e) $g(1)$

(f) $g(4)$

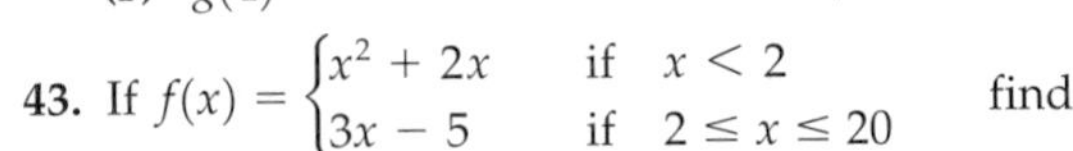
43. If $f(x) = \begin{cases} x^2 + 2x & \text{if } x < 2 \\ 3x - 5 & \text{if } 2 \leq x \leq 20 \end{cases}$ find

(a) The domain of f
(b) $f(-3)$ **(c)** $f(-1)$ **(d)** $f(2)$ **(e)** $f(7/3)$

44. If $g(x) = \begin{cases} 2x - 3 & \text{if } x < -1 \\ |x| - 5 & \text{if } -1 \leq x \leq 2 \\ x^2 & \text{if } x > 2 \end{cases}$ find

(a) The domain of g
(b) $g(-2.5)$ **(c)** $g(-1)$ **(d)** $g(2)$ **(e)** $g(4)$

45. In a certain state the sales tax $T(p)$ on an item of price p dollars is 5% of p. Which of the following formulas give the correct sales tax in all cases?

(i) $T(p) = p + 5$
(ii) $T(p) = 1 + 5p$
(iii) $T(p) = p/20$
(iv) $T(p) = p + (5/100)p = p + .05p$
(v) $T(p) = (5/100)p = .05p$

46. Let T be the sales tax function of Exercise 45, and find $T(3.60)$, $T(4.80)$, $T(.60)$, and $T(0)$.

In Exercises 47–60, determine the domain of the function according to the usual convention.

47. $f(x) = x^2$

48. $g(x) = \dfrac{1}{x^2} + 2$

49. $h(t) = |t| - 1$

50. $k(u) = \sqrt{u}$

51. $k(x) = |x| + \sqrt{x} - 1$

52. $h(x) = \sqrt{(x+1)^2}$

53. $g(u) = \dfrac{|u|}{u}$

54. $h(x) = \dfrac{\sqrt{x-1}}{x^2 - 1}$

55. $g(y) = [-y]$

56. $f(t) = \sqrt{-t}$

57. $g(u) = \dfrac{u^2 + 1}{u^2 - u - 6}$

58. $f(t) = \sqrt{4 - t^2}$

59. $f(x) = -\sqrt{9 - (x-9)^2}$

60. $f(x) = \sqrt{-x} + \dfrac{2}{x+1}$

61. Give an example of two different functions f and g that have all of the following properties:

$$f(-1) = 1 = g(-1) \quad \text{and} \quad f(0) = 0 = g(0)$$
$$\text{and} \quad f(1) = 1 = g(1).$$

62. Give an example of a function g with the property that $g(x) = g(-x)$ for every real number x.

In Exercises 63–66, the rule of a function f is given. Write an algebraic formula for $f(x)$.

63. Double the input, subtract 5, and take the square root of the result.

64. Square the input, multiply by 3, and subtract the result from 8.

65. Cube the input, add 6, and divide the result by 5.

66. Take the square root of the input, add 7, divide the result by 8, and add this result to the original input.

67. Jack and Jill are salespersons in the suit department of a clothing store. Jack is paid $200 per week plus $5 for each suit he sells, whereas Jill is paid $10 for every suit she sells.

(a) Let $f(x)$ denote Jack's weekly income and $g(x)$ Jill's weekly income from selling x suits. Find the rules of the functions f and g.

(b) Use algebra or a table to find: $f(20)$ and $g(20)$; $f(35)$ and $g(35)$; $f(50)$ and $g(50)$.

(c) If Jack sells 50 suits a week, how many must Jill sell to have the same income as Jack?

68. The table shows the 1999 federal income tax rates for a single person. Write the rule of a piecewise defined function T such that $T(x)$ is the tax due on a 1999 taxable income of x dollars.

Taxable Income	Tax
Not over $25,750	15% of income
Over $25,750, but not over $62,450	$3862.50 + 28% of amount over $25,750
Over $62,450, but not over $130,250	$14,138.50 + 31% of amount over $62,450
Over $130,250, but not over $283,150	$35,156.50 + 36% of amount over $130,250
Over $283,150	$90,200.50 + 39.6% of amount over $283,150

69. Suppose a car travels at a constant rate of 55 mph for 2 hours and travels at 45 mph thereafter. Show that distance traveled is a function of time, and find the rule of the function.

70. A man walks for 45 minutes at a rate of 3 mph, then jogs for 75 minutes at a rate of 5 mph, then sits and rests for 30 minutes, and finally walks for $1\frac{1}{2}$ hours. Find the rule of the function that expresses his distance traveled as a function of time. [*Caution:* Don't mix up the units of time; use either minutes or hours, not both.]

71. The list price of a workbook is $12. But if ten or more copies are purchased, then the price per copy is reduced by 25¢ for every copy above ten. (That is, $11.75 per copy for 11 copies, $11.50 per copy for 12 copies, and so on.)

(a) The price per copy is a function of the number of copies purchased. Find the rule of this function.

(b) The total cost of a quantity purchase is

(number of copies) × (price per copy).

Show that the total cost is a function of the number of copies and find the rule of the function.

72. A potato chip factory has a daily overhead from salaries and building costs of $1800. The cost of ingredients and packaging to produce a pound of potato chips is 50¢. A pound of potato chips sells for $1.20. Show that the factory's daily profit is a function of the number of pounds of potato chips sold and find the rule of this function. (Assume that the factory sells all the potato chips it produces each day.)

73. A rectangular region of 6000 square feet is to be fenced in on three sides with fencing costing $3.75 per foot and on the fourth side with fencing costing $2.00 per foot. Express the cost of the fence as a function of the length x of the fourth side.

74. A box with a square base measuring $t \times t$ ft is to be made of three kinds of wood. The cost of the wood for the base is 85¢ per square foot; the wood for the sides costs 50¢ per square foot, and the wood for the top $1.15 per square foot. The volume of the box is to be 10 cubic feet. Express the total cost of the box as a function of the length t.

75. Average tuition and fees in private four-year colleges in recent years were as follows. (*Source:* U.S. Department of Education and the College Board)

Year	Tuition & Fees
1992	$10,294
1993	$10,952
1994	$11,481

Year	Tuition & Fees
1995	$12,216
1996	$12,994
1997	$13,664

(a) Use linear regression to find the rule of a function f that gives the approximate average tuition in year x, where $x = 0$ corresponds to 1990.

(b) Find $f(3)$, $f(5)$, and $f(7)$. How do they compare with the actual figures?

(c) Use f to estimate tuition in 1999.

76. The table shows the national debt (in billions of dollars) in selected years. (*Source:* U.S. Department of Treasury)

Year	Debt
1980	907.7
1982	1142
1984	1572.3
1986	2125.3
1988	2602.3

Year	Debt
1990	3233.3
1992	4064.6
1994	4692.8
1996	5224.8

(a) Use linear regression to find the rule of a function f that gives the approximate national debt (in billions) in year x, where $x = 0$ corresponds to 1980.

(b) Find $f(6)$, $f(10)$, and $f(14)$. How do they compare with the actual figures?

(c) Use f to estimate the national debt in 2000.

3.3 Graphs of Functions

The graph of a function f is the graph of the *equation* $y = f(x)$. Hence

The graph of the function f consists of all points $(x, f(x))$, where x is any number in the domain of f.

The graphs of most functions whose rules are given by algebraic formulas are easily obtained on a calculator. However, there are situations in which a calculator-generated graph may be incomplete or misleading. So the emphasis here is on using your algebraic knowledge *before* reaching for a calculator. Doing this will often tell you that a calculator is inappropriate, or help you to interpret screen images when a calculator is used.

Example 1 The graph of $f(x) = 3x - 2$ is the graph of the equation $y = 3x - 2$. As we saw in Section 1.4, this graph is a straight line with slope 3 and y-intercept -2, which is easily graphed by hand (Figure 3–8). On a calculator, the graph of f will look a bit bumpy and jagged, rather than the smooth line in Figure 3–8 (try it!). ■

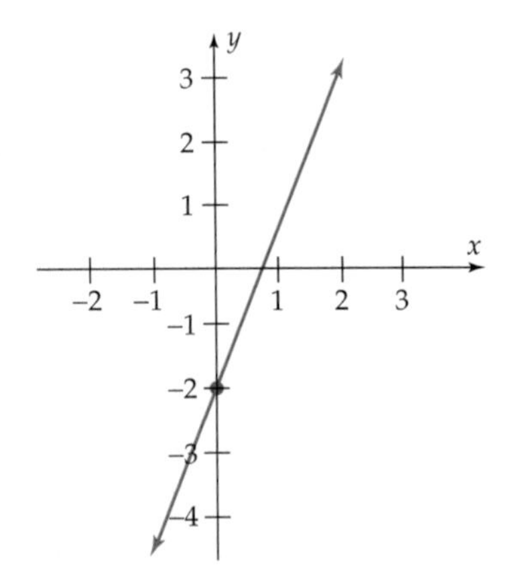

Figure 3–8

Example 2 The greatest integer function $f(x) = [x]$ was introduced in Example 4 of Section 3.1. It can easily be graphed by hand, by considering the values of the function between each two consecutive integers. For instance,

x	$-2 \le x < -1$	$-1 \le x < 0$	$0 \le x < 1$	$1 \le x < 2$	$2 \le x < 3$
$[x]$	-2	-1	0	1	2

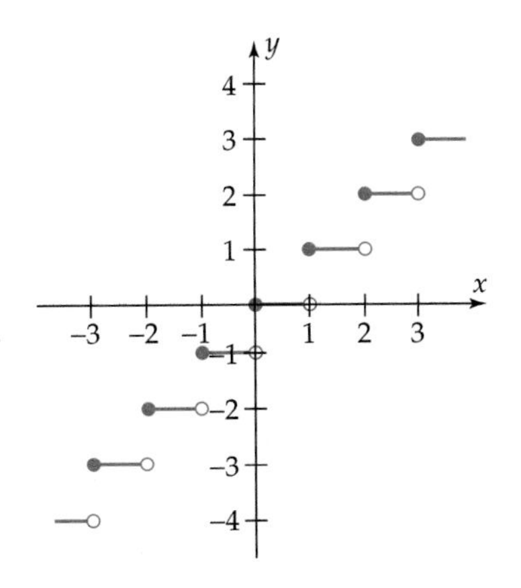

Figure 3–9

Thus, between $x = -2$ and $x = -1$, the value of $f(x) = [x]$ is always -2, so that the graph there is a horizontal line segment, all of whose points have second coordinate -2. The rest of the graph is obtained similarly (Figure 3–9). An open circle in Figure 3–9 indicates that the endpoint of the segment is *not* on the graph, whereas a closed circle indicates that the endpoint is on the graph. ■

A function whose graph consists of horizontal line segments, such as Figure 3–9, is called a **step function.** Graphing step functions with reasonable accuracy on a calculator requires some care. Even then, some features of the graph may not be shown.

GRAPHING EXPLORATION

Graph the greatest integer function $f(x) = [x]$ on your calculator (see the Technology Tip on page 120). Does your graph look like Figure 3–9, or does it include vertical segments? Now change the graphing mode of your calculator to "dot" rather than "connected" (see the Technology Tip in the margin), and graph again. How does this graph compare with Figure 3–9? Can you tell from the graph which endpoints are included and which are excluded?

Technology Tip

Directions for switching to dot graphing mode are in Calculator Investigation 5 at the end of Section 2.1.

Example 3 An overnight delivery service charges \$18 for a package weighing less than 1 pound, \$21 for one weighing at least 1 pound, but less than 2 pounds, \$24 for one weighing at least 2 pounds, but less than 3 pounds, and so on. Verify that the cost $c(x)$ of shipping a package weighing x pounds is given by $c(x) = 18 + 3[x]$. For example,

$$\text{If } 2 \leq x < 3, \text{ then } [x] = 2 \text{ and } c(x) = 18 + 3[x] = 18 + 3(2) = 24.$$

Although this rule makes sense for all real numbers, the domain of this cost function consists of positive numbers (why?). The graph of c is in Figure 3–10. ■

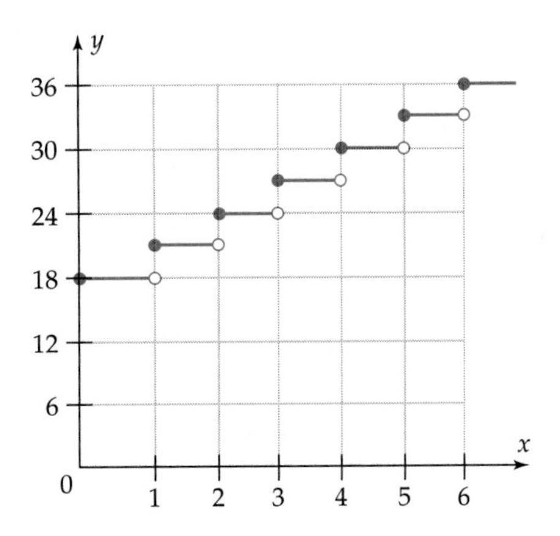

Figure 3–10

Example 4 The graph of the piecewise-defined function

$$f(x) = \begin{cases} x^2 & \text{if } x \le 1 \\ x + 2 & \text{if } 1 < x \le 4 \end{cases}$$

is made up of *parts* of two graphs, corresponding to the different parts of the rule of the function:

$x \le 1$ — For these values of x, the graph of f coincides with the graph of $y = x^2$, which was sketched in Figure 2–2 on page 60.

$1 < x \le 4$ — For these values of x, the graph of f coincides with the graph of $y = x + 2$, which is a straight line.

Therefore, we must graph

$$y = x^2 \quad \text{when } x \le 1 \qquad \text{and} \qquad y = x + 2 \quad \text{when } 1 < x \le 4.$$

Combining these partial graphs produces the graph of f in Figure 3–11. ■

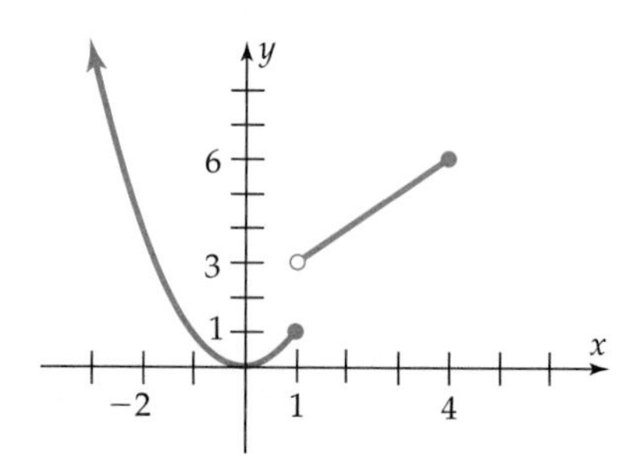

Figure 3–11

Piecewise-defined functions can be graphed on a calculator, provided that you use the correct syntax. Once again, however, the screen does not show which endpoints are included or excluded from the graph.

Technology Tip

Inequality symbols are in the TEST menu of TI-83/86, in the TESTS submenu of the HP-38 MATH menu, and in the INEQ submenu of the Sharp 9600 MATH menu. TI-89 has the symbols <, >, and | on the keyboard; other inequality symbols and logical symbols (such as "and") are in the TEST submenu of the MATH menu.

GRAPHING EXPLORATION

Graph the function f of Example 4 on a calculator, as follows. On Sharp 9600 or HP-38 or TI-83/86 calculators, use the Tip in the margin to graph these two equations on the same screen:

$$y_1 = x^2/(x \le 1)$$

$$y_2 = (x + 2)/((x > 1)(x \le 4)).$$

On TI-89/92, use the Tip to graph these equations on the same screen:

$$y_1 = x^2 \,|\, x \le 1$$

$$y_2 = x + 2 \,|\, x > 1 \quad \text{and} \quad x \le 4.$$

To graph f on Casio 9850, with the viewing window of Figure 3–11, graph these equations on the same screen (including commas and square brackets):

$$y_1 = x^2, [-6, 1]$$

$$y_2 = x + 2, [1, 4].$$

How does your graph compare with Figure 3–11?

Example 5 The absolute value function $f(x) = |x|$ is also a piecewise-defined function, since by definition

$$|x| = \begin{cases} x & \text{if } x \geq 0 \\ -x & \text{if } x < 0. \end{cases}$$

Its graph can be obtained by drawing the part of the line $y = x$ to the right of the origin and the part of the line $y = -x$ to the left of the origin (Figure 3–12) or by graphing $y = \text{ABS } x$ on a calculator (Figure 3–13). ■

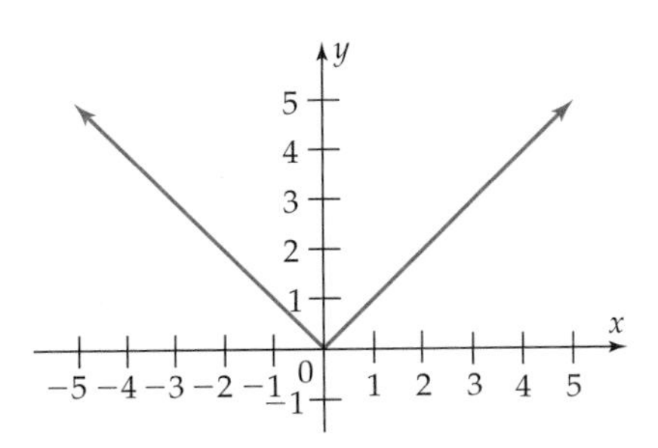

Figure 3–12

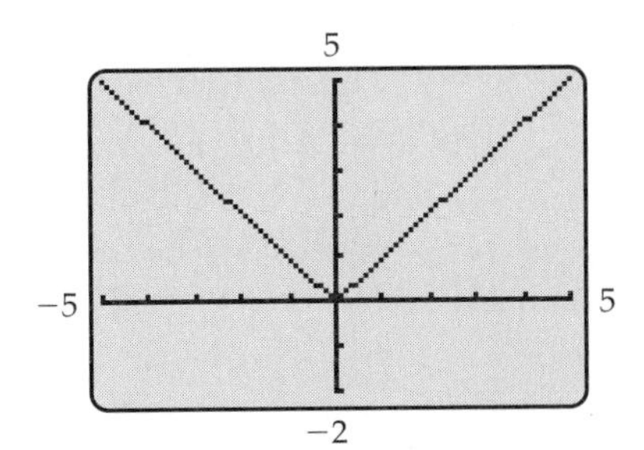

Figure 3–13

The Vertical Line Test

The following fact, which distinguishes graphs of functions from other graphs, can also be used to interpret some calculator-generated graphs.

Vertical Line Test

The graph of a function $y = f(x)$ has this property:

No vertical line intersects the graph more than once.

Conversely, any graph with this property is the graph of a function.

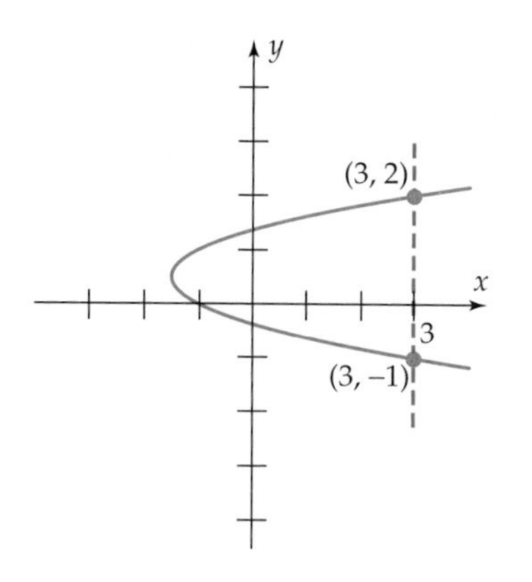

Figure 3–14

To see why this is true, consider Figure 3–14, in which the graph intersects the vertical line at two points. If this were the graph of a function f, then we would have $f(3) = 2$ [because (3, 2) is on the graph] *and* $f(3) = -1$ [because (3, −1) is on the graph]. This means that the input 3 produces two different outputs, which is impossible for a function. Therefore, Figure 3–14 is not the graph of a function. A similar argument works in the general case.

Example 6 The rule $g(x) = x^{15} + 2$ does give a function (each input produces exactly one output), but its graph in Figure 3–15 appears to be vertical near $x = 1$. To see that this is not actually true, change the viewing window. Figure 3–16 shows that the graph is not vertical between $x = 1$ and $x = 1.2$. This detail is lost in Figure 3–15 because all the x values in Figure 3–16 occupy only a single pixel width in Figure 3–15. ■

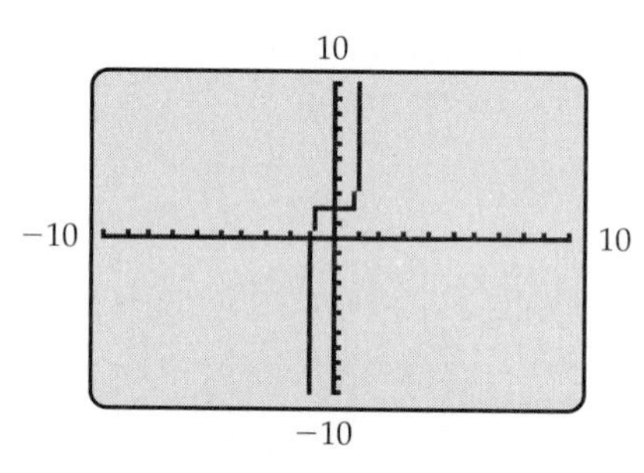

Figure 3–15

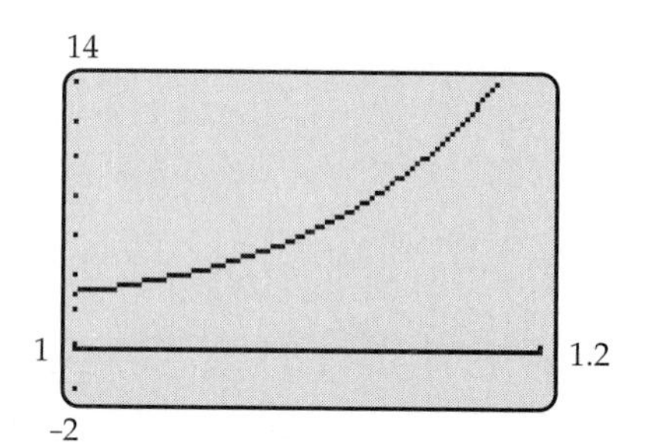

Figure 3–16

GRAPHING EXPLORATION

Find a viewing window that shows that the graph of the function g in Example 6 is not actually vertical near $x = -1$.

Local Maxima and Minima

The graph of a function may include some peaks and valleys (Figure 3–17). A peak is not necessarily the highest point on the graph, but it is the highest point in its neighborhood. Similarly, a valley is the lowest point in the neighborhood, but not necessarily the lowest point on the graph.

More formally, we say that a function f has a **local maximum** at $x = c$ if the graph of f has a peak at the point $(c, f(c))$. This means that all nearby points $(x, f(x))$ have smaller y-coordinates, that is,

$$f(x) \leq f(c) \quad \text{for all } x \text{ near } c.$$

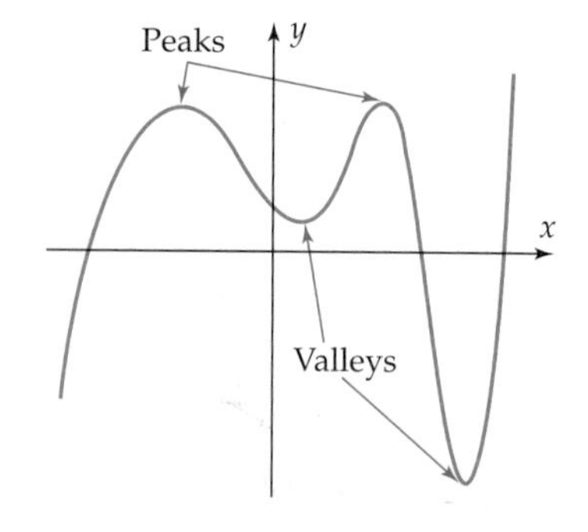

Figure 3–17

Similarly, a function has a **local minimum** at $x = d$ provided that

$$f(x) \geq f(d) \quad \text{for all } x \text{ near } d.$$

In other words, the graph of f has a valley at $(d, f(d))$ because all nearby points $(x, f(x))$ have larger y-coordinates.

Calculus is usually needed to find the exact location of local maxima and minima (the plural forms of maximum and minimum). However, they can be accurately approximated by the maximum finder or minimum finder of a calculator.

Example 7 The graph of $f(x) = x^3 - 1.8x^2 + x + 1$ in Figure 3–18 does not appear to have any local maxima or minima. However, if you use the trace feature to move along the flat segment to the right of the y-axis, you find that the y-coordinates increase, then decrease, then increase (try it!). To see what's really going on, we change viewing windows (Figure 3–19) and see that the function actually has a local maximum and a local minimum (Figure 3–20). The calculator's minimum finder shows that the local minimum occurs when $x \approx .7633$. ■

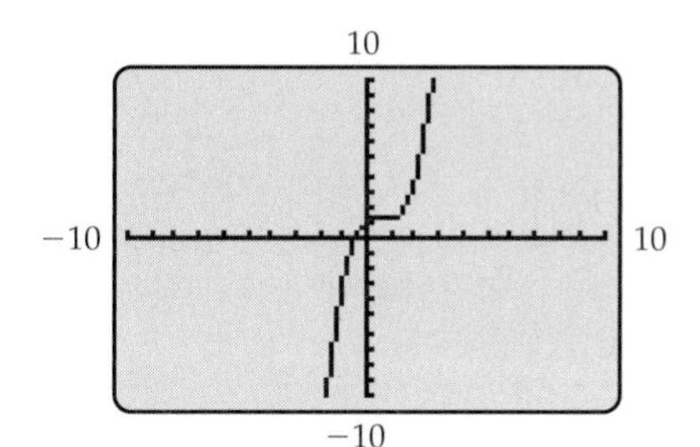

Figure 3–18

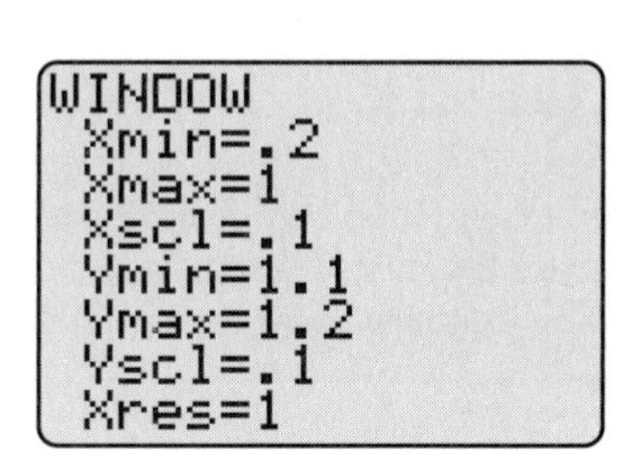

Figure 3–19

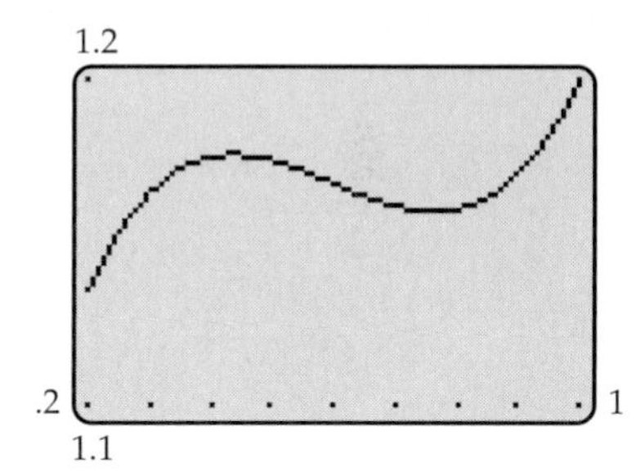

Figure 3–20

GRAPHING EXPLORATION

Graph the function in Example 7 in the viewing window of Figure 3–19. Use the maximum finder to approximate the location of the local maximum.

Example 8 The number of radio stations that primarily play country music is approximated by the function

$$g(x) = 1.2x^3 - 31.4x^2 + 192x + 2300 \quad (0 \leq x \leq 10),$$

where $x = 0$ corresponds to 1990. When was the number of country stations the largest during the 1990s?

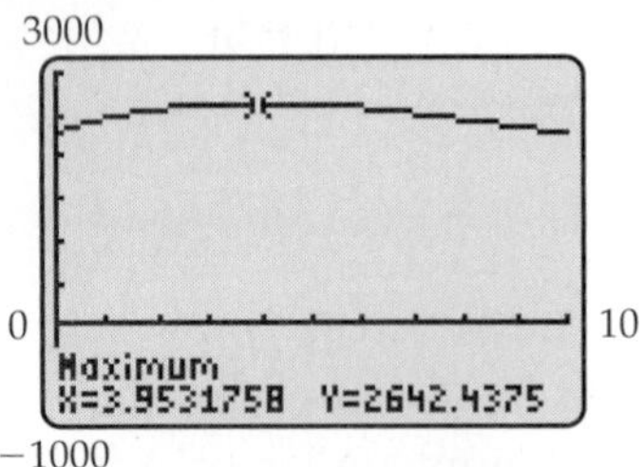

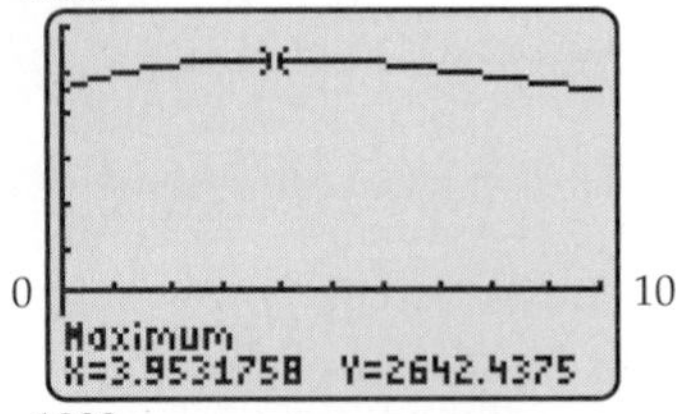

Figure 3–21

Solution We graph the function and use the maximum finder to determine that the local maximum occurs when $x \approx 3.95$ and $g(3.95) \approx 2642$, as shown in Figure 3–21. Therefore the maximum number of country stations was about 2642, and this occurred very late in 1993. ■

Increasing and Decreasing Functions

A function is said to be **increasing on an interval** if its graph always rises as you move from left to right over the interval. It is **decreasing on an interval** if its graph always falls as you move from left to right over the interval. A function is said to be **constant on an interval** if its graph is horizontal over the interval.

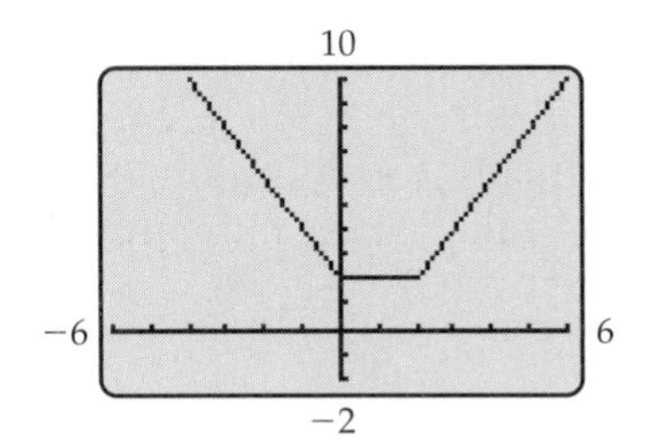

Figure 3–22

Example 9 Figure 3–22 suggests that $f(x) = |x| + |x - 2|$ is decreasing on the interval $(-\infty, 0)$, increasing on $(2, \infty)$, and constant on $[0, 2]$. You can confirm that the function is actually constant between 0 and 2 by using the trace feature to move along the graph there (the y-coordinates remain the same, as they should on a horizontal segment). For an algebraic proof that f is constant on $[0, 2]$, see Exercise 20. ■

> **CAUTION**
>
> A horizontal segment on a calculator graph does not always mean that the function is constant there. There may be **hidden behavior,** as was the case in Example 7. When in doubt, use the trace feature to see if the y-coordinates remain constant as you move along the 'horizontal" segment, or change the viewing window.

Example 10 On what (approximate) intervals is the function $g(x) = .5x^3 - 3x$ increasing and decreasing?

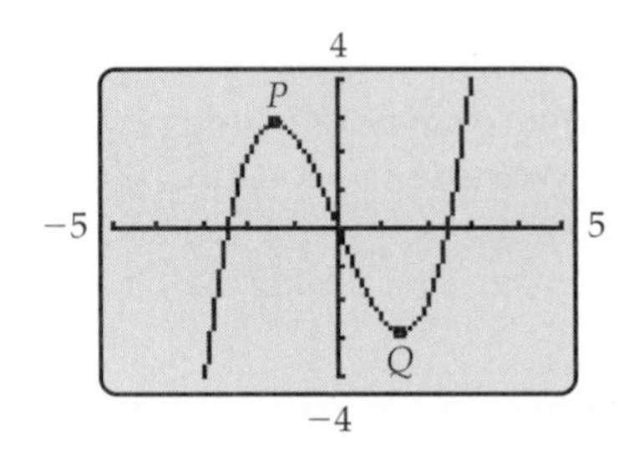

Figure 3–23

Solution The (complete) graph of g in Figure 3–23 shows that g has a local maximum at P and a local minimum at Q. The maximum and minimum finders show that the approximate coordinates of P and Q are

$$P = (-1.4142, 2.8284) \quad \text{and} \quad Q = (1.4142, -2.8284).$$

Therefore, f is increasing on $(-\infty, -1.4142)$ and $(1.4142, \infty)$. It is decreasing on $(-1.4142, 1.4142)$. ■

Graph Reading

Until now we have concentrated on translating statements into functional notation and functional notation into graphs. It is just as important, however, to be able to translate graphical information into equivalent statements in English or functional notation.

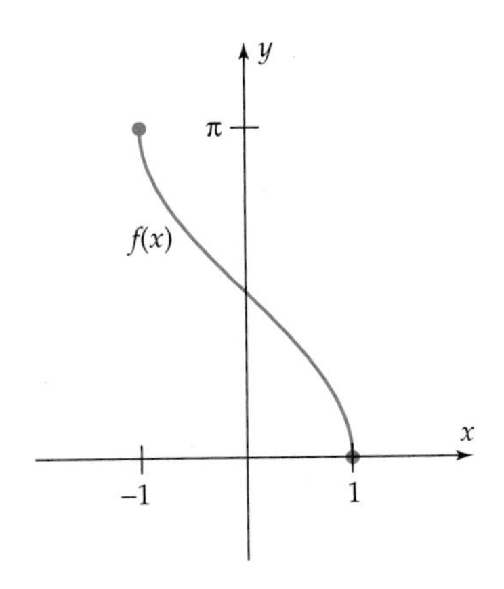

Figure 3–24

Example 11 The entire graph of a function f is shown in Figure 3–24. Find the domain and range of f.

Solution The graph of f consists of all points of the form $(x, f(x))$. Thus, the first coordinates of points on the graph are the inputs (numbers in the domain of f), and the second coordinates are the outputs (the numbers in the range of f). Figure 3–24 shows that the first coordinates of points on the graph all satisfy $-1 \le x \le 1$, so the domain of f is the interval $[-1, 1]$. Similarly, the second coordinates satisfy $0 \le y \le \pi$, so the range of f is the interval $[0, \pi]$. ■

Example 12 Use the graphs of the functions g and h in Figure 3–25 to find:

(a) All numbers x such that $h(x) < 0$;

(b) All numbers x in the interval $[-3, 3]$ such that $g(x) = 2$;

(c) The largest interval over which g is increasing and h is decreasing and $h(x) \ge g(x)$ for every x in the interval.

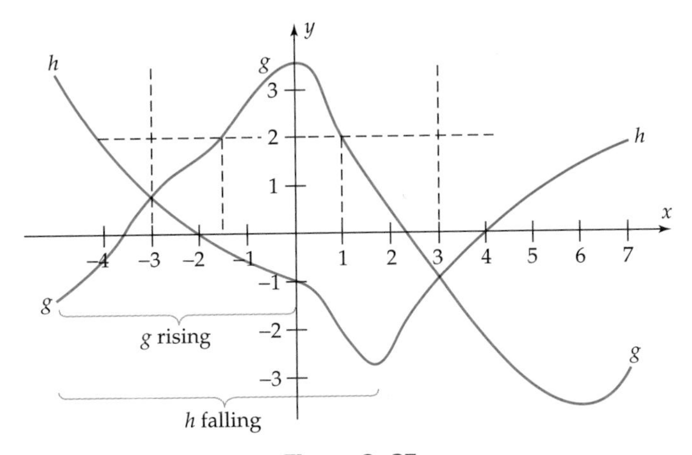

Figure 3–25

Solution

(a) The graph of h consists of all points $(x, h(x))$. The numbers x such that $h(x) < 0$ correspond to the points on the graph with negative second coordinates, that is, the points that lie below the x-axis. The graph of h shows that these are the points whose first coordinates satisfy $-2 < x < 4$. So the answer is all numbers in the interval $(-2, 4)$.

(b) The graph of g consists of the points $(x, g(x))$. The points on the graph that lie *between* the vertical lines through -3 and 3 and *on* the horizontal line through 2 have first coordinate x in the interval $[-3, 3]$ and second coordinate $g(x) = 2$. Figure 3–25 shows that the only such points are $(-1.5, 2)$ and $(1, 2)$. So the answer is $x = -1.5$ and $x = 1$.

(c) Figure 3–25 shows that $[-5, 0]$ is the only interval over which the graph of h is falling *and* the graph of g is rising. Clearly, $h(x) \ge g(x)$

exactly when the point $(x, h(x))$ lies above the point $(x, g(x))$. The only time this occurs in the interval $[-5, 0]$ is when $-5 \le x \le -3$. Therefore, the answer is the interval $[-5, -3]$. ■

Parametric Graphing*

Most of the functions graphed above can be described by equations in which y is a function of x, and hence, are easily graphed on a calculator. When you must graph an equation that expresses x as a function of y, a different calculator graphing technique is needed.

Example 13 Graph $x = y^3 - 3y^2 - 4y + 7$.

Solution Let t be any real number. If $y = t$, then

$$x = y^3 - 3y^2 - 4y + 7 = t^3 - 3t^2 - 4t + 7.$$

Thus, the graph consists of all points (x, y) such that

$$x = t^3 - 3t^2 - 4t + 7 \quad \text{and} \quad y = t \quad (t \text{ any real number}).$$

When written in this form, the equation can be graphed as follows.

Change the graphing mode to **parametric mode** (see the Tip in the margin), and enter the equations

$$x_{1t} = t^3 - 3t^2 - 4t + 7$$
$$y_{1t} = t$$

in the equation memory.† Next set the viewing window so that

$$-6 \le t \le 6, \qquad -10 \le x \le 10, \qquad -6 \le y \le 6,$$

as partially shown in Figure 3–26 (scroll down to see the rest). We use the same range for t and y here because $y = t$. Note that we must also set "t-step" (or "t pitch"), which determines how much t changes each time a point is plotted. A t-step between .05 and .15 usually produces a relatively smooth graph in a reasonable amount of time. Finally, pressing GRAPH (or PLOT or DRAW) produces the graph in Figure 3–27. ■

Technology Tip

To change to parametric graphing mode, choose PAR (or PARAM or PARAMETRIC or PARM) in the TI MODE menu, or the HP-38 LIB menu, or the COORD submenu of the Sharp 9600 SETUP menu, or in the TYPE submenu of the Casio 9850 GRAPH menu (on the main menu).

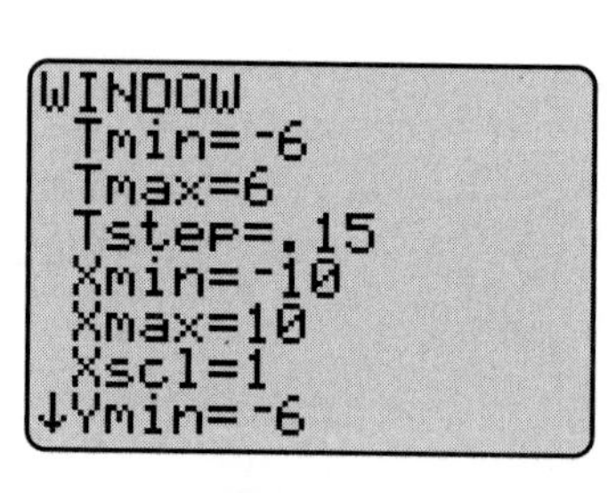

Figure 3–26

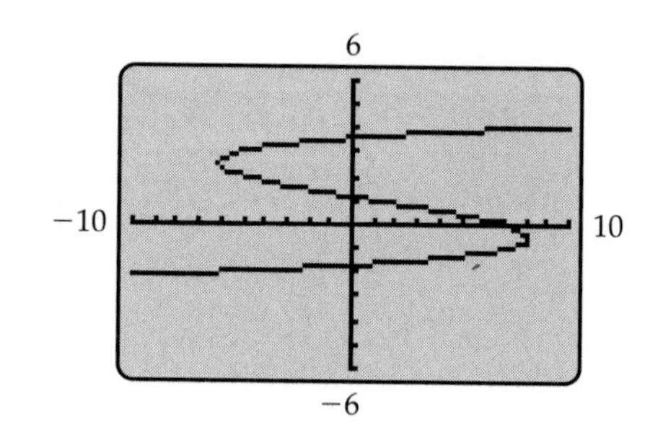

Figure 3–27

*This material will be used only in Sections 3.7 and 6.2, Chapter 10, and occasional exercises. It may be omitted or postponed until then, if desired.

†On some calculators x_{1t} is denoted x_{t1} or $x_1(T)$, and similarly for y_{1t}.

Technology Tip

If you have trouble finding appropriate ranges for t, x, and y, it may help to use the TABLE feature to display a table of t-x-y values produced by the parametric equations.

As illustrated in Example 13, the underlying idea of **parametric graphing** is to express both x and y as functions of a third variable t. The equations that define x and y are called **parametric equations,** and the variable t is called the **parameter.** Example 13 illustrates just one of the many applications of parametric graphing. It can also be used to graph curves that are not graphs of a single equation in x and y.

Example 14 Graph the curve given by

$$x = t^2 - t - 1 \quad \text{and} \quad y = t^3 - 4t - 6 \quad (-2 \le t \le 3).$$

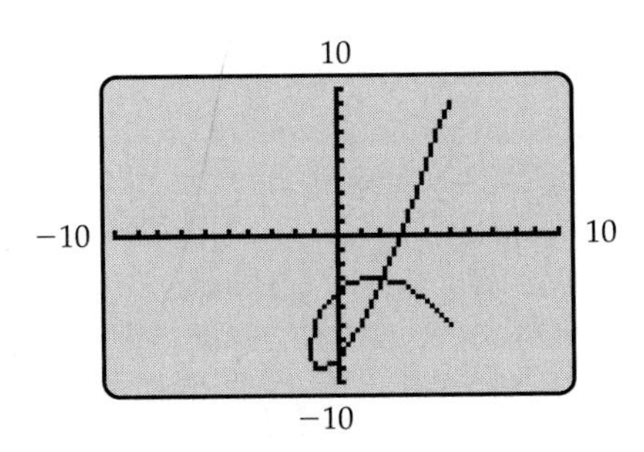

Figure 3–28

Solution Using the standard viewing window, we obtain the graph in Figure 3–28. Note that the graph crosses over itself at one point and that it does not extend forever to the left and right, but has "endpoints."

GRAPHING EXPLORATION

Graph these same parametric equations, but set the range of t values so that $-4 \le t \le 4$. What happens to the graph? Now change the range of t values so that $-10 \le t \le 10$. Find a viewing window large enough to show the entire graph, including endpoints. ■

Any function of the form $y = f(x)$ can be expressed in terms of parametric equations and graphed that way. For instance, to graph $f(x) = x^2 + 1$, let $x = t$ and $y = f(x) = t^2 + 1$. Parametric graphing will be used hereafter whenever it is convenient and will be studied more thoroughly in Section 10.1.

Exercises 3.3

In Exercises 1–4, sketch the graph of the function. You should be able to do this without using a calculator. Regardless of how you obtain these graphs, memorize their shapes.

1. $f(x) = x$ **2.** $g(x) = x^2$ **3.** $h(x) = x^3$

4. $p(x) = \sqrt{x}$

In Exercises 5–12, sketch the graph of the function, being sure to indicate which endpoints are included and which ones are excluded.

5. $f(x) = 2[x]$

6. $f(x) = -[x]$

7. $g(x) = [-x]$ [This is *not* the same function as in Exercise 6.]

8. $h(x) = [x] + [-x]$

9. $f(x) = \begin{cases} x^2 & \text{if } x \ge -1 \\ 2x + 3 & \text{if } x < -1 \end{cases}$

10. $g(x) = \begin{cases} |x| & \text{if } x < 1 \\ -3x + 4 & \text{if } x \ge 1 \end{cases}$

11. $k(u) = \begin{cases} -2u - 2 & \text{if } u < -3 \\ u - [u] & \text{if } -3 \le u \le 1 \\ 2u^2 & \text{if } u > 1 \end{cases}$

12. $f(x) = \begin{cases} x^2 & \text{if } x < -2 \\ x & \text{if } -2 \le x < 4 \\ \sqrt{x} & \text{if } x \ge 4 \end{cases}$

13. At this writing, first-class postage rates are 33¢ for the first ounce or fraction thereof, plus 23¢ for each additional ounce or fraction thereof. Assume that each first-class letter carries one 33¢ stamp and as many 23¢ stamps as are necessary. Then the *number* of stamps required for a first-class letter is a function of the weight of the letter in ounces. Call this function the *postage stamp function.*

(a) Describe the rule of the postage stamp function algebraically.

(b) Sketch the graph of the postage stamp function.

(c) Sketch the graph of the function whose rule is $f(x) = p(x) - [x]$, where p is the postage stamp function.

14. A common mistake is to graph the function f in Example 4 by graphing both $y = x^2$ and $y = x + 2$ on the same screen (with no restrictions on x). Explain why this graph could not possibly be the graph of a function.

In Exercises 15–19: (a) Use the fact that the absolute value function is piecewise-defined (see Example 5) to write the rule of the given function as a piecewise-defined function whose rule does not include any absolute value bars. (b) Graph the function.

15. $f(x) = |x| + 2$

16. $g(x) = |x| - 4$

17. $h(x) = |x|/2 - 2$

18. $g(x) = |x + 3|$

19. $f(x) = |x - 5|$

20. Show that the function $f(x) = |x| + |x - 2|$ is constant on the interval $[0, 2]$. [*Hint:* Use the definition of absolute value (see Example 5) to compute $f(x)$ when $0 \le x \le 2$.]

In Exercises 21–26, find the approximate location of all local maxima and minima of the function.

21. $f(x) = x^3 - x$

22. $g(t) = -\sqrt{16 - t^2}$

23. $h(x) = \dfrac{x}{x^2 + 1}$

24. $k(x) = x^3 - 3x + 1$

25. $f(x) = x^3 - 1.8x^2 + x + 2$

26. $g(x) = 2x^3 + x^2 + 1$

27. (a) A rectangle has perimeter 100 inches and one side has length x. Express the area of the rectangle as a function of x.

(b) Use the function in part (a) to find the dimensions of the rectangle with perimeter 100 inches and largest possible area.

28. (a) A rectangle has area 240 square inches and one side has length x. Express the perimeter of the rectangle as a function of x.

(b) Use the function in part (a) to find the dimensions of the rectangle with area 240 square inches and smallest possible perimeter.

29. (a) A box with a square base has a volume of 867 cubic inches. Express the surface area of the box as a function of the length x of a side of the base. [Be sure to include the top of the box.]

(b) Use the function in part (a) to find the dimensions of the box with volume 867 cubic inches and smallest possible surface area.

30. (a) A cylindrical can has a surface area of 60 square inches. Express the volume of the can as a function of the radius r.

(b) Use the function in part (a) to find the radius and height of the can with surface area 60 square inches and the largest possible volume.

In Exercises 31–32, find the approximate intervals on which the function whose graph is shown is increasing and those on which it is decreasing.

31.

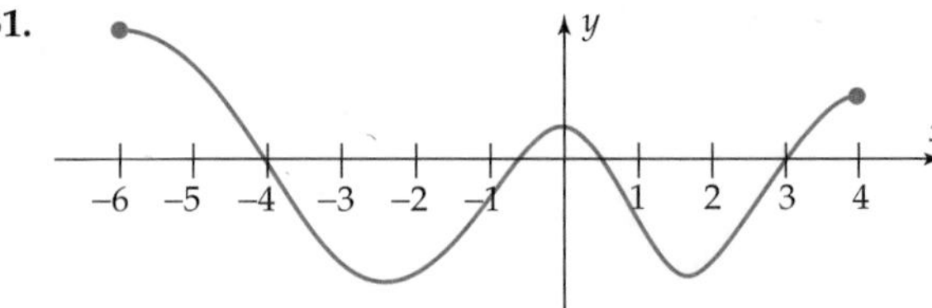

32.

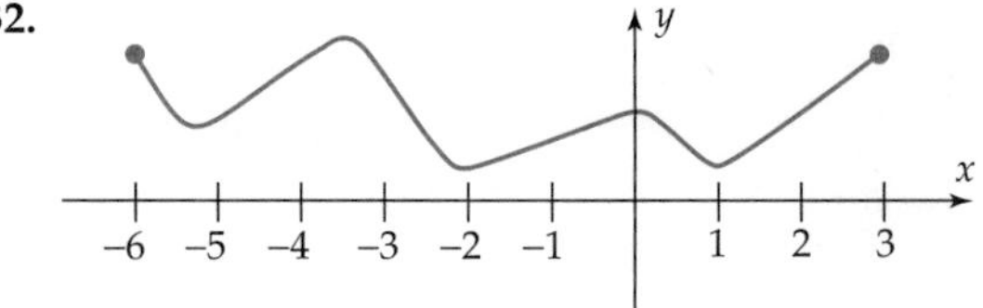

In Exercises 33–38, find the approximate intervals on which the function is increasing, those on which it is decreasing, and those on which it is constant.

33. $f(x) = |x - 1| - |x + 1|$

34. $g(x) = |x - 1| + |x + 2|$

35. $f(x) = -x^3 - 8x^2 + 8x + 5$

36. $f(x) = x^4 - .7x^3 - .6x^2 + 1$

37. $g(x) = .2x^4 - x^3 + x^2 - 2$

38. $g(x) = x^4 + x^3 - 4x^2 + x - 1$

39. Match each of the following functions with the graph that best fits the situation.

(a) The phases of the moon as a function of time

(b) The demand for a product as a function of its price

(c) The height of a ball thrown from the top of a building as a function of time

(d) The distance a woman runs at constant speed as a function of time

(e) The temperature of an oven turned on and set to 350° as a function of time

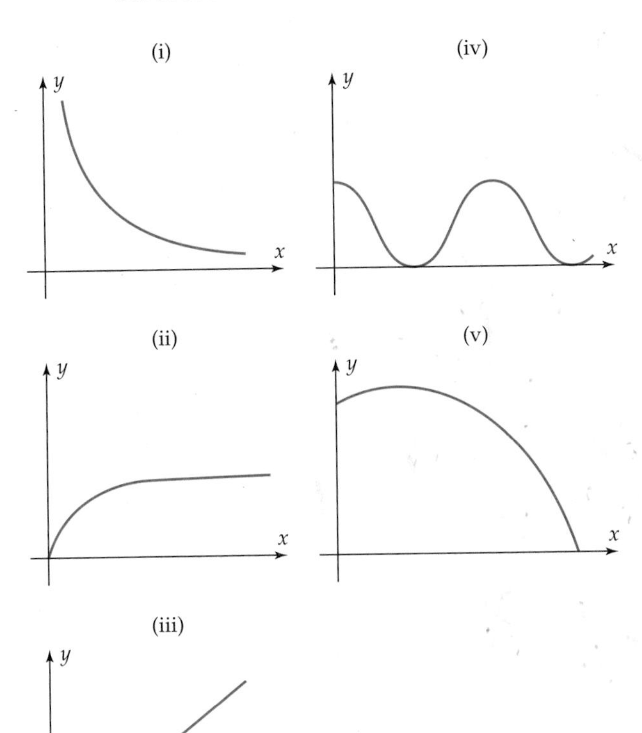

In Exercises 40–42, sketch a plausible graph of the given function. Label the axes, and specify a reasonable domain and range. Many correct answers are possible.

40. The distance from the top of your head to the ground as you jump on a trampoline as a function of time

41. The amount you spend on gas each week as a function of the number of gallons you put in your car

42. The temperature of an oven that is turned on, set to 350° and 45 minutes later turned off as a function of time

43. A bacteria population in a laboratory culture contains about a million bacteria at 8 A.M. The culture grows very rapidly until noon, when a bactericide is introduced and the bacteria population plunges. By 4 P.M. the bacteria have adapted to the bactericide and the culture slowly increases in population until 9 P.M. when the culture is accidentally destroyed by the clean-up crew. Let $g(t)$ denote the bacteria population at time t, and draw a plausible graph of the function g. [Many correct answers are possible.]

44. A plane flies from Austin, Texas, to Cleveland, Ohio, a distance of 1200 miles. Let f be the function whose rule is $f(t)$ = distance (in miles) from Austin at time t hours. Draw a plausible graph of f under the given circumstances. [There are many possible correct answers for each part.]

(a) The flight is nonstop and takes less than 4 hours.

(b) Bad weather forces the plane to land in Dallas (about 200 miles from Austin), remain overnight (for 8 hours), and continue the next day.

(c) The flight is nonstop, but due to heavy traffic the plane must fly in a holding pattern over Cincinnati (about 200 miles from Cleveland) for an hour before going on to Cleveland.

In Exercises 45–46, the graph of a function f is shown. Find and label the given points on the graph.

45. **(a)** $(k, f(k))$
(b) $(-k, f(-k))$
(c) $(k, -f(k))$

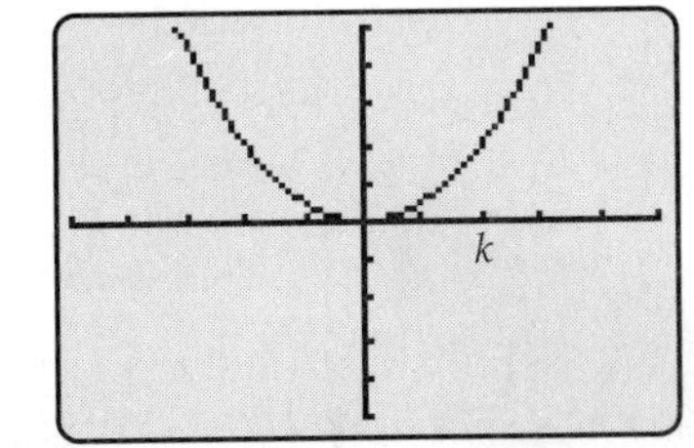

46. **(a)** $(k, f(k))$
(b) $(k, .5f(k))$
(c) $(.5k, f(.5k))$
(d) $(2k, f(2k))$

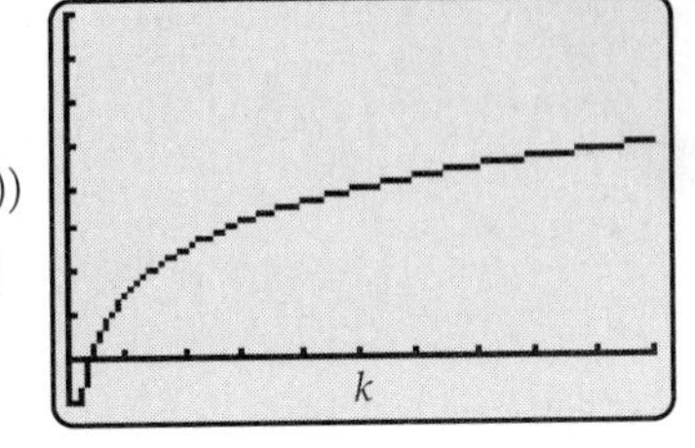

In Exercises 47–50, use your calculator to determine the domain and range of the function by reading its graph as best you can (the trace feature may be helpful). Careful approximate answers are acceptable.

47. $f(x) = 3x - 2$

48. $g(x) = x^2 - 4$

49. $h(x) = \sqrt{x^2 - 4}$

50. $k(x) = \sqrt{x^2 + 4}$

Exercises 51–62 deal with the function g whose entire graph is shown in the figure.

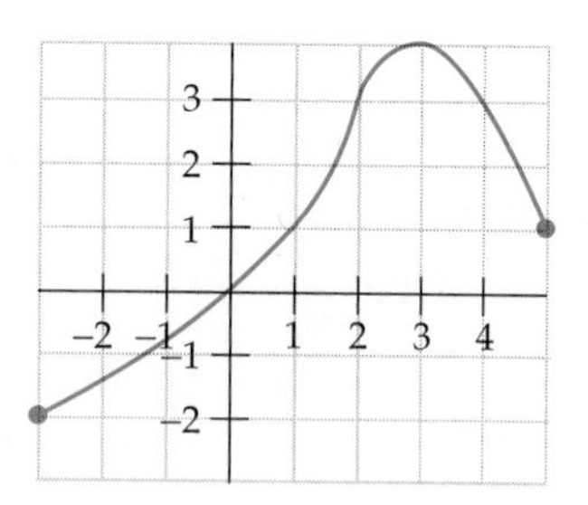

51. What is the domain of g?
52. What is the range of g?
53. If $t = 1.5$, then $g(2t) = ?$
54. If $t = 1.5$, then $2g(t) = ?$
55. If $y = 2$, then $g(y + 1.5) = ?$
56. If $y = 2$, then $g(y) + g(1.5) = ?$
57. If $y = 2$, then $g(y) + 1.5 = ?$
58. For what values of x is $g(x) < 0$?
59. If $v = 1.5$, then $g(3v - 1.5) = ?$
60. If $s = 2$, then $g(-s) = ?$
61. For what values of z is $g(z) = 1$?
62. For what values of z is $g(z) = -1$?

Exercises 63–70 deal with the function f whose entire graph is shown in the figure.

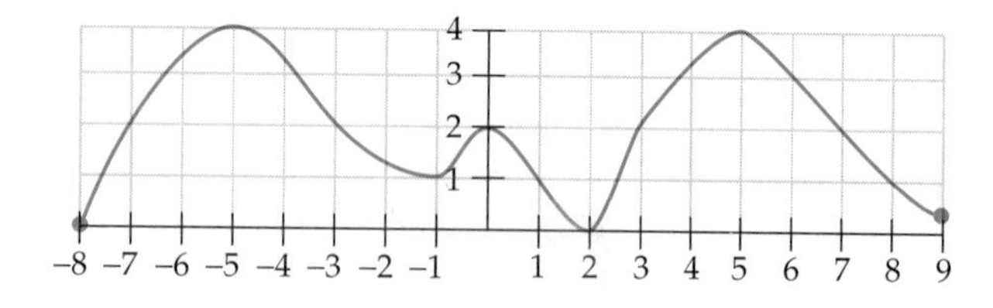

63. What is the domain of f?
64. What is the range of f?
65. Find all numbers x such that $f(x) = 2$.
66. Find all numbers x such that $f(x) > 2$.
67. Find at least three numbers x such that $f(x) = f(-x)$.
68. Find all numbers x such that $f(x) = f(7)$.
69. Find a number x such that $f(x + 1) = 0$.
70. Find two numbers x such that $f(x - 2) = 4$.

Exercises 71–78 deal with the two functions f and g whose entire graphs are shown in the figure.

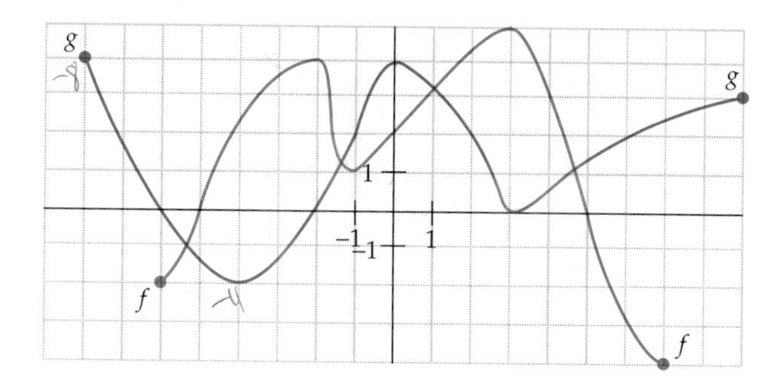

71. What is the domain of f? The domain of g?
72. What is the range of f? The range of g?
73. Find all numbers x in the interval $[-3, 1]$ such that $f(x) = 2$.
74. Find all numbers x in the interval $[-3, 3]$ such that $g(x) \geq 2$.
75. Find the number x for which $f(x) - g(x)$ is largest.
76. For how many values of x is it true that $f(x) = g(x)$?
77. Find all intervals over which both functions are defined, f is decreasing, and g is increasing.
78. Find all intervals over which g is decreasing.

Exercises 79–81 deal with this situation: The owners of the Melville & Pluth Hammer Factory have determined that both their weekly manufacturing expenses and their weekly sales income are functions of the number of hammers manufactured each week. The figure shows the graphs of these two functions.

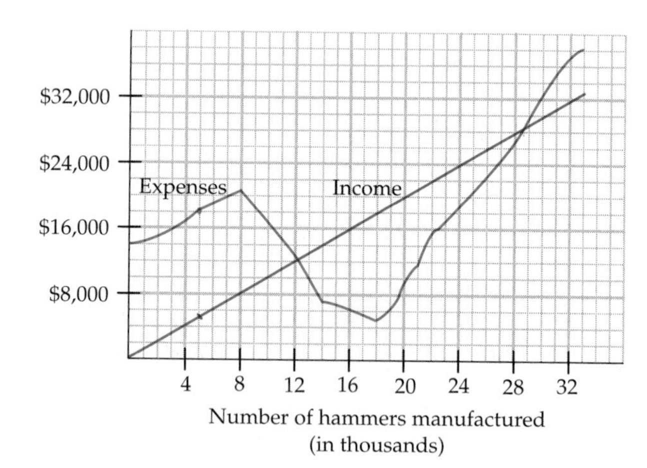

79. Use careful measurement on the graph and the fact that profit = income − expenses to

determine the weekly profit if 5000 hammers are manufactured.

80. Do the same if 10,000, 14,000, 18,000, or 22,000 hammers are manufactured.

81. **(a)** What is the smallest number of hammers that can be manufactured each week without losing money?

(b) What is the largest number of hammers that can be manufactured without losing money?

82. The graph of the function f, whose rule is $f(x)$ = average interest rate on a 30-year fixed-rate mortgage in year x, is shown in the figure. Use it to answer these questions (reasonable approximations are OK).

(a) $f(1973) = ?$ **(b)** $f(1982) = ?$ **(c)** $f(1995) = ?$

(d) In what year between 1990 and 1996 were rates the lowest? The highest?

(e) During what three-year period were rates changing the fastest? How do you determine this from the graph?

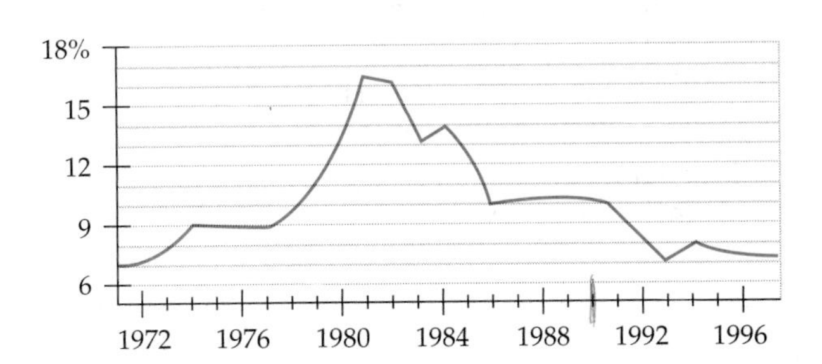

83. Draw the graph of a function f that satisfies the following four conditions:

(i) domain $f = [-2, 4]$

(ii) range $f = [-5, 6]$

(iii) $f(-1) = f(3)$

(iv) $f\left(\frac{1}{2}\right) = 0$

84. Sketch the graph of a function f that satisfies these five conditions:

(i) $f(-1) = 2$

(ii) $f(x) \geq 2$ when x is in the interval $(-1, \frac{1}{2})$

(iii) $f(x)$ starts decreasing when $x = 1$

(iv) $f(3) = 3 = f(0)$

(v) $f(x)$ starts increasing when $x = 5$

[*Note:* The function whose graph you sketch need not be given by an algebraic formula.]

In Exercises 85–90, use parametric graphing. Find a viewing window that shows a complete graph of the equation.

85. $x = y^3 + 5y^2 - 4y - 5$

86. $\sqrt[3]{y^2 - y + 1} - x + 2 = 0$

87. $xy^2 + xy + x = y^3 - 2y^2 + 4$
[*Hint:* First solve for x.]

88. $2y = xy^2 + 180x$

89. $x - \sqrt{y} + y^2 + 8 = 0$

90. $y^2 - x - \sqrt{y + 5} + 4 = 0$

In Exercises 91–96, find a viewing window that shows a complete graph of the curve determined by the parametric equations.

91. $x = 3t^2 - 5$ and $y = t^2$ $(-4 \leq t \leq 4)$

92. The Zorro curve: $x = .1t^3 - .2t^2 - 2t + 4$ and $y = 1 - t$ $(-5 \leq t \leq 6)$

93. $x = t^2 - 3t + 2$ and $y = 8 - t^3$ $(-4 \leq t \leq 4)$

94. $x = t^2 - 6t$ and $y = \sqrt{t + 7}$ $(-5 \leq t \leq 9)$

95. $x = 1 - t^2$ and $y = t^3 - t - 1$ $(-4 \leq t \leq 4)$

96. $x = t^2 - t - 1$ and $y = 1 - t - t^2$

Thinkers

97. A jogger begins her daily run from her home. The graph shows her distance from home at time t minutes. The graph shows, for example, that she ran at a slow but steady pace for 10 minutes, then increased her pace for 5 minutes, all the time moving farther from home. Describe the rest of her run.

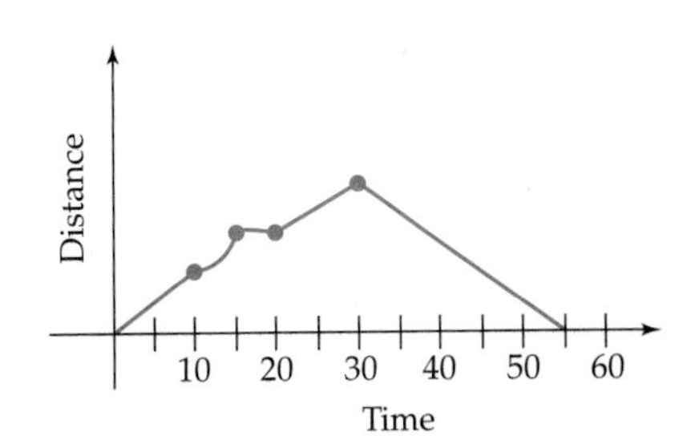

98. The graph shows the speed (in mph) at which a driver is going at time t minutes. Describe his journey.

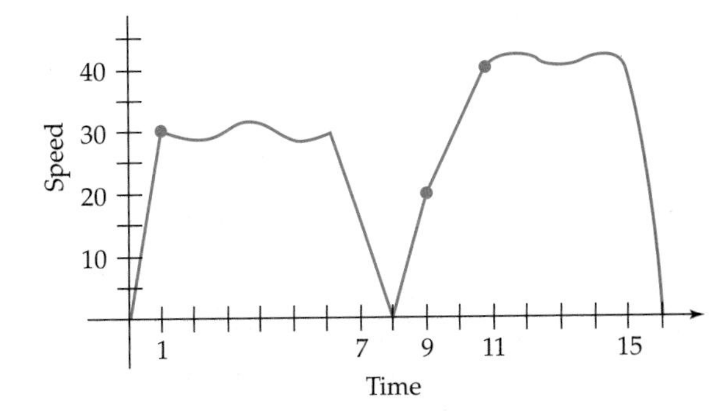

99. Graph the curve given by

$$x = (t^2 - 1)(t^2 - 4)(t + 5) + t + 3$$
$$y = (t^2 - 1)(t^2 - 4)(t^3 + 4) + t - 1$$
$$(-2.5 \le t \le 2.5)$$

How many times does this curve cross itself?

100. Use parametric equations to describe a curve that crosses itself more times than the curve in Exercise 99. [Many correct answers are possible.]

In Exercises 101 and 102, sketch the graph of the equation.

101. $|x| + |y| = 1$

102. $|y| = x^2$

3.4 Graphs and Transformations

In this section we shall see that when the rule of a function is algebraically changed in certain ways, so as to produce a new function, then the graph of the new function can be obtained from the graph of the original function by a simple geometric transformation. The same format will be used for each topic:

First You will be asked to assemble some evidence by doing a graphing exploration.

Next General conclusions deduced from the evidence will be summarized in the boxes.

Last There may be a discussion of how these conclusions can be proved in particular cases and possibly some additional examples.

Vertical Shifts

GRAPHING EXPLORATION

Using the standard viewing window, graph these three functions on the same screen:

$$f(x) = x^2 \qquad g(x) = x^2 + 5 \qquad h(x) = x^2 - 7,$$

and answer these questions:

Do the graphs of g and h look very similar to the graph of f in *shape*?

How do their vertical positions differ?

Where would you predict that the graph of $k(x) = x^2 - 9$ is located relative to the graph $f(x) = x^2$, and what is its shape?

Confirm your prediction by graphing k on the same screen as f, g, and h.

The results of this Exploration should make the following statement plausible:

Vertical Shifts

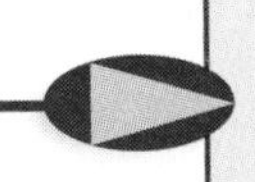

> **If $c > 0$, then the graph of $g(x) = f(x) + c$ is the graph of f shifted upward c units.**
>
> **If $c > 0$, then the graph $h(x) = f(x) - c$ is the graph of f shifted downward c units.**

To see why the first statement is true, suppose $f(x) = x^2$, $c = 5$, and $g(x) = x^2 + 5$. The points $P = (x, x^2)$ on the graph of f and $Q = (x, x^2 + 5)$ on the graph of g lie on the same vertical line (because they have the same first coordinate). Since the second coordinate of Q is larger than the second coordinate of P, the point Q lies 5 units directly above P. Since this is true for every number x, the graph of $g(x) = f(x) + 5$ is just the graph of f shifted 5 units upward.

Example 1 A calculator was used to obtain a complete graph of $f(x) = .04x^3 - x - 3$ in Figure 3–29. The graph of

$$h(x) = f(x) - 4 = (.04x^3 - x - 3) - 4 = .04x^3 - x - 7$$

is the graph of f shifted 4 units downward, as shown in Figure 3–30.

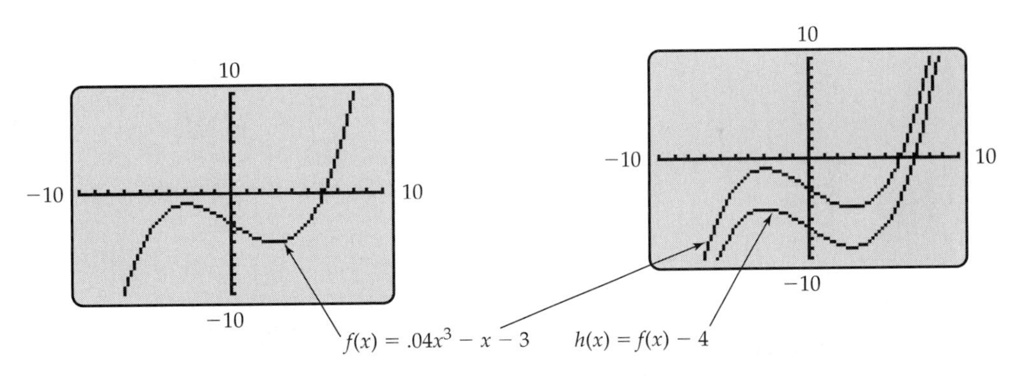

Figure 3–29 **Figure 3–30**

Although it may appear that the graph of h is closer to the graph of f at the outer edges of Figure 3–30 than in the center, this is an optical illusion.

The *vertical* distance between the graphs is always 4 units.

GRAPHING EXPLORATION

Use the trace feature of your calculator as follows to confirm that the vertical distance is always 4:*

Move the cursor to any point on the graph of f, and note its coordinates.

Use the down arrow to drop the cursor to the graph of h, and note the coordinates of the cursor in its new position.

The x-coordinates will be the same in both cases, and the new y-coordinate will be 4 less than the original y-coordinate. ■

Horizontal Shifts

GRAPHING EXPLORATION

Using the standard viewing window, graph these three functions on the same screen:

$$f(x) = 2x^3 \qquad g(x) = 2(x + 6)^3 \qquad h(x) = 2(x - 8)^3,$$

and answer these questions:

Do the graphs of g and h look very similar to the graph of f in *shape*?

How do their horizontal positions differ?

Where would you predict that the graph of $k(x) = 2(x + 2)^3$ is located relative to the graph of $f(x) = 2x^3$ and what is its shape?

Confirm your prediction by graphing k on the same screen as f, g, and h.

The results of this Exploration should make the following statement plausible:

Horizontal Shifts

Let f be a function and c a positive constant.

The graph of $g(x) = f(x + c)$ is the graph of f shifted horizontally c units to the left.

The graph of $h(x) = f(x - c)$ is the graph of f shifted horizontally c units to the right.

*The trace cursor can be moved vertically from graph to graph by using the up and down arrows.

To see why the first statement is true, suppose $c = 4$, and $g(x) = f(x + 4)$. Then the value of g at x is the same as the value of f at $x + 4$. So the graph of g at x looks like the graph of f at $x + 4$, which is 4 units to the right of x on the horizontal axis. So the graph of f is the graph of g shifted 4 units to the *right*, which means that the graph of g is the graph of f shifted 4 units to the *left*. An analogous argument works for the second statement in the box.

Technology Tip

If the function f of Example 2 is entered as $y_1 = x^2 - 7$, then the functions g and h can be entered as $y_2 = y_1(x + 5)$ and $y_3 = y_1(x - 4)$ on TI-82/83/86/89, Sharp 9600, and HP-38 (but *not* on TI-85 or Casio 9850). See the Tip on page 128 for how to enter y_1.

Example 2 In some cases, shifting the graph of a function f horizontally may produce a graph that overlaps the graph of f. For instance, a complete graph of $f(x) = x^2 - 7$ is shown in Figure 3–31. The graph of

$$g(x) = f(x + 5) = (x + 5)^2 - 7 = x^2 + 10x + 25 - 7 = x^2 + 10x + 18$$

is the graph of f shifted 5 units to the left, and the graph of

$$h(x) = f(x - 4) = (x - 4)^2 - 7 = x^2 - 8x + 16 - 7 = x^2 - 8x + 9$$

is the graph of f shifted 4 units to the right, as shown in Figure 3–31. ■

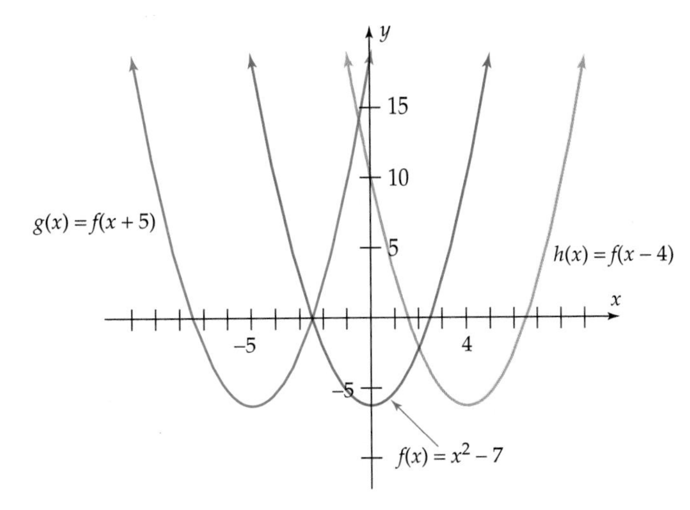

Figure 3–31

Expansions and Contractions

GRAPHING EXPLORATION

In the viewing window with $-5 \le x \le 5$ and $-15 \le y \le 15$, graph these functions on the same screen:

$$f(x) = x^2 - 4 \qquad g(x) = 3f(x) = 3(x^2 - 4).$$

Technology Tip

On TI calculators (except TI-81) you can graph both functions in the Exploration at the same time by keying in

$y = \{1, 3\}(x^2 - 4).$

One way to understand the relationship between the two graphs is to imagine that the graph of f is nailed to the x-axis at its intercepts (± 2) and that you can vertically "stretch" the graph by pulling from the top and bottom away from the x-axis (with the nails holding the x-intercepts in place) so that it fits onto the graph of g. In this process (Figure 3–32), the point $(0, -4)$ on the graph of f is stretched down to the point $(0, -12)$ on the graph of g, that is, it is stretched away from the x-axis by a factor of 3. Similarly, the point $(3, 5)$ on the graph of f is stretched up by a factor of 3 to the point $(3, 15)$ on the graph of g, shown in Figure 3–32.

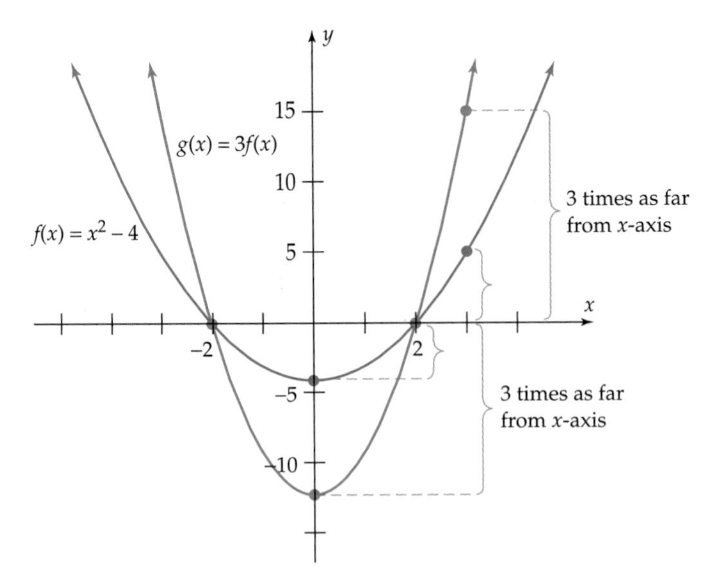

Figure 3–32

> **GRAPHING EXPLORATION**
>
> In the viewing window with $-5 \le x \le 5$ and $-5 \le y \le 10$, graph these functions on the same screen:
>
> $$f(x) = x^2 - 4 \qquad h(x) = \frac{1}{4}(x^2 - 4).$$
>
> Your screen should suggest that the graph of h is the graph of f "shrunk" vertically toward the x-axis by a factor of $1/4$.

Analogous facts are true in the general case:

Expansions and Contractions

> **If $c > 1$, then the graph of $g(x) = cf(x)$ is the graph of f stretched vertically away from the x-axis by a factor of c.**
>
> **If $0 < c < 1$, then the graph of $h(x) = cf(x)$ is the graph of f shrunk vertically toward the x-axis by a factor of c.**

Reflections

> **GRAPHING EXPLORATION**
>
> In the standard viewing window, graph these functions on the same screen.
>
> $$f(x) = .04x^3 - x \qquad g(x) = -f(x) = -(.04x^3 - x).$$
>
> If your trace cursor can be moved from graph to graph, verify that for every point on the graph of f there is a point on the graph of g with the same first coordinate that is on the opposite side of the x-axis, the same distance from the x-axis.

This Exploration shows that the graph of g is the mirror image (reflection) of the graph of f, with the x-axis being the (two-way) mirror. The same thing is true in the general case:

Reflections

> **Let f be a function.**
>
> **The graph of $g(x) = -f(x)$ is the graph of f reflected in the x-axis.**

Example 3 If $f(x) = x^2 - 3$, then the graph of

$$g(x) = -f(x) = -(x^2 - 3)$$

is the reflection of the graph of f in the x-axis, as shown in Figure 3–33. ■

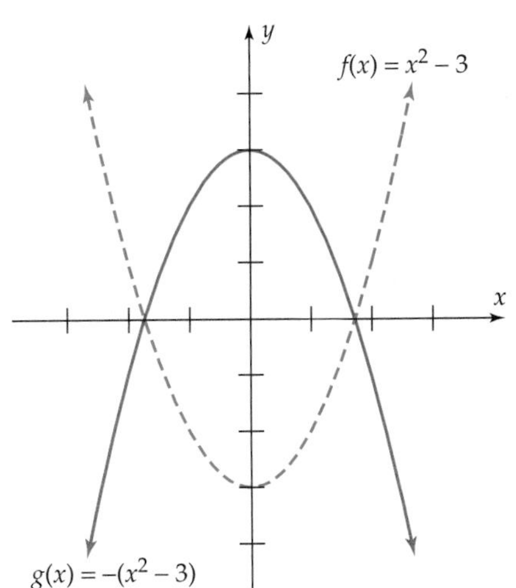

Figure 3–33

> **GRAPHING EXPLORATION**
>
> In the standard viewing window, graph these functions on the same screen:
>
> $$f(x) = \sqrt{5x + 10} \qquad \text{and} \qquad h(x) = f(-x) = \sqrt{5(-x) + 10}.$$
>
> Reflect carefully: How are the two graphs related to the y-axis? Now graph these two functions on the same screen:
>
> $$f(x) = x^2 + 3x - 3 \qquad \text{and}$$
>
> $$h(x) = f(-x) = (-x)^2 + 3(-x) - 3 = x^2 - 3x - 3.$$
>
> Are the graphs of f and h related in the same way as the first pair?

This Exploration shows that the graph of h in each case is the mirror image (reflection) of the graph of f, with the y-axis as the mirror. The same thing is true in the general case.

Reflections

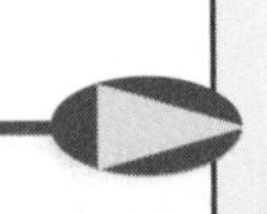

Let f be a function.

The graph of $h(x) = f(-x)$ is the graph of f reflected in the y-axis.

To see why this statement is true, let a be a number. Then $(a, f(a))$ is on the graph of f. On the other hand, note that $h(-a) = f(-(-a)) = f(a)$ so that $(-a, h(-a)) = (-a, f(a))$ is on the graph of h. The points $(a, f(a))$ and $(-a, f(a))$ are on the same horizontal line (because they have the same second coordinate) and lie on opposite sides of the y-axis at the same distance from the y-axis (because their first coordinates are negatives of each other). Thus every point on the graph of f has a "mirror image point" on the graph of h, with the y-axis being the mirror.

For other types of algebraic operations on the rule of a function and their effects on its graph, see Exercises 37–48.

Combining Transformations

The transformations described above may be used in sequence to analyze the graphs of functions whose rules are algebraically complicated.

Example 4 To understand the graph of $g(x) = 2(x-3)^2 - 1$, note that the rule of g may be obtained from the rule of $f(x) = x^2$ in three steps:

$$f(x) = x^2 \xrightarrow{\text{Step 1}} (x-3)^2 \xrightarrow{\text{Step 2}} 2(x-3)^2 \xrightarrow{\text{Step 3}} 2(x-3)^2 - 1 = g(x)$$

Step 1 shifts the graph of f horizontally 3 units to the right; step 2 stretches the resulting graph away from the x-axis by a factor of 2; step 3 shifts this graph 1 unit downward, thus producing the graph of g in Figure 3–34. ■

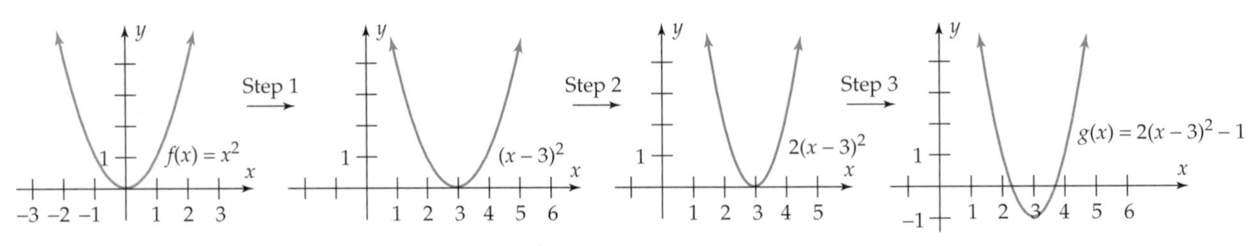

Figure 3–34

Calculator Investigations

Investigations 1 and 2 show that, because of its small screen size, a calculator may not always display clearly what it should.

1. Graph $f(x) = x^2 - x - 6$ in the standard viewing window. Describe verbally what the graph of $h(x) = f(x - 1000)$ should look like (see the box on page 152). Now find an appropriate viewing window and graph h. Can you find a viewing window that clearly displays both the graph of f and the graph of h?

2. Graph $f(x) = x^2 - x - 6$ in the standard viewing window. Describe verbally what the graph of $g(x) = 1000f(x)$ should look like (see the box on page 154). Now find an appropriate viewing window and graph g. Can you find a viewing window that clearly displays both the graph of f and the graph of g?

Exercises 3.4

In Exercises 1–4, find a single viewing window that shows complete graphs of the functions f, g, h.

1. $f(x) = .25x^3 - 9x + 5;\quad g(x) = f(x) + 15;$
   $h(x) = f(x) - 20$
2. $f(x) = \sqrt{x^2 - 9} - 5;\quad g(x) = 3f(x);$
   $h(x) = .5f(x)$
3. $f(x) = |x^2 - 5|;\quad g(x) = f(x + 8);$
   $h(x) = f(x - 6)$
4. $f(x) = .125x^3 - .25x^2 - 1.5x + 5;$
   $g(x) = f(x) - 5;\quad h(x) = 5 - f(x)$

In Exercises 5 and 6, find complete graphs of the functions f and g in the same viewing window.

5. $f(x) = \dfrac{4 - 5x^2}{x^2 + 1};\quad g(x) = -f(x)$
6. $f(x) = x^4 - 4x^3 + 2x^2 + 3;\quad g(x) = f(-x)$

In Exercises 7–12, describe a sequence of transformations that will transform the graph of the function f into the graph of the function g.

7. $f(x) = x^2 + x;\quad g(x) = (x - 3)^2 + (x - 3) + 2$
8. $f(x) = x^2 + 5;\quad g(x) = (x + 2)^2 + 10$
9. $f(x) = \sqrt{x^3 + 5};\quad g(x) = -\dfrac{1}{2}\sqrt{x^3 + 5} - 6$
10. $f(x) = \sqrt{x^4 + x^2 + 1};$
    $g(x) = 10 - \sqrt{4x^4 + 4x^2 + 4}$
11. $f(x) = \dfrac{3x}{x^2 + 10};\quad g(x) = \dfrac{-6x + 12}{(x - 2)^2 + 10}$
12. $f(x) = \dfrac{1}{2x^2 + 2};\quad g(x) = \dfrac{-1}{(x - 1)^2 + 1}$

In Exercises 13–18, write the rule of a function g whose graph can be obtained from the graph of the function f by performing the transformations in the order given.

13. $f(x) = x^2 + 2$; shift the graph horizontally 5 units to the left and then vertically upward 4 units.
14. $f(x) = x^2 - x + 1$; reflect the graph in the x-axis, then shift it vertically upward 3 units.
15. $f(x) = \sqrt{x}$; shift the graph horizontally 6 units to the right, stretch it away from the x-axis by a factor of 2, and shift it vertically downward 3 units.
16. $f(x) = \sqrt{-x}$; shift the graph horizontally 3 units to the left, then reflect it in the x-axis, and shrink it toward the x-axis by a factor of 1/2.
17. $f(x) = x^2 + 3x + 1$; stretch the graph away from the x-axis by a factor of 2, shift it horizontally 2 units to the right, and shift it vertically upward 2 units.

18. $f(x) = x^2 + 3x + 1$; shift the graph horizontally 2 units to the right, then shift it vertically upward 2 units, and stretch it away from the x-axis by a factor of 2. Compare the result with Exercise 17.

In Exercises 19–22, use the graph of the function f in the figure to sketch the graph of the function g.

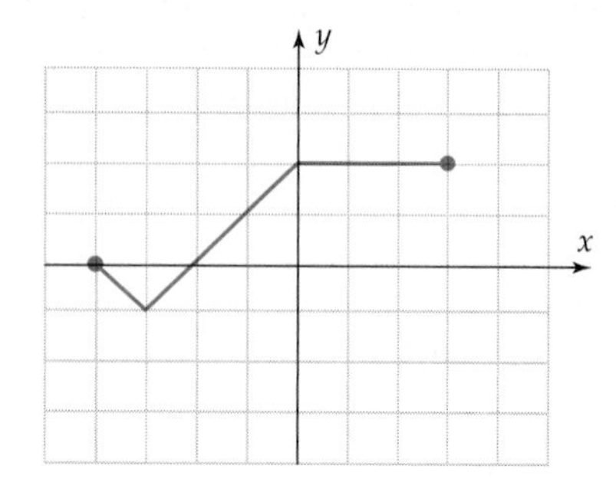

19. $g(x) = f(x) + 3$ **20.** $g(x) = f(x) - 1$

21. $g(x) = 3f(x)$ **22.** $g(x) = .25f(x)$

In Exercises 23–26, use the graph of the function f in the figure to sketch the graph of the function h.

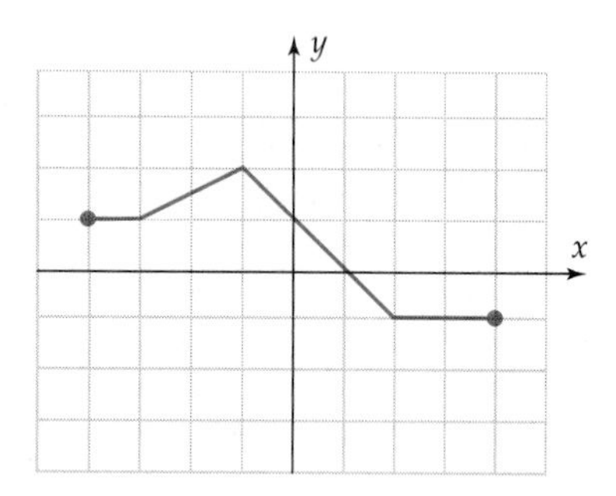

23. $h(x) = -f(x)$ **24.** $h(x) = -4f(x)$

25. $h(x) = f(-x)$ **26.** $h(x) = f(-x) + 2$

In Exercises 27–32, use the graph of the function f in the figure to sketch the graph of the function g.

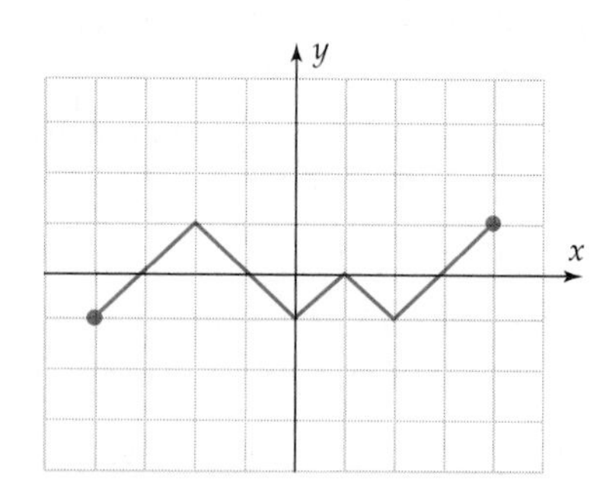

27. $g(x) = f(x + 3)$ **28.** $g(x) = f(x - 2)$

29. $g(x) = f(x - 2) + 3$ **30.** $g(x) = f(x + 1) - 3$

31. $g(x) = 2 - f(x)$ **32.** $g(x) = f(-x) + 2$

In Exercises 33–35, use the standard viewing window to graph the function f and the function $g(x) = |f(x)|$ on the same screen. Exercise 36 may be helpful for interpreting the results.

33. $f(x) = .5x^2 - 5$ **34.** $f(x) = x^3 - 4x^2 + x + 3$

35. $f(x) = x + 3$

36. (a) Let f be a function, and let g be the function defined by $g(x) = |f(x)|$. Use the definition of absolute value (page 8) to explain why the following statement is true:

$$g(x) = \begin{cases} f(x) & \text{if } f(x) \geq 0 \\ -f(x) & \text{if } f(x) < 0 \end{cases}$$

(b) Use part (a) and your knowledge of transformations to explain why the graph of g consists of those parts of the graph of f that lie above the x-axis together with the reflection in the x-axis of those parts of the graph of f that lie below the x-axis.

In Exercises 37–39, assume $f(x) = (.2x)^6 - 4$. Use the standard viewing window to graph the functions f and g on the same screen.

37. $g(x) = f(2x)$ **38.** $g(x) = f(3x)$ **39.** $g(x) = f(4x)$

40. Based on the results of Exercises 37–39, describe the transformation that transforms the graph of a function $f(x)$ into the graph of the function $f(cx)$, where c is a constant with $c > 1$. [*Hint:* How are the two graphs related to the y-axis? Stretch your mind.]

In Exercises 41–43, assume $f(x) = x^2 - 3$. Use the standard viewing window to graph the functions f and g on the same screen.

41. $g(x) = f\left(\frac{1}{2}x\right)$ **42.** $g(x) = f\left(\frac{1}{3}x\right)$

43. $g(x) = f\left(\frac{1}{4}x\right)$

44. Based on the results of Exercises 41–43, describe the transformation that transforms the graph of a function $f(x)$ into the graph of the function $f(cx)$, where c is a constant with $0 < c < 1$. [*Hint:* How are the two graphs related to the y-axis?]

In Exercises 45–47, use the standard viewing window to graph the function f and the function $g(x) = f(|x|)$ on the same screen.

45. $f(x) = x - 4$

46. $f(x) = x^3 - 3$

47. $f(x) = .5(x - 4)^2 - 9$

48. Based on the results of Exercises 45–47, describe the relationship between the graph of a function $f(x)$ and the graph of the function $f(|x|)$.

49. A factory has a linear cost function $f(x) = ax + b$, where b represents fixed costs and a represents the labor and material costs of making one item, both in thousands of dollars.

(a) If property taxes (part of the fixed costs) are increased by $28,000 per year, what effect does this have on the graph of the cost function?

(b) If labor and material costs for making 100,000 items increase by $12,000, what effect does this have on the graph of the cost function?

3.4.A EXCURSION Symmetry

A graph is **symmetric with respect to the y-axis** if the part of the graph on the right side of the y-axis is the mirror image of the part on the left side of the y-axis (with the y-axis being the mirror), as shown in Figure 3–35.

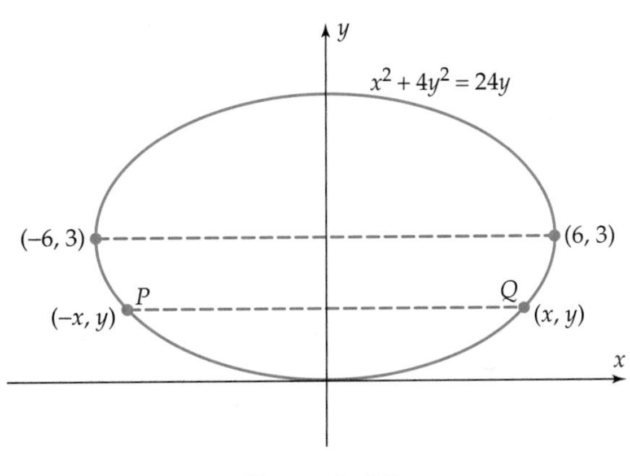

Figure 3–35

Each point P on the left side of the graph has a mirror image point Q on the right side of the graph, as indicated by the dashed lines. Note that

Their second coordinates are the same (P and Q are on the same side of the x-axis and the same distance from it);

Their first coordinates are negatives of each other (P and Q lie on opposite sides of the y-axis and the same distance from it).

Thus, a graph is symmetric with respect to the y-axis provided that

Whenever (x, y) is on the graph, then $(-x, y)$ is also on it.

In algebraic terms, this means that replacing x by $-x$ in the equation leads to the same number y. In other words, replacing x by $-x$ produces an equivalent equation.

Example 1 Replacing x by $-x$ in the equation $y = x^4 - 5x^2 + 3$ produces $(-x)^4 - 5(-x)^2 + 3$, which is the same equation because $(-x)^2 = x^2$ and $(-x)^4 = x^4$. Therefore, the graph is symmetric with respect to the y-axis.

GRAPHING EXPLORATION

Confirm this fact by graphing the equation. ■

x-Axis Symmetry

A graph is **symmetric with respect to the x-axis** if the part of the graph above the x-axis is the mirror image of the part below the x-axis (with the x-axis being the mirror), as shown in Figure 3–36.

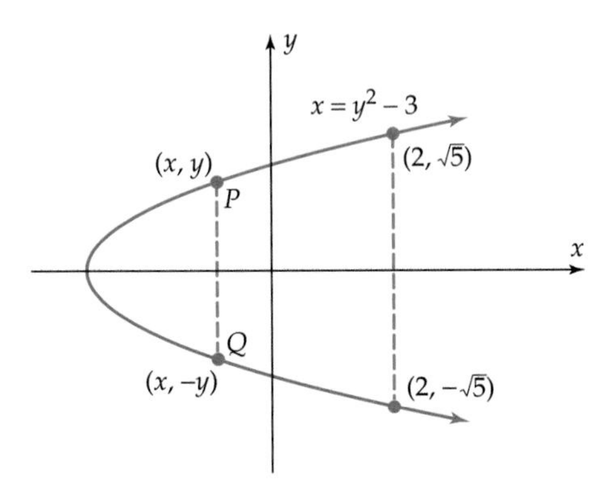

Figure 3–36

Using Figure 3–36 and argument analogous to the one preceding Example 1, we see that a graph is symmetric with respect to the x-axis provided that

Whenever (x, y) is on the graph, then $(x, -y)$ is also on it.

In algebraic terms, this means that replacing y by $-y$ in the equation leads to the same number x. In other words, replacing y by $-y$ produces an equivalent equation.

Example 2 Replacing y by $-y$ in the equation $y^2 = 4x - 12$ produces $(-y)^2 = 4x - 12$, which is the same equation, so the graph is symmetric with respect to the x-axis.

GRAPHING EXPLORATION

Confirm this fact by graphing the equation. In order to do this, note that every point on the graph of $y^2 = 4x - 12$ is also on the graph of either $y = \sqrt{4x - 12}$ or $y = -\sqrt{4x - 12}$. Each of these latter equations defines a function; graph them both on the same screen. ■

Origin Symmetry

A graph is **symmetric with respect to the origin** if a straight line through the origin and any point P on the graph also intersects the graph at a point Q such that the origin is the midpoint of segment PQ, as shown in Figure 3–37.

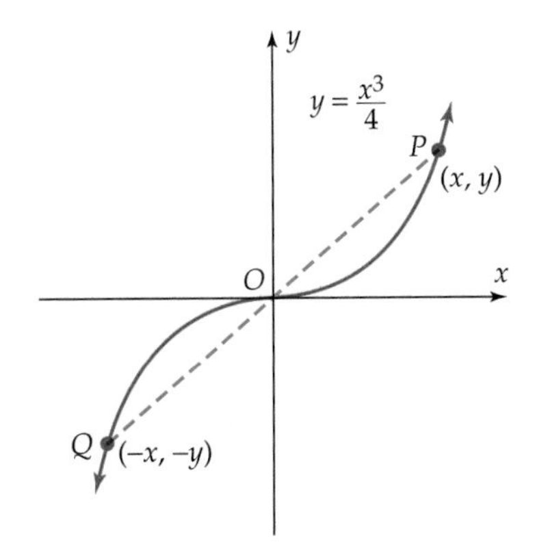

Figure 3–37

Using Figure 3–37, we can also describe symmetry with respect to the origin in terms of coordinates and equations (Exercise 34):

Whenever (x, y) is on the graph, then $(-x, -y)$ is also on it.

In algebraic terms, this means that replacing x by $-x$ and y by $-y$ in the equation produces an equivalent equation.

Example 3 Replacing x by $-x$ and y by $-y$ in the equation $y = \frac{x^3}{10} - x$ yields

$$-y = \frac{(-x)^3}{10} - (-x), \qquad \text{that is,} \qquad -y = \frac{-x^3}{10} + x.$$

This equation is equivalent to $y = \frac{x^3}{10} - x$ since it can be obtained from it by multiplying by -1. Therefore, the graph of $y = \frac{x^3}{10} - x$ is symmetric with respect to the origin.

GRAPHING EXPLORATION

Confirm this fact by graphing the equation. ■

Here is a summary of the various tests for symmetry:

Symmetry Tests

Symmetry with Respect to	*Coordinate Test for Symmetry*	*Algebraic Test for Symmetry*
y-axis	(x, y) on graph implies $(-x, y)$ on graph.	Replacing x by $-x$ produces an equivalent equation.
x-axis	(x, y) on graph implies $(x, -y)$ on graph.	Replacing y by $-y$ produces an equivalent equation.
origin	(x, y) on graph implies $(-x, -y)$ on graph.	Replacing x by $-x$ and y by $-y$ produces an equivalent equation.

Even and Odd Functions

For *functions,* the algebraic description of symmetry takes a different form. A function f whose graph is symmetric with respect to the y-axis is called an **even function.** To say that the graph of $y = f(x)$ is symmetric with respect to the y-axis means that replacing x by $-x$ produces the same y value. In other words, the function takes the same value at both x and $-x$. Therefore,

Even Functions

A function f is even provided that

$$f(x) = f(-x) \text{ for every number } x \text{ in the domain of } f.$$

The graph of an even function is symmetric with respect to the y-axis.

For example, $f(x) = x^4 + x^2$ is even because

$$f(-x) = (-x)^4 + (-x)^2 = x^4 + x^2 = f(x).$$

Thus, the graph of f is symmetric with respect to the y-axis, as you can easily verify with your calculator (do it!).

Except for zero functions ($f(x) = 0$ for every x in the domain), *the graph of a function is never symmetric with respect to the x-axis.* The reason is the Vertical Line Test: The graph of a function never contains two points with the same first coordinate. If both $(5, 3)$ and $(5, -3)$, for instance, were on the graph, this would say that $f(5) = 3$ and $f(5) = -3$, which is impossible when f is a function.

A function whose graph is symmetric with respect to the origin is called an **odd function.** If both (x, y) and $(-x, -y)$ are on the graph of such a function f, then we must have both

$$y = f(x) \qquad \text{and} \qquad -y = f(-x)$$

so that $f(-x) = -y = -f(x)$. Therefore,

Odd Functions

> **A function f is odd provided that**
>
> $$f(-x) = -f(x) \text{ for every number } x \text{ in the domain of } f.$$
>
> **The graph of an odd function is symmetric with respect to the origin.**

For example, $f(x) = x^3$ is an odd function because

$$f(-x) = (-x)^3 = -x^3 = -f(x).$$

Hence, the graph of f is symmetric with respect to the origin (verify this with your calculator).

Exercises 3.4.A

In Exercises 1–4, find the graph of the equation. If the graph is symmetric with respect to the x-axis, the y-axis, or the origin, say so.

1. $y = x^2 + 2$

2. $x = (y - 3)^2$

3. $y = x^3 + 2$

4. $y = (x + 2)^3$

In Exercises 5–14, determine whether the given function is even, odd, or neither.

5. $f(x) = 4x$

6. $k(t) = -5t$

7. $f(x) = x^2 - |x|$

8. $h(u) = |3u|$

9. $k(t) = t^4 - 6t^2 + 5$

10. $f(x) = x(x^4 - x^2) + 4$

11. $f(t) = \sqrt{t^2 - 5}$

12. $h(x) = \sqrt{7 - 2x^2}$

13. $f(x) = \dfrac{x^2 + 2}{x - 7}$

14. $g(x) = \dfrac{x^2 + 1}{x^2 - 1}$

In Exercises 15–18, determine algebraically whether or not the graph of the given equation is symmetric with respect to the x-axis.

15. $x^2 - 6x + y^2 + 8 = 0$

16. $x^2 + 8x + y^2 = -15$

17. $x^2 - 2x + y^2 + 2y = 2$

18. $x^2 - x + y^2 - y = 0$

In Exercises 19–24, determine whether the given graph is symmetric with respect to the y-axis, the x-axis, or the origin.

19.

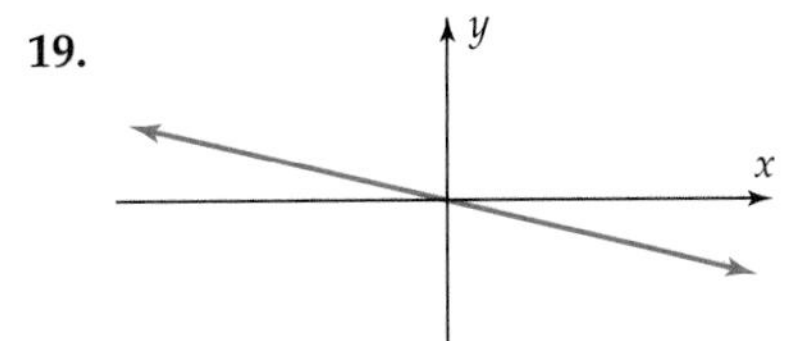

20.

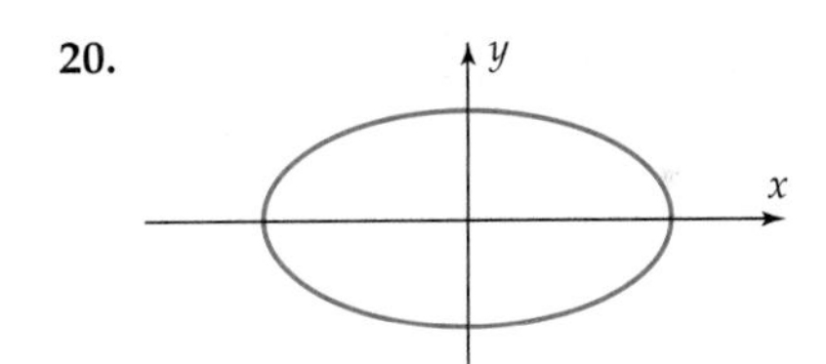

21.

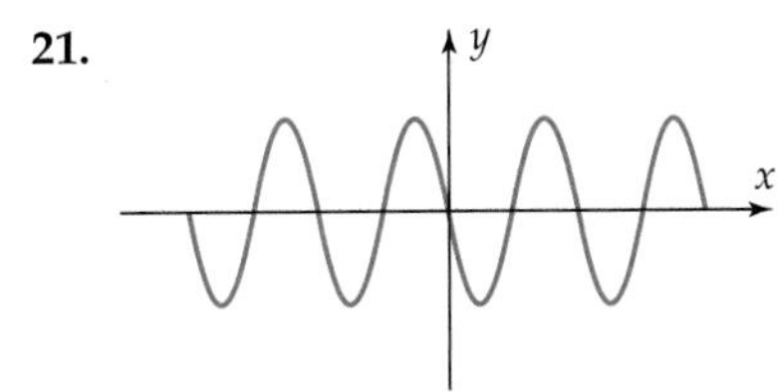

22.

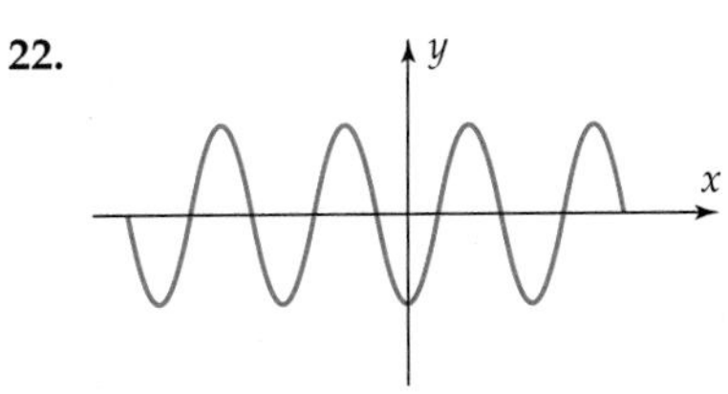

23.

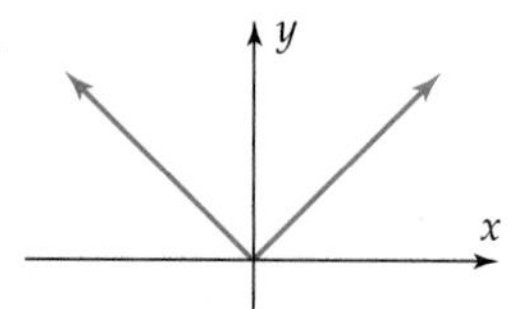

24.

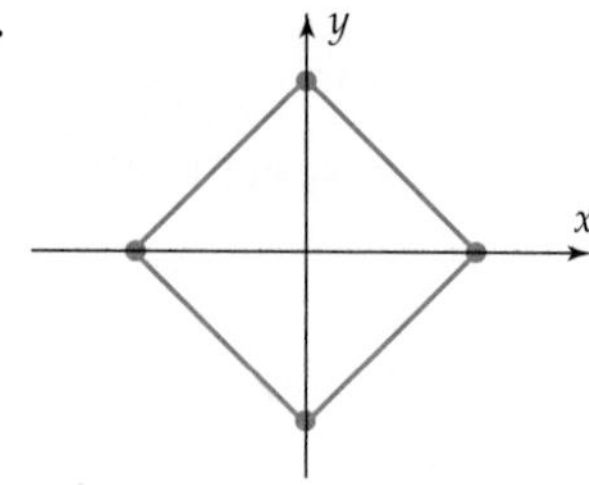

In Exercises 25–28, complete the graph of the given function, assuming that it satisfies the given symmetry condition.

25. Even

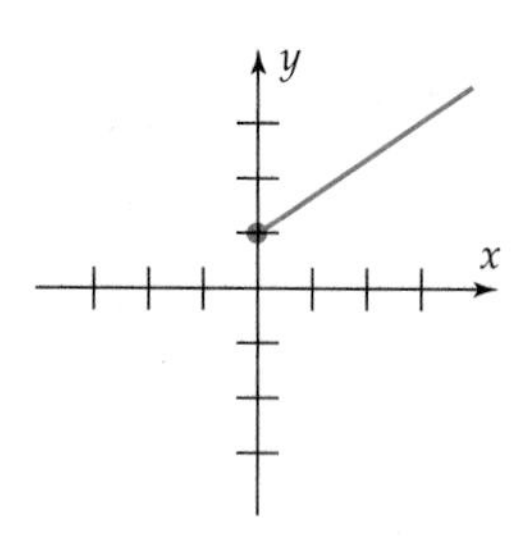

26. Even

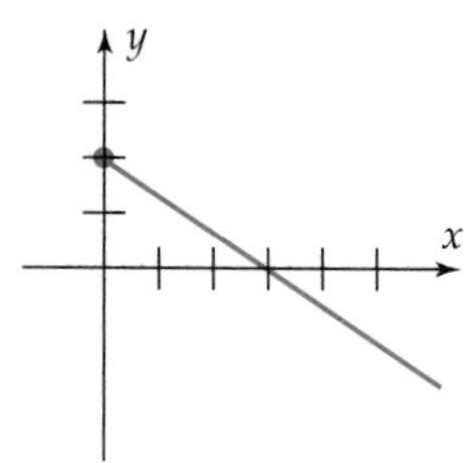

27. Odd

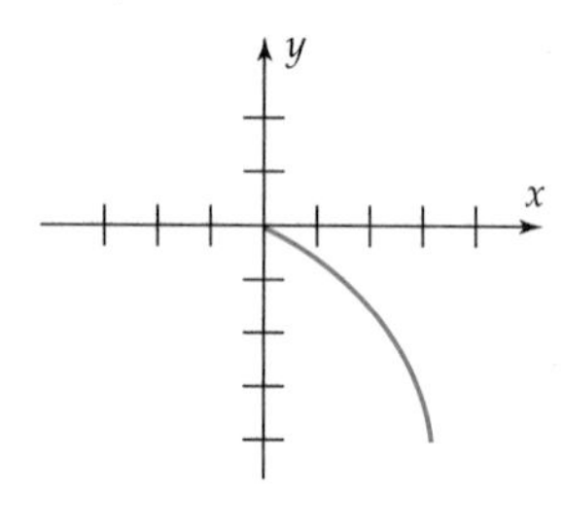

28. Odd

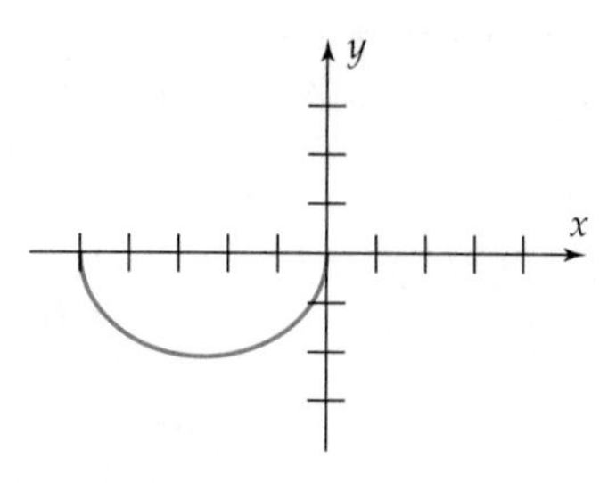

29. (a) Draw some coordinate axes, and plot the points $(0, 1)$, $(1, -3)$, $(-5, 2)$, $(-3, 5)$, $(2, 3)$, and $(4, 1)$.

(b) Suppose the points in part (a) lie on the graph of an *even* function f. Plot the points $(0, f(0))$, $(-1, f(-1))$, $(5, f(5))$, $(3, f(3))$, $(-2, f(-2))$, and $(-4, f(-4))$.

30. Draw the graph of an *even* function that includes the points $(0, -3)$, $(-3, 0)$, $(2, 0)$, $(1, -4)$, $(2.5, -1)$, $(-4, 3)$, and $(-5, 3)$.*

31. (a) Plot the points $(0, 0)$; $(2, 3)$; $(3, 4)$; $(5, 0)$; $(7, -3)$; $(-1, -1)$; $(-4, -1)$; $(-6, 1)$.

(b) Suppose the points in part (a) lie on the graph of an *odd* function f. Plot the points $(-2, f(-2))$; $(-3, f(-3))$; $(-5, f(-5))$; $(-7, f(-7))$; $(1, f(1))$; $(4, f(4))$; $(6, f(6))$.

(c) Draw the graph of an odd function f that includes all the points plotted in parts (a) and (b).*

32. Draw the graph of an odd function that includes the points $(-3, 5)$, $(-1, 1)$, $(2, -6)$, $(4, -9)$, and $(5, -5)$.*

33. Show that any graph that has two of the three types of symmetry (x-axis, y-axis, origin) necessarily has the third type also.

34. Use the midpoint formula to show that $(0, 0)$ is the midpoint of the segment joining (x, y) and $(-x, -y)$. Conclude that the coordinate test for symmetry with respect to the origin (page 162) is correct.

*There are many correct answers.

3.5 Operations on Functions

We now examine ways in which two or more given functions can be used to create new functions. If f and g are functions, then their **sum** is the function h defined by the rule

$$h(x) = f(x) + g(x).$$

For example, if $f(x) = 3x^2 + x$ and $g(x) = 4x - 2$, then

$$h(x) = f(x) + g(x) = (3x^2 + x) + (4x - 2) = 3x^2 + 5x - 2.$$

Instead of using a different letter h for the sum function, we shall usually denote it by $f + g$. Thus, the sum $f + g$ is defined by the rule

$$\mathbf{(f + g)(x) = f(x) + g(x).}$$

This rule is *not* just a formal manipulation of symbols. If x is a number, then so are $f(x)$ and $g(x)$. The plus sign in $f(x) + g(x)$ is addition of *numbers,* and the result is a number. But the plus sign in $f + g$ is addition of *functions,* and the result is a new function.

The **difference** $f - g$ is the function defined by the rule

$$\mathbf{(f - g)(x) = f(x) - g(x).}$$

The domain of the sum and difference functions is the set of all real numbers that are in both the domain of f and the domain of g.

Technology Tip

If you have two functions entered in the equation memory as y_1 and y_2 [or $y_1(x)$, $y_2(x)$ or $f_1(x)$, $f_2(x)$], you can graph their sum by entering $y_1 + y_2$ as y_3 in the equation memory and graphing y_3. Differences, products, and quotients are graphed similarly.

You can find y_1 and y_2 in the FUNCTION submenu of the TI-83 VARS or Y-VARS menu, in the EQVARS submenu of the Sharp 9600 VARS menu, and in the GRPH submenu of the Casio 9850 VARS menu. On other calculators, type them in from the keyboard.

Example 1 If $f(x) = \sqrt{9 - x^2}$ and $g(x) = \sqrt{x - 2}$, then

$$(f + g)(x) = \sqrt{9 - x^2} + \sqrt{x - 2}$$
$$(f - g)(x) = \sqrt{9 - x^2} - \sqrt{x - 2}.$$

The domain of f consists of all x such that $9 - x^2 \geq 0$ (so that the square root will be defined), that is, all x with $-3 \leq x \leq 3$. Similarly, the domain of g consists of all x such that $x \geq 2$. The domain of $f + g$ and $f - g$ consists of all real numbers in both the domain of f and the domain of g, namely, all x such that $2 \leq x \leq 3$. ■

The product and quotient of functions f and g are the functions defined by the rules

$$\mathbf{(fg)(x) = f(x)g(x)} \qquad \text{and} \qquad \mathbf{\left(\frac{f}{g}\right)(x) = \frac{f(x)}{g(x)}}.$$

The domain of fg consists of all real numbers in both the domain of f and the domain of g. The domain of f/g consists of all real numbers x in both the domain of f and the domain of g such that $g(x) \neq 0$.

Example 2 If $f(x) = \sqrt{3x}$ and $g(x) = x^2 - 1$, then

$$(fg)(x) = \sqrt{3x}(x^2 - 1) = \sqrt{3x}x^2 - \sqrt{3x}$$

$$\left(\frac{f}{g}\right)(x) = \frac{\sqrt{3x}}{x^2 - 1}.$$

The domain of fg consists of all numbers x in both the domain of f (all nonnegative real numbers) and the domain of g (all real numbers), that is, all $x \geq 0$. The domain of f/g consists of all these x for which $g(x) \neq 0$, that is, all nonnegative real numbers *except* $x = 1$. ■

If c is a real number and f is a function, then the product of f and the constant function $g(x) = c$ is usually denoted cf. For example, if the function $f(x) = x^3 - x + 2$, and $c = 5$, then $5f$ is the function given by

$$(5f)(x) = 5 \cdot f(x) = 5(x^3 - x + 2) = 5x^3 - 5x + 10.$$

Composition of Functions

Another way of combining functions is illustrated by the function $h(x) = \sqrt{x^3}$. To compute $h(4)$, for example, you first find $4^3 = 64$ and then take the square root $\sqrt{64} = 8$. So the rule of h may be rephrased as:

First apply the function $f(x) = x^3$,

Then apply the function $g(t) = \sqrt{t}$ to the result.

The same idea can be expressed in functional notation like this:

$$x \xrightarrow{\textit{first apply } f} f(x) \xrightarrow{\textit{then apply } g \textit{ to the result}} g(f(x))$$

$$x \qquad x^3 \qquad \sqrt{x^3}$$

apply h (from x to $\sqrt{x^3}$)

So the rule of h may be written as $h(x) = g(f(x))$, where $f(x) = x^3$ and $g(t) = \sqrt{t}$. We can think of h as being made up of two simpler functions f and g, or we can think of f and g being "composed" to create the function h. Both viewpoints are useful.

Example 3 Suppose $f(x) = 4x^2 + 1$ and $g(t) = \dfrac{1}{t + 2}$. Define a new function h whose rule is "first apply f; then apply g to the result." In functional notation

$$x \xrightarrow{\textit{first apply } f} f(x) \xrightarrow{\textit{then apply } g \textit{ to the result}} g(f(x))$$

So the rule of the function h is $h(x) = g(f(x))$. Evaluating $g(f(x))$ means that whenever t appears in the formula for $g(t)$, we must replace it by $f(x) = 4x^2 + 1$:

$$h(x) = g(f(x)) = \frac{1}{f(x) + 2} = \frac{1}{(4x^2 + 1) + 2} = \frac{1}{4x^2 + 3}. \quad ■$$

The function h in Example 3 is an illustration of the following definition.

Composite Functions

> **Let f and g be functions. The *composite function* of f and g is given by:**
>
> **For input x, the output is $g(f(x))$.**
>
> **This composite function is denoted $g \circ f$.**

The symbol "$g \circ f$" is read "g circle f" or "f followed by g." (Note the order carefully; the functions are applied *right* to *left*.) So the rule of the composite function is:

$$(g \circ f)(x) = g(f(x)).$$

Example 4 If $f(x) = 2x + 5$ and $g(t) = 3t^2 + 2t + 4$, then

$$(f \circ g)(2) = f(g(2)) = f(3 \cdot 2^2 + 2 \cdot 2 + 4) = f(20) = 2 \cdot 20 + 5 = 45.$$

Similarly,

$$(g \circ f)(-1) = g(f(-1)) = g(2(-1) + 5) = g(3) = 3 \cdot 3^2 + 2 \cdot 3 + 4 = 37.$$

The value of a composite function can also be computed like this:

$$(g \circ f)(5) = g(f(5)) = 3(f(5)^2) + 2(f(5)) + 4 = 3(15^2) + 2(15) + 4 = 709.$$

■

The domain of $g \circ f$ is determined by the usual convention:

Domain of $g \circ f$

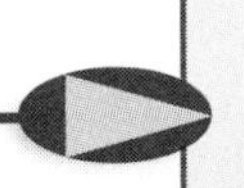

> **The domain of the composite function $g \circ f$ is the set of all real numbers x such that x is in the domain of f and $f(x)$ is in the domain of g.**

Example 5 If $f(x) = \sqrt{x}$ and $g(t) = t^2 - 5$, then

$$(g \circ f)(x) = g(f(x)) = (f(x))^2 - 5 = (\sqrt{x})^2 - 5 = x - 5.$$

Although $x - 5$ is defined for every real number x, the domain of $g \circ f$ is *not* the set of all real numbers. The domain of g is the set of all real numbers, but the function $f(x) = \sqrt{x}$ is defined only when $x \geq 0$. So the domain of $g \circ f$ is the set of nonnegative real numbers, that is, the interval $[0, \infty)$. ■

Technology Tip

Evaluating composite functions is easy on TI-82/83/86/89, Sharp 9600, and HP-38. If the functions are entered in the equation memory as $y_1 = g(x)$ and $y_2 = h(x)$ (with f in place of y on HP-38), then keying in $y_2(y_1(5))$ ENTER produces the number $h(g(5))$.

On other calculators (including TI-85) this syntax does *not* produce $h(g(5))$; it produces $h(x)\cdot g(x)\cdot 5$ for whatever number is stored in the x-memory.

Example 6 If $h(x) = \sqrt{3x^2 + 1}$, then h may be considered as the composite $g \circ f$, where $f(x) = 3x^2 + 1$ and $g(u) = \sqrt{u}$ because

$$(g \circ f)(x) = g(f(x)) = g(3x^2 + 1) = \sqrt{3x^2 + 1} = h(x).$$

There are other ways to consider $h(x) = \sqrt{3x^2 + 1}$ as a composite function. For instance, h is also the composite $j \circ k$, where $j(x) = \sqrt{x + 1}$ and $k(x) = 3x^2$:

$$(j \circ k)(x) = j(k(x)) = j(3x^2) = \sqrt{3x^2 + 1} = h(x). \quad \blacksquare$$

Example 7 If $k(x) = (x^2 - 2x + \sqrt{x})^3$, then k is $g \circ f$, where $f(x) = x^2 - 2x + \sqrt{x}$ and $g(t) = t^3$ because

$$(g \circ f)(x) = g(f(x)) = g\left(x^2 - 2x + \sqrt{x}\right) = \left(x^2 - 2x + \sqrt{x}\right)^3 = k(x). \quad \blacksquare$$

As you may have noticed, there are two possible ways to form a composite function from two given functions. If f and g are functions, we can consider either

$$(g \circ f)(x) = g(f(x)), \qquad \text{[the composite of } f \text{ and } g\text{]}$$

$$(f \circ g)(x) = f(g(x)), \qquad \text{[the composite of } g \text{ and } f\text{]}$$

The *order is important,* as we shall now see:

$g \circ f$ and $f \circ g$ usually are *not* the same function.

Example 8 If $f(x) = x^2$ and $g(x) = x + 3$,* then

$$(g \circ f)(x) = g(f(x)) = g(x^2) = x^2 + 3$$

but

$$(f \circ g)(x) = f(g(x)) = f(x + 3) = (x + 3)^2 = x^2 + 6x + 9.$$

Obviously, $g \circ f \neq f \circ g$ since, for example, they have different values at $x = 0$. $\blacksquare$

CAUTION

Don't confuse the product function fg with the composite function $f \circ g$ (g followed by f). For instance, if $f(x) = 2x^2$ and $g(x) = x - 3$, then the product fg is given by:

$$(fg)(x) = f(x)g(x) = 2x^2(x - 3) = 2x^3 - 6x^2.$$

It is *not* the same as the composite $f \circ g$ because

$$(f \circ g)(x) = f(g(x)) = f(x - 3) = 2(x - 3)^2 = 2x^2 - 12x + 18.$$

*Now that you have the idea of composite functions, we'll use the same letter for the variable in both functions.

By using the operations above, a complicated function may be considered as being built up from simple parts.

Example 9 The function $f(x) = \sqrt{\dfrac{3x^2 - 4x + 5}{x^3 + 1}}$ may be considered as the composite $f = g \circ h$, where

$$h(x) = \frac{3x^2 - 4x + 5}{x^3 + 1} \quad \text{and} \quad g(x) = \sqrt{x}$$

since

$$(g \circ h)(x) = g(h(x)) = g\left(\frac{3x^2 - 4x + 5}{x^3 + 1}\right) = \sqrt{\frac{3x^2 - 4x + 5}{x^3 + 1}} = f(x).$$

The function $h(x) = \dfrac{3x^2 - 4x + 5}{x^3 + 1}$ is the quotient $\dfrac{p}{q}$, where

$$p(x) = 3x^2 - 4x + 5 \quad \text{and} \quad q(x) = x^3 + 1.$$

The function $p(x) = 3x^2 - 4x + 5$ may be written $p = k - s + r$, where

$$k(x) = 3x^2, \qquad s(x) = 4x, \qquad r(x) = 5.$$

The function k, in turn, can be considered as the product $3I^2$, where I is the *identity function* [whose rule is $I(x) = x$]:

$$(3I^2)(x) = 3(I^2(x)) = 3(I(x)I(x)) = 3 \cdot x \cdot x = 3x^2 = k(x).$$

Similarly, $s(x) = (4I)(x) = 4I(x) = 4x$. The function $q(x) = x^3 + 1$ may be "decomposed" in the same way.

Thus the complicated function f is just the result of performing suitable operations on the identity function I and various constant functions. ■

Applications

Composition of functions arises in applications involving several functional relationships simultaneously. In such cases one quantity may have to be expressed as a function of another.

Example 10 A circular puddle of liquid is evaporating and slowly shrinking in size. After t minutes, the radius r of the puddle measures $\dfrac{18}{2t + 3}$ inches; in other words, the radius is a function of time. The area A of the puddle is given by $A = \pi r^2$, that is, area is a function of the radius r. We can express the area as a function of time by substituting $r = \dfrac{18}{2t + 3}$ in the area equation:

$$A = \pi r^2 = \pi\left(\frac{18}{2t + 3}\right)^2.$$

This amounts to forming the composite function $f \circ g$, where $f(r) = \pi r^2$ and $g(t) = \dfrac{18}{2t + 3}$:

$$(f \circ g)(t) = f(g(t)) = f\left(\frac{18}{2t + 3}\right) = \pi\left(\frac{18}{2t + 3}\right)^2.$$

When area is expressed as a function of time, it is easy to compute the area of the puddle at any time. For instance, after 12 minutes the area of the puddle is

$$A = \pi\left(\frac{18}{2t + 3}\right)^2 = \pi\left(\frac{18}{2 \cdot 12 + 3}\right)^2 = \frac{4\pi}{9} \approx 1.396 \text{ square inches.}$$ ■

Example 11 At noon a car leaves Podunk on a straight road, heading south at 45 mph, and a plane 3 miles above the ground passes over Podunk heading east at 350 mph.

(a) Express the distance r traveled by the car and the distance s traveled by the plane as functions of time.

(b) Express the distance d between the plane and the car in terms of r and s.

(c) Express d as a function of time.

(d) How far apart were the plane and the car at 1:30 P.M.?

Solution

(a) Traveling at 45 mph for t hours the car will go a distance of $45t$ miles. Hence the equation $r = 45t$ expresses the distance r as a function of the time t. Similarly, the equation $s = 350t$ expresses the distance s as a function of the time t.

(b) In order to express the distance d as a function of r and s, consider Figure 3–38.

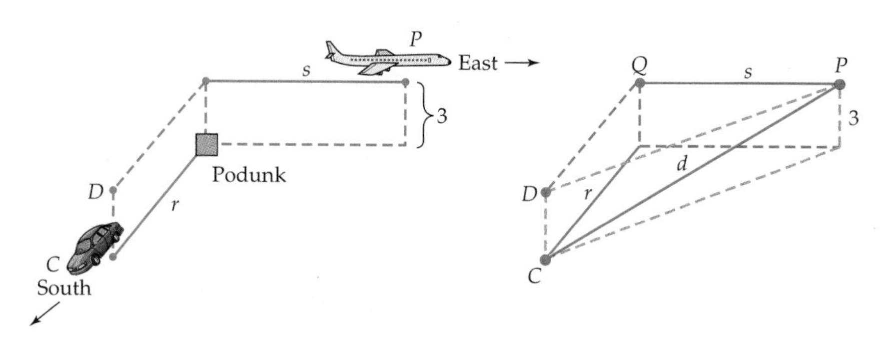

Figure 3–38

Right triangle PQD and the Pythagorean Theorem show that $(PD)^2 = r^2 + s^2$; hence $PD = \sqrt{r^2 + s^2}$. Applying the Pythagorean Theorem to right triangle PDC, we have

$$d^2 = 3^2 + (PD)^2$$
$$d^2 = 3^2 + \left(\sqrt{r^2 + s^2}\right)^2$$
$$d^2 = 9 + r^2 + s^2$$
$$d = \sqrt{9 + r^2 + s^2}$$

(c) The preceding equation expresses d in terms of r and s. By substituting $r = 45t$ and $s = 350t$ in this equation, we can express d as a function of the time t:

$$d = \sqrt{9 + r^2 + s^2}$$
$$d = \sqrt{9 + (45t)^2 + (350t)^2}$$
$$d = \sqrt{9 + 2025t^2 + 122{,}500t^2} = \sqrt{9 + 124{,}525t^2}.$$

(d) At 1:30 P.M. we have $t = 1.5$ (since noon is $t = 0$). At this time

$$d = \sqrt{9 + 124{,}525t^2} = \sqrt{9 + 124{,}525(1.5)^2} = \sqrt{280{,}190.25}$$
$$\approx 529.33 \text{ miles.}$$ ■

Exercises 3.5

In Exercises 1–4, find $(f + g)(x)$, $(f - g)(x)$, *and* $(g - f)(x)$.

1. $f(x) = -3x + 2, \quad g(x) = x^3$
2. $f(x) = x^2 + 2, \quad g(x) = -4x + 7$
3. $f(x) = 1/x, \quad g(x) = x^2 + 2x - 5$
4. $f(x) = \sqrt{x}, \quad g(x) = x^2 + 1$

In Exercises 5–8, find $(fg)(x)$, $(f/g)(x)$, *and* $(g/f)(x)$.

5. $f(x) = -3x + 2, \quad g(x) = x^3$
6. $f(x) = 4x^2 + x^4, \quad g(x) = \sqrt{x^2 + 4}$
7. $f(x) = x^2 - 3, \quad g(x) = \sqrt{x - 3}$
8. $f(x) = \sqrt{x^2 - 1}, \quad g(x) = \sqrt{x - 1}$

In Exercises 9–12, find the domains of fg *and* f/g.

9. $f(x) = x^2 + 1, \quad g(x) = 1/x$
10. $f(x) = x + 2, \quad g(x) = \dfrac{1}{x + 2}$
11. $f(x) = \sqrt{4 - x^2}, \quad g(x) = \sqrt{3x + 4}$
12. $f(x) = 3x^2 + x^4 + 2, \quad g(x) = 4x - 3$

In Exercises 13–16, find the indicated values, where $g(t) = t^2 - t$ *and* $f(x) = 1 + x$.

13. $g(f(0))$
14. $(f \circ g)(3)$
15. $g(f(2) + 3)$
16. $f(2g(1))$

In Exercises 17–20, find $(g \circ f)(3)$, $(f \circ g)(1)$, *and* $(f \circ f)(0)$.

17. $f(x) = 3x - 2, \quad g(x) = x^2$
18. $f(x) = |x + 2|, \quad g(x) = -x^2$
19. $f(x) = x, \quad g(x) = -3$
20. $f(x) = x^2 - 1, \quad g(x) = \sqrt{x}$

In Exercises 21–24, find the rule of the function $f \circ g$, *the domain of* $f \circ g$, *the rule of* $g \circ f$, *and the domain of* $g \circ f$.

21. $f(x) = x^2, \quad g(x) = x + 3$
22. $f(x) = -3x + 2, \quad g(x) = x^3$
23. $f(x) = 1/x, \quad g(x) = \sqrt{x}$
24. $f(x) = \dfrac{1}{2x + 1}, \quad g(x) = x^2 - 1$

In Exercises 25–28, find the rules of the functions ff *and* $f \circ f$.

25. $f(x) = x^3$
26. $f(x) = (x - 1)^2$
27. $f(x) = 1/x$
28. $f(x) = \dfrac{1}{x - 1}$

In Exercises 29–32, verify that $(f \circ g)(x) = x$ and $(g \circ f)(x) = x$ for every x.

29. $f(x) = 9x + 2, \quad g(x) = \dfrac{x - 2}{9}$

30. $f(x) = \sqrt[3]{x - 1}, \quad g(x) = x^3 + 1$

31. $f(x) = \sqrt[3]{x} + 2, \quad g(x) = (x - 2)^3$

32. $f(x) = 2x^3 - 5, \quad g(x) = \sqrt[3]{\dfrac{x + 5}{2}}$

Exercises 33 and 34 refer to the function f whose graph is shown in the figure.

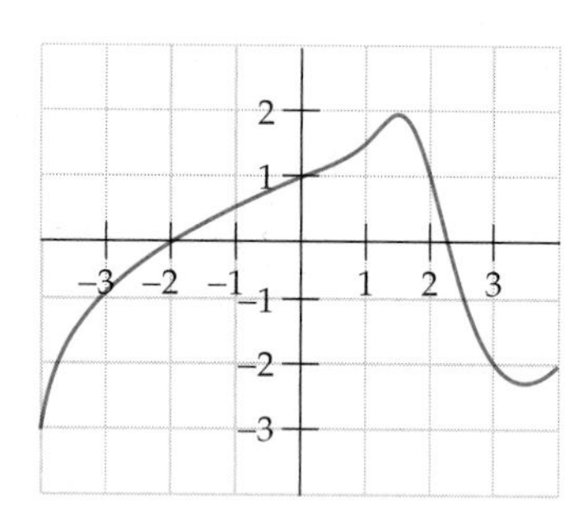

33. Let g be the composite function $f \circ f$ [that is, $g(x) = (f \circ f)(x) = f(f(x))$]. Use the graph of f to fill in the following table (approximate where necessary).

x	$f(x)$	$g(x) = f(f(x))$
−4		
−3		
−2	0	1
−1		
0		
1		
2		
3		
4		

34. Use the information obtained in Exercise 33 to sketch the graph of the function g.

In Exercises 35–38, fill the blanks in the given table. In each case the values of the functions f and g are given by these tables:

x	$f(x)$
1	3
2	5
3	1
4	2
5	3

t	$g(t)$
1	5
2	4
3	4
4	3
5	2

35.

x	$(g \circ f)(x)$
1	4
2	2
3	5
4	4
5	4

36.

t	$(f \circ g)(t)$
1	3
2	2
3	2
4	1
5	5

37.

x	$(f \circ f)(x)$
1	1
2	3
3	3
4	5
5	1

38.

t	$(g \circ g)(t)$
1	2
2	3
3	3
4	4
5	4

In Exercises 39–44, write the given function as the composite of two functions, neither of which is the identity function, as in Examples 6 and 7. (There may be more than one way to do this.)

39. $f(x) = \sqrt[3]{x^2 + 2}$

40. $g(x) = \sqrt{x + 3} - \sqrt[3]{x + 3}$

41. $h(x) = (7x^3 - 10x + 17)^7$

42. $k(x) = \sqrt[3]{(7x - 3)^2}$

43. $f(x) = \dfrac{1}{3x^2 + 5x - 7}$

44. $g(t) = \dfrac{3}{\sqrt{t} - 3} + 7$

45. If $f(x) = x + 1$ and $g(t) = t^2$, then

$$(g \circ f)(x) = g(f(x)) = g(x + 1) = (x + 1)^2$$
$$= x^2 + 2x + 1$$

Find two other functions $h(x)$ and $k(t)$ such that $(k \circ h)(x) = x^2 + 2x + 1$.

46. If f is any function and I is the identity function, what are $f \circ I$ and $I \circ f$?

In Exercises 47–50, determine whether the functions $f \circ g$ and $g \circ f$ are defined. If a composite is defined, find its domain.

47. $f(x) = x^3, \quad g(x) = \sqrt{x}$

48. $f(x) = x^2 + 1, \quad g(x) = \sqrt{x}$

49. $f(x) = \sqrt{x + 10}, \quad g(x) = 5x$

50. $f(x) = -x^2, \quad g(x) = \sqrt{x}$

51. **(a)** If $f(x) = 2x^3 + 5x - 1$, find $f(x^2)$.
(b) If $f(x) = 2x^3 + 5x - 1$, find $(f(x))^2$.
(c) Are the answers in parts (a) and (b) the same? What can you conclude about $f(x^2)$ and $(f(x))^2$?

52. Give an example of a function f such that $f\left(\dfrac{1}{x}\right) \neq \dfrac{1}{f(x)}$.

In Exercises 53 and 54, graph both $f \circ g$ and $g \circ f$ on the same screen. Use the graphs to determine whether $f \circ g$ is the same function as $g \circ f$.

53. $f(x) = x^5 - x^3 - x; \quad g(x) = x - 2$

54. $f(x) = x^3 + x; \quad g(x) = \sqrt[3]{x - 1}$

55. **(a)** What is the area of the puddle in Example 10 after one day? After a week? After a month?
(b) Does the puddle ever totally evaporate? Is this realistic? Under what circumstances might this area function be an accurate model of reality?

56. In a laboratory culture, the number $N(d)$ of bacteria (in thousands) at temperature d degrees Celsius is given by the function

$$N(d) = \frac{-90}{d + 1} + 20 \quad (4 \le d \le 32).$$

The temperature $D(t)$ at time t hours is given by the function $D(t) = 2t + 4 \quad (0 \le t \le 14)$.
(a) What does the composite function $N \circ D$ represent?
(b) How many bacteria are in the culture after 4 hours? After 10 hours?

57. A certain fungus grows in a circular shape. Its diameter after t weeks is $6 - \dfrac{50}{t^2 + 10}$ inches.
(a) Express the area covered by the fungus as a function of time.
(b) What is the area covered by the fungus when $t = 0$? What area does it cover at the end of 8 weeks?
(c) When is its area 25 square inches?

58. Tom left point P at 6 A.M. walking south at 4 mph. Anne left point P at 8 A.M. walking west at 3.2 mph.
(a) Express the distance between Tom and Anne as a function of the time t elapsed since 6 A.M.
(b) How far apart are Tom and Anne at noon?
(c) At what time are they 35 miles apart?

59. As a weather balloon is inflated its radius increases at the rate of 4 centimeters per second. Express the volume of the balloon as a function of time and determine the volume of the balloon after 4 seconds. [*Hint:* The volume of a sphere of radius r is $4\pi r^3/3$.]

60. Express the surface area of the weather balloon in Exercise 59 as a function of time. [*Hint:* The surface area of a sphere of radius r is $4\pi r^2$.]

61. Charlie, who is 6 feet tall, walks away from a streetlight that is 15 feet high at a rate of 5 feet per second, as shown in the figure. Express the length s of Charlie's shadow as a function of time. [*Hint:* First use similar triangles to express s as a function of the distance d from the streetlight to Charlie.]

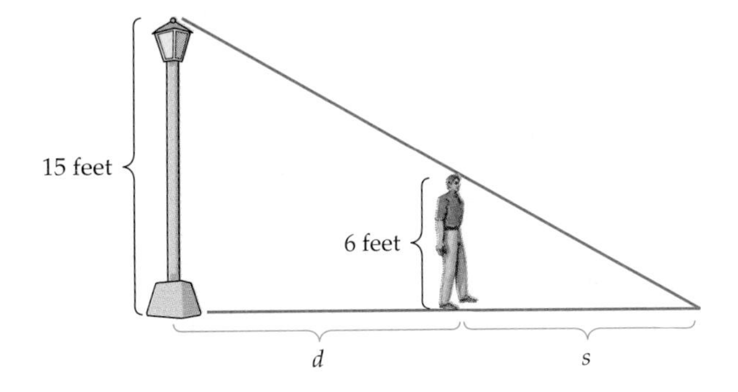

62. A water-filled balloon is dropped from a window 120 feet above the ground. Its height above the ground after t seconds is $120 - 16t^2$ feet. Laura is standing on the ground 40 feet from the point where the balloon will hit the ground, as shown in the figure on the next page.

(a) Express the distance d between Laura and the balloon as a function of time.
(b) When is the balloon exactly 90 feet from Laura?

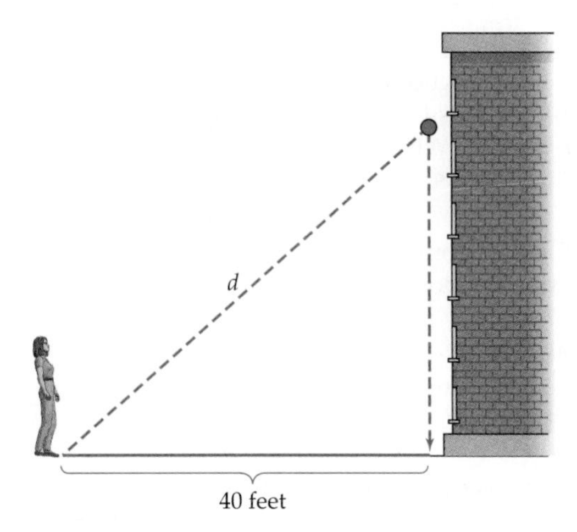

Thinker

63. Find a function f (other than the identity function) such that $(f \circ f \circ f)(x) = x$ for every x in the domain of f. [Several correct answers are possible.]

3.6 Rates of Change*

Rates of change play a central role in calculus. They also have an important connection with the difference quotient of a function, which was introduced in Section 3.2. We begin with a familiar example that illustrates the basic idea of average rate of change.

Recall from Sections 3.1 and 3.2 that when a rock is dropped from a high place, then the distance the rock travels (ignoring wind resistance) is given by the function

$$d(t) = 16t^2$$

with distance $d(t)$ measured in feet and time t in seconds. The following table shows the distance the rock has fallen at various times:

Time t	0	1	2	3	3.5	4	4.5	5
Distance $d(t)$	0	16	64	144	196	256	324	400

To find the distance the rock falls from time $t = 1$ to $t = 3$, we note that at the end of three seconds, the rock has fallen $d(3) = 144$ feet, whereas

*This section may be omitted or postponed. It will regularly be used in clearly identifiable exercises, but not elsewhere in the text.

it had only fallen $d(1) = 16$ feet at the end of one second. So during this time interval the rock traveled

$$d(3) - d(1) = 144 - 16 = 128 \text{ feet.}$$

The distance traveled by the rock during other time intervals can be found similarly:

Time Interval	Distance Traveled
$t = 1$ to $t = 4$	$d(4) - d(1) = 256 - 16 = 240$
$t = 2$ to $t = 3.5$	$d(3.5) - d(2) = 196 - 64 = 132$
$t = 2$ to $t = 4.5$	$d(4.5) - d(2) = 324 - 64 = 260$

The same procedure works in general:

The distance traveled from time $t = a$ to time $t = b$ is $d(b) - d(a)$ feet.

In the preceding chart, the length of each time interval can be computed by taking the difference between the two times. For example, from $t = 1$ to $t = 4$ is a time interval of length $4 - 1 = 3$ seconds. Similarly, the interval from $t = 2$ to $t = 3.5$ is of length $3.5 - 2 = 1.5$ seconds and in general,

The time interval from $t = a$ to $t = b$ is an interval of $b - a$ seconds.

Since distance = average speed × time,

$$\text{average speed} = \frac{\text{distance traveled}}{\text{time interval}}.$$

Hence, the average speed over the time interval from $t = a$ to $t = b$ is

$$\text{average speed} = \frac{\text{distance traveled}}{\text{time interval}} = \frac{d(b) - d(a)}{b - a}.$$

For example, to find the average speed from $t = 1$ to $t = 4$, apply the preceding formula with $a = 1$ and $b = 4$:

$$\text{average speed} = \frac{d(4) - d(1)}{4 - 1} = \frac{256 - 16}{4 - 1} = \frac{240}{3} = 80 \text{ feet per second.}$$

Similarly, the average speed from $t = 2$ to $t = 4.5$ is

$$\frac{d(4.5) - d(2)}{4.5 - 2} = \frac{324 - 64}{4.5 - 2} = \frac{260}{2.5} = 104 \text{ feet per second}$$

The units in which average speed is measured here (feet per second) indicate the number of units of distance traveled during each unit of time, that is, the *rate of change* of distance (feet) with respect to time (seconds).

The preceding discussion can be summarized by saying that the average speed (rate of change of distance with respect to time) as time changes from $t = a$ to $t = b$ is given by

$$\text{average speed} = \text{average rate of change}$$
$$= \frac{\text{change in distance}}{\text{change in time}} = \frac{d(b) - d(a)}{b - a}.$$

Although speed is the most familiar example, rates of change play a role in many other situations as well, as illustrated in Examples 1–3 below. Consequently, we define the average rate of change of any function as follows.

Average Rate of Change

Let f be a function. The *average rate of change of $f(x)$* with respect to x as x changes from a to b is the number

$$\frac{\text{change in } f(x)}{\text{change in } x} = \frac{f(b) - f(a)}{b - a}.$$

Example 1 A large heavy-duty balloon is being filled with water. Its approximate volume (in gallons) is given by

$$V(x) = \frac{x^3}{55}$$

where x is the radius of the balloon (in inches). The average rate of change of the volume of the balloon as the radius increases from 5 to 10 inches is

$$\frac{\text{change in volume}}{\text{change in radius}} = \frac{V(10) - V(5)}{10 - 5} \approx \frac{18.18 - 2.27}{10 - 5} = \frac{15.91}{5}$$
$$= 3.182 \text{ gallons per inch.}$$ ■

Example 2 A small manufacturing company makes specialty office desks. The cost (in thousands of dollars) of producing x desks is given by the function

$$c(x) = .0009x^3 - .06x^2 + 1.6x + 5.$$

For example, you can readily verify that $c(10) = 15.9$ and $c(30) = 23.3$. Since $c(x)$ is measured in thousands, this means that the cost of making 10 desks is \$15,900 and the cost of making 30 is \$23,300.

GRAPHING EXPLORATION

Graph the cost function in the viewing window with scale marks 5 units apart on each axis). As you can see, the $0 \le x \le 50$ and $0 \le y \le 50$ (for convenient reading, set the graph rises from $x = 0$ to $x = 10$, continues rising somewhat more slowly from $x = 10$ to $x = 35$, and then begins to rise quite sharply. This reflects the fact that costs continually rise, but not always at the same rate.*

As production increases from 0 to 10 desks, the average rate of change of cost is

$$\frac{\text{change in cost}}{\text{change in production}} = \frac{c(10) - c(0)}{10 - 0} = \frac{15.9 - 5}{10} = \frac{10.9}{10} = 1.09.$$

This means that costs are rising at an average rate of 1.09 thousand dollars (that is, \$1090) per desk. As production goes from 10 to 30 desks, the average rate of change of cost is

$$\frac{c(30) - c(10)}{30 - 10} = \frac{23.3 - 15.9}{30 - 10} = \frac{7.4}{20} = .37$$

so that costs are rising at an average rate of only \$370 per desk. The rate increases as production goes from 30 to 50:

$$\frac{c(50) - c(30)}{50 - 30} = \frac{47.5 - 23.3}{50 - 30} = \frac{24.2}{20} = 1.21, \quad \text{that is,} \quad \$1210 \text{ per desk.}$$ ■

Example 3 Figure 3–39 is the graph of the temperature function f during a particular day; $f(x)$ is the temperature at x hours after midnight. What is the average rate of change of the temperature (a) from 4 A.M. to noon? (b) from 3 P.M. to 8 P.M.?

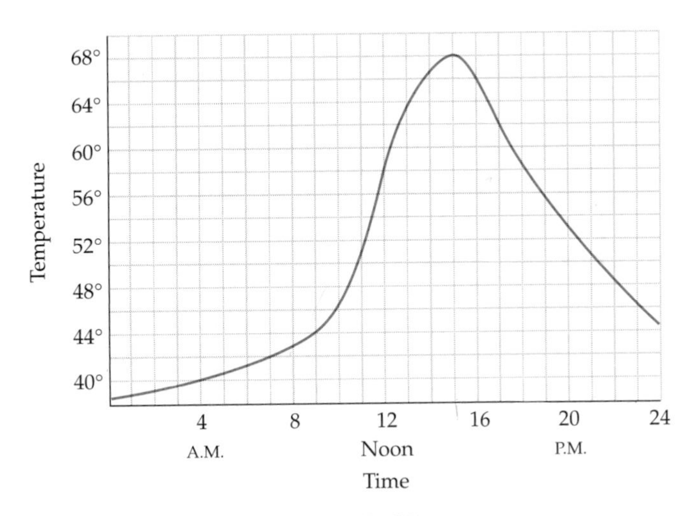

Figure 3–39

*Costs are high at the beginning because of initial setup costs (rent, equipment, etc.); in fact, it costs \$5000 even if no desks are manufactured ($c(0) = 5$). As more desks are produced, costs increase more slowly because of efficiencies of scale. Then they climb again (increasing production past a certain point might require a new building, or more machines, or a second shift of workers, etc.).

Solution

(a) The graph shows that the temperature at 4 A.M. is $f(4) = 40°$ and the temperature at noon is $f(12) = 58°$. The average rate of change of temperature is

$$\frac{\text{change in temperature}}{\text{change in time}} = \frac{f(12) - f(4)}{12 - 4} = \frac{58 - 40}{12 - 4} = \frac{18}{8}$$
$$= 2.25° \text{ per hour.}$$

The rate of change is positive because the temperature is increasing at an average rate of 2.25° per hour.

(b) Now 3 P.M. corresponds to $x = 15$ and 8 P.M. to $x = 20$. The graph shows that $f(15) = 68°$ and $f(20) = 53°$. Hence the average rate of change of temperature is

$$\frac{\text{change in temperature}}{\text{change in time}} = \frac{f(20) - f(15)}{20 - 15} = \frac{53 - 68}{20 - 15} = \frac{-15}{5}$$
$$= -3° \text{ per hour.}$$

The rate of change is negative because the temperature is decreasing at an average rate of 3° per hour. ■

Geometric Interpretation of Average Rate of Change

If P and Q are points on the graph of a function f, then the straight line determined by P and Q is called a **secant line.** Figure 3–40 shows the secant line joining the points (4, 40) and (12, 58) on the graph of the temperature function f of Example 3.

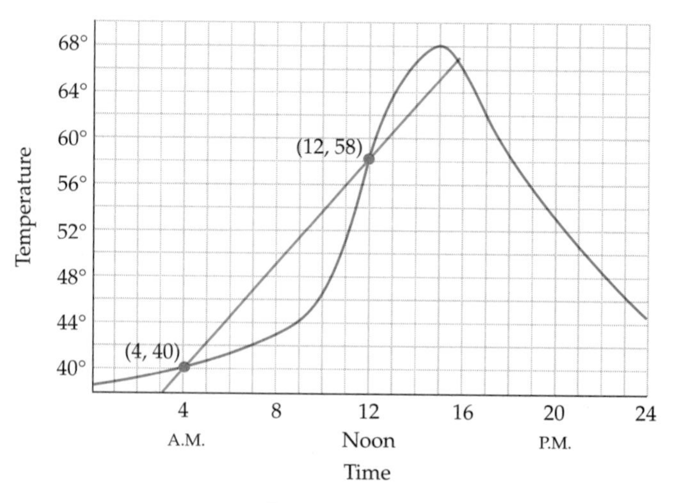

Figure 3–40

Using the points (4, 40) and (12, 58), we see that the slope of this secant line is $\frac{58 - 40}{12 - 4} = \frac{18}{8} = 2.25$. To say that (4, 40) and (12, 58) are on the graph of f means that $f(4) = 40$ and $f(12) = 58$. Thus,

$$\text{slope of secant line} = 2.25 = \frac{58 - 40}{12 - 4} = \frac{f(12) - f(4)}{12 - 4}$$

$$= \text{average rate of change as } x \text{ goes from 4 to 12.}$$

The same thing happens in the general case:

Secant Lines and Average Rates of Change

If f is a function, then the average rate of change of $f(x)$ with respect to x as x changes from $x = a$ to $x = b$ is the slope of the secant line joining the points $(a, f(a))$ and $(b, f(b))$ on the graph of f.

The Difference Quotient

Average rates of change are often computed for very small intervals. For instance, we might compute the rate from 4 to 4.01 or from 4 to 4.001. Since $4.01 = 4 + .01$ and $4.001 = 4 + .001$, we are doing essentially the same thing in both cases: computing the rate of change over the interval from 4 to $4 + h$ for some small nonzero quantity h. Furthermore, it's often possible to use a single calculation to determine the average rate for all possible values of h.

Example 4 Consider the falling rock with which this section began. The distance the rock has traveled at time t is given by $d(t) = 16t^2$ and its average speed (rate of change) from $t = 4$ to $t = 4 + h$ is:

$$\text{average speed} = \frac{d(4 + h) - d(4)}{(4 + h) - 4} = \frac{16(4 + h)^2 - 16 \cdot 4^2}{h}$$

$$= \frac{16(16 + 8h + h^2) - 256}{h} = \frac{256 + 128h + 16h^2 - 256}{h}$$

$$= \frac{128h + 16h^2}{h} = \frac{h(128 + 16h)}{h} = 128 + 16h.$$

Thus, we can quickly compute the average speed over the interval from 4 to $4 + h$ seconds for any value of h by using the formula

$$\text{average speed} = 128 + 16h.$$

For example, the average speed from 4 seconds to 4.001 seconds (here $h = .001$) is

$$128 + 16h = 128 + 16(.001) = 128 + .016 = 128.016 \text{ feet per second.} \quad ■$$

Similar calculations can be done with any number in place of 4. In each such case, we are dealing with an interval from x to $x + h$ for some number x. As in Example 4, a single computation can often be used for all possible x and h.

Example 5 The average speed of the falling rock of Example 4 from time x to time $x + h$ is:*

$$\begin{aligned}\text{average speed} &= \frac{d(x+h) - d(x)}{(x+h) - x} = \frac{16(x+h)^2 - 16x^2}{h} \\ &= \frac{16(x^2 + 2xh + h^2) - 16x^2}{h} = \frac{16x^2 + 32xh + 16h^2 - 16x^2}{h} \\ &= \frac{32xh + 16h^2}{h} = \frac{h(32x + 16h)}{h} = 32x + 16h.\end{aligned}$$

When $x = 4$, then this result states that the average speed from 4 to $4 + h$ is $32(4) + 16h = 128 + 16h$, which is exactly what we found in Example 4. To find the average speed from 3 to 3.1 seconds, apply the formula

$$\text{average speed} = 32x + 16h$$

with $x = 3$ and $h = .1$:

$$\text{average speed} = 32 \cdot 3 + 16(.1) = 96 + 1.6 = 97.6 \text{ feet per second.} \quad ■$$

More generally, we can compute the average rate of change of any function f over the interval from x to $x + h$ just as we did in Example 5: Apply the definition of average rate of change in the box on page 176 with x in place of a and $x + h$ in place of b:

$$\begin{aligned}\textbf{average rate of change} &= \frac{f(b) - f(a)}{b - a} = \frac{f(x+h) - f(x)}{(x+h) - x} \\ &= \frac{f(x+h) - f(x)}{h}.\end{aligned}$$

This last quantity is just the difference quotient of f (see page 130). Therefore,

*Note that this calculation is the same as in Example 4, except that 4 has been replaced by x.

Difference Quotients and Rates of Change

If f is a function, then the average rate of change of f over the interval from x to $x + h$ is given by the difference quotient

$$\frac{f(x+h) - f(x)}{h}.$$

Example 6 Find the difference quotient of $V(x) = x^3/55$ and use it to find the average rate of change of V as x changes from 8 to 8.01.

Solution Use the definition of the difference quotient and algebra:

$$\frac{V(x+h) - V(x)}{h} = \frac{\overbrace{\frac{(x+h)^3}{55}}^{V(x+h)} - \overbrace{\frac{x^3}{55}}^{V(x)}}{h} = \frac{\frac{1}{55}[(x+h)^3 - x^3]}{h}$$

$$= \frac{1}{55} \cdot \frac{(x+h)^3 - x^3}{h} = \frac{1}{55} \cdot \frac{x^3 + 3x^2h + 3xh^2 + h^3 - x^3}{h}$$

$$= \frac{1}{55} \cdot \frac{3x^2h + 3xh^2 + h^3}{h} = \frac{1}{55} \cdot \frac{h(3x^2 + 3xh + h^2)}{h}$$

$$= \frac{3x^2 + 3xh + h^2}{55}.$$

When x changes from 8 to $8.01 = 8 + .01$, we have $x = 8$ and $h = .01$. So the average rate of change is

$$\frac{3x^2 + 3xh + h^2}{55} = \frac{3 \cdot 8^2 + 3 \cdot 8(.01) + (.01)^2}{55} \approx 3.495. \quad \blacksquare$$

Instantaneous Rate of Change

Rates of change are a major theme in calculus—not just the average rate of change discussed above, but also the *instantaneous rate of change* of a function (that is, its rate of change at a particular instant). Even without calculus, however, we can obtain quite accurate approximations of instantaneous rates of change by using average rates appropriately.

Example 7 A rock is dropped from a high place. What is its speed exactly 3 seconds after it is dropped?

Solution The distance the rock has fallen at time t is given by the function $d(t) = 16t^2$. The exact speed at $t = 3$ can be approximated by finding the average speed over very small time intervals, say, 3 to 3.01 or even shorter intervals. Over a very short time span, such as a hundredth of a second, the rock cannot change speed very much so these average speeds should be a reasonable approximation of its speed at the instant $t = 3$. Example 5 shows that the average speed is given by the difference quotient $32x + 16h$. When $x = 3$, the difference quotient is $32 \cdot 3 + 16h = 96 + 16h$ and we have:

Change in Time **3 to 3 + h**	h	**Average Speed** **[Difference Quotient at $x = 3$]** **$96 + 16h$**
3 to 3.1	.1	$96 + 16(.1) = 97.6$ ft per sec
3 to 3.01	.01	$96 + 16(.01) = 96.16$ ft per sec
3 to 3.005	.005	$96 + 16(.005) = 96.08$ ft per sec
3 to 3.00001	.00001	$96 + 16(.00001) = 96.00016$ ft per sec

The table suggests that the exact speed of the rock at the instant $t = 3$ seconds is very close to 96 feet per second. ■

Example 8 A balloon is being filled with water in such a way that when its radius is x inches, then its volume is $V(x) = x^3/55$ gallons. In Example 1 we saw that the average rate of change of the volume as the radius increases from 5 inches to 10 inches is 3.182 gallons per inch. What is the rate of change at the instant when the radius is 7 inches?

Solution The average rate of change when the radius goes from x to $x + h$ inches is given by the difference quotient of $V(x)$, which was found in Example 6:

$$\frac{V(x + h) - V(x)}{h} = \frac{3x^2 + 3xh + h^2}{55}.$$

Therefore, when $x = 7$ the difference quotient is

$$\frac{3 \cdot 7^2 + 3 \cdot 7 \cdot h + h^2}{55} = \frac{147 + 21h + h^2}{55}$$

and we have these average rates of change over small intervals near 7:

Change in Radius **7 to 7 + h**	h	**Average Rate of Change of Volume** **[Difference Quotient at $x = 7$]** $\dfrac{147 + 21h + h^2}{55}$
7 to 7.01	.01	2.6765 gallons per inch
7 to 7.001	.001	2.6731 gallons per inch
7 to 7.0001	.0001	2.6728 gallons per inch
7 to 7.00001	.00001	2.6727 gallons per inch

The chart suggests that at the instant the radius is 7 inches, the volume is changing at a rate of approximately 2.673 gallons per inch. ■

Exercises 3.6

1. A car moves along a straight test track. The distance traveled by the car at various times is shown in this table:

Time (seconds)	0	5	10	15	20	25	30
Distance (feet)	0	20	140	400	680	1400	1800

Find the average speed of the car over the interval from
(a) 0 to 10 seconds **(b)** 10 to 20 seconds
(c) 20 to 30 seconds **(d)** 15 to 30 seconds

2. The yearly profit of a small manufacturing firm is shown in the following tables.

Year	1986	1987	1988	1989
Profit	$5000	$6000	$6500	$6800

Year	1990	1991	1992	1993
Profit	$7200	$6700	$6500	$7000

What is the average rate of change of profits over the given time span?
(a) 1986–1990 **(b)** 1986–1993
(c) 1989–1992 **(d)** 1988–1992

3. Find the average rate of change of the volume of the balloon in Example 1 as the radius increases from
(a) 2 to 5 inches **(b)** 4 to 8 inches

4. Find the average rate of change of cost for the company in Example 2 when production decreases from
(a) 5 to 25 desks **(b)** 0 to 40 desks

5. The graph in the figure shows the monthly sales of floral pattern ties (in thousands of ties) made by Neckwear, Inc., over a 48-month period. Sales are very low when the ties are first introduced, increase significantly, hold steady for a while, and then drop off as the ties go out of fashion. Find the average rate of change of sales (in ties per month) over the interval:
(a) 0 to 12 **(b)** 8 to 24 **(c)** 12 to 24
(d) 20 to 28 **(e)** 28 to 36 **(f)** 32 to 44
(g) 36 to 40 **(h)** 40 to 48

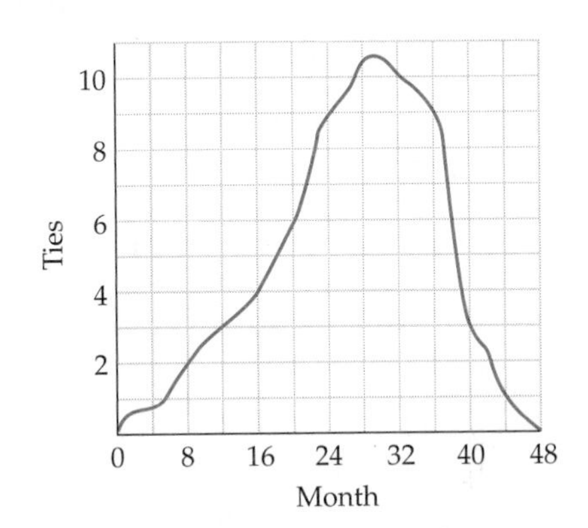

6. The XYZ Company has found that its sales are related to the amount of advertising it does in trade magazines. The graph in the figure shows the sales (in thousands of dollars) as a function of the amount of advertising (in number of magazine ad pages). Find the average rate of change of sales when the number of ad pages increases from
(a) 10 to 20 **(b)** 20 to 60
(c) 60 to 100 **(d)** 0 to 100
(e) Is it worthwhile to buy more than 70 pages of ads, if the cost of a one-page ad is $2000? If the cost is $5000? If the cost is $8000?

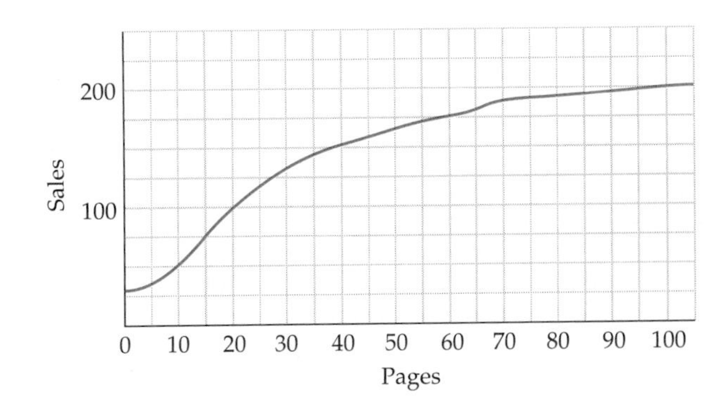

7. When blood flows through an artery (which can be thought of as a cylindrical tube) its velocity is greatest at the center of the artery. Because of friction along the walls of the tube, the blood's velocity decreases as the distance r from the center of the artery increases, finally becoming 0 at the wall of the artery. The velocity (in centimeters per second) is given by the function $v = 18{,}500(.000065 - r^2)$, where r is measured in centimeters. Find the average rate of change of the velocity as the distance from the center changes from
(a) $r = .001$ to $r = .002$ **(b)** $r = .002$ to $r = .003$
(c) $r = 0$ to $r = .025$

8. A car is stopped at a traffic light and begins to move forward along a straight road when the light turns green. The distance (in feet) traveled by the car in t seconds is given by $s(t) = 2t^2$ $(0 \le t \le 30)$. What is the average speed of the car from
(a) $t = 0$ to $t = 5$? **(b)** $t = 5$ to $t = 10$?
(c) $t = 10$ to $t = 30$? **(d)** $t = 10$ to $t = 10.1$?

In Exercises 9–14, find the average rate of change of the function f over the given interval.

9. $f(x) = 2 - x^2$ from $x = 0$ to $x = 2$

10. $f(x) = .25x^4 - x^2 - 2x + 4$ from $x = -1$ to $x = 4$

11. $f(x) = x^3 - 3x^2 - 2x + 6$ from $x = -1$ to $x = 3$

12. $f(x) = -\sqrt{x^4 - x^3 + 2x^2 - x + 4}$ from $x = 0$ to $x = 3$

13. $f(x) = \sqrt{x^3 + 2x^2 - 6x + 5}$ from $x = 1$ to $x = 2$

14. $f(x) = \dfrac{x^2 - 3}{2x - 4}$ from $x = 3$ to $x = 6$

In Exercises 15–22, compute the difference quotient of the function.

15. $f(x) = x + 5$ **16.** $f(x) = 7x + 2$

17. $f(x) = x^2 + 3$ **18.** $f(x) = x^2 + 3x - 1$

19. $f(t) = 160{,}000 - 8000t + t^2$ **20.** $V(x) = x^3$

21. $A(r) = \pi r^2$ **22.** $V(p) = 5/p$

23. Water is draining from a large tank. After t minutes there are $160{,}000 - 8000t + t^2$ gallons of water in the tank.
(a) Use the results of Exercise 19 to find the average rate at which the water runs out in the interval from 10 to 10.1 minutes.
(b) Do the same for the interval from 10 to 10.01 minutes.
(c) Estimate the rate at which the water runs out after exactly 10 minutes.

24. Use the results of Exercise 20 to find the average rate of change of the volume of a cube whose side has length x as x changes from
(a) 4 to 4.1 **(b)** 4 to 4.01 **(c)** 4 to 4.001
(d) Estimate the rate of change of the volume at the instant when $x = 4$.

25. Use the results of Exercise 21 to find the average rate of change of the area of a circle of radius r as r changes from
(a) 3 to 3.5 **(b)** 3 to 3.2 **(c)** 3 to 3.1
(d) Estimate the rate of change at the instant when $r = 3$.
(e) How is your answer in part (d) related to the circumference of a circle of radius 3?

26. Under certain conditions, the volume V of a quantity of air is related to the pressure p (which is measured in kilopascals) by the equation $V = 5/p$. Use the results of Exercise 22 to estimate the rate at which the volume is changing at the instant when the pressure is 50 kilopascals.

27. Two cars race on a straight track, beginning from a dead stop. The distance (in feet) each car has covered at each time during the first 16 seconds is shown in the figure.

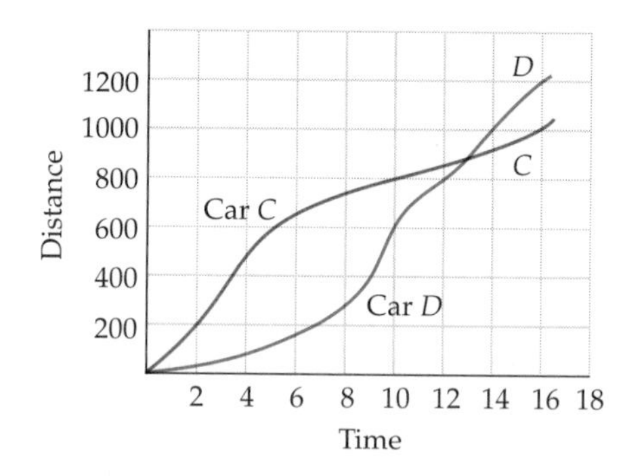

(a) What is the average speed of each car during this 16-second interval?
(b) Find an interval beginning at $t = 4$ during which the average speed of car D was approximately the same as the average speed of car C from $t = 2$ to $t = 10$.
(c) Use secant lines and slopes to justify the statement "car D traveled at a higher average speed than car C from $t = 4$ to $t = 10$."

28. The figure shows the profits earned by a certain company during the last quarters of three consecutive years.

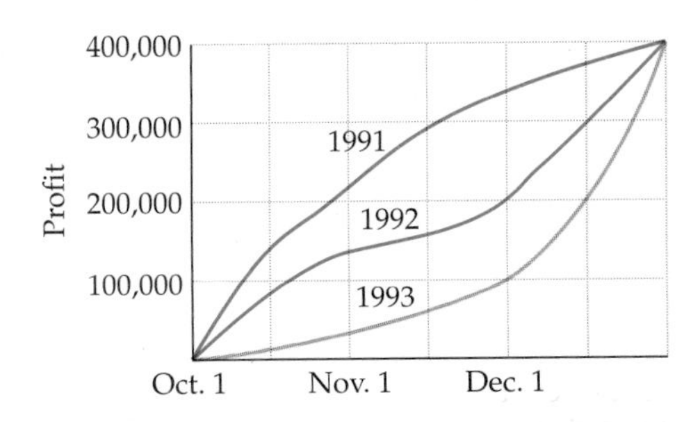

(a) Explain why the average rate of change of profits from October 1 to December 31 was the same in all three years.

(b) During what month in what year was the average rate of change of profits the greatest?

29. The graph in the figure shows the chipmunk population in a certain wilderness area. The population increases as the chipmunks reproduce, but then decreases sharply as predators move into the area.

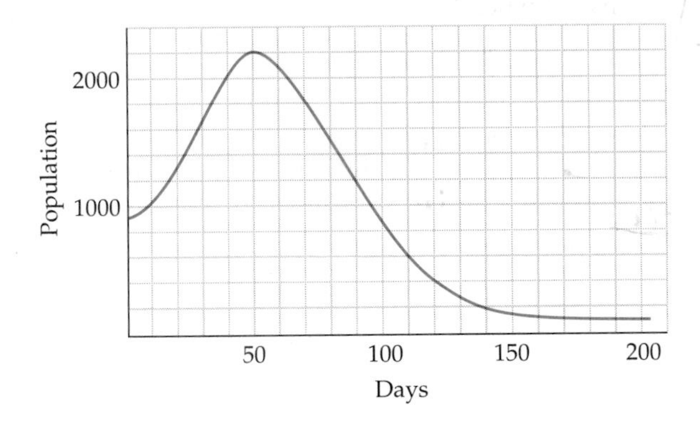

(a) During what approximate time period (beginning on day 0) is the average growth rate of the chipmunk population positive?

(b) During what approximate time period, beginning on day 0, is the average growth rate of the chipmunk population 0?

(c) What is the average growth rate of the chipmunk population from day 50 to day 100? What does this number mean?

(d) What is the average growth rate from day 45 to day 50? From day 50 to day 55? What is the approximate average growth rate from day 49 to day 51?

30. Lucy has a viral flu. How bad she feels depends primarily on how fast her temperature is rising at that time. The figure shows her temperature during the first day of the flu.

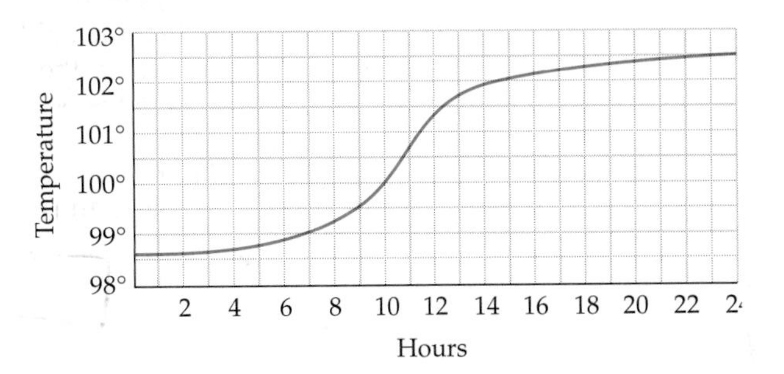

(a) At what average rate does her temperature rise during the entire day?

(b) During what 2-hour period during the day does she feel worst?

(c) Find two time intervals, one in the morning and one in the afternoon, during which she feels about the same (that is, during which her temperature is rising at the same average rate).

3.7 Inverse Functions*

We begin with a simplified example that illustrates the basic ideas to be studied in this section. Suppose the function f is given by the following table.

f-input	-2	-1	0	1	2
f-output	-3	-2	1	4	5

The table shows, for example, that $f(1) = 4$ and $f(-2) = -3$. The domain of f (set of inputs) consists of $-2, -1, 0, 1, 2$ and the range of f (set of outputs) consists of $-3, -2, 1, 4, 5$.

Now define a new function g by the following table (which simply *reverses* the rows in the f table).

g-input	-3	-2	1	4	5
g-output	-2	-1	0	1	2

The table shows, for example, that $g(4) = 1$ and $g(-3) = -2$. The domain of g (set of inputs) consists of $-3, -2, 1, 4, 5$ (which are the *outputs* of f) and the range of g (set of outputs) consists of $-2, -1, 0, 1, 2$ (which are the *inputs* of f). The rule of g simply *reverses* the action of f by taking each output of f back to the input it came from:

$$g(4) = 1 \quad \text{and} \quad f(1) = 4$$

$$g(-3) = -2 \quad \text{and} \quad f(-2) = -3$$

and, in general

$$g(y) = x \quad \text{exactly when} \quad f(x) = y.$$

We say that g is the *inverse function* of f.

Some functions do not have inverse functions. Consider, for example, the function h given by the following table.

h-input	1	2	3	4	5
h-output	-1	3	0	3	2

If we try to define a function k by reversing inputs and outputs, we run into trouble. Since $h(2) = 3$, we must have $k(3) = 2$, but since $h(4) = 3$, we must also have $k(3) = 4$. Thus, the input 3 has two different outputs under k, which means that k is not a function. So the first step in dealing

*This section is used only in Section 5.3, Excursion 5.3.A, and Section 8.5. It may be postponed until then.

with inverse functions in the general case is to determine which functions have inverses.

One-to-One Functions

The reason the function h in the previous paragraph did not have an inverse was that two different inputs (2 and 4) produced the same output 3. This was not the case with the function f above; its table shows that different inputs always produce different outputs. Consequently, f is an example of the following definition.

A function f is said to be **one-to-one** if different inputs always produce different outputs, that is,

$$\text{if } a \neq b, \text{ then } f(a) \neq f(b).$$

In graphical terms this means that two points on the graph, $(a, f(a))$ and $(b, f(b))$, that have different x-coordinates $[a \neq b]$ must also have different y-coordinates $[f(a) \neq f(b)]$. Consequently, these points cannot lie on the same horizontal line because all points on a horizontal line have the same y-coordinate. Therefore, we have this geometric test to determine if a function is one-to-one.

The Horizontal Line Test

If a function f is one-to-one, then it has this property:

No horizontal line intersects the graph of f more than once.

Conversely, if the graph of a function has this property, then the function is one-to-one.

Example 1 Which of the following functions are one-to-one?

(a) $f(x) = 7x^5 + 3x^4 - 2x^3 + 2x + 1$

(b) $g(x) = x^3 - 3x - 1$

(c) $h(x) = 1 - .2x^3$

Solution Complete graphs of each function are shown in Figure 3–41.

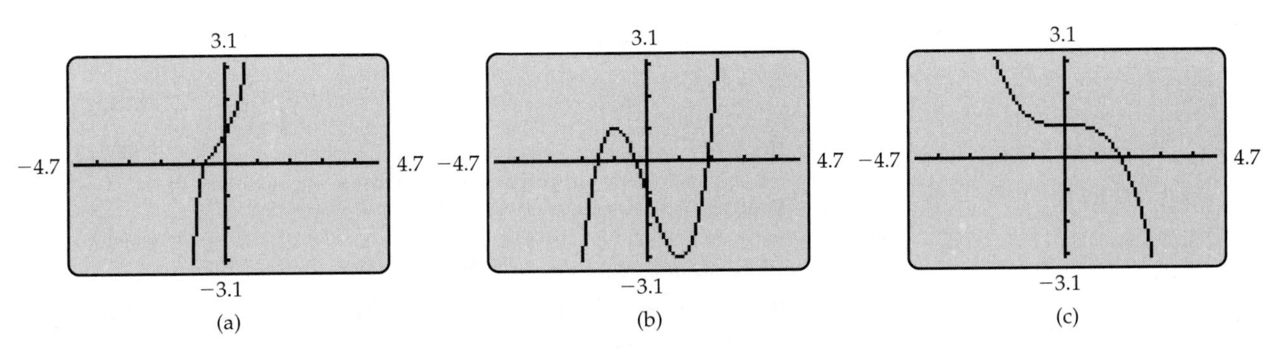

Figure 3–41

(a) The graph of f in Figure 3–41(a) passes the Horizontal Line Test since no horizontal line intersects the graph more than once. Hence, f is one-to-one.

(b) The graph of g in Figure 3–41(b) obviously fails the Horizontal Line Test because many horizontal lines (including the x-axis) intersect the graph more than once. Therefore, g is not one-to-one.

(c) The graph of h in Figure 3–41(c) appears to contain a horizontal line segment. So, h appears to fail the Horizontal Line Test because the horizontal line through (0, 1) seems to intersect the graph infinitely many times. But appearances are deceiving.

Technology Tip

Although a horizontal segment may appear on a calculator screen when the graph is actually rising or falling, there is another possibility. The graph may have a tiny wiggle (less than the height of a pixel) and thus fail the Horizontal Line Test:

You can usually detect such a wiggle by zooming in to magnify that portion of the graph, or by using the trace feature to see if the y-coordinates increase, then decrease (or vice versa) along the "horizontal" segment.

GRAPHING EXPLORATION

Graph $h(x) = 1 - .2x^3$ and use the trace feature to move from left to right along the "horizontal" segment. Do the y-coordinates stay the same, or do they decrease?

The Exploration shows that the graph is actually falling from left to right, so that each horizontal line intersects it only once. (It appears to have a horizontal segment because the amount the graph falls there is less than the height of a pixel on the screen.) Therefore, h is a one-to-one function. ■

The function f in Example 1 is an **increasing function** (its graph is always rising from left to right) and the function h is a **decreasing function** (its graph is always falling from left to right). Every increasing or decreasing function is necessarily one-to-one because its graph can never touch the same horizontal line twice (it would have to change from rising to falling, or vice versa, to do so).

Inverse Functions

With a one-to-one function f, each output comes from exactly one input (because different inputs must produce different outputs). Consequently, we can define a new function g that reverses the action of f by sending each output back to the unique input it came from. For instance, if $f(7) = 11$, then $g(11) = 7$. Thus, the outputs of f become the inputs of g and we have this definition.

Inverse Functions

Let f be a one-to-one function. Then the *inverse function* of f is the function g whose rule is

$$g(y) = x \qquad \text{exactly when} \qquad f(x) = y.$$

The domain of g is the range of f and the range of g is the domain of f.

Example 2 The graph of $f(x) = 3x - 2$ is a straight line that certainly passes the Horizontal Line Test, so f is one-to-one and has an inverse function g. From the definition of g we know that

$$g(y) = x \qquad \text{exactly when} \qquad f(x) = y$$

that is,

$$g(y) = x \qquad \text{exactly when} \qquad 3x - 2 = y.$$

In order to find the rule of g we need only solve this last equation for x:

$$3x - 2 = y$$

Add 2 to both sides: $$3x = y + 2$$

Divide both sides by 3: $$x = \frac{y+2}{3}$$

Since $g(y) = x$, we see that the rule of g is $g(y) = \dfrac{y+2}{3}$. ■

Recall that the letter used for the variable of a function doesn't matter. For instance, $h(x) = x^2$ and $h(t) = t^2$ and $h(u) = u^2$ all describe the same function, whose rule is "square the input." When dealing with inverse functions, it is customary to use the same variable for both f and its inverse g. Consequently, the inverse function in Example 2 would normally be written as $g(x) = \dfrac{x+2}{3}$.

Example 3 Use your calculator to verify that the function $f(x) = x^3 + 5$ passes the Horizontal Line Test and hence is one-to-one. Its inverse can be found by solving for x in the equation $y = x^3 + 5$:

Subtract 5 from both sides: $$x^3 = y - 5$$

Take cube roots on both sides: $$x = \sqrt[3]{y - 5}$$

By the Round-Trip Theorem, $g(y) = \sqrt[3]{y - 5}$ is the inverse function of f. Using the same variable as f, we write its rule as $g(x) = \sqrt[3]{x - 5}$. ■

Example 4 The function $f(x) = \sqrt{x - 3}$ is one-to-one, as you can verify with your calculator. To find its inverse we solve the equation:

$$y = \sqrt{x - 3}$$

Square both sides: $$y^2 = x - 3$$

Add 3 to both sides: $$x = y^2 + 3$$

Although this last equation is defined for all real numbers y, the original equation $y = \sqrt{x - 3}$ has $y \geq 0$ (since square roots are nonnegative). In other words, the range of the function f (the possible values of y) consists of all nonnegative real numbers. Consequently, the domain of the inverse function g is the set of all nonnegative real numbers and its rule is:

$$g(y) = y^2 + 3 \quad (y \geq 0).$$

Once again, it's customary to use the same variable to describe both f and its inverse function, so we write the rule of g as $g(x) = x^2 + 3$ $(x \geq 0)$. ■

The Round-Trip Properties

The inverse function g of a function f was designed to send each output of f back to the input it came from, that is,

$$g(d) = c \qquad \text{exactly when} \qquad f(c) = d.$$

Consequently, if you first apply f and then apply g to the result, you obtain the number you started with:

$$\begin{aligned} g(f(c)) &= g(d) \quad &&(\text{because } f(c) = d) \\ &= c &&(\text{because } g(d) = c). \end{aligned}$$

A similar argument shows that $f(g(d)) = d$.

Example 5 As we saw in Example 2, the inverse function of $f(x) = 3x - 2$ is $g(x) = \dfrac{x + 2}{3}$. If we start with a number c and apply f we obtain $f(c) = 3c - 2$. If we now apply g to this result, we obtain

$$g(f(c)) = g(3c - 2) = \frac{(3c - 2) + 2}{3} = c.$$

So we are back where we started. Similarly, if we first apply g and then apply f to a number, we end up where we started:

$$f(g(c)) = f\left(\frac{c + 2}{3}\right) = 3\left(\frac{c + 2}{3}\right) - 2 = c.$$

The function $f(x) = x^3 + 5$ of Example 3 and its inverse function $g(x) = \sqrt[3]{x - 5}$ also have these "round-trip" properties. If you apply one function and then the other, you wind up at the number you started with:

$$g(f(x)) = g(x^3 + 5) = \sqrt[3]{(x^3 + 5) - 5} = \sqrt[3]{x^3} = x$$

and

$$f(g(x)) = f(\sqrt[3]{x - 5}) = (\sqrt[3]{x - 5})^3 + 5 = (x - 5) + 5 = x. \quad ■$$

Not only do a function and its inverse have the round-trip properties illustrated in Example 5, but somewhat more is true (as is proved in Exercise 53).

Round-Trip Theorem

A one-to-one function f and its inverse function g have these properties:

$$g(f(x)) = x \quad \text{for every } x \text{ in the domain of } f;$$

$$f(g(x)) = x \quad \text{for every } x \text{ in the domain of } g.$$

Conversely, if f and g are functions having these properties, then f is one-to-one and its inverse is g.

Example 6 If $f(x) = \dfrac{5}{2x-4}$ and $g(x) = \dfrac{4x+5}{2x}$, then for every x in the domain of f (that is, all $x \neq 2$),

$$g(f(x)) = g\left(\frac{5}{2x-4}\right) = \frac{4\left(\dfrac{5}{2x-4}\right)+5}{2\left(\dfrac{5}{2x-4}\right)} = \frac{\dfrac{20+5(2x-4)}{2x-4}}{\dfrac{10}{2x-4}}$$

$$= \frac{20+5(2x-4)}{10} = \frac{20+10x-20}{10} = \frac{10x}{10} = x$$

and for every x in the domain of g (all $x \neq 0$),

$$f(g(x)) = f\left(\frac{4x+5}{2x}\right) = \frac{5}{2\left(\dfrac{4x+5}{2x}\right)-4} = \frac{5}{\dfrac{4x+5}{x}-4}$$

$$= \frac{5}{\dfrac{4x+5-4x}{x}} = \frac{5}{\dfrac{5}{x}} = x.$$

By the Round-Trip Theorem, f is a one-to-one function with inverse g. ■

Graphs of Inverse Functions

Finding the rule of the inverse function g of a one-to-one function f by solving the equation $y = f(x)$ for x, as in the preceding examples, is not always possible (some equations are hard to solve). But even if you don't know the rule of g, you can always find its graph, as shown below.

Suppose f is a one-to-one function and g is its inverse function. If (a, b) is on the graph of f, then by definition $f(a) = b$. Since the inverse function g takes each output of f back to its corresponding input, we know that $g(b) = a$. Hence, (b, a) is on the graph of g. A similar argument works in the other direction and leads to this conclusion:

(*) **(a, b) is on the graph of f exactly when (b, a) is on the graph of the inverse function g.**

This fact makes it very easy to graph an inverse function by using parametric graphing mode.

Example 7 The function $f(x) = .7x^5 + .3x^4 - .2x^3 + 2x + .5$ can be graphed in parametric mode by letting $x = t$ and $y = f(t) = .7t^5 + .3t^4 - .2t^3 + 2t + .5$. Its complete graph in Figure 3–42 shows that f has an inverse function g (why?), but it would be difficult to find its rule algebraically. According to statement (*) above, the graph of g can be obtained by taking each point on the graph of f and reversing its coordinates. In other words, g can be graphed parametrically by letting

$$x = f(t) = .7t^5 + .3t^4 - .2t^3 + 2t + 5 \quad \text{and} \quad y = t.$$

Figure 3–43 shows the graphs of g and f on the same screen. ■

Technology Tip

Inverse functions can be graphed directly on some calculators. When the function has been entered in the equation memory as y_1, use DRAWINV y_1 or DRINV y_1 in the TI-83 and Sharp 9600 DRAW menu or in the DRAW submenu of the TI-86/89 GRAPH menu.

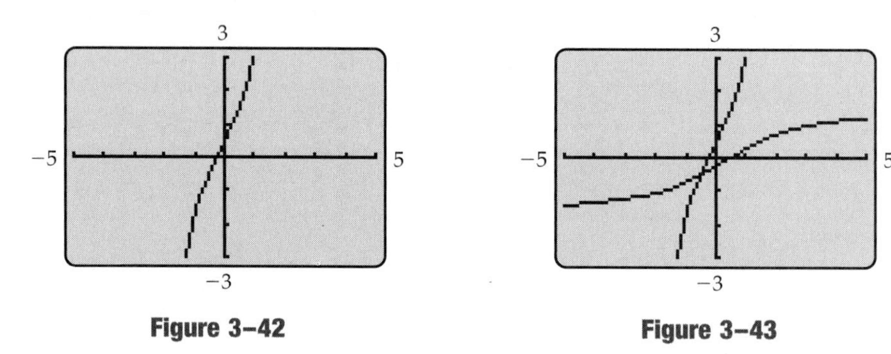

Figure 3–42 **Figure 3–43**

Suppose f is a function with inverse function g. Then we have seen that

(a, b) is on the graph of f exactly when (b, a) is on the graph of g.

Exercise 51 shows that the line $y = x$ is the perpendicular bisector of the line through points (a, b) and (b, a). Thus, (a, b) and (b, a) lie on opposite sides of $y = x$, the same distance from it.* If you think of the line $y = x$ as a mirror, the points (a, b) and (b, a) are mirror images of each other. Consequently, the graph of the inverse function g is the mirror image of the graph of f, with the line $y = x$ being the mirror.

GRAPHING EXPLORATION

Illustrate this fact by graphing the line $y = x$, the function $f(x) = x^3 + 5$ of Example 3, and its inverse $g(x) = \sqrt[3]{x - 5}$ on the same screen (use a square viewing window so that the mirror effect won't be distorted).

In formal terms:

Inverse Function Graphs

If g is the inverse function of f, then the graph of g is the reflection of the graph of f in the line $y = x$.

*In technical terms, (a, b) and (b, a) are said to be **symmetric** with respect to the line $y = x$.

NOTE In many texts the inverse function of a function f is denoted f^{-1}. In this notation, for instance, the inverse of the function $f(x) = x^3 + 5$ in Example 3 would be written as $f^{-1}(x) = \sqrt[3]{x - 5}$. Similarly, the reversal properties of inverse functions become

$$f^{-1}(f(x)) = x \text{ for every } x \text{ in the domain of } f\text{; and}$$
$$f(f^{-1}(x)) = x \text{ for every } x \text{ in the domain of } f^{-1}.$$

In this context, f^{-1} does *not* mean $1/f$ (see Exercise 47).

Exercises 3.7

In Exercises 1–8, use a calculator and the Horizontal Line Test to determine whether or not the function f is one-to-one.

1. $f(x) = x^4 - 4x^2 + 3$
2. $f(x) = x^4 - 4x + 3$
3. $f(x) = x^3 + x - 5$
4. $f(x) = \begin{cases} x - 3 & \text{for } x \le 3 \\ 2x - 6 & \text{for } x > 3 \end{cases}$
5. $f(x) = x^5 + 2x^4 - x^2 + 4x - 5$
6. $f(x) = x^3 - 4x^2 + x - 10$
7. $f(x) = .1x^3 - .1x^2 - .005x + 1$
8. $f(x) = .1x^3 + .005x + 1$

In Exercises 9–22, use algebra to find the inverse of the given one-to-one function.

9. $f(x) = -x$
10. $f(x) = -x + 1$
11. $f(x) = 5x - 4$
12. $f(x) = -3x + 5$
13. $f(x) = 5 - 2x^3$
14. $f(x) = (x^5 + 1)^3$
15. $f(x) = \sqrt{4x - 7}$
16. $f(x) = 5 + \sqrt{3x - 2}$
17. $f(x) = 1/x$
18. $f(x) = 1/\sqrt{x}$
19. $f(x) = \dfrac{1}{2x + 1}$
20. $f(x) = \dfrac{x}{x + 1}$
21. $f(x) = \dfrac{x^3 - 1}{x^3 + 5}$
22. $f(x) = \sqrt[5]{\dfrac{3x - 1}{x - 2}}$

In Exercises 23–28, use the Round-Trip Theorem on page 191 to show that g is the inverse of f.

23. $f(x) = x + 1, \quad g(x) = x - 1$
24. $f(x) = 2x - 6, \quad g(x) = \dfrac{x}{2} + 3$
25. $f(x) = \dfrac{1}{x + 1}, \quad g(x) = \dfrac{1 - x}{x}$
26. $f(x) = \dfrac{-3}{2x + 5}, \quad g(x) = \dfrac{-3 - 5x}{2x}$
27. $f(x) = x^5, \quad g(x) = \sqrt[5]{x}$
28. $f(x) = x^3 - 1, \quad g(x) = \sqrt[3]{x + 1}$
29. Show that the inverse function of the function f whose rule is $f(x) = \dfrac{2x + 1}{3x - 2}$ is f itself.
30. List three different functions (other than the one in Exercise 29), each of which is its own inverse. [Many correct answers are possible.]

In Exercises 31 and 32, the graph of a function f is given. Sketch the graph of the inverse function of f. [Reflect carefully.]

31.

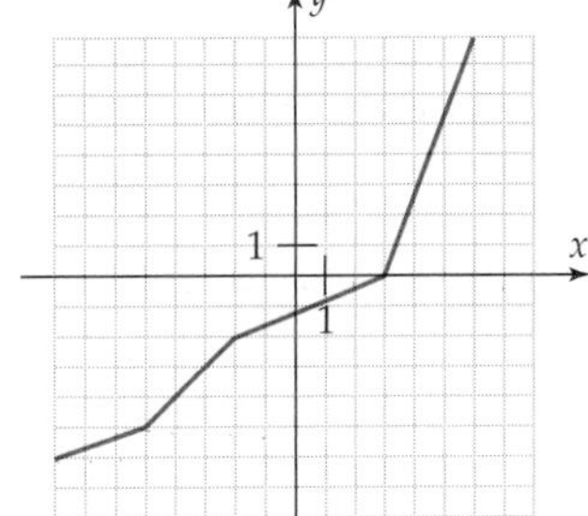

32.
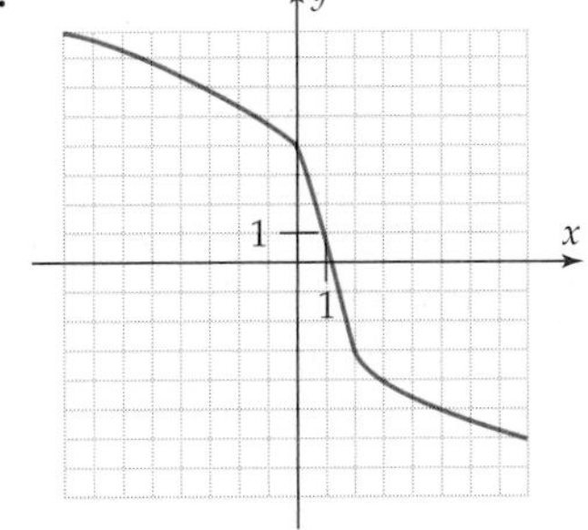

In Exercises 33–38, each given function has an inverse function. Sketch the graph of the inverse function.

33. $f(x) = \sqrt{x + 3}$ **34.** $f(x) = \sqrt{3x - 2}$

35. $f(x) = .3x^5 + 2$ **36.** $f(x) = \sqrt[3]{x + 3}$

37. $f(x) = \sqrt[5]{x^3 + x - 2}$

38. $f(x) = \begin{cases} x^2 - 1 & \text{for } x \le 0 \\ -.5x - 1 & \text{for } x > 0 \end{cases}$

In Exercises 39–46, none of the functions has an inverse. State at least one way of restricting the domain of the function (that is, find a function with the same rule and a smaller domain) so that the restricted function has an inverse. Then find the rule of the inverse function.

Example: *$f(x) = x^2$ has no inverse. But the function h with domain all $x \ge 0$ and rule $h(x) = x^2$ is increasing (its graph is the right half of the graph of f—see Figure 2–2 on page 60)—and therefore has an inverse.*

39. $f(x) = |x|$ **40.** $f(x) = |x - 3|$

41. $f(x) = -x^2$ **42.** $f(x) = x^2 + 4$

43. $f(x) = \dfrac{x^2 + 6}{2}$ **44.** $f(x) = \sqrt{4 - x^2}$

45. $f(x) = \dfrac{1}{x^2 + 1}$ **46.** $f(x) = 3(x + 5)^2 + 2$

47. (a) Using the f^{-1} notation for inverse functions, find $f^{-1}(x)$ when $f(x) = 3x + 2$.
(b) Find $f^{-1}(1)$ and $1/f(1)$. Conclude that f^{-1} is not the same function as $1/f$.

48. Let C be the temperature in degrees Celsius. Then the temperature in degrees Fahrenheit is given by $f(C) = \frac{9}{5}C + 32$. Let g be the function that converts degrees Fahrenheit to degrees Celsius. Show that g is the inverse function of f and find the rule of g.

Thinkers

49. Let m and b be constants with $m \neq 0$. Show that the function $f(x) = mx + b$ has an inverse function g and find the rule of g.

50. Prove that the function $f(x) = 1 - .2x^3$ of Example 1(c) is one-to-one by showing that it satisfies the definition:

If $a \neq b$, then $f(a) \neq f(b)$.

[*Hint:* Use the rule of f to show that when $f(a) = f(b)$, then $a = b$. If this is the case, then it is impossible to have $f(a) = f(b)$ when $a \neq b$.]

51. Show that the points $P = (a, b)$ and $Q = (b, a)$ are symmetric with respect to the line $y = x$ as follows.
(a) Find the slope of the line through P and Q.
(b) Use slopes to show that the line through P and Q is perpendicular to $y = x$.
(c) Let R be the point where the line $y = x$ intersects line segment PQ. Since R is on $y = x$, it has coordinates (c, c) for some number c, as shown in the figure. Use the distance formula to show that segment PR has the same length as segment RQ. Conclude that the line $y = x$ is the perpendicular bisector of segment PQ. Therefore, P and Q are symmetric with respect to the line $y = x$.

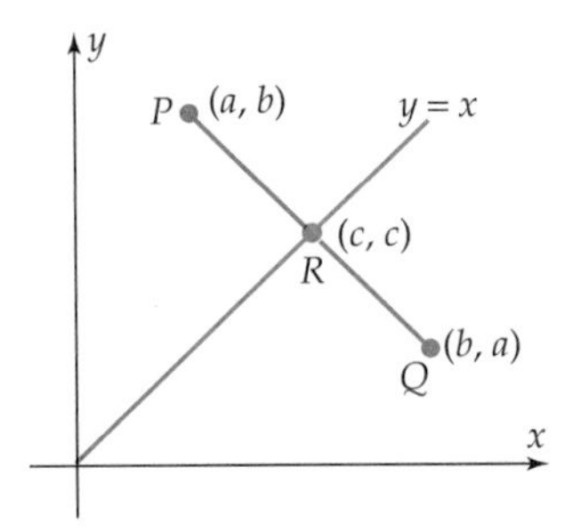

52. (a) Experiment with your calculator or use some of the preceding exercises to find four different increasing functions. For each function, sketch the graph of the function and the graphs of its inverse on the same set of axes.
(b) Based on the evidence in part (a), do you think the following statement true or false: The inverse function of every increasing function is also an increasing function.
(c) Do parts (a) and (b) with "increasing" replaced by "decreasing."

53. Prove the Round-Trip Theorem (page 191) as follows. By hypothesis, f and g have these properties:
(1) $g(f(x)) = x$ for every number x in the domain of f;
(2) $f(g(y)) = y$ for every number y in the domain of g.
(a) Prove that f is one-to-one by showing that

if $a \neq b$, then $f(a) \neq f(b)$.

[*Hint:* If $f(a) = f(b)$, apply g to both sides and use **(1)** to show that $a = b$. Consequently, if $a \neq b$, it is impossible to have $f(a) = f(b)$.]
(b) If $g(y) = x$, show that $f(x) = y$. [*Hint:* Use **(2)**.]
(c) If $f(x) = y$, show that $g(y) = x$. [*Hint:* Use **(1)**.]
Parts (b) and (c) prove that

$g(y) = x$ exactly when $f(x) = y$.

Hence, g is the inverse function of f (see page 188).

Chapter 3 Review

Important Concepts

Section	Concept	Page
Section 3.1	Function	119
	Domain	119
	Range	119
	Rule	119
	Greatest integer function	120
	Functions defined by equations and graphs	120
Section 3.2	Functional notation	127–129
	Difference quotient	130
	Common mistakes with functional notation	130
	Domain convention	131
	Piecewise-defined functions	131
Section 3.3	Graphs of functions	136
	Step functions	137
	Graphs of piecewise-defined functions	138
	Vertical Line Test	139
	Local maximum and local minimum	140–141
	Increasing and decreasing functions	142
	Graph reading	143
	Parametric graphing	144
Section 3.4	Vertical and horizontal shifts	150–152
	Expansions and contractions	154
	Reflections in the x- and y-axis	155–156
Excursion 3.4.A	Symmetry with respect to the y-axis	159
	Symmetry with respect to the x-axis	160
	Symmetry with respect to the origin	161
	Coordinate and algebraic tests for symmetry	162
	Even and odd functions	162–163
Section 3.5	Sums, differences, products, and quotients of functions	165
	Composition of functions	166
Section 3.6	Average rate of change	176
	Secant lines and average rate of change	179
	Difference quotient	181
	Estimating instantaneous rates of change	181
Section 3.7	One-to-one functions	187
	Horizontal Line Test	187
	Inverse functions	188
	Round-Trip Theorem	191
	Graphs of inverse functions	192

Important Facts and Formulas

- The average rate of change of a function f as x changes from a to b is the number

$$\frac{f(b) - f(a)}{b - a}.$$

- The difference quotient of the function f is the quantity

$$\frac{f(x+h)-f(x)}{h}.$$

- The average rate of change of a function f as x changes from a to b is the slope of the secant line joining the points $(a, f(a))$ and $(b, f(b))$.

Review Questions

1. Let $[x]$ denote the greatest integer function and evaluate
   (a) $[-5/2] =$ ____. (b) $[1755] =$ ____.
   (c) $[18.7] + [-15.7] =$ ____. (d) $[-7] - [7] =$ ____.
2. If $f(x) = x + |x| + [x]$, then find $f(0)$, $f(-1)$, $f(1/2)$, and $f(-3/2)$.
3. Let f be the function given by the rule $f(x) = 7 - 2x$. Complete this table:

x	0	1	2	-4	t	k	$b-1$	$1-b$	$6-2u$
$f(x)$	7								

4. What is the domain of the function g given by
$$g(t) = \frac{\sqrt{t-2}}{t-3}?$$
5. In each case give a *specific* example of a function and numbers a, b to show that the given statement may be *false*.
   (a) $f(a+b) = f(a) + f(b)$ (b) $f(ab) = f(a)f(b)$
6. If $f(x) = |3 - x|\sqrt{x-3} + 7$, then $f(7) - f(4) =$ ____.
7. What is the domain of the function given by
$$g(r) = \sqrt{r-4} + \sqrt{r} - 2?$$
8. What is the domain of the function $f(x) = \sqrt{-x+2}$?
9. If $h(x) = x^2 - 3x$, then $h(t+2) =$ ____.
10. Which of the following statements about the greatest integer function $f(x) = [x]$ is true for *every* real number x?
    (a) $x - [x] = 0$ (b) $x - [x] \le 0$
    (c) $[x] + [-x] \le 0$ (d) $[-x] \ge [x]$
    (e) $3[x] = [3x]$
11. If $f(x) = 2x^3 + x + 1$, then $f(x/2) =$ ____.
12. If $g(x) = x^2 - 1$, then $g(x-1) - g(x+1) =$ ____.
13. The radius of an oil spill (in meters) is 50 times the square root of the time t (in hours).
    (a) Write the rule of a function f that gives the radius of the spill at time t.
    (b) Write the rule of a function g that gives the area of the spill at time t.
    (c) What are the radius and area of the spill after 9 hours?
    (d) When will the spill have an area of 100,000 square meters?
14. The cost of renting a limousine for 24 hours is given by
$$C(x) = \begin{cases} 150 & \text{if } 0 < x \le 25 \\ 1.75x + 150 & \text{if } x > 25 \end{cases},$$
    where x is the number of miles driven.
    (a) What is the cost if the limo is driven 20 miles? 30 miles?
    (b) If the cost is \$218.25, how many miles were driven?

15. Sketch the graph of the function f given by

$$f(x) = \begin{cases} x^2 & \text{if } x \leq 0 \\ x + 1 & \text{if } 0 < x < 4 \\ \sqrt{x} & \text{if } x \geq 4 \end{cases}$$

16. U.S. Express Mail rates in 1998 are shown in the following table. Sketch the graph of the function e, whose rule is $e(x)$ = cost of sending a package weighing x pounds by Express Mail.

Express Mail

Letter Rate—Post Office to Addressee Service

Up to 8 ounces	$10.75
Over 8 ounces to 2 pounds	15.00
Up to 3 pounds	17.25
Up to 4 pounds	19.40
Up to 5 pounds	21.55
Up to 6 pounds	25.40
Up to 7 pounds	26.45

17. Which of the following are graphs of functions of x?

(a)

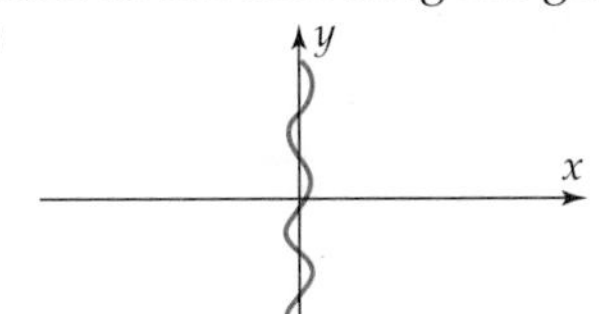

(b)

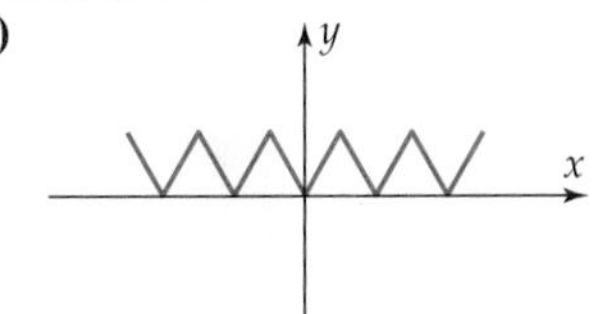

18. The function whose graph is shown gives the amount of money (in millions of dollars) spent on tickets for major concerts in selected years. (*Source:* Pollstar)

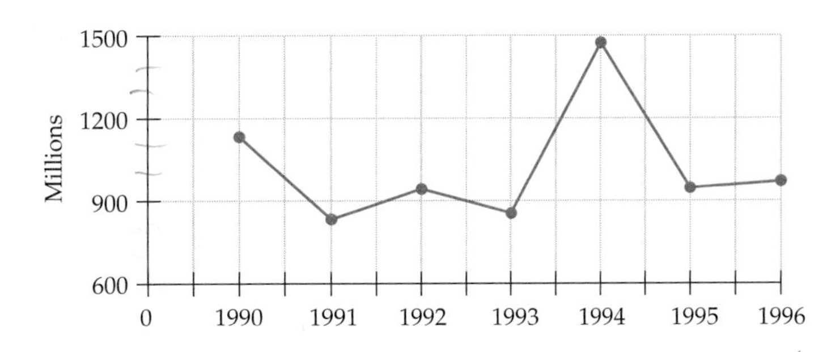

(a) What is the domain of the function?
(b) What is the approximate range of the function?
(c) Over what one-year interval is the rate of change the largest?

In Questions 19–22, determine the local maxima and minima of the function, the intervals on which the function is increasing, and the intervals on which it is decreasing.

19. $g(x) = \sqrt{x^2 + x + 1}$

20. $f(x) = 2x^3 - 5x^2 + 4x - 3$

21. $g(x) = x^3 + 8x^2 + 4x - 3$

22. $f(x) = .5x^4 + 2x^3 - 6x^2 - 16x + 2$

In Questions 23 and 24, sketch the graph of the curve given by the parametric equations.

23. $x = t^2 - 4$ and $y = 2t + 1 \quad (-3 \leq t \leq 3)$

24. $x = t^3 + 3t^2 - 1$ and $y = t^2 + 1 \quad (-3 \leq t \leq 2)$

25. Sketch a graph that is symmetric with respect to both the x-axis and the y-axis. (There are many correct answers and your graph need not be the graph of an equation.)

26. Sketch the graph of a function that is symmetric with respect to the origin. (There are many correct answers and you don't have to state the rule of your function.)

In Questions 27 and 28, determine algebraically whether the graph of the given equation is symmetric with respect to the x-axis, and y-axis, or the origin.

27. $x^2 = y^2 + 2$

28. $5y = 7x^2 - 2x$

In Questions 29–31, determine whether the given function is even, odd, or neither.

29. $g(x) = 9 - x^2$

30. $f(x) = |x|x + 1$

31. $h(x) = 3x^5 - x(x^4 - x^2)$

32. **(a)** Draw some coordinate axes and plot the points $(-2, 1)$, $(-1, 3)$, $(0, 1)$, $(3, 2)$, $(4, 1)$.

(b) Suppose the points plotted in part (a) lie on the graph of an *even* function f. Plot these points: $(2, f(2))$, $(1, f(1))$, $(0, f(0))$, $(-3, f(-3))$, $(-4, f(-4))$.

33. Determine whether the circle with equation $x^2 + y^2 + 6y = -5$ is symmetric with respect to the x-axis, the y-axis, or the origin.

34. Sketch the graph of a function f that satisfies all of these conditions:

(i) domain of $f = [-3, 4]$ **(ii)** range of $f = [-2, 5]$

(iii) $f(-2) = 0$ **(iv)** $f(1) > 2$

[*Note:* There are many possible correct answers and the function whose graph you sketch need *not* have an algebraic rule.]

35. Sketch the graph of $g(x) = 5 + \dfrac{4}{x - 5}$.

Use the graph of the function f in the figure to answer Questions 36–39.

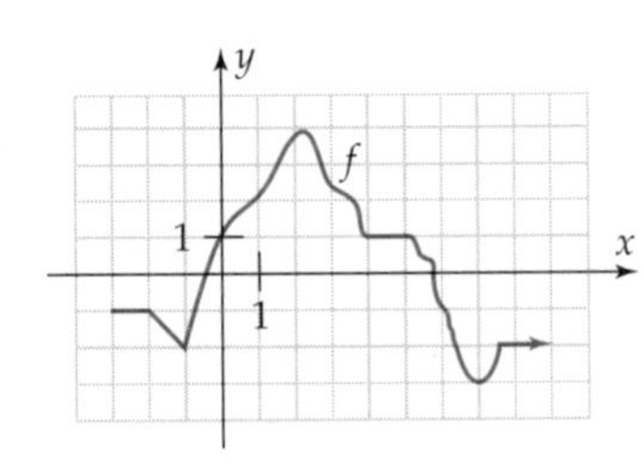

36. What is the domain of f?

37. What is the range of f?

38. Find all numbers x such that $f(x) = 1$.

39. Find a number x such that $f(x + 1) < f(x)$. (Many correct answers are possible.)

Use the graph of the function f in the figure to answer Questions 40–46.

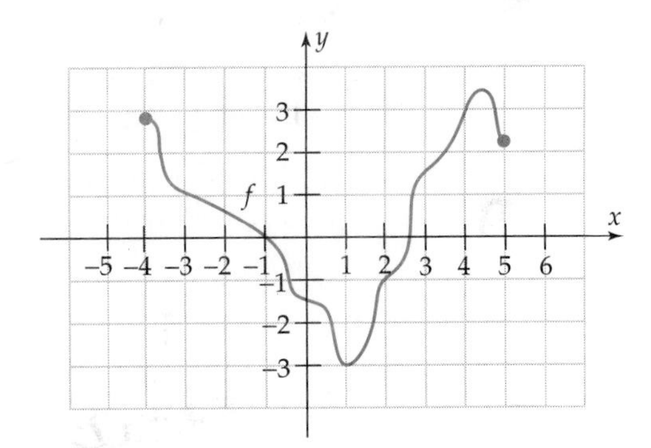

40. What is the domain of f?

41. $f(-3) =$ ____.

42. $f(2 + 2) =$ ____.

43. $f(-1) + f(1) =$ ____.

44. True or false: $2f(2) = f(4)$.

45. True or false: $3f(2) = -f(4)$.

46. True or false: $f(x) = 3$ for exactly one number x.

Use the graphs of the functions f and g in the figure to answer Questions 47–52.

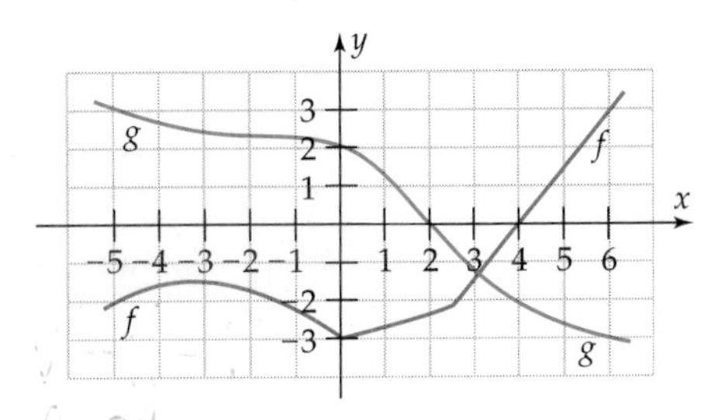

47. For which values of x is $f(x) = 0$?

48. True or false: If a and b are numbers such that $-5 \leq a < b \leq 6$, then $g(a) < g(b)$.

49. For which values of x is $g(x) \geq f(x)$?

50. Find $f(0) - g(0)$.

51. For which values of x is $f(x + 1) < 0$?

52. What is the distance from the point $(-5, g(-5))$ to the point $(6, g(6))$?

53. Fireball Bob and King Richard are two NASCAR racers. The graph shows their distance traveled in a recent race as a function of time.

(a) Which car made the most pit stops?

(b) Which car started out the fastest?

(c) Which car won the race?

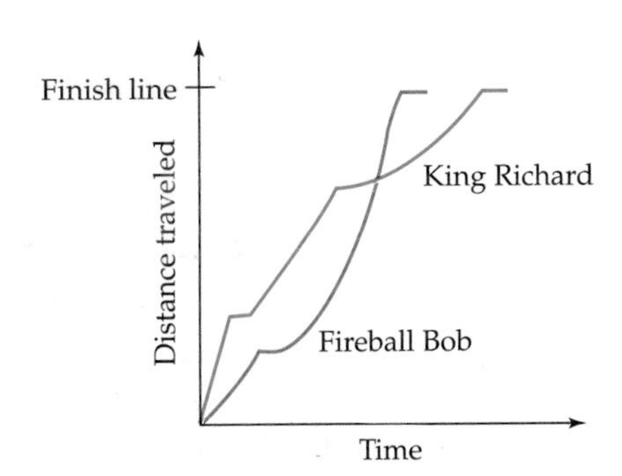

In Questions 54–57, list the transformations, in the order they should be performed on the graph of $g(x) = x^2$, so as to produce a complete graph of the function f.

54. $f(x) = (x - 2)^2$

55. $f(x) = .25x^2 + 2$

56. $f(x) = -(x + 4)^2 - 5$

57. $f(x) = -3(x - 7)^2 + 2$

58. The graph of a function f is shown in the figure. On the same coordinate plane, carefully draw the graphs of the functions g and h whose rules are:

$$g(x) = -f(x) \qquad \text{and} \qquad h(x) = 1 - f(x)$$

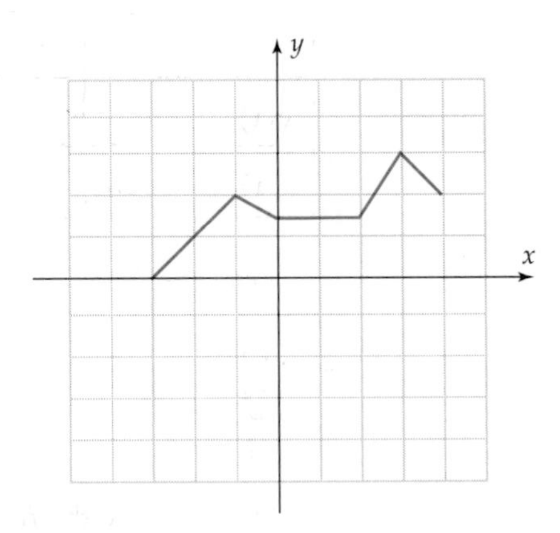

59. The figure shows the graph of a function f. If g is the function given by $g(x) = f(x + 2)$, then which of these statements about the graph of g is true?

(a) It does not cross the x-axis.
(b) It does not cross the y-axis.
(c) It crosses the y-axis at $y = 4$.
(d) It crosses the y-axis at the origin.
(e) It crosses the x-axis at $x = -3$.

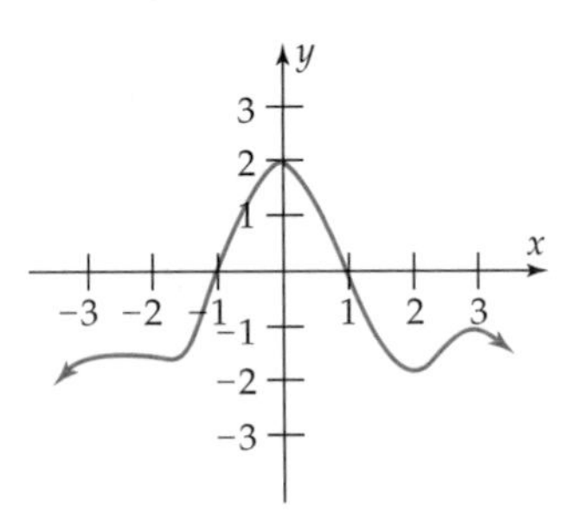

60. If $f(x) = 3x + 2$ and $g(x) = x^3 + 1$, find:

(a) $(f + g)(-1)$ **(b)** $(f - g)(2)$ **(c)** $(fg)(0)$

61. If $f(x) = \dfrac{1}{x - 1}$ and $g(x) = \sqrt{x^2 + 5}$, find:

(a) $(f/g)(2)$ **(b)** $(g/f)(x)$
(c) $(fg)(c + 1)$ $(c \neq 1)$

62. Find two functions f and g such that neither is the identity function and

$$(f \circ g)(x) = (2x + 1)^2$$

63. Use the graph of the function g in the figure to fill in the following table, in which h is the composite function $g \circ g$.

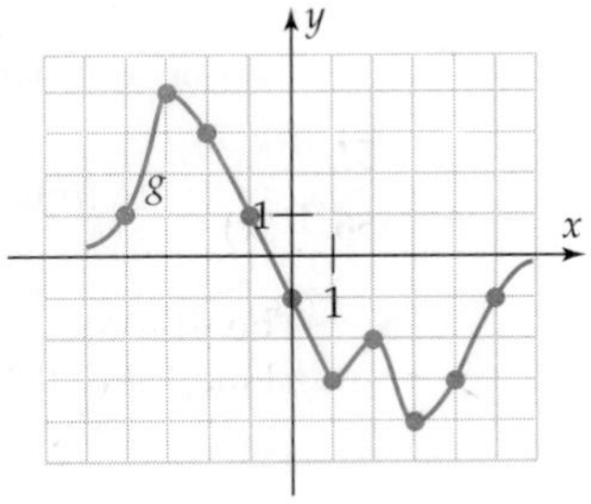

x	-4	-3	-2	-1	0	1	2	3	4
$g(x)$					-1				
$h(x) = g(g(x))$									

Questions 64–69 refer to the functions $f(x) = \dfrac{1}{x+1}$ *and* $g(t) = t^3 + 3$.

64. $(f \circ g)(1) =$ _____.

65. $(g \circ f)(2) =$ _____.

66. $g(f(-2)) =$ _____.

67. $(g \circ f)(x - 1) =$ _____.

68. $g(2 + f(0)) =$ _____.

69. $f(g(1) - 1) =$ _____.

70. Let f and g be the functions given by

$$f(x) = 4x + x^4 \quad \text{and} \quad g(x) = \sqrt{x^2 + 1}$$

(a) $(f \circ g)(x) =$ _____. **(b)** $(g - f)(x) =$ _____.

71. If $f(x) = \dfrac{1}{x}$ and $g(x) = x^2 - 1$, then

$$(f \circ g)(x) = \_\_\_\_\_ . \quad \text{and} \quad (g \circ f)(x) = \_\_\_\_\_ .$$

72. Let $f(x) = x^2$. Give an example of a function g with domain all real numbers such that $g \circ f \neq f \circ g$.

73. If $f(x) = \dfrac{1}{1 - x}$ and $g(x) = \sqrt{x}$, then find the domain of the composite function $f \circ g$.

74. These tables show the values of the functions f and g at certain numbers:

x	-1	0	1	2	3
$f(x)$	1	0	1	3	5

and

t	0	1	2	3	4
$g(t)$	-1	0	1	2	5

Which of the following statements are *true*?

(a) $(g - f)(1) = 1$

(b) $(f \circ g)(2) = (f - g)(0)$

(c) $f(1) + f(2) = f(3)$

(d) $(g \circ f)(2) = 1$

(e) None of the above is true.

75. Find the average rate of change of the function $g(x) = \dfrac{x^3 - x + 1}{x + 2}$ as x changes from

(a) -1 to 1 **(b)** 0 to 2

76. Find the average rate of change of the function $f(x) = \sqrt{x^2 - x + 1}$ as x changes from

(a) -3 to 0 **(b)** -3 to 3.5 **(c)** -3 to 5

77. If $f(x) = 2x + 1$ and $g(x) = 3x - 2$, find the average rate of change of the composite function $f \circ g$ as x changes from 3 to 5.

78. If $f(x) = x^2 + 1$ and $g(x) = x - 2$, find the average rate of change of the composite function $f \circ g$ as x changes from -1 to 1.

In Questions 79–82, find the difference quotient of the function.

79. $f(x) = 3x + 4$

80. $g(x) = \sqrt{x}$

81. $g(x) = x^2 - 1$

82. $f(x) = x^2 + x$

83. The profit (in hundreds of dollars) from selling x tons of Wonderchem is given by $P(x) = .2x^2 + .5x - 1$. What is the average rate of change of profit when the number of tons of Wonderchem sold increases from

(a) 4 to 8 tons? **(b)** 4 to 5 tons? **(c)** 4 to 4.1 tons?

84. On the planet Mars, the distance traveled by a falling rock (ignoring atmospheric resistance) in t seconds is $6.1t^2$ feet. How far must a rock fall in order to have an average speed of 25 feet per second over that time interval?

85. The graph in the figure shows the population of fruit flies during a 50-day experiment in a controlled atmosphere.

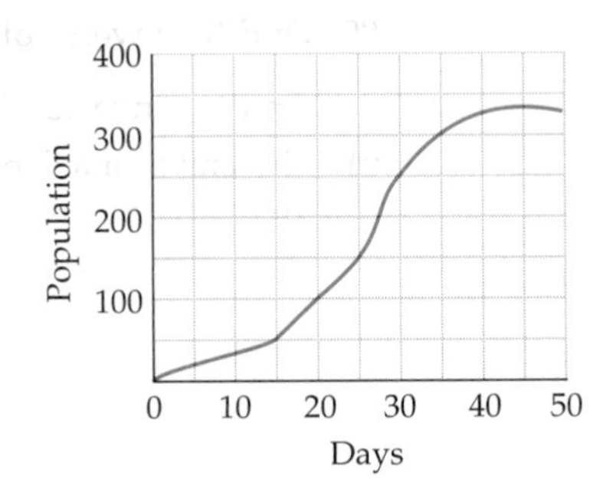

(a) During what 5-day period is the average rate of population growth the slowest?

(b) During what 10-day period is the average rate of population growth the fastest?

(c) Find an interval beginning at the 30th day during which the average rate of population growth is the same as the average rate from day 10 to day 20.

86. The graph of the function g in the figure consists of straight line segments. Find an interval over which the average rate of change of g is

(a) 0 **(b)** -3 **(c)** .5

(d) Explain why the average rate of change of g is the same from -3 to -1 as it is from -2.5 to 0.

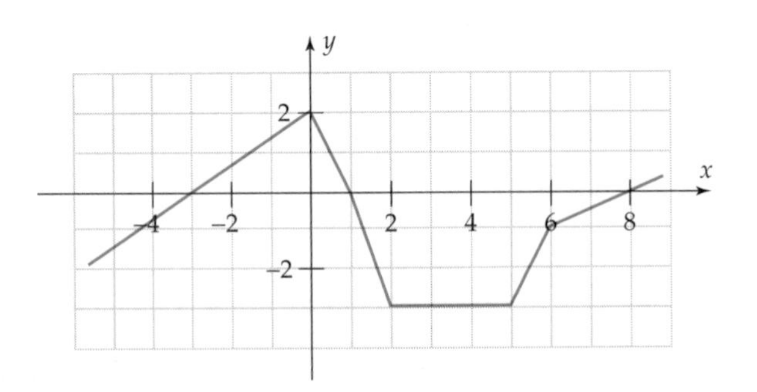

87. Use the table to answer the questions below.

Teen Births (age 15–19)		
	1991 (%)	1996 (%)
White	4.3	3.9
Black	11.6	9.2
Hispanic	10.7	10.2

(a) What has been the average rate of change (of percentage points per year) of teen births from 1991 to 1996 for Whites? for Blacks? for Hispanics?

(b) If these rates of change continue, in what year would there be no teen births for Whites? for Blacks? for Hispanics? Do these projections make sense? Why or why not?

88. Find the inverse of the function $f(x) = 2x + 1$.

89. Find the inverse of the function $f(x) = \sqrt{5 - x} + 7$.

90. Find the inverse of the function $f(x) = \sqrt[5]{x^3 + 1}$.

91. The graph of a function f is shown in the figure. Sketch the graph of the inverse function of f.

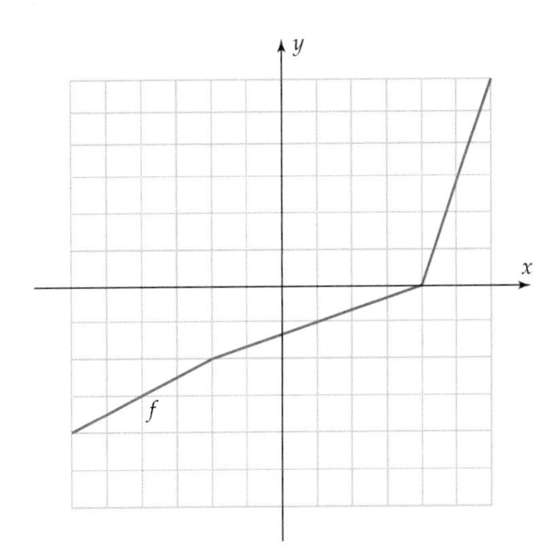

92. Which of the following functions have inverse functions (give reasons for your answers):

(a) $f(x) = x^3$ **(b)** $f(x) = 1 - x^2, \quad x \leq 0$

(c) $f(x) = |x|$

In Exercises 93–95, determine whether or not the given function has an inverse function (give reasons for your answer). If it does, find the graph of the inverse function.

93. $f(x) = 1/x$ **94.** $f(x) = .02x^3 - .04x^2 + .6x - 4$

95. $f(x) = .2x^3 - 4x^2 + 6x - 15$

Discovery Project 3

Building an Odometer

Simple mathematical functions often occur in everyday life in ways that are transparent to the casual observer. Consider the construction of a simple device for measuring the distance a bicycle has traveled. In this device, a pin is placed on one of the spokes of the bicycle wheel and a device that counts the number of times the pin passes by is fixed to the frame of the bicycle. The counter device can be either mechanical, incrementing when the pin strokes the counter, or electronic, sensing the passage of the pin with some sort of electromagnetic radiation.

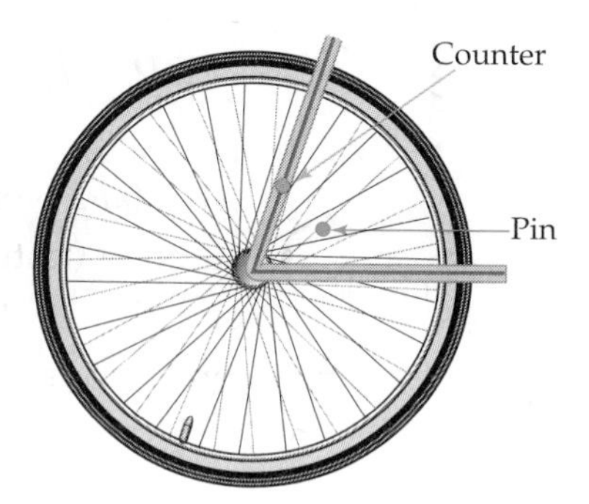

1. How many times will the pin pass by the counter over the course of 1 mile if the wheel has a diameter of 27 inches?
2. Write a function $d(x)$ where x is the number of times that the pin passes by the counter and $d(x)$ tells you the distance traveled in miles.
3. What is the domain of $d(x)$? What *kind* of numbers are there in the domain?
4. The electronic version of the device described can be turned into a speedometer. Write a function $s_1(x)$, where x is the number of times that the pin passes by the counter *each second* and $s_1(x)$ is the speed of the bicycle in miles per hour.
5. What is the domain of $s_1(x)$? Calculate $s_1(1)$ and $s_1(2)$. Why is this a problem?
6. A more efficient method of using the device as a speedometer is to measure the time interval between clicks of the counter. Write a function $s_2(x)$, where x is the time interval between clicks and $s_2(x)$ is the speed of the bicycle in miles per hour. Why is $s_2(x)$ better than $s_1(x)$?

Polynomial and Rational Functions

Can you afford to go to college?

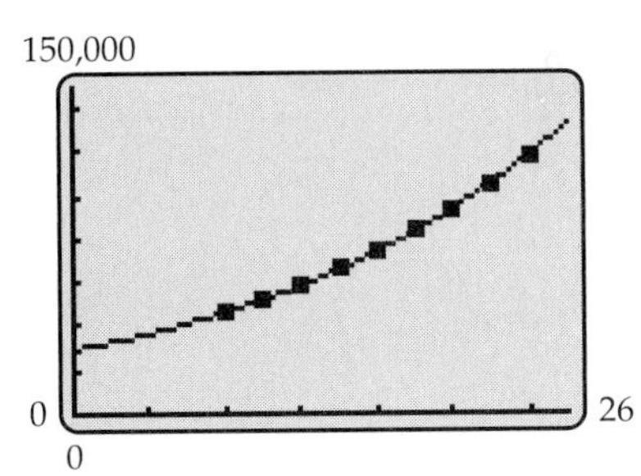

As you (and your parents) know, the cost of a four-year college education (tuition, fees, room, board, books) has steadily increased. It is expected to continue to do so in the foreseeable future, according to projections by a large insurance company. This growth can be modeled by a fourth-degree polynomial function. See Exercise 46 on page 299.

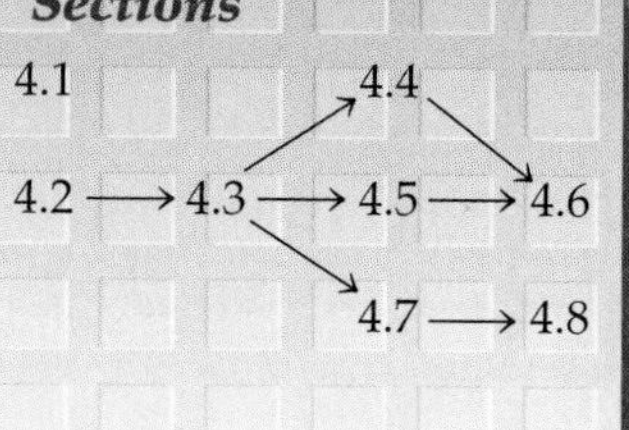

Chapter Outline

Polynomial functions arise naturally in many applications. Many complicated functions in applied mathematics can be approximated by polynomial functions or their quotients (rational functions).

4.1 Quadratic Functions*

A **quadratic function** is a function whose rule can be written in the form

$$f(x) = ax^2 + bx + c$$

for some constants a, b, c, with $a \neq 0$. The graph of a quadratic function is called a **parabola.**

GRAPHING EXPLORATION

Using the standard viewing window, graph the following quadratic functions on the same screen:

$$f(x) = x^2, \qquad f(x) = 3x^2 + 30x + 77, \qquad f(x) = -x^2 + 4x,$$
$$f(x) = -.2x^2 + 1.5x - 5$$

As the Exploration illustrates, all parabolas have the same basic "cup" shape, though the cup may be broad or narrow. The parabola opens upward when the coefficient of x^2 is positive and downward when this coefficient is negative.

*This section may be omitted or postponed if desired. Section 3.4 (Graphs and Transformations) is a prerequisite.

If a parabola opens upward, its **vertex** is the lowest point on the graph. For instance, (0, 0) is the vertex of $f(x) = x^2$ since every other point on the graph has a positive y-coordinate. If a parabola opens downward, its **vertex** is the highest point on the graph. Every parabola is symmetric with respect to the vertical line through its vertex; this line is called the **axis** of the parabola.

The vertex of a parabola can always be located approximately by graphing it and using trace or a minimum/maximum finder on a calculator. However, there are algebraic techniques for finding the vertex precisely.

Example 1 The function $g(x) = 2(x - 3)^2 + 1$ is quadratic because its rule can be written in the required form:

$$g(x) = 2(x - 3)^2 + 1 = 2(x^2 - 6x + 9) + 1 = 2x^2 - 12x + 19$$

Graphing the function in the standard viewing window (Figure 4–1) and using the trace feature shows that the coordinates of the vertex are approximately (3.0526, 1.0055).

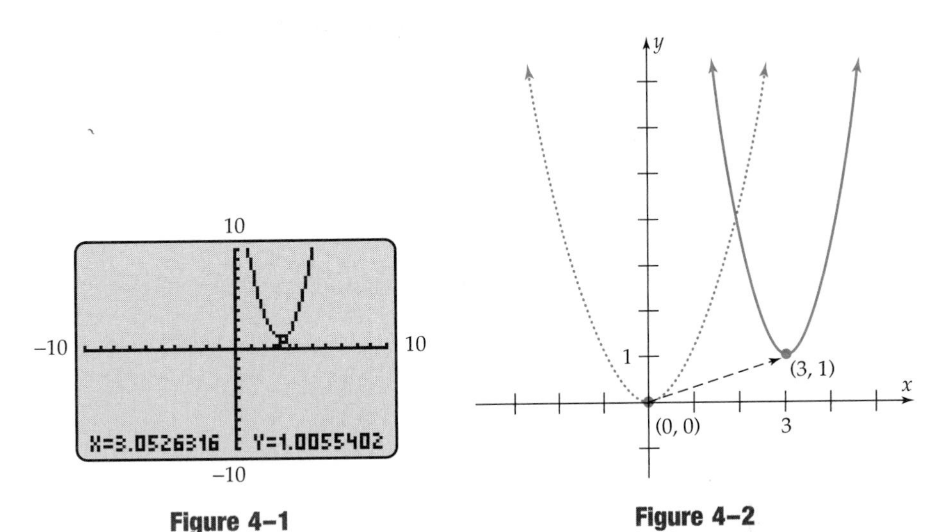

Figure 4–1 **Figure 4–2**

As we saw in Section 3.4, the graph of $g(x) = 2(x - 3)^2 + 1$ is the graph of $f(x) = x^2$ shifted horizontally 3 units to the right, stretched by a factor of 2, and shifted vertically 1 unit upward, as shown in Figure 4–2. When the vertex of f, namely (0, 0), is shifted 3 units right and 1 unit up, it moves to (3, 1). Thus, (3, 1) is the vertex of $g(x) = 2(x - 3)^2 + 1$. Note how the coordinates of the vertex of g are related to the constants in its rule:

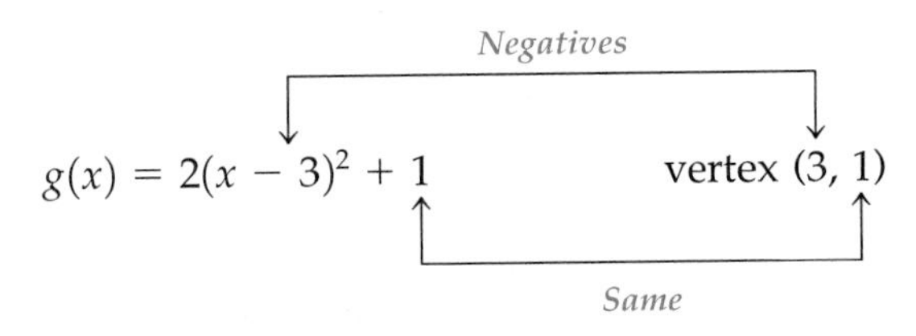

■

The vertex of the function g in Example 1 was easily determined because the rule of g had a special algebraic form. The vertex of the graph of any quadratic function can be determined in a similar fashion by first rewriting its rule.

Example 2 To find the vertex of the graph $g(x) = 3x^2 + 30x + 77$, we first rewrite its rule as $g(x) = 3(x^2 + 10x) + 77$. The next step is to complete the square in the expression in parentheses by adding 25 (the square of half the coefficient of x).* In order not to change the rule of the function, we must also *subtract* 25:

$$g(x) = 3(x^2 + 10x + 25 - 25) + 77.$$

Using the distributive law and factoring, we have:

$$g(x) = 3(x^2 + 10x + 25) - 3 \cdot 25 + 77$$

$$g(x) = 3(x + 5)^2 + 2.$$

As we saw in Section 3.4, the graph of $g(x) = 3(x + 5)^2 + 2$ is the graph of $f(x) = x^2$ shifted horizontally 5 units to the left, stretched by a factor of 3, then shifted vertically 2 units upward, as shown in Figure 4–3. In this process, the vertex (0, 0) of f moves to $(-5, 2)$, which is therefore the vertex of g. Once again, note how the coordinates of the vertex are related to the rule of the function:

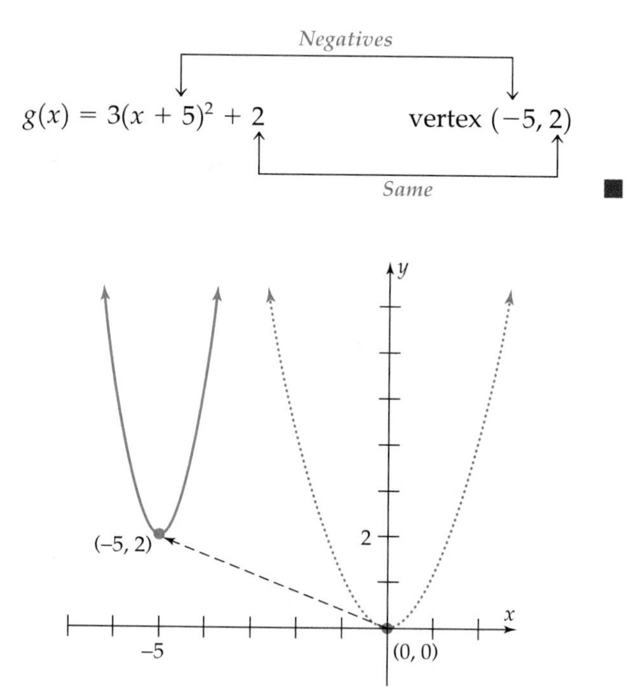

Figure 4–3

*Completing the square is discussed on page 20.

The technique in Example 2 works for any quadratic function $f(x) = ax^2 + bx + c$.* First, rewrite the rule as $f(x) = a\left(x^2 + \frac{b}{a}x\right) + c$. Then complete the square in the expression in parentheses by adding the square of half the coefficient of x, namely $\frac{b^2}{4a^2}$. In order not to change the rule of the function, subtract the same quantity:

$$f(x) = a\left(x^2 + \frac{b}{a}x + \frac{b^2}{4a^2} - \frac{b^2}{4a^2}\right) + c.$$

Then use the distributive law and factor:

$$f(x) = a\left(x^2 + \frac{b}{a}x + \frac{b^2}{4a^2}\right) - a\frac{b^2}{4a^2} + c$$
$$= a\left(x + \frac{b}{2a}\right)^2 + \left(c - \frac{b^2}{4a}\right).$$

As in the preceding examples, the graph of f is just the graph of x^2 shifted horizontally, stretched by a factor of a, and shifted vertically. As above, the vertex of this parabola can be read from the rule of the function:

Negatives

$$f(x) = a\left(x + \frac{b}{2a}\right)^2 + \left(c - \frac{b^2}{4a}\right) \qquad \text{vertex}\left(\frac{-b}{2a}, c - \frac{b^2}{4a}\right)$$

Same

Consequently, the preceding discussion can be summarized as follows:

Quadratic Functions

> **The graph of the quadratic function $f(x) = ax^2 + bx + c$ is a parabola that opens upward if $a > 0$ and downward if $a < 0$. The vertex of this parabola has x-coordinate $-b/2a$.**

It isn't necessary to memorize the y-coordinate of the vertex here because you can always compute $f(-b/2a)$ to find it.

Example 3 The graph of $f(x) = -4x^2 + 12x - 8$ is a downward-opening parabola because the coefficient of x^2 is negative. Its vertex has x-coordinate

$$-\frac{b}{2a} = -\frac{12}{2(-4)} = \frac{-12}{-8} = \frac{3}{2}$$

*The following argument is exactly the one used in Example 2, with a in place of 3, b in place of 30, c in place of 77, and $b^2/4a^2$ in place of 25.

and y-coordinate

$$f\left(\frac{3}{2}\right) = -4\left(\frac{3}{2}\right)^2 + 12\left(\frac{3}{2}\right) - 8 = 1. \quad \blacksquare$$

Applications

The solution of many applied problems depends on finding the vertex of a parabola.

Example 4 Find the area and dimensions of the largest rectangular field that can be enclosed with 3000 feet of fence.

Solution Let x denote the length and y the width of the field, as shown in Figure 4–4.

$$\text{Perimeter} = x + y + x + y$$
$$= 2x + 2y$$
$$\text{Area} = xy$$

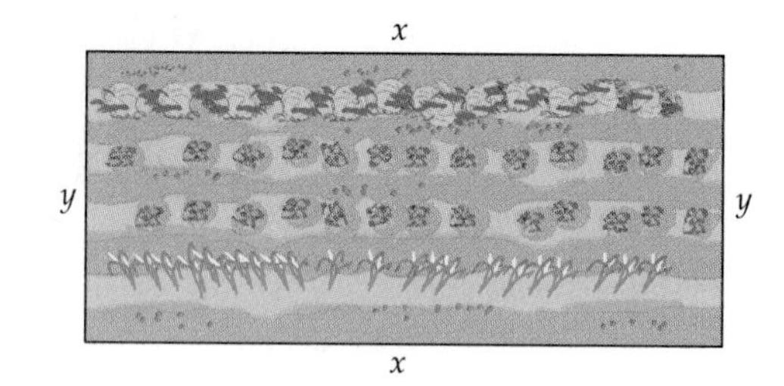

Figure 4–4

Since the perimeter is the length of the fence, $2x + 2y = 3000$. Hence, $2y = 3000 - 2x$ and $y = 1500 - x$. Consequently, the area is

$$A = xy = x(1500 - x) = 1500x - x^2 = -x^2 + 1500x.$$

The largest possible area is just the maximum value of the quadratic function $A(x) = -x^2 + 1500x$. This maximum occurs at the vertex of the graph of $A(x)$ (which is a downward-opening parabola because the coefficient of x^2 is negative). The vertex may be found by using the fact in the box on page 209 (with $a = -1$ and $b = 1500$):

$$\text{The } x\text{-coordinate of the vertex is } -\frac{1500}{2(-1)} = 750 \text{ feet.}$$

Hence, the y-coordinate of the vertex, the maximum value of $A(x)$, is

$$A(750) = -750^2 + 1500 \cdot 750 = 562{,}500 \text{ square feet.}$$

It occurs when the length is $x = 750$. In this case the width is $y = 1500 - x = 1500 - 750 = 750$. $\blacksquare$

Example 5 The owner of a 20-unit apartment complex has found that each \$50 increase in monthly rent results in a vacant apartment. All units are now rented at \$400 per month. How many \$50 increases in rent will produce the largest possible income for the owner?

Solution Let x represent the number of \$50 increases. Then the monthly rent will be $400 + 50x$ dollars. Since one apartment goes vacant for each increase, the number of occupied apartments will be $20 - x$. Then the owner's monthly income $R(x)$ is given by

$$R(x) = (\text{number of apartments rented}) \times (\text{rent per apartment})$$

$$R(x) = (20 - x)(400 + 50x)$$

$$R(x) = -50x^2 + 600x + 8000$$

Since there are only 20 apartments, there are two practical ways to find the maximum possible income.

Table Method: Make a table of values of $R(x)$ for $0 \leq x \leq 20$, as in Figure 4–5. It shows that the maximum income of \$9800 occurs when $x = 6$. In this case, there will be $20 - 6 = 14$ apartments rented at a monthly rent of $400 + 6(50) = \$700$.

X	Y1
0	8000
1	8550
2	9000
3	9350
4	9600
5	9750
6	9800

X=0

X	Y1
7	9750
8	9600
9	9350
10	9000
11	8550
12	8000
13	7350

X=7

X	Y1
14	6600
15	5750
16	4800
17	3750
18	2600
19	1350
20	0

X=14

Figure 4–5

Algebraic Method: The graph of $R(x) = -50x^2 + 600x + 8000$ is a downward-opening parabola (why?). Maximum income occurs at the vertex of this parabola, that is, when

$$x = \frac{-b}{2a} = \frac{-600}{2(-50)} = \frac{-600}{-100} = 6.$$

Therefore, 6 increases of \$50 will produce maximum income. ■

Exercises 4.1

In Exercises 1–16, without graphing, *determine the vertex of the given parabola and state whether it opens upward or downward.*

1. $f(x) = 3(x - 5)^2 + 2$

2. $g(x) = -6(x - 2)^2 - 5$

3. $y = -(x - 1)^2 + 2$

4. $h(x) = -x^2 + 1$

5. $f(x) = x^2 - 6x + 3$

6. $g(x) = x^2 + 8x - 1$

7. $h(x) = x^2 + 3x + 6$

8. $f(x) = x^2 - 5x - 7$

9. $y = 2x^2 + 12x - 3$

10. $y = 3x^2 + 6x + 1$

11. $f(x) = -x^2 + 8x - 2$

12. $g(x) = -x^2 - 6x + 4$

13. $f(x) = -3x^2 + 4x + 5$

14. $g(x) = 2x^2 - x - 1$

15. $y = -x^2 + x$

16. $y = -2x^2 + 2x - 1$

17. The graph of the quadratic function g is obtained from the graph of $f(x) = x^2$ by vertically stretching it by a factor of 2 and then shifting vertically 5 units downward. What is the rule of the function g? What is the vertex of its graph?

18. The graph of the quadratic function g is obtained from the graph of $f(x) = x^2$ by shifting it horizontally 4 units to the left, then vertically stretching it by a factor of 3, and then shifting vertically 2 units upward. What is the rule of the function g? What is the vertex of its graph?

19. If the graph of the quadratic function h is shifted vertically 4 units downward, then shrunk by a factor of $1/2$, and then shifted horizontally 3 units to the left, the resulting graph is the parabola $f(x) = x^2$. What is the rule of the function h? What is the vertex of its graph?

20. If the graph of the quadratic function h is shifted vertically 3 units upward, then reflected in the x-axis, and then shifted horizontally 5 units to the right, the resulting graph is the parabola $f(x) = x^2$. What is the rule of the function h? What is the vertex of its graph?

21. Find the rule of the quadratic function whose graph is the parabola with vertex at the origin that passes through $(2, 12)$.

22. Find the rule of the quadratic function whose graph is the parabola with vertex $(0, 1)$ that passes through $(2, -7)$.

23. Find the number b such that the vertex of the parabola $y = x^2 + bx + c$ lies on the y-axis.

24. Find the number c such that the vertex of the parabola $y = x^2 + 8x + c$ lies on the x-axis.

25. If the vertex of the parabola $f(x) = x^2 + bx + c$ is at $(2, 4)$, find b and c.

26. If the vertex of the parabola $f(x) = -x^2 + bx + 8$ has second coordinate 17 and is in the second quadrant, find b.

27. The braking distance (in meters) for a car with excellent brakes on a good road with an alert driver can be modeled by the quadratic function $B(s) = .01s^2 + .7s$, where s is the car's speed in kilometers per hour.

(a) What is the braking distance for a car traveling 30 kilometers per hour? For one traveling 100 kilometers per hour?

(b) If the car takes 60 meters to come to a complete stop, what was its speed?

28. The median sale price of existing single family homes in the United States from 1970 to 1996 can be approximated by the quadratic function $P(t) = -5.006t^2 + 3921.13t + 19{,}275.39$, where $t = 0$ corresponds to 1970.

(a) Assuming the trend continues, estimate the median sales price in the year 2000.

(b) Explain why this model needs a restricted domain.

(c) In what year was the median sale price \$100,000?

29. What is the minimum product of two numbers whose difference is 4? What are the numbers?

30. Find numbers c and d whose sum is -18 and whose product is as large as possible.

31. The sum of the height h and the base b of a triangle is 30. What height and base will produce a triangle of maximum area?

32. A trough is to be made by bending a long, flat piece of tin 10 inches wide into a rectangular shape. What depth should the trough be in order to have the maximum possible cross-sectional area?

33. A field bounded on one side by a river is to be fenced on three sides so as to form a rectangular enclosure. If the total length of fence to be used is 200 feet, what dimensions will yield an enclosure of the largest possible area?

34. A rectangular box (with top) has a square base. The sum of the lengths of its 12 edges is 8 feet. What dimensions should the box have so that its surface area is as large as possible?

35. At Middleton Place, a plantation near Charleston, South Carolina, there is a "joggling board" that was once used for courting. A young girl would sit at one end, her suitor at the other end, and her mother in the center. The mother would bounce on the board, thus causing the girl and her suitor to move closer together. A joggling board is 8 feet long and an average mother sitting at its center causes the board to deflect 2 inches, as shown in the figure. The shape of the deflected board is parabolic.

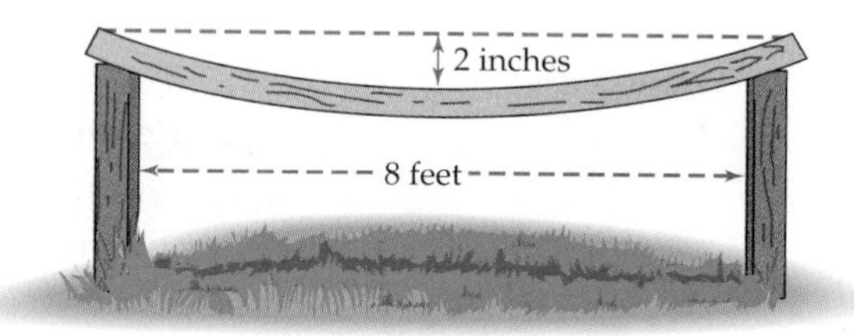

(a) Find the equation of the parabola, assuming that the joggling board lies on the x-axis with its center at the origin.

(b) How far from the center of the board is the deflection 1 inch?

36. A salesperson finds that her sales average 40 cases per store when she visits 20 stores a week. Each time she visits an additional store per week, the average sales per store decreases by 1 case. How many stores should she visit each week if she wants to maximize her sales?

37. A potter can sell 120 bowls per week at \$4 per bowl. For each 50¢ decrease in price 20 more bowls are sold. What price should be charged in order to maximize sales income?

38. A vendor can sell 200 souvenirs per day at a price of \$2 each. Each 10¢ price increase decreases the number of sales by 25 per day. Souvenirs cost the vendor \$1.50 each. What price should be charged in order to maximize the profit?

39. When a basketball team charges \$4 per ticket, average attendance is 500 people. For each 20¢ decrease in ticket price, average attendance increases by 30 people. What should the ticket price be to ensure maximum income?

40. A ballpark concessions manager finds that each salesperson sells an average of 40 boxes of popcorn per game when 20 salespeople are working. When an additional salesperson is employed, each salesperson averages 1 less box per game. How many salespeople should be hired to ensure maximum income?

In Exercises 41–44, use the formula for the height h of an object that is traveling vertically (subject only to gravity) at time t: $h = -16t^2 + v_0t + h_0$, *where* h_0 *is the initial height and* v_0 *is the initial velocity.*

41. A ball is thrown upward from the top of a 96-foot-high tower with an initial velocity of 80 feet per second. When does the ball reach its maximum height and how high is it at that time?

42. A rocket is fired upward from ground level with an initial velocity of 1600 feet per second. When does it attain its maximum height and what is that height?

43. A ball is thrown upward from a height of 6 feet with an initial velocity of 32 feet per second. Find its maximum height.

44. A bullet is fired upward from ground level with an initial velocity of 1500 feet per second. How high does it go?

45. A projectile is fired at an angle of 45° upward. Exactly t seconds after firing, its vertical height above the ground is $500t - 16t^2$. What is the greatest height the projectile reaches and at what times does this occur?

Thinker

46. The *discriminant* of a quadratic function $f(x) = ax^2 + bx + c$ is the number $b^2 - 4ac$. For each of the discriminants listed, state which graphs could possibly be the graph of f.
   (a) $b^2 - 4ac = 25$
   (b) $b^2 - 4ac = 0$
   (c) $b^2 - 4ac = -49$
   (d) $b^2 - 4ac = 72$

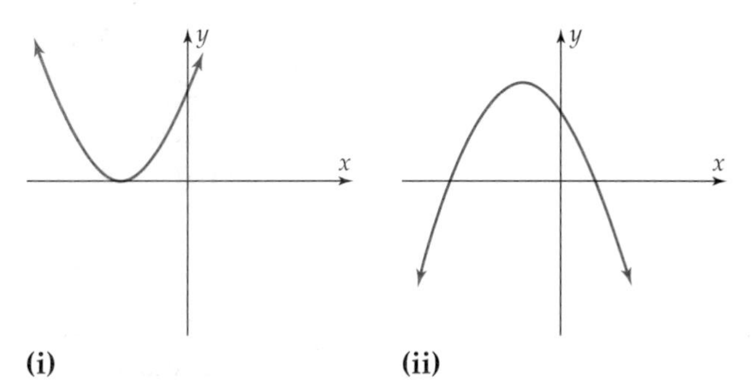

(i) **(ii)**

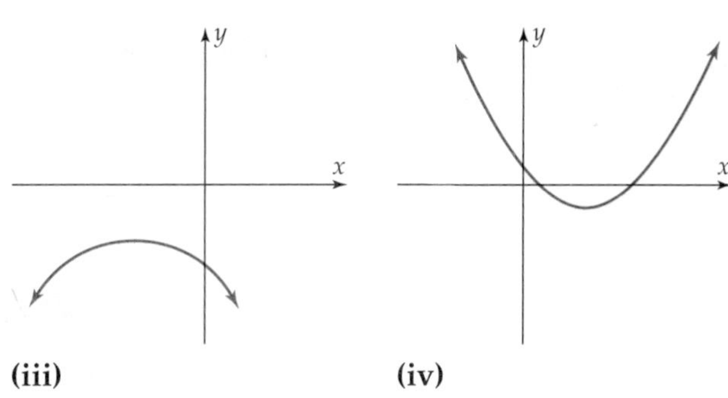

(iii) **(iv)**

4.2 Polynomial Functions

Informally, a **polynomial** is an algebraic expression such as

$$x^3 - 6x^2 + \tfrac{1}{2} \quad \text{or} \quad x^{15} + x^{10} + 7 \quad \text{or} \quad x - 6.7 \quad \text{or} \quad 12.$$

Formally, a **polynomial in** x is an algebraic expression that can be written in the form

$$a_nx^n + a_{n-1}x^{n-1} + \cdots + a_3x^3 + a_2x^2 + a_1x + a_0,$$

where n is a nonnegative integer, x is a variable, and each of $a_0, a_1, \ldots, a_n$ is a constant, called a **coefficient.*** The coefficient a_0 is called the **constant**

*Any letter may be used as the variable in a polynomial.

term. A polynomial that consists only of a constant term, such as 12, is called a **constant polynomial.** The **zero polynomial** is the constant polynomial 0.

The *exponent* of the highest power of x that appears with *nonzero* coefficient is the **degree** of the polynomial, and the nonzero coefficient of this highest power of x is the **leading coefficient.** For example,

Polynomial	Degree	Leading Coefficient	Constant Term
$6x^7 + 4x^3 + 5x^2 - 7x + 10$	7	6	10
x^3	3	1	0
12 (think of this as $12x^0$)	0	12	12
$0x^9 + 2x^6 + 3x^7 + x^8 - 2x - 4$	8	1	−4

The degree of the zero polynomial is *not defined* since no exponent of x occurs with nonzero coefficient.

A **polynomial function** is a function whose rule is given by a polynomial, such as $f(x) = x^5 + 3x^2 - 2$. First-degree polynomial functions, such as $g(x) = 3x - 4$, are called **linear functions,** and, as we saw in Section 4.1, second-degree polynomial functions are called **quadratic functions.**

Polynomial Arithmetic

You should be familiar with addition, subtraction, and multiplication of polynomials, which are presented in the Algebra Review Appendix. Long division of polynomials is quite similar to long division of numbers, as we now see.

Example 1 To divide $2x^5 + 5x^4 - 4x^3 + 8x^2 + 1$ by $2x^2 - x + 1$, we first write:

$$\textit{Divisor} \rightarrow \quad 2x^2 - x + 1 \overline{\big)\, 2x^5 + 5x^4 - 4x^3 + 8x^2 + 1} \quad \leftarrow \textit{Dividend}$$

We begin by dividing the first term of the divisor $(2x^2)$ into the first term of the dividend $(2x^5)$ and putting the result $\left(\text{namely, } \frac{2x^5}{2x^2} = x^3\right)$ on the top line, as shown below. Then multiply x^3 times the entire divisor, put the result on the third line, and subtract:

$$\begin{array}{rll} & x^3 & \leftarrow \textit{Partial quotient} \\ 2x^2 - x + 1 \big) & 2x^5 + 5x^4 - 4x^3 + 8x^2 + 1 & \\ & 2x^5 - x^4 + x^3 & \leftarrow x^3 \cdot (2x^2 - x + 1) \\ \hline & 6x^4 - 5x^3 + 8x^2 + 1 & \leftarrow \textit{Subtraction*} \end{array}$$

*If this subtraction is confusing, write it out horizontally and watch the signs carefully:
$(2x^5 + 5x^4 - 4x^3 + 8x^2 + 1) - (2x^5 - x^4 + x^3)$

$$= 2x^5 + 5x^4 - 4x^3 + 8x^2 + 1 - 2x^5 + x^4 - x^3$$
$$= 6x^4 - 5x^3 + 8x^2 + 1$$

Next, divide the first term of the divisor ($2x^2$) into $6x^4$ and put the result $\left(\frac{6x^4}{2x^2} = 3x^2\right)$ on the top line, as shown below. Then multiply $3x^2$ times the entire divisor, put the result on the fifth line, and subtract. Continuing this procedure, we obtain:

$$
\begin{array}{rrl}
 & x^3 + 3x^2 - x + 2 & \leftarrow \textit{Quotient} \\
2x^2 - x + 1 \,\big) & 2x^5 + 5x^4 - 4x^3 + 8x^2 \qquad + 1 & \\
 & 2x^5 - x^4 + x^3 \qquad\qquad\qquad & \leftarrow x^3 \cdot (2x^2 - x + 1) \\
 & 6x^4 - 5x^3 + 8x^2 \qquad + 1 & \leftarrow \textit{Subtraction} \\
 & 6x^4 - 3x^3 + 3x^2 \qquad\qquad & \leftarrow 3x^2 \cdot (2x^2 - x + 1) \\
 & -2x^3 + 5x^2 \qquad + 1 & \leftarrow \textit{Subtraction} \\
 & -2x^3 + x^2 - x \qquad & \leftarrow (-x)(2x^2 - x + 1) \\
 & 4x^2 + x + 1 & \leftarrow \textit{Subtraction} \\
 & 4x^2 - 2x + 2 & \leftarrow 2 \cdot (2x^2 - x + 1) \\
 & \textit{Remainder} \rightarrow 3x - 1 & \leftarrow \textit{Subtraction}
\end{array}
$$

The division process stops when the remainder is zero or has *smaller degree* than the divisor (here the divisor $2x^2 - x + 1$ has degree 2 and the remainder $3x - 1$ has degree 1). ■

When the divisor in polynomial division is a first-degree polynomial such as $x - 2$ or $x + 5$, there is a convenient shorthand method of doing the division called **synthetic division.** See Excursion 4.2.A for details.

Recall how you check a long division problem with numbers:

$$
\begin{array}{rr}
 & 145 \\
31 \,\big) & 4509 \\
 & 31 \\
 & 140 \\
 & 124 \\
 & 169 \\
 & 155 \\
 & 14
\end{array}
\qquad \textit{Check:} \qquad
\begin{array}{rl}
145 & \leftarrow \textit{Quotient} \\
\times\ 31 & \leftarrow \textit{Divisor} \\
145 & \\
435 & \\
4495 & \\
+\ 14 & \leftarrow \textit{Remainder} \\
4509 & \leftarrow \textit{Dividend}
\end{array}
$$

We can summarize this process in one line:

$$\text{Dividend} = \text{Divisor} \cdot \text{Quotient} + \text{Remainder}.$$

The same process can be used with polynomial division. In Example 1, you can easily verify that the divisor times the quotient is

$$(2x^2 - x + 1)(x^3 + 3x^2 - x + 2) = 2x^5 + 5x^4 - 4x^3 + 8x^2 - 3x + 2.$$

Adding the remainder $3x - 1$ to this result yields the original dividend:

$$(2x^5 + 5x^4 - 4x^3 + 8x^2 - 3x + 2) + (3x - 1) = 2x^5 + 5x^4 - 4x^3 + 8x^2 + 1.$$

So just as with division of numbers we have:

$$\text{Dividend} = \text{Divisor} \cdot \text{Quotient} + \text{Remainder}.$$

This fact is so important that it is given a special name and a formal statement:

The Division Algorithm

If a polynomial $f(x)$ is divided by a nonzero polynomial $h(x)$, then there is a quotient polynomial $q(x)$ and a remainder polynomial $r(x)$ such that

$$f(x) = h(x)q(x) + r(x)$$

where either $r(x) = 0$ or $r(x)$ has degree less than the degree of the divisor $h(x)$.

The Division Algorithm can be used to determine if one polynomial is a factor of another polynomial.

Technology Tip

The TI-89 does polynomial division (use PROPFRAC in the ALGEBRA menu). It displays the answer as the sum of a fraction and a polynomial:

$$\frac{\text{Remainder}}{\text{Divisor}} + \text{Quotient}.$$

Example 2 To determine if $2x^2 + 1$ is a factor of $6x^3 - 4x^2 + 3x - 2$, we divide:

$$\begin{array}{r|l}
 & 3x \quad - 2 \\ \hline
2x^2 + 1 & 6x^3 - 4x^2 + 3x - 2 \\
 & \underline{6x^3 \qquad\quad + 3x \quad\;\;} \\
 & \quad\; - 4x^2 \qquad\quad - 2 \\
 & \quad\; \underline{- 4x^2 \qquad\quad - 2} \\
 & \qquad\qquad\qquad\quad 0
\end{array}$$

Since the remainder is 0, the Division Algorithm tells us that:

$$\begin{aligned}
\text{Dividend} &= \text{Divisor} \cdot \text{Quotient} + \text{Remainder}. \\
6x^3 - 4x^2 + 3x - 2 &= (2x^2 + 1)(3x - 2) + 0 \\
&= (2x^2 + 1)(3x - 2).
\end{aligned}$$

Therefore, $2x^2 + 1$ is a factor of $6x^3 - 4x^2 + 3x - 2$, and the other factor is the quotient $3x - 2$. ■

The same argument works in the general case:

Remainders and Factors

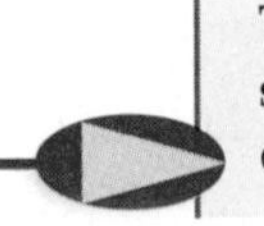

The remainder in polynomial division is 0 exactly when the divisor is a factor of the dividend. In this case the other factor is the quotient.

Remainders and Roots

When a polynomial $f(x)$ is divided by a first-degree polynomial, such as $x - 3$ or $x + 5$, the remainder is a constant (because constants are the only polynomials of degree less than 1, the degree of the divisor). For example, you can verify that when $f(x) = x^3 - 2x^2 - 4x + 5$ is divided by

$x - 3$, the quotient is $x^2 + x - 1$ and the remainder is 2. Hence, by the Division Algorithm

$$\text{Dividend} = \text{Divisor} \cdot \text{Quotient} + \text{Remainder}$$

$$f(x) = x^3 - 2x^2 - 4x + 5 = (x - 3)(x^2 + x - 1) + 2$$

Observe that

$$f(3) = (3 - 3)(3^2 + 3 - 1) + 2 = 0 + 2 = 2.$$

Thus, value of $f(x)$ at 3 (namely, $f(3) = 2$) is the same as the remainder when $f(x)$ is divided by $x - 3$. A similar argument works in the general case and proves:

Remainder Theorem

If a polynomial $f(x)$ is divided by $x - c$, then the remainder is the number $f(c)$.

Example 3 To find the remainder when $f(x) = x^{79} + 3x^{24} + 5$ is divided by $x - 1$, we apply the Remainder Theorem with $c = 1$. The remainder is

$$f(1) = 1^{79} + 3 \cdot 1^{24} + 5 = 1 + 3 + 5 = 9. \quad ■$$

Example 4 To find the remainder when $f(x) = 3x^4 - 8x^2 + 11x + 1$ is divided by $x + 2$, we must apply the Remainder Theorem *carefully*. The divisor in the theorem is $x - c$, not $x + c$. So we rewrite $x + 2$ as $x - (-2)$ and apply the theorem with $c = -2$. The remainder is

$$f(-2) = 3(-2)^4 - 8(-2)^2 + 11(-2) + 1 = 48 - 32 - 22 + 1 = -5. \quad ■$$

If $f(x)$ is a polynomial, then a solution of the equation $f(x) = 0$ is called a **root** or **zero** of $f(x)$. Thus, a number c is a root of $f(x)$ if $f(c) = 0$. A root that is a real number is called a **real root.** Graphical methods for finding the real roots of polynomials were considered in Section 2.2, and algebraic methods will be considered in Section 4.3. For now we examine the connection between roots and factors.

Example 5 By the Remainder Theorem, the remainder when $f(x) = x^3 - 4x^2 + 2x + 3$ is divided by $x - 3$ is $f(3)$. We have

$$f(3) = 3^3 - 4(3^2) + 2(3) + 3 = 0.$$

Therefore, 3 is a root of $f(x)$. Since the remainder is 0, $x - 3$ is a factor of $f(x)$. ■

Example 5 is an illustration of:

Factor Theorem

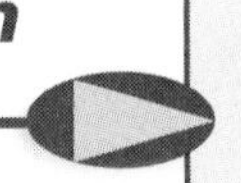

The number c is a root of the polynomial $f(x)$ exactly when $x - c$ is a factor of $f(x)$.

Proof of the Factor Theorem Divide $f(x)$ by $x - c$. The Remainder Theorem shows that the remainder is $f(c)$. Hence, by the Division Algorithm,

$$f(x) = (x - c)q(x) + f(c).$$

If c is a root of $f(x)$, then $f(c) = 0$ and we have $f(x) = (x - c)q(x)$. Hence, $x - c$ is a factor. Conversely, if $x - c$ is a factor, then the remainder $f(c)$ must be 0, so c is a root. ■

The Factor Theorem and a calculator can sometimes be used to factor polynomials.

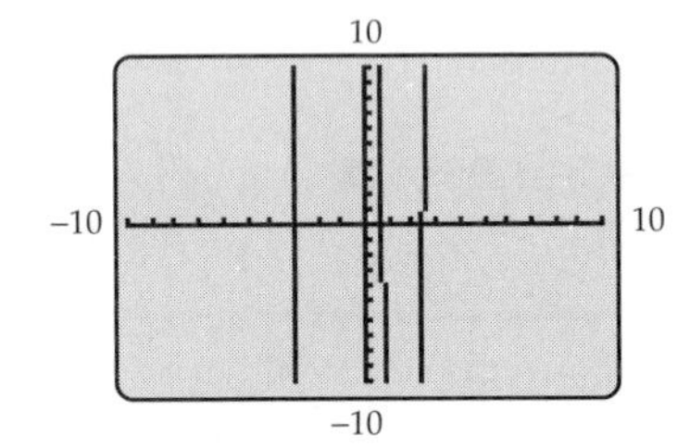

Figure 4–6

Example 6 The graph of $f(x) = 15x^3 - x^2 - 114x + 72$ in the standard viewing window (Figure 4–6) is obviously not complete, but suggests that -3 is an x-intercept, and hence a root of $f(x)$. It is easy to verify that this is indeed the case:

$$f(-3) = 15(-3)^3 - (-3)^2 - 114(-3) + 72 = -405 - 9 + 342 + 72 = 0.$$

Since -3 is a root, $x - (-3) = x + 3$ is a factor of $f(x)$. Use synthetic or long division to verify that the other factor is $15x^2 - 46x + 24$. By factoring this quadratic, we obtain a complete factorization of $f(x)$:

$$f(x) = (x + 3)(15x^2 - 46x + 24) = (x + 3)(3x - 2)(5x - 12).$$ ■

Example 7 Find three polynomials of different degrees that have 1, 2, 3, and -5 as roots.

Solution A polynomial that has 1, 2, 3, and -5 as roots must have $x - 1$, $x - 2$, $x - 3$, and $x - (-5) = x + 5$ as factors. Many polynomials satisfy these conditions, such as

$$g(x) = (x - 1)(x - 2)(x - 3)(x + 5) = x^4 - x^3 - 19x^2 + 49x - 30$$

$$h(x) = 8(x - 1)(x - 2)(x - 3)^2(x + 5)$$

$$k(x) = 2(x + 4)^2(x - 1)(x - 2)(x - 3)(x + 5)(x^2 + x + 1).$$

Note that g has degree 4. When h is multiplied out, its leading term is $8x^5$, so h has degree 5. Similarly, k has degree 8 since its leading term is $2x^8$. ■

If a polynomial $f(x)$ has four roots, say a, b, c, d, then by the same argument used in Example 7, it must have

$$(x - a)(x - b)(x - c)(x - d)$$

as a factor. Since $(x - a)(x - b)(x - c)(x - d)$ has degree 4 (multiply it out—its leading term is x^4), $f(x)$ must have degree at least 4. In particular, this means that no polynomial of degree 3 can have four or more roots. A similar argument works in the general case.

Number of Roots

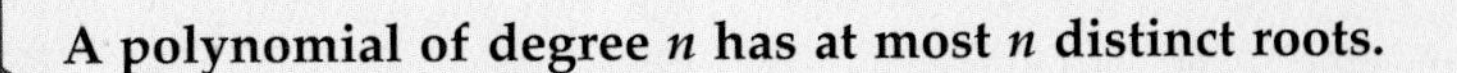

A polynomial of degree n has at most n distinct roots.

Exercises 4.2

In Exercises 1–8, determine whether the given algebraic expression is a polynomial. If it is, list its leading coefficient, constant term, and degree.

1. $1 + x^3$ **2.** -7 **3.** $(x - 1)(x^2 + 1)$

4. $7^x + 2x + 1$ **5.** $(x + \sqrt{3})(x - \sqrt{3})$

6. $4x^2 + 3\sqrt{x} + 5$ **7.** $\frac{7}{x^2} + \frac{5}{x} - 15$

8. $(x - 1)^k$ (where k is a fixed positive integer)

In Exercises 9–14, state the quotient and remainder when the first polynomial is divided by the second. Check your division by calculating (Divisor)(Quotient) + Remainder.

9. $3x^4 + 2x^2 - 6x + 1;\quad x + 1$

10. $x^5 - x^3 + x - 5;\quad x - 2$

11. $x^5 + 2x^4 - 6x^3 + x^2 - 5x + 1;\quad x^3 + 1$

12. $3x^4 - 3x^3 - 11x^2 + 6x - 1;\quad x^3 + x^2 - 2$

13. $5x^4 + 5x^2 + 5;\quad x^2 - x + 1$

14. $x^5 - 1;\quad x - 1$

In Exercises 15–18, determine whether the first polynomial is a factor of the second.

15. $x^2 + 3x - 1;\quad x^3 + 2x^2 - 5x - 6$

16. $x^2 + 9;\quad x^5 + x^4 - 81x - 81$

17. $x^2 + 3x - 1;\quad x^4 + 3x^3 - 2x^2 - 3x + 1$

18. $x^2 - 5x + 7;\quad x^3 - 3x^2 - 3x + 9$

In Exercises 19–22, determine which of the given numbers are roots of the given polynomial.

19. $2, 3, 0, -1;\quad g(x) = x^4 + 6x^3 - x^2 - 30x$

20. $1, 1/2, 2, -1/2, 3;\quad f(x) = 6x^2 + x - 1$

21. $2\sqrt{2}, \sqrt{2}, -\sqrt{2}, 1, -1;\quad h(x) = x^3 + x^2 - 8x - 8$

22. $\sqrt{3}, -\sqrt{3}, 1, -1;\quad k(x) = 8x^3 - 12x^2 - 6x + 9$

In Exercises 23–32, find the remainder when $f(x)$ is divided by $g(x)$, without using division.

23. $f(x) = x^{10} + x^8;\quad g(x) = x - 1$

24. $f(x) = x^6 - 10;\quad g(x) = x - 2$

25. $f(x) = 3x^4 - 6x^3 + 2x - 1;\quad g(x) = x + 1$

26. $f(x) = x^5 - 3x^2 + 2x - 1;\quad g(x) = x - 2$

27. $f(x) = x^3 - 2x^2 + 5x - 4;\quad g(x) = x + 2$

28. $f(x) = 10x^{75} - 8x^{65} + 6x^{45} + 4x^{32} - 2x^{15} + 5;$
$g(x) = x - 1$

29. $f(x) = 2x^5 - 3x^4 + x^3 - 2x^2 + x - 8;$
$g(x) = x - 10$

30. $f(x) = x^3 + 8x^2 - 29x + 44;\quad g(x) = x + 11$

31. $f(x) = 2x^5 - 3x^4 + 2x^3 - 8x - 8;$
$g(x) = x - 20$

32. $f(x) = x^5 - 10x^4 + 20x^3 - 5x - 95;$
$g(x) = x + 10$

In Exercises 33–38, use the Factor Theorem to determine whether or not $h(x)$ is a factor of $f(x)$.

33. $h(x) = x - 1$; $f(x) = x^5 + 1$
34. $h(x) = x - 1/2$; $f(x) = 2x^4 + x^3 + x - 3/4$
35. $h(x) = x + 2$; $f(x) = x^3 - 3x^2 - 4x - 12$
36. $h(x) = x + 1$; $f(x) = x^3 - 4x^2 + 3x + 8$
37. $h(x) = x - 1$; $f(x) = 14x^{99} - 65x^{56} + 51$
38. $h(x) = x - 2$; $f(x) = x^3 + x^2 - 4x + 4$

In Exercises 39–42, use the Factor Theorem and a calculator to factor the polynomial, as in Example 6.

39. $f(x) = 6x^3 - 7x^2 - 89x + 140$
40. $g(x) = x^3 - 5x^2 - 5x - 6$
41. $h(x) = 4x^4 + 4x^3 - 35x^2 - 36x - 9$
42. $f(x) = x^5 - 5x^4 - 5x^3 + 25x^2 + 6x - 30$

In Exercises 43–46, each graph is of a polynomial function $f(x)$ of degree 5 whose leading coefficient is 1. The graph is not drawn to scale. Use the Factor Theorem to find the polynomial. [Hint: What are the roots of $f(x)$? What does the Factor Theorem tell you?]

43.

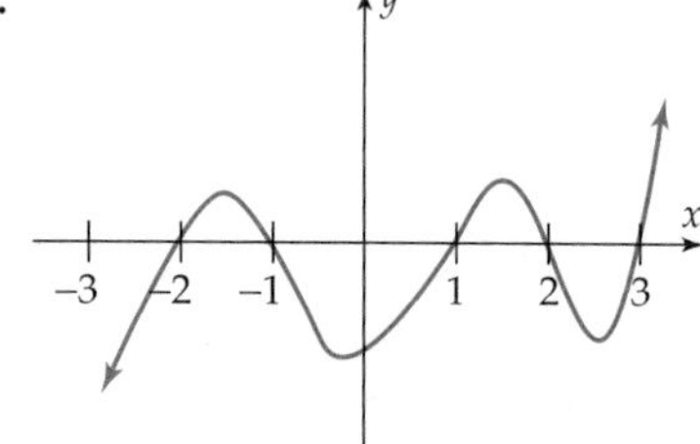

44.

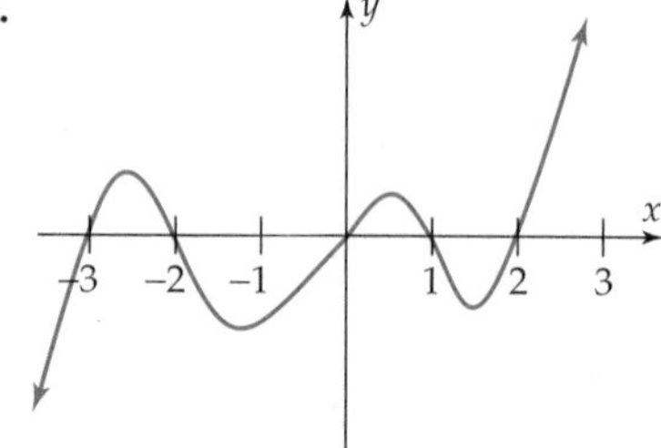

45.

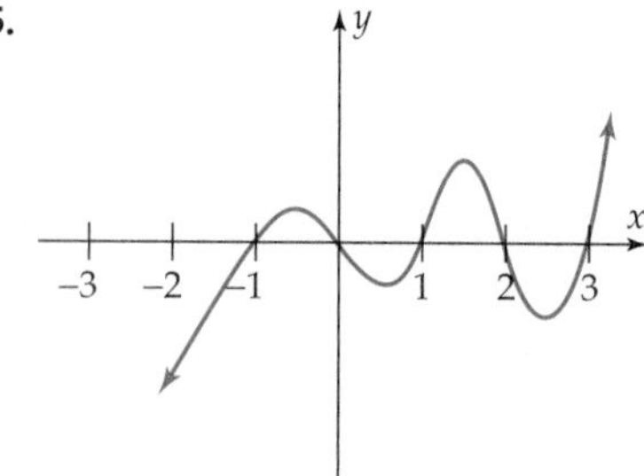

46.

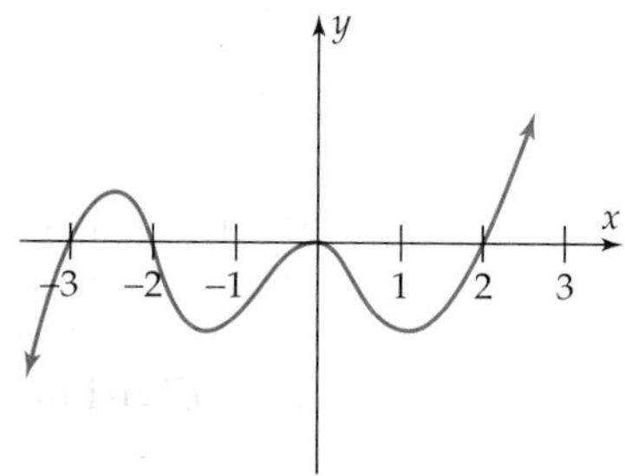

In Exercises 47–50, find a polynomial with the given degree n, the given roots, and no other roots.

47. $n = 3$; roots, 1, 7, −4
48. $n = 3$; roots 1, −1
49. $n = 6$; roots 1, 2, π
50. $n = 5$; root 2

51. Find a polynomial function f of degree 3 such that $f(10) = 17$ and the roots of $f(x)$ are 0, 5, and 8.
52. Find a polynomial function g of degree 4 such that the roots of g are 0, −1, 2, −3 and $g(3) = 288$.

In Exercises 53–56, find a number k satisfying the given condition.

53. $x + 2$ is a factor of $x^3 + 3x^2 + kx - 2$.
54. $x - 3$ is a factor of $x^4 - 5x^3 - kx^2 + 18x + 18$.
55. $x - 1$ is a factor of $k^2x^4 - 2kx^2 + 1$.
56. $x + 2$ is a factor of $x^3 - kx^2 + 3x + 7k$.

57. Use the Factor Theorem to show that for every real number c, $x - c$ is *not* a factor of $x^4 + x^2 + 1$.
58. Let c be a real number and n a positive integer.
    (a) Show that $x - c$ is a factor of $x^n - c^n$.
    (b) If n is even, show that $x + c$ is a factor of $x^n - c^n$. [*Remember:* $x + c = x - (-c)$.]
59. **(a)** If c is a real number and n an odd positive integer, give an example to show that $x + c$ may not be a factor of $x^n - c^n$.
    (b) If c and n are as in part (a), show that $x + c$ is a factor of $x^n + c^n$.

Thinkers

60. For what value of k is the difference quotient of $g(x) = kx^2 + 2x + 1$ equal to $7x + 2 - (3.5)h$?
61. For what value of k is the difference quotient of $f(x) = x^2 + kx$ equal to $2x + 5 + h$?

4.2.A

EXCURSION Synthetic Division

Synthetic division is a fast method of doing polynomial division when the divisor is a first-degree polynomial of the form $x - c$ for some real number c. To see how it works, we first consider an example of ordinary long division:

$$
\begin{array}{r|lllll l}
\multicolumn{1}{r}{} & 3x^3 & +\,6x^2 & +\,4x & -\,3 & & \leftarrow \textit{Quotient}\\ \cline{2-6}
\textit{Divisor} \rightarrow \underline{x-2} & 3x^4 & & -\,8x^2 & -\,11x & +\,1 & \leftarrow \textit{Dividend}\\
\multicolumn{1}{r}{} & \underline{3x^4} & \underline{-\,6x^3} & & & & \\
\multicolumn{1}{r}{} & & 6x^3 & -\,8x^2 & & & \\
\multicolumn{1}{r}{} & & \underline{6x^3} & \underline{-\,12x^2} & & & \\
\multicolumn{1}{r}{} & & & 4x^2 & -\,11x & & \\
\multicolumn{1}{r}{} & & & \underline{4x^2} & \underline{-\,8x} & & \\
\multicolumn{1}{r}{} & & & & -\,3x & +\,1 & \\
\multicolumn{1}{r}{} & & & & \underline{-\,3x} & \underline{+\,6} & \\
\multicolumn{1}{r}{} & & & & & -\,5 & \leftarrow \textit{Remainder}
\end{array}
$$

This calculation obviously involves a lot of repetitions. If we insert 0 coefficients for terms that don't appear above and keep the various coefficients in the proper columns, we can eliminate the repetition and all the x's:

$$
\begin{array}{r|rrrrr l}
\multicolumn{1}{r}{} & 3 & 6 & 4 & -3 & & \leftarrow \textit{Quotient}\\ \cline{2-6}
\textit{Divisor} \rightarrow \underline{1-2} & 3 & 0 & -8 & -11 & 1 & \leftarrow \textit{Dividend}\\
\multicolumn{1}{r}{} & & \underline{-6} & & & & \\
\multicolumn{1}{r}{} & & 6 & & & & \\
\multicolumn{1}{r}{} & & & \underline{-12} & & & \\
\multicolumn{1}{r}{} & & & 4 & & & \\
\multicolumn{1}{r}{} & & & & \underline{-8} & & \\
\multicolumn{1}{r}{} & & & & -3 & & \\
\multicolumn{1}{r}{} & & & & & \underline{+6} & \\
\multicolumn{1}{r}{} & & & & & -5 & \leftarrow \textit{Remainder}
\end{array}
$$

We can save space by moving the lower lines upward and writing 2 in the divisor position (since that's enough to remind us that the divisor is $x - 2$):

$$
\begin{array}{r|rrrrr l}
\multicolumn{1}{r}{} & 3 & 6 & 4 & -3 & & \leftarrow \textit{Quotient}\\ \cline{2-6}
\textit{Divisor} \rightarrow 2 & 3 & 0 & -8 & -11 & 1 & \leftarrow \textit{Dividend}\\
\multicolumn{1}{r}{} & & -6 & -12 & -8 & 6 & \\ \cline{2-6}
\multicolumn{1}{r}{} & & 6 & 4 & -3 & \multicolumn{1}{|r}{-5} & \leftarrow \textit{Remainder}
\end{array}
$$

Since the last line contains most of the quotient line, we can save more space and still preserve the essential information by inserting a 3 in the last line and omitting the top line:

$$
\begin{array}{r|rrrrr l}
\textit{Divisor} \rightarrow 2 & 3 & 0 & -8 & -11 & 1 & \leftarrow \textit{Dividend}\\
\multicolumn{1}{r}{} & & -6 & -12 & -8 & 6 & \\ \cline{2-6}
\multicolumn{1}{r}{} & \multicolumn{4}{l}{\underbrace{3 \quad 6 \quad 4 \quad -3}_{\textit{Quotient}}} & \multicolumn{1}{|r}{-5} & \leftarrow \textit{Remainder}
\end{array}
$$

Synthetic division is a quick method for obtaining the last row of this array. Here is a step-by-step explanation of the division of $3x^4 - 8x^2 - 11x + 1$ by $x - 2$:

Step 1 In the first row list the 2 from the divisor and the coefficients of the dividend in order of decreasing powers of x (insert 0 coefficients for missing powers of x).

$$\begin{array}{r|rrrrr} 2 & 3 & 0 & -8 & -11 & 1 \end{array}$$

Step 2 Bring down the first dividend coefficient (namely, 3) to the third row

$$\begin{array}{r|rrrrr} 2 & 3 & 0 & -8 & -11 & 1 \\ & & & & & \\ \hline & 3 & & & & \end{array}$$

Step 3 Multiply $2 \cdot 3$ and insert the answer 6 in the second row, in the position shown here.

$$\begin{array}{r|rrrrr} 2 & 3 & 0 & -8 & -11 & 1 \\ & & 6 & & & \\ \hline & 3 & & & & \end{array}$$

Step 4 Add $0 + 6$ and write the answer 6 in the third row.

$$\begin{array}{r|rrrrr} 2 & 3 & 0 & -8 & -11 & 1 \\ & & 6 & & & \\ \hline & 3 & 6 & & & \end{array}$$

Step 5 Multiply $2 \cdot 6$ and insert the answer 12 in the second row.

$$\begin{array}{r|rrrrr} 2 & 3 & 0 & -8 & -11 & 1 \\ & & 6 & 12 & & \\ \hline & 3 & 6 & & & \end{array}$$

Step 6 Add $-8 + 12$ and write the answer 4 in the third row.

$$\begin{array}{r|rrrrr} 2 & 3 & 0 & -8 & -11 & 1 \\ & & 6 & 12 & & \\ \hline & 3 & 6 & 4 & & \end{array}$$

Step 7 Multiply $2 \cdot 4$ and insert the answer 8 in the second row.

$$\begin{array}{r|rrrrr} 2 & 3 & 0 & -8 & -11 & 1 \\ & & 6 & 12 & 8 & \\ \hline & 3 & 6 & 4 & & \end{array}$$

Step 8 Add $-11 + 8$ and write the answer -3 in the third row.

$$\begin{array}{r|rrrrr} 2 & 3 & 0 & -8 & -11 & 1 \\ & & 6 & 12 & 8 & \\ \hline & 3 & 6 & 4 & -3 & \end{array}$$

Step 9 Multiply $2 \cdot (-3)$ and insert the answer -6 in the second row.

$$\begin{array}{r|rrrrr} 2 & 3 & 0 & -8 & -11 & 1 \\ & & 6 & 12 & 8 & -6 \\ \hline & 3 & 6 & 4 & -3 & \end{array}$$

Step 10 Add $1 + (-6)$ and write the answer -5 in the third row.

$$\begin{array}{r|rrrrr} 2 & 3 & 0 & -8 & -11 & 1 \\ & & 6 & 12 & 8 & -6 \\ \hline & 3 & 6 & 4 & -3 & \boxed{-5} \end{array}$$

Except for the signs in the second row, this last array is the same as the array obtained from the long division process, and we can read off the quotient and remainder:

The last number in the third row is the remainder.

The other numbers in the third row are the coefficients of the quotient (arranged in order of decreasing powers of x).

Since we are dividing the *fourth*-degree polynomial $3x^4 - 8x^2 - 11x + 1$ by the *first*-degree polynomial $x - 2$, the quotient must be a polynomial of degree *three* with coefficients 3, 6, 4, -3, namely, $3x^3 + 6x^2 + 4x - 3$. The remainder is -5.

> **CAUTION**
>
> Synthetic division can be used *only* when the divisor is a first-degree polynomial of the form $x - c$. In the example above, $c = 2$. If you want to use synthetic division with a divisor such as $x + 3$, you must write it as $x - (-3)$, which is of the form $x - c$ with $c = -3$.

Example 1 To divide $x^5 + 5x^4 + 6x^3 - x^2 + 4x + 29$ by $x + 3$, we write the divisor as $x - (-3)$ and proceed as above:

$$\begin{array}{r|rrrrrr} -3 & 1 & 5 & 6 & -1 & 4 & 29 \\ & & -3 & -6 & 0 & 3 & -21 \\ \hline & 1 & 2 & 0 & -1 & 7 & \boxed{8} \end{array}$$

> **Technology Tip**
>
> Synthetic division programs are in the Program Appendix.

The last row shows that the quotient is $x^4 + 2x^3 - x + 7$ and the remainder is 8. ■

Example 2 Show that $x - 7$ is a factor of $8x^5 - 52x^4 + 2x^3 - 198x^2 - 86x + 14$ and find the other factor.

Solution $x - 7$ is a factor exactly when division by $x - 7$ leaves remainder 0, in which case the quotient is the other factor. Using synthetic division we have:

$$\begin{array}{r|rrrrrr} 7 & 8 & -52 & 2 & -198 & -86 & 14 \\ & & 56 & 28 & 210 & 84 & -14 \\ \hline & 8 & 4 & 30 & 12 & -2 & \boxed{0} \end{array}$$

Since the remainder is 0, the divisor $x - 7$ and the quotient $8x^4 + 4x^3 + 30x^2 + 12x - 2$ are factors:

$$8x^5 - 52x^4 + 2x^3 - 198x^2 - 86x + 14$$
$$= (x - 7)(8x^4 + 4x^3 + 30x^2 + 12x - 2). \quad ■$$

Exercises 4.2.A

In Exercises 1–8, use synthetic division to find the quotient and remainder.

1. $(3x^4 - 8x^3 + 9x + 5) \div (x - 2)$
2. $(4x^3 - 3x^2 + x + 7) \div (x - 2)$
3. $(2x^4 + 5x^3 - 2x - 8) \div (x + 3)$
4. $(3x^3 - 2x^2 - 8) \div (x + 5)$
5. $(5x^4 - 3x^2 - 4x + 6) \div (x - 7)$
6. $(3x^4 - 2x^3 + 7x - 4) \div (x - 3)$
7. $(x^4 - 6x^3 + 4x^2 + 2x - 7) \div (x - 2)$
8. $(x^6 - x^5 + x^4 - x^3 + x^2 - x + 1) \div (x + 3)$

In Exercises 9–12, use synthetic division to find the quotient and the remainder. In each divisor $x - c$, the number c is not an integer, but the same technique will work.

9. $(3x^4 - 2x^2 + 2) \div \left(x - \frac{1}{4}\right)$

10. $(2x^4 - 3x^2 + 1) \div \left(x - \frac{1}{2}\right)$

11. $(2x^4 - 5x^3 - x^2 + 3x + 2) \div \left(x + \frac{1}{2}\right)$

12. $\left(10x^5 - 3x^4 + 14x^3 + 13x^2 - \frac{4}{3}x + \frac{7}{3}\right) \div \left(x + \frac{1}{5}\right)$

In Exercises 13–16, use synthetic division to show that the first polynomial is a factor of the second and find the other factor.

13. $x + 4$; $3x^3 + 9x^2 - 11x + 4$

14. $x - 5$; $x^5 - 8x^4 + 17x^2 + 293x - 15$

15. $x - 1/2$; $2x^5 - 7x^4 + 15x^3 - 6x^2 - 10x + 5$

16. $x + 1/3$; $3x^6 + x^5 - 6x^4 + 7x^3 + 3x^2 - 15x - 5$

In Exercises 17 and 18, use a calculator and synthetic division to find the quotient and remainder.

17. $(x^3 - 5.27x^2 + 10.708x - 10.23) \div (x - 3.12)$

18. $(2.79x^4 + 4.8325x^3 - 6.73865x^2 + .9255x - 8.125) \div (x - 1.35)$

Thinkers

19. When $x^3 + cx + 4$ is divided by $x + 2$, the remainder is 4. Find c.

20. If $x - d$ is a factor of $2x^3 - dx^2 + (1 - d^2)x + 5$, what is d?

4.3 Real Roots of Polynomials

Finding the real roots of polynomials is the same as solving polynomial equations. The root of a first-degree polynomial, such as $5x - 3$, can always be found by solving the equation $5x - 3 = 0$. Similarly, the roots of any second-degree polynomial can be found by using the quadratic formula (Section 1.2). Although the roots of higher degree polynomials can always be approximated graphically as in Section 2.2, it is better to find exact solutions, if possible.

Rational Roots

When a polynomial has integer coefficients, all of its **rational roots** (roots that are rational numbers) can be found exactly by using the following result.

The Rational Root Test

If a rational number r/s (in lowest terms) is a root of the polynomial

$$a_nx^n + \cdots + a_1x + a_0,$$

where the coefficients $a_n, \ldots, a_1, a_0$ are integers with $a_n \neq 0$, $a_0 \neq 0$, then

r is a factor of the constant term a_0 and

s is a factor of the leading coefficient a_n.

The test states that every rational root must satisfy certain conditions.* By finding all the numbers that satisfy these conditions, we produce a list of *possible* rational roots. Then we must evaluate the polynomial at each number on the list to see if the number actually is a root. This testing process can be considerably shortened by using a calculator, as in the next example.

Example 1 Find the rational roots of

$$f(x) = 2x^4 + x^3 - 17x^2 - 4x + 6.$$

Solution If $f(x)$ has a rational root r/s, then by the Rational Root Test r must be a factor of the constant term 6. Therefore, r must be one of ± 1, $\pm 2, \pm 3$, or ± 6 (the only factors of 6). Similarly, s must be a factor of the leading coefficient 2, so s must be one of ± 1 or ± 2 (the only factors of 2). Consequently, the only *possibilities* for r/s are

$$\frac{\pm 1}{\pm 1}, \frac{\pm 2}{\pm 1}, \frac{\pm 3}{\pm 1}, \frac{\pm 6}{\pm 1}, \frac{\pm 1}{\pm 2}, \frac{\pm 2}{\pm 2}, \frac{\pm 3}{\pm 2}, \frac{\pm 6}{\pm 2}.$$

Eliminating duplications from this list, we see that the only *possible* rational roots are

$$1, -1, 2, -2, 3, -3, 6, -6, \frac{1}{2}, -\frac{1}{2}, \frac{3}{2}, -\frac{3}{2}.$$

Now graph $f(x)$ in a viewing window that includes all of these numbers on the x-axis, say $-7 \le x \le 7$ and $-5 \le y \le 5$ (Figure 4–7). A complete graph isn't necessary since we are interested only in the x-intercepts.

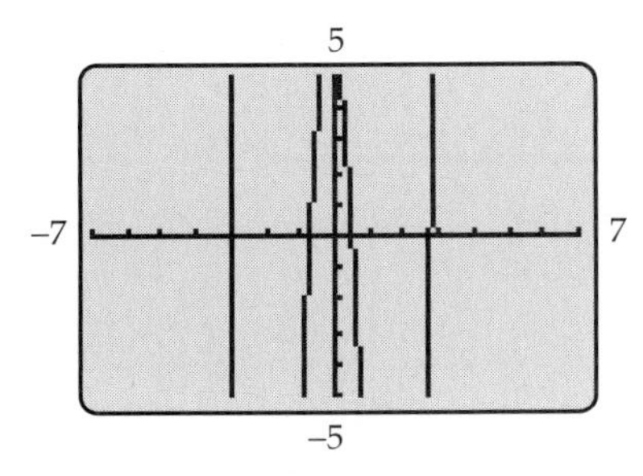

Figure 4–7

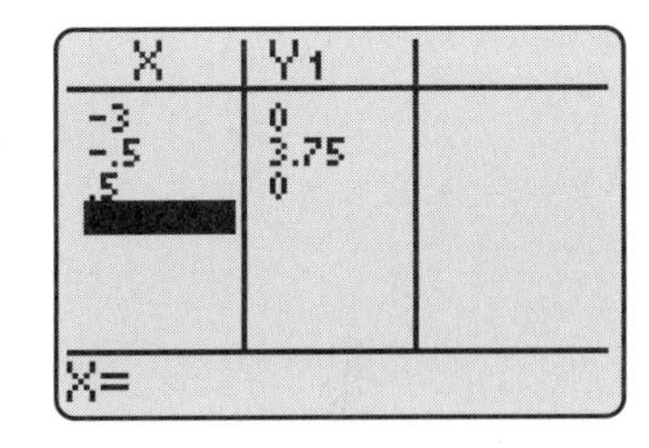

Figure 4–8

Figure 4–7 shows that the only numbers on our list that could possibly be roots (x-intercepts) are -3, $-1/2$, and $1/2$, so these are the only ones that need be tested. We use the table feature to evaluate $f(x)$ at these three numbers (Figure 4–8). The table shows that -3 and $1/2$ are the only rational roots of $f(x)$. Its other roots (x-intercepts) in Figure 4–7 must be irrational numbers. ■

*Since the proof of the Rational Root Test sheds no light on how the test is actually used to solve equations, it will be omitted.

Roots and the Factor Theorem

Once some roots of a polynomial have been found, the Factor Theorem can be used to factor the polynomial, which may lead to additional roots.

Example 2 Find all the roots of $f(x) = 2x^4 + x^3 - 17x^2 - 4x + 6$.

Solution In Example 1 we saw that -3 and $1/2$ are the rational roots of $f(x)$. By the Factor Theorem $x - (-3) = x + 3$ and $x - 1/2$ are factors of $f(x)$. Using synthetic or long division twice, we have

$$\begin{aligned} 2x^4 + x^3 - 17x^2 - 4x + 6 &= (x + 3)(2x^3 - 5x^2 - 2x + 2) \\ &= (x + 3)(x - .5)(2x^2 - 4x - 4) \end{aligned}$$

The remaining roots of $f(x)$ are the roots of $2x^2 - 4x - 4$, that is, the solutions of

$$2x^2 - 4x - 4 = 0$$

$$x^2 - 2x - 2 = 0.$$

They are easily found by the quadratic formula.

$$\begin{aligned} x &= \frac{-(-2) \pm \sqrt{(-2)^2 - 4\cdot 1\cdot(-2)}}{2\cdot 1} \\ &= \frac{2 \pm \sqrt{12}}{2} = \frac{2 \pm 2\sqrt{3}}{2} = 1 \pm \sqrt{3}. \end{aligned}$$

Therefore, $f(x)$ has rational roots -3 and $1/2$, and irrational roots $1 + \sqrt{3}$ and $1 - \sqrt{3}$. ■

Example 3 Factor $f(x) = 2x^5 - 10x^4 + 7x^3 + 13x^2 + 3x + 9$ completely.

Solution We begin by finding as many roots of $f(x)$ as we can. By the Rational Root Test, every rational root is of the form r/s, where $r = \pm 1$, ± 3, or ± 9 and $s = \pm 1$ or ± 2. Thus, the possible rational roots are

$$\pm 1, \pm 3, \pm 9, \pm\frac{1}{2}, \pm\frac{3}{2}, \pm\frac{9}{2}.$$

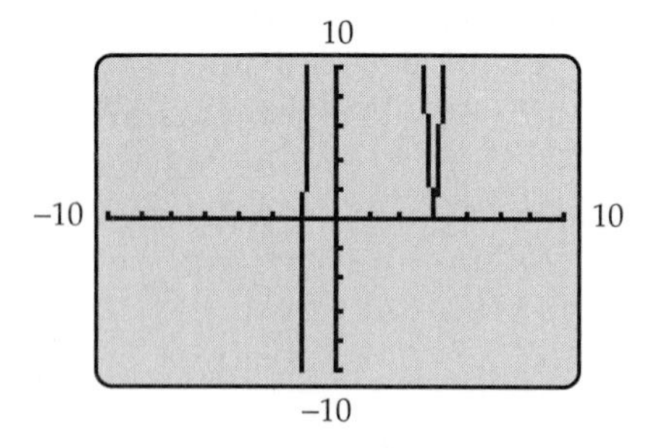

Figure 4–9

The partial graph of $f(x)$ in Figure 4–9 shows that the only possible roots (x-intercepts) are -1 and 3. You can easily verify that both -1 and 3 are roots of $f(x)$.

Since -1 and 3 are roots, $x - (-1) = x + 1$ and $x - 3$ are factors of $f(x)$ by the Factor Theorem. Division shows that

$$\begin{aligned} f(x) &= 2x^5 - 10x^4 + 7x^3 + 13x^2 + 3x + 9 \\ &= (x + 1)(2x^4 - 12x^3 + 19x^2 + 3x + 9) \\ &= (x + 1)(x - 3)(2x^3 - 6x^2 + x - 3) \end{aligned}$$

The other roots of $f(x)$ are the roots of $g(x) = 2x^3 - 6x^2 + x - 3$. We first check for rational roots of $g(x)$. Since every root of $g(x)$ is also a root of $f(x)$ (why?), the only possible rational roots of $g(x)$ are -1 and 3 [the rational roots of $f(x)$]. We have

$$g(-1) = 2(-1)^3 - 6(-1)^2 + (-1) - 3 = -12;$$
$$g(3) = 2(3^3) - 6(3^2) + 3 - 3 = 0.$$

So -1 is not a root, but 3 is a root of $g(x)$. By the Factor Theorem, $x - 3$ is a factor of $g(x)$. Division shows that

$$\begin{aligned} f(x) &= (x+1)(x-3)(2x^3 - 6x^2 + x - 3) \\ &= (x+1)(x-3)(x-3)(2x^2+1). \end{aligned}$$

Since $2x^2 + 1$ has no real roots, it cannot be factored. So the factorization of $f(x)$ is complete. ■

Bounds

The polynomial $f(x)$ in Example 2 had degree 4 and had four real roots. Since a polynomial of degree n has at most n roots, we know that we found all the roots of $f(x)$. In other cases, however, special techniques may be needed to guarantee that we have found all the roots.

Example 4 Prove that all the real roots of $g(x) = x^5 - 2x^4 - x^3 + 3x + 1$ lie between -1 and 3. Then find all the real roots of $g(x)$.

Solution We first prove that $g(x)$ has no root larger than 3, as follows. Use synthetic division to divide $g(x)$ by $x - 3$:*

$$\begin{array}{r|rrrrrr} 3 & 1 & -2 & -1 & 0 & 3 & 1 \\ & & 3 & 3 & 6 & 18 & 63 \\ \hline & 1 & 1 & 2 & 6 & 21 & \boxed{64} \end{array}$$

Thus, the quotient is $x^4 + x^3 + 2x^2 + 6x + 21$ and the remainder is 64. Applying the Division Algorithm, we have:

$$f(x) = (x-3)(x^4 + x^3 + 2x^2 + 6x + 21) + 64.$$

When $x > 3$, then the factor $x - 3$ is positive and the quotient $x^4 + x^3 + 2x^2 + 6x + 21$ is also positive (because all its coefficients are). The remainder 64 is also positive. Therefore, $f(x)$ is positive whenever $x > 3$. In particular, $f(x)$ is never zero when $x > 3$ and so there are no roots of $f(x)$ greater than 3.

Now we show that $g(x)$ has no root less than -1. Divide $g(x)$ by $x - (-1) = x + 1$:

$$\begin{array}{r|rrrrrr} -1 & 1 & -2 & -1 & 0 & 3 & 1 \\ & & -1 & 3 & -2 & 2 & -5 \\ \hline & 1 & -3 & 2 & -2 & 5 & \boxed{-4} \end{array}$$

*If you haven't read Excursion 4.2.A, use long division to find the quotient and remainder.

Read off the quotient and remainder and apply the Division Algorithm:

$$f(x) = (x + 1)(x^4 - 3x^3 + 2x^2 - 2x + 5) - 4.$$

When $x < -1$, then the factor $x + 1$ is negative. When x is negative, its odd powers are negative and its even powers are positive. Consequently, the quotient $x^4 - 3x^3 + 2x^2 - 2x + 5$ is positive (because the odd powers of x are multiplied by negative coefficients). The product of the positive quotient with the negative factor $x + 1$ is negative. The remainder -4 is also negative. Hence, $f(x)$ is negative whenever $x < -1$. So there are no real roots less than -1. Therefore, all the real roots of $g(x)$ lie between -1 and 3.

Finally, we find the roots of $g(x) = x^5 - 2x^4 - x^3 + 3x + 1$. The only possible rational roots are ± 1 (why?) and it is easy to verify that neither is actually a root. The graph of $g(x)$ in Figure 4–10 shows that there are exactly three real roots (x-intercepts) between -1 and 3. Since all the real roots of $g(x)$ lie between -1 and 3, $g(x)$ has only these three real roots. They are readily approximated by a root finder:

$$x \approx -.3361, \qquad x \approx 1.4268, \qquad \text{and} \qquad x \approx 2.2012. \quad ■$$

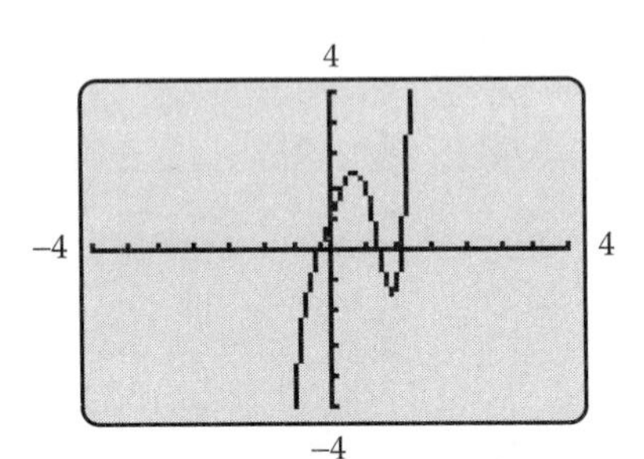

Figure 4–10

Suppose $f(x)$ is a polynomial and r and s are real numbers with $r < s$. If all the real roots of $f(x)$ are between r and s, we say that r is a **lower bound** and s is an **upper bound** for the real roots of $f(x)$.* Example 4 shows that -1 is a lower bound and 3 is an upper bound for the real roots of $g(x) = x^5 - 2x^4 - x^3 + 3x + 1$.

If you know lower and upper bounds for the real roots of a polynomial, you can usually determine the number of real roots the polynomial has, as we did in Example 4. The technique used in Example 4 to test possible lower and upper bounds works in the general case:

Bounds Test

Let $f(x)$ be a polynomial with positive leading coefficient.

If $d > 0$ and every number in the last row in the synthetic division of $f(x)$ by $x - d$ is nonnegative,† then d is an upper bound for the real roots of $f(x)$.

If $c < 0$ and the numbers in the last row of the synthetic division of $f(x)$ by $x - c$ are alternately positive and negative [with 0 considered as either,‡ then c is a lower bound for the real roots of $f(x)$.]

*The bounds are not unique. Any number smaller than r is also a lower bound and any number larger than s is also an upper bound.

†Equivalently, all the coefficients of the quotient and the remainder are nonnegative.

‡Equivalently, the coefficients of the quotient are alternatively positive and negative, with the last one and the remainder having opposite signs.

Example 5 Find all real roots of

$$f(x) = x^7 - 6x^6 + 9x^5 + 7x^4 - 28x^3 + 33x^2 - 36x + 20.$$

Solution By the Rational Root Test, the only possible roots are

$$\pm 1, \pm 2, \pm 4, \pm 5, \pm 10, \text{and } \pm 20.$$

The graph of $f(x)$ in Figure 4–11 is hard to read, but shows that the possible roots are quite close to the origin. Changing the window (Figure 4–12), we see that the only numbers on the list that could possibly be roots (x-intercepts) are 1 and 2. You can easily verify that both 1 and 2 are roots of $f(x)$.

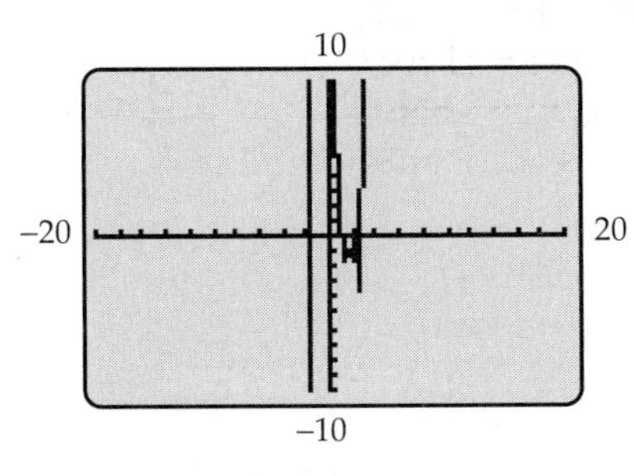

Figure 4–11

Figure 4–12

In Figures 4–11 and 4–12 all the real roots of $f(x)$ lie between -2 and 6, which suggests that these numbers might be lower and upper bounds for the real roots of $f(x)$. The Bounds Test shows that this is indeed the case:

−2	1	−6	9	7	−28	33	−36	20
		−2	16	−50	86	−116	172	−272
	1	−8	25	−43	58	−86	136	−252

Alternating Signs
−2 is a lower bound

6	1	−6	9	7	−28	33	−36	20
		6	0	54	366	2028	12,366	73,980
	1	0	9	61	338	2061	12,330	74,000

All Nonnegative
6 is an upper bound

Therefore, the four x-intercepts in Figure 4–12 are the only real roots of $f(x)$. We have seen that two of these are the rational roots, 1 and 2. A root finder shows that the other roots are $x \approx -1.7913$ and $x \approx 2.7913$. ■

Summary

The examples above illustrate the following guidelines for finding all the real roots of a polynomial $f(x)$.

1. Use the Rational Root Test to find all the rational roots of $f(x)$. [*Examples 1, 3, 5*]
2. Write $f(x)$ as the product of linear factors (one for each rational root) and another factor $g(x)$. [*Examples 2, 3*]

Technology Tip

The polynomial solvers on TI-86/89, Casio FX 2.0, and HP-38 can find or approximate all the roots of a polynomial simultaneously. Those on Sharp 9600 and Casio 9850 are limited to polynomials of degree 2 or 3. See the Calculator Investigation on page 24 for details.

3. If $g(x)$ has degree 2, find its roots by factoring or the quadratic formula. [*Example 2*]
4. If $g(x)$ has degree 3 or more, use the Bounds Test, if possible, to find lower and upper bounds for the roots of $g(x)$ and approximate the remaining roots graphically. [*Examples 4, 5*]

Shortcuts and variations are always possible. For instance, if the graph of a cubic shows three x-intercepts, then it has three real roots (the maximum possible) and there is no point in finding bounds on the roots. In order to find as many roots as possible exactly in guideline 4, check to see if the rational roots of $f(x)$ are also roots of $g(x)$ and factor $g(x)$ accordingly, as in Example 3.

Exercises 4.3

Directions: *When asked to find the roots of a polynomial, find exact roots whenever possible and approximate the other roots.*

In Exercises 1–12, find all the rational roots of the polynomial.

1. $x^3 + 3x^2 - x - 3$

2. $x^3 - x^2 - 3x + 3$

3. $x^3 + 5x^2 - x - 5$

4. $3x^3 + 8x^2 - x - 20$

5. $f(x) = 2x^5 + 5x^4 - 11x^3 + 4x^2$ [*Hint:* The Rational Root Test can only be used on polynomials with nonzero constant terms. Factor $f(x)$ as a product of a power of x and a polynomial $g(x)$ with nonzero constant term. Then use the Rational Root Test on $g(x)$.]

6. $2x^6 - 3x^5 - 7x^4 - 6x^3$

7. $f(x) = \frac{1}{12}x^3 - \frac{1}{12}x^2 - \frac{2}{3}x + 1$ [*Hint:* The Rational Root Test can only be used on polynomials with integer coefficients. Note that $f(x)$ and $12f(x)$ have the same roots. (Why?)]

8. $\frac{2}{3}x^4 + \frac{1}{2}x^3 - \frac{5}{4}x^2 - x - \frac{1}{6}$

9. $\frac{1}{3}x^4 - x^3 - x^2 + \frac{13}{3}x - 2$

10. $\frac{1}{3}x^7 - \frac{1}{2}x^6 - \frac{1}{6}x^5 + \frac{1}{6}x^4$

11. $.1x^3 - 1.9x + 3$

12. $.05x^3 + .45x^2 - .4x + 1$

In Exercises 13–18, factor the polynomial as a product of linear factors and a factor $g(x)$ such that $g(x)$ is either a constant or a polynomial that has no rational roots.

13. $2x^3 - 4x^2 + x - 2$

14. $6x^3 - 5x^2 + 3x - 1$

15. $x^6 + 2x^5 + 3x^4 + 6x^3$

16. $x^5 - 2x^4 + 2x^3 - 3x + 2$

17. $x^5 - 4x^4 + 8x^3 - 14x^2 + 15x - 6$

18. $x^5 + 4x^3 + x^2 + 6x$

In Exercises 19–22, use the Bounds Test to find lower and upper bounds for the real roots of the polynomial.

19. $x^3 + 2x^2 - 7x + 20$

20. $x^3 - 15x^2 - 16x + 12$

21. $-x^5 - 5x^4 + 9x^3 + 18x^2 - 68x + 176$ [*Hint:* The Bounds Test applies only to polynomials with positive leading coefficient. The polynomial $f(x)$ has the same roots as $-f(x)$. (Why?)]

22. $-.002x^3 - 5x^2 + 8x - 3$

In Exercises 23–36, find all real roots of the polynomial.

23. $2x^3 - 5x^2 + x + 2$

24. $t^4 - t^3 + 2t^2 - 4t - 8$

25. $6x^3 - 11x^2 + 6x - 1$

26. $z^3 + z^2 + 2z + 2$

27. $x^4 + x^3 - 19x^2 + 32x - 12$

28. $3x^5 + 2x^4 - 7x^3 + 2x^2$

29. $2x^5 - x^4 - 10x^3 + 5x^2 + 12x - 6$

30. $x^5 - x^3 + x$

31. $x^6 - 4x^5 - 5x^4 - 9x^2 + 36x + 45$

32. $x^5 + 3x^4 - 4x^3 - 11x^2 - 3x + 2$

33. $3x^4 + 2x^3 - 4x^2 + 4x - 1$

34. $x^5 + 8x^4 + 20x^3 + 9x^2 - 27x - 27$

35. $x^4 - 48x^3 - 101x^2 + 49x + 50$

36. $3x^7 + 8x^6 - 13x^5 - 36x^4 - 10x^3 + 21x^2 + 41x + 10$

37. **(a)** Show that $\sqrt{2}$ is an irrational number. [*Hint:* $\sqrt{2}$ is a root of $x^2 - 2$. Does this polynomial have any rational roots?]
(b) Show that $\sqrt{3}$ is irrational.

38. Graph $f(x) = .001x^3 - .199x^2 - .23x + 6$ in the standard viewing window.

(a) How many roots does $f(x)$ appear to have? Without changing the viewing window, explain why $f(x)$ must have an additional root. [*Hint:* Each root corresponds to a factor of $f(x)$. What does the rest of the factorization consist of?]

(b) Find all the roots of $f(x)$.

39. According to data from the FBI, the number of people murdered each year per 100,000 population can be approximated by the polynomial function

$$f(x) = .0011x^4 - .0233x^3 + .1144x^2 + .0126x + 8.1104 \quad (0 \le x \le 10),$$

where $x = 0$ corresponds to 1987.

(a) What was the murder rate in 1990?

(b) In what year was the rate 8 people per 100,000?

(c) In what year was the rate the highest?

40. During the first 150 hours of an experiment, the growth rate of a bacteria population at time t hours is $g(t) = -.0003t^3 + .04t^2 + .3t + .2$ bacteria per hour.

(a) What is the growth rate at 50 hours? At 100 hours?

(b) What is the growth rate at 145 hours? What does this mean?

(c) At what time is the growth rate 0?

(d) At what time is the growth rate -50 bacteria per hour?

(e) Approximately at what time does the highest growth rate occur?

41. An open-top reinforced box is to be made from a 12-by-36-inch piece of cardboard by cutting along the marked lines, discarding the shaded pieces, and folding as shown in the figure. If the box must be less than 2.5 inches high, what size squares should be cut from the corners in order for the box to have a volume of 448 cubic inches?

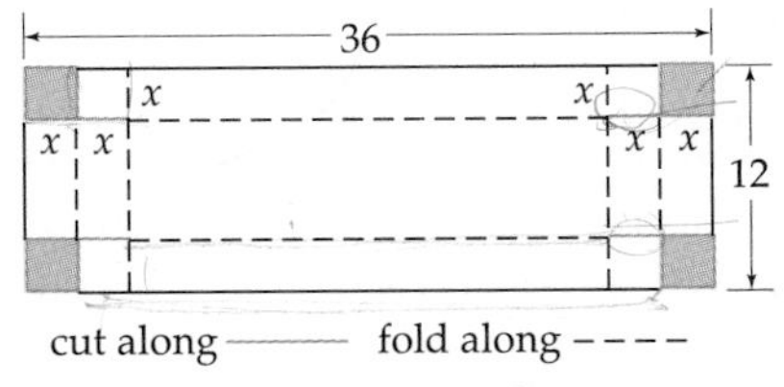

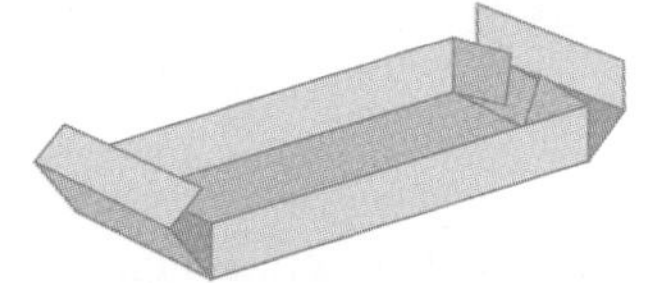

42. A box with a lid is to be made from a 48-by-24-inch piece of cardboard by cutting and folding, as shown in the figure. If the box must be at least 6 inches high, what size squares should be cut from the two corners in order for the box to have a volume of 1000 cubic inches?

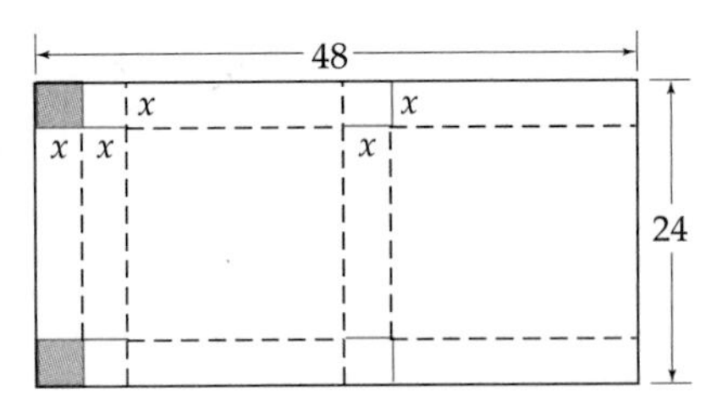

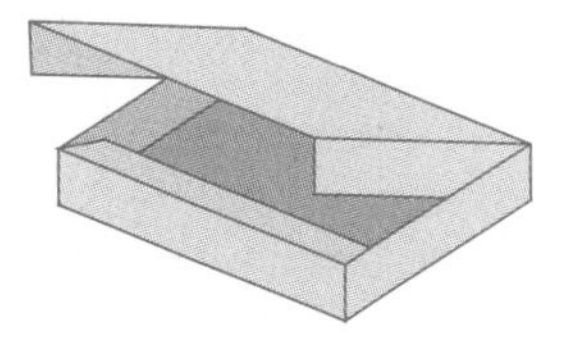

43. In a sealed chamber where the temperature varies, the instantaneous rate of change of temperature with respect to time over an 11-day period is given by $F(t) = .0035t^4 - .4t^2 - .2t + 6$, where time is measured in days and temperature in degrees Fahrenheit (so that rate of change is in degrees per day).

(a) At what rate is the temperature changing at the beginning of the period ($t = 0$)? At the end of the period ($t = 11$)?

(b) When is the temperature increasing at a rate of 4°F per day?

(c) When is the temperature decreasing at a rate of 3°F per day?

(d) When is the temperature decreasing at the fastest rate?

44. (a) If c is a root of

$$f(x) = 5x^4 - 4x^3 + 3x^2 - 4x + 5,$$

show that $1/c$ is also a root.

(b) Do part (a) with $f(x)$ replaced by $g(x) = 2x^6 + 3x^5 + 4x^4 - 5x^3 + 4x^2 + 3x + 2$.

(c) Let $f(x) = a_{12}x^{12} + a_{11}x^{11} + \cdots + a_2x^2 + a_1x + a_0$. What conditions must the coefficients a_i satisfy in order that this statement be true: If c is a root of $f(x)$, then $1/c$ is also a root?

4.4 Graphs of Polynomial Functions

The graphs of first- and second-degree polynomial functions are straight lines and parabolas respectively (Sections 1.4 and 4.1). The emphasis here will be on higher degree polynomial functions.

The simplest ones are those of the form $f(x) = ax^n$ (where a is a constant). Their graphs are of four types, as shown in the following chart.

Graph of $f(x) = ax^n$

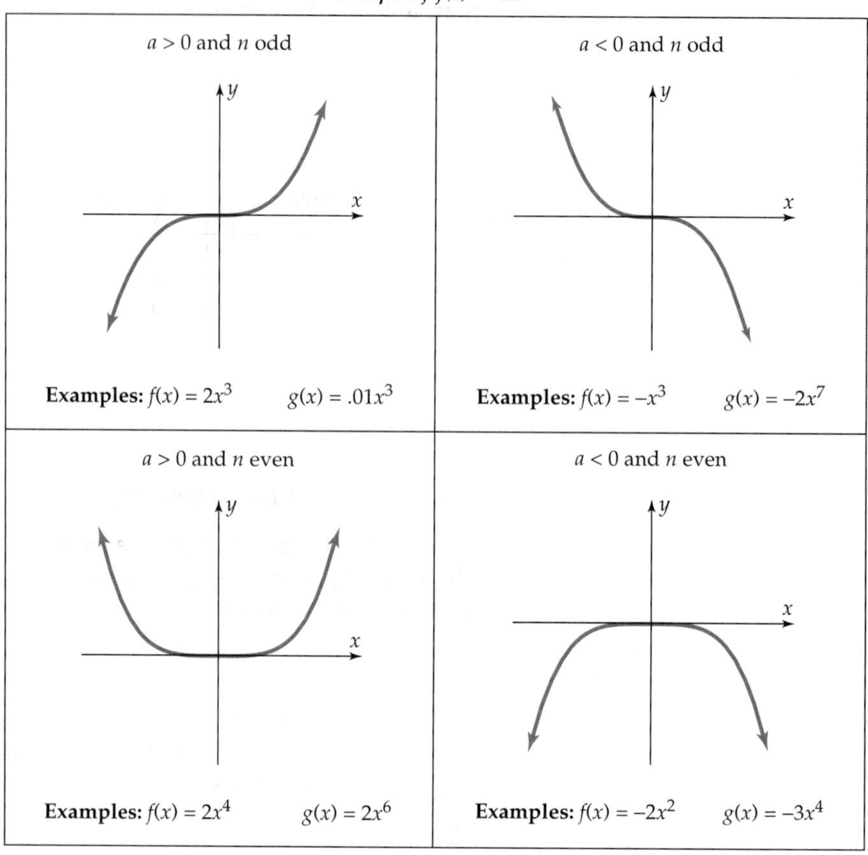

GRAPHING EXPLORATION

Verify the accuracy of the preceding summary by graphing each of the examples in the window with $-5 \le x \le 5$ and $-30 \le y \le 30$.

Properties of Polynomial Graphs

The graphs of more complicated polynomial functions can vary considerably in shape. Understanding the properties discussed below should assist you to interpret screen images correctly and to determine when a polynomial graph is complete.

Continuity

Every polynomial graph is **continuous,** meaning that it is an unbroken curve, with no jumps gaps, or holes. Furthermore, polynomial graphs have no sharp corners. Thus, neither of the graphs in Figure 4–13 is the graph of a polynomial function. On a calculator screen, however, a polynomial graph may look like a series of juxtaposed line segments, rather than a smooth, continuous curve.

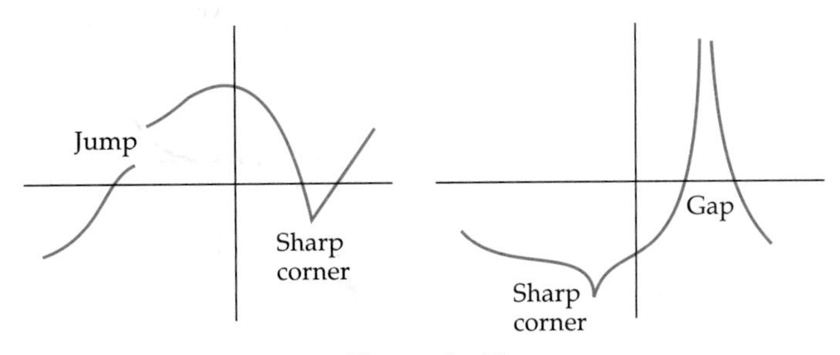

Figure 4–13

Shape of the Graph When $|x|$ Is Large

The shape of a polynomial graph at the far left and far right of the coordinate plane is easily determined by using our knowledge of graphs of functions of the form $f(x) = ax^n$.

Example 1 Consider the function $f(x) = 2x^3 + x^2 - 6x$ and the function determined by its leading term $g(x) = 2x^3$.

GRAPHING EXPLORATION

Using the standard viewing window, graph f and g on the same screen.

Do the graphs look different? Now graph f and g in the viewing window with $-20 \leq x \leq 20$ and $-10{,}000 \leq y \leq 10{,}000$. Do the graphs look almost the same?

Finally, graph f and g in the viewing window with $-100 \leq x \leq 100$ and $-1{,}000{,}000 \leq y \leq 1{,}000{,}000$. Do the graphs look virtually identical?

The reason the answer to the last question is "yes" can be understood from this table:

x	-100	-50	70	100
$-6x$	600	300	-420	-600
x^2	10,000	2,500	4,900	10,000
$g(x) = 2x^3$	$-2{,}000{,}000$	$-250{,}000$	686,000	2,000,000
$f(x) = 2x^3 + x^2 - 6x$	$-1{,}989{,}400$	$-247{,}200$	690,480	2,009,400

It shows that when $|x|$ is large, the terms x^2 and $-6x$ are insignificant compared with $2x^3$ and play a very minor role in determining the value of $f(x)$. Hence the values of $f(x)$ and $g(x)$ are relatively close. ■

Example 1 is typical of what happens in every case: When $|x|$ is very large, the highest power of x totally overwhelms all lower powers and plays the greatest role in determining the value of the function.

Behavior When $|x|$ Is Large

When $|x|$ is very large, the graph of a polynomial function closely resembles the graph of its highest degree term.

In particular, when the polynomial function has odd degree, one end of its graph shoots upward and the other end downward. When the polynomial function has even degree, both ends of its graph shoot upward or both ends shoot downward.

x-Intercepts

As we saw in Section 4.3, the x-intercepts of the graph of a polynomial function are the real roots of the polynomial. Since a polynomial of degree n has at most n distinct roots (page 219), we have:

x-Intercepts

The graph of a polynomial function of degree n meets the x-axis at most n times.

There is another connection between roots and graphs. For example, it is easy to see that the roots of $f(x) = (x + 3)^2(x + 1)(x - 1)^3$ are -3, -1, and 1. We say that

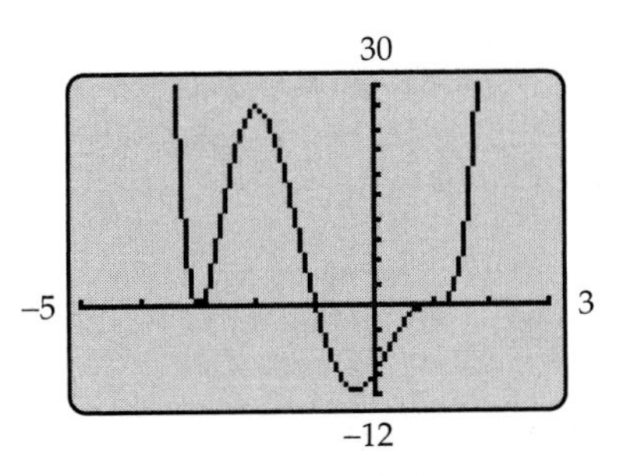

Figure 4–14

-3 is a root of multiplicity 2;

-1 is a root of multiplicity 1;

1 is a root of multiplicity 3.

Observe that the graph of $f(x)$ in Figure 4–14 does not cross the x-axis at -3 (a root whose multiplicity is an *even* number), but does cross the x-axis at -1 and 1 (roots of *odd* multiplicity). More generally, a number c is a **root of multiplicity** k of a polynomial $f(x)$ if $(x - c)^k$ is a factor of $f(x)$ and no higher power of $(x - c)$ is a factor and we have this fact:

Multiplicity and Graphs

Let c be a root of multiplicity k of a polynomial function f.

If k is odd, the graph of f crosses the x-axis at c.

If k is even, the graph of f touches, but does not cross, the x-axis at c.

Local Extrema

The term **local extremum** (plural, extrema) refers to either a local maximum or a local minimum, that is, a point where the graph as a peak or a valley.

GRAPHING EXPLORATION

Graph $f(x) = x^3 + 2x^2 - 4x - 3$ in the standard viewing window. What is the total number of peaks and valleys on the graph? What is the degree of $f(x)$?

Now graph $g(x) = x^4 - 3x^3 - 2x^2 + 4x + 5$ in the standard viewing window. What is the total number of peaks and valleys on the graph? What is the degree of $g(x)$? ■

The two polynomials you have just graphed are illustrations of the following fact, which is proved in calculus:

Local Extrema

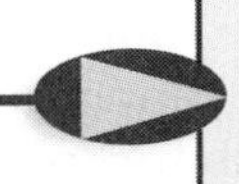

A polynomial function of degree n has at most $n - 1$ local extrema. In other words, the total number of peaks and valleys on the graph is at most $n - 1$.

Bending

A polynomial graph may bend upward or downward as indicated here by the vertical arrows (Figure 4–15):

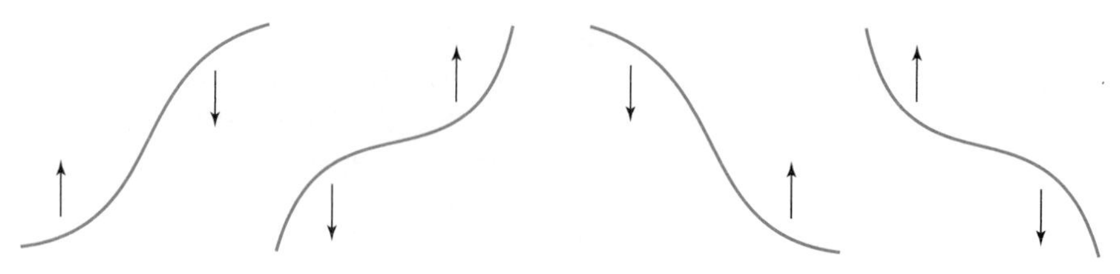

Figure 4–15

A point at which the graph changes from bending downward to bending upward (or vice versa) is called a **point of inflection.** The direction in which a graph bends may not always be clear on a calculator screen, and calculus is usually required to determine the exact location of points of inflection. The number of inflection points and hence the amount of bending in the graph are governed by these facts, which are proved in calculus:

Points of Inflection

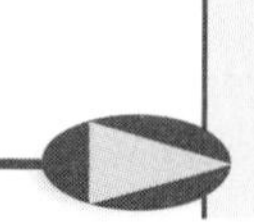

The graph of a polynomial function of degree n (with $n \geq 2$) has at most $n - 2$ points of inflection. The graph of a polynomial function of odd degree has at least one point of inflection.

Technology Tip

Points of inflection may be found by using INFLC in the TI-86/89 GRAPH MATH menu or INFLEC in the Sharp 9600 CALC menu.

Thus, the graph of a quadratic function (degree 2) has no points of inflection ($n - 2 = 2 - 2 = 0$), and the graph of a cubic has exactly one (since it has at least one and at most $3 - 2 = 1$).

Complete Graphs of Polynomial Functions

By using the facts discussed earlier, you can often determine whether or not the graph of a polynomial function is complete (that is, shows all the important features).

Example 2 Find a complete graph of

$$f(x) = x^4 + 10x^3 + 21x^2 - 40x - 80.$$

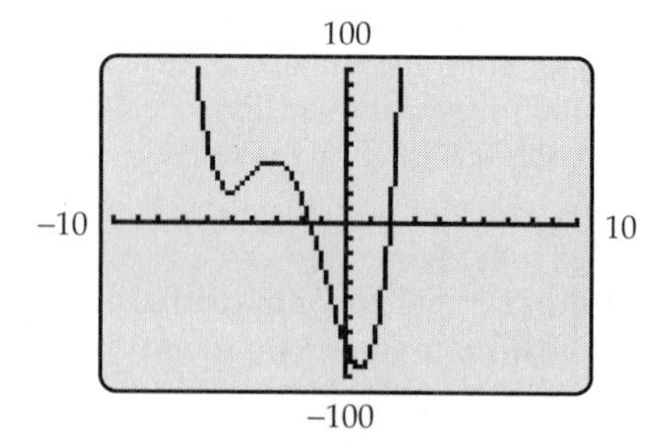

Figure 4–16

Solution Since $f(0) = -80$, the standard viewing window probably won't show a complete graph, so we try the window with $-10 \leq x \leq 10$ and $-100 \leq y \leq 100$ and obtain Figure 4–16. The three peaks and valleys shown here are the only ones because a fourth-degree polynomial graph has at most three local extrema. There cannot be more x-intercepts than the two shown here because if the graph turned toward the x-axis farther out, there would be an additional peak, which is impossible. Finally, the outer ends of the graph resemble the graph of x^4, the highest degree term (see the chart on page 232). Hence, Figure 4–16 includes all the important features of the graph and is therefore complete. ■

Example 3 The graph of $f(x) = x^3 - 1.8x^2 + x + 2$ in Figure 4–17 is similar to the graph of its leading term $y = x^3$, but does not appear to have any local extrema. However, if you use the trace feature on the flat portion of the graph to the right of the x-axis, you see that the y-coordinates increase, then decrease, then increase (try it!). Zooming in on the portion of the graph between 0 and 1 (Figure 4–18), we see that the graph actually has a tiny peak and valley (the maximum possible number of local extrema for a cubic). So Figures 4–17 and 4–18 together provide a complete graph of f. ■

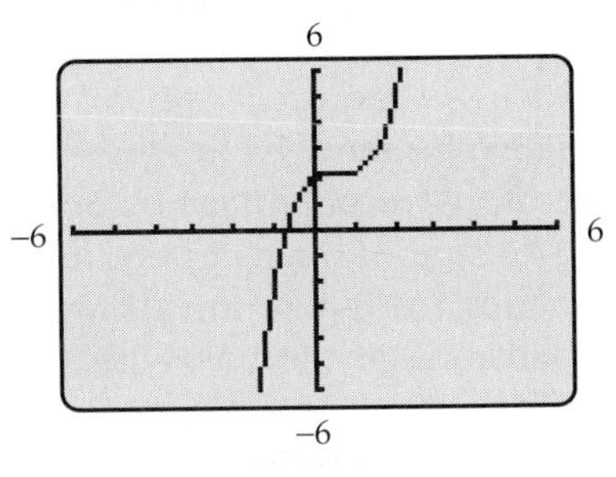

Figure 4–17

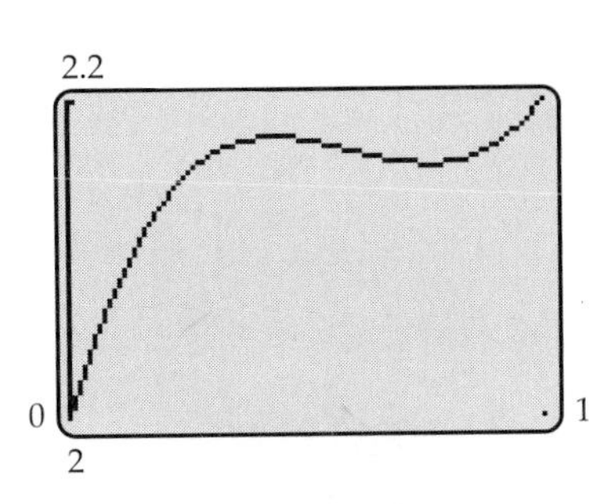

Figure 4–18

Exercise 63 shows that *no polynomial graph contains any horizontal line segments.* However, a calculator may erroneously show some, as in Figure 4–17. So always investigate such segments by using trace or zoom-in to determine any hidden behavior, as in Example 3.

Example 4 In the standard viewing window, the graph of $f(x) = .01x^5 + x^4 - x^3 - 6x^2 + 5x + 4$ looks like Figure 4–19. This cannot be a complete graph because, when $|x|$ is large, the graph of $f(x)$ must resemble the graph of $g(x) = .01x^5$, whose left end goes downward (see the chart on page 232). So the graph of $f(x)$ must turn downward and cross the x-axis somewhere to the left of the origin. Even without graphing, we can see that there must be one more peak (where the graph turns downward), making a total of four local extrema (the most a fifth-degree polynomial can have), and another x-intercept for a total of five. When these additional features are shown, we will have a complete graph.

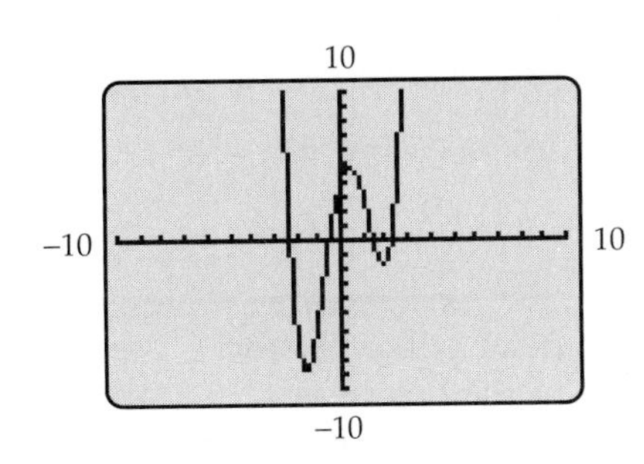

Figure 4–19

GRAPHING EXPLORATION

Find a viewing window that includes the local maximum and x-intercept not shown in Figure 4–19. When you do, the scale will be such that the local extrema and x-intercepts shown in Figure 4–19 will no longer be visible.

Consequently, a complete graph of $f(x)$ requires several viewing windows in order to see all the important features. ■

The graphs obtained in Examples 2–4 were known to be complete because in each case they included the maximum possible number of local extrema. In many cases, however, a graph may not have the largest possible number of peaks and valleys. In such cases, use any available information and try several viewing windows to obtain the most likely complete graph.

Applications

The solution of many applied problems reduces to finding a local extremum of a polynomial function.

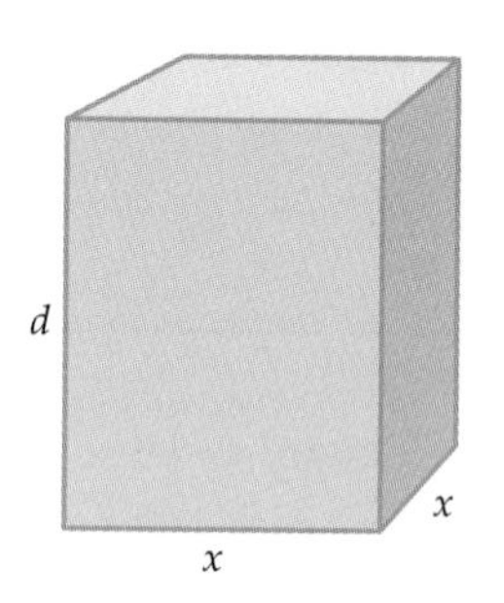

Figure 4–20

Example 5 A rectangular box with a square base (Figure 4–20) is to be mailed. The sum of the height of the box and the perimeter of the base is to be 84 inches, the maximum allowable under postal regulations. What are the dimensions of the box with largest possible volume that meets these conditions?

Solution If the length of one side of the base is x, then the perimeter of the base (the sum of the length of its four sides) is $4x$. If the height of the box is d, then $4x + d = 84$, so that $d = 84 - 4x$ and hence the volume is

$$V = x \cdot x \cdot d = x \cdot x \cdot (84 - 4x) = 84x^2 - 4x^3.$$

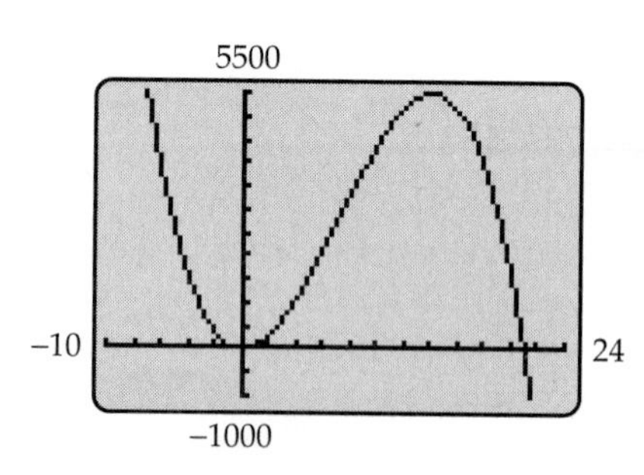

Figure 4–21

The graph of the polynomial function $V(x) = 84x^2 - 4x^3$ in Figure 4–21 is complete. (Why?) However, the only relevant part of the graph in this situation is the portion with x and $V(x)$ positive (because x is a length and $V(x)$ is a volume). The graph of $V(x)$ has a local maximum between 10 and 20 and this local maximum value is the largest possible volume for the box.

GRAPHING EXPLORATION

Use a maximum finder to find the x-value at which the local maximum occurs. State the dimensions of the box in this case. ■

Exercises 4.4

In Exercises 1–6, decide whether the given graph could possibly be the graph of a polynomial function.

1.

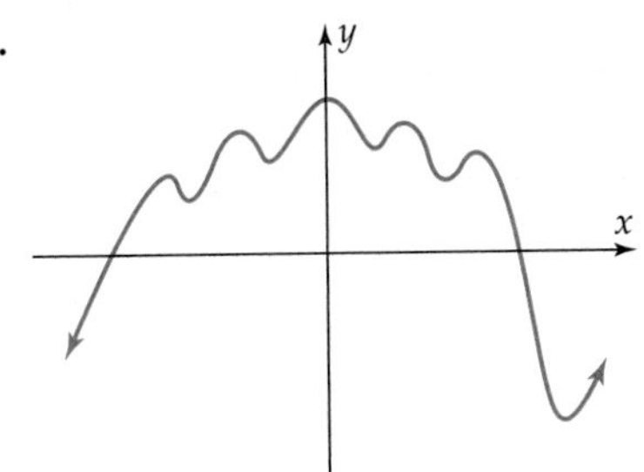

2.

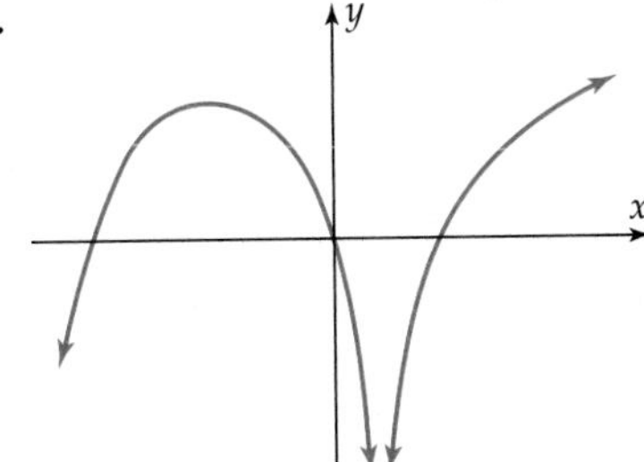

3.

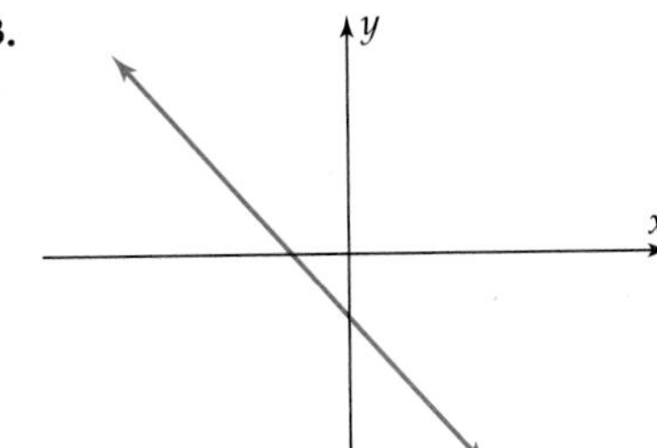

4.

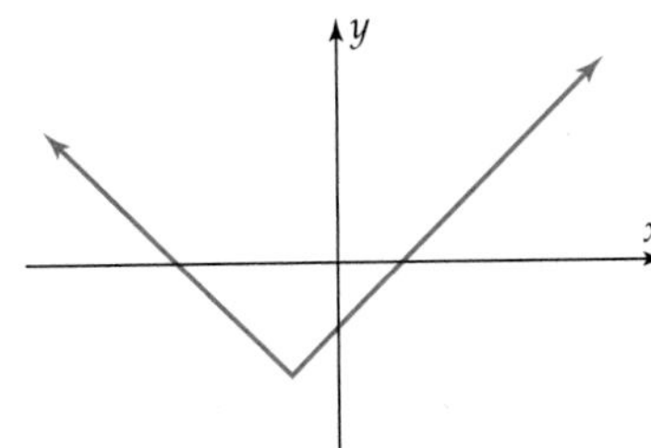

5.

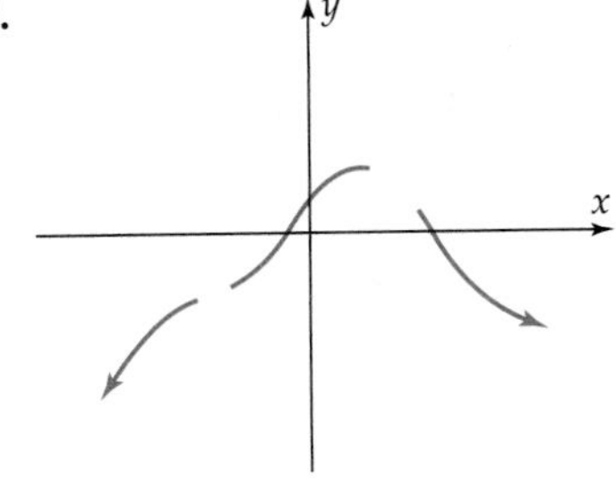

6.

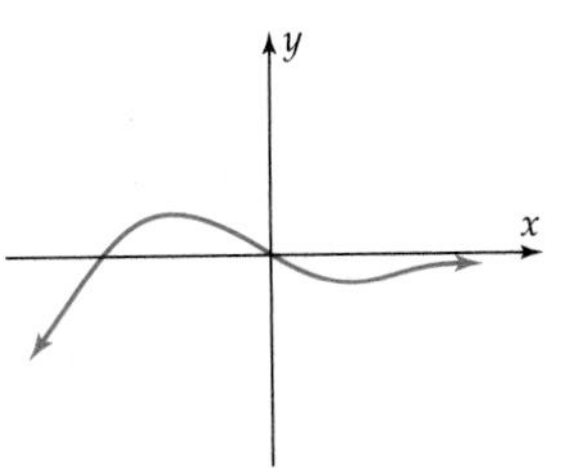

In Exercises 7–12, determine whether the given graph could possibly be the graph of a polynomial function of degree 3, of degree 4, or of degree 5.

7.

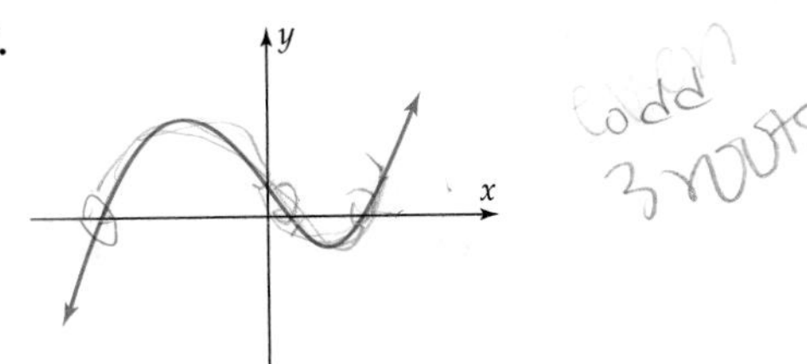

8.

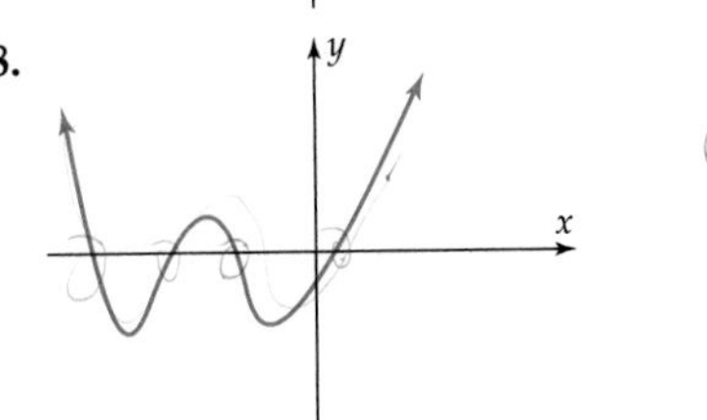

9.

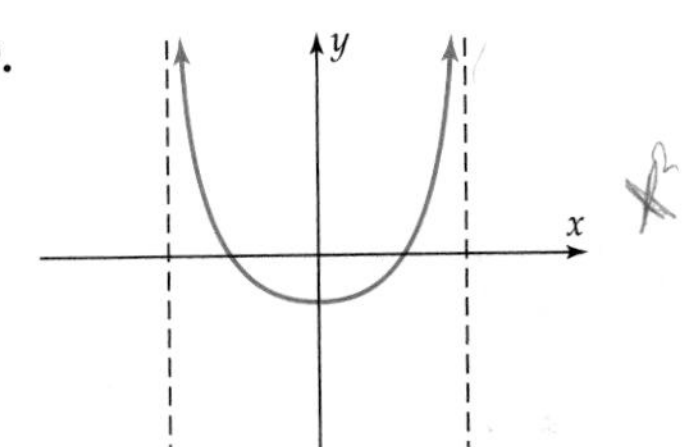

10.

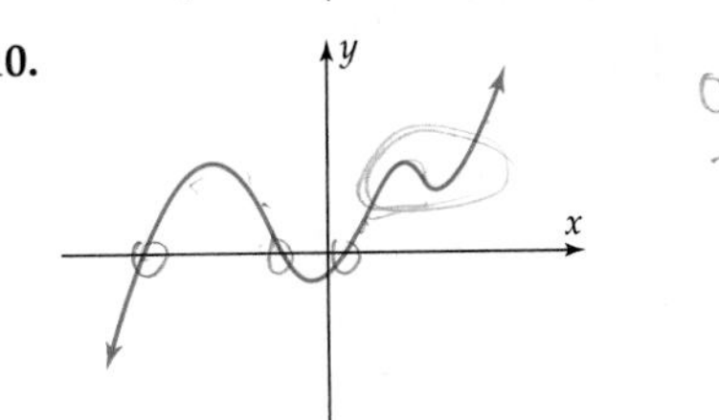

11.

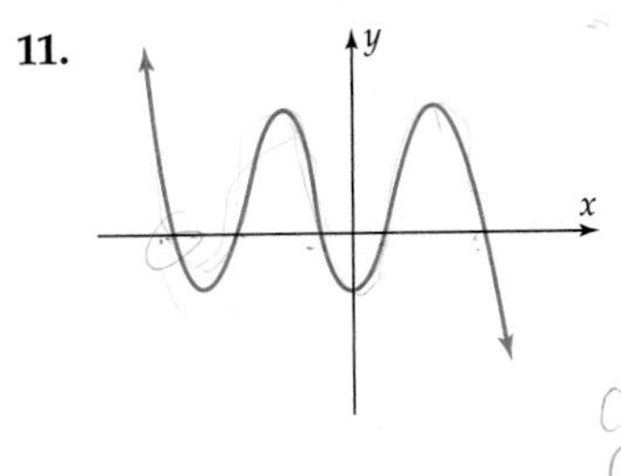

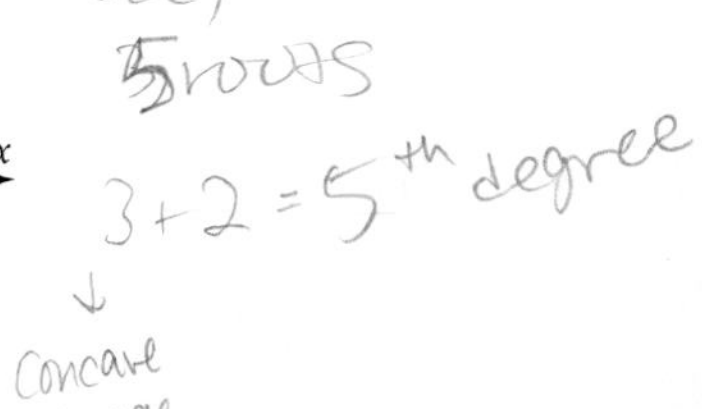

12.

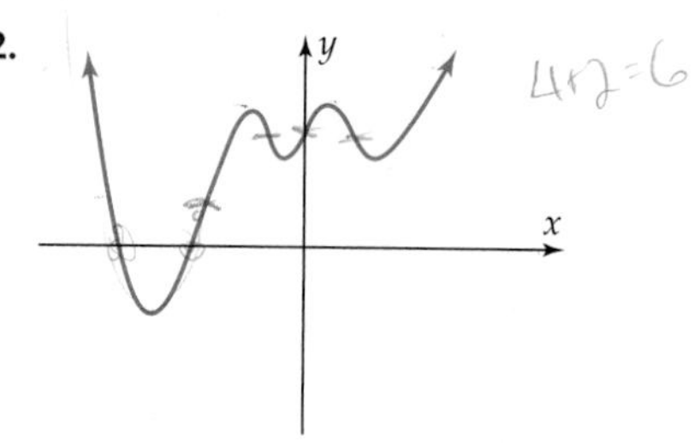

In Exercises 13 and 14, find a viewing window in which the graph of the given polynomial function f appears to have the same general shape as the graph of its leading term.

13. $f(x) = x^4 - 6x^3 + 9x^2 - 3$

14. $f(x) = x^3 - 5x^2 + 4x - 2$

In Exercises 15–18, the graph of a polynomial function is shown. List each root of the polynomial and state whether its multiplicity is even or odd.

15.

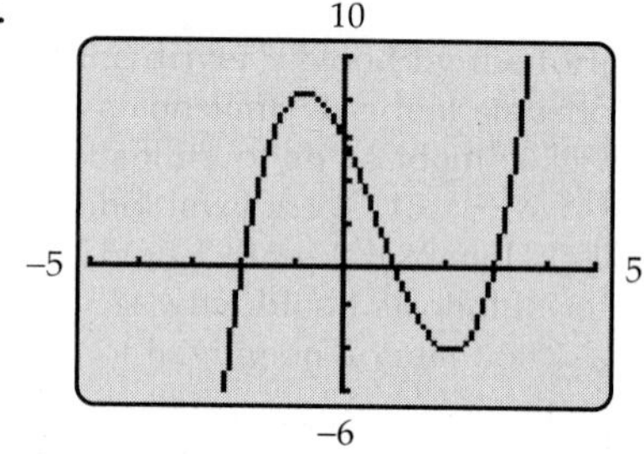

16.

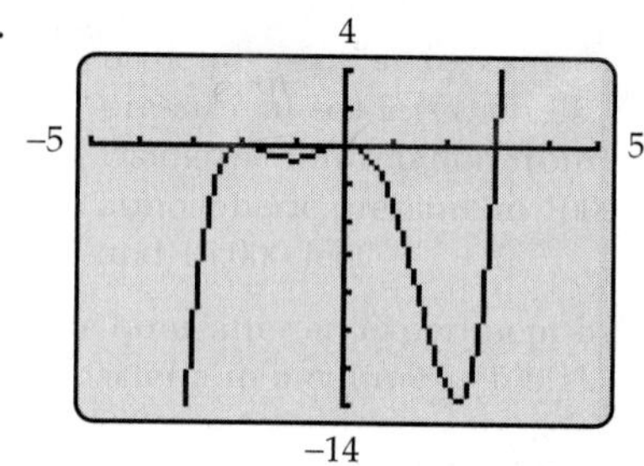

17.

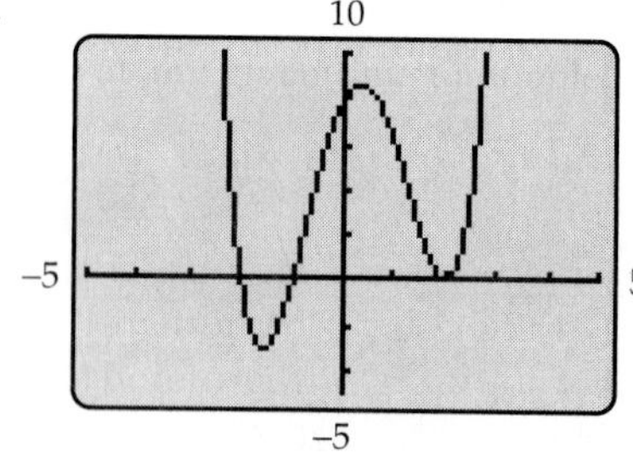

18.

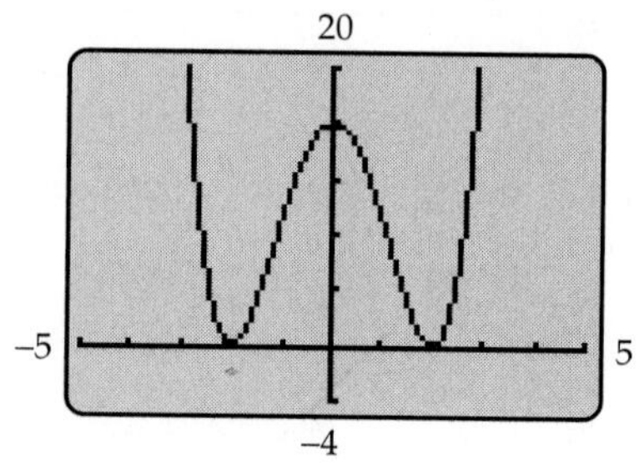

In Exercises 19–24, use your knowledge of polynomial graphs, not *a calculator, to match the given function with its graph, which is one of those shown here.*

(a)

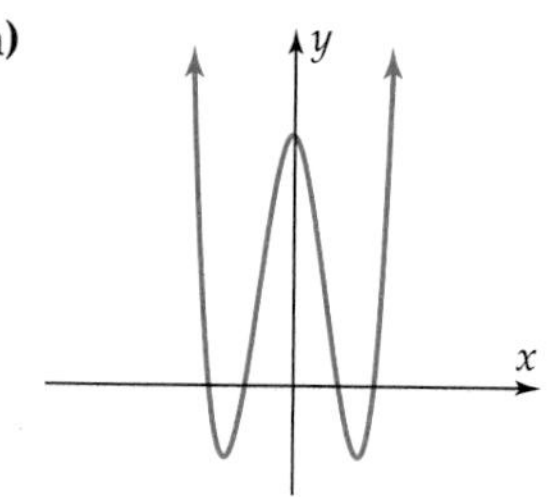

(b)

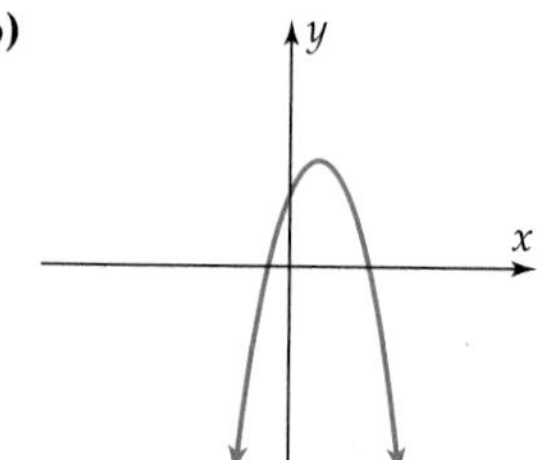

(c)

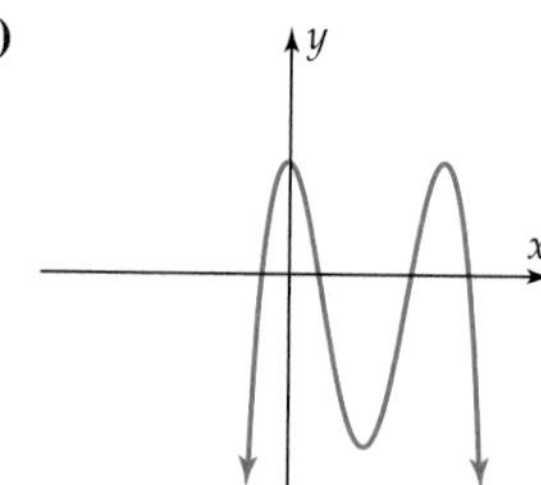

(d)

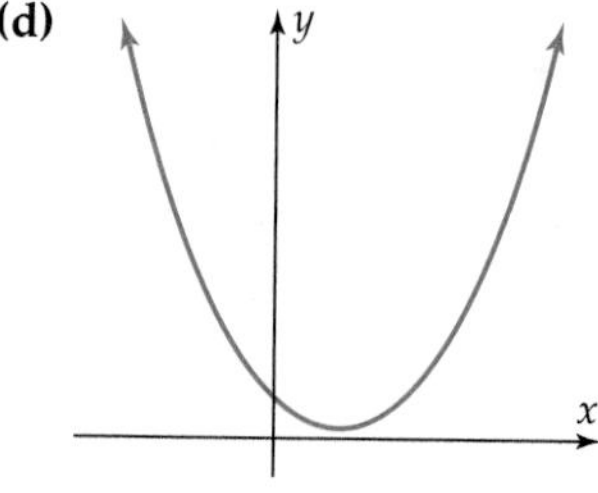

(e)

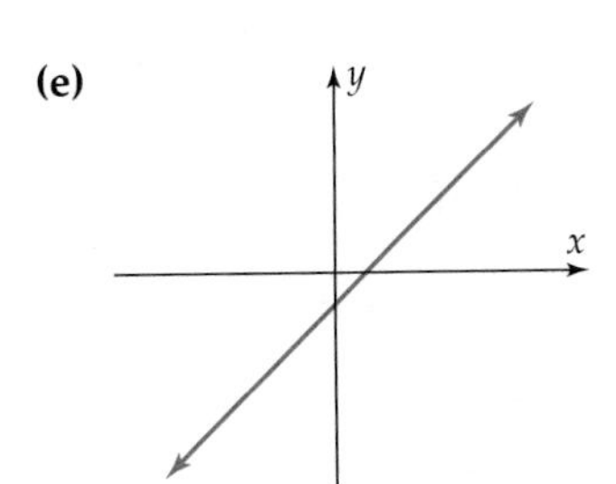

(f)

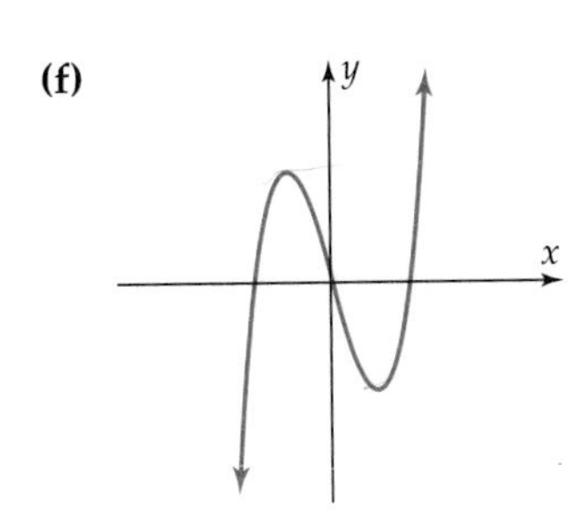

19. $f(x) = 2x - 3$

20. $g(x) = x^2 - 4x + 7$

21. $g(x) = x^3 - 4x$

22. $f(x) = x^4 - 5x^2 + 4$

23. $f(x) = -x^4 + 6x^3 - 9x^2 + 2$

24. $g(x) = -2x^2 + 3x + 1$

In Exercises 25–28, graph the function in the standard viewing window and explain why that graph cannot possibly be complete.

25. $f(x) = .01x^3 - .2x^2 - .4x + 7$

26. $g(x) = .01x^4 + .1x^3 - .8x^2 - .7x + 9$

27. $h(x) = .005x^4 - x^2 + 5$

28. $f(x) = .001x^5 - .01x^4 - .2x^3 + x^2 + x - 5$

In Exercises 29–34, find a single viewing window that shows a complete graph of the function.

29. $f(x) = x^3 + 8x^2 + 5x - 14$

30. $g(x) = x^3 - 3x^2 - 4x - 5$

31. $g(x) = -x^4 - 3x^3 + 24x^2 + 80x + 15$

32. $f(x) = x^4 - 10x^3 + 35x^2 - 50x + 24$

33. $f(x) = 2x^5 - 3.5x^4 - 10x^3 + 5x^2 + 12x + 6$

34. $g(x) = x^5 + 8x^4 + 20x^3 + 9x^2 - 27x - 7$

In Exercises 35–40, find a complete graph of the function and list the viewing window(s) that show this graph.

35. $f(x) = .1x^5 + 3x^4 - 4x^3 - 11x^2 + 3x + 2$

36. $g(x) = x^4 - 48x^3 - 101x^2 + 49x + 50$

37. $g(x) = .03x^3 - 1.5x^2 - 200x + 5$

38. $f(x) = .25x^6 + .25x^5 - 35x^4 - 7x^3 + 823x^2 + 25x - 2750$

39. $g(x) = 2x^3 - .33x^2 - .006x + 5$

40. $f(x) = .3x^5 + 2x^4 - 7x^3 + 2x^2$

41. (a) Explain why the graph of a cubic polynomial function has either two local extrema or none at all. [*Hint:* If it had only one, what would the graph look like when $|x|$ is very large?]

(b) Explain why the general shape of the graph of a cubic polynomial function must be one of the following:

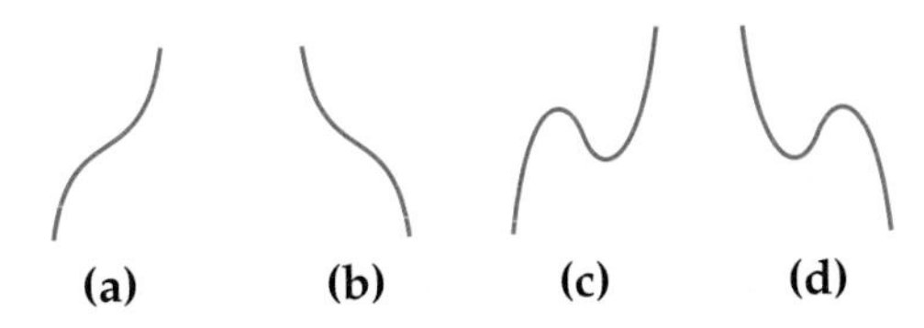

42. The figure shows an incomplete graph of a fourth-degree even polynomial function f. (Even functions were defined in Excursion 3.4.A.)

(a) Find the roots of f.

(b) Explain why

$$f(x) = k(x - a)(x - b)(x - c)(x - d),$$

where a, b, c, d are the roots of f.

(c) Experiment with your calculator to find the value of k that produces the graph in the figure.

(d) Find all local extrema of f.

(e) List the approximate intervals on which f is increasing and those on which it is decreasing.

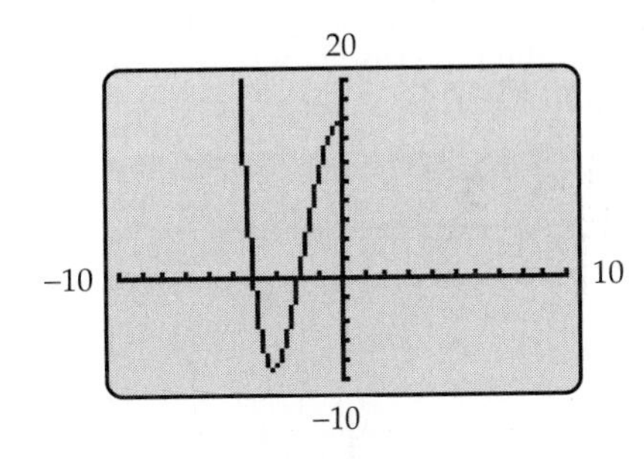

43. A complete graph of a polynomial function g is shown on the next page.

(a) Is the degree of $g(x)$ even or odd?

(b) Is the leading coefficient of $g(x)$ positive or negative?

(c) What are the real roots of $g(x)$?

(d) What is the smallest possible degree of $g(x)$?

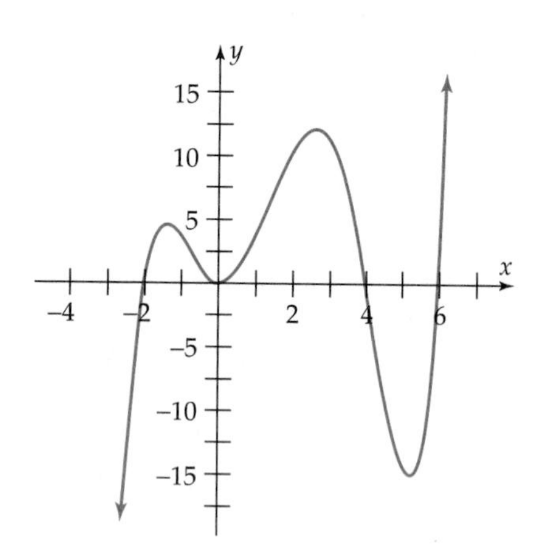

44. Do Exercise 43 for the polynomial function g whose complete graph is shown here.

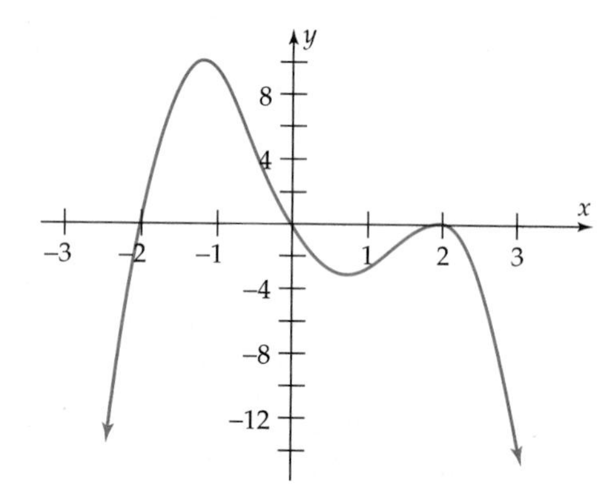

45. The figure is a partial view of the graph of a cubic polynomial whose leading coefficient is negative. Which of the patterns shown in Exercise 41 does this graph have?

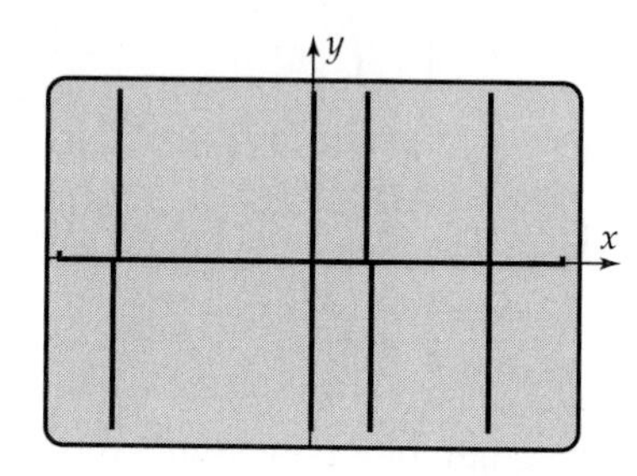

46. The figure is a partial view of the graph of a fourth-degree polynomial. Sketch the general shape of the graph and state whether the leading coefficient is positive or negative.

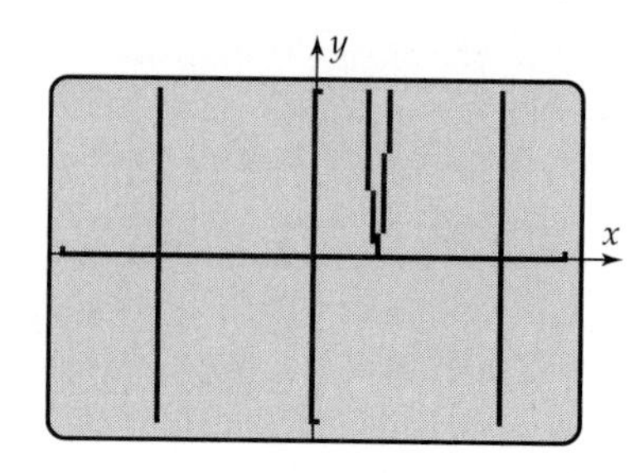

In Exercises 47–56, sketch a complete graph of the function. Label each x-intercept and the coordinates of each local extremum; find intercepts and coordinates exactly when possible and otherwise approximate them.

47. $f(x) = x^3 - 3x^2 + 4$ **48.** $g(x) = 4x - 4x^3/3$

49. $h(x) = .25x^4 - 2x^3 + 4x^2$

50. $f(x) = .25x^4 - 2x^3/3$

51. $g(x) = 3x^3 - 18.5x^2 - 4.5x - 45$

52. $h(x) = 2x^3 + x^2 - 4x - 2$

53. $f(x) = x^5 - 3x^3 + x + 1$

54. $g(x) = .25x^4 - x^2 + .5$

55. $h(x) = 8x^4 + 22.8x^3 - 50.6x^2 - 94.8x + 138.6$

56. $f(x) = 32x^6 - 48x^4 + 18x^2 - 1$

57. Name tags can be sold for $29 per thousand. The cost of manufacturing x thousand tags is $.001x^3 + .06x^2 - 1.5x$ dollars. Assuming that all tags manufactured are sold,

(a) What number of tags should be made to guarantee a maximum profit? What will that profit be?

(b) What is the largest number of tags that can be made without losing money?

58. When there are 22 apple trees per acre, the average yield has been found to be 500 apples per tree. For each additional tree planted per acre, the yield per tree decreases by 15 apples per tree. How many additional trees per acre should be planted to maximize the yield?

59. An auto parts manufacturer makes radiators that sell for $350 each. The cost of producing x radiators is approximated by the function $C(x) = 600{,}000 - 25x + .01x^2$.

(a) What is the revenue function in this situation? What is the profit function?

(b) How many radiators must be sold in order to make any profit?

(c) What is the maximum possible number of radiators that can be sold without losing money?

(d) What number of radiators will produce the largest possible profit?

60. The top of a 12-ounce can of soda pop is three times thicker than the sides and bottom (so that the flip top opener will work properly) and the can has a volume of 355 cubic centimeters. What should the radius and height of the can be in order to use the least possible amount of metal? [Assume that the entire can is made from a single sheet of metal, with three layers being used for the top. Example 3 in Section 2.4 may be helpful.]

61. An open-top reinforced box is to be made from a 12-by-36-inch piece of cardboard as in Exercise 41 of Section 4.3. What size squares should be cut from the corners in order to have a box with maximum volume?

62. A box with a lid is to be made from a 48-by-24-inch piece of cardboard by cutting and folding, as in Exercise 42 of Section 4.3. What size squares should be cut from the two corners in order to have a box of maximum volume?

Thinkers

63. (a) Graph $g(x) = .01x^3 - .06x^2 + .12x + 3.92$ in the viewing window with $-3 \le x \le 3$ and $0 \le y \le 6$ and verify that the graph appears to coincide with the horizontal line $y = 4$ between $x = 1$ and $x = 3$. In other words, it appears that every x with $1 \le x \le 3$ is a solution of the equation

$$.01x^3 - .06x^2 + .12x + 3.92 = 4.$$

Explain why this is impossible. Conclude that the actual graph is not horizontal between $x = 1$ and $x = 3$.

(b) Use the trace feature to verify that the graph is actually rising from left to right between $x = 1$ and $x = 3$. Find a viewing window that shows this.

(c) Show that it is not possible for the graph of a polynomial $f(x)$ to contain a horizontal segment. [*Hint:* A horizontal line segment is part of the horizontal line $y = k$ for some constant k. Adapt the argument in part (a), which is the case $k = 4$.]

64. (a) Let $f(x)$ be a polynomial of odd degree. Explain why $f(x)$ must have at least one real root. [*Hint:* Why must the graph of f cross the x-axis, and what does this mean?]

(b) Let $g(x)$ be a polynomial of even degree, with a negative leading coefficient and a positive constant term. Explain why $g(x)$ must have at least one positive and at least one negative root.

65. For x-values in a particular interval, a nonpolynomial function $g(x)$ can often be approximated by a polynomial function, meaning that there is some polynomial $f(x)$ such that $g(x) \approx f(x)$ for every x in the interval.

(a) In the standard viewing window, graph both $g(x) = \sqrt{x}$ (which is not a polynomial function) and the polynomial function

$$f(x) = .26705x^3 - .78875x^2 + 1.3021x + .22033.$$

Are the graphs similar?

(b) Graph $f(x)$ and $g(x)$ in the viewing window with $0.26 \le x \le 1$ and $0 \le y \le 1$. Does it now appear that $f(x)$ is a good approximation of $g(x)$ over the interval [.26, 1]?

(c) For any particular value of x, the error in this approximation is the difference between $f(x)$ and $g(x)$. In other words, $h(x) = f(x) - g(x)$ measures the error in the approximation. Graph the function $h(x)$ in the viewing window with $.26 \le x \le 1$ and $-.001 \le y \le .001$ and use the trace feature to determine the maximum error in the approximation.

66. (a) Graph $f(x) = x^3 - 4x$ in the viewing window with $-3 \le x \le 3$ and $-5 \le y \le 5$.

(b) Graph the difference quotient of $f(x)$ (with $h = .01$) on the same screen.

(c) Find the x-coordinates of the relative extrema of $f(x)$. How do these numbers compare with the x-intercepts of the difference quotient?

67. The graph of

$$f(x) = (x + 18)(x^2 - 20)(x - 2)^2(x - 10)$$

has x-intercepts at each of its roots, that is, at $x = -18, \pm\sqrt{20} \approx \pm 4.472$, 2, and 10. It is also true that $f(x)$ has a relative minimum at $x = 2$.

(a) Draw the x-axis and mark the roots of $f(x)$. Then use the fact that $f(x)$ has degree 6 (why?) to sketch the general shape of the graph (as was done for cubics in Exercise 41).

(b) Now graph $f(x)$ in the standard viewing window. Does the graph resemble your sketch? Does it even show all the x-intercepts between -10 and 10?

(c) Graph $f(x)$ in the viewing window with $-19 \le x \le 11$ and $-10 \le y \le 10$. Does this window include all the x-intercepts as it should?

(d) List viewing windows that give a complete graph of $f(x)$.

EXCURSION Polynomial Models*

Linear regression was used in Section 2.5 to construct a linear function that modeled a set of data points. When the scatter plot of the data points looks more like a higher degree polynomial graph than a straight line, similar least squares regression procedures are available on most calculators for constructing quadratic, cubic, and quartic (fourth-degree) polynomial functions to model the data.

Example 1 The table below, which is based on statistics from the Department of Health and Human Services, gives the cumulative number of reported cases of AIDS in the United States from 1982 through 1996. It shows, for example, that 41,662 cases were reported from 1982 through 1986.

Year	Cases	Year	Cases	Year	Cases
1982	1,563	1987	70,222	1992	278,189
1983	4,647	1988	105,489	1993	380,601
1984	10,845	1989	147,170	1994	457,789
1985	22,620	1990	188,872	1995	529,282
1986	41,662	1991	232,383	1996	598,433

Technology Tip

Quadratic, cubic, and quartic regression are denoted by QuadReg, CubicReg, QuartReg on the CALC submenus of the TI-83 STAT menu and TI-89 Data Editor (APPS menu); by P2Reg, P3Reg, P4Reg on the CALC menu of the TI-86 STAT menu; by x^2, x^3, x^4 on the CALC REG submenu of the Casio 9850 STAT menu; and by RG_x^2, Rg_x^3, Rg_x^4 on the REG submenu of the Sharp 9600 STAT menu. On HP-38, select STAT in the LIB menu and choose quadratic or cubic on the SYMB SETUP menu; quartic regression is not available.

Letting $x = 0$ correspond to 1980 and plotting the data points (2, 1563), (3, 4647), etc., we obtain the scatter plot in Figure 4–22. The points are not in a straight line, but could be part of a polynomial graph of degree 2 or more.

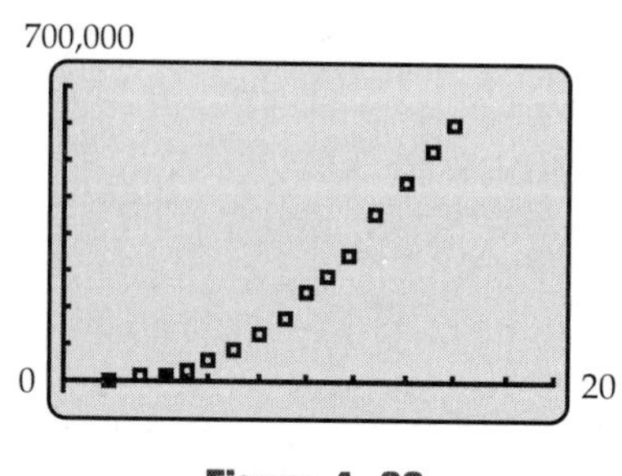

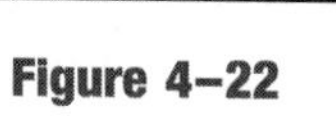

Figure 4–22

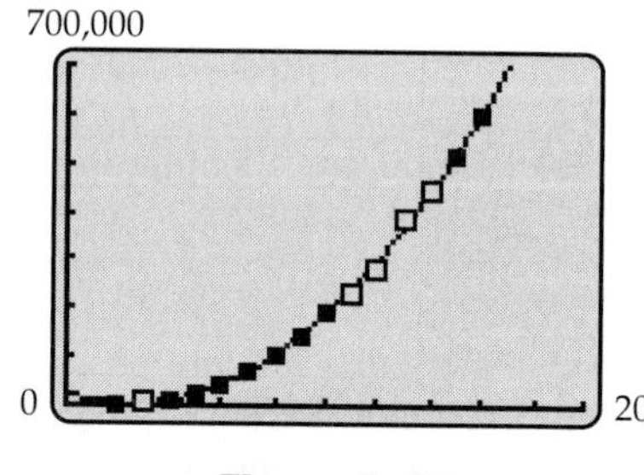

Figure 4–23

We use the regression feature of a calculator to find three possible models for this data (the procedure is the same as for linear regression; see the Technology Tip on page 101 and the one in the margin here):

*This section is optional; its prerequisite is Section 2.5. This material will be used in the optional Section 5.5 and in clearly identifiable exercises, but not elsewhere in the text.

Quadratic: $f(x) = 3362.1x^2 - 17{,}270.3x + 24{,}977.1$*

Cubic: $g(x) = 2.18x^3 + 3303.21x^2 - 16{,}811.16x + 24{,}042.78$

Quartic: $h(x) = -13.17x^4 + 476.33x^3 - 2474.98x^2 + 10{,}384.48x - 15{,}338.41.$

The graph of the quadratic model f in Figure 4–23 appears to be a good fit. In fact the graphs of f, g, and h are virtually identical in this viewing window, as you can easily verify.

Although any one of f, g, or h provides a reasonable model for the given data, our knowledge of polynomial graphs suggests that the quartic model h should not be used for predicting future results. As x gets larger, the graph of h will resemble that of $y = -13.25x^4$, which turns downward. However, the cumulative number of cases can't decrease (even when there are no new cases, the cumulative total stays the same).

GRAPHING EXPLORATION

Graph the functions f, g, and h in the window with $0 \le x \le 26$ and $0 \le y \le 1{,}800{,}000$. In this window can you distinguish the graphs of f and g? Assuming no medical breakthroughs or changes in the current social situation, does the graph of f seem to be a plausible model for the next few years? What about the graph of h? ■

Example 2 The average yearly price of Pepsico stock in selected years is given in the table.

Year	1981	1983	1985	1987	1989	1991	1993	1995	1997
Price	1.88	2.02	3.15	5.74	8.79	15.30	19.60	22.94	35.75

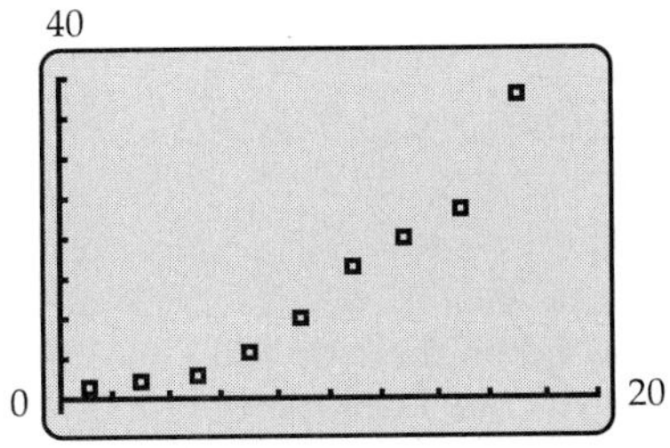

40

0 20

Figure 4–24

The scatter plot of the data in Figure 4–24 suggests a parabola. However, the data points climb quite steeply at the right side, so a fourth-degree polynomial graph might fit them better. Quartic regression produces this model for the data:

$$f(x) = .0018x^4 - .0626x^3 + .825x^2 - 3.0741x + 4.5459.$$

The graph of f in Figure 4–25 appears to be a reasonably good fit for the data.

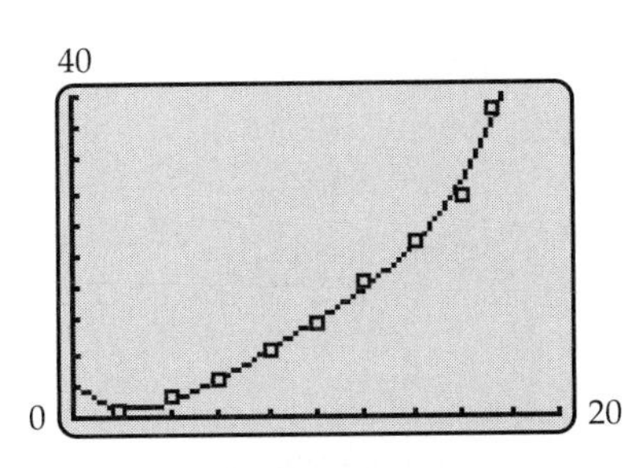

Figure 4–25

*Here and below, coefficients are rounded for convenient reading, but the full coefficients are used to produce the graphs.

The function f may be used to estimate the average price of the stock in other years. For instance,

$$f(10) \approx 11.89 \qquad \text{and} \qquad f(16) \approx 29.38,$$

meaning that the average price was approximately $11.89 in 1990 and $29.38 in 1996. These estimates compare favorably with the actual average prices of $11.75 and $31.01. Once again, however, a model may not be accurate when applied outside the range of points used to construct it. For instance, $f(18.75) \approx 49.19$, suggesting that the average price in late 1998 was about $49.19. In fact, Pepsico stock was selling at $29.44 on September 30, 1998. ■

NOTE You must have at least three data points for quadratic regression, at least four for cubic regression, and at least 5 for quartic regression. If you have exactly the required minimum number of data points, no two of them can have the same first coordinate. In this case, the polynomial regression function will pass through all of the data points (an exact fit). When you have more than the minimum number of data points required, the fit will generally be approximate rather than exact.

Exercises 4.4.A

In Exercises 1–4, a scatter plot of data is shown. State the type of polynomial model that seems most appropriate for the data (linear, quadratic, cubic, or quartic). If none of them is likely to provide a reasonable model, say so.

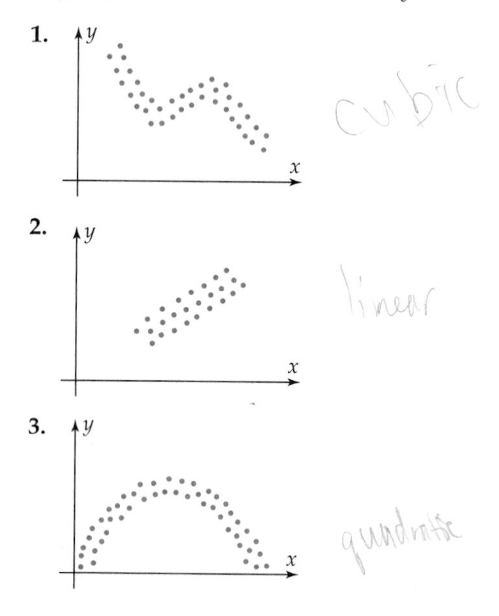

4.

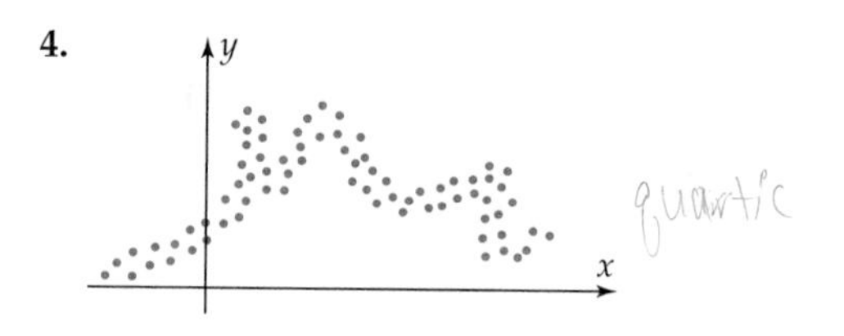

5. The table, which is based on the 1996 FBI Uniform Crime Report, shows the rate of property crimes per 100,000 population.

Year	Crimes	Year	Crimes
1980	5553.3	1990	5088.5
1982	5032.5	1992	4902.7
1984	4492.1	1994	4660
1986	4862.6	1996	4444.8
1988	5027.1		

(a) Use cubic regression to find a polynomial function that models this data, with $x = 0$ corresponding to 1980.

(b) According to this model, what was the property crime rate in 1987 and 1995?
(c) The actual crime rate was about 4250 in 1997. What does the model predict?
(d) For how many years in the future is this model likely to be a reasonable one?

6. The table, which is based on the FBI Uniform Crime Report, shows the rate of violent crimes per 100,000 population.

Year	Crimes	Year	Crimes
1980	596.6	1990	731.8
1982	571.1	1992	757.5
1984	539.2	1994	713.6
1986	617.7	1996	634.1
1988	637.2		

(a) Use cubic regression to find a polynomial function that models this data, with $x = 0$ corresponding to 1980.
(b) According to this model, what was the violent crime rate in 1983 and 1993?
(c) According to the model, in what year were violent crimes at their highest level?
(d) Do you think this model could accurately predict the crime rate in 2002?

7. The table shows the air temperature at various times during a spring day in Gainesville, Florida.

Time	Temp (F°)	Time	Temp (F°)
6 A.M.	52	1 P.M.	82
7 A.M.	56	2 P.M.	86
8 A.M.	61	3 P.M.	85
9 A.M.	67	4 P.M.	83
10 A.M.	72	5 P.M.	78
11 A.M.	77	6 P.M.	72
noon	80		

(a) Sketch a scatter plot of the data, with $x = 0$ corresponding to midnight.
(b) Find a quadratic polynomial model for the data.
(c) What is the predicted temperature for noon? For 9 A.M.? For 2 P.M.?

8. The table, which is based on data from the U.S. Education Department, Office of Special Education and Rehabilitation Services, shows the number of disabled children (in thousands) in federally supported programs.

School Year Beginning in Fall of	Number of Children
1976	3692
1980	4142
1985	4317
1987	4446
1988	4544
1989	4641
1990	4771
1991	4949
1992	5125
1993	5309
1994	5378
1995	5573

(a) Sketch a scatter plot of the data, with $x = 1$ corresponding to 1976.
(b) Find a cubic polynomial model for this data.

Use the following table for Exercises 9–10. It shows the median income of U.S. households in 1995 dollars. (Source: U.S. Census Bureau)

Year	Median Income	Year	Median Income
1980	$32,795	1988	$35,073
1981	32,263	1989	35,526
1982	32,155	1990	34,914
1983	31,957	1991	33,709
1984	32,878	1992	33,278
1985	33,452	1993	32,949
1986	34,620	1994	33,178
1987	34,962	1995	34,076

9. **(a)** Sketch a scatter plot of the data from 1980 to 1995, with $x = 0$ representing 1980.
(b) Decide whether a quadratic or quartic model seems more appropriate.
(c) Find an appropriate polynomial model.
(d) Use the model to predict the median income in 2000. Does your answer seem reasonable?
(e) According to this model, in what year will the median income exceed $50,000?

10. **(a)** Sketch a scatter plot of the data from 1989 to 1995, with $x = 0$ representing 1989.
(b) Decide whether a quadratic or quartic model seems more appropriate.
(c) Find an appropriate polynomial model.
(d) Use the model to predict the median income in 2000. Does your answer seem reasonable?
(e) According to this model, in what year will the median income exceed $50,000?
(f) Which model [the one in part (c) or the one in Exercise 9] is the better predictor? Justify your answer.

11. The table shows the percent of eighth graders who had smoked within 30 days of the time the data was collected each year.

Year	Percent Who Smoked
1991	14.3
1992	15.5
1993	16.7
1994	18.6
1995	19.1
1996	21.0

(a) Sketch a scatter plot of the data, with $x = 0$ corresponding to 1990.
(b) Find a polynomial model of the data and justify your choice.
(c) Do you think your model will remain accurate into the future? For how long?

12. The table shows the U.S. public debt per person (in dollars). (*Source:* U.S. Department of Treasury, Bureau of Public Debt)

Year	Debt	Year	Debt
1980	$3,985	1989	$11,545
1981	4,338	1990	13,000
1982	4,913	1991	14,436
1983	5,870	1992	15,846
1984	6,640	1993	17,105
1985	7,598	1994	18,025
1986	8,774	1995	18,930
1987	9,615	1996	19,805
1988	10,534		

(a) Sketch a scatter plot of the data.
(b) Find a polynomial model of the data and justify your choice.

13. **(a)** Find both a cubic and a quartic model for the data on the number of unemployed people in the labor force in Example 4 of Section 2.5.
(b) Does either model seem likely to be accurate in the future?

14. The table shows the number of people murdered each year per 100,000 population. (*Source:* FBI)

Year	Number	Year	Number
1987	8.12	1993	8.74
1989	8.35	1994	8.47
1990	8.71	1995	8.12
1991	8.88	1996	7.71
1992	8.65		

(a) Sketch a scatter plot of the data with $x = 0$ corresponding to 1987.
(b) Find a cubic model for the data.
(c) Compare the model in part (b) with the quartic model for the same data in Exercise 39 of Section 4.3. Which model seems likely to be accurate for years after 1996? For how long do you think it is likely to remain accurate?

4.5 Rational Functions

A **rational function** is a function whose rule is the quotient of two polynomials, such as

$$f(x) = \frac{1}{x}, \qquad t(x) = \frac{4x - 3}{2x + 1}, \qquad k(x) = \frac{2x^3 + 5x + 2}{x^2 - 7x + 6}.$$

A polynomial function is defined for every real number, but the rational function $f(x) = g(x)/h(x)$ is defined only when its denominator is nonzero. Hence,

Domain

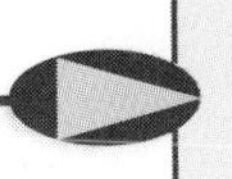

The domain of the rational function $f(x) = \frac{g(x)}{h(x)}$ is the set of all real numbers that are *not* roots of the denominator $h(x)$.

For instance, the domain of $f(x) = \frac{x^2 + 3x + 1}{x^2 - x - 6}$ is the set of all real numbers except -2 and 3 [the roots of $x^2 - x - 6 = (x + 2)(x - 3)$].

Calculators sometimes do a poor job of graphing rational functions. Hence, the emphasis here is on the algebraic analysis of rational functions, so that you will be able to interpret misleading screen images. The key to understanding the behavior of rational functions is this fact from arithmetic:

The Big-Little Principle

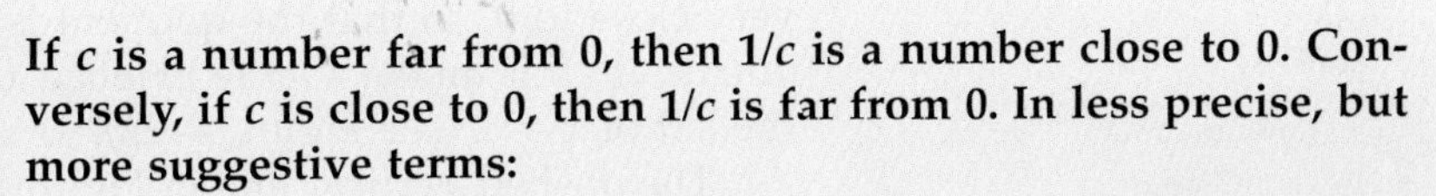

If c is a number far from 0, then $1/c$ is a number close to 0. Conversely, if c is close to 0, then $1/c$ is far from 0. In less precise, but more suggestive terms:

$$\frac{1}{\textbf{big}} = \textbf{little} \qquad \textbf{and} \qquad \frac{1}{\textbf{little}} = \textbf{big}.$$

For example, 5000 is big (far from 0) and $1/5000$ is little (close to 0). Similarly, $-1/1000$ is very close to 0, but $\frac{1}{-1/1000} = -1000$ is far from 0. To see the role played by the Big-Little Principle, we consider a typical example.

Example 1 Without using a calculator graph, describe the graph of $f(x) = \frac{x + 1}{2x - 4}$ near $x = 2$ and far from $x = 2$. Then sketch the graph.

Solution The function is not defined when $x = 2$ because the denominator is 0 there. When $x > 2$ and is very close to 2, then

The numerator $x + 1$ is very close to $2 + 1 = 3$;

The denominator $2x - 4$ is positive and very close to $2 \cdot 2 - 4 = 0$.

Therefore,

$$f(x) = \frac{x+1}{2x-4} \approx \frac{3}{\text{little}} = 3 \cdot \frac{1}{\text{little}} = 3 \cdot \text{big} = \text{BIG!}$$

You can confirm this fact by making a table of values for $f(x)$ when $x = 2.1$, 2.01, 2.001, etc., as in Figure 4–26. In graphical terms, the points with x-coordinates slightly larger than 2 have gigantic y-coordinates, so that the graph shoots upward just to the right of $x = 2$ (Figure 4–27).

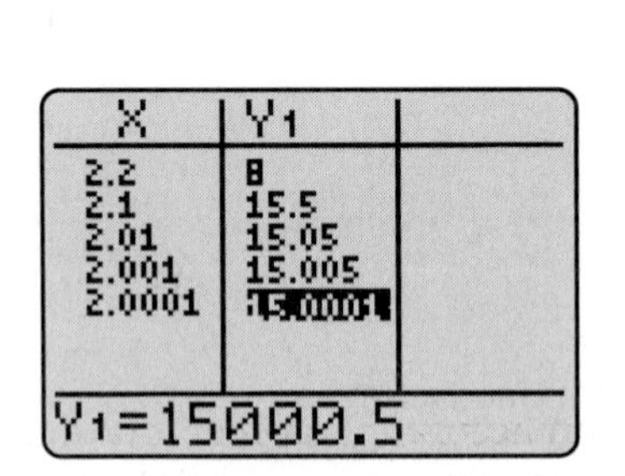

Figure 4–26

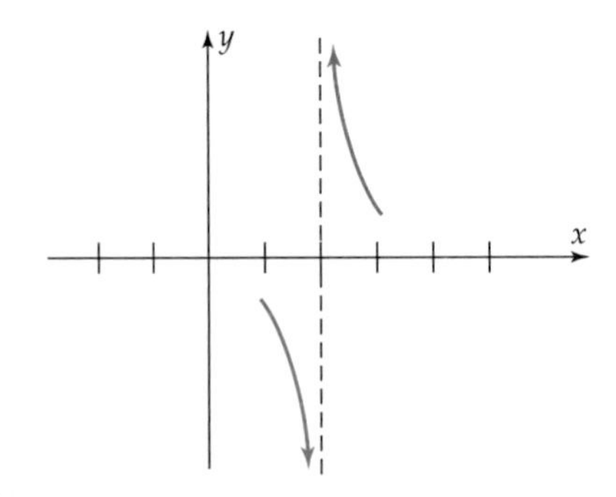

Figure 4–27

A similar analysis when $x < 2$ and very close to 2 shows that the numerator of $f(x)$ is very close to 3 and the denominator is negative and very close to 0, so that the quotient is a negative number far from 0. (If you doubt this, make a table.) Hence, the graph shoots downward just to the left of $x = 2$ (see Figure 4–27).

The dashed vertical line in Figure 4–27 is included for easier visualization; it is *not* part of the graph, but is called a **vertical asymptote** of the graph. The graph gets closer and closer to the vertical asymptote, but never crosses it because $f(x)$ is not defined when $x = 2$.

To see what the graph looks like far from $x = 2$, we rewrite the rule of f like this:

$$f(x) = \frac{x+1}{2x-4} = \frac{\dfrac{x+1}{x}}{\dfrac{2x-4}{x}} = \frac{1 + \dfrac{1}{x}}{2 - \dfrac{4}{x}}$$

As x gets larger in absolute value (far from 0), both $1/x$ and $4/x$ get very close to 0 by the Big-Little Principle. Consequently, $f(x) = \dfrac{1 + (1/x)}{2 - (4/x)}$ gets very close to $\dfrac{1+0}{2-0} = \dfrac{1}{2}$. So when $|x|$ is large, the graph gets closer and closer to the horizontal line $y = 1/2$, but never touches it, as shown in

Figure 4–28. The line $y = 1/2$ is called a **horizontal asymptote** of the graph.

The preceding information, together with a few hand-plotted points, produces the graph in Figure 4–28. ■

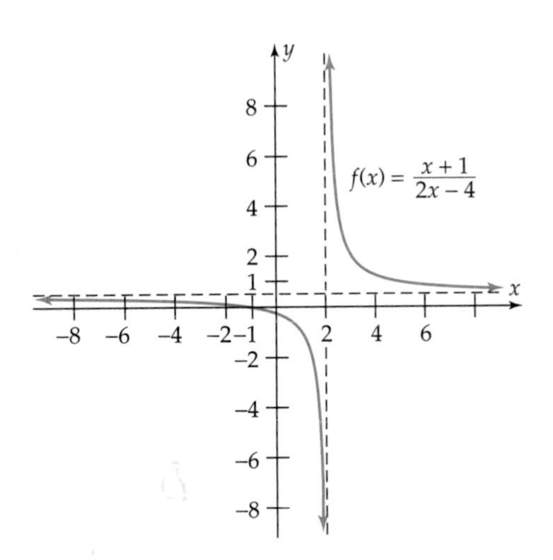

Figure 4–28

Getting an accurate graph of a rational function on a calculator often depends on choosing an appropriate viewing window. For example, a TI-83 produced the following graphs of $f(x) = \dfrac{x+1}{2x-4}$ in Figure 4–29, two of which do not look like Figure 4–28 as they should.

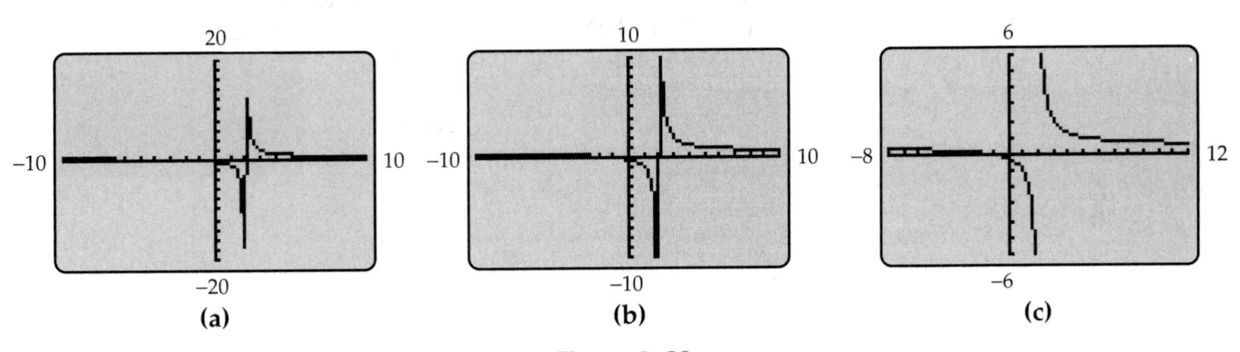

Figure 4–29

The vertical segments in graphs (a) and (b) are *not* representations of the vertical asymptote. They are a result of the calculator evaluating $f(x)$ just to the left of $x = 2$ and just to the right of $x = 2$, but not at $x = 2$, and then erroneously connecting these points with a near vertical segment that looks like an asymptote.

In the accurate graph (c) the calculator attempted to plot a point with $x = 2$ and when it found that $f(2)$ was not defined, skipped a pixel and did not join the points on either side of the skipped one.

Technology Tip

To avoid erroneous vertical lines, use a window with the vertical asymptote in the center, as in Figure 4–29(c) (where the asymptote at $x = 2$ is half-way between -8 and 12). Also see the Tip on page 257.

GRAPHING EXPLORATION

Find a viewing window on your calculator (other than the one in Figure 4–29) that displays the graph of $f(x)$ without any erroneous vertical line segments being shown. The Tip in the margin may be helpful.

The analysis in Example 1 works in the general case:

Linear Rational Functions

The graph of $f(x) = \dfrac{ax + b}{cx + d}$ (with $c \neq 0$ and $ad \neq bc$) has two asymptotes:

> **The vertical asymptote occurs at the root of the denominator.**
>
> **The horizontal asymptote is the line $y = a/c$.**

Example 1 is the case where $a = 1$, $b = 1$, $c = 2$, and $d = -4$. Figure 4–30 shows some additional examples, in which the asymptotes are indicated by dashed lines that are not part of the graph.

$$f(x) = \frac{-5x + 12}{2x - 4}$$

Vertical asymptote $x = 2$
Horizontal asymptote $y = -\frac{5}{2}$

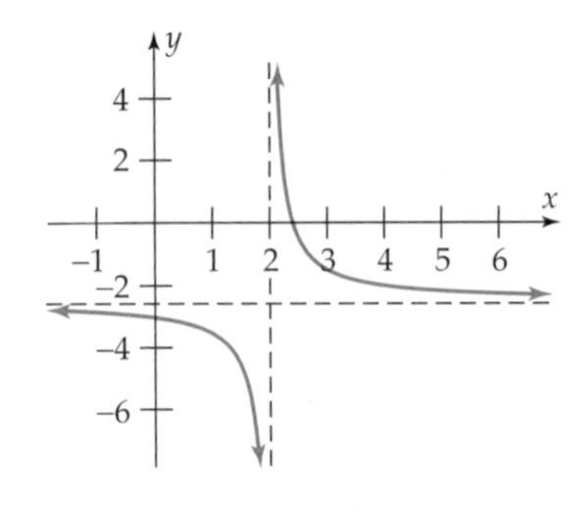

(a)

$$k(x) = \frac{3x + 6}{x} = \frac{3x + 6}{1x + 0}$$

Vertical asymptote $x = 0$
Horizontal asymptote $y = \frac{3}{1} = 3$

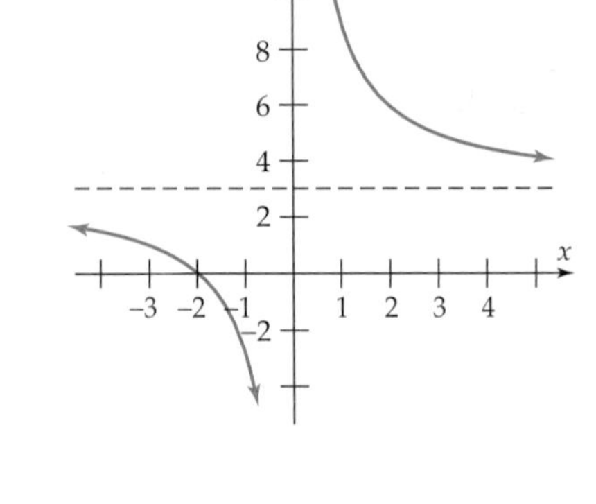

(b)

$$f(x) = \frac{1}{x} = \frac{0x + 1}{1x + 0}$$

Vertical asymptote $x = 0$
Horizontal asymptote $y = \frac{0}{1} = 0$

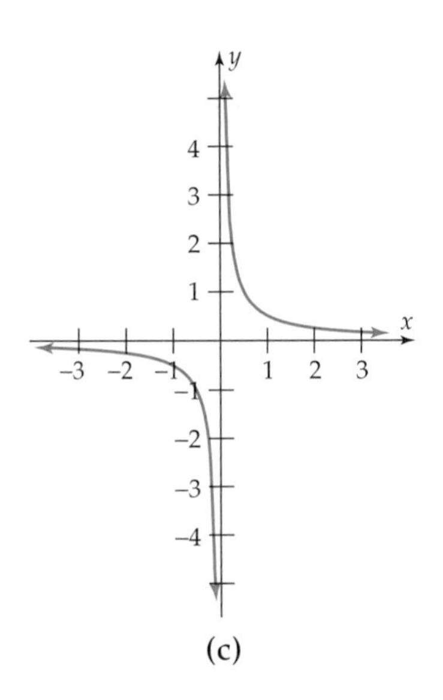

(c)

Figure 4–30

Properties of Rational Graphs

Here is a summary of the important characteristics of graphs of more complicated rational functions.

Continuity

There will be breaks in the graph of a rational function wherever the function is not defined. Except for breaks at these undefined points, the graph is a continuous unbroken curve. In addition, the graph has no sharp corners.

Local Maxima and Minima

The graph may have some local extrema (peaks and valleys) and calculus is needed to determine their exact location. There are no simple rules for the possible number of peaks and valleys as there were with polynomial functions.

Points of Inflection

There may be points of inflection, where the graph changes from bending downward to bending upward (or vice versa). However, there is no easy algebraic way to determine the maximum number of inflection points.

Intercepts

As with any function, the y-intercept of the graph of a rational function f occurs at $f(0)$, provided that f is defined at $x = 0$. The x-intercepts of the graph of any function f occur at each number c for which $f(c) = 0$. Now a fraction is 0 only when its numerator is 0 and its denominator nonzero (since division by 0 is not defined). Thus,

Intercepts

The x-intercepts of the graph of the rational function $f(x) = \dfrac{g(x)}{h(x)}$ occur at the numbers that are roots of the numerator $g(x)$ but *not* of the denominator $h(x)$. If f has a y-intercept, it occurs at $f(0)$.

For example, the graph of $f(x) = \dfrac{x^2 - x - 2}{x - 5}$ has x-intercepts at $x = -1$ and $x = 2$ [which are the roots of $x^2 - x - 2 = (x + 1)(x - 2)$, but not of $x - 5$] and y-intercept at $y = 2/5$ (the value of f at $x = 0$).

Vertical Asymptotes

In Example 1 we saw that the graph of $f(x) = \dfrac{x + 1}{2x - 4}$ had a vertical asymptote at $x = 2$. Note that $x = 2$ is a root of the denominator $2x - 4$, but not of the numerator $x + 1$. The same thing occurs in the general case:

Vertical Asymptotes

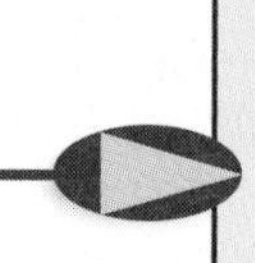

The function $f(x) = \dfrac{g(x)}{h(x)}$ has a vertical asymptote at every number that is a root of the denominator $h(x)$, but *not* of the numerator $g(x)$.

Near a vertical asymptote, the graph of a rational fraction may look like the graph in Example 1, or like one of these graphs (Figure 4–31):

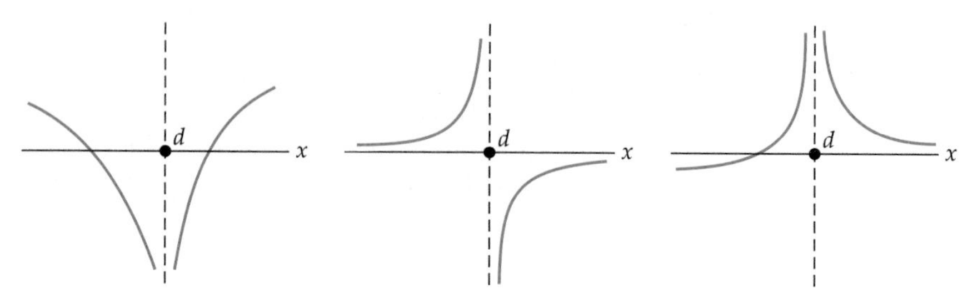

Vertical asymptotes at $x = d$

Figure 4–31

Holes

You have often cancelled factors, as here:

$$\frac{x^2 - 4}{x - 2} = \frac{(x + 2)(x - 2)}{x - 2} = x + 2.$$

But the functions given by

$$p(x) = \frac{x^2 - 4}{x - 2} \qquad \text{and} \qquad q(x) = x + 2$$

are *not* the same, because when $x = 2$,

$$q(2) = 2 + 2 = 4, \qquad \text{but} \qquad p(2) = \frac{2^2 - 2}{2 - 2} = \frac{0}{0}, \text{ which is not defined.}$$

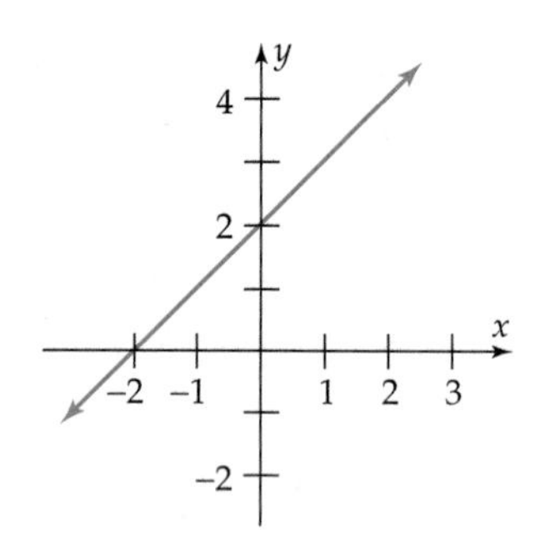

Figure 4–32

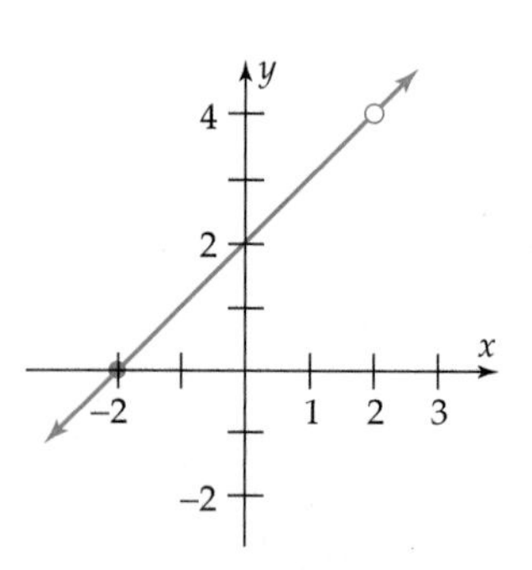

Figure 4–33

For any number other than 2, the two functions do have the same value, and hence, the same graph. The graph of $q(x) = x + 2$ is a straight line that includes the point (2, 4), as shown in Figure 4–32. The graph of $p(x)$ is the same straight line, but with the point (2, 4) omitted, so that the graph of p has a **hole** at $x = 2$ (indicated by an open circle in Figure 4–33). Note that the hole occurs at $x = 2$, which is a root of multiplicity 1 in both the numerator and the denominator of

$$p(x) = \frac{x^2 - 4}{x - 2} = \frac{(x + 2)(x - 2)}{x - 2}.^*$$

*Multiplicity of roots was discussed on page 235.

Similarly, the graph of $g(x) = \frac{x^2}{x^3}$ is the same as the graph of $f(x) = \frac{1}{x}$, except at $x = 0$, where neither function is defined. In this case, however, there is a vertical asymptote rather than a hole at $x = 0$ (see Figure 4–30(c)). Note that the vertical asymptote occurs at $x = 0$, which is a root of multiplicity 2 in the numerator, but of larger multiplicity 3 in the denominator. In general,

Holes

Let $f(x) = \frac{g(x)}{h(x)}$ be a rational function and d a root of both $g(x)$ and $h(x)$. If the multiplicity of d as a root of $g(x)$ is greater than or equal to its multiplicity as a root of $h(x)$, then the graph of f has a hole at $x = d$. Otherwise, the graph has a vertical asymptote at $x = d$.

A calculator-drawn graph may not show holes where it should. If the calculator actually attempts to compute an undefined quantity, it indicates a hole by skipping a pixel; otherwise, it may erroneously show a continuous graph with no hole.

Behavior When $|x|$ Is Large

The shape of a rational graph at the far left and far right (that is, when $|x|$ is large) can usually be found by algebraic analysis.

Example 2 Determine the shape of the graph when $|x|$ is large for the following functions.

(a) $f(x) = \frac{7x^4 - 6x^3 + 4}{2x^4 + x^2}$ (b) $g(x) = \frac{x^2 - 2}{x^3 - 3x^2 + x - 3}$

Solution

(a) When $|x|$ is very large, a polynomial function behaves in essentially the same way as its highest degree term, as we saw on page 234. Consequently, we have this approximation

$$f(x) = \frac{7x^4 - 6x^3 + 4}{2x^4 + x^2} \approx \frac{7x^4}{2x^4} = \frac{7}{2} = 3.5.$$

Thus, when $|x|$ is large, the graph of $f(x)$ is very close to the horizontal line $y = 3.5$, which is a horizontal asymptote of the graph. This means that at the far left and far right, the graph of $f(x)$ is almost flat and very close to the line $y = 3.5$.

GRAPHING EXPLORATION

Graph $f(x)$ in the standard viewing window. Use the trace feature to determine how far the left and right ends of the graph are from the horizontal asymptote $y = 3.5$. Now find a wider viewing window in which the ends of the graph are within .1 of the horizontal asymptote.

(b) When $|x|$ is large, the graph of g closely resembles the graph of $y = \frac{x^2}{x^3} = \frac{1}{x}$. By the Big-Little Principle, $1/x$ is very close to 0 when $|x|$ is large. So, the line $y = 0$ (that is, the x-axis) is the horizontal asymptote.

GRAPHING EXPLORATION

Find a viewing window that provides the answer to this question: as you move to the left from the origin does the graph of $g(x)$ approach its horizontal asymptote from above or from below? Does the graph ever cross the horizontal asymptote? ■

Arguments similar to those in the preceding example, using the highest degree terms in the numerator and denominator, carry over to the general case and lead to this conclusion:

Horizontal Asymptotes

Let $f(x) = \frac{ax^n + \cdots}{cx^k + \cdots}$ be a rational function whose numerator has degree n and whose denominator has degree k.

If $n = k$, then the line $y = a/c$ is a horizontal asymptote.

If $n < k$, then the x-axis (the line $y = 0$) is a horizontal asymptote.

The asymptotes of rational functions in which the denominator has smaller degree than the numerator are discussed in Excursion 4.5.A.

Graphs of Rational Functions

The facts presented above can be used in conjunction with a calculator to find accurate, complete graphs of rational functions whose numerators have degree less than or equal to the degree of their denominators. The basic procedure is as follows.

Graphing $f(x) = \dfrac{g(x)}{h(x)}$ When Degree $g(x) \leq$ Degree $h(x)$

1. **Analyze the function algebraically to determine its vertical asymptotes, holes, and intercepts.**
2. **Determine the horizontal asymptote of the graph when $|x|$ is large by using the facts in the box on page 256.**
3. **Use the preceding information to select an appropriate viewing window (or windows), to interpret the calculator's version of the graph (if necessary), and to sketch an accurate graph.**

Example 3 If you ignore the preceding advice and simply graph $f(x) = \dfrac{x-1}{x^2 - x - 6}$ in the standard viewing window, you get garbage (Figure 4–34). So let's try analyzing the function. We begin by factoring:

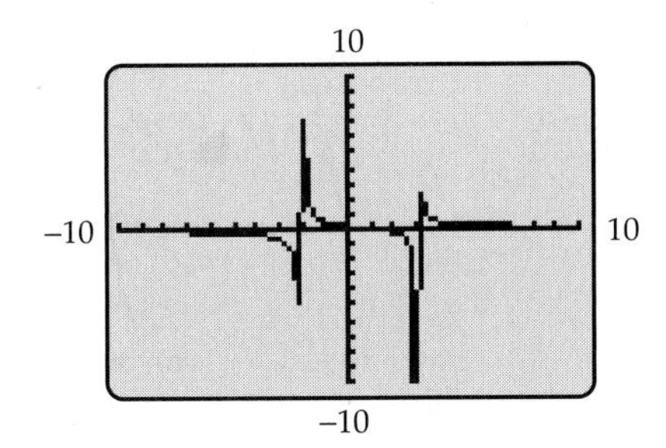

Figure 4–34

$$f(x) = \frac{x-1}{x^2 - x - 6} = \frac{x-1}{(x+2)(x-3)}.$$

The factored form allows us to read off the necessary information:

Vertical Asymptotes: $x = -2$ and $x = 3$ (roots of the denominator but not of the numerator).

Horizontal Asymptote: x-axis (because denominator has larger degree than the numerator).

Intercepts: y-intercept at $f(0) = \dfrac{0-1}{0^2 - 0 - 6} = \dfrac{1}{6}$; x-intercept at $x = 1$ (root of the numerator but not of the denominator).

Interpreting Figure 4–34 in the light of this information suggests that a complete graph of f looks something like Figure 4–35.

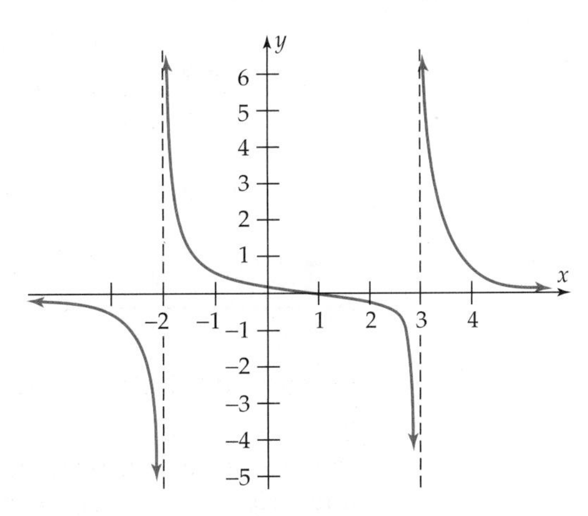

Figure 4–35

Technology Tip

When the vertical asymptotes of a rational function occur at numbers such as $-2.1, -2, -1.9, \dots, 2.9, 3, 3.1$, a decimal window normally produces an accurate graph because the calculator actually evaluates the function at the asymptotes, finds that it is undefined, and skips a pixel.

GRAPHING EXPLORATION

Find a viewing window in which the graph of f looks similar to Figure 4–35. The Tip in the margin may be helpful. ■

NOTE The graph of a rational function never touches a horizontal asymptote when x is large in absolute value. For smaller values of x, however, the graph may cross the asymptote, as in Example 3.

Example 4 To graph $f(x) = \dfrac{2x^2}{x^2 + x - 2}$ we factor and then read off the necessary information:

$$f(x) = \frac{2x^2}{x^2 + x - 2} = \frac{2x^2}{(x + 2)(x - 1)}.$$

Vertical Asymptotes: $x = -2$ and $x = 1$ (roots of denominator).

Horizontal Asymptote: $y = 2/1 = 2$ (because numerator and denominator have the same degree; see the box on page 256).

Intercepts: x-intercept at $x = 0$ (root of numerator); y-intercept at $f(0) = 0$.

Using this information and selecting a decimal viewing window that will accurately portray the graph near the vertical asymptotes, we obtain what seems to be a reasonably complete graph in Figure 4–36. The graph appears to be falling to the right of $x = 1$, but this is deceptive.

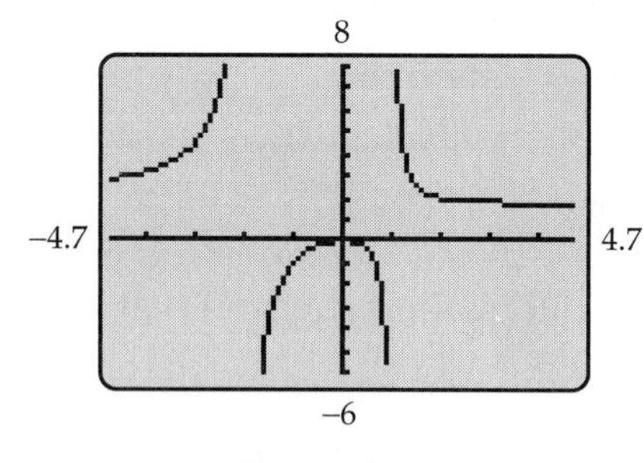

Figure 4–36

GRAPHING EXPLORATION

Graph f in this same viewing window and use the trace feature, beginning at approximately $x = 1.1$ and moving to the right. For what values of x is the graph above the horizontal asymptote $y = 2$? For what values of x is the graph below the horizontal asymptote?

This use of the trace feature indicates that there is some *hidden behavior* of the graph that is not visible in Figure 4–36.

GRAPHING EXPLORATION

To see this hidden behavior, graph both f and the line $y = 2$ in the viewing window with $1 \le x \le 50$ and $1.7 \le y \le 2.1$.

This Exploration shows that the graph has a local minimum near $x = 4$ and then stays below the asymptote, moving closer and closer to it as x takes larger values. ■

Applications

Several applications of rational functions were considered in Section 2.4. Here is another one.

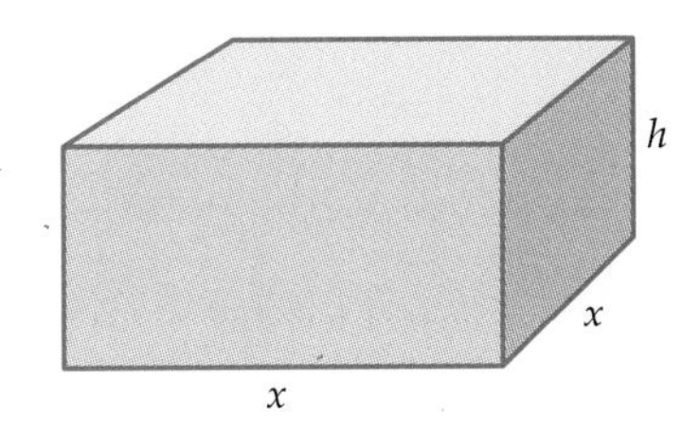

Figure 4–37

Example 5 A cardboard box with a square base and a volume of 1000 cubic inches is to be constructed (Figure 4–37). The box must be at least 2 inches in height.

(a) What are the possible lengths for a side of the base if no more than 1100 square inches of cardboard can be used to construct the box?

(b) What is the least possible amount of cardboard that can be used?

(c) What are the dimensions of the box that uses the least possible amount of cardboard?

Solution The amount of cardboard needed to construct the box is given by the surface area of the box. Since the top and bottom each have area x^2 (why?) and each of the four sides has area xh, the surface area S is given by

$$S = x^2 + x^2 + xh + xh + xh + xh = 2x^2 + 4xh$$

Since the volume of the box is given by:

$$\text{Length} \times \text{Width} \times \text{Height} = \text{Volume}$$

we have

$$x \cdot x \cdot h = 1000 \qquad \text{or equivalently,} \qquad h = \frac{1000}{x^2}.$$

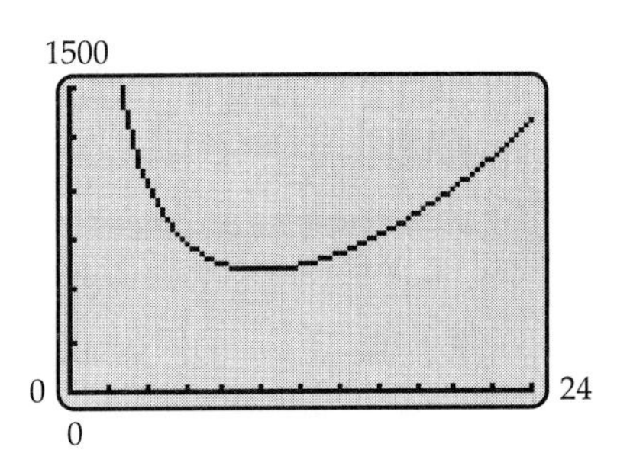

Figure 4–38

Substituting this into the surface area formula allows us to express the surface area as a function of x:

$$S(x) = 2x^2 + 4xh = 2x^2 + 4xh\left(\frac{1000}{x^2}\right) = 2x^2 + \frac{4000}{x} = \frac{2x^3 + 4000}{x}.$$

Although the rational function $S(x)$ is defined for all nonzero real numbers, x is a length here and must be positive. Furthermore, $x^2 \leq 500$ because if $x^2 > 500$, then $h = \frac{1000}{x^2}$ would be less than 2, contrary to specifications. Hence, the only values of x that make sense in this context are those with $0 < x \leq \sqrt{500}$. Since $\sqrt{500} \approx 22.4$, we choose the viewing window in Figure 4–38. For each point (x, y) on the graph, x is a possible side length for the base of the box and y is the corresponding surface area.

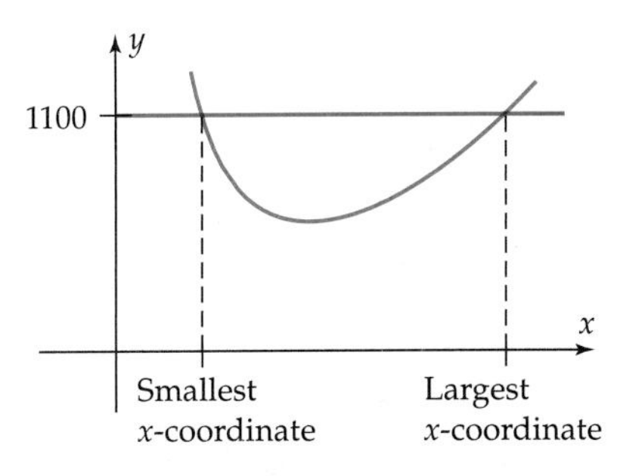

Figure 4–39

(a) The points on the graph corresponding to the requirement that no more than 1100 square inches of cardboard be used are those whose y-coordinates are less than or equal to 1100. The x-coordinates of these points are the possible side lengths. The x-coordinates of the points where the graph of S meets the horizontal line $y = 1100$ are the smallest and largest possible values for x, as indicated schematically in Figure 4–39.

> **GRAPHING EXPLORATION**
>
> Graph $S(x)$ and $y = 1100$ on the same screen. Use an intersection finder to show that the possible side lengths that use no more than 1100 square inches of cardboard are those with $3.73 \le x \le 21.36$.

(b) The least possible amount of cardboard corresponds to the point on the graph of $S(x)$ with the smallest y-coordinate.

> **GRAPHING EXPLORATION**
>
> Show that the graph of S has a local minimum at the point (10.00, 600.00). Consequently, the least possible amount of cardboard is 600 square inches and this occurs when $x = 10$.

(c) When $x = 10$, $h = 1000/10^2 = 10$. So the dimensions of the box using the least amount of cardboard are $10 \times 10 \times 10$. ■

Exercises 4.5

In Exercises 1–6, find the domain of the function. You may need to use some of the techniques of Section 4.3.

1. $f(x) = \dfrac{-3x}{2x + 5}$

2. $g(x) = \dfrac{x^3 + x + 1}{2x^2 - 5x - 3}$

3. $h(x) = \dfrac{6x - 5}{x^2 - 6x + 4}$

4. $g(x) = \dfrac{x^3 - x^2 - x - 1}{x^5 - 36x}$

5. $f(x) = \dfrac{x^5 - 2x^3 + 7}{x^3 - x^2 - 2x + 2}$

6. $h(x) = \dfrac{x^5 - 5}{x^4 + 12x^3 + 60x^2 + 50x - 125}$

In Exercises 7 – 12, use algebra to determine the location of the vertical asymptotes and holes in the graph of the function.

7. $f(x) = \dfrac{x^2 + 4}{x^2 - 5x - 6}$

8. $g(x) = \dfrac{x - 5}{x^3 + 7x^2 + 2x}$

9. $f(x) = \dfrac{x}{x^3 + 2x^2 + x}$

10. $g(x) = \dfrac{x}{x^3 + 5x}$

11. $f(x) = \dfrac{x^2 - 4x + 4}{(x + 2)(x - 2)^3}$

12. $h(x) = \dfrac{x - 3}{x^2 - x - 6}$

In Exercises 13–18, find the horizontal asymptote of the graph of the function when $|x|$ is large and find a viewing window in which the ends of the graph are within .1 of this asymptote.

13. $f(x) = \dfrac{3x - 2}{x + 3}$

14. $g(x) = \dfrac{3x^2 + x}{2x^2 - 2x + 4}$

15. $h(x) = \dfrac{5 - x}{x - 2}$

16. $f(x) = \dfrac{4x^2 - 5}{2x^3 - 3x^2 + x}$

17. $g(x) = \dfrac{5x^3 - 8x^2 + 4}{2x^3 + 2x}$

18. $h(x) = \dfrac{8x^5 - 6x^3 + 2x - 1}{.5x^5 + x^4 + 3x^2 + x}$

In Exercises 19–38, analyze the function algebraically: List its vertical asymptotes, holes, and horizontal asymptote. Then, sketch a complete graph of the function.

19. $f(x) = \dfrac{1}{x + 5}$

20. $q(x) = \dfrac{-7}{x - 6}$

21. $k(x) = \dfrac{-3}{2x + 5}$

22. $g(x) = \dfrac{-4}{2 - x}$

23. $f(x) = \dfrac{3x}{x - 1}$

24. $p(x) = \dfrac{x - 2}{x}$

25. $f(x) = \dfrac{2 - x}{x - 3}$

26. $g(x) = \dfrac{3x - 2}{x + 3}$

27. $f(x) = \dfrac{1}{x(x + 1)^2}$

28. $g(x) = \dfrac{x}{2x^2 - 5x - 3}$

29. $f(x) = \dfrac{x - 3}{x^2 + x - 2}$

30. $g(x) = \dfrac{x + 2}{x^2 - 1}$

31. $h(x) = \dfrac{(x^2 + 6x + 5)(x + 5)}{(x + 5)^3(x - 1)}$

32. $f(x) = \dfrac{x^2 - 1}{x^3 - 2x^2 + x}$

33. $f(x) = \dfrac{-4x^2 + 1}{x^2}$

34. $k(x) = \dfrac{x^2 + 1}{x^2 - 1}$

35. $q(x) = \dfrac{x^2 + 2x}{x^2 - 4x - 5}$

36. $F(x) = \dfrac{x^2 + x}{x^2 - 2x + 4}$

37. $p(x) = \dfrac{(x + 3)(x - 3)}{(x - 5)(x + 4)(x + 3)}$

38. $p(x) = \dfrac{x^3 + 3x^2}{x^4 - 4x^2}$

In Exercises 39–48, find a viewing window, or windows, that shows a complete graph of the function (if possible, with no erroneous vertical line segments). Be alert for hidden behavior, such as that in Example 4.

39. $f(x) = \dfrac{x^3 + 4x^2 - 5x}{(x^2 - 4)(x^2 - 9)}$

40. $g(x) = \dfrac{x^2 + x - 6}{x^3 - 19x + 30}$

41. $h(x) = \dfrac{2x^2 - x - 6}{x^3 + x^2 - 6x}$

42. $f(x) = \dfrac{x^3 - x + 1}{x^4 - 2x^3 - 2x^2 + x - 1}$

43. $f(x) = \dfrac{2x^4 - 3x^2 + 1}{3x^4 - x^2 + x - 1}$

44. $g(x) = \dfrac{x^4 + 2x^3}{x^5 - 25x^3}$

45. $h(x) = \dfrac{3x^2 + x - 4}{2x^2 - 5x}$

46. $f(x) = \dfrac{2x^2 - 1}{3x^3 + 2x + 1}$

47. $g(x) = \dfrac{x - 4}{2x^3 - 5x^2 - 4x + 12}$

48. $h(x) = \dfrac{x^2 - 9}{x^3 + 2x^2 - 23x - 60}$

49. (a) Graph $f(x) = 1/x$ in the viewing window with $-6 \le x \le 6$ and $-6 \le y \le 6$.

(b) Without using a calculator, describe how the graph of $g(x) = 2/x$ can be obtained from the graph of $f(x)$. [*Hint:* $g(x) = 2f(x)$; see Section 3.4.]

(c) Without using a calculator, describe how the graphs of each of the following functions can be obtained from the graph of $f(x)$:

$$h(x) = \frac{1}{x} + 4, \quad k(x) = \frac{1}{x - 3}, \quad t(x) = \frac{1}{x + 2}.$$

[*Hint:* $h(x) = f(x) + 4$; $k(x) = f(x - 3)$; and $t(x) = f(x + 2)$.]

(d) Without using a calculator, describe how the graph of $p(x) = \dfrac{2}{x - 3} + 4$ can be obtained from the graph of $f(x) = 1/x$.

(e) Show that the function $p(x)$ of part (d) is a rational function by rewriting its rule as the quotient of two first-degree polynomials.

(f) If r, s, t are constants, describe how the graph of $q(x) = \dfrac{r}{x + s} + t$ can be obtained from the graph of $f(x) = 1/x$.

(g) Show that the function $q(x)$ of part (f) is a rational function by rewriting its rule as the quotient of two first-degree polynomials.

50. The graph of $f(x) = \dfrac{2x^3 - 2x^2 - x + 1}{3x^3 - 3x^2 + 2x - 1}$ has a vertical asymptote. Find a viewing window that demonstrates this fact.

51. (a) Find the difference quotient of the function $f(x) = 1/x$ and express it as a single fraction in lowest terms. [*Hint:* Section 3.6]

(b) Use the difference quotient in part (a) to determine the average rate of change of $f(x)$ as x changes from 2 to 2.1, from 2 to 2.01, and from 2 to 2.001. Estimate the instantaneous rate of change of $f(x)$ at $x = 2$.

(c) Use the difference quotient in part (a) to determine the average rate of change of $f(x)$ as x changes from 3 to 3.1, from 3 to 3.01, and from 3 to 3.001. Estimate the instantaneous rate of change of $f(x)$ at $x = 3$.

(d) How are the estimated instantaneous rates of change of $f(x)$ at $x = 2$ and $x = 3$ related to the values of the function $g(x) = -1/x^2$ at $x = 2$ and $x = 3$?

52. Do Exercise 51 for the functions $f(x) = 1/x^2$ and $g(x) = -2/x^3$.

53. (a) When $x \ge 0$, what rational function has the same graph as $f(x) = \dfrac{x - 1}{|x| - 2}$? [*Hint:* Use the definition of absolute value on page 8.]

(b) When $x < 0$, what rational function has the same graph as $f(x) = \frac{x-1}{|x|-2}$? [See the hint for part (a).]

(c) Use parts (a) and (b) to explain why the graph of $f(x) = \frac{x-1}{|x|-2}$ has two vertical asymptotes. What are they? Confirm your answer by graphing the function.

54. The percentage c of a drug in a person's bloodstream t hours after its injection is approximated by

$$c(t) = \frac{5t}{4t^2 + 5}.$$

(a) Approximately what percentage of the drug is in the person's bloodstream after four and a half hours?

(b) Graph the function c in an appropriate window for this situation.

(c) What is the horizontal asymptote of the graph? What does it tell you about the amount of the drug in the bloodstream?

(d) At what time is the percentage the highest? What is the percentage at that time?

55. It costs 2.5 cents per square inch to make the top and bottom of the box in Example 5. The sides cost 1.5 cents per square inch. What are the dimensions of the cheapest possible box?

56. A box with a square base and a volume of 1000 cubic inches is to be constructed. The material for the top and bottom of the box costs \$3 per 100 square inches and the material for the sides costs \$1.25 per 100 square inches.

(a) If x is the length of a side of the base, express the cost of constructing the box as a function of x.

(b) If the side of the base must be at least 6 inches long, for what value of x will the cost of the box be \$7.50?

57. A truck traveling at a constant speed on a reasonably straight, level road burns fuel at the rate of $g(x)$ gallons per mile, where x is the speed of the truck (in miles per hour) and $g(x)$ is given by $g(x) = \frac{800 + x^2}{200x}$.

(a) If fuel costs \$1.40 per gallon, find the rule of the cost function $c(x)$ that expresses the cost of fuel for a 500-mile trip as a function of the speed. [*Hint:* $500g(x)$ gallons of fuel are needed to go 500 miles. (Why?)]

(b) What driving speed will make the cost of fuel for the trip \$250?

(c) What driving speed will minimize the cost of fuel for the trip?

58. Pure alcohol is being added to 50 gallons of a coolant mixture that is 40% alcohol.

(a) Find the rule of the concentration function $c(x)$ that expresses the percentage of alcohol in the resulting mixture as a function of the number x of gallons of pure alcohol that are added. [*Hint:* The final mixture contains $50 + x$ gallons. (Why?) So $c(x)$ is the amount of alcohol in the final mixture divided by the total amount $50 + x$. How much alcohol is in the original 50-gallon mixture? How much is in the final mixture?]

(b) How many gallons of pure alcohol should be added to produce a mixture that is at least 60% alcohol and no more than 80% alcohol?

(c) Determine algebraically the exact amount of pure alcohol that must be added to produce a mixture that is 70% alcohol.

59. A rectangular garden with an area of 250 square meters is to be located next to a building and fenced on three sides, with the building acting as a fence on the fourth side.

(a) If the side of the garden parallel to the building has length x meters, express the amount of fencing needed as a function of x.

(b) For what values of x will less than 60 meters of fencing be needed?

(c) What value of x will result in the least possible amount of fencing being used? What are the dimensions of the garden in this case?

60. A certain company has fixed costs of \$40,000 and variable costs of \$2.60 per unit.

(a) Let x be the number of units produced. Find the rule of the average cost function. [The average cost is the cost of the units divided by the number of units.]

(b) Graph the average cost function in a window with $0 \le x \le 100{,}000$ and $0 \le y \le 20$.

(c) Find the horizontal asymptote of the average cost function. Explain what the asymptote means in this situation. [How low can the average cost possibly be?]

61. Radioactive waste is stored in a cylindrical tank, whose exterior has radius r and height h as shown in the figure. The sides, top, and bottom of the tank are one foot thick and the tank has a volume of 150 cubic feet (including top, bottom, and walls).

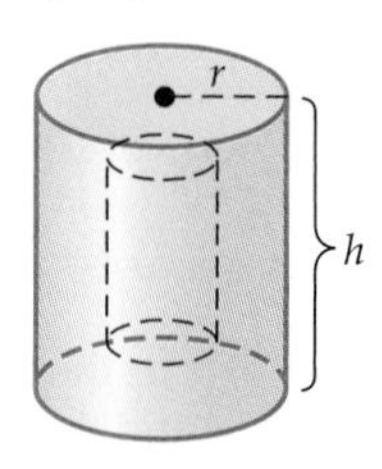

(a) Express the interior height h_1 (that is, the height of the storage area) as a function of h.
(b) Express the interior height as a function of r.
(c) Express the volume of the interior as a function of r.
(d) Explain why r must be greater than 1.
(e) What should the dimensions of the tank be in order for it to hold as much as possible?

62. The relationship between the fixed focal length F of a camera, the distance u from the object being photographed to the lens, and the distance v from the lens to the film is given by $\frac{1}{F} = \frac{1}{u} + \frac{1}{v}$.

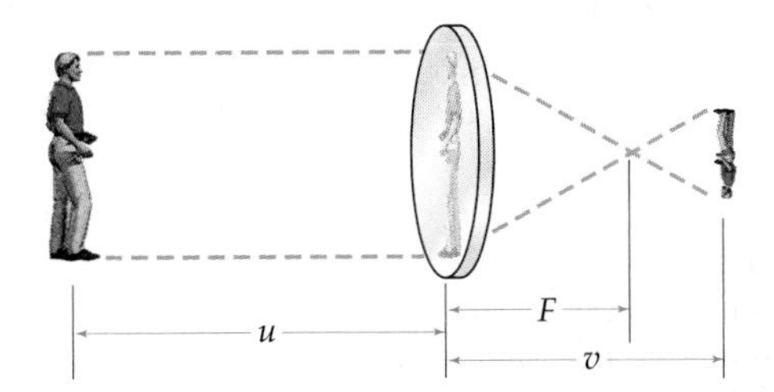

(a) If the focal length is 50 millimeters, express v as a function of u.
(b) What is the horizontal asymptote of the graph of the function in part (a)?
(c) Graph the function in part (a) when 50 millimeters $< u <$ 35,000 millimeters.
(d) When you focus the camera on an object, the distance between the lens and the film is changed. If the distance from the lens to the camera changes by less than .1 millimeter, the object will remain in focus. Explain why you have more latitude in focusing on distant objects than on very close ones.

63. The formula for the gravitational acceleration (in units of meters per second squared) of an object relative to the earth is

$$g(r) = \frac{3.987 \times 10^{14}}{(6.378 \times 10^6 + r)^2}$$

where r is the distance in meters above the earth's surface.
(a) What is the gravitational acceleration at the earth's surface?
(b) Graph the function $g(r)$ for $r \geq 0$.
(c) Can you ever escape the pull of gravity? [Does the graph have any r-intercepts?]

4.5.A EXCURSION Other Rational Functions

We now examine the graphs of rational functions in which the degree of the denominator is smaller than the degree of the numerator. Such a graph has no horizontal asymptote. However, it does have some polynomial curve as an asymptote, which means that the graph will get very close to this curve when $|x|$ is very large.

Example 1 To graph $f(x) = \frac{x^3 + 3x^2 + x + 1}{x^2 + 2x - 1}$, we begin by finding the vertical asymptotes and the x- and y-intercepts. The quadratic formula can be used to find the roots of the denominator:

$$x = \frac{-2 \pm \sqrt{2^2 - 4 \cdot 1(-1)}}{2 \cdot 1} = \frac{-2 \pm \sqrt{8}}{2} = \frac{-2 \pm 2\sqrt{2}}{2} = -1 \pm \sqrt{2}.$$

It is easy to verify that neither of these numbers is a root of the numerator, so the graph has vertical asymptotes at $x = -1 - \sqrt{2}$ and $x = -1 + \sqrt{2}$. The y-intercept is $f(0) = -1$. The x-intercepts are the roots of the numerator.

> **GRAPHING EXPLORATION**
>
> Use a calculator to verify that $x^3 + 3x^2 + x + 1$ has exactly one real root, located between -3 and -2.

Therefore, the graph of $f(x)$ has one x-intercept. Using this information and the calculator graph in Figure 4–40 (which erroneously shows some vertical segments), we conclude that the graph looks approximately like Figure 4–41.

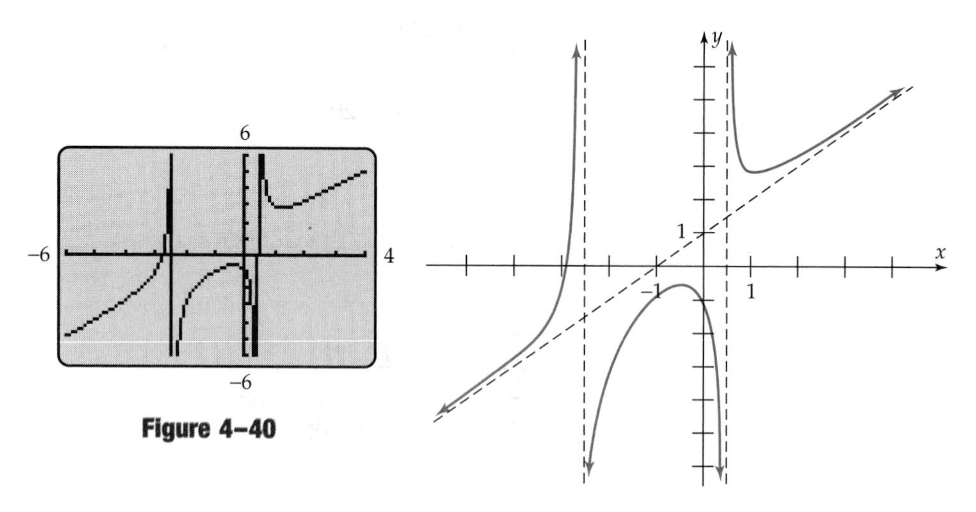

Figure 4–40

Figure 4–41

At the left and right ends, the graph moves away from the x-axis. To understand the behavior of the graph when $|x|$ is large, divide the numerator by $f(x)$ by its denominator:

$$\begin{array}{r|l} & x + 1 \\ \hline x^2 + 2x - 1 & x^3 + 3x^2 + x + 1 \\ & \underline{x^3 + 2x^2 - x} \\ & \qquad x^2 + 2x + 1 \\ & \qquad \underline{x^2 + 2x - 1} \\ & \qquad\qquad\qquad 2 \end{array}$$

By the Division Algorithm

$$x^3 + 3x^2 + x + 1 = (x^2 + 2x - 1)(x + 1) + 2.$$

Dividing both sides by $x^2 + 2x - 1$, we have

$$\frac{x^3 + 3x^2 + x + 1}{x^2 + 2x - 1} = \frac{(x^2 + 2x - 1)(x + 1) + 2}{x^2 + 2x - 1}$$

$$f(x) = (x + 1) + \frac{2}{x^2 + 2x - 1}.$$

Now when x is very large in absolute value, so is $x^2 + 2x - 1$. Hence, $2/(x^2 + 2x - 1)$ is very close to 0 by the Big-Little Principle and $f(x)$ is very close to $(x + 1) + 0$. Therefore, as x gets larger in absolute value, the graph of $f(x)$ gets closer and closer to the line $y = x + 1$ (the dashed slanted line in Figure 4–41) and this line is an asymptote of the graph.* Note that $x + 1$ is just the quotient obtained in the long division above. ■

It is instructive to examine the graph in Example 1 further to see that the asymptote accurately indicates the behavior of the function when $|x|$ is large.

GRAPHING EXPLORATION

Using the viewing window with $-20 \leq x \leq 20$ and $-20 \leq y \leq 20$, graph both $f(x)$ and $y = x + 1$ on the same screen.

Except near the vertical asymptotes of $f(x)$, the two graphs are virtually identical.

GRAPHING EXPLORATION

Now change the range values, so that the viewing window has $-100 \leq x \leq 100$ and $-100 \leq y \leq 100$.

In this viewing window, the vertical asymptotes of $f(x)$ are no longer visible and the graph is indistinguishable from the graph of the asymptote $y = x + 1$.

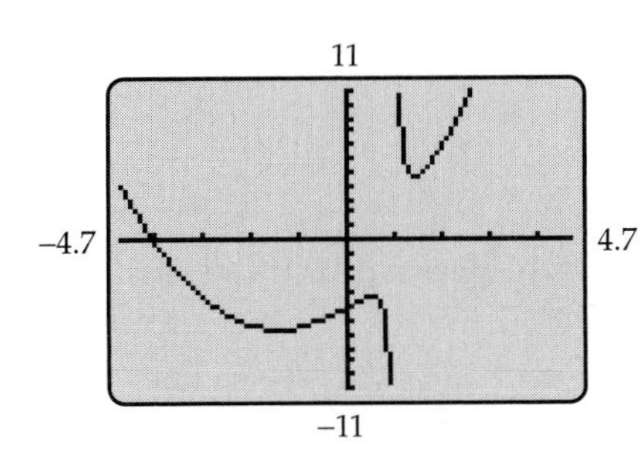

Figure 4–42

Example 2 To graph $g(x) = \dfrac{x^3 + 2x^2 - 7x + 5}{x - 1}$, we first note that there is a vertical asymptote at $x = 1$ (root of the denominator, but not the numerator). The y-intercept is at $g(0) = -5$. By carefully choosing a viewing window that accurately portrays the behavior of $g(x)$ near its vertical asymptote, we obtain Figure 4–42.

GRAPHING EXPLORATION

Verify that the x-intercept near $x = -4$ is the only one by showing graphically that the numerator of $g(x)$ has exactly one real root.

*An asymptote that is a nonvertical and nonhorizontal straight line is called an **oblique asymptote.**

To confirm that Figure 4–42 is a complete graph, we find its asymptote when $|x|$ is large. Divide the denominator by the numerator:

$$\begin{array}{r|l} & x^2 + 3x - 4 \\ \hline x - 1 & x^3 + 2x^2 - 7x + 5 \\ & \underline{x^3 - x^2} \\ & \quad 3x^2 - 7x + 5 \\ & \quad \underline{3x^2 - 3x} \\ & \qquad -4x + 5 \\ & \qquad \underline{-4x + 4} \\ & \qquad\qquad 1 \end{array}$$

Hence, by the Division Algorithm,

$$x^3 + 2x^2 - 7x + 5 = (x - 1)(x^2 + 3x - 4) + 1$$

$$\frac{x^3 + 2x^2 - 7x + 5}{x - 1} = \frac{(x - 1)(x^2 + 3x - 4) + 1}{x - 1}$$

$$g(x) = (x^2 + 3x - 4) + \frac{1}{x - 1}.$$

When $|x|$ is large, $1/(x - 1)$ is very close to 0 (why?), so that $y = x^2 + 3x - 4$ is the asymptote. Once again, the asymptote is given by the quotient of the division.

GRAPHING EXPLORATION

Graph $g(x)$ and $y = x^2 + 3x - 4$ on the same screen to show that the graph of $g(x)$ does get very close to the asymptote when $|x|$ is large. Then find a large enough viewing window so that the two graphs appear to be identical. ■

The procedures used in the preceding examples may be summarized as follows.

Graphing $f(x) = \frac{g(x)}{h(x)}$ When Degree $g(x)$ > Degree $h(x)$

1. **Analyze the function algebraically to determine its vertical asymptotes, holes, and intercepts.**
2. **Divide the numerator $g(x)$ by the denominator $h(x)$. The quotient $q(x)$ is the nonvertical asymptote of the graph, which describes the behavior of the graph when $|x|$ is large.**
3. **Use the preceding information to select an appropriate viewing window (or windows), to interpret the calculator's version of the graph (if necessary), and to sketch an accurate graph.**

Exercises 4.5.A

In Exercises 1–4, find the nonvertical asymptote of the graph of the function when $|x|$ is large and find a viewing window in which the ends of the graph are within .1 of this asymptote.

1. $f(x) = \dfrac{x^3 - 1}{x^2 - 4}$

2. $g(x) = \dfrac{x^3 - 4x^2 + 6x + 5}{x - 2}$

3. $h(x) = \dfrac{x^3 + 3x^2 - 4x + 1}{x + 4}$

4. $f(x) = \dfrac{x^3 + 3x^2 - 4x + 1}{x^2 - x}$

In Exercises 5–12, analyze the function algebraically. List its vertical asymptotes and holes, and determine its nonvertical asymptote. Then sketch a complete graph of the function.

5. $f(x) = \dfrac{x^2 - x - 6}{x - 2}$

6. $k(x) = \dfrac{x^2 + x - 2}{x}$

7. $Q(x) = \dfrac{4x^2 + 4x - 3}{2x - 5}$

8. $K(x) = \dfrac{3x^2 - 12x + 15}{3x + 6}$

9. $f(x) = \dfrac{x^3 - 2}{x - 1}$

10. $p(x) = \dfrac{x^3 + 8}{x + 1}$

11. $q(x) = \dfrac{x^3 - 1}{x - 2}$

12. $f(x) = \dfrac{x^4 - 1}{x^2}$

In Exercises 13–18, find a viewing window (or windows) that shows a complete graph of the function (if possible, with no erroneous vertical line segments). Be alert for hidden behavior.

13. $f(x) = \dfrac{2x^2 + 5x + 2}{2x + 7}$

14. $g(x) = \dfrac{2x^3 + 1}{x^2 - 1}$

15. $h(x) = \dfrac{x^3 - 2x^2 + x - 2}{x^2 - 1}$

16. $f(x) = \dfrac{3x^3 - 11x - 1}{x^2 - 4}$

17. $g(x) = \dfrac{2x^4 + 7x^3 + 7x^2 + 2x}{x^3 - x + 50}$

18. $h(x) = \dfrac{2x^3 + 7x^2 - 4}{x^2 + 2x - 3}$

19. (a) Show that when $0 < x < 4$, the rational function

$$r(x) = \frac{4096x^3 + 34{,}560x^2 + 19{,}440x + 729}{18{,}432x^2 + 34{,}560x + 5832}$$

is a good approximation of the function $s(x) = \sqrt{x}$ by graphing both functions in the viewing window with $0 \le x \le 4$ and $0 \le y \le 2$.

(b) For what values of x is $r(x)$ within .01 of $s(x)$?

20. Find a rational function f that has these properties:

(i) The curve $y = x^3 - 8$ is an asymptote of the graph of f.

(ii) $f(2) = 1$.

(iii) The line $x = 1$ is a vertical asymptote of the graph of f.

4.6 Polynomial and Rational Inequalities

Inequalities may be solved using algebraic or geometric methods, both of which are discussed here. Whenever possible we shall use algebra to obtain exact solutions. When algebraic methods are too difficult, approximate graphical solutions will be found. The basic tools for working with inequalities are the following principles.

Basic Principles for Solving Inequalities

Performing any of the following operations on an inequality produces an equivalent inequality:*

1. **Add or subtract the same quantity on both sides of the inequality.**
2. **Multiply or divide both sides of the inequality by the same *positive* quantity.**
3. **Multiply or divide both sides of the inequality by the same *negative* quantity and *reverse the direction of the inequality.***

Note principle 3 carefully. It says, for example, that if you multiply both sides of $-3 < 5$ by -2, the equivalent inequality is $6 > -10$ (direction of inequality is reversed).

Linear Inequalities

Example 1 To solve $5x + 3 \le 6 + 7x$ we use the Basic Principles to transform it into an inequality whose solutions are obvious:

Subtract 7x from both sides: $\quad -2x + 3 \ge 6$

Subtract 3 from both sides: $\quad -2x \le 3$

Divide both sides by −2 and reverse the direction of the inequality: $\quad x \ge -3/2$

Therefore, the solutions are all real numbers greater than or equal to $-3/2$, that is, the interval $[-3/2, \infty)$, as shown in Figure 4–43. ■

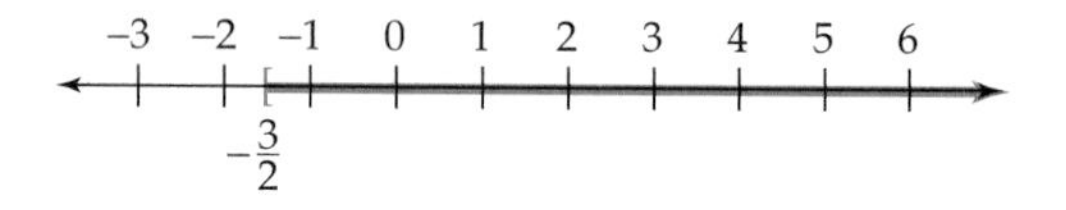

Figure 4–43

Example 2 A solution of the inequality $2 \le 3x + 5 < 2x + 11$ is any number that is a solution of *both* of these inequalities:

$$2 \le 3x + 5 \qquad \text{and} \qquad 3x + 5 < 2x + 11$$

*Two inequalities are **equivalent** if they have the same solutions.

Each of these inequalities can be solved by the methods used earlier. For the first one we have:

$$2 \le 3x + 5$$

Subtract 5 from both sides: $-3 \le 3x$

Divide both sides by 3: $-1 \le x$

The second inequality is solved similarly:

$$3x + 5 < 2x + 11$$

Subtract 5 from both sides: $3x < 2x + 6$

Subtract 2x from both sides: $x < 6$

The solutions of the original inequality are the numbers x that satisfy *both* $-1 \le x$ *and* $x < 6$, that is, all x with $-1 \le x < 6$. Thus, the solutions are precisely the numbers in the interval $[-1, 6)$, as shown in Figure 4–44. ■

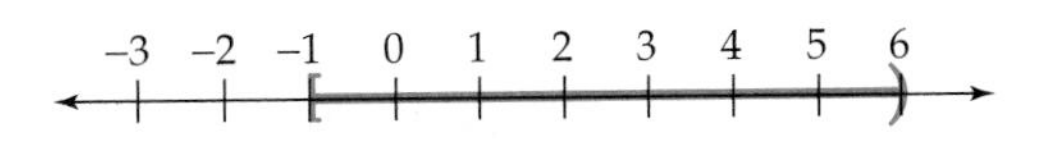

Figure 4–44

Example 3 When solving the inequality $4 < 3 - 5x < 18$, in which the variable appears only in the middle part, you can proceed as follows:

> **CAUTION**
> All inequality signs in an inequality should point in the same direction. *Don't* write things like $4 < x > 2$ or $-3 \ge x < 5$.

$$4 < 3 - 5x < 18$$

Subtract 3 from each part: $1 < -5x < 15$

Divide each part by −5 and reverse the direction of the inequalities: $-\frac{1}{5} > x > -3$

Reading this last inequality from right to left we see that

$$-3 < x < -1/5$$

so that the solutions are precisely the numbers in the interval $(-3, -1/5)$. ■

Polynomial Inequalities

Although the Basic Principles play a role in the solution of nonlinear inequalities, the key to solving such inequalities is this geometric fact:

> **The graph of $y = f(x)$ lies above the x-axis exactly when $f(x) > 0$ and below the x-axis exactly when $f(x) < 0$.**

Consequently, the solutions of $f(x) > 0$ are the numbers x for which the graph of f lies above the x-axis and the solutions $f(x) < 0$ are the numbers x for which the graph of f lies below the x-axis.

Example 4 To solve $2x^3 - 15x < x^2$, replace it by the equivalent inequality

$$2x^3 - x^2 - 15x < 0$$

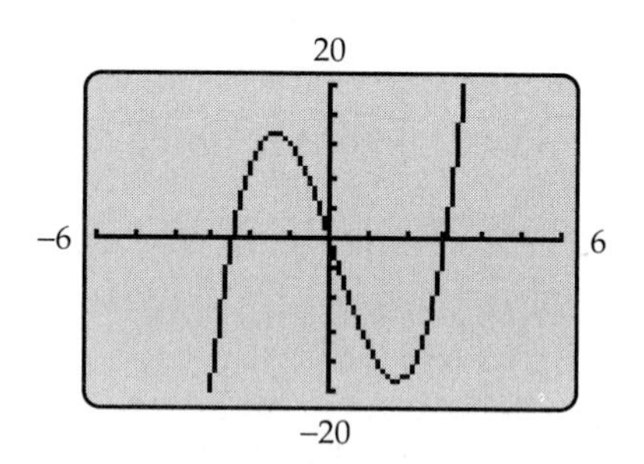

Figure 4–45

and consider the graph of the function $f(x) = 2x^3 - x^2 - 15x$ (Figure 4–45). Since $f(x)$ factors as

$$f(x) = 2x^3 - x^2 - 15x = x(2x^2 - x - 15) = x(2x + 5)(x - 3)$$

its roots (the x-intercepts of its graph) are $x = 0$, $x = -5/2$, and $x = 3$. The graph of $f(x) = 2x^3 - x^2 - 15x$ in Figure 4–45 is complete (why?) and lies below the x-axis when $x < -5/2$ or $0 < x < 3$. Therefore, the solutions of

$$2x^3 - x^2 - 15x < 0,$$

and hence of the original inequality, are all numbers x such that $x < -5/2$ or $0 < x < 3$. ■

Example 5 Solve $2x^3 - x^2 - 15x \geq 0$.

Solution Figure 4–45 shows that the solutions of $2x^3 - x^2 - 15x > 0$ (that is, the numbers x for which the graph of $f(x) = 2x^3 - x^2 - 15x$ lies above the x-axis) are all x such that $-5/2 < x < 0$ or $x > 3$. The solutions of the equation $2x^3 - x^2 - 15x = 0$ are the roots of $f(x) = 2x^3 - x^2 - 15x$, namely, 0, $-5/2$, and 3 as we saw in Example 4. Therefore, the solutions of the given inequality are all numbers x such that $-5/2 \leq x \leq 0$ or $x \geq 3$. ■

When the roots of a polynomial $f(x)$ cannot be determined exactly, zoom-in or root finder can be used to approximate them and to find approximate solutions of the inequalities $f(x) > 0$ and $f(x) < 0$.

Example 6 To solve $x^4 + 10x^3 + 21x^2 + 8 > 40x + 88$, we note that this inequality is equivalent to

$$x^4 + 10x^3 + 21x^2 - 40x - 80 > 0.$$

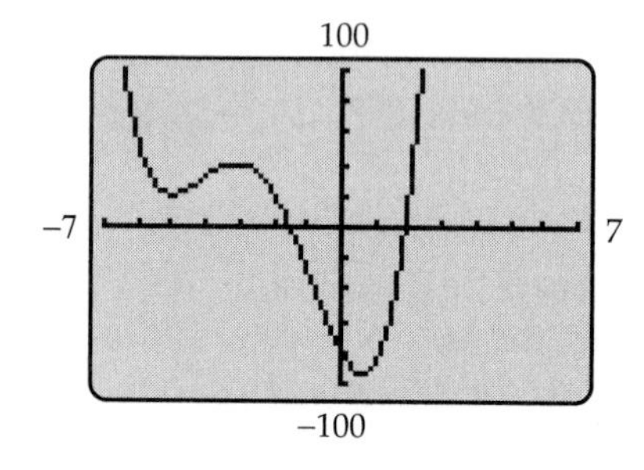

Figure 4–46

The graph $f(x) = x^4 + 10x^3 + 21x^2 - 40x - 80$ in Figure 4–46 is complete (why?) and shows that $f(x)$ has two roots, one between -2 and -1 and the other near 2.

GRAPHING EXPLORATION

Use zoom-in or a root finder to show that the approximate roots of $f(x)$ are -1.53 and 1.89.

Therefore, the approximate solutions of the inequality (the numbers x for which the graph is above the x-axis) are all numbers x such that $x < -1.53$ or $x > 1.89$. ■

> **CAUTION**
>
> Do not attempt to write the solution in Example 6, namely, "$x < -1.53$ or $x > 1.89$" as a single inequality. If you do, the result will be a *nonsense statement* such as $-1.53 > x > 1.89$ (which says, among other things, that $-1.53 > 1.89$).

Quadratic and Factorable Inequalities

The preceding examples show that solving a polynomial inequality depends only on knowing the roots of a polynomial and the places where its graph is above or below the x-axis. In the case of quadratic inequalities or completely factored polynomial inequalities a calculator is not needed to determine this information.

Example 7 The solutions of $2x^2 + 3x - 4 \le 0$ are the numbers x at which the graph of $f(x) = 2x^2 + 3x - 4$ lies on or below the x-axis. The points where the graph meets the x-axis are the roots of $f(x) = 2x^2 + 3x - 4$, which can be found by means of the quadratic formula:

$$x = \frac{-3 \pm \sqrt{3^2 - 4 \cdot 2(-4)}}{2 \cdot 2} = \frac{-3 \pm \sqrt{41}}{4}.$$

From Section 4.1 we know that the graph of $f(x)$ is an upward opening parabola, so the graph must have the general shape shown in Figure 4–47:

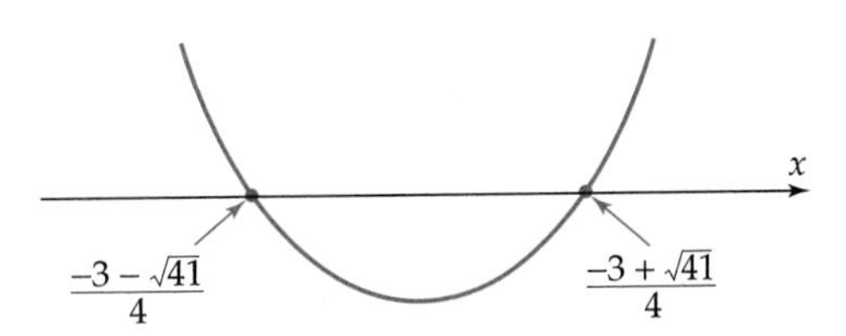

Figure 4–47

The graph lies below the x-axis between the two roots. Therefore, the solutions of the original inequality are all numbers x such that

$$\frac{-3 - \sqrt{41}}{4} \le x \le \frac{-3 + \sqrt{41}}{4}. \quad ■$$

Example 8 Solve $(x + 15)(x - 2)^6(x - 10) \le 0$.

Solution The roots of $f(x) = (x + 15)(x - 2)^6(x - 10)$ are easily read from the factored form: -15, 2, and 10. So we need only determine where the graph of $f(x)$ is on or below the x-axis. To do this without a calculator, note that the three roots of $f(x)$ divide the x-axis into four intervals:

$$x < -15, \qquad -15 < x < 2, \qquad 2 < x < 10, \qquad x > 10.$$

For each of these intervals, we shall determine whether the graph is above or below the x-axis.

Consider, for example, the interval between the roots 2 and 10. The graph of $f(x)$ touches the x-axis at $x = 2$ and $x = 10$, but does not touch the axis at any point in between since the only other root (x-intercept) is -15. Furthermore, the graph of $f(x)$ is continuous, which means that between the points (2, 0) and (10, 0) it can be drawn without lifting the pencil from the paper. If you experiment with Figure 4–48 you will find that the only way to draw a graph from (2, 0) to (10, 0) without lifting your pencil and without touching the x-axis is to stay entirely on one side of the x-axis (try it!). In other words, between the two adjacent roots 2 and 10 the graph of $f(x)$ is either entirely above the x-axis or entirely below it.

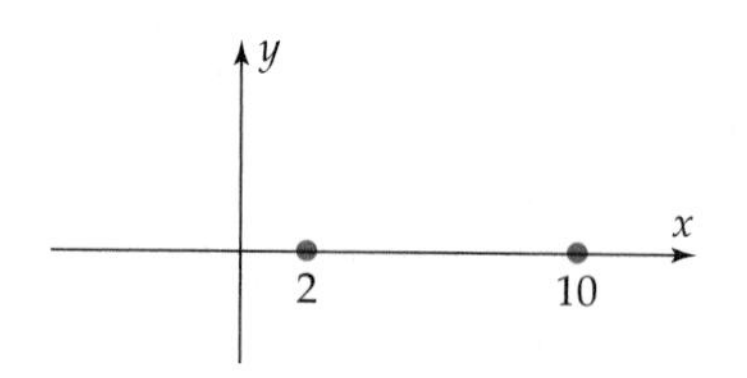

Figure 4–48

In order to determine which is the case, choose any number between 2 and 10, say $x = 4$, and evaluate $f(4)$:

$$f(4) = (4 + 15)(4 - 2)^6(4 - 10) = 19(2^6)(-6).$$

You don't even have to finish the computation to see that $f(4)$ is a negative number. Therefore the point $(4, f(4))$ on the graph of $f(x)$ lies below the x-axis. Since one point of the graph between 2 and 10 lies below the x-axis, the entire graph must be below the x-axis between 2 and 10.

The location of the graph on the other intervals can be determined similarly, by choosing a "test number" in each interval, as summarized in this chart:

Interval	$x < -15$	$-15 < x < 2$	$2 < x < 10$	$x > 10$
Test number in this interval	-20	0	4	11
Value of $f(x)$ at test number	$(-5)(-22)^6(-30)$	$15(-2)^6(-10)$	$19(2^6)(-6)$	$26(9^6)(1)$
Sign of $f(x)$ at test number	+	−	−	+
Graph	Above x-axis	Below x-axis	Below x-axis	Above x-axis

The last line of the chart shows that the intervals where the graph is below the x-axis are $-15 < x < 2$ and $2 < x < 10$. Since the graph touches the x-axis at the roots -15, 2, and 10, the solutions of the original inequality (the numbers x for which the graph is on or below the x-axis) are all numbers x such that $-15 \leq x \leq 10$. ■

The procedures used in Examples 4–8 may be summarized as follows.

Solving Polynomial Inequalities

1. **Write the inequality in one of these forms:**

$$f(x) > 0, \qquad f(x) \geq 0, \qquad f(x) < 0, \qquad f(x) \leq 0.$$

2. **Determine the roots of $f(x)$, exactly if possible, approximately otherwise.**
3. **Use a calculator (as in Examples 4–6), your knowledge of quadratic functions (as in Example 7), or a sign chart (as in Example 8) to determine whether the graph of $f(x)$ is above or below the x-axis on each of the intervals determined by the roots.**
4. **Use the information in step 3 to find the solutions of the inequality.**

Rational Inequalities

Rational inequalities are solved in essentially the same way that polynomial inequalities are solved, with one difference. The graph of a rational function may cross the x-axis at an x-intercept, but there is another possibility: The graph may be above the x-axis on one side of a vertical asymptote and below it on the other side (see, for instance, Examples 3–4 in Section 4.5).* Since the x-intercepts of the graph of the rational function $g(x)/h(x)$ are determined by the roots of its numerator $g(x)$ and the vertical asymptotes by the roots of its denominator $h(x)$, all of these roots must be considered in determining the solution of an inequality involving $g(x)/h(x)$.

Example 9 Solve $\dfrac{x}{x-1} > -6$.

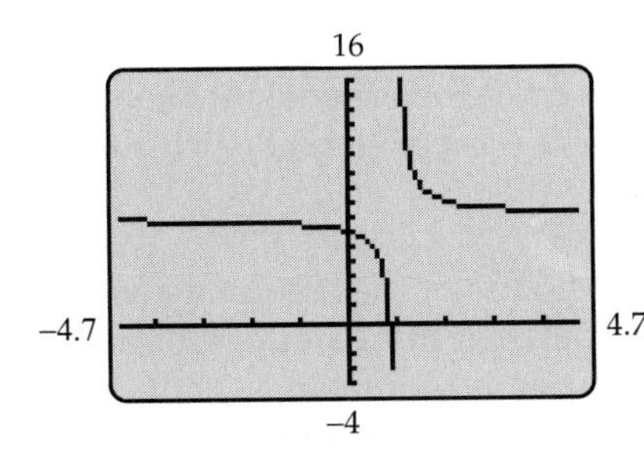

Figure 4–49

Solution There are three ways to solve this inequality.

Geometric: The fastest way to get an approximate solution is to replace the given inequality by an equivalent one

$$(*) \qquad \frac{x}{x-1} + 6 > 0$$

and graph the function $f(x) = \dfrac{x}{x-1} + 6$ as in Figure 4–49.

*These are the only ways that the graph of a rational function can change from one side of the x-axis to the other because a rational function is continuous everywhere that it is defined. On any interval on which the function is defined, its graph can be drawn without lifting pencil from paper; see the discussion in the second paragraph on page 272.

The graph is above the x-axis everywhere except between the x-intercept and the vertical asymptote $x = 1$. Using zoom-in or a root finder, we see that the x-intercept is approximately .857. Therefore, the approximate solutions of the original inequality are all numbers x such that $x < .857$ or $x > 1$.

Algebraic/Geometric: Proceed as above, but rewrite the rule of the function f as a single rational expression before graphing:

$$f(x) = \frac{x}{x-1} + 6 = \frac{x}{x-1} + \frac{6(x-1)}{x-1} = \frac{x + 6x - 6}{x-1} = \frac{7x-6}{x-1}.$$

When the rule of f is written in this form, it is easy to see that the x-intercept of the graph (the root of the numerator) is $x = 6/7$ (whose decimal approximation begins .857). Therefore, the exact solutions of the original inequality (the numbers x for which the graph in Figure 4–49 is above the x-axis) are all numbers x such that $x < 6/7$ or $x > 1$.

Algebraic: Write the rule of the function f as a single rational expression $f(x) = \frac{7x-6}{x-1}$. The roots of the numerator and denominator (6/7 and 1) divide the x-axis into three intervals. Use test numbers and a sign chart instead of graphing to determine the location of the graph on each interval:*

Interval	$x < 6/7$	$6/7 < x < 1$	$x > 1$
Test number in this interval	0	.9	2
Value of $f(x)$ at test number	$\frac{7 \cdot 0 - 6}{0 - 1}$	$\frac{7(.9) - 6}{.9 - 1}$	$\frac{7 \cdot 2 - 6}{2 - 1}$
Sign of $f(x)$ at test number	+	−	+
Graph	Above x-axis	Below x-axis	Above x-axis

The last line of the chart shows that the solutions of the original inequality (the numbers x for which the graph is above the x-axis) are all such that $x < 6/7$ or $x > 1$. ■

The algebraic technique of writing the left side of the inequality as a single rational expression is useful whenever the resulting numerator has low degree (so that its roots can be found exactly), but can usually be omitted when the roots of the numerator must be approximated.

*The justification for this approach is essentially the same as that in Example 8: Because f is continuous everywhere that it is defined, the graph can change from one side of the x-axis to the other only at x-intercepts or vertical asymptotes, so testing one number in each interval is sufficient to determine the side on which the graph lies.

CAUTION

Don't treat rational inequalities as if they are equations, as in this *incorrect* "solution" of the preceding example:

$$\frac{x}{x-1} > -6$$

$$x > -6(x-1) \qquad \text{[Both sides multiplied by } x - 1\text{]}$$

$$x > -6x + 6$$

$$7x > 6$$

$$x > \frac{6}{7}$$

According to this, the inequality has no negative solution and $x = 1$ is a solution, but as we saw in Example 9, *every* negative number is a solution and $x = 1$ is not.*

Applications

Example 10 A computer store has determined that the cost C of ordering and storing x laser printers is given by

$$C = 2x + \frac{300{,}000}{x}.$$

If the delivery truck can bring at most 450 printers per order, how many printers should be ordered at a time to keep the cost below \$1600?

Solution To find the values of x that make C less than 1600, we must solve the inequality

$$2x + \frac{300{,}000}{x} < 1600 \qquad \text{or equivalently,} \qquad 2x + \frac{300{,}000}{x} - 1600 < 0.$$

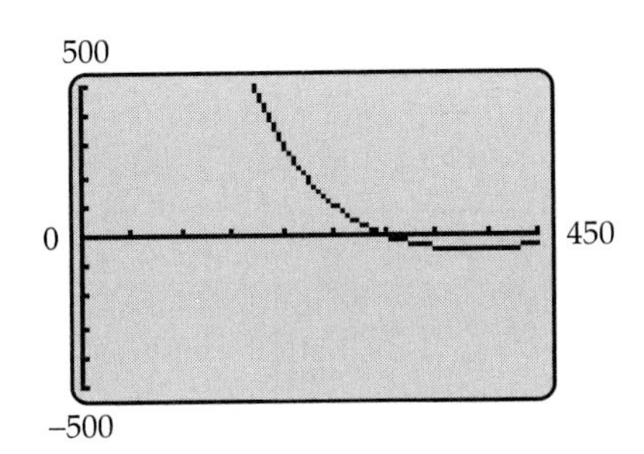

Figure 4–50

We shall solve this inequality graphically, although it can also be solved algebraically. In this context, the only solutions that make sense are those between 0 and 450. So we choose the viewing window in Figure 4–50 and graph

$$f(x) = 2x + \frac{300{,}000}{x} - 1600.$$

Figure 4–50 is consistent with the fact that $f(x)$ has a vertical asymptote at $x = 0$ and shows that the desired solutions (numbers where the graph is below the x-axis) are all numbers x between the root and 450. Zoom-in or a root finder shows that the root is $x \approx 300$. In fact this is the exact root since a simple computation shows that $f(300) = 0$. (Do it!) Therefore, to keep costs under \$1600, x printers should be ordered each time, with $300 < x \leq 450$. ■

*The source of the error is multiplying by $x - 1$. This quantity is negative for some values of x and positive for others. To do this calculation correctly, you must consider two separate cases and reverse the direction of the inequality when $x - 1$ is negative.

Exercises 4.6

In Exercises 1–20, solve the inequality and express your answer in interval notation.

1. $2x + 4 \le 7$
2. $3x - 5 > -6$
3. $3 - 5x < 13$
4. $2 - 3x < 11$
5. $6x + 3 \le x - 5$
6. $5x + 3 \le 2x + 7$
7. $5 - 7x < 2x - 4$
8. $5 - 3x > 7x - 3$
9. $2 < 3x - 4 < 8$
10. $1 < 5x + 6 < 9$
11. $0 < 5 - 2x \le 11$
12. $-4 \le 7 - 3x < 0$
13. $2x + 7(3x - 2) < 2(x - 1)$
14. $x + 3(x - 5) \ge 3x + 2(x + 1)$
15. $\frac{x+1}{2} - 3x \le \frac{x+5}{3}$
16. $\frac{x-1}{4} + 2x \ge \frac{2x-1}{3} + 2$
17. $2x + 3 \le 5x + 6 < -3x + 7$
18. $4x - 2 < x + 8 < 9x + 1$
19. $3 - x < 2x + 1 \le 3x - 4$
20. $2x + 5 \le 4 - 3x < 1 - 4x$

In Exercises 21–24, a, b, c, and d are positive constants. Solve the inequality for x.

21. $ax - b < c$
22. $d - cx > a$
23. $0 < x - c < a$
24. $-d < x - c < d$

In Exercises 25–54, solve the inequality. Find exact solutions when possible and approximate ones otherwise.

25. $x^2 - 4x + 3 \le 0$
26. $x^2 - 7x + 10 \le 0$
27. $x^2 + 9x + 15 \ge 0$
28. $x^2 + 8x + 20 \ge 0$
29. $8 + x - x^2 \le 0$
30. $4 - 3x - x^2 \ge 0$
31. $x^3 - x \ge 0$
32. $x^3 + 2x^2 + x > 0$
33. $x^3 - 2x^2 - 3x < 0$
34. $x^4 - 14x^3 + 48x^2 \ge 0$
35. $x^4 - 5x^2 + 4 < 0$
36. $x^4 - 10x^2 + 9 \le 0$
37. $x^3 - 2x^2 - 5x + 7 \ge 2x + 1$
38. $x^4 - 6x^3 + 2x^2 < 5x - 2$
39. $2x^4 + 3x^3 < 2x^2 + 4x - 2$
40. $x^5 + 5x^4 > 4x^3 - 3x^2 + 2$
41. $\frac{3x+1}{2x-4} > 0$
42. $\frac{2x-1}{5x+3} \ge 0$
43. $\frac{x^2+x-2}{x^2-2x-3} < 0$
44. $\frac{2x^2+x-1}{x^2-4x+4} \ge 0$
45. $\frac{x-2}{x-1} < 1$
46. $\frac{-x+5}{2x+3} \ge 2$
47. $\frac{x-3}{x+3} \le 5$
48. $\frac{2x+1}{x-4} > 3$
49. $\frac{2}{x+3} \ge \frac{1}{x-1}$
50. $\frac{1}{x-1} < \frac{-1}{x+2}$
51. $\frac{x^3 - 3x^2 + 5x - 29}{x^2 - 7} > 3$
52. $\frac{x^4 - 3x^3 + 2x^2 + 2}{x - 2} > 15$
53. $\frac{2x^2 + 6x - 8}{2x^2 + 5x - 3} < 1$ [Be alert for hidden behavior.]
54. $\frac{1}{x^2 + x - 6} + \frac{x-2}{x+3} > \frac{x+3}{x-2}$

In Exercises 55–57, read the solution of the inequality from the given graph.

55. $3 - 2x < .8x + 7$

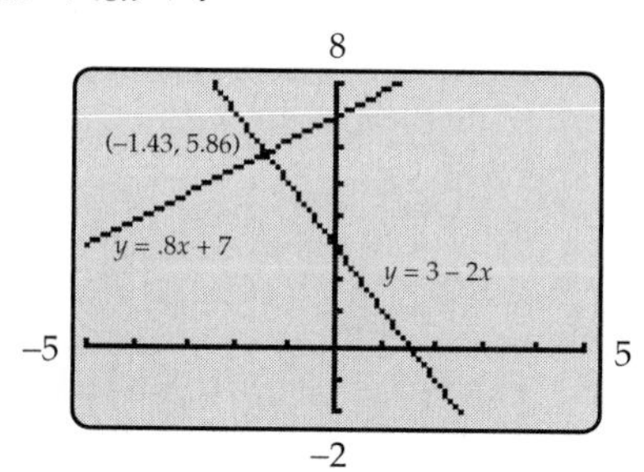

56. $8 - |7 - 5x| > 3$

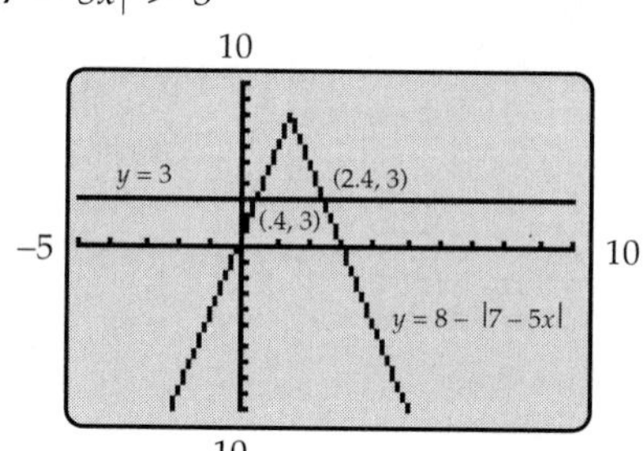

57. $x^2 + 3x + 1 \ge 4$

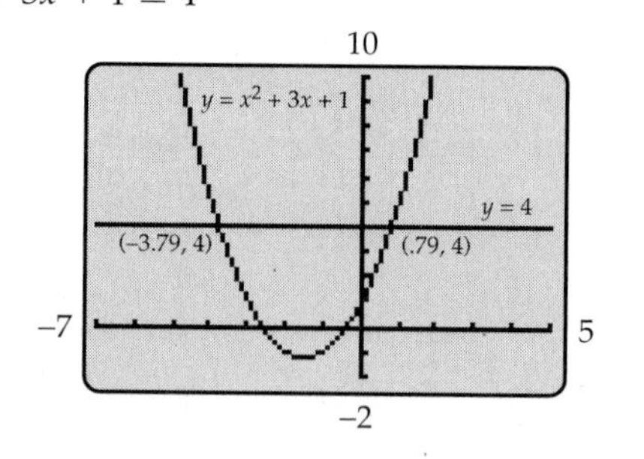

58. The graphs of the revenue and cost functions for a manufacturing firm are shown in the figure.
 (a) What is the break-even point?
 (b) Shade in the region representing profit.

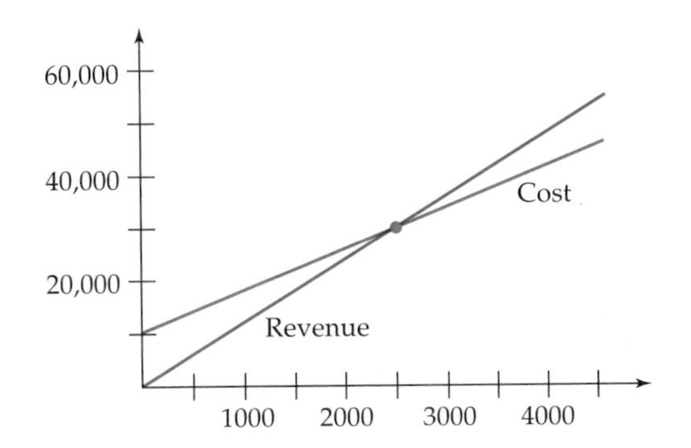

59. One freezer costs $623.95 and uses 90 kilowatt hours (kwh) of electricity each month. A second freezer costs $500 and uses 100 kwh of electricity each month. The expected life of each freezer is 12 years. What is the minimum electric rate (in *cents* per kwh) for which the 12-year total cost (purchase price + electricity costs) will be less for the first freezer?

60. A business executive leases a car for $300 per month. She decides to lease another brand for $250 per month, but has to pay a penalty of $1000 for breaking the first lease. How long must she keep the second car in order to come out ahead?

61. One salesperson is paid a salary of $1000 per month plus a commission of 2% of her total sales. A second salesperson receives no salary, but is paid a commission of 10% of her total sales. What dollar amount of sales must the second salesperson have in order to earn more per month than the first?

62. A developer subdivided 60 acres of a 100-acre tract, leaving 20% of the 60 acres as a park. Zoning laws require that at least 25% of the total tract be set aside for parks. For financial reasons the developer wants to have no more than 30% of the tract as parks. How many one-quarter-acre lots can the developer sell in the remaining 40 acres and still meet the requirements for the whole tract?

63. If $5000 is invested at 8%, how much more should be invested at 10% in order to guarantee a total annual interest income between $800 and $940?

64. How many gallons of a 12% salt solution should be added to 10 gallons of an 18% salt solution in order to produce a solution whose salt content is between 14% ad 16%?

65. Find all pairs of numbers that satisfy these two conditions: Their sum is 20 and the sum of their squares is less than 362.

66. The length of a rectangle is 6 inches longer than its width. What are the possible widths if the area of the rectangle is at least 667 square inches?

67. It costs a craftsman $5 in materials to make a medallion. He has found that if he sells the medallions for $50 - x$ dollars each, where x is the number of medallions produced each week, then he can sell all that he makes. His fixed costs are $350 per week. If he wants to sell all he makes and show a profit each week, what are the possible numbers of medallions he should make?

68. A retailer sells file cabinets for $80 - x$ dollars each, where x is the number of cabinets she receives from the supplier each week. She pays $10 for each file cabinet and has fixed costs of $600 per week. How many file cabinets should she order from the supplier each week in order to guarantee that she makes a profit?

In Exercises 69–72, you will need the formula for the height h of an object above the ground at time t seconds: $h = -16t^2 + v_0t + h_0$; this formula was explained on page 213.

69. A toy rocket is fired straight up from ground level with an initial velocity of 80 feet per second. During what time interval will it be at least 64 feet above the ground?

70. A projectile is fired straight up from ground level with an initial velocity of 72 feet per second. During what time interval is it at least 37 feet above the ground?

71. A ball is dropped from the roof of a 120-foot-high building. During what time period will it be strictly between 56 feet and 39 feet above the ground?

72. A ball is thrown straight up from a 40-foot-high tower with an initial velocity of 56 feet per second.
 (a) During what time interval is the ball at least 8 feet above the ground?
 (b) During what time interval is the ball between 53 feet and 80 feet above the ground?

73. **(a)** Solve the inequalities $x^2 < x$ and $x^2 > x$.
 (b) Use the results of part (a) to show that for any nonzero real number c with $|c| < 1$, it is always true that $c^2 < |c|$.
 (c) Use the results of part (a) to show that for any nonzero real number c with $|c| > 1$, it is always true that $c^2 > c$.

74. **(a)** If $0 < a \leq b$, prove that $1/a \geq 1/b$.
 (b) If $a \leq b < 0$, prove that $1/a \geq 1/b$.
 (c) If $a < 0 < b$, how are $1/a$ and $1/b$ related?

4.6.A

EXCURSION Absolute Value Inequalities

Polynomial and rational inequalities involving absolute value can be solved graphically just as was done earlier: Rewrite the inequality in an equivalent form that has 0 on the right side of the inequality sign; then graph the function whose rule is given by the left side and determine where the graph is above or below the x-axis.

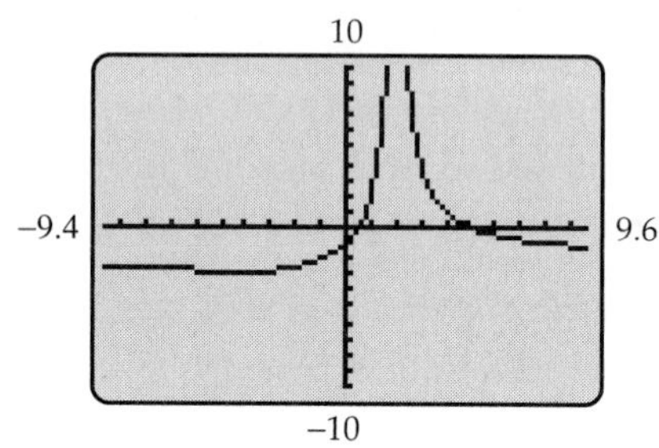

Figure 4–51

Example 1 To solve $\left|\dfrac{x+4}{x-2}\right| > 3$, use the equivalent inequality

$$\left|\frac{x+4}{x-2}\right| - 3 > 0$$

and graph the function $f(x) = \left|\dfrac{x+4}{x-2}\right| - 3$ (Figure 4–51). The graph is above the x-axis between the two x-intercepts, which can be found algebraically or graphically.

GRAPHING EXPLORATION

Verify that the x-intercepts are $x = 1/2$ and $x = 5$.

Since $f(x)$ is not defined at $x = 2$ (where the graph has a vertical asymptote), the solutions of the original inequality are all x such that $1/2 < x < 2$ or $2 < x < 5$. ■

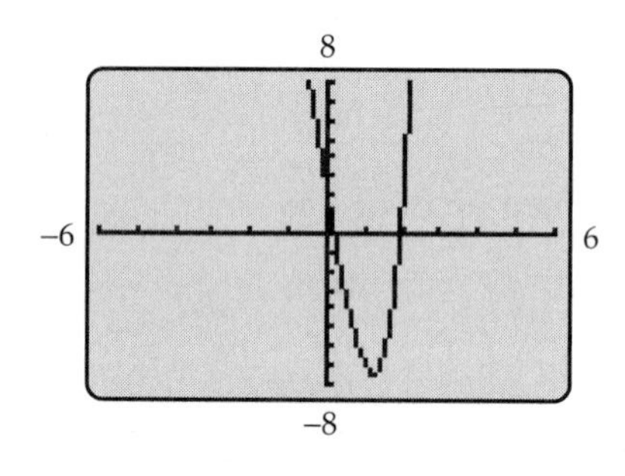

Figure 4–52

Example 2 The solutions of $|x^4 + 2x^2 - x + 2| < 11x$ can be found by determining the numbers for which the graph of

$$f(x) = |x^4 + 2x^2 - x + 2| - 11x$$

lies below the x-axis. (Why?) Convince yourself that the graph of $f(x)$ in Figure 4–52 is complete.

A root finder shows that the approximate x-intercepts are $x = .17$ and $x = 1.92$. Therefore, the approximate solutions of the original inequality (the numbers where the graph is below the x-axis) are all x such that $.17 < x < 1.92$. ■

Algebraic Methods

Most linear and quadratic inequalities involving absolute values can be solved exactly by algebraic means. In fact, this is often the easiest way to solve such inequalities. The key to the algebraic method is the fact that the absolute value of a number can be interpreted as distance on the number line. For example, the inequality $|r| \le 5$ states that the distance from

r to 0 (namely, $|r|$) is 5 units or less. A glance at the number line in Figure 4–53 shows that these are the numbers r with $-5 \le r \le 5$:

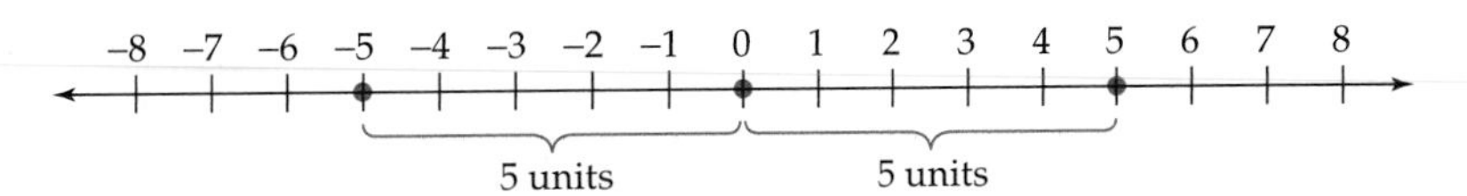

Figure 4–53

Similarly, the numbers r such that $|r| \ge 5$ are those whose distance to 0 is 5 or more units, that is, the numbers r with $r \le -5$ or $r \ge 5$. This argument works with any positive number k in place of 5 and proves the following facts (which are also true with $<$ and $>$ in place of $\le$ and $\ge$):

Absolute Value Inequalities

Let k be a positive number and r any real number.

$|r| \le k$ **is equivalent to** $-k \le r \le k$.

$|r| \ge k$ **is equivalent to** $r \le -k$ **or** $r \ge k$.

Example 3 To solve $|3x - 7| \le 11$, apply the first fact in the box, with $3x - 7$ in place of r and 11 in place of k, and obtain this equivalent inequality $-11 \le 3x - 7 \le 11$. Then

Add 7 to each part: $-4 \le 3x \le 18$

Divide each part by 3: $-4/3 \le x \le 6$.

Therefore, the solutions of the original inequality are all numbers in the interval $[-4/3, 6]$, as shown in Figure 4–54. ■

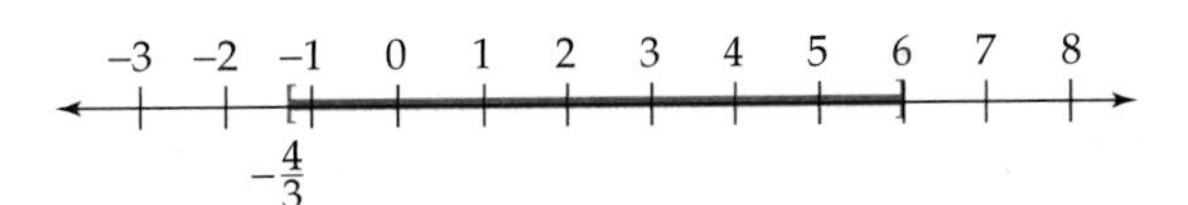

Figure 4–54

Example 4 To solve $|5x + 2| > 3$, apply the second fact in the box, with $5x + 2$ in place of r, and 3 in place of k, and $>$ in place of $\ge$. This produces the equivalent statement:

$$5x + 2 < -3 \quad \text{or} \quad 5x + 2 > 3$$
$$5x < -5 \qquad\qquad 5x > 1$$
$$x < -1 \quad \text{or} \quad x > 1/5.$$

Therefore, the solutions of the original inequality are the numbers in *either* of the intervals $(-\infty, -1)$ or $(1/5, \infty)$. ■

Example 5 If a and δ are real numbers with δ positive, then the inequality $|x - a| < \delta$ is equivalent to $-\delta < x - a < \delta$. Adding a to each part shows that $a - \delta < x < a + \delta$. ■

Example 6 To solve $|x^2 - x - 4| \geq 2$, we use the fact in the box on the preceding page to replace it by an equivalent inequality:

$$x^2 - x - 4 \leq -2 \quad \text{or} \quad x^2 - x - 4 \geq 2$$

which is the same as

$$x^2 - x - 2 \leq 0 \quad \text{or} \quad x^2 - x - 6 \geq 0.$$

The solutions are all numbers that are solutions of *either one* of the two inequalities.

To solve the first of these inequalities, note that the graph of $f(x) = x^2 - x - 2$ is an upward-opening parabola, which crosses the x-axis at the roots of $f(x)$. The roots are -1 and 2, as can be seen from the factorization: $x^2 - x - 2 = (x + 1)(x - 2)$. Therefore, the solutions of

$$x^2 - x - 2 \leq 0$$

(the numbers for which the graph of $f(x)$ is on or below the x-axis) are all x with $-1 \leq x \leq 2$. The second inequality above, $x^2 - x - 6 \geq 0$, is solved similarly.

GRAPHING EXPLORATION

What is the shape of the graph of $g(x) = x^2 - x - 6$ and what are its x-intercepts?

This Exploration shows that the solutions of the second inequality (the numbers for which the graph of $g(x)$ is on or above the x-axis) are all x with $x \leq -2$ or $x \geq 3$.

Consequently, the solutions of the original inequality are all numbers x such that $x \leq -2$ or $-1 \leq x \leq 2$ or $x \geq 3$, as shown in Figure 4–55. ■

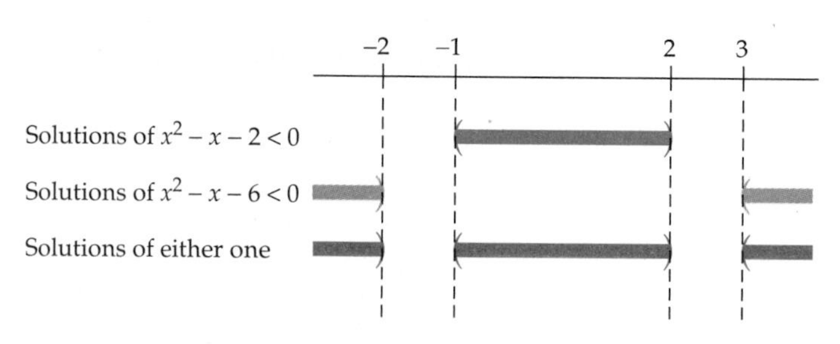

Figure 4–55

Exercises 4.6.A

In Exercises 1–32, solve the inequality. Find exact solutions when possible and approximate ones otherwise.

1. $|3x + 2| \le 2$ **2.** $|5x - 1| < 3$

3. $|3 - 2x| < 2/3$ **4.** $|4 - 5x| \le 4$

5. $|2x + 3| > 1$ **6.** $|3x - 1| \ge 2$

7. $|5x + 2| \ge \dfrac{3}{4}$ **8.** $|2 - 3x| > 4$

9. $\left|\dfrac{12}{5} + 2x\right| > \dfrac{1}{4}$ **10.** $\left|\dfrac{5}{6} + 3x\right| < \dfrac{7}{6}$

11. $\left|\dfrac{x - 1}{x + 2}\right| \le 3$ **12.** $\left|\dfrac{x + 1}{3x + 5}\right| < 2$

13. $\left|\dfrac{2x - 1}{x + 5}\right| > 1$ **14.** $\left|\dfrac{x + 1}{x + 2}\right| \ge 2$

15. $\left|\dfrac{1 - 4x}{2 + 3x}\right| < 1$ **16.** $\left|\dfrac{3x + 1}{1 - 2x}\right| \ge 2$

17. $|x^2 - 2| < 1$ **18.** $|x^2 - 4| \le 3$

19. $|x^2 - 2| > 4$ **20.** $\left|\dfrac{1}{x^2 - 1}\right| \le 2$

21. $|x^2 + x - 1| \ge 1$ **22.** $|x^2 + x - 4| \le 2$

23. $|3x^2 - 8x + 2| < 2$ **24.** $|x^2 + 3x - 4| < 6$

25. $|x^5 - x^3 + 1| < 2$ **26.** $|4x - x^3| > 1$

27. $|x^4 - x^3 + x^2 - x + 1| > 4$

28. $|x^3 - 6x^2 + 4x - 5| < 3$ **29.** $\dfrac{x + 2}{|x - 3|} \le 4$

30. $\dfrac{x^2 - 9}{|x^2 - 4|} < -2$

31. $\left|\dfrac{2x^2 + 2x - 12}{x^3 - x^2 + x - 2}\right| > 2$ **32.** $\left|\dfrac{x^2 - x - 2}{x^2 + x - 2}\right| > 3$

Thinkers

33. Let E be a fixed real number. Show that every solution of $|x - 3| < E/5$ is also a solution of $|(5x - 4) - 11| < E$.

34. Let a and b be fixed real numbers with $a < b$. Show that the solutions of

$$\left|x - \frac{a + b}{2}\right| < \frac{b - a}{2}$$

are all x with $a < x < b$.

4.7 Complex Numbers

If you are restricted to nonnegative integers, you can't solve the equation $x + 5 = 0$. Enlarging the number system to include negative integers makes it possible to solve this equation ($x = -5$). Enlarging it again, to include rational numbers, makes it possible to solve equations like $3x = 7$, which have no integer solutions. Similarly, the equation $x^2 = 2$ has no solutions in the rational number system, but has $\sqrt{2}$ and $-\sqrt{2}$ as solutions in the real number system. So the idea of enlarging a number system in order to solve an equation that can't be solved in the present system is a natural one.

Equations such as $x^2 = -1$ and $x^2 = -4$ have no solutions in the real number system because $\sqrt{-1}$ and $\sqrt{-4}$ are not real numbers. In order to solve such equations (or equivalently, in order to find square roots of negative numbers) we must enlarge the number system again. We claim that there is a number system, called the **complex number system,** with these properties:

Properties of the Complex Number System

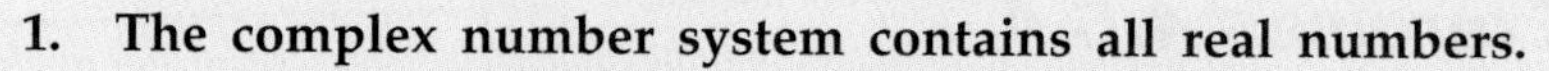

1. **The complex number system contains all real numbers.**
2. **Addition, subtraction, multiplication, and division of complex numbers obey the same rules of arithmetic that hold in the real number system, with one exception: the exponent laws hold for *integer* exponents, but not necessarily for fractional ones.**
3. **The complex number system contains a number (usually denoted by i) such that $i^2 = -1$.**
4. **Every complex number can be written in the *standard form* $a + bi$, where a and b are real numbers.***
5. **Two complex numbers $a + bi$ and $c + di$ are equal exactly when $a = c$ and $b = d$.**

In view of our past experience with enlarging the number system, this claim *ought* to appear plausible. But the mathematicians who invented the complex numbers in the seventeenth century were very uneasy about a number i such that $i^2 = -1$ (that is, $i = \sqrt{-1}$). Consequently, they called numbers of the form bi (b any real number), such as $5i$ and $-\frac{1}{4}i$, **imaginary numbers.** The old familiar numbers (integers, rationals, irrationals) were called **real numbers.** Sums of real and imaginary numbers, numbers of the form $a + bi$, such as

$$5 + 2i, \qquad 7 - 4i, \qquad 18 + \frac{3}{2}i, \qquad \sqrt{3} - 12i$$

were called **complex numbers.**†

Every real number is a complex number; for instance, $7 = 7 + 0i$. Similarly, every imaginary number bi is a complex number since $bi = 0 + bi$. Since the usual laws of arithmetic still hold, it's easy to add, subtract, and multiply complex numbers. As the following examples demonstrate, *all symbols can be treated as if they were real numbers, provided that i^2 is replaced by -1.* Unless directed otherwise, express your answers in the standard form $a + bi$.

Example 1

(a) $(1 + i) + (3 - 7i) = 1 + i + 3 - 7i$
$\qquad = (1 + 3) + (i - 7i) = 4 - 6i.$

(b) $(4 + 3i) - (8 - 6i) = 4 + 3i - 8 - (-6i)$
$\qquad = (4 - 8) + (3i + 6i) = -4 + 9i.$

*Hereafter whenever we write $a + bi$ or $c + di$, it is assumed that a, b, c, d are real numbers and $i^2 = -1$.

†This terminology is still used, even though there is nothing complicated, unreal, or imaginary about complex numbers—they are just as valid mathematically as are real numbers. See Exercise 80 for a formal construction of the complex numbers and proofs of the claims made here.

(c) $4i\left(2 + \frac{1}{2}i\right) = 4i \cdot 2 + 4i\left(\frac{1}{2}i\right) = 8i + 4 \cdot \frac{1}{2} \cdot i^2$

$$= 8i + 2i^2 = 8i + 2(-1) = -2 + 8i.$$

(d) $(2 + i)(3 - 4i) = 2 \cdot 3 + 2(-4i) + i \cdot 3 + i(-4i)$

$$= 6 - 8i + 3i - 4i^2 = 6 - 8i + 3i - 4(-1)$$

$$= (6 + 4) + (-8i + 3i) = 10 - 5i. \quad \blacksquare$$

The familiar multiplication patterns and exponent laws for integer exponents hold in the complex number system.

Example 2

(a) $(3 + 2i)(3 - 2i) = 3^2 - (2i)^2$

$$= 9 - 4i^2 = 9 - 4(-1) = 9 + 4 = 13.$$

(b) $(4 + i)^2 = 4^2 + 2 \cdot 4 \cdot i + i^2 = 16 + 8i + (-1) = 15 + 8i.$

(c) To find i^{54}, we first note that $i^4 = i^2 i^2 = (-1)(-1) = 1$ and that $54 = 52 + 2 = 4 \cdot 13 + 2$. Consequently,

$$i^{54} = i^{52+2} = i^{52}i^2 = i^{4 \cdot 13}i^2 = (i^4)^{13}i^2 = 1^{13}(-1) = -1. \quad \blacksquare$$

Technology Tip

You can do complex arithmetic directly by using the special i key on the TI-83/89 and Sharp 9600 keyboards and in the CPLX submenu of the Casio 9850 OPTN menu. On TI-86 and HP-38, enter $a + bi$ as (a, b). Other calculators can do complex arithmetic by means of matrices; see the Calculator Investigation on page 286.

The **conjugate** of the complex number $a + bi$ is the number $a - bi$, and the conjugate of $a - bi$ is $a + bi$. For example, the conjugate of $3 + 4i$ is $3 - 4i$ and the conjugate of $-3i = 0 - 3i$ is $0 + 3i = 3i$. *Every real number is its own conjugate;* for instance, the conjugate of $17 = 17 + 0i$ is $17 - 0i = 17$.

For any complex number $a + bi$, we have

$$(a + bi)(a - bi) = a^2 - (bi)^2 = a^2 - b^2i^2 = a^2 - b^2(-1) = a^2 + b^2.$$

Since a^2 and b^2 are nonnegative real numbers, so is $a^2 + b^2$. Therefore *the product of a complex number and its conjugate is a nonnegative real number.* This fact enables us to express quotients of complex numbers in standard form.

Example 3 To express $\frac{3 + 4i}{1 + 2i}$ in the form $a + bi$, *multiply both numerator and denominator by the conjugate of the denominator,* namely, $1 - 2i$:

$$\frac{3 + 4i}{1 + 2i} = \frac{3 + 4i}{1 + 2i} \cdot \frac{1 - 2i}{1 - 2i} = \frac{(3 + 4i)(1 - 2i)}{(1 + 2i)(1 - 2i)}$$

$$= \frac{3 + 4i - 6i - 8i^2}{1^2 - (2i)^2} = \frac{3 + 4i - 6i - 8(-1)}{1 - 4i^2} = \frac{11 - 2i}{1 - 4(-1)}$$

$$= \frac{11 - 2i}{5} = \frac{11}{5} - \frac{2}{5}i.$$

This is the form $a + bi$ with $a = 11/5$ and $b = -2/5$. $\blacksquare$

Example 4 To express $\dfrac{1}{1-i}$ in standard form, note that the conjugate of the denominator is $1 + i$ and therefore:

$$\frac{1}{1-i} = \frac{1\cdot(1+i)}{(1-i)(1+i)} = \frac{1+i}{1^2 - i^2} = \frac{1+i}{1-(-1)} = \frac{1+i}{2} = \frac{1}{2} + \frac{1}{2}i.$$

We can check this result by multiplying $\frac{1}{2} + \frac{1}{2}i$ by $1 - i$ to see if the product is 1 $\left(\text{which it should be if } \frac{1}{2} + \frac{1}{2}i = \frac{1}{1-i}\right)$:

$$\left(\frac{1}{2} + \frac{1}{2}i\right)(1-i) = \frac{1}{2}\cdot 1 - \frac{1}{2}i + \frac{1}{2}i\cdot 1 - \frac{1}{2}i^2 = \frac{1}{2} - \frac{1}{2}(-1) = 1. \quad ■$$

Since $i^2 = -1$, we define $\sqrt{-1}$ to be the complex number i. Similarly, since $(5i)^2 = 5^2 i^2 = 25(-1) = -25$, we define $\sqrt{-25}$ to be $5i$. In general,

Square Roots of Negative Numbers

Let b be a positive real number.

$$\sqrt{-b} \text{ is defined to be } \sqrt{b}\, i$$

because $(\sqrt{b}\, i)^2 = (\sqrt{b})^2 i^2 = b(-1) = -b$.

CAUTION

$\sqrt{b}\, i$ is *not* the same as $\sqrt{bi}$. To avoid confusion it may help to write $\sqrt{b}\, i$ as $i\sqrt{b}$.

Example 5

(a) $\sqrt{-3} = \sqrt{3}i = i\sqrt{3}$.

(b) $\dfrac{1 - \sqrt{-7}}{3} = \dfrac{1 - \sqrt{7}i}{3} = \dfrac{1}{3} - \dfrac{\sqrt{7}}{3}i.$ ■

CAUTION

The property $\sqrt{cd} = \sqrt{c}\sqrt{d}$ (or equivalently in exponential notation, $(cd)^{1/2} = c^{1/2}d^{1/2}$), which is valid for positive real numbers, *does not hold* when both c and d are negative.

$$\sqrt{-20}\sqrt{-5} = \sqrt{20}i\cdot\sqrt{5}i = \sqrt{20}\,\sqrt{5}\cdot i^2 = \sqrt{20\cdot 5}(-1)$$
$$= \sqrt{100}(-1) = -10$$

But $\sqrt{(-20)(-5)} = \sqrt{100} = 10$, so that

$$\sqrt{(-20)(-5)} \neq \sqrt{-20}\sqrt{-5}.$$

To avoid difficulty, *always write square roots of negative numbers in terms of i before doing any simplification.*

Technology Tip

Most calculators that do complex number arithmetic automatically return a complex number when asked for the square root of a negative number. On TI-83/89, however, the MODE must be set to "rectangular" or "$a + bi$."

Example 6

$$(7 - \sqrt{-4})(5 + \sqrt{-9}) = (7 - \sqrt{4}i)(5 + \sqrt{9}i)$$
$$= (7 - 2i)(5 + 3i)$$
$$= 35 + 21i - 10i - 6i^2$$
$$= 35 + 11i - 6(-1) = 41 + 11i. \quad ■$$

Since every negative real number has a square root in the complex number system, we can now find complex solutions for equations that have no real solutions. For example, the solutions of $x^2 = -25$ are $x = \pm\sqrt{-25} = \pm 5i$. In fact,

Every quadratic equation with real coefficients has solutions in the complex number system.

Example 7 To solve the equation $2x^2 + x + 3 = 0$, we apply the quadratic formula:

$$x = \frac{-1 \pm \sqrt{1^2 - 4 \cdot 2 \cdot 3}}{2 \cdot 2} = \frac{-1 \pm \sqrt{-23}}{4}.$$

Since $\sqrt{-23}$ is not a real number, this equation has no real number solutions. But $\sqrt{-23}$ *is* a complex number, namely, $\sqrt{-23} = \sqrt{23}i$. Thus the equation does have solutions in the complex number system:

$$x = \frac{-1 \pm \sqrt{-23}}{4} = \frac{-1 \pm \sqrt{23}i}{4} = -\frac{1}{4} \pm \frac{\sqrt{23}}{4}i.$$

Note that the two solutions, $-\frac{1}{4} + \frac{\sqrt{23}}{4}i$ and $-\frac{1}{4} - \frac{\sqrt{23}}{4}i$, are conjugates of each other. ■

Technology Tip

The polynomial solvers on TI-86, HP-38, Sharp 9600, and Casio 9850 and FX 2.0 produce all real and complex solutions of any polynomial equation that they can solve. See Calculator Investigation 1 on page 24 for details. On TI-89, use cSOLVE in the COMPLEX submenu of the ALGEBRA menu to find all solutions.

Example 8 To find *all* solutions of $x^3 = 1$, we rewrite the equation and use the Difference of Cubes pattern (see the Algebra Review Appendix) to factor:

$$x^3 = 1$$
$$x^3 - 1 = 0$$
$$(x - 1)(x^2 + x + 1) = 0$$
$$x - 1 = 0 \qquad \text{or} \qquad x^2 + x + 1 = 0.$$

The solution of the first equation is $x = 1$. The solutions of the second can be obtained from the quadratic formula:

$$x = \frac{-1 \pm \sqrt{1^2 - 4 \cdot 1 \cdot 1}}{2 \cdot 1} = \frac{-1 \pm \sqrt{-3}}{2} = \frac{-1 \pm \sqrt{3}i}{2} = -\frac{1}{2} \pm \frac{\sqrt{3}}{2}i.$$

Therefore, the equation $x^3 = 1$ has one real solution ($x = 1$) and two nonreal complex solutions [$x = -1/2 + (\sqrt{3}/2)i$ and $x = -1/2 - (\sqrt{3}/2)i$]. Each of these solutions is said to be a **cube root of one** or a **cube root of unity.** Observe that the two nonreal complex cube roots of unity are conjugates of each other. ■

The preceding examples illustrate this useful fact (whose proof is discussed in Section 4.8):

Conjugate Solutions

If $a + bi$ is a solution of a polynomial equation with *real* coefficients, then its conjugate $a - bi$ is also a solution of this equation.

Calculator Investigation

The following investigation of complex number arithmetic is for use with TI-81/82 and other calculators that do not do complex number arithmetic, but do have matrix capabilities. Before doing it, look up "matrix" or "matrices" in your instruction manual and learn how to enter and store 2 by 2 matrices and how to do addition, subtraction, and multiplication with them.

1. The complex number $a + bi$ is expressed in matrix notation as the matrix $\begin{pmatrix} a & b \\ -b & a \end{pmatrix}$. For example, $-3 + 6i$ is written as $\begin{pmatrix} -3 & 6 \\ -6 & -3 \end{pmatrix}$.

(a) Write $3 + 4i$, $1 + 2i$, and $1 - i$ in matrix form and enter them in your calculator as $[A]$, $[B]$, $[C]$.

(b) We know that $(3 + 4i) + (1 + 2i) = 4 + 6i$. Verify that $[A] + [B]$ is $\begin{pmatrix} 4 & 6 \\ -6 & 4 \end{pmatrix}$, which represents the complex number $4 + 6i$.

(c) Use matrix addition, subtraction, and multiplication to find the following. Interpret the answers as complex numbers: $[A] - [C]$, $[B] + [C]$, $[A][B]$, $[B][C]$.

(d) In Example 3 we saw that $\dfrac{3 + 4i}{1 + 2i} = \dfrac{11}{5} - \dfrac{2}{5}i = 2.2 - .4i$. Do this problem in matrix form by computing $[A]\cdot[B]^{-1}$ (use the x^{-1} key for the exponent).

(e) Do each of the following calculations and interpret the problem in terms of complex numbers: $[A]\cdot[C]^{-1}$, $[B][A]^{-1}$, $[B][C]^{-1}$.

Exercises 4.7

In Exercises 1–54, perform the indicated operation and write the result in the form $a + bi$.

1. $(2 + 3i) + (6 - i)$
2. $(-5 + 7i) + (14 + 3i)$
3. $(2 - 8i) - (4 + 2i)$
4. $(3 + 5i) - (3 - 7i)$
5. $\frac{5}{4} - \left(\frac{7}{4} + 2i\right)$
6. $(\sqrt{3} + i) + (\sqrt{5} - 2i)$
7. $\left(\frac{\sqrt{2}}{2} + i\right) - \left(\frac{\sqrt{3}}{2} - i\right)$
8. $\left(\frac{1}{2} + \frac{\sqrt{3}i}{2}\right) + \left(\frac{3}{4} - \frac{5\sqrt{3}i}{2}\right)$
9. $(2 + i)(3 + 5i)$
10. $(2 - i)(5 + 2i)$
11. $(-3 + 2i)(4 - i)$
12. $(4 + 3i)(4 - 3i)$
13. $(2 - 5i)^2$
14. $(1 + i)(2 - i)i$
15. $(\sqrt{3} + i)(\sqrt{3} - i)$
16. $\left(\frac{1}{2} - i\right)\left(\frac{1}{4} + 2i\right)$
17. i^{15}
18. i^{26}
19. i^{33}
20. $(-i)^{53}$
21. $(-i)^{107}$
22. $(-i)^{213}$
23. $\frac{1}{5 - 2i}$
24. $\frac{1}{i}$
25. $\frac{1}{3i}$
26. $\frac{i}{2 + i}$
27. $\frac{3}{4 + 5i}$
28. $\frac{2 + 3i}{i}$
29. $\frac{1}{i(4 + 5i)}$
30. $\frac{1}{(2 - i)(2 + i)}$
31. $\frac{2 + 3i}{i(4 + i)}$
32. $\frac{2}{(2 + 3i)(4 + i)}$
33. $\frac{2 + i}{1 - i} + \frac{1}{1 + 2i}$
34. $\frac{1}{2 - i} + \frac{3 + i}{2 + 3i}$
35. $\frac{i}{3 + i} - \frac{3 + i}{4 + i}$
36. $6 + \frac{2i}{3 + i}$
37. $\sqrt{-36}$
38. $\sqrt{-81}$
39. $\sqrt{-14}$
40. $\sqrt{-50}$
41. $-\sqrt{-16}$
42. $-\sqrt{-12}$
43. $\sqrt{-16} + \sqrt{-49}$
44. $\sqrt{-25} - \sqrt{-9}$
45. $\sqrt{-15} - \sqrt{-18}$
46. $\sqrt{-12}\sqrt{-3}$
47. $\sqrt{-16}/\sqrt{-36}$
48. $-\sqrt{-64}/\sqrt{-4}$
49. $(\sqrt{-25} + 2)(\sqrt{-49} - 3)$
50. $(5 - \sqrt{-3})(-1 + \sqrt{-9})$
51. $(2 + \sqrt{-5})(1 - \sqrt{-10})$
52. $\sqrt{-3}(3 - \sqrt{-27})$
53. $1/(1 + \sqrt{-2})$
54. $(1 + \sqrt{-4})/(3 - \sqrt{-9})$

In Exercises 55–58, find x and y. Remember that $a + bi = c + di$ exactly when $a = c$ and $b = d$.

55. $3x - 4i = 6 + 2yi$
56. $8 - 2yi = 4x + 12i$
57. $3 + 4xi = 2y - 3i$
58. $8 - xi = \frac{1}{2}y + 2i$

In Exercises 59–70, solve the equation and express each solution in the form $a + bi$.

59. $3x^2 - 2x + 5 = 0$
60. $5x^2 + 2x + 1 = 0$
61. $x^2 + x + 2 = 0$
62. $5x^2 - 6x + 2 = 0$
63. $2x^2 - x = -4$
64. $x^2 + 1 = 4x$
65. $2x^2 + 3 = 6x$
66. $3x^2 + 4 = -5x$
67. $x^3 - 8 = 0$
68. $x^3 + 125 = 0$
69. $x^4 - 1 = 0$
70. $x^4 - 81 = 0$

71. Simplify: $i + i^2 + i^3 + \cdots + i^{15}$

72. Simplify: $i - i^2 + i^3 - i^4 + i^5 - \cdots + i^{15}$

Thinkers

If $z = a + bi$ is a complex number, then its conjugate is usually denoted $\bar{z}$, that is, $\bar{z} = a - bi$. In Exercises 73–77, prove that for any complex numbers $z = a + bi$ and $w = c + di$:

73. $\overline{z + w} = \bar{z} + \bar{w}$
74. $\overline{zw} = \bar{z} \cdot \bar{w}$
75. $\overline{\left(\frac{z}{w}\right)} = \frac{\bar{z}}{\bar{w}}$
76. $\bar{\bar{z}} = z$

77. z is a real number exactly when $\bar{z} = z$.

78. The **real part** of the complex number $a + bi$ is defined to be the real number a. The **imaginary part** of $a + bi$ is defined to the real number b (*not* bi).

(a) Show that the real part of $z = a + bi$ is $\frac{z + \bar{z}}{2}$.

(b) Show that the imaginary part of $z = a + bi$ is $\frac{z - \bar{z}}{2i}$.

79. If $z = a + bi$ (with a, b real numbers, not both 0), express $1/z$ in standard form.

80. Construction of the Complex Numbers. We assume that the real number system is known. In order to construct a new number system with the desired properties, we must do the following:

(i) Define a set C (whose elements will be called complex numbers).

(ii) Ensure that the set C contains the real numbers or at least a copy of them.

(iii) Define addition and multiplication in the set C in such a way that the usual laws of arithmetic are valid.

(iv) Show that C has the other properties listed on page 282.

We begin by defining C to be the set of all ordered pairs of real numbers. Thus, $(1, 5)$, $(-6, 0)$, $(4/3, -17)$, and $(\sqrt{2}, 12/5)$ are some of the elements of the set C. More generally, a complex number (= element of C) is any pair (a, b) where a and b are real numbers. By definition, two complex numbers are *equal* exactly when they have the same first and the same second coordinate.

(a) *Addition in* C is defined by this rule:

$$(a, b) + (c, d) = (a + c, b + d)$$

For example,

$$(3, 2) + (5, 4) = (3 + 5, 2 + 4) = (8, 6).$$

Verify that this addition has the following properties. For any complex numbers (a, b), (c, d), (e, f) in C:

(i) $(a, b) + (c, d) = (c, d) + (a, b)$

(ii) $[(a, b) + (c, d)] + (e, f) = (a, b) + [(c, d) + (e, f)]$

(iii) $(a, b) + (0, 0) = (a, b)$

(iv) $(a, b) + (-a, -b) = (0, 0)$

(b) *Multiplication in* C is defined by this rule:

$$(a, b)(c, d) = (ac - bd, bc + ad)$$

For example,

$$(3, 2)(4, 5) = (3\cdot 4, -2\cdot 5, 2\cdot 4 + 3\cdot 5) = (12 - 10, 8 + 15) = (2, 23).$$

Verify that this multiplication has the following properties. For any complex numbers (a, b), (c, d), (e, f) in C:

(i) $(a, b)(c, d) = (c, d)(a, b)$

(ii) $[(a, b)(c, d)](e, f) = (a, b)[(c, d)(e, f)]$

(iii) $(a, b)(1, 0) = (a, b)$

(iv) $(a, b)(0, 0) = (0, 0)$

(c) Verify that for any two elements of C with second coordinate zero:

(i) $(a, 0) + (c, 0) = (a + c, 0)$

(ii) $(a, 0)(c, 0) = (ac, 0)$

Identify $(t, 0)$ with the real number t. Statements (i) and (ii) show that when addition or multiplication in C is performed on two real numbers (that is, elements of C with second coordinate 0), the result is the usual sum or product of real numbers. Thus, C contains (a copy of) the real number system.

(d) *New Notation.* Since we are identifying the complex number $(a, 0)$ with the real number a, we shall hereafter denote $(a, 0)$ simply by the symbol a. Also, let i denote the complex number $(0, 1)$.

(i) Show that $i^2 = -1$ [that is, $(0, 1)(0, 1) = (-1, 0)$].

(ii) Show that for any complex number $(0, b)$, $(0, b) = bi$ [that is, $(0, b) = (b, 0)(0, 1)$].

(iii) Show that any complex number (a, b) can be written: $(a, b) = a + bi$ [that is, $(a, b) = (a, 0) + (b, 0)(0, 1)$].

In this new notation, every complex number is of the form $a + bi$ with a, b real and $i^2 = -1$, and our construction is finished.

4.8 Theory of Equations*

The complex numbers were constructed in order to obtain a solution for the equation $x^2 = -1$, that is, a root of the polynomial $x^2 + 1$. In Section 4.7 we saw that *every* quadratic polynomial with real coefficients has roots in the complex number system. A natural question now arises:

Do we have to enlarge the complex number system (perhaps many times) to find roots for higher degree polynomials?

In this section we shall see that the somewhat surprising answer is no.

In order to give the full answer, we shall consider not just polynomials with real coefficients, but also those with complex coefficients, such as

$$x^3 - ix^2 + (4 - 3i)x + 1 \quad \text{or} \quad (-3 + 2i)x^6 - 3x + (5 - 4i).$$

*Section 4.7 is a prerequisite for this section.

The discussion of polynomial division in Section 4.2 can easily be extended to include polynomials with complex coefficients. In fact, *all of the results in Section 4.2 are valid for polynomials with complex coefficients.* For example, you can check that i is a root of $f(x) = x^2 + (i - 1)x + (2 + i)$ and that $x - i$ is a factor of $f(x)$:

$$f(x) = x^2 + (i - 1)x + (2 + i) = (x - i)[x - (1 - 2i)].$$

Since every real number is also a complex number, polynomials with real coefficients are just special cases of polynomials with complex coefficients. So in the rest of this section, "polynomial" means "polynomial with complex (possibly real) coefficients" unless specified otherwise. We can now answer the question posed in the first paragraph.

Fundamental Theorem of Algebra

Every nonconstant polynomial has a root in the complex number system.

Although this is obviously a powerful result, neither the Fundamental Theorem nor its proof provides a practical method for *finding* a root of a given polynomial.* The proof of the Fundamental Theorem is beyond the scope of this book, but we shall explore some of the useful implications of the theorem, such as this one:

Factorization over the Complex Numbers

Let $f(x)$ be a polynomial of degree $n > 0$ with leading coefficient d. Then there are (not necessarily distinct) complex numbers $c_1, c_2, \ldots, c_n$ such that

$$f(x) = d(x - c_1)(x - c_2)(x - c_3) \cdots (x - c_n)$$

Furthermore, $c_1, c_2, \ldots, c_n$ are the only roots of $f(x)$.

Proof By the Fundamental Theorem, $f(x)$ has a complex root c_1. The Factor Theorem shows that $x - c_1$ must be a factor of $f(x)$, say,

$$f(x) = (x - c_1)g(x)$$

where $g(x)$ has degree $n - 1$.† If $g(x)$ is nonconstant, then it has a complex root c_2 by the Fundamental Theorem. Hence $x - c_2$ is a factor of $g(x)$, so that

$$f(x) = (x - c_1)(x - c_2)h(x)$$

*It may seem strange that you can prove that a root exists without actually exhibiting one. But such "existence theorems" are quite common. A rough analogy is the situation that occurs when someone is killed by a sniper's bullet. The police know that there *is* a killer, but *finding* the killer may be impossible.

†The degree of $g(x)$ is 1 less than the degree n of $f(x)$ because $f(x)$ is the product of $g(x)$ and $x - c_1$ (which has degree 1).

for some $h(x)$ of degree $n - 2$ [1 less than the degree of $g(x)$]. If $h(x)$ is nonconstant, then it has a complex root c_3 and the argument can be repeated. Continuing in this way, with the degree of the last factor going down by 1 at each step, we reach a factorization in which the last factor is a constant (degree 0 polynomial):

$$(*) \qquad f(x) = (x - c_1)(x - c_2)(x - c_3)\cdots(x - c_n)d.$$

If the right side were multiplied out, it would look like

$$dx^n + \text{lower degree terms.}$$

So the constant factor d is the leading coefficient of $f(x)$.

It is easy to see from the factored form (*) that the numbers $c_1, c_2, \ldots, c_n$ are roots of $f(x)$. If k is *any* root of $f(x)$, then

$$0 = f(k) = d(k - c_1)(k - c_2)(k - c_3)\cdots(k - c_n).$$

The product on the right is 0 only when one of the factors is 0. Since the leading coefficient d is nonzero, we must have

$$\begin{array}{ccccccc} k - c_1 = 0 & \text{or} & k - c_2 = 0 & \text{or} & \cdots & k - c_n = 0 \\ k = c_1 & \text{or} & k = c_2 & \text{or} & \cdots & k = c_n. \end{array}$$

Therefore, k is one of the c's and $c_1, \ldots, c_n$ are the only roots of $f(x)$. This completes the proof. ■

Technology Tip

To find all the roots of a polynomial, see the second Tip on page 285.

To factor a polynomial as a product of linear factors; use cFACTOR in the COMPLEX submenu of the TI-89 ALGEBRA menu.

Since the n roots $c_1, \ldots, c_n$ of $f(x)$ may not all be distinct, we see that:

Number of Roots

Every polynomial of degree $n > 0$ has at most n different roots in the complex number system.

Suppose $f(x)$ has repeated roots, meaning that some of the $c_1, c_2, \ldots, c_n$ are the same in factorization (*). Recall that a root c is said to have multiplicity k if $(x - c)^k$ is a factor of $f(x)$, but no higher power of $(x - c)$ is a factor. Consequently, if every root is counted as many times as its multiplicity, then the statement in the preceding box implies that

A polynomial of degree n has exactly n roots.

Example 1 Find a polynomial $f(x)$ of degree 5 such that 1, -2, and 5 are roots, 1 is a root of multiplicity 3 and $f(2) = -24$.

Solution Since 1 is a root of multiplicity 3, $(x - 1)^3$ must be a factor of $f(x)$. There are at least two other factors corresponding to the roots -2 and 5: $x - (-2) = x + 2$ and $x - 5$. The product of these factors $(x - 1)^3(x + 2)(x - 5)$ has degree 5, as does $f(x)$, so $f(x)$ must look like this:

$$f(x) = d(x - 1)^3(x + 2)(x - 5)$$

where d is the leading coefficient. Since $f(2) = -24$ we have:

$$d(2-1)^3(2+2)(2-5) = f(2) = -24$$

which reduces to $-12d = -24$. Therefore, $d = (-24)/(-12) = 2$ and

$$\begin{aligned} f(x) &= 2(x-1)^3(x+2)(x-5) \\ &= 2x^5 - 12x^4 + 4x^3 + 40x^2 - 54x + 20. \quad \blacksquare \end{aligned}$$

Polynomials with Real Coefficients

Recall that the **conjugate** of the complex number $a + bi$ is the number $a - bi$ (see page 283). We usually write a complex number as a single letter, say z, and indicate its conjugate by $\bar{z}$ (sometimes read "z bar"). For instance, if $z = 3 + 7i$, then $\bar{z} = 3 - 7i$. Conjugates play a role whenever a quadratic polynomial with real coefficients has complex roots.

Example 2 The quadratic formula shows that $x^2 - 6x + 13$ has two complex roots:

$$\frac{-(-6) \pm \sqrt{(-6)^2 - 4 \cdot 1 \cdot 13}}{2 \cdot 1} = \frac{6 \pm \sqrt{-16}}{2} = \frac{6 \pm 4i}{2} = 3 \pm 2i.$$

The complex roots are $z = 3 + 2i$ and its conjugate $\bar{z} = 3 - 2i$. ■

The preceding example is a special case of a more general theorem, whose proof is outlined in Exercises 59 and 60:

Conjugate Roots Theorem

Let $f(x)$ be a polynomial with *real* coefficients. If the complex number z is a root of $f(x)$, then its conjugate $\bar{z}$ is also a root of $f(x)$.

Example 3 Find a polynomial with real coefficients whose roots include the numbers 2 and $3 + i$.

Solution Since $3 + i$ is a root, its conjugate $3 - i$ must also be a root. Consider the polynomial

$$f(x) = (x-2)[(x-(3+i)][x-(3-i)].$$

Obviously 2, $3 + i$, and $3 - i$ are roots of $f(x)$. Multiplying out this factored form shows that $f(x)$ *does* have real coefficients:

$$\begin{aligned} f(x) &= (x-2)[x^2 - (3-i)x - (3+i)x + (3+i)(3-i)] \\ &= (x-2)(x^2 - 3x + ix - 3x - ix + 9 - i^2) \\ &= (x-2)(x^2 - 6x + 10) \\ &= x^3 - 8x^2 + 22x - 20. \end{aligned}$$

The next-to-last line of this calculation also shows that $f(x)$ can be factored as a product of a linear and a quadratic polynomial, each with *real* coefficients. ■

The technique in Example 3 works because the polynomial

$$[x - (3 + i)][x - (3 - i)]$$

turns out to have real coefficients. The proof of the following result shows why this must always be the case:

Factorization over the Real Numbers

Every nonconstant polynomial with real coefficients can be factored as a product of linear and quadratic polynomials with real coefficients in such a way that the quadratic factors, if any, have no real roots.

Proof The box on page 289 shows that

$$f(x) = d(x - c_1)(x - c_2)\cdots(x - c_n)$$

where $c_1, \ldots, c_n$ are the roots of $f(x)$. If some c_i is a real number, then the factor $x - c_i$ is a linear polynomial with real coefficients.* If some c_j is a nonreal complex root, then its conjugate must also be a root. Thus some c_k is the conjugate of c_j, say, $c_j = a + bi$ (with a, b real) and $c_k = a - bi$.† In this case,

$$\begin{aligned}(x - c_j)(x - c_k) &= [x - (a + bi)][x - (a - bi)] \\ &= x^2 - (a - bi)x - (a + bi)x + (a + bi)(a - bi) \\ &= x^2 - ax + bix - ax - bix + a^2 - (bi)^2 \\ &= x^2 - 2ax + (a^2 + b^2).\end{aligned}$$

Therefore, the factor $(x - c_j)(x - c_k)$ of $f(x)$ is a quadratic with real coefficients (because a and b are real numbers). Its roots (c_j and c_k) are nonreal. By taking the real roots of $f(x)$ one at a time and the nonreal ones in conjugate pairs in this fashion, we obtain the desired factorization of $f(x)$. ■

Example 4 Given that $1 + i$ is a root of $f(x) = x^4 - 2x^3 - x^2 + 6x - 6$, factor $f(x)$ completely over the real numbers.

*In Example 3, for instance, 2 is a real root and $x - 2$ a linear factor.

†In Example 3, for instance, $c_j = 3 + i$ and $c_k = 3 - i$ are conjugate roots.

Solution Since $1 + i$ is a root of $f(x)$, so is its conjugate $1 - i$, and hence $f(x)$ has this quadratic factor:

$$[x - (1 + i)][x - (1 - i)] = x^2 - 2x + 2.$$

Dividing $f(x)$ by $x^2 - 2x + 2$ shows that the other factor is $x^2 - 3$, which factors as $(x + \sqrt{3})(x - \sqrt{3})$. Therefore

$$f(x) = (x + \sqrt{3})(x - \sqrt{3})(x^2 - 2x + 2). \quad \blacksquare$$

Exercises 4.8

In Exercises 1–6, find the remainder when $f(x)$ is divided by $g(x)$ without using synthetic or long division.

1. $f(x) = x^{10} + x^8; \quad g(x) = x - 1$
2. $f(x) = x^6 - 10; \quad g(x) = x - 2$
3. $f(x) = 3x^4 - 6x^3 + 2x - 1; \quad g(x) = x + 1$
4. $f(x) = x^5 - 3x^2 + 2x - 1; \quad g(x) = x - 2$
5. $f(x) = x^3 - 2x^2 + 5x - 4; \quad g(x) = x + 2$
6. $f(x) = 10x^{75} - 8x^{65} + 6x^{45} + 4x^{32} - 2x^{15} + 5;$ $g(x) = x - 1$

In Exercises 7–10, list the roots of the polynomial and state the multiplicity of each root.

7. $f(x) = x^{54}\left(x + \frac{4}{5}\right)$
8. $g(x) = 3\left(x + \frac{1}{6}\right)\left(x - \frac{1}{5}\right)\left(x + \frac{1}{4}\right)$
9. $h(x) = 2x^{15}(x - \pi)^{14}[x - (\pi + 1)]^{13}$
10. $k(x) = (x - \sqrt{7})^7(x - \sqrt{5})^5(2x - 1)$

In Exercises 11–22, find all the roots of $f(x)$ in the complex number system; then write $f(x)$ as a product of linear factors.

11. $f(x) = x^2 - 2x + 5$
12. $f(x) = x^2 - 4x + 13$
13. $f(x) = 3x^2 + 2x + 7$
14. $f(x) = 3x^2 - 5x + 2$
15. $f(x) = x^3 - 27$ [*Hint:* Factor first.]
16. $f(x) = x^3 + 125$
17. $f(x) = x^3 + 8$
18. $f(x) = x^6 - 64$ [*Hint:* Let $u = x^3$ and factor $u^2 - 64$ first.]
19. $f(x) = x^4 - 1$
20. $f(x) = x^4 - x^2 - 6$
21. $f(x) = x^4 - 3x^2 - 10$
22. $f(x) = 2x^4 - 7x^2 - 4$

In Exercises 23–44, find a polynomial $f(x)$ with real coefficients that satisfies the given conditions. Some of these problems have many correct answers.

23. Degree 3; only roots are $1, 7, -4$.
24. Degree 3; only roots are 1 and -1.
25. Degree 6; only roots are $1, 2, \pi$.
26. Degree 5; only root is 2.
27. Degree 3; roots $-3, 0, 4$; $f(5) = 80$.
28. Degree 3; roots $-1, 1/2, 2$; $f(0) = 2$.
29. Roots include $2 + i$ and $2 - i$.
30. Roots include $1 + 3i$ and $1 - 3i$.
31. Roots include 2 and $2 + i$.
32. Roots include 3 and $4i - 1$.
33. Roots include $-3, 1 - i, 1 + 2i$.
34. Roots include $1, 2 + i, 3i - 1$.
35. Degree 2; roots $1 + 2i$ and $1 - 2i$.
36. Degree 4; roots $3i$ and $-3i$, each of multiplicity 2.
37. Degree 4; only roots are $4, 3 + i$, and $3 - i$.
38. Degree 5; roots 2 (of multiplicity 3), i, and $-i$.
39. Degree 6; roots 0 (of multiplicity 3) and $3, 1 + i, 1 - i$, each of multiplicity 1.
40. Degree 6; roots include i (of multiplicity 2) and 3.
41. Degree 2; roots include $1 + i$; $f(0) = 6$.
42. Degree 2; roots include $3 + i$; $f(2) = 3$.
43. Degree 3; roots include i and 1; $f(-1) = 8$.
44. Degree 3; roots include $2 + 3i$ and -2; $f(2) = -3$.

In Exercises 45–48, find a polynomial with complex coefficients that satisfies the given conditions.

45. Degree 2; roots i and $1 - 2i$.
46. Degree 2; roots $2i$ and $1 + i$.
47. Degree 3; roots $3, i$, and $2 - i$.
48. Degree 4; roots $\sqrt{2}, -\sqrt{2}, 1 + i$, and $1 - i$.

In Exercises 49–56, one root of the polynomial is given; find all the roots.

49. $x^3 - 2x^2 - 2x - 3$; root 3.

50. $x^3 + x^2 + x + 1$; root i.

51. $x^4 + 3x^3 + 3x^2 + 3x + 2$; root i.

52. $x^4 - x^3 - 5x^2 - x - 6$; root i.

53. $x^4 - 2x^3 + 5x^2 - 8x + 4$; root 1 of multiplicity 2.

54. $x^4 - 6x^3 + 29x^2 - 76x + 68$; root 2 of multiplicity 2.

55. $x^4 - 4x^3 + 6x^2 - 4x + 5$; root $2 - i$.

56. $x^4 - 5x^3 + 10x^2 - 20x + 24$; root $2i$.

57. Let $z = a + bi$ and $w = c + di$ be complex numbers (a, b, c, d are real numbers). Prove the given equality by computing each side and comparing the results:

(a) $\overline{z + w} = \overline{z} + \overline{w}$ (The left side says: First find $z + w$ and then take the conjugate. The right side says: First take the conjugates of z and w and then add.)

(b) $\overline{z \cdot w} = \overline{z} \cdot \overline{w}$

58. Let $g(x)$ and $h(x)$ be polynomials of degree n and assume that there are $n + 1$ numbers $c_1, c_2, \ldots, c_n, c_{n+1}$ such that

$$g(c_i) = h(c_i) \quad \text{for every } i.$$

Prove that $g(x) = h(x)$. [*Hint:* Show that each c_i is a root of $f(x) = g(x) - h(x)$. If $f(x)$ is nonzero, what is its largest possible degree? To avoid a contradiction, conclude that $f(x) = 0$.]

59. Suppose $f(x) = ax^3 + bx^2 + cx + d$ has real coefficients and z is a complex root of $f(x)$.

(a) Use Exercise 57 and the fact that $\overline{r} = r$, when r is a real number, to show that

$$\overline{f(z)} = \overline{az^3 + bz^2 + cz + d}$$
$$= a\overline{z}^3 + b\overline{z}^2 + c\overline{z} + d = f(\overline{z}).$$

(b) Conclude that $\overline{z}$ is also a root of $f(x)$. [*Note:* $f(\overline{z}) = \overline{f(z)} = \overline{0} = 0$.]

60. Let $f(x)$ be a polynomial with real coefficients and z a complex root of $f(x)$. Prove that the conjugate $\overline{z}$ is also a root of $f(x)$. [*Hint:* Exercise 59 is the case when $f(x)$ has degree 3; the proof in the general case is similar.]

61. Use the statement in the box on page 292 to show that every polynomial with real coefficients and *odd* degree must have at least one real root.

62. Give an example of a polynomial $f(x)$ with complex, nonreal coefficients and a complex number z such that z is a root of $f(x)$, but its conjugate is not. Hence, the conclusion of the Conjugate Roots Theorem (page 291) may be *false* if $f(x)$ doesn't have real coefficients.

Chapter 4 Review

Important Concepts

Section 4.1	Quadratic function 206–209 Parabola 206 Vertex 207 Applications of quadratic functions 210
Section 4.2	Polynomial 213 Coefficient 213 Constant term 213 Zero polynomial 214 Degree of a polynomial 214 Leading coefficient 214 Polynomial function 214 Division of polynomials 214 Division Algorithm 216 Remainder Theorem 217 Root or zero of a polynomial 217 Real root 217 Factor Theorem 218 Number of roots of a polynomial 219
Excursion 4.2.A	Synthetic division 221
Section 4.3	The Rational Root Test 224 Factoring polynomials 226 Bounds Test 228
Section 4.4	Graph of $y = ax^n$ 232 Continuity 233 Behavior when $\|x\|$ is large 233 x-Intercepts 234 Multiplicity 235 Local extrema 235 Points of inflection 236
Excursion 4.4.A	Quadratic, cubic, and quartic regression 244
Section 4.5	Rational function 249 Domain 249 The Big-Little Principle 249 Linear rational functions 252 Intercepts 253 Vertical asymptotes 254 Holes 255 Horizontal asymptotes 256
Excursion 4.5.A	Asymptote when $\|x\|$ is large 263
Section 4.6	Basic principles for solving inequalities 268 Linear inequalities 268 Polynomial inequalities 269 Rational inequalities 273
Excursion 4.6.A	Absolute value inequalities 278
Section 4.7	Complex number 282 Imaginary number 282 Real numbers 282 Conjugate 283 Square roots of negative numbers 284

Important Facts and Formulas

- The graph of $f(x) = ax^2 + bx + c$ is a parabola whose vertex has x-coordinate $-b/2a$.

Review Questions

In Questions 1–5, find the vertex of the graph of the quadratic function.

1. $f(x) = (x - 2)^2 + 3$ **2.** $f(x) = 2(x + 1)^2 - 1$

3. $f(x) = x^2 - 8x + 12$ **4.** $f(x) = x^2 - 7x + 6$ **5.** $f(x) = 3x^2 - 9x + 1$

6. Which of the following statements about the functions

$$f(x) = 3x^2 + 2 \qquad \text{and} \qquad g(x) = -3x^2 + 2$$

is *false*?

(a) The graphs of f and g are parabolas.

(b) The graphs of f and g have the same vertex.

(c) The graphs of f and g open in opposite directions.

(d) The graph of f is the graph of $y = 3x^2$ shifted 2 units to the right.

7. A preschool wants to construct a fenced playground. The fence will be attached to the building at two corners, as shown in the figure. There is 400 feet of fencing available, all of it to be used.

(a) Write an equation in x and y that gives the amount of fencing to be used. Solve the equation for y.

(b) Write the area of the playground as a function of x. [Part (a) may be helpful.]

(c) What are the dimensions of the playground with the largest possible area?

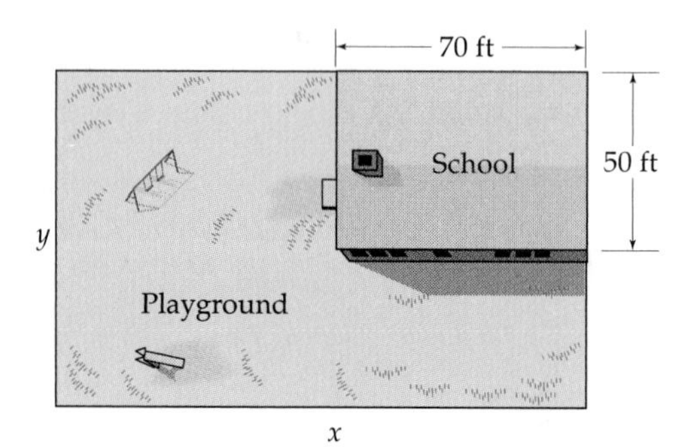

8. A model rocket is launched straight up from a platform at time $t = 0$ (where t is time measured in seconds). The altitude $h(t)$ of the rocket above the ground at given time (t) is given by $h(t) = 10 + 112t - 16t^2$ (where $h(t)$ is measured in feet).

(a) What is the altitude of the rocket the instant it is launched?

(b) What is the altitude of the rocket 2 seconds after launching?

(c) What is the maximum *altitude* attained by the rocket?

(d) At what *time* does the rocket return to the altitude at which it was launched?

9. A rectangular garden next to a building is to be fenced with 120 feet of fencing. The side against the building will not be fenced. What should the lengths of the other three sides be in order to ensure the largest possible area?

10. A factory offers 100 calculators to a retailer at a price of $20 each. The price per calculator on the entire order will be reduced 5¢ for each additional calculator over 100. What number of calculators will produce the largest possible sales revenue for the factory?

11. Which of the following are polynomials?

(a) $2^3 + x^2$ **(b)** $x + \frac{1}{x}$

(c) $x^3 - \frac{1}{\sqrt{2}}$ **(d)** $\sqrt[3]{x^4}$

(e) $\pi^3 - x$ **(f)** $\sqrt{2} + 2x^2$

(g) $\sqrt{x} + 2x^2$ **(h)** $|x|$

12. What is the remainder when $x^4 + 3x^3 + 1$ is divided by $x^2 + 1$?

13. What is the remainder when $x^{112} - 2x^8 + 9x^5 - 4x^4 + x - 5$ is divided by $x - 1$?

14. Is $x - 1$ a factor of $f(x) = 14x^{87} - 65x^{56} + 51$? Justify your answer.

15. Use synthetic division to show that $x - 2$ is a factor of $x^6 - 5x^5 + 8x^4 + x^3 - 17x^2 + 16x - 4$ and find the other factor.

16. List the roots of this polynomial and the multiplicity of each root:

$$f(x) = 5(x - 4)^3(x - 2)(x + 17)^3(x^2 - 4).$$

17. Find a polynomial f of degree 3 such that $f(-1) = 0$, $f(1) = 0$, and $f(0) = 5$.

18. Find the root(s) of $2\left(\frac{x}{5} + 7\right) - 3x - \frac{x + 2}{5} + 4$.

19. Find the roots of $3x^2 - 2x - 5$.

20. Factor the polynomial $x^3 - 8x^2 + 9x + 6$. [*Hint:* 2 is a root.]

21. Find all real roots of $x^6 - 4x^3 + 4$.

22. Find all real roots of $9x^3 - 6x^2 - 35x + 26$. [*Hint:* Try $x = -2$].

23. Find all real roots of $3y^3(y^4 - y^2 - 5)$.

24. Find the rational roots of $x^4 - 2x^3 - 4x^2 + 1$.

25. Consider the polynomial $2x^3 - 8x^2 + 5x + 3$.

(a) List the only *possible* rational roots.

(b) Find one rational root.

(c) Find all the roots of the polynomial.

26. **(a)** Find all rational roots of $x^3 + 2x^2 - 2x - 2$.

(b) Find two consecutive integers such that an irrational root of $x^3 + 2x^2 - 2x - 2$ lies between them.

27. How many distinct real roots does $x^3 + 4x$ have?

28. How many distinct real roots does $x^3 - 6x^2 + 11x - 6$ have?

29. Find the roots of $x^4 - 11x^2 + 18$.

30. The polynomial $x^3 - 2x + 1$ has

(a) no real roots.

(b) only one real root.

(c) three rational roots.

(d) only one rational root.

(e) none of the above.

31. Show that 5 is an upper bound for the real roots of $x^4 - 4x^3 + 16x - 16$.

32. Show that -1 is a lower bound for the real roots of $x^4 - 4x^3 + 15$.

In Questions 33 and 34, find the real roots of the polynomial.

33. $x^6 - 2x^5 - x^4 + 3x^3 - x^2 - x + 1$

34. $x^5 - 3x^4 - 2x^3 - x^2 - 23x - 20$

In Questions 35 and 36, compute and simplify the difference quotient of the function.

35. $f(x) = x^2 + x$

36. $g(x) = x^3 - x + 1$

37. Draw the graph of a function that could not possibly be the graph of a polynomial function and explain why.

38. Draw a graph that could be the graph of a polynomial function of degree 5. You need not list a specific polynomial, nor do any computation.

39. Which of the statements below is *not* true about the polynomial function f whose graph is shown in the figure?

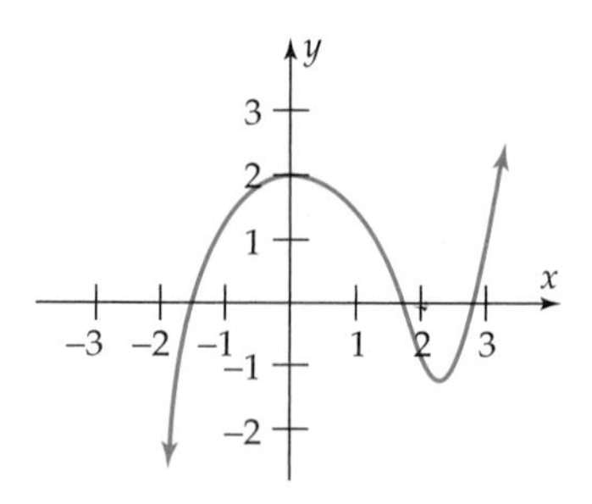

(a) f has three roots between -2 and 3.
(b) $f(x)$ could possibly be a fifth-degree polynomial.
(c) $(f \circ f)(0) > 0$.
(d) $f(2) - f(-1) < 3$.
(e) $f(x)$ is positive for all x in the interval $[-1, 0]$.

40. Which of the statements (i)–(v) about the polynomial function f whose graph is shown in the figure are *false*?

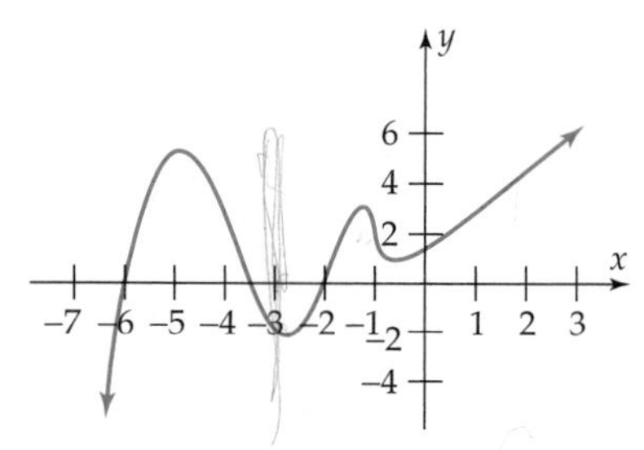

(i) f as 2 roots in the interval $[-6, -3)$.
(ii) $f(-3) - f(-6) < 0$.
(iii) $f(0) < f(1)$.
(iv) $f(2) - 2 = 0$.
(v) f has degree ≤ 4.

In Questions 41 – 44, find a viewing window (or windows) that shows a complete graph of the function. Be alert for hidden behavior.

41. $f(x) = .5x^3 - 4x^2 + x + 1$

42. $g(x) = .3x^5 - 4x^4 + x^3 - 4x^2 + 5x + 1$

43. $h(x) = 4x^3 - 100x^2 + 600x$

44. $f(x) = 32x^3 - 99x^2 + 100x + 2$

45. HomeArt makes plastic replicas of famous statues. Their total cost to produce copies of a particular statue are shown in the table.

(a) Sketch a scatter plot of the data.

(b) Use cubic regression to find a function $C(x)$ that models the data [that is, $C(x)$ is the cost of making x statues]. Assume C is reasonably accurate when $x \leq 100$.

(c) Use C to estimate the cost of making the seventy-first statue.

(d) Use C to approximate the average cost per statue when 35 are made and when 75 are made. [Recall that the average cost of x statues is $C(x)/x$.]

Number of Statues	Total Cost
0	$2,000
10	2,519
20	2,745
30	2,938
40	3,021
50	3,117
60	3,269
70	3,425

46. The table gives the estimated cost of a college education at a public institution. Costs include tuition, fees, books, and room and board for four years. (*Source:* Teachers Insurance and Annuity Association College Retirement Equities Fund)

Enrollment Year	Costs
1998	$46,691
2000	52,462
2002	58,946
2004	66,232
2006	74,418

Enrollment Year	Costs
2008	$ 83,616
2010	93,951
2012	105,564
2014	118,611

(a) Sketch a scatter plot of the data (with $x = 0$ corresponding to 1990).

(b) Use quartic regression to find a function C that models the data.

(c) Estimate the cost of a college education in 2007 and in 2015.

In Questions 47–54, sketch a complete graph of the function. Label the x-intercepts, all local extrema, holes, and asymptotes.

47. $f(x) = x^3 - 9x$

48. $g(x) = x^3 - 2x^2 + 3$

49. $h(x) = x^4 - x^3 - 4x^2 + 4x + 2$

50. $f(x) = x^4 - 3x - 2$

51. $g(x) = \dfrac{-2}{x + 4}$

52. $h(x) = \dfrac{3 - x}{x - 2}$

53. $k(x) = \dfrac{4x + 10}{3x - 9}$

54. $f(x) = \dfrac{x + 1}{x^2 - 1}$

In Questions 55 and 56, list all asymptotes of the graph of the function.

55. $f(x) = \dfrac{x^2 - 1}{x^3 - 2x^2 - 5x + 6}$

56. $g(x) = \dfrac{x^4 - 6x^3 + 2x^2 - 6x + 2}{x^2 - 3}$

In Questions 57–60, find a viewing window (or windows) that shows a complete graph of the function. Be alert for hidden behavior.

57. $f(x) = \dfrac{x - 3}{x^2 + x - 2}$

58. $g(x) = \dfrac{x^2 - x - 6}{x^3 - 3x^2 + 3x - 1}$

59. $h(x) = \dfrac{x^4 + 4}{x^4 - 99x^2 - 100}$

60. $k(x) = \dfrac{x^3 - 2x^2 - 4x + 8}{x - 10}$

61. It costs the Junkfood Company 50¢ to produce a bag of Munchies. There are fixed costs of \$500 per day for building, equipment, etc. The company has found that if the price of a bag of Munchies is set at $1.95 - \dfrac{x}{2000}$ dollars, where x is the number of bags produced per day, then all the bags that are produced will be sold. What number of bags can be produced each day if all are to be sold and the company is to make a profit? What are the possible prices?

62. Highway engineers have found that a good model of the relationship between the density of automobile traffic and the speed at which it moves along a particular section of highway is given by the function

$$s = \frac{100}{1 + d^2} \quad (0 \le d \le 3)$$

where the density of traffic d is measured in hundreds of cars per mile and the speed s at which traffic moves is measured in miles per hour. Then the traffic flow q is given by the product of the density and the speed, that is, $q = ds$, with q being measured in hundreds of cars per hour.

(a) Express traffic flow as a function of traffic density.

(b) For what densities will traffic flow be at least 3000 cars per hour?

(c) What traffic density will maximize traffic flow? What is the maximum flow?

63. Charlie lives 150 miles from the city. He drives 40 miles to the station and catches a train to the city. The average speed of the train is 25 mph faster than the average speed of the car.

(a) Express the total time for the journey as a function of the speed of the car. What speeds make sense in this context?

(b) How fast should Charlie drive if the entire journey is to take no more than two hours?

64. The survival rate s of seedlings in the vicinity of a parent tree is given by

$$s = \frac{.5x}{1 + .4x^2}$$

where x is the distance from the seedling to the tree (in meters) and $0 < x \le 10$.

(a) For what distances is the survival rate at least .21?

(b) What distance produces the maximum survival rate?

In Questions 65 and 66, find the average rate of change of the function between x and $x + h$.

65. $f(x) = \dfrac{x}{x + 1}$

66. $g(x) = \dfrac{1}{x^2 + 1}$

67. Which of these statements about the graph of $f(x) = \dfrac{(x - 1)(x + 3)}{(x^2 + 1)(x^2 - 1)}$ is *true*?

(a) The graph has two vertical asymptotes.

(b) The graph touches the x-axis at $x = 3$.

(c) The graph lies above the x-axis when $x < -1$.

(d) The graph has a hole at $x = 1$.

(e) The graph has no horizontal asymptotes.

68. Solve for x: $-3(x - 4) \le 5 + x$.

69. Solve for y: $\left|\dfrac{y + 2}{3}\right| \ge 5$.

70. Solve for x: $-4 < 2x + 5 < 9$.

71. On which intervals is $\dfrac{2x - 1}{3x + 1} < 1$?

72. On which intervals is $\dfrac{2}{x + 1} < x$?

73. Solve for x: $\left|\dfrac{1}{1 - x^2}\right| \ge \dfrac{1}{2}$.

74. Solve for x: $x^2 + x > 12$.

75. Solve for x: $(x - 1)^2(x^2 - 1)x \le 0$.

76. If $0 < r \le s - t$, then which of these statements is *false*?

(a) $s \ge r + t$

(b) $t - s \le -r$

(c) $-r \ge s - t$

(d) $\dfrac{s - t}{r} > 0$

(e) $s - r \ge t$

77. If $\dfrac{x + 3}{2x + 3} > 1$, then which of these statements is *true*?

(a) $\dfrac{x - 3}{2x + 3} < -1$

(b) $\dfrac{2x - 3}{x + 3} < -1$

(c) $\dfrac{3 - 2x}{x + 3} > 1$

(d) $2x + 3 < x - 3$

(e) None of these

78. Solve and express your answer in interval notation:

$$2x - 3 \le 5x + 9 < -3x + 4.$$

In Questions 79–86, solve the inequality.

79. $|3x + 2| \geq 2$

80. $x^2 + x - 20 > 0$

81. $\dfrac{x - 2}{x + 4} \leq 3$

82. $(x + 1)^2(x - 3)^4(x + 2)^3(x - 7)^5 > 0$

83. $\dfrac{x^2 + x - 9}{x + 3} < 1$

84. $\dfrac{x^2 - x - 6}{x - 3} > 1$

85. $\dfrac{x^2 - x - 5}{x^2 + 2} > -2$

86. $\dfrac{x^4 - 3x^2 + 2x - 3}{x^2 - 4} < -1$

In Questions 87–94, solve the equation in the complex number system.

87. $x^2 + 3x + 10 = 0$

88. $x^2 + 2x + 5 = 0$

89. $5x^2 + 2 = 3x$

90. $-3x^2 + 4x - 5 = 0$

91. $3x^4 + x^2 - 2 = 0$

92. $8x^4 + 10x^2 + 3 = 0$

93. $x^3 + 8 = 0$

94. $x^3 - 27 = 0$

95. One root of $x^4 - x^3 - x^2 - x - 2$ is i. Find all the roots.

96. One root of $x^4 + x^3 - 5x^2 + x - 6$ is i. Find all the roots.

97. Give an example of a fourth-degree polynomial with real coefficients whose roots include 0 and $1 + i$.

98. Find a fourth-degree polynomial f whose only roots are $2 + i$ and $2 - i$, such that $f(-1) = 50$.

Discovery Project 4

Architectural Arches

You can see arches almost everywhere you look—in windows, entryways, tunnels, and bridges. Common arch shapes are semicircles, semi-ellipses, and parabolas. When constructed on a level base, arches are symmetric left to right. This means that a mathematical function which describes an arch must be an even function.

A *semicircular arch* always has the property that its base is twice as wide as its height. This ratio can be modified by placing a rectangular area under the semicircle, giving a shape known as a *Norman arch.* This approach gives a tunnel or room a vaulted ceiling. *Parabolic arches* can also be created to give a more vaulted appearance.

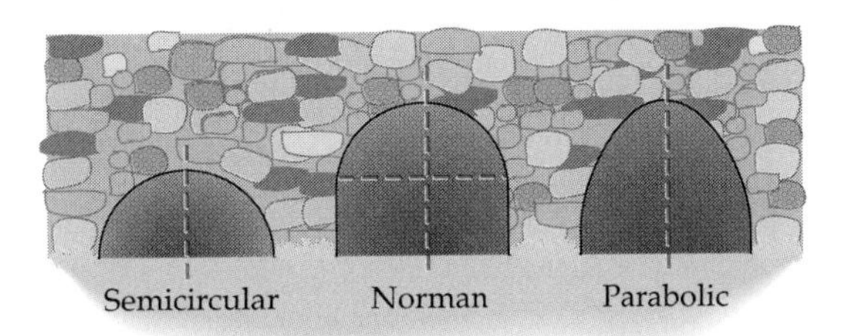

For the following exercises in arch modeling, you should always set the origin of your coordinate system to be the center of the base of the arch.

1. Show that the function that models a semicircular arch of radius r is $h(x) = \sqrt{r^2 - x^2}$.
2. Write a function $h(x)$ that models a semicircular arch that is 15 feet tall. How wide is the arch?
3. Write a function $n(x)$ that models a Norman arch that is 15 feet tall and 16 feet wide at the base.
4. Parabolic arches are typically modeled using the function $p(x) = H - ax^2$, where H is the height of the arch. Write a function $p(x)$ for an arch that is 15 feet tall and 16 feet wide at the base.
5. Would a truck that is 12 feet tall and 9 feet wide fit through all three arches? How could you fix any of the arches that is too small so that the truck would fit through?

Exponential and Logarithmic Functions

How old is that dinosaur?

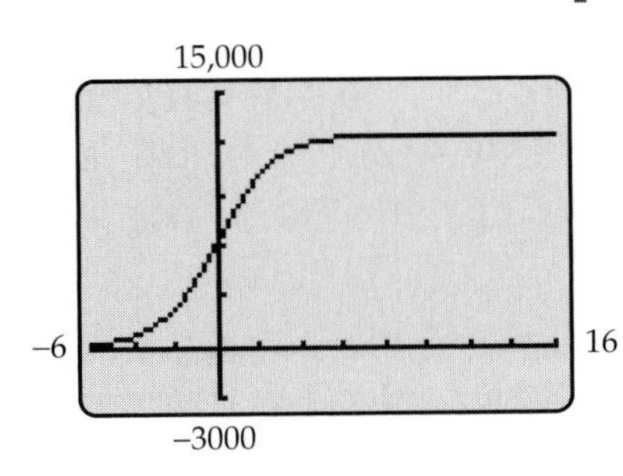

Population growth (of humans, fish, bacteria, etc.), compound interest, radioactive decay, and a host of other phenomena can be mathematically described by exponential functions (see Exercises 46 and 51 on page 332). Archeologists sometimes use carbon-14 dating to determine the approximate age of an artifact (such as a dinosaur skeleton, a mummy, or a wooden statue). This involves using logarithms to solve an appropriate exponential equation. See Exercise 55 on page 367.

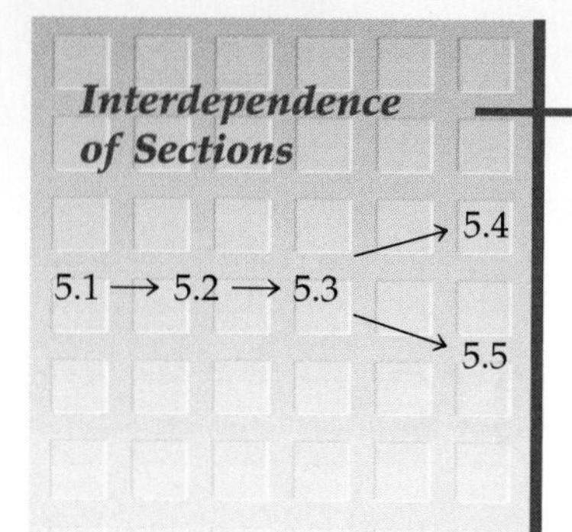

Chapter Outline

Exponential and logarithmic functions are essential for the mathematical description of a variety of phenomena in the physical sciences, engineering, and economics. Although a calculator is necessary to evaluate these functions at most numbers, you won't be able to use your calculator efficiently or interpret its answers unless you understand the properties of these functions. When calculations can readily be done by hand, you will be expected to do them without a calculator.

5.1 Radicals and Rational Exponents

Recall that when $c \geq 0$, the square root of c is a nonnegative solution of the equation $x^2 = c$. Cube roots, fourth roots, and higher roots are defined in a similar fashion as solutions of the equation $x^n = c$. This equation can be solved graphically by finding the x-coordinate of the intersection point of the graphs of $y = x^n$ and $y = c$. (Recall that the graph of $y = x^n$ was discussed at the beginning of Section 4.4.) Depending on whether n is even or odd and whether c is positive or negative, $x^n = c$ may have two, one, or no solutions, as shown here.

$x^n = c$ n **odd**	$x^n = c$ n **even**		
Exactly one solution for any c	$c > 0$ One positive and one negative solution	$c = 0$ One solution $x = 0$	$c < 0$ No solution
$y = x^n$, $y = c$	$y = x^n$, $y = c$	$y = x^n$, $y = 0$	$y = x^n$, $y = c$

Consequently, we have the following definition.

nth Roots

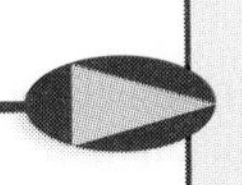

Let c be a real number and n a positive integer. The *nth root of c* is denoted by either of the symbols

$$\sqrt[n]{c} \quad \text{or} \quad c^{1/n}$$

and is defined to be

The solution of $x^n = c$, when n is odd;

The positive solution of $x^n = c$, when n is even and $c \geq 0$.

Thus, for example,

$$\sqrt[3]{-8} = (-8)^{1/3} = -2 \text{ because } -2 \text{ is the solution of } x^3 = -8$$

and

$$\sqrt[4]{81} = 81^{1/4} = 3 \text{ because 3 is the positive solution of } x^4 = 81.$$

As before, square roots are denoted by $\sqrt{\ }$ rather than $\sqrt[2]{\ }$. Expressions involving nth roots can often be written in a variety of ways by using this basic fact, which is justified in Exercise 66:

$$\sqrt[n]{cd} = \sqrt[n]{c}\,\sqrt[n]{d}, \quad \text{or equivalently,} \quad (cd)^{1/n} = c^{1/n}d^{1/n}.$$

Example 1

(a) $\sqrt{8}\sqrt{12} = \sqrt{8 \cdot 12} = \sqrt{96} = \sqrt{16 \cdot 6} = \sqrt{16}\sqrt{6} = 4\sqrt{6}.$

(b) $\sqrt{12} - \sqrt{75} = \sqrt{4\cdot 3} - \sqrt{25\cdot 3} = \sqrt{4}\sqrt{3} - \sqrt{25}\sqrt{3}$
$= 2\sqrt{3} - 5\sqrt{3} = -3\sqrt{3}.$

(c) $\sqrt[3]{8x^6y^4} = \sqrt[3]{8}\sqrt[3]{x^6}\sqrt[3]{y^3y} = 2x^2\sqrt[3]{y^3}\sqrt[3]{y} = 2x^2y\sqrt[3]{y}.$

(d) If $c > 0$, then $(5 + \sqrt{c})(5 - \sqrt{c}) = 5^2 - (\sqrt{c})^2 = 25 - c.$ ■

Exponent notation for the nth roots is usually more convenient than radical notation when using a calculator.

Example 2 Use a calculator to approximate

(a) $40^{1/5}$ (b) $225^{1/11}$

```
40^.2
        2.091279105
225^(1/11)
        1.636193919
225^.0909
         1.63611336
```

Figure 5–1

Solution

(a) Since $1/5 = .2$, you can compute $40^{.2}$, as in Figure 5–1.

(b) $1/11$ is the infinite decimal $.090909\cdots$. In such cases it is best to leave the exponent in fractional form and use parentheses, as shown in Figure 5–1. If you round off the decimal equivalent of $1/11$, say as .0909, you will not get the same answer, as Figure 5–1 shows. ■

Rational Exponents

We have now defined fractional exponents of the form $1/n$. The next step is to define fractional exponents of the form m/n, as in $4^{3/2}$ for example. We would like to do this in such a way that the various exponent laws continue to hold. If we want $c^{rs} = (c^r)^s$ to be valid, then since $\frac{3}{2} = 3\cdot\left(\frac{1}{2}\right) = \left(\frac{1}{2}\right)\cdot 3$, it might be reasonable to define $4^{3/2}$ to be either $(4^3)^{1/2}$ or $(4^{1/2})^3$. Observe that

$$(4^3)^{1/2} = 64^{1/2} = \sqrt{64} = 8 \qquad \text{and} \qquad (4^{1/2})^3 = (\sqrt{4})^3 = 2^3 = 8.$$

This illustrates a fact that can be proved more generally, namely, that for any nonnegative real number c and any integers t, k, with k positive:

$$(c^t)^{1/k} = (c^{1/k})^t, \qquad \text{or in radical notation,} \qquad \sqrt[k]{c^t} = (\sqrt[k]{c})^t.$$

Consequently, we make this definition.

Rational Exponents

Let c be a positive real number and let t/k be a rational number with positive denominator. Then,

$$c^{t/k} \text{ is defined to be the number } (c^t)^{1/k} = (c^{1/k})^t.$$

Since every terminating decimal is a rational number, expressions such as $13^{3.78}$ now have a meaning, namely $13^{378/100}$. (Actually, we used this fact earlier when we computed $225^{.0909}$ in Figure 5–1.) Although the preceding definition requires c to be positive, it remains valid when c is negative, *provided that* the exponent is in lowest terms with an odd denominator, such as $(-8)^{2/3}$. To understand why these restrictions are necessary when c is negative, see Exercise 65.

CAUTION

We know that

$$(-8)^{2/3} = \sqrt[3]{(-8)^2} = \sqrt[3]{64} = 4,$$

but entering $(-8)^{2/3}$ on TI-82/85, Casio 9850, HP-38, or Sharp 9600 produces either an error message or a complex number (which the calculator indicates by an ordered pair or an expression involving i). However, you can get the correct answer by keying in one of

$$[(-8)^2]^{1/3} \quad \text{or} \quad [(-8)^{1/3}]^2,$$

each of which is equal to $(-8)^{2/3}$. TI-89 will produce the correct answer provided that the COMPLEX FORMAT in the MODE menu is set to "Real."

Rational exponents were defined in a way that guaranteed that one of the familiar exponent laws would remain valid. In fact, all exponent laws developed for integer exponents are valid for rational exponents, as summarized here and illustrated in Examples 3 through 6.

Exponent Laws

Let c and d be nonnegative real numbers and let r and s be any rational numbers. Then

1. $c^r c^s = c^{r+s}$
2. $\dfrac{c^r}{c^s} = c^{r-s} \quad (c \neq 0)$
3. $(c^r)^s = c^{rs}$
4. $(cd)^r = c^r d^r$
5. $\left(\dfrac{c}{d}\right)^r = \dfrac{c^r}{d^r} \quad (d \neq 0)$
6. $\dfrac{1}{c^{-r}} = c^r \quad (c \neq 0)$

Example 3

$$(8r^{3/4}s^{-3})^{2/3} = 8^{2/3}(r^{3/4})^{2/3}(s^{-3})^{2/3} = \sqrt[3]{8^2}(r^{2/4})(s^{-2})$$

$$= \sqrt[3]{64}r^{1/2}s^{-2} = \frac{4r^{1/2}}{s^2}.$$

We can leave the answer in exponential form, or if it is more convenient, write it as $4\sqrt{r}/s^2$. ■

CAUTION

The exponent laws deal only with products and quotients. There are no analogous properties for sums. In particular,

$(c + d)^r$ may **not** be equal to $c^r + d^r$.

Example 4

$$x^{1/2}(x^{3/4} - x^{3/2}) = x^{1/2}x^{3/4} - x^{1/2}x^{3/2}$$
$$= x^{1/2+3/4} - x^{1/2+3/2} = x^{5/4} - x^2. \quad \blacksquare$$

Example 5

$$(x^{5/2}y^4)(xy^{7/4})^{-2} = (x^{5/2}y^4)x^{-2}(y^{7/4})^{-2}$$
$$= x^{5/2}y^4x^{-2}y^{(7/4)(-2)} = x^{5/2}x^{-2}y^4y^{-7/2}$$
$$= x^{(5/2)-2}y^{4-(7/2)} = x^{1/2}y^{1/2}. \quad \blacksquare$$

Example 6 Let k be a positive rational number and express $\sqrt[10]{c^{5k}}\sqrt{(c^{-k})^{1/2}}$ without radicals, using only positive exponents.

Solution

$$\sqrt[10]{c^{5k}}\sqrt{(c^{-k})^{1/2}} = (c^{5k})^{1/10}[(c^{-k})^{1/2}]^{1/2} = c^{k/2}c^{-k/4} = c^{k/2-k/4} = c^{k/4}. \quad \blacksquare$$

Rationalizing Denominators and Numerators

When dealing with fractions in the days before calculators, it was customary to *rationalize the denominators*, that is, write equivalent fractions with no radicals in the denominator, because this made many computations easier. With calculators, of course, there is no computational advantage to rationalizing denominators. Nevertheless, rationalizing denominators or numerators is sometimes needed to simplify expressions and to derive useful formulas.

Example 7

(a) To rationalize the denominator of $\frac{7}{\sqrt{5}}$, multiply it by 1, with 1 written as a suitable radical fraction:

$$\frac{7}{\sqrt{5}} = \frac{7}{\sqrt{5}}\cdot 1 = \frac{7}{\sqrt{5}}\cdot\frac{\sqrt{5}}{\sqrt{5}} = \frac{7\sqrt{5}}{5}.$$

(b) To rationalize the denominator of $\frac{2}{3 + \sqrt{6}}$, multiply by 1 written as a radical fraction and use the multiplication pattern $(a + b)(a - b) = a^2 - b^2$.

$$\frac{2}{3+\sqrt{6}} = \frac{2}{3+\sqrt{6}}\cdot 1$$

$$= \frac{2}{3+\sqrt{6}}\cdot\frac{3-\sqrt{6}}{3-\sqrt{6}} = \frac{2(3-\sqrt{6})}{(3+\sqrt{6})(3-\sqrt{6})}$$

$$= \frac{6-2\sqrt{6}}{3^2-(\sqrt{6})^2} = \frac{6-2\sqrt{6}}{9-6} = \frac{6-2\sqrt{6}}{3}. \quad ■$$

Example 8 Assume $h \neq 0$ and rationalize the numerator of $\dfrac{\sqrt{x+h}-\sqrt{x}}{h}$; that is, write an equivalent fraction with no radicals in the numerator.

Solution Again, the technique is to multiply the fraction by 1, with 1 written as a suitable radical fraction:

$$\frac{\sqrt{x+h}-\sqrt{x}}{h} = \frac{\sqrt{x+h}-\sqrt{x}}{h}\cdot 1$$

$$= \frac{\sqrt{x+h}-\sqrt{x}}{h}\cdot\frac{\sqrt{x+h}+\sqrt{x}}{\sqrt{x+h}+\sqrt{x}} = \frac{(\sqrt{x+h})^2-(\sqrt{x})^2}{h(\sqrt{x+h}+\sqrt{x})}$$

$$= \frac{x+h-x}{h(\sqrt{x+h}+\sqrt{x})} = \frac{h}{h(\sqrt{x+h}+\sqrt{x})} = \frac{1}{\sqrt{x+h}+\sqrt{x}}. \quad ■$$

Irrational Exponents

An example will illustrate how a^t is defined when t is an irrational number.* To compute $10^{\sqrt{2}}$ we use the infinite decimal expansion $\sqrt{2} \approx 1.414213562\cdots$ (see Excursion 1.1.A). Each of

$$1.4, 1.41, 1.414, 1.4142, 1.41421, \cdots$$

is a rational number approximation of $\sqrt{2}$, and each is a more accurate approximation than the preceding one. We know how to raise 10 to each of these rational numbers:

$$10^{1.4} \approx 25.1189 \qquad 10^{1.4142} \approx 25.9537$$
$$10^{1.41} \approx 25.7040 \qquad 10^{1.41421} \approx 25.9543$$
$$10^{1.414} \approx 25.9418 \qquad 10^{1.414213} \approx 25.9545$$

*This example is not a proof, but should make the idea plausible. Calculus is required for a rigorous proof.

It appears that as the exponent r gets closer and closer to $\sqrt{2}$, 10^r gets closer and closer to a real number whose decimal expansion begins $25.954\cdots$. We define $10^{\sqrt{2}}$ to be this number.

Similarly, for any $a > 0$.

a^t is a well-defined *positive* number for each real exponent t.

We shall also assume this fact:

The exponent laws (page 309) are valid for *all* real exponents.

Exercises 5.1

Note: *Unless directed otherwise, assume all letters represent positive real numbers.*

In Exercises 1–12, simplify the expression without using a calculator.

1. $\sqrt{.0081}$
2. $\sqrt{.000169}$
3. $\sqrt{(.08)^{12}}$
4. $\sqrt{(-11)^{28}}$
5. $\sqrt{6}\sqrt{12}$
6. $\sqrt{8}\sqrt{96}$
7. $\dfrac{\sqrt{10}}{\sqrt{8}\sqrt{5}}$
8. $\dfrac{\sqrt{6}}{\sqrt{14}\sqrt{63}}$
9. $(1+\sqrt{3})(2-\sqrt{3})$
10. $(3+\sqrt{2})(3-\sqrt{2})$
11. $(4-\sqrt{3})(5+2\sqrt{3})$
12. $(2\sqrt{5}-4)(3\sqrt{5}+2)$

In Exercises 13–18, write the given expression without using radicals.

13. $\sqrt[3]{a^2+b^2}$
14. $\sqrt[4]{a^3-b^3}$
15. $\sqrt[4]{\sqrt[4]{a^3}}$
16. $\sqrt{\sqrt[3]{a^3b^4}}$
17. $\sqrt[5]{t}\sqrt{16t^5}$
18. $\sqrt{x}(\sqrt[3]{x^2})(\sqrt[4]{x^3})$

In Exercises 19–38, simplify the given expression.

19. $\sqrt{16a^8b^{-2}}$
20. $\sqrt{24x^6y^{-4}}$
21. $\dfrac{\sqrt{c^2d^6}}{\sqrt{4c^3d^{-4}}}$
22. $\dfrac{\sqrt{a^{-10}b^{-12}}}{\sqrt{a^{14}d^{-4}}}$
23. $5\sqrt{20}-\sqrt{45}+2\sqrt{80}$
24. $\sqrt[3]{40}+2(\sqrt[3]{135})-5(\sqrt[3]{320})$
25. $\sqrt[4]{(4x+2y)^8}$
26. $(\sqrt[3]{a+b})(\sqrt[3]{-(a+b)^2})+\sqrt[3]{a+b}$
27. $\dfrac{2^{11/2}\cdot 2^{-7}\cdot 2^{-5}}{2^3\cdot 2^{1/2}\cdot 2^{-10}}$
28. $\dfrac{(3^2)^{-1/2}(9^4)^{-1}}{27^{-3}}$
29. $\sqrt{x^7}\cdot x^{5/2}\cdot x^{-3/2}$
30. $(x^{1/2}y^3)(x^0y^7)^{-2}$
31. $(c^{2/5}d^{-2/3})(c^6d^3)^{4/3}$
32. $\left(\dfrac{r^{2/3}}{s^{1/5}}\right)^{15/9}$
33. $\dfrac{(7a)^2(5b)^{3/2}}{(5a)^{3/2}(7b)^4}$
34. $\dfrac{(6a)^{1/2}\sqrt{ab}}{a^2b^{3/2}}$
35. $\dfrac{(2a)^{1/2}(3b)^{-2}(4a)^{3/5}}{(4a)^{-3/2}(3b)^2(2a)^{1/5}}$
36. $\dfrac{(a^{3/4}b)^2(ab^{1/4})^3}{(ab)^{1/2}(bc)^{-1/4}}$
37. $(a^{x^2})^{1/x}$
38. $\dfrac{(b^x)^{x-1}}{b^{-x}}$

In Exercises 39–42, write the given expression without radicals, using only positive exponents.

39. $(\sqrt[3]{xy^2})^{-3/5}$
40. $(\sqrt[4]{r^{14}s^{-21/5}})^{-3/7}$
41. $\dfrac{c}{(c^{5/6})^{42}(c^{51})^{-2/3}}$
42. $(c^{5/6}-c^{-5/6})^2$

In Exercises 43–48, compute and simplify.

43. $x^{1/2}(x^{2/3}-x^{4/3})$
44. $x^{1/2}(3x^{3/2}+2x^{-1/2})$
45. $(x^{1/2}+y^{1/2})(x^{1/2}-y^{1/2})$
46. $(x^{1/3}+y^{1/2})(2x^{1/3}-y^{3/2})$
47. $(x+y)^{1/2}[(x+y)^{1/2}-(x+y)]$
48. $(x^{1/3}+y^{1/3})(x^{2/3}-x^{1/3}y^{1/3}+y^{2/3})$

In Exercises 49–54, factor the given expression. For example,

$$x-x^{1/2}-2=(x^{1/2}-2)(x^{1/2}+1).$$

49. $x^{2/3}+x^{1/3}-6$
50. $x^{2/5}+11x^{1/5}+30$
51. $x+4x^{1/2}+3$
52. $x^{1/3}+7x^{1/6}+10$
53. $x^{4/5}-81$
54. $x^{2/3}-6x^{1/3}+9$

In Exercises 55–60, rationalize the denominator and simplify your answer.

55. $\dfrac{3}{\sqrt{8}}$
56. $\dfrac{2}{\sqrt{6}}$
57. $\dfrac{3}{2+\sqrt{12}}$
58. $\dfrac{1+\sqrt{3}}{5+\sqrt{10}}$
59. $\dfrac{2}{\sqrt{x}+2}$
60. $\dfrac{\sqrt{x}}{\sqrt{x}-\sqrt{c}}$

In Exercises 61–64, rationalize the numerator and simplify your answer.

61. $\dfrac{\sqrt{x+h+1}-\sqrt{x+1}}{h}$

62. $\dfrac{2\sqrt{x+h+3}-2\sqrt{x+3}}{h}$

63. $\dfrac{\sqrt{(x+h)^2+1}-\sqrt{x^2+1}}{h}$

64. $\dfrac{\sqrt{(x+h)^2-(x+h)}-\sqrt{x^2-x}}{h}$

65. Here are some of the reasons why restrictions are necessary when defining fractional powers of a negative number.
(a) Explain why the equations $x^2 = -4$, $x^4 = -4$, $x^6 = -4$, etc., have no real solutions. Hence we cannot define $c^{1/2}, c^{1/4}, c^{1/6}$ when $c = -4$.
(b) Since $1/3$ is the same as $2/6$, it should be true that $c^{1/3} = c^{2/6}$, that is, that $\sqrt[3]{c} = \sqrt[6]{c^2}$. Show that this is false when $c = -8$.

66. (a) Suppose r is a solution of the equation $x^n = c$ and s is a solution of $x^n = d$. Verify that rs is a solution of $x^n = cd$.
(b) Explain why part (a) shows that $\sqrt[n]{cd} = \sqrt[n]{c}\sqrt[n]{d}$.

67. Write exponent laws 3, 4, and 5 in radical notation in the case when $r = 1/m$ and $s = 1/n$.

68. Write the caution in the margin on page 310 in radical notation in the case when $r = 1/n$.

69. (a) Graph $f(x) = x^5$ and explain why this function has an inverse function.
(b) Show algebraically that the inverse function is $g(x) = x^{1/5}$.

70. If n is an odd positive integer, show that $f(x) = x^n$ has an inverse function and find the rule of the inverse function. [*Hint:* Exercise 69 is the case when $n = 5$.]

71. Using the viewing window with $0 \le x \le 4$ and $0 \le y \le 2$, graph the following functions on the same screen:

$$f(x) = x^{1/2}, \qquad g(x) = x^{1/4}, \qquad h(x) = x^{1/6}.$$

In each of the following cases, arrange $x^{1/2}$, $x^{1/4}$, and $x^{1/6}$ in order of increasing size and justify your answer by using the graphs.
(a) $0 < x < 1$ (b) $x > 1$

72. Using the viewing window with $-3 \le x \le 3$ and $-1.5 \le y \le 1.5$, graph the following functions on the same screen:

$$f(x) = x^{1/3}, \qquad g(x) = x^{1/5}, \qquad h(x) = x^{1/7}.$$

In each of the following cases, arrange $x^{1/3}$, $x^{1/5}$, and $x^{1/7}$ in order of increasing size and justify your answer by using the graphs.
(a) $x < -1$ **(b)** $-1 < x < 0$ **(c)** $0 < x < 1$ **(d)** $x > 1$

73. Graph $f(x) = \sqrt{x}$ in the standard viewing window. Then, without doing any more graphing, describe the graphs of these functions:
(a) $g(x) = \sqrt{x+3}$ [*Hint:* $g(x) = f(x+3)$; see Section 3.4.]
(b) $h(x) = \sqrt{x} - 2$ **(c)** $k(x) = \sqrt{x+3} - 2$

74. Do Exercise 73 with $\sqrt[3]{\ }$ in place of $\sqrt{\ }$.

5.1.A EXCURSION Radical Equations

Equations involving radicals and rational exponents can be solved either graphically or algebraically.

Example 1 There are two ways to solve

$$\sqrt[5]{x^2 - 6x + 2} = x - 4$$

graphically. One is to rewrite the equation as

$$\sqrt[5]{x^2 - 6x + 2} - x + 4 = 0$$

and graph the function $h(x) = \sqrt[5]{x^2 - 6x + 2} - x + 4$ (Figure 5–2 on the next page). The x-intercept of the graph is the solution of the equation.

Alternatively, you can graph both $f(x) = \sqrt[5]{x^2 - 6x + 2}$ and $g(x) = x - 4$ on the same screen (Figure 5–3). The x-coordinate of a point of intersection is a solution of the original equation.

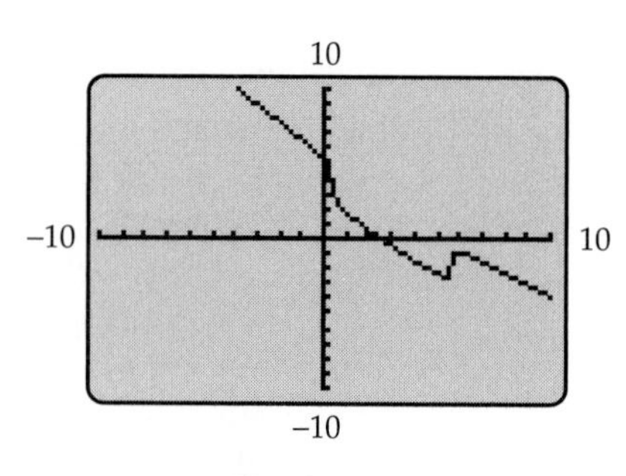

Figure 5–2

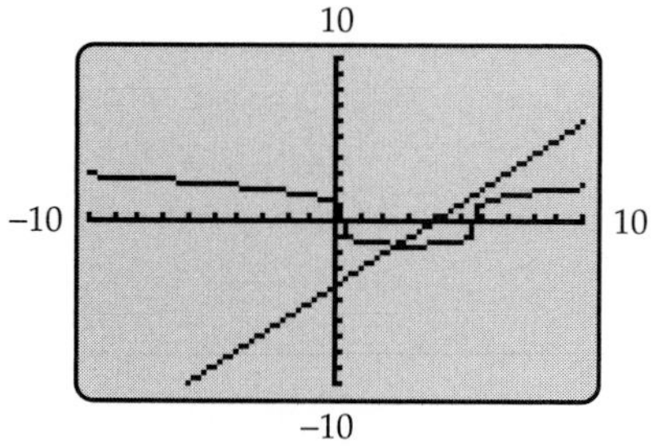

Figure 5–3

GRAPHING EXPLORATION

Verify that the solution of the equation is $x \approx 2.534$ by using a root finder in Figure 5–2 or an intersection finder in Figure 5–3. ■

Although the (approximate) solutions of any radical equation can be found graphically, exact solutions can be found algebraically in many cases. The algebraic solution of radical equations depends on the following fact. If two numbers are equal, say $a = b$, then $a^r = b^r$ for every positive integer r. The same is true of algebraic expressions. For example,

$$\text{if} \quad x - 2 = 3, \quad \text{then} \quad (x - 2)^2 = 3^2 = 9.$$

Thus, every solution of $x - 2 = 3$ is also a solution of $(x - 2)^2 = 9$. But *be careful:* This only works in *one* direction. For instance, -1 is a solution of $(x - 2)^2 = 9$, but not of $x - 2 = 3$. Similarly, 1 is the only solution of $x = 1$, but $x^3 = 1^3$ has two additional complex solutions, as shown in Example 8 on page 285. Therefore,

Power Principle

If both sides of an equation are raised to the same positive integer power, every solution of the original equation is also a solution of the new equation. But the new equation may have solutions that are *not* solutions of the original one.*

Consequently, if you raise both sides of an equation to a power, you must *check your solutions* in the *original* equation. Graphing provides a quick way to eliminate most extraneous solutions. But only an algebraic computation can confirm an exact solution.

*Such solutions are called **extraneous solutions.**

Example 2 To solve $\sqrt[3]{2x^2 + 7x - 6} = 3^{2/3}$, we first cube both sides, and then solve the resulting equation:

$$\left(\sqrt[3]{2x^2 + 7x - 6}\right)^3 = (3^{2/3})^3$$
$$2x^2 + 7x - 6 = 9$$
$$2x^2 + 7x - 15 = 0$$
$$(2x - 3)(x + 5) = 0$$
$$2x - 3 = 0 \quad \text{or} \quad x + 5 = 0$$
$$2x = 3 \qquad\qquad x = -5$$
$$x = \frac{3}{2}$$

Substituting 3/2 and −5 in the left side of the original equation shows that:

$$\sqrt[3]{2\left(\frac{3}{2}\right)^2 + 7\left(\frac{3}{2}\right) - 6} = \sqrt[3]{\frac{30}{2} - 6} = \sqrt[3]{9} = \sqrt[3]{3^2} = 3^{2/3}$$
$$\sqrt[3]{2(-5)^2 + 7(-5) - 6} = \sqrt[3]{9} = \sqrt[3]{3^2} = 3^{2/3}$$

Therefore, both 3/2 and −5 are solutions. ■

Example 3 To solve $5 + \sqrt{3x - 11} = x$, we first rearrange terms to get the radical expression alone on one side:

$$\sqrt{3x - 11} = x - 5.$$

Then, square both sides and solve the resulting equation:

$$\left(\sqrt{3x - 11}\right)^2 = (x - 5)^2$$
$$3x - 11 = x^2 - 10x + 25$$
$$0 = x^2 - 13x + 36$$
$$0 = (x - 4)(x - 9)$$
$$x - 4 = 0 \quad \text{or} \quad x - 9 = 0$$
$$x = 4 \quad \text{or} \quad x = 9$$

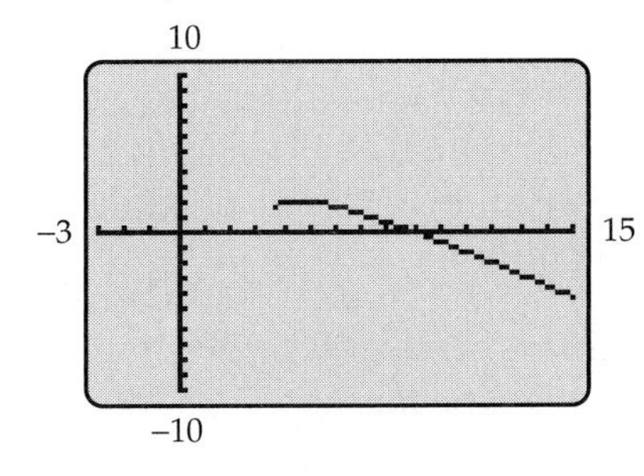

Figure 5–4

If these numbers are solutions of the original equation, they should be x-intercepts of the graph of $f(x) = \sqrt{3x - 11} - x + 5$. (Why?) But the graph of f in Figure 5–4 doesn't have an x-intercept at $x = 4$, so 4 is not a solution of the original equation. The graph appears to show that $x = 9$ is a solution of the original equation, and direct calculation confirms this:

Left side: $5 + \sqrt{3 \cdot 9 - 11} = 5 + \sqrt{16} = 9$ *Right side:* 9

Hence, 9 is the only solution of the original equation. ■

Example 4 To solve $\sqrt{2x-3} - \sqrt{x+7} = 2$, we first rearrange terms so that one side contains only a single radical term:

$$\sqrt{2x-3} = \sqrt{x+7} + 2.$$

Then, square both sides and simplify:

$$\begin{aligned}
(\sqrt{2x-3})^2 &= (\sqrt{x+7} + 2)^2 \\
2x-3 &= (\sqrt{x+7})^2 + 2 \cdot 2 \cdot \sqrt{x+7} + 2^2 \\
2x-3 &= x+7+4\sqrt{x+7}+4 \\
x-14 &= 4\sqrt{x+7}.
\end{aligned}$$

Now, square both sides and solve the resulting equation:

$$\begin{aligned}
(x-14)^2 &= (4\sqrt{x+7})^2 \\
x^2 - 28x + 196 &= 4^2 \cdot (\sqrt{x+7})^2 \\
x^2 - 28x + 196 &= 16(x+7) \\
x^2 - 28x + 196 &= 16x + 112 \\
x^2 - 44x + 84 &= 0 \\
(x-2)(x-42) &= 0
\end{aligned}$$

$$x - 2 = 0 \quad \text{or} \quad x - 42 = 0$$

$$x = 2 \qquad\qquad x = 42$$

Substituting 2 and 42 in the left side of the original equation shows that

$$\sqrt{2 \cdot 2 - 3} - \sqrt{2+7} = \sqrt{1} - \sqrt{9} = 1 - 3 = -2$$
$$\sqrt{2 \cdot 42 - 3} - \sqrt{42+7} = \sqrt{81} - \sqrt{49} = 9 - 7 = 2$$

Therefore, 42 is the only solution of the equation. ■

Example 5 Stella, who is standing at point A on the bank of a 2.5-kilometer-wide river wants to reach point B, 15 kilometers downstream on the opposite bank. She plans to row to a point C on the opposite shore and then run to B, as shown in Figure 5–5. She can row at a rate of 4 kilometers per hour and can run at 8 kilometers per hour.

(a) If her trip is to take 3 hours, how far from B should she land?

(b) How far from B should she land to make the time for the trip as short as possible?

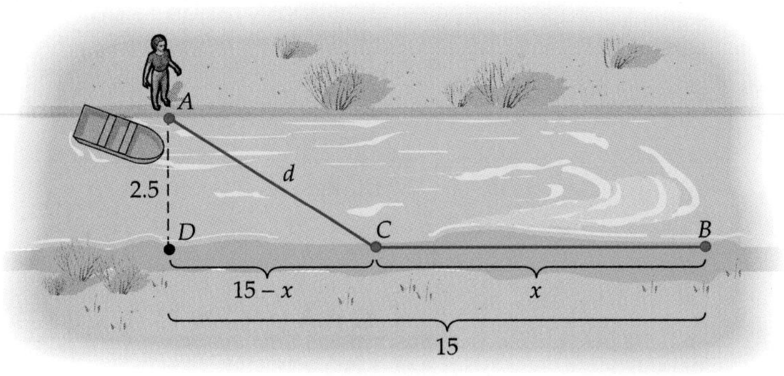

Figure 5–5

Solution Let x be the distance that Stella must run from C to B. Using the basic formula for distance, we have:

$$\text{Rate} \times \text{Time} = \text{Distance}$$

$$\text{Time} = \frac{\text{Distance}}{\text{Rate}} = \frac{x}{8}.$$

Similarly, the time required to row distance d is

$$\text{Time} = \frac{\text{Distance}}{\text{Rate}} = \frac{d}{4}.$$

Since $15 - x$ is the distance from D to C, the Pythagorean Theorem applied to right triangle ADC shows that

$$d^2 = (15 - x)^2 + 2.5^2 \qquad \text{or equivalently} \qquad d = \sqrt{(15 - x)^2 + 6.25}.$$

Therefore, the total time for the trip is given by

$$T(x) = \text{Rowing time} + \text{Running time} = \frac{d}{4} + \frac{x}{8} = \frac{\sqrt{(x - 15)^2 + 6.25}}{4} + \frac{x}{8}.$$

(a) If the trip is to take 3 hours, then $T(x) = 3$ and we must solve the equation

$$\frac{\sqrt{(x - 15)^2 + 6.25}}{4} + \frac{x}{8} = 3.$$

GRAPHING EXPLORATION

Using the viewing window with $0 \le x \le 15$ and $-2 \le y \le 2$, graph the function

$$f(x) = \frac{\sqrt{(x - 15)^2 + 6.25}}{4} + \frac{x}{8} - 3$$

and use a root finder to find its x-intercept (the solution of the equation).

This exploration shows that Stella should land approximately 6.74 kilometers from B in order to make the trip in 3 hours.

(b) To find the shortest possible time, we must find the value of x that makes $T(x) = \dfrac{\sqrt{(x-15)^2 + 6.25}}{4} + \dfrac{x}{8}$ as small as possible.

GRAPHING EXPLORATION

Using the viewing window with $0 \le x \le 15$ and $0 \le y \le 4$, graph $T(x)$ and use a minimum finder to verify that the lowest point on the graph (that is, the point with the y-coordinate $T(x)$ as small as possible) is approximately (13.56, 2.42).

Therefore, the shortest time for the trip will be 2.42 hours and will occur if Stella rows to a point 13.56 kilometers from B. ■

Equations involving rational exponents can be solved graphically, provided that the function to be graphed is entered properly. Many such equations can also be solved algebraically by making an appropriate substitution.

Example 6 Solve $x^{2/3} - 2x^{1/3} - 15 = 0$ algebraically and graphically.

Solution *Algebraic:* Let $u = x^{1/3}$, rewrite the equation, and solve:

$$x^{2/3} - 2x^{1/3} - 15 = 0$$

$$(x^{1/3})^2 - 2x^{1/3} - 15 = 0$$

$$u^2 - 2u - 15 = 0$$

$$(u+3)(u-5) = 0$$

$$u + 3 = 0 \qquad \text{or} \qquad u - 5 = 0$$

$$u = -3 \qquad\qquad u = 5$$

$$x^{1/3} = -3 \qquad\qquad x^{1/3} = 5$$

Cubing both sides of these last equations shows that

$$(x^{1/3})^3 = (-3)^3 \qquad \text{or} \qquad (x^{1/3})^3 = 5^3$$

$$x = -27 \qquad\qquad x = 125.$$

Since we cubed both sides, we must check these numbers in the original equation. Verify that both *are* solutions.

Geometric: Graph the function $f(x) = x^{2/3} - 2x^{1/3} - 15$ and find the x-intercepts, namely $x = -27$ and $x = 125$. The only difficulty is the one mentioned in the Caution box on page 309. Depending on your calculator, you may have to enter the function f in one of these forms:

$$f(x) = (x^2)^{1/3} - 2x^{1/3} - 15 \quad \text{or} \quad f(x) = (x^{1/3})^2 - 2x^{1/3} - 15.$$

Otherwise, the calculator may not produce a graph when x is negative. ■

Exercises 5.1.A

In Exercises 1–30, find all real solutions of each equation. Find exact solutions when possible and approximate ones otherwise.

1. $\sqrt{x+2} = 3$
2. $\sqrt{x-7} = 4$
3. $\sqrt{4x+9} = 5$
4. $\sqrt{3x-2} = 7$
5. $\sqrt[3]{5-11x} = 3$
6. $\sqrt[3]{6x-10} = 2$
7. $\sqrt[3]{x^2-1} = 2$
8. $(x+1)^{2/3} = 4$
9. $\sqrt{x^2-x-1} = 1$
10. $\sqrt{x^2-5x+4} = 2$
11. $\sqrt{x+7} = x-5$
12. $\sqrt{x+5} = x-1$
13. $\sqrt{3x^2+7x-2} = x+1$
14. $\sqrt{4x^2-10x+5} = x-3$
15. $\sqrt[3]{x^3+x^2-4x+5} = x+1$
16. $\sqrt[3]{x^3-6x^2+2x+3} = x-1$
17. $\sqrt[5]{9-x^2} = x^2+1$
18. $\sqrt[4]{x^3-x+1} = x^2-1$
19. $\sqrt[3]{x^5-x^3-x} = x+2$
20. $\sqrt{x^3+2x^2-1} = x^3+2x-1$
21. $\sqrt{x^2+3x-6} = x^4-3x^2+2$
22. $\sqrt[3]{x^4+x^2+1} = x^2-x-5$
23. $\sqrt{5x+6} = 3+\sqrt{x+3}$
24. $\sqrt{3y+1} - 1 = \sqrt{y+4}$
25. $\sqrt{2x-5} = 1+\sqrt{x-3}$
26. $\sqrt{x-3} + \sqrt{x+5} = 4$
27. $\sqrt{3x+5} + \sqrt{2x+3} + 1 = 0$
28. $\sqrt{20-x} = \sqrt{9-x} + 3$
29. $\sqrt{6x^2+x+7} - \sqrt{3x+2} = 2$
30. $\sqrt{x^3+x^2-3} = \sqrt{x^3-x+3} - 1$

In Exercises 31–34, assume that all letters represent positive numbers and solve each equation for the required letter.

31. $A = \sqrt{1+\dfrac{a^2}{b^2}}$ for b
32. $T = 2\pi\sqrt{\dfrac{m}{g}}$ for g
33. $K = \sqrt{1-\dfrac{x^2}{u^2}}$ for u
34. $R = \sqrt{d^2+k^2}$ for d

In Exercises 35–46, solve each equation algebraically.

35. $x - 4x^{1/2} + 4 = 0$ [*Hint:* Let $u = x^{1/2}$.]
36. $x - x^{1/2} - 12 = 0$
37. $2x - \sqrt{x} - 6 = 0$
38. $3x - 11\sqrt{x} - 4 = 0$
39. $x^{2/3} + 3x^{1/3} + 2 = 0$ [*Hint:* Let $u = x^{1/3}$.]
40. $x^{2/3} - 4x^{1/3} + 3 = 0$
41. $\sqrt[3]{x^2} + 2\sqrt[3]{x} - 8 = 0$
42. $2\sqrt[3]{x^2} - \sqrt[3]{x} - 6 = 0$
43. $x^{1/2} - x^{1/4} - 2 = 0$ [*Hint:* Let $u = x^{1/4}$.]
44. $x^{1/3} + x^{1/6} - 2 = 0$
45. $x^{-2} - x^{-1} - 6 = 0$ [*Hint:* Let $u = x^{-1}$.]
46. $x^{-2} - 6x^{-1} + 5 = 0$

In Exercises 47–50, solve each equation graphically.

47. $x^{3/5} - 2x^{2/5} + x^{1/5} - 6 = 0$
48. $x^{5/3} + x^{4/3} - 3x^{2/3} + x = 5$
49. $x^{-3} + 2x^{-2} - 4x^{-1} + 5 = 0$
50. $x^{-2/3} - 3x^{-1/3} = 4$

51. A rope is to be stretched at uniform height from a tree to a 35-foot-long fence, which is 20 feet from the tree, and then to the side of a building at a point 30 feet from the fence, as shown in the figure.

(a) If 63 feet of rope is to be used, how far from the building wall should the rope meet the fence?

(b) How far from the building wall should the rope meet the fence if as little rope as possible is to be used?

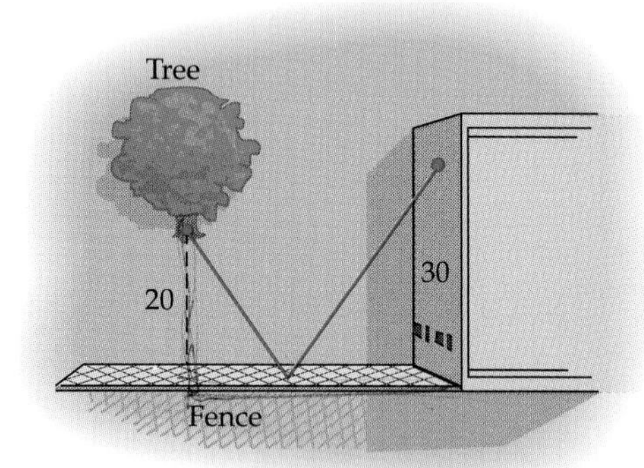

52. Anne is standing on a straight road and wants to reach her helicopter, which is located 2 miles down the road from her, a mile from the road in a field (see figure). She can run 5 miles per hour on the road and 3 miles per hour in the field. She plans to run down the road, then cut diagonally across the field to reach the helicopter.

(a) Where should she leave the road in order to reach the helicopter in exactly 42 minutes (.7 hour)?

(b) Where should she leave the road in order to reach the helicopter as soon as possible?

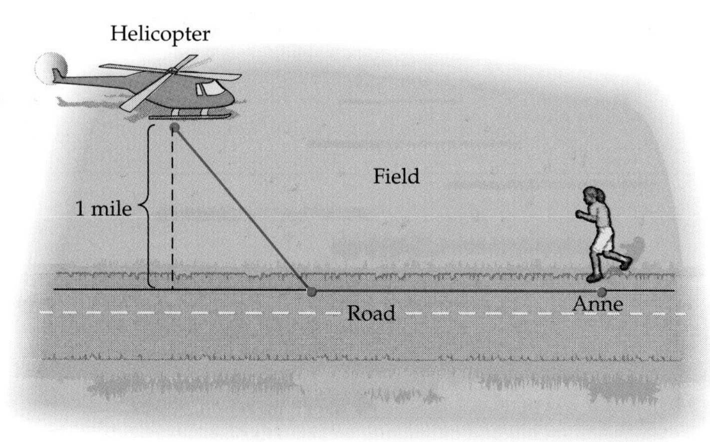

53. A spotlight is to be placed on a building wall to illuminate a bench that is 32 feet from the base of the wall. The intensity I of the light at the bench is known to be x/d^3, where x is the height of the spotlight above the ground and d is the distance from the bench to the spotlight.

(a) Express I as a function of x. [It may help to draw a picture.]

(b) How high should the spotlight be in order to provide maximum illumination at the bench?

5.2 Exponential Functions

For each positive real number a there is a function (called the **exponential function with base** $\boldsymbol{a}$) whose domain is all real numbers and whose rule is $f(x) = a^x$. For example,

$$f(x) = 10^x, \qquad g(x) = 2^x, \qquad h(x) = \left(\frac{1}{2}\right)^x, \qquad k(x) = \left(\frac{3}{2}\right)^x.$$

The shape of the graph of the exponential function $f(x) = a^x$ depends only on the size of a, as shown in the following figures.

Graph of $f(x) = a^x$

$a > 1$	$0 < a < 1$
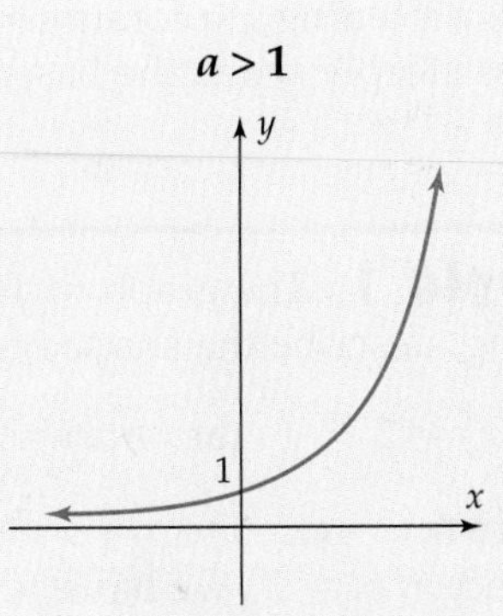	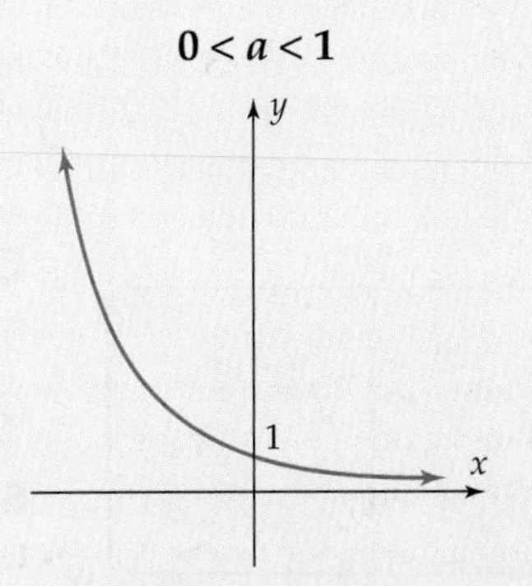
Graph is above x-axis. y-intercept is 1. $f(x)$ is increasing. Negative x-axis is a horizontal asymptote.	Graph is above x-axis. y-intercept is 1. $f(x)$ is decreasing. Positive x-axis is a horizontal asymptote.

GRAPHING EXPLORATION

(a) Using the viewing window with $-3 \le x \le 7$ and graph

$$f(x) = 1.3^x, \qquad g(x) = 2^x \qquad h(x) = 10^x$$

on the same screen and observe their behavior to the *right* of the y-axis.

Which one rises least steeply? Which one most steeply?

How does the steepness of the graph of $f(x) = a^x$ seem to be related to the size of the base a?

(b) To see what's going on to the *left* of the y-axis, graph the same four functions in the viewing window with $-4 \le x \le 2$ and $-.5 \le y \le 2$.

As you move to the left, how does size of the base a seem to be related to how quickly the graph of $f(x) = a^x$ falls toward the x-axis?

GRAPHING EXPLORATION

Using the viewing window with $-4 \le x \le 4$ and $-1 \le y \le 4$, graph

$$f(x) = .2^x, \qquad g(x) = .4^x, \qquad h(x) = .6^x, \qquad k(x) = .8^x$$

on the same screen. Note that the bases of these exponential functions are increasing in size: $0 < .2 < .4 < .6 < .8 < 1$.

How is the size relationship of the bases related to the graphs? Which graph falls least steeply? Which one falls most steeply?

The preceding Explorations show that the graph of $f(x) = a^x$ rises or falls less steeply when the base a is close to 1.

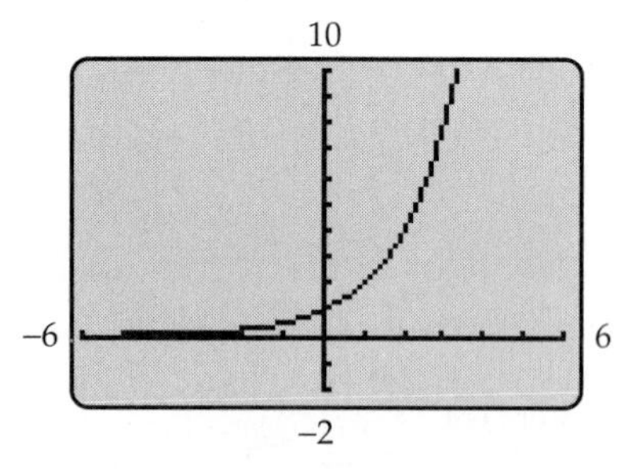

Figure 5–6

Example 1 The graph of $f(x) = 2^x$ is shown in Figure 5–6. Without graphing, describe the shape of each of the following graphs:

(a) $g(x) = 2^{x+3}$ (b) $h(x) = 2^{x-3} - 4$

Solution

(a) Since $f(x) = 2^x$, we have $f(x + 3) = 2^{x+3} = g(x)$. So the graph of $g(x)$ is just the graph of $f(x) = 2^x$ shifted horizontally 3 units to the left. [See page 152.]

GRAPHING EXPLORATION

Verify the preceding statement graphically by graphing $f(x)$ and $g(x)$ on the same screen.

(b) In this case $f(x - 3) - 4 = 2^{x-3} - 4 = h(x)$. So the graph of $h(x)$ is the graph of $f(x) = 2^x$ shifted horizontally 3 units to the right and vertically 4 units downward. [See pages 151 and 152.]

GRAPHING EXPLORATION

Verify the preceding statement graphically by graphing $f(x)$ and $h(x)$ on the same screen. ■

Calculator screens don't really show how explosive exponential graph is. To see what this means, consider the graph of $f(x) = 2^x$ in Figure 5–6. Take a pencil and extend the x-axis to the right, keeping the same scale. Then $x = 50$ will be at the right edge of the page (try it). At this point the graph of $f(x) = 2^x$ is 2^{50} units high. Now the y-axis scale in Figure 5–6 is approximately 12 units per inch, which is equivalent to 144 units per foot or 760,320 units per mile. Therefore the height of the graph at $x = 50$ is

$$\frac{2^{50}}{760{,}320} = 1{,}480{,}823{,}741 \text{ MILES},$$

which would put that part of the graph well beyond the planet Saturn!

Since most quantities that grow exponentially do not change as dramatically as the graph of $y = 2^x$, exponential functions that model real-life growth or decay are usually modified by the insertion of appropriate constants. These functions are generally of the form $f(x) = Pa^{kx}$, such as

$$f(x) = 5 \cdot 2^{.45x}, \qquad g(x) = 3.5(10^{-.03x}), \qquad h(x) = (-6)(1.076^{2x}).$$

Their graphs have the same shape as the graph of $f(x) = a^x$, but may rise or fall more or less steeply, depending on the constants P, k, and a.

Example 2 Figure 5–7 shows the graphs of

$$f(x) = 3^x, \quad g(x) = 3^{.15x}, \quad h(x) = 3^{.35x}, \quad k(x) = 3^{-x}, \quad p(x) = 3^{-.4x}.$$

Note how the coefficient of x determines the steepness of the graph. When this coefficient is positive, the graph rises and when it is negative, the graph falls from left to right.

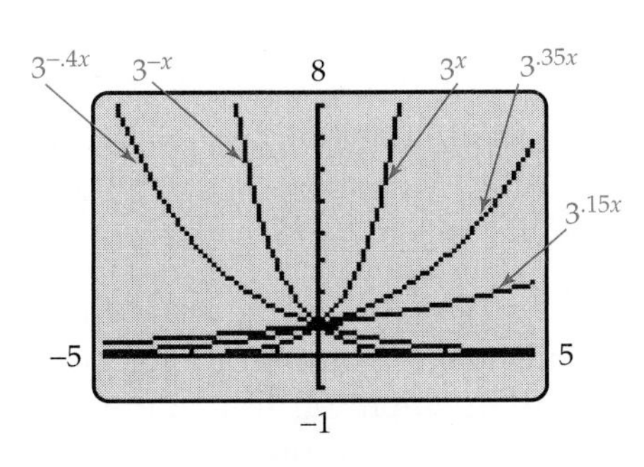

Figure 5–7

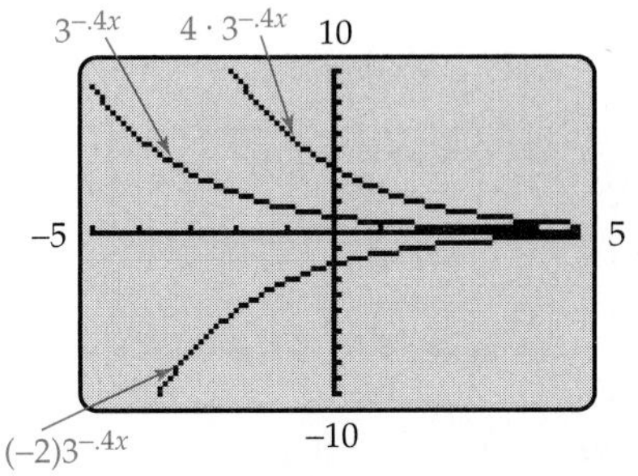

Figure 5–8

Figure 5–8 shows the graphs of

$$p(x) = 3^{-.4x}, \quad q(x) = 4 \cdot 3^{-.4x}, \quad r(x) = (-2)3^{-.4x}$$

As we saw in Section 3.4, the graph of $q(x) = 4 \cdot 3^{-.4x}$ is the graph of $p(x) = 3^{-.4x}$ stretched away from the x-axis by a factor of 4. The graph of $r(x) = (-2)3^{-.4x}$ is the graph of $p(x) = 3^{-.4x}$ stretched away from the x-axis by a factor of 2 *and* reflected in the x-axis. ■

Exponential Growth and Decay

Exponential functions are useful for modeling situations in which a quantity increases or decreases by a fixed factor.

Example 3 **Compound Interest.**

If you deposit \$5000 in a savings account that pays 3% interest, compounded annually,

(a) How much money is in the account after nine years?

(b) When will the balance reach \$50,000?

Solution

(a) After one year the account balance is

$$5000 + 3\% \text{ of } 5000 = 5000 + .03 \cdot 5000 = 5000(1 + .03) = 5000(1.03).$$

In other words, the account balance has changed by a factor of 1.03, from \$5000 to \$5000(1.03) = \$5150. If you leave the \$5150 in the account for another year, your new balance is

$$5150 + 3\% \text{ of } 5150 = 5150 + .03 \cdot 5150 = 5150(1 + .03) = 5150(1.03).$$

Since 5150 = 5000(1.03), we see that the account balance at the end of two years is

$$5150(1.03) = 5000(1.03)(1.03) = 5000(1.03)^2.$$

Similarly, your balance will change by a factor of 1.03 every year so that the balance in the account at the end of year x is given by

$$B(x) = 5000 \cdot (1.03)^x.$$

To find the account balance after nine years evaluate $B(x)$ at $x = 9$:

$$B(9) = 5000 \cdot (1.03)^9 = \$6523.87.$$

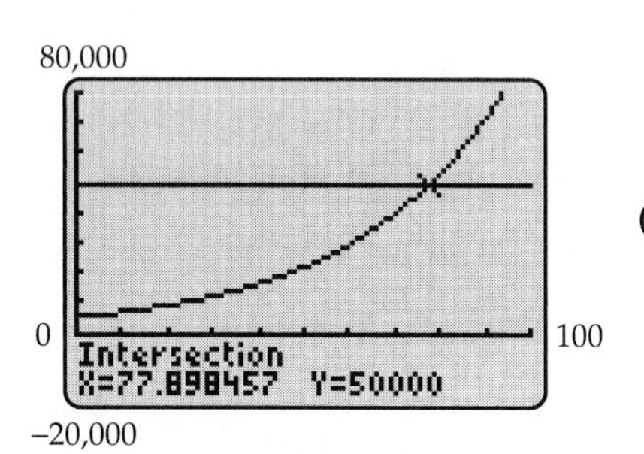

Figure 5–9

(b) In the graph of $B(x)$ in Figure 5–9, the x-coordinate of each point represents the year, and the y-coordinate the balance at that time. So we must find the point on the graph with y-coordinate 50,000, that is, the intersection of the graph of $B(x)$ with the horizontal line $y = 50{,}000$. A graphical intersection finder shows that the approximate coordinates of the point of intersection are (77.90, 50,000). Therefore the balance will reach \$50,000 in about 77.9 years. ■

Example 4 **Population Growth.**

The world population in 1950 was about 2.5 billion people and has been increasing at approximately 1.85% per year.

(a) Find the world population in 2000.

(b) In what year will the population be double what it is in 2000?

Solution

(a) The world population in 1951 was

$$2.5 + 1.85\% \text{ of } 2.5 = 2.5 + .0185(2.5) = 2.5(1 + .0185) = 2.5(1.0185).$$

Similarly, in each successive year the population changed by a factor of 1.0185 so that the population in billions in year x is given by the function

$$S(x) = 2.5(1.0185)^x,$$

where $x = 0$ corresponds to 1950, $x = 1$ to 1951, etc. The year 2000 corresponds to $x = 50$, so the population is

$$S(50) = 2.5(1.0185)^{50} \approx 6.25 \text{ billion people.}$$

(b) Twice the population in 2000 is 2(6.25) = 12.5 billion. We must find the number x such that $S(x) = 12.5$, that is, solve the equation

$$2.5(1.0185)^x = 12.5.$$

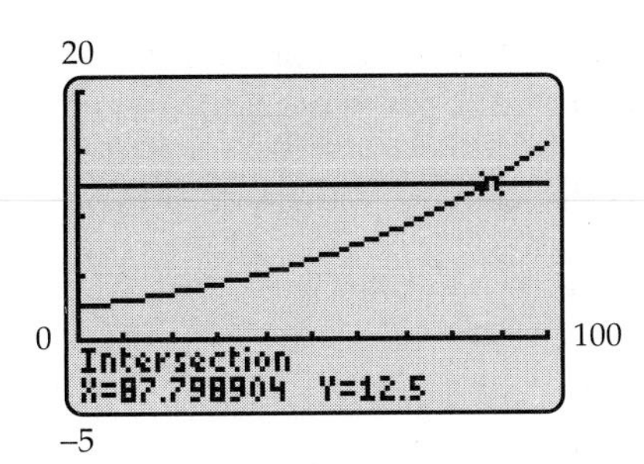

Figure 5–10

This can be done with an equation solver or by either of the following graphical methods:

(i) Write the equation as $2.5(1.0185)^x - 12.5 = 0$ and find the x-intercept of the graph of $f(x) = 2.5(1.0185)^x - 12.5$.

(ii) Find the x-coordinate of the intersection point of the graph of $S(x) = 2.5(1.0185)^x$ and the line $y = 12.5$ (Figure 5–10).

Both methods show that $x \approx 87.8$, or 88 when rounded to the nearest year. This corresponds to the year 2038. In other words, the world population will double in just 38 years. This is what is meant by the population explosion. ■

The preceding examples illustrate **exponential growth.** The general case is similar. Suppose some quantity P is increasing at a rate of r per period ($r = .08$ in Example 1 and $r = .0185$ in Example 2 and in each case the period was a year). At the end of the period the amount will be $P + rP = P(1 + r)$, that is,

> The amount at the end of the period is the starting amount multiplied by $(1 + r)$.

At the beginning of the second period, the amount is $P(1 + r)$ and by the end of the second period it has been multiplied by a factor of $(1 + r)$, so the total is

$$[P(1 + r)](1 + r) = P(1 + r)^2.$$

The total continues to increase by a factor of $1 + r$ each time period, leading to this conclusion:

Exponential Growth

Exponential growth can be described by a function of the form

$$f(x) = Pa^x,$$

where $f(x)$ is the quantity at time x, P is the initial quantity (the amount when $x = 0$), and $a > 1$ is the factor by which the quantity changes when x increases by 1. If the quantity is growing at rate r per time period, then $a = 1 + r$ and

$$f(x) = Pa^x = P(1 + r)^x.$$

Example 5 At the beginning of an experiment a culture contains 1000 bacteria. Five hours later there are 7600 bacteria. Assuming that the bacteria grow exponentially, how many will there be after 24 hours?

Solution The bacterial population is given by $f(x) = Pa^x$, where P is the initial population, a is the change factor, and x is the time in hours.

We are given that $P = 1000$, so that $f(x) = 1000a^x$. The next step is to determine a. Since there are 7600 bacteria when $x = 5$, we have

$$7600 = f(5) = 1000a^5$$

so that

$$1000a^5 = 7600$$
$$a^5 = 7.6$$
$$a = \sqrt[5]{7.6} = (7.6)^{.2}.$$

Therefore, the population function is $f(x) = 1000(7.6^{.2})^x = 1000 \cdot (7.6)^{.2x}$. After 24 hours the bacteria population will be

$$f(24) = 1000(7.6)^{.2(24)} \approx 16{,}900{,}721. \quad \blacksquare$$

In some situations a quantity decreases by a fixed factor as time goes on, as in the next example.

Example 6 When tap water is filtered through a layer of charcoal and other purifying agents, 30% of the chemical impurities in the water are removed and 70% remain. If the water is filtered through a second purifying layer, then the amount of impurities remaining is 70% of 70%, that is, $(.7)(.7) = .7^2 = .49$ or 49%. A third layer results in $.7^3$ of the impurities remaining. Thus, the function

$$f(x) = .7^x$$

gives the percentage of impurities remaining in the water after it passes through x layers of purifying material. How many layers are needed to ensure that 95% of the impurities are removed from the water?

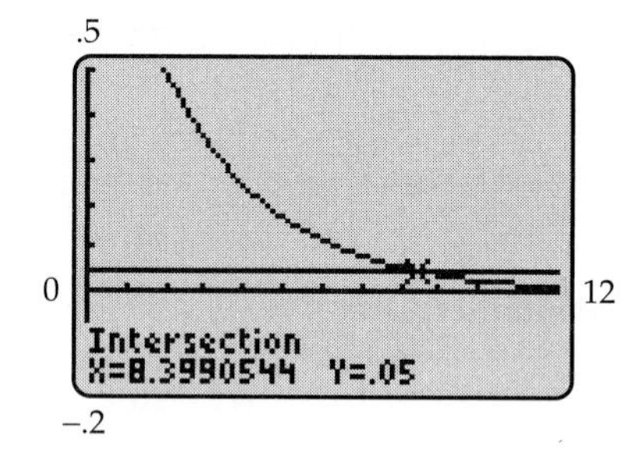

Figure 5–11

Solution If 95% of the impurities are removed, then 5% will remain. Hence, we must find x such that $f(x) = .05$, that is, we must solve the equation $.7^x = .05$. This can be done numerically or graphically. Figure 5–11 shows that the solution is $x \approx 8.4$, so 8.4 layers of material are needed.

GRAPHING EXPLORATION

How many layers are needed to ensure that 99% of the impurities are removed? $\blacksquare$

Example 6 illustrates **exponential decay.** Note that the impurities were removed at a rate of $30\% = .3$ and that the amount of impurities remaining in the water was changing by a factor of $1 - .30 = .7$. The same thing is true in the general case:

Exponential Decay

Exponential decay can be described by a function of the form

$$f(x) = Pa^x,$$

where $f(x)$ is the quantity at time x, P is the initial quantity (the amount when $x = 0$), and $0 < a < 1$; here a is the factor by which the quantity changes when x increases by 1. If the quantity is decaying at rate r per period, then $a = 1 - r$ and

$$f(x) = Pa^x = P(1 - r)^x.$$

Exponential functions are particularly useful for dealing with radioactive decay. The **half-life** of a radioactive element is the time it takes a given quantity to decay to one-half of its original mass. The half-life depends only on the substance and does not depend on the size of the sample. Exercise 68 shows that the function

$$M(x) = c\left(\frac{1}{2}\right)^{x/h} = c(.5)^{x/h},$$

gives the mass at time x, where h is the half-life and c is the original mass of the element.

Example 7 Radioactive Decay.

Plutonium (^{239}Pu) has a half-life of 24,360 years. So the amount remaining from 1 kilogram after x years is given by the function $M(x) = 1(.5)^{x/24360}$. The rule of M can be written as

$$M(x) = (.5)^{x/24360} = (.5^{1/24360})^x \approx (.99997)^x.$$

Since M is an exponential function with base smaller than but very close to 1, its graph falls *very slowly* from left to right.

GRAPHING EXPLORATION

Verify that in a viewing window with $0 \le x \le 400$, the graph of M looks like a horizontal straight line. Find a viewing window in which you can actually see the graph falling to the right of the y-axis.

The fact that the graph falls so slowly as x gets large means that even after an extremely long time, a substantial amount of plutonium will remain. Most of the original kilogram is still there after *ten thousand* years because

$$M(10{,}000) \approx .75 \text{ kilogram}.$$

This is the reason that nuclear waste disposal is such a serious problem. ■

The Number *e* and the Natural Exponential Function

There is an irrational number, denoted e, that arises naturally in a variety of phenomena and plays a central role in the mathematical description of the physical universe.* Its decimal expansion begins

$$e = 2.718281828459045\cdots.$$

Your calculator has an e^x key that can be used to evaluate the **natural exponential function** $f(x) = e^x$. If you key in e^1, the calculator will display the first part of the decimal expansion of e. At the end of Section 5.3 we shall show that every exponential function can be written in the form $f(x) = Pe^{kx}$ for suitable constants P and k.

The graph of $f(x) = e^x$ has the same shape as the graph of $g(x) = 2^x$ in Figure 5–6, but climbs more steeply.

Technology Tip

On most calculators you use the e^x key, but not the x^y or $\wedge$ keys to enter the function $f(x) = e^x$.

GRAPHING EXPLORATION

Graph $f(x) = e^x$, $g(x) = 2^x$, and $h(x) = 3^x$ on the same screen in a window with $-5 \le x \le 5$. The Tip in the margin may be helpful.

Example 8 **Population Growth.**

If the population of the United States continues to grow as it has recently, then the approximate population of the United States (in millions) in year t will be given by the function

$$P(t) = 227e^{.0093t},$$

where 1980 corresponds to $t = 0$.

(a) Estimate the population in 2015.

(b) When will the population reach half a billion?

Solution

(a) The population in 2015 (that is, $t = 35$) will be approximately

$$P(35) = 227e^{.0093(35)} \approx 314.3 \text{ million people.}$$

(b) Half a billion is 500 million people. So we must find the value of t for which $P(t) = 500$, that is, we must solve the equation

$$227e^{.0093t} = 500.$$

This can be done graphically by finding the intersection of the graph of $P(t)$ and the horizontal line $y = 500$, which occurs when $t \approx 84.9$ (Figure

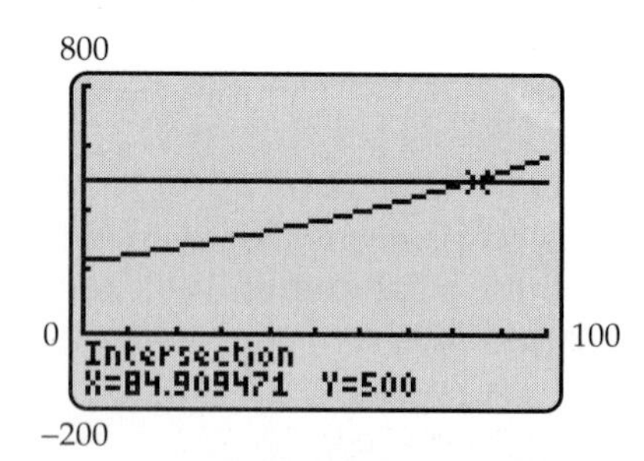

Figure 5–12

*See Example 5 in Excursion 5.2.A.

5–12). Therefore, the population will reach half a billion late in the year 2064. ■

Other Exponential Functions

The population growth models in earlier examples do not take into account factors that may limit population growth in the future (wars, new diseases, etc.). Example 9 illustrates a function, called a **logistic model,** which is designed to model such situations more accurately.

Example 9 **Inhibited Population Growth.**

There is an upper limit on the fish population in a certain lake due to the oxygen supply, available food, etc. The population of fish in this lake at time t months is given by the function

$$p(t) = \frac{20{,}000}{1 + 24e^{-t/4}} \quad (t \geq 0).$$

What is the upper limit on the fish population?

Solution The graph of $p(t)$ in Figure 5–13 suggests that the horizontal line $y = 20{,}000$ is a horizontal asymptote of the graph.

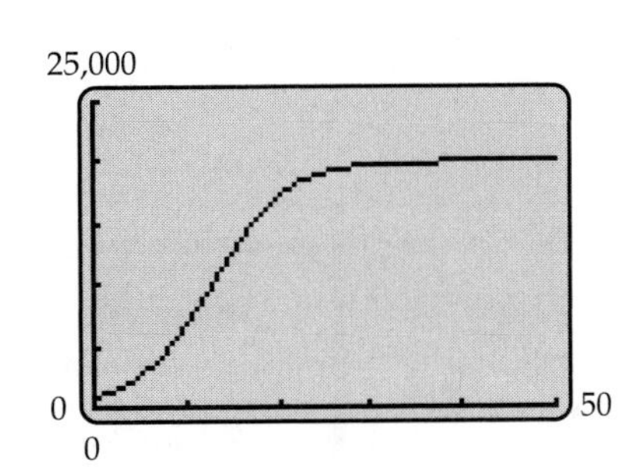

Figure 5–13

In other words, the fish population never goes above 20,000. You can confirm this algebraically by rewriting the rule of p in this form:

$$p(t) = \frac{20{,}000}{1 + 24e^{-t/4}} = \frac{20{,}000}{1 + \dfrac{24}{e^{t/4}}}.$$

When t is very large, so is $t/4$, which means that $e^{t/4}$ is huge. Hence, by the Big-Little Principle (page 249), $\dfrac{24}{e^{t/4}}$ is very close to 0 and $p(t)$ is very close to $\dfrac{20{,}000}{1 + 0} = 20{,}000$. Since $e^{t/4}$ is positive, the denominator of $p(t)$ is slightly bigger than 1, so that $p(t)$ is always less than 20,000. ■

When a cable, such as a power line, is suspended between towers of equal height, it forms a curve called a **catenary,** which is the graph of a function of the form

$$f(x) = A(e^{kx} + e^{-kx})$$

for suitable constants A and k. The Gateway Arch in St. Louis (Figure 5–14) has the shape of an inverted catenary, which was chosen because it evenly distributes the internal structural forces.

Figure 5–14

> **GRAPHING EXPLORATION**
>
> Graph each of the following functions in the window with $-5 \le x \le 5$ and $-10 \le y \le 80$:
>
> $y_1 = 10(e^{.4x} + e^{-.4x})$, $y_2 = 10(e^{2x} + e^{-2x})$, $y_3 = 10(e^{3x} + e^{-3x})$.
>
> How does the coefficient of x affect the shape of the graph? Predict the shape of the graph of $y = -y_1 + 80$. Confirm your answer by graphing.

Exercises 5.2

In Exercises 1–6, sketch a complete graph of the function.

1. $f(x) = 4^{-x}$

2. $f(x) = (5/2)^{-x}$

3. $f(x) = 2^{3x}$

4. $g(x) = 3^{x/2}$

5. $f(x) = 2^{x^2}$

6. $g(x) = 2^{-x^2}$

In Exercises 7–12, list the transformations needed to transform the graph of $h(x) = 2^x$ into the graph of the given function. [Section 3.4 may be helpful.]

7. $f(x) = 2^x - 5$

8. $g(x) = -(2^x)$

9. $k(x) = 3(2^x)$

10. $g(x) = 2^{x-1}$

11. $f(x) = 2^{x+2} - 5$

12. $g(x) = -5(2^{x-1}) + 7$

In Exercises 13 and 14, match the functions to the graphs. Assume $a > 1$ and $c > 1$.

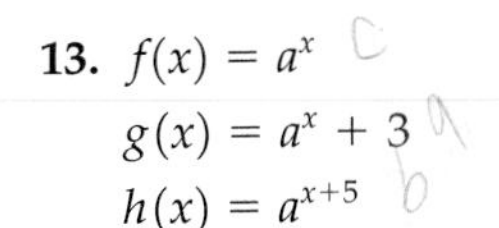

13. $f(x) = a^x$
$g(x) = a^x + 3$
$h(x) = a^{x+5}$

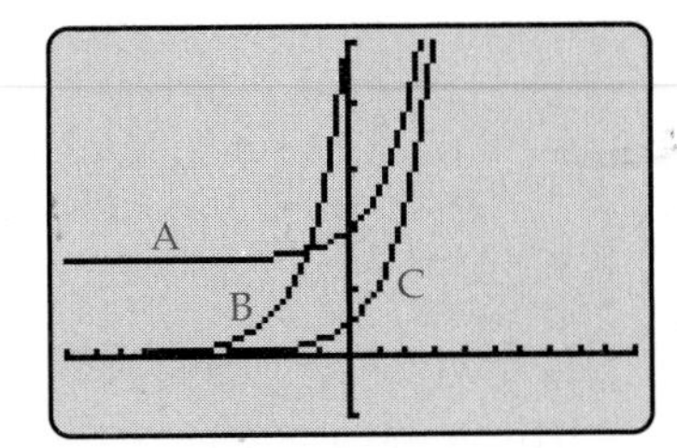

14. $f(x) = c^x$
$g(x) = -3c^x$
$h(x) = c^{x+5}$
$k(x) = -3c^x - 2$

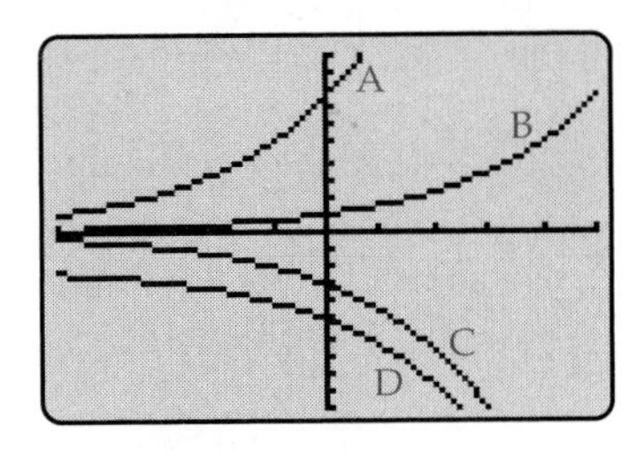

In Exercises 15–19, determine whether the function is even, odd, or neither (see Excursion 3.4.A).

15. $f(x) = 10^x$ **16.** $g(x) = 2^x - x$

17. $f(x) = \dfrac{e^x + e^{-x}}{2}$ **18.** $f(x) = \dfrac{e^x - e^{-x}}{2}$

19. $f(x) = e^{-x^2}$

20. Use the Big-Little Principle to explain why $e^x + e^{-x}$ is approximately equal to e^x when x is large.

In Exercises 21–24, find the average rate of change of the function.

21. $f(x) = x2^x$ as x goes from 1 to 3

22. $g(x) = 3^{x^2-x}$ as x goes from -1 to 1

23. $h(x) = 5^{-x^2}$ as x goes from -1 to 0

24. $f(x) = e^x - e^{-x}$ as x goes from -3 to -1

In Exercises 25–28, find the difference quotient of the function.

25. $f(x) = 10^x$ **26.** $g(x) = 5^{x^2}$

27. $f(x) = 2^x + 2^{-x}$ **28.** $f(x) = e^x - e^{-x}$

In Exercises 29–36, find a viewing window (or windows) that shows a complete graph of the function.

29. $k(x) = e^{-x}$ **30.** $f(x) = e^{-x^2}$

31. $f(x) = \dfrac{e^x + e^{-x}}{2}$ **32.** $h(x) = \dfrac{e^x - e^{-x}}{2}$

33. $g(x) = 2^x - x$ **34.** $k(x) = \dfrac{2}{e^x + e^{-x}}$

35. $f(x) = \dfrac{5}{1 + e^{-x}}$ **36.** $g(x) = \dfrac{10}{1 + 9e^{-x/2}}$

In Exercises 37–42, list all asymptotes of the graph of the function and the approximate coordinates of each local extremum.

37. $f(x) = x2^x$ **38.** $g(x) = x2^{-x}$ **39.** $h(x) = e^{x^2/2}$

40. $k(x) = 2^{x^2-6x+2}$ **41.** $f(x) = e^{-x^2}$

42. $g(x) = -xe^{x^2/20}$

43. (a) A genetic engineer is growing cells in a fermenter. The cells multiply by splitting in half every 15 minutes. The new cells have the same DNA as the original ones. Complete the following table.

Time (hours)	Number of Cells
0	1
.25	2
.5	4
.75	
1	

(b) Write the rule of the function that gives the number of C cells at time t hours.

44. Do Exercise 43, using the following table:

Time (hours)	Number of Cells
0	300
.25	600
.5	1200
.75	
1	

45. The Gateway Arch (Figure 5–14) is 630 feet high and 630 feet wide at ground level. Suppose it were placed on a coordinate plane with the x-axis at ground level and the y-axis going through the center of the arch. Find a catenary function

$g(x) = A(e^{kx} + e^{-kx})$ and a constant C such that the graph of the function $f(x) = g(x) + C$ provides a model of the arch. [*Hint:* Experiment with various values of A, k, C as in the Graphing Exploration on page 330.]

46. If you deposit $750 at 2.2% interest, compounded annually and paid from the day of deposit to the day of withdrawal, your balance at time t is given by $B(t) = 750(1.022)^t$. How much will you have after 2 years? After 3 years and 9 months?

47. The population of a colony of fruit flies t days from now is given by the function $p(t) = 100 \cdot 3^{t/10}$.

(a) What will the population be in 15 days? In 25 days?

(a) When will the population reach 2500?

48. A certain type of bacteria grows according to the function $f(x) = 5000e^{.4055x}$, where the time x is measured in hours.

(a) What will the population be in 8 hours?

(b) When will the population reach one million?

49. According to data from the National Center for Health Statistics, the life expectancy at birth for a person born in a year x is approximated by the function

$$D(x) = \frac{79.257}{1 + 9.7135 \times 10^{24} \cdot e^{-.0304x}} \quad (1900 \le x \le 2050).$$

(a) What is the life expectancy of someone born in 1980? In 2000?

(b) In what year was life expectancy at birth 60 years?

50. Based on data from 1989–1994, the number of babies born each year in the United States through assisted reproductive technology (ART), such as *in vitro* fertilization, is approximated by the function

$$K(x) = \frac{12{,}439}{1 + 4.76 \cdot e^{-.4713x}}$$

where $x = 1$ corresponds to 1989. (*Source:* American Society for Reproductive Medicine)

(a) In what year did the number of babies born through ART first exceed 8000?

(b) If this model remains accurate in the future, will the number of babies born through ART ever reach 13,000 per year?

51. The population of Mexico was 67.4 million in 1980 and has been increasing by approximately 2.6% each year.

(a) If $g(x)$ is the population of Mexico (in millions) in year x (with $x = 0$ being 1980), find the rule of the function g. [See Example 4.]

(b) Estimate the population of Mexico in the year 2000.

52. There are now 3.2 million people who play bridge and the number increases by 3.5% a year.

(a) Write the rule of a function that gives the number of bridge players in year x.

(b) How many people will be playing bridge in 15 years?

(c) When will there be 10 million bridge players?

53. The number of dandelions in your lawn increases by 5% a week and there are 75 dandelions now.

(a) If $f(x)$ is the number of dandelions in week x, find the rule of the function f.

(b) How many dandelions will there be in 16 weeks?

54. An eccentric billionaire offers you a job for the month of September. She says that she will pay you 2¢ on the first day, 4¢ on the second day, 8¢ on the third day, and so on, doubling your pay on each successive day.

(a) Let $P(x)$ denote your salary in *dollars* on day x. Find the rule of the function P.

(b) Would you be better off financially if instead you were paid $10,000 per day? [*Hint:* Consider $P(30)$.]

55. Kerosene is passed through a pipe filled with clay in order to remove various pollutants. Each foot of pipe removes 25% of the pollutants.

(a) Write the rule of a function that gives the percentage of pollutants remaining in the kerosene after it has passed through x feet of pipe. [See Example 6.]

(b) How many feet of pipe are needed to ensure that 90% of the pollutants have been removed from the kerosene?

56. If inflation runs at a steady 3% per year, then the amount a dollar is worth decreases by 3% each year.

(a) Write the rule of a function that gives the value of a dollar in year x.

(b) How much will the dollar be worth in 5 years? In 10 years?

(c) How many years will it take before today's dollar is worth only a dime?

57. A weekly census of the tree-frog population in Frog Hollow State Park produces the following results:

y = P(a)x
y = 6(3)x

Week	1	2	3	4	5	6
Population	18	54	162	486	1458	4374

(a) Find a function of the form $f(x) = Pa^x$ that describes the frog population at time x weeks.
(b) What is the growth factor in this situation (that is, by what number must this week's population be multiplied to obtain next week's population)?
(c) Each tree frog requires 10 square feet of space and the park has an area of 6.2 square miles. Will the space required by the frog population exceed the size of the park in 12 weeks? In 14 weeks? [Remember: 1 square mile $= 5280^2$ square feet.]

58. The fruit fly population in a certain laboratory triples every day. Today there are 200 fruit flies.
(a) Make a table showing the number of fruit flies present for the first 4 days (today is day 0, tomorrow is day 1, etc.).
(b) Find a function of the form $f(x) = Pa^x$ that describes the fruit fly population at time x days.
(c) What is the growth factor here (that is, by what number must each day's population be multiplied to obtain the next day's population)?
(d) How many fruit flies will there be a week from now?

59. Take an ordinary piece of typing paper and fold it in half; then the folded sheet is twice as thick as the single sheet was. Fold it in half again, so that it is twice as thick as before. Keep folding it in half as long as you can. Soon the folded paper will be so thick and small that you will be unable to continue, but suppose you could keep folding the paper as long as you wanted. Assume the paper is .002 inches thick.
(a) Make a table showing the thickness of the folded paper for the first four folds (with fold 0 being the thickness of the original unfolded paper).
(b) Find a function of the form $f(x) = Pa^x$ that describes the thickness of the folded paper after x folds.
(c) How thick would the paper be after 20 folds?
(d) How many folds would it take to reach the moon (which is 243,000 miles from the earth)? [*Hint:* One mile is 5280 feet.]

60. The figure is the graph of an exponential growth function $f(x) = Pa^x$.
(a) In this case, what is P? [*Hint:* What is $f(0)$?]
(b) Find the rule of the function f by finding a. [*Hint:* What is $f(2)$?]

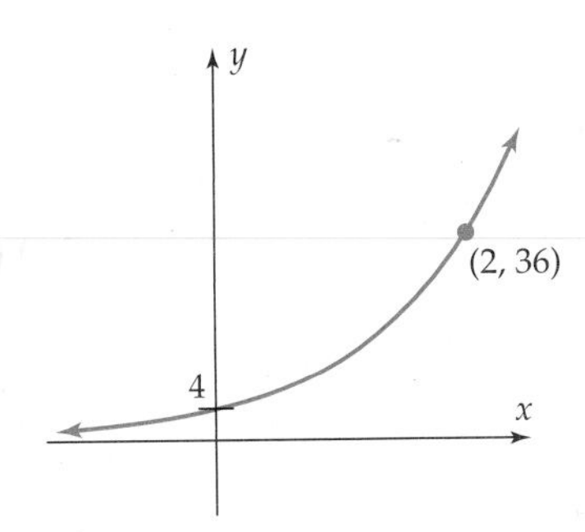

61. At the beginning of an experiment a culture contains 200 *h-pylori* bacteria. An hour later there are 205 bacteria. Assuming that the *h-pylori* bacteria grow exponentially, how many will there be after 10 hours? After 2 days? [See Example 5.]

62. If the population of India was 650 million a decade ago and is now 790 million people and continues to grow exponentially at the same rate, what will the population be in 5 years?

63. You have 5 grams of carbon-14, whose half-life is 5730 years.
(a) Write the rule of the function that gives the amount of carbon-14 remaining after x years. [See the discussion preceding Example 7.]
(b) How much carbon-14 will be left after 4000 years? After 8000 years?
(c) When will there be just one gram left?

64. (a) The half-life of radium is 1620 years. If you start with 100 milligrams of radium, what is the rule of the function that gives the amount remaining after t years?
(b) How much radium is left after 800 years? After 1600 years? After 3200 years?

65. The estimated number of units that will be sold by a certain company t months from now is given by $N(t) = 100{,}000e^{-.09t}$.
(a) What are the current sales ($t = 0$)? What will sales be in 2 months? In 6 months?
(b) Will sales ever start to increase again? (What does the graph of $N(t)$ look like?)

66. (a) The function $g(t) = 1 - e^{-.0479t}$ gives the percentage of the population (expressed as a decimal) that has seen a new TV show t weeks after it goes on the air. What percentage of people have seen the show after 24 weeks?
(b) Approximately when will 90% of the people have seen it?

67. (a) The beaver population near a certain lake in year t is approximately $p(t) = \dfrac{2000}{1 + 199e^{-.5544t}}$.

What is the population now ($t = 0$) and what will it be in 5 years?

(b) Approximately when will there be 1000 beavers?

Thinkers

68. This exercise provides a justification for the claim that the function $M(x) = c(.5)^{x/h}$ gives the mass after x years of a radioactive element with half-life h years. Suppose we have c grams of an element that has a half-life of 50 years. Then after 50 years we would have $c(\frac{1}{2})$ grams. After another 50 years we would have half of that, namely $c(\frac{1}{2})(\frac{1}{2}) = c(\frac{1}{2})^2$.

(a) How much remains after a third 50-year period? After a fourth 50-year period?

(b) How much remains after t 50-year periods?

(c) If x is the number of years, then $x/50$ is the number of 50-year periods. By replacing the number of periods t in part (b) by $x/50$, you obtain the amount remaining after x years. This gives the function $M(x)$ when $h = 50$. The same argument works in the general case (just replace 50 by h).

69. Find a function $f(x)$ with the property $f(r + s) = f(r)f(s)$ for all real numbers r and s. [*Hint:* Think exponential.]

70. Find a function $g(x)$ with the property $g(2x) = (g(x))^2$ for every real number x.

71. (a) Using the viewing window with $-4 \le x \le 4$ and $-1 \le y \le 8$, graph $f(x) = (\frac{1}{2})^x$ and $g(x) = 2^x$ on the same screen. If you think of the y-axis as a mirror, how would you describe the relationship between the two graphs?

(b) Without graphing, explain how the graphs of $g(x) = 2^x$ and $k(x) = 2^{-x}$ are related.

72. Look back at Section 4.4, where the basic properties of graphs of polynomial functions were discussed. Then review the basic properties of the graph of $f(x) = a^x$ discussed in this section. Using these various properties, give an argument to show that for any fixed positive number $a\ (\neq 1)$, it is *not* possible to find a polynomial function $g(x) = c_n x^n + \cdots + c_1 x + c_0$ such that $a^x = g(x)$ for *all* numbers x. In other words, *no exponential function is a polynomial function.* However, see Exercise 73.

73. *Approximating exponential functions by polynomials.* For each positive integer n, let f_n be the polynomial function whose rule is

$$f_n(x) = 1 + x + \frac{x^2}{2!} + \frac{x^3}{3!} + \frac{x^4}{4!} + \cdots + \frac{x^n}{n!}.$$

(a) Using the viewing window with $-4 \le x \le 4$ and $-5 \le y \le 55$ graph $g(x) = e^x$ and $f_4(x)$ on the same screen. Do the graphs appear to coincide?

(b) Replace the graph of $f_4(x)$ by that of $f_5(x)$, then by $f_6(x)$, $f_7(x)$, and so on until you find a polynomial $f_n(x)$ whose graph appears to coincide with the graph of $g(x) = e^x$ in this viewing window. Use the trace feature to move from graph to graph at the same value of x to see how accurate this approximation is.

(c) Change the viewing window so that $-6 \le x \le 6$ and $-10 \le y \le 400$. Is the polynomial you found in part (b) a good approximation for $g(x)$ in this viewing window? What polynomial is?

5.2.A EXCURSION Compound Interest and the Number *e*

When money earns compound interest, as in Example 3 on page 323, the exponential growth function can be described as follows.

Compound Interest Formula

If P dollars is invested at interest rate r per time period (expressed as a decimal), then the amount A after t periods is

$$A = P(1 + r)^t$$

Interest is often paid from day of deposit to day of withdrawal, regardless of the period used for compounding the interest. So the formula is used even when t is not an integer.

Example 1 **Compound Interest.**

If $7500 is invested at 12% interest compounded yearly, how much is in the account **(a)** after 5 years? **(b)** after 9 years and 3 months?

Solution

(a) Apply the compound interest formula with $P = 7500$, $r = .12$, and $t = 5$. Then $A = 7500(1.12)^5 = \$13{,}217.56$.

(b) Since 9 years and 3 months is 9.25 years, the amount in the account then is $A = 7500(1.12)^{9.25} = \$21{,}395.77$. ■

Example 2 **Compound Interest.**

If $9000 is invested at 8% annual interest, compounded monthly, how much will the investment be worth after 6 years?

Solution Interest is compounded 12 times per year, so the time period is 1/12 of a year and the interest rate per period is $r = .08/12$. The number of periods in 6 years is $t = 6 \cdot 12 = 72$. Using these numbers in the compound interest formula shows that

$$A = 9000\left(1 + \frac{.08}{12}\right)^{72} \approx 9000(1.0067)^{72} = \$14{,}521.52.$$

This is more money than there would be if the interest were compounded annually ($r = .08$ and $t = 6$ in the formula):

$$A = 9000(1 + .08)^6 = 9000(1.08)^6 = \$14{,}281.87. \quad ■$$

Example 3 **Compound Interest.**

If $5,000 is invested at 6.5% annual interest compounded monthly, how long will it take for the investment to double in value?

Solution The compound interest formula (with $P = 5000$ and $r = .065/12$) shows that the amount in the account at time t months is $5000(1 + .065/12)^t$. We must find the value of t such that

$$5000\left(1 + \frac{.065}{12}\right)^t = 10{,}000.$$

GRAPHING EXPLORATION

Graph

$$f(t) = 5000\left(1 + \frac{.065}{12}\right)^t - 10{,}000$$

in a viewing window with $0 \le t \le 240$ (that's 20 years). Show that the t-intercept is approximately $t = 128.3$.

Therefore, it will take 128.3 months (approximately 10.7 years) for the investment to double. ■

Example 4 Compound Interest.

What interest rate, compounded annually, is needed in order that a $16,000 investment grow to $50,000 in 18 years?

Solution In the compound interest formula, we have $A = 50{,}000$, $P = 16{,}000$, and $t = 18$ and must find r. The equation

$$16{,}000(1 + r)^{18} = 50{,}000$$

may be solved either graphically by finding the r-intercept of the graph of $f(r) = 16{,}000(1 + r)^{18} - 50{,}000$ or algebraically as follows:

Divide both sides by 16,000: $$(1 + r)^{18} = \frac{50{,}000}{16{,}000} = 3.125$$

Take 18th roots on both sides: $$\sqrt[18]{(1 + r)^{18}} = \sqrt[18]{3.125}$$

$$1 + r = \sqrt[18]{3.125}$$

$$r = \sqrt[18]{3.125} - 1 \approx .06535$$

So, the necessary interest rate is about 6.535%. ■

As a general rule, the more often your interest is compounded, the better off you are. But there is, alas, a limit.

Example 5 The Number *e*.

You have $1 to invest for 1 year. The Exponential Bank offers to pay 100% annual interest, compounded n times per year and rounded to the nearest penny. You may pick any value you want for n. Can you choose n so large that your $1 will grow to some huge amount?

Solution Since the interest rate 100% (=1.00) is compounded n times per year, the interest rate per period is $r = 1/n$ and the number of periods in 1 year is n. According to the formula, the amount at the end of the year will be $A = \left(1 + \frac{1}{n}\right)^n$. Here's what happens for various values of n:

Interest Is Compounded	$n =$	$\left(1 + \frac{1}{n}\right)^n =$
Annually	1	$(1 + \frac{1}{1})^1 = 2$
Semiannually	2	$(1 + \frac{1}{2})^2 = 2.25$
Quarterly	4	$(1 + \frac{1}{4})^4 \approx 2.4414$
Monthly	12	$(1 + \frac{1}{12})^{12} \approx 2.6130$
Daily	365	$(1 + \frac{1}{365})^{365} \approx 2.71457$
Hourly	8760	$(1 + \frac{1}{8760})^{8760} \approx 2.718127$
Every minute	525,600	$(1 + \frac{1}{525,600})^{525,600} \approx 2.7182792$
Every second	31,536,000	$(1 + \frac{1}{31,536,000})^{31,536,000} \approx 2.7182818$

Since interest is rounded to the nearest penny, your dollar will grow no larger than \$2.72, no matter how big n is. ■

The last entry in the preceding table, 2.7182818, is the number e to seven decimal places. This is just one example of how e arises naturally in real-world situations. In calculus, it is provided that e is the *limit* of $\left(1 + \frac{1}{n}\right)^n$, meaning that as n gets larger and larger, $\left(1 + \frac{1}{n}\right)^n$ gets closer and closer to e.

GRAPHING EXPLORATION

Confirm this fact graphically by graphing the function $f(x) = \left(1 + \frac{1}{x}\right)^x$ and the horizontal line $y = e$ in the viewing window with $0 \le x \le 5000$ and $-1 \le y \le 4$ and noting that the two graphs appear identical.

When interest is compounded n times per year for larger and larger values of n, as in Example 5, we say that the interest is **continuously compounded.** In this terminology, Example 5 says that \$1 will grow to \$2.72 in 1 year at an interest rate of 100% compounded continuously.

Now consider a more realistic interest rate. Suppose P dollars is invested at 8% interest compounded continuously. If interest is compounded m times per year, then the interest rate per period is $.08/m$ and there are m periods in 1 year. Therefore at the end of 1 year, P dollars will have grown to

$$A = P\left(1 + \frac{.08}{m}\right)^m. \qquad [1]$$

To see what happens to this amount as m gets larger, let n denote the number $\frac{m}{.08}$. Algebraic manipulation shows that $n = \frac{m}{.08}$ is equivalent to

$$m = .08n \qquad \text{and} \qquad \frac{1}{n} = \frac{.08}{m}$$

and, therefore, equation [1] may be rewritten as:

$$A = P\left(1 + \frac{.08}{m}\right)^m = P\left(1 + \frac{1}{n}\right)^{.08n} = P\left[\left(1 + \frac{1}{n}\right)^n\right]^{.08}.$$

As m gets larger and larger, so does $.08m = n$. Example 5 shows that when n is very large,

$$\left(1 + \frac{1}{n}\right)^n \text{ is very close to the number } e.$$

Therefore, as m gets larger and larger,

$$A = P\left[\left(1 + \frac{1}{n}\right)^n\right]^{.08} \text{ gets closer and closer to } Pe^{.08}.$$

Since dollar amounts are rounded, we can say that after 1 year P dollars grows to $Pe^{.08}$ dollars. In other words,

The amount at the end of a year is the beginning amount multiplied by $e^{.08}$. Applying this fact repeatedly, we have:

Amount at end of year 1	$Pe^{.08}$
Amount at end of year 2	$(Pe^{.08})e^{.08} = P(e^{.08})^2 = Pe^{(.08)2}$
Amount at end of year 3	$(Pe^{(.08)2})e^{.08} = P(e^{.08})^2e^{.08} = P(e^{.08})^3 = Pe^{(.08)3}$

Continuing in this manner, we see that P dollars invested at 8% compounded continuously for 6 years will grow to

$$Pe^{(.08)6} \text{ dollars.}$$

When $P = 250$, for example, this says that \$250 invested at 8% compounded continuously for 6 years will grow to

$$250e^{(.08)6} = 250e^{.48} \approx \$404.02.$$

If we use interest rate r in place of 8% and t years in place of 6 years in the preceding discussion, then a virtually identical argument leads to this conclusion:

Continuous Compounding

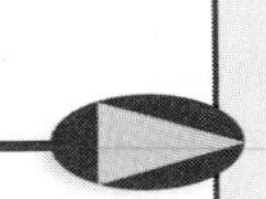

If P dollars is invested at interest rate r, compounded continuously, then the amount A after t years is

$$A = Pe^{rt}$$

Example 6 **Continuous Compounding.**

If $3800 is invested at 9.2% compounded continuously, then after seven and a half years the amount in the account will be

$$A = 3800e^{(.092)7.5} = 3800e^{.69} \approx \$7576.12.^*$$

To find out how long it will take for the value of the investment to reach $5,000, you need only solve the equation

$$3800e^{.092t} = 5000, \quad \text{or equivalently,} \quad 3800e^{.092t} - 5000 = 0.$$

GRAPHING EXPLORATION

Solve the equation graphically and verify that it will take just a few days less than 3 years for the investment to be worth $5,000. ■

Exercises 5.2.A

1. If $1,000 is invested at 8%, find the value of the investment after 5 years if interest is compounded
   (a) annually. (b) quarterly. (c) monthly.
   (d) weekly.
2. If $2500 is invested at 11.5%, what is the value of the investment after 10 years if interest is compounded
   (a) annually? (b) monthly? (c) daily?

In Exercises 3–12, determine how much money will be in a savings account if the initial deposit was $500 and the interest rate is:

3. 2% compounded annually for 8 years.
4. 2% compounded annually for 10 years.
5. 2% compounded quarterly for 10 years.
6. 2.3% compounded monthly for 9 years.
7. 2.9% compounded daily for 8.5 years.
8. 3.5% compounded weekly for 7 years and 7 months.
9. 3% compounded continuously for 4 years.
10. 3.5% compounded continuously for 10 years.
11. 2.45% compounded continuously for 6.2 years.
12. 3.25% compounded continuously for 11.6 years.
13. A typical credit card company charges 18% annual interest, compounded monthly, on the unpaid balance. If your current balance is $520 and you don't make any payments for 6 months, how much will you owe (assuming they don't sue you in the meantime)?
14. How long will it take to double an investment of $100 if the interest rate is 8% compounded annually? Answer the same question for $500 investment.
15. How long will it take to double an investment of $100 if the interest rate is 7% compounded continuously?

*This is the continuous compounding formula with $P = 3800$, $r = .092$, and $t = 7.5$.

16. How long will it take to triple an investment of \$5000 if the interest rate is 8% compounded continuously?

17. If an investment of \$1000 grows to \$1407.10 in seven years with interest compounded annually, what is the interest rate?

18. If an investment of \$2000 grows to \$2700 in three and a half years, with an annual interest rate that is compounded quarterly, what is the annual interest rate?

19. If you put \$3000 in a savings account today, what interest rate (compounded annually) must you receive in order to have \$4000 after five years?

20. If interest is compounded continuously, what annual rate must you receive if your investment of \$1500 is to grow to \$2100 in six years?

5.3 Common and Natural Logarithmic Functions*

> ***Roadmap***
> We begin with the only logarithms that are in widespread use, common and natural logarithms. Natural logarithms are emphasized because of their central role in calculus. Those who prefer to begin with logarithms to an arbitrary base b should cover Excursion 5.3.A before reading this section.

From their invention in the seventeenth century until the development of computers and calculators, logarithms were the only effective tool for numerical computation in astronomy, chemistry, physics, and engineering. Although they are no longer needed for computation, logarithmic functions still play an important role in the sciences and engineering. In this section we examine the two most important types of logarithms, those to base 10 and those to base e. Logarithms to other bases are considered in Excursion 5.3.A.

Common Logarithms

The exponential function $f(x) = 10^x$, whose graph is shown in Figure 5–15, is an increasing function and hence is one-to-one (as explained on page 188). Therefore, f has an inverse function g whose graph is the reflection of the graph of f in the line $y = x$ (see p 192), as shown in Figure 5–16.†

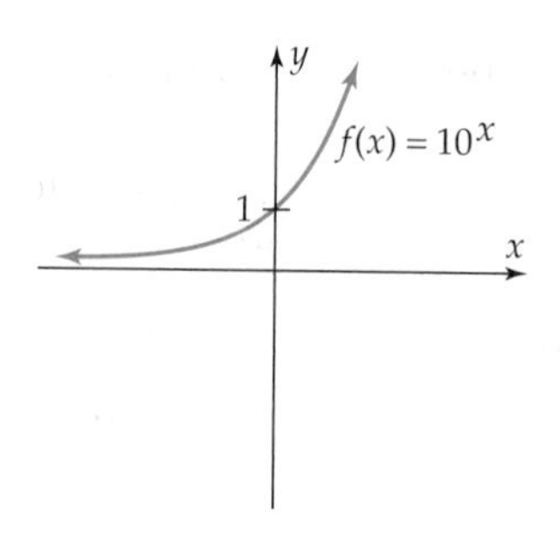

Figure 5–15

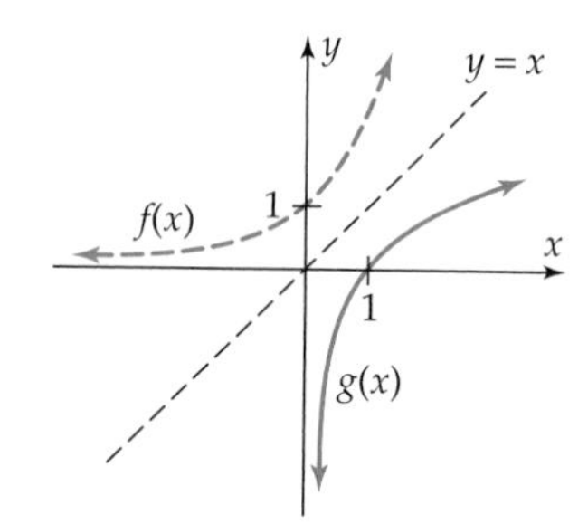

Figure 5–16

*Section 3.7 (Inverse Functions) is a prerequisite for this section.

†The graph of $f(x) = 10^x$ can be obtained in parametric mode by letting

$$x = t \quad \text{and} \quad y = 10^t \quad (t \text{ any real number}).$$

As explained on page 192, the graph of the inverse function g can then be obtained by letting

$$x = 10^t \quad \text{and} \quad y = t \quad (t \text{ any real number}).$$

This inverse function g is called the **common logarithmic function.** The value of $g(x)$ of this function at the number x is denoted **log x** and called the **common logarithm** of the number x. Every calculator has a LOG key for evaluating the function $g(x) = \log x$. For instance,

$$\log .6 = -.2218 \qquad \text{and} \qquad \log 327 = 2.5145.\ *$$

Although many properties of the logarithmic function $g(x) = \log x$ can be read from its graph, we also need an algebraic description of this function. On page 188 we saw that the relationship between a function f and its inverse function g is given by:

$$g(v) = u \qquad \text{exactly when} \qquad f(u) = v.$$

When $f(x) = 10^x$ and $g(x) = \log x$, this statement says

$$\mathbf{\log v = u \qquad \text{exactly when} \qquad 10^u = v.}$$

In other words,

log v is the exponent to which 10 must be raised to produce v.

Example 1 Without using a calculator, find

(a) $\log 1000$ (b) $\log 1$ (c) $\log \sqrt{10}$

Solution

(a) To find log 1000, ask yourself "what power of 10 equals 1000?" The answer is 3 because $10^3 = 1000$. Therefore, $\log 1000 = 3$.

(b) What power must 10 be raised to in order to produce 1? Since $10^0 = 1$, we conclude that $\log 1 = 0$.

(c) $\text{Log}\sqrt{10} = 1/2$ because $1/2$ is the exponent to which 10 must be raised to produce $\sqrt{10}$, that is $10^{1/2} = \sqrt{10}$. ■

A calculator is necessary to find most logarithms, but even then you should first get a rough estimate by hand. For instance, log 795 is the exponent to which 10 must be raised to produce 795. Since $10^2 = 100$ and $10^3 = 1000$, this exponent must be between 2 and 3, that is, $2 < \log 795 < 3$.

Since logarithms are a special kind of exponent, every statement about logarithms is equivalent to a statement about exponents. For instance,

Logarithmic Statement	Equivalent Exponential Statement
$\log v = u$	$10^u = v$
$\log 29 = 1.4624$	$10^{1.4624} = 29$
$\log 378 = 2.5775$	$10^{2.5775} = 378$

*Here and below all logarithms are rounded to four decimal places and an equal sign is used rather than the more correct "approximately equal." The word "common" will be omitted except when it is necessary to distinguish these logarithms from other types that are introduced below.

Example 2 To solve the equation $\log x = 2$, note that the equation is equivalent to the exponential statement $10^2 = x$. So the solution is $x = 100$. ■

Natural Logarithms

The exponential function $f(x) = 10^x$ and the resulting common logarithms played an important role in mathematics and still do. With the advent of calculus, however, it became clear that the exponential function $f(x) = e^x$ was even more useful in science and engineering. Consequently, a new type of logarithm, based on the number e instead of 10, was developed.* This development is essentially a carbon copy of what was done above.

The exponential function $f(x) = e^x$ whose graph is shown in Figure 5–17 is increasing and hence one-to-one, so f has an inverse function g whose graph is the reflection of the graph of f in the line $y = x$, as shown in Figure 5–18.

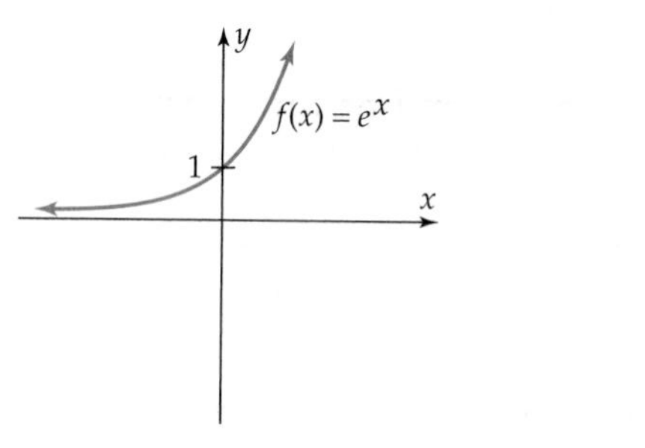

Figure 5–17

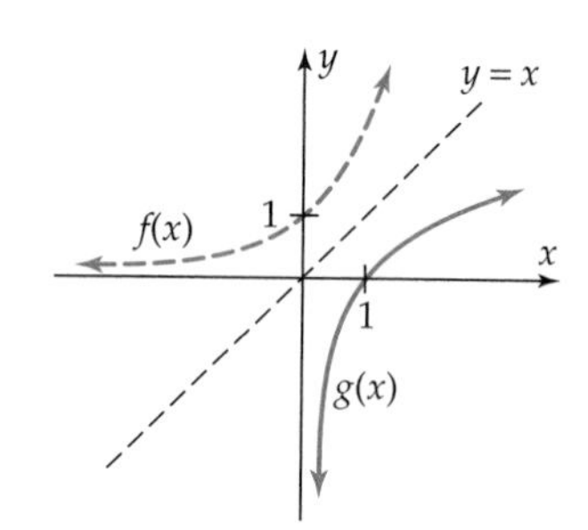

Figure 5–18

This inverse function g is called the **natural logarithmic function.** The value $g(x)$ of this function at a number x is denoted **ln x** and called the **natural logarithm** of the number x. When the relationship of inverse functions (page 188)

$$g(v) = u \qquad \text{exactly when} \qquad f(u) = v$$

is applied to the function $f(x) = e^x$ and its inverse $g(x) = \ln x$, it says

$$\mathbf{\ln v = u \qquad \text{exactly when} \qquad e^u = v.}$$

In other words,

ln v is the exponent to which e must be raised to produce v.

Example 3 Every calculator has an LN key for evaluating the function $g(x) = \ln x$. For instance,

$$\ln .15 = -1.8971 \qquad \text{and} \qquad \ln 186 = 5.2257.$$

*The number e was introduced on page 328 and discussed further in Excursion 5.2.A.

In a few cases you can determine $\ln x$ without a calculator. For example,

$$\ln 1 = 0$$

because $e^0 = 1$, that is, 0 is the exponent to which e must be raised to produce 1. Similarly, $\ln e = 1$ because 1 is the exponent to which e must be raised to produce e. ■

Once again, (natural) logarithms are a special kind of exponent and every statement about logarithms is equivalent to a statement about exponents:

Logarithmic Statement	**Equivalent Exponential Statement**
$\ln v = u$	$e^u = v$
$\ln 14 = 2.6391$	$e^{2.6391} = 14$
$\ln 158 = 5.0626$	$e^{5.0626} = 158$

Properties of Logarithms

Since common and natural logarithms have almost identical definitions (just replace 10 by e), it is not surprising that they share the same essential properties. You don't need a calculator to understand these properties. You need only translate logarithmic statements into equivalent exponential ones (or vice versa).

Example 4 What is $\ln(-10)$?

Translation: To what power must e be raised to produce -10?

Answer: The graph of $f(x) = e^x$ in Figure 5–17 shows that every power of e is *positive.* So e^x can *never* be -10 or any negative number or zero and hence $\ln(-10)$ is not defined. Similarly, $\log(-10)$ is not defined because every power of 10 is positive. Therefore,

$\ln v$ and $\log v$ are defined only when $v > 0$. ■

Example 5 What is $\log 1$?

Translation: To what power must 10 be raised to produce 1?

Answer: 0 because $10^0 = 1$. So $\log 1 = 0$. Example 3 shows that the same thing is true for natural logarithms: $\ln 1 = 0$. ■

Example 6 What is $\ln e^9$?

Translation: To what power must e be raised to produce e^9?

Answer: Obviously, the answer is 9. So $\ln e^9 = 9$ and in general

$$\ln e^k = k \qquad \text{for every real number } k.$$

Similarly,

$$\log 10^k = k \qquad \text{for every real number } k$$

because k is the exponent to which 10 must be raised to produce 10^k. ■

Example 7 What are $10^{\log 678}$ and $e^{\ln 678}$? By definition log 678 is the exponent to which 10 must be raised to produce 678, so if you raise 10 to this exponent the answer will be 678, that is, $10^{\log 678} = 678$. Similarly, ln 678 is the exponent to which e must be raised to produce 678 and hence $e^{\ln 678} = 678$. Similarly, for every $v > 0$

$$e^{\ln v} = v \qquad \text{and} \qquad 10^{\log v} = v. \quad ■$$

The facts presented in the preceding examples may be summarized as follows.

Properties of Logarithms

Natural Logarithms	Common Logarithms
1. $\ln v$ is defined only when $v > 0$;	$\log v$ is defined only when $v > 0$.
2. $\ln 1 = 0$ and $\ln e = 1$;	$\log 1 = 0$ and $\log 10 = 1$.
3. $\ln e^k = k$ for every real number k;	$\log 10^k = k$ for every real number k.
4. $e^{\ln v} = v$ for every $v > 0$;	$10^{\log v} = v$ for every $v > 0$.

Properties 3 and 4 (for natural logarithms) are restatements of the fact that the exponential function $f(x) = e^x$ and its inverse function $g(x) = \ln x$ have the usual "round-trip properties" (see page 191):

$$(g \circ f)(x) = g(f(x)) = g(e^x) = \ln e^x = x \quad \text{for all } x;$$

$$(f \circ g)(x) = f(g(x)) = f(\ln x) = e^{\ln x} = x \quad \text{for all } x > 0.^*$$

Analogous statements hold for common logarithms.

The properties of logarithms can be used to simplify expressions and solve equations.

Example 8 Applying Property 3 with $k = 2x^2 + 7x + 9$ shows that

$$\ln e^{2x^2+7x+9} = 2x^2 + 7x + 9. \quad ■$$

*A calculator provides a visual demonstration of these facts. For instance, if you key in LN e^x 52 ENTER the calculator displays 52 (that is, $g(f(52)) = 52$). Similarly, keying in e^x LN 167 ENTER produces 167 (that is $f(g(167)) = 167$).

Example 9 Solve the equation $\ln(x + 1) = 2$.

Solution Since $\ln(x + 1) = 2$, we have:

$$e^{\ln(x+1)} = e^2.$$

Applying Property 4 with $v = x + 1$ shows that

$$x + 1 = e^{\ln(x+1)} = e^2$$

$$x = e^2 - 1 \approx 6.3891. \quad ■$$

Logarithm Laws

The first law of exponents states that $b^m b^n = b^{m+n}$, or in words,

The exponent of a product is the sum of the exponents of the factors.

Since logarithms are just particular kinds of exponents, this statement translates as:

The logarithm of a product is the sum of the logarithms of the factors.

Here is the same statement in symbolic language:

Product Law for Logarithms

For all $v, w > 0$,

$$\ln(vw) = \ln v + \ln w$$

and

$$\log(vw) = \log v + \log w.$$

Proof According to Property 4 of logarithms,

$$e^{\ln v} = v \quad \text{and} \quad e^{\ln w} = w.$$

Therefore, by the first law of exponents (with $m = \ln v$ and $n = \ln w$):

$$vw = e^{\ln v} e^{\ln w} = e^{\ln v + \ln w}.$$

So raising e to the exponent $(\ln v + \ln w)$ produces vw. But the definition of logarithm says that $\ln vw$ is the exponent to which e must be raised to produce vw. Therefore, we must have $\ln vw = \ln v + \ln w$. A similar argument works for common logarithms (just replace e by 10 and "ln" by "log"). ■

Example 10 A calculator shows that $\ln 7 = 1.9459$ and $\ln 9 = 2.1972$. Therefore,

$$\ln 63 = \ln(7 \cdot 9) = \ln 7 + \ln 9 = 1.9459 + 2.1972 = 4.1341.$$

Similarly, you can readily verify that

$$\log 33 = \log(3 \cdot 11) = \log 3 + \log 11 = .4771 + 1.0414 = 1.5185. \quad ■$$

CAUTION

A common error in applying the Product Law for Logarithms is to write the *false* statement

$$\ln 7 + \ln 9 = \ln (7 + 9) = \ln 16$$

instead of the correct statement

$$\ln 7 + \ln 9 = \ln (7 \cdot 9) = \ln 63.$$

GRAPHING EXPLORATION

Illustrate the caution in the margin graphically by graphing both

$$f(x) = \ln x + \ln 9 \qquad \text{and} \qquad g(x) = \ln (x + 9)$$

in the standard viewing window and verifying that the graphs are not the same. In particular, the functions have different values at $x = 7$.

The second law of exponents, namely, $b^m/b^n = b^{m-n}$, may be roughly stated in words as

The exponent of the quotient is the difference of exponents.

When the exponents are logarithms, this says

The logarithm of a quotient is the difference of the logarithms.

In other words,

Quotient Law for Logarithms

For all $v, w > 0$,

$$\ln\left(\frac{v}{w}\right) = \ln v - \ln w$$

and

$$\log\left(\frac{v}{w}\right) = \log v - \log w.$$

The proof of the Quotient Law is very similar to the proof of the Product Law (see Exercise 92).

Example 11

$$\ln\left(\frac{17}{44}\right) = \ln 17 - \ln 44 \qquad \text{and} \qquad \log\left(\frac{297}{39}\right) = \log 297 - \log 39. \quad \blacksquare$$

Example 12 For any $w > 0$,

$$\ln\left(\frac{1}{w}\right) = \ln 1 - \ln w = 0 - \ln w = -\ln w$$

and

$$\log\left(\frac{1}{w}\right) = \log 1 - \log w = 0 - \log w = -\log w. \quad \blacksquare$$

CAUTION

Do not confuse $\ln\left(\frac{v}{w}\right)$ with the quotient $\frac{\ln v}{\ln w}$. They are *different* numbers. For example,

$$\ln\left(\frac{36}{3}\right) = \ln(12) = 2.4849, \quad \text{but} \quad \frac{\ln 36}{\ln 3} = \frac{3.5835}{1.0986} = 3.2619.$$

GRAPHING EXPLORATION

Illustrate the preceding Caution graphically by graphing both $f(x) = \ln(x/3)$ and $g(x) = (\ln x)/(\ln 3)$ and verifying that the graphs are not the same at $x = 36$ (or anywhere else, for that matter).

The third law of exponents, namely $(b^m)^k = b^{mk}$, can also be translated into logarithmic language:

Power Law for Logarithms

For all k and all $v > 0$,

$$\ln(v^k) = k(\ln v)$$

and

$$\log(v^k) = k(\log v).$$

Proof Since $v = 10^{\log v}$ (why?), the third law of exponents (with $b = 10$ and $m = \log v$) shows that

$$v^k = (10^{\log v})^k = 10^{(\log v)k} = 10^{k(\log v)}.$$

So raising 10 to the exponent $k(\log v)$ produces v^k. But the exponent to which 10 must be raised to produce v^k is, by definition, $\log(v^k)$. Therefore $\log(v^k) = k(\log v)$, and the proof is complete. A similar argument with e in place of 10 and "ln" in place of "log" works for natural logarithms. ■

Example 13 $\ln\sqrt{19} = \ln 19^{1/2} = \frac{1}{2}(\ln 19)$. ■

The logarithm laws can be used to simplify various expressions.

Example 14

$$\ln 3x + 4\cdot\ln x - \ln 3xy = \ln 3x + \ln x^4 - \ln 3xy \quad \text{[Power Law]}$$

$$= \ln(3x\cdot x^4) - \ln 3xy \quad \text{[Product Law]}$$

$$= \ln \frac{3x^5}{3xy} \qquad \text{[Quotient Law]}$$

$$= \ln \frac{x^4}{y} \qquad \text{[Cancel } 3x\text{]} \quad ■$$

Example 15 To simplify $\ln(\sqrt{x}/x) + \ln \sqrt[4]{ex^2}$, we begin by changing to exponential notation:

$$\ln\left(\frac{x^{1/2}}{x}\right) + \ln(ex^2)^{1/4} = \ln(x^{-1/2}) + \ln(ex^2)^{1/4}$$

$$= -\frac{1}{2}\cdot\ln x + \frac{1}{4}\cdot\ln ex^2 \qquad \text{[Power Law]}$$

$$= -\frac{1}{2}\cdot\ln x + \frac{1}{4}(\ln e + \ln x^2) \qquad \text{[Product Law]}$$

$$= -\frac{1}{2}\cdot\ln x + \frac{1}{4}(\ln e + 2\cdot\ln x) \qquad \text{[Power Law]}$$

$$= -\frac{1}{2}\cdot\ln x + \frac{1}{4}\cdot\ln e + \frac{1}{2}\cdot\ln x$$

$$= \frac{1}{4}\cdot\ln e = \frac{1}{4} \qquad [\ln e = 1] \quad ■$$

Applications

Because logarithmic growth is slow, measurements on a logarithmic scale (that is, on a scale determined by a logarithmic function) can sometimes be deceptive.

Example 16 **Earthquakes**

The magnitude $R(i)$ of an earthquake on the Richter scale is given by $R(i) = \log(i/i_0)$, where i is the amplitude of the ground motion of the earthquake and i_0 is the amplitude of the ground motion of the so-called zero earthquake.* A moderate earthquake might have 1000 times the ground motion of the zero earthquake (that is, $i = 1000i_0$). So its magnitude would be

$$\log(1000i_0/i_0) = \log 1000 = \log 10^3 = 3.$$

An earthquake with 10 times this ground motion (that is, $i = 10\cdot1000i_0 = 10{,}000i_0$) would have a magnitude of

$$\log(10{,}000i_0/i_0) = \log 10{,}000 = \log 10^4 = 4.$$

*The zero earthquake has ground motion amplitude of less than 1 micron on a standard seismograph 100 kilometers from the epicenter.

So a *tenfold* increase in ground motion produces only a 1-point change on the Richter scale. In general,

Increasing the ground motion by a factor of 10^k increases the Richter magnitude by k units.*

For instance, the 1989 World Series earthquake in San Francisco measured 7.0 on the Richter scale, and the great earthquake of 1906 measured 8.3. The difference of 1.3 points means that the 1906 quake was $10^{1.3} \approx 20$ times more intense than the 1989 one in terms of ground motion. ■

Exponential Functions

If a is any positive number, then $e^{\ln a} = a$ by the fourth property of natural logarithms. Hence, a^x may be written as $a^x = (e^{\ln a})^x = e^{(\ln a)x}$. For example, $2^x = e^{(\ln 2)x} \approx e^{.6931x}$. Thus, we have this useful result.

Exponential Functions

Every exponential growth or decay function can be written in the form

$$f(x) = Pe^{kx},$$

where $f(x)$ is the amount at time x, P is the initial quantity, and k is positive for growth and negative for decay.

Exercises 5.3

Unless stated otherwise, all letters represent positive numbers.

In Exercises 1–4, find the logarithm, without using a calculator.

1. $\log 10{,}000$ **2.** $\log .001$

3. $\log \dfrac{\sqrt{10}}{1000}$ **4.** $\log \sqrt[3]{.01}$

In Exercises 5–14, translate the given logarithmic statement into an equivalent exponential statement.

5. $\log 1000 = 3$ **6.** $\log .001 = -3$

7. $\log 750 = 2.88$ **8.** $\log .8 = -.097$

9. $\ln 3 = 1.0986$ **10.** $\ln 10 = 2.3026$

11. $\ln .01 = -4.6052$ **12.** $\ln s = r$

13. $\ln (x^2 + 2y) = z + w$ **14.** $\log (a + c) = d$

In Exercises 15–24, translate the given exponential statement into an equivalent logarithmic one.

15. $10^{-2} = .01$ **16.** $10^3 = 1000$

17. $10^{.4771} = 3$ **18.** $10^{7k} = r$

19. $e^{3.25} = 25.79$ **20.** $e^{-4} = .0183$

21. $e^{12/7} = 5.5527$ **22.** $e^k = t$

23. $e^{2/r} = w$ **24.** $e^{4uv} = m$

In Exercises 25–36, evaluate the given expression without using a calculator.

25. $\log 10^{\sqrt{43}}$ **26.** $\log 10^{\sqrt{x^2+y^2}}$ **27.** $\ln e^{15}$

**Proof:* If one quake has ground motion amplitude i and the other $10^k i$, then

$$R(10^k i) = \log 10^k i/i_0 = \log 10^k + \log (i/i_0) = k + \log (i/i_0) = k + R(i).$$

28. $\ln e^{3.78}$ **29.** $\ln \sqrt{e}$ **30.** $\ln \sqrt[5]{e}$

31. $e^{\ln 931}$ **32.** $e^{\ln 34.17}$ **33.** $\ln e^{x+y}$

34. $\ln e^{x^2+2y}$ **35.** $e^{\ln x^2}$ **36.** $e^{\ln \sqrt{x+3}}$

In Exercises 37–46, write the given expression as a single logarithm, as in Example 14.

37. $\ln x^2 + 3 \ln y$

38. $\ln 2x + 2(\ln x) - \ln 3y$

39. $\log (x^2 - 9) - \log (x + 3)$

40. $\log 3x - 2[\log x - \log (2 + y)]$

41. $2(\ln x) - 3(\ln x^2 + \ln x)$

42. $\ln \dfrac{e}{\sqrt{x}} - \ln \sqrt{ex}$

43. $3 \ln (e^2 - e) - 3$ **44.** $2 - 2 \log (20)$

45. $\log (10x) + \log (20y) - 1$

46. $\ln (e^2x) + \ln (ey) - 3$

In Exercises 47–52, let $u = \ln x$ and $v = \ln y$. Write the given expression in terms of u and v. For example, $\ln x^3y = \ln x^3 + \ln y = 3(\ln x) + \ln y = 3u + v$

47. $\ln (x^2y^5)$ **48.** $\ln (x^3y^2)$ **49.** $\ln \left(\sqrt{x} \cdot y^2\right)$

50. $\ln \left(\sqrt{x/y}\right)$ **51.** $\ln \left(\sqrt[3]{x^2\sqrt{y}}\right)$ **52.** $\ln \left(\sqrt{x^2y} / \sqrt[3]{y}\right)$

In Exercises 53–56, find the domain of the given function (that is, the largest set of real numbers for which the rule produces well-defined real numbers).

53. $f(x) = \ln (x + 1)$ **54.** $g(x) = \ln (x + 2)$

55. $h(x) = \log (-x)$ **56.** $k(x) = \log (2 - x)$

57. **(a)** Graph $y = x$ and $y = e^{\ln x}$ in separate viewing windows (or use a split-screen, if your calculator has that feature). For what values of x are the graphs identical?

(b) Use the properties of logarithms to explain your answer in part (a).

58. **(a)** Graph $y = x$ and $y = \ln (e^x)$ in separate viewing windows (or a split-screen, if your calculator has that feature). For what values of x are the graphs identical?

(b) Use the properties of logarithms to explain your answer in part (a).

In Exercises 59–64, use graphical or algebraic means to determine whether the statement is true or false.

59. $\ln |x| = |\ln x|$? **60.** $\ln \left(\dfrac{1}{x}\right) = \dfrac{1}{\ln x}$?

61. $\log x^5 = 5(\log x)$? **62.** $e^{x \ln x} = x^x \quad (x > 0)$?

63. $\ln x^3 = (\ln x)^3$? **64.** $\log \sqrt{x} = \sqrt{\log x}$?

In Exercises 65–70, list the transformations that will change the graph of $g(x) = \ln x$ into the graph of the given function. [Section 3.4 may be helpful.]

65. $f(x) = 2 \cdot \ln x$ **66.** $f(x) = \ln x - 7$

67. $h(x) = \ln (x - 4)$ **68.** $k(x) = \ln (x + 2)$

69. $h(x) = \ln (x + 3) - 4$ **70.** $k(x) = \ln (x - 2) + 2$

In Exercises 71–74, sketch the graph of the function.

71. $f(x) = \log (x - 3)$ **72.** $g(x) = 2 \ln x + 3$

73. $h(x) = -2 \log x$ **74.** $f(x) = \ln (-x) - 3$

In Exercises 75–80, find a viewing window (or windows) that shows a complete graph of the function.

75. $f(x) = \dfrac{x}{\ln x}$ **76.** $g(x) = \dfrac{\ln x}{x}$

77. $h(x) = \dfrac{\ln x^2}{x}$ **78.** $k(x) = e^{2/\ln x}$

79. $f(x) = 10 \log x - x$ **80.** $f(x) = \dfrac{\log x}{x}$

In Exercises 81–84, find the average rate of change of the function.

81. $f(x) = \ln (x - 2)$, as x goes from 3 to 5.

82. $g(x) = x - \ln x$, as x goes from .5 to 1.

83. $g(x) = \log (x^2 + x + 1)$, as x goes from -5 to -3.

84. $f(x) = x \log|x|$, as x goes from 1 to 4.

85. **(a)** What is the average rate of change of $f(x) = \ln x$, as x goes from 3 to $3 + h$?

(b) What is the value of h when the average rate of $f(x) = \ln x$, as x goes from 3 to $3 + h$, is .25?

86. **(a)** Find the average rate of change of $f(x) = \ln x^2$, as x goes from .5 to 2.

(b) Find the average rate of change of $g(x) = \ln (x - 3)^2$, as x goes from 3.5 to 5.

(c) What is the relationship between your answers in parts (a) and (b) and why is this so?

87. The concentration of hydrogen ions in a given solution is denoted $[H^+]$ and is measured in moles per liter. For example, $[H^+] = .00008$ for beer and $[H^+] = .0004$ for wine. Chemists define the pH of the solution to be the number $\text{pH} = -\log [H^+]$. The solution is said to be an *acid* if $\text{pH} < 7$ and a *base* if $\text{pH} > 7$.

(a) Is beer an acid or a base? What about wine?

(b) If a solution has a pH of 2, what is its $[H^+]$?
(c) For hominy, $[H^+] = 5 \cdot 10^{-8}$. Is hominy a base?

88. The doubling function $D(x) = \dfrac{\ln 2}{\ln (1 + x)}$ gives the years required to double your money when it is invested at interest rate x (expressed as a decimal), compounded annually.
(a) Find the time it takes to double your money at each of these interest rates: 4%, 6%, 8%, 12%, 18%, 24%, 36%.
(b) Round the answers in part (a) to the nearest year and compare them with these numbers: 72/4, 72/6, 72/8, 72/12, 72/18, 72/24, 72/36. Use this evidence to state a "rule of thumb" for determining approximate doubling time, without using the function D. This rule of thumb, which has long been used by bankers, is called the **rule of 72.**

89. Suppose $f(x) = A \ln x + B$, where A and B are constants. If $f(1) = 10$ and $f(e) = 1$, what are A and B?

90. If $f(x) = A \ln x + B$ and $f(e) = 5$ and $f(e^2) = 8$, what are A and B?

91. Show that $g(x) = \ln\left(\dfrac{x}{1 - x}\right)$ is the inverse function of $f(x) = \dfrac{1}{1 + e^{-x}}$. (See Section 3.7.)

92. Prove the Quotient Law for Logarithms: for v, $w > 0$, $\ln\left(\dfrac{v}{w}\right) = \ln v - \ln w$. (Use properties of exponents and the fact that $v = e^{\ln v}$ and $w = e^{\ln w}$.)

In Exercises 93–96, state the magnitude on the Richter scale of an earthquake that satisfies the given condition.

93. 100 times stronger than the zero quake.

94. $10^{4.7}$ times stronger than the zero quake.

95. 350 times stronger than the zero quake.

96. 2500 times stronger than the zero quake.

Exercises 97–100 deal with the energy intensity i of a sound, which is related to the loudness of the sound by the function $L(i) = 10 \cdot \log(i/i_0)$, where i_0 is the minimum intensity detectable by the human ear and L(i) is measured in decibels. Find the decibel measure of the sound.

97. Ticking watch (intensity is 100 times i_0).

98. Soft music (intensity is 10,000 times i_0).

99. Loud conversation (intensity is 4 million times i_0).

100. Victoria Falls in Africa (intensity is 10 billion times i_0).

101. The height h above sea level (in meters) is related to air temperature t (in degrees Celsius), the atmospheric pressure p (in centimeters of mercury at height h), and the atmosphere pressure c at sea level by

$$h = (30t + 8000) \ln (c/p).$$

If the pressure at the top of Mount Rainier is 44 centimeters on a day when sea level pressure is 75.126 centimeters and the temperature is 7°, what is the height of Mount Rainier?

102. Mount Everest is 8850 meters high. What is the atmospheric pressure at the top of the mountain on a day when the temperature is −25° and the atmospheric pressure at sea level is 75 centimeters? [See Exercise 101.]

103. A class in elementary Sanskrit is tested at the end of the semester and weekly thereafter on the same material. The average score on the exam taken after t weeks is given by the "forgetting function"

$$g(t) = 77 - 10 \cdot \ln (t + 1).$$

(a) What was the average score on the original exam?
(b) What was the average score after 2 weeks? After 5 weeks?
(c) When did the average score drop below 50?

104. Students in a precalculus class were given a final exam. Each month thereafter, they took an equivalent exam. The class average on the exam taken after t months is given by $F(t) = 82 - 8 \cdot \ln (t + 1)$.
(a) What was the class average after 6 months?
(b) After a year?
(c) When did the class average drop below 55?

105. One person with a flu virus visited the campus. The number T of days it took for the virus to infect x people was given by:

$$T = -.93 \ln \left[\frac{7000 - x}{6999x}\right].$$

(a) How many days did it take for 6000 people to become infected?
(b) After two weeks, how many people were infected?

106. *Approximating logarithmic functions by polynomials.* For each positive integer n, let f_n be the polynomial function whose rule is

$$f_n(x) = x - \frac{x^2}{2} + \frac{x^3}{3} - \frac{x^4}{4} + \frac{x^5}{5} - \cdots \pm \frac{x^n}{n}$$

where the sign of the last term is + if n is odd and − if n is even. In the viewing window with $-1 \le x \le 1$ and $-4 \le y \le 1$, graph $g(x) = \ln(1 + x)$ and $f_4(x)$ on the same screen. For what values of x does f_4 appear to be a good approximation of g?

107. Using the viewing window in Exercise 106, find a value of n for which the graph of the function f_n (as defined in Exercise 106) appears to coincide with the graph of $g(x) = \ln(1 + x)$. Use the trace feature to move from graph to graph to see how good this approximation actually is.

108. The perceived loudness L of a sound of intensity I is given by $L = k \cdot \ln I$, where k is a certain constant. By how much must the intensity be increased to double the loudness? (That is, what must be done to I to produce $2L$?)

109. A bicycle store finds that the number N of bikes sold is related to the number d of dollars spent on advertising by $N = 51 + 100 \cdot \ln(d/100 + 2)$.

(a) How many bikes will be sold if nothing is spent on advertising? If \$1000 is spent? If \$10,000 is spent?

(b) If the average profit is \$25 per bike, is it worthwhile to spend \$1000 on advertising? What about \$10,000?

(c) What are the answers in part (b) if the average profit per bike is \$35?

110. The number N of days of training needed for a factory worker to produce x tools per day is given by

$$N = -25 \cdot \ln\left(1 - \frac{x}{60}\right).$$

(a) How many training days are needed for the worker to be able to produce 40 tools a day?

(b) It costs \$135 to train 1 worker for 1 day. If the profit on 1 tool is \$1.85, how many work days does it take before the factory breaks even on the training costs for a worker who can produce 40 tools a day?

5.3.A EXCURSION Logarithmic Functions to Other Bases*

Common and natural logarithms were defined by considering the inverse functions of the exponential functions $f(x) = 10^x$ and $f(x) = e^x$. We now show that a similar procedure can be carried out with any positive number b in place of 10 and e.

Throughout this excursion, b is a fixed positive number with $b > 1$.†

The exponential function $f(x) = b^x$, whose graph is shown in Figure 5–19, is an increasing function and hence is one-to-one (as explained on page 188). Therefore, f has an inverse function g whose graph is the reflection of the graph of f in the line $y = x$ (see page 192), as shown in Figure 5–20.

*This material is not needed in the sequel and may be read before Section 5.3 if desired. Section 3.7 (Inverse Functions) is a prerequisite for this section, which replicates the discussion of Section 5.3 in a more general context.

†The discussion is also valid when $0 < b < 1$, but in that case the graphs have a different shape.

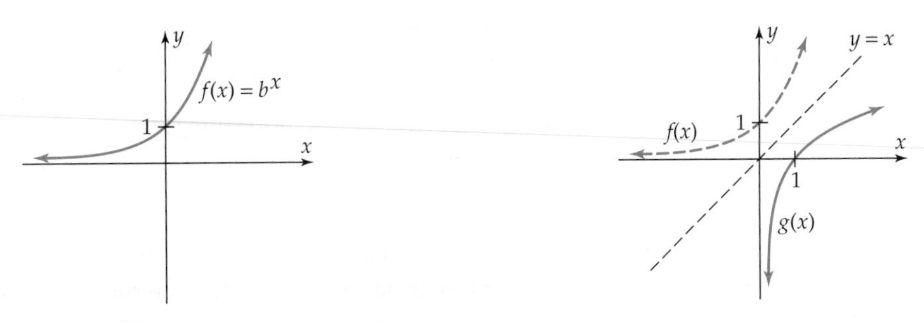

Figure 5–19 **Figure 5–20**

This inverse function g is called the **logarithmic function to the base b.** The value $g(x)$ of this function at a number x is denoted $\log_b x$ and called the **logarithm to the base b** of the number x.

Although many properties of the logarithmic function $g(x) = \log_b x$ can be read from its graph, we also need an algebraic description of this function. On page 188 we saw that the relationship between a function f and its inverse function g is given by:

$$g(v) = u \qquad \text{exactly when} \qquad f(u) = v.$$

When $f(x) = b^x$ and $g(x) = \log_b x$, this statement says

$$\log_b v = u \qquad \textbf{exactly when} \qquad b^u = v.$$

In other words, the number $\log_b v$ is the answer to the question

To what exponent must b be raised to produce v?

Example 1 To find $\log_2 16$, ask yourself "what power of 2 equals 16?" Since $2^4 = 16$, we see that $\log_2 16 = 4$. Similarly, $\log_2 (1/8) = -3$ because $2^{-3} = 1/8$. ■

Example 2 Since logarithms are just exponents, every logarithmic statement can be translated into exponential language:

Logarithmic Statement	Equivalent Exponential Statement
$\log_3 81 = 4$	$3^4 = 81$
$\log_4 64 = 3$	$4^3 = 64$
$\log_{125} 5 = \dfrac{1}{3}$	$125^{1/3} = 5$*
$\log_8 \left(\dfrac{1}{4}\right) = -\dfrac{2}{3}$	$8^{-2/3} = \dfrac{1}{4}$ (verify!) ■

*Because $125^{1/3} = \sqrt[3]{125} = 5$.

Example 3 Solve $\log_5 x = 3$.

Solution The equation $\log_5 x = 3$ is equivalent to the exponential statement $5^3 = x$, so the solution is $x = 125$. ■

Example 4 Logarithms to the base 10 are called **common logarithms.** It is customary to write $\log v$ instead of $\log_{10} v$. Then,

$$\log 100 = 2 \qquad \text{because} \qquad 10^2 = 100;$$

$$\log .001 = -3 \qquad \text{because} \qquad 10^{-3} = \frac{1}{10^3} = \frac{1}{1000} = .001.$$

Calculators have a LOG key for evaluating common logarithms. For instance,*

$$\log .4 = -0.3979, \qquad \log 45.3 = 1.6561, \qquad \log 685 = 2.8357.$$ ■

Example 5 The most frequently used base for logarithms in modern applications is the number e ($\approx 2.71828\cdots$). Logarithms to the base e are called **natural logarithms** and use a different notation: We write $\ln v$ instead of $\log_e v$. Scientific calculators also have an LN key for evaluating natural logarithms. For example,

$$\ln .5 = -0.6931, \qquad \ln 65 = 4.1744, \qquad \ln 158 = 5.0626.$$ ■

You *don't* need a calculator to understand the essential properties of logarithms. You need only translate logarithmic statements into exponential ones (or vice versa).

Example 6 What is $\log(-25)$?

Translation: To what power must 10 be raised to produce -25? The graph of $f(x) = 10^x$ lies entirely above the x-axis (see Figure 5–15 on page 340 or use your calculator), which means that *every* power of 10 is *positive.* So 10^x can *never* be -25, or any negative number, or zero. The same argument works for any base b:

$\log_b v$ is defined only when $v > 0$. ■

*Here and below, all logarithms are rounded to four decimal places. So strictly speaking, the equal sign should be replaced by an "approximately equal" sign ($\approx$).

Example 7 What is $\log_5 1$?

Translation: To what power must 5 be raised to produce 1? The answer, of course, is $5^0 = 1$. So $\log_5 1 = 0$. Similarly, $\log_5 5 = 1$ because 1 is the answer to "what power of 5 equals 5?" In general,

$$\log_b 1 = 0 \qquad \text{and} \qquad \log_b b = 1. \quad \blacksquare$$

Example 8 What is $\log_2 2^9$?

Translation: To what power must 2 be raised to produce 2^9? Obviously, the answer is 9. So $\log_2 2^9 = 9$ and, in general,

$$\log_b b^k = k \qquad \text{for every real number } k.$$

This property holds even when k is a complicated expression. For instance, if x and y are positive, then

$$\log_6 6^{\sqrt{3x+y}} = \sqrt{3x+y} \quad \left(\text{here } k = \sqrt{3x+y}\right). \quad \blacksquare$$

Example 9 What is $10^{\log 439}$? Well, $\log 439$ is the power to which 10 must be raised to produce 439, that is $10^{\log 439} = 439$. Similarly,

$$b^{\log_b v} = v \qquad \text{for every } v > 0. \quad \blacksquare$$

Here is a summary of the facts illustrated in the preceding examples.

Properties of Logarithms

1. **$\log_b v$ is defined only when $v > 0$.**
2. **$\log_b 1 = 0$ and $\log_b b = 1$.**
3. **$\log_b (b^k) = k$ for every real number k.**
4. **$b^{\log_b v} = v$ for every $v > 0$.**

Property 1 is simply a restatement of the fact that the domain of the logarithmic function $g(x) = \log_b x$ is the set of all positive real numbers, as can be seen from the fact that its graph (Figure 5–19 on page 353) lies entirely to the right of the y-axis. Properties 3 and 4 are a restatement of the fact that the exponential function $f(x) = b^x$ and its inverse function $g(x) = \log_b x$ have the usual "round-trip properties" (see page 191):

$$(g \circ f)(x) = g(f(x)) = g(b^x) = \log_b b^x = x \qquad \text{for all } x;$$

$$(f \circ g)(x) = f(g(x)) = f(\log_b x) = b^{\log_b x} = x \qquad \text{for all } x > 0.\text{*}$$

*A calculator provides a visual demonstration of these facts for base 10 logarithms. For instance, if you key in LOG 10^x 52 ENTER the calculator displays 52 (that is, $g(f(52)) = 52$). Similarly, keying in 10^x LOG 167 ENTER produces 167 (that is, $f(g(167)) = 167$).

Logarithm Laws

The first law of exponents states that $b^m b^n = b^{m+n}$, or in words,

The exponent of a product is the sum of the exponents of the factors.

Since logarithms are just particular kinds of exponents, this statement translates as:

The logarithm of a product is the sum of the logarithms of the factors.

The second and third laws of exponents, namely, $b^m/b^n = b^{m-n}$ and $(b^m)^k = b^{mk}$, can also be translated into logarithmic language:

Logarithm Laws

Let b, v, w, k be real numbers, with b, v, and w positive and $b \neq 1$.

Product Law: $\log_b (vw) = \log_b v + \log_b w.$

Quotient Law: $\log_b\left(\frac{v}{w}\right) = \log_b v - \log_b w.$

Power Law: $\log_b (v^k) = k(\log_b v).$

Proof According to Property 4 in the box on page 355,

$$b^{\log_b v} = v \qquad \text{and} \qquad b^{\log_b w} = w.$$

Therefore, by the second law of exponents (with $m = \log_b v$ and $n = \log_b w$) we have:

$$\frac{v}{w} = \frac{b^{\log_b v}}{b^{\log_b w}} = b^{\log_b v - \log_b w}$$

Since $\log_b (v/w)$ is the exponent to which b must be raised to produce v/w, we must have $\log_b (v/w) = \log_b v - \log_b w$. This proves the Quotient Law. The Product and Power Laws are proved in a similar fashion. ■

Example 10 Given that $\log_7 2 = .3562$, $\log_7 3 = .5646$, and $\log_7 5 = .8271$, find: (a) $\log_7 10$; (b) $\log_7 2.5$; (c) $\log_7 48$.

Solution

(a) By the Product Law,

$$\log_7 10 = \log_7(2 \cdot 5) = \log_7 2 + \log_7 5 = .3562 + .8271 = 1.1833.$$

(b) By the Quotient Law,

$$\log_7 2.5 = \log_7\left(\frac{5}{2}\right) = \log_7 5 - \log_7 2 = .8271 - .3562 = .4709.$$

(c) By the Product and Power Laws,

$$\log_7 48 = \log_7 (3 \cdot 16) = \log_7 3 + \log_7 16 = \log_7 3 + \log_7 2^4$$
$$= \log_7 3 + 4 \cdot \log_7 2 = .5646 + 4(.3562)$$
$$= 1.9894. \quad \blacksquare$$

Example 11 Simplify and write as a single logarithm:

(a) $\log_3 (x + 2) + \log_3 y - \log_3 (x^2 - 4)$

(b) $3 - \log_5 (125z)$

Solution

(a) $\log_3 (x + 2) + \log_3 y - \log_3 (x^2 - 4)$

$$= \log_3[(x + 2)y] - \log_3 (x^2 - 4) \qquad \textit{[Product Law]}$$

$$= \log_3\left[\frac{(x + 2)y}{x^2 - 4}\right] \qquad \textit{[Quotient Law]}$$

$$= \log_3\left[\frac{(x + 2)y}{(x + 2)(x - 2)}\right] \qquad \textit{[Factor denominator]}$$

$$= \log_3\left(\frac{y}{x - 2}\right) \qquad \textit{[Cancel common factor]}$$

(b) $3 - \log_5 (125x)$

$$= 3 - (\log_5 125 + \log_5 x) \qquad \textit{[Product Law]}$$

$$= 3 - \log_5 125 - \log_5 x$$

$$= 3 - 3 - \log_5 x \qquad [\log_5 125 - 3 \textit{ because } 5^3 - 125]$$

$$= -\log_5 x$$

$$= \log_5 x^{-1} = \log_5\left(\frac{1}{x}\right) \qquad \textit{[Power Law]} \quad \blacksquare$$

> **CAUTION**
>
> 1. A common error in using the Product Law is to write something like $\log 6 + \log 7 = \log (6 + 7) = \log 13$ instead of the correct statement $\log 6 + \log 7 = \log (6 \cdot 7) = \log 42$.
>
> 2. Do not confuse $\log_b\left(\frac{v}{w}\right)$ with the quotient $\frac{\log_b v}{\log_b w}$. They are *different* numbers. For example, when $b = 10$
>
> $$\log\left(\frac{48}{4}\right) = \log 12 = 1.0792, \text{ but } \frac{\log 48}{\log 4} = \frac{1.6812}{0.6021} = 2.7922.$$

For graphic illustrations of the errors mentioned in the Caution, see Exercises 82 and 83.

Example 10 worked because we were *given* several logarithms to base 7. But there's no $\log_7$ key on the calculator, so how do you find logarithms to base 7, or to any base other than e or 10? *Answer:* Use the LN key on the calculator and the following formula:

Change of Base Formula

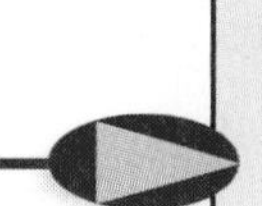

For any positive number v,

$$\log_b v = \frac{\ln v}{\ln b} = \frac{1}{\ln b} \cdot \ln v.$$

Proof By Property 4 in the box on page 355, $b^{\log_b v} = v$. Take the natural logarithm of each side of this equation:

$$\ln (b^{\log_b v}) = \ln v.$$

Apply the Power Law for natural logarithms on the left side:

$$(\log_b v)(\ln b) = \ln v.$$

Dividing both sides by $\ln b$ finishes the proof:

$$\log_b v = \frac{\ln v}{\ln b}. \quad \blacksquare$$

Example 12 To find $\log_7 3$, apply the change of base formula with $b = 7$:

$$\log_7 3 = \frac{\ln 3}{\ln 7} = \frac{1.0986}{1.9459} = .5646. \quad \blacksquare$$

Exercises 5.3.A

Note: *Unless stated otherwise, all letters represent positive numbers and $b \neq 1$.*

In Exercises 1–8, fill in the missing entries in each table.

1.

x	0	1	2	4
$f(x) = \log_4 x$				

2.

x	1/25	5	25	$\sqrt{5}$
$g(x) = \log_5 x$				

3.

x		1/6	1	216
$h(x) = \log_6 x$	−2			

4.

x	10/3	4	6	12
$k(x) = \log_3 (x - 3)$				

5.

x	0	1/7	$\sqrt{7}$	49
$f(x) = 2 \log_7 x$				

6.

x			100	1000
$g(x) = 3 \log x$	6	3		

7.

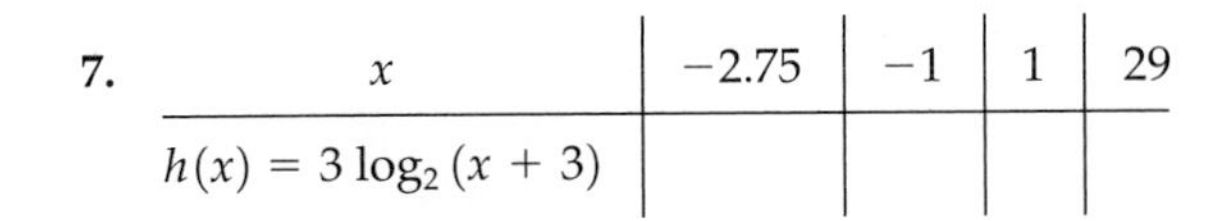

x	-2.75	-1	1	29
$h(x) = 3 \log_2 (x + 3)$				

8.

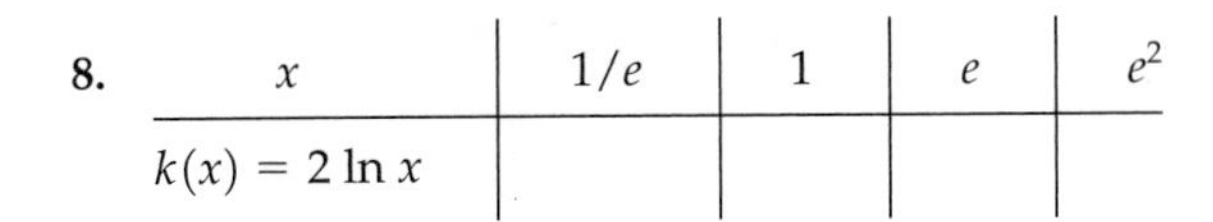

x	$1/e$	1	e	e^2
$k(x) = 2 \ln x$				

In Exercises 9–18, translate the given exponential statement into an equivalent logarithmic one.

9. $10^{-2} = .01$

10. $10^3 = 1000$

11. $\sqrt[3]{10} = 10^{1/3}$

12. $10^{.4771} \approx 3$

13. $10^{7k} = r$

14. $10^{(a+b)} = c$

15. $7^8 = 5{,}764{,}801$

16. $2^{-3} = 1/8$

17. $3^{-2} = 1/9$

18. $b^{14} = 3379$

In Exercises 19–28, translate the given logarithmic statement into an equivalent exponential one.

19. $\log 10{,}000 = 4$

20. $\log .001 = -3$

21. $\log 750 \approx 2.88$

22. $\log (.8) = -.097$

23. $\log_5 125 = 3$

24. $\log_8 (1/4) = -2/3$

25. $\log_2 (1/4) = -2$

26. $\log_2 \sqrt{2} = 1/2$

27. $\log (x^2 + 2y) = z + w$

28. $\log (a + c) = d$

In Exercises 29–36, evaluate the given expression without using a calculator.

29. $\log 10^{\sqrt{43}}$

30. $\log_{17} (17^{17})$

31. $\log 10^{\sqrt{x^2+y^2}}$

32. $\log_{3.5} (3.5^{(x^2-1)})$

33. $\log_{16} 4$

34. $\log_2 64$

35. $\log_{\sqrt{3}} (27)$

36. $\log_{\sqrt{3}} (1/9)$

In Exercises 37–40, a graph or a table of values for the function $f(x) = \log_b x$ is given. Find b.

37.

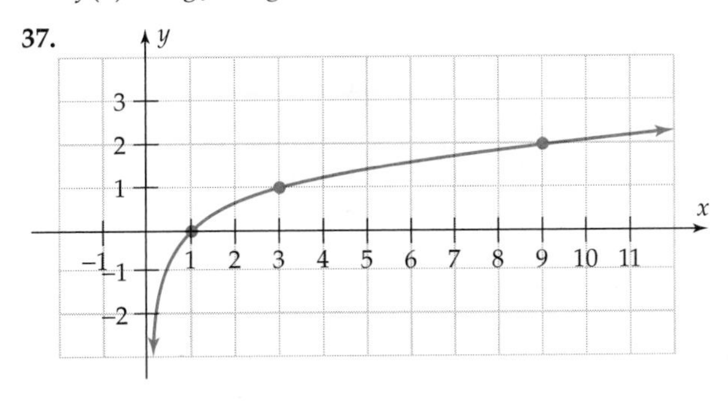

38.

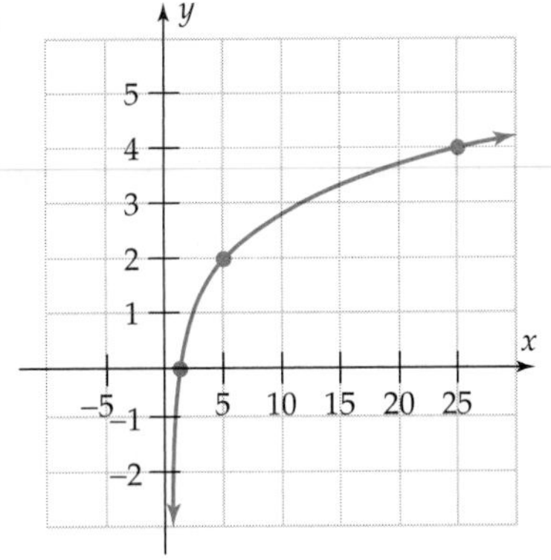

39.

x	.05	1	400	$2\sqrt{5}$
$f(x)$	-1	0	2	$1/2$

40.

x	$1/25$	1	5	125
$f(x)$	-4	0	2	6

In Exercises 41–46, find x.

41. $\log_3 243 = x$

42. $\log_{81} 27 = x$

43. $\log_{27} x = 1/3$

44. $\log_5 x = -4$

45. $\log_x 64 = 3$

46. $\log_x (1/9) = -2/3$

In Exercises 47–60, write the given expression as the logarithm of a single quantity, as in Example 11.

47. $2 \log x + 3 \log y - 6 \log z$

48. $5 \log_8 x - 3 \log_8 y + 2 \log_8 z$

49. $\log x - \log (x + 3) + \log (x^2 - 9)$

50. $\log_3 (y + 2) + \log_3 (y - 3) - \log_3 y$

51. $\frac{1}{2} \log_2 (25c^2)$

52. $\frac{1}{3} \log_2 (27b^6)$

53. $-2 \log_4 (7c)$

54. $\frac{1}{3} \log_5 (x + 1)$

55. $2 \ln (x + 1) - \ln (x + 2)$

56. $\ln (z - 3) + 2 \ln (z + 3)$

57. $\log_2 (2x) - 1$

58. $2 - \log_5 (25z)$

59. $2 \ln (e^2 - e) - 2$

60. $4 - 4 \log_5 (20)$

In Exercises 61–68, use a calculator and the change of base formula to find the logarithm.

61. $\log_2 10$
62. $\log_2 22$
63. $\log_7 5$
64. $\log_5 7$
65. $\log_{500} 1000$
66. $\log_{500} 250$
67. $\log_{12} 56$
68. $\log_{12} 725$

In Exercises 69–74, answer true *or* false *and give reasons for your answer.*

69. $\log_b (r/5) = \log_b r - \log_b 5$
70. $\dfrac{\log_b a}{\log_b c} = \log_b\left(\dfrac{a}{c}\right)$
71. $(\log_b r)/t = \log_b (r^{1/t})$
72. $\log_b (cd) = \log_b c + \log_b d$
73. $\log_5 (5x) = 5(\log_5 x)$
74. $\log_b (ab)^t = t(\log_b a) + t$

75. Which is larger: 397^{398} or 398^{397}? [*Hint:* $\log 397 \approx 2.5988$ and $\log 398 \approx 2.5999$ and $f(x) = 10^x$ is an increasing function.]
76. If $\log_b 9.21 = 7.4$ and $\log_b 359.62 = 19.61$, then what is $\log_b 359.62/\log_b 9.21$?

In Exercises 77–80, assume that a and b are positive, with $a \neq 1$ and $b \neq 1$.

77. Express $\log_b u$ in terms of logarithms to the base a.
78. Show that $\log_b a = 1/\log_a b$.
79. How are $\log_{10} u$ and $\log_{100} u$ related?
80. Show that $a^{\log b} = b^{\log a}$.

81. If $\log_b x = \dfrac{1}{2}\log_b v + 3$, show that $x = (b^3)\sqrt{v}$.
82. Graph the functions $f(x) = \log x + \log 7$ and $g(x) = \log (x + 7)$ on the same screen. For what values of x is it true that $f(x) = g(x)$? What do you conclude about the statement "$\log 6 + \log 7 = \log (6 + 7)$"?
83. Graph the functions $f(x) = \log (x/4)$ and $g(x) = (\log x)/(\log 4)$. Are they the same? What does this say about a statement such as "$\log\left(\dfrac{48}{4}\right) = \dfrac{\log 48}{\log 4}$"?

In Exercises 84–86, sketch a complete graph of the function, labeling any holes, asymptotes, or local extrema.

84. $f(x) = \log_5 x + 2$
85. $h(x) = x \log x^2$
86. $g(x) = \log_{20} x^2$

5.4 Algebraic Solutions of Exponential and Logarithmic Equations

Earlier in this chapter, many exponential and logarithmic equations were solved by graphical means. Most of them could also have been solved algebraically by the techniques presented in this section, which depend primarily on the properties of logarithms.

The easiest exponential equations to solve are those in which both sides are powers of the same base.

Example 1 The equation $8^x = 2^{x+1}$ can be rewritten $(2^3)^x = 2^{x+1}$ or, equivalently, $2^{3x} = 2^{x+1}$. Since the powers of 2 are the same, the exponents must be equal, that is,

$$3x = x + 1$$

$$x = \frac{1}{2}. \quad ■$$

When different bases are involved in an exponential equation, a different solution technique is needed.

Example 2 To solve $5^x = 2$,

*Take logarithms on each side:** $\ln 5^x = \ln 2$

Use the Power Law: $x(\ln 5) = \ln 2$

Divide both sides by ln 5: $x = \dfrac{\ln 2}{\ln 5} \approx \dfrac{.6931}{1.6094} \approx .4307.$

Remember: $\dfrac{\ln 2}{\ln 5}$ is *not* $\ln \dfrac{2}{5}$ or $\ln 2 - \ln 5$. ■

Example 3 To solve $2^{4x-1} = 3^{1-x}$,

Take logarithms of each side: $\ln 2^{4x-1} = \ln 3^{1-x}$

Use the Power Law: $(4x - 1)(\ln 2) = (1 - x)(\ln 3)$

Multiply out both sides: $4x(\ln 2) - \ln 2 = \ln 3 - x(\ln 3)$

Rearrange terms: $4x(\ln 2) + x(\ln 3) = \ln 2 + \ln 3$

Factor left side: $(4 \cdot \ln 2 + \ln 3)x = \ln 2 + \ln 3$

Divide both sides by (4·ln 2 + ln 3): $x = \dfrac{\ln 2 + \ln 3}{4 \cdot \ln 2 + \ln 3} \approx .4628.$ ■

Example 4 To solve $e^x - e^{-x} = 4$, we first multiply both sides by e^x. Since $e^x > 0$ for every x, we get an equivalent equation:

$$e^x e^x - e^{-x} e^x = 4\,e^x$$

$$(e^x)^2 - 4\,e^x - 1 = 0$$

Let $u = e^x$ so that the equation becomes

$$u^2 - 4u - 1 = 0$$

The solutions are given by the quadratic formula:

$$u = \frac{-(-4) \pm \sqrt{(-4)^2 - 4 \cdot 1 \cdot (-1)}}{2 \cdot 1} = \frac{4 \pm \sqrt{20}}{2} = \frac{4 \pm 2\sqrt{5}}{2} = 2 \pm \sqrt{5}.$$

Since $u = e^x$, we have

$$e^x = 2 - \sqrt{5} \qquad \text{or} \qquad e^x = 2 + \sqrt{5}.$$

*We shall use natural logarithms, but the same techniques are valid for logarithms to other bases (Exercise 34).

But e^x is always positive and $2 - \sqrt{5} < 0$, so the first equation has no solutions. The second can be solved as above:

$$\ln e^x = \ln\left(2 + \sqrt{5}\right)$$
$$x(\ln e) = \ln\left(2 + \sqrt{5}\right)$$
$$x(1) = \ln\left(2 + \sqrt{5}\right) \qquad [\ln e = 1]$$
$$x = \ln\left(2 + \sqrt{5}\right) \approx 1.4436. \quad \blacksquare$$

Applications of Exponential Equations

Example 5 **Radiocarbon Dating.**

When a living organism dies, its carbon-14 decays. The half-life of carbon-14 is 5730 years. As explained before Example 7 on page 327, the amount left at time t is given by $M(t) = c(.5)^{t/5730}$, where c is the mass of carbon-14 that was present initially.

The skeleton of a mastodon has lost 58% of its original carbon-14.* When did the mastodon die?

Solution Time is measured from the death of the mastodon. The present mass of carbon-14 is $.58c$ less than the original mass c. So the present value of $M(t)$ is $c - .58c = .42c$ and we have:

$$M(t) = c(.5)^{t/5730}$$
$$.42c = c(.5)^{t/5730}$$
$$.42 = (.5)^{t/5730}$$

The solution of this equation is the time elapsed from the mastodon's death to the present. It can be solved as above.

$$\ln .42 = \ln (.5)^{t/5730}$$
$$\ln .42 = \frac{t}{5730}(\ln .5)$$
$$5730(\ln .42) = t(\ln .5)$$
$$t = \frac{5730(\ln .42)}{\ln .5} \approx 7171.32$$

Therefore, the mastodon died approximately 7200 years ago. ■

*Archeologists can determine how much carbon-14 has been lost by a technique that involves measuring the ratio of carbon-14 to carbon-12 in the skeleton.

Example 6 **Compound Interest.***

\$3000 is to be invested at 8% per year, compounded quarterly. In how many years will the investment be worth \$10,680?

Solution The interest rate per quarter is $.08/4 = .02$. The compound interest formula (page 334) shows that the value of the investment after t quarters is $A = 3000(1 + .02)^t = 3000(1.02)^t$. So we must solve the equation

$$\begin{aligned} 3000(1.02)^t &= 10{,}680 \\ 1.02^t &= \frac{10{,}680}{3000} = 3.56 \\ \ln 1.02^t &= \ln 3.56 \\ t(\ln 1.02) &= \ln 3.56 \\ t &= \frac{\ln 3.56}{\ln 1.02} \approx 64.12 \text{ quarters} \end{aligned}$$

Therefore, it will take $\frac{64.12}{4} = 16.03$ years. ■

Example 7 **Population Growth.**

A biologist knows that if there are no inhibiting or stimulating factors, a certain type of bacteria will increase exponentially, with the population at time t given by a function of the form $S(t) = Pe^{kt}$, where P is the original population and k is the continuous growth rate. The biologist has a culture that contains 1000 bacteria. Seven hours later there are 5000 bacteria.

(a) Find the rule of the population function S.

(b) When will the bacteria population reach one billion?

Solution

(a) The original population is $P = 1000$, so the population function is $S(t) = 1000e^{kt}$. To determine the growth rate k we use the fact that $S(7) = 5000$, that is,

$$1000e^{k7} = 5000.$$

This equation can be solved for k as above; first, divide both sides by 1000:

$$\begin{aligned} e^{7k} &= 5 \\ \ln e^{7k} &= \ln 5 \\ 7k \cdot \ln e &= \ln 5 \\ k &= \frac{\ln 5}{7 \cdot \ln e} = \frac{\ln 5}{7 \cdot 1} \approx .2299. \end{aligned} \qquad [\ln e = 1]$$

*Skip this example if you haven't read Excursion 5.2.A.

Therefore, the population function is $S(t) = 1000e^{.2299t}$.

(b) To find the value of t for which $S(t)$ is one billion, we must solve:

$$\begin{aligned}
1000e^{.2299t} &= 1{,}000{,}000{,}000 \\
e^{.2299t} &= 1{,}000{,}000 \\
\ln e^{.2299t} &= \ln 1{,}000{,}000 \\
.2299t \cdot \ln e &= \ln 1{,}000{,}000 \\
t &= \frac{\ln 1{,}000{,}000}{.2299 \cdot \ln e} = \frac{\ln 1{,}000{,}000}{.2299} \approx 60.09 \text{ hours.} \quad \blacksquare
\end{aligned}$$

Example 8 **Inhibited Population Growth.**

The population of fish in a lake at time t months is given by the function

$$p(t) = \frac{20{,}000}{1 + 24e^{-t/4}}.$$

How long will it take for the population to reach 15,000?

Solution We must solve this equation for t:

$$\begin{aligned}
15{,}000 &= \frac{20{,}000}{1 + 24e^{-t/4}} \\
15{,}000(1 + 24e^{-t/4}) &= 20{,}000 \\
1 + 24e^{-t/4} &= \frac{20{,}000}{15{,}000} = \frac{4}{3} \\
24e^{-t/4} &= \frac{1}{3} \\
e^{-t/4} &= \frac{1}{3} \cdot \frac{1}{24} = \frac{1}{72} \\
\ln e^{-t/4} &= \ln\left(\frac{1}{72}\right) \\
\left(-\frac{t}{4}\right)(\ln e) &= \ln 1 - \ln 72 \\
-\frac{t}{4} &= -\ln 72 \qquad [\ln e = 1 \text{ and } \ln 1 = 0] \\
t &= 4(\ln 72) \approx 17.1067.
\end{aligned}$$

So the population reaches 15,000 in a little over 17 months. ■

Logarithmic Equations

Equations that involve only logarithmic terms may be solved by using this fact, which is proved in Exercise 33:

$$\text{If } \ln u = \ln v, \text{ then } u = v.$$

Example 9 To solve $\ln(x - 3) + \ln(2x + 1) = 2 \cdot \ln x$, use the Product Law on the left side and the Power Law on the right side:

$$\ln[(x - 3)(2x + 1)] = \ln x^2$$
$$\ln(2x^2 - 5x - 3) = \ln x^2$$

Since the logarithms are equal, we must have:

$$2x^2 - 5x - 3 = x^2$$
$$x^2 - 5x - 3 = 0$$
$$x = \frac{-(-5) \pm \sqrt{(-5)^2 - 4 \cdot 1 \cdot (-3)}}{2 \cdot 1} = \frac{5 \pm \sqrt{37}}{2}$$

But $x - 3$ is negative when $x = (5 - \sqrt{37})/2$, so $\ln(x - 3)$ is not defined in that case. Therefore the only solution of the original equation is $x = (5 + \sqrt{37})/2 \approx 5.5414$. ■

Equations that involve both logarithmic and constant terms may be solved by using the basic property of logarithms that was discussed on pages 344 and 355:

(*) $$b^{\log_b v} = v.$$

Example 10 To solve $\ln(x - 3) = 5 - \ln(x - 3)$, start by getting all the logarithmic terms on one side and the constant on the other.

$$2\ln(x - 3) = 5$$

Divide both sides by 2: $$\ln(x - 3) = \frac{5}{2}$$

Exponentiate both sides: $$e^{\ln(x-3)} = e^{5/2}$$

Use the basic property of logarithms (*): $$x - 3 = e^{5/2}$$

Add 3 to both sides: $$x = e^{5/2} + 3 \approx 15.1825$$

This is the only possibility for a solution. A calculator shows that it actually is a solution of the original equation. ■

Example 11 To solve $\log(x - 16) = 2 - \log(x - 1)$,

Rearrange terms: $$\log(x - 16) + \log(x - 1) = 2$$

Use the Product Law: $$\log[(x - 16)(x - 1)] = 2$$
$$\log(x^2 - 17x + 16) = 2$$

Exponentiate both sides: $$10^{\log(x^2-17x+16)} = 10^2$$

Use the basic logarithm property (*): $$x^2 - 17x + 16 = 100$$

Subtract 10 from both sides: $$x^2 - 17x - 84 = 0$$

Factor: $(x + 4)(x - 21) = 0$

$x + 4 = 0$ or $x - 21 = 0$

$x = -4$ or $x = 21$

You can easily verify that 21 is a solution of the original equation, but -4 is not (when $x = -4$, then $\log(x - 16) = \log(-20)$, which is not defined). ■

Example 12 To solve $\log(x + 5) = 1 - \log(x - 2)$,

Rearrange terms: $\log(x + 5) + \log(x - 2) = 1$

Use the Product Law: $\log[(x + 5)(x - 2)] = 1$

$\log(x^2 + 3x - 10) = 1$

Exponentiate both sides: $10^{\log(x^2+3x-10)} = 10^1$

Use the basic logarithm property (*): $x^2 + 3x - 10 = 10$

$x^2 + 3x - 20 = 0$

This equation can be solved with the quadratic formula:

$$x = \frac{-3 \pm \sqrt{3^2 - 4\cdot 1\cdot(-20)}}{2\cdot 1} = \frac{-3 \pm \sqrt{89}}{2} \approx \begin{cases} -6.217 \\ 3.217 \end{cases}$$

The original equation is not defined when $x = -6.217$ (why?), so this is not a solution. Verify that $x = (3 + \sqrt{89})/2$ is a solution. ■

Exercises 5.4

In Exercises 1–8, solve the equation without using logarithms.

1. $3^x = 81$
2. $3^x + 3 = 30$
3. $3^{x+1} = 9^{5x}$.
4. $4^{5x} = 16^{2x-1}$
5. $3^{5x}9^{x^2} = 27$
6. $2^{x^2+5x} = 1/16$
7. $9^{x^2} = 3^{-5x-2}$
8. $4^{x^2-1} = 8^x$

In Exercises 9–29, solve the equation. First express your answer in terms of natural logarithms (for instance, $x = (2 + \ln 5)/(\ln 3)$). Then use a calculator to find an approximation for the answer.

9. $3^x = 5$
10. $5^x = 4$
11. $2^x = 3^{x-1}$
12. $4^{x+2} = 2^{x-1}$
13. $3^{1-2x} = 5^{x+5}$
14. $4^{3x-1} = 3^{x-2}$
15. $2^{1-3x} = 3^{x+1}$
16. $3^{z+3} = 2^z$
17. $e^{2x} = 5$
18. $e^{-3x} = 2$
19. $6e^{-1.4x} = 21$
20. $3.4e^{-x/3} = 5.6$
21. $2.1e^{(x/2)\ln 3} = 5$
22. $7.8e^{(x/3)\ln 5} = 14$
23. $9^x - 4\cdot 3^x + 3 = 0$ [*Hint:* Note that $9^x = (3^x)^2$; let $u = 3^x$.]
24. $4^x - 6\cdot 2^x = -8$
25. $e^{2x} - 5e^x + 6 = 0$ [*Hint:* Let $u = e^x$.]
26. $2e^{2x} - 9e^x + 4 = 0$
27. $6e^{2x} - 16e^x = 6$
28. $8e^{2x} + 8e^x = 6$
29. $4^x + 6\cdot 4^{-x} = 5$

In Exercises 30–32, solve the equation for x.

30. $\dfrac{e^x + e^{-x}}{e^x - e^{-x}} = t$
31. $\dfrac{e^x - e^{-x}}{2} = t$
32. $\dfrac{e^x - e^{-x}}{e^x + e^{-x}} = t$
33. Prove that if $\ln u = \ln v$, then $u = v$. [*Hint:* Property (*) on page 365.]

34. **(a)** Solve $7^x = 3$, using natural logarithms. Leave your answer in logarithmic form; don't approximate with a calculator.
(b) Solve $7^x = 3$, using common (base 10) logarithms. Leave your answer in logarithmic form.
(c) Use the change of base formula in Excursion 5.3.A to show that your answers in parts (a) and (b) are the same.

In Exercises 35–44, solve the equation as in Example 9.

35. $\ln(3x - 5) = \ln 11 + \ln 2$

36. $\log(4x - 1) = \log(x + 1) + \log 2$

37. $\log(3x - 1) + \log 2 = \log 4 + \log(x + 2)$

38. $\ln(x + 6) - \ln 10 = \ln(x - 1) - \ln 2$

39. $2 \ln x = \ln 36$

40. $2 \log x = 3 \log 4$

41. $\ln x + \ln(x + 1) = \ln 3 + \ln 4$

42. $\ln(6x - 1) + \ln x = \frac{1}{2}\ln 4$

43. $\ln x = \ln 3 - \ln(x + 5)$

44. $\ln(2x + 3) + \ln x = \ln e$

In Exercises 45–52, solve the equation.

45. $\ln(x + 9) - \ln x = 1$

46. $\ln(2x + 1) - 1 = \ln(x - 2)$

47. $\log x + \log(x - 3) = 1$

48. $\log(x - 1) + \log(x + 2) = 1$

49. $\log \sqrt{x^2 - 1} = 2$

50. $\log \sqrt[3]{x^2 + 21x} = 2/3$

51. $\ln(x^2 + 1) - \ln(x - 1) = 1 + \ln(x + 1)$

52. $\dfrac{\ln(x + 1)}{\ln(x - 1)} = 2$

Exercises 53–62 deal with the half-life function $M(x) = c(.5)^{x/h}$, which was discussed on page 327 and used in Example 5 of this section.

53. How old is a piece of ivory that has lost 36% of its carbon-14?

54. How old is a mummy that has lost 49% of its carbon-14?

55. A Native American mummy was found recently. If it has lost 26.4% of its carbon-14, approximately how long ago did the Native American die?

56. How old is a wooden statue that has only one-third of its original carbon-14?

57. A quantity of uranium decays to two-thirds of its original mass in .26 billion years. Find the half-life of uranium.

58. A certain radioactive substance loses one-third of its original mass in 5 days. Find its half-life.

59. Krypton-85 loses 6.44% of its mass each year. What is its half-life?

60. Strontium-90 loses 2.5% of its mass each year. What is its half-life?

61. The half-life of a certain substance is 3.6 days. How long will it take for 20 grams to decay to 3 grams?

62. The half-life of cobalt-60 is 4.945 years. How long will it take for 25 grams to decay to 15 grams?

Exercises 63–68 deal with the compound interest formula $A = P(1 + r)^t$, which was discussed in Excursion 5.2.A and used in Example 6 of this section.

63. At what annual rate of interest should \$1000 be invested so that it will double in 10 years, if interest is compounded quarterly?

64. How long does it take \$500 to triple if it is invested at 6% compounded: **(a)** annually, **(b)** quarterly, **(c)** daily?

65. **(a)** How long will it take to triple your money if you invest \$500 at a rate of 5% per year compounded annually?
(b) How long will it take at 5% compounded quarterly?

66. At what rate of interest (compounded annually) should you invest \$500 if you want to have \$1500 in 12 years?

67. How much money should be invested at 5% interest, compounded quarterly, so that 9 years later the investment will be worth \$5000? This amount is called the **present value** of \$5000 at 5% interest.

68. Find a formula that gives the time needed for an investment of P dollars to double, if the interest rate is r% compounded annually. [*Hint:* Solve the compound interest formula for t, when $A = 2P$.]

Exercises 69–76 deal with functions of the form $f(x) = Pe^{kx}$, where k is the continuous exponential growth rate (see Example 7).

69. The present concentration of carbon dioxide in the atmosphere is 364 parts per million (ppm) and is increasing exponentially at a continuous yearly rate of .4% (that is, $k = .004$). How many years will it take for the concentration to reach 500 ppm?

70. The amount P of ozone in the atmosphere is currently decaying exponentially each year at a continuous rate of $\frac{1}{4}\%$ (that is, $k = -.0025$). How long will it take for half the ozone to disappear (that is, when will the amount be $P/2$)? [Your answer is the half-life of ozone.]

71. The population of Brazil increased from 122 million in 1980 to 158 million in 1992.

(a) At what continuous rate was the population growing during this period?

(b) Assuming that Brazil's population continues to increase at this rate, when will it reach 250 million?

72. A colony of 1000 weevils grows exponentially to 1750 in one week.

(a) At what continuous rate is the population growing?

(b) How many weeks does it take for the weevil population to reach 3000?

73. The probability P percent of having an accident while driving a car is related to the alcohol level of the driver's blood by the formula $P = e^{kt}$, where k is a constant. Accident statistics show that the probability of an accident is 25% when the blood alcohol level is $t = .15$.

(a) Find k. [Use $P = 25$, not .25.]

(b) At what blood alcohol level is the probability of having an accident 50%?

74. Under normal conditions, the atmospheric pressure (in millibars) at height h feet above sea level is given by $P(h) = 1015e^{-kh}$, where k is a positive constant.

(a) If the pressure at 18,000 feet is half the pressure at sea level, find k.

(b) Using the information from part (a), find the atmospheric pressure at 1000 feet, 5000 feet, and 15,000 feet.

75. One hour after an experiment begins, the number of bacteria in a culture is 100. An hour later there are 500.

(a) Find the number of bacteria at the beginning of the experiment and the number 3 hours later.

(b) How long does it take the number of bacteria at any given time to double?

76. If the population at time t is given by $S(t) = ce^{kt}$, find a formula that gives the time it takes for the population to double.

77. The spread of a flu virus in a community of 45,000 people is given by the function $f(t) = \dfrac{45{,}000}{1 + 224e^{-.899t}}$, where $f(t)$ is the number of people infected in week t.

(a) How many people had the flu at the outbreak of the epidemic? After 3 weeks?

(b) When will half the town be infected?

78. The beaver population near a certain lake in year t is approximately $p(t) = \dfrac{2000}{1 + 199e^{-.5544t}}$.

(a) When will the beaver population reach 1000?

(b) Will the population ever reach 2000? Why?

Thinkers

79. According to one theory of learning, the number of words per minute N that a person can type after t weeks of practice is given by $N = c(1 - e^{-kt})$, where c is an upper limit that N cannot exceed and k is a constant that must be determined experimentally for each person.

(a) If a person can type 50 wpm (words per minute) after 4 weeks of practice and 70 wpm after 8 weeks, find the values of k and c for this person. According to the theory, this person will never type faster than c wpm.

(b) Another person can type 50 wpm after 4 weeks of practice and 90 wpm after 8 weeks. How many weeks must this person practice to be able to type 125 wpm?

80. Wendy has been offered two jobs, each with the same starting salary of $24,000 and identical benefits. Assuming satisfactory performance, she will receive a $1200 raise each year at the Great Gizmo Company, whereas the Wonder Widget Company will give her a 4% raise each year.

(a) In what year (after the first year) would her salary be the same at either company? Until then, which company pays better? After that, which company pays better?

(b) Answer the questions in part (a) assuming that the annual raise at Great Gizmo is $1800.

5.5 Exponential, Logarithmic, and Other Models*

Many data sets can be modeled by suitable exponential, logarithmic, and related functions. Most calculators have regression procedures for constructing the following models.

Model	Equation	Examples	
Power	$y = ax^r$	$y = 5x^{2.7}$	$y = 3.5x^{-.045}$
Exponential	$y = ab^x$ or $y = ae^{kx}$	$y = 2(1.64)^x$	$y = 2\cdot e^{.4947x}$
Logistic	$y = \dfrac{a}{1 + be^{-kx}}$	$y = \dfrac{20{,}000}{1 + 24e^{-.25x}}$	$y = \dfrac{650}{1 + 6e^{.3x}}$
Logarithmic	$y = a + b \ln x$	$y = 5 + 4.2 \ln x$	$y = 2 - 3 \ln x$

We begin by examining exponential models, such as $y = 3\cdot 2^x$. A table of values for this model is shown below. Look carefully at the ratio of successive entries (that is, each entry divided by its predecessor).

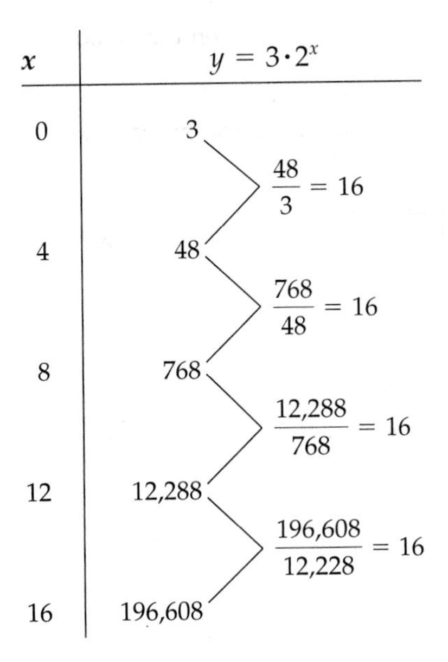

x	$y = 3\cdot 2^x$
0	3
4	48
8	768
12	12,288
16	196,608

It should not be a surprise that the ratio of successive entries is constant. For at each step, x changes from x to $x + 4$ (from 0 to 4, from 4 to 8, and so on) and y changes from $3\cdot 2^x$ to $3\cdot 2^{x+4}$. Hence, the ratio of successive terms is always

$$\frac{3\cdot 2^{x+4}}{3\cdot 2^x} = \frac{3\cdot 2^x\cdot 2^4}{3\cdot 2^x} = 2^4 = 16.$$

*This section is optional; its prerequisites are Section 2.5 and Excursion 4.4.A. It will be used in clearly identifiable exercises, but not elsewhere in the text.

A similar argument applies to any exponential model $y = ab^x$ and shows that if x changes by a fixed amount k, then the ratio of the corresponding y values is the constant b^k (in our example b was 2 and k was 4). This suggests that

when the ratio of successive entries in a table of data is approximately constant, an exponential model is appropriate.

Example 1 In the years before the Civil War, the population of the United States grew rapidly, as shown in the following table from the U.S. Bureau of the Census. Find a model for this growth.

Year	Population in Millions	Year	Population in Millions
1790	3.93	1830	12.86
1800	5.31	1840	17.07
1810	7.24	1850	23.19
1820	9.64	1860	31.44

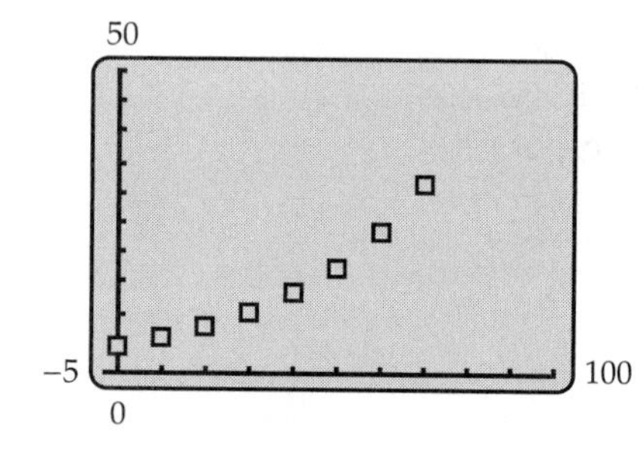

Figure 5–21

Solution The data points (with $x = 0$ corresponding to 1790) are shown in Figure 5–21. Their shape suggests either a polynomial graph of even degree or an exponential graph. Since populations generally grow exponentially, an exponential model is likely to be a good choice. We can confirm this by looking at the ratios of successive entries in the table.

Year	Population	Ratio
1790	3.93	
		$\frac{5.31}{3.93} \approx 1.351$
1800	5.31	
		$\frac{7.24}{5.31} \approx 1.363$
1810	7.24	
		$\frac{9.64}{7.24} \approx 1.331$
1820	9.64	
		$\frac{12.86}{9.64} \approx 1.334$
1830	12.86	

Year	Population	Ratio
1830	12.86	
		$\frac{17.07}{12.86} \approx 1.327$
1840	17.07	
		$\frac{23.19}{17.07} \approx 1.359$
1850	23.19	
		$\frac{31.44}{23.19} \approx 1.356$
1860	31.44	

The ratios are almost constant, as they would be in an exponential model. So we use regression to find such an exponential model. The procedure

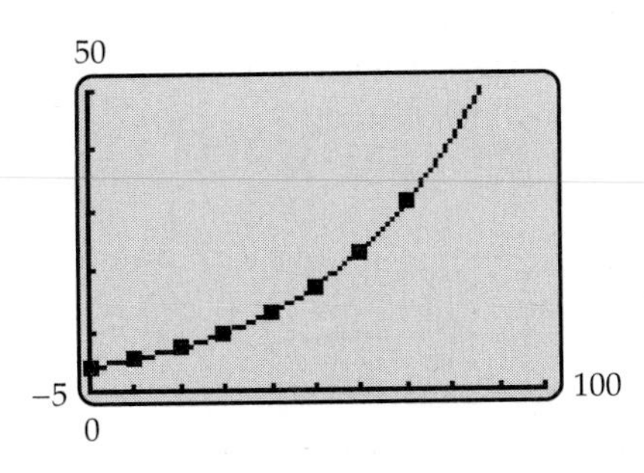

Figure 5–22

is the same as for linear and polynomial regression (see the Tips on pages 101 and 244). It produces this model:*

$$y = 3.9572\,(1.0299^x).$$

The graph in Figure 5–22 appears to fit the data quite well. In fact, you can readily verify that the model has an error of less than 1% for each of the data points. Furthermore, as discussed before the example, when x changes by 10, the value of y changes by approximately $1.0299^{10} \approx 1.343$, which is very close to the successive ratios of the data that were computed above. ■

Example 2 After the Civil War, the U.S. population continued to increase, as shown in this table:

Year	Population in Millions	Year	Population in Millions	Year	Population in Millions
1870	38.56	1920	106.02	1960	179.32
1880	50.19	1930	123.20	1970	202.30
1890	62.98	1940	132.16	1980	226.54
1900	76.21	1950	151.33	1990	248.72
1910	92.23				

However, the model from Example 1 does not remain valid, as can be seen in Figure 5–23, which shows its graph together with all the data points from 1790 through 1990 ($x = 0$ corresponds to 1790).

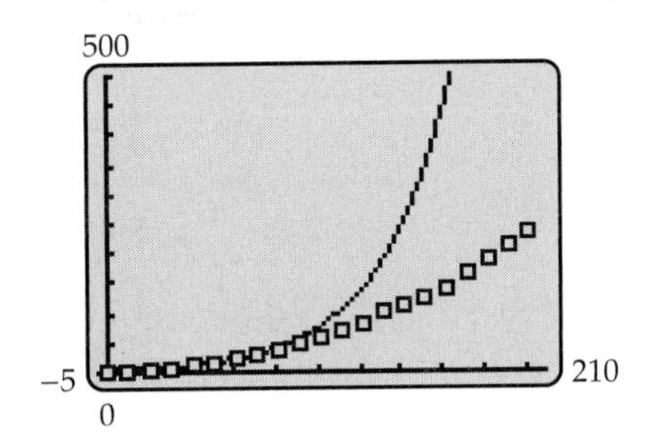

Figure 5–23

The problem is that the *rate* of growth has steadily decreased since the Civil War. For instance, the ratio of the first two entries is $\dfrac{50.19}{38.56} \approx 1.302$ and the ratio of the last two is $\dfrac{248.72}{226.54} \approx 1.098$. So an exponential model may not be the best choice now. Other possibilities are polynomial models (which grow at a slower rate) or logistic models (in which the growth rate decreases with time). Figure 5–24 shows three possible models, each obtained by using the appropriate regression program on a calculator, with all the data points from 1790 through 1990.

*Throughout this section, coefficients are rounded for convenient reading, but the full expansion is used for calculations and graphs.

Exponential Model

$y = 5.5381 \cdot 1.02098^x$

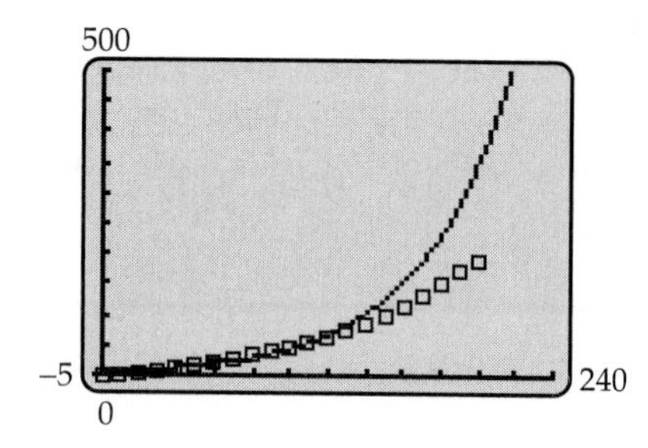

Polynomial Model

$$y = (5.61 \times 10^{-8})x^4 - (1.92 \times 10^{-5})x^3 + .0084x^2 - .1276x + 5.247$$

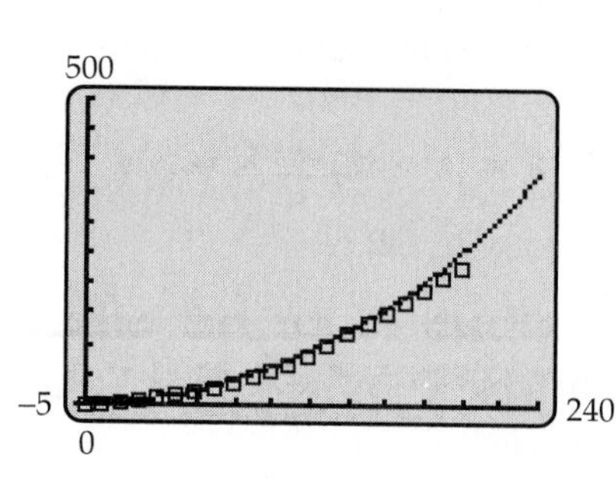

Logistic Model

$$y = \frac{384.57}{1 + 54e^{-.0228x}}$$

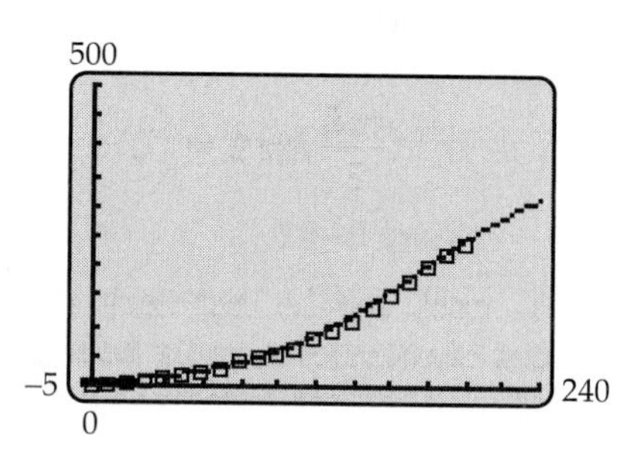

Figure 5–24

As expected, the polynomial and logistic models fit the data better than does the exponential model. The main difference between them is that the polynomial model indicates unlimited future growth, whereas the logistic model has the population growing more slowly in the future (and eventually leveling off—see Exercise 13). ■

In Example 1 we used the ratios of successive entries of the data table to determine that an exponential model was appropriate. Here is another way to make that determination. Consider the exponential function $y = ab^x$. Taking natural logarithms of both sides and using the logarithm laws on the right side shows that

$$\ln y = \ln (ab^x) = \ln a + \ln b^x = \ln a + x \ln b.$$

Now $\ln a$ and $\ln b$ are constants, say $k = \ln a$ and $m = \ln b$, so that

$$\ln y = mx + k.$$

Thus, the points $(x, \ln y)$ lie on the straight line with slope m and y-intercept k. Consequently, we have this guideline:

> **If (x, y) are data points and if the points $(x, \ln y)$ are approximately linear, then an exponential model may be appropriate for the data.**

Similarly, if $y = ax^r$ is a power function, then

$$\ln y = \ln (ax^r) = \ln a + r \ln x.$$

Since $\ln a$ is a constant, say $k = \ln a$, we have

$$\ln y = r \ln x + k,$$

which means that the points $(\ln x, \ln y)$ lie on a straight line with slope r and y-intercept k. Consequently, we have this guideline:

> **If (x, y) are data points and if the points $(\ln x, \ln y)$ are approximately linear, then a power model may be appropriate for the data.**

Example 3 The length of time that a planet takes to make one complete rotation around the sun is its year. The table shows the length (in earth years) of each planet's year and the distance of that planet from the sun (in millions of miles).* Find a model for this data in which x is the length of the year and y the distance from the sun.

Planet	Year	Distance
Mercury	.24	36.0
Venus	.62	67.2
Earth	1	92.9
Mars	1.88	141.6
Jupiter	11.86	483.6
Saturn	29.46	886.7
Uranus	84.01	1783.0
Neptune	164.79	2794.0
Pluto	247.69	3674.5

Solution Figure 5–25 shows the data points for the five planets with the shortest years. Figure 5–26 shows all the data points, but on this scale, the first four points look like a single large one near the origin.

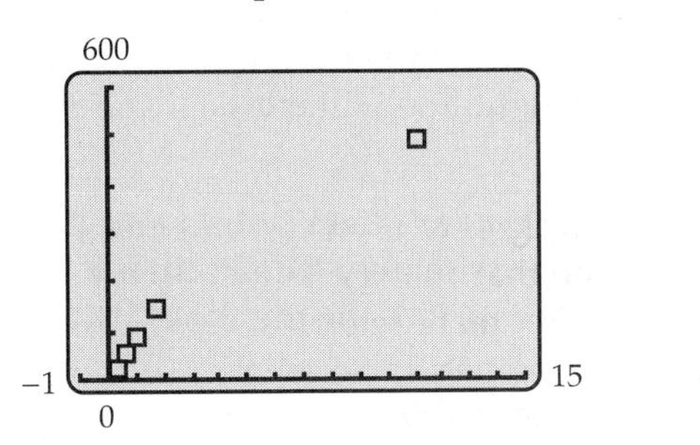

Figure 5–25

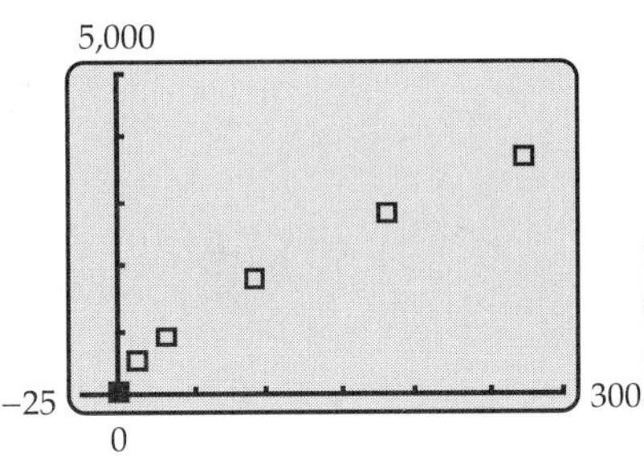

Figure 5–26

Plotting the point $(x, \ln y)$ for each data point (x, y) (see the Tip in the margin) produces Figure 5–27. Its points do not form a linear pattern (four of them are almost vertical near the y-axis and the other five almost horizontal), so an exponential function is not an appropriate model. On the other hand, the points $(\ln x, \ln y)$ in Figure 5–28 do form a linear pattern, which suggests that a power model will work.

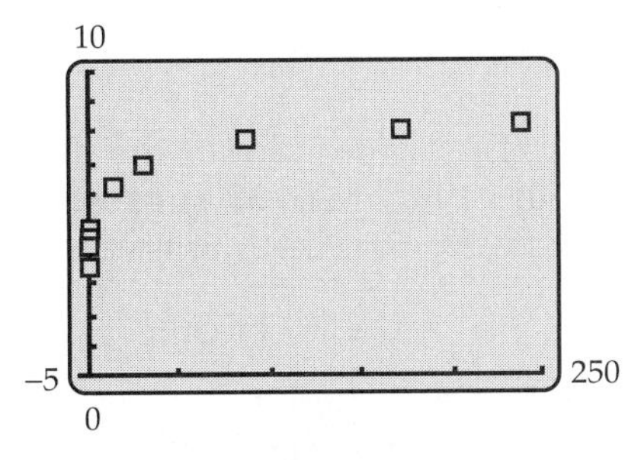

Figure 5–27

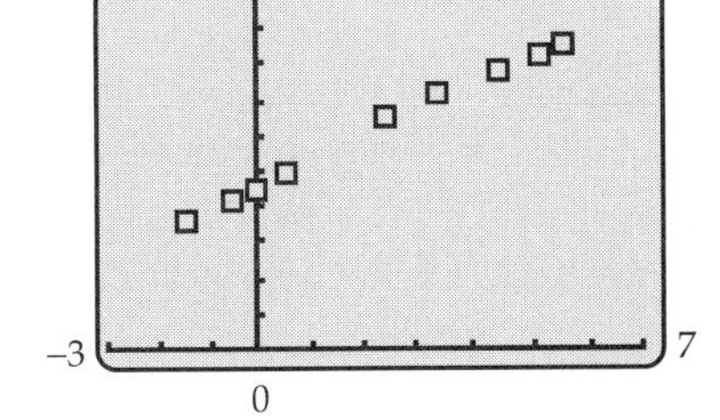

Figure 5–28

Technology Tip

Suppose the x- and y-coordinates of the data points are stored in lists L_1 and L_2, respectively. On calculators other than TI-89, keying in

$\ln L_2$ STO $\rightarrow L_4$

produces the list L_4, whose entries are the natural logarithms of the numbers in list L_2, and stores it in the statistics editor. You can then use lists L_1 and L_4 to plot the points $(x, \ln y)$. For TI-89, check your instruction manual.

*Since the orbit of a planet around the sun is not circular, its distance from the sun varies through the year. The number given here is the average of its maximum and minimum distances from the sun.

A calculator's power regression feature produces this model:

$$y = 92.8935\,x^{.6669}.$$

Its graph in Figure 5–29 shows that it fits the original data points quite well. ■

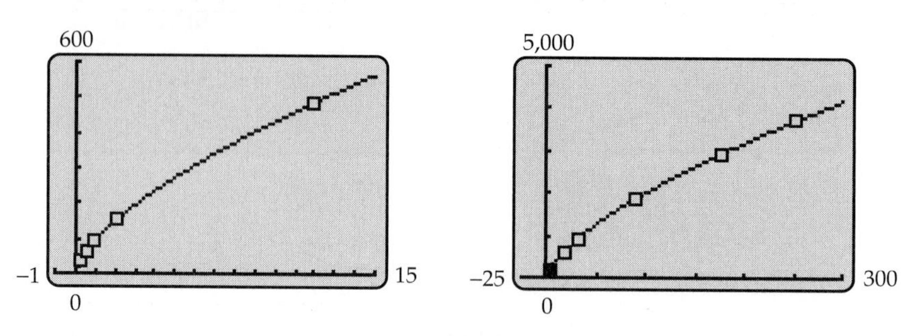

Figure 5–29

If $y = a + b \ln x$ is a logarithmic model, then the points $(\ln x, y)$ lie on the straight line with slope b and y-intercept a. (Why?) Thus we have this guideline:

> **If (x, y) are data points and if the points $(\ln x, y)$ are approximately linear, then a logarithmic model may be appropriate for the data.**

Example 4 Find a model for population growth in Anaheim, California, given the following data. (*Source:* U.S. Bureau of the Census)

Year	1950	1970	1980	1990	1994
Population	14,556	166,408	219,494	266,406	282,133

Solution The scatter plot of the data points (with $x = 50$ corresponding to 1950) in Figure 5–30 suggests a logarithmic curve with a very slight bend. So we plot the points

$$(\ln 50, 14556), \ldots, (\ln 94, 282133)$$

in Figure 5–31. Since these points appear to lie on a straight line, a logarithmic model is appropriate. Using logarithmic regression on a calculator, we obtain this model:

$$y = -1{,}643{,}983.42 + 424{,}768.97 \ln x.$$

Its graph is a good fit for the data (Figure 5–32). ■

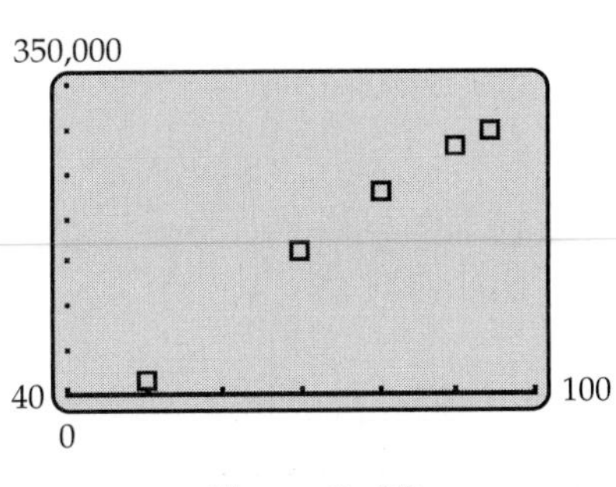

Figure 5–30

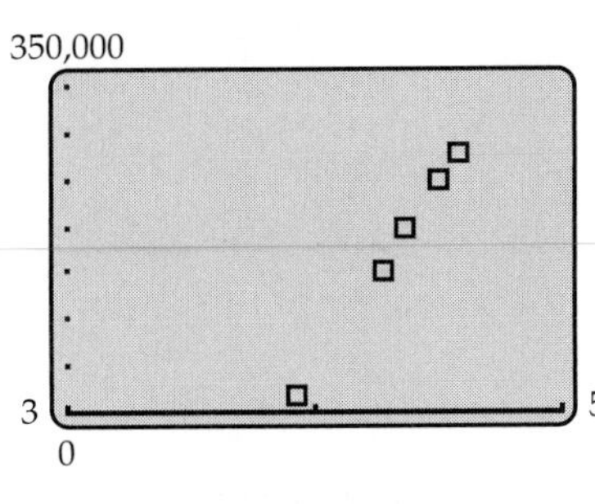

Figure 5–31

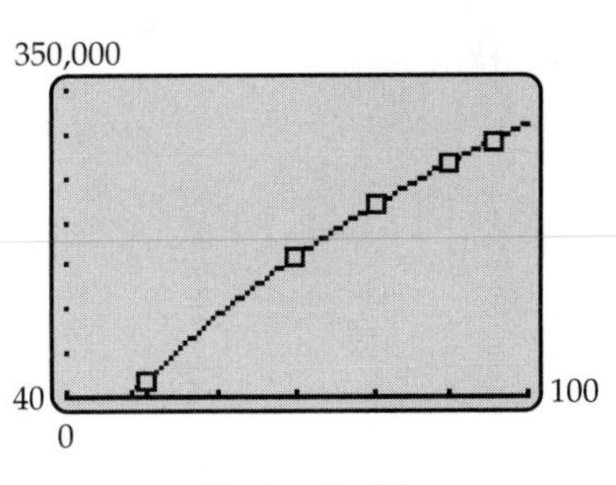

Figure 5–32

> **CAUTION**
> When using logarithmic models, you must have data points with positive first coordinates (since logarithms of negative numbers and 0 are not defined).

Exercises 5.5

In Exercises 1–10, state which of the following models might be appropriate for the given scatter plot of data (more than one model may be appropriate).

Model	Corresponding Function
A. Linear	$y = ax + b$
B. Quadratic	$y = ax^2 + bx + c$
C. Power	$y = ax^r$
D. Cubic	$y = ax^3 + bx^2 + cx + d$
E. Exponential	$y = ab^x$
F. Logarithmic	$y = a + b \ln x$
G. Logistic	$y = \dfrac{a}{1 + be^{-kx}}$

1.

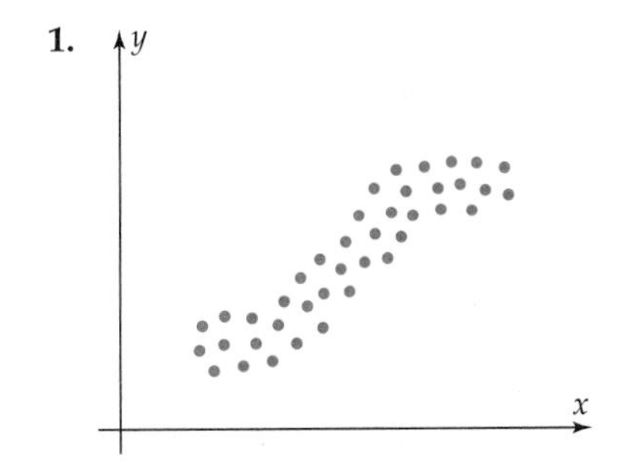

2.

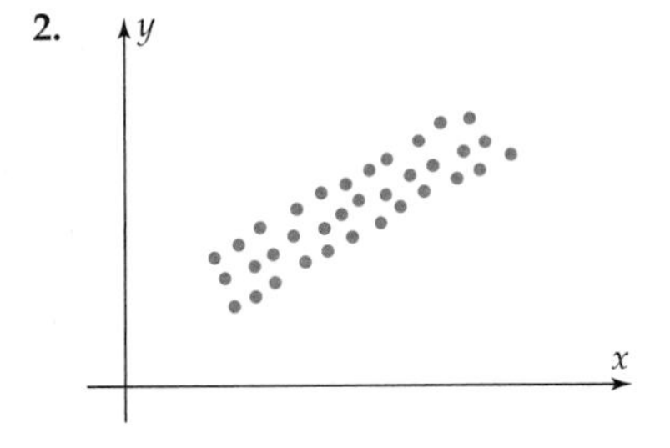

3.

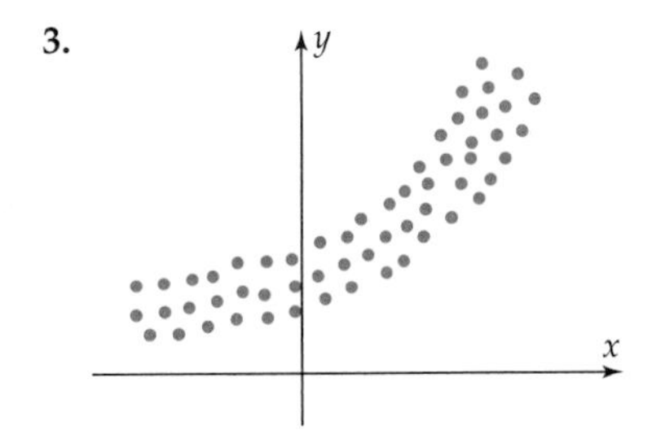

4.

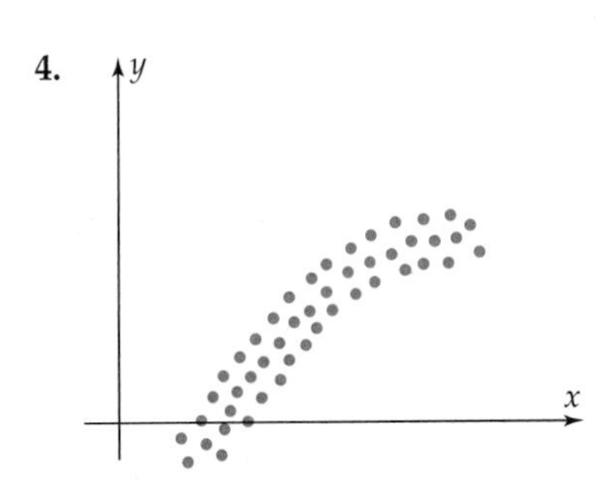

5.

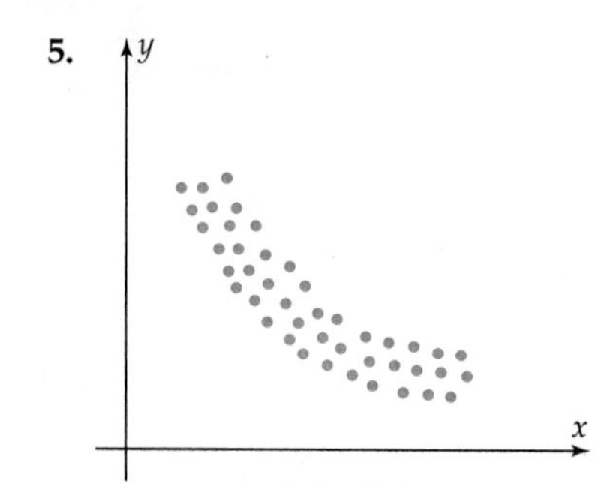

6.

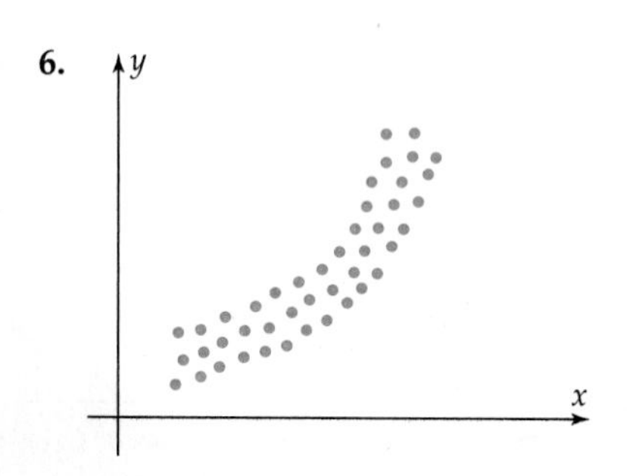

7.

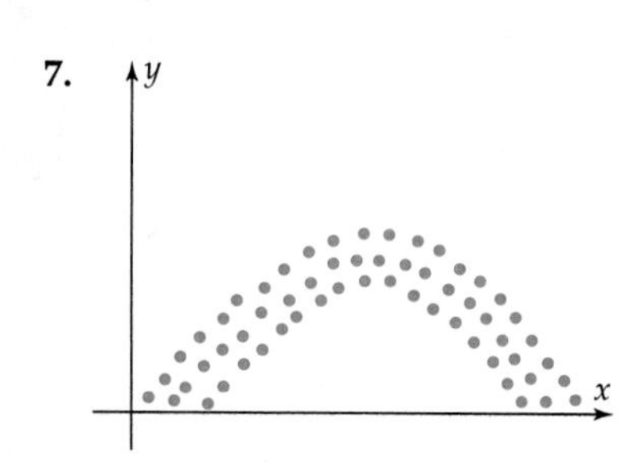

8.

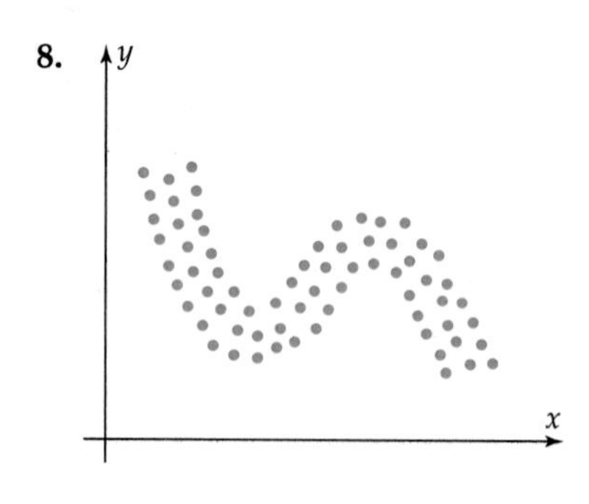

9.

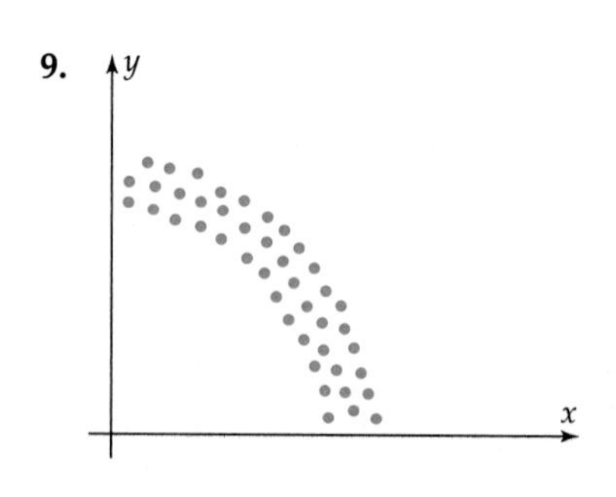

10.

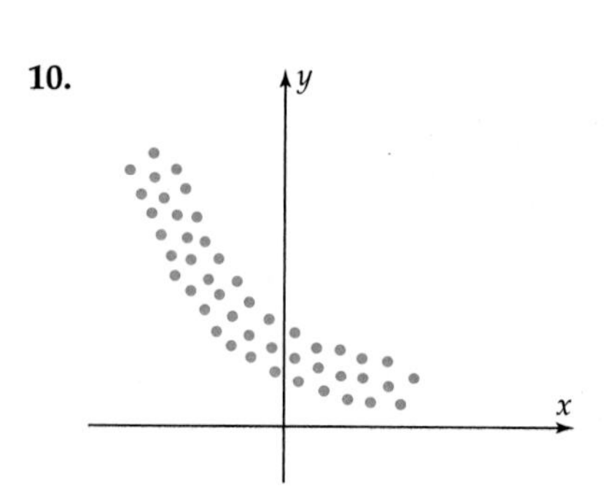

In Exercises 11 and 12, compute the ratios of successive entries in the table to determine whether or not an exponential model is appropriate for the data.

11.

x	0	2	4	6	8	10
y	3	15.2	76.9	389.2	1975.5	9975.8

12.

x	1	3	5	7	9	11
y	3	21	55	105	171	253

13. (a) Show algebraically that in the logistic model for the U.S. population in Example 2, the population can never exceed 384.57 million people.

(b) Confirm your answer in part (a) by graphing the logistic model in a window that includes the next three centuries.

14. According to estimates by the U.S. Bureau of the Census, the U.S. population was 267.6 million in 1997. Based on this information, which of the models in Example 2 appears to be the most accurate predictor?

15. Graph each of the following power functions in a window with $0 \leq x \leq 20$.

(a) $f(x) = x^{-1.5}$ **(b)** $g(x) = x^{.75}$ **(c)** $h(x) = x^{2.4}$

16. Based on your graphs in Exercise 15, describe the general shape of the graph of $y = ax^r$, when $a > 0$ and

(a) $r < 0$ **(b)** $0 < r < 1$ **(c)** $r > 1$

In Exercises 17–20, determine whether an exponential, power, or logarithmic model (or none or several of these) is appropriate for the data by determining which (if any) of the following sets of points are approximately linear:

$$\{(x, \ln y)\}, \qquad \{(\ln x, \ln y)\}, \qquad \{(\ln x, y)\},$$

where the given data set consists of the points $\{(x, y)\}$.

17.

x	1	3	5	7	9	11
y	2	25	81	175	310	497

18.

x	3	6	9	12	15	18
y	385	74	14	2.75	.5	.1

19.

x	5	10	15	20	25	30
y	17	27	35	40	43	48

20.

x	5	10	15	20	25	30
y	2	110	460	1200	2500	4525

21. The table shows the number of babies born as twins, triplets, quadruplets, etc., in recent years.

Year	Multiple Births
1989	92,916
1990	96,893
1991	98,125
1992	99,255
1993	100,613
1994	101,658
1995	101,709

(a) Sketch a scatter plot of the data, with $x = 1$ corresponding to 1989.

(b) Plot each of the following models on the same screen as the scatter plot:

$$f(x) = 93{,}201.973 + 4{,}545.977 \ln x;$$

$$g(x) = \frac{102{,}519.98}{1 + .1536e^{-.4263x}}.$$

(c) Use the table feature to estimate the number of multiple births in 2000 and 2005.

(d) Over the long run which model do you think is the better predictor?

22. The graph shows the Census Bureau estimates of future U.S. population.

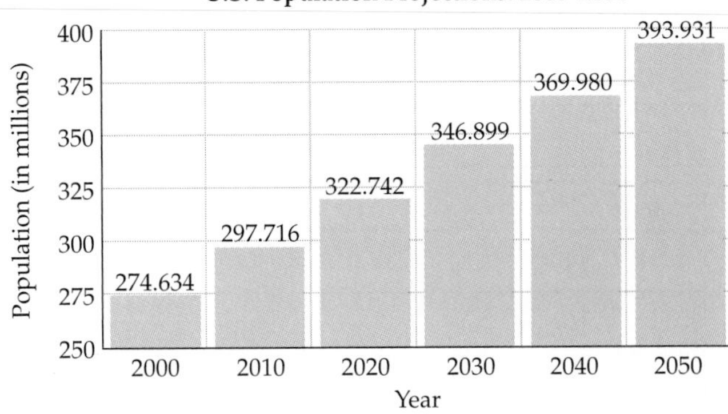

(a) How well do the projections in the graph compare with those given by the logistic model in Example 2?

(b) Find a logistic model of the U.S. population, using the data given in Example 2 for the years from 1900 to 1990 and the estimated 1997 population of 267.2 million

(c) How well do the projections in the graph compare with those given by the model in part (b)?

23. Infant mortality rates in the United States are shown in the table.

Year	Infant Mortality Rate*
1920	76.7
1930	60.4
1940	47.0
1950	29.2
1960	26.0
1970	20.0

Year	Infant Mortality Rate*
1980	12.6
1985	10.6
1990	9.2
1995	7.6
1996	7.2

*Rates are infant (under 1 year) deaths per 1000 live births

(a) Sketch a scatter plot of the data, with $x = 0$ corresponding to 1900.

(b) Verify that the set of points $(x, \ln y)$, where (x, y) are the original data points, is approximately linear.

(c) Based on part (b), what type of model would be appropriate for this data? Find such a model.

24. The average number of students per computer in the U.S. public schools (elementary through high school) is shown in the table.

Fall of School Year	Students/Computer
1983	125
1984	75
1985	50
1986	37
1987	32
1988	25
1989	22
1990	20
1991	18
1992	16
1993	14
1994	10.5
1995	10

(a) Sketch a scatter plot of the data, with $x = 1$ corresponding to 1983.
(b) Find an exponential model for the data.
(c) Use the model to estimate the number of students per computer in 2000.
(d) In what year, according to this model, will each student have his or her own computer in school?
(e) What are the limitations of this model?

25. The number of children who were home schooled in the United States in selected years is shown in the table at the top of the next column. (*Source:* National Home Education Research Institute)
(a) Sketch a scatter plot of the data, with $x = 0$ corresponding to 1980.
(b) Find a quadratic model for the data.
(c) Find a logistic model for the data.
(d) What is the number of home-schooled children predicted by each model for the year 2000?
(e) What are the limitations of each model?

Fall of School Year	Number of Children (in thousands)
1983	92.5
1985	183
1988	225
1990	301
1992	470
1993	588
1994	735
1995	800
1996	920
1997	1100

26. (a) Find an exponential model for the federal debt, based on the data in the table. Let $x = 0$ correspond to 1960.

Accumulated Gross Federal Debt	
Year	*Amount (in billions of dollars)*
1960	291
1965	322
1970	381
1975	542
1980	909
1985	1,818
1990	3,207
1995	4,921
1996	5,182

(b) Use the model to predict the federal debt in 2000.

27. Although the number of farms in the United States has been decreasing (see Exercise 28), the average size of farms (in acres) has increased, as shown in the table. (*Source:* U.S. Department of Agriculture)

Year	Average Farm Size (in acres)
1940	174
1950	213
1960	297
1970	374
1980	426
1990	460
1997	470

(a) Sketch a scatter plot of the data, with $x = 0$ corresponding to 1940.

(b) Find a cubic model for the data. Explain why this is *not* a reasonable model for making predictions about the future.

(c) Find a logistic model for the data (which may be more reasonable because the amount of farming land is limited).

(d) Based on the logistic model, what is the estimated size of the average farm in 2000?

(e) According to the logistic model, when will the average farm size be 1000 acres?

28. (a) Find an exponential model for the number of farms in the United States, based on the table below. Let $x = 0$ correspond to 1950. (*Source:* U.S. Department of Agriculture)

Year	Number of Farms (in millions)
1950	5.647800
1955	4.653800
1960	3.962520
1965	3.356170
1970	2.949140
1975	2.521420
1980	2.439510
1985	2.292530
1990	2.145820
1995	2.071520
1996	2.063910
1997	2.058910

(b) Use the model to predict the number of farms in 2000.

(c) Is this model realistic for the next millennium?

29. Worldwide production of computers has grown dramatically, as shown in the first two columns of the following table. (*Source:* Dataquest)

Year	Worldwide Shipments (thousands)	Predicted Number of Shipments (thousands)	Ratio
1985	14.7		
1986	15.1		
1987	16.7		
1988	18.1		
1989	21.3		
1990	23.7		
1991	27		
1992	32.4		
1993	38.9		
1994	47.9		
1995	60.2		
1996	70.9		
1987	84.3		

(a) Sketch a scatter plot of the data, with $x = 1$ corresponding to 1985.

(b) Find an exponential model for the data.

(c) Use the model to complete column 3 of the table.

(d) Fill in column 4 of the table by dividing each entry in column 2 by the preceding one.

(e) What does column 4 tell you about the appropriateness of the model?

30. A teacher invested $2000 in an Individual Retirement Account (IRA) in 1986. Its value has increased since then, as shown in the table.

(a) Find an exponential model for this data, with $x = 0$ corresponding to 1980.

(b) Based on your model, what is the effective rate of return on this investment?

(c) Assuming this rate of return continues, how much will be in the account in 2010?

Year	Value of Account
1986	$2,000.00
1987	2,387.54
1988	3,770.55
1989	4,651.03
1990	5,639.20
1991	6,014.69

Year	Value of Account
1992	$ 6,687.36
1993	7,410.43
1994	7,383.54
1995	9,513.97
1996	10,980.12
1997	13,424.73
1998	15,207.44

Chapter 5 Review

Important Concepts

Important Facts and Formulas

- *Laws of Exponents:*

$$c^r c^s = c^{r+s} \qquad (cd)^r = c^r d^r$$

$$\frac{c^r}{c^s} = c^{r-s} \qquad \left(\frac{c}{d}\right)^r = \frac{c^r}{d^r}$$

$$(c^r)^s = c^{rs} \qquad c^{-r} = \frac{1}{c^r}$$

- $g(x) = \log x$ is the inverse function of $f(x) = 10^x$:

$$10^{\log v} = v \text{ for all } v > 0 \qquad \text{and} \qquad \log 10^u = u \text{ for all } u.$$

- $g(x) = \ln x$ is the inverse function of $f(x) = e^x$:

$$e^{\ln v} = v \text{ for all } v > 0 \qquad \text{and} \qquad \ln(e^u) = u \text{ for all } u.$$

- $h(x) = \log_b x$ is the inverse function of $k(x) = b^x$:

$$b^{\log_b v} = v \text{ for all } v > 0 \qquad \text{and} \qquad \log_b(b^u) = u \text{ for all } u.$$

- *Logarithm Laws:* For all $v, w > 0$ and any k:

$$\ln(vw) = \ln v + \ln w \qquad \log_b(vw) = \log_b v + \log_b w$$

$$\ln\left(\frac{v}{w}\right) = \ln v - \ln w \qquad \log_b\left(\frac{v}{w}\right) = \log_b v - \log_b w$$

$$\ln(v^k) = k(\ln v) \qquad \log_b(v^k) = k(\log_b v)$$

- *Exponential Growth Functions:*

$$f(x) = P(1+r)^x \quad (0 < r < 1),$$
$$f(x) = Pa^x \quad (a > 1),$$
$$f(x) = Pe^{kx} \quad (k > 0)$$

- *Exponential Decay Functions:*

$$f(x) = P(1-r)^x \quad (0 < r < 1),$$
$$f(x) = Pa^x \quad (0 < a < 1),$$
$$f(x) = Pe^{kx} \quad (k < 0)$$

- *Compound Interest Formula:* $A = P(1+r)^t$
- *Change of Base Formula:* $\log_b v = \dfrac{\ln v}{\ln b}$

Review Questions

In Questions 1–6, simplify the expression.

1. $\sqrt{\sqrt[3]{c^{12}}}$
2. $(\sqrt[3]{4c^3d^2})^3(c\sqrt{d})^2$
3. $(a^{-2/3}b^{2/5})(a^3b^6)^{4/3}$
4. $\dfrac{(3c)^{3/5}(2d)^{-2}(4c)^{1/2}}{(4c)^{1/5}(2d)^4(2c)^{-3/2}}$
5. $(u^{1/4} - v^{1/4})(u^{1/4} + v^{1/4})$
6. $c^{3/2}(2c^{1/2} + 3c^{-3/2})$

In Exercises 7 and 8, simplify and write the expression without radicals or negative exponents:

7. $\dfrac{\sqrt[3]{6c^4d^{14}}}{\sqrt[3]{48c^{-2}d^2}}$
8. $\dfrac{(8u^5)^{1/4}2^{-1}u^{-3}}{2u^8}$

9. Rationalize the numerator and simplify: $\dfrac{\sqrt{2x+2h+1} - \sqrt{2x+1}}{h}$.

10. Rationalize the denominator: $\dfrac{5}{\sqrt{x}-3}$.

In Questions 11–16, find all real solutions of the equation.

11. $\sqrt{x-1} = 2 - x$
12. $\sqrt[3]{1-t^2} = -2$
13. $\sqrt{x+1} + \sqrt{x-1} = 1$
14. $x^{1/4} - 5x^{1/2} + 6 = 0$
15. $\sqrt[3]{x^4 - 2x^3 + 6x - 7} = x + 3$
16. $x^{4/3} + x - 2x^{2/3} + x^{1/3} = 4$

In Questions 17 and 18, find a viewing window (or windows) that shows a complete graph of the function.

17. $f(x) = 2^{x^3 - x - 2}$
18. $g(x) = \ln\left(\dfrac{x}{x-2}\right)$

In Questions 19–22, sketch a complete graph of the function. Indicate all asymptotes clearly.

19. $g(x) = 2^x - 1$
20. $f(x) = 2^{x-1}$

21. $h(x) = \ln(x + 4) - 2$

22. $k(x) = \ln\left(\frac{x + 4}{x}\right)$

23. A computer software company claims the following function models the "learning curve" for their mathematical software:

$$P(t) = \frac{100}{1 + 48.2e^{-.52t}}$$

where t is measured in months and $P(t)$ is the average percent of the software program's capabilities mastered after t months.

(a) Initially what percent of the program is mastered?

(b) After 6 months what percent of the program is mastered?

(c) Roughly, when can a person expect to "learn the most in the least amount of time"?

(d) If the company's claim is true, how many months will it take to have completely mastered the program?

24. Compunote has offered you a starting salary of $60,000 with $1000 yearly raises. Calcuplay offers you an initial salary of $30,000 and a guaranteed 6% raise each year.

(a) Complete the following table for each company.

Year	Compunote
1	$60,000
2	$61,000
3	
4	
5	

Year	Calcuplay
1	$30,000
2	$31,800
3	
4	
5	

(b) For each company write a function that gives your salary in terms of years employed.

(c) If you plan on staying with the company for only five years, which job should you take to earn the most money?

(d) If you plan on staying with the company for 20 years, which is your best choice?

(e) In what year does the salary at Calcuplay exceed the salary at Compunote?

In Questions 25–30, translate the given exponential statement into an equivalent logarithmic one.

25. $e^{6.628} = 756$

26. $e^{5.8972} = 364$

27. $e^{r^2-1} = u + v$

28. $e^{a-b} = c$

29. $10^{2.8785} = 756$

30. $10^{c+d} = t$

In Questions 31–36, translate the given logarithmic statement into an equivalent exponential one.

31. $\ln 1234 = 7.118$

32. $\ln(ax + b) = y$

33. $\ln(rs) = t$

34. $\log 1234 = 3.0913$

35. $\log_5(cd - k) = u$

36. $\log_d(uv) = w$

In Questions 37–40, evaluate the given expression without using a calculator.

37. $\ln e^3$

38. $\ln \sqrt[3]{e}$

39. $e^{\ln 3/4}$

40. $e^{\ln (x+2y)}$

41. Simplify: $3 \ln \sqrt{x} + (1/2)\ln x$

42. Simplify: $\ln (e^{4e})^{-1} + 4e$

In Questions 43–45, write the given expression as a single logarithm.

43. $\ln 3x - 3 \ln x + \ln 3y$

44. $\log_7 7x + \log_7 y - 1$

45. $4 \ln x - 2(\ln x^3 + 4 \ln x)$

46. $\log (-.01) = ?$

47. $\log_{20} 400 = ?$

48. You are conducting an experiment about memory. The people who participate agree to take a test at the end of your course and every month thereafter for a period of two years. The average score for the group is given by the model $M(t) = 91 - 14 \ln (t + 1)$, $0 \le t \le 24$, where t is time in months after the first test.

(a) What is the average score on the initial exam?
(b) What is the average score after three months?
(c) When will the average drop below 50%?
(d) Is the magnitude of the rate of memory loss greater in the first month after the course (from $t = 0$ to $t = 1$) or after the first year (from $t = 12$ to $t = 13$)?
(e) Hypothetically, if the model could be extended past $t = 24$ months, would it be possible for the average score to be 0%?

49. Which of the following statements is *true*?

(a) $\ln 10 = (\ln 2)(\ln 5)$
(b) $\ln (e/6) = \ln e + \ln 6$
(c) $\ln (1/7) + \ln 7 = 0$
(d) $\ln (-e) = -1$
(e) None of the above is true.

50. Which of the following statements is *false*?

(a) $10 (\log 5) = \log 50$
(b) $\log 100 + 3 = \log 10^5$
(c) $\log 1 = \ln 1$
(d) $\log 6/\log 3 = \log 2$
(e) All of the above are false.

Use the six graphs here and on the next page for Questions 51 and 52.

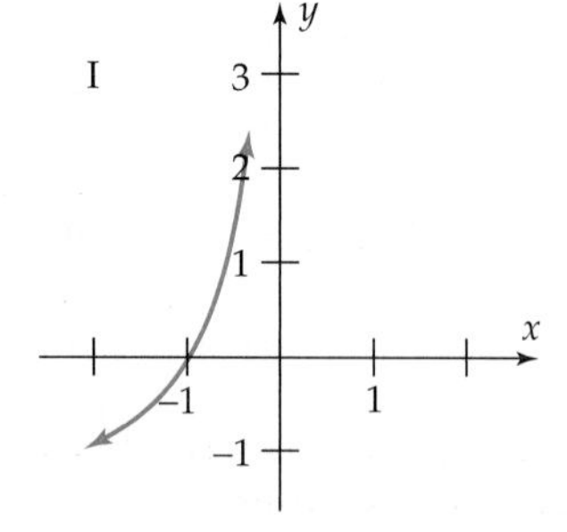

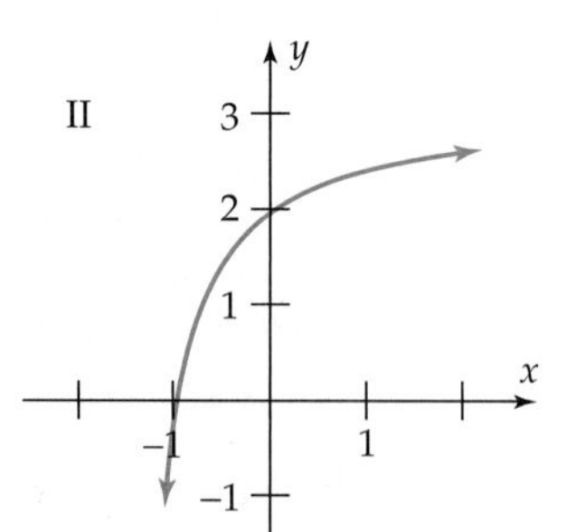

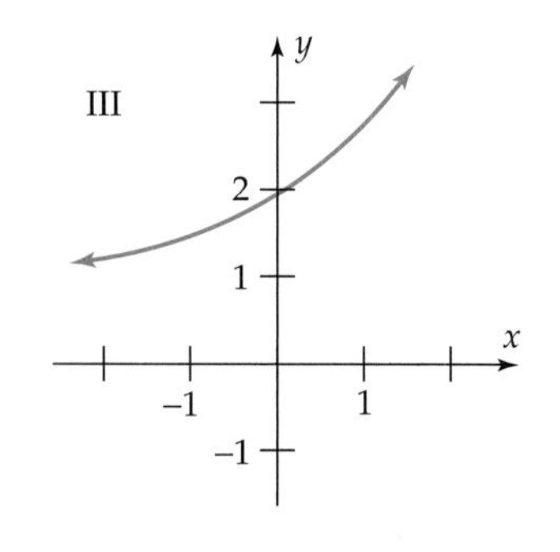

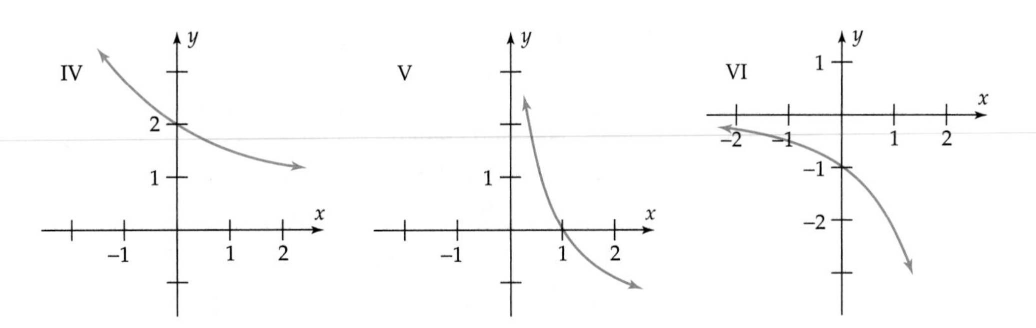

51. If $b > 1$, then the graph of $f(x) = -\log_b x$ could possibly be:
(a) I
(b) IV
(c) V
(d) VI
(e) none of these

52. If $0 < b < 1$, then the graph of $g(x) = b^x + 1$ could possibly be:
(a) II
(b) III
(c) IV
(d) VI
(e) none of these

53. If $\log_3 9^{x^2} = 4$, what is x?

54. What is the domain of the function $f(x) = \ln\left(\dfrac{x}{x-1}\right)$?

In Questions 55–63, solve the equation for x.

55. $8^x = 4^{x^2-3}$

56. $e^{3x} = 4$

57. $2 \cdot 4^x - 5 = -4$

58. $725e^{-4x} = 1500$

59. $u = c + d \ln x$

60. $2^x = 3^{x+3}$

61. $\ln x + \ln(3x - 5) = \ln 2$

62. $\ln(x + 8) - \ln x = 1$

63. $\log(x^2 - 1) = 2 + \log(x + 1)$

64. At a small community college the spread of a rumor through the population of 500 faculty and students can be modeled by:

$$\ln(n) - \ln(1000 - 2n) = .65t - \ln 998$$

where n is the number of people who have heard the rumor after t days.

(a) How many people know the rumor initially? (at $t = 0$)
(b) How many people have heard the rumor after four days?
(c) Roughly, in how many weeks will the entire population have heard the rumor?
(d) Use the properties of logarithms to write n as a function of t; in other words solve the model above for n in terms of t.
(e) Enter the function you found in part (d) into your calculator and use the table feature to check your answers to parts (a), (b), and (c). Do they agree?
(f) Now graph the function. Roughly over what time interval does the rumor seem to "spread" the fastest?

65. The half-life of polonium (^{210}Po) is 140 days. If you start with 10 milligrams, how much will be left at the end of a year?

66. An insect colony grows exponentially from 200 to 2000 in 3 months' time. How long will it take for the insect population to reach 50,000?

67. Hydrogen-3 decays at a rate of 5.59% per year. Find its half-life.

68. The half-life of radium-88 is 1590 years. How long will it take for 10 grams to decay to 1 gram?

69. How much money should be invested at 8% per year, compounded quarterly, in order to have $1000 in 10 years?

70. At what annual interest rate should you invest your money if you want to double it in 6 years?

71. One earthquake measures 4.6 on the Richter scale. A second earthquake is 1000 times more intense than the first. What does it measure on the Richter scale?

72. The table gives the population of Austin, TX.

Year	1950	1970	1980	1990	1994
Population	132,459	253,539	345,890	465,648	514,013

(a) Sketch a scatter plot of the data, with $x = 0$ corresponding to 1950.
(b) Find an exponential model for the data.
(c) Use the model to estimate the population of Austin in 1960 and 1985.

73. The wind-chill factor is the temperature that would produce the same cooling effect on a person's skin if there were no wind. The table shows the wind-chill factors for various wind speeds when the temperature is 35°F. (*Source:* National Weather Service)

Wind Speed (mph)	0	5	10	15	20	25	30	35	40	45
Wind Chill Temperature (in °F)	35	33	22	16	12	8	6	4	3	2

(a) What does a 20-mph wind make 35°F feel like?
(b) Sketch a scatter plot of the data.
(c) Explain why an exponential model would be appropriate.
(d) Find an exponential model for the data.
(e) According to the model, what is the wind-chill factor for a 22-mph wind?

Discovery Project 5

Exponential and Logistic Modeling of Diseases

Diseases that are contagious and are transmitted homogeneously through a population often appear to be spreading exponentially. That is, the rate of spread is proportional to the number of people in the population who are already infected. This is a reasonable model as long as the number of infected people is relatively small compared with the number of people in the population who can be infected. The standard exponential model looks like this:

$$f(t) = Y_0 e^{rt}.$$

Y_0 is the initial number of infected people, (the number on the arbitrarily decided day 0) and r is the rate by which the disease spreads through the population. If the time t is measured in days, then r is the ratio of new infections to current infections each day.

Suppose that in Big City, population 3855, there is an outbreak of dingbat disease. On the first Monday after the outbreak was discovered (day 0), 72 people have dingbat disease. On the following Monday (day 7), 193 people have dingbat disease.

1. Using the exponential model, Y_0 is clearly 72. Calculate the value of r.
2. Using your values of Y_0 and r, predict the number of cases of dingbat disease that will be reported on day 14.

It turns out that eventually, the spread of disease must slow as the number of infected people approaches the number of susceptible people. What happens is that some of the people to whom the disease would spread are already infected. As time goes on, the spread of the disease becomes proportional to the number of susceptible and uninfected people. The disease then follows the logistic model:

$$g(t) = \frac{rY_0}{aY_0 + (r - aY_0)e^{-rt}}.$$

Y_0 is still the initial value, and r serves the same function as before, at least at the initial time. The extra parameter a is not so obvious, but it is inversely related to the number of people susceptible to the disease. Unfortunately, the algebra to solve for a is quite complicated. It is much easier to approximate a using the same r from the exponential model.

3. On day 14 in Big City, 481 people have dingbat disease. Using the values of Y_0 and r from Exercise 1 in the rule of the function g, determine the value of a. [*Hint:* $g(14) = 481$.] Does g overestimate or underestimate the number of people with dingbat disease on day 7?

4. Use the function g from Exercise 3 to approximate the number of people in Big City who are susceptible to the disease. Does this model make sense? [Remember, as time goes on, the number of people infected approaches the number of people susceptible.]

5. In the logistic model, the rate at which the disease spreads tends to fall over time. This means that the value of r you calculated in Exercise 1 is a little low. Raise the value of r and find the new value of a as in Exercise 3. Experiment until you find a value of r for which $g(7) = 193$ (meaning that the model g matches the data on day 7).

6. Using the function g from Exercise 5, repeat Exercise 4.

Chapter 6

Trigonometric Functions

Don't touch that dial!

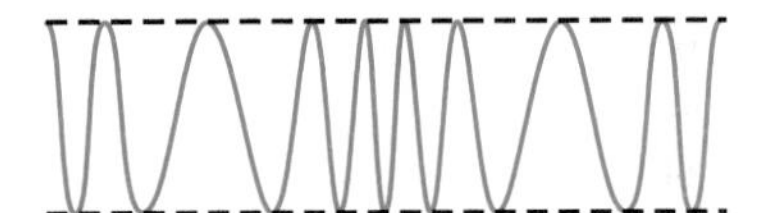

AM signal

FM signal

Radio stations transmit by sending out a signal in the form of an electromagnetic wave that can be described by a trigonometric function. The shape of this signal is modified by the sounds being transmitted. AM radio signals are modified by varying the "height," or amplitude, of the waves, whereas FM signals are modified by varying the frequency of the waves. See Exercise 41 on page 438.

Interdependence of Sections

$$6.1 \to 6.2 \to 6.3 \to 6.4 \nearrow 6.5, \searrow 6.6^*$$

Chapter Outline

The ancient Greeks developed trigonometry for measuring angles and sides of triangles in order to solve problems in astronomy, navigation, and surveying.[†] But with the invention of calculus in the 17th century and the subsequent explosion of knowledge in the physical sciences, a different viewpoint toward trigonometry arose.

Whereas the ancients dealt only with *angles,* the classical trigonometric concepts of sine and cosine are now considered as *functions* with domain the set of all *real numbers.* The advantage of this switch in viewpoint is that almost any phenomena involving rotation or vibration can be described in terms of trigonometric functions, including light rays, sound waves, electron orbitals, planetary orbits, radio transmission, vibrating strings, pendulums, and many more.

The presentation of trigonometry here reflects this modern viewpoint. Nevertheless, angles still play an important role in defining the trigonometric functions, so the chapter begins with them.

6.1 Angles and Their Measurement

In trigonometry an **angle** is formed by rotating a half-line around its endpoint (the **vertex**), as shown in Figure 6–1, where the arrow indicates the direction of rotation. The position of the half-line at the beginning is the **initial side** and its final position is the **terminal side** of the angle.

*Parts of Section 6.6 may be covered much earlier; see the table at the beginning of Section 6.2.

[†]In fact, "trigonometry" means "triangle measurement."

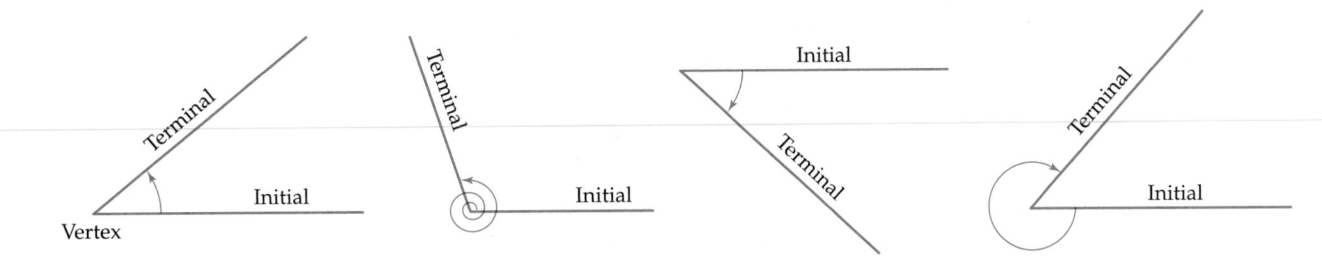

Figure 6–1

Figure 6–2 shows that different angles (that is, angles obtained by different rotations) may have the same initial and terminal side.* Such angles are said to be **coterminal.**

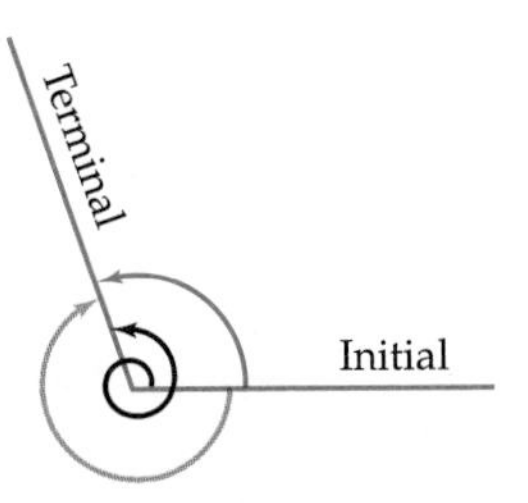

Figure 6–2

An angle in the coordinate plane is said to be in **standard position** if its vertex is at the origin and its initial side on the positive x-axis, as in Figure 6–3. When measuring angles in standard position, we use positive numbers for angles obtained by counterclockwise rotation (**positive angles**) and negative numbers for ones obtained by clockwise rotation (**negative angles**).

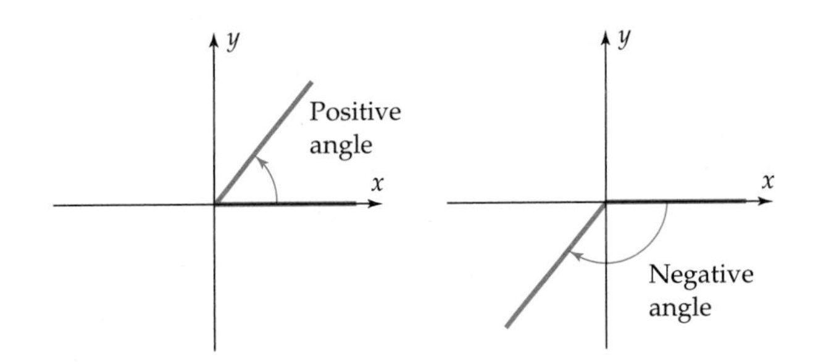

Figure 6–3

The classical unit for angle measurement is the **degree** (in symbols °), as explained in the Geometry Review Appendix. You should be familiar with the positive angles in standard position shown in Figure 6–4 on the next page. Note that a 360° angle corresponds to one full revolution and thus is coterminal with an angle of 0°.

*They are *not* the same angle, however. For instance, both $\frac{1}{2}$ turn and $1\frac{1}{2}$ turns put a circular faucet handle in the same position, but the water flow is quite different.

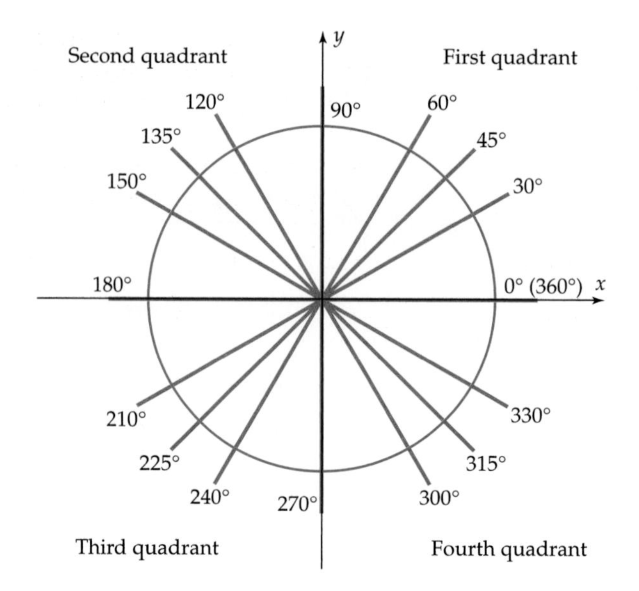

Figure 6–4

Radian Measure

Because it simplifies many formulas in calculus and physics, a different unit of angle measurement is used in mathematical and scientific applications. Recall that the unit circle is the circle of radius 1 with center at the origin, whose equation is $x^2 + y^2 = 1$. When a positive angle θ is in standard position, its initial side passes through (1, 0) and its terminal side intersects the unit circle, as shown in Figure 6–5. The **radian** measure of such an angle is defined to be

> **the distance along the unit circle in the counterclockwise direction from the point (1, 0) to the point where the terminal side of the angle intersects the unit circle.**

The radian measure of a negative angle in standard position is found in the same way, except that you move clockwise along the unit circle. Figure 6–5 shows angles of 3.75 and -2 radians, respectively.

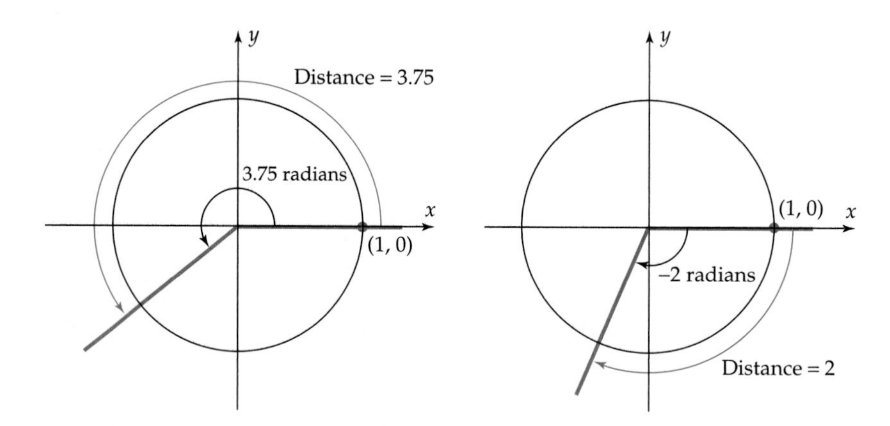

Figure 6–5

To become comfortable with radian measure, think of the terminal side of the angle revolving around the origin: when it makes one full revolution it produces an angle of 2π radians (because the circumference of the unit circle is 2π). When it makes half a revolution it forms an angle whose radian measure is 1/2 of 2π, that is, π radians, and so on, as illustrated in Figure 6–6 and the table below.

1 revolution

2π radians

3/4 revolution

$\frac{3}{4} \cdot 2\pi = \frac{3\pi}{2}$ radians

1/2 revolution

$\frac{1}{2} \cdot 2\pi = \pi$ radians

1/4 revolution

$\frac{1}{4} \cdot 2\pi = \frac{\pi}{2}$ radians

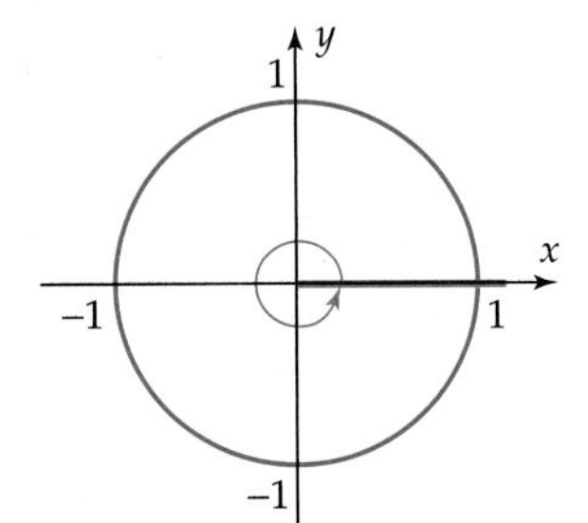

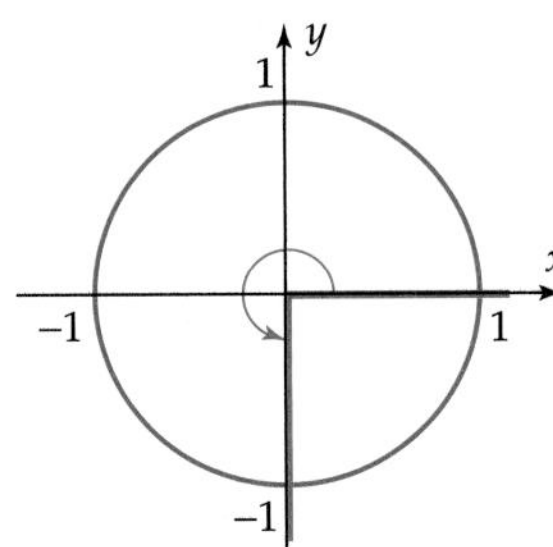

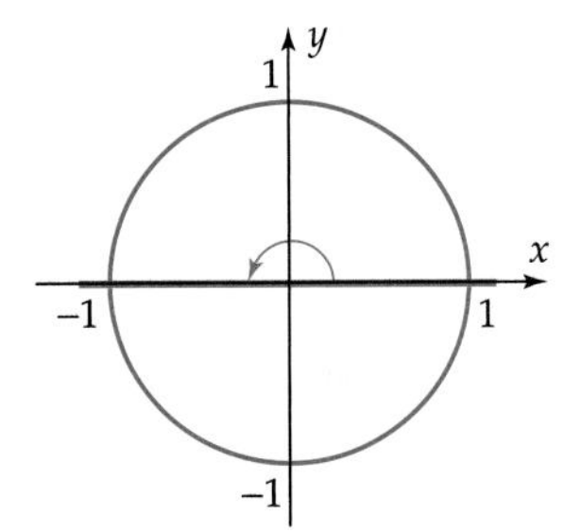

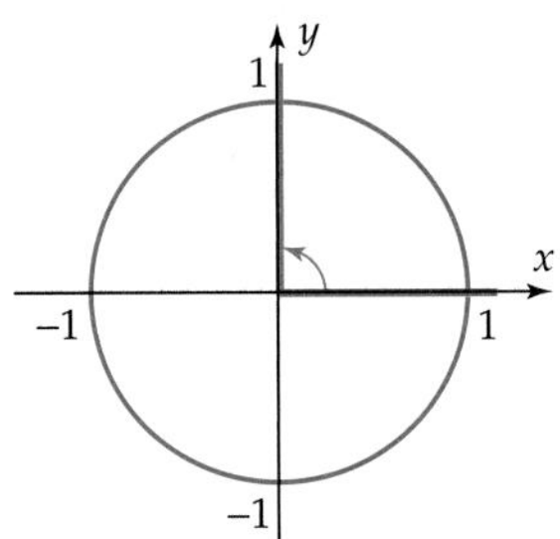

Figure 6–6

Terminal Side	**Radian Measure of Angle**	**Equivalent Degree Measure**
1 revolution	2π	360°
$\frac{7}{8}$ revolution	$\frac{7}{8} \cdot 2\pi = \frac{7\pi}{4}$	$\frac{7}{8} \cdot 360 = 315°$
$\frac{3}{4}$ revolution	$\frac{3}{4} \cdot 2\pi = \frac{3\pi}{2}$	$\frac{3}{4} \cdot 360 = 270°$
$\frac{2}{3}$ revolution	$\frac{2}{3} \cdot 2\pi = \frac{4\pi}{3}$	$\frac{2}{3} \cdot 360 = 240°$
$\frac{1}{2}$ revolution	$\frac{1}{2} \cdot 2\pi = \pi$	$\frac{1}{2} \cdot 360 = 180°$
$\frac{1}{3}$ revolution	$\frac{1}{3} \cdot 2\pi = \frac{2\pi}{3}$	$\frac{1}{3} \cdot 360 = 120°$
$\frac{1}{4}$ revolution	$\frac{1}{4} \cdot 2\pi = \frac{\pi}{2}$	$\frac{1}{4} \cdot 360 = 90°$
$\frac{1}{6}$ revolution	$\frac{1}{6} \cdot 2\pi = \frac{\pi}{3}$	$\frac{1}{6} \cdot 360 = 60°$
$\frac{1}{8}$ revolution	$\frac{1}{8} \cdot 2\pi = \frac{\pi}{4}$	$\frac{1}{8} \cdot 360 = 45°$
$\frac{1}{12}$ revolution	$\frac{1}{12} \cdot 2\pi = \frac{\pi}{6}$	$\frac{1}{12} \cdot 360 = 30°$

Although equivalent degree measures are given in the table, you should learn to "think radian" as much as possible, rather than mentally translating from degrees to radians.

Example 1 To construct an angle of $16\pi/3$ radians in standard position, note that

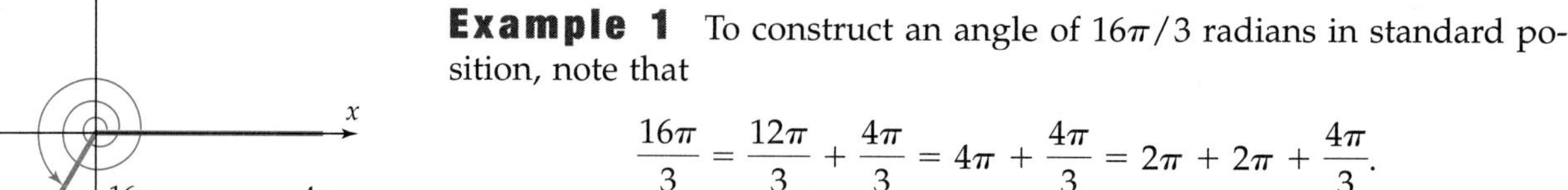

$$\frac{16\pi}{3} = \frac{12\pi}{3} + \frac{4\pi}{3} = 4\pi + \frac{4\pi}{3} = 2\pi + 2\pi + \frac{4\pi}{3}.$$

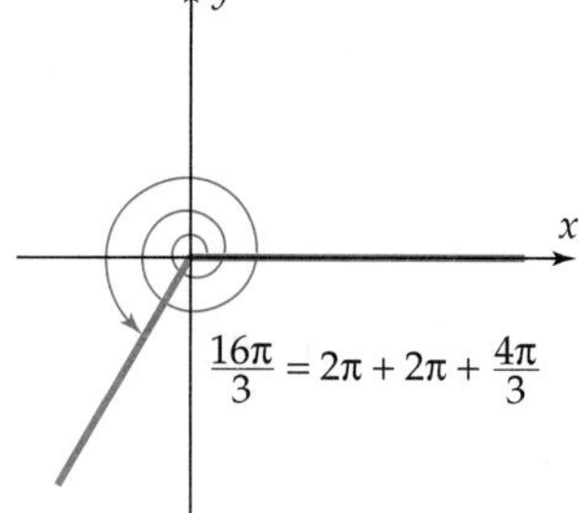

Figure 6–7

So, the terminal side must be rotated counterclockwise through two complete revolutions (each full-circle revolution is 2π radians) and then rotated an additional 2/3 of a revolution (since $4\pi/3$ is 2/3 of a complete revolution of 2π radians), as shown in Figure 6–7. ■

Example 2 Since $-5\pi/4 = -\pi - \pi/4$, an angle of $-5\pi/4$ radians in standard position is obtained by rotating the terminal side *clockwise* for half a revolution (π radians) plus an additional 1/8 of a revolution (since $\pi/4$ is 1/8 of a full-circle revolution of 2π radians), as shown in Figure 6–8. ■

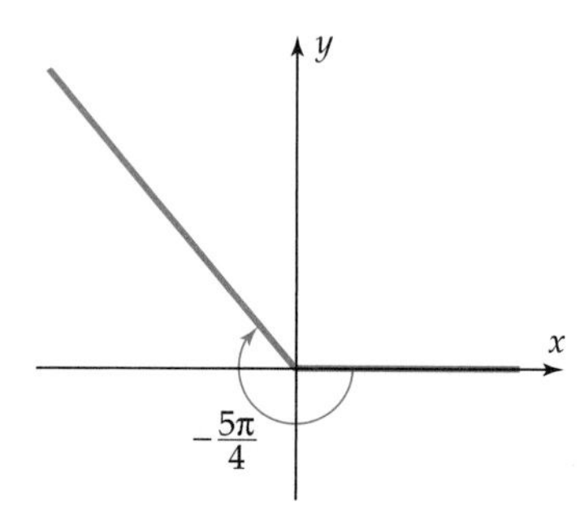

Figure 6–8

Consider an angle of t radians in standard position (Figure 6–9). Since 2π radians corresponds to a full revolution of the terminal side, this angle has the same terminal side as an angle of $t + 2\pi$ radians or $t - 2\pi$ radians or $t + 4\pi$ radians:

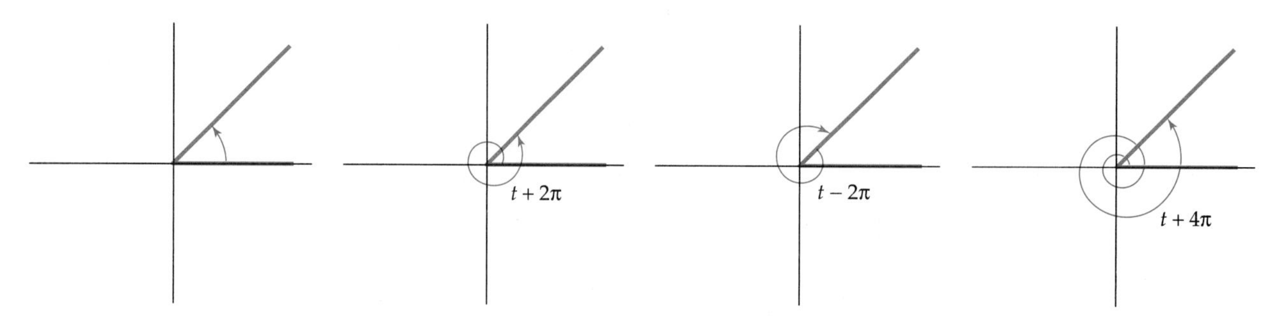

Figure 6–9

The same thing is true in general:

Coterminal Angles

Increasing or decreasing the radian measure of an angle by an integer multiple of 2π results in a coterminal angle.

Example 3 Find angles in standard position that are coterminal with an angle of: **(a)** $23\pi/5$ radians **(b)** $-\pi/12$ radians.

Solution

(a) We can subtract 2π to obtain a coterminal angle whose measure is

$$\frac{23\pi}{5} - 2\pi = \frac{23\pi}{5} - \frac{10\pi}{5} = \frac{13\pi}{5} \text{ radians,}$$

or we can subtract 4π to obtain a coterminal angle of measure

$$\frac{23\pi}{5} - 4\pi = \frac{3\pi}{5} \text{ radians.}$$

Subtracting 6π produces a coterminal angle of

$$\frac{23\pi}{5} - 6\pi = -\frac{7\pi}{5} \text{ radians.}$$

(b) An angle of $\frac{-\pi}{12}$ radians is coterminal with an angle of

$$-\frac{\pi}{12} + 2\pi = \frac{23\pi}{12} \text{ radians}$$

and with an angle of

$$-\frac{\pi}{12} - 2\pi = -\frac{25\pi}{12} \text{ radians.} \quad \blacksquare$$

Radian/Degree Conversion

Although we shall generally work with radians, it may occasionally be necessary to convert from radian to degree measure or vice versa. The key to doing this is the fact that

$$(*) \qquad \pi \textit{ radians} = 180°.$$

Dividing both sides of $(*)$ by π shows that

$$1 \text{ radian} = \frac{180}{\pi} \text{ degrees} \approx 57.3°,$$

and dividing both sides of $(*)$ by 180 shows that

$$1° = \frac{\pi}{180} \text{ radians} \approx .0175 \text{ radians.}$$

Consequently, we have these rules:

Radian/Degree Conversion

To convert radians to degrees, multiply by $\frac{180}{\pi}$.

To convert degrees to radians, multiply by $\frac{\pi}{180}$.

Example 4

(a) To find the degree measure of an angle of 2.4 radians, multiply by $\frac{180}{\pi}$:

$$(2.4)\left(\frac{180}{\pi}\right) = \frac{432}{\pi} \approx 137.51°.$$

(b) An angle of $-.3$ radians has degree measure

$$(-.3)\left(\frac{180}{\pi}\right) = \frac{-54}{\pi} \approx -17.19°. \quad \blacksquare$$

Angle conversions from radian to degree measure or vice versa can be done directly on a calculator. Since procedures vary from one model to another, check your instruction manual.

Arc Length and Angular Speed*

Consider a circle of radius r, with center at the origin, and an angle of θ radians in standard position, as shown in Figure 6–10. As you can see, the sides of the angle of θ radians determine an arc length s along the circle. We say that the **central angle** of θ radians **subtends an arc** of length s on the circle. It can be shown that the ratio of the arc length s to the circumference of the entire circle (namely, $2\pi r$) is the same as the ratio of the angle of θ radians to the full-circle angle of 2π radians; that is, $\frac{s}{2\pi r} = \frac{\theta}{2\pi}$. Solving this equation for s, we obtain this fact:

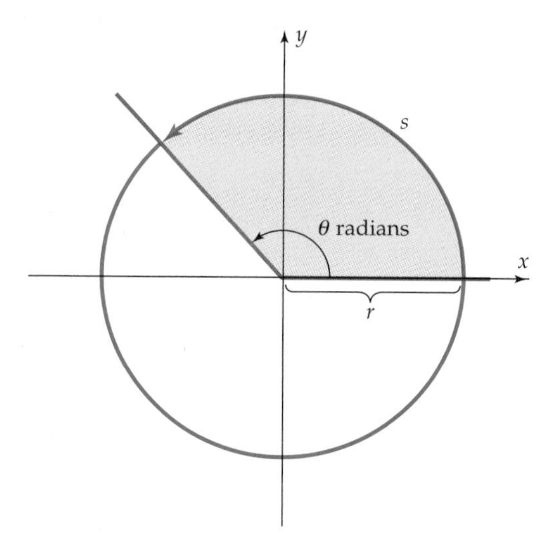

Figure 6–10

*The remainder of this section is not needed in the sequel and may be omitted if desired.

Arc Length

A central angle of θ radians in a circle of radius r subtends an arc of length

$$s = \theta r,$$

that is, the length s of the arc is the product of the radian measure of the angle and the radius of the circle.

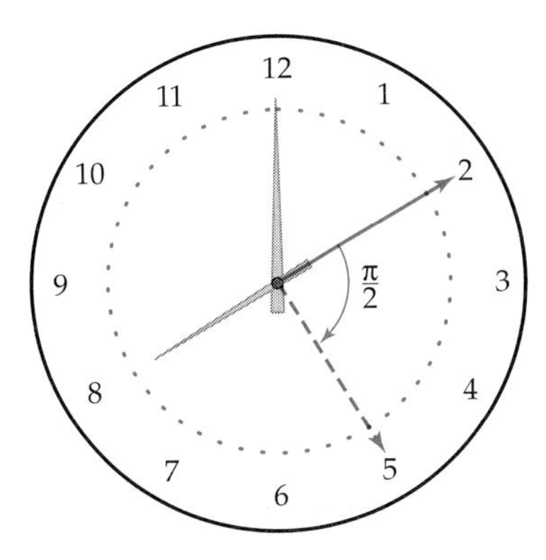

Figure 6–11

Example 5 The second hand on a large clock is 6 inches long. How far does the tip of the second hand move in 15 seconds?

Solution The second hand makes a full revolution every 60 seconds, that is, it moves through an angle of 2π radians. During a 15-second interval it will make $\frac{15}{60} = \frac{1}{4}$ of a revolution, moving through an angle of $\pi/2$ radians (Figure 6–11). If we think of the second hand as the radius of a circle, then during a 15-second interval its tip travels along the arc subtended by an angle of $\pi/2$ radians. Therefore the distance (arc length) traveled by the tip of the second hand is

$$s = \theta r = \frac{\pi}{2} \cdot 6 = 3\pi \approx 9.425 \text{ inches.} \quad \blacksquare$$

Example 6 Suppose the circle in Figure 6–10 has radius 6 and the arc length s is 14.64. We can find the measure of the central angle that subtends the arc s by solving $s = \theta r$ for θ:

$$\theta = \frac{s}{r} = \frac{14.64}{6} = 2.44 \text{ radians.} \quad \blacksquare$$

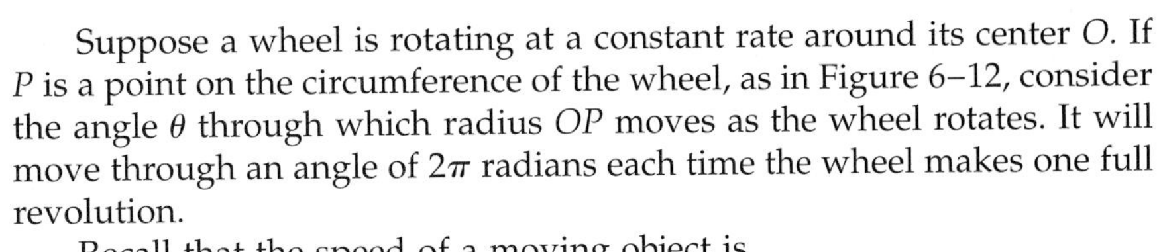
Suppose a wheel is rotating at a constant rate around its center O. If P is a point on the circumference of the wheel, as in Figure 6–12, consider the angle θ through which radius OP moves as the wheel rotates. It will move through an angle of 2π radians each time the wheel makes one full revolution.

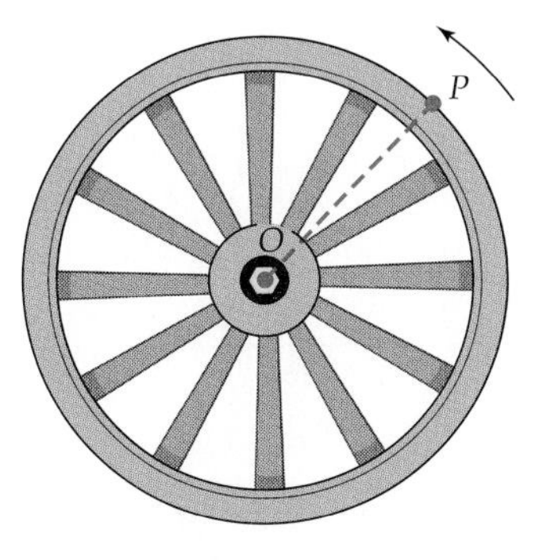

Figure 6–12

Recall that the speed of a moving object is

$$\frac{\text{distance traveled}}{\text{time elapsed}}, \quad \text{or more briefly,} \quad \frac{\text{distance}}{\text{time}}.$$

Thus, if an object moves s feet in t seconds, its speed is s/t feet per second. Similarly, the **angular speed** of the wheel is the radian measure of the angle through which OP travels divided by the time, that is,

$$\textbf{angular speed} = \frac{\textbf{angle}}{\textbf{time}} = \frac{\theta}{t}.$$

Example 7 A merry-go-round makes 8 revolutions per minute.

(a) What is the angular speed of the merry-go-round in radians per minute?

(b) How fast is a horse 12 feet from the center traveling?

(c) How fast is a horse 4 feet from the center traveling?

Solution

(a) Each revolution of the merry-go-round corresponds to a central angle of 2π radians, so it travels through an angle of $8 \cdot 2\pi = 16\pi$ radians in one minute. Therefore, its

$$\text{angular speed} = \frac{\text{angle}}{\text{time}} = \frac{16\pi}{1} = 16\pi \text{ radians per minute.}$$

(b) The horse 12 feet from the center travels along a circle of radius 12. As we saw in part (a), the angle through which the horse travels in one minute is 16π radians. By the arc length formula, the distance traveled by the horse is $r\theta = 12(16\pi) = 192\pi$ ft. Consequently, its linear speed is 192π ft/min (approximately 6.9 miles per hour).

(c) Proceeding as in part (b), with 4 in place of 12, we see that the linear speed of the horse is

$$\frac{\text{distance}}{\text{time}} = \frac{4(16\pi)}{1} = 64\pi \text{ ft/min}, \qquad \text{or equivalently,} \qquad 2.28 \text{ mph.}$$ ■

In Example 7(a), note that the angular speed is $8 \cdot 2\pi$, that is, the number of revolutions per minute times 2π. In Example 7(b), the linear speed of the horse is $12(16\pi)$, that is, the radius times the angular speed. The same relationships are valid for the angular speed of any rotating disc and the linear speed of a point on its circumference:

$$\textbf{angle of rotation} = \textbf{(number of revolutions)} \cdot (2\pi)$$

$$\textbf{angular speed} = \textbf{(revolutions per unit of time)} \cdot (2\pi)$$

$$\textbf{linear speed} = \textbf{(radius)} \cdot \textbf{(angular speed).}$$

Exercises 6.1

In Exercises 1–5, find the radian measure of the angle in standard position formed by rotating the terminal side the given amount.

1. 1/9 of a circle

2. 1/24 of a circle

3. 1/18 of a circle

4. 1/72 of a circle

5. 1/36 of a circle

6. State the radian measure of *every* standard position angle in the figure on the next page.*

*This is the same diagram that appears in Figure 6–4 on page 392, showing positive angles in standard position.

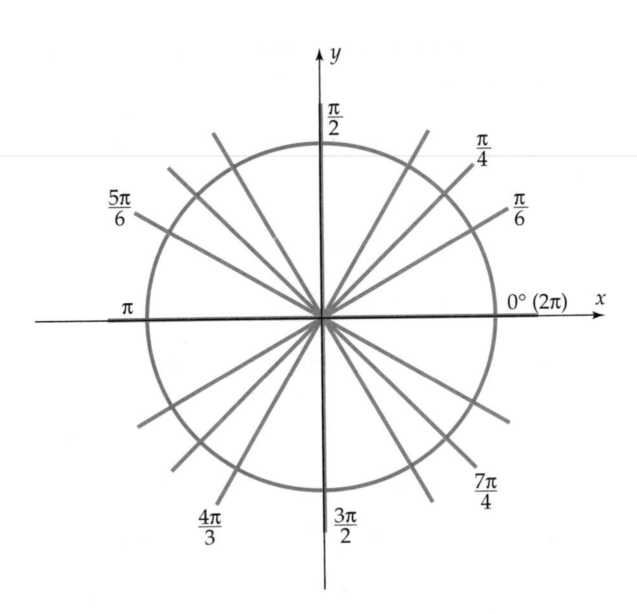

In Exercises 7–10, find the radian measure of four angles in standard position that are coterminal with the angle in standard position whose measure is given.

7. $\pi/4$ **8.** $7\pi/5$

9. $-\pi/6$ **10.** $-9\pi/7$

In Exercises 11–18, find the radian measure of an angle in standard position that has measure between 0 and 2π and is coterminal with the angle in standard position whose measure is given.

11. $-\pi/3$ **12.** $-3\pi/4$ **13.** $19\pi/4$

14. $16\pi/3$ **15.** $-7\pi/5$ **16.** $45\pi/8$

17. 7 **18.** 18.5

In Exercises 19–30, convert the given degree measure to radians.

19. 6° **20.** −10° **21.** −12° **22.** 36°

23. 75° **24.** −105° **25.** 135° **26.** −165°

27. −225° **28.** 252° **29.** 930° **30.** −585°

In Exercises 31–42, convert the given radian measure to degrees.

31. $\pi/5$ **32.** $-\pi/6$ **33.** $-\pi/10$

34. $2\pi/5$ **35.** $3\pi/4$ **36.** $-5\pi/3$

37. $\pi/45$ **38.** $-\pi/60$ **39.** $-5\pi/12$

40. $7\pi/15$ **41.** $27\pi/5$ **42.** $-41\pi/6$

In Exercises 43–48, determine the positive radian measure of the angle that the second hand of a clock traces out in the given time.

43. 40 seconds **44.** 50 seconds **45.** 35 seconds

46. 2 minutes and 15 seconds

47. 3 minutes and 25 seconds

48. 1 minute and 55 seconds

49. The second hand on a clock is 6 centimeters long. How far does its tip travel in 40 seconds? (See Exercise 43.)

50. The second hand on a clock is 5 centimeters long. How far does its tip travel in 2 minutes and 15 seconds? (See Exercise 46.)

51. If the radius of the circle in the figure is 20 centimeters and the length of arc s is 85 centimeters, what is the radian measure of the angle θ?

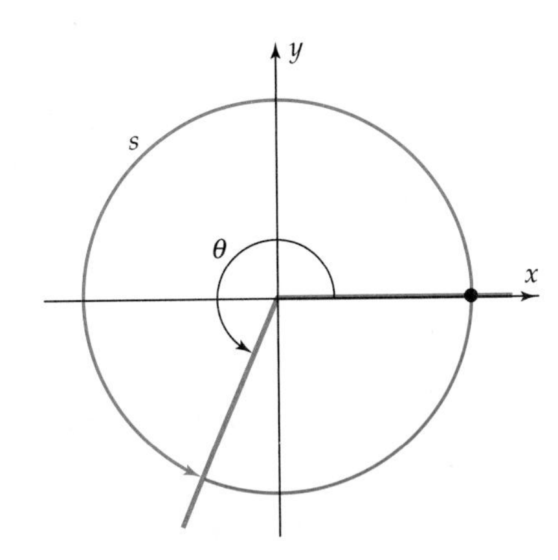

52. Find the radian measure of the angle θ in the preceding figure if the *diameter* of the circle is 150 centimeters and s has length 360 centimeters.

In Exercises 53–56, assume that a wheel on a car has radius 36 centimeters. Find the angle (in radians) that the wheel turns while the car travels the given distance.

53. 2 meters (= 200 centimeters) **54.** 5 meters

55. 720 meters **56.** 1 kilometer (= 1000 meters)

In Exercises 57–60, find the length of the circular arc subtended by the central angle whose radian measure is given. Assume the circle has diameter 10.

57. 1 radian **58.** 2 radians **59.** 1.75 radians

60. 2.2 radians

In Exercises 61–64, the latitudes of a pair of cities are given. Assume that one city is directly south of the other and that the earth is a perfect sphere of radius 4000 miles. Find the distance between the two cities. [The ***latitude*** *of a point P on the earth is the degree measure of the angle θ between the point and the plane of the equator (with the center of the earth being the vertex), as shown in the figure on the next page. Remember that angles are measured in radians in the arc length formula.]*

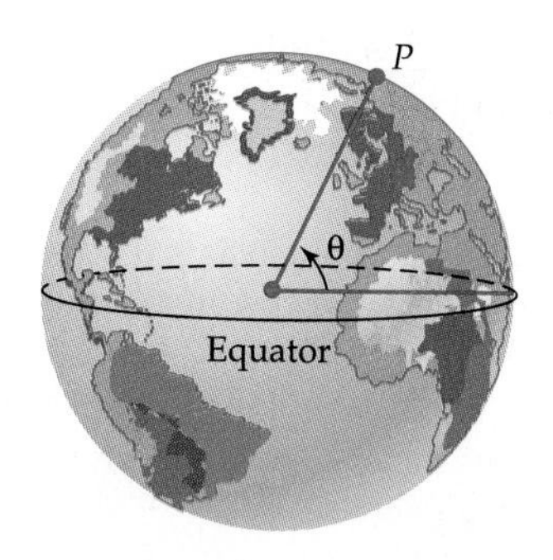

61. The North Pole (latitude 90° north) and Springfield, Illinois (latitude 40° north).

62. San Antonio, Texas (latitude 29.5° north), and Mexico City, Mexico (latitude 20° north).

63. Cleveland, Ohio (latitude 41.5° north), and Tampa, Florida (latitude 28° north).

64. Copenhagen, Denmark (latitude 54.3° north), and Rome, Italy (latitude 42° north).

In Exercises 65–72, a wheel is rotating around its axle. Find the angle (in radians) through which the wheel turns in the given time when it rotates at the given number of revolutions per minute (rpm). Assume $t > 0$ and $k > 0$.

65. 3.5 minutes, 1 rpm	**66.** t minutes, 1 rpm
67. 1 minute, 2 rpm	**68.** 3.5 minutes, 2 rpm
69. 4.25 minutes, 5 rpm	**70.** t minutes, 5 rpm
71. 1 minute, k rpm	**72.** t minutes, k rpm

73. One end of a rope is attached to a winch (circular drum) of radius 2 feet and the other to a steel beam on the ground. When the winch is rotated, the rope wraps around the drum and pulls the object upward (see figure). Through what angle must the winch be rotated in order to raise the beam 6 feet above the ground?

74. A circular saw blade has an angular speed of 15,000 radians per minute.
    (a) How many revolutions per minute does the saw make?
    (b) How long will it take the saw to make 6000 revolutions?

75. A circular gear rotates at the rate of 200 revolutions per minute (rpm).
    (a) What is the angular speed of the gear in radians per minute?
    (b) What is the linear speed of a point on the gear 2 inches from the center in inches per minute and in feet per minute?

76. A wheel in a large machine is 2.8 feet in diameter and rotates at 1200 rpm.
    (a) What is the angular speed of the wheel?
    (b) How fast is a point on the circumference of the wheel traveling in feet per minute? In miles per hour?

77. A riding lawn mower has wheels that are 15 inches in diameter. If the wheels are making 2.5 revolutions per second
    (a) What is the angular speed of a wheel?
    (b) How fast is the lawn mower traveling in miles per hour?

78. A bicycle has wheels that are 26 inches in diameter. If the bike is traveling at 14 mph, what is the angular speed of each wheel?

79. A merry-go-round horse is traveling at 10 feet per second and the merry-go-round is making 6 revolutions per minute. How far is the horse from the center of the merry-go-round?

80. The pedal sprocket of a bicycle has radius 4.5 inches and the rear wheel sprocket has radius 1.5 inches (see figure). If the rear wheel has a radius of 13.5 inches and the cyclist is pedaling at the rate of 80 rpm, how fast is the bicycle traveling in feet per minute? In miles per hour?

6.2 The Sine, Cosine, and Tangent Functions

Roadmap

Instructors who wish to cover all six trigonometric functions simultaneously should incorporate Section 6.6 into Sections 6.2–6.4, as follows.

Subsection of Section 6.6	Cover at the end of
Definitions; Alternative Descriptions	Section 6.2
Algebra and Identities	Section 6.3
Graphs	Section 6.4

Instructors who prefer to introduce triangle trigonometry early should cover Section 7.1 between Sections 6.2 and 6.3.

Unlike most of the functions seen thus far, the definitions of the sine and cosine functions do not involve algebraic formulas. Instead, these functions are defined geometrically, using the unit circle.* Recall that the unit circle is the circle of radius 1 with center at the origin, whose equation is $x^2 + y^2 = 1$.

Both the sine and cosine functions have the set of all real numbers as domain. Their rules are given by the following three-step geometric process:

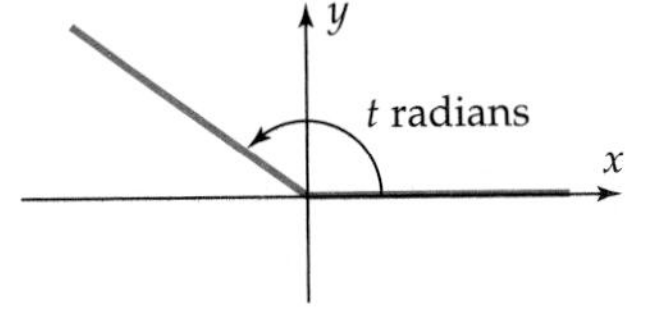

1. Given a real number t, construct an angle of t radians in standard position.
2. Find the coordinates of the point P where the terminal side of this angle meets the unit circle $x^2 + y^2 = 1$, say $P = (a,b)$.
3. The value of the **cosine function** at t (denoted $\cos t$) is the x-coordinate of P:

$$\cos t = a.$$

The value of the **sine function** at t (denoted $\sin t$) is the y-coordinate of P:

$$\sin t = b.$$

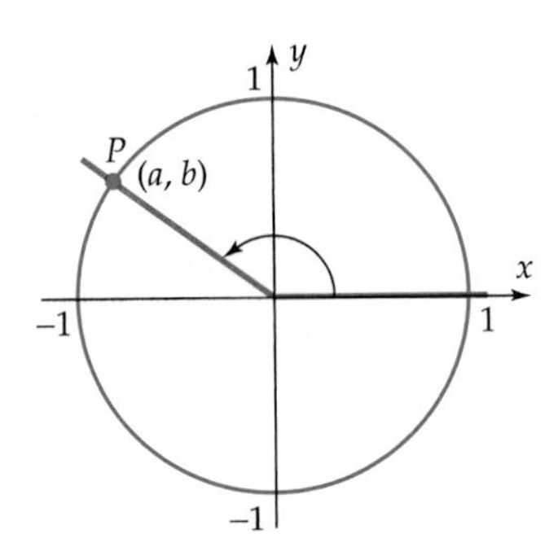

In other words,

Sine and Cosine

If P is the point where the terminal side of an angle of t radians in standard position meets the unit circle, then

P has coordinates $(\cos t, \sin t)$.

*If you have previously studied the trigonometry of triangles, the definition given here may not look familiar. If so, just concentrate on this definition and don't worry about relating it to any definition you remember from the past. The connection between this definition and the trigonometry of triangles will be explained in Chapter 7.

GRAPHING EXPLORATION

With your calculator in radian mode and parametric graphing mode, set the range values as follows:

$$0 \le t \le 2\pi, \qquad -1.8 \le x \le 1.8, \qquad -1.2 \le y \le 1.2^*$$

Then, graph the curve given by the parametric equations

$$x = \cos t \quad \text{and} \quad y = \sin t.$$

The graph is the unit circle. Use the trace to move around the circle. At each point, the screen will display three numbers: the values of t, x, and y. For each t, the cursor is on the point where the terminal side of an angle of t radians meets the unit circle, so the corresponding x is the number $\cos t$ and the corresponding y is the number $\sin t$.

The **tangent function** is defined as the quotient of the sine and cosine functions. Its value at the number t, denoted $\tan t$, is given by

$$\tan t = \frac{\sin t}{\cos t}$$

Technology Tip

Throughout this chapter, be sure your calculator is set for RADIAN mode. Use the MODE(S) menu on TI and HP and the SETUP menu on Sharp and Casio.

Every calculator has SIN, COS, and TAN keys for evaluating the sine, cosine, and tangent functions. For instance, a calculator (in radian mode) gives these approximations (Figure 6–13).

```
sin(3.5)
          -.3507832277
cos(5)
           .2836621855
tan(12.7)
           .1344305066
```

Figure 6–13

Nevertheless, you should know how to use the definition to find exact values of these functions when possible, as illustrated in Examples 1–4.

Example 1 Evaluate the three trigonometric functions at

(a) $t = \pi$ (b) $t = \pi/2$.

*These settings give a square viewing window on calculators with a screen measuring approximately 95 by 63 pixels, and hence the unit circle will look like a circle. For wider screens, adjust the x range settings to obtain a square window.

Solution

(a) Construct an angle of π radians, as in Figure 6–14. Its terminal side lies on the negative x-axis and intersects the unit circle at $P = (-1, 0)$. Hence,

$$\sin \pi = y\text{-coordinate of } P = 0$$
$$\cos \pi = x\text{-coordinate of } P = -1$$
$$\tan \pi = \frac{\sin \pi}{\cos \pi} = \frac{0}{-1} = 0$$

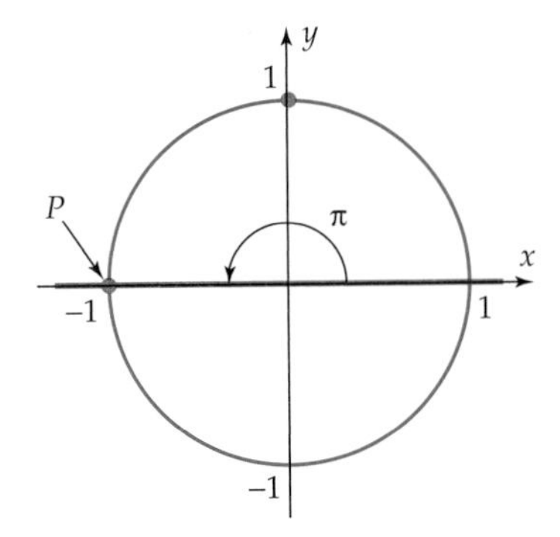

Figure 6–14

(b) An angle of $\pi/2$ radians (Figure 6–15) has its terminal side on the positive y-axis and intersects the unit circle at $P = (0, 1)$.

$$\cos \frac{\pi}{2} = x\text{-coordinate of } P = 0$$
$$\sin \frac{\pi}{2} = y\text{-coordinate of } P = 1$$
$$\tan \frac{\pi}{2} = \frac{\sin(\pi/2)}{\cos(\pi/2)} = \frac{1}{0} \text{ undefined} \quad \blacksquare$$

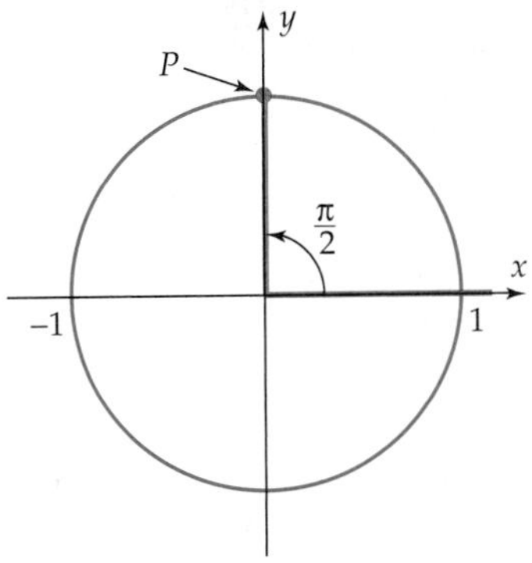

Figure 6–15

Example 1(b) shows that the domain of the tangent function must exclude numbers for which $\cos t = 0$. These occur when the point P has coordinates (0, 1) or (0, −1). Therefore,

> **the domain of the tangent function consists of all real numbers except $t = \pm\pi/2, \pm 3\pi/2, \pm 5\pi/2, \ldots$.**

Special Values

The trigonometric functions can be evaluated exactly at $t = \pi/3$, $t = \pi/4$, $t = \pi/6$, and any integer multiples of these numbers by using the following facts (which are explained in the Geometry Review Appendix):

A right triangle with hypotenuse 1 and angles of $\pi/6$ and $\pi/3$ radians has sides of lengths 1/2 (opposite the angle of $\pi/6$) and $\sqrt{3}/2$ (opposite the angle of $\pi/3$).

A right triangle with hypotenuse 1 and two angles of $\pi/4$ radians has two sides of length $\sqrt{2}/2$.

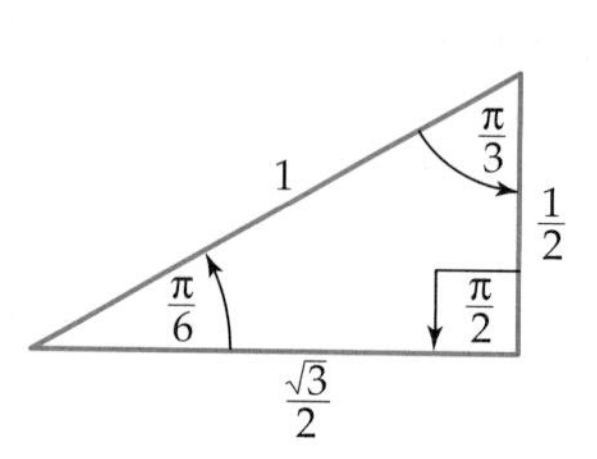

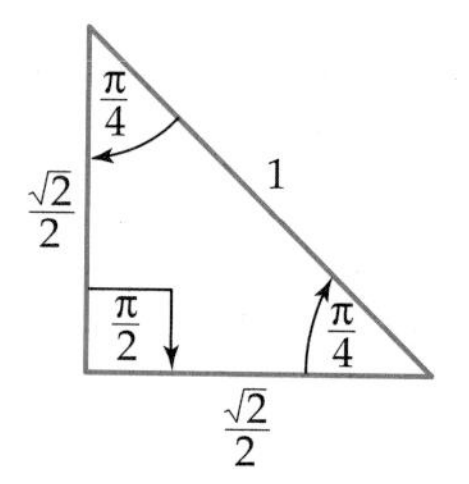

Figure 6–16

Example 2 Evaluate the three trigonometric functions at $t = \pi/6$.

Solution Construct an angle of $\pi/6$ radians in standard position and let P be the point where its terminal side intersects the unit circle (Figure 6–17). Draw a vertical line from P to the x-axis, forming a right triangle with hypotenuse 1, angles of $\pi/6$ and $\pi/3$ radians, and sides of lengths $1/2$ and $\sqrt{3}/2$, as described above. Then P has coordinates $(\sqrt{3}/2, 1/2)$ and by definition

$$\sin\frac{\pi}{6} = y\text{-coordinate of } P = \frac{1}{2}$$

$$\cos\frac{\pi}{6} = x\text{-coordinate of } P = \frac{\sqrt{3}}{2}$$

$$\tan\frac{\pi}{6} = \frac{\sin(\pi/6)}{\cos(\pi/6)} = \frac{1/2}{\sqrt{3}/2} = \frac{1}{\sqrt{3}} = \frac{\sqrt{3}}{3} \quad \blacksquare$$

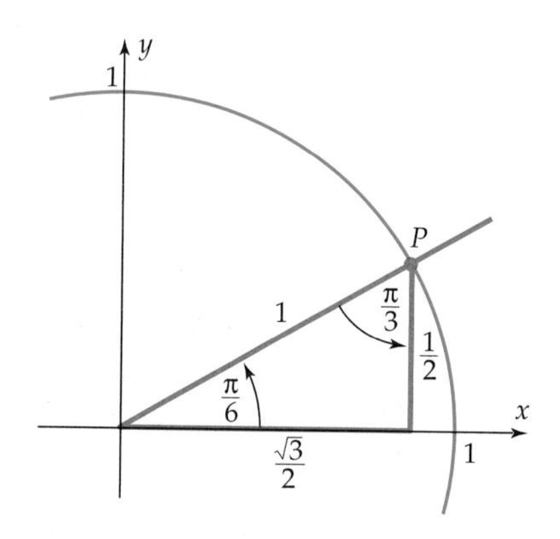

Figure 6–17

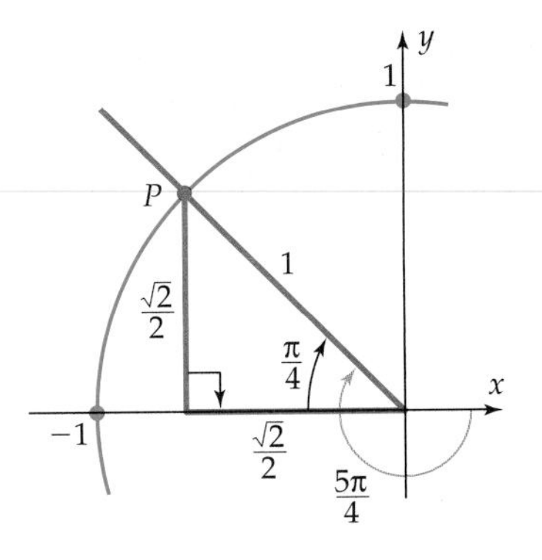

Figure 6–18

Example 3 Evaluate the trigonometric functions at $-5\pi/4$.

Solution Construct an angle of $-5\pi/4$ radians in standard position and let P be the point where the terminal side intersects the unit circle. Draw a vertical line from P to the x-axis, forming a right triangle with hypotenuse 1 and two angles of $\pi/4$ radians, as shown in Figure 6–18.

Each side of this triangle has length $\sqrt{2}/2$, as explained above. Therefore, the coordinates of P are $(-\sqrt{2}/2, \sqrt{2}/2)$, so that

$$\sin\frac{-5\pi}{4} = y\text{-coordinate of } P = \frac{\sqrt{2}}{2}$$

$$\cos\frac{-5\pi}{4} = x\text{-coordinate of } P = -\frac{\sqrt{2}}{2}$$

$$\tan\frac{-5\pi}{4} = \frac{\sin t}{\cos t} = \frac{\sqrt{2}/2}{-\sqrt{2}/2} = -1. \quad \blacksquare$$

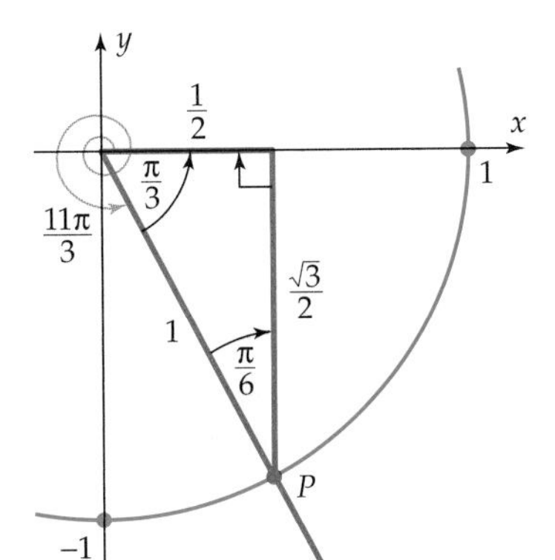

Figure 6–19

Example 4 Evaluate the trigonometric functions at $11\pi/3$.

Solution Construct an angle of $11\pi/3$ radians in standard position and draw a vertical line from the x-axis to the point P where the terminal side of the angle meets the unit circle, as shown in Figure 6–19. The right triangle formed in this way has hypotenuse 1 and angles of $\pi/6$ and $\pi/3$ radians. The sides of the triangle must have lengths $1/2$ and $\sqrt{3}/2$, so the coordinates of P are $(1/2, -\sqrt{3}/2)$ and

$$\sin t = y\text{-coordinate of } P = -\frac{\sqrt{3}}{2}$$

$$\cos t = x\text{-coordinate of } P = \frac{1}{2}$$

$$\tan t = \frac{\sin t}{\cos t} = \frac{-\sqrt{3}/2}{1/2} = -\sqrt{3}. \quad \blacksquare$$

Point-in-the-Plane Description of Trigonometric Functions

In evaluating $\sin t$, $\cos t$ and $\tan t$, from the definition, we use the point where the unit circle intersects the terminal side of an angle of t radians in standard position. Here is an alternative method of evaluating the trigonometric functions that uses *any* point on the terminal side of the angle (except the origin).

Point-in-the-Plane Description

Let t be a real number. Let (x, y) be any point (except the origin) on the terminal side of an angle of t radians in standard position. Then,

$$\sin t = \frac{y}{r} \qquad \cos t = \frac{x}{r} \qquad \tan t = \frac{y}{x}$$

where $r = \sqrt{x^2 + y^2}$ is the distance from (x, y) to the origin.

Proof Let Q be the point on the terminal side of the standard position angle of t radians and let P be the point where the terminal side meets the unit circle, as in Figure 6–20. The definition of sine and cosine shows that P has coordinates $(\cos t, \sin t)$. The distance formula shows that the segment OQ has length $\sqrt{x^2 + y^2}$, which we denote by r.

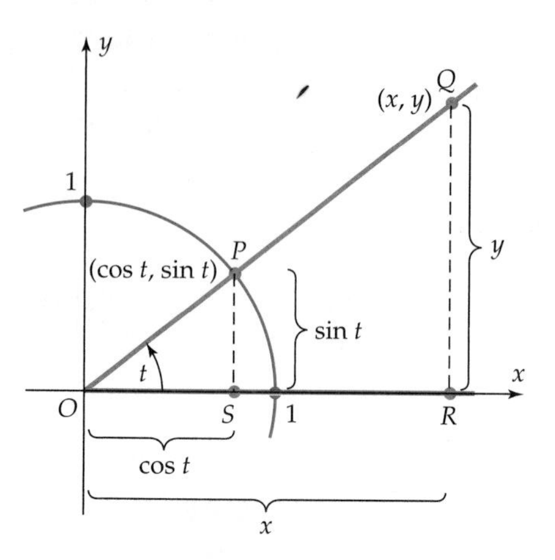

Figure 6–20

Both triangles QOR and POS are right triangles containing an angle of t radians. Therefore, these triangles are *similar.** Consequently,

$$\frac{\text{length } OP}{\text{length } OQ} = \frac{\text{length } PS}{\text{length } QR} \quad \text{and} \quad \frac{\text{length } OP}{\text{length } OQ} = \frac{\text{length } OS}{\text{length } OR}.$$

Figure 6–20 shows what each of these lengths is. Hence,

$$\frac{1}{r} = \frac{\sin t}{y} \quad \text{and} \quad \frac{1}{r} = \frac{\cos t}{x}$$

$$r \sin t = y \qquad\qquad r \cos t = x$$

$$\sin t = \frac{y}{r} \qquad\qquad \cos t = \frac{x}{r}$$

Similar arguments work when the terminal side is not in the first quadrant. In every case, $\tan t = \dfrac{\sin t}{\cos t} = \dfrac{y/r}{x/r} = \dfrac{y}{x}$. This completes the proof of the statements in the box on page 405. ■

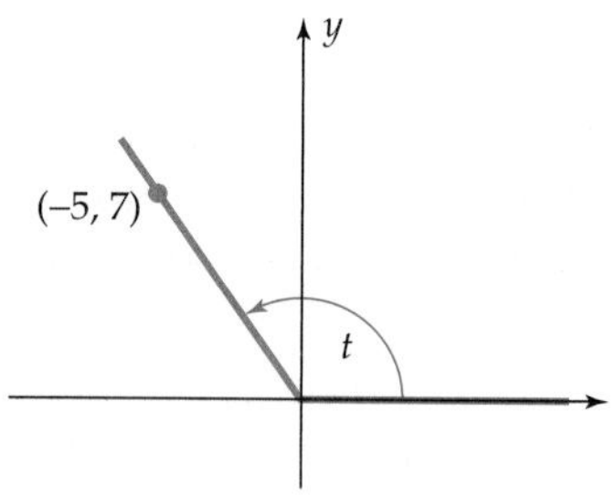

Example 5 Figure 6–21 shows an angle of t radians in standard position. Evaluate the three trigonometric functions at t.

Figure 6–21

*See the Geometry Review Appendix for the basic facts about similar triangles.

Solution Apply the facts in the box above with $(x, y) = (5, 7)$ and $r = \sqrt{x^2 + y^2} = \sqrt{(-5)^2 + 7^2} = \sqrt{74}$:

$$\sin t = \frac{y}{r} = \frac{7}{\sqrt{74}} \qquad \cos t = \frac{x}{r} = \frac{-5}{\sqrt{74}} \qquad \tan t = \frac{y}{x} = \frac{7}{-5} = -\frac{7}{5}. \quad ■$$

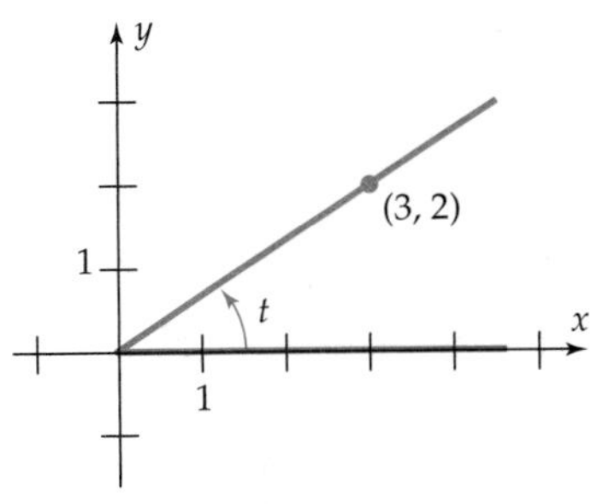

Figure 6–22

Example 6 The terminal side of a first-quadrant angle of t radians in standard position lies on the line with equation $2x - 3y = 0$. Evaluate the three trigonometric functions at t.

Solution Verify that the point (3, 2) satisfies the equation and hence lies on the terminal side of the angle (Figure 6–22). Now we have $(x, y) = (3, 2)$ and $r = \sqrt{x^2 + y^2} = \sqrt{3^2 + 2^2} = \sqrt{13}$. Therefore,

$$\sin t = \frac{y}{r} = \frac{2}{\sqrt{13}}, \qquad \cos t = \frac{x}{r} = \frac{3}{\sqrt{13}}, \qquad \tan t = \frac{y}{x} = \frac{2/\sqrt{13}}{3/\sqrt{13}} = \frac{2}{3}. \quad ■$$

Exercises 6.2

Note: Unless stated otherwise, all angles are in standard position.

In Exercises 1–10, use the definition (not a calculator) to find the function value.

1. $\sin 3\pi/2$ **2.** $\sin(-\pi)$ **3.** $\cos 3\pi/2$

4. $\cos(-\pi/2)$ **5.** $\tan 4\pi$ **6.** $\tan(-\pi)$

7. $\cos(-3\pi/2)$ **8.** $\sin 9\pi/2$ **9.** $\cos(-11\pi/2)$

10. $\tan(-13\pi)$

In Exercises 11–14, assume that the terminal side of an angle of t radians passes through the given point on the unit circle. Find sin t, cos t, tan t.

11. $(-2/\sqrt{5}, 1/\sqrt{5})$ **12.** $(1/\sqrt{10}, -3/\sqrt{10})$

13. $(-3/5, -4/5)$ **14.** $(.6, -.8)$

In Exercises 15–29, find the exact value of the sine, cosine, and tangent of the number, without using a calculator.

15. $5\pi/6$ **16.** $7\pi/6$ **17.** $7\pi/3$

18. $17\pi/3$ **19.** $11\pi/4$ **20.** $5\pi/4$

21. $-3\pi/2$ **22.** 3π **23.** $-23\pi/6$

24. $11\pi/6$ **25.** $-19\pi/3$ **26.** $-10\pi/3$

27. $-15\pi/4$ **28.** $-25\pi/4$ **29.** $-17\pi/2$

30. Fill the blanks in the following table. Write each entry as a fraction with denominator 2 and with a radical in the numerator. For example,

$$\sin \frac{\pi}{2} = 1 = \frac{\sqrt{4}}{2}.$$

Some students find the resulting pattern an easy way to remember these functional values.

t	0	$\pi/6$	$\pi/4$	$\pi/3$	$\pi/2$
sin t					
cos t					

In Exercises 31–36, write the expression as a single real number. Do not use decimal approximations.

31. $\sin(\pi/3)\cos\pi + \sin\pi\cos(\pi/3)$

32. $\sin(\pi/6)\cos(\pi/2) - \cos(\pi/6)\sin(\pi/2)$

33. $\cos(\pi/2)\cos(\pi/4) - \sin(\pi/2)\sin(\pi/4)$

34. $\cos(2\pi/3)\cos\pi + \sin(2\pi/3)\sin\pi$

35. $\sin(3\pi/4)\cos(5\pi/6) - \cos(3\pi/4)\sin(5\pi/6)$

36. $\sin(-7\pi/3)\cos(5\pi/4) + \cos(-7\pi/3)\sin(5\pi/4)$

In Exercises 37–42, find sin t, cos t, tan t when the terminal side of an angle of t radians in standard position passes through the given point.

37. $(2, 7)$ **38.** $(-3, 2)$ **39.** $(-5, -6)$

40. $(4, -3)$ **41.** $(\sqrt{3}, -10)$ **42.** $(-\pi, 2)$

In Exercises 43–48, assume that the terminal side of an angle of t radians in standard position lies in the given quadrant on the given straight line. Find sin t, cos t, tan t. [Hint: *Find a point on the terminal side of the angle.*]

43. Quadrant IV; line with equation $y = -3x$.

44. Quadrant III; line with equation $2y - 4x = 0$.

45. Quadrant IV; line through $(-3, 5)$ and $(-9, 15)$.

46. Quadrant III; line through the origin parallel to
$$7x - 2y = -6.$$

47. Quadrant III; line through the origin parallel to
$$2y + x = 6.$$

48. Quadrant I; line through the origin perpendicular to
$$3y + x = 6.$$

49. The values of $\sin t$, $\cos t$, and $\tan t$ are determined by the point (x, y) where the terminal side of an angle of t radians in standard position intersects the unit circle. The coordinates x and y are positive or negative, depending on what quadrant (x, y) lies in. For instance, in the second quadrant x is negative and y is positive, so that $\cos t$ (which is x by definition) is negative. Fill the blanks in this chart with the appropriate sign (+ or −).

Quadrant II $\pi/2 < t < \pi$		**Quadrant I** $0 < t < \pi/2$	
$\sin t$		$\sin t$	+
$\cos t$	−	$\cos t$	
$\tan t$		$\tan t$	

Quadrant III $\pi < t < 3\pi/2$		**Quadrant IV** $3\pi/2 < t < 2\pi$	
$\sin t$		$\sin t$	
$\cos t$		$\cos t$	
$\tan t$		$\tan t$	

50. (a) Find two numbers c and d such that
$$\sin(c + d) \neq \sin c + \sin d.$$

(b) Find two numbers c and d such that
$$\cos(c + d) \neq \cos c + \cos d.$$

In Exercises 51–56, draw a rough sketch to determine if the given number is positive.

51. $\sin 1$ [*Hint:* The terminal side of an angle of 1 radian lies in the first quadrant (why?), so any point on it will have a positive y-coordinate.]

52. $\cos 2$ **53.** $\tan 3$ **54.** $(\cos 2)(\sin 2)$

55. $\tan 1.5$ **56.** $\cos 3 + \sin 3$

In Exercises 57–62, find all the solutions of the equation.

57. $\sin t = 1$ **58.** $\cos t = -1$ **59.** $\tan t = 0$

60. $\sin t = -1$ **61.** $|\sin t| = 1$ **62.** $|\cos t| = 1$

Thinkers

63. Using only the definition and no calculator, determine which number is larger: $\sin(\cos 0)$ or $\cos(\sin 0)$.

64. With your calculator in radian mode and function graphing mode, graph the following functions on the same screen, using the viewing window with $0 \leq x \leq 2\pi$ and $-3 \leq y \leq 3$: $f(x) = \cos x^3$ and $g(x) = (\cos x)^3$. Are the graphs the same? What do you conclude about the statement $\cos x^3 = (\cos x)^3$?

65. Figure R is a diagram of a merry-go-round that includes horses A through F. The distance from the center P to A is 1 unit and the distance from P to D is 5 units. Define six functions as follows:

$A(t)$ = vertical distance from horse A to the x-axis at time t minutes;

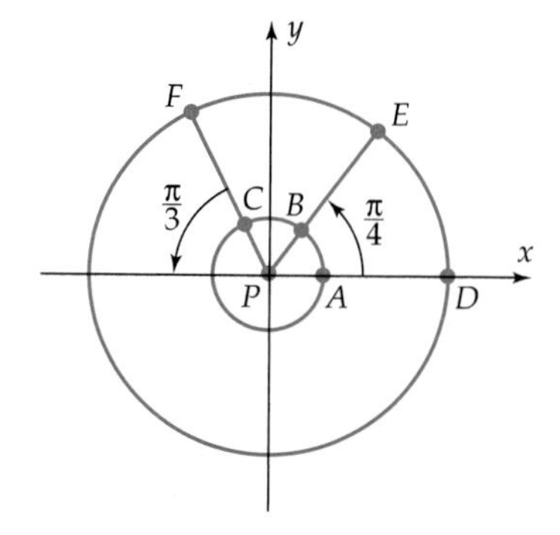

Figure R

and similarly for $B(t)$, $C(t)$, $D(t)$, $E(t)$, $F(t)$. The merry-go-round rotates counterclockwise at a rate of 1 revolution per minute and the horses are in the positions shown in Figure R at the starting time $t = 0$. As the merry-go-round rotates, the horses move around the circles shown in Figure R.

(a) Show that $B(t) = A(t + 1/8)$ for every t.
(b) In a similar manner, express $C(t)$ in terms of the function $A(t)$.
(c) Express $E(t)$ and $F(t)$ in terms of the function $E(t)$.
(d) Explain why Figure S is valid and use it and similar triangles to express $D(t)$ in terms of $A(t)$.
(e) In a similar manner, express $E(t)$ and $F(t)$ in terms of $A(t)$.
(f) Show that $A(t) = \sin(2\pi t)$ for every t. [*Hint:* Exercises 65–72 in Section 6.1 may be helpful.]
(g) Use parts (a), (b), and (f) to express $B(t)$ and $C(t)$ in terms of the sine function.
(h) Use parts (d), (e), and (f) to express $D(t)$, $E(t)$, and $F(t)$ in terms of the sine function.

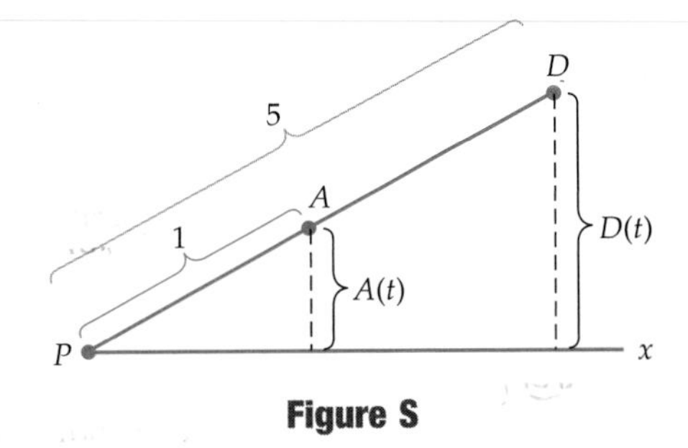

Figure S

6.3 Algebra and Identities

> ***Roadmap***
> Section 7.1 may be covered at this point by instructors who prefer to introduce right triangle trigonometry early.

In the previous section we concentrated on evaluating the trigonometric functions. In this section the emphasis is on the algebra of such functions. When dealing with trigonometric functions two notational conventions are normally observed:

1. **Parentheses are omitted whenever no confusion can result.** For example,

$$\sin(t) \quad \text{is written} \quad \sin t$$
$$-(\cos(5t)) \quad \text{is written} \quad -\cos 5t$$
$$4(\tan t) \quad \text{is written} \quad 4 \tan t.$$

On the other hand, parentheses *are* needed to distinguish

$$\cos(t + 3) \qquad \text{from} \qquad \cos t + 3,$$

because the first one says "add 3 to t and take the cosine of the result," but the second one says "take the cosine of t and add 3 to the result." When $t = 5$, for example, these are different numbers:

> **Technology Tip**
>
> TI-83 and HP-38 automatically insert an opening parenthesis when the COS key is pushed. The display COS(5 + 3 is interpreted as COS(5 + 3). If you want cos 5 + 3, you must insert a parenthesis after the 5: COS(5) + 3.

Figure 6–23

2. **When dealing with powers of trigonometric functions, exponents (other than -1) are written between the function symbol and the variable.**

For example,

$$(\cos t)^3 \quad \text{is written} \quad \cos^3 t$$
$$(\sin t)^4(\tan 7t)^2 \quad \text{is written} \quad \sin^4 t \tan^2 7t.$$

Furthermore,

$$\sin t^3 \quad \text{means} \quad \sin(t^3) \quad [\textit{not } (\sin t)^3 \text{ or } \sin^3 t]$$

For instance, when $t = 4$, we have:

Technology Tip

Calculators do not use convention 2. In order to obtain $\sin^3 4$, you must enter $(\sin 4)^3$.

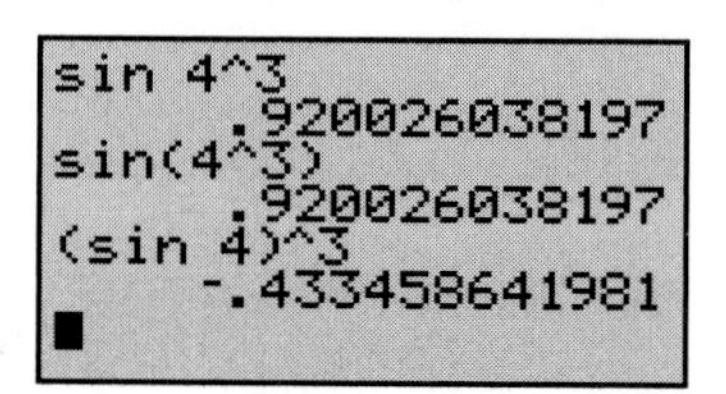

Figure 6–24

CAUTION

Convention 2 is not used when the exponent is -1. For example, $\sin^{-1} t$ does *not* mean $(\sin t)^{-1}$ or $1/(\sin t)$; it has an entirely different meaning that will be discussed in Section 8.5. Similar remarks apply to $\cos^{-1} t$ and $\tan^{-1} t$.

Except for these two conventions, the algebra of trigonometric functions is just like the algebra of other functions. They may be added, subtracted, multiplied, composed, etc.

Example 1 If $f(t) = \sin^2 t + \tan t$ and $g(t) = \tan^3 t + 5$, then the product function fg is given by the rule

$$\begin{aligned}(fg)(t) = f(t)g(t) &= (\sin^2 t + \tan t)(\tan^3 t + 5)\\ &= \sin^2 t \tan^3 t + 5\sin^2 t + \tan^4 t + 5\tan t. \quad \blacksquare\end{aligned}$$

Example 2 Factor $2\cos^2 t - 5\cos t - 3$.

Solution You can do this directly, but it may be easier to understand if you make a substitution. Let $u = \cos t$; then,

$$\begin{aligned}2\cos^2 t - 5\cos t - 3 &= 2(\cos t)^2 - 5\cos t - 3\\ &= 2u^2 - 5u - 3\\ &= (2u + 1)(u - 3)\\ &= (2\cos t + 1)(\cos t - 3). \quad \blacksquare\end{aligned}$$

Example 3 If $f(t) = \cos^2 t - 9$ and $g(t) = \cos t + 3$, then the quotient function f/g is given by the rule

$$\left(\frac{f}{g}\right)(t) = \frac{f(t)}{g(t)} = \frac{\cos^2 t - 9}{\cos t + 3} = \frac{(\cos t + 3)(\cos t - 3)}{\cos t + 3} = \cos t - 3. \quad ■$$

CAUTION

You are dealing with *functional notation* here, so the symbol sin t is a *single entity*, as are cos t and tan t. Don't try some nonsensical "canceling" operation, such as

$$\frac{\sin t}{\cos t} = \frac{\sin}{\cos} \quad \text{or} \quad \frac{\cos t^2}{\cos t} = \frac{\cos t}{\cos} = t.$$

Example 4 If $f(t) = \sin t$ and $g(t) = t^2 + 3$, then the composite function $g \circ f$ is given by the rule

$$(g \circ f)(t) = g(f(t)) = g(\sin t) = \sin^2 t + 3.$$

The composite function $f \circ g$ is given by the rule

$$(f \circ g)(t) = f(g(t)) = f(t^2 + 3) = \sin(t^2 + 3).$$

The parentheses are absolutely necessary here because $\sin(t^2 + 3)$ is *not* the same function as $\sin t^2 + 3$. For instance, a calculator in radian mode shows that for $t = 5$,

$$\sin(5^2 + 3) = \sin(25 + 3) = \sin 28 \approx .2709$$

whereas

$$\sin 5^2 + 3 = \sin 25 + 3 \approx (-.1324) + 3 = 2.8676. \quad ■$$

The Pythagorean Identity

Trigonometric functions have numerous interrelationships that are usually expressed as *identities*. An **identity** is an equation that is true for all values of the variable for which every term of the equation is defined. Here is one of the most important trigonometric identities:

Pythagorean Identity

For every real number t,

$$\sin^2 t + \cos^2 t = 1.$$

GRAPHING EXPLORATION

Recall that the graph of $y = 1$ is a horizontal line through $(0, 1)$. Verify the Pythagorean identity by graphing the equation $y = (\sin x)^2 + (\cos x)^2$ in the window with $-10 \le x \le 10$ and $-3 \le y \le 3$ and using the trace feature.

The Pythagorean identity can be proved algebraically by using the definition of sine and cosine. For each real number t, $\cos t$ is the first coordinate and $\sin t$ the second coordinate of the point P where the unit circle intersects the terminal side of an angle of t radians in standard position (Figure 6–25). Since P lies on the unit circle, its coordinates $(\cos t, \sin t)$ must satisfy the equation of the unit circle: $x^2 + y^2 = 1$, that is, $\cos^2 t + \sin^2 t = 1$.

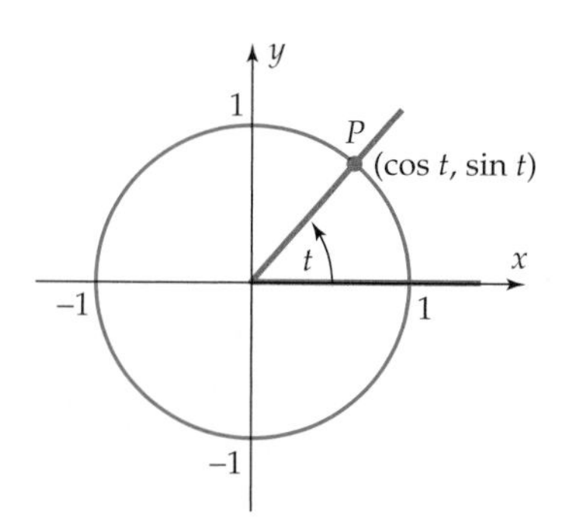

Figure 6–25

Example 5 If $\pi/2 < t < \pi$ and $\sin t = 2/3$, find $\cos t$ and $\tan t$.

Solution By the Pythagorean identity

$$\cos^2 t = 1 - \sin^2 t = 1 - \left(\frac{2}{3}\right)^2 = 1 - \frac{4}{9} = \frac{5}{9}.$$

So there are two possibilities:

$$\cos t = \sqrt{5/9} = \sqrt{5}/3 \quad \text{or} \quad \cos t = -\sqrt{5/9} = -\sqrt{5}/3.$$

Since $\pi/2 < t < \pi$, $\cos t$ is negative (see Exercise 49 on page 408). Therefore, $\cos t = -\sqrt{5}/3$ and

$$\tan t = \frac{\sin t}{\cos t} = \frac{2/3}{-\sqrt{5}/3} = \frac{-2}{\sqrt{5}} = \frac{-2\sqrt{5}}{5}. \quad ■$$

Example 6 The Pythagorean identity is valid for *any* number t. For instance, if $t = 3k + 7$, then $\sin^2(3k + 7) + \cos^2(3k + 7) = 1$. ■

Example 7 To simplify the expression $\tan^2 t \cos^2 t + \cos^2 t$, we use the definition of tangent and the Pythagorean identity:

$$\tan^2 t \cos^2 t + \cos^2 t = \frac{\sin^2 t}{\cos^2 t}\cos^2 t + \cos^2 t$$

$$= \sin^2 t + \cos^2 t = 1. \quad ■$$

Periodicity Identities

Let t be any real number and construct two angles in standard position of measure t and $t + 2\pi$ radians, respectively, as shown in Figure 6–26. As we saw in Section 6.1, both of these angles have the same terminal side. Therefore, the point P where the terminal side meets the unit circle is the *same* in both cases:

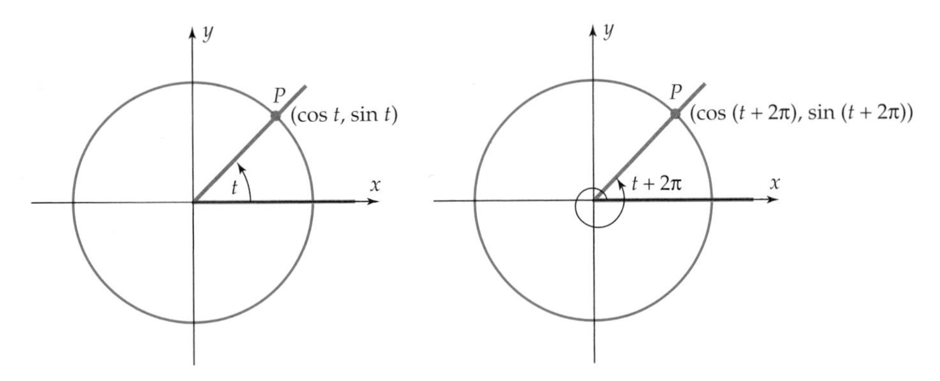

Figure 6–26

Since the second coordinate of P is the value of the sine function in each case, we see that $\sin t = \sin(t + 2\pi)$. Furthermore, since an angle of t radians in standard position has the same terminal side as angles of radian measure $t \pm 2\pi$, $t \pm 4\pi$, $t \pm 6\pi$, and so forth, essentially the same argument shows that for every number t

$$\sin t = \sin(t \pm 2\pi) = \sin(t \pm 4\pi) = \sin(t \pm 6\pi) = \cdots$$

Similarly, the first coordinate of P is the value of the cosine function in each case so that

$$\cos t = \cos(t \pm 2\pi) = \cos(t \pm 4\pi) = \cos(t \pm 6\pi) = \cdots$$

CALCULATOR EXPLORATION

Illustrate the preceding discussion by computing each of the following numbers: $\sin 3$, $\sin(3 + 2\pi)$, $\sin(3 - 4\pi)$. Do the same for $\cos 4$, $\cos(4 + 2\pi)$, and $\cos(4 + 6\pi)$.

There is a special name for functions that repeat their values at regular intervals. A function f is said to be **periodic** if there is a positive constant k such that $f(t) = f(t + k)$ for every number t in the domain of f. There will be more than one constant k with this property; the smallest one is called the **period** of the function f. We have just seen that sine and cosine are periodic with $k = 2\pi$. Exercises 65 and 66 show that 2π is the smallest such positive constant k. Therefore,

Period of Sine and Cosine

The sine and cosine functions are periodic with period 2π: For every real number t,

$$\sin t = \sin(t \pm 2\pi) \quad \text{and} \quad \cos t = \cos(t \pm 2\pi)$$

Example 8 Find $\sin \dfrac{13\pi}{6}$.

Solution The periodicity identity shows that

$$\sin \frac{13\pi}{6} = \sin\left(\frac{\pi}{6} + \frac{12\pi}{6}\right) = \sin\left(\frac{\pi}{6} + 2\pi\right) = \sin\frac{\pi}{6} = \frac{1}{2}. \quad \blacksquare$$

The tangent function is also periodic (see Exercise 36), but its period is π rather than 2π, that is,

$$\tan(t + \pi) = \tan t \quad \text{for every real number } t,$$

as we shall see in Section 6.4.

Negative Angle Identities

GRAPHING EXPLORATION

(a) In a viewing window with $-2\pi \leq x \leq 2\pi$, graph $y = \sin x$ and $y = \sin(-x)$ on the same screen. Use trace to move along $y = \sin x$. Stop at a point and note its y-coordinate. Use the up or down arrow to move vertically to the graph of $y = \sin(-x)$. The x-coordinate remains the same, but the y-coordinate is different. How are the two y-coordinates related? Is one the negative of the other? Repeat the procedure for other points. Are the results the same?

(b) In a viewing window with $-2\pi \leq x \leq 2\pi$, graph $y = \cos x$ and $y = \cos(-x)$ on the same screen. How do the graphs compare?

The preceding exploration suggests the truth of the following statement:

Negative Angle Identities for Sine and Cosine

For every real number t,

$$\sin(-t) = -\sin t$$

and

$$\cos(-t) = \cos t.$$

These identities can be proved by using the definitions of sine and cosine. Consider two angles in standard position, one of t radians and the other of $-t$ radians, as in Figure 6–27. By the definitions of sine and cosine, the point P, where the terminal side of the angle of t radians meets the unit circle, has coordinates $(\cos t, \sin t)$. Similarly, the point Q, where the terminal side of the angle of $-t$ radians meets the unit circle, has coordinates $(\cos(-t), \sin(-t))$:

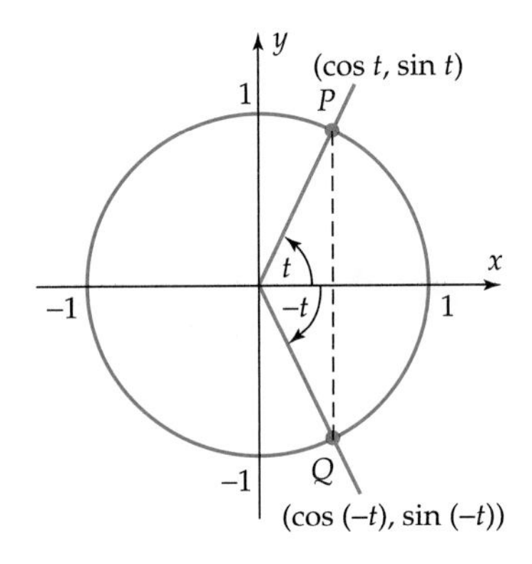

Figure 6–27

As the figure suggests, P and Q lie on the same vertical line. Therefore, they have the same first coordinate, that is, $\cos(-t) = \cos t$. Furthermore, as indicated in the figure, P and Q lie at equal distances from the x-axis.* So the y-coordinate of Q must be the negative of the y-coordinate of P, that is, $\sin(-t) = -\sin t$.

Example 9 On page 404, we saw that $\sin \pi/6 = 1/2$ and that $\cos \pi/6 = \sqrt{3}/2$. Therefore,

$$\sin\left(-\frac{\pi}{6}\right) = -\sin\frac{\pi}{6} = -\frac{1}{2}$$

and

$$\cos\left(-\frac{\pi}{6}\right) = \cos\frac{\pi}{6} = \frac{\sqrt{3}}{2}. \quad \blacksquare$$

*These facts can be proved by considering basic facts about congruent triangles and right angles.

Example 10 To simplify $(1 + \sin t)(1 + \sin(-t))$, we use the negative angle identity and the Pythagorean identity:

$$\begin{aligned}(1 + \sin t)(1 + \sin(-t)) &= (1 + \sin t)(1 - \sin t)\\ &= 1 - \sin^2 t\\ &= \cos^2 t.\end{aligned}$$

■

The negative angle identity for the tangent function is now easy to prove:

$$\tan(-t) = \frac{\sin(-t)}{\cos(-t)} = \frac{-\sin t}{\cos t} = -\tan t.$$

Hence,

Negative Angle Identity for Tangent

For every real number t,

$$\tan(-t) = -\tan t.$$

Exercises 6.3

In Exercises 1–4, find the rule of the product function fg.

1. $f(t) = 3 \sin t$; $g(t) = \sin t + 2 \cos t$
2. $f(t) = 5 \tan t$; $g(t) = \tan^3 t - 1$
3. $f(t) = 3 \sin^2 t$; $g(t) = \sin t + \tan t$
4. $f(t) = \sin 2t + \cos^4 t$; $g(t) = \cos 2t + \cos^2 t$

In Exercises 5–14, factor the given expression.

5. $\cos^2 t - 4$
6. $25 - \tan^2 t$
7. $\sin^2 t - \cos^2 t$
8. $\sin^3 t - \sin t$
9. $\tan^2 t + 6 \tan t + 9$
10. $\cos^2 t - \cos t - 2$
11. $6 \sin^2 t - \sin t - 1$
12. $\tan t \cos t + \cos^2 t$
13. $\cos^4 t + 4 \cos^2 t - 5$
14. $3 \tan^2 t + 5 \tan t - 2$

In Exercises 15–18, find the rules of the composite functions $f \circ g$ and $g \circ f$.

15. $f(t) = \cos t$; $g(t) = 2t + 4$
16. $f(t) = \sin t + 2$; $g(t) = t^2$
17. $f(t) = \tan(t + 3)$; $g(t) = t^2 - 1$
18. $f(t) = \cos^2(t - 2)$; $g(t) = 5t + 2$

In Exercises 19–24, determine if it is possible for a number t to satisfy the given conditions. [Hint: *Think Pythagorean.*]

19. $\sin t = 5/13$ and $\cos t = 12/13$
20. $\sin t = -2$ and $\cos t = 1$
21. $\sin t = -1$ and $\cos t = 1$
22. $\sin t = 1/\sqrt{2}$ and $\cos t = -1/\sqrt{2}$
23. $\sin t = 1$ and $\tan t = 1$
24. $\cos t = 8/17$ and $\tan t = 15/8$

In Exercises 25–28, use the Pythagorean identity to find sin t.

25. $\cos t = -.5$ and $\pi < t < 3\pi/2$
26. $\cos t = -3/\sqrt{10}$ and $\pi/2 < t < \pi$
27. $\cos t = 1/2$ and $0 < t < \pi/2$
28. $\cos t = 2/\sqrt{5}$ and $3\pi/2 < t < 2\pi$

In Exercises 29–35, assume that $\sin t = 3/5$ *and* $0 < t < \pi/2$. *Use identities in the text to find the number.*

29. $\sin(-t)$ **30.** $\sin(t + 10\pi)$ **31.** $\sin(2\pi - t)$

32. $\cos t$ **33.** $\tan t$ **34.** $\cos(-t)$

35. $\tan(2\pi - t)$

36. (a) Show that $\tan(t + 2\pi) = \tan t$ for every t in the domain of $\tan t$. [*Hint:* Use the definition of tangent and some identities proved in the text.]

(b) Verify that it appears true that $\tan(x + \pi) = \tan x$ for every t in the domain by using your calculator's table feature to make a table of values for $y_1 = \tan(x + \pi)$ and $y_2 = \tan x$.

In Exercises 37–42, assume that $\cos t = -2/5$ *and* $\pi < t < 3\pi/2$. *Use identities to find the number.*

37. $\sin t$ **38.** $\tan t$ **39.** $\cos(2\pi - t)$

40. $\cos(-t)$ **41.** $\sin(4\pi + t)$ **42.** $\tan(4\pi - t)$

In Exercises 43–46, assume that $\sin(\pi/8) = \dfrac{\sqrt{2 - \sqrt{2}}}{2}$ *and use identities to find the exact functional value.*

43. $\cos(\pi/8)$ **44.** $\tan(\pi/8)$

45. $\sin(17\pi/8)$ **46.** $\tan(-15\pi/8)$

In Exercises 47–58, use algebra and identities in the text to simplify the expression. Assume all denominators are nonzero.

47. $(\sin t + \cos t)(\sin t - \cos t)$

48. $(\sin t - \cos t)^2$ **49.** $\tan t \cos t$ **50.** $(\sin t)/(\tan t)$

51. $\sqrt{\sin^3 t \cos t}\sqrt{\cos t}$

52. $(\tan t + 2)(\tan t - 3) - (6 - \tan t) + 2 \tan t$

53. $\left(\dfrac{4 \cos^2 t}{\sin^2 t}\right)\left(\dfrac{\sin t}{4 \cos t}\right)^2$

54. $\dfrac{5 \cos t}{\sin^2 t} \cdot \dfrac{\sin^2 t - \sin t \cos t}{\sin^2 t - \cos^2 t}$

55. $\dfrac{\cos^2 t + 4 \cos t + 4}{\cos t + 2}$

56. $\dfrac{\sin^2 t - 2 \sin t + 1}{\sin t - 1}$

57. $\dfrac{1}{\cos t} - \sin t \tan t$

58. $\dfrac{1 - \tan^2 t}{1 + \tan^2 t} + 2 \sin^2 t$

In Exercises 59–63, show that the given function is periodic with period less than 2π. [Hint: *Find a positive number k with* $k < 2\pi$, *such that* $f(t + k) = f(t)$ *for every t in the domain of f.*]

59. $f(t) = \sin 2t$ **60.** $f(t) = \cos 3t$

61. $f(t) = \sin 4t$ **62.** $f(t) = \cos(3\pi t/2)$

63. $f(t) = \tan 2t$

64. Fill the blanks with "even" or "odd" so that the resulting statement is true. Then prove the statement by using an appropriate identity. [Excursion 3.4.A may be helpful.]

(a) $f(t) = \sin t$ is an _____ function.

(b) $g(t) = \cos t$ is an _____ function.

(c) $h(t) = \tan t$ is an _____ function.

(d) $f(t) = t \sin t$ is an _____ function.

(e) $g(t) = t + \tan t$ is an _____ function.

65. Here is a proof that the cosine function has period 2π. We saw in the text that $\cos(t + 2\pi) = \cos t$ for every t. We must show that there is no positive number smaller than 2π with this property. Do this as follows:

(a) Find all numbers k such that $0 < k < 2\pi$ and $\cos k = 1$. [*Hint:* Draw a picture and use the definition of the cosine function.]

(b) Suppose k is a number such that $\cos(t + k) = \cos t$ for every number t. Show that $\cos k = 1$. [*Hint:* Consider $t = 0$.]

(c) Use parts (a) and (b) to show that there is no positive number k less than 2π with the property that $\cos(t + k) = \cos t$ for *every* number t. Therefore, $k = 2\pi$ is the smallest such number and the cosine function has period 2π.

66. Here is proof that the sine function has period 2π. We saw in the text that $\sin(t + 2\pi) = \sin t$ for every t. We must show that there is no positive number smaller than 2π with this property. Do this as follows:

(a) Find a number t such that $\sin(t + \pi) \neq \sin t$.

(b) Find all numbers k such that $0 < k < 2\pi$ and $\sin k = 0$. [*Hint:* Draw a picture and use the definition of the sine function.]

(c) Suppose k is a number such that $\sin(t + k) = \sin t$ for every number t. Show that $\sin k = 0$. [*Hint:* Consider $t = 0$.]

(d) Use parts (a)–(c) to show that there is no positive number k less than 2π with the property that $\sin(t + k) = \sin t$ for *every* number t. Therefore, $k = 2\pi$ is the smallest such number and the sine function has period 2π.

6.4 Basic Graphs

Although a graphing calculator will quickly sketch the graphs of the sine, cosine, and tangent functions, it will not give you much insight into why these graphs have the shape they do and why these shapes are important. So the emphasis here is on the connection between the definition of these functions and their graphs.

If P is the point where the unit circle meets the terminal side of an angle of t radians, then the y-coordinate of P is the number $\sin t$. We can get a rough sketch of the graph of $f(t) = \sin t$ by watching the y-coordinate of P:

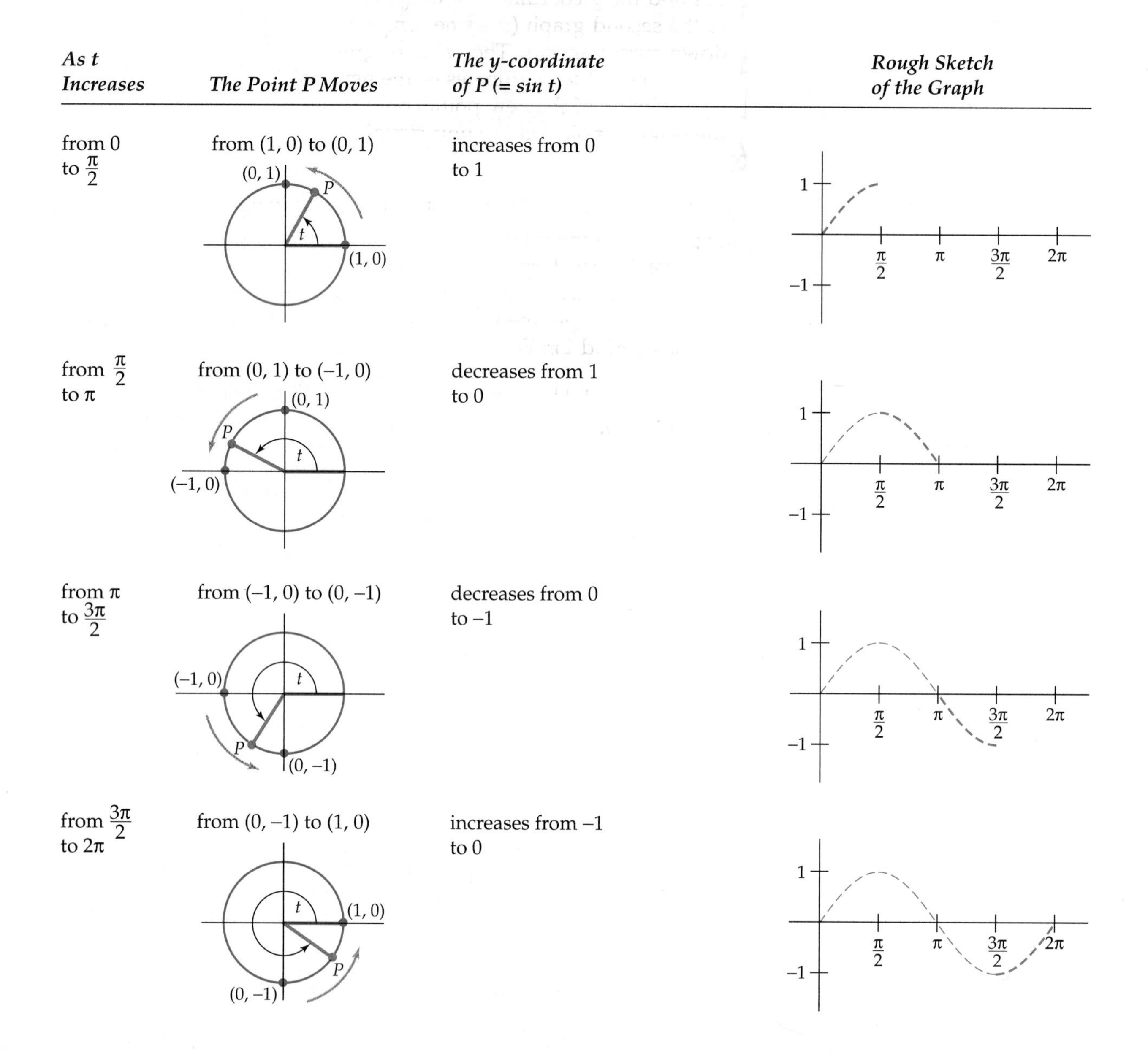

As t Increases	*The Point P Moves*	*The y-coordinate of P (= sin t)*	*Rough Sketch of the Graph*
from 0 to $\frac{\pi}{2}$	from (1, 0) to (0, 1)	increases from 0 to 1	
from $\frac{\pi}{2}$ to π	from (0, 1) to (−1, 0)	decreases from 1 to 0	
from π to $\frac{3\pi}{2}$	from (−1, 0) to (0, −1)	decreases from 0 to −1	
from $\frac{3\pi}{2}$ to 2π	from (0, −1) to (1, 0)	increases from −1 to 0	

GRAPHING EXPLORATION

Your calculator can provide a dynamic simulation of this process. Put it in parametric graphing mode and set the range values as follows:

$$0 \le t \le 6.28 \qquad -1 \le x \le 6.28 \qquad -2.5 \le y \le 2.5.$$

On the same screen, graph the two functions given by

$$x_1 = \cos t, \quad y_1 = \sin t \qquad \text{and} \qquad x_2 = t, \quad y_2 = \sin t.$$

Using the trace feature, move the cursor along the first graph (the unit circle). Stop at a point on the circle, note the value of t and the y-coordinate of the point. Then switch the trace to the second graph (the sine function) by using the up or down cursor arrows. The value of t remains the same. What are the x- and y-coordinates of the new point? How does the y-coordinate of the new point compare with the y-coordinate of the original point on the unit circle?

To complete the graph of the sine function, note that as t goes from 2π to 4π, the point P on the unit circle *retraces* the path it took from 0 to 2π, so *the same wave shape will repeat* on the graph. The same thing happens when t goes from 4π to 6π, or from -2π to 0, and so on. This repetition of the same pattern is simply the graphical expression of the fact that the sine function has period 2π: For any number t, the points

$$(t, \sin t) \qquad \text{and} \qquad (t + 2\pi, \sin(t + 2\pi))$$

on the graph have the same second coordinate.

A graphing calculator or some point plotting with an ordinary calculator now produces the graph of $f(t) = \sin t$ (Figure 6–28):

Technology Tip

Calculators have built-in windows for trigonometric functions, in which the x-axis tick marks are at intervals of $\pi/2$. Choose TRIG or ZTRIG in the ZOOM menu of TI and Sharp 9600, or the VIEWS menu of HP-38, or the V-WINDOW menu of Casio 9850.

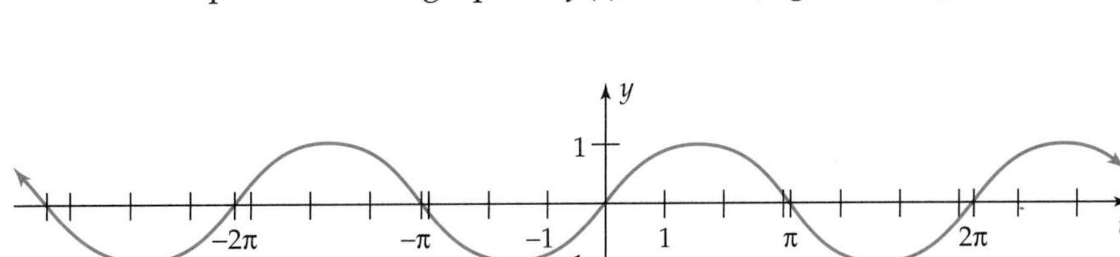

Figure 6–28

NOTE Throughout this chapter, we use t as the variable for trigonometric functions, to avoid any confusion with the x's and y's that are part of the definition of these functions. For calculator graphing in "function mode," however, you must use x as the variable: $f(x) = \sin x$, $g(x) = \cos x$, etc.

The graph of the sine function and the techniques of Section 3.4 can be used to graph other trigonometric functions.

Example 1 The graph of $h(t) = 3 \sin t$ is the graph of $f(t) = \sin t$ stretched away from the horizontal axis by a factor of 3, as shown in Figure 6–29. ■

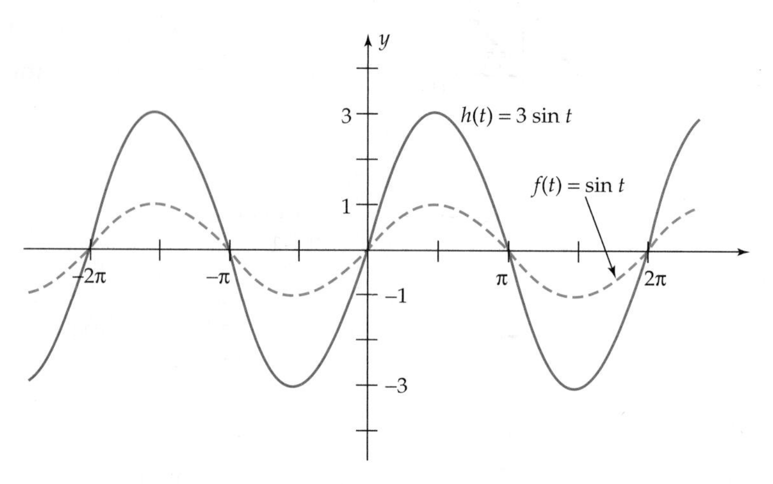

Figure 6–29

Example 2 The graph of $k(t) = -\frac{1}{2} \sin t$ is the graph of $f(t) = \sin t$ shrunk by a factor of 1/2 toward the horizontal axis and then reflected in the horizontal axis, as shown in Figure 6–30. ■

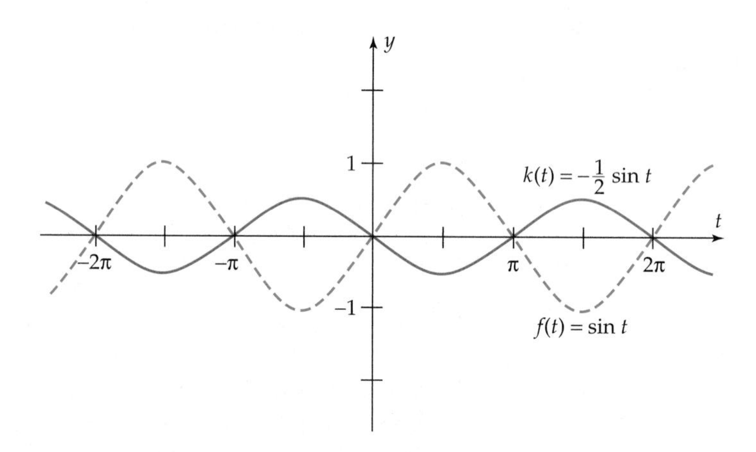

Figure 6–30

Graph of the Cosine Function

To obtain the graph of $g(t) = \cos t$, we follow the same procedure, except that we now watch the x-coordinate of P (which is $\cos t$).

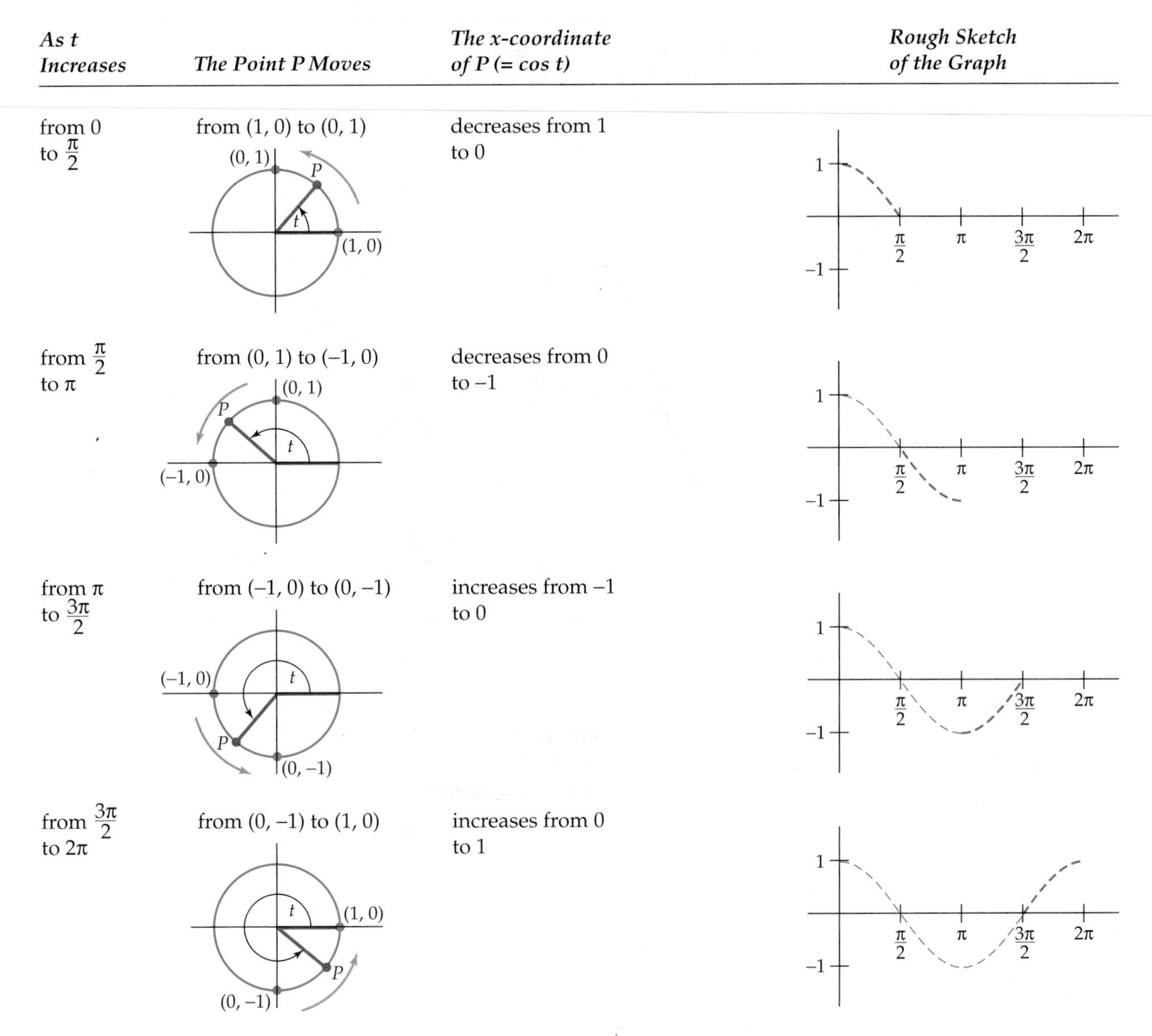

As t Increases	*The Point P Moves*	*The x-coordinate of P (= cos t)*	*Rough Sketch of the Graph*
from 0 to $\frac{\pi}{2}$	from (1, 0) to (0, 1)	decreases from 1 to 0	
from $\frac{\pi}{2}$ to π	from (0, 1) to (−1, 0)	decreases from 0 to −1	
from π to $\frac{3\pi}{2}$	from (−1, 0) to (0, −1)	increases from −1 to 0	
from $\frac{3\pi}{2}$ to 2π	from (0, −1) to (1, 0)	increases from 0 to 1	

As t takes larger values, P begins to retrace its path around the unit circle, so the graph of $g(t) = \cos t$ repeats the same wave pattern, and similarly for negative values of t. So the graph looks like Figure 6–31:

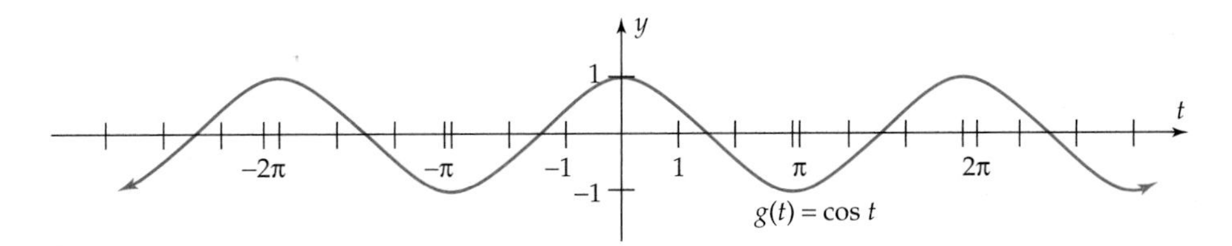

Figure 6–31

For a dynamic simulation of the cosine graphing process described earlier, see Exercise 51.

The graphs of the cosine function in Figure 6–31 and the sine function in Figure 6–28 lie between -1 and 1 on the y-axis. This is a graphical illustration of the following fact:

Range of Sine and Cosine

For every real number t

$$-1 \leq \sin t \leq 1$$

and

$$-1 \leq \cos t \leq 1.$$

The techniques of Section 3.4 can be used to graph variations of the cosine function.

Example 3 The graph of $h(t) = 4 \cos t$ is the graph of $g(t) = \cos t$ stretched away from the horizontal axis by a factor of 4, as shown in Figure 6–32. ■

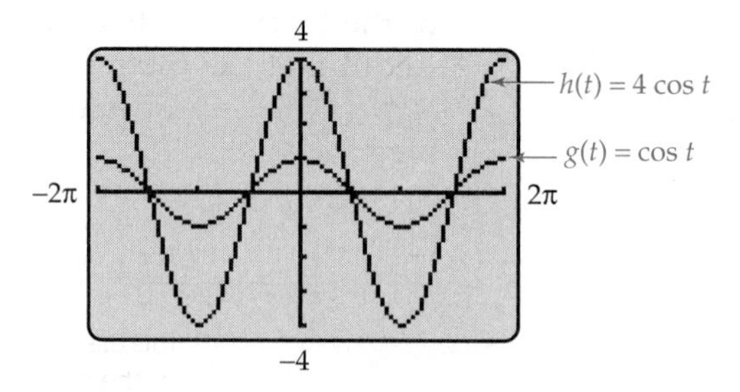

Figure 6–32

Example 4 The graph of $k(t) = -2 \cos t + 3$ is the graph of $g(t) = \cos t$ stretched away from the horizontal axis by a factor of 2, reflected in the horizontal axis, and shifted vertically 3 units upward as shown in Figure 6–33. ■

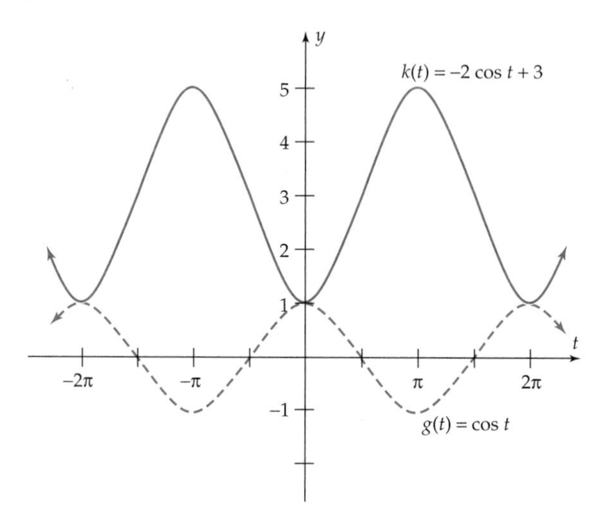

Figure 6–33

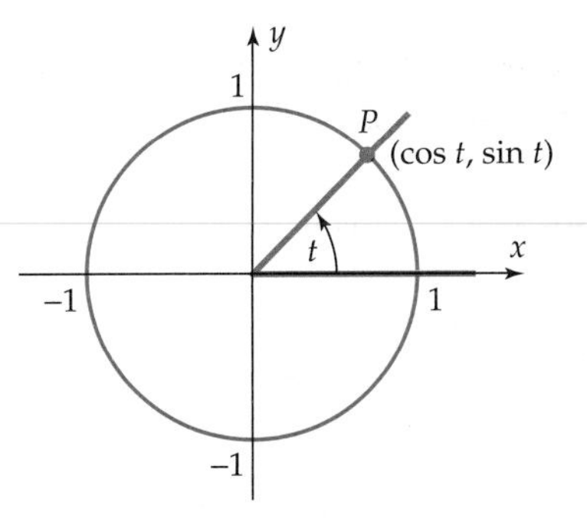

Figure 6–34

Graph of the Tangent Function

To determine the shape of the graph of $h(t) = \tan t$, we use an interesting connection between the tangent function and straight lines. As shown in Figure 6–34, the point P where the terminal side of an angle of t radians in standard position meets the unit circle has coordinates $(\cos t, \sin t)$. We can use this point and the point $(0, 0)$ to compute the *slope* of the terminal side:

$$\text{slope} = \frac{\sin t - 0}{\cos t - 0} = \frac{\sin t}{\cos t} = \tan t$$

Therefore,

Slope and Tangent

> **The slope of the terminal side of an angle of t radians in standard position is the number $\tan t$.**

The graph of $h(x) = \tan t$ can now be sketched by watching the slope of the terminal side of an angle of t radians, as t takes different values. Recall that the more steeply a line rises from left to right, the larger its slope. Similarly, lines that fall from left to right have negative slopes that increase in absolute value as the line falls more steeply.

As t Changes	*The Terminal Side of the Angle Moves*	*Its Slope (tan t)*	*Rough Sketch of the Graph*
from 0 to $\frac{\pi}{2}$	from horizontal upward toward vertical	increases from 0 in the positive direction and keeps getting larger	
from 0 to $-\frac{\pi}{2}$	from horizontal downward toward vertical 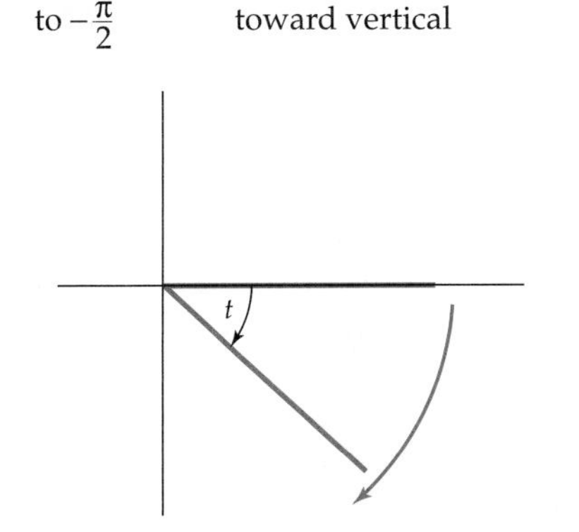	decreases from 0 in the negative direction and keeps getting larger in absolute value	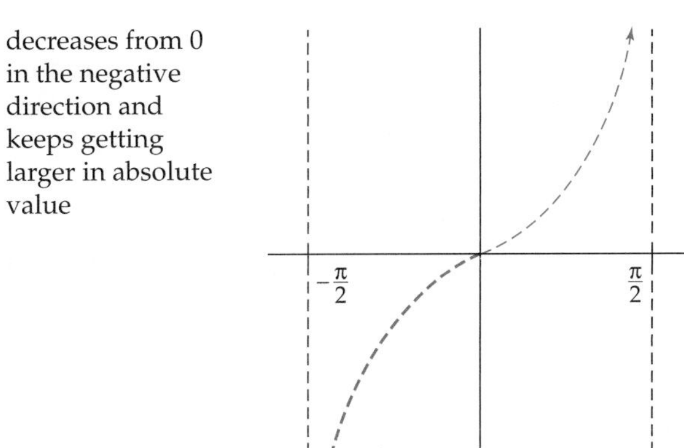

When $t = \pm\pi/2$, the terminal side of the angle is vertical and hence its slope is not defined. This corresponds to the fact that the tangent function is not defined when $t = \pm\pi/2$. The vertical lines through $\pm\pi/2$ are vertical asymptotes of the graph: It gets closer and closer to these lines, but never touches them.

When t is slightly larger than $\pi/2$, the terminal side falls from left to right and has negative slope (draw a picture). As t goes from $\pi/2$ to $3\pi/2$, the terminal side goes from almost vertical with negative slope to horizontal to almost vertical with positive slope, exactly as it does between $-\pi/2$ and $\pi/2$. So, the graph repeats the same pattern. The same thing happens between $3\pi/2$ and $5\pi/2$, between $-3\pi/2$ and $-\pi/2$, etc. Therefore, the entire graph looks like this (Figure 6–35):

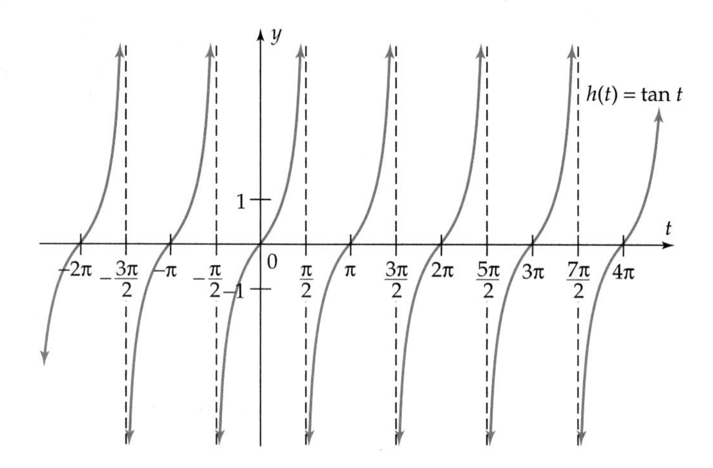

Figure 6–35

Because calculators sometimes do not graph accurately across vertical asymptotes, the graph may look slightly different on a calculator screen (with vertical line segments where the asymptotes should be).

The graph of the tangent function repeats the same pattern at intervals of length π. This means that the tangent function repeats its values at intervals of π.

Period of Tangent

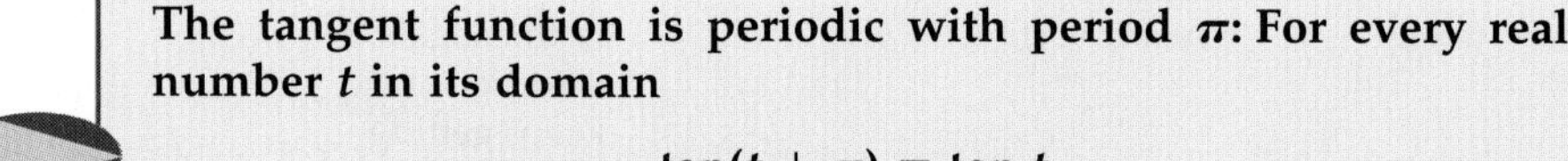

The tangent function is periodic with period π: For every real number t in its domain

$$\tan(t \pm \pi) = \tan t.$$

Example 5 As we saw in Section 3.4, the graph of

$$k(t) = \tan\left(t - \frac{\pi}{2}\right)$$

is the graph of $h(t) = \tan t$ shifted horizontally $\pi/2$ units to the right (Figure 6–36). ■

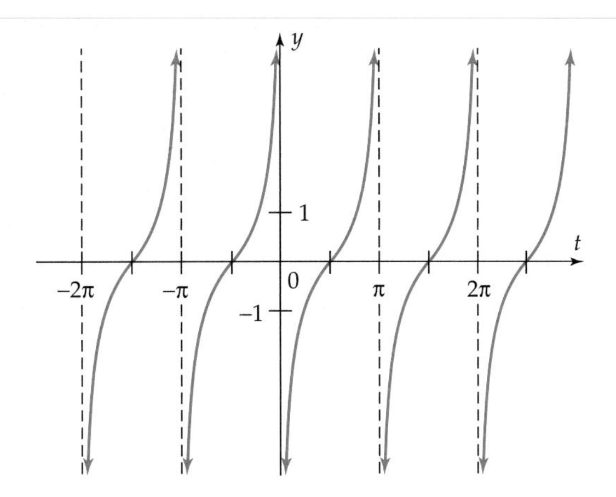

Figure 6–36

Graphs and Identities

Graphing calculators can be used to identify equations that could possibly be identities. A calculator cannot *prove* that such an equation is an identity; but it can provide evidence that it *might* be one. On the other hand, a calculator *can* prove that a particular equation is not an identity.

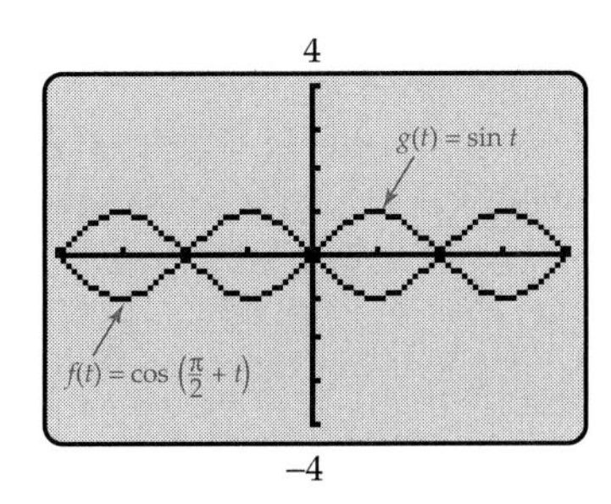

Figure 6–37

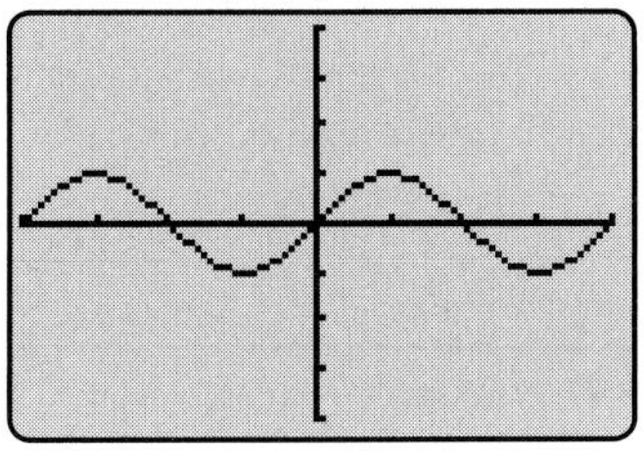

4

−4

Figure 6–38

Example 6 Which of the following equations could possibly be an identity?

(a) $\cos\left(\dfrac{\pi}{2} + t\right) = \sin t$ (b) $\cos\left(\dfrac{\pi}{2} - t\right) = \sin t$

Solution

(a) Consider the functions $f(t) = \cos\left(\dfrac{\pi}{2} + t\right)$ and $g(t) = \sin t$, whose rules are given by the two sides of the equation $\cos\left(\dfrac{\pi}{2} + t\right) = \sin t$. If this equation is an identity, then $f(t) = g(t)$ for every real number t, and hence f and g have the same graph. But the graphs of f and g on the interval $[-2\pi, 2\pi]$ (Figure 6–37) are obviously different. Therefore, this equation is *not* an identity.

(b) We can test this equation in the same manner. The graph of the left side, that is, the graph of $h(t) = \cos\left(\dfrac{\pi}{2} - t\right)$, in Figure 6–38 appears to be the same as the graph of $g(t) = \sin t$ on the interval $[-2\pi, 2\pi]$ (Figure 6–37). So let's explore further.

GRAPHING EXPLORATION

Graph $h(t) = \cos\left(\frac{\pi}{2} - t\right)$ and $g(t) = \sin t$ on the same screen and use the trace feature to confirm that the graphs appear to be identical.

The fact that the graphs appear identical means that the two functions have the same value at every number t that the calculator computed in making the graphs (at least 95 numbers). This evidence strongly suggests that the equation $\cos\left(\frac{\pi}{2} - t\right) = \sin t$ is an identity, but does not prove it. All we can say at this point is that the equation could possibly be an identity. ■

CAUTION

Do not assume that two graphs that look the same on a calculator screen actually are the same. Depending on the viewing window, two graphs that are actually quite different may appear identical. See Exercises 43–50 for some examples.

Exercises 6.4

In Exercises 1–6, use the graphs of the sine and cosine functions to find all the solutions of the equation.

1. $\sin t = 0$ **2.** $\cos t = 0$ **3.** $\sin t = 1$

4. $\sin t = -1$ **5.** $\cos t = -1$ **6.** $\cos t = 1$

In Exercises 7–10, find tan t, where the terminal side of an angle of t radians lies on the given line.

7. $y = 11x$ **8.** $y = 1.5x$ **9.** $y = 1.4x$

10. $y = .32x$

In Exercises 11–22, list the transformations needed to change the graph of f(t) into the graph of g(t). [See Section 3.4.]

11. $f(t) = \sin t;$ $g(t) = \sin t + 3$

12. $f(t) = \cos t;$ $g(t) = \cos t - 2$

13. $f(t) = \cos t;$ $g(t) = -\cos t$

14. $f(x) = \sin t;$ $g(t) = -3 \sin t$

15. $f(t) = \tan t;$ $g(t) = \tan t + 5$

16. $f(t) = \tan t;$ $g(t) = -\tan t$

17. $f(t) = \cos t;$ $g(t) = 3 \cos t$

18. $f(t) = \sin t;$ $g(t) = -2 \sin t$

19. $f(t) = \sin t;$ $g(t) = 3 \sin t + 2$

20. $f(t) = \cos t;$ $g(t) = 5 \cos t + 3$

21. $f(t) = \sin t;$ $g(t) = \sin (t - 2)$

22. $f(t) = \cos t;$ $g(t) = 3 \cos (t + 2) - 3$

In Exercises 23–30, use the graphs of the trigonometric functions to determine the number *of solutions of the equation between 0 and 2π.*

23. $\sin t = 3/5$ [*Hint:* How many points on the graph of $f(t) = \sin t$ between $t = 0$ and $t = 2\pi$ have second coordinate $3/5$?]

24. $\cos t = -1/4$ **25.** $\tan t = 4$ **26.** $\cos t = 2/3$

27. $\sin t = -1/2$

28. $\sin t = k$, where k is a nonzero constant such that $-1 < k < 1$.

29. $\cos t = k$, where k is a constant such that $-1 < k < 1$.

30. $\tan t = k$, where k is any constant.

In Exercises 31–42, use graphs to determine whether the equation could possibly be an identity or definitely is not an identity.

31. $\sin(-t) = -\sin t$

32. $\cos(-t) = \cos t$

33. $\sin^2 t + \cos^2 t = 1$

34. $\sin(t + \pi) = -\sin t$

35. $\sin t = \cos(t - \pi/2)$

36. $\sin^2 t - \tan^2 t = -(\sin^2 t)(\tan^2 t)$

37. $\dfrac{\sin t}{1 + \cos t} = \tan t$

38. $\dfrac{\cos t}{1 - \sin t} = \dfrac{1}{\cos t} + \tan t$

39. $\cos\left(\dfrac{\pi}{2} + t\right) = -\sin t$

40. $\sin\left(\dfrac{\pi}{2} + t\right) = -\cos t$

41. $(1 + \tan t)^2 = \dfrac{1}{\cos t}$

42. $(\cos^2 t - 1)(\tan^2 t + 1) = -\tan^2 t$

Thinkers

Exercises 43–46 explore various ways in which a calculator can produce inaccurate graphs of trigonometric functions. These exercises also provide examples of two functions, with different graphs, whose graphs appear identical in certain viewing windows.

43. Choose a viewing window with $-3 \le y \le 3$ and $0 \le x \le k$, where k is chosen as follows.

Width of Screen	k
95 pixels (TI-83)	188π
127 pixels (TI-86, Sharp 9600, Casio 9850)	252π
131 pixels (HP-38)	260π
159 pixels (TI-89)	316π

(a) Graph $y = \cos x$ and the constant function $y = 1$ on the same screen. Do the graphs look identical? Are the functions the same?

(b) Use the trace feature to move the cursor along the graph of $y = \cos x$, starting at $x = 0$. For what values of x did the calculator plot points? [*Hint:* $2\pi \approx 6.28$.] Use this information to explain why the two graphs look identical.

44. Using the viewing window in Exercise 43, graph $y = \tan x + 2$ and $y = 2$ on the same screen. Explain why the graphs look identical even though the functions are not the same.

45. The graph of $g(x) = \cos x$ is a series of repeated waves (see Figure 6–31). A full wave (from the peak, down to the trough, and up to the peak again) starts at $x = 0$ and finishes at $x = 2\pi$.

(a) How many full waves will the graph make between $x = 0$ and $x = 502.65$ ($\approx 80 \cdot 2\pi$)?

(b) Graph $g(t) = \cos t$ in a viewing window with $0 \le t \le 502.65$. How many full waves are shown on the graph? Is your answer the same as in part (a)? What's going on?

46. Find a viewing window in which the graphs of $y = \cos x$ and $y = .54$ appear identical. [*Hint:* See the chart in Exercise 43 and note that $\cos 1 \approx .54$.]

Exercises 47–50 provide further examples of functions with different graphs, whose graphs appear identical in certain viewing windows.

47. *Approximating trigonometric functions by polynomials.* For each odd positive integer n, let f_n be the function whose rule is

$$f_n(t) = t - \frac{t^3}{3!} + \frac{t^5}{5!} - \frac{t^7}{7!} + \cdots - \frac{t^n}{n!}.$$

Since the signs alternate, the sign of the last term might be + instead of −, depending on what n is. Recall that $n!$ is the product of all integers from 1 to n; for instance, $5! = 1\cdot2\cdot3\cdot4\cdot5 = 120$.

(a) Graph $f_7(t)$ and $g(t) = \sin t$ on the same screen in a viewing window with $-2\pi \le t \le 2\pi$. For what values of t does f appear to be a good approximation of g?

(b) What is the smallest value of n for which the graphs of f_n and g appear to coincide in this window? In this case, determine how accurate the approximation is by finding $f_n(2)$ and $g(2)$.

48. For each even positive integer n, let f_n be the function whose rule is

$$f_n(t) = 1 - \frac{t^2}{2!} + \frac{t^4}{4!} - \frac{t^6}{6!} + \frac{t^8}{8!} - \cdots + \frac{t^n}{n!}.$$

(The sign of the last term may be − instead of +, depending on what n is.)

(a) In a viewing window with $-2\pi \le t \le 2\pi$, graph f_6, f_{10}, and f_{12}.

(b) Find a value of n for which the graph of f_n appears to coincide (in this window) with the graph of a well-known trigonometric function. What is the function?

49. Find a rational function whose graph appears to coincide with the graph of $h(t) = \tan t$ when

$$-2\pi \le t \le 2\pi.$$

[*Hint:* Exercises 47 and 48.]

50. Find a periodic function whose graph consists of "square waves." [*Hint:* Consider the sum

$$\sin \pi t + \frac{1}{3}\sin 3\pi t + \frac{1}{5}\sin 5\pi t + \frac{1}{7}\sin 7\pi t + \cdots.]$$

51. With your calculator in parametric graphing mode and these range values

$0 \le t \le 6.28 \qquad -1 \le x \le 6.28 \qquad -2.5 \le y \le 2.5,$

graph the following two functions on the same screen:

$x_1 = \cos t,\ y_1 = \sin t \qquad$ and $\qquad x_2 = t,\ y_2 = \cos t.$

Using the trace feature, move the cursor along the first graph (the unit circle). Stop at a point on the circle, note the value of t and the x-coordinate of the point. Then switch the trace to the second graph (the cosine function) by using the up or down cursor arrows. The value of t remains the same. How does the y-coordinate of the new point compare with the x-coordinate of the original point on the unit circle? Explain what's going on.

52. (a) Judging from their graphs, which of the functions $f(t) = \sin t$, $g(t) = \cos t$, and $h(t) = \tan t$ appear to be even functions? Which appear to be odd functions?

(b) Confirm your answers in part (a) algebraically by using appropriate identities from Section 6.3.

6.5 Periodic Graphs and Simple Harmonic Motion

Functions whose graphs are of the form

$$f(t) = A\sin(bt + c) \qquad \text{or} \qquad g(t) = A\cos(bt + c)$$

with A, b, c constants, are used to describe many periodic physical phenomena. Figure 6–39 shows the graphs of three such functions:

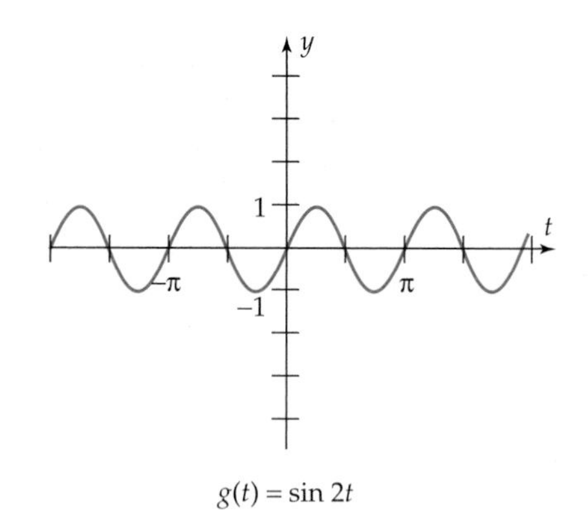

$g(t) = \sin 2t$

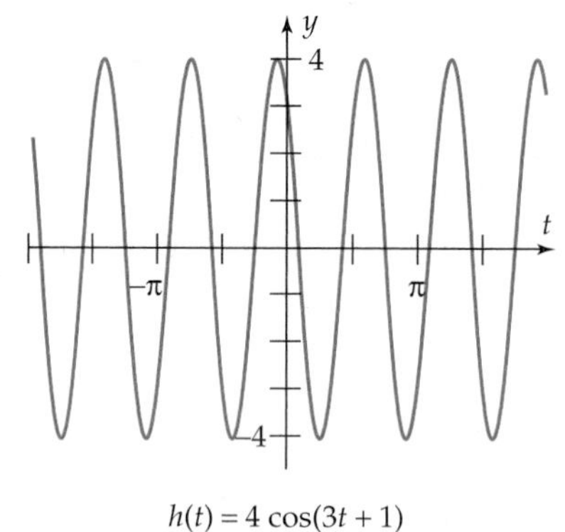

$h(t) = 4\cos(3t + 1)$

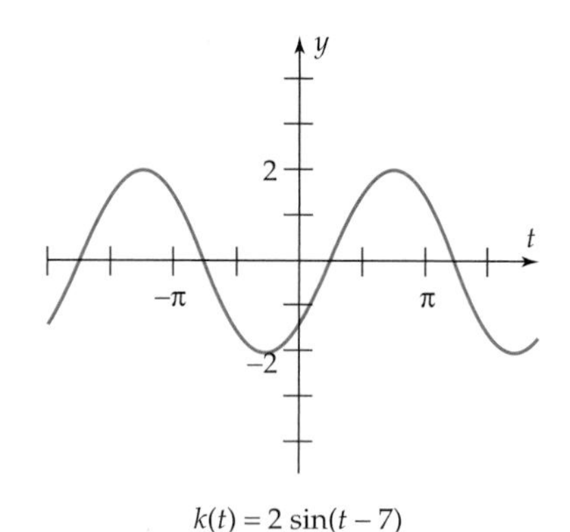

$k(t) = 2\sin(t - 7)$

Figure 6–39

In addition to having different periods, these graphs also differ in the heights of their waves and in the place where the first wave begins. We now analyze each of these factors.

Figure 6–40

Period

The graph of $f(t) = \sin t$ makes one complete wave between 0 and 2π: It begins on the horizontal axis, rises to height 1, falls to -1, and returns to the axis (see Figure 6–40). This corresponds to the fact that the sine function has period 2π. The graph of $g(t) = \sin 2t$ in Figure 6–39 makes two complete waves between $t = 0$ and $t = 2\pi$. So its period (the length of one wave) is $2\pi/2 = \pi$.

GRAPHING EXPLORATION

Graph each of the following functions, one at a time, in a viewing window with $0 \leq t \leq 2\pi$ and determine the number of complete waves in each graph and the period of each function (the length of one wave):

$$f(t) = \sin 3t, \qquad g(t) = \cos 4t, \qquad h(t) = \sin 5t.$$

Keep in mind that a complete wave of the cosine function starts at height 1, falls to -1, then rises to height 1 again.

This exploration suggests

Period

If $b > 0$, then the graph of either

$$f(t) = \sin bt \quad \text{or} \quad g(t) = \cos bt$$

makes b complete waves between 0 and 2π. Hence, each function has period $2\pi/b$.

Although we arrived at this statement by generalizing from several graphs, it can also be explained algebraically.

Example 1 The graph of $g(t) = \cos t$ makes one complete wave as t takes values from 0 to 2π. Similarly, the graph of $k(t) = \cos 3t$ will complete one wave as the quantity $3t$ takes values from 0 to 2π. However,

$$3t = 0 \text{ when } t = 0 \qquad \text{and} \qquad 3t = 2\pi \text{ when } t = 2\pi/3.$$

So the graph of $k(t) = \cos 3t$ makes one complete wave between $t = 0$ and $t = 2\pi/3$, as shown in Figure 6–41, and hence k has period $2\pi/3$. Similarly, the graph makes a complete wave from $t = 2\pi/3$ to $t = 4\pi/3$ and another one from $t = 4\pi/3$ to $t = 2\pi$, as shown in Figure 6–41. ■

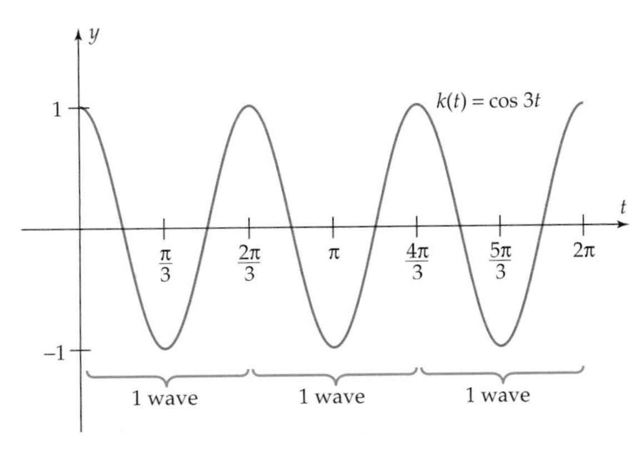

Figure 6–41

Example 2 According to the box above, the function $f(t) = \sin \frac{1}{2}t$ has period $\dfrac{2\pi}{1/2} = 4\pi$. Its graph makes *half* a wave from $t = 0$ to $t = 2\pi$ (just as $\sin t$ does from $t = 0$ to $t = \pi$) and the other half of the wave from $t = 2\pi$ to $t = 4\pi$, as shown in Figure 6–42. ■

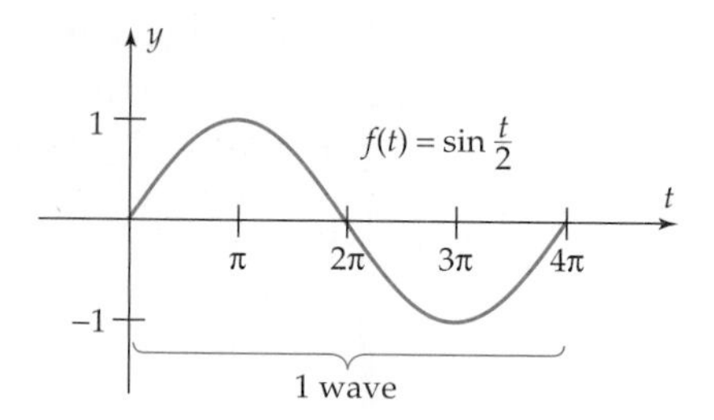

Figure 6–42

> **CAUTION**
>
> A calculator may produce a wildly inaccurate graph of sin *bt* or cos *bt* for some values of *b*. For instance, the graph of
>
> $$f(t) = \sin 50t$$
>
> has 50 complete waves of the same height between 0 and 2π, but that's not what your calculator screen will show (try it!). Similarly, the graph of $g(t) = \cos 100t$ has 100 complete waves between 0 and 2π. How many does your calculator screen show? See Exercises 53–54 for an explanation.

Amplitude

As we saw in Section 3.4, multiplying the rule of a function by a positive constant has the effect of stretching its graph away from or shrinking it toward the horizontal axis.

Example 3 The function $g(t) = 7 \cos 3t$ is just the function $k(t) = \cos 3t$ multiplied by 7. Consequently, the graph of g is just the graph of k (which was obtained in Example 1) stretched away from the horizontal axis by a factor of 7:

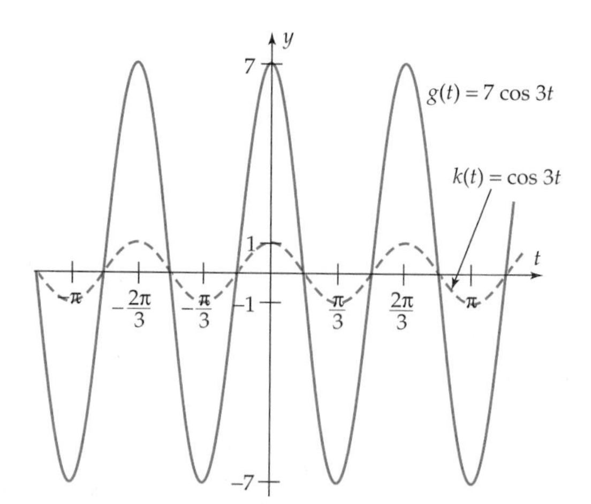

Figure 6–43

As Figure 6–43 shows, stretching the graph affects only the height of the waves, not the period of the function: Both graphs have period $2\pi/3$ and each full wave has length $2\pi/3$. The "stretching factor" 7 is the maximum height of the wave, that is, the maximum vertical distance from the graph to the horizontal axis. ■

The waves of the graph of $g(t) = 7 \cos 3t$ in Figure 6–43 rise 7 units above the t-axis and drop 7 units below the axis. More generally, the waves of the graph of $f(t) = A \sin bt$ or $g(t) = A \cos bt$ move a distance of $|A|$ units above and below the t-axis and we say that these functions have **amplitude** $|A|$. In summary,

Amplitude and Period

> **If $A \neq 0$ and $b > 0$, then each of the functions**
>
> $$f(t) = A \sin bt \qquad \text{or} \qquad g(t) = A \cos bt$$
>
> **has amplitude $|A|$ and period $2\pi/b$.**

Example 4 The function $f(t) = -2 \sin 4t$ has amplitude $|-2| = 2$ and period $2\pi/4 = \pi/2$. So the graph consists of waves of length $\pi/2$ that rise and fall between -2 and 2. But be careful: The waves in the graph of 2 sin 4t (like the waves of sin t) begin at height 0, rise, and then fall. But the graph of $f(t) = -2 \sin 4t$ is the graph of 2 sin 4t reflected in the horizontal axis (see page 155). So its waves start at height 0, move *downward,* and then rise, as shown in Figure 6–44. ■

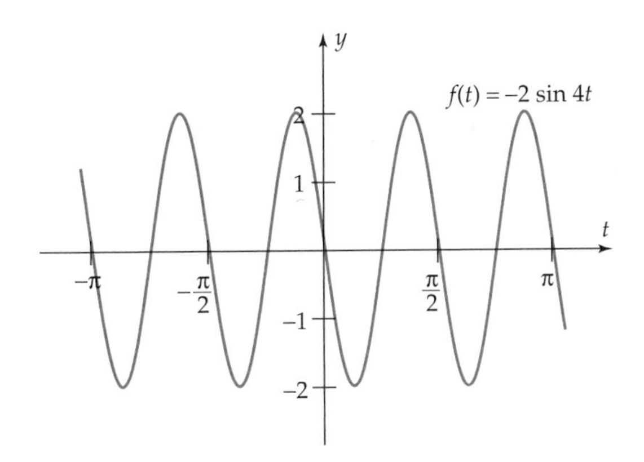

Figure 6–44

Phase Shift

In Section 3.4 we saw that replacing t by $t \pm 3$ in the rule of a function $f(t)$ shifts the graph horizontally (see page 152). Similarly, the graph of $\sin(t - 3)$ is the graph of sin t shifted 3 units to the right and the graph of $\sin(t + 3)$ is the graph of sin t shifted 3 units to the left.

Example 5

(a) Find a sine function whose graph looks like Figure 6–45.

(b) Find a cosine function whose graph looks like Figure 6–45.

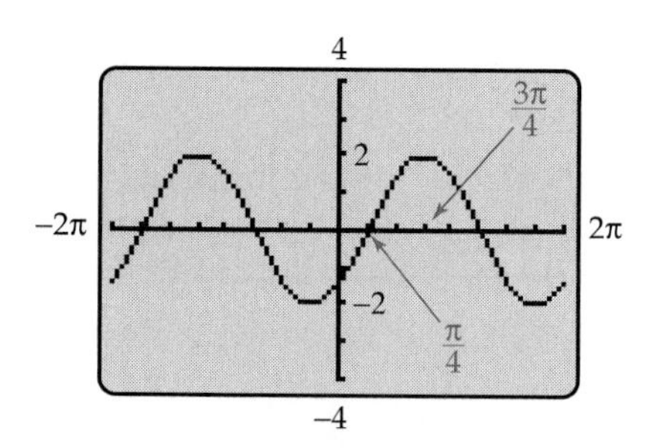

Figure 6–45

Solution

(a) Since each wave has height 2, Figure 6–45 looks like the graph of $2 \sin t$ shifted $\pi/4$ units to the right (so that a sine wave starts at $t = \pi/4$). Since the graph of $2 \sin(t - \pi/4)$ is the graph of $2 \sin t$ shifted $\pi/4$ units to the right (see page 152), we conclude that Figure 6–45 closely resembles the graph of $f(t) = 2 \sin(t - \pi/4)$.

(b) Figure 6–45 also looks like the graph of $2 \cos t$ shifted $3\pi/4$ units to the right (so that a cosine wave starts at $t = 3\pi/4$). Hence Figure 6–45 could also be the graph of $g(t) = 2 \cos(t - 3\pi/4)$. ■

Example 6

(a) Find the amplitude and the period of

$$f(t) = 3 \sin(2t + 5).$$

(b) Do the same for the function $f(t) = A \sin(bt + c)$, where A, b, c are constants.

Solution The analysis of $f(t) = 3 \sin(2t + 5)$ is in the left-hand column below and the analysis of the general case $f(t) = A \sin(bt + c)$ is in the right-hand column. Observe that exactly the same procedure is used in both cases: just change 3 to A, 2 to b, and 5 to c.

(a) Rewrite the rule of $f(t) = 3 \sin(2t + 5)$ as

$$f(t) = 3 \sin(2t + 5) = 3 \sin\left(2\left(t + \frac{5}{2}\right)\right).$$

Thus, the rule of f can be obtained from the rule of the function $k(t) = 3 \sin 2t$ by replacing t with $t + \frac{5}{2}$. Therefore, the graph of f is just the graph of k shifted horizontally $5/2$ units to the left, as shown in Figure 6–46.

(b) Rewrite the rule of $f(t) = A \sin(bt + c)$ as

$$f(t) = A \sin(bt + c) = A \sin\left(b\left(t + \frac{c}{b}\right)\right).$$

Thus, the rule of f can be obtained from the rule of the function $k(t) = A \sin bt$ by replacing t with $t + \frac{c}{b}$. Therefore, the graph of f is just the graph of k shifted horizontally by c/b units.

Hence, $f(t) = 3\sin(2t + 5)$ has the same amplitude as $k(t) = 3\sin 2t$, namely 3, and the same period, namely $2\pi/2 = \pi$.

On the graph of $k(t) = 3\sin 2t$ a wave begins when $t = 0$. On the graph of

$$f(t) = 3\sin 2\left(t + \frac{5}{2}\right)$$

the shifted wave begins when $t + 5/2 = 0$, that is, when $t = -5/2$.

Hence, $f(t) = A\sin(bt + c)$ has the same amplitude as $k(t) = A\sin bt$, namely $|A|$, and the same period, namely $2\pi/b$.

On the graph of $k(t) = A\sin bt$, a wave begins when $t = 0$. On the graph of

$$f(t) = A\sin b\left(t + \frac{c}{b}\right)$$

the shifted wave begins when $t + c/b = 0$, that is, when $t = -c/b$. ■

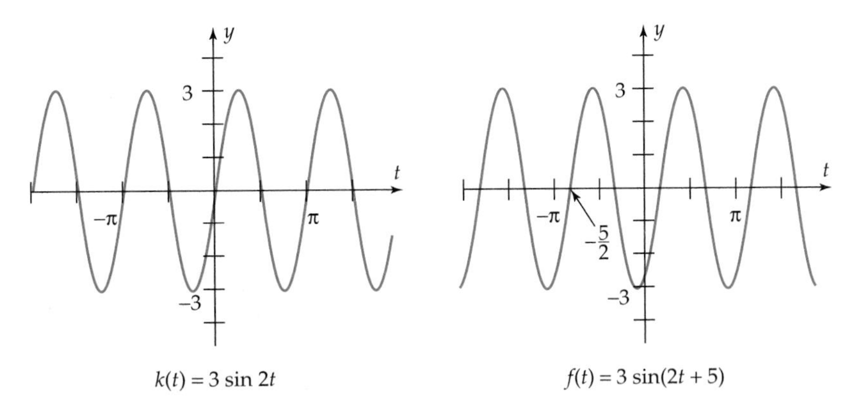

Figure 6–46

We say that the function $f(t) = A\sin(bt + c)$ has **phase shift** $-c/b$. A similar analysis applies to the function $g(t) = \cos(bt + c)$ and leads to this conclusion:

Amplitude, Period, and Phase Shift

If $A \neq 0$ and $b > 0$, then each of the functions

$$f(t) = A\sin(bt + c) \qquad \text{and} \qquad g(t) = A\cos(bt + c)$$

has

amplitude $|A|$, period $2\pi/b$, phase shift $-c/b$.

A wave of the graph begins at $t = -c/b$.

Example 7 Describe the graph of $g(t) = 2\cos(3t - 4)$.

Solution The rule of g can be rewritten as

$$g(t) = 2\cos(3t + (-4)).$$

This is the case described in the preceding box where $A = 2$, $b = 3$, and $c = -4$. Therefore, the function g has

$$\text{amplitude } |A| = |2| = 2, \qquad \text{period } \frac{2\pi}{b} = \frac{2\pi}{3},$$

$$\text{phase shift } -\frac{c}{b} = -\frac{-4}{3} = \frac{4}{3}.$$

Hence, the graph of g consists of waves of length of $2\pi/3$ that run vertically between 2 and -2. A wave begins at $t = 4/3$.

GRAPHING EXPLORATION

Verify the accuracy of this analysis by graphing $y = 2\cos(3t - 4)$ in the viewing window with $-2\pi \leq t \leq 2\pi$ and $-3 \leq y \leq 3$. Keeping in mind that a wave of the cosine graph begins at the maximum height above the horizontal axis, use the trace feature to show that a wave begins at $t = 4/3$. ■

Many other types of trigonometric graphs, including those consisting of waves of varying height and length, are considered in Excursion 6.5.A.

Applications

The sine and cosine functions, or variations of them, can be used to describe many different phenomena.

Example 8 A wheel of radius 2 centimeters is rotating counterclockwise at 3 radians per second. A free-hanging rod 10 centimeters long is connected to the edge of the wheel at point P and remains vertical as the wheel rotates (Figure 6–47). Assuming that the center of the wheel is at the origin and that P is at (2, 0) at time $t = 0$, find a function that describes the y-coordinate of the tip E of the rod at time t.

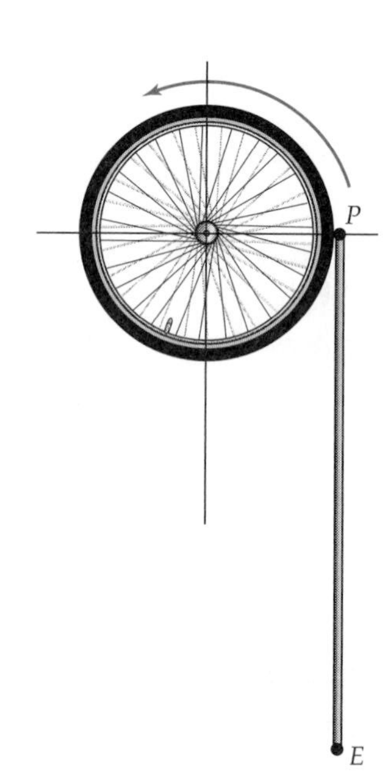

Figure 6–47

Solution The wheel is rotating at 3 radians per second, so after t seconds the point P has moved through an angle of $3t$ radians and is 2 units from the origin. Using the point-in-the-plane description, we see that the coordinates (x, y) of P satisfy

$$\frac{x}{2} = \cos 3t \qquad \frac{y}{2} = \sin 3t$$

$$x = 2\cos 3t \qquad y = 2\sin 3t.$$

Since E lies 10 centimeters directly below P, its y-coordinate is 10 less than the y-coordinate of P. Hence, the function giving the y-coordinate of E at time t is

$$f(t) = y - 10 = 2\sin 3t - 10. \quad ■$$

Example 9 Suppose that a weight hanging from a spring is set in motion by an upward push (Figure 6–48) and that it takes 5 seconds for it to move from its equilibrium position to 8 centimeters above, then drop to 8 centimeters below, and finally return to its equilibrium position. [We consider an idealized situation in which the spring has perfect elasticity and friction, air resistance, etc., are negligible.]

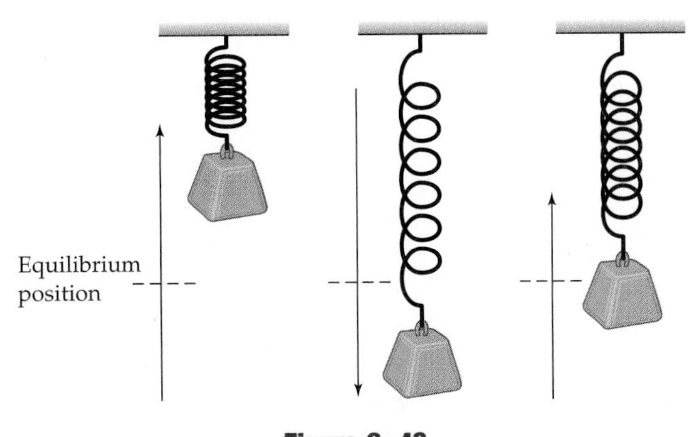

Figure 6–48

Let $h(t)$ denote the distance of the weight above $(+)$ or below $(-)$ its equilibrium position at time t. Then $h(t)$ is 0 when $t = 0$. As t runs from 0 to 5, $h(t)$ increases from 0 to 8, decreases to -8, and increases again to 0. In the next 5 seconds it repeats the same pattern, and so on. Thus, the graph of h has some kind of wave shape. Two possibilities are shown in Figure 6–49.

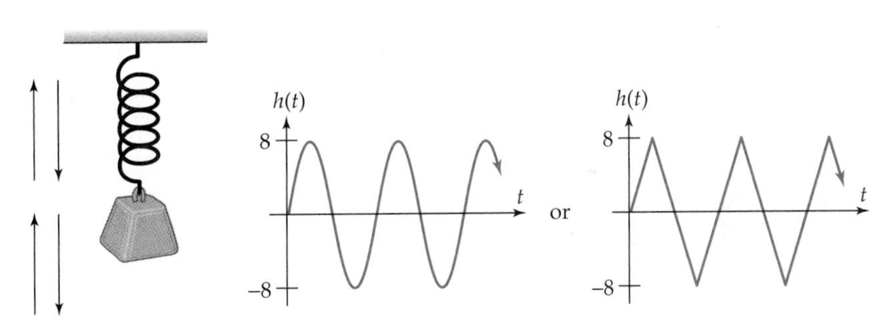

Figure 6–49

Careful physical experiment suggests that the left-hand curve in Figure 6–49, which resembles the sine graphs studied above is a reasonably accurate model of this process. Facts from physics, calculus, and differential equations show that the rule of the function h is the form $h(t) = A\sin(bt + c)$ for some constants A, b, c. Since the amplitude of h is 8, its period is 5, and its phase shift is 0, the constants A, b, and c must satisfy

$$A = 8, \qquad \frac{2\pi}{b} = 5, \qquad -\frac{c}{b} = 0$$

or equivalently,

$$A = 8, \qquad b = \frac{2\pi}{5}, \qquad c = 0.$$

Therefore, the motion of the moving spring can be described by the function

$$h(t) = A\sin(bt + c) = 8\sin\left(\frac{2\pi}{5}t + 0\right) = 8\sin\frac{2\pi t}{5}. \quad ■$$

Motion that can be described by a function of the form $f(t) = A\sin(bt + c)$ or $f(t) = A\cos(bt + c)$ is called **simple harmonic motion.** Many kinds of physical motion are simple harmonic motions. Other periodic phenomena, such as sound waves, are more complicated to describe. Their graphs consist of waves of varying amplitude. Such graphs are discussed in Excursion 6.5.A.

Example 10* The table shows the average monthly temperature in Cleveland, OH, based on data from 1961 to 1990. (*Source:* National Climatic Data Center.) Since average temperatures are not likely to vary much from year to year, the data essentially repeats the same pattern in subsequent years. So a periodic model is appropriate.

Month	Temperature (°F)
Jan	24.8
Feb	27.3
Mar	37.2
Apr	47.5
May	57.9
Jun	67.5

Month	Temperature (°F)
Jul	71.8
Aug	70.3
Sep	63.9
Oct	52.7
Nov	42.4
Dec	30.9

The data for a two-year period is plotted in Figure 6–50 (with $x = 1$ corresponding to January, $x = 2$ to February, and so on). The sine regression feature on a calculator produces this model from the 24 data points:

$$y = 23.0799\sin(.5212x - 2.1943) + 49.4178.$$

The period of this function is $2\pi/.5212 \approx 12.06$, slightly off from the 12-month period we would expect. However, its graph in Figure 6–51 appears to fit the data well. ■

*Skip this example if you haven't read Sections 2.5 and 5.5 on regression.

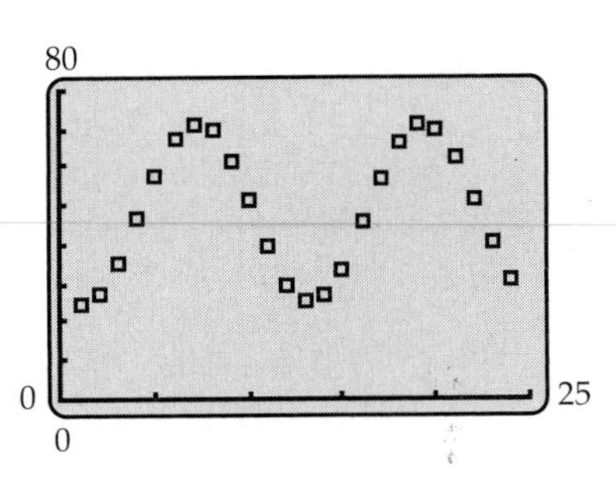

Figure 6–50

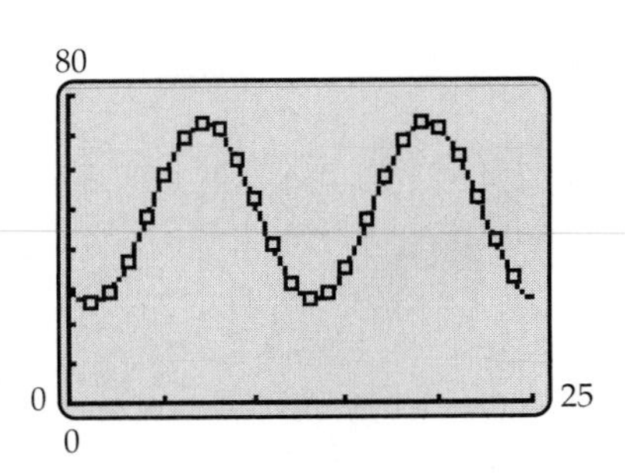

Figure 6–51

Exercises 6.5

In Exercises 1–7, state the amplitude, period, and phase shift of the function.

1. $g(t) = 3\sin(2t - \pi)$
2. $h(t) = -4\cos(3t - \pi/6)$
3. $q(t) = -7\sin(7t + 1/7)$
4. $g(t) = 97\cos(14t + 5)$
5. $f(t) = \cos 2\pi t$
6. $k(t) = \cos(2\pi t/3)$
7. $p(t) = 6\cos(3\pi t + 1)$

8. **(a)** What is the period of $f(t) = \sin 2\pi t$?
   (b) For what values of t (with $0 \le t \le 2\pi$) is $f(t) = 0$?
   (c) For what values of t (with $0 \le t \le 2\pi$) is $f(t) = 1$? or $f(t) = -1$?

In Exercises 9–14, give the rule of a periodic function with the given numbers as amplitude, period, and phase shift (in this order).

9. $3, \pi/4, \pi/5$ **10.** $2, 3, 0$ **11.** $2/3, 1, 0$
12. $4/5, 2, 3$ **13.** $7, 5/3, -\pi/2$ **14.** $19, 4, -5$

In Exercises 15–18, state the rule of a function of the form $f(t) = A\sin bt$ or $g(t) = A\cos bt$ whose graph appears to be identical to the given graph.

15.

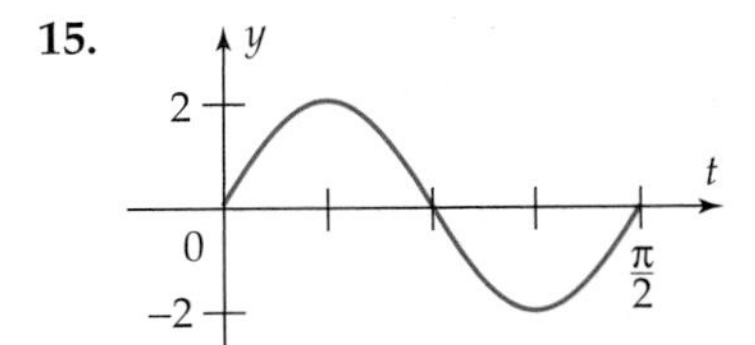

16.

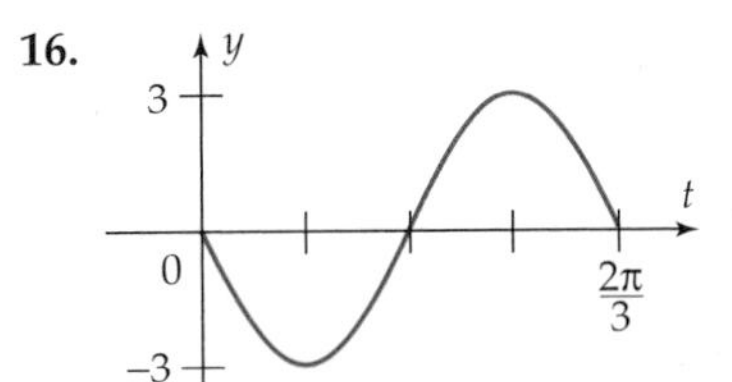

17.

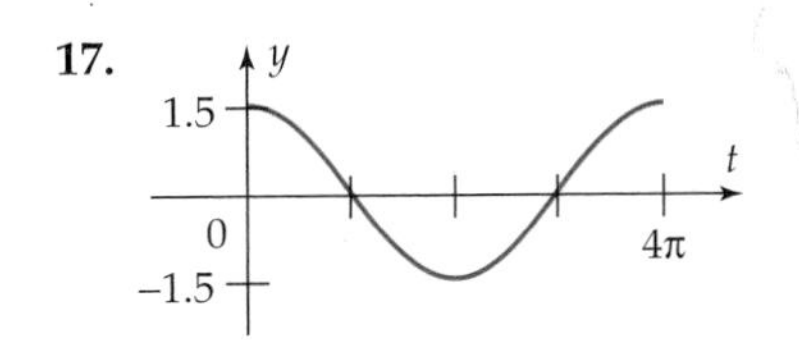

18.

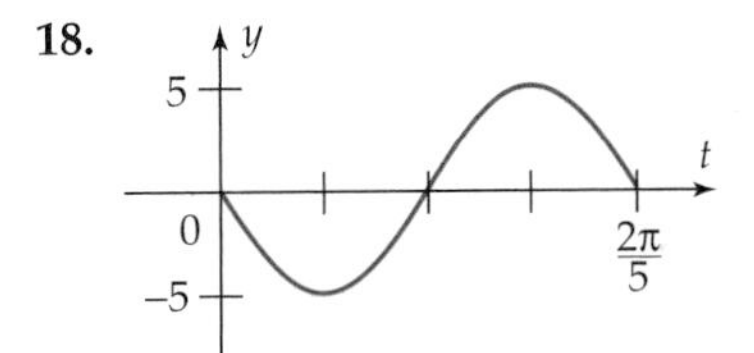

In Exercises 19–24,
(a) State the rule of a function of the form

$$f(t) = A\sin(bt + c)$$

whose graph appears to be identical with the given graph.
(b) State the rule of a function of the form

$$g(t) = A\cos(bt + c)$$

whose graph appears to be identical with the given graph.

19.

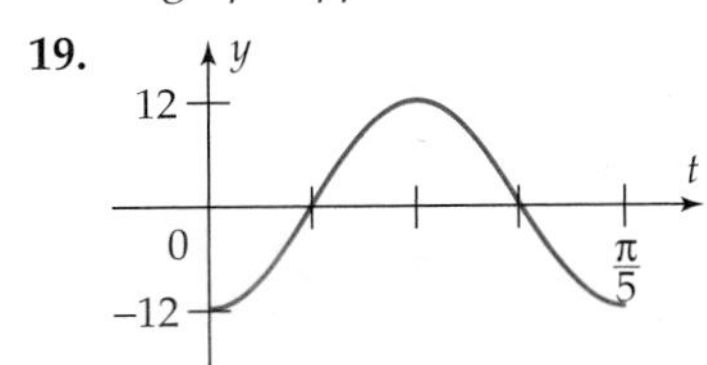

20.

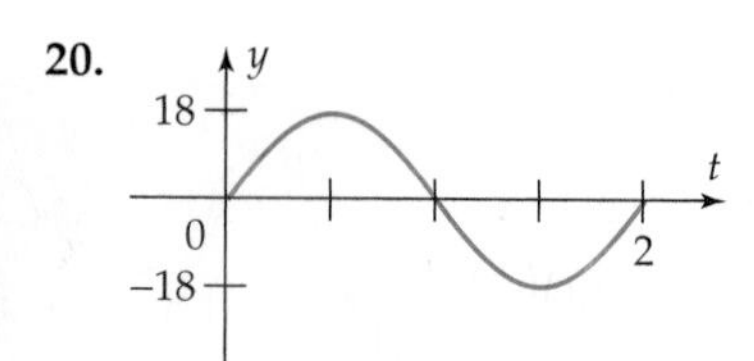

21.

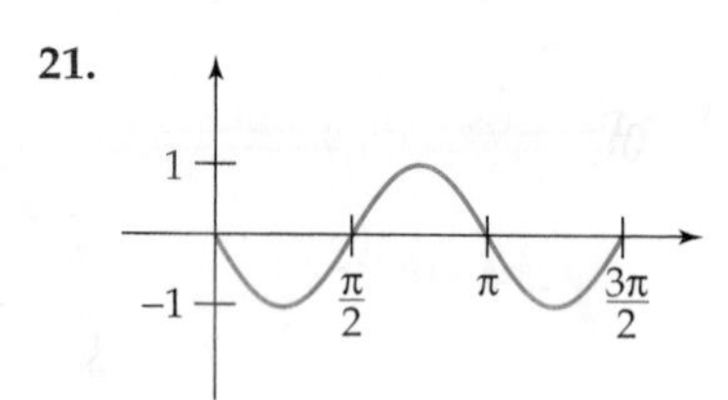

22.

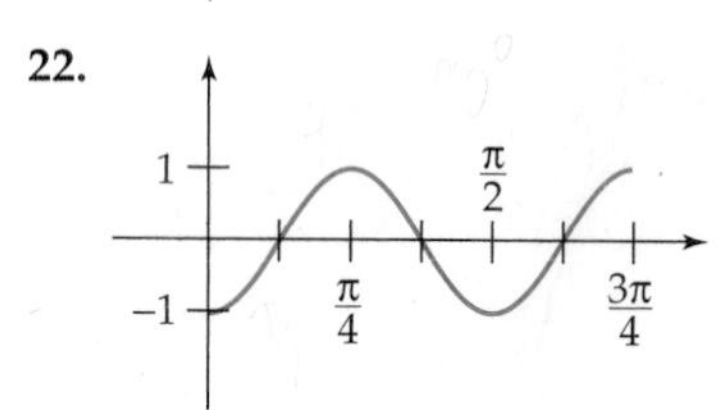

23.

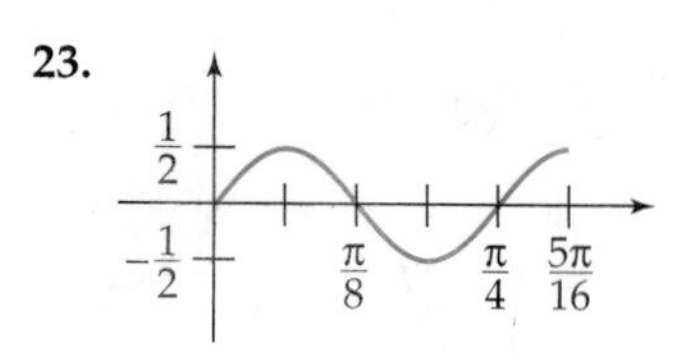

24.

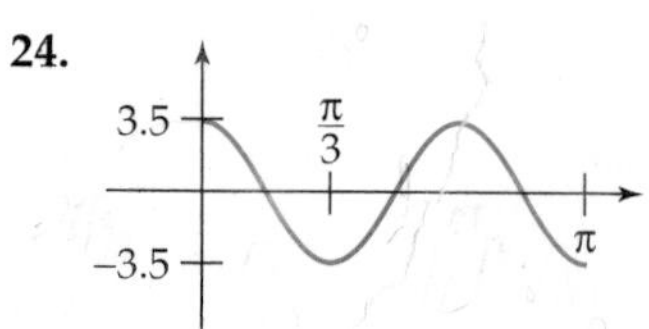

In Exercises 25–30, sketch a complete graph of the function.

25. $k(t) = -3 \sin t$

26. $y(t) = -2 \cos 3t$

27. $p(t) = -\dfrac{1}{2} \sin 2t$

28. $q(t) = \dfrac{2}{3} \cos \dfrac{3}{2} t$

29. $h(t) = 3 \sin(2t + \pi/2)$

30. $p(t) = 3 \cos(3t - \pi)$

In Exercises 31–34, graph the function over the interval $[0, 2\pi)$ and determine the location of all local maxima and minima. [This can be done either graphically or algebraically.]

31. $f(t) = \dfrac{1}{2} \sin\left(t - \dfrac{\pi}{3}\right)$

32. $g(t) = 2 \sin(2t/3 - \pi/9)$

33. $f(t) = -2 \sin(3t - \pi)$

34. $h(t) = \dfrac{1}{2} \cos\left(\dfrac{\pi}{2} t - \dfrac{\pi}{8}\right) + 1$

In Exercises 35–38, graph $f(t)$ in a viewing window with $-2\pi \le t \le 2\pi$. Use the trace feature to determine constants A, b, c such that the graph of $f(t)$ appears to coincide with the graph of $g(t) = A \sin(bt + c)$.

35. $f(t) = 2 \sin t + 5 \cos t$

36. $f(t) = -3 \sin t + 2 \cos t$

37. $f(t) = 3 \sin(4t + 2) + 2 \cos(4t - 1)$

38. $f(t) = 2 \sin(3t - 5) - 3 \cos(3t + 2)$

In Exercises 39 and 40, explain why there could not possibly be constants A, b, and c such that the graph of $g(t) = A \sin(bt + c)$ coincides with the graph of $f(t)$.

39. $f(t) = \sin 2t + \cos 3t$

40. $f(t) = 2 \sin(3t - 1) + 3 \cos(4t + 1)$

41. The current generated by an AM radio transmitter is given by a function of the form $f(t) = A \sin 2000\pi m t$, where $550 \le m \le 1600$ is the location on the broadcast dial and t is measured in seconds. For example, a station at 980 on the AM dial has a function of the form

$$f(t) = A \sin 2000\pi(980)t = A \sin 1{,}960{,}000\pi t.$$

Sound information is added to this signal by varying (modulating) A, that is, by changing the amplitude of the waves being transmitted. (AM means amplitude modulation.) For a station at 980 on the dial, what is the period of function f? What is the frequency (number of complete waves per second)?

42. Find the function f (as in Exercise 41), its period, and its frequency for a station at 1440 on the dial.

43. The original Ferris wheel, built by George Ferris for the Columbian Exposition of 1893, was much larger and slower than its modern counterparts: It had a diameter of 250 feet and contained 36 cars, each of which held 40 people; it made one

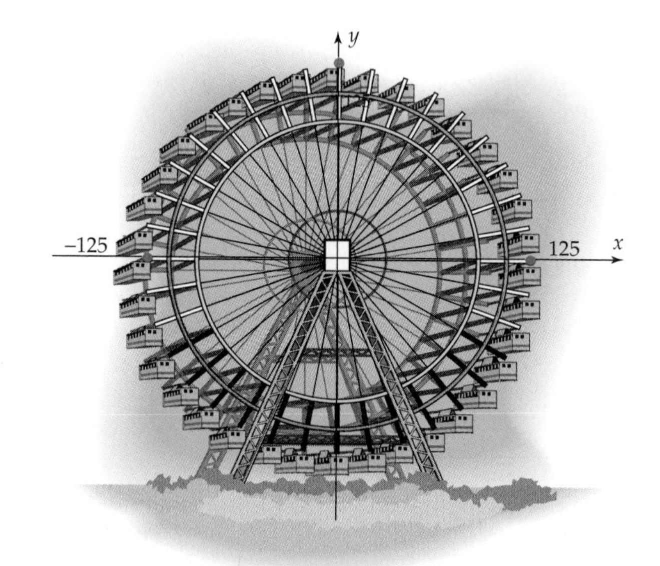

revolution every 10 minutes. Imagine that the Ferris wheel revolves counterclockwise in the x-y plane with its center at the origin. Car D in the figure had coordinates (125, 0) at time $t = 0$. Find the rule of a function that gives the y-coordinate of car D at time t.

44. Do Exercise 43 if the wheel turns at 2 radians per minute and car D is at $(0, -125)$ at time $t = 0$.

45. A circular wheel of radius 1 foot rotates counterclockwise. A 4-foot rod has one end attached to the edge of this wheel and the other end to the base of a piston (see figure). It transfers the rotary motion of the wheel into a back-and-forth linear motion of the piston. If the wheel is rotating at 10 revolutions per second, point W is at (1, 0) at time $t = 0$, and point P is always on the x-axis, find the rule of a function that gives the x-coordinate of P at time t.

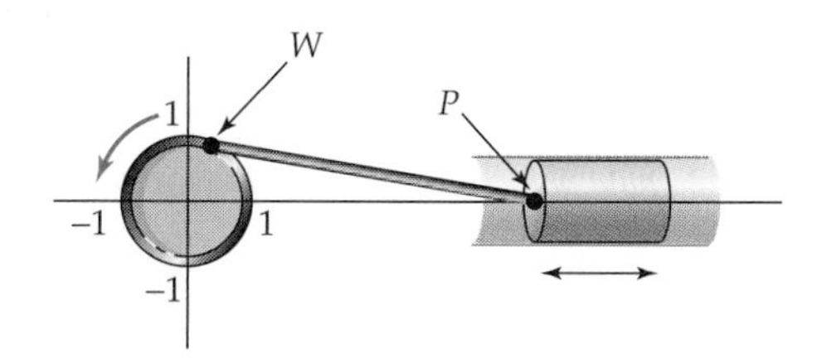

46. Do Exercise 45 if the wheel has a radius of 2 feet, rotates at 50 revolutions per second, and is at (2, 0) when $t = 0$.

In Exercises 47–50, suppose there is a weight hanging from a spring (under the same idealized conditions described in Example 9). The weight is given a push to start it moving. At any time t, let h(t) be the height (or depth) of the weight above (or below) its equilibrium point. Assume that the maximum distance the weight moves in either direction from the equilibrium point is 6 centimeters and that it moves through a complete cycle every 4 seconds. Express h(t) in terms of the sine or cosine function under the stated conditions.

47. Initial push is *upward* from the equilibrium.

48. Initial push is *downward* from the equilibrium point. [*Hint:* What does the graph of $A \sin bt$ look like when $A < 0$?]

49. Weight is pulled 6 centimeters above equilibrium, and the initial movement (at $t = 0$) is downward. [*Hint:* Think cosine.]

50. Weight is pulled 6 centimeters below equilibrium, and the initial movement is upward.

51. A pendulum swings uniformly back and forth, taking 2 seconds to move from the position directly above point A to the position directly above point B.

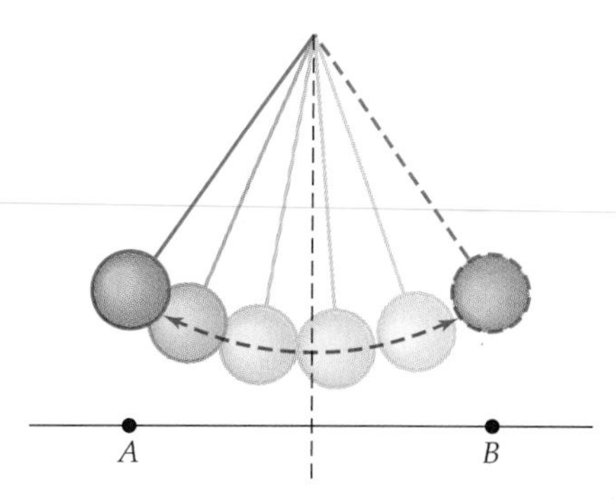

The distance from A to B is 20 centimeters. Let $d(t)$ be the horizontal distance from the pendulum to the (dashed) center line at time t seconds (with distances to the right of the line measured by positive numbers and distances to the left by negative ones). Assume that the pendulum is on the center line at time $t = 0$ and moving to the right. Assume the motion of the pendulum is simple harmonic motion. Find the rule of the function $d(t)$.

52. The diagram shows a merry-go-round that is turning counterclockwise at a constant rate, making 2 revolutions in 1 minute. On the merry-go-round are horses A, B, C, and D at 4 meters from the center and horses E, F, and G at 8 meters from the center. There is a function $a(t)$ that gives the distance the horse A is from the y-axis (this is the x-coordinate of the position A is in) as a function of time t (measured in minutes). Similarly, $b(t)$ gives the x-coordinate for B as a function of time, and so on. Assume the diagram shows the situation at time $t = 0$.

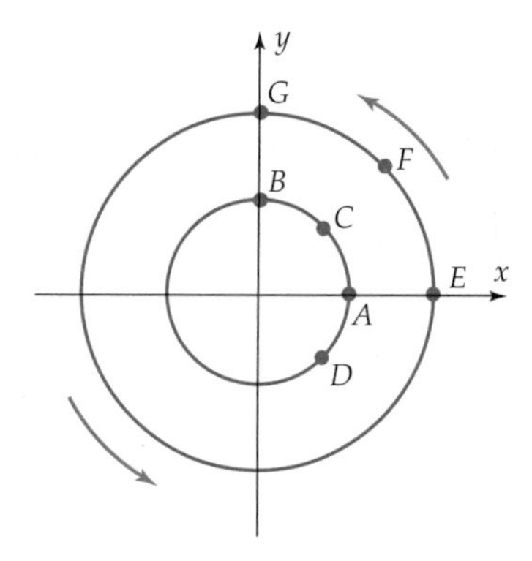

(a) Which of the following functions does $a(t)$ equal:

$$4 \cos t, \quad 4 \cos \pi t, \quad 4 \cos 2t, \quad 4 \cos 2\pi t,$$
$$4 \cos \left(\tfrac{1}{2}t\right), \quad 4 \cos ((\pi/2)t), \quad 4 \cos 4\pi t?$$

Explain.

(b) Describe the functions $b(t)$, $c(t)$, $d(t)$, and so on using the cosine function:

$b(t) =$ ____, $c(t) =$ ____, $d(t) =$ ____.
$e(t) =$ ____, $f(t) =$ ____, $g(t) =$ ____.

(c) Suppose the x-coordinate of a horse S is given by the function $4 \cos ((4\pi t - (5\pi/6))$ and the x-coordinate of another horse T is given by $8 \cos (4\pi t - (\pi/3))$. Where are these horses located in relation to the rest of the horses? Mark the positions of T and S at $t = 0$ into the figure.

Exercises 53–54 explore various ways in which a calculator can produce inaccurate or misleading graphs of trigonometric functions.

53. (a) If you were going to draw a rough picture of a full wave of the sine function by plotting some points and connecting them with straight-line segments, approximately how many points would you have to plot?

(b) If you were drawing a rough sketch of the graph of $f(t) = \sin 100t$ when $0 \leq t \leq 2\pi$, according to the method in part (a), approximately how many points would have to be plotted?

(c) How wide (in pixels) is your calculator screen? Your answer to this question is the maximum number of points that your calculator plots when graphing any function.

(d) Use parts (a)–(c) to explain why your calculator cannot possibly produce an accurate graph of $f(t) = \sin 100t$ in any viewing window with $0 \leq t \leq 2\pi$.

54. (a) Using a viewing window with $0 \leq t \leq 2\pi$, use the trace feature to move the cursor along the horizontal axis. [On some calculators it may be necessary to graph $y = 0$ in order to do this.] What is the distance between one pixel and the next (to the nearest hundredth)?

(b) What is the period of $f(t) = \sin 300t$? Since the period is the length of one full wave of the graph, approximately how many waves should there be between two adjacent pixels? What does this say about the possibility of your calculator's producing an accurate graph of this function between 0 and 2π?

55. The table at the top of the next column shows the number of unemployed people in the labor force (in millions) for 1984–1995.

(a) Sketch a scatter plot of the data, with $x = 0$ corresponding to 1980.

(b) Does the data appear to be periodic? If so, find an appropriate model.

(c) Do you think this model is likely to be accurate much beyond the 1995? Why?

Year	Unemployed
1984	8.539
1985	8.312
1986	8.237
1987	7.425
1988	6.701
1989	6.528

Year	Unemployed
1990	7.047
1991	8.628
1992	9.613
1993	8.940
1994	7.996
1995	7.404

In Exercises 56 and 57, do the following:

(a) Use 12 data points (with $x = 1$ corresponding to January) to find a periodic model of the data.

(b) What is the period of the function found in part (a)? Is this reasonable?

(c) Plot 24 data points (two years) and graph the function from part (a) on the same screen. Is the function a good model in the second year?

(d) Use the 24 data points in part (c) to find another periodic model for the data.

(e) What is the period of the function in part (d)? Does its graph fit the data well?

56. The table shows the average monthly temperature in Chicago, IL, based on data from 1961 to 1990. (*Source:* National Climatic Data Center.)

sin

Month	Temperature (°F)
Jan	21.2
Feb	25.7
Mar	36.7
Apr	48.6
May	59.0
Jun	68.4
Jul	73.0
Aug	71.8
Sep	64.2
Oct	52.5
Nov	39.7
Dec	27.3

57. The table shows the average monthly precipitation (in inches) in San Francisco, CA, based on data from 1961 to 1990. (*Source:* National Climatic Data Center.)

Month	Precipitation
Jan	4.4
Feb	3.2
Mar	3.1
Apr	1.4
May	.2
Jun	.1
Jul	0
Aug	.1
Sep	.2
Oct	1.2
Nov	2.9
Dec	3.1

Thinkers

58. Based on the results of Exercises 35–38, under what conditions on the constants a, k, h, d, r, s does it appear that the graph of

$$f(t) = a\sin(kt + h) + d\cos(rt + s)$$

coincides with the graph of the function $g(t) = A\sin(bt + c)$?

59. A grandfather clock has a pendulum length of k meters and its swing is given (as in Exercise 51) by the function $f(t) = .25\sin(\omega t)$, where $\omega = \sqrt{\frac{9.8}{k}}$.

(a) Find k such that the period of the pendulum is 2 seconds.

(b) The temperature in the summer months causes the pendulum to increase its length by .01%. How much time will the clock lose in June, July, and August? [*Hint:* These three months have a total of 92 days (7,948,800 seconds). If k is increased by .01%, what is $f(2)$?]

EXCURSION Other Trigonometric Graphs

A graphing calculator enables you to explore with ease a wide variety of trigonometric functions. Some of the possibilities are considered here.

GRAPHING EXPLORATION

Graph

$$g(t) = \cos t \quad \text{and} \quad f(t) = \sin(t + \pi/2)$$

on the same screen. Is there any apparent difference between the two graphs?

This exploration suggests that the equation $\cos t = \sin(t + \pi/2)$ is an identity and hence that the graph of the cosine function can be obtained by horizontally shifting the graph of the sine function. This is indeed the case, as will be proved in Section 8.2. Consequently, every graph in Section 6.5 is actually the graph of a function of the form $f(t) = A\sin(bt + c)$. In fact, considerably more is true.

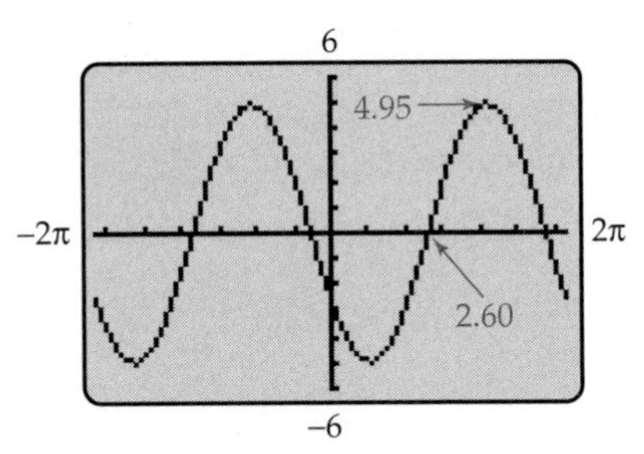

Figure 6–52

Example 1 The function $g(t) = -2\sin(t + 7) + 3\cos(t + 2)$ has period 2π because this is the period of both $\sin(t + 7)$ and $\cos(t + 2)$. Its graph in Figure 6–52 consists of repeating waves of uniform height, as do graphs of functions of the form $f(t) = A\sin(bt + c)$.

By using a maximum finder and root finder, we see that the maximum height of a wave is approximately 4.95 and that a wave similar to a sine wave begins at approximately $t = 2.60$, as indicated in Figure 6–52. Thus, the graph looks very much like a sine wave with amplitude 4.95 and phase shift 2.60. As we saw in Section 6.5, the function

$$f(t) = 4.95\sin(t - 2.60)$$

has amplitude of 4.95, period $2\pi/1 = 2\pi$, and phase shift $-(-2.60)/1 = 2.60$.

GRAPHING EXPLORATION

Graph

$$g(t) = -2\sin(t + 7) + 3\cos(t + 2)$$

and

$$f(t) = 4.95\sin(t - 2.60)$$

on the same screen. Do the graphs look identical? ■

Example 1 is an illustration (but not a proof) of the following fact.

Sinusoidal Graphs

If b, D, E, r, s are constants, then the graph of the function

$$g(t) = D\sin(bt + r) + E\cos(bt + s)$$

is a sine curve: there exist constants A and c such that

$$D\sin(bt + r) + E\cos(bt + s) = A\sin(bt + c).$$

Example 2 Estimate the constants A, b, c such that

$$A\sin(bt + c) = 4\sin(3t + 2) + 2\cos(3t - 4).$$

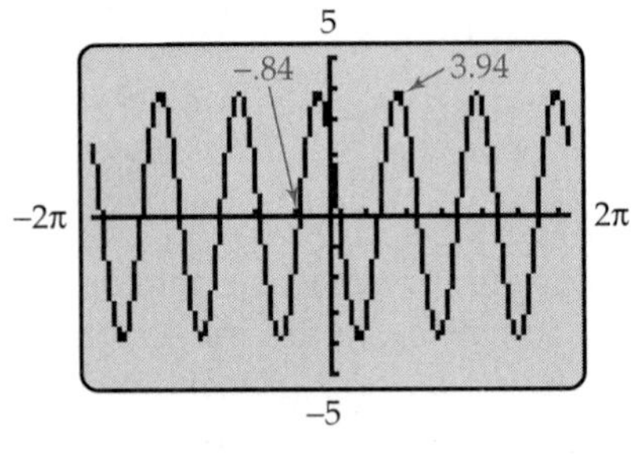

Figure 6–53

Solution The function $g(t) = 4\sin(3t + 2) + 2\cos(3t - 4)$ has period $2\pi/3$ because this is the period of both $\sin(3t + 2)$ and $\cos(3t - 4)$. The function $f(t) = A\sin(bt + c)$ has period $2\pi/b$. So we must have

$$\frac{2\pi}{b} = \frac{2\pi}{3}, \qquad \text{or equivalently,} \qquad b = 3.$$

Using a maximum finder and a root finder on the graph of $g(t) = 4\sin(3t + 2) + 2\cos(3t - 4)$ in Figure 6–53, we see that the maximum height (amplitude) of a wave is approximately 3.94 and that a sine wave begins at approximately $t = -.84$. Therefore, the graph has (approximate) amplitude 3.94 and phase shift $-.84$. Since $b = 3$ and $f(t) = A\sin(bt + c)$ has amplitude $|A|$ and phase shift $-c/b = -c/3$, we have $A \approx 3.94$ and

$$-\frac{c}{3} \approx -.84, \qquad \text{or equivalently,} \qquad c \approx 3(.84) = 2.52.$$

Therefore,

$$3.94\sin(3t + 2.52) \approx 4\sin(3t + 2) + 2\cos(3t - 4).$$

GRAPHING EXPLORATION

Graphically confirm this fact by graphing

$$f(t) = 3.94\sin(3t + 2.52)$$

and

$$g(t) = 4\sin(3t + 2) + 2\cos(3t - 4)$$

on the same screen. Do the graphs appear identical? ■

In the preceding examples, the variable t had the same coefficient in both the sine and cosine term of the function's rule. When this is not the case, the graph will consist of waves of varying size and shape, as you can readily illustrate.

GRAPHING EXPLORATION

Graph each of the following functions separately in the viewing window with $-2\pi \le t \le 2\pi$ and $-6 \le y \le 6$:

$$f(t) = \sin 3t + \cos 2t \qquad g(t) = -2\sin(3t + 5) + 4\cos(t + 2),$$
$$h(t) = 2\sin 2t - 3\cos 3t.$$

Example 3 Find a complete graph of

$$f(t) = 4\sin 100\pi t + 2\cos 40\pi t.$$

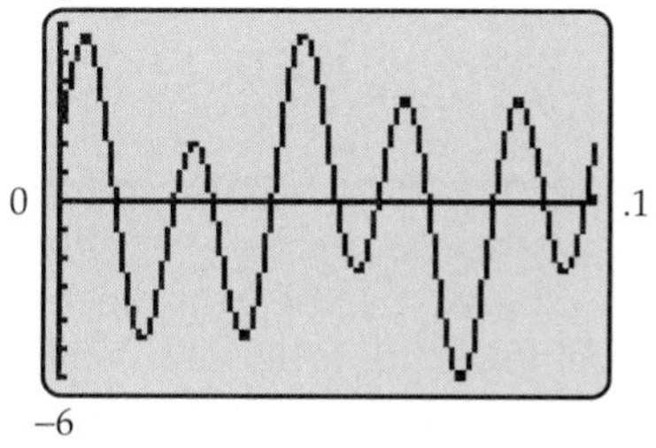

Solution If you graph f in a window with $-2\pi \le t \le 2\pi$, you will get "garbage" on the screen (try it!). Trial and error may lead to a viewing window that shows a readable graph, but the graph may not be accurate. A better procedure is to note that this is a periodic function. Hence we need only graph it over one period to have a complete graph. The period of $4\sin 100\pi t$ is $2\pi/100\pi = 1/50 = .02$ and the period of $2\cos 40\pi t$ is $2\pi/40\pi = 1/20 = .05$. So the period of their sum will be the least common multiple of .02 and .05, which is .1.* By graphing f in the viewing window with $0 \le t \le .1$ and $-6 \le y \le 6$, we obtain the complete graph in Figure 6–54. ■

6
0
.1
–6

Figure 6–54

*The multiples of .02 are .02, .04, .06, .08, .10, . . . and the multiples of .05 are .05, .10, ... Hence, the smallest common multiple is .10.

Damped and Compressed Trigonometric Graphs

Suppose a weight hanging from a spring is set in motion by an upward push. No spring is perfectly elastic and friction acts to slow the motion of the weight as time goes on.* Consequently, the graph showing the height of the spring (above or below its equilibrium point) at time t will consist of waves that get smaller and smaller as t gets larger. Many other physical situations can be described by functions whose graphs consist of waves of diminishing or increasing heights. Other situations (for instance, sound waves in FM radio transmission) are modeled by functions whose graphs consist of waves of uniform height and varying frequency. Here are some examples of such functions.

Example 4 Graph $f(g) = t \cos t$ in the viewing window with $-35 \le t \le 35$ and $-35 \le y \le 35$. You will see that the graph consists of waves that are quite small near the origin and get larger as you move away from the origin to the left or right. Some algebraic analysis may help you to explain just what's going on. We know that

$$-1 \le \cos t \le 1 \quad \text{for every } t.$$

If we multiply each term of this inequality by t and remember the rules for changing the direction of inequalities when multiplying by negatives, we see that

$$-t \le t \cos t \le t \quad \text{when } t \ge 0$$

and

$$-t \ge t \cos t \ge t \quad \text{when } t < 0.$$

In graphical terms this means that the graph of $f(x) = t \cos t$ lies between the straight lines $y = t$ and $y = -t$, with the waves growing larger or smaller to fill this space. The graph touches the lines $y = \pm t$ exactly when $t \cos t = \pm t$, that is, when $\cos t = \pm 1$. This occurs when $t = 0, \pm\pi, \pm 2\pi, \pm 3\pi, \cdots$.

GRAPHING EXPLORATION

Illustrate this analysis by graphing $f(t) = t \cos t, y = t,$ and $y = -t$ on the same screen. ■

Example 5 No single viewing window gives a completely readable graph of $g(t) = .5^t \sin t$ (try some). To the left of the y-axis, the graph gets quite large, but to the right, it almost coincides with the horizontal axis. To get a better mental picture, note that $.5^t > 0$ for every t. Multiplying each term of the known inequality $-1 \le \sin t \le 1$ by $.5^t$ we see that

$$-.5^t \le .5^t \sin t \le .5^t \quad \text{for every } t.$$

*These factors were ignored in Example 9 of Section 6.5.

Hence, the graph of g lies between the graphs of the exponential functions $y = -.5^t$ and $y = .5^t$, which are shown in Figure 6–55. The graph of g will consist of sine waves rising and falling between those exponential graphs, as indicated in the sketch in Figure 6–56 (which is not to scale).

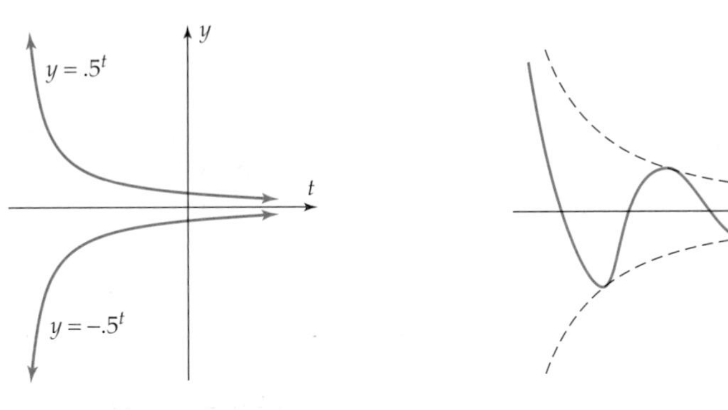

Figure 6–55 **Figure 6–56**

The best you can do with a calculator is to look at various viewing windows in which a portion of the graph is readable.

> **GRAPHING EXPLORATION**
>
> Find viewing windows that clearly show the graph of $g(t) = .5^t \sin t$ in each of these ranges:
>
> $$-2\pi \le t \le 0, \qquad 0 \le t \le 2\pi, \qquad 2\pi \le t \le 4\pi.$$ ■

Example 6 If you graph $f(t) = \sin(\pi/t)$ in a wide viewing window such as Figure 6–57, it is clear that the horizontal axis is an asymptote of the graph.* Near the origin, however, the graph is not very readable, even in a very narrow viewing window like Figure 6–58.

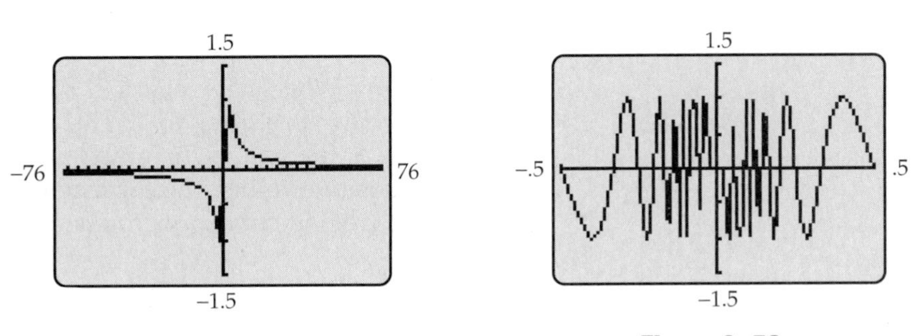

Figure 6–57 **Figure 6–58**

*This can also be demonstrated algebraically: When t is very large in absolute value, then π/t is very close to 0 by the Big-Little Principle and hence $\sin(\pi/t)$ is very close to 0 as well.

To understand the behavior of f near the origin, consider what happens as you move left from $t = 1/2$ to $t = 0$:

$$\text{As } t \text{ goes from } \frac{1}{2} \text{ to } \frac{1}{4}, \text{ then } \frac{\pi}{t} \text{ goes from } \frac{\pi}{1/2} = 2\pi \text{ to } \frac{\pi}{1/4} = 4\pi.$$

As π/t takes all values from 2π to 4π, the graph of $f(t) = \sin(\pi/t)$ makes one complete sine wave. Similarly,

$$\text{As } t \text{ goes from } \frac{1}{4} \text{ to } \frac{1}{6}, \text{ then } \frac{\pi}{t} \text{ goes from } \frac{\pi}{1/4} = 4\pi \text{ to } \frac{\pi}{1/6} = 6\pi.$$

As π/t takes all values from 4π to 6π, the graph of $f(t) = \sin(\pi/t)$ makes another complete sine wave. The same pattern continues, so that the graph of f makes a complete wave from $t = 1/2$ to $t = 1/4$, another from $t = 1/4$ to $t = 1/6$, another from $t = 1/6$ to $t = 1/8$, and so on. A similar phenomenon occurs as t takes values between $-1/2$ and 0. Consequently, the graph of f near 0 oscillates infinitely often between -1 and 1, with the waves becoming more and more compressed as t gets closer to 0, as indicated in Figure 6–59. Since the function is not defined at $t = 0$, the left and right halves of the graph are not connected. ■

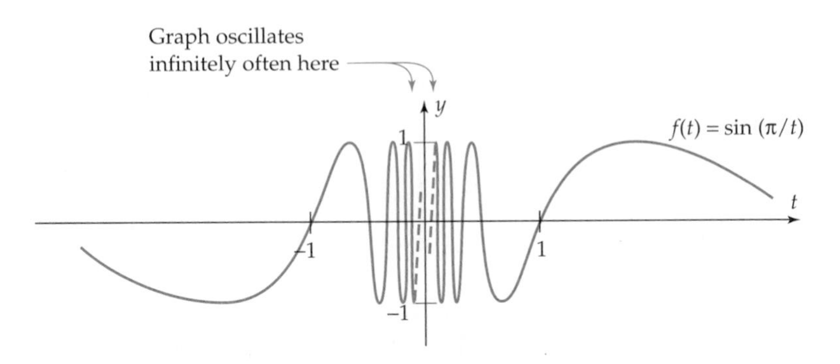

Figure 6–59

Example 7 Describe the graph of $g(t) = \cos e^t$.

Solution When t is negative, then e^t is very close to 0 (why?) and hence $\cos e^t$ is very close to 1. Therefore the horizontal line $y = 1$ is an asymptote of the half of the graph to the left of the origin. As t takes increasing positive values, the corresponding values of e^t increase at a much faster rate (remember exponential growth). For instance, as t goes from 0 to 2π,

$$e^t \text{ goes from } e^0 = 1 \text{ to } e^{2\pi} \approx 535.5 \approx 107\pi = 85(2\pi).$$

Consequently, $\cos e^t$ runs through 85 periods, that is, the graph of g makes 85 full waves between 0 and 2π. As t gets larger, the graph of g makes waves at a faster and faster rate.

GRAPHING EXPLORATION

To see how compressed the waves become, graph $g(t)$ in three viewing windows, with

$$0 \le t \le 3.5, \qquad 4.5 \le t \le 5, \qquad 6 \le t \le 6.2$$

and note how the number of waves increases in each succeeding window, even though the widths of the windows are getting smaller. ■

Exercises 6.5.A

In Exercises 1–6, estimate constants A, b, c such that

$$f(t) = A\sin(bt + c).$$

1. $f(t) = \sin t + 2\cos t$
2. $f(t) = 3\sin t + 2\cos t$
3. $f(t) = 2\sin 4t - 5\cos 4t$
4. $f(t) = 3\sin(2t - 1) + 4\cos(2t + 3)$
5. $f(t) = -5\sin(3t + 2) + 2\cos(3t - 1)$
6. $f(t) = .3\sin(2t + 4) - .4\cos(2t - 3)$

In Exercises 7–16, find a viewing window that shows a complete graph of the function.

7. $g(t) = (5\sin 2t)(\cos 5t)$
8. $h(t) = e^{\sin t}$
9. $f(t) = t/2 + \cos 2t$
10. $g(t) = \sin\left(\frac{t}{3} - 2\right) + 2\cos\left(\frac{t}{4} - 2\right)$
11. $h(t) = \sin 300t + \cos 500t$
12. $f(t) = 3\sin(200t + 1) - 2\cos(300t + 2)$
13. $g(t) = -5\sin(250\pi t + 5) + 2\cos(400\pi t - 7)$
14. $h(t) = 4\sin(600\pi t + 3) - 6\cos(500\pi t - 3)$
15. $f(t) = 4\sin .2\pi t - 5\cos .4\pi t$
16. $g(t) = 6\sin .05\pi t + 2\cos .04\pi t$

In Exercises 17–24, describe the graph of the function verbally (including such features as asymptotes, undefined points, amplitude and number of waves between 0 and 2π, etc.) as in Examples 5–7. Find viewing windows that illustrate the main features of the graph.

17. $g(t) = \sin e^t$
18. $h(t) = \dfrac{\cos 2t}{1 + t^2}$
19. $f(t) = \sqrt{|t|}\cos t$
20. $g(t) = e^{-t^2/8}\sin 2\pi t$
21. $h(t) = \dfrac{1}{t}\sin t$
22. $f(t) = t\sin\dfrac{1}{t}$
23. $g(t) = \ln|\cos t|$
24. $h(t) = \ln|\sin t + 1|$

6.6 Other Trigonometric Functions

This section is divided into three parts, each of which may be covered earlier, as shown in the table.

Subsection of Section 6.6	May be covered at the end of
Definitions, Alternate Descriptions	Section 6.2
Algebra and Identities	Section 6.3
Graphs	Section 6.4

Definitions

The three remaining trigonometric functions are defined in terms of sine and cosine, as follows:

Definition of Cotangent, Secant, and Cosecant Functions

Name of Function	Value of Function at t Is Denoted	Rule of Function
cotangent	$\cot t$	$\cot t = \dfrac{\cos t}{\sin t}$
secant	$\sec t$	$\sec t = \dfrac{1}{\cos t}$
cosecant	$\csc t$	$\csc t = \dfrac{1}{\sin t}$

The domain of each function consists of all real numbers for which the denominator is not 0. The graphs of the sine and cosine function in Section 6.4 show that $\sin t = 0$ only when t is an integer multiple of π and that $\cos t = 0$ only when t is an odd integer multiple of $\pi/2$. Therefore,

The domain of the cotangent and cosecant functions consists of all real numbers *except* $0, \ \pm\pi, \ \pm 2\pi, \ \pm 3\pi$, and so on.

The domain of the secant function consists of all real numbers *except* $\pm\pi/2, \ \pm 3\pi/2, \ \pm 5\pi/2$, and so on.

The values of the secant and cosecant functions may be approximated on a calculator by using the SIN and COS keys. For instance,

$$\sec 7 = \frac{1}{\cos 7} \approx 1.3264 \quad \text{and} \quad \csc 18.5 = \frac{1}{\sin 18.5} \approx -2.9199.$$

To evaluate the cotangent function, either use the SIN or COS keys, or use the TAN key and this identity (which is proved below): $\cot t = 1/\tan t$. For example,

$$\cot(-5) = \frac{\cos(-5)}{\sin(-5)} \approx .2958 \quad \text{or} \quad \cot(-5) = \frac{1}{\tan(-5)} \approx .2958.$$

CAUTION

The calculator keys labeled SIN^{-1}, COS^{-1}, and TAN^{-1} do *not* denote the functions $1/\sin t$, $1/\cos t$, and $1/\tan t$. For instance, if you key in

COS^{-1} 7 ENTER

you will get an error message, not the number sec 7 and if you key in

TAN^{-1} -5 ENTER

you will obtain -1.3734, which is *not* $\cot(-5)$.

These new trigonometric functions may be evaluated exactly at any integer multiple of $\pi/3$, $\pi/4$, or $\pi/6$.

Example 1 Evaluate the cotangent, secant, and cosecant functions at $t = \pi/3$.

Solution Let P be the point where the terminal side of an angle of $\pi/3$ radians in standard position meets the unit circle (Figure 6–60). Draw the vertical line from P to the x-axis, forming a right triangle with hypotenuse 1, angles of $\pi/3$ and $\pi/6$ radians, and sides of lengths of $1/2$ and $\sqrt{3}/2$, as explained on page 404. Then P has coordinates $(1/2, \sqrt{3}/2)$ and by definition

$$\sin\frac{\pi}{3} = y\text{-coordinate of } P = \sqrt{3}/2$$

$$\cos\frac{\pi}{3} = x\text{-coordinate of } P = 1/2.$$

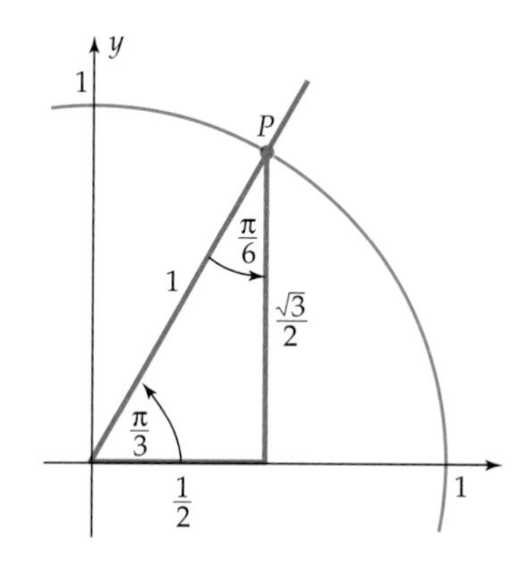

Figure 6–60

Therefore,

$$\csc\frac{\pi}{3} = \frac{1}{\sin(\pi/3)} = \frac{1}{\sqrt{3}/2} = \frac{2}{\sqrt{3}} = \frac{2\sqrt{3}}{3}$$

$$\sec\frac{\pi}{3} = \frac{1}{\cos(\pi/3)} = \frac{1}{1/2} = 2$$

$$\cot\frac{\pi}{3} = \frac{\cos(\pi/3)}{\sin(\pi/3)} = \frac{1/2}{\sqrt{3}/2} = \frac{1}{\sqrt{3}} = \frac{\sqrt{3}}{3}. \quad ■$$

Alternate Descriptions

The point-in-the-plane description of sine, cosine, and tangent readily extends to these new functions.

Point-in-the-Plane Description

Let t be a real number and (x, y) any point (except the origin) on the terminal side of an angle of t radians in standard position. Let

$$r = \sqrt{x^2 + y^2}.$$

Then,

$$\cot t = \frac{x}{y} \qquad \sec t = \frac{r}{x} \qquad \csc t = \frac{r}{y}$$

for each number t in the domain of the given function.

These statements are proved by using the similar descriptions of sine and cosine. For instance,

$$\cot t = \frac{\cos t}{\sin t} = \frac{x/r}{y/r} = \frac{x}{y}.$$

The proofs of the other statements are similar.

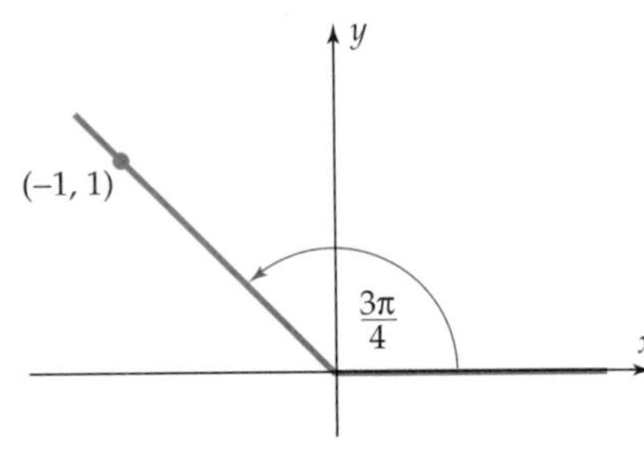

Figure 6–61

Example 2 Evaluate all six trigonometric functions at $t = 3\pi/4$.

Solution The terminal side of an angle of $3\pi/4$ radians in standard position lies on the line $y = -x$, as shown in Figure 6–61. We shall use the point $(-1, 1)$ on this line to compute the function values. In this case $r = \sqrt{x^2 + y^2} = \sqrt{(-1)^2 + 1^2} = \sqrt{2}$. Therefore,

$$\sin\frac{3\pi}{4} = \frac{y}{r} = \frac{1}{\sqrt{2}} = \frac{\sqrt{2}}{2} \qquad \cos\frac{3\pi}{4} = \frac{x}{r} = \frac{-1}{\sqrt{2}} = \frac{-\sqrt{2}}{2}$$

$$\tan\frac{3\pi}{4} = \frac{y}{x} = \frac{1}{-1} = -1 \qquad \csc\frac{3\pi}{4} = \frac{r}{y} = \frac{\sqrt{2}}{1} = \sqrt{2}$$

$$\sec\frac{3\pi}{4} = \frac{r}{x} = \frac{\sqrt{2}}{-1} = -\sqrt{2} \qquad \cot\frac{3\pi}{4} = \frac{x}{y} = \frac{-1}{1} = -1.$$ ■

Algebra and Identities

We begin by noting the relationship between the cotangent and tangent functions.

Reciprocal Identities

The cotangent and tangent functions are reciprocals; that is,

$$\cot t = \frac{1}{\tan t} \qquad \tan t = \frac{1}{\cot t}$$

for every number t in the domain of both functions.

The proof of these facts comes directly from the definitions; for instance,

$$\cot t = \frac{\cos t}{\sin t} = \frac{1}{\frac{\sin t}{\cos t}} = \frac{1}{\tan t}.$$

Period of Secant, Cosecant, Cotangent

The secant and cosecant functions are periodic with period 2π and the cotangent function is periodic with period π. In symbols,

$$\sec(t + 2\pi) = \sec t \qquad \csc(t + 2\pi) = \csc t$$

$$\cot(t + \pi) = \cot t$$

for every number t in the domain of the given function.

The proof of these statements uses the fact that each of these functions is the reciprocal of a function whose period is known. For instance,

$$\csc(t + 2\pi) = \frac{1}{\sin(t + 2\pi)} = \frac{1}{\sin t} = \csc t$$

$$\cot(t + \pi) = \frac{1}{\tan(t + \pi)} = \frac{1}{\tan t} = \cot t.$$

The other details are left as an exercise.

Pythagorean Identities

For every number t in the domain of both functions,

$$\mathbf{1 + \tan^2 t = \sec^2 t}$$

and

$$\mathbf{1 + \cot^2 t = \csc^2 t.}$$

The proof of these identities uses the definitions of the functions and the Pythagorean identity $\sin^2 t + \cos^2 t = 1$:

$$1 + \tan^2 t = 1 + \frac{\sin^2 t}{\cos^2 t} = \frac{\cos^2 t + \sin^2 t}{\cos^2 t} = \frac{1}{\cos^2 t} = \left(\frac{1}{\cos t}\right)^2 = \sec^2 t.$$

The second identity is proved similarly.

Example 3 Simplify the expression $\dfrac{30\cos^3 t \sin t}{6\sin^2 t \cos t}$, assuming $\sin t \neq 0$, $\cos t \neq 0$.

Solution

$$\frac{30\cos^3 t \sin t}{6\sin^2 t \cos t} = \frac{5\cos^3 t \sin t}{\cos t \sin^2 t} = \frac{5\cos^2 t}{\sin t} = 5\frac{\cos t}{\sin t}\cos t = 5\cot t \cos t. \quad ■$$

Example 4 Assume $\cos t \neq 0$ and simplify $\cos^2 t + \cos^2 t \tan^2 t$.

Solution

$$\cos^2 t + \cos^2 t \tan^2 t = \cos^2 t(1 + \tan^2 t) = \cos^2 t \sec^2 t = \cos^2 t \cdot \frac{1}{\cos^2 t} = 1. \quad ■$$

Example 5 If $\tan t = 3/4$ and $\sin t < 0$, find $\cot t$, $\cos t$, $\sin t$, $\sec t$, $\csc t$.

Solution First we have $\cot t = 1/\tan t = 1/(3/4) = 4/3$. Next we use the Pythagorean identity to obtain:

$$\sec^2 t = 1 + \tan^2 t = 1 + \left(\frac{3}{4}\right)^2 = 1 + \frac{9}{16} = \frac{25}{16}$$

$$\sec t = \pm\sqrt{\frac{25}{16}} = \pm\frac{5}{4}$$

$$\frac{1}{\cos t} = \pm\frac{5}{4} \qquad \text{or equivalently} \qquad \cos t = \pm\frac{4}{5}.$$

Since $\sin t$ is given as negative and $\tan t = \sin t/\cos t$ is positive, $\cos t$ must be negative. Hence, $\cos t = -4/5$. Consequently,

$$\frac{3}{4} = \tan t = \frac{\sin t}{\cos t} = \frac{\sin t}{(-4/5)}$$

so that

$$\sin t = \left(-\frac{4}{5}\right)\left(\frac{3}{4}\right) = -\frac{3}{5}.$$

Therefore,

$$\sec t = \frac{1}{\cos t} = \frac{1}{(-4/5)} = -\frac{5}{4} \quad \text{and} \quad \csc t = \frac{1}{\sin t} = \frac{1}{(-3/5)} = -\frac{5}{3}. \quad ■$$

Graphs

The general shape of the graph of $g(t) = \sec t$ can be determined by using the graph of the cosine function and the fact that $\sec t = 1/\cos t$. First of all, $\sec t$ is not defined whenever $\cos t = 0$, that is, when $t = \pm\pi/2$, $\pm 3\pi/2$, $\pm 5\pi/2$, etc. When $\cos t$ is a number near 1 or -1, then so is $\sec t = 1/\cos t$ and their graphs are close together. When $\cos t$ is near 0 (so that its graph is close to the x-axis), then $\sec t = 1/\cos t$ is very large in absolute value* (so that its graph is far from the x-axis), as shown in Figure 6–62. The graph of $g(t) = \sec t$ has vertical asymptotes at those

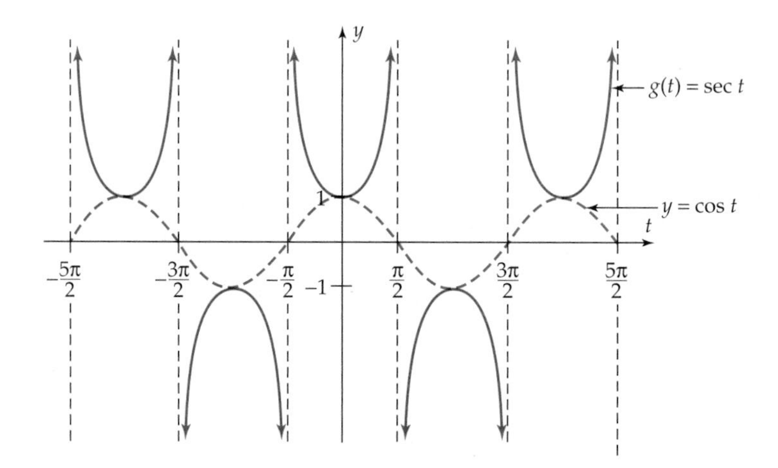

Figure 6–62

*See the Big-Little Principle on page 249.

points where the function is not defined, that is $x = \pi/2$, $x = -\pi/2$, $x = 3\pi/2$, $x = -3\pi/2$, etc.

The graphs of $h(t) = \csc t = 1/\sin t$ and $f(t) = \cot t = 1/\tan t$ can be obtained in a similar fashion (Figure 6–63).

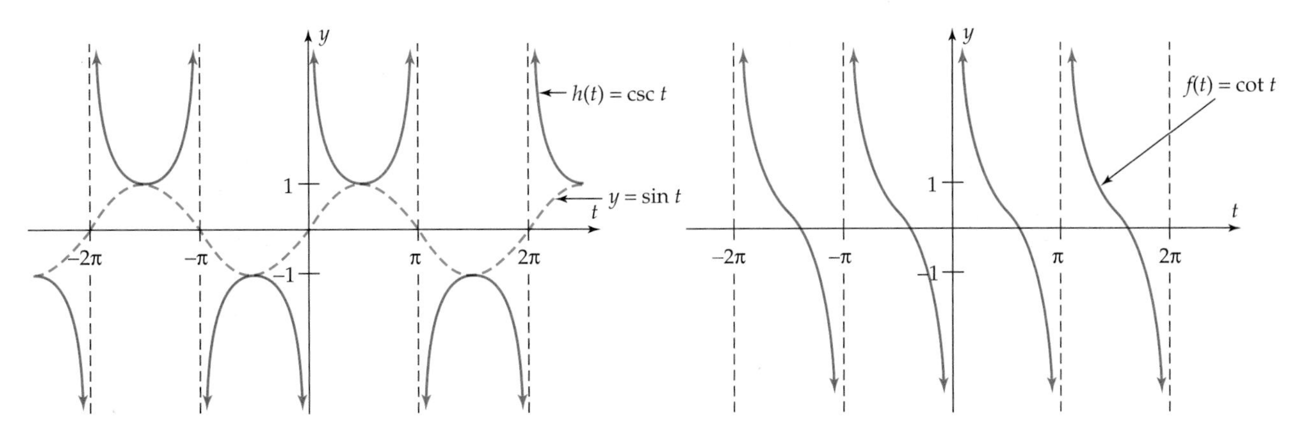

Figure 6–63

Exercises 6.6

Note: The arrangement of the exercises corresponds to the subsections of this section.

Definitions and Alternate Descriptions

In Exercises 1–6, determine the quadrant containing the terminal side of an angle of t radians in standard position under the given conditions.

1. $\cos t > 0$ and $\sin t < 0$
2. $\sin t < 0$ and $\tan t > 0$
3. $\sec t < 0$ and $\cot t < 0$
4. $\csc t < 0$ and $\sec t > 0$
5. $\sec t > 0$ and $\cot t < 0$
6. $\sin t > 0$ and $\sec t < 0$

In Exercises 7–16, evaluate all six trigonometric functions at t, where the given point lies on the terminal side of an angle of t radians in standard position.

7. $(3, 4)$
8. $(0, 6)$
9. $(-5, 12)$
10. $(-2, -3)$
11. $(-1/5, 1)$
12. $(4/5, -3/5)$
13. $(\sqrt{2}, \sqrt{3})$
14. $(-2\sqrt{3}, \sqrt{3})$
15. $(1 + \sqrt{2}, 3)$
16. $(1 + \sqrt{3}, 1 - \sqrt{3})$

In Exercises 17–19, evaluate all six trigonometric functions at the given number without using a calculator.

17. $\frac{4\pi}{3}$
18. $-\frac{7\pi}{6}$
19. $\frac{7\pi}{4}$

20. Fill in the missing entries in the following table. Give exact answers, not decimal approximations.

t	0	$\frac{\pi}{6}$	$\frac{\pi}{4}$	$\frac{\pi}{3}$	$\frac{\pi}{2}$	$\frac{2\pi}{3}$	$\frac{3\pi}{4}$	$\frac{5\pi}{6}$	π	$\frac{3\pi}{2}$
$\sin t$										
$\cos t$										
$\tan t$					—					—
$\cot t$	—								—	
$\sec t$					—					—
$\csc t$	—								—	

Algebra and Identities

21. Find the average rate of change of $f(t) = \cot t$ from $t = 1$ to $t = 3$.

22. Find the average rate of change of $g(t) = \csc t$ from $t = 2$ to $t = 3$.

23. (a) Find the average rate of change of $f(t) = \tan t$ from $t = 2$ to $t = 2 + h$, for each of these values of h: .01, .001, .0001, and .00001.

(b) Compare your answers in part (a) with the number $(\sec 2)^2$. What would you guess that the instantaneous rate of change of $f(t) = \tan t$ is at $t = 2$?

In Exercises 24–30, perform the indicated operations, then simplify your answers by using appropriate definitions and identities.

24. $\tan t(\cos t - \csc t)$

25. $\cos t \sin t(\csc t + \sec t)$

26. $(1 + \cot t)^2$

27. $(1 - \sec t)^2$

28. $(\sin t - \csc t)^2$

29. $(\cot t - \tan t)(\cot^2 t + 1 + \tan^2 t)$

30. $(\sin t + \csc t)(\sin^2 t + \csc^2 t - 1)$

In Exercises 31–36, factor and simplify the given expression.

31. $\sec t \csc t - \csc^2 t$

32. $\tan^2 t - \cot^2 t$

33. $\tan^4 t - \sec^4 t$

34. $4\sec^2 t + 8\sec t + 4$

35. $\cos^3 t - \sec^3 t$

36. $\csc^4 t + 4\csc^2 t - 5$

In Exercises 37–42, simplify the given expression. Assume all denominators are nonzero.

37. $\dfrac{\cos^2 t \sin t}{\sin^2 t \cos t}$

38. $\dfrac{\sec^2 t + 2\sec t + 1}{\sec t}$

39. $\dfrac{4\tan t \sec t + 2\sec t}{6\sin t \sec t + 2\sec t}$

40. $\dfrac{\sec^2 t \csc t}{\csc^2 t \sec t}$

41. $(2 + \sqrt{\tan t})(2 - \sqrt{\tan t})$

42. $\dfrac{6\tan t \sin t - 3\sin t}{9\sin^2 t + 3\sin t}$

In Exercises 43–46, prove the given identity.

43. $1 + \cot^2 t = \csc^2 t$ [*Hint:* Look at the proof of the similar identity on page 451]

44. $\cot(-t) = -\cot t$ [*Hint:* Express the left side in terms of sine and cosine; then use the negative angle identities and express the result in terms of cotangent.]

45. $\sec(-t) = \sec t$ [Adapt the hint for Exercise 44.]

46. $\csc(-t) = -\csc t$

In Exercises 47–52, find the values of all six trigonometric functions at t if the given conditions are true.

47. $\cos t = -1/2$ and $\sin t > 0$
[*Hint:* $\sin^2 t + \cos^2 t = 1$.]

48. $\cos t = \dfrac{1}{2}$ and $\sin t < 0$

49. $\cos t = 0$ and $\sin t = 1$

50. $\sin t = -2/3$ and $\sec t > 0$

51. $\sec t = -13/5$ and $\tan t < 0$

52. $\csc t = 8$ and $\cos t < 0$

Graphs

In Exercises 53–56, use graphs to determine whether the equation could possibly be an identity or is definitely not an identity.

53. $\tan t = \cot\left(\frac{\pi}{2} - t\right)$

54. $\dfrac{\cos t}{\cos(t - \pi/2)} = \cot t$

55. $\dfrac{\sin t}{1 - \cos t} = \cot t$

56. $\dfrac{\sec t + \csc t}{1 + \tan t} = \csc t$

57. Show graphically that the equation $\sec t = t$ has infinitely many solutions, but none between $-\pi/2$ and $\pi/2$.

Thinkers

58. In the diagram of the unit circle in the figure, find six line segments whose respective lengths are $\sin t$, $\cos t$, $\tan t$, $\cot t$, $\sec t$, $\csc t$. [*Hint:* $\sin t$ = length CA. Why? Note that OC has length 1 and various right triangles in the figure are similar.]

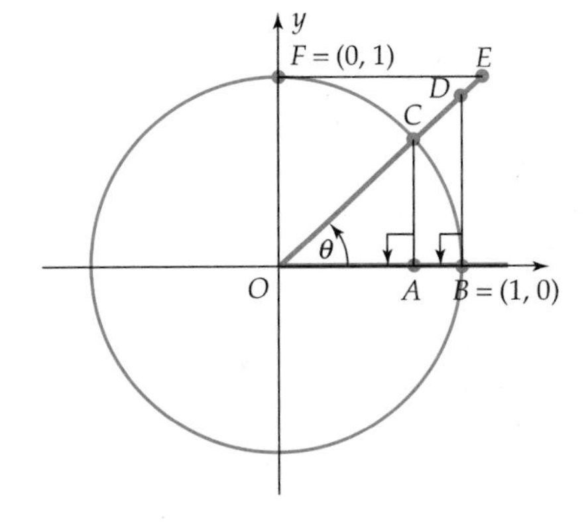

59. In the figure for Exercise 58, find the following areas in terms of θ.

(a) triangle OCA

(b) triangle ODB

(c) circular segment OCB

Chapter 6 Review

Important Concepts

Section 6.1	Angle 390
	Vertex 390
	Initial side 390
	Terminal side 390
	Coterminal angles 391, 395
	Standard position 391
	Degree measure 391
	Radian measure 392
	Arc length 397
	Angular speed 397
Section 6.2	Sine function 401
	Cosine function 401
	Tangent function 402
	Special values 403–405
	Point-in-the-plane description 405
Section 6.3	Algebra with trigonometric functions 409–410
	Pythagorean identity 411
	Periodicity 414
	Negative angle identities 415–416
Section 6.4	Graphs of sine, cosine, and tangent functions 418, 421, 423
	Graphs and identities 425
Section 6.5	Period 429
	Amplitude 431
	Phase shift 433
	Simple harmonic motion 436
Excursion 6.5.A	Sinusoidal graphs 442
	Damped and compressed graphs 444
Section 6.6	Cotangent function 448
	Secant function 448
	Cosecant function 448
	Alternative descriptions 449
	Identities 450
	Graphs of cotangent, secant, and cosecant functions 452–453

Important Facts and Formulas

- *Conversion Rules:* To convert radians to degrees, multiply by $180/\pi$. To convert degrees to radians, multiply by $\pi/180$.
- *Definition of Trigonometric Functions:* If P is the point where the terminal side of an angle of t radians in standard position meets the unit circle, then

$$\sin t = y\text{-coordinate of } P$$

$$\cos t = x\text{-coordinate of } P$$

$$\tan t = \frac{\sin t}{\cos t} \qquad \cot t = \frac{\cos t}{\sin t} \qquad \sec t = \frac{1}{\cos t} \qquad \csc t = \frac{1}{\sin t}$$

- *Point-in-the-Plane Description:* If (x, y) is any point other than the origin on the terminal side of an angle of t radians in standard position and

$r = \sqrt{x^2 + y^2}$, then

$$\sin t = \frac{y}{r} \qquad \cos t = \frac{x}{r}$$

$$\tan t = \frac{y}{x} \qquad \cot t = \frac{x}{y}$$

$$\sec t = \frac{r}{x} \qquad \csc t = \frac{r}{y}$$

- *Basic Identities:*

$$\sin^2 t + \cos^2 t = 1 \qquad \sin(-t) = -\sin t$$

$$1 + \tan^2 t = \sec^2 t \qquad \cos(-t) = \cos t$$

$$1 + \cot^2 t = \csc^2 t \qquad \tan(-t) = -\tan t$$

$$\tan t = \frac{1}{\cot t} \qquad \cot t = \frac{1}{\tan t}$$

$$\sin(t \pm 2\pi) = \sin t \qquad \csc(t \pm 2\pi) = \csc t$$

$$\cos(t \pm 2\pi) = \cos t \qquad \sec(t \pm 2\pi) = \sec t$$

$$\tan(t \pm \pi) = \tan t \qquad \cot(t \pm \pi) = \cot t$$

- If $A \neq 0$ and $b > 0$, then each of the $f(t) = A\sin(bt + c)$ and $g(t) = A\cos(bt + c)$ has

$$\text{amplitude } |A|, \qquad \text{period } 2\pi/b, \qquad \text{phase shift } -c/b.$$

Review Questions

1. Find a number t between 0 and 2π such that an angle of t radians in standard position is coterminal with an angle of $-23\pi/3$ radians in standard position.
2. Through how many radians does the second hand of a clock move in 2 minutes and 40 seconds?
3. $\dfrac{9\pi}{5}$ radians = ____ degrees.
4. 36 degrees = ____ radians.
5. $220°$ = ____ radians.
6. $\dfrac{17\pi}{12}$ radians = ____ degrees.
7. $-\dfrac{11\pi}{4}$ radians = ____ degrees.
8. $-135°$ = ____ radians.
9. If an angle of v radians has its terminal side in the second quadrant and $\sin v = \sqrt{8/9}$, then find $\cos v$.
10. $\cos\dfrac{47\pi}{2} = ?$
11. $\sin(-13\pi) = ?$
12. Simplify: $\dfrac{\tan(t + \pi)}{\sin(t + 2\pi)}$

Use the figure in Questions 13–17.

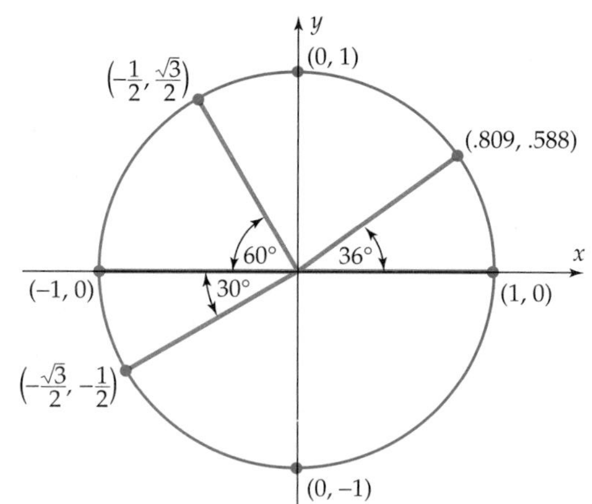

13. $\cos \frac{\pi}{5} = ?$

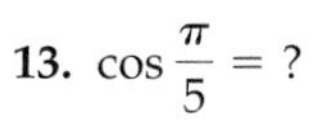

14. $\sin\left(\frac{7\pi}{6}\right) = ?$

15. $\cos\left(\frac{-5\pi}{6}\right) = ?$

16. $\sin\left(\frac{16\pi}{6}\right) = ?$

17. $\sin\left(\frac{-4\pi}{3}\right) = ?$

18. $\left[3 \sin\left(\frac{\pi}{5^{500}}\right)\right]^2 + \left[3 \cos\left(\frac{\pi}{5^{500}}\right)\right]^2 = ?$

19. Fill in the blanks (approximations not allowed):

t	0	$\frac{\pi}{6}$	$\frac{\pi}{4}$	$\frac{\pi}{3}$	$\frac{\pi}{2}$
sin t					
cos t					

20. Express as a single real number:

$$\cos \frac{3\pi}{4} \sin \frac{5\pi}{6} - \sin \frac{3\pi}{4} \cos \frac{5\pi}{6}$$

21. $\left(\sin \frac{\pi}{6} + 1\right)^2 = ?$

22. $\sin(\pi/2) + \sin 0 + \cos 0 = ?$

23. If $f(x) = \log_{10} x$ and $g(t) = -\cos t$, then $(f \circ g)(\pi) = ?$

24. Cos t is negative when the terminal side of an angle of t radians in standard position lies in which quadrants?

25. If $\sin t = 1/\sqrt{3}$ and the terminal side of an angle of t radians in standard position lies in the second quadrant, then $\cos t = ?$

26. Which of the following could possibly be a true statement about a real number t?

(a) $\sin t = -2$ and $\cos t = 1$

(b) $\sin t = 1/2$ and $\cos t = \sqrt{2}/2$

(c) $\sin t = -1$ and $\cos t = 1$

(d) $\sin t = \pi/2$ and $\cos t = 1 - (\pi/2)$

(e) $\sin t = 3/5$ and $\cos t = 4/5$

27. If $\sin t = -4/5$ and the terminal side of an angle of t radians in standard position lies in the third quadrant, then $\cos t =$ _____.

28. If $\sin(-101\pi/2) = -1$, then $\sin(-105\pi/2) = ?$

29. If $\pi/2 < t < \pi$ and $\sin t = 5/13$, then $\cos t = ?$

30. $\cos\left(-\frac{\pi}{6}\right) = ?$

31. $\cos\left(\frac{2\pi}{3}\right) = ?$

32. $\sin\left(-\frac{11\pi}{6}\right) = ?$ **33.** $\sin\left(\frac{\pi}{3}\right) = ?$ **34.** $\tan(5\pi/3) = ?$

35. Which of the following is *not* true about the graph of $f(t) = \sin t$?

(a) It has no sharp corners.
(b) It crosses the horizontal axis more than once.
(c) It rises higher and higher as t gets larger.
(d) It is periodic.
(e) It has no vertical asymptotes.

36. Which of the following functions has the graph in the figure between $-\pi$ and π?

(a) $f(x) = \begin{cases} \sin x, & \text{if } x \geq 0 \\ \cos x, & \text{if } x < 0 \end{cases}$

(b) $g(x) = \cos x - 1$

(c) $h(x) = \begin{cases} \sin x, & \text{if } x \geq 0 \\ \sin(-x), & \text{if } x < 0 \end{cases}$

(d) $k(x) = |\cos x|$

(e) $p(x) = \sqrt{1 - \sin^2 x}$

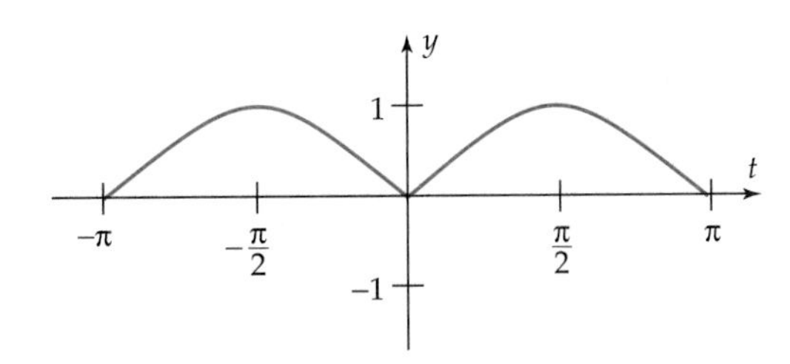

The point $(-3/\sqrt{50}, 7/\sqrt{50})$ *lies on the terminal side of an angle of t radians (in standard position). Find:*

37. $\sin t$ **38.** $\cos t$

39. $\tan t$ **40** $\sec t$

41. Find the equation of the straight line containing the terminal side of an angle of $5\pi/3$ radians (in standard position).

Suppose that an angle of w radians has its terminal side in the fourth quadrant and $\cos w = 2/\sqrt{13}$. *Find:*

42. $\sin w$ **43.** $\tan w$

44. $\csc w$ **45.** $\cos(-w)$

46. Fill in the blanks (approximations not allowed):

t	$\sin t$	$\tan t$	$\sec t$
$\pi/4$			
$2\pi/3$			
$5\pi/6$			

47. Sketch the graphs of $f(t) = \sin t$ and $h(t) = \csc t$ on the same set of coordinate axes ($-2\pi \le t \le 2\pi$).

48. Let θ be the angle shown in the figure. Which of the following statements is true?

(a) $\sin \theta = \frac{\sqrt{2}}{2}$

(b) $\cos \theta = \frac{\sqrt{2}}{2}$

(c) $\tan \theta = 1$

(d) $\cos \theta = \sqrt{2}$

(e) $\tan \theta = -1$

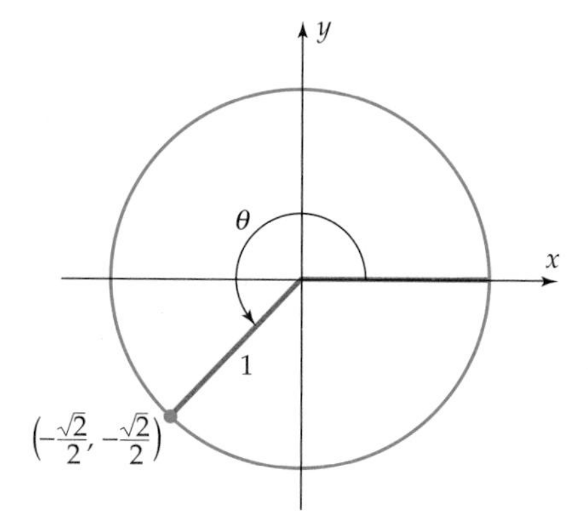

49. Let θ be as indicated in the figure. Which of the statements **(i)–(iii)** are *true*?

(i) $\cos \theta = -\frac{1}{3}$

(ii) $\tan \theta = \frac{2\sqrt{2}}{9}$

(iii) $\sin \theta = -\frac{2\sqrt{2}}{3}$

(a) only ii
(b) only ii and iii
(c) all of them
(d) only i and iii
(e) none of them

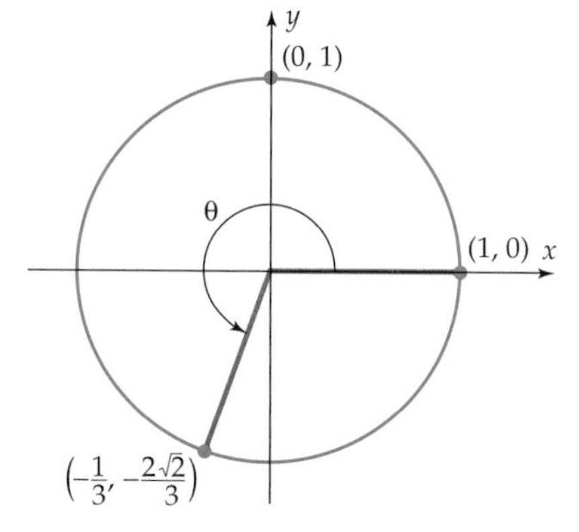

50. Between (and including) 0 and 2π, the function $h(t) = \tan t$ has:
(a) 3 roots and is undefined at 2 places.
(b) 2 roots and is undefined at 3 places.
(c) 2 roots and is undefined at 2 places.
(d) 3 roots and is defined everywhere.
(e) no roots and is undefined at 3 places.

51. If the terminal side of an angle of θ radians in standard position passes through the point $(-2, 3)$, and then $\tan \theta =$ _____.

52. Which of the statements **(i)–(iii)** are true?

(i) $\sin(-x) = -\sin x$
(ii) $\cos(-x) = -\cos x$
(iii) $\tan(-x) = -\tan x$

(a) (i) and (ii) only
(b) (ii) only
(c) (i) and (iii) only
(d) all of them
(e) none of them

53. If $\sec x = 1$ and $-\pi/2 < x < \pi/2$, then $x =$?

54. If $\tan t = 4/3$ and $0 < t < \pi$, what is $\cos t$?

55. Which of the following is true about sec t?
(a) $\sec(0) = 0$
(b) $\sec t = 1/\sin t$
(c) Its graph has no asymptotes.
(d) It is a periodic function.
(e) It is never negative.

56. If $\cot t = 0$ and $0 < t \leq \pi$, then $t =$ ____.

57. What is $\cot\left(\frac{2\pi}{3}\right)$?

58. Which of the following functions has the graph in the figure?
(a) $f(t) = \tan t$
(b) $g(t) = \tan\left(t + \frac{\pi}{2}\right)$
(c) $h(t) = 1 + \tan t$
(d) $k(t) = 3\tan t$
(e) $p(t) = -\tan t$

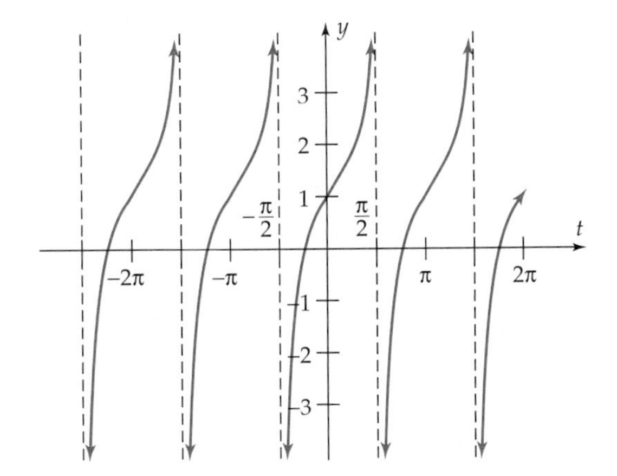

59. Let $f(t) = \frac{3}{2}\sin 5t$.
(a) What is the largest possible value of $f(x)$?
(b) Find the smallest positive number t such that $f(t) = 0$.

60. Sketch the graph of $g(t) = -2\cos t$.

61. Sketch the graph of $f(t) = -\frac{1}{2}\sin 2t\ (-2\pi \leq t \leq 2\pi)$.

62. Sketch the graph of $f(t) = \sin 4t\ (0 \leq t \leq 2\pi)$.

In Questions 63–66, determine graphically whether the given equation could possibly be an identity.

63. $\cos t = \sin\left(t - \frac{\pi}{2}\right)$

64. $\tan\frac{t}{2} = \frac{\sin t}{1 + \cos t}$

65. $\frac{\sin t - \sin 3t}{\cos t + \cos 3t} = -\tan t$

66. $\cos 2t = \frac{1}{1 - 2\sin^2 t}$

67. What is the period of the function $g(t) = \sin 4\pi t$?

68. What are the amplitude, period, and phase shift of the function

$$h(t) = 13\cos(14t + 15)?$$

69. State the rule of a periodic function whose graph from $t = 0$ to $t = 2\pi$ closely resembles the one in the figure.

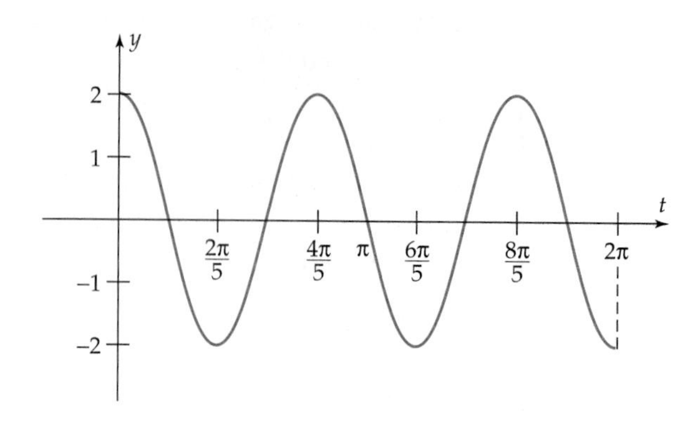

70. State the rule of a periodic function with amplitude 3, period π, and phase shift $\pi/3$.

71. State the rule of a periodic function with amplitude 8, period 5, and phase shift 14.

72. If $g(t) = 20\sin(200t)$, for how many values of t, with $0 \leq t \leq 2\pi$, is it true that $g(t) = 1$?

In Exercises 73 and 74, estimate constants A, b, c such that $f(t) = A\sin(bt + c)$.

73. $f(t) = 6\sin(4t + 7) - 5\cos(4t + 8)$

74. $f(t) = -5\sin(5t - 3) + 2\cos(5t + 2)$

In Exercises 75 and 76, find a viewing window that shows a complete graph of the function.

75. $f(t) = 3\sin(300t + 5) - 2\cos(500t + 8)$

76. $g(t) = -5\sin(400\pi t + 1) + 2\cos(150\pi t - 6)$

Discovery Project 6

Pistons and Flywheels

A common and well-proven piece of technology is the piston and flywheel combination. It is clearly visible in photographs of steam locomotives from the middle of the last century. The structure involves a wheel (or crankshaft) connected to a sliding plug in a cylinder by a rigid arm. The axis of rotation of the wheel is perpendicular to the central axis of the cylinder. The sliding plug, the piston, moves in one dimension, in and out of the cylinder. The motion of the piston is periodic, like the basic trigonometric functions, but doesn't have the same elegant symmetry.

It is quite easy to superimpose a coordinate plane on the flywheel-piston system so that the center of the flywheel is the origin, the flywheel rotates counterclockwise and the piston moves along the x-axis. As you can see in the diagram, the flywheel typically has a larger radius than the radial distance from the center to the point where the arm attaches. In this particular figure, the radius of the flywheel is 50 centimeters and the radial distance to the attachment point Q is 45 centimeters. The arm is 150 centimeters long measured from the base of the piston to the attachment point.

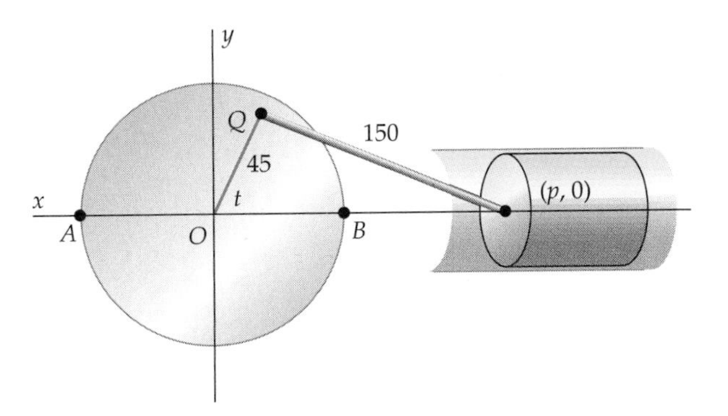

1. What are the coordinates of A and B in the figure? How close does the base of the piston come to the flywheel? What is the length of the piston stroke?
2. Show that Q has coordinates $(45 \cos t, 45 \sin t)$, where t is the radian measure of angle BOQ. [*Hint:* If Q has coordinates (x, y), use the point-in-the-plane description to compute $\cos t$ and solve for x; find y similarly.]
3. Let p be the x-coordinate of the center point of the base of the piston. Express p as a function of t. [*Hint:* Use the distance formula to express the distance from Q to $(p, 0)$ in terms of p and t. Set this expression equal to 150 (why?) and solve for p.]

4. Let $p(t)$ be the function found in Question 3 (that is, $p(t)$ = the x-coordinate of the base of the piston when angle BOQ measures t radians). What is the range of this function? How does this relate to Question 1?
5. Approximate the values of $p(0)$, $p(\pi/2)$, and $p(\pi)$. Note that $p(\pi/2)$ is not half way between $p(0)$ and $p(\pi)$. Which side of the half-way point is it on? Does this mean that the piston moves faster on the average when it is to the right or left of the half-way point? Find the value of t that places the base of the piston at the half-way point of its motion.
6. If the flywheel spins at a constant speed, does the piston move back and forth at a constant speed? How do you know?

Chapter

7 Triangle Trigonometry

Where are we?

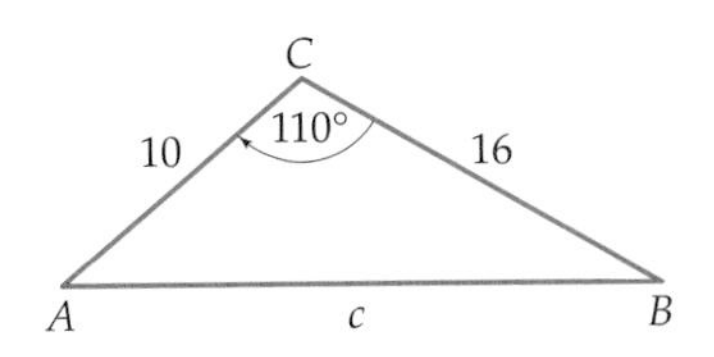

Navigators at sea must determine their location. Surveyors need to determine the height of a mountain or the width of a canyon when direct measurement isn't feasible. A fighter plane's computer must set the course of a missile so that it will hit a moving target. These and many similar problems can be solved by using triangle trigonometry. See Exercise 33 on page 487 and Exercise 41 on page 497.

Interdependence of Sections

7.1 → 7.2
7.1 → 7.3

Roadmap

Chapters 7–9 are independent of each other and may be read in any order.

Chapter Outline

Trigonometry was first used by the ancients to solve practical problems in astronomy, navigation, and surveying that involved triangles. Trigonometric functions, as presented in Chapter 6, came much later. The early mathematicians took a somewhat different viewpoint than we have used up to now, but their approach is often the best way to deal with problems involving triangles.

7.1 Right Triangle Trigonometry

The process of evaluating the sine function may be summarized like this:

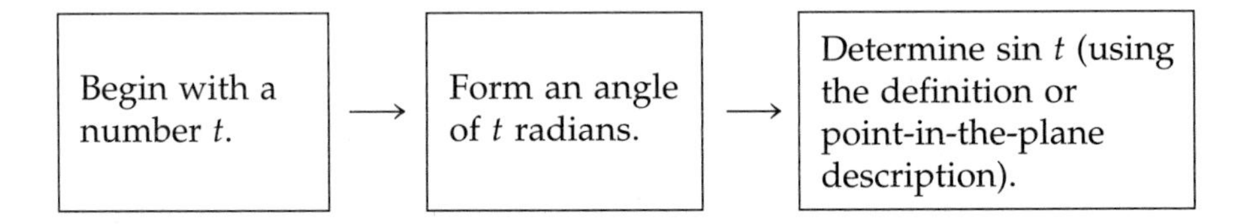

The classical approach begins at the second step, with *angles* rather than numbers. Although the ancients wouldn't have used these terms, their approach amounts to defining the sine function as a function whose domain consists of all *angles* instead of all real numbers. Analogous remarks apply to the other trigonometric functions.

In this chapter (and hereafter, whenever convenient) we shall take this classical approach. Instead of starting with numbers and *then* passing to angles, we shall just begin with the angles. From there on everything is essentially the same. The *values* of the various trigonometric functions are still *numbers* and are obtained as before. For example, the point-in-the-plane method yields this:

Point-in-the-Plane Description

Let θ be an angle in standard position and let (x, y) be any point (except the origin) on the terminal side of θ. Let $r = \sqrt{x^2 + y^2}$. Then, the values of the six trigonometric functions of the angle θ are given by

$$\sin\theta = \frac{y}{r} \qquad \cos\theta = \frac{x}{r} \qquad \tan\theta = \frac{y}{x}$$

$$\csc\theta = \frac{r}{y} \qquad \sec\theta = \frac{r}{x} \qquad \cot\theta = \frac{x}{y}$$

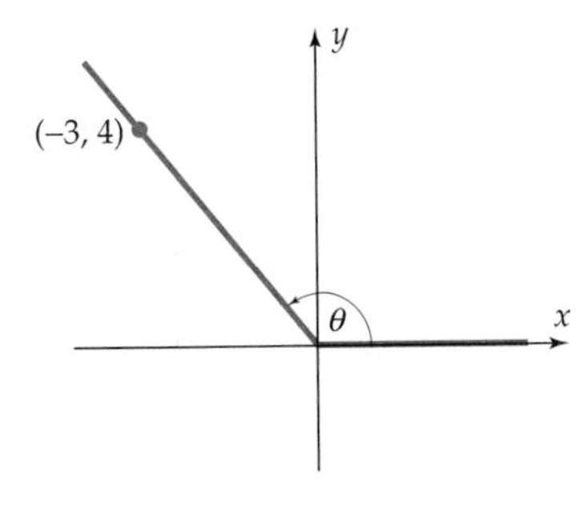

Figure 7–1

Example 1 Evaluate the six trigonometric functions at the angle θ shown in Figure 7–1.

Solution We use $(-3, 4)$ as the point (x, y), so that

$$r = \sqrt{x^2 + y^2} = \sqrt{9 + 16} = \sqrt{25} = 5.$$

Thus,

$$\sin\theta = \frac{y}{r} = \frac{4}{5} \qquad \cos\theta = \frac{x}{r} = \frac{-3}{5} \qquad \tan\theta = \frac{y}{x} = \frac{4}{-3}$$

$$\csc\theta = \frac{r}{y} = \frac{5}{4} \qquad \sec\theta = \frac{r}{x} = \frac{5}{-3} \qquad \cot\theta = \frac{x}{y} = \frac{-3}{4} \quad ■$$

Degrees and Radians

Angles can be measured in either degrees or radians. If radian measure is used (as was the case in Chapter 6), then everything is the same as before. For example, sin 30 denotes the sine of an angle of 30 radians.

But when angles are measured in degrees (as will be done in the rest of this chapter), new notation is needed. In order to denote the value of the sine function at an angle of 30 *degrees,* we write

$$\sin 30^\circ \qquad \text{[note the degree symbol]}$$

The degree symbol here is absolutely essential to avoid error.

Example 2 Since an angle of 30 degrees is the same as one of $\pi/6$ radians, $\sin 30^\circ$ is the same number as $\sin \pi/6$. Hence,

$$\sin 30^\circ = \sin \pi/6 = 1/2.$$

This is *different* from sin 30 (the sine of an angle of 30 *radians*); a calculator in radian mode shows that $\sin 30 = -.988$. ■

The various identities proved in earlier sections are valid for angles measured in degrees, provided that π radians is replaced by 180°. For any angle θ measured in degrees for which the functions are defined, the following identities hold.

Identities for Angles Measured in Degrees

Periodicity Identities

$\sin(\theta + 360°) = \sin\theta$ $\qquad$ $\csc(\theta + 360°) = \csc\theta$

$\cos(\theta + 360°) = \cos\theta$ $\qquad$ $\sec(\theta + 360°) = \sec\theta$

$\tan(\theta + 180°) = \tan\theta$ $\qquad$ $\cot(\theta + 180°) = \cot\theta$

Pythagorean Identities

$\sin^2\theta + \cos^2\theta = 1 \qquad 1 + \tan^2\theta = \sec^2\theta \qquad 1 + \cot^2\theta = \csc^2\theta$

Negative Angle Identities

$\sin(-\theta) = -\sin\theta \qquad \cos(-\theta) = \cos\theta \qquad \tan(-\theta) = -\tan\theta$

Right Triangle Description of Trigonometric Functions

For angles between 0° and 90°, the trigonometric functions may be evaluated by using right triangles as follows. Suppose θ is an angle in a right triangle. Place the triangle so that angle θ is in standard position, with the hypotenuse as its terminal side, as shown in Figure 7–2.

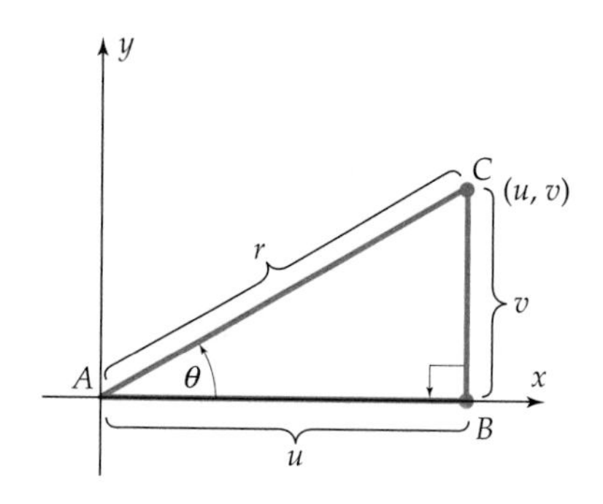

Figure 7–2

Denote the length of the side AB (the one *adjacent* to angle θ) by u and the length of side BC (the one *opposite* angle θ) by v. Then the coordinates of C are (u, v). Let r be the length of the *hypotenuse* AC (the distance from (u, v) to the origin). Then the point-in-the-plane description shows that

$$\sin\theta = \frac{v}{r} = \frac{\text{length of opposite side}}{\text{length of hypotenuse}} \qquad \cos\theta = \frac{u}{r} = \frac{\text{length of adjacent side}}{\text{length of hypotenuse}}$$

$$\tan\theta = \frac{v}{u} = \frac{\text{length of opposite side}}{\text{length of adjacent side}},$$

and similarly for the other trigonometric functions. These facts can be summarized as follows.

Right Triangle Description

Consider a right triangle containing an angle θ:

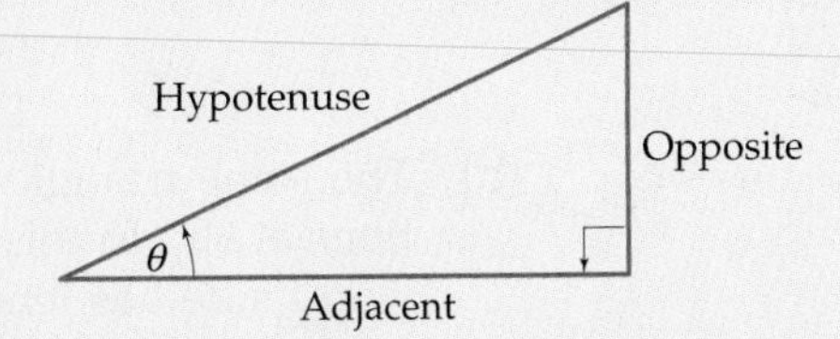

The values of the six trigonometric functions of the angle θ are given by:

$$\sin\theta = \frac{\text{opposite}}{\text{hypotenuse}} \qquad \cos\theta = \frac{\text{adjacent}}{\text{hypotenuse}} \qquad \tan\theta = \frac{\text{opposite}}{\text{adjacent}}$$

$$\csc\theta = \frac{\text{hypotenuse}}{\text{opposite}} \qquad \sec\theta = \frac{\text{hypotenuse}}{\text{adjacent}} \qquad \cot\theta = \frac{\text{adjacent}}{\text{opposite}}$$

This is the description of the trigonometric functions that is usually presented first when trigonometry is studied from the viewpoint of angles and triangles, rather than from the viewpoint of functions. It has the advantage of being independent of both the unit circle and the coordinate system in the plane.

Figure 7–3

Example 3 Evaluate the six trigonometric functions at the angle θ shown in Figure 7–3.

Solution

$$\sin\theta = \frac{\text{opposite}}{\text{hypotenuse}} = \frac{5}{13} \qquad \csc\theta = \frac{\text{hypotenuse}}{\text{opposite}} = \frac{13}{5}$$

$$\cos\theta = \frac{\text{adjacent}}{\text{hypotenuse}} = \frac{12}{13} \qquad \sec\theta = \frac{\text{hypotenuse}}{\text{adjacent}} = \frac{13}{12}$$

$$\tan\theta = \frac{\text{opposite}}{\text{adjacent}} = \frac{5}{12} \qquad \cot\theta = \frac{\text{adjacent}}{\text{opposite}} = \frac{12}{5}$$

■

Example 4 Evaluate $\sin\theta$, $\cos\theta$, $\tan\theta$ when (a) $\theta = 30°$ (b) $\theta = 60°$.

Solution

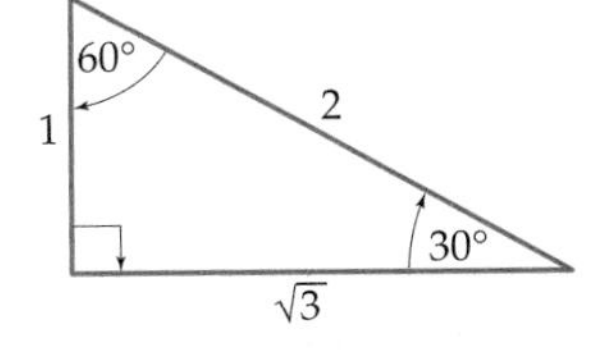

Figure 7–4

(a) Consider a 30°-60°-90° triangle whose hypotenuse has length 2. As explained in Example 3 of the Geometry Review Appendix, the side opposite the 30° angle must have length 1 (half the hypotenuse) and the side adjacent to this angle must have length $\sqrt{3}$, as shown in Figure 7–4. According to the right triangle description,

$$\sin 30° = \frac{\text{opposite}}{\text{hypotenuse}} = \frac{1}{2} \qquad \cos 30° = \frac{\text{adjacent}}{\text{hypotenuse}} = \frac{\sqrt{3}}{2}$$

$$\tan 30° = \frac{\text{opposite}}{\text{adjacent}} = \frac{1}{\sqrt{3}} = \frac{\sqrt{3}}{3}.$$

(b) The same triangle can be used to evaluate the trigonometric functions at 60°. In this case, the opposite side has length $\sqrt{3}$ and the adjacent side has length 1. Therefore,

$$\sin 60° = \frac{\text{opposite}}{\text{hypotenuse}} = \frac{\sqrt{3}}{2} \qquad \cos 60° = \frac{\text{adjacent}}{\text{hypotenuse}} = \frac{1}{2}$$

$$\tan 60° = \frac{\text{opposite}}{\text{adjacent}} = \frac{\sqrt{3}}{1} = \sqrt{3}. \quad ■$$

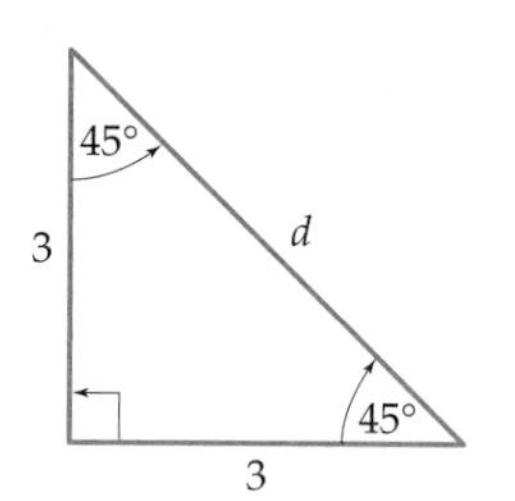

Figure 7–5

Example 5 Evaluate sin 45°, cos 45°, tan 45°.

Solution Consider a 45°-45°-90° triangle whose sides each have length 3 (Figure 7–5). According to the Pythagorean Theorem, the hypotenuse d satisfies

$$d^2 = 3^2 + 3^2 = 18$$

so that

$$d = \sqrt{18} = \sqrt{9 \cdot 2} = \sqrt{9}\sqrt{2} = 3\sqrt{2}.$$

Therefore,

$$\sin 45° = \frac{\text{opposite}}{\text{hypotenuse}} = \frac{3}{3\sqrt{2}} = \frac{1}{\sqrt{2}} = \frac{\sqrt{2}}{2}$$

$$\cos 45° = \frac{\text{adjacent}}{\text{hypotenuse}} = \frac{3}{3\sqrt{2}} = \frac{1}{\sqrt{2}} = \frac{\sqrt{2}}{2}$$

$$\tan 45° = \frac{\text{opposite}}{\text{adjacent}} = \frac{3}{3} = 1. \quad ■$$

You should memorize the values of the trigonometric functions of angles of 30°, 45°, and 60° that were found in Examples 4 and 5. Except for these special angles (and integer multiples of them), a calculator in *degree mode* should be used to evaluate trigonometric functions of angles. For instance,

Technology Tip

Throughout this chapter be sure your calculator is set for DEGREE mode. Use the MODE(S) menu on TI and HP and the SETUP menus on Sharp and Casio.

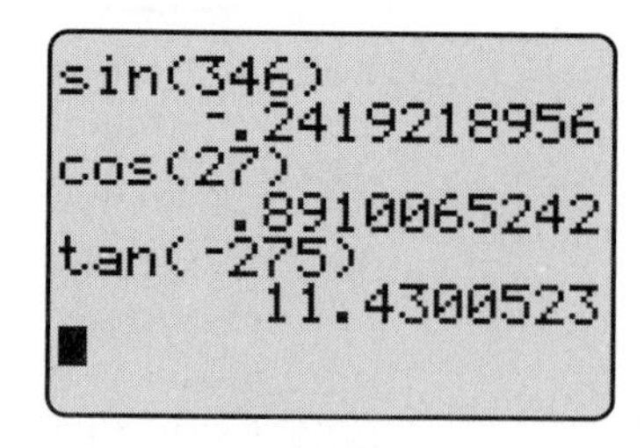

Figure 7–6

Solving Right Triangles

Many applications of trigonometry involve **"solving a triangle."** This means finding the lengths of all three sides and the measures of all three angles when only some of these quantities are given. Solving the right triangles depends on this fact:

> **The right angle description of a trigonometric function (such as $\sin\theta$ = opposite/hypotenuse) relates three quantities: the angle θ and two sides of the right triangle.**

When two of these three quantities are known, then the third can always be found.

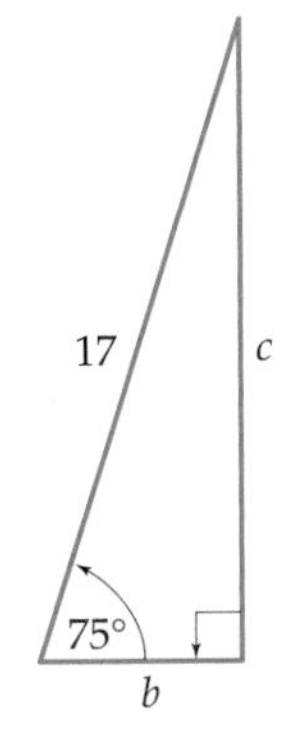

Figure 7–7

Example 6 Find the lengths of sides b and c in the right triangle shown in Figure 7–7.

Solution Since the side c is opposite the 75° angle and the hypotenuse is 17, we have:

$$\sin 75^\circ = \frac{\text{opposite}}{\text{hypotenuse}} = \frac{c}{17}.$$

Solving the equation $\sin 75^\circ = c/17$ for c and using a calculator in degree mode, we have:

$$c = 17\sin 75^\circ \approx 17(.9659) \approx 16.42.$$

Side b can now be found by the Pythagorean Theorem or by using the cosine function:

$$\cos 75^\circ = \frac{\text{adjacent}}{\text{hypotenuse}} = \frac{b}{17}$$

or equivalently, $b = 17\cos 75^\circ \approx 4.40.$ ■

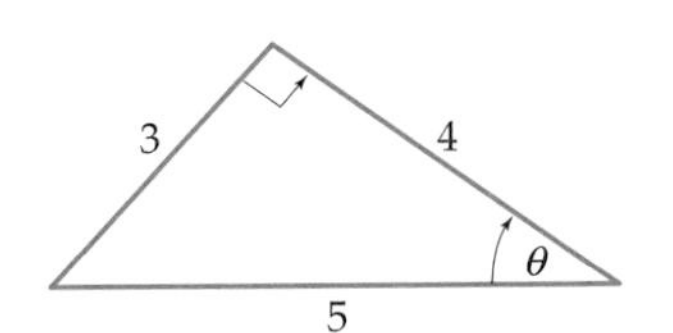

Figure 7–8

Example 7 Find the degree measure of angle θ in Figure 7–8.

Solution We first note that

$$\cos\theta = \frac{\text{adjacent}}{\text{hypotenuse}} = \frac{4}{5} = .8.$$

Figure 7–9

Although you could graphically solve the equation $\cos\theta = .8$, it is easier to use the COS^{-1} key on your calculator. When you key in $\text{COS}^{-1}.8$, as in Figure 7–9, the calculator produces an angle between 0° and 90° whose cosine is .8, namely, $\theta \approx 36.8699°$. (You can think of this as the electronic equivalent of searching through a table of cosine values until you find an angle whose cosine is .8.) ■

NOTE In this chapter we shall use the COS^{-1} key, and the analogous keys SIN^{-1} and TAN^{-1}, as they were used in the last example—as a way to find an angle θ in a triangle, when $\sin\theta$ or $\cos\theta$ or $\tan\theta$ is known.* The other uses of these keys are discussed in Section 8.5, which deals with the inverse functions of sine, cosine, and tangent.

Applications

The following examples illustrate a variety of practical applications of the right triangle description of trigonometric functions.

Example 8 A flagpole casts a 60-foot-long shadow and the angle at the tip of the shadow measures 35° (Figure 7–10). Find the length of the flagpole.

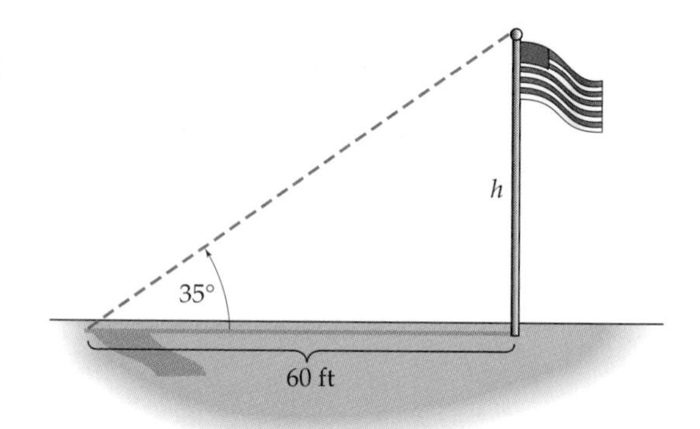

Figure 7–10

Solution Using the right triangle whose legs are the flagpole and its shadow, we see that

$$\tan 35^\circ = \frac{\text{opposite}}{\text{adjacent}} = \frac{h}{60}$$

Solving this equation for h, we have

$$h = 60 \tan 35^\circ \approx 42.01 \text{ feet.}$$

So the flagpole is about 42 feet long. ■

Example 9 A straight road leads from an ocean beach into the nearby hills. The road has a constant upward grade of 3°. After taking this road from the beach for 1 mile, how high above sea level are you?

*On HP calculators, the COS^{-1}, SIN^{-1}, and TAN^{-1} keys are labeled ACOS, ASIN, and ATAN, respectively.

Solution Figure 7–11 shows the situation [*Remember:* 1 mile = 5280 feet]:

Figure 7–11

Figure 7–11 shows that

$$\sin 3° = \frac{\text{opposite}}{\text{hypotenuse}} = \frac{h}{5280}$$

$$h = 5280 \sin 3° \approx 276.33 \text{ feet.} \quad ■$$

In many practical applications, one uses the angle between the horizontal and some other line (for instance, the line of sight from an observer to a distant object). This angle is called the **angle of elevation** or the **angle of depression,** depending on whether the line is above or below the horizontal, as shown in Figure 7–12.

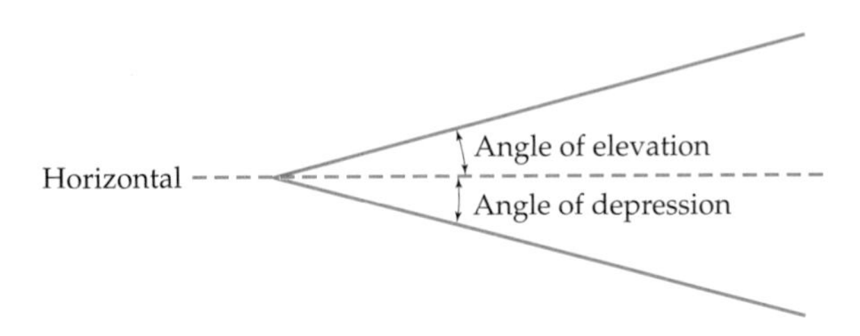

Figure 7–12

Example 10 A wire is to be stretched from the top of a 10-meter-high building to a point on the ground. From the top of the building the angle of depression to the ground point is 22°. How long must the wire be?

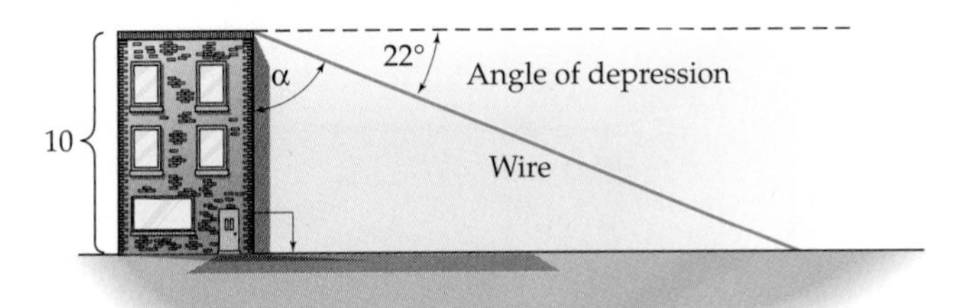

Figure 7–13

Solution Figure 7–13 shows that the sum of the angle of depression and the angle α is 90°. Hence α measures $90° - 22° = 68°$. We know the length of the side of the triangle adjacent to the angle α and must find the hypotenuse w (the length of the wire). Using the cosine function we see that

$$\cos 68° = \frac{\text{adjacent}}{\text{hypotenuse}} = \frac{10}{w}.$$

$$w = \frac{10}{\cos 68°} \approx 26.7 \text{ meters.} \quad \blacksquare$$

Example 11 A person standing on the edge of one bank of a canal observes a lamp post on the edge of the other bank of the canal. The person's eye level is 152 centimeters above the ground (approximately 5 feet). The angle of elevation from eye level to the top of the lamp post is 12°, and the angle of depression from eye level to the bottom of the lamp post is 7°, as shown in Figure 7–14. How wide is the canal? How high is the lamp post?

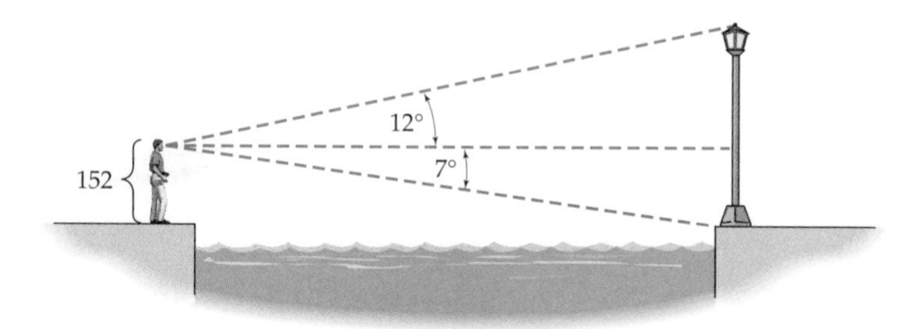

Figure 7–14

Solution Abstracting the essential information, we obtain the diagram in Figure 7–15.

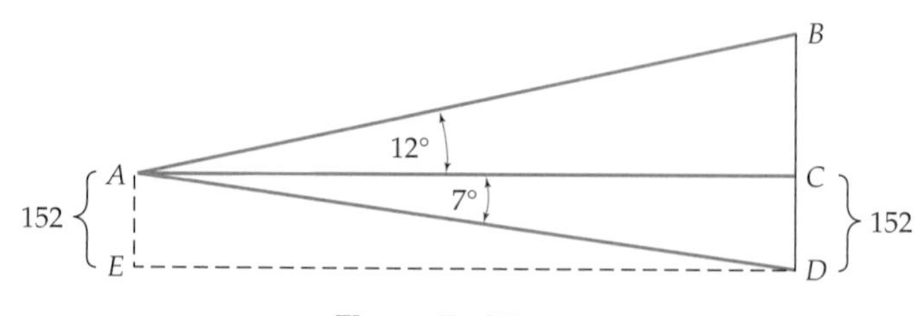

Figure 7–15

We must find the height of the lamp post BD and the width of the canal AC (or ED). The eye level height AE of the observer is 152 centimeters. Since AC and ED are parallel, CD also has length 152 centimeters. In right triangle ACD we know the angle of 7° and the side CD opposite it. We must find the adjacent side AC. The tangent function is needed:

$$\tan 7° = \frac{\text{opposite}}{\text{adjacent}} = \frac{152}{AC}, \quad \text{or equivalently,} \quad AC = \frac{152}{\tan 7°}$$

$$AC = \frac{152}{\tan 7°} \approx 1237.94 \text{ centimeters}$$

So, the canal is approximately 12.3794 meters* wide (about 40.6 feet). Now using right triangle ACB, we see that

$$\tan 12° = \frac{\text{opposite}}{\text{adjacent}} = \frac{BC}{AC} \approx \frac{BC}{1237.94}$$

or equivalently,

$$BC \approx 1237.94(\tan 12°) \approx 263.13 \text{ centimeters.}$$

Therefore, the height of the lamp post BD is $BC + CD \approx 263.13 + 152 = 415.13$ centimeters. ■

Exercises 7.1

Directions: *When solving triangles here, all decimal approximations should be rounded off to one decimal place at the end of the computation.*

In Exercises 1–6, evaluate all six trigonometric functions at the angle (in standard position) whose terminal side contains the given point.

1. $(2, 3)$ **2.** $(4, -2)$ **3.** $(-5, 6)$

4. $(\sqrt{2}, \sqrt{3})$ **5.** $(-3, -\sqrt{2})$ **6.** $(-5, 3)$

In Exercises 7–12, find $\sin\theta$, $\cos\theta$, $\tan\theta$.

7.

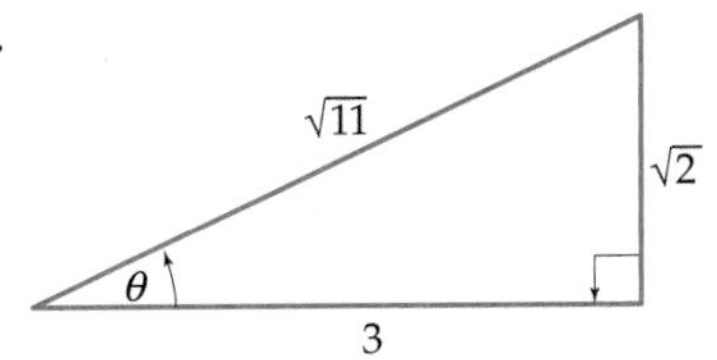

8.

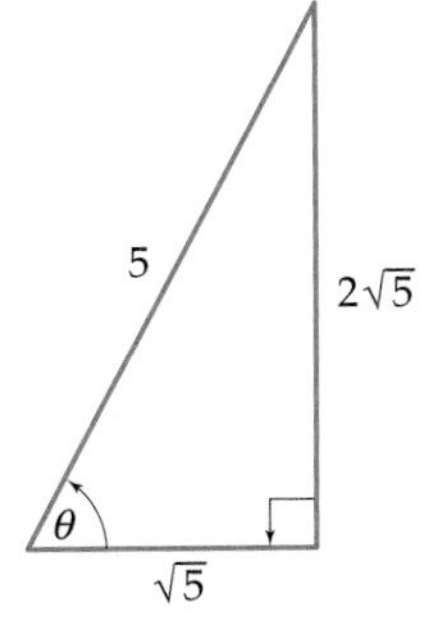

9.

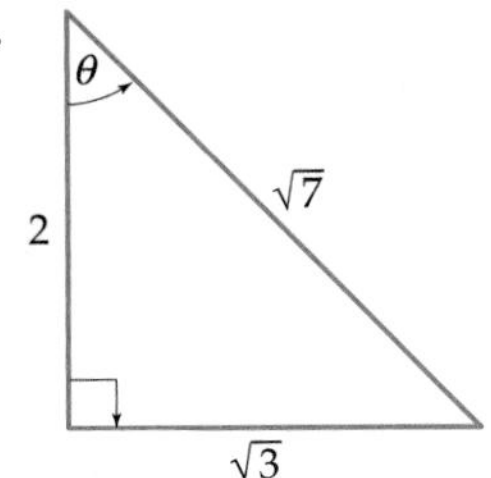

10.

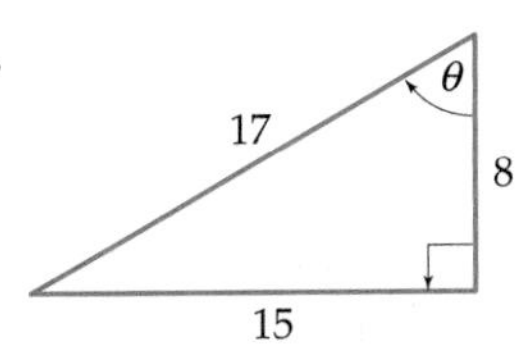

11.

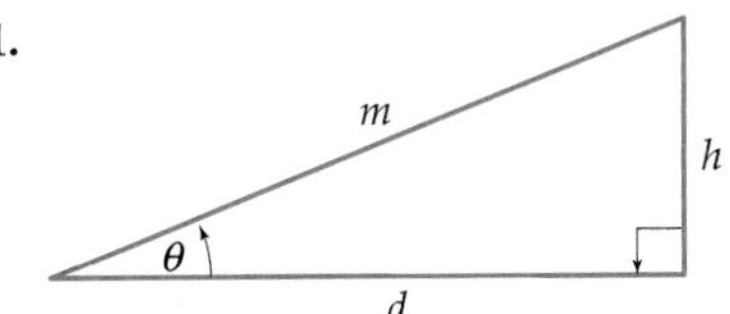

12.

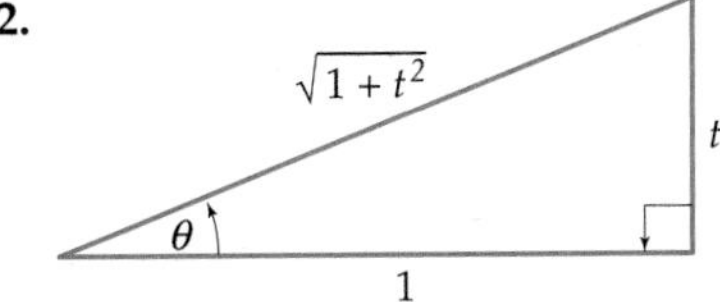

*Remember, 100 centimeters = 1 meter.

In Exercises 13–18, find side c of the right triangle in the figure under the given conditions.

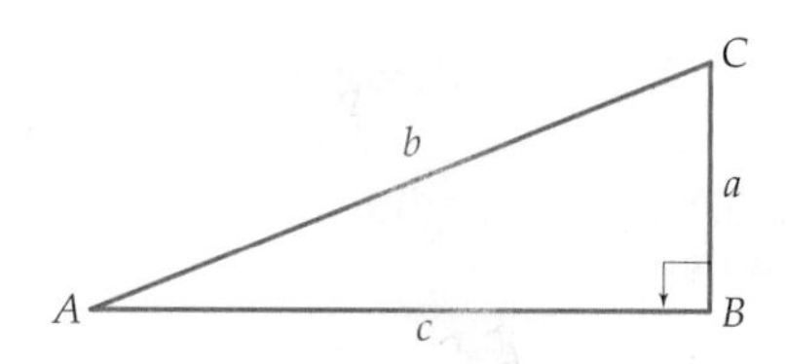

13. $\cos A = 12/13$ and $b = 39$
14. $\sin C = 3/4$ and $b = 12$
15. $\tan A = 5/12$ and $a = 15$
16. $\sec A = 2$ and $b = 8$
17. $\cot A = 6$ and $a = 1.4$
18. $\csc C = 1.5$ and $b = 4.5$

In Exercises 19–24, find the length h of the side of the right triangle, without using a calculator.

19.

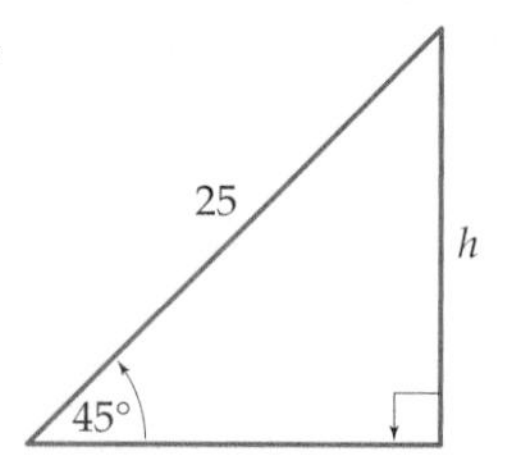

20.

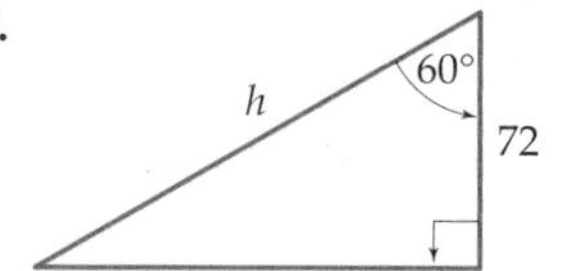

21.

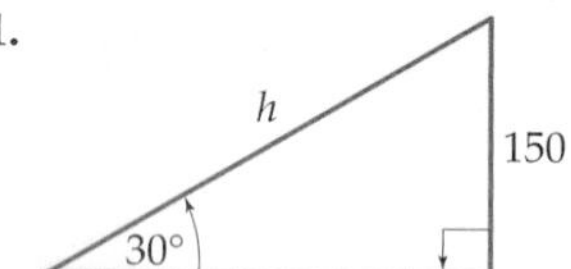

22.

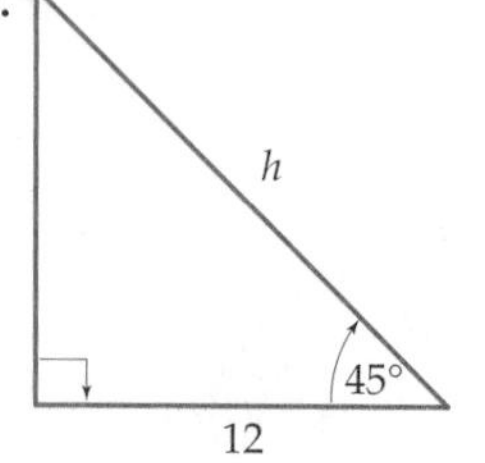

23.

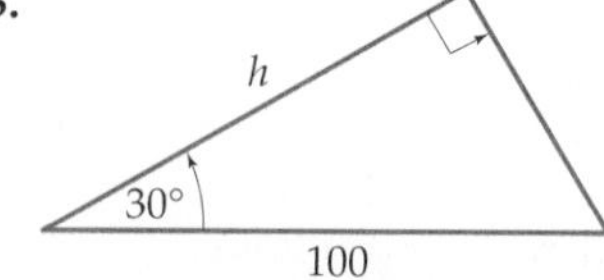

24.

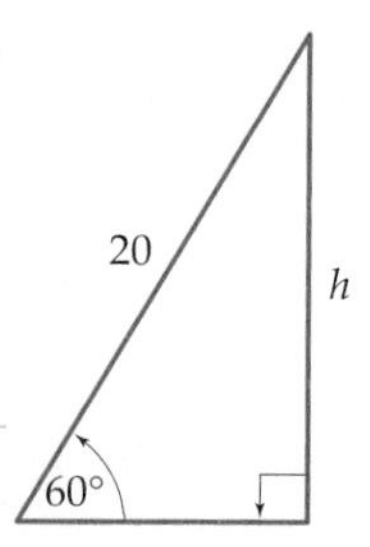

In Exercises 25–28, find the required side without using a calculator.

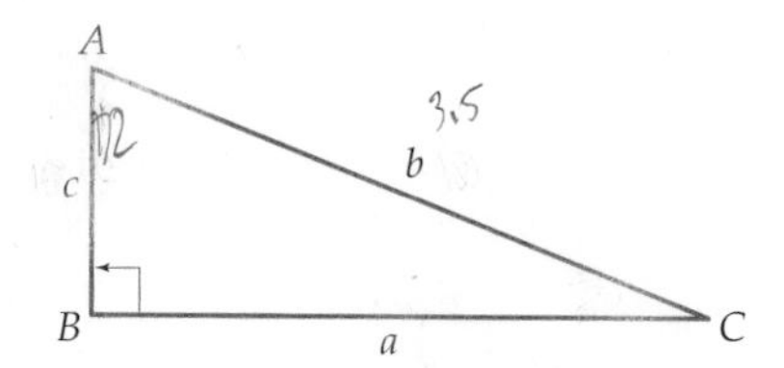

25. $a = 4$ and angle A measures 60°; find c.
26. $c = 5$ and angle A measures 60°; find a.
27. $c = 10$ and angle A measures 30°; find a.
28. $a = 12$ and angle A measures 30°; find c.

In Exercises 29–36, use the figure for Exercises 25–28. Solve the right triangle under the given conditions.

29. $b = 10$ and $\measuredangle C = 50°$
30. $c = 12$ and $\measuredangle C = 37°$
31. $a = 6$ and $\measuredangle A = 14°$
32. $a = 8$ and $\measuredangle A = 40°$
33. $c = 5$ and $\measuredangle A = 65°$
34. $c = 4$ and $\measuredangle C = 28°$
35. $b = 3.5$ and $\measuredangle A = 72°$
36. $a = 4.2$ and $\measuredangle C = 33°$

In Exercises 37–40, find angle θ.

37.

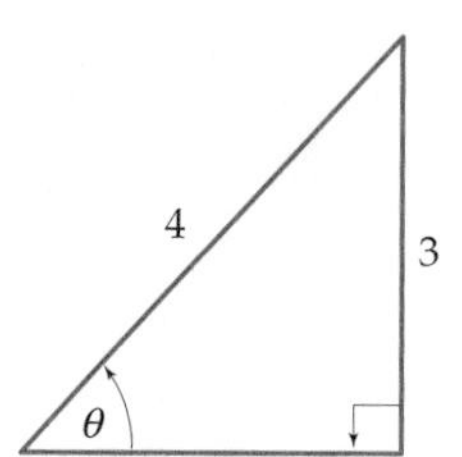

38.

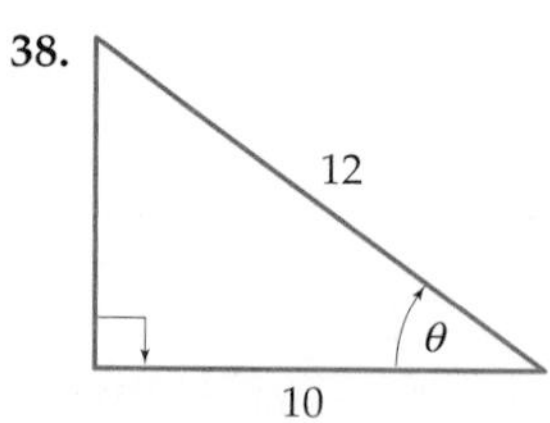

39.

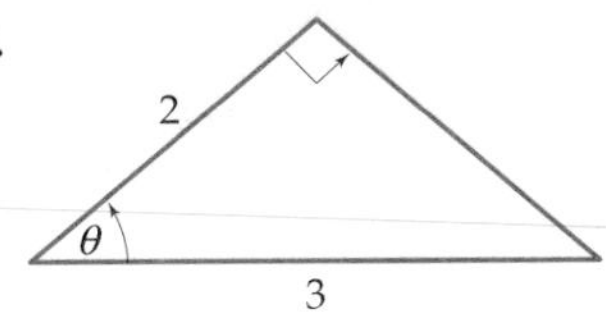

40.

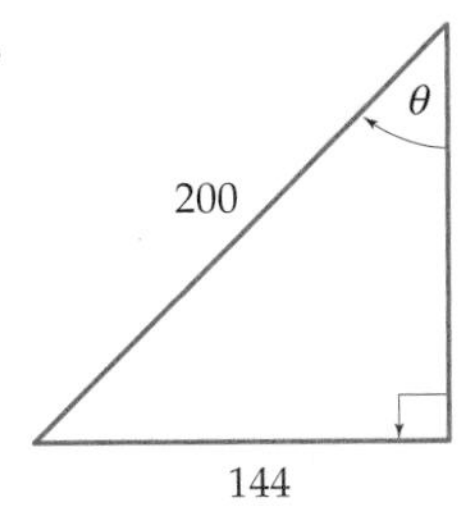

In Exercises 41–48, use the figure for Exercises 25–28 to find angles A and C under the given conditions.

41. $a = 4$ and $c = 6$

42. $b = 14$ and $c = 5$

43. $a = 7$ and $b = 10$

44. $a = 5$ and $c = 3$

45. $b = 18$ and $c = 12$

46. $a = 4$ and $b = 9$

47. $a = 2.5$ and $c = 1.4$

48. $b = 3.7$ and $c = 2.2$

49. A 20-foot-long ladder leans on a wall of a building. The foot of the ladder makes an angle of 50° with the ground. How far above the ground does the top of the ladder touch the wall?

50. A guy wire stretches from the top of an antenna tower to a point on level ground 18 feet from the base of the tower. The angle between the wire and the ground is 63°. How high is the tower?

51. A 150-foot-long ramp connects a ground-level parking lot with the entrance of a building. If the entrance is 8 feet above the ground, what angle does the ramp make with the ground?

52. For maximum safety, the distance from the base of a ladder to the building wall should be one fourth of the length of the ladder. If a ladder is in this position, what angle does it make with the ground?

53. Consider a 16-foot-long drawbridge on a medieval castle, as shown in the figure. The royal army is engaged in ignominious retreat. The king would like to raise the end of the drawbridge 8 feet off the ground so that Sir Rodney can jump onto the drawbridge and scramble into the castle, while the enemy's cavalry are held at bay. Through how much of an angle must the drawbridge be raised in order for the end of it to be 8 feet off the ground?

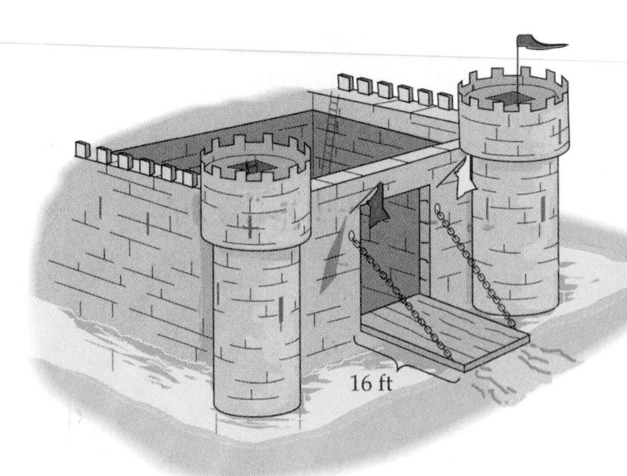

54. Through what angle must the drawbridge in Exercise 53 be raised in order that its end be directly above the center of the moat?

55. A plane takes off at an angle of 5°. After traveling 1 mile along this flight path, how high (in feet) is the plane above the ground?

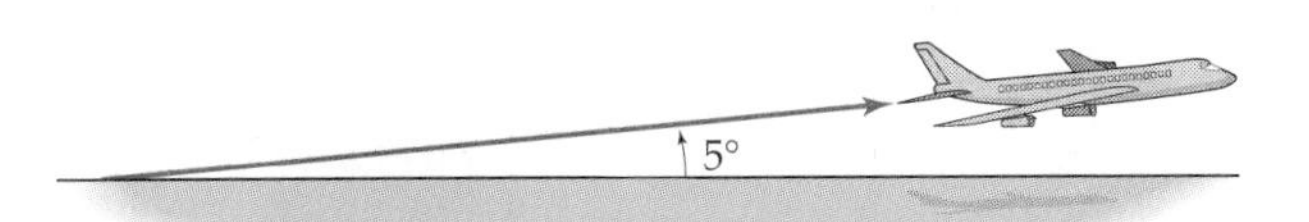

56. A plane takes off at an angle of 6° traveling at the rate of 200 feet/second. If it continues on this flight path at the same speed, how many minutes will it take to reach an altitude of 8000 feet?

57. It is claimed that the Ohio Turnpike never has an uphill grade of more than 3°. How long must a straight uphill segment of the road be in order to allow a vertical rise of 450 feet?

58. Ruth is flying a kite. Her hand is 3 feet above ground level and is holding the end of a 300-foot-long kite string, which makes an angle of 57° with the horizontal. How high is the kite above the ground?

59. If you stand upright on a mountainside that makes a 62° angle with the horizontal and stretch your arm straight out at shoulder height, you may be able to touch the mountain (as shown in the figure on the next page). Can a person with an arm reach of 27 inches, whose shoulder is 5 feet above the ground, touch the mountain?

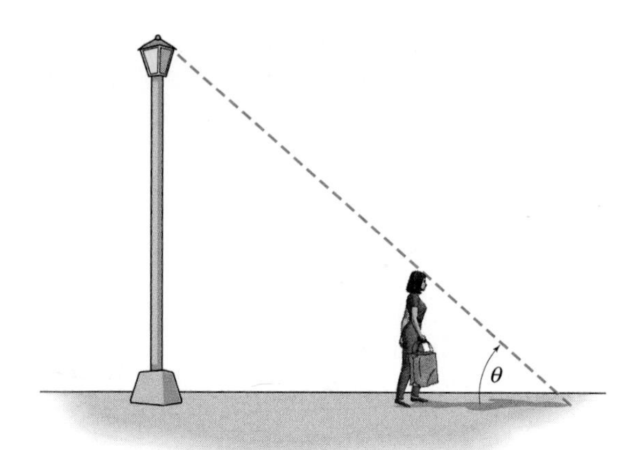

60. A swimming pool is 3 feet deep in the shallow end. The bottom of the pool has a steady downward drop of 12°. If the pool is 50 feet long, how deep is it at the deep end?

61. Batman is on the edge of a 200-foot-deep chasm and wants to jump to the other side. A tree on the edge of the chasm is directly across from him. He walks 20 feet to his right and notes that the angle to the tree is 54°. His jet belt enables him to jump a maximum of 24 feet. How wide is the chasm and is it safe for Batman to jump?

62. A wire from the top of a TV tower to the ground makes an angle of 49.5° with the ground and touches ground 225 feet from the base of the tower. How high is the tower?

63. A woman 5.5 feet tall stands 10 feet from a streetlight and casts a 4-foot-long shadow. How tall is the streetlight? What is angle θ?

64. A plane flies a straight course. On the ground directly below the flight path, observers 2 miles apart spot the plane at the same time. The plane's angle of elevation is 46° from one observation point and 71° from the other. How high is the plane?

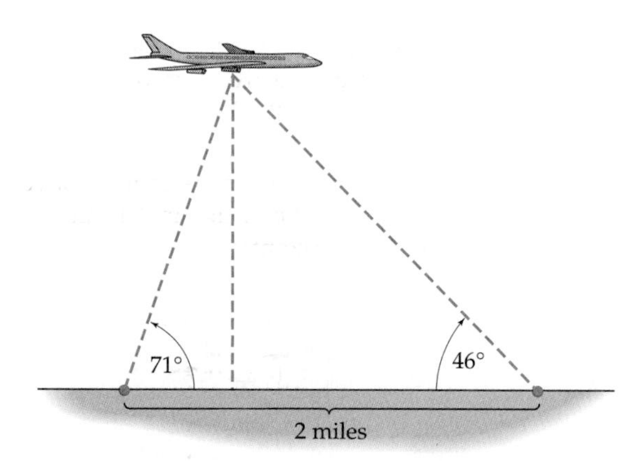

65. A buoy in the ocean is observed from the top of a 40-meter-high radar tower on shore. The angle of depression from the top of the tower to the base of the buoy is 6.5°. How far is the buoy from the base of the radar tower?

66. A plane passes directly over your head at an altitude of 500 feet. Two seconds later you observe that its angle of elevation is 42°. How far did the plane travel during those 2 seconds?

67. A man stands 12 feet from a statue. The angle of elevation from eye level to the top of the statue is 30°, and the angle of depression to the base of the statue is 15°. How tall is the statue?

68. Two boats lie on a straight line with the base of a lighthouse. From the top of the lighthouse (21 meters above water level) it is observed that the

angle of depression of the nearest boat is 53° and the angle of depression of the farthest boat is 27°. How far apart are the boats?

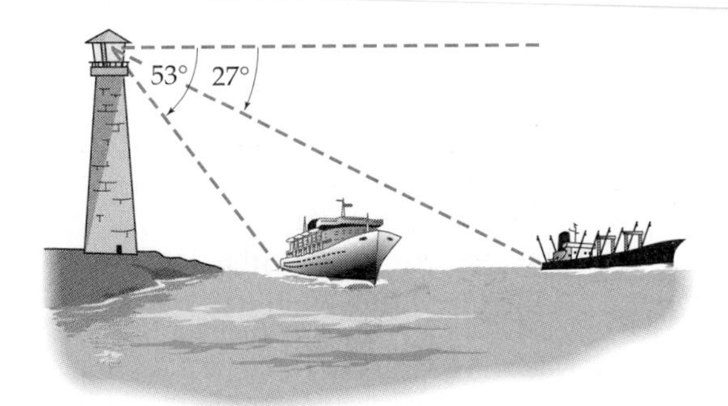

69. A rocket shoots straight up from the launchpad. Five seconds after lift-off an observer 2 miles away notes that the rocket's angle of elevation is 3.5°. Four seconds later the angle of elevation is 41°. How far did the rocket rise during those 4 seconds?

70. From a 35-meter-high window, the angle of depression to the top of a nearby streetlight is 55°. The angle of depression to the base of the streetlight is 57.8°. How high is the streetlight?

71. A closed 60-foot-long drawbridge is 24 feet above water level. When open, the bridge makes an angle of 33° with the horizontal.

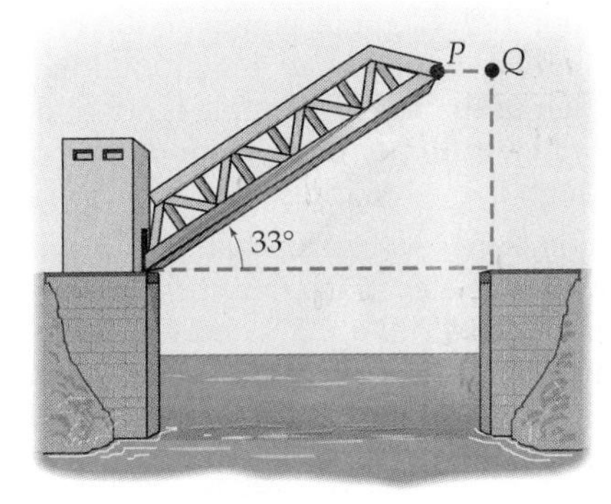

(a) How high is the tip P of the open bridge above the water?

(b) When the bridge is open, what is the distance from P to Q?

72. One plane flies straight east at an altitude of 31,000 feet. A second plane is flying west at an altitude of 14,000 feet on a course that lies directly below that of the first plane and directly above the straight road from Thomasville to Johnsburg. As the first plane passes over Thomasville, the second is passing over Johnsburg. At that instant both planes spot a beacon next to the road between Thomasville to Johnsburg. The angle of depression from the first plane to the beacon is 61° and the angle of depression from the second plane to the beacon is 34°. How far is Thomasville from Johnsburg?

73. A schematic diagram of a pedestrian overpass is shown in the figure. If you walk on the overpass from one end to the other, how far have you walked?

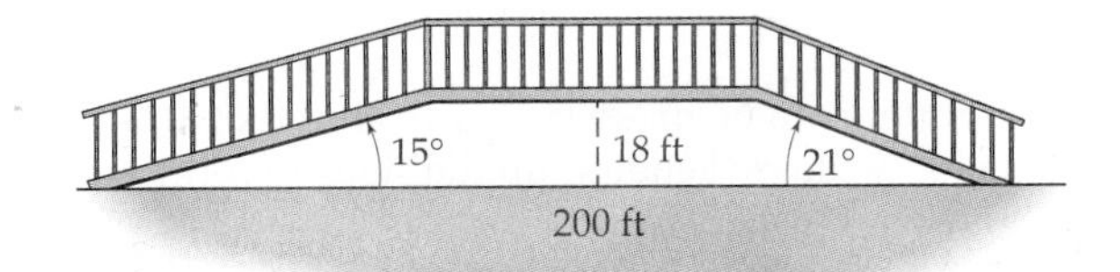

74. A 5-inch-high plastic beverage glass has a 2.5-inch diameter base. Its sides slope outward at a 4° angle as shown. What is the diameter of the top of the glass?

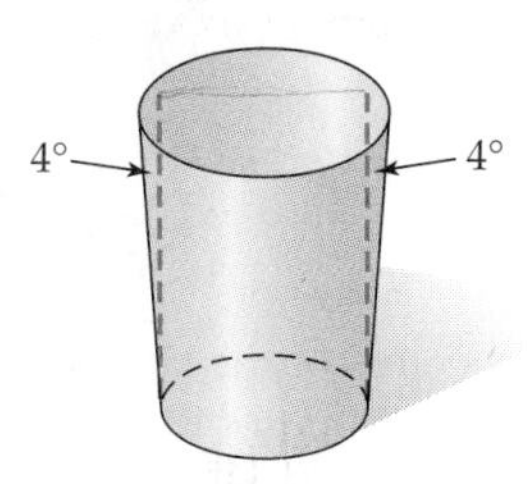

75. In aerial navigation, directions are given in degrees clockwise from north. Thus east is 90°, south is 180°, and so on, as shown below. A plane travels from an airport for 200 miles in the direction 300°. How far west of the airport is the plane then?

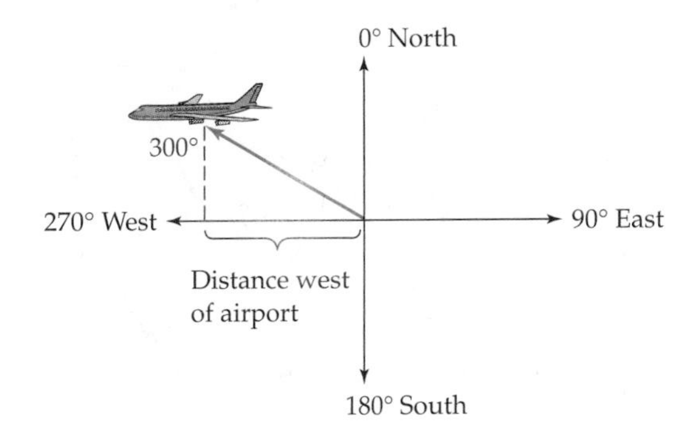

76. A plane travels at a constant 300 mph in the direction 65° (see Exercise 75).
(a) How far east of its starting point is the plane after half an hour?
(b) How far north of its starting point is the plane after 2 hours and 24 minutes?

77. A car on a straight road passes under a bridge. Two seconds later an observer on the bridge, 20 feet above the road, notes that the angle of depression to the car is 7.4°. How fast (in miles per hour) is the car traveling? [*Note:* 60 mph is equivalent to 88 feet/second.]

Thinkers

78. A gutter is to be made from a strip of metal 24 inches wide by bending up the sides to form a trapezoid.

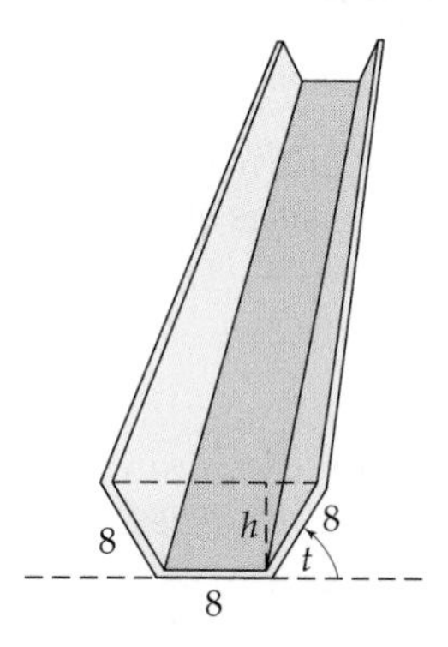

(a) Express the area of the cross section of the gutter as a function of the angle t. [*Hint:* The area of a trapezoid with bases b and b' and height h is $h(b + b')/2$.]
(b) For what value of t will this area be as large as possible?

79. The cross section of a tunnel is a semicircle with radius 10 meters. The interior walls of the tunnel form a rectangle.

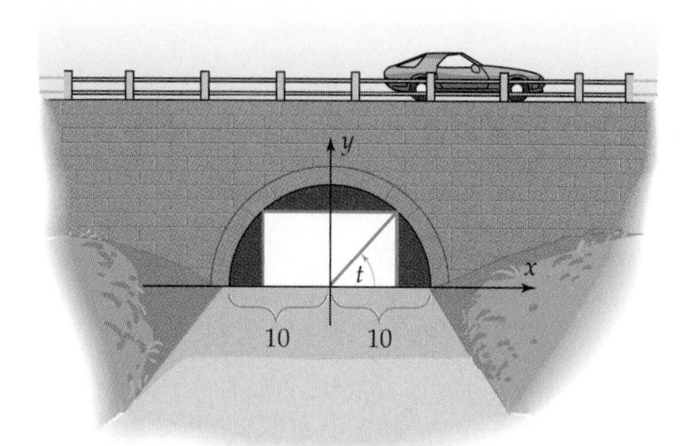

(a) Express the area of the rectangular cross section of the tunnel opening as a function of angle t.
(b) For what value of t is the cross-sectional area of the tunnel opening as large as possible? What are the dimensions of the tunnel opening in this case?

80. A spy plane on a practice run over the Midwest takes a picture that shows Cleveland, Ohio, on the eastern horizon and St. Louis, Missouri, 520 miles away, on the western horizon (the figure is not to scale). Assuming that the radius of the earth is 3950 miles, how high was the plane when the picture was taken? [*Hint:* The sight lines from the plane to the horizons are tangent to the earth and a tangent line to a circle is perpendicular to the radius at that point. The arc of the earth between St. Louis and Cleveland is 520 miles long. Use this fact and the arc length formula to find angle θ (your answers will be in radians). Note that $\alpha = \theta/2$ (why?).]

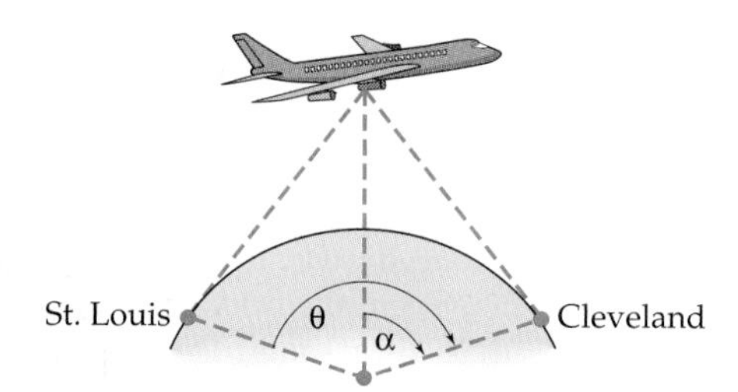

81. A 50-foot-high flagpole stands on top of a building. From a point on the ground the angle of elevation of the top of the pole is 43° and the angle of elevation of the bottom of the pole is 40°. How high is the building?

82. Two points on level ground are 500 meters apart. The angles of elevation from these points to the top of a nearby hill are 52° and 67°, respectively. The two points and the ground level point directly below the top of the hill lie on a straight line. How high is the hill?

7.2 The Law of Cosines

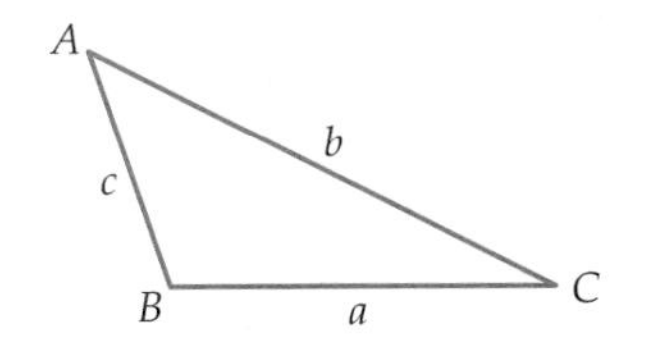

Figure 7–16

We now consider the solution of *oblique* triangles (ones that don't contain a right angle). We shall use **standard notation** for triangles: Each vertex is labeled with a capital letter, and the length of the side opposite that vertex is denoted by the same letter in lower case, as shown in Figure 7–16. The letter A will also be used to label the *angle* at vertex A, and similarly for B and C. So we shall make statements such as $A = 37°$ or $\cos B = .326$.

The first fact needed to solve oblique triangles is the Law of Cosines, whose proof is given at the end of this section.

Law of Cosines

In any triangle ABC, with sides of lengths a, b, c, as in Figure 7–16,

$$a^2 = b^2 + c^2 - 2bc \cos A$$

$$b^2 = a^2 + c^2 - 2ac \cos B$$

$$c^2 = a^2 + b^2 - 2ab \cos C$$

You need only memorize one of these equations since each of them provides essentially the same information: a description of one side of a triangle in terms of the angle opposite it and the other two sides.

NOTE When C is a right angle, then c is the hypotenuse and

$$\cos C = \cos 90° = 0,$$

so that the third equation in the Law of Cosines becomes the Pythagorean Theorem:

$$c^2 = a^2 + b^2$$

Solving the first equation in the Law of Cosines for $\cos A$, we obtain

Law of Cosines: Alternate Form

In any triangle ABC, with sides of lengths a, b, c, as in Figure 7–16,

$$\cos A = \frac{b^2 + c^2 - a^2}{2bc}.$$

The other two equations can be similarly rewritten. In this form, the Law of Cosines provides a description of each angle of a triangle in terms of the three sides. Consequently, the Law of Cosines can be used to solve triangles in these cases:

1. Two sides and the angle between them are known (SAS).
2. Three sides are known (SSS).

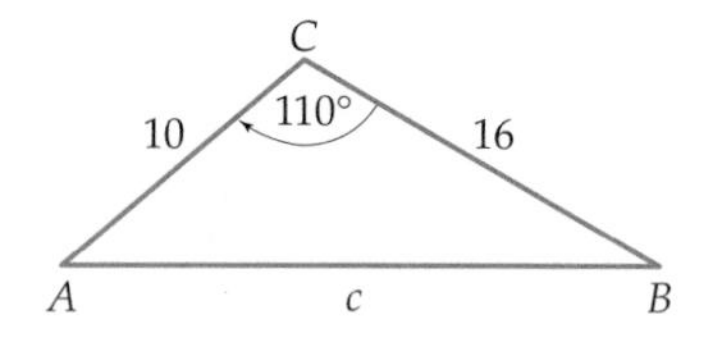

Figure 7–17

Example 1 (SAS) Solve triangle *ABC* in Figure 7–17.

Solution We have $a = 16$, $b = 10$, and $C = 110°$. The right side of the third equation in the Law of Cosines involves only these known quantities. Hence,

$$\begin{aligned} c^2 &= a^2 + b^2 - 2ab \cos C \\ &= 16^2 + 10^2 - 2 \cdot 16 \cdot 10 \cos 110° \\ &\approx 256 + 100 - 320(-.342) \approx 465.4.^* \end{aligned}$$

Therefore, $c \approx \sqrt{465.4} \approx 21.6$. Now use the alternate form of the Law of Cosines:

$$\cos A = \frac{b^2 + c^2 - a^2}{2bc} \approx \frac{10^2 + (21.6)^2 - 16^2}{2 \cdot 10 \cdot 21.6} \approx .7172.$$

The COS^{-1} key (in degree mode) produces an angle with cosine .7172: $A \approx 44.2°$. Hence, $B \approx 180° - (44.2° + 110°) = 25.8°$. ■

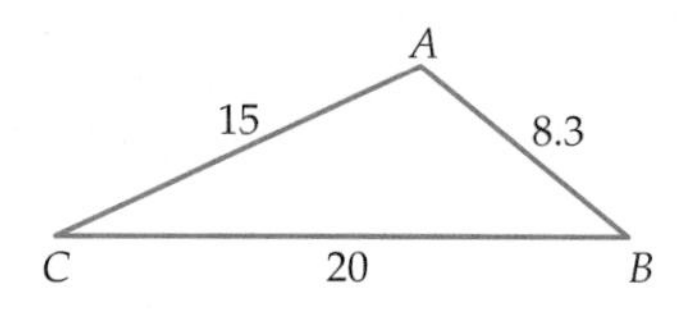

Figure 7–18

Example 2 (SSS) Find the angles of triangle *ABC* in Figure 7–18.

Solution In this case, $a = 20$, $b = 15$, and $c = 8.3$. By the alternate form of the Law of Cosines:

$$\cos A = \frac{b^2 + c^2 - a^2}{2bc} = \frac{15^2 + 8.3^2 - 20^2}{2 \cdot 15 \cdot 8.3} = \frac{-106.11}{249} \approx -.4261.$$

The COS^{-1} key shows that $A \approx 115.2°$. Similarly, the alternate form of the Law of Cosines yields:

$$\cos B \approx \frac{a^2 + c^2 - b^2}{2ac} = \frac{20^2 + 8.3^2 - 15^2}{2 \cdot 20 \cdot 8.3} = \frac{243.89}{332} \approx .7346$$

$$B \approx 42.7°.$$

Therefore, $C \approx 180° - (115.2° + 42.7°) = 180° - 157.9° = 22.1°$. ■

Example 3 Two trains leave a station on different tracks. The tracks make an angle of 125° with the station as vertex. The first train travels at

*Throughout this chapter all decimals are printed in rounded-off form for reading convenience, but no rounding is done in the actual computation until the final answer is obtained.

an average speed of 100 kilometers per hour and the second an average of 65 kilometers per hour. How far apart are the trains after 2 hours?

Solution The first train A traveling at 100 kilometers per hour for 2 hours goes a distance of $100 \times 2 = 200$ kilometers. The second train B travels a distance of $65 \times 2 = 130$ kilometers. So, we have the situation shown in Figure 7–19.

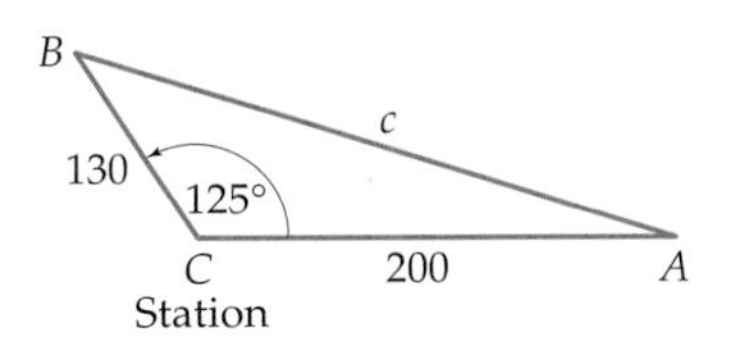

Figure 7–19

By the Law of Cosines

$$\begin{aligned} c^2 &= a^2 + b^2 - 2ab\cos C \\ &= 130^2 + 200^2 - 2\cdot 130\cdot 200\cos 125^\circ \\ &= 56{,}900 - 52{,}000\cos 125^\circ \approx 86{,}725.97 \\ c &\approx \sqrt{86{,}725.97} = 294.5 \text{ kilometers.} \end{aligned}$$

The trains are 294.5 kilometers apart after 2 hours. ■

Example 4 A 100-foot-tall antenna tower is to be placed on a hillside that makes an angle of 12° with the horizontal. It is to be anchored by two cables from the top of the tower to points 85 feet uphill and 95 feet downhill from the base. How much cable is needed?

Solution The situation is shown in Figure 7–20, where AB represents the tower and AC and AD the cables.

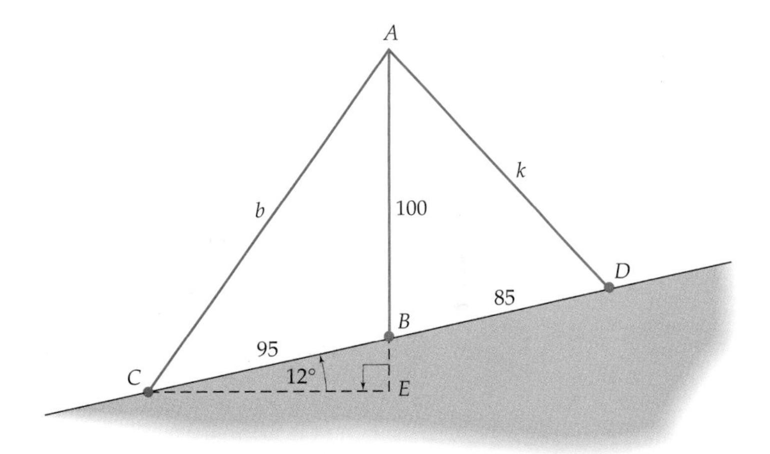

Figure 7–20

In triangle BEC, angle E is a right angle and by hypothesis, angle C measures 12°. Since the sum of the angles of a triangle is 180°, we must have

$$\measuredangle CBE = 180° - (90° + 12°) = 78°.$$

As shown in the figure, the sum of angles CBE and CBA is a straight angle (180°). Hence,

$$\measuredangle CBA = 180° - 78° = 102°.$$

Apply the Law of Cosines to triangle ABC:

$$\begin{aligned} b^2 &= 95^2 + 100^2 - 2 \cdot 95 \cdot 100 \cos 102° \\ &= 9025 + 10{,}000 - 19{,}000 \cos 102° \\ &\approx 22{,}975.32. \end{aligned}$$

Therefore, the length of the downhill cable is $b \approx \sqrt{22{,}975.32} \approx$ 151.58 feet.

To find the length of the uphill cable, note that the sum of angles CBA and DBA is a straight angle, so that

$$\measuredangle DBA = 180° - \measuredangle CBA = 180° - 102° = 78°.$$

Applying the Law of Cosines to triangle DBA, we have:

$$\begin{aligned} k^2 &= 85^2 + 100^2 - 2 \cdot 85 \cdot 100 \cos 78° \\ &= 7225 + 10{,}000 - 17{,}000 \cos 78° \approx 13{,}690.50. \end{aligned}$$

Hence, the length of the uphill cable is $k = \sqrt{13{,}690.50} \approx 117.01$ feet. ■

Proof of the Law of Cosines

Given triangle ABC, position it on a coordinate plane so that angle A is in standard position with initial side c and terminal side b. Depending on the size of angle A, there are two possibilities, as shown in Figure 7–21.

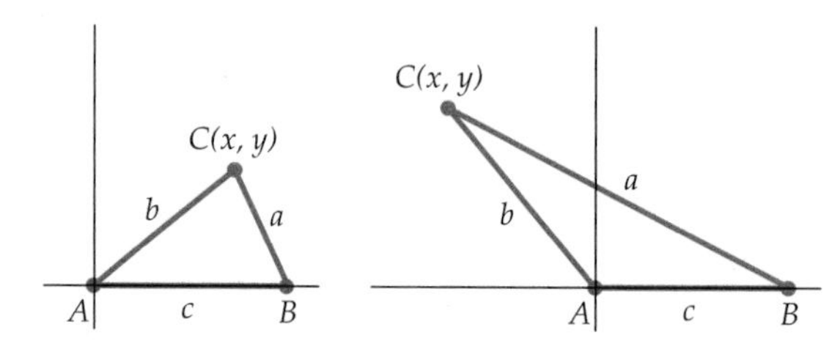

Figure 7–21

The coordinates of B are $(c, 0)$. Let (x, y) be the coordinates of C. Now C is a point on the terminal side of angle A, and the distance from C to the origin A is obviously b. Therefore, according to the point-in-the-plane description of sine and cosine, we have

$$\frac{x}{b} = \cos A, \qquad \text{or equivalently,} \qquad x = b\cos A$$

$$\frac{y}{b} = \sin A, \qquad \text{or equivalently,} \qquad y = b\sin A$$

Using the distance formula on the coordinates of B and C, we have

$$\begin{aligned} a &= \text{distance from } C \text{ to } B \\ &= \sqrt{(x-c)^2 + (y-0)^2} = \sqrt{(b\cos A - c)^2 + (b\sin A - 0)^2}. \end{aligned}$$

Squaring both sides of this last equation and simplifying, using the Pythagorean identity, yields:

$$\begin{aligned} a^2 &= (b\cos A - c)^2 + (b\sin A)^2 \\ a^2 &= b^2\cos^2 A - 2bc\cos A + c^2 + b^2\sin^2 A \\ a^2 &= b^2(\sin^2 A + \cos^2 A) + c^2 - 2bc\cos A \\ a^2 &= b^2 + c^2 - 2bc\cos A. \end{aligned}$$

This proves the first equation in the Law of Cosines. Similar arguments beginning with angles B or C in standard position prove the other two equations.

Exercises 7.2

Directions: *Standard notation for triangle ABC is used throughout. Use a calculator and round off your answers to one decimal place at the end of the computation.*

In Exercises 1–16, solve the triangle ABC under the given conditions.

1. $A = 20°$, $b = 10$, $c = 7$
2. $B = 40°$, $a = 12$, $c = 20$
3. $C = 118°$, $a = 6$, $b = 10$
4. $C = 52.5°$, $a = 6.5$, $b = 9$
5. $A = 140°$, $b = 12$, $c = 14$
6. $B = 25.4°$, $a = 6.8$, $c = 10.5$
7. $C = 78.6°$, $a = 12.1$, $b = 20.3$
8. $A = 118.2°$, $b = 16.5$, $c = 10.7$
9. $a = 7$, $b = 3$, $c = 5$
10. $a = 8$, $b = 5$, $c = 10$
11. $a = 16$, $b = 20$, $c = 32$
12. $a = 5.3$, $b = 7.2$, $c = 10$
13. $a = 7.2$, $b = 6.5$, $c = 11$
14. $a = 6.8$, $b = 12.4$, $c = 15.1$
15. $a = 12$, $b = 16.5$, $c = 21.3$
16. $a = 5.7$, $b = 20.4$, $c = 16.8$
17. Find the angles of the triangle whose vertices are $(0, 0)$, $(5, -2)$, $(1, -4)$.
18. Find the angles of the triangle whose vertices are $(-3, 4)$, $(6, 1)$, $(2, -1)$.
19. Two trains leave a station on different tracks. The tracks make a 112° angle with the station as vertex. The first train travels at an average speed of 90 kilometers/hour and the second at an average speed of 55 kilometers/hour. How far apart are the trains after 2 hours and 45 minutes?
20. One plane flies west from Cleveland at 350 mph. A second plane leaves Cleveland at the same time and flies southeast at 200 mph. How far apart are the planes after 1 hour and 36 minutes?
21. The pitcher's mound on a standard baseball diamond (which is actually a square) is 60.5 feet from home plate (see figure on next page). How far is the pitcher's mound from first base?

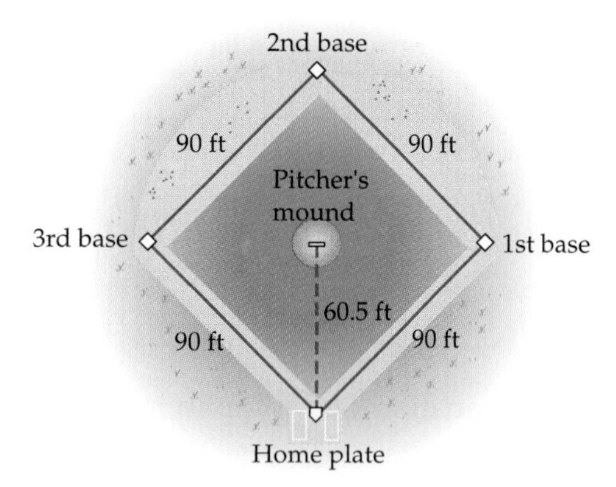

22. If the straight line distance from home plate over second base to the center field wall in a baseball stadium is 400 feet, how far is it from first base to the same point in center field? [Adapt the figure from Exercise 21.]
23. A stake is located 10.8 feet from the end of a closed gate that is 8 feet long. The gate swings open, and its end hits the stake. Through what angle did the gate swing?

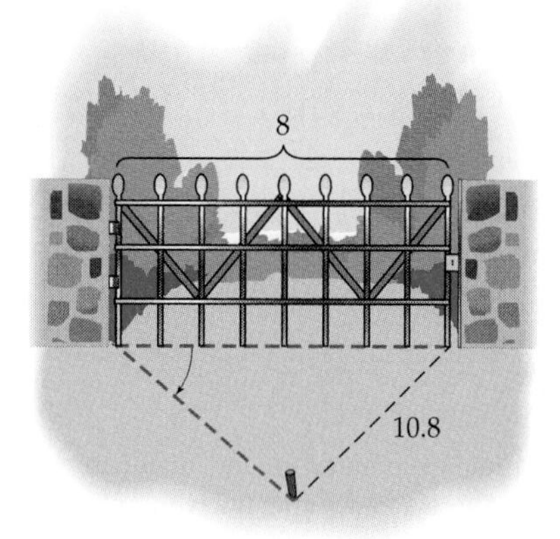

24. The distance from Chicago to St. Louis is 440 kilometers, from St. Louis to Atlanta 795 kilometers, and from Atlanta to Chicago 950 kilometers. What are the angles in the triangle with these three cities as vertices?
25. A boat runs in a straight line for 3 kilometers, then makes a 45° turn and goes for another 6 kilometers (see figure). How far is the boat from its starting point?

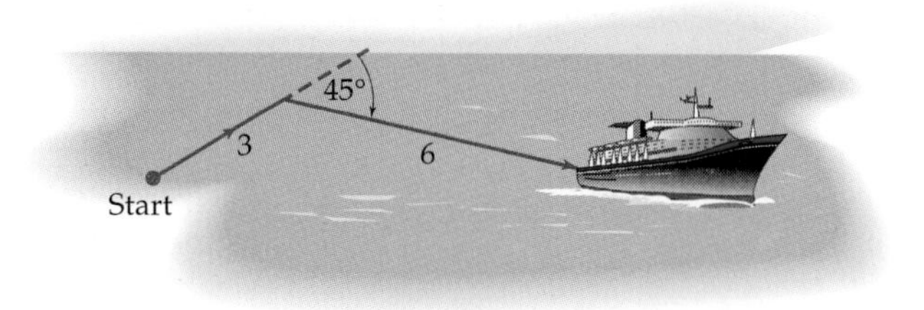

26. A plane flies in a straight line at 400 mph for 1 hour and 12 minutes. It makes a 15° turn and flies at 375 mph for 2 hours and 27 minutes. How far is it from its starting point?
27. The side of a hill makes an angle of 12° with the horizontal. A wire is to be run from the top of a 175-foot tower on the top of the hill to a stake located 120 feet down the hillside from the base of the tower. How long a wire is needed?
28. Two ships leave port, one traveling in a straight course at 22 mph and the other traveling a straight course at 31 mph. Their courses diverge by 38°. How far apart are they after 3 hours?
29. An engineer wants to measure the width *CD* of a sinkhole. So he places a stake *B* and determines the measurements shown in the figure. How wide is the sinkhole?

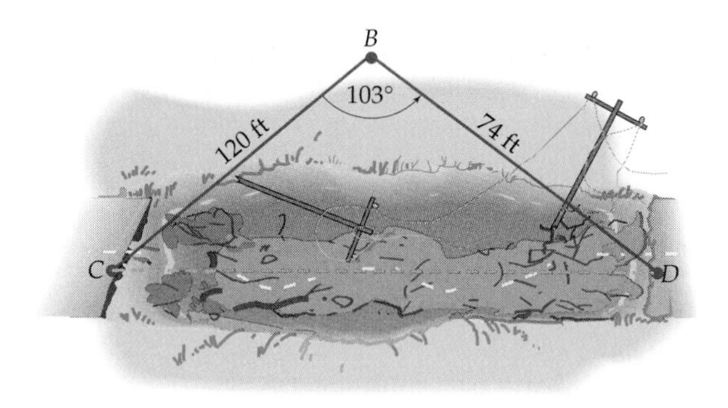

30. A straight tunnel is to be dug through a hill. Two people stand on opposite sides of the hill where the tunnel entrances are to be located. Both can see a stake located 530 meters from the first person and 755 meters from the second. The angle determined by the two people and the stake (vertex) is 77°. How long must the tunnel be?
31. One diagonal of a parallelogram is 6 centimeters long, and the other is 13 centimeters long. They form an angle of 42° with each other. How long

are the sides of the parallelogram? [*Hint:* The diagonals of a parallelogram bisect each other.]

32. A parallelogram has diagonals of lengths 12 and 15 inches that intersect at an angle of 63.7°. How long are the sides of the parallelogram?

33. A ship is traveling at 18 mph from Corsica to Barcelona, a distance of 350 miles. To avoid bad weather, the ship leaves Corsica on a route 22° south of the direct route (see figure). After 7 hours the bad weather has been bypassed. Through what angle should the ship now turn to head directly to Barcelona?

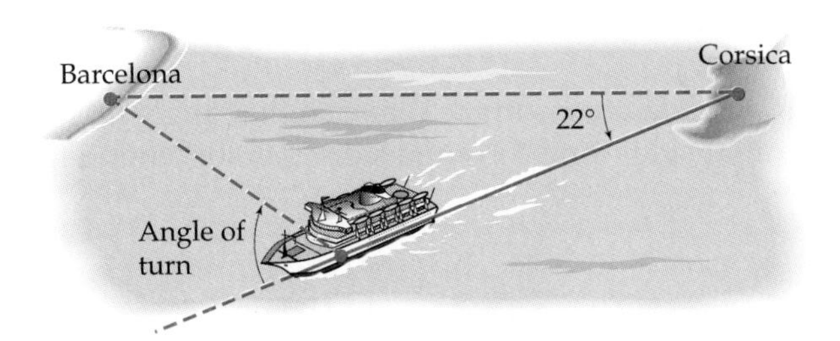

34. A plane leaves South Bend for Buffalo, 400 miles away, intending to fly a straight course in the direction 70° (aerial navigation is explained in Exercise 75 of Section 7.1). After flying 180 miles, the pilot realizes that an error has been made and that he has actually been flying in the direction 55°.

(a) At that time, how far is the plane from Buffalo?

(b) In what direction should the plane now go to reach Buffalo?

35. Assume that the earth is a sphere of radius 3980 miles. A satellite travels in a circular orbit around the earth, 900 miles above the equator, making one full orbit every 6 hours. If it passes directly over a tracking station at 2 P.M., what is the distance from the satellite to the tracking station at 2:05 P.M.?

36. Two planes at the same altitude approach an airport. One plane is 16 miles from the control tower and the other is 22 miles from the tower. The angle determined by the planes and the tower, with the tower as vertex, is 11°. How far apart are the planes?

37. Assuming that the circles in the figure are mutually tangent, find the lengths of the sides and the measures of the angles in triangle ABC.

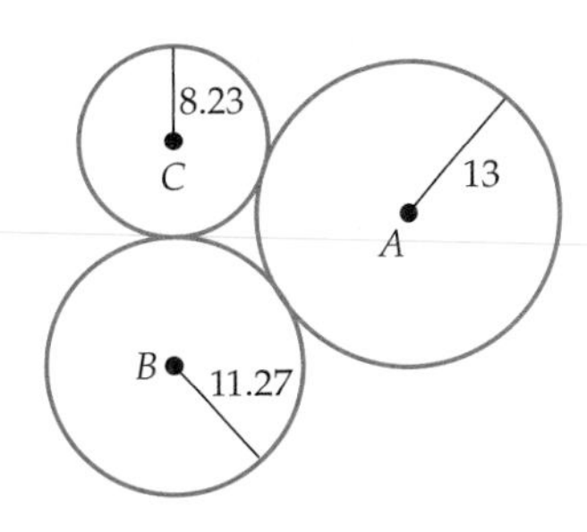

38. Assuming that the circles in the figure are mutually tangent, find the lengths of the sides and the measures of the angles in triangle ABC.

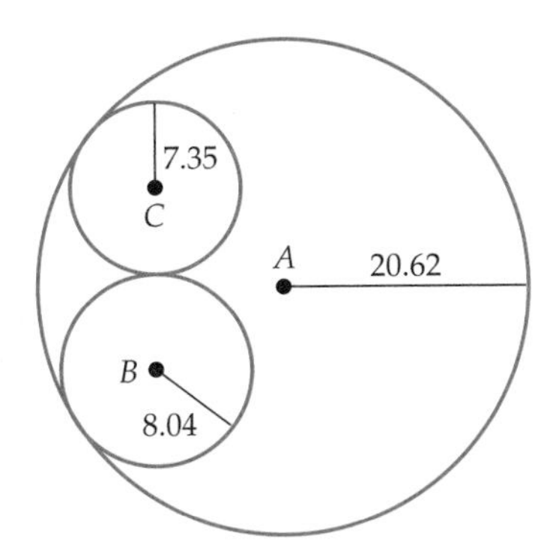

Thinkers

39. A rope is attached at points A and B and taut around a pulley whose center is at C, as shown in the figure (in which AC has length 8 and BC length 7). The rope lies on the pulley from D to E and the radius of the pulley is 1 meter. How long is the rope?

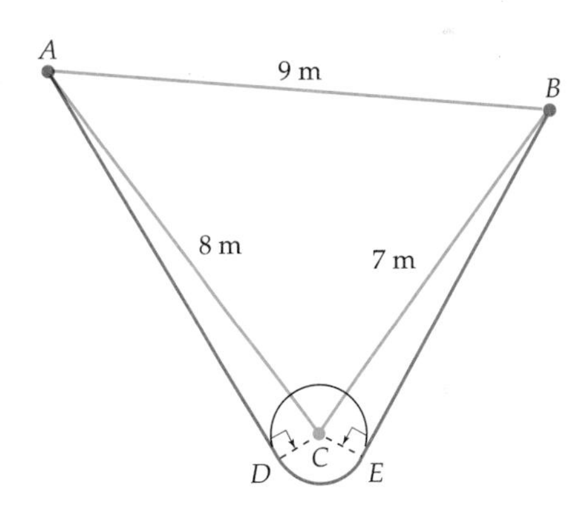

40. Use the Law of Cosines to prove that the sum of the squares of the lengths of the two diagonals of a parallelogram equals the sum of the squares of the lengths of the four sides.

7.3 The Law of Sines

In order to solve oblique triangles in cases where the Law of Cosines cannot be used, we need this fact:

Law of Sines

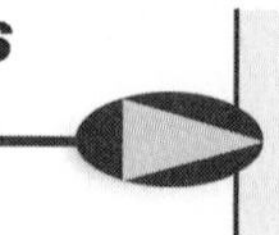

In any triangle *ABC* (in standard notation)

$$\frac{a}{\sin A} = \frac{b}{\sin B} = \frac{c}{\sin C}.^*$$

Proof Position triangle *ABC* on a coordinate plane so that angle *C* is in standard position, with initial side *b* and terminal side *a*, as shown in Figure 7–22.

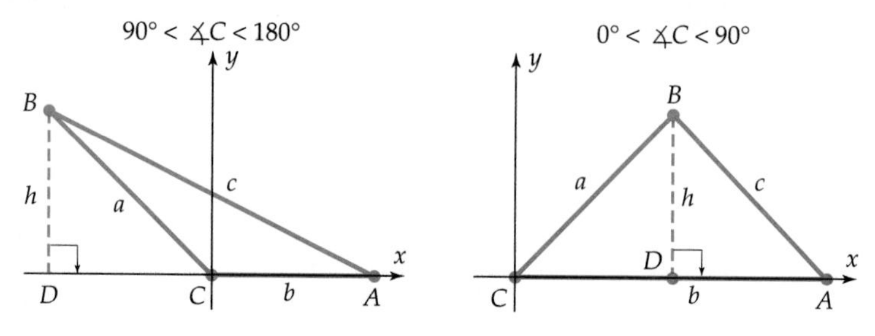

Figure 7–22

In each case we can compute sin *C* by using the point *B* on the terminal side of angle *C*. The second coordinate of *B* is *h* and the distance from *B* to the origin is *a*. Therefore, by the point-in-the-plane description of sine,

$$\sin C = \frac{h}{a}, \qquad \text{or equivalently,} \qquad h = a \sin C.$$

In each case, right triangle *ADB* shows that

$$\sin A = \frac{\text{opposite}}{\text{hypotenuse}} = \frac{h}{c}, \qquad \text{or equivalently,} \qquad h = c \sin A.$$

Combining this with the fact that $h = a \sin C$, we have

$$a \sin C = c \sin A.$$

Since angles in a triangle are nonzero, $\sin A \neq 0$ and $\sin C \neq 0$. Dividing both sides of the last equation by $\sin A \sin C$ yields

*An equality of the form $u = v = w$ is shorthand for the statement $u = v$ and $v = w$ and $w = u$.

$$\frac{a}{\sin A} = \frac{c}{\sin C}.$$

This proves one equation in the Law of Sines. Similar arguments beginning with angles A or B in standard position prove the other equations. ■

The Law of Sines can be used to solve triangles in these cases:

1. Two angles and one side are known (AAS).
2. Two sides and the angle opposite one of them are known (SSA).

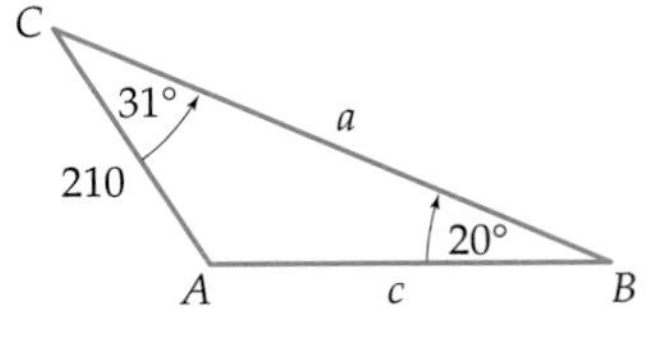

Figure 7–23

Example 1 (AAS) If $B = 20°$, $C = 31°$, and $b = 210$ in Figure 7–23, find the other angles and sides.

Solution Since the sum of the angles of a triangle is 180°,

$$A = 180° - (20° + 31°) = 180° - 51° = 129°.$$

In order to find side a, we observe that we know three of the four quantities in one of the equations given by the Law of Sines:

$$\frac{a}{\sin A} = \frac{b}{\sin B}$$

$$\frac{a}{\sin 129°} = \frac{210}{\sin 20°}.$$

Multiplying both sides by sin 129°, we obtain:

$$a = \frac{210(\sin 129°)}{\sin 20°} \approx 477.2.$$

Side c is found similarly. Beginning with an equation from the Law of Sines involving c and three known quantities, we have:

$$\frac{c}{\sin C} = \frac{b}{\sin B}$$

$$\frac{c}{\sin 31°} = \frac{210}{\sin 20°}$$

$$c = \frac{210 \sin 31°}{\sin 20°} \approx 316.2. \quad ■$$

In the AAS case, there is exactly one triangle satisfying the given data.* But when two sides of a triangle and the angle opposite one of them are known (SSA), there may be one, two, or no triangles that satisfy the given

*Once you know two angles, you know all three (their sum must be 180°). Hence, you know two angles and the included side. Any two triangles satisfying these conditions will be congruent by the ASA Theorem of plane geometry.

data (the **ambiguous case**). To see why this can happen, suppose sides a and b and angle A are given. Place angle A in standard position with terminal side b. If angle A is less than 90°, then there are four possibilities for side a:

(i) Side a is too short to reach the third side: *no solution.*

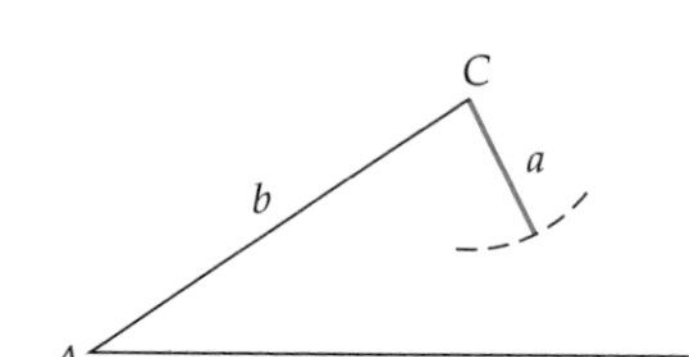

(ii) Side a just reaches the third side and is perpendicular to it: *one solution.*

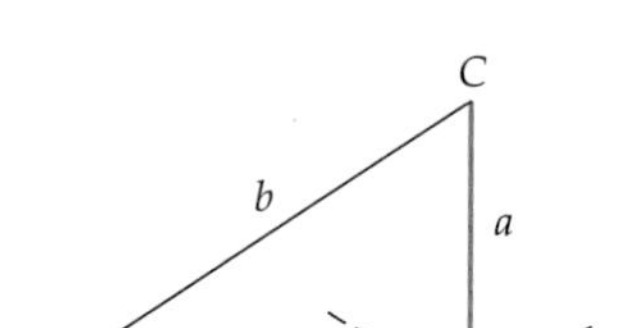

(iii) An arc of radius a meets the third side at 2 points to the right of A: *two solutions.*

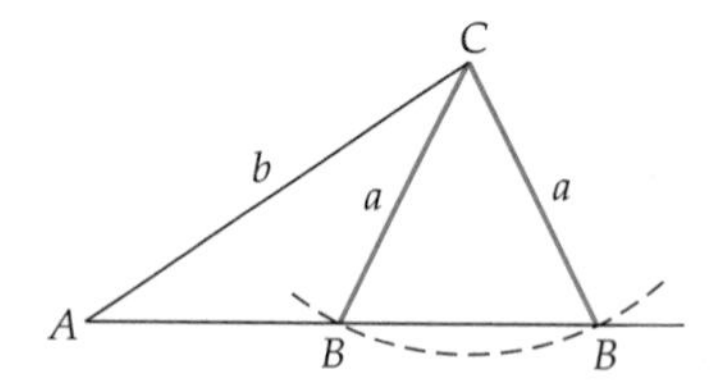

(iv) $a \geq b$, so that an arc of radius a meets the third side at just one point to the right of A: *one solution.*

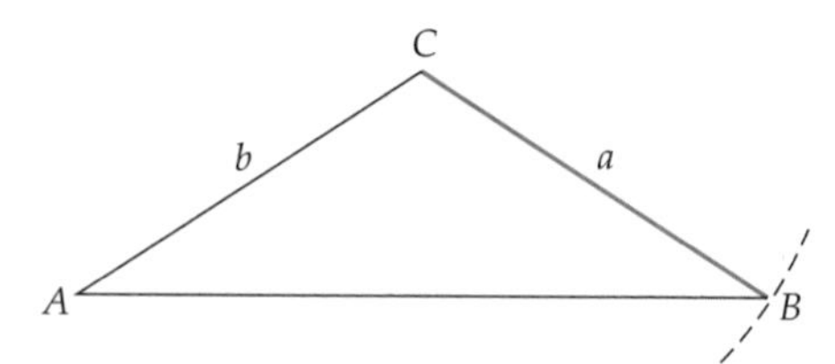

When angle A is more than 90°, then there are only two possibilities:

(i) $a \leq b$, so that side a is too short to reach the third side: *no solution.*

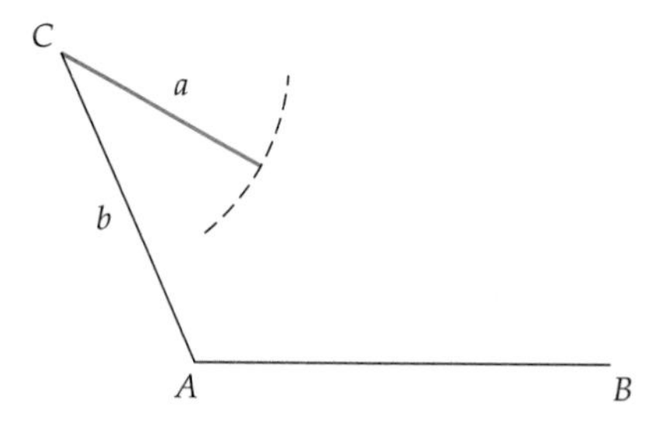

(ii) $a > b$, so that an arc of radius a meets the third side at just one point to the right of A: one *solution.*

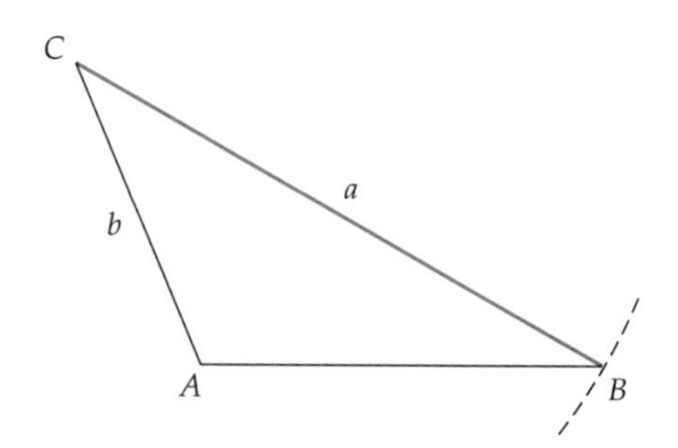

In order to deal with the case of two solutions, we need this identity:

Supplementary Angle Identity

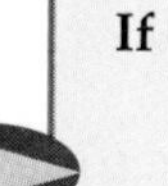

If $0° \leq \theta \leq 90°$, then

$$\sin \theta = \sin(180° - \theta).$$

Proof* Place the angle $180° - \theta$ in standard position and choose a point D on its terminal side. Let r be the distance from D to the origin. The situation looks like this (Figure 7–24):

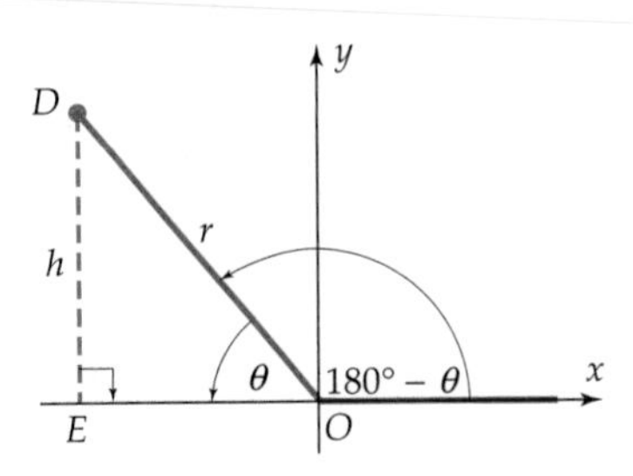

Figure 7–24

Since h is the second coordinate of D, we have $\sin(180° - \theta) = h/r$. Right triangle OED shows that

$$\sin\theta = \frac{\text{opposite}}{\text{hypotenuse}} = \frac{h}{r} = \sin(180° - \theta). \quad \blacksquare$$

Example 2 (SSA) Given triangle ABC with $a = 6$, $b = 7$, and $A = 65°$, find angle B.

Solution We use an equation from the Law of Sines involving the known quantities:

$$\frac{b}{\sin B} = \frac{a}{\sin A}$$

$$\frac{7}{\sin B} = \frac{6}{\sin 65°}$$

$$\sin B = \frac{7\sin 65°}{6} \approx 1.06.$$

There is no angle B whose sine is greater than 1. Therefore, there is no triangle satisfying the given data. ■

Example 3 An airplane A takes off from carrier B and flies in a straight line for 12 kilometers. At that instant, an observer on destroyer C, located 5 kilometers from the carrier, notes that the angle determined by the carrier, the destroyer (vertex), and the plane is 37°. How far is the plane from the destroyer?

*You may skip this proof if you've read Example 2 in Section 8.2, where the identity [in the form $\sin y = \sin(\pi - y)$] is proved for every angle.

Solution The given data provide Figure 7–25:

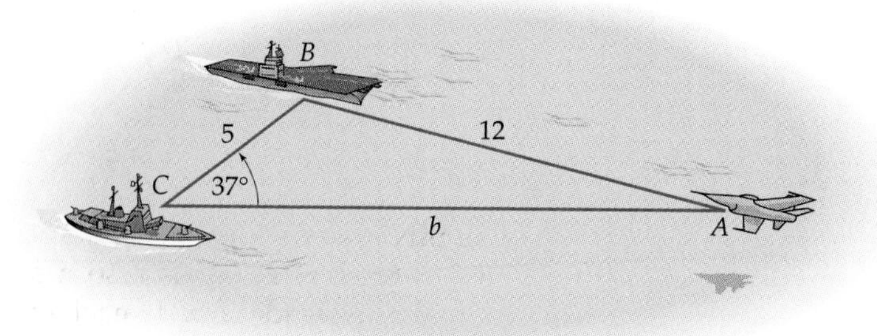

Figure 7–25

We must find side b. To do this, we first use the Law of Sines to find angle A:

$$\frac{a}{\sin A} = \frac{c}{\sin C}$$

$$\frac{5}{\sin A} = \frac{12}{\sin 37°}$$

$$\sin A = \frac{5 \sin 37°}{12} \approx .2508.$$

The SIN^{-1} key on a calculator in degree mode shows that 14.5° is an angle whose sine is .2508. The supplementary angle identity shows that $180° - 14.5° = 165.5°$ is also an angle with sine .2508. But if $A = 165.5°$ and $C = 37°$, the sum of angles A, B, C would be greater than 180°. Since this is impossible, $A = 14.5°$ is the only solution here. Therefore,

$$B = 180° - (37° + 14.5°) = 180° - 51.5° = 128.5°.$$

Using the Law of Sines again, we have

$$\frac{b}{\sin B} = \frac{c}{\sin C}$$

$$\frac{b}{\sin 128.5°} = \frac{12}{\sin 37°}$$

$$b = \frac{12 \sin 128.5°}{\sin 37°} \approx 15.6.$$

Thus, the plane is approximately 15.6 kilometers from the destroyer. ■

Example 4 (SSA) Solve triangle ABC when $a = 7.5$, $b = 12$, and $A = 35°$.

Solution The Law of Sines shows that

$$\frac{b}{\sin B} = \frac{a}{\sin A}$$

$$\frac{12}{\sin B} = \frac{7.5}{\sin 35^\circ}$$

$$\sin B = \frac{12 \sin 35^\circ}{7.5} \approx .9177.$$

The SIN^{-1} key shows that 66.6° is a solution of $\sin B = .9177$. Therefore, $180^\circ - 66.6^\circ = 113.4^\circ$ is also a solution of $\sin B = .9177$ by the supplementary angle identity. In each case the sum of angles A and B is less than 180°, so there are two triangles ABC satisfying the given data, as shown in Figure 7–26.

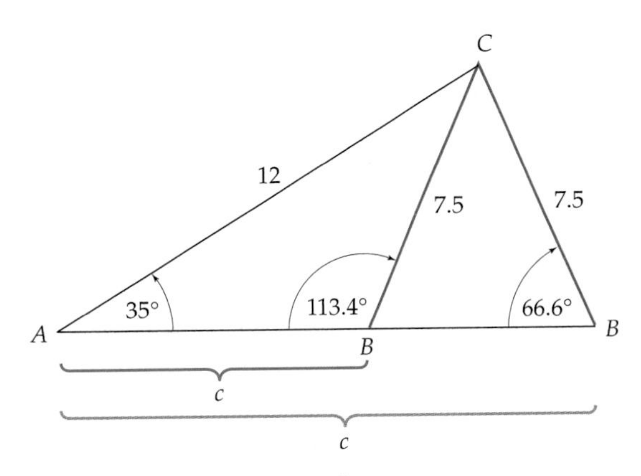

Figure 7–26

Case 1. If $B = 66.6^\circ$, then $C = 180^\circ - (35^\circ + 66.6^\circ) = 78.4^\circ$. By the Law of Sines,

$$\frac{c}{\sin C} = \frac{a}{\sin A}$$

$$\frac{c}{\sin 78.4^\circ} = \frac{7.5}{\sin 35^\circ}$$

$$c = \frac{7.5 \sin 78.4^\circ}{\sin 35^\circ} \approx \frac{7.5(.9796)}{.5736} \approx 12.8.$$

Case 2. If $B = 113.4^\circ$, then $C = 180^\circ - (35^\circ + 113.4^\circ) = 31.6^\circ$. Consequently,

$$\frac{c}{\sin C} = \frac{a}{\sin A}$$

$$\frac{c}{\sin 31.6^\circ} = \frac{7.5}{\sin 35^\circ}$$

$$c = \frac{7.5 \sin 31.6^\circ}{\sin 35^\circ} \approx \frac{7.5(.5240)}{.5736} \approx 6.9. \quad \blacksquare$$

Example 5 A plane flying in a straight line passes directly over point A on the ground and later directly over point B, which is 3 miles from A. A few minutes after the plane passes over B, the angle of elevation from A to the plane is 43° and the angle of elevation from B to the plane is 67°. How high is the plane at that moment?

Solution If C represents the plane, then the situation is represented in Figure 7–27. We must find the length of h.

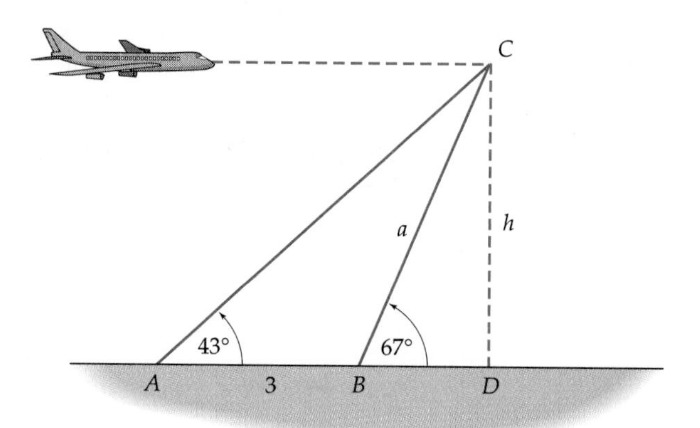

Figure 7–27

Note that angle ABC measures $180° - 67° = 113°$ and hence

$$\measuredangle BCA = 180° - (43° + 113°) = 24°.$$

Use the Law of Sines to find side a of triangle ABC:

$$\frac{a}{\sin 43°} = \frac{3}{\sin 24°}$$

$$a = \frac{3 \sin 43°}{\sin 24°} \approx 5.03.$$

Now in the right triangle CBD, we have

$$\sin 67° = \frac{\text{opposite}}{\text{hypotenuse}} = \frac{h}{a} \approx \frac{h}{5.03}.$$

Therefore, $h \approx 5.03 \sin 67° \approx 4.63$ miles. ■

The Area of a Triangle

The proof of the Law of Sines leads to this fact:

Area of a Triangle

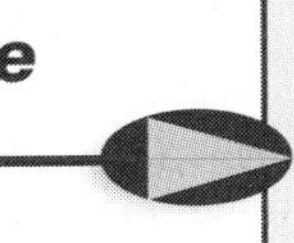

The area of a triangle containing an angle C with sides of lengths a and b is

$$\frac{1}{2}ab \sin C.$$

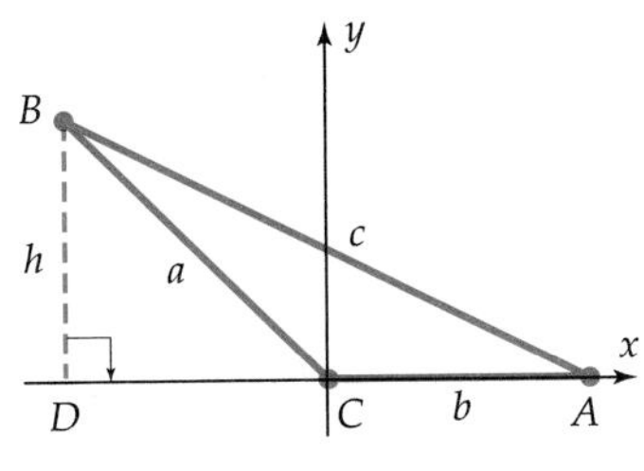

Figure 7–28

Proof Place the vertex of angle C at the origin, with side b on the positive x-axis (Figure 7–28).* Then b is the base and h is the altitude of the triangle so that

$$\text{area of triangle } ABC = \frac{1}{2} \times \text{base} \times \text{altitude} = \frac{1}{2} \cdot b \cdot h.$$

The proof of the Law of Sines on page 488 shows that $h = a \sin C$. Therefore,

$$\text{area of triangle } ABC = \frac{1}{2} \cdot b \cdot h = \frac{1}{2} \cdot b \cdot a \sin C = \frac{1}{2}ab \sin C. \quad \blacksquare$$

Example 6 Find the area of the triangle shown in Figure 7–29.

Solution

$$\frac{1}{2} \cdot 8 \cdot 13 \sin 130° \approx 39.83 \text{ square centimeters.} \quad \blacksquare$$

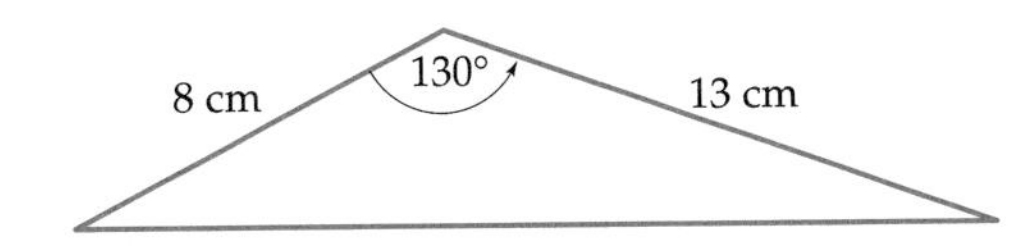

Figure 7–29

Here is a useful formula for the area of a triangle in terms of its sides:

Heron's Formula

The area of a triangle with sides a, b, c is

$$\sqrt{s(s-a)(s-b)(s-c)}$$

where $s = \frac{1}{2}(a + b + c)$.

*Figure 7–28 is the case when C is larger than 90°; the argument when C is less than 90° is similar.

Proof The preceding area formula and the Pythagorean identity

$$\sin^2 C = 1 - \cos^2 C = (1 + \cos C)(1 - \cos C)$$

show that the area of triangle ABC (standard notation) is

$$\begin{aligned} \frac{1}{2}ab\sin C &= \sqrt{\left(\frac{1}{2}ab\sin C\right)^2} = \sqrt{\frac{1}{4}a^2b^2\sin^2 C} \\ &= \sqrt{\frac{1}{4}a^2b^2(1 - \cos^2 C)} \\ &= \sqrt{\frac{1}{2}ab(1 + \cos C)\frac{1}{2}ab(1 - \cos C)}. \end{aligned}$$

Exercise 62 uses the Law of Cosines to show that

$$\begin{aligned} \frac{1}{2}ab(1 + \cos C) &= \frac{(a + b)^2 - c^2}{4} = \frac{(a + b) + c}{2} \cdot \frac{(a + b) - c}{2} \\ &= s(s - c) \end{aligned}$$

and

$$\begin{aligned} \frac{1}{2}ab(1 - \cos C) &= \frac{c^2 - (a - b)^2}{4} = \frac{c - (a - b)}{2} \cdot \frac{c + (a - b)}{2} \\ &= (s - a)(s - b). \end{aligned}$$

Combining these facts completes the proof:

$$\begin{aligned} \text{Area} = \frac{1}{2}ab\sin C &= \sqrt{\frac{1}{2}ab(1 + \cos C)\frac{1}{2}ab(1 - \cos C)} \\ &= \sqrt{s(s - a)(s - b)(s - c)}. \end{aligned}$$

■

Example 7 Find the area of the triangle whose sides have lengths 7, 9, and 12.

Solution Apply Heron's Formula with $a = 7$, $b = 9$, $c = 12$, and

$$s = \frac{1}{2}(a + b + c) = \frac{1}{2}(7 + 9 + 12) = 14.$$

The area is

$$\begin{aligned} \sqrt{s(s - a)(s - b)(s - c)} &= \sqrt{14(14 - 7)(14 - 9)(14 - 12)} \\ &= \sqrt{980} \approx 31.3 \text{ square units.} \end{aligned}$$

■

Exercises 7.3

Directions: *Standard notation for triangle ABC is used throughout. Use a calculator and round off your answers to one decimal place at the end of the computation.*

In Exercises 1–8, solve triangle ABC under the given conditions.

1. $A = 48°, B = 22°, a = 5$
2. $B = 33°, C = 46°, b = 4$
3. $A = 116°, C = 50°, a = 8$
4. $A = 105°, B = 27°, b = 10$
5. $B = 44°, C = 48°, b = 12$
6. $A = 67°, C = 28°, a = 9$
7. $A = 102.3°, B = 36.2°, a = 16$
8. $B = 97.5°, C = 42.5°, b = 7$

In Exercises 9–16, find the area of triangle ABC under the given conditions.

9. $a = 4, b = 8, C = 27°$
10. $b = 10, c = 14, A = 36°$
11. $c = 7, a = 10, B = 68°$
12. $a = 9, b = 13, C = 75°$
13. $a = 11, b = 15, c = 18$
14. $a = 4, b = 12, c = 14$
15. $a = 7, b = 9, c = 11$
16. $a = 17, b = 27, c = 40$

In Exercises 17–36, solve the triangle. The Law of Cosines may be needed in Exercises 27–36.

17. $b = 15, c = 25, B = 47°$
18. $b = 30, c = 50, C = 60°$
19. $a = 12, b = 5, B = 20°$
20. $b = 12.5, c = 20.1, B = 37.3°$
21. $a = 5, c = 12, A = 102°$
22. $a = 9, b = 14, B = 95°$
23. $b = 11, c = 10, C = 56°$
24. $a = 12.4, c = 6.2, A = 72°$
25. $A = 41°, B = 67° \; a = 10.5$
26. $a = 30, b = 40, A = 30°$
27. $b = 4, c = 10, A = 75°$
28. $a = 50, c = 80, C = 45°$
29: $a = 6, b = 12, c = 16$
30. $B = 20.67°, C = 34°, b = 185$
31. $a = 16.5, b = 18.2, C = 47°$
32. $a = 21, c = 15.8, B = 71°$
33. $b = 17.2, c = 12.4, B = 62.5°$
34. $b = 24.1, c = 10.5, C = 26.3°$
35. $a = 10.1, b = 18.2, A = 50.7°$
36. $b = 14.6, c = 7.8, B = 40.4°$

In Exercises 37 and 38, find the area of the triangle with the given vertices.

37. (0, 0), (2, −5), (−3, 1) **38.** (−4, 2), (5, 7), (3, 0)

In Exercises 39 and 40, find the area of the polygonal region. [Hint: *Divide the region into triangles.*]

39. 

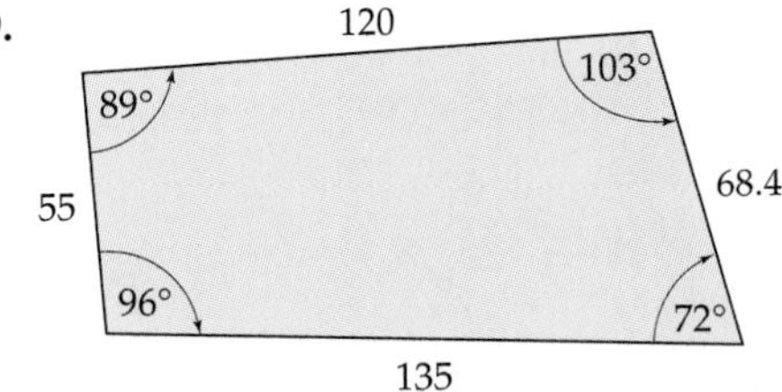

40. 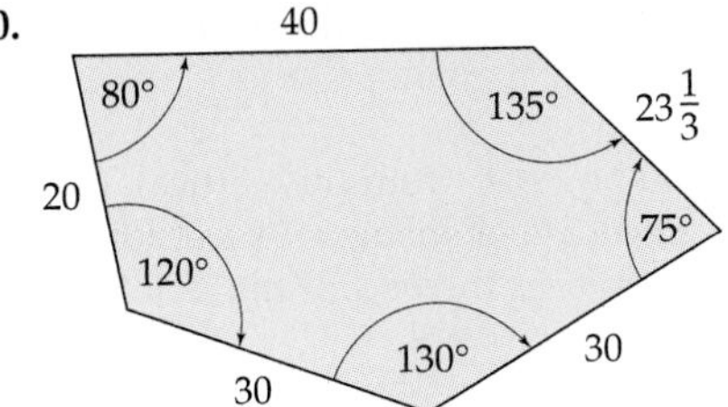

41. A surveyor marks points *A* and *B* 200 meters apart on one bank of a river. She sights a point *C* on the opposite bank and determines the angles shown in the figure. What is the distance from *A* to *C*?

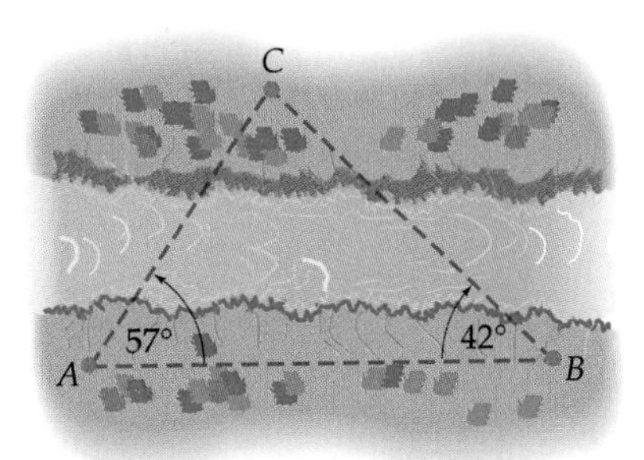

42. A forest fire is spotted from two fire towers. The triangle determined by the two towers and the fire has angles of 28° and 37° at the tower vertices. If the towers are 3000 meters apart, which one is closest to the fire?

43. A visitor to the Leaning Tower of Pisa observed that the tower's shadow was 40 meters long and that the angle of elevation from the tip of the shadow to the top of the tower was 57°. The tower is now 54 meters tall (measured from the ground to the top along the center line of the tower). Approximate the angle α that the center line of the tower makes with the vertical.

44. A pole tilts at an angle 9° from the vertical, away from the sun, and casts a shadow 24 feet long. The angle of elevation from the end of the pole's shadow to the top of the pole is 53°. How long is the pole?

45. A side view of a bus shelter is shown in the figure. The brace d makes an angle of 37.25° with the back and an angle of 34.85° with the top of the shelter. How long is this brace?

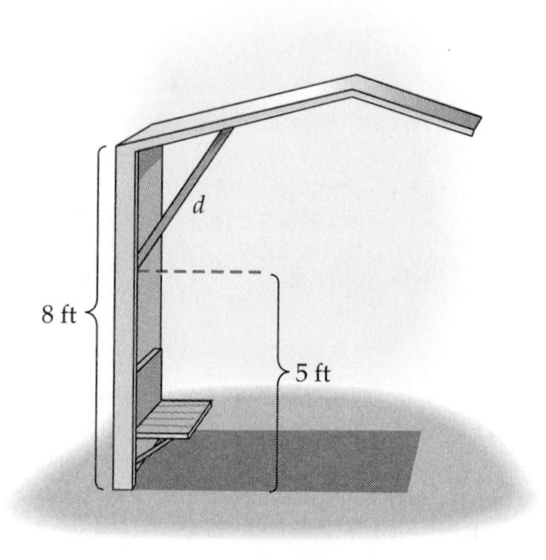

46. A straight path makes an angle of 6° with the horizontal. A statue at the higher end of the path casts a 6.5-meter-long shadow straight down the path. The angle of elevation from the end of the shadow to the top of the statue is 32°. How tall is the statue?

47. A vertical statue 6.3 meters high stands on top of a hill. At a point on the side of the hill 35 meters from the statue's base, the angle between the hillside and a line from the top of the statue is 10°. What angle does the side of the hill make with the horizontal?

48. A fence post is located 36 feet from one corner of a building and 40 feet from the adjacent corner. Fences are put up between the post and the building corners to form a triangular garden area. The 40-foot fence makes a 58° angle with the building. What is the area of the garden?

49. Two straight roads meet at an angle of 40° in Harville, one leading to Eastview and the other to Wellston. Eastview is 18 kilometers from Harville and 20 kilometers from Wellston. What is the distance from Harville to Wellston?

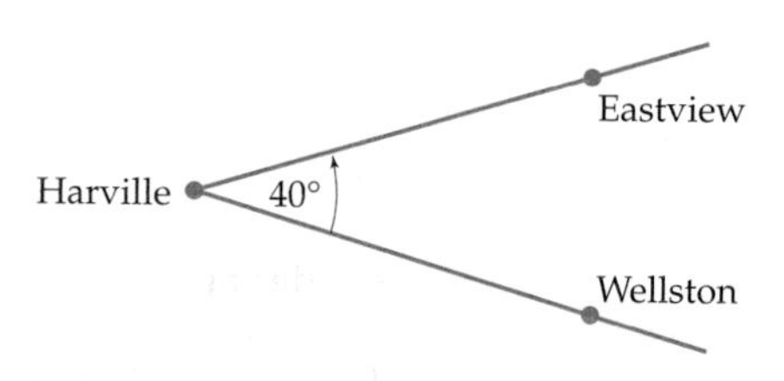

50. Each of two observers 400 feet apart measures the angle of elevation to the top of a tree that sits on the straight line between them. These angles are 51° and 65°, respectively. How tall is the tree? How far is the base of its trunk from each observer?

51. From the top of the 800-foot-tall Cartalk Tower, Tom sees a plane; the angle of elevation is 67°. At the same instant, Ray, who is on the ground, 1 mile from the building, notes that his angle of elevation to the plane is 81° and that his angle of elevation to the top of Cartalk Tower is 8.6°. Assuming that Tom and Ray and the airplane are in a plane perpendicular to the ground, how high is the airplane?

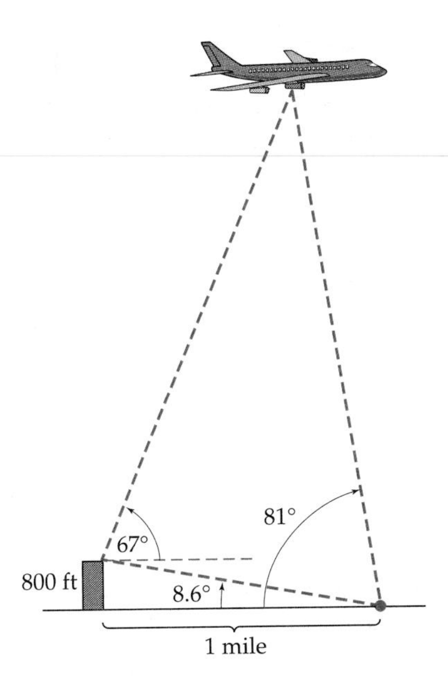

52. A plane flies in a direction of 105° from airport A. After a time, it turns and proceeds in a direction of 267°. Finally, it lands at airport B, 120 miles directly south of airport A. How far has the plane traveled? [*Note:* Aerial navigation directions are explained in Exercise 75 of Section 7.1.]

53. Charlie is afraid of water; he can't swim and refuses to get in a boat. However, he must measure the width of a river for his geography class. He has a long tape measure, but no way to measure angles. While pondering what to do, he paces along the side of the river using the five paths joining points A, B, C, and D. If he can't determine the width of the river, he will flunk the course.

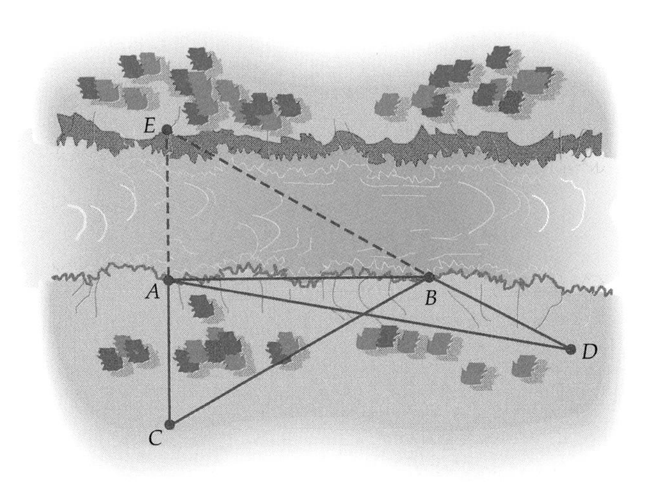

(a) Save Charlie from disaster by explaining how he can determine the width AE simply by measuring the lengths AB, AC, AD, BC, and BD and using trigonometry.

(b) Charlie determines that $AB = 75$ feet, $AC = 25$ feet, $AD = 90$ feet, $BC = 80$ feet, and $BD = 22$ feet. How wide is the river between A and E?

54. A plane flies in a direction of 85° from Chicago. It then turns and flies in the direction of 200° for 150 miles. It is then 195 miles from its starting point. How far did the plane fly in the direction of 85°? (See the note in Exercise 52.)

55. A hinged crane makes an angle of 50° with the ground. A malfunction causes the lock on the hinge to fail and the top part of the crane swings down. How far from the base of the crane does the top hit the ground?

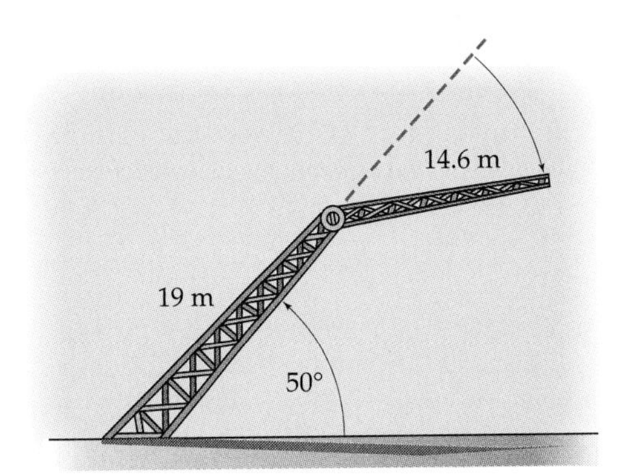

56. A triangular lot has sides of length 120 feet and 160 feet. The angle between these sides is 42°. Adjacent to this lot is a rectangular lot whose longest side has length 200 feet and whose shortest side is the same length as the shortest side of the triangular lot. What is the total area of both lots?

57. If a gallon of paint covers 400 square feet, how many gallons are needed to paint a triangular deck with sides of lengths 65 feet, 72 feet, and 88 feet?

58. Find the volume of the prism in the figure. The volume is given by the formula $V = \frac{1}{3}Bh$, where B is the area of the base and h is the height.

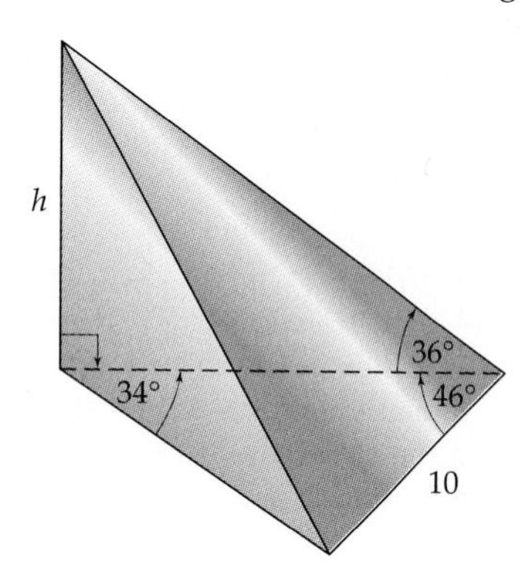

59. A rigid plastic triangle ABC rests on three vertical rods, as shown in the figure. What is its area?

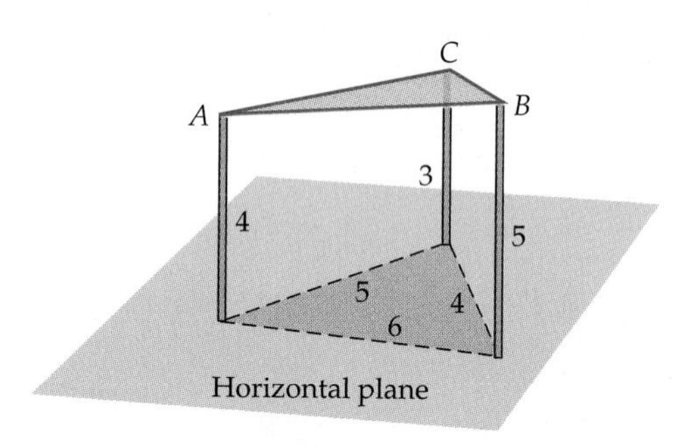

60. Prove that the area of triangle ABC (standard notation) is given by $\dfrac{a^2 \sin B \sin C}{2 \sin A}$.

61. What is the area of a triangle whose sides have lengths 12, 20, and 36? (If your answer turns out strangely, try drawing a picture.)

62. Complete the proof of Heron's Formula as follows. Let $s = \frac{1}{2}(a + b + c)$.

(a) Show that

$$\begin{aligned}\frac{1}{2}ab(1 + \cos C) &= \frac{(a + b)^2 - c^2}{4}\\ &= \frac{(a + b) + c}{2} \cdot \frac{(a + b) - c}{2}\\ &= s(s - c).\end{aligned}$$

[*Hint:* Use the Law of Cosines to express $\cos C$ in terms of a, b, c; then simplify.]

(b) Show that

$$\begin{aligned}\frac{1}{2}ab(1 - \cos C) &= \frac{c^2 - (a - b)^2}{4}\\ &= \frac{c - (a - b)}{2} \cdot \frac{c + (a - b)}{2}\\ &= (s - a)(s - b).\end{aligned}$$

Chapter 7 Review

Important Concepts

Important Facts and Formulas

- *Right Triangle Description:* In a right triangle containing an angle θ,

$$\sin\theta = \frac{\text{opposite}}{\text{hypotenuse}} \qquad \cos\theta = \frac{\text{adjacent}}{\text{hypotenuse}}$$

$$\tan\theta = \frac{\text{opposite}}{\text{adjacent}} \qquad \cot\theta = \frac{\text{adjacent}}{\text{opposite}}$$

$$\sec\theta = \frac{\text{hypotenuse}}{\text{adjacent}} \qquad \csc\theta = \frac{\text{hypotenuse}}{\text{opposite}}$$

- *Law of Cosines:* $a^2 = b^2 + c^2 - 2bc\cos A$
- *Law of Cosines—Alternate Form:* $\cos A = \dfrac{b^2 + c^2 - a^2}{2bc}$
- *Law of Sines:* $\dfrac{a}{\sin A} = \dfrac{b}{\sin B} = \dfrac{c}{\sin C}$
- $\sin D = \sin(180° - D)$
- Area of triangle $ABC = \dfrac{ab\sin C}{2}$
- *Heron's Formula:*

$$\text{Area of triangle } ABC = \sqrt{s(s-a)(s-b)(s-c)}$$

where $s = \dfrac{1}{2}(a + b + c)$.

Review Questions

Note: *Standard notation is used for triangles.*

1. Which of the following statements about the angle θ is true?

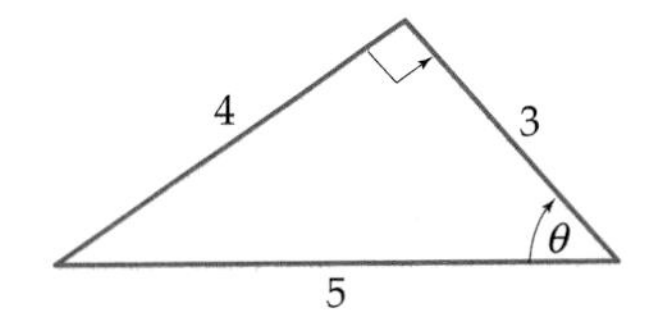

(a) $\sin\theta = 3/4$
(b) $\cos\theta = 5/4$
(c) $\tan\theta = 3/5$
(d) $\sin\theta = 4/5$
(e) $\sin\theta = 4/3$

2. Suppose θ is a real number. Consider the right triangle with sides as shown in the figure. Then:
   (a) $x = 1$
   (b) $x = 2$
   (c) $x = 4$
   (d) $x = 2(\cos\theta + \sin\theta)$
   (e) none of the above

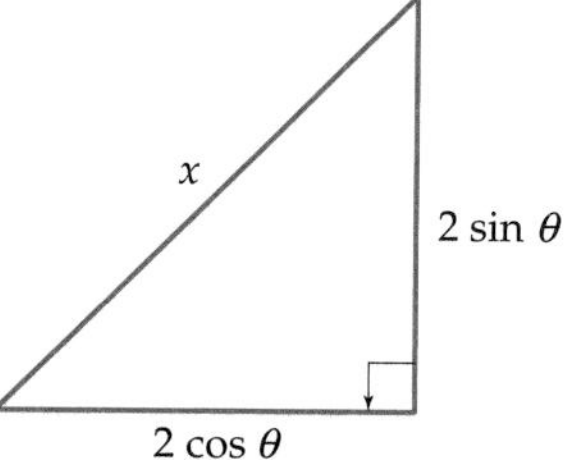

3. Use the right triangle in the figure to find $\sec\theta$.
   (a) $\sec\theta = 7/4$
   (b) $\sec\theta = 4/\sqrt{65}$
   (c) $\sec\theta = 7/\sqrt{65}$
   (d) $\sec\theta = \sqrt{65}/7$
   (e) $\sec\theta = \sqrt{65}/4$

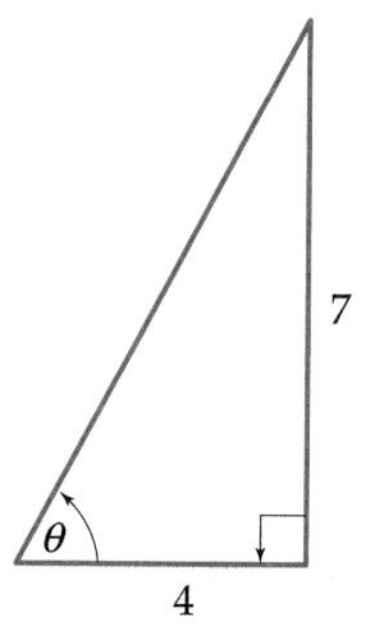

4. Find the length of side h in the triangle, where angle A measures 40° and the distance from C to A is 25.

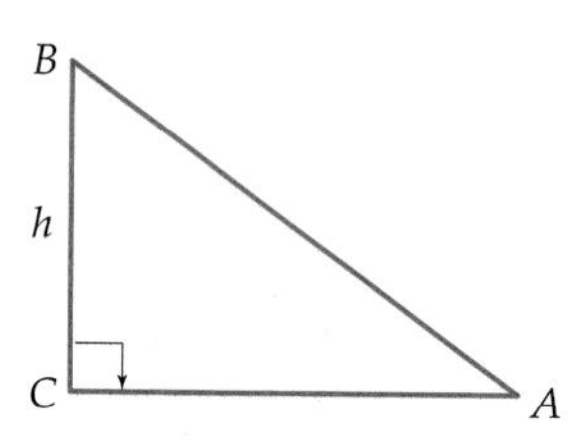

In Questions 5–8, angle B is a right angle. Solve triangle ABC.

5. $a = 12$, $c = 13$
6. $A = 40°$, $b = 10$
7. $C = 35°$, $a = 12$
8. $A = 56°$, $a = 11$

9. From a point on level ground 145 feet from the base of a tower, the angle of elevation to the top of the tower is 57.3°. How high is the tower?
10. A pilot in a plane at an altitude of 22,000 feet observes that the angle of depression to a nearby airport is 26°. How many *miles* is the airport from a point on the ground directly below the plane?
11. A road rises 140 feet per horizontal mile. What angle does the road make with the horizontal?
12. A lighthouse keeper 100 feet above the water sees a boat sailing in a straight line directly toward her. As she watches, the angle of depression to the boat changes from 25° to 40°. How far has the boat traveled during this time?

In Questions 13–16, use the Law of Cosines to solve triangle ABC.

13. $a = 12$, $b = 10$, $c = 15$
14. $a = 7.5$, $b = 3.2$, $c = 6.4$
15. $a = 10$, $c = 14$, $B = 130°$
16. $a = 7$, $b = 8.6$, $C = 72.4°$

17. Two trains depart simultaneously from the same station. The angle between the two tracks on which they leave is 120°. One train travels at an average

speed of 45 mph and the other at 70 mph. How far apart are the trains after 3 hours?

18. A 40-foot-high flagpole sits on the side of a hill. The hillside makes a 17° angle with the horizontal. How long is a wire that runs from the top of the pole to a point 72 feet downhill from the base of the pole?

In Questions 19–24, use the Law of Sines to solve triangle ABC.

19. $B = 124°, C = 31°, c = 3.5$

20. $A = 96°, B = 44°, b = 12$

21. $a = 75, c = 84, C = 62°$

22. $a = 5, c = 2.5, C = 30°$

23. $a = 3.5, b = 4, A = 60°$

24. $a = 3.8, c = 2.8, C = 41°$

25. Find the area of triangle ABC if $b = 24$, $c = 15$, and $A = 55°$.

26. Find the area of triangle ABC if $a = 10$, $c = 14$, and $B = 75°$.

27. A boat travels for 8 kilometers in a straight line from the dock. It is then sighted from a lighthouse which is 6.5 kilometers from the dock. The angle determined by the dock, the lighthouse (vertex), and the boat is 25°. How far is the boat from the lighthouse?

28. A pole tilts 12° from the vertical, away from the sun, and casts a 34-foot-long shadow on level ground. The angle of elevation from the end of the shadow to the top of the pole is 64°. How long is the pole?

In Questions 29–32, solve triangle ABC.

29. $A = 48°, B = 57°, b = 47$

30. $A = 67°, c = 125, a = 100$

31. $a = 5, c = 8, B = 76°$

32. $a = 90, b = 70, c = 40$

33. Two surveyors, Joe and Alice, are 240 meters apart on a riverbank. Each sights a flagpole on the opposite bank. The angle from the pole to Joe (vertex) to Alice is 63°. The angle from the pole to Alice (vertex) to Joe is 54°. How far are Joe and Alice from the pole?

34. A surveyor stakes out points A and B on opposite sides of a building. Point C on the side of the building is 300 feet from A and 440 feet from B. Angle ACB measures 38°. What is the distance from A to B?

35. A woman on the top of a 448-foot-high building spots a small plane. As she views the plane, its angle of elevation is 62°. At the same instant a man at the ground-level entrance to the building sees the plane and notes that its angle of elevation is 65°.

(a) How far is the woman from the plane?
(b) How far is the man from the plane?
(c) How high is the plane?

36. A straight road slopes at an angle of 10° with the horizontal. When the angle of elevation of the sun (from horizontal) is 62.5°, a telephone pole at the side of the road casts a 15-foot shadow downhill, parallel to the road. How high is the telephone pole?

37. Find angle ABC.

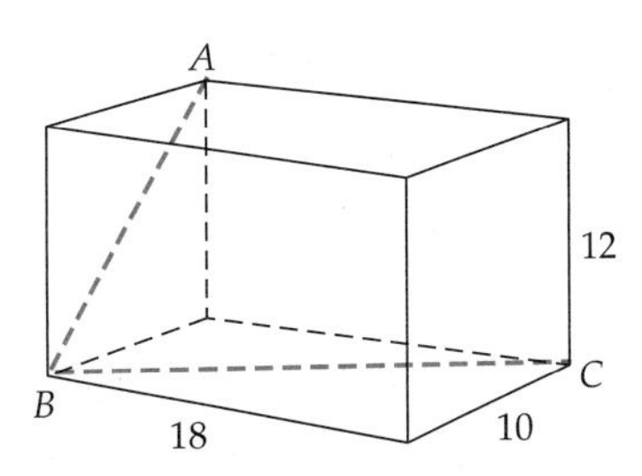

38. Use the Law of Sines to prove Engelsohn's equations: For any triangle ABC (standard notation),

$$\frac{a+b}{c} = \frac{\sin A + \sin B}{\sin C}$$

and

$$\frac{a-b}{c} = \frac{\sin A - \sin B}{\sin C}.$$

In Questions 39–42, find the area of triangle ABC under the given conditions.

39. There is an angle of 30°, the sides of which have lengths 5 and 8.

40. There is an angle of 40°, the sides of which have lengths 3 and 12.

41. The sides have lengths 7, 11, and 14.

42. The sides have lengths 4, 8, and 10.

Discovery Project 7

Life on a Sphere

Although we experience the surface of the earth as a level surface, it is not. The earth, of course, is a sphere whose radius is about 6370 kilometers. When you stand on a sphere, strange things happen, especially when the surface is particularly smooth.

Granted, the horizon is often interfered with by hills, forests, buildings, and such. However, if you are out in an area of flat land like that found in eastern Kansas or northern Ontario, the hills are very slight and the ground is very much like a sphere. The same is true of the large lakes and the ocean.

Generally speaking, you are not so interested in seeing the horizon as you are objects which the horizon may hide. On the ocean, for instance, you may want to see another ship. On land, you might be looking for a particular building or vehicle.

1. Suppose that your average human is standing on a highway and a 3.5-meter-high truck passes by. Assume also that your average human has eyes that are 1.55 meters off the ground. If the road is flat and straight, how far away is the truck when it disappears below the horizon?

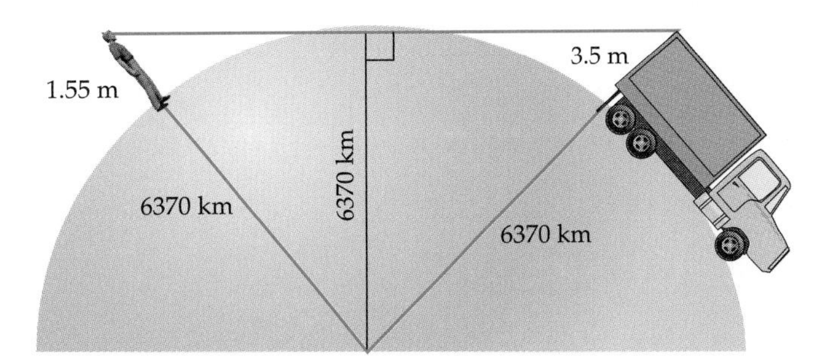

2. Engineers building large bridges must also account for the curvature of the earth. In the figure, a bridge with towers 200 meters tall and 900 meters apart (straight line distance) at the base is constructed. How much farther apart are the tops of the towers than the bases?

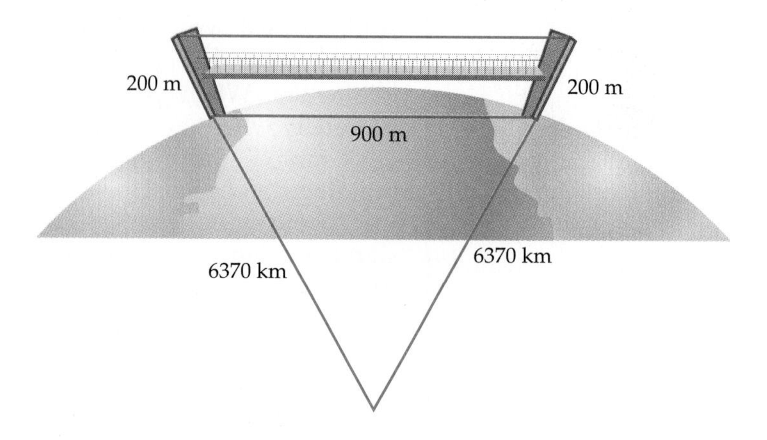

3. How much difference does it make if the distance between the towers is measured along the curve of the earth?

Trigonometric Identities and Equations

Beam me up, Scotty!

When a light beam passes from one medium to another (for instance, from air to water), it changes both its speed and direction. If you know what some of these numbers are (say the speed of light in air, or the angle at which a light beam hits the water), then you can determine the unknown ones by solving a trigonometric equation. See Exercises 41–44 on page 547.

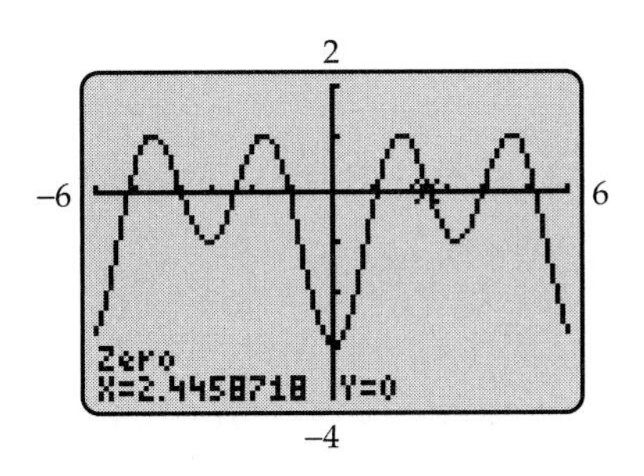

Interdependence of Chapters

Chapters 8 and 9 are independent of each other and may be read in either order. The interdependence of the sections of this chapter is shown below. Sections 8.1, 8.4, and 8.5 are independent of one another and may be read in any order.

8.1 ⟶ 8.2 ⟶ 8.3

8.4

8.5

Chapter Outline

Until now the variable t has been used for trigonometric functions, to avoid confusion with the x's and y's that appear in their definitions. Now that you are comfortable with these functions, we shall usually use the letter x (or occasionally y) for the variable. Unless stated otherwise, all trigonometric functions in this chapter are considered as functions of real numbers, rather than functions of angles in degree measure.

Two kinds of trigonometric equations are considered here. *Identities* (Sections 8.1–8.3) are equations that are valid for all values of the variable for which the equation is defined, such as

$$\sin^2 x + \cos^2 x = 1 \qquad \text{and} \qquad \cot x = \frac{1}{\tan x}.$$

Conditional equations (Section 8.4) are valid only for certain values of the variable, such as $\sin x = 0$ and $\cos x = \frac{1}{2}$. Inverse trigonometric functions are discussed in Section 8.5.

8.1 Basic Identities and Proofs

Trigonometric identities can be used for simplifying expressions, rewriting the rule of a trigonometric function, performing numerical calculations, and in other ways. There are no hard and fast rules for dealing with identities, but some suggestions are given below. The phrases "prove the identity" and "verify the identity" mean "prove that the given equation is an identity."

Graphical Testing

When presented with a trigonometric equation that *might* be an identity, it is a good idea to determine graphically whether or not this is possible. For instance, in Example 6 on page 425 we tested the equation $\cos\left(\frac{\pi}{2} + t\right) = \sin t$ by graphing the functions $f(t) = \cos\left(\frac{\pi}{2} + t\right)$ and

$g(t) = \sin t$ on the same screen. Since the graphs were different, we concluded that the equation was not an identity.

Any equation can be tested in the same way, by simultaneously graphing the two functions whose rules are given by the left and right sides of the equation. If the graphs are different, the equation is not an identity. If the graphs appear to be the same, then it is *possible* that the equation is an identity. However,

The fact that the graphs appear identical does not prove that the equation is an identity, as the following exploration demonstrates.

GRAPHING EXPLORATION

In the viewing window with $-\pi \le x \le \pi$ and $-2 \le y \le 2$, graph both sides of the equation

$$\cos x = 1 - \frac{x^2}{2} + \frac{x^4}{24} - \frac{x^6}{720} + \frac{x^8}{40{,}320}.$$

Do the graphs appear identical? Now change the viewing window so that $-2\pi \le x \le 2\pi$. Is the equation an identity?

Example 1 Is either of the following equations an identity?

(a) $2\sin^2 x - \cos x = 2\cos^2 x + \sin x$

(b) $\dfrac{1 + \sin x - \sin^2 x}{\cos x} = \cos x + \tan x$

Solution

(a) Test the equation graphically to see if it might be an identity.

GRAPHING EXPLORATION

Graph the functions $f(x) = 2\sin^2 x - \cos x$ and $g(x) = 2\cos^2 x + \sin x$ on the same screen, using a viewing window with $-2\pi \le x \le 2\pi$. Do the graphs appear identical? Is the equation an identity?

(b) Test the second equation graphically.

GRAPHING EXPLORATION

Graph the functions $f(x) = \dfrac{1 + \sin x - \sin^2 x}{\cos x}$ and $g(x) = \cos x + \tan x$ on the same screen, using a viewing window with $-2\pi \le x \le 2\pi$. Do the graphs appear identical?

The exploration suggests that the equations *may* be an identity, but the proof that it actually is an identity must be done algebraically. ■

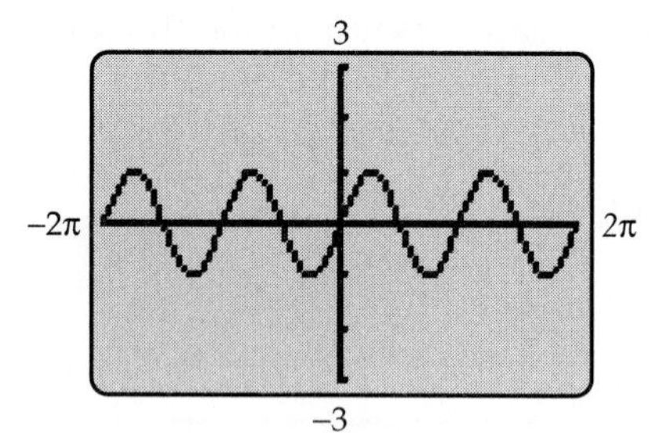

Figure 8–1

Example 2 Find an equation involving $2 \sin x \cos x$ that could possibly be an identity.

Solution First graph $y = 2\sin x \cos x$ and examine the graph (Figure 8–1). Does it look familiar? At first glance it looks like the graph of $\sin x$, but there's an important difference. The function graphed in Figure 8–1 has period π (a complete wave runs from $x = 0$ to $x = \pi$). As we saw in Section 6.5, the graph of $y = \sin 2x$ has waves like the sine graph and period π (see the box on page 429).

GRAPHING EXPLORATION

Graph $y = 2\sin x \cos x$ and $y = \sin 2x$ on the same screen. Do the graphs appear identical? What identity does this suggest? ■

Proving Identities

In the proofs presented below, we shall assume the elementary identities that were proved in Chapter 6 and are summarized here.

Basic Trigonometric Identities

Reciprocal Identities

$$\sec x = \frac{1}{\cos x} \qquad \csc x = \frac{1}{\sin x}$$

$$\tan x = \frac{\sin x}{\cos x} \qquad \cot x = \frac{\cos x}{\sin x}$$

$$\cot x = \frac{1}{\tan x} \qquad \tan x = \frac{1}{\cot x}$$

Periodicity Identities

$$\sin(x + 2\pi) = \sin x \qquad \cos(x + 2\pi) = \cos x$$

$$\csc(x + 2\pi) = \csc x \qquad \sec(x + 2\pi) = \sec x$$

$$\tan(x + \pi) = \tan x \qquad \cot(x + \pi) = \cot x$$

Pythagorean Identities

$$\sin^2 x + \cos^2 x = 1 \qquad 1 + \tan^2 x = \sec^2 x \qquad 1 + \cot^2 x = \csc^2 x$$

Negative Angle Identities

$$\sin(-x) = -\sin x \qquad \cos(-x) = \cos x \qquad \tan(-x) = -\tan x$$

There are no cut-and-dried rules for simplifying trigonometric expressions or proving identities, but there are some common strategies that are often helpful. Four of these strategies are illustrated in the following examples. There are often a variety of ways to proceed and it will take some practice before you can easily decide which strategies are likely to be the most efficient in a particular case.

Strategy 1

Express everything in terms of sine and cosine.

Example 3 Simplify $(\csc x + \cot x)(1 - \cos x)$.

Solution Using Strategy 1, we have

$$(\csc x + \cot x)(1 - \cos x)$$

$$= \left(\frac{1}{\sin x} + \frac{\cos x}{\sin x}\right)(1 - \cos x) \qquad \textit{[Reciprocal identities]}$$

$$= \frac{(1 + \cos x)}{\sin x}(1 - \cos x)$$

$$= \frac{(1 + \cos x)(1 - \cos x)}{\sin x}$$

$$= \frac{1 - \cos^2 x}{\sin x} = \frac{\sin^2 x}{\sin x} \qquad \textit{[Pythagorean identity]}$$

$$= \sin x. \quad \blacksquare$$

Strategy 2

Use algebra and identities to transform the expression on one side of the equal sign into the expression on the other side.*

Example 4 In Example 1 we verified graphically that the equation:

$$\frac{1 + \sin x - \sin^2 x}{\cos x} = \cos x + \tan x$$

might be an identity. Prove that it is.

*That is, start with expression A on one side and use identities and algebra to produce a sequence of equalities $A = B$, $B = C$, $C = D$, $D = E$, where E is the other side of the identity to be proved; conclude that $A = E$.

Solution We use Strategy 2, beginning with the left side of the equation:

$$\frac{1 + \sin x - \sin^2 x}{\cos x} = \frac{(1 - \sin^2 x) + \sin x}{\cos x}$$

$$= \frac{\cos^2 x + \sin x}{\cos x} \qquad \textit{[Pythagorean identity]}$$

$$= \frac{\cos^2 x}{\cos x} + \frac{\sin x}{\cos x}$$

$$= \cos x + \frac{\sin x}{\cos x}$$

$$= \cos x + \tan x. \quad \blacksquare$$

The strategies presented above and those to be considered below are "plans of attack." By themselves they are not much help unless you also have some techniques for carrying out these plans. In the previous examples we used the techniques of basic algebra and the use of known identities to change trigonometric expressions to equivalent ones. Here is another technique that is often useful when dealing with fractions:

Rewrite a fraction in equivalent form by multiplying its numerator and denominator by the same quantity.

Example 5 Prove that $\dfrac{\sin x}{1 + \cos x} = \dfrac{1 - \cos x}{\sin x}$.

Solution We shall use Strategy 2, beginning with the left side, whose denominator is $1 + \cos x$. Multiply its numerator and denominator by $1 - \cos x$:*

$$\frac{\sin x}{1 + \cos x} = \frac{\sin x}{1 + \cos x} \cdot \frac{1 - \cos x}{1 - \cos x} = \frac{\sin x(1 - \cos x)}{(1 + \cos x)(1 - \cos x)}$$

$$= \frac{\sin x(1 - \cos x)}{1 - \cos^2 x}$$

$$= \frac{\sin x(1 - \cos x)}{\sin^2 x} \qquad \textit{[Pythagorean identity]}$$

$$= \frac{1 - \cos x}{\sin x}.$$

*This is analogous to the process used to rationalize the denominator of a fraction by multiplying its numerator and denominator by the conjugate of the denominator, as in this example:

$$\frac{1}{3 + \sqrt{2}} = \frac{1}{3 + \sqrt{2}} \cdot \frac{3 - \sqrt{2}}{3 - \sqrt{2}} = \frac{3 - \sqrt{2}}{3^2 - (\sqrt{2})^2} = \frac{3 - \sqrt{2}}{7}.$$

Alternate Solution The numerators of the given equation look similar to the Pythagorean identity—with the squares missing. So we begin with the left side and introduce some squares by multiplying it by $\frac{\sin x}{\sin x} = 1$:

$$\frac{\sin x}{1+\cos x} = 1 \cdot \frac{\sin x}{1+\cos x} = \frac{\sin x}{\sin x} \cdot \frac{\sin x}{1+\cos x} = \frac{\sin^2 x}{\sin x(1+\cos x)}$$

$$= \frac{1-\cos^2 x}{\sin x(1+\cos x)} \qquad \textit{[Pythagorean identity]}$$

$$= \frac{(1-\cos x)(1+\cos x)}{\sin x(1+\cos x)} \qquad \textit{[Factor numerator]}$$

$$= \frac{1-\cos x}{\sin x}. \quad \blacksquare$$

Strategy 3

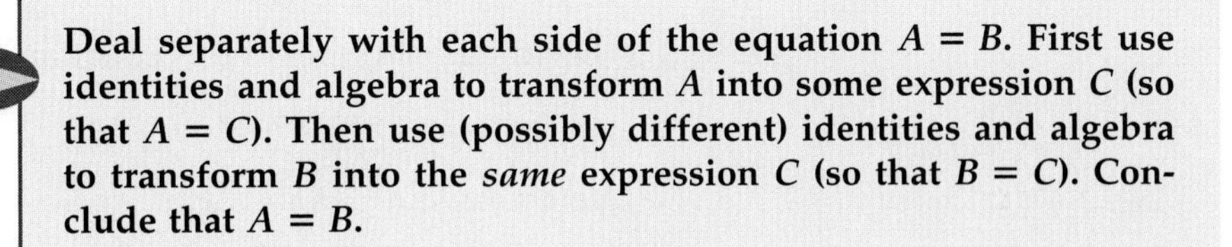

Deal separately with each side of the equation $A = B$. First use identities and algebra to transform A into some expression C (so that $A = C$). Then use (possibly different) identities and algebra to transform B into the *same* expression C (so that $B = C$). Conclude that $A = B$.

Example 6 Prove that $\csc x - \cot x = \frac{\sin x}{1+\cos x}$.

Solution We use Strategy 3, together with Strategy 1, beginning with the left side:

$$(*) \qquad \csc x - \cot x = \frac{1}{\sin x} - \frac{\cos x}{\sin x} = \frac{1-\cos x}{\sin x}.$$

Example 5 shows that the right side of the identity to be proved can also be transformed into this same expression:

$$(**) \qquad \frac{\sin x}{1+\cos x} = \frac{1-\cos x}{\sin x}.$$

Combining the equalities (*) and (**) proves the identity:

$$\csc x - \cot x = \frac{1-\cos x}{\sin x} = \frac{\sin x}{1+\cos x}. \quad \blacksquare$$

Proving identities involving fractions can sometimes be quite complicated. It often helps to approach a fractional identity indirectly, as in the following example.

Example 7 Prove these identities:

(a) $\sec x(\sec x - \cos x) = \tan^2 x$ (b) $\dfrac{\sec x}{\tan x} = \dfrac{\tan x}{\sec x - \cos x}$.

Solution

(a) Beginning with the left side and using Strategy 1, we have

$$\begin{aligned}\sec x(\sec x - \tan x) &= \sec^2 x - \sec x \cos x\\ &= \sec^2 x - \frac{1}{\cos x}\cos x\\ &= \sec^2 x - 1\\ &= \tan^2 x \qquad \textit{[Pythagorean identity]}\end{aligned}$$

Therefore, $\sec x(\sec x - \cos x) = \tan^2 x$.

(b) By part (a), we know that

$$\sec x(\sec x - \cos x) = \tan^2 x$$

Dividing both sides of this equation by $\tan x(\sec x - \cos x)$ show that

$$\frac{\sec x(\sec x - \cos x)}{\tan x(\sec x - \cos x)} = \frac{\tan^2 x}{\tan x(\sec x - \cos x)}$$

$$\frac{\sec x(\sec x - \cos x)}{\tan x(\sec x - \cos x)} = \frac{\tan x \tan x}{\tan x(\sec x - \cos x)}$$

$$\frac{\sec x}{\tan x} = \frac{\tan x}{\sec x - \cos x}. \quad ■$$

Look carefully at how identity (b) was proved in Example 7. We first proved identity (a), which is of the form $AD = BC$ (with $A = \sec x$, $B = \tan x$, $C = \tan x$, and $D = \sec x - \cos x$). Then we divided both sides by BD (that is, by $\tan x(\sec x - \cos x)$ to conclude that $A/B = C/D$. The same argument works in general and provides the following useful strategy for dealing with identities involving fractions.

Strategy 4

If you can prove that $AD = BC$, with $B \neq 0$ and $D \neq 0$, then you can conclude that

$$\frac{A}{B} = \frac{C}{D}.$$

Many students misunderstand Strategy 4: It does *not* say that you begin with a fractional equation $A/B = C/D$ and cross multiply to eliminate the fractions. If you did that, you'd be assuming what has to be proved. What the strategy says is that to prove an identity involving fractions, you need only prove a different identity that does not involve fractions. In other words, if you prove that $AD = BD$ whenever $B \neq 0$ and $D \neq 0$, then you can conclude that $A/B = C/D$. Note that you do not *assume* that $AD = BC$; you use Strategy 2 or 3 or some other means to *prove* this statement.

Example 8 Prove that $\dfrac{\cot x - 1}{\cot x + 1} = \dfrac{1 - \tan x}{1 + \tan x}$.

Solution We use Strategy 4, with $A = \cot x - 1$, $B = \cot x + 1$, $C = 1 - \tan x$, and $D = 1 + \tan x$. We must prove that this equation is an identity:

$$AD = BC$$

$$(***) \qquad (\cot x - 1)(1 + \tan x) = (\cot x + 1)(1 - \tan x).$$

Strategy 3 will be used. Multiplying out the left side shows that

$$\begin{aligned}(\cot x - 1)(1 + \tan x) &= \cot x - 1 + \cot x \tan x - \tan x \\ &= \cot x - 1 + \frac{1}{\tan x}\tan x - \tan x \\ &= \cot x - 1 + 1 - \tan x \\ &= \cot x - \tan x.\end{aligned}$$

Similarly, on the right side of (***):

$$\begin{aligned}(\cot x + 1)(1 - \tan x) &= \cot x + 1 - \cot x \tan x - \tan x \\ &= \cot x + 1 - 1 - \tan x \\ &= \cot x - \tan x.\end{aligned}$$

Since the left and right sides are equal to the same expression, we have proved that (***) is an identity. Therefore, by Strategy 4, we conclude that

$$\frac{\cot x - 1}{\cot x + 1} = \frac{1 - \tan x}{1 + \tan x}$$

is also an identity. ■

Technology Tip

Using SOLVE in the TI-89 ALGEBRA menu to solve an equation that might be an identity produces one of three responses. "True" means the equation *probably* is an identity [algebraic proof is required for certainty]. "False" means the equation is not an identity. A numerical answer is inconclusive [the equation may or may not be an identity].

It takes a good deal of practice, as well as *much* trial and error, to become proficient in proving identities. The more practice you have, the easier it will get. Since there are many correct methods, your proofs may be quite different from those of your instructor or the text answers.

If you don't see what to do immediately, try something and see where it leads: Multiply out or factor or multiply numerator and denominator by the same nonzero quantity. Even if this doesn't lead anywhere, it may give you some ideas on other things to try. When you do obtain a proof, check to see if it can be done more efficiently. In your final proof, don't include the side trips that may have given you some ideas but aren't themselves part of the proof.

Exercises 8.1

In Exercises 1–4, test the equation graphically to determine whether it might be an identity. You need not prove those equations that seem to be identities.

1. $\dfrac{\sec x - \cos x}{\sec x} = \sin^2 x$
2. $\tan x + \cot x = (\sin x)(\cos x)$
3. $\dfrac{1 - \cos(2x)}{2} = \sin^2 x$
4. $\dfrac{\tan x + \cot x}{\csc x} = \sec x$

In Exercises 5–8, insert one of A–F on the right of the equal sign so that the resulting equation appears to be an identity when you test it graphically. You need not prove the identity.

A. $\cos x$ **B.** $\sec x$ **C.** $\sin^2 x$

D. $\sec^2 x$ **E.** $\sin x - \cos x$ **F.** $\dfrac{1}{\sin x \cos x}$

5. $\csc x \tan x =$ ____
6. $\dfrac{\sin x}{\tan x} =$ ____
7. $\dfrac{\sin^4 x - \cos^4 x}{\sin x + \cos x} =$ ____
8. $\tan^2(-x) - \dfrac{\sin(-x)}{\sin x} =$ ____

In Exercises 9–18, prove the identity.

9. $\tan x \cos x = \sin x$
10. $\cot x \sin x = \cos x$
11. $\cos x \sec x = 1$
12. $\sin x \csc x = 1$
13. $\tan x \csc x = \sec x$
14. $\sec x \cot x = \csc x$
15. $\dfrac{\tan x}{\sec x} = \sin x$
16. $\dfrac{\cot x}{\csc x} = \cos x$
17. $(1 + \cos x)(1 - \cos x) = \sin^2 x$
18. $(\csc x - 1)(\csc x + 1) = \cot^2 x$

In Exercises 19–48, state whether or not the equation is an identity. If it is an identity, prove it.

19. $\sin x = \sqrt{1 - \cos^2 x}$
20. $\cot x = \dfrac{\csc x}{\sec x}$
21. $\dfrac{\sin(-x)}{\cos(-x)} = -\tan x$
22. $\tan x = \sqrt{\sec^2 x - 1}$
23. $\cot(-x) = -\cot x$
24. $\sec(-x) = \sec x$
25. $1 + \sec^2 x = \tan^2 x$
26. $\sec^4 x - \tan^4 x = 1 + 2\tan^2 x$
27. $\sec^2 x - \csc^2 x = \tan^2 x - \cot^2 x$
28. $\sec^2 x + \csc^2 x = \sec^2 x \csc^2 x$
29. $\sin^2 x(\cot x + 1)^2 = \cos^2 x(\tan x + 1)^2$
30. $\cos^2 x(\sec x + 1)^2 = (1 + \cos x)^2$
31. $\sin^2 x - \tan^2 x = -\sin^2 x \tan^2 x$
32. $\cot^2 x - 1 = \csc^2 x$
33. $(\cos^2 x - 1)(\tan^2 x + 1) = -\tan^2 x$
34. $(1 - \cos^2 x)\csc x = \sin x$
35. $\tan x = \dfrac{\sec x}{\csc x}$
36. $\dfrac{\cos(-x)}{\sin(-x)} = -\cot x$
37. $\cos^4 x - \sin^4 x = \cos^2 x - \sin^2 x$
38. $\cot^2 x - \cos^2 x = \cos^2 x \cot^2 x$
39. $(\sin x + \cos x)^2 = \sin^2 x + \cos^2 x$
40. $(1 + \tan x)^2 = \sec^2 x$
41. $\dfrac{\sec x}{\csc x} + \dfrac{\sin x}{\cos x} = 2\tan x$
42. $\dfrac{1 + \cos x}{\sin x} + \dfrac{\sin x}{1 + \cos x} = 2\csc x$
43. $\dfrac{\sec x + \csc x}{1 + \tan x} = \csc x$
44. $\dfrac{\cot x - 1}{1 - \tan x} = \dfrac{\csc x}{\sec x}$
45. $\dfrac{1}{\csc x - \sin x} = \sec x \tan x$
46. $\dfrac{1 + \csc x}{\csc x} = \dfrac{\cos^2 x}{1 - \sin x}$
47. $\dfrac{\sin x - \cos x}{\tan x} = \dfrac{\tan x}{\sin x + \cos x}$
48. $\dfrac{\cot x}{\csc x - 1} = \dfrac{\csc x + 1}{\cot x}$

In Exercises 49–52, half of an identity is given. Graph this half in a viewing window with $-2\pi \le x \le 2\pi$ and make a conjecture as to what the right side of the identity is. Then prove your conjecture.

49. $1 - \dfrac{\sin^2 x}{1 + \cos x} = ?$ [*Hint:* What familiar function has a graph that looks like this?]
50. $\dfrac{1 + \cos x - \cos^2 x}{\sin x} - \cot x = ?$
51. $(\sin x + \cos x)(\sec x + \csc x) - \cot x - 2 = ?$
52. $\cos^3 x(1 - \tan^4 x + \sec^4 x) = ?$

In Exercises 53–66, prove the identity.

53. $\dfrac{1 - \sin x}{\sec x} = \dfrac{\cos^3 x}{1 + \sin x}$

54. $\dfrac{\sin x}{1 - \cot x} + \dfrac{\cos x}{1 - \tan x} = \cos x + \sin x$

55. $\dfrac{\cos x}{1 - \sin x} = \sec x + \tan x$

56. $\dfrac{1 + \sec x}{\tan x + \sin x} = \csc x$

57. $\dfrac{\cos x \cot x}{\cot x - \cos x} = \dfrac{\cot x + \cos x}{\cos x \cot x}$

58. $\dfrac{\cos^3 x - \sin^3 x}{\cos x - \sin x} = 1 + \sin x \cos x$

59. $\log_{10}(\cot x) = -\log_{10}(\tan x)$

60. $\log_{10}(\sec x) = -\log_{10}(\cos x)$

61. $\log_{10}(\csc x + \cot x) = -\log_{10}(\csc x - \cot x)$

62. $\log_{10}(\sec x + \tan x) = -\log_{10}(\sec x - \tan x)$

63. $\tan x - \tan y = -\tan x \tan y(\cot x - \cot y)$

64. $\dfrac{\tan x - \tan y}{\cot x - \cot y} = -\tan x \tan y$

65. $\dfrac{\cos x - \sin y}{\cos y - \sin x} = \dfrac{\cos y + \sin x}{\cos x + \sin y}$

66. $\dfrac{\tan x + \tan y}{\cot x + \cot y} = \dfrac{\tan x \tan y - 1}{1 - \cot x \cot y}$

8.2 Addition and Subtraction Identities

A common student ERROR is to write

$$\sin\left(x + \frac{\pi}{6}\right) = \sin x + \sin\frac{\pi}{6} = \sin x + \frac{1}{2}.$$

GRAPHING EXPLORATION

Verify graphically that the equation above is NOT an identity by graphing $y = \sin(x + \pi/6)$ and $y = \sin x + 1/2$ on the same screen.

The exploration shows that $\sin(x + y) = \sin x + \sin y$ is NOT an identity (because it's false when $y = \pi/6$). So, is there an identity involving $\sin(x + y)$ and $\sin x$ and $\sin y$?

GRAPHING EXPLORATION

Graph $y = \sin\left(x + \dfrac{\pi}{6}\right)$ and $y = \dfrac{\sqrt{3}}{2}\sin x + \dfrac{1}{2}\cos x$ on the same screen. Do the graphs appear identical?

The exploration suggests that

$$\sin\left(x + \frac{\pi}{6}\right) = \frac{\sqrt{3}}{2}\sin x + \frac{1}{2}\cos x$$

is an identity. Furthermore, note that the coefficients on the right side can be expressed in terms of $\pi/6$: $\frac{\sqrt{3}}{2} = \cos\frac{\pi}{6}$ and $\frac{1}{2} = \sin\frac{\pi}{6}$. In other words, the following equation appears to be an identity

$$\sin\left(x + \frac{\pi}{6}\right) = \sin x\left(\cos\frac{\pi}{6}\right) + \cos x\left(\sin\frac{\pi}{6}\right).$$

Is there something special about $\pi/6$, or would we get the same result with another number?

GRAPHING EXPLORATION

Graph $y = \sin(x + 5)$ and $y = (\sin x)(\cos 5) + (\cos x)(\sin 5)$ on the same screen. Do the graphs appear identical? What identity does this suggest? Repeat the process with some other number in place of 5. Are the results the same?

The equations examined in the discussion and exploration above are examples of the first identity listed below (they are the cases when $y = \pi/6$ and when $y = 5$).

Addition and Subtraction Identities

$$\sin(x + y) = \sin x \cos y + \cos x \sin y$$

$$\sin(x - y) = \sin x \cos y - \cos x \sin y$$

$$\cos(x + y) = \cos x \cos y - \sin x \sin y$$

$$\cos(x - y) = \cos x \cos y + \sin x \sin y$$

GRAPHING EXPLORATION

Confirm the last identity listed above graphically in the case when $y = 3$ by graphing $\cos(x - 3)$ and $(\cos x)(\cos 3) + (\sin x)(\sin 3)$ on the same screen.

The addition and subtraction identities are probably the most important of all the trigonometric identities. Before reading their proofs at the end of this section, you should become familiar with the examples and special cases below.

Example 1 Use the addition identities to find the *exact* value of $\sin(5\pi/12)$ and $\cos(5\pi/12)$.

Solution Since

$$\frac{5\pi}{12} = \frac{2\pi}{12} + \frac{3\pi}{12} = \frac{\pi}{6} + \frac{\pi}{4},$$

we apply the addition identities with $x = \pi/6$ and $y = \pi/4$:

$$\sin\frac{5\pi}{12} = \sin\left(\frac{\pi}{6} + \frac{\pi}{4}\right) = \sin\frac{\pi}{6}\cos\frac{\pi}{4} + \cos\frac{\pi}{6}\sin\frac{\pi}{4}$$

$$= \frac{1}{2}\cdot\frac{\sqrt{2}}{2} + \frac{\sqrt{3}}{2}\cdot\frac{\sqrt{2}}{2} = \frac{\sqrt{2}(\sqrt{3}+1)}{4}$$

$$\cos\frac{5\pi}{12} = \cos\left(\frac{\pi}{6} + \frac{\pi}{4}\right) = \cos\frac{\pi}{6}\cos\frac{\pi}{4} - \sin\frac{\pi}{6}\sin\frac{\pi}{4}$$

$$= \frac{\sqrt{3}}{2}\cdot\frac{\sqrt{2}}{2} - \frac{1}{2}\cdot\frac{\sqrt{2}}{2} = \frac{\sqrt{2}(\sqrt{3}-1)}{4}. \quad ■$$

Example 2 Find $\sin(\pi - y)$.

Solution Apply the subtraction identity with $x = \pi$:

$$\sin(\pi - y) = \sin\pi\cos y - \cos\pi\sin y$$

$$= (0)(\cos y) - (-1)(\sin y)$$

$$= \sin y. \quad ■$$

Example 3 Show that for the function $f(x) = \sin x$ and any number $h \neq 0$,

$$\frac{f(x+h) - f(x)}{h} = \sin x\left(\frac{\cos h - 1}{h}\right) + \cos x\left(\frac{\sin h}{h}\right).$$

[This fact is needed in calculus.]

Solution Use the addition identity for $\sin(x + y)$ with $y = h$:

$$\frac{f(x+h) - f(x)}{h} = \frac{\sin(x+h) - \sin x}{h}$$

$$= \frac{\sin x\cos h + \cos x\sin h - \sin x}{h}$$

$$= \frac{\sin x(\cos h - 1) + \cos x\sin h}{h}$$

$$= \sin x\left(\frac{\cos h - 1}{h}\right) + \cos x\left(\frac{\sin h}{h}\right). \quad ■$$

Example 4 Prove that

$$\cos x\cos y = \frac{1}{2}[\cos(x+y) + \cos(x-y)].$$

Solution We begin with the more complicated right side and use the addition and subtraction identities for cosine to transform it into the left side:

$$\begin{aligned}\frac{1}{2}[\cos(x+y)+\cos(x-y)] &= \frac{1}{2}[(\cos x\cos y-\sin x\sin y)\\ &\qquad +(\cos x\cos y+\sin x\sin y)]\\ &= \frac{1}{2}(\cos x\cos y+\cos x\cos y)\\ &= \frac{1}{2}(2\cos x\cos y)=\cos x\cos y. \quad \blacksquare\end{aligned}$$

The addition and subtraction identities for sine and cosine can be used to obtain

Addition and Subtraction Identities for Tangent

$$\tan(x+y)=\frac{\tan x+\tan y}{1-\tan x\tan y}$$

$$\tan(x-y)=\frac{\tan x-\tan y}{1+\tan x\tan y}$$

A proof of these identities is outlined in Exercise 36.

Example 5 Suppose x and y are numbers such that $0 < x < \pi/2$ and $\pi < y < 3\pi/2$. If $\sin x = 3/4$ and $\cos y = -1/3$, find the exact values of $\sin(x+y)$ and $\tan(x+y)$ and determine in which of the following intervals $x+y$ lies: $(0, \pi/2)$, $(\pi/2, \pi)$, $(\pi, 3\pi/2)$, $(3\pi/2, 2\pi)$.

Solution Using the Pythagorean identity and the fact that $\cos x$ and $\tan x$ are positive when $0 < x < \pi/2$, we have

$$\cos x=\sqrt{1-\sin^2 x}=\sqrt{1-\left(\frac{3}{4}\right)^2}=\sqrt{1-\frac{9}{16}}=\sqrt{\frac{7}{16}}=\frac{\sqrt{7}}{4}$$

$$\tan x=\frac{\sin x}{\cos x}=\frac{3/4}{\sqrt{7}/4}=\frac{3}{4}\cdot\frac{4}{\sqrt{7}}=\frac{3}{\sqrt{7}}=\frac{3\sqrt{7}}{7}.$$

Since y lies between π and $3\pi/2$, its sine is negative; hence,

$$\sin y=-\sqrt{1-\cos^2 y}=-\sqrt{1-\left(-\frac{1}{3}\right)^2}=-\sqrt{\frac{8}{9}}=-\frac{\sqrt{8}}{3}=-\frac{2\sqrt{2}}{3}$$

$$\tan y=\frac{\sin y}{\cos y}=\frac{-2\sqrt{2}/3}{-1/3}=\frac{-2\sqrt{2}}{3}\cdot\frac{3}{-1}=2\sqrt{2}.$$

The addition identities for sine and tangent now show that

$$\sin(x+y) = \sin x \cos y + \cos x \sin y$$
$$= \frac{3}{4}\cdot\frac{-1}{3} + \frac{\sqrt{7}}{4}\cdot\frac{-2\sqrt{2}}{3} = \frac{-1}{4} - \frac{2\sqrt{14}}{12} = \frac{-3-2\sqrt{14}}{12}$$

$$\tan(x+y) = \frac{\tan x + \tan y}{1 - \tan x \tan y}$$
$$= \frac{\frac{3\sqrt{7}}{7} + 2\sqrt{2}}{1 - \left(\frac{3\sqrt{7}}{7}\right)(2\sqrt{2})} = \frac{\frac{3\sqrt{7}+14\sqrt{2}}{7}}{\frac{7-6\sqrt{14}}{7}} = \frac{3\sqrt{7}+14\sqrt{2}}{7-6\sqrt{14}}.$$

So both the sine and tangent of $x + y$ are negative numbers. Therefore, $x + y$ must be in the interval $(3\pi/2, 2\pi)$ since the sign chart in Exercise 49 on page 408 shows that this is the only one of the four intervals in which both sine and tangent are negative. ■

Cofunction Identities

Other special cases of the addition and subtraction identities are the cofunction identities:

Cofunction Identities

$$\sin x = \cos\left(\frac{\pi}{2} - x\right) \qquad \cos x = \sin\left(\frac{\pi}{2} - x\right)$$
$$\tan x = \cot\left(\frac{\pi}{2} - x\right) \qquad \cot x = \tan\left(\frac{\pi}{2} - x\right)$$
$$\sec x = \csc\left(\frac{\pi}{2} - x\right) \qquad \csc x = \sec\left(\frac{\pi}{2} - x\right)$$

The first cofunction identity is proved by using the identity for $\cos(x - y)$ with $\pi/2$ in place of x and x in place of y:

$$\cos\left(\frac{\pi}{2} - x\right) = \cos\frac{\pi}{2}\cos x + \sin\frac{\pi}{2}\sin x = (0)(\cos x) + (1)(\sin x) = \sin x.$$

Since the first cofunction identity is valid for *every* number x, it is also valid with the number $\pi/2 - x$ in place of x:

$$\sin\left(\frac{\pi}{2} - x\right) = \cos\left[\frac{\pi}{2} - \left(\frac{\pi}{2} - x\right)\right] = \cos x.$$

Thus, we have proved the second cofunction identity. The others now follow from these two. For instance,

$$\tan\left(\frac{\pi}{2} - x\right) = \frac{\sin[(\pi/2) - x]}{\cos[(\pi/2) - x]} = \frac{\cos x}{\sin x} = \cot x.$$

Example 6 Verify that $\dfrac{\cos(x - \pi/2)}{\cos x} = \tan x$.

Solution Beginning on the left side, we see that the term $\cos(x - \pi/2)$ looks almost, but not quite, like the term $\cos(\pi/2 - x)$ in the cofunction identity. But note that $-(x - \pi/2) = \pi/2 - x$. Therefore,

$$\frac{\cos\left(x - \frac{\pi}{2}\right)}{\cos x} = \frac{\cos\left[-\left(x - \frac{\pi}{2}\right)\right]}{\cos x} \qquad \textit{[Negative angle identity with } x - \tfrac{\pi}{2} \textit{ in place of } x\textit{]}$$

$$= \frac{\cos\left(\frac{\pi}{2} - x\right)}{\cos x}$$

$$= \frac{\sin x}{\cos x} \qquad \textit{[Cofunction identity]}$$

$$= \tan x. \qquad \textit{[Reciprocal identity]}$$ ■

Proof of the Addition and Subtraction Identities

We first prove that

$$\cos(x - y) = \cos x \cos y + \sin x \sin y$$

If $x = y$, then this is true by the Pythagorean identity:

$$\cos(x - x) = \cos 0 = 1 = \cos^2 x + \sin^2 x = \cos x \cos x + \sin x \sin x.$$

Next we prove the identity in the case when $x > y$. Let P be the point where the terminal side of an angle of x radians in standard position meets the unit circle and let Q be the point where the terminal side of an angle of y radians in standard position meets the circle, as shown in Figure 8–2. According to the definitions of sine and cosine, P has coordinates $(\cos x, \sin x)$ and Q has coordinates $(\cos y, \sin y)$.

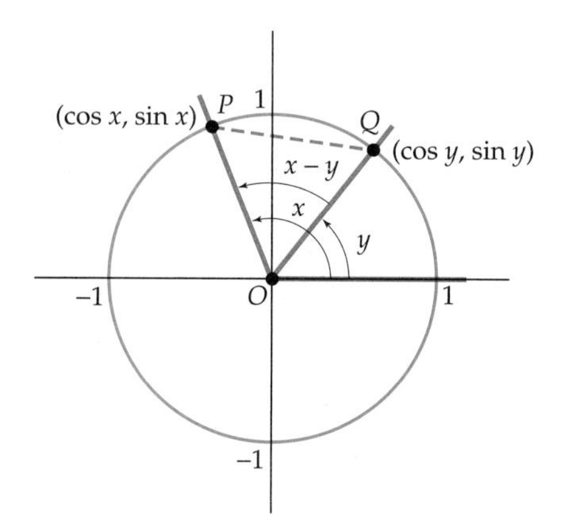

Figure 8–2

The angle QOP formed by the two terminal sides has radian measure $x - y$. Rotate this angle clockwise until side OQ lies on the horizontal axis, as shown in Figure 8–3. Angle QOP is now in standard position, and its terminal side meets the unit circle at P. Since angle QOP has radian measure $x - y$, the definitions of sine and cosine show that the point P, in this new location, has coordinates $(\cos(x - y), \sin(x - y))$. Q now has coordinates $(1, 0)$.

Using the coordinates of P and Q *before* the angle was rotated and the distance formula, we have:

distance from P to Q

$$= \sqrt{(\cos x - \cos y)^2 + (\sin x - \sin y)^2}$$

$$= \sqrt{\cos^2 x - 2\cos x \cos y + \cos^2 y + \sin^2 x - 2\sin x \sin y + \sin^2 y}$$

$$= \sqrt{(\cos^2 x + \sin^2 x) + (\cos^2 y + \sin^2 y) - 2\cos x \cos y - 2\sin x \sin y}$$

$$= \sqrt{1 + 1 - 2\cos x \cos y - 2\sin x \sin y}$$

$$= \sqrt{2 - 2\cos x \cos y - 2\sin x \sin y}.$$

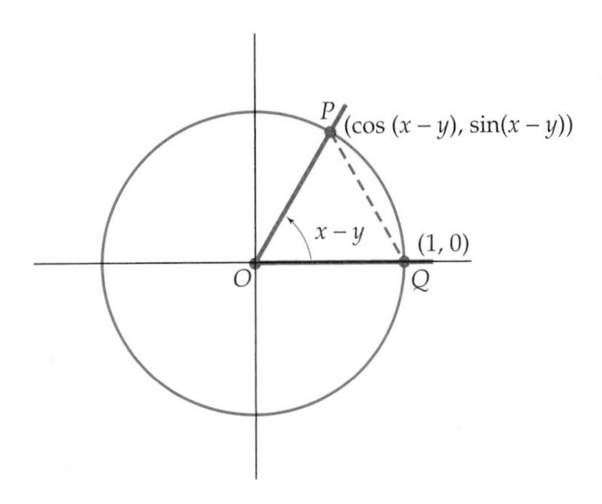

Figure 8–3

But using the coordinates of P and Q *after* the angle is rotated shows that

$$\text{distance from } P \text{ to } Q = \sqrt{[\cos(x - y) - 1]^2 + [\sin(x - y) - 0]^2}$$

$$= \sqrt{\cos^2(x - y) - 2\cos(x - y) + 1 + \sin^2(x - y)}$$

$$= \sqrt{\cos^2(x - y) + \sin^2(x - y) - 2\cos(x - y) + 1}$$

$$= \sqrt{1 - 2\cos(x - y) + 1} \qquad \textit{[Pythagorean identity]}$$

$$= \sqrt{2 - 2\cos(x - y)}.$$

The two expressions for the distance from P to Q must be equal. Hence,

$$\sqrt{2 - 2\cos(x - y)} = \sqrt{2 - 2\cos x \cos y - 2\sin x \sin y}.$$

Squaring both sides of this equation and simplifying the result yields:

$$\begin{aligned} 2 - 2\cos(x - y) &= 2 - 2\cos x \cos y - 2\sin x \sin y \\ -2\cos(x - y) &= -2(\cos x \cos y + \sin x \sin y) \\ \cos(x - y) &= \cos x \cos y + \sin x \sin y. \end{aligned}$$

This completes the proof of the last addition identity when $x > y$. If $y > x$, then the proof just given is valid with the roles of x and y interchanged; it shows that

$$\begin{aligned} \cos(y - x) &= \cos y \cos x + \sin y \sin x \\ &= \cos x \cos y + \sin x \sin y. \end{aligned}$$

The negative angle identity with $x - y$ in place of x shows that

$$\cos(x - y) = \cos[-(x - y)] = \cos(y - x).$$

Combining this fact with the previous one shows that

$$\cos(x - y) = \cos x \cos y + \sin x \sin y$$

in this case also. Therefore, the last addition identity is proved.

The identity for $\cos(x + y)$ now follows from one for $\cos(x - y)$ by using the negative angle identities for sine and cosine:

$$\begin{aligned} \cos(x + y) &= \cos[x - (-y)] = \cos x \cos(-y) + \sin x \sin(-y) \\ &= \cos x \cos y + (\sin x)(-\sin y) = \cos x \cos y - \sin x \sin y. \end{aligned}$$

The proof of the first two cofunction identities on page 521 depended only on the addition identity for $\cos(x - y)$. Since that has been proved, we can validly use the first two cofunction identities in the remainder of the proof. In particular:

$$\sin(x - y) = \cos\left[\frac{\pi}{2} - (x - y)\right] = \cos\left[\left(\frac{\pi}{2} - x\right) + y\right].$$

Applying the proven identity for $\cos(x + y)$ with $(\pi/2) - x$ in place of x and the two cofunction identities now yields

$$\begin{aligned} \sin(x - y) &= \cos\left[\left(\frac{\pi}{2} - x\right) + y\right] \\ &= \cos\left(\frac{\pi}{2} - x\right)\cos y - \sin\left(\frac{\pi}{2} - x\right)\sin y \\ &= \sin x \cos y - \cos x \sin y. \end{aligned}$$

This proves the second of the addition and subtraction identities. The first is obtained from the second in the same way the third was obtained from the last.

Exercises 8.2

In Exercises 1–12, find the exact value.

1. $\sin \frac{\pi}{12}$ **2.** $\cos \frac{\pi}{12}$ **3.** $\tan \frac{\pi}{12}$

4. $\sin \frac{5\pi}{12}$ **5.** $\cot \frac{5\pi}{12}$ **6.** $\cos \frac{7\pi}{12}$

7. $\tan \frac{7\pi}{12}$ **8.** $\cos \frac{11\pi}{12}$ **9.** $\cot \frac{11\pi}{12}$

10. $\sin 75°$ [*Hint:* $75° = 45° + 30°$.]

11. $\sin 105°$ **12.** $\cos 165°$

In Exercises 13–18, rewrite the given expression in terms of $\sin x$ and $\cos x$.

13. $\sin\left(\frac{\pi}{2} + x\right)$ **14.** $\cos\left(x + \frac{\pi}{2}\right)$

15. $\cos\left(x - \frac{3\pi}{2}\right)$ **16.** $\csc\left(x + \frac{\pi}{2}\right)$

17. $\sec(x - \pi)$ **18.** $\cot(x + \pi)$

In Exercises 19–24, simplify the given expression.

19. $\sin 3 \cos 5 - \cos 3 \sin 5$

20. $\sin 37° \sin 53° - \cos 37° \cos 53°$

21. $\cos(x + y) \cos y + \sin(x + y) \sin y$

22. $\sin(x - y) \cos y + \cos(x - y) \sin y$

23. $\cos(x + y) - \cos(x - y)$

24. $\sin(x + y) - \sin(x - y)$

25. If $\sin x = \frac{1}{3}$ and $0 < x < \frac{\pi}{2}$, then $\sin\left(\frac{\pi}{4} + x\right) = ?$

26. If $\cos x = -\frac{1}{4}$ and $\frac{\pi}{2} < x < \pi$, then $\cos\left(\frac{\pi}{6} - x\right) = ?$

27. If $\cos x = -\frac{1}{5}$ and $\pi < x < \frac{3\pi}{2}$, then $\sin\left(\frac{\pi}{3} - x\right) = ?$

28. If $\sin x = -\frac{3}{4}$ and $\frac{3\pi}{2} < x < 2\pi$, then $\cos\left(\frac{\pi}{4} + x\right) = ?$

In Exercises 29–34, assume $\sin x = .8$ and $\sin y = \sqrt{.75}$ and that x, y lie between 0 and $\pi/2$. Evaluate the given expressions.

29. $\cos(x + y)$ **30.** $\sin(x + y)$ **31.** $\cos(x - y)$

32. $\sin(x - y)$ **33.** $\tan(x + y)$ **34.** $\tan(x - y)$

35. If $f(x) = \cos x$ and h is a fixed nonzero number, prove that:

$$\frac{f(x + h) - f(x)}{h} = \cos x\left(\frac{\cos h - 1}{h}\right) - \sin x\left(\frac{\sin h}{h}\right).$$

36. Prove the addition and subtraction identities for the tangent function (page 520). [*Hint:* $\tan(x + y) = \frac{\sin(x + y)}{\cos(x + y)}$. Use the addition identities on the numerator and denominator; then divide both numerator and denominator by $\cos x \cos y$ and simplify.]

37. If x is in the first and y is in the second quadrant, $\sin x = 24/25$, and $\sin y = 4/5$, find the exact value of $\sin(x + y)$ and $\tan(x + y)$ and the quadrant in which $x + y$ lies.

38. If x and y are in the second quadrant, $\sin x = 1/3$, and $\cos y = -3/4$, find the exact value of $\sin(x + y)$, $\cos(x + y)$, $\tan(x + y)$, and find the quadrant in which $x + y$ lies.

39. If x is in the first and y is in the second quadrant, $\sin x = 4/5$, and $\cos y = -12/13$, find the exact value of $\cos(x + y)$ and $\tan(x + y)$ and the quadrant in which $x + y$ lies.

40. If x is in the fourth and y is in the first quadrant, $\cos x = 1/3$, and $\cos y = 2/3$, find the exact value of $\sin(x - y)$ and $\tan(x - y)$ and the quadrant in which $x - y$ lies.

41. Express $\sin(u + v + w)$ in terms of sines and cosines of u, v, and w. [*Hint:* First apply the addition identity with $x = u + v$ and $y = w$.]

42. Express $\cos(x + y + z)$ in terms of sines and cosines of x, y, and z.

43. If $x + y = \pi/2$, show that $\sin^2 x + \sin^2 y = 1$.

44. Prove that $\cot(x + y) = \frac{\cot x \cot y - 1}{\cot x + \cot y}$.

In Exercises 45–56, prove the identity.

45. $\sin(x - \pi) = -\sin x$ **46.** $\cos(x - \pi) = -\cos x$

47. $\cos(\pi - x) = -\cos x$ **48.** $\tan(\pi - x) = -\tan x$

49. $\sin(x + \pi) = -\sin x$ **50.** $\cos(x + \pi) = -\cos x$

51. $\tan(x + \pi) = \tan x$

52. $\sin x \cos y = \frac{1}{2}[\sin(x + y) + \sin(x - y)]$

53. $\sin x \sin y = \frac{1}{2}[\cos(x - y) - \cos(x + y)]$

54. $\cos x \sin y = \frac{1}{2}[\sin(x + y) - \sin(x - y)]$

55. $\cos(x + y)\cos(x - y) = \cos^2 x \cos^2 y - \sin^2 x \sin^2 y$

56. $\sin(x + y)\sin(x - y) = \sin^2 x \cos^2 y - \cos^2 x \sin^2 y$

In Exercises 57–66, determine graphically whether the equation could possibly be an identity (by choosing a numerical value for y and graphing both sides). If it could, prove that it is.

57. $\dfrac{\cos(x - y)}{\sin x \cos y} = \cot x + \tan y$

58. $\dfrac{\cos(x + y)}{\sin x \cos y} = \cot x - \tan y$

59. $\sin(x - y) = \sin x - \sin y$

60. $\cos(x + y) = \cos x + \cos y$

61. $\dfrac{\sin(x + y)}{\sin(x - y)} = \dfrac{\tan x + \tan y}{\tan x - \tan y}$

62. $\dfrac{\sin(x + y)}{\sin(x - y)} = \dfrac{\cot y + \cot x}{\cot y - \cot x}$

63. $\dfrac{\cos(x + y)}{\cos(x - y)} = \dfrac{\cot x + \tan y}{\cot x - \tan y}$

64. $\dfrac{\cos(x - y)}{\cos(x + y)} = \dfrac{\cot y + \tan x}{\cot y - \tan x}$

65. $\tan(x + y) = \tan x + \tan y$

66. $\cot(x - y) = \cot x - \cot y$

8.2.A EXCURSION Lines and Angles

Several interesting concepts dealing with lines are defined in terms of trigonometry. They lead to some useful facts whose proofs are based on the addition and subtraction identities for sine, cosine, and tangent.

If L is a nonhorizontal straight line, the **angle of inclination** of L is the angle θ formed by the part of L above the x-axis and the x-axis in the positive direction, as shown in Figure 8–4.

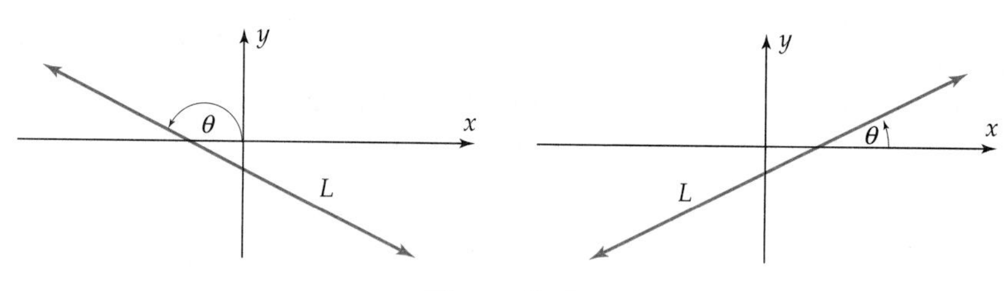

Figure 8–4

The angle of inclination of a horizontal line is defined to be $\theta = 0$. Thus, the radian measure of the angle of inclination of any line satisfies $0 \le \theta < \pi$. Furthermore,

Angle of Inclination

If L is a nonvertical line with angle of inclination θ, then

$$\tan \theta = \text{slope of } L.$$

Proof If L is horizontal, then L has slope 0 and angle of inclination $\theta = 0$. Hence, $\tan \theta = \tan 0 = 0 = \text{slope } L$. If L is not horizontal, then it intersects the x-axis at some point $(x_1, 0)$, as shown for two possible cases in Figure 8–5. Let (x_2, y_2) be any point on L above the x-axis.

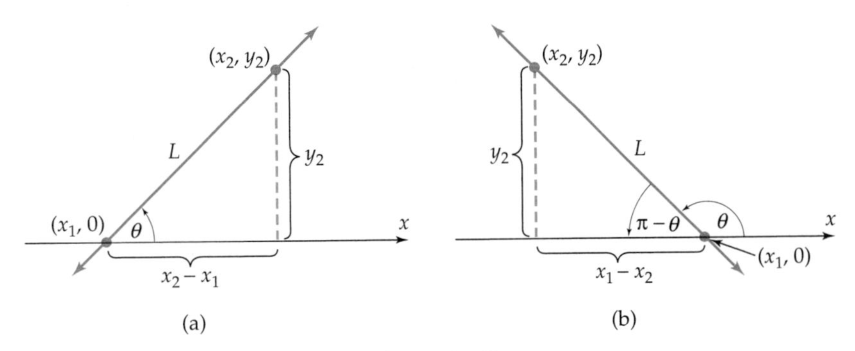

Figure 8–5

The right triangle in Figure 8–5(a) shows that

$$\text{slope of } L = \frac{y_2 - 0}{x_2 - x_1} = \frac{y_2}{x_2 - x_1} = \frac{\text{opposite}}{\text{adjacent}} = \tan\theta.$$

The right triangle in Figure 8–5(b) shows that

$$(*) \quad \text{slope of } L = \frac{0 - y_2}{x_1 - x_2} = -\frac{y_2}{x_1 - x_2} = -\frac{\text{opposite}}{\text{adjacent}} = -\tan(\pi - \theta).$$

Using the fact that the tangent function has period π and the negative angle identity for tangent, we have

$$-\tan(\pi - \theta) = -\tan(-\theta) = -(-\tan\theta) = \tan\theta.$$

Combining this fact with $(*)$ shows that slope of $L = \tan\theta$ in this case also. ■

Example 1 Find the angle of inclination θ of a line of slope $5/3$.

Solution We know that $\tan\theta = 5/3$. The TAN^{-1} key on a calculator shows that $\theta \approx 1.03$ radians (equivalently, $59.04°$). ■

A line with negative slope moves downward from left to right, and hence its angle of inclination lies between $\pi/2$ and π radians (as illustrated in Figure 8–5(b)).

Example 2 Find the angle of inclination of a line L with slope -2.

Solution Since a line L has slope -2, then its angle of inclination is a solution of $\tan\theta = -2$ that lies between $\pi/2$ and π. A calculator shows that an approximate solution is -1.10715 and we know that $\tan t = \tan(t + \pi)$ for every t (see page 424). Hence, $\tan(-1.10715 + \pi) \approx -2$, Since $-1.10715 + \pi \approx 2.03444$ is between $\pi/2$ and π, the angle of inclination of L is $\theta = 2.03444$ radians (equivalently, $116.565°$). ■

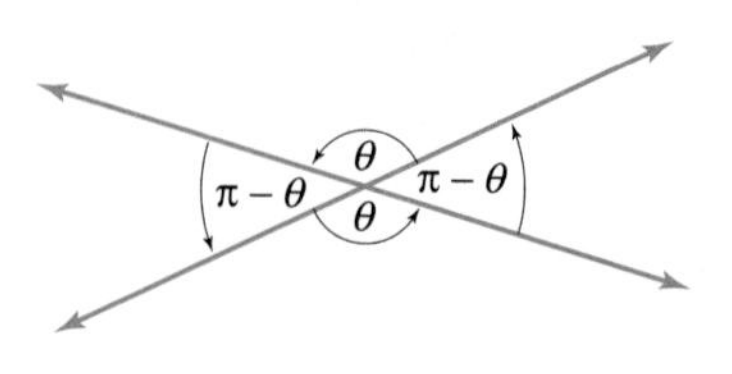

Figure 8–6

Angles Between Two Lines

If two lines intersect, then they determine four angles with vertex at the point of intersection, as shown in Figure 8–6. If one of these angles measures θ radians, then each of the two angles adjacent to it measures $\pi - \theta$ radians. (Why?) The fourth angle also measures θ radians by the vertical angle theorem from plane geometry.

The angles between intersecting lines can be determined from the angles of inclination of the lines. Suppose L and M have angles of inclination α and β, respectively, such that $\beta \geq \alpha$. Basic facts about parallel lines, as illustrated in Figure 8–7, show that $\beta - \alpha$ is one angle between L and M and $\pi - (\beta - \alpha)$ is the other one.*

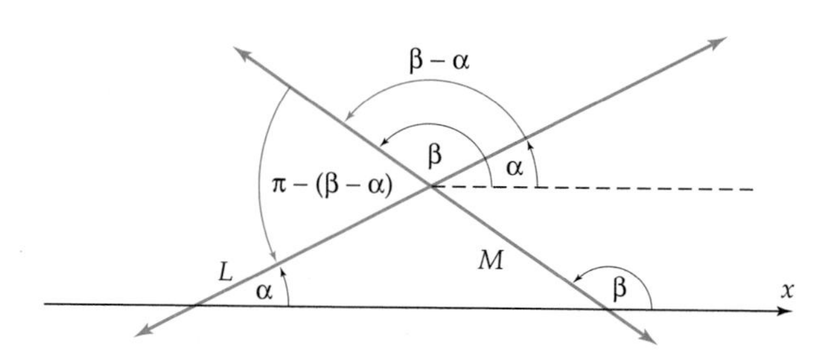

Figure 8–7

The angle between two lines can also be found from their slopes by using this fact:

Angle Between Two Lines

If two nonvertical, nonperpendicular lines have slopes m and k, then one angle θ between them satisfies

$$\tan\theta = \left|\frac{m-k}{1+mk}\right|.$$

Proof Suppose L has slope k and angle of inclination α and that M has slope m and angle of inclination β. By the definition of absolute value

$$\left|\frac{m-k}{1+mk}\right| = \frac{m-k}{1+mk} \qquad \text{or} \qquad \left|\frac{m-k}{1+mk}\right| = -\frac{m-k}{1+mk}$$

whichever is positive. We shall show that one angle between L and M has tangent $\frac{m-k}{1+mk}$ and that the other one has tangent $-\frac{m-k}{1+mk}$. Thus, one

*Figure 8–7 illustrates one case. Analogous figures in the other possible cases lead to the same conclusion.

of them necessarily has tangent $\left|\frac{m-k}{1+mk}\right|$. If $\beta \geq \alpha$, then $\beta - \alpha$ is one angle between L and M. By the subtraction identity for tangent,

$$\tan(\beta - \alpha) = \frac{\tan\beta - \tan\alpha}{1 + \tan\beta\tan\alpha} = \frac{m-k}{1+mk}.$$

the other angle between L and M is $\pi - (\beta - \alpha)$ and by periodicity, the negative angle identity, and the addition identity, we have:

$$\begin{aligned}\tan[\pi - (\beta - \alpha)] &= \tan[-(\beta - \alpha)] \\ &= -\tan(\beta - \alpha) \\ &= -\frac{\tan\beta - \tan\alpha}{1 + \tan\beta\tan\alpha} = -\frac{m-k}{1+mk}.\end{aligned}$$

This completes the proof when $\beta \geq \alpha$. The proof in the case $\alpha \geq \beta$ is similar. ■

Example 3 If L and M are lines of slopes 8 and -3, respectively, then the angle between them satisfies

$$\tan\theta = \left|\frac{8-(-3)}{1+8(-3)}\right| = \left|\frac{11}{-23}\right| = \frac{11}{23}.$$

A calculator shows that $\theta \approx .446$ radians, or 25.56°. ■

The Slope Theorem for Perpendicular Lines

We can now prove the following fact, which was first presented in Section 1.4.

Let L be a line with slope k and M a line with slope m. Then L and M are perpendicular exactly when $km = -1$.

First, suppose L and M are perpendicular. We must show that $km = -1$. If α and β (with $\beta \geq \alpha$) are the angles of inclination of L and M, then $\beta - \alpha$ is the angle between L and M, so that $\beta - \alpha = \pi/2$, or equivalently, $\beta = \alpha + \pi/2$. Therefore, by the addition identities for sine and cosine,

$$\begin{aligned}m = \tan\beta = \tan\left(\alpha + \frac{\pi}{2}\right) &= \frac{\sin[\alpha + (\pi/2)]}{\cos[\alpha + (\pi/2)]} \\ &= \frac{\sin\alpha\cos(\pi/2) + \cos\alpha\sin(\pi/2)}{\cos\alpha\cos(\pi/2) - \sin\alpha\sin(\pi/2)} \\ &= \frac{\sin\alpha\,(0) + \cos\alpha\,(1)}{\cos\alpha\,(0) - \sin\alpha\,(1)} \\ &= -\frac{\cos\alpha}{\sin\alpha} = -\cot\alpha = \frac{-1}{\tan\alpha} = \frac{-1}{k}.\end{aligned}$$

Thus, $m = -1/k$ and hence $mk = -1$.

Now suppose that $mk = -1$. We must show that L and M are perpendicular. If L and M are *not* perpendicular, then neither of the angles between them is $\pi/2$. In this case, if θ is either of the angles between L and M, then $\tan\theta$ is a well-defined real number. But we know that one of these angles must satisfy

$$\tan\theta = \left|\frac{m-k}{1+mk}\right| = \left|\frac{m-k}{1+(-1)}\right| = \frac{|m-k|}{0}$$

which is *not* defined. This contradiction shows that L and M must be perpendicular.

Exercises 8.2.A

In Exercises 1–6, find the angles of inclination of the straight line through the given points.

1. $(-1, 2), (3, 5)$
2. $(0, 4), (5, -1)$
3. $(1, 4), (6, 0)$
4. $(4, 2), (-3, -2)$
5. $(3, -7), (3, 5)$
6. $(0, 0), (-4, -5)$

In Exercises 7–12, find one of the angles between the straight lines L and M.

7. L has slope $3/2$ and M has slope -1.
8. L has slope 1 and M has slope 3.
9. L has slope -1 and M has slope 0.
10. L has slope -2 and M has slope -3.
11. $(3, 2)$ and $(5, 6)$ are on L; $(0, 3)$ and $(4, 0)$ are on M.
12. $(-1, 2)$ and $(3, -3)$ are on L; $(3, -3)$ and $(6, 1)$ are on M.

8.3 Other Identities

We now present a variety of identities that are special cases of the addition and subtraction identities of Section 8.2, beginning with

Double-Angle Identities

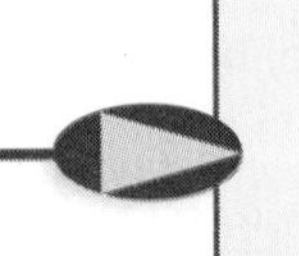

$$\sin 2x = 2\sin x\cos x$$

$$\cos 2x = \cos^2 x - \sin^2 x$$

$$\tan 2x = \frac{2\tan x}{1-\tan^2 x}$$

To prove these identities, just let $x = y$ in the addition identities:

$$\sin 2x = \sin(x+x) = \sin x\cos x + \cos x\sin x = 2\sin x\cos x$$

$$\cos 2x = \cos(x+x) = \cos x\cos x - \sin x\sin x = \cos^2 x - \sin^2 x$$

$$\tan 2x = \tan(x+x) = \frac{\tan x + \tan x}{1 - \tan x\tan x} = \frac{2\tan x}{1-\tan^2 x}.$$

Example 1 If $\pi < x < 3\pi/2$ and $\cos x = -8/17$, find $\sin 2x$ and $\cos 2x$, and show that $5\pi/2 < 2x < 3\pi$.

Solution In order to use the double-angle identities, we first must determine $\sin x$. Now $\sin x$ can be found by using the Pythagorean identity:

$$\sin^2 x = 1 - \cos^2 x = 1 - \left(-\frac{8}{17}\right)^2 = 1 - \frac{64}{289} = \frac{225}{289}.$$

Since $\pi < x < 3\pi/2$, we know $\sin x$ is negative. Therefore,

$$\sin x = -\sqrt{\frac{225}{289}} = -\frac{15}{17}.$$

We now substitute these values in the double-angle identities:

$$\sin 2x = 2 \sin x \cos x = 2\left(-\frac{15}{17}\right)\left(-\frac{8}{17}\right) = \frac{240}{289} \approx .83$$

$$\begin{aligned}\cos 2x = \cos^2 x - \sin^2 x &= \left(-\frac{8}{17}\right)^2 - \left(-\frac{15}{17}\right)^2 \\ &= \frac{64}{289} - \frac{225}{289} = -\frac{161}{289} \approx -.56.\end{aligned}$$

Since $\pi < x < 3\pi/2$, we know that $2\pi < 2x < 3\pi$. The calculations above show that at $2x$, sine is positive and cosine is negative. This can occur only if $2x$ lies between $5\pi/2$ and 3π. ■

Example 2 Express the rule of the function $f(x) = \sin 3x$ in terms of $\sin x$ and constants.

Solution We first use the addition identity for $\sin(x + y)$ with $y = 2x$:

$$f(x) = \sin 3x = \sin(x + 2x) = \sin x \cos 2x + \cos x \sin 2x.$$

Next apply the double-angle identities for $\cos 2x$ and $\sin 2x$:

$$\begin{aligned}f(x) = \sin 3x &= \sin x \cos 2x + \cos x \sin 2x \\ &= \sin x(\cos^2 x - \sin^2 x) + \cos x(2 \sin x \cos x) \\ &= \sin x \cos^2 x - \sin^3 x + 2 \sin x \cos^2 x \\ &= 3 \sin x \cos^2 x - \sin^3 x.\end{aligned}$$

Finally, use the Pythagorean identity:

$$\begin{aligned}f(x) = \sin 3x = 3 \sin x \cos^2 x - \sin^3 x &= 3 \sin x(1 - \sin^2 x) - \sin^3 x \\ &= 3 \sin x - 3 \sin^3 x - \sin^3 x = 3 \sin x - 4 \sin^3 x.\end{aligned}$$ ■

The double-angle identity for $\cos 2x$ can be rewritten in several useful ways. For instance, we can use the Pythagorean identity in the form of $\cos^2 x = 1 - \sin^2 x$ to obtain:

$$\cos 2x = \cos^2 x - \sin^2 x = (1 - \sin^2 x) - \sin^2 x = 1 - 2\sin^2 x.$$

Similarly, using the Pythagorean identity in the form $\sin^2 x = 1 - \cos^2 x$, we have:

$$\cos 2x = \cos^2 x - \sin^2 x = \cos^2 x - (1 - \cos^2 x) = 2\cos^2 x - 1.$$

In summary:

More Double-Angle Identities

$$\mathbf{\cos 2x = 1 - 2\sin^2 x}$$

$$\mathbf{\cos 2x = 2\cos^2 x - 1}$$

Example 3 Prove that $\dfrac{1 - \cos 2x}{\sin 2x} = \tan x$.

Solution The first identity in the preceding box and the double-angle identity for sine show that

$$\frac{1 - \cos 2x}{\sin 2x} = \frac{1 - (1 - 2\sin^2 x)}{2\sin x \cos x} = \frac{2\sin^2 x}{2\sin x \cos x} = \frac{\sin x}{\cos x} = \tan x. \quad ■$$

If we solve the first equation in the preceding box for $\sin^2 x$ and the second one for $\cos^2 x$, we obtain a useful alternate form for these identities:

Power-Reducing Identities

$$\mathbf{\sin^2 x = \frac{1 - \cos 2x}{2}}$$

$$\mathbf{\cos^2 x = \frac{1 + \cos 2x}{2}}$$

Example 4 Express the rule of the function $f(x) = \sin^4 x$ in terms of constants and first powers of the cosine function.

Solution We begin by applying the power-reducing identity

$$f(x) = \sin^4 x = \sin^2 x \sin^2 x = \frac{1 - \cos 2x}{2} \cdot \frac{1 - \cos 2x}{2}$$

$$= \frac{1 - 2\cos 2x + \cos^2 2x}{4}.$$

Next we apply the power-reducing identity for cosine to $\cos^2 2x$. Note that this means using $2x$ in place of x in the identity:

$$\cos^2 2x = \frac{1 + \cos 2(2x)}{2} = \frac{1 + \cos 4x}{2}.$$

Finally, we substitute this last result in the expression for $\sin^4 x$ above:

$$f(x) = \sin^4 x = \frac{1 - 2\cos 2x + \cos^2 2x}{4} = \frac{1 - 2\cos 2x + \dfrac{1 + \cos 4x}{2}}{4}$$

$$= \frac{1}{4} - \frac{1}{2}\cos 2x + \frac{1}{8}(1 + \cos 4x)$$

$$= \frac{3}{8} - \frac{1}{2}\cos 2x + \frac{1}{8}\cos 4x. \quad \blacksquare$$

Half-Angle Identities

If we use the power-reducing identity with $x/2$ in place of x, we obtain

$$\sin^2\left(\frac{x}{2}\right) = \frac{1 - \cos 2\left(\dfrac{x}{2}\right)}{2} = \frac{1 - \cos x}{2}.$$

Consequently, we must have:

$$\sin\left(\frac{x}{2}\right) = \pm\sqrt{\frac{1 - \cos x}{2}}.$$

This proves the first of the half-angle identities.

Half-Angle Identities

$$\sin\frac{x}{2} = \pm\sqrt{\frac{1 - \cos x}{2}} \qquad \cos\frac{x}{2} = \pm\sqrt{\frac{1 + \cos x}{2}}$$

$$\tan\frac{x}{2} = \pm\sqrt{\frac{1 - \cos x}{1 + \cos x}}$$

The half-angle identity for cosine is derived from a power-reducing identity, as was the half-angle identity for sine. The half-angle identity for tangent then follows immediately since $\tan(x/2) = \sin(x/2)/\cos(x/2)$. In all cases, *the sign in front of the radical depends on the quadrant in which x/2 lies.*

Example 5 Find the exact value of

(a) $\cos 5\pi/8$ (b) $\sin \pi/12$.

Solution

(a) Since $5\pi/8 = \frac{1}{2}(5\pi/4)$, we use the half-angle identity with $x = 5\pi/4$ and the fact that $\cos(5\pi/4) = -\sqrt{2}/2$. The sign chart in Exercise 49 on page 408 shows that $\cos(5\pi/8)$ is negative because $5\pi/8$ is in the second quadrant. So we use the negative sign in front of the radical:

$$\cos\frac{5\pi}{8} = \cos\frac{5\pi/4}{2} = -\sqrt{\frac{1+\cos(5\pi/4)}{2}}$$

$$= -\sqrt{\frac{1+\left(-\sqrt{2}/2\right)}{2}} = -\sqrt{\frac{(2-\sqrt{2})/2}{2}}$$

$$= -\sqrt{\frac{2-\sqrt{2}}{4}}$$

$$= \frac{-\sqrt{2-\sqrt{2}}}{2}.$$

(b) Since $\pi/12 = \frac{1}{2}(\pi/6)$ and $\pi/12$ is in the first quadrant, where sine is positive, we have:

$$\sin\frac{\pi}{12} = \sin\frac{\pi/6}{2} = \sqrt{\frac{1-\cos(\pi/6)}{2}}$$

$$= \sqrt{\frac{1-\sqrt{3}/2}{2}} = \sqrt{\frac{(2-\sqrt{3})/2}{2}} = \sqrt{\frac{2-\sqrt{3}}{4}}$$

$$= \frac{\sqrt{2-\sqrt{3}}}{2}. \quad ■$$

The problem of determining signs in the half-angle formulas can be eliminated with tangent by using these:

Half-Angle Identities for Tangent

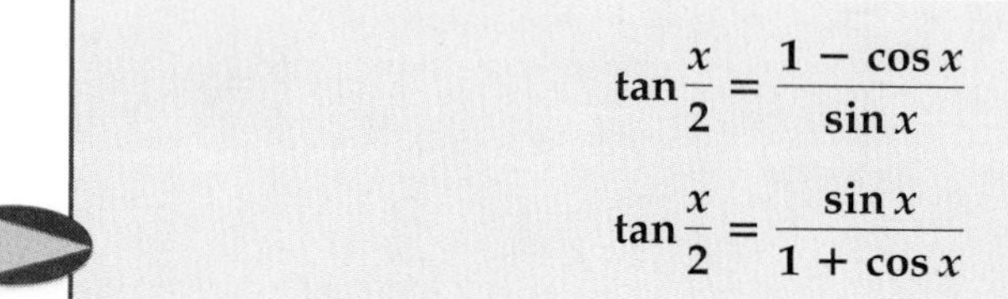

$$\tan\frac{x}{2} = \frac{1-\cos x}{\sin x}$$

$$\tan\frac{x}{2} = \frac{\sin x}{1+\cos x}$$

The proof of the first of these identities follows from the identity $\tan x = \dfrac{1-\cos 2x}{\sin 2x}$, which was proved in Example 3; simply replace x by $x/2$:

$$\tan\left(\frac{x}{2}\right) = \frac{1-\cos 2(x/2)}{\sin 2(x/2)} = \frac{1-\cos x}{\sin x}.$$

The second identity in the box is proved in Exercise 71.

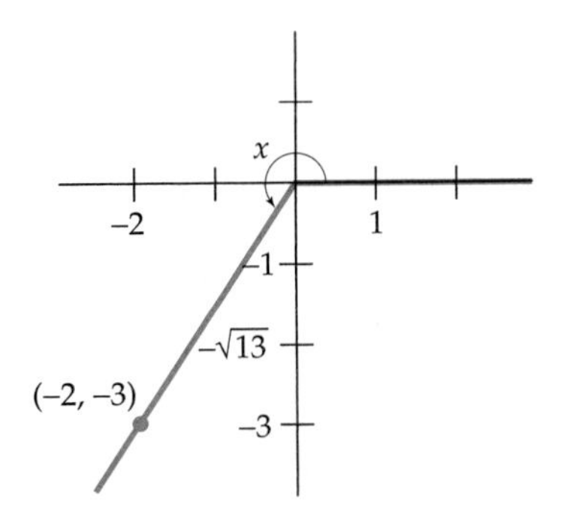

Figure 8–8

Example 6 If $\tan x = \frac{3}{2}$ and $\pi < x < \frac{3\pi}{2}$, find $\tan\frac{x}{2}$.

Solution The terminal side of an angle of x radians in standard position lies in the third quadrant, as shown in Figure 8–8. The tangent of the angle in standard position whose terminal side passes through the point $(-2, -3)$ is $\frac{-3}{-2} = \frac{3}{2}$. Since there is only one angle in the third quadrant with tangent $3/2$, the point $(-2, -3)$ must lie on the terminal side of the angle of x radians.

Since the distance from $(-2, -3)$ to the origin is

$$\sqrt{(-2-0)^2 + (-3-0)^2} = \sqrt{13},$$

we have

$$\sin x = \frac{-3}{\sqrt{13}} \quad \text{and} \quad \cos x = \frac{-2}{\sqrt{13}}.$$

Therefore, by the first of the half-angle identities for tangent

$$\tan\frac{x}{2} = \frac{1-\cos x}{\sin x} = \frac{1-\left(\frac{-2}{\sqrt{13}}\right)}{\frac{-3}{\sqrt{13}}} = \frac{\frac{\sqrt{13}+2}{\sqrt{13}}}{\frac{-3}{\sqrt{13}}} = -\frac{\sqrt{13}+2}{3}. \quad \blacksquare$$

Product and Factoring Identities

Using the addition and subtraction identities to compute

$$\sin(x+y) + \sin(x-y),$$

we see that

$$\begin{aligned}\sin(x+y) + \sin(x-y) &= \sin x \cos y + \cos x \sin y + \sin x \cos y - \cos x \sin y \\ &= 2 \sin x \cos y.\end{aligned}$$

Dividing both sides of this last equation by 2 produces the first of the following identities:

Product Identities

$$\sin x \cos y = \frac{1}{2}[\sin(x+y) + \sin(x-y)]$$

$$\sin x \sin y = \frac{1}{2}[\cos(x-y) - \cos(x+y)]$$

$$\cos x \cos y = \frac{1}{2}[\cos(x+y) + \cos(x-y)]$$

$$\cos x \sin y = \frac{1}{2}[\sin(x+y) - \sin(x-y)]$$

The proofs of the second and fourth product identities are similar to the proof of the first. The third product identity was proved in Example 4 of Section 8.2.

If we use the first product identity with $\frac{1}{2}(x + y)$ in place of x and $\frac{1}{2}(x - y)$ in place of y, we obtain:

$$\sin\left[\frac{1}{2}(x + y)\right]\cos\left[\frac{1}{2}(x - y)\right] = \frac{1}{2}\left[\sin\left(\frac{1}{2}(x + y) + \frac{1}{2}(x - y)\right)\right.$$

$$\left. + \sin\left(\frac{1}{2}(x + y) - \frac{1}{2}(x - y)\right)\right]$$

$$= \frac{1}{2}(\sin x + \sin y).$$

Multiplying both sides of the last equation by 2 produces the first of the following identities:

Factoring Identities

$$\sin x + \sin y = 2\sin\left(\frac{x + y}{2}\right)\cos\left(\frac{x - y}{2}\right)$$

$$\sin x - \sin y = 2\cos\left(\frac{x + y}{2}\right)\sin\left(\frac{x - y}{2}\right)$$

$$\cos x + \cos y = 2\cos\left(\frac{x + y}{2}\right)\cos\left(\frac{x - y}{2}\right)$$

$$\cos x - \cos y = -2\sin\left(\frac{x + y}{2}\right)\sin\left(\frac{x - y}{2}\right)$$

The last three factoring identities are proved in the same way as the first.

Example 7 Prove the identity

$$\frac{\sin t + \sin 3t}{\cos t + \cos 3t} = \tan 2t.$$

Solution Using the first factoring identity with $x = t$ and $y = 3t$ yields:

$$\sin t + \sin 3t = 2\sin\left(\frac{t + 3t}{2}\right)\cos\left(\frac{t - 3t}{2}\right) = 2\sin 2t\cos(-t).$$

Similarly,

$$\cos t + \cos 3t = 2\cos\left(\frac{t + 3t}{2}\right)\cos\left(\frac{t - 3t}{2}\right) = 2\cos 2t\cos(-t)$$

so that

$$\frac{\sin t + \sin 3t}{\cos t + \cos 3t} = \frac{2\sin 2t\cos(-t)}{2\cos 2t\cos(-t)} = \frac{\sin 2t}{\cos 2t} = \tan 2t. \quad ■$$

Exercises 8.3

In Exercises 1–12, use the half-angle identities to evaluate the given expression exactly.

1. $\cos\frac{\pi}{8}$ **2.** $\tan\frac{\pi}{8}$ **3.** $\sin\frac{3\pi}{8}$ **4.** $\cos\frac{3\pi}{8}$

5. $\tan\frac{\pi}{12}$ **6.** $\sin\frac{5\pi}{8}$ **7.** $\cos\frac{\pi}{12}$ **8.** $\tan\frac{5\pi}{8}$

9. $\sin\frac{7\pi}{8}$ **10.** $\cos\frac{7\pi}{8}$ **11.** $\tan\frac{7\pi}{8}$ **12.** $\cot\frac{\pi}{8}$

In Exercises 13–18, write each expression as a sum or difference.

13. $\sin 4x \cos 6x$ **14.** $\sin 5x \sin 7x$

15. $\cos 2x \cos 4x$ **16.** $\sin 3x \cos 5x$

17. $\sin 17x \sin(-3x)$ **18.** $\cos 13x \cos(-5x)$

In Exercises 19–22, write each expression as a product.

19. $\sin 3x + \sin 5x$ **20.** $\cos 2x + \cos 6x$

21. $\sin 9x - \sin 5x$ **22.** $\cos 5x - \cos 7x$

In Exercises 23–30, find sin 2x, cos 2x, and tan 2x under the given conditions.

23. $\sin x = \frac{5}{13}\quad \left(0 < x < \frac{\pi}{2}\right)$

24. $\sin x = -\frac{4}{5}\quad \left(\pi < x < \frac{3\pi}{2}\right)$

25. $\cos x = -\frac{3}{5}\quad \left(\pi < x < \frac{3\pi}{2}\right)$

26. $\cos x = -\frac{1}{3}\quad \left(\frac{\pi}{2} < x < \pi\right)$

27. $\tan x = \frac{3}{4}\quad \left(\pi < x < \frac{3\pi}{2}\right)$

28. $\tan x = -\frac{3}{2}\quad \left(\frac{\pi}{2} < x < \pi\right)$

29. $\csc x = 4\quad \left(0 < x < \frac{\pi}{2}\right)$

30. $\sec x = -5\quad \left(\pi < x < \frac{3\pi}{2}\right)$

In Exercises 31–36, find $\sin\frac{x}{2}$, $\cos\frac{x}{2}$, and $\tan\frac{x}{2}$ under the given conditions.

31. $\cos x = .4\quad \left(0 < x < \frac{\pi}{2}\right)$

32. $\sin x = .6\quad \left(\frac{\pi}{2} < x < \pi\right)$

33. $\sin x = -\frac{3}{5}\quad \left(\frac{3\pi}{2} < x < 2\pi\right)$

34. $\cos x = .8\quad \left(\frac{3\pi}{2} < x < 2\pi\right)$

35. $\tan x = \frac{1}{2}\quad \left(\pi < x < \frac{3\pi}{2}\right)$

36. $\cot x = 1\quad \left(-\pi < x < -\frac{\pi}{2}\right)$

In Exercises 37–42, assume $\sin x = .6$ and $0 < x < \pi/2$ and evaluate the given expression.

37. $\sin 2x$ **38.** $\cos 4x$ **39.** $\cos 2x$ **40.** $\sin 4x$

41. $\sin\frac{x}{2}$ **42.** $\cos\frac{x}{2}$

43. Express $\cos 3x$ in terms of $\cos x$.

44. **(a)** Express the rule of the function $f(x) = \cos^3 x$ in terms of constants and first powers of the cosine function as in Example 4.
(b) Do the same for $f(x) = \cos^4 x$.

In Exercises 45–50, simplify the given expression.

45. $\frac{\sin 2x}{2\sin x}$ **46.** $1 - 2\sin^2\left(\frac{x}{2}\right)$

47. $2\cos 2y \sin 2y$ (Think!)

48. $\cos^2\left(\frac{x}{2}\right) - \sin^2\left(\frac{x}{2}\right)$

49. $(\sin x + \cos x)^2 - \sin 2x$

50. $2\sin x\cos^3 x - 2\sin^3 x\cos x$

In Exercises 51–64, determine graphically whether the equation could possibly be an identity. If it could, prove that it is.

51. $\sin 16x = 2\sin 8x\cos 8x$

52. $\cos 8x = \cos^2 4x - \sin^2 4x$

53. $\cos^4 x - \sin^4 x = \cos 2x$

54. $\sec 2x = \frac{1}{1 - 2\sin^2 x}$

55. $\cos 4x = 2\cos 2x - 1$

56. $\sin^2 x = \cos^2 x - 2\sin x$

57. $\frac{1 + \cos 2x}{\sin 2x} = \cot x$

58. $\sin 2x = \dfrac{2\cot x}{\csc^2 x}$

59. $\sin 3x = (\sin x)(3 - 4\sin^2 x)$

60. $\sin 4x = (4\cos x \sin x)(1 - 2\sin^2 x)$

61. $\cos 2x = \dfrac{2\tan x}{\sec^2 x}$

62. $\cos 3x = (\cos x)(3 - 4\cos^2 x)$

63. $\csc^2\left(\dfrac{x}{2}\right) = \dfrac{2}{1 - \cos x}$

64. $\sec^2\left(\dfrac{x}{2}\right) = \dfrac{2}{1 + \cos x}$

In Exercises 65–70, prove the identity.

65. $\dfrac{\sin x - \sin 3x}{\cos x + \cos 3x} = -\tan x$

66. $\dfrac{\sin x - \sin 3x}{\cos x - \cos 3x} = -\cot 2x$

67. $\dfrac{\sin 4x + \sin 6x}{\cos 4x - \cos 6x} = \cot x$

68. $\dfrac{\cos 8x + \cos 4x}{\cos 8x - \cos 4x} = -\cot 6x \cot 2x$

69. $\dfrac{\sin x + \sin y}{\cos x - \cos y} = -\cot\left(\dfrac{x - y}{2}\right)$

70. $\dfrac{\sin x - \sin y}{\cos x + \cos y} = \tan\left(\dfrac{x - y}{2}\right)$

71. **(a)** Prove that $\dfrac{1 - \cos x}{\sin x} = \dfrac{\sin x}{1 + \cos x}$.

(b) Use part (a) and the half-angle identity proved in the text to prove that $\tan\dfrac{x}{2} = \dfrac{\sin x}{1 + \cos x}$.

8.4 Trigonometric Equations

Any equation involving trigonometric functions can be solved graphically and many can be solved algebraically. We begin with the graphical method, which is the same as that used to solve other equations. Unlike the equations solved previously, however, trigonometric equations typically have an infinite number of solutions. In many cases, these solutions can be systematically determined by using periodicity, as we now see.

Example 1 Solve: $3\sin^2 x - \cos x - 2 = 0$.

Solution Both sine and cosine have period 2π, so the period of $f(x) = 3\sin^2 x - \cos x - 2$ is at most 2π. The graph of f, which is shown in two viewing windows in Figure 8–9, does not repeat its pattern over any interval of less than 2π, so we conclude that f has period 2π.

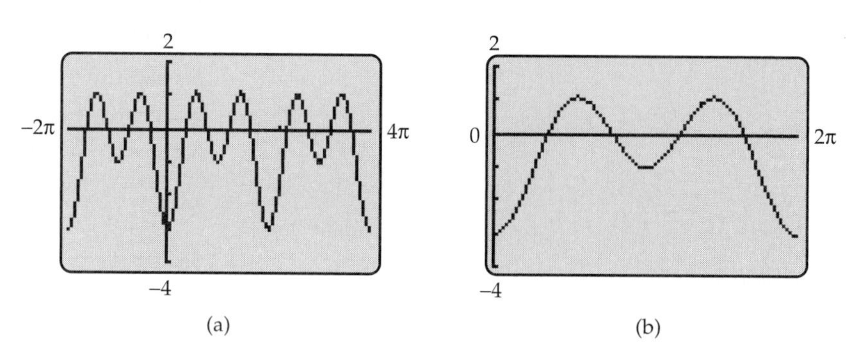

Figure 8–9

The function f makes one complete period in the interval $[0, 2\pi)$, as shown in Figure 8–9(b). The equation has four solutions between 0 and 2π, namely, the four x-intercepts of the graph in that interval, as explained in Section 2.2. A graphical root finder shows that these four solutions are

$$x \approx 1.12, \qquad x \approx 2.45, \qquad x \approx 3.84, \qquad x \approx 5.16.$$

Since the graph repeats its pattern to the left and right, the other x-intercepts (solutions) will differ from these four by multiples of 2π. For instance, in addition to the x-intercept at 1.12, there will be x-intercepts (solutions) at

$$1.12 \pm 2\pi, \qquad 1.12 \pm 4\pi, \qquad 1.12 \pm 6\pi, \text{ etc.,}$$

the first two of which can be seen in Figure 8–9(a). These solutions are customarily written like this:

$$x = 1.12 + 2k\pi \qquad (k = 0, \pm 1, \pm 2, \pm 3, \ldots).$$

A similar analysis applies to the other solutions between 0 and 2π. Hence, all solutions of the equation are given by:

$$x \approx 1.12 + 2k\pi, \quad x \approx 2.45 + 2k\pi, \quad x \approx 3.84 + 2k\pi, \quad x \approx 5.16 + 2k\pi,$$

$$\text{where } k = 0, \pm 1, \pm 2, \pm 3, \ldots. \quad \blacksquare$$

Although the graph in Figure 8–9(a) of Example 1 shows three periods of the function f, only one complete period, the interval $(0, 2\pi)$, was actually used to solve the equation. A similar procedure can be used to solve any trigonometric equation graphically:

Graphical Method for Solving Trigonometric Equations

1. **Write the equation in the form $f(x) = 0$.**
2. **Determine the period of p of $f(x)$.**
3. **Graph $f(x)$ over an interval of length p.**
4. **Use a graphical root finder to determine the x-intercepts of the graph in this interval.**
5. **For each x-intercept u, all of the numbers**

$$u + kp \qquad (k = 0, \pm 1, \pm 2, \pm 3, \ldots)$$

are solutions of the equation.

In Example 1, for instance, p was 2π. However, many trigonometric functions have periods other than 2π.

Example 2 To solve the equation $\tan x = 3 \sin 2x$, we first rewrite it as

$$\tan x - 3 \sin 2x = 0.$$

Both $\tan x$ and $\sin 2x$ have period π (see pages 424 and 429). Hence the function given by the left side of the equation, $f(x) = \tan x - 3\sin 2x$, also has period π. The graph of f on the interval $[0, \pi)$ (Figure 8–10) shows erroneous vertical line segments instead of the vertical asymptote at $x = \pi/2$, where tangent is not defined, as well as x-intercepts at the endpoints of the interval. Consequently, we use the more easily read graph f in Figure 8–11, which uses the interval $(-\pi/2, \pi/2)$.

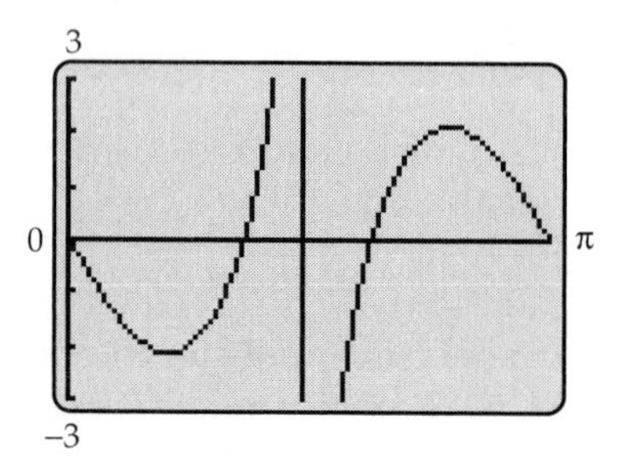

Figure 8–10

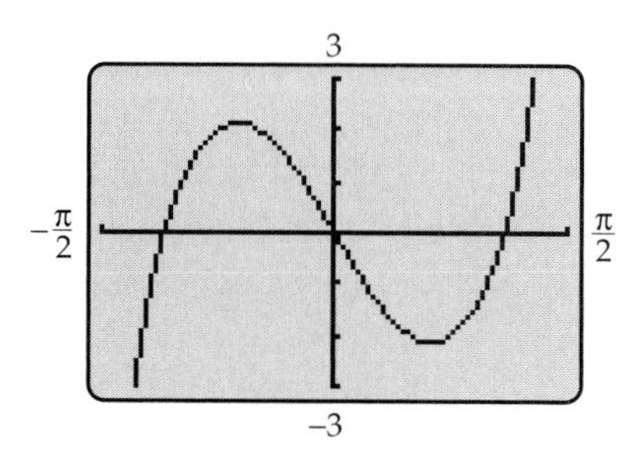

Figure 8–11

Even without the graph, we can verify that there is an x-intercept at the origin because $f(0) = \tan 0 - 3\sin(2\cdot 0) = 0$. Using zoom-in or a root finder on the other two x-intercepts in Figure 8–11 shows that they are $x \approx -1.15$ and $x \approx 1.15$. All other solutions differ from these by multiples of π (the period of f) and are given by:

$$x \approx -1.15 + k\pi, \qquad x \approx 0 + k\pi, \qquad x \approx 1.15 + k\pi$$
$$(k = 0, \pm 1, \pm 2, \pm 3, \ldots). \quad \blacksquare$$

Basic Equations

We shall call equations such as

$$\sin x = .39, \qquad \cos x = .2, \qquad \tan x = -3$$

basic equations. The following examples show how they can be quickly solved algebraically.

Example 3 Solve $\tan x = 2$.

Solution The graph of $f(x) = \tan x$ in Figure 8–12 shows that there are infinitely many points with second coordinate 2 (each point where the graph intersects the horizontal line through 2). The first coordinate of every such point is a number whose tangent is 2, that is, a solution of $\tan x = 2$.

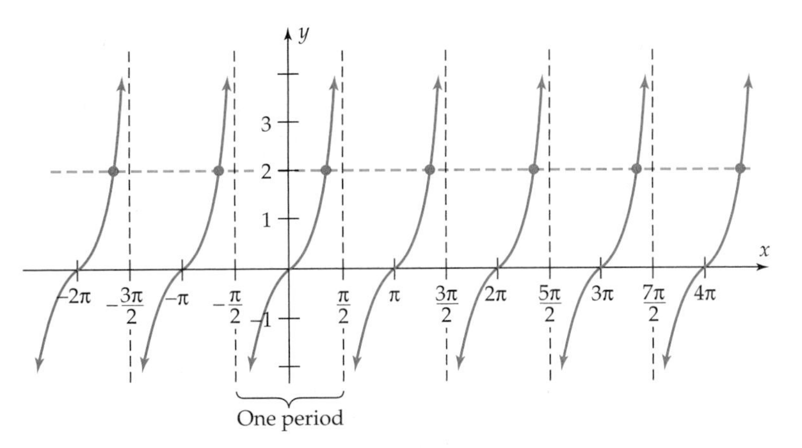

Figure 8–12

There is exactly one solution between $-\pi/2$ and $\pi/2$. Although this solution can be found graphically, it is faster to key in TAN^{-1} 2 ENTER.* The calculator then displays the number between $-\pi/2$ and $\pi/2$ whose tangent is 2, namely, 1.1071. In other words, $x = 1.1071$ is a solution of $\tan x = 2$.† Since the interval $(-\pi/2, \pi/2)$ is one full period of the tangent function and $x = 1.1071$ is the only solution of $\tan x = -2$ in that interval, all solutions are given by

$$x = 1.1071 + k\pi \quad (k = 0, \pm 1, \pm 2, \pm 3, \ldots). \quad \blacksquare$$

With only a slight modification, the procedure used in Example 3 also works for basic sine and cosine equations.

Example 4 Solve $\sin x = -.75$.

Solution There are infinitely many points on the graph of $f(x) = \sin x$ that have second coordinate $-.75$ (Figure 8–13). The first coordinate of every such point is a solution of the equation $\sin x = -.75$. (Why?)

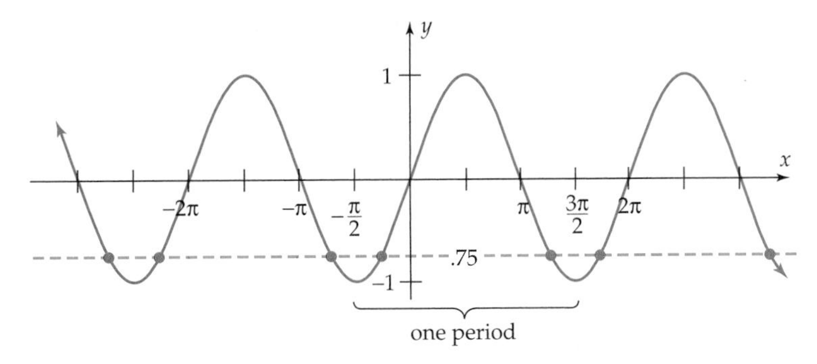

Figure 8–13

*Unless stated otherwise, radian mode is used throughout this section.

† For convenient reading, all solutions in the text are rounded to four decimal places, but the full decimal expansion given by the calculator is used in all computations. We write = rather than ≈ even though these calculator solutions are approximations of the actual solutions.

There is one solution between $-\pi/2$ and $\pi/2$, which can be found by keying in SIN^{-1} $-.75$ ENTER. Verify that this solution is $x = -.8481$. Note that there is a second solution in the period of the sine function from $-\pi/2$ to $3\pi/2$. It can be found by using the identity that was proved in Example 2 of Section 8.2:

$$\sin(\pi - y) = \sin y.$$

Applying this identity with $y = -.8481$ shows that

$$\sin[\pi - (-.8481)] = \sin(-.8481) = -.75.$$

Hence, $x = \pi - (-.8481) = 3.9897$ is also a solution of $\sin x = -.75$. Since the interval $[-\pi/2, 3\pi/2)$ is one full period of the sine function, all solutions of the equation $\sin x = -.75$ are given by

$$x = -.8481 + 2k\pi \qquad \text{and} \qquad x = 3.9897 + 2k\pi$$
$$(k = 0, \pm 1, \pm 2, \pm 3, \ldots). \quad ■$$

Example 5 Solve $\cos x = .2$.

Solution A solution between 0 and π can be found by keying in COS^{-1} .2 ENTER, namely, $x = 1.3694$.

NOTE The reason why the COS^{-1} key produces a solution in the interval from 0 to π, whereas the SIN^{-1} and TAN^{-1} keys produce solutions in the interval from $-\pi/2$ to $\pi/2$, is explained in Section 8.5.

GRAPHING EXPLORATION

Verify that the equation $\cos x = .2$ has another solution between $-\pi$ and π by showing that the graph of $g(x) = \cos x$ intersects the horizontal line $y = .2$ at two points in that interval. Note that these points appear to be at equal distances from the y-axis, which suggests that their first coordinates are negatives of each other.

The second solution in the interval $[-\pi, \pi)$ is indeed the negative of the first one, because by the negative angle identity, $\cos(-x) = \cos x$, we have

$$\cos(-1.3694) = \cos 1.3694 = .2.$$

Thus, $x = 1.3694$ and $x = -1.3694$ are the solutions of $\cos x = .2$ in the interval $[-\pi, \pi)$. Since this interval is one full period of the cosine function, all solutions of the equation are given by

$$x = 1.3694 + 2k\pi \qquad \text{and} \qquad x = -1.3694 + 2k\pi$$
$$(k = 0, \pm 1, \pm 2, \pm 3, \ldots). \quad ■$$

The procedures in Examples 3–5 work in the general case and lead to the following conclusion.*

*See Exercises 13 and 14 for the equations $\sin x = \pm 1$ and $\cos x = \pm 1$.

Solution Algorithm for Basic Trigonometric Equations

If c is any real number, then the equation $\tan x = c$ has one solution between $-\pi/2$ and $\pi/2$, which can be found by using the TAN^{-1} key. All solutions can be found by adding or subtracting integer multiples of π to this one.

If $|c| < 1$, then the equation $\sin x = c$ has a solution between $-\pi/2$ and $\pi/2$, which can be found by using the SIN^{-1} key. A second solution is $\pi - u$, where u is the first solution. All solutions can be found by adding or subtracting integer multiples of 2π to these two.

If $|c| < 1$, then the equation $\cos x = c$ has a solution between 0 and π, which can be found by using the COS^{-1} key. A second solution is the negative of this one. All solutions can be found by adding or subtracting integer multiples of 2π to these two.

NOTE The solution algorithm for basic equations can be used to find solutions in degrees by replacing π by 180° and using a calculator in degree mode.

Example 6 Find all solutions of $\sec x = 8$ in the interval $[0, 2\pi)$.

Solution Note that $\sec x = 8$ exactly when

$$\frac{1}{\cos x} = 8, \qquad \text{or equivalently,} \qquad \cos x = \frac{1}{8} = .125.$$

Since $\cos^{-1}(.125) = 1.4455$, the solution of $\cos x = .125$, and hence of $\sec x = 8$ are

$$x = 1.4455 + 2k\pi \qquad \text{and} \qquad x = -1.4455 + 2k\pi$$
$$(k = 0, \pm 1, \pm 2, \pm 3, \ldots).$$

Of these solutions, the two between 0 and 2π are

$$x = 1.4455 \qquad \text{and} \qquad x = -1.4455 + 2\pi = 4.8377. \quad \blacksquare$$

Example 7 Solve $\sin u = \sqrt{2}/2$ exactly, without using a calculator.

Solution Our knowledge of special values shows that $u = \pi/4$ is one solution. As above, the identity $\sin(\pi - x) = \sin x$ shows that $u = \pi - \pi/4 = 3\pi/4$ is a second solution between $-\pi/2$ and $3\pi/2$. Therefore, the exact solutions of $\sin x = \sqrt{2}/2$ are

$$u = \frac{\pi}{4} + 2k\pi \qquad \text{and} \qquad u = \frac{3\pi}{4} + 2k\pi$$
$$(k = 0, \pm 1, \pm 2, \pm 3, \ldots). \quad \blacksquare$$

Example 8 Solve exactly: $\sin 2x = \sqrt{2}/2$.

Solution First, let $u = 2x$ and solve the basic equation $\sin u = \sqrt{2}/2$. As we saw in Example 7, the solutions are

$$u = \frac{\pi}{4} + 2k\pi \quad \text{and} \quad u = \frac{3\pi}{4} + 2k\pi \quad (k = 0, \pm 1, \pm 2, \pm 3, \ldots).$$

Since $u = 2x$, each of these solutions leads to a solution of the original equation:

$$2x = u = \frac{\pi}{4} + 2k\pi, \quad \text{or equivalently,} \quad x = \frac{1}{2}\left(\frac{\pi}{4} + 2k\pi\right) = \frac{\pi}{8} + k\pi.$$

Similarly,

$$2x = u = \frac{3\pi}{4} + 2k\pi, \quad \text{or equivalently,} \quad x = \frac{1}{2}\left(\frac{3\pi}{4} + 2k\pi\right) = \frac{3\pi}{8} + k\pi.$$

Therefore, all solutions of $\sin 2x = \sqrt{2}/2$ are given by

$$x = \frac{\pi}{8} + k\pi \quad \text{and} \quad x = \frac{3\pi}{8} + k\pi \quad (k = 0, \pm 1, \pm 2, \pm 3, \ldots).$$

The fact that the solutions are obtained by adding multiples of π rather than 2π is a reflection of the fact that the period of $\sin 2x$ is π. ■

Algebraic Solution of Other Trigonometric Equations

Many trigonometric equations can be solved algebraically by using factoring, the quadratic formula, and identities to reduce the problem to an equivalent one that involves only basic equations.

Example 9 Solve $-10\cos^2 x - 3\sin x + 9 = 0$.

Solution We first use the Pythagorean identity to rewrite the equation in terms of the sine function:

$$\begin{aligned} -10\cos^2 x - 3\sin x + 9 &= 0 \\ -10(1 - \sin^2 x) - 3\sin x + 9 &= 0 \\ -10 + 10\sin^2 x - 3\sin x + 9 &= 0 \\ 10\sin^2 x - 3\sin x - 1 &= 0. \end{aligned}$$

Now factor the left side:*

$$(2\sin x - 1)(5\sin x + 1) = 0$$

$$\begin{aligned} 2\sin x - 1 &= 0 & \quad \text{or} \quad 5\sin x + 1 &= 0 \\ 2\sin x &= 1 & 5\sin x &= -1 \\ \sin &= 1/2 & \sin x &= -1/5 = -.2 \end{aligned}$$

*The factorization may be easier to see if you first substitute v for $\sin x$, so that $10\sin^2 x - 3\sin x - 1$ becomes $10v^2 - 3v - 1 = (2v - 1)(5v + 1)$.

Each of these basic equations is readily solved. We note that $\sin(\pi/6) = 1/2$, so that $x = \pi/6$ and $x = \pi - \pi/6 = 5\pi/6$ are solutions of the first one. Since $\sin^{-1}(-.2) = -.2014$, both $x = -.2014$ and $x = \pi - (-.2014) = 3.3430$ are solutions of the second equation. Therefore, all solutions of the original equation are given by:

$$x = \frac{\pi}{6} + 2k\pi, \qquad x = \frac{5\pi}{6} + 2k\pi, \qquad x = -.2014 + 2k\pi,$$

$$x = 3.3430 + 2k\pi$$

where $k = 0, \pm 1, \pm 2, \pm 3, \ldots.$ ■

Example 10 Solve $\sec^2 x + 5\tan x = -2$.

Solution We use the Pythagorean identity $\sec^2 x = 1 + \tan^2 x$ to obtain an equivalent equation:

$$\sec^2 x + 5\tan x = -2$$

$$\sec^2 x + 5\tan x + 2 = 0$$

$$(1 + \tan^2 x) + 5\tan x + 2 = 0$$

$$\tan^2 x + 5\tan x + 3 = 0$$

If we let $u = \tan x$, this last equation becomes $u^2 + 5u + 3 = 0$. Since the left side does not readily factor, we use the quadratic formula to solve the equation:

$$u = \frac{-5 \pm \sqrt{5^2 - 4\cdot 1\cdot 3}}{2} = \frac{-5 \pm \sqrt{13}}{2}.$$

Since $u = \tan x$, the original equation is equivalent to

$$\tan x = \frac{-5 + \sqrt{13}}{2} \approx -.6972 \qquad \text{or} \qquad \tan x = \frac{-5 - \sqrt{13}}{2} \approx -4.3028.$$

Solving these basic equations as above, we find that $x = -.6089$ is a solution of the first and $x = -1.3424$ is a solution of the second. Hence, the solutions of the original equation are

$$x = -.6089 + k\pi \qquad \text{and} \qquad x = -1.3424 + k\pi$$

$$(k = 0, \pm 1, \pm 2, \pm 3, \ldots).$$ ■

Example 11 Solve $5\cos x + 3\cos 2x = 3$.

Solution We use the double-angle identity: $\cos 2x = 2\cos^2 x - 1$ as follows.

$$5\cos x + 3\cos 2x = 3$$

Use double-angle identity: $$5\cos x + 3(2\cos^2 x - 1) = 3$$

Multiply out left side: $$5\cos x + 6\cos^2 x - 3 = 3$$

Rearrange terms: $6\cos^2 x + 5\cos x - 6 = 0$

Factor left side: $(2\cos x + 3)(3\cos x - 2) = 0$

$$2\cos x + 3 = 0 \quad \text{or} \quad 3\cos x - 2 = 0$$
$$2\cos x = -3 \qquad 3\cos x = 2$$
$$\cos x = -\frac{3}{2} \qquad \cos x = \frac{2}{3}$$

The equation $\cos x = -3/2$ has no solutions because $\cos x$ always lies between -1 and 1. A calculator shows that the solutions of $\cos x = 2/3$ are

$$x = .8411 + 2k\pi \quad \text{and} \quad x = -.8411 + 2k\pi$$
$$(k = 0, \pm1, \pm2, \pm3, \ldots).$$ ■

Exercises 8.4

In all exercises, find exact solutions if possible (as in Examples 7 and 8) and approximate ones otherwise. *When a calculator is used, round your answers (but not any intermediate results) to four decimal places.*

In Exercises 1–12, solve the equation graphically.

1. $4\sin 2x - 3\cos 2x = 2$
2. $5\sin 3x + 6\cos 3x = 1$
3. $3\sin^3 2x = 2\cos x$
4. $\sin^2 2x - 3\cos 2x + 2 = 0$
5. $\tan x + 5\sin x = 1$
6. $2\cos^2 x + \sin x + 1 = 0$
7. $\cos^3 x - 3\cos x + 1 = 0$
8. $\tan x = 3\cos x$
9. $\cos^4 x - 3\cos^3 x + \cos x = 1$
10. $\sec x + \tan x = 3$
11. $\sin^3 x + 2\sin^2 x - 3\cos x + 2 = 0$
12. $\csc^2 x + \sec x = 1$

13. Use the graph of the sine function to show the following:
   (a) The solutions of $\sin x = 1$ are
   $$x = \frac{\pi}{2}, \frac{5\pi}{2}, \frac{9\pi}{2}, \ldots \quad \text{and}$$
   $$x = \frac{-3\pi}{2}, \frac{-7\pi}{2}, \frac{-11\pi}{2}, \ldots.$$
   (b) The solutions of $\sin x = -1$ are
   $$x = \frac{3\pi}{2}, \frac{7\pi}{2}, \frac{11\pi}{2}, \ldots \quad \text{and}$$
   $$x = \frac{-\pi}{2}, \frac{-5\pi}{2}, \frac{-9\pi}{2}, \ldots.$$

14. Use the graph of the cosine function to show the following:
   (a) The solutions of $\cos x = 1$ are
   $$x = 0, \pm 2\pi, \pm 4\pi, \pm 6\pi, \ldots.$$
   (b) The solutions of $\cos x = -1$ are
   $$\pm\pi, \pm 3\pi, \pm 5\pi, \ldots.$$

In Exercises 15–22, use your knowledge of special angles to find the exact solutions of the equation.

15. $\sin x = \sqrt{3}/2$
16. $2\cos x = \sqrt{2}$
17. $\tan x = -\sqrt{3}$
18. $\tan x = 1$
19. $2\cos x = -\sqrt{3}$
20. $\sin x = 0$
21. $2\sin x + 1 = 0$
22. $\csc x = \sqrt{2}$

In Exercises 23–26, approximate all solutions in $[0, 2\pi)$ of the given equation.

23. $\sin x = .119$
24. $\cos x = .958$
25. $\tan x = 5$
26. $\tan x = 17.65$

In Exercises 27–36, find all angles θ with $0° \le \theta < 360°$ that are solutions of the given equation. [Hint: Put your calculator in degree mode and replace π by 180° in the solution algorithms for basic equations.]

27. $\tan\theta = 7.95$
28. $\tan\theta = 69.4$
29. $\cos\theta = -.42$
30. $\cot\theta = -2.4$
31. $2\sin^2\theta + 3\sin\theta + 1 = 0$
32. $4\cos^2\theta + 4\cos\theta - 3 = 0$
33. $\tan^2\theta - 3 = 0$
34. $2\sin^2\theta = 1$
35. $4\cos^2\theta + 4\cos\theta + 1 = 0$
36. $\sin^2\theta - 3\sin\theta = 10$

At the instant you hear a sonic boom from an airplane overhead, your angle of elevation α to the plane is given by the equation $\sin \alpha = 1/m$, where m is the Mach number for the speed of the plane (Mach 1 is the speed of sound, Mach 2.5 is 2.5 times the speed of sound, etc.). In Exercises 37–40, find the angle of elevation (in degrees) for the given Mach number. Remember that an angle of elevation must be between 0° and 90°.

37. $m = 1.1$ **38.** $m = 1.6$ **39.** $m = 2$

40. $m = 2.4$

When a light beam passes from one medium to another (for instance, from air to water), it changes both its speed and direction. According to Snell's Law of Refraction,

$$\frac{\sin \theta_1}{\sin \theta_2} = \frac{v_1}{v_2},$$

where v_1 is the speed of light in the first medium, v_2 its speed in the second medium, θ_1 the angle of incidence, and θ_2 the angle of refraction, as shown in the figure. The number v_1/v_2 is called the index of refraction. *Use this information to do Exercises 41–44.*

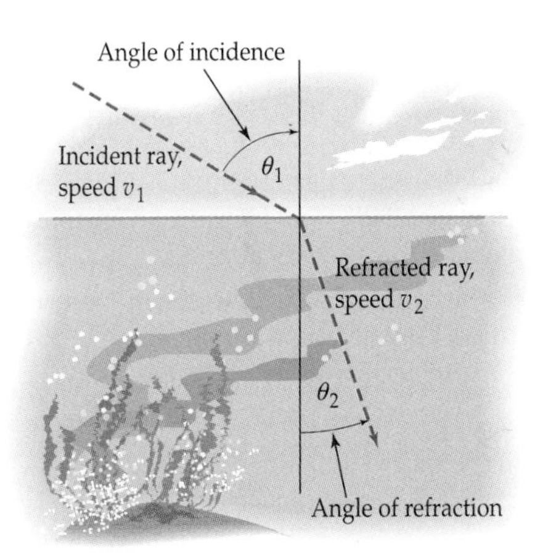

41. The index of refraction of light passing from air to water is 1.33. If the angle of incidence is 38°, find the angle of refraction.

42. The index of refraction of light passing from air to ordinary glass is 1.52. If the angle of incidence is 17°, find the angle of refraction.

43. The index of refraction of light passing from air to dense glass is 1.66. If the angle of incidence is 24°, find the angle of refraction.

44. The index of refraction of light passing from air to quartz is 1.46. If the angle of incidence is 50°, find the angle of refraction.

In Exercises 45–64, find all the solutions of the equation.

45. $\sin x = -.465$ **46.** $\sin x = -.682$

47. $\cos x = -.564$ **48.** $\cos x = -.371$

49. $\tan x = -.237$ **50.** $\tan x = -12.45$

51. $\cot x = 2.3$ [*Remember:* $\cot x = 1/\tan x$.]

52. $\cot x = -3.5$ **53.** $\sec x = -2.65$

54. $\csc x = 5.27$ **55.** $\sin 2x = -\sqrt{3}/2$

56. $\cos 2x = \sqrt{2}/2$ **57.** $2 \cos \frac{x}{2} = \sqrt{2}$

58. $2 \sin \frac{x}{3} = 1$ **59.** $\tan 3x = -\sqrt{3}$

60. $5 \sin 2x = 2$ **61.** $5 \cos 3x = -3$

62. $2 \tan 4x = 16$ **63.** $4 \tan \frac{x}{2} = 8$

64. $5 \sin \frac{x}{4} = 4$

In Exercises 65–90, use factoring, the quadratic formula, or identities to solve the equation. Find all solutions in the interval $[0, 2\pi)$,

65. $3 \sin^2 x - 8 \sin x - 3 = 0$

66. $5 \cos^2 x + 6 \cos x = 8$

67. $2 \tan^2 x + 5 \tan x + 3 = 0$

68. $3 \sin^2 x + 2 \sin x = 5$

69. $\cot x \cos x = \cos x$ (Be careful; see Exercise 99.)

70. $\tan x \cos x = \cos x$

71. $\cos x \csc x = 2 \cos x$

72. $\tan x \sec x + 3 \tan x = 0$

73. $4 \sin x \tan x - 3 \tan x + 20 \sin x - 15 = 0$
[*Hint:* One factor is $\tan x + 5$.]

74. $25 \sin x \cos x - 5 \sin x + 20 \cos x = 4$

75. $\sin^2 x + 2 \sin x - 2 = 0$

76. $\cos^2 x + 5 \cos x = 1$

77. $\tan^2 x + 1 = 3 \tan x$

78. $4 \cos^2 x - 2 \cos x = 1$

79. $2 \tan^2 x - 1 = 3 \tan x$

80. $6 \sin^2 x + 4 \sin x = 1$

81. $\sin^2 x + 3 \cos^2 x = 0$

82. $\sec^2 x - 2 \tan^2 x = 0$

83. $\sin 2x + \cos x = 0$

84. $\cos 2x - \sin x = 1$

85. $9 - 12 \sin x = 4 \cos^2 x$

86. $\sec^2 x + \tan x = 3$

87. $\cos^2 x - \sin^2 x + \sin x = 0$

88. $2\tan^2 x + \tan x = 5 - \sec^2 x$

89. $\sin\dfrac{x}{2} = 1 - \cos x$

90. $4\sin^2\left(\dfrac{x}{2}\right) + \cos^2 x = 2$

91. The number of hours of daylight in Detroit on day t of a non–leap year (with $t = 0$ being January 1) is given by the function
$d(t) = 3\sin\left[\dfrac{2\pi}{365}(t - 80)\right] + 12.$

(a) On what days of the year is there exactly 11 hours of daylight?

(b) What day has the maximum amount of daylight?

92. A weight hanging from a spring is set into motion (see Figure 6–48 on page 435), moving up and down. Its distance (in centimeters) above or below the equilibrium point at time t seconds is given by $d = 5(\sin 6t - 4\cos 6t)$. At what times during the first two seconds is the weight at the equilibrium position ($d = 0$)?

In Exercises 93–96, use the following fact: When a projectile (such as a ball or a bullet) leaves its starting point at angle of elevation θ with velocity v, the horizontal distance d it travels is given by the equation $d = \dfrac{v^2}{32}\sin 2\theta$, where d is measured in feet and v in feet per second. Note that the horizontal distance traveled may be the same for two different angles of elevation, so that some of these exercises may have more than one correct answer.

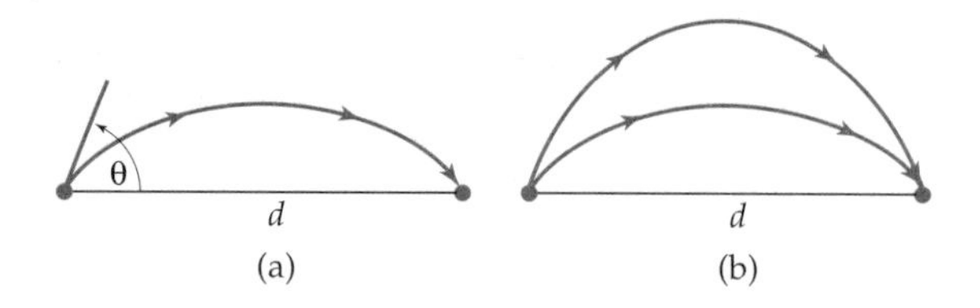

(a) (b)

93. If muzzle velocity of a rifle is 300 feet/second, at what angle of elevation (in radians) should it be aimed in order for the bullet to hit a target 2500 feet away?

94. Is it possible for the rifle in Exercise 93 to hit a target that is 3000 feet away? [At what angle of elevation would it have to be aimed?]

95. A fly ball leaves the bat at a velocity of 98 mph and is caught by an outfielder 288 feet away. At what angle of elevation (in degrees) did the ball leave that bat?

96. An outfielder throws the ball at a speed of 75 mph to the catcher who is 200 feet away. At what angle of elevation was the ball thrown?

Thinkers

97. Under what conditions (on the constant) does a basic equation involving the sine and cosine function have *no* solutions?

98. Do Exercise 97 for the secant and cosecant functions.

99. What is wrong with this so-called solution?

$$\sin x \tan x = \sin x$$
$$\tan x = 1$$
$$x = \frac{\pi}{4} \quad \text{or} \quad \frac{5\pi}{4}.$$

[*Hint:* Solve the original equation by moving all terms to one side and factoring. Compare your answers with the ones above.]

100. Let n be a fixed positive integer. Describe *all* solutions of the equation $\sin nx = 1/2$. [*Hint:* See Exercises 55–64.]

8.5 Inverse Trigonometric Functions

You should review the concept of inverse functions (Section 3.7) before reading this section. As explained there, a function cannot have an inverse function unless its graph has this property: No horizontal line intersects the graph more than once. The graphs of the sine, cosine, and tangent functions certainly don't have this property, as you can readily verify with your calculator. However, functions that are very closely related to these functions (same rules, but smaller domains) *do* have inverse functions.

We begin with the *restricted* sine function whose rule is $f(x) = \sin x$, but whose domain is restricted to the interval $[-\pi/2, \pi/2]$.* Its graph in Figure 8–14 shows that for each number v between -1 and 1, there is exactly one number u between $-\pi/2$ and $\pi/2$ such that $\sin u = v$.

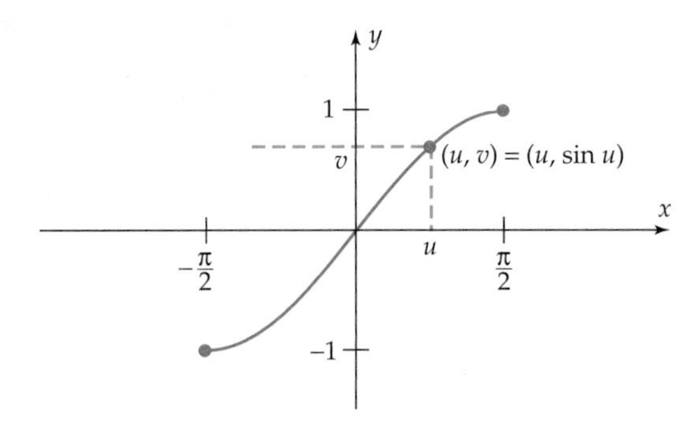

Figure 8–14

Since the graph of the restricted sine function passes the horizontal line test, we know that it has an inverse function. This inverse function is called the **inverse sine** (or **arcsine**) **function** and is denoted by $g(x) = \sin^{-1}x$ or $g(x) = \arcsin x$. The domain of the inverse sine function is the interval $[-1, 1]$ and its rule is

Inverse Sine Function

For each v with $-1 \le v \le 1$,

$$\sin^{-1} v = \text{the unique number } u \text{ between } -\pi/2 \text{ and } \pi/2 \text{ whose sine is } v;$$

that is,

$$\sin^{-1} v = u \qquad \text{exactly when} \qquad \sin u = v.$$

*Other ways of restricting the domain are possible. Those presented here for sine, cosine, and tangent are the ones universally agreed on by mathematicians.

Example 1 Find **(a)** $\sin^{-1}(1/2)$ **(b)** $\sin^{-1}(-\sqrt{2}/2)$.

Solution

(a) $\text{Sin}^{-1}(1/2)$ is the one number between $-\pi/2$ and $\pi/2$ whose sine is $1/2$. From our study of special values, we know that $\sin \pi/6 = 1/2$, *and* $\pi/6$ is between $-\pi/2$ and $\pi/2$. Hence, $\sin^{-1}(1/2) = \pi/6$.

(b) $\text{Sin}^{-1}(-\sqrt{2}/2) = -\pi/4$ because $\sin(-\pi/4) = -\sqrt{2}/2$ and $-\pi/4$ is between $-\pi/2$ and $\pi/2$. ■

Example 2 Except for special values you should use the SIN^{-1} key (labeled ASIN on some calculators) in *radian mode* to evaluate the inverse sine function. For instance,

$$\sin^{-1}(-.67) = -.7342 \qquad \text{and} \qquad \sin^{-1}(.42) = .4334. \quad ■$$

Example 3 If you key in SIN^{-1} 2 ENTER you will get an error message, because 2 is not in the domain of the inverse sine function.* ■

CAUTION

The notation $\sin^{-1}x$ is *not* exponential notation. It does *not* mean either $(\sin x)^{-1}$ or $\frac{1}{\sin x}$. For instance, Example 1 shows that $\sin^{-1}(1/2) = \pi/6 \approx .5236$, but

$$\left(\sin\frac{1}{2}\right)^{-1} = \frac{1}{\sin\frac{1}{2}} \approx \frac{1}{.4794} \approx 2.0858.$$

Suppose $-1 \le v \le 1$ and $\sin^{-1}v = u$. Then by the definition of the inverse sine function we know that $-\pi/2 \le u \le \pi/2$ and $\sin u = v$. Therefore,

$$\sin^{-1}(\sin u) = \sin^{-1}(v) = u \qquad \text{and} \qquad \sin(\sin^{-1}v) = \sin u = v.$$

This shows that the restricted sine function and the inverse sine function have the usual "round-trip properties" of inverse functions. In summary,

Properties of Inverse Sine

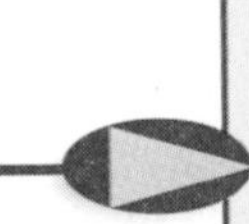

$$\sin^{-1}(\sin u) = u \quad \text{if} \quad -\frac{\pi}{2} \le u \le \frac{\pi}{2}$$

$$\sin(\sin^{-1}v) = v \quad \text{if} \quad -1 \le v \le 1$$

*TI-85/86 and HP-38 display the complex number $(1.5707\cdots, -1.3169\cdots)$ for $\sin^{-1}(2)$. For our purposes this is equivalent to an error message since we only deal with functions whose values are real numbers.

Example 4 Since $\sin \pi/6 = 1/2$, we see that

$$\sin^{-1}(\sin \pi/6) = \sin^{-1}\left(\frac{1}{2}\right) = \pi/6$$

because $\pi/6$ is between $-\pi/2$ and $\pi/2$. On the other hand, $\sin 5\pi/6$ is also $1/2$, so an expression such as $\sin^{-1}(\sin 5\pi/6)$ *is* defined. But

$$\sin^{-1}\left(\sin \frac{5\pi}{6}\right) = \sin^{-1}\left(\frac{1}{2}\right) = \frac{\pi}{6}, \qquad \textit{not} \qquad \frac{5\pi}{6}.$$

The identity in the box is valid only when u ranges from $-\pi/2$ to $\pi/2$. ■

Example 5 A calculator can visually demonstrate the identities in the preceding box. For instance, for any number v between -1 and 1 key in $\sin(\sin^{-1}v)$. The display will show the number you started with (except for minor round-off errors). ■

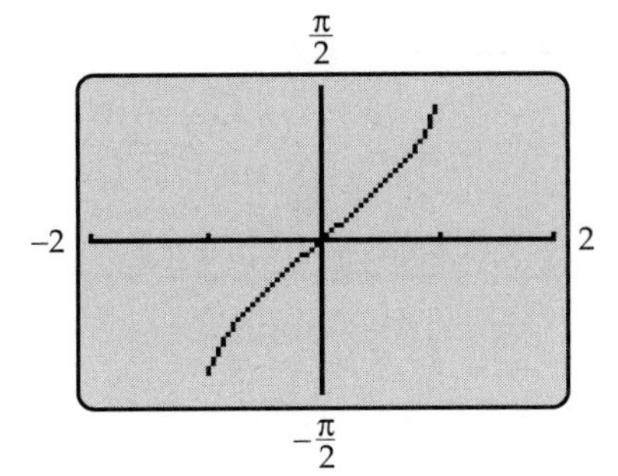

Figure 8–15

The graph of the inverse sine function is readily obtained from a calculator (Figure 8–15). Since the inverse sine function is the inverse of the restricted sine function, its graph is the reflection of the graph of the restricted sine function (Figure 8–14) in the line $y = x$ (as explained on page 192).

The Inverse Cosine Function

The rule of the *restricted* cosine function is $f(x) = \cos x$, and its domain is the interval $[0, \pi]$. Its graph in Figure 8–16 shows that for each v between -1 and 1 there is exactly one number u between 0 and π such that $\cos u = v$.

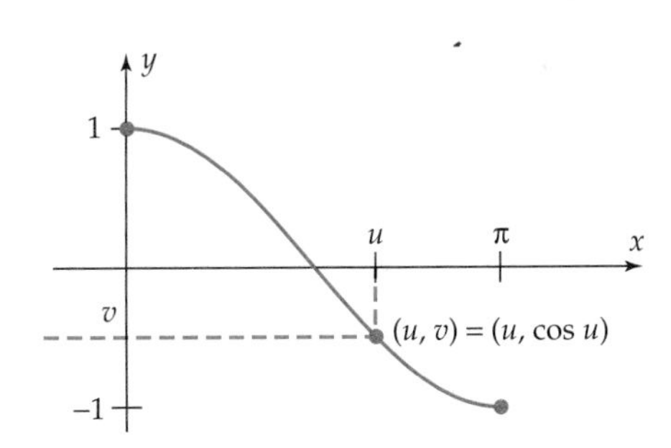

Figure 8–16

Since the graph of the restricted cosine function passes the horizontal line test, we know that it has an inverse function. This inverse function is called the **inverse cosine** (or **arccosine**) **function** and is denoted by $h(x) = \cos^{-1}x$ or $h(x) = \arccos x$. The domain of the inverse cosine function is the interval $[-1, 1]$ and its rule is as follows.

Inverse Cosine Function

For each v with $-1 \le v \le 1$,

$$\cos^{-1} v = \text{the unique number } u \text{ ranging from 0 to } \pi \text{ whose cosine is } v;$$

that is,

$$\cos^{-1} v = u \quad \text{exactly when} \quad \cos u = v.$$

The inverse cosine function has these properties:

$$\cos^{-1}(\cos u) = u \quad \text{if} \quad 0 \le u \le \pi;$$

$$\cos(\cos^{-1} v) = v \quad \text{if} \quad -1 \le v \le 1.$$

Example 6 Find **(a)** $\cos^{-1}(1/2)$ **(b)** $\cos^{-1}(0)$ **(c)** $\cos^{-1}(-.63)$.

Solution

(a) $\text{Cos}^{-1}(1/2) = \pi/3$ since $\pi/3$ is the unique number between 0 and π whose cosine is $1/2$.

(b) $\text{Cos}^{-1}(0) = \pi/2$ because $\cos \pi/2 = 0$ and $0 \le \pi/2 \le \pi$.

(c) The COS^{-1} key on a calculator in *radian mode* shows that $\cos^{-1}(-.63) = 2.2523$. ■

CAUTION

$\text{Cos}^{-1} x$ does *not* mean $(\cos x)^{-1}$, or $1/\cos x$.

Example 7 Write $\sin(\cos^{-1} v)$ as an algebraic expression in v.

Solution $\text{Cos}^{-1} v = u$, where $\cos u = v$ and $0 \le u \le \pi$. Hence, $\sin u$ is nonnegative, and by the Pythagorean identity, $\sin u = \sqrt{\sin^2 u} = \sqrt{1 - \cos^2 u}$. Also, $\cos^2 u = v^2$. Therefore,

$$\sin(\cos^{-1} v) = \sin u = \sqrt{1 - \cos^2 u} = \sqrt{1 - v^2}. \quad ■$$

Example 8 Prove the identity $\sin^{-1} x + \cos^{-1} x = \pi/2$.

Solution The identity is defined only when $-1 \le x \le 1$ and is true for $x = 0$ since $\sin^{-1} 0 = 0$ and $\cos^{-1} 0 = \pi/2$. Assume that $0 < x \le 1$ and consider a right triangle with hypotenuse 1, side x, and acute angles that measure u radians and w radians, respectively (Figure 8–17). We have

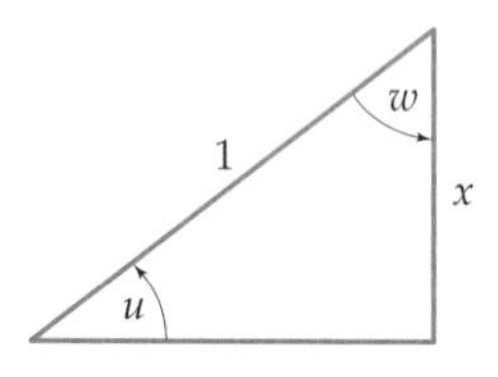

Figure 8–17

$$\sin u = \frac{\text{opposite}}{\text{hypotenuse}} = \frac{x}{1} = x \quad \text{and} \quad \cos w = \frac{\text{adjacent}}{\text{hypotenuse}} = \frac{x}{1} = x.$$

Therefore,

$$\sin^{-1} x = u \quad \text{and} \quad \cos^{-1} x = w.$$

But $u + w = \pi/2$; hence,

$$\sin^{-1} x + \cos^{-1} x = u + w = \pi/2.$$

The proof when $-1 \le x < 0$ is given in Exercise 62. ■

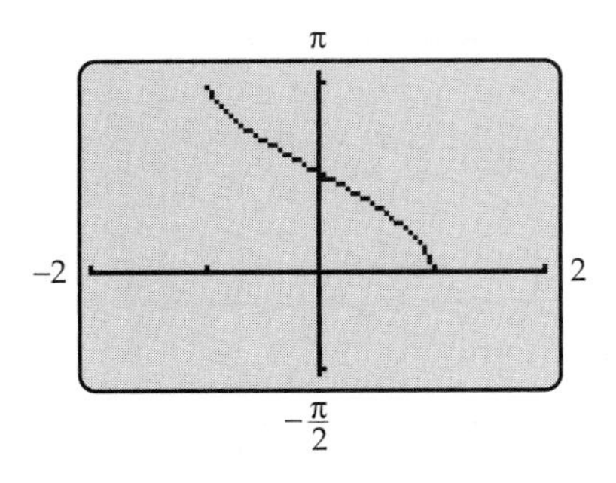

Figure 8–18

The graph of the inverse cosine function, which is the reflection of the graph of the restricted cosine function (Figure 8–16) in the line $y = x$, is shown in Figure 8–18.

The Inverse Tangent Function

The rule of the *restricted* tangent function is $f(x) = \tan x$, and its domain is the open interval $(-\pi/2, \pi/2)$. Its graph in Figure 8–19 shows that for every real number v, there is exactly one number u between $-\pi/2$ and $\pi/2$ such that $\tan u = v$.

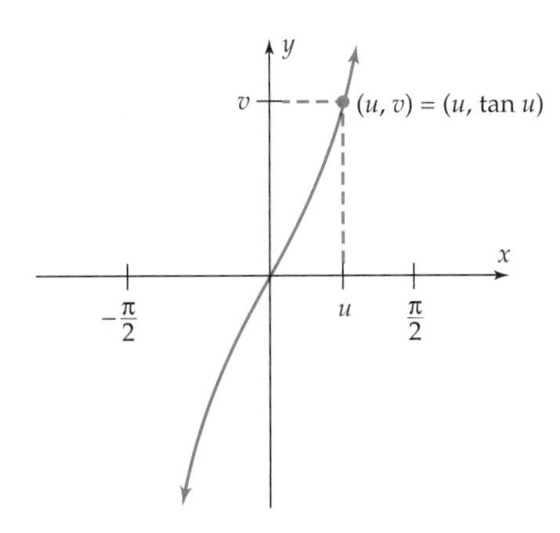

Figure 8–19

Since the graph of the restricted tangent function passes the horizontal line test, we know that it has an inverse function. This inverse function is called the **inverse tangent** (or **arctangent**) **function** and is denoted by $g(x) = \tan^{-1} x$ or $g(x) = \arctan x$. The domain of the inverse tangent function is the set of all real numbers and its rule is as follows.

Inverse Tangent Function

For each real number v,

$$\tan^{-1} v = \text{the unique number } u \text{ between } -\pi/2 \text{ and } \pi/2 \text{ whose tangent is } v;$$

that is,

$$\tan^{-1} v = u \quad \text{exactly when} \quad \tan u = v.$$

The inverse tangent function has these properties:

$$\tan^{-1}(\tan u) = u \quad \text{if} \quad -\frac{\pi}{2} < u < \frac{\pi}{2};$$

$$\tan(\tan^{-1} v) = v \quad \text{for every number } v.$$

CAUTION
$\text{Tan}^{-1}x$ does *not* mean $(\tan x)^{-1}$, or $1/\tan x$.

Example 9 $\text{Tan}^{-1}1 = \pi/4$ because $\pi/4$ is the unique number between $-\pi/2$ and $\pi/2$ such that $\tan \pi/4 = 1$. A calculator in *radian mode* shows that $\tan^{-1}(136) = 1.5634$. ■

Example 10 Find the exact value of $\cos\left[\tan^{-1}(\sqrt{5}/2)\right]$.

Solution Let $\tan^{-1}(\sqrt{5}/2) = u$; then $\tan u = \sqrt{5}/2$ and $-\pi/2 < u < \pi/2$. Since $\tan u$ is positive, u must be between 0 and $\pi/2$. Construct a right triangle containing an angle of u radians whose tangent is $\sqrt{5}/2$ (Figure 8–20):

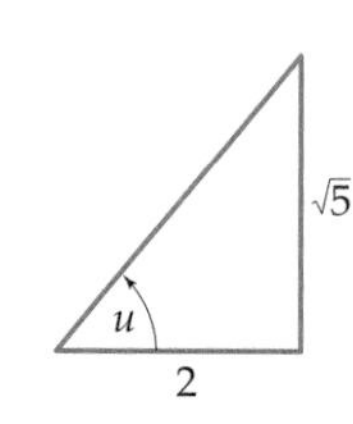

$$\tan u = \frac{\text{opposite}}{\text{adjacent}} = \frac{\sqrt{5}}{2}.$$

Figure 8–20

The hypotenuse has length $\sqrt{2^2 + (\sqrt{5})^2} = \sqrt{4+5} = 3$. Therefore,

$$\cos\left[\tan^{-1}\left(\frac{\sqrt{5}}{2}\right)\right] = \cos u = \frac{\text{adjacent}}{\text{hypotenuse}} = \frac{2}{3}. \quad ■$$

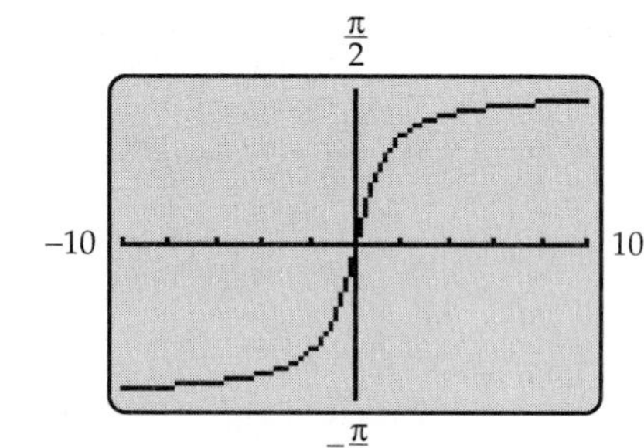

Figure 8–21

The graph of the inverse tangent function (Figure 8–21) is the reflection in the line $y = x$ of the graph of the restricted tangent function.

Exercises 8.5

In Exercises 1–14, find the exact functional value without using a calculator.

1. $\sin^{-1}1$
2. $\cos^{-1}0$
3. $\tan^{-1}(-1)$
4. $\sin^{-1}(-1)$
5. $\cos^{-1}1$
6. $\tan^{-1}1$
7. $\tan^{-1}(\sqrt{3}/3)$
8. $\cos^{-1}(\sqrt{3}/2)$
9. $\sin^{-1}(-\sqrt{2}/2)$
10. $\sin^{-1}(\sqrt{3}/2)$
11. $\tan^{-1}(-\sqrt{3})$
12. $\cos^{-1}(-\sqrt{2}/2)$
13. $\cos^{-1}\left(-\frac{1}{2}\right)$
14. $\sin^{-1}\left(-\frac{1}{2}\right)$

In Exercises 15–24, use a calculator in radian mode to approximate the functional value.

15. $\sin^{-1}.35$
16. $\cos^{-1}.76$
17. $\tan^{-1}(-3.256)$
18. $\sin^{-1}(-.795)$
19. $\sin^{-1}(\sin 7)$ [The answer is *not* 7.]
20. $\cos^{-1}(\cos 3.5)$
21. $\tan^{-1}[\tan(-4)]$
22. $\sin^{-1}[\sin(-2)]$
23. $\cos^{-1}[\cos(-8.5)]$
24. $\tan^{-1}(\tan 12.4)$
25. Given that $u = \sin^{-1}(-\sqrt{3}/2)$, find the exact value of $\cos u$ and $\tan u$.
26. Given that $u = \tan^{-1}(4/3)$, find the exact value of $\sin u$ and $\sec u$.

In Exercises 27–42, find the exact functional value without using a calculator.

27. $\sin^{-1}(\cos 0)$
28. $\cos^{-1}(\sin \pi/6)$
29. $\cos^{-1}(\sin 4\pi/3)$
30. $\tan^{-1}(\cos \pi)$
31. $\sin^{-1}(\cos 7\pi/6)$
32. $\cos^{-1}(\tan 7\pi/4)$

33. $\sin^{-1}(\sin 2\pi/3)$ (See Exercise 19.)

34. $\cos^{-1}(\cos 5\pi/4)$ **35.** $\cos^{-1}[\cos(-\pi/6)]$

36. $\tan^{-1}[\tan(-4\pi/3)]$

37. $\sin[\cos^{-1}(3/5)]$ (See Example 10.)

38. $\tan[\sin^{-1}(3/5)]$ **39.** $\cos[\tan^{-1}(-3/4)]$

40. $\cos\left[\sin^{-1}\left(\sqrt{3}/5\right)\right]$ **41.** $\tan[\sin^{-1}(5/13)]$

42. $\sin\left[\cos^{-1}\left(3/\sqrt{13}\right)\right]$

In Exercises 43–46, write the expression as an algebraic expression in v, as in Example 7.

43. $\cos(\sin^{-1}v)$ **44.** $\cot(\cos^{-1}v)$

45. $\tan(\sin^{-1}v)$ **46.** $\sin(2\sin^{-1}v)$

In Exercises 47–50, graph the function.

47. $f(x) = \cos^{-1}(x+1)$ **48.** $g(x) = \tan^{-1}x + \pi$

49. $h(x) = \sin^{-1}(\sin x)$ **50.** $k(x) = \sin(\sin^{-1}x)$

51. In an alternating current circuit, the voltage is given by the formula

$$V = V_{max} \cdot \sin(2\pi ft + \phi),$$

where V_{max} is the maximum voltage, f is the frequency (in cycles per second), t is the time in seconds, and ϕ is the phase angle.

(a) If the phase angle is 0, solve the voltage equation for t.

(b) If $\phi = 0$, $V_{max} = 20$, $V = 8.5$, and $f = 120$, find the smallest positive value of t.

52. A model plane 40 feet above the ground is flying away from an observer.

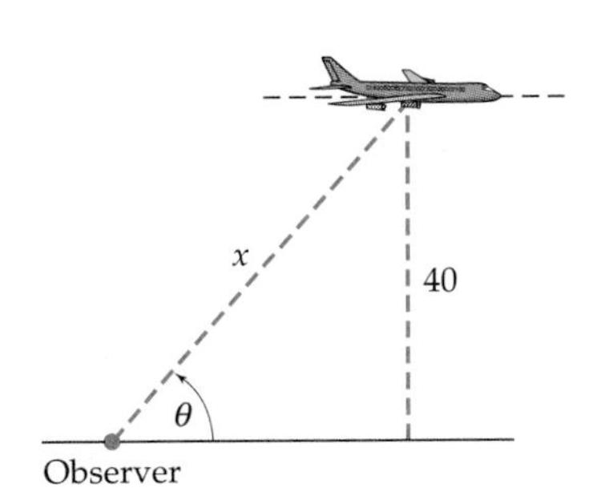

(a) Express the angle of elevation θ of the plane as a function of the distance x from the observer to the plane.

(b) What is θ when the plane is 250 feet from the observer?

53. A 15-foot-wide highway sign is placed 10 feet from a road, perpendicular to the road (see figure). A spotlight at the edge of the road is aimed at the sign.

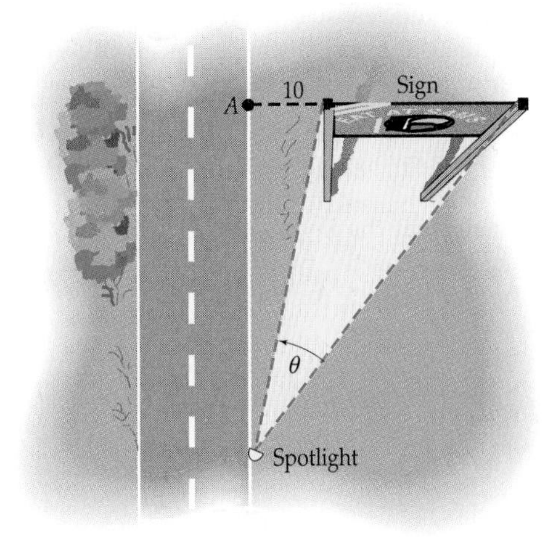

(a) Express θ as a function of the distance x from point A to the spotlight.

(b) How far from point A should the spotlight be placed so that the angle θ is as large as possible?

54. A camera on a 5-foot-high tripod is placed in front of a 6-foot-high picture that is mounted 3 feet above the floor.

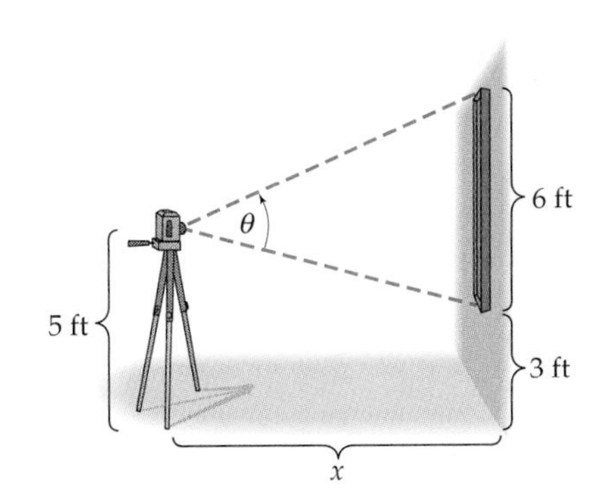

(a) Express angle θ as a function of the distance x from the camera to the wall.

(b) The photographer wants to use a particular lens, for which $\theta = 36°$ ($\pi/5$ radians). How far should she place the camera from the wall to be sure the entire picture will show in the photograph?

55. Show that the restricted secant function, whose domain consists of all numbers x such that $0 \le x \le \pi$ and $x \ne \pi/2$, has an inverse function. Sketch its graph.

56. Show that the restricted cosecant function, whose domain consists of all numbers x such that $-\pi/2 \le x \le \pi/2$ and $x \ne 0$, has an inverse function. Sketch its graph.

57. Show that the restricted cotangent function, whose domain is the interval $(0, \pi)$, has an inverse function. Sketch its graph.

58. (a) Show that the inverse cosine function actually has the two properties listed in the box on page 552.

(b) Show that the inverse tangent function actually has the two properties listed in the box on page 553.

In Exercises 59–67, prove the identity.

59. $\sin^{-1}(-x) = -\sin^{-1}x$ [*Hint:* Let $u = \sin^{-1}(-x)$ and show that $\sin^{-1}x = -u$.]

60. $\tan^{-1}(-x) = -\tan^{-1}x$

61. $\cos^{-1}(-x) = \pi - \cos^{-1}x$ [*Hint:* Let $u = \cos^{-1}(-x)$ and show that $0 \le \pi - u \le \pi$; use the identity $\cos(\pi - u) = -\cos u$.]

62. $\sin^{-1}x + \cos^{-1}x = \pi/2$ [*Hint:* The case when $0 \le x \le 1$ was demonstrated in Example 8. If $-1 < x < 0$, then $0 < -x \le 1$; use Exercises 59 and 61.]

63. $\tan^{-1}(\cot x) = \pi/2 - x \quad (0 < x < \pi)$

64. $\sin^{-1}(\cos x) = \pi/2 - x \quad (0 \le x \le \pi)$

65. $\sin^{-1}x = \tan^{-1}\left(\dfrac{x}{\sqrt{1 - x^2}}\right) \quad (-1 < x < 1)$

[*Hint:* Let $u = \sin^{-1}x$ and show that $\tan u = x/\sqrt{1 - x^2}$. Since $\sin u = x$, $\cos u = \pm\sqrt{1 - x^2}$. Show that in this case, $\cos u = \sqrt{1 - x^2}$.]

66. $\cos^{-1}x = \dfrac{\pi}{2} - \tan^{-1}\left(\dfrac{x}{\sqrt{1 - x^2}}\right) \quad (-1 < x < 1)$
[*Hint:* See Exercises 62 and 65.]

67. $\tan^{-1}x + \tan^{-1}\left(\dfrac{1}{x}\right) = \dfrac{\pi}{2}$

68. Using the viewing window with $-2\pi \le x \le 2\pi$ and $-4 \le y \le 4$ graph the functions $f(x) = \cos(\cos^{-1}x)$ and $g(x) = \cos^{-1}(\cos x)$. How do you explain the shapes of the two graphs?

69. Is it true that $\tan^{-1}x = \dfrac{\sin^{-1}x}{\cos^{-1}x}$? Justify your answer.

Chapter 8 Review

Important Concepts

Section 8.1	Reciprocal identities 510 Periodicity identities 510 Pythagorean identities 510 Negative angle identities 510 Strategies and proof techniques for identities 511–514
Section 8.2	Addition and subtraction identities 518, 520 Cofunction identities 521
Section 8.2.A	Angle of inclination 526 Angle between two lines 528 Slope theorem for perpendicular lines 529
Section 8.3	Double-angle identities 530, 532 Power-reducing identities 532 Half-angle identities 533–534 Product identities 535 Factoring identities 536
Section 8.4	Graphical solution of trigonometric equations 539 Solution algorithms for basic trigonometric equations 543 Algebraic solution of trigonometric equations 544
Section 8.5	Inverse sine function 549–550 Inverse cosine function 552 Inverse tangent function 553

Important Facts and Formulas

- All identities in the Chapter 6 Review
- *Addition and Subtraction Identities:*

$$\sin(x + y) = \sin x \cos y + \cos x \sin y$$
$$\sin(x - y) = \sin x \cos y - \cos x \sin y$$
$$\cos(x + y) = \cos x \cos y - \sin x \sin y$$
$$\cos(x - y) = \cos x \cos y + \sin x \sin y$$
$$\tan(x + y) = \frac{\tan x + \tan y}{1 - \tan x \tan y}$$
$$\tan(x - y) = \frac{\tan x - \tan y}{1 + \tan x \tan y}$$

- *Cofunction Identities:*

$$\sin x = \cos\left(\frac{\pi}{2} - x\right) \qquad \cos x = \sin\left(\frac{\pi}{2} - x\right)$$
$$\tan x = \cot\left(\frac{\pi}{2} - x\right) \qquad \cot x = \tan\left(\frac{\pi}{2} - x\right)$$
$$\sec x = \csc\left(\frac{\pi}{2} - x\right) \qquad \csc x = \sec\left(\frac{\pi}{2} - x\right)$$

- *Double Angle Identities:*

$$\sin 2x = 2\sin x\cos x$$
$$\cos 2x = \cos^2 x - \sin^2 x$$
$$\cos 2x = 2\cos^2 x - 1 \qquad \cos 2x = 1 - 2\sin^2 x$$
$$\tan 2x = \frac{2\tan x}{1 - \tan^2 x}$$

- *Half-Angle Identities:*

$$\sin\frac{x}{2} = \pm\sqrt{\frac{1-\cos x}{2}}$$
$$\cos\frac{x}{2} = \pm\sqrt{\frac{1+\cos x}{2}}$$
$$\tan\frac{x}{2} = \frac{1-\cos x}{\sin x} \qquad \tan\frac{x}{2} = \frac{\sin x}{1+\cos x}$$

- $\sin^{-1} v = u$ exactly when $\sin u = v$ $\left(-\frac{\pi}{2} \le u \le \frac{\pi}{2}, -1 \le v \le 1\right)$
- $\cos^{-1} v = u$ exactly when $\cos u = v$ $(0 \le u \le \pi, -1 \le v \le 1)$
- $\tan^{-1} v = u$ exactly when $\tan u = v$ $\left(-\frac{\pi}{2} < u < \frac{\pi}{2}, \text{any } v\right)$

Review Questions

In Questions 1–4, simplify the given expression.

1. $\dfrac{\sin^2 t + (\tan^2 t + 2\tan t - 4) + \cos^2 t}{3\tan^2 t - 3\tan t}$
2. $\dfrac{\sec^2 t\csc t}{\csc^2 t\sec t}$
3. $\dfrac{\tan^2 x - \sin^2 x}{\sec^2 x}$
4. $\dfrac{(\sin x + \cos x)(\sin x - \cos x) + 1}{\sin^2 x}$

In Questions 5–12, determine graphically whether the equation could possibly be an identity. If it could, prove that it is.

5. $\sin^4 t - \cos^4 t = 2\sin^2 t - 1$
6. $1 + 2\cos^2 t + \cos^4 t = \sin^4 t$
7. $\dfrac{\sin t}{1 - \cos t} = \dfrac{1 + \cos t}{\sin t}$
8. $\dfrac{\sin^2 t}{\cos^2 t} + 1 = \dfrac{1}{\cos^2 t}$
9. $\dfrac{\cos^2(\pi + t)}{\sin^2(\pi + t)} - 1 = \dfrac{1}{\sin^2 t}$
10. $\tan x + \cot x = \sec x\csc x$
11. $(\sin x + \cos x)^2 - \sin 2x = 1$
12. $\dfrac{1 - \cos 2x}{\tan x} = \sin 2x$

In Questions 13–22, prove the given identity.

13. $\dfrac{\tan x - \sin x}{2\tan x} = \sin^2\left(\dfrac{x}{2}\right)$
14. $2\cos x - 2\cos^3 x = \sin x\sin 2x$
15. $\cos(x + y)\cos(x - y) = \cos^2 x - \sin^2 y$

16. $\dfrac{\cos(x - y)}{\cos x \cos y} = 1 + \tan x \tan y$

17. $\dfrac{\sec x + 1}{\tan x} = \dfrac{\tan x}{\sec x - 1}$

18. $\dfrac{\cos^4 x - \sin^4 x}{1 - \tan^4 x} = \cos^4 x$

19. $\dfrac{1 + \tan^2 x}{\tan^2 x} = \csc^2 x$

20. $\sec x - \cos x = \sin x \tan x$

21. $\tan^2 x - \sec^2 x = \cot^2 x - \csc^2 x$

22. $\sin 2x = \dfrac{1}{\tan x + \cot 2x}$

23. If $\tan x = 5/12$ and $\sin x > 0$, find $\sin 2x$.

24. If $\cos x = 15/17$ and $0 < x < \pi/2$, find $\sin(x/2)$.

25. If $\tan x = 4/3$ and $\pi < x < 3\pi/2$, and $\cot y = -5/12$ with $3\pi/2 < y < 2\pi$, find $\sin(x - y)$.

26. If $\sin x = -12/13$ with $\pi < x < 3\pi/2$, and $\sec y = 13/12$ with $3\pi/2 < y < 2\pi$, find $\cos(x + y)$.

27. If $\sin x = 1/4$ and $0 < x < \pi/2$, then $\sin(\pi/3 + x) = ?$

28. If $\sin x = -2/5$ and $3\pi/2 < x < 2\pi$, then $\cos(\pi/4 + x) = ?$

29. If $\sin x = 0$, is it true that $\sin 2x = 0$? Justify your answer.

30. If $\cos x = 0$, is it true that $\cos 2x = 0$? Justify your answer.

31. Show that $\sqrt{2 + \sqrt{3}} = \dfrac{\sqrt{2} + \sqrt{6}}{2}$ by computing $\cos(\pi/12)$ in two ways, using the half-angle identity and the subtraction identity for cosine.

32. True or false: $2 \sin x = \sin 2x$. Justify your answer.

33. $\sin(5\pi/12) = ?$

34. Express $\sec(x - \pi)$ in terms of $\sin x$ and $\cos x$.

35. $\sqrt{\dfrac{1 - \cos^2 x}{1 - \sin^2 x}} =$ ____.

(a) $|\tan x|$
(b) $|\cot x|$
(c) $\sqrt{\dfrac{1 - \sin^2 x}{1 - \cos^2 x}}$
(d) $\sec x$
(e) undefined

36. $\dfrac{1}{(\csc x)(\sec^2 x)} =$ ____.

(a) $\dfrac{1}{(\sin x)(\cos^2 x)}$
(b) $\sin x - \sin^3 x$
(c) $\dfrac{1}{(\sin x)(1 + \tan^2 x)}$
(d) $\sin x - \dfrac{1}{1 + \tan^2 x}$
(e) $1 + \tan^3 x$

37. If $\sin x = .6$ and $0 < x < \pi/2$, find $\sin 2x$.

38. If $\sin x = .6$ and $0 < x < \pi/2$, find $\sin(x/2)$.

39. Find the angle of inclination of the straight line through the points (2, 6) and (−2, 2).

40. Find one of the angles between the line L through the points $(-3, 2)$ and $(5, 1)$ and the line M, which has slope 2.

In Questions 41–44, solve the equation graphically.

41. $5 \tan x = 2 \sin 2x$

42 $\sin^3 x + \cos^2 x - \tan x = 2$

43. $\sin x + \sec^2 x = 3$

44. $\cos^2 x - \csc^2 x + \tan(x - \pi/2) + 5 = 0$

In Questions 45–60, solve the equation by any means. Find exact solutions when possible and approximate ones otherwise.

45. $2 \sin x = 1$

46. $\cos x = \sqrt{3}/2$

47. $\tan x = -1$

48. $\sin 3x = -\sqrt{3}/2$

49. $\sin x = .7$

50. $\cos x = -.8$

51. $\tan x = 13$

52. $\cot x = .4$

53. $2 \sin^2 x + 5 \sin x = 3$

54. $4 \cos^2 x - 2 = 0$

55. $2 \sin^2 x - 3 \sin x = 2$

56. $\cos 2x = \cos x$
[*Hint:* First use an identity.]

57. $\sec^2 x + 3 \tan^2 x = 13$

58. $\sec^2 x = 4 \tan x - 2$

59. $2 \sin^2 x + \sin x - 2 = 0$

60. $\cos^2 x - 3 \cos x - 2 = 0$

61. Find all angles θ with $0° \leq \theta \leq 360°$ such that $\sin \theta = -.7133$.

62. Find all angles θ with $0° \leq \theta \leq 360°$ such that $\tan \theta = 3.7321$.

63. A cannon has a muzzle velocity of 600 feet per second. At what angle of elevation should it be fired in order to hit a target 3500 feet away? [*Hint:* Use the projectile equation preceding Exercise 93 of Section 8.4.]

64. A weight hanging from a spring is set into motion (see Figure 6–48 on page 435), moving up and down. Its distance (in centimeters) above or below the equilibrium point at time t seconds is given by $d = 5 \sin 3t - 3 \cos 3t$. At what times during the first two seconds is the weight at the equilibrium position $(d = 0)$?

65. $\cos^{-1}(\sqrt{2}/2) = ?$

66. $\sin^{-1}(\sqrt{3}/2) = ?$

67. $\tan^{-1}\sqrt{3} = ?$

68. $\sin^{-1}(\cos 11\pi/6) = ?$

69. $\cos^{-1}(\sin 5\pi/3) = ?$

70. $\tan^{-1}(\cos 7\pi/2) = ?$

71. $\sin^{-1}(\sin .75) = ?$

72. $\cos^{-1}(\cos 2) = ?$

73. $\sin^{-1}(\sin 8\pi/3) = ?$

74. $\cos^{-1}(\cos 13\pi/4) = ?$

75. Sketch the graph of $f(x) = \tan^{-1} x - \pi$.

76. Sketch the graph of $g(x) = \sin^{-1}(x - 2)$.

77. Find the exact value of $\sin[\cos^{-1}(1/4)]$.

78. Find the exact value of $\sin[\tan^{-1}(1/2) - \cos^{-1}(4/5)]$.

Discovery Project 8

The Sun and the Moon

It has long been known that the cycles of the sun and the moon are periodic; that is, the moon is full at regular intervals and it is new at regular intervals. The same is true of the sun; the interval between the summer and winter solstices is also regular. It is therefore quite natural to use the sun and the moon to keep track of time and also to study the interaction between the two in order to predict events like full moons, new moons, solstices, equinoxes, and eclipses. Indeed, these solar and lunar events have consequences on earth, including the succession of the seasons and the severity of tides.

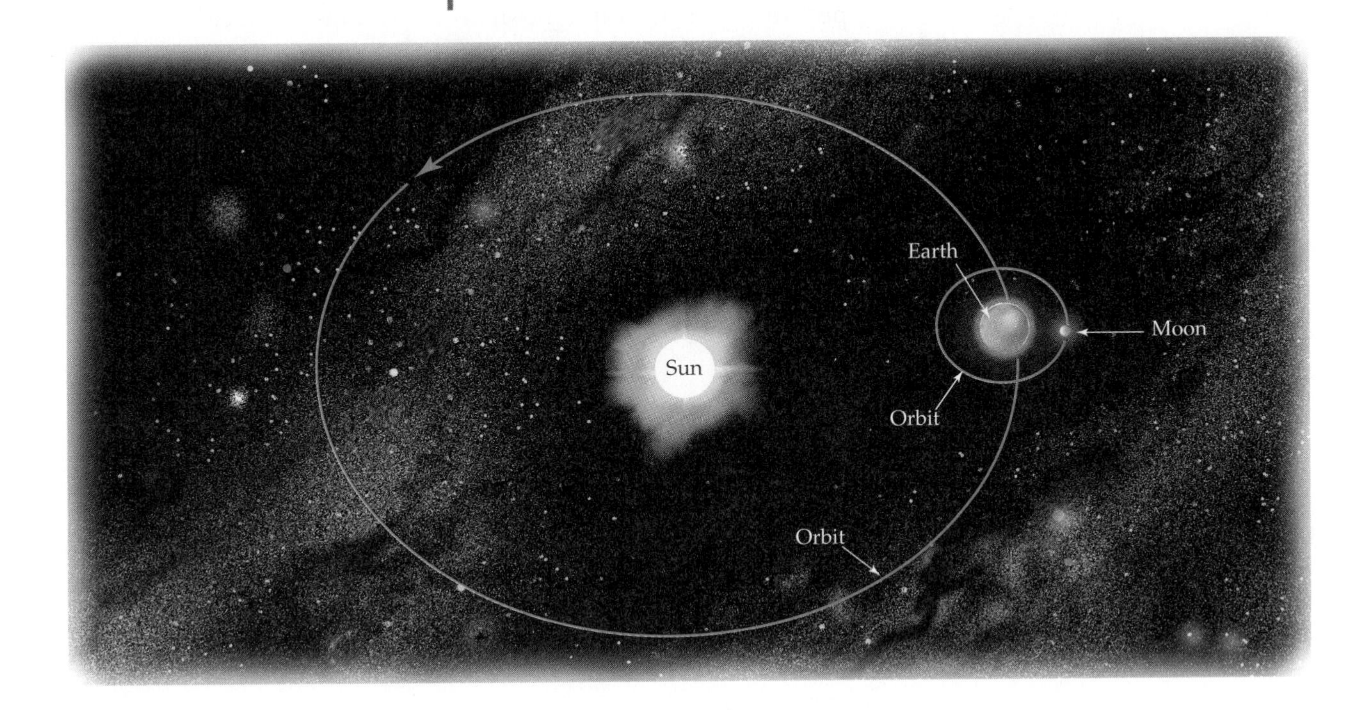

1. The following is a list of the days in 1999 when the moon was full. The length of the lunar month is the length of time between full moons. Use the data to approximate the length of the lunar month.

January 2	May 30	October 24
January 31	June 28	November 23
March 2	July 28	December 22
March 31	August 26	
April 30	September 25	

2. Write a function which has value 1 when the moon is full and 0 when the moon is new. Measure the independent variable in days with January 2, 1999, set as time 0. Use a function of the form $m(x) = \dfrac{\cos kt + 1}{2}$ with period equal to the length of the lunar month.
3. Use your function to predict the date of the first full moon of the 21st century (in January of the year 2001). Does your function agree with the actual date of January 9? If not, what could have caused the discrepancy?
4. The solar year is approximately 365.24 days long. Write a function $s(x)$ with period equal to the length of the solar year so that on the date of the summer solstice, $s(x) = 1$ and on the day of the winter solstice, $s(x) = 0$. The summer solstice in 1999 was June 21 and the winter solstice falls midway between summer solstices.
5. If your functions $s(x)$ and $m(x)$ were accurate, when would you expect to see the next full moon on the summer solstice?
6. How would you go about making your models $s(x)$ and $m(x)$ more accurate?

Chapter

9 Applications of Trigonometry

Is this bridge safe?

y
$(a - b, c - d)$
(a, b)
$\mathbf{u} - \mathbf{v}$
$\mathbf{u}$
$-\mathbf{v}$
$\mathbf{w}$
x
$\mathbf{v}$
(c, d)

When planning a bridge or a building, architects and engineers must determine the stress on cables and other parts of the structure to be sure that all parts are adequately supported. Problems like these can be modeled and solved by using vectors. See Exercise 73 on page 592.

Interdependence of Chapters

The chapter is divided into two independent parts, each consisting of two sections. The parts may be read in either order:

9.1 ⟶ 9.2

9.3 ⟶ 9.4

Chapter Outline

Trigonometry has a variety of useful applications in geometry, algebra, and the physical sciences, several of which are discussed in this chapter.

9.1 The Complex Plane and Polar Form for Complex Numbers*

The real number system is represented geometrically by the number line. The complex number system can be represented geometrically by the coordinate plane:

The complex number $a + bi$ corresponds to the point (a, b) in the plane.

For example, the point (2, 3) in Figure 9–1 is labeled by $2 + 3i$; and similarly for the other points shown:

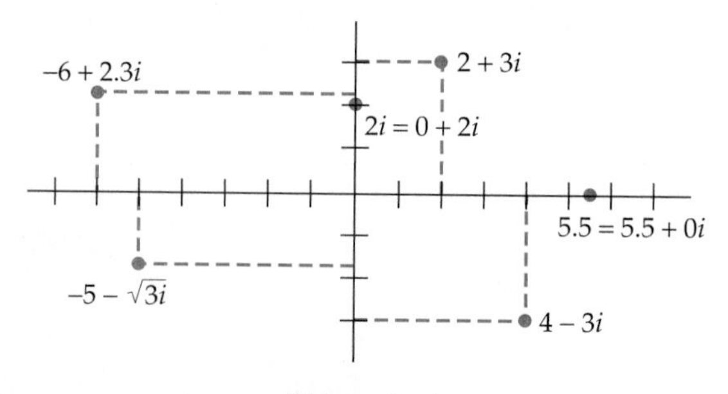

Figure 9–1

When the coordinate plane is labeled by complex numbers in this way, it is called the **complex plane.** Each real number $a = a + 0i$ corresponds to the point $(a, 0)$ on the horizontal axis; so this axis is called the **real axis.** The vertical axis is called the **imaginary axis** because every imaginary number $bi = 0 + bi$ corresponds to the point $(0, b)$ on the vertical axis.

Technology Tip

Recall that complex numbers are entered by using the special i key on the TI-83/89 and Sharp 9600 keyboards or in the CPLX submenu of the Casio 9850 OPTN menu. On TI-86 and HP-38, $a + bi$ is entered as (a, b). TI-82 does not handle complex numbers.

*Section 4.7 is a prerequisite for this section.

The absolute value of a real number c is the distance from c to 0 on the number line (see page 10). So we define the **absolute value** (or **modulus**) of the complex number $a + bi$ to be the distance from $a + bi$ to the origin in the complex plane. Hence,

$$|a + bi| = \text{distance from } (a, b) \text{ to } (0, 0) = \sqrt{(a - 0)^2 + (b - 0)^2}.$$

Therefore,

Absolute Value

The *absolute value* (or *modulus*) of the complex number $a + bi$ is

$$|a + bi| = \sqrt{a^2 + b^2}.$$

Technology Tip

Use the ABS key to find the absolute value of a complex number. It is on the HP-38 keyboard, in the CPLX menu of TI-86, in the CPX or NUM submenus of the TI-83/89 and Sharp 9600 MATH menus, and in the CPLX submenu of the Casio 9850 OPTN menu.

Example 1

(a) $|3 + 2i| = \sqrt{3^2 + 2^2} = \sqrt{13}.$

(b) $|4 - 5i| = \sqrt{4^2 + (-5)^2} = \sqrt{41}.$ ■

Absolute values and trigonometry now lead to a useful way of representing complex numbers. Let $a + bi$ be a nonzero complex number and denote $|a + bi|$ by r. Then r is the length of the line segment joining (a, b) and $(0, 0)$ in the plane. Let θ be the angle in standard position with this line segment as terminal side (Figure 9–2).

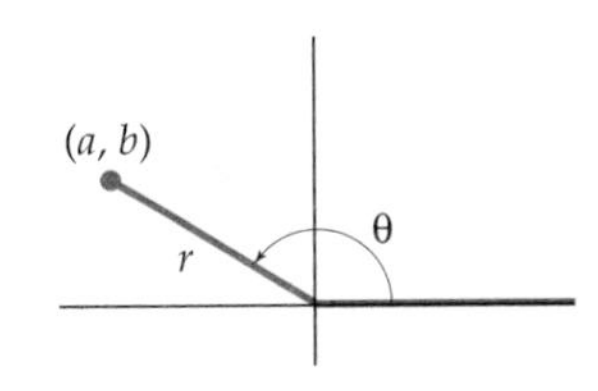

Figure 9–2

Technology Tip

Use the ARG or ANGLE key to find the argument θ. It is in the CPLX menu of TI-86, in the CPX or COMPLEX submenu of the TI-83/89, Sharp 9600, and HP-38 MATH menus, and in the CPLX submenu of the Casio 9850 OPTN menu.

According to the point-in-the-plane description of sine and cosine

$$\cos\theta = \frac{a}{r} \quad \text{and} \quad \sin\theta = \frac{b}{r}$$

so that

$$a = r\cos\theta \quad \text{and} \quad b = r\sin\theta.$$

Consequently,

$$a + bi = r\cos\theta + (r\sin\theta)i = r(\cos\theta + i\sin\theta).^*$$

When a complex number $a + bi$ is written in this way, it is said to be in **polar form** or **trigonometric form.** The angle θ is called the **argument** and

*It is customary to place i in front of $\sin\theta$ rather than after it. Some books abbreviate $r(\cos\theta + i\sin\theta)$ as $r \operatorname{cis} \theta$.

is usually expressed in radian measure. The number $r = |a + bi|$ is sometimes called the **modulus** (plural, moduli). The number 0 can also be written in polar notation by letting $r = 0$ and θ be any angle. Thus,

Polar Form

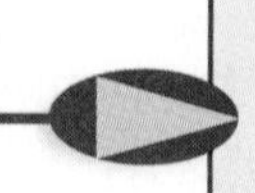

Every complex number $a + bi$ can be written in polar form

$$r(\cos\theta + i\sin\theta)$$

where $r = |a + bi| = \sqrt{a^2 + b^2}$, $a = r\cos\theta$, and $b = r\sin\theta$.

When a complex number is written in polar form, the argument θ is not uniquely determined since θ, $\theta \pm 2\pi$, $\theta \pm 4\pi$, etc., all satisfy the conditions in the box.

Technology Tip

The complex number $r(\cos\theta + i\sin\theta)$ is entered as $r\angle\theta$ on TI-86/89 and Sharp 9600 ($\angle$ is on the keyboard) and as $re^{\theta i}$ on TI-83 (use the special i key). It can also be entered as $re^{i\theta}$ on TI-89.

Example 2 Express $-\sqrt{3} + i$ in polar form.

Solution Here $a = -\sqrt{3}$ and $b = 1$, so that

$$r = \sqrt{a^2 + b^2} = \sqrt{(-\sqrt{3})^2 + 1^2} = \sqrt{3 + 1} = 2.$$

The angle θ must satisfy

$$\cos\theta = \frac{a}{r} = \frac{-\sqrt{3}}{2} \quad \text{and} \quad \sin\theta = \frac{b}{r} = \frac{1}{2}.$$

Since $-\sqrt{3} + i$ lies in the second quadrant (Figure 9–3), θ must be a second-quadrant angle. Our knowledge of special angles and Figure 9–3 show that $\theta = 5\pi/6$ satisfies these conditions. Hence,

$$-\sqrt{3} + i = 2\left(\cos\frac{5\pi}{6} + i\sin\frac{5\pi}{6}\right). \quad ■$$

Technology Tip

Most calculators can convert complex numbers from rectangular to polar form or vice versa. Check your instruction manual. On TI calculators, use ▶POL or ▶RECT in the CMPLX menu of TI-86, the CPX submenu of the TI-83 MATH menu, or the VECTOR OPS submenu of the MATRIX submenu of the TI-89 MATH menu.

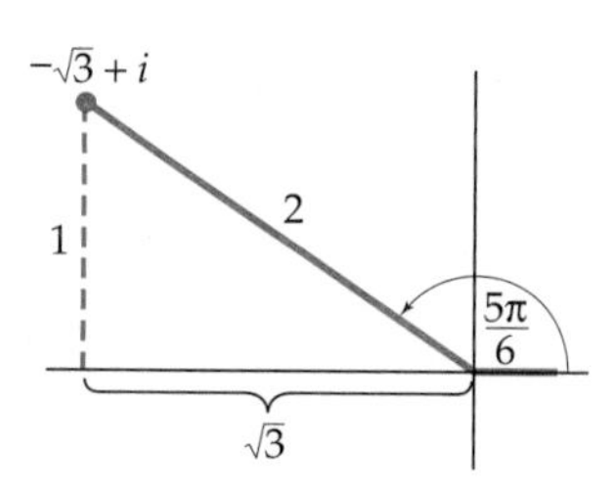

Figure 9–3

Example 3* Express $-2 + 5i$ in polar form.

Solution Since $a = -2$ and $b = 5$, $r = \sqrt{(-2)^2 + 5^2} = \sqrt{29}$. The angle θ must satisfy

$$\cos\theta = \frac{a}{r} = \frac{-2}{\sqrt{29}} \quad \text{and} \quad \sin\theta = \frac{b}{r} = \frac{5}{\sqrt{29}}$$

so that

$$\tan\theta = \frac{\sin\theta}{\cos\theta} = \frac{5/\sqrt{29}}{-2/\sqrt{29}} = -\frac{5}{2} = -2.5.$$

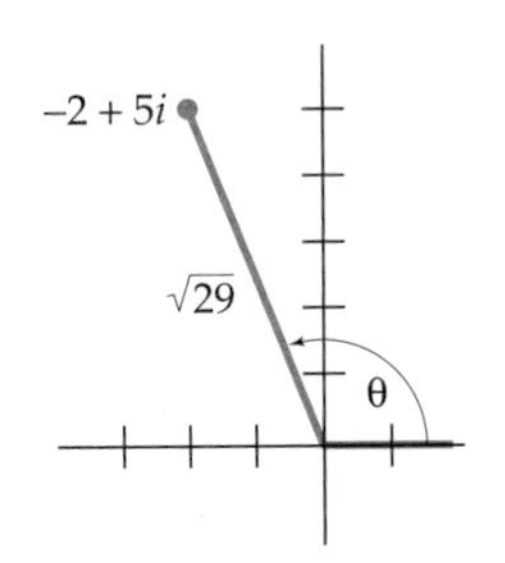

Figure 9–4

Since $-2 + 5i$ lies in the second quadrant (Figure 9–4), θ lies between $\pi/2$ and π. As we saw in Section 8.4, the only solution of the equation $\tan\theta = -2.5$ that lies between $\pi/2$ and π is $\theta \approx -1.1903 + \pi = 1.9513$. Therefore,

$$-2 + 5i \approx \sqrt{29}(\cos 1.9513 + i\sin 1.9513). \quad \blacksquare$$

Multiplication and division of complex numbers in polar form are done by the following rules, which are proved at the end of the section:

Polar Multiplication and Division Rules

If $z_1 = r_1(\cos\theta_1 + i\sin\theta_1)$ and $z_2 = r_2(\cos\theta_2 + i\sin\theta_2)$ are any two complex numbers, then

$$z_1z_2 = r_1r_2[\cos(\theta_1 + \theta_2) + i\sin(\theta_1 + \theta_2)]$$

and

$$\frac{z_1}{z_2} = \frac{r_1}{r_2}[\cos(\theta_1 - \theta_2) + i\sin(\theta_1 - \theta_2)] \quad (z_2 \neq 0).$$

In other words, to multiply two numbers in polar form, just *multiply the moduli and add the arguments.* To divide, just *divide the moduli and subtract the arguments.* Before proving the statements in the box, we will illustrate them with some examples.

Example 4 Find z_1z_2, when

$$z_1 = 2[\cos(5\pi/6) + i\sin(5\pi/6)] \quad \text{and} \quad z_2 = 3[\cos(7\pi/4) + i\sin(7\pi/4)].$$

*Omit this example if you haven't read Section 8.4.

Technology Tip

Complex arithmetic can be done with numbers in polar form on TI calculators (except TI-82) and Sharp 9600. Some answers may be expressed in rectangular form.

Solution Here r_1 is the number 2 and $\theta_1 = 5\pi/6$; similarly, $r_2 = 3$ and $\theta_2 = 7\pi/4$ and we have:

$$\begin{aligned} z_1z_2 &= r_1r_2[\cos(\theta_1 + \theta_2) + i\sin(\theta_1 + \theta_2)] \\ &= 2\cdot 3\left[\cos\left(\frac{5\pi}{6} + \frac{7\pi}{4}\right) + i\sin\left(\frac{5\pi}{6} + \frac{7\pi}{4}\right)\right] \\ &= 6\left[\cos\left(\frac{10\pi}{12} + \frac{21\pi}{12}\right) + i\sin\left(\frac{10\pi}{12} + \frac{21\pi}{12}\right)\right] \\ &= 6\left(\cos\frac{31\pi}{12} + i\sin\frac{31\pi}{12}\right). \quad \blacksquare \end{aligned}$$

Example 5 Find z_1/z_2, where

$$z_1 = 10[\cos(\pi/3) + i\sin(\pi/3)] \quad \text{and} \quad z_2 = 2[\cos(\pi/4) + i\sin(\pi/4)].$$

Solution

$$\begin{aligned} \frac{z_1}{z_2} &= \frac{10\left(\cos\frac{\pi}{3} + i\sin\frac{\pi}{3}\right)}{2\left(\cos\frac{\pi}{4} + i\sin\frac{\pi}{4}\right)} = \frac{10}{2}\left[\cos\left(\frac{\pi}{3} - \frac{\pi}{4}\right) + i\sin\left(\frac{\pi}{3} - \frac{\pi}{4}\right)\right] \\ &= 5\left(\cos\frac{\pi}{12} + i\sin\frac{\pi}{12}\right). \quad \blacksquare \end{aligned}$$

Proof of the Polar Multiplication Rule

If $z_1 = r_1(\cos\theta_1 + i\sin\theta_1)$ and $z_2 = r_2(\cos\theta_2 + i\sin\theta_2)$, then

$$\begin{aligned} z_1z_2 &= r_1(\cos\theta_1 + i\sin\theta_1)r_2(\cos\theta_2 + i\sin\theta_2) \\ &= r_1r_2(\cos\theta_1 + i\sin\theta_1)(\cos\theta_2 + i\sin\theta_2) \\ &= r_1r_2(\cos\theta_1\cos\theta_2 + i\sin\theta_1\cos\theta_2 + i\cos\theta_1\sin\theta_2 + i^2\sin\theta_1\sin\theta_2) \\ &= r_1r_2[(\cos\theta_1\cos\theta_2 - \sin\theta_1\sin\theta_2) + i(\sin\theta_1\cos\theta_2 + \cos\theta_1\sin\theta_2)]. \end{aligned}$$

But the addition identities for sine and cosine (page 518) show that

$$\cos\theta_1\cos\theta_2 - \sin\theta_1\sin\theta_2 = \cos(\theta_1 + \theta_2)$$

and

$$\sin\theta_1\cos\theta_2 + \cos\theta_1\sin\theta_2 = \sin(\theta_1 + \theta_2).$$

Therefore,

$$\begin{aligned} z_1z_2 &= r_1r_2[(\cos\theta_1\cos\theta_2 - \sin\theta_1\sin\theta_2) + i(\sin\theta_1\cos\theta_2 + \cos\theta_1\sin\theta_2)] \\ &= r_1r_2[\cos(\theta_1 + \theta_2) + i\sin(\theta_1 + \theta_2)]. \end{aligned}$$

This completes the proof of the multiplication rule. The division rule is proved similarly (Exercise 51).

Exercises 9.1

In Exercises 1–8, plot the point in the complex plane corresponding to the number.

1. $3 + 2i$ **2.** $-7 + 6i$ **3.** $-\frac{8}{3} - \frac{5}{3}i$

4. $\sqrt{2} - 7i$ **5.** $(1 + i)(1 - i)$

6. $(2 + i)(1 - 2i)$ **7.** $2i\left(3 - \frac{5}{2}i\right)$

8. $\frac{4i}{3}(-6 - 3i)$

In Exercises 9–14, find the absolute value.

9. $|5 - 12i|$ **10.** $|2i|$ **11.** $|1 + \sqrt{2}i|$

12. $|2 - 3i|$ **13.** $|-12i|$ **14.** $|i^7|$

15. Give an example of complex numbers z and w such that $|z + w| \neq |z| + |w|$.

16. If $z = 3 - 4i$, find $|z|^2$ and $z\bar{z}$, where $\bar{z}$ is the conjugate of z (see page 283).

In Exercises 17–24, sketch the graph of the equation in the complex plane (z denotes a complex number of the form $a + bi$).

17. $|z| = 4$ [*Hint:* The graph consists of all points that lie 4 units from the origin.]

18. $|z| = 1$

19. $|z - 1| = 10$ [*Hint:* 1 corresponds to (1, 0) in the complex plane. What does the equation say about the distance from z to 1?]

20. $|z + 3| = 1$

21. $|z - 2i| = 4$

22. $|z - 3i + 2| = 9$
[*Hint:* Rewrite it as $|z - (-2 + 3i)| = 9$.]

23. $\text{Re}(z) = 2$ [The **real part** of the complex number $z = a + bi$ is defined to be the number a and is denoted $\text{Re}(z)$.]

24. $\text{Im}(z) = -5/2$ [The **imaginary part** of $z = a + bi$ is defined to be the number b *(not bi)* and is denoted $\text{Im}(z)$.]

In Exercises 25–32, express the number in polar form.

25. $3 + 4i$ **26.** $-4 + 3i$ **27.** $5 - 12i$

28. $-\sqrt{7} - 3i$ **29.** $1 + 2i$ **30.** $3 - 5i$

31. $-\frac{5}{2} + \frac{7}{2}i$ **32.** $\sqrt{5} + \sqrt{11}i$

In Exercises 33–38, perform the indicated multiplication or division; express your answer in both rectangular form $a + bi$ and polar form $r(\cos\theta + i\sin\theta)$.

33. $\left(\cos\frac{\pi}{12} + i\sin\frac{\pi}{12}\right)\cdot 2\left(\cos\frac{7\pi}{12} + i\sin\frac{7\pi}{12}\right)$

34. $3\left(\cos\frac{\pi}{8} + i\sin\frac{\pi}{8}\right)\cdot 12\left(\cos\frac{3\pi}{8} + i\sin\frac{3\pi}{8}\right)$

35. $12\left(\cos\frac{11\pi}{12} + i\sin\frac{11\pi}{12}\right)\cdot\frac{7}{2}\left(\cos\frac{\pi}{4} + i\sin\frac{\pi}{4}\right)$

36. $\dfrac{8\left(\cos\frac{5\pi}{18} + i\sin\frac{5\pi}{18}\right)}{4\left(\cos\frac{\pi}{9} + i\sin\frac{\pi}{9}\right)}$

37. $\dfrac{6\left(\cos\frac{7\pi}{20} + i\sin\frac{7\pi}{20}\right)}{4\left(\cos\frac{\pi}{10} + i\sin\frac{\pi}{10}\right)}$

38. $\dfrac{\sqrt{54}\left(\cos\frac{9\pi}{4} + i\sin\frac{9\pi}{4}\right)}{\sqrt{6}\left(\cos\frac{7\pi}{12} + i\sin\frac{7\pi}{12}\right)}$

In Exercises 39–46, convert to polar form and then multiply or divide. Express your answer in polar form.

39. $(1 + i)(1 + \sqrt{3}i)$ **40.** $(1 - i)(3 - 3i)$

41. $\dfrac{1 + i}{1 - i}$ **42.** $\dfrac{2 - 2i}{-1 - i}$

43. $3i(2\sqrt{3} + 2i)$ **44.** $\dfrac{-4i}{\sqrt{3} + i}$

45. $i(i + 1)(-\sqrt{3} + i)$

46. $(1 - i)(2\sqrt{3} - 2i)(-4 - 4\sqrt{3}i)$

47. Explain what is meant by saying that multiplying a complex number $z = r(\cos\theta + i\sin\theta)$ by i amounts to rotating z 90° counterclockwise around the origin. [*Hint:* Express i and iz in polar form; what are their relative positions in the complex plane?]

48. Describe what happens geometrically when you multiply a complex number by 2.

Thinkers

49. The sum of two distinct complex numbers, $a + bi$ and $c + di$, can be found geometrically by means of the so-called **parallelogram rule:** Plot the points $a + bi$ and $c + di$ in the complex plane and form the parallelogram, three of whose vertices are 0, $a + bi$, and $c + di$, as in the figure. Then the fourth vertex of the parallelogram is the point whose coordinate is the sum

$$(a + bi) + (c + di) = (a + c) + (b + d)i.$$

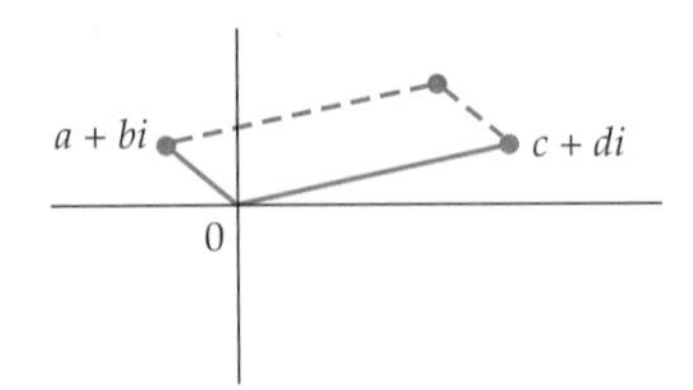

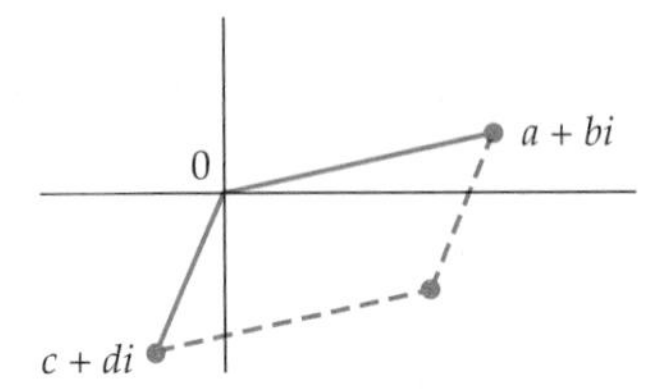

Complete the following *proof* of the parallelogram rule when $a \neq 0$ and $c \neq 0$.

(a) Find the *slope* of the line K from 0 to $a + bi$. [*Hint:* K contains the points $(0, 0)$ and (a, b).]

(b) Find the *slope* of the line N from 0 to $c + di$.

(c) Find the *equation* of the line L, through $a + bi$ and parallel to line N of part (b). [*Hint:* The point (a, b) is on L; find the slope of L by using part (b) and facts about the slope of parallel lines.]

(d) Find the *equation* of the line M, through $c + di$ and parallel to line K of part (a).

(e) Label the lines K, L, M, N in the figure.

(f) Show by using substitution that the point $(a + c, b + d)$ satisfies both the equation of line L and the equation of line M. Therefore, $(a + c, b + d)$ lies on both L and M. Since the only point on both L and M is the fourth vertex of the parallelogram (see the figure), this vertex must be $(a + c, b + d)$. Hence, this vertex has coordinate

$$(a + c) + (b + d)i = (a + bi) + (c + di).$$

50. Let $z = a + bi$ be a complex number and denote its conjugate $a - bi$ by $\bar{z}$. Prove that $|z|^2 = z\bar{z}$.

51. *Proof of the polar division rule.* Let $z_1 = r_1(\cos\theta_1 + i\sin\theta_1)$ and $z_2 = r_2(\cos\theta_2 + i\sin\theta_2)$. Then,

$$\begin{aligned}\frac{z_1}{z_2} &= \frac{r_1(\cos\theta_1 + i\sin\theta_1)}{r_2(\cos\theta_2 + i\sin\theta_2)} \\ &= \frac{r_1(\cos\theta_1 + i\sin\theta_1)}{r_2(\cos\theta_2 + i\sin\theta_2)}\cdot\frac{\cos\theta_2 - i\sin\theta_2}{\cos\theta_2 - i\sin\theta_2}.\end{aligned}$$

(a) Multiply out the denominator on the right side and use the Pythagorean identity to show that it is just the number r_2.

(b) Multiply out the numerator on the right side; use the subtraction identities for sine and cosine (page 518) to show that it is

$$r_1[\cos(\theta_1 - \theta_2) + i\sin(\theta_1 - \theta_2)].$$

Therefore,

$$\frac{z_1}{z_2} = \left(\frac{r_1}{r_2}\right)[\cos(\theta_1 - \theta_2) + i\sin(\theta_1 - \theta_2)].$$

52. (a) If $s(\cos\beta + i\sin\beta) = r(\cos\theta + i\sin\theta)$, explain why we must have $s = r$. [*Hint:* Think distance.]

(b) If $r(\cos\beta + i\sin\beta) = r(\cos\theta + i\sin\theta)$, explain why $\cos\beta = \cos\theta$ and $\sin\beta = \sin\theta$. [*Hint:* See property 5 of the complex numbers on page 282.]

(c) If $\cos\beta = \cos\theta$ and $\sin\beta = \sin\theta$, show that angles β and θ in standard position have the same terminal side. [*Hint:* $(\cos\beta, \sin\beta)$ and $(\cos\theta, \sin\theta)$ are points on the unit circle.]

(d) Use parts (a)–(c) to prove this **equality rule for polar form:**

$$s(\cos\beta + i\sin\beta) = r(\cos\theta + i\sin\theta)$$

exactly when $s = r$ and $\beta = \theta + 2k\pi$ for some integer k. [*Hint:* Angles with the same terminal side must differ by an integer multiple of 2π.]

9.2 DeMoivre's Theorem and *n*th Roots of Complex Numbers

Polar form provides a convenient way to calculate both powers and roots of complex numbers. If $z = r(\cos\theta + i\sin\theta)$, then the multiplication formula on page 567 shows that

$$\begin{aligned}z^2 = z\cdot z &= r\cdot r[\cos(\theta + \theta) + i\sin(\theta + \theta)] \\ &= r^2(\cos 2\theta + i\sin 2\theta).\end{aligned}$$

Similarly,

$$z^3 = z^2 \cdot z = r^2 \cdot r[\cos(2\theta + \theta) + i\sin(2\theta + \theta)]$$
$$= r^3(\cos 3\theta + i\sin 3\theta).$$

Repeated application of the multiplication formula proves:

DeMoivre's Theorem

For any complex number $z = r(\cos\theta + i\sin\theta)$ and any positive integer n,

$$z^n = r^n(\cos n\theta + i\sin n\theta).$$

Example 1 Compute $\left(-\sqrt{3} + i\right)^5$.

Solution We first express $-\sqrt{3} + i$ in polar form (as in Example 2 on page 566):

$$-\sqrt{3} + i = 2\left(\cos\frac{5\pi}{6} + i\sin\frac{5\pi}{6}\right).$$

By DeMoivre's Theorem,

$$\left(-\sqrt{3} + i\right)^5 = 2^5\left[\cos\left(5\cdot\frac{5\pi}{6}\right) + i\sin\left(5\cdot\frac{5\pi}{6}\right)\right] = 32\left(\cos\frac{25\pi}{6} + i\sin\frac{25\pi}{6}\right).$$

Since $25\pi/6 = (\pi/6) + (24\pi/6) = (\pi/6) + 4\pi$, we have

$$\left(-\sqrt{3} + i\right)^5 = 32\left(\cos\frac{25\pi}{6} + i\sin\frac{25\pi}{6}\right) = 32\left(\cos\frac{\pi}{6} + i\sin\frac{\pi}{6}\right)$$
$$= 32\left(\frac{\sqrt{3}}{2} + \frac{1}{2}i\right) = 16\sqrt{3} + 16i. \quad \blacksquare$$

Example 2 Find $(1 + i)^{10}$.

Solution First verify that the polar form of $1 + i$ is $1 + i = \sqrt{2}[\cos(\pi/4) + i\sin(\pi/4).]$ Therefore, by DeMoivre's Theorem,

$$(1 + i)^{10} = \left(\sqrt{2}\right)^{10}\left(\cos\frac{10\pi}{4} + i\sin\frac{10\pi}{4}\right)$$
$$= (2^{1/2})^{10}\left(\cos\frac{5\pi}{2} + i\sin\frac{5\pi}{2}\right) = 2^5(0 + i\cdot 1) = 32i. \quad \blacksquare$$

If $a + bi$ is a complex number and n a positive integer, the equation $z^n = a + bi$ may have n different solutions in the complex numbers, as we shall see below. Furthermore, there is no obvious way to designate one of these solutions as *the* nth root of $a + bi$.* Consequently, *any* solution of the equation $z^n = a + bi$ is called an **nth root** of $a + bi$.

Every real number is, of course, a complex number. When the definition of an nth root of a complex number is applied to a real number, we must change our previous terminology. For instance, 16 now has *four* fourth roots, since each of 2, -2, $2i$, and $-2i$ is a solution of $z^4 = 16$, whereas we previously said that 2 was *the* fourth root of 16. In the context of complex numbers, this change will not cause any confusion.

Although nth roots are no longer unique, the radical symbol will be used only for nonnegative real numbers and will have the same meaning as before: If r is a nonnegative real number, then $\sqrt[n]{r}$ denotes the unique nonnegative real number whose nth power is r.

All nth roots of a complex number $a + bi$ can easily be found if $a + bi$ is written in polar form, as illustrated in the next example.

Example 3 Find the fourth roots of $-8 + 8\sqrt{3}i$.

Solution To solve $z^4 = -8 + 8\sqrt{3}i$, first verify that the polar form of $-8 + 8\sqrt{3}i$ is $16\left(\cos\frac{2\pi}{3} + i\sin\frac{2\pi}{3}\right)$. We must find numbers s and β such that

$$[s(\cos\beta + i\sin\beta)]^4 = 16\left(\cos\frac{2\pi}{3} + i\sin\frac{2\pi}{3}\right).$$

By DeMoivre's Theorem we must have:

$$s^4(\cos 4\beta + i\sin 4\beta) = 16\left(\cos\frac{2\pi}{3} + i\sin\frac{2\pi}{3}\right).$$

The equality rules for complex numbers (Exercise 52 in Section 9.1) show that this can happen only when

$$s^4 = 16 \qquad \text{and} \qquad 4\beta = \frac{2\pi}{3} + 2k\pi \quad (k \text{ an integer})$$

$$s = \sqrt[4]{16} = 2 \qquad\qquad \beta = \frac{2\pi/3 + 2k\pi}{4}$$

Substituting these values in $s(\cos\beta + i\sin\beta)$ shows that the solutions of $z^4 = 16\left(\cos\frac{2\pi}{3} + i\sin\frac{2\pi}{3}\right)$ are

$$z = 2\left(\cos\frac{2\pi/3 + 2k\pi}{4} + i\sin\frac{2\pi/3 + 2k\pi}{4}\right) \quad (k = 0, \pm1, \pm2, \pm3, \ldots)$$

*You can't just choose the positive one, as we did in the real numbers, since "positive" and "negative" aren't meaningful terms in the complex numbers. For instance, should $3 - 2i$ be called positive or negative?

which can be simplified as

$$z = 2\left[\cos\left(\frac{\pi}{6} + \frac{k\pi}{2}\right) + i\sin\left(\frac{\pi}{6} + \frac{k\pi}{2}\right)\right] \quad (k = 0, \pm 1, \pm 2, \pm 3, \ldots).$$

Letting $k = 0, 1, 2, 3,$ produces four distinct solutions:

$$k = 0: \quad z = 2\left(\cos\frac{\pi}{6} + i\sin\frac{\pi}{6}\right) = \sqrt{3} + i.$$

$$k = 1: \quad z = 2\left[\cos\left(\frac{\pi}{6} + \frac{\pi}{2}\right) + i\sin\left(\frac{\pi}{6} + \frac{\pi}{2}\right)\right] = 2\left(\cos\frac{2\pi}{3} + i\sin\frac{2\pi}{3}\right)$$
$$= -1 + \sqrt{3}i.$$

$$k = 2: \quad z = 2\left[\cos\left(\frac{\pi}{6} + \pi\right) + i\sin\left(\frac{\pi}{6} + \pi\right)\right] = 2\left(\cos\frac{7\pi}{6} + i\sin\frac{7\pi}{6}\right)$$
$$= -\sqrt{3} - i.$$

$$k = 3: \quad z = 2\left[\cos\left(\frac{\pi}{6} + \frac{3\pi}{2}\right) + i\sin\left(\frac{\pi}{6} + \frac{3\pi}{2}\right)\right] = 2\left(\cos\frac{5\pi}{3} + i\sin\frac{5\pi}{3}\right)$$
$$= 1 - \sqrt{3}i.$$

Any other value of k produces an angle β with the same terminal side as one of the four angles used above, and hence leads to the same solution. For instance, when $k = 4$, then $\beta = \frac{\pi}{6} + \frac{4\pi}{2} = \frac{\pi}{6} + 2\pi$ and β has the same terminal side as $\pi/6$. Therefore, we have found *all* the solutions—the four fourth roots of $-8 + 8\sqrt{3}i$.* ■

The general equation $z^n = r(\cos\theta + i\sin\theta)$ can be solved by exactly the same method used in the preceding example—just substitute n for 4, r for 16, and θ for $2\pi/3$, as follows. A solution is a number $s(\cos\beta + i\sin\beta)$ such that

$$[s(\cos\beta + i\sin\beta)]^n = r(\cos\theta + i\sin\theta)$$
$$s^n(\cos n\beta + i\sin n\beta) = r(\cos\theta + i\sin\theta).$$

Therefore,

$$s^n = r \qquad \text{and} \qquad n\beta = \theta + 2k\pi \quad (k \text{ any integer})$$
$$s = \sqrt[n]{r} \qquad\qquad \beta = \frac{\theta + 2k\pi}{n}$$

Taking $k = 0, 1, 2, \ldots, n - 1$ produces n distinct angles β. Any other value of k leads to an angle β with the same terminal side as one of these. Hence,

*Alternatively, page 290 shows that a fourth-degree equation, such as $z^4 = -8 + 8\sqrt{3}i$, has at most four distinct solutions.

Formula for nth Roots

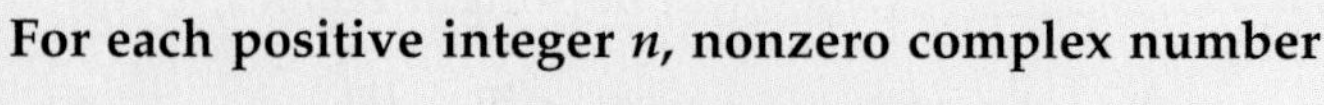

For each positive integer n, nonzero complex number

$$r(\cos\theta + i\sin\theta)$$

has exactly n distinct nth roots. They are given by:

$$\sqrt[n]{r}\left[\cos\left(\frac{\theta + 2k\pi}{n}\right) + i\sin\left(\frac{\theta + 2k\pi}{n}\right)\right]$$

where $k = 0, 1, 2, 3, \ldots n - 1$.

Technology Tip

To solve polynomial equations with complex coefficients, such as $z^n = 4 + 4i$, use POLY on the TI-86 keyboard, C-SOLVE in the COMPLEX submenu of the TI-89 ALGEBRA menu, or POLYROOT in the POLYNOMIAL submenu of the HP-38 MATH menu.

Example 4 Find the fifth roots of $4 + 4i$.

Solution First write it in polar form as $4\sqrt{2}\left(\cos\frac{\pi}{4} + i\sin\frac{\pi}{4}\right)$. Now apply the root formula with $n = 5$, $r = 4\sqrt{2}$, $\theta = \pi/4$, and $k = 0, 1, 2, 3, 4$. Note that

$$\sqrt[5]{r} = \sqrt[5]{4\sqrt{2}} = (4\sqrt{2})^{1/5} = (2^2 2^{1/2})^{1/5} = (2^{5/2})^{1/5} = 2^{5/10} = 2^{1/2} = \sqrt{2}.$$

Therefore, the fifth roots are

$$\sqrt{2}\left[\cos\left(\frac{\pi/4 + 2k\pi}{5}\right) + i\sin\left(\frac{\pi/4 + 2k\pi}{5}\right)\right] \quad k = 0, 1, 2, 3, 4,$$

that is,

$$k = 0{:}\ \sqrt{2}\left[\cos\left(\frac{\pi/4 + 0}{5}\right) + i\sin\left(\frac{\pi/4 + 0}{5}\right)\right] = \sqrt{2}\left(\cos\frac{\pi}{20} + i\sin\frac{\pi}{20}\right).$$

$$k = 1{:}\ \sqrt{2}\left[\cos\left(\frac{\pi/4 + 2\pi}{5}\right) + i\sin\left(\frac{\pi/4 + 2\pi}{5}\right)\right] = \sqrt{2}\left(\cos\frac{9\pi}{20} + i\sin\frac{9\pi}{20}\right).$$

$$k = 2{:}\ \sqrt{2}\left[\cos\left(\frac{\pi/4 + 4\pi}{5}\right) + i\sin\left(\frac{\pi/4 + 4\pi}{5}\right)\right] = \sqrt{2}\left(\cos\frac{17\pi}{20} + i\sin\frac{17\pi}{20}\right).$$

$$k = 3{:}\ \sqrt{2}\left[\cos\left(\frac{\pi/4 + 6\pi}{5}\right) + i\sin\left(\frac{\pi/4 + 6\pi}{5}\right)\right] = \sqrt{2}\left(\cos\frac{25\pi}{20} + i\sin\frac{25\pi}{20}\right).$$

$$k = 4{:}\ \sqrt{2}\left[\cos\left(\frac{\pi/4 + 8\pi}{5}\right) + i\sin\left(\frac{\pi/4 + 8\pi}{5}\right)\right] = \sqrt{2}\left(\cos\frac{33\pi}{20} + i\sin\frac{33\pi}{20}\right). \quad \blacksquare$$

Roots of Unity

The n distinct nth roots of 1 (the solutions of $z^n = 1$) are called the ***n*th roots of unity.** Since $\cos 0 = 1$ and $\sin 0 = 0$, the polar form of the number 1 is $\cos 0 + i\sin 0$. Applying the root formula with $r = 1$ and $\theta = 0$ shows that

Roots of Unity

For each positive integer n, there are n distinct nth roots of unity:

$$\cos\frac{2k\pi}{n} + i\sin\frac{2k\pi}{n} \quad (k = 0, 1, 2, \ldots, n - 1).$$

Example 5 Find the cube roots of unity.

Solution Apply the formula with $n = 3$ and $k = 0, 1, 2$:

$$k = 0: \qquad \cos 0 + i \sin 0 = 1$$

$$k = 1: \qquad \cos \frac{2\pi}{3} + i \sin \frac{2\pi}{3} = -\frac{1}{2} + \frac{\sqrt{3}}{2} i$$

$$k = 2: \qquad \cos \frac{4\pi}{3} + i \sin \frac{4\pi}{3} = -\frac{1}{2} - \frac{\sqrt{3}}{2} i. \quad \blacksquare$$

Denote by ω the first complex cube root of unity obtained in Example 5:

$$\omega = \cos \frac{2\pi}{3} + i \sin \frac{2\pi}{3}.$$

If we use DeMoivre's Theorem to find ω^2 and ω^3, we see that these numbers are the other two cube roots of unity found in Example 5:

$$\omega^2 = \left(\cos \frac{2\pi}{3} + i \sin \frac{2\pi}{3}\right)^2 = \cos \frac{4\pi}{3} + i \sin \frac{4\pi}{3}$$

$$\omega^3 = \left(\cos \frac{2\pi}{3} + i \sin \frac{2\pi}{3}\right)^3 = \cos \frac{6\pi}{3} + i \sin \frac{6\pi}{3} = \cos 2\pi + i \sin 2\pi$$

$$= 1 + 0 \cdot i = 1.$$

In other words, all the cube roots of unity are powers of ω. The same thing is true in the general case.

Roots of Unity

Let n be a positive integer with $n > 1$. Then the number

$$z = \cos \frac{2\pi}{n} + i \sin \frac{2\pi}{n}$$

is an nth root of unity and all the nth roots of unity are

$$z, z^2, z^3, z^4, \ldots z^{n-1}, z^n = 1.$$

The nth roots of unity have an interesting geometric interpretation. Every nth root of unity has absolute value 1 by the Pythagorean identity:

$$\left|\cos \frac{2k\pi}{n} + i \sin \frac{2k\pi}{n}\right| = \left(\cos \frac{2k\pi}{n}\right)^2 + \left(\sin \frac{2k\pi}{n}\right)^2$$

$$= \cos^2\left(\frac{2k\pi}{n}\right) + \sin^2\left(\frac{2k\pi}{n}\right) = 1.$$

Therefore, in the complex plane, every nth root of unity is exactly 1 unit from the origin. In other words, the nth roots of unity all lie on the unit circle.

Example 6 Find the fifth roots of unity.

Solution They are

$$\cos\frac{2k\pi}{5} + i\sin\frac{2k\pi}{5} \quad (k = 0, 1, 2, 3, 4)$$

that is,

$$\cos 0 + i\sin 0 = 1, \qquad \cos\frac{2\pi}{5} + i\sin\frac{2\pi}{5}, \qquad \cos\frac{4\pi}{5} + i\sin\frac{4\pi}{5},$$

$$\cos\frac{6\pi}{5} + i\sin\frac{6\pi}{5}, \qquad \cos\frac{8\pi}{5} + i\sin\frac{8\pi}{5}.$$

These five roots can be plotted in the complex plane by starting at $1 = 1 + 0i$ and moving counterclockwise around the unit circle, moving through an angle of $2\pi/5$ at each step, as shown in Figure 9–5. If you connect these five roots, they form the vertices of a regular pentagon (Figure 9–6). ■

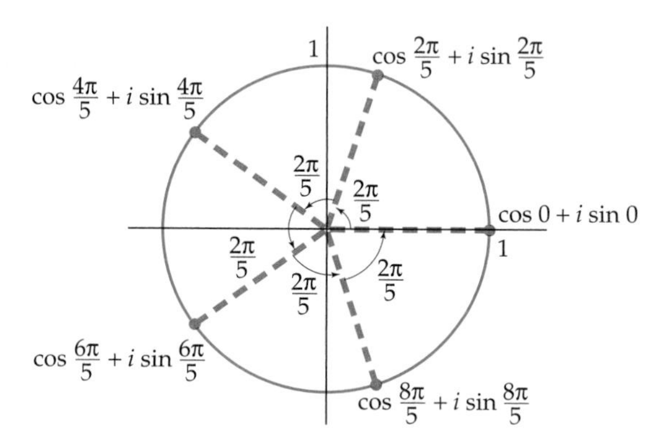

Figure 9–5

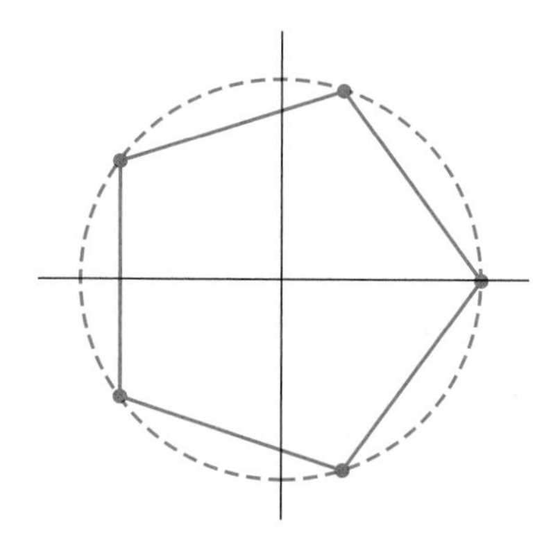

Figure 9–6

GRAPHING EXPLORATION

With your calculator in parametric graphing mode, set these range values:

$$0 \le t \le 2\pi, \qquad t\text{-step} \approx .067,$$
$$-1.5 \le x \le 1.5, \qquad -1 \le y \le 1$$

and graph the unit circle, whose parametric equations are

$$x = \cos t \qquad \text{and} \qquad y = \sin t.^*$$

Reset the t-step to be $2\pi/5$ and graph again. Your screen now looks exactly like the solid lines in Figure 9–6 because the calculator plotted only the five points corresponding to $t = 0$, $2\pi/5$, $4\pi/5$, $6\pi/5$, $8\pi/5$† and connected them with the shortest possible segments. Use the trace feature to move along the graph. The cursor will jump from vertex to vertex, that is, it will move from one fifth root of unity to the next.

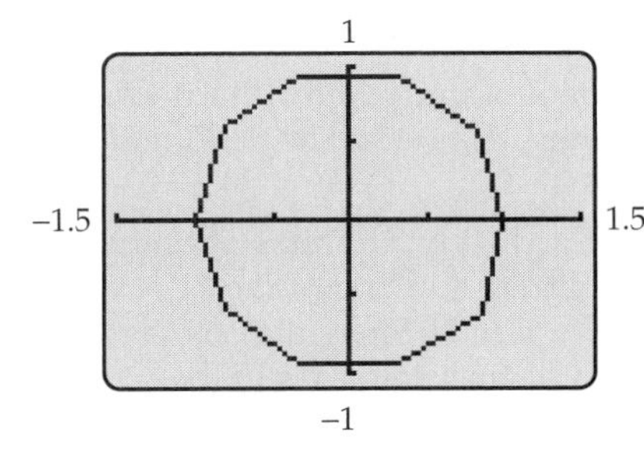

Figure 9–7

Example 7 Find the tenth roots of unity graphically.

Solution Graph the unit circle as in the preceding exploration, but use $2\pi/10$ as the t-step. The result (Figure 9–7) is a regular decagon whose vertices are the tenth roots of unity. By using the trace feature you can approximate each of them.

GRAPHING EXPLORATION

Verify that the two tenth roots of unity in the first quadrant are (approximately) $.8090 + .5878i$ and $.3090 + .9511i$. ■

*On wide-screen calculators, use $-2 \le x \le 2$ or $-1.7 \le x \le 1.7$ so that the unit circle looks like a circle.

†The point corresponding to $t = 10\pi/5 = 2\pi$ is the same as the one corresponding to $t = 0$.

Exercises 9.2

In Exercises 1–10, calculate the given product and express your answer in the form $a + bi$.

1. $\left(\cos\frac{\pi}{12} + i\sin\frac{\pi}{12}\right)^6$

2. $\left(\cos\frac{\pi}{5} + i\sin\frac{\pi}{5}\right)^{20}$

3. $\left[3\left(\cos\frac{7\pi}{30} + i\sin\frac{7\pi}{30}\right)\right]^5$

4. $\left[\sqrt[3]{4}\left(\cos\frac{7\pi}{36} + i\sin\frac{7\pi}{36}\right)\right]^{12}$

5. $(1 - i)^{12}$ [*Hint:* Use polar form and DeMoivre's Theorem.]

6. $(2 + 2i)^8$

7. $\left(\frac{\sqrt{3}}{2} + \frac{1}{2}i\right)^{10}$

8. $\left(-\frac{1}{2} + \frac{\sqrt{3}}{2}i\right)^{20}$ **9.** $\left(\frac{-1}{\sqrt{2}} + \frac{i}{\sqrt{2}}\right)^{14}$

10. $(-1 + \sqrt{3}i)^8$

In Exercises 11 and 12, find the indicated roots of unity and express your answers in the form $a + bi$.

11. Fourth roots of unity **12.** Sixth roots of unity

In Exercises 13–22, find the nth roots of the given number in polar form.

13. $64\left(\cos\frac{\pi}{5} + i\sin\frac{\pi}{5}\right); \quad n = 3$

14. $8\left(\cos\frac{\pi}{10} + i\sin\frac{\pi}{10}\right); \quad n = 3$

15. $81\left(\cos\frac{\pi}{12} + i\sin\frac{\pi}{12}\right); \quad n = 4$

16. $16\left(\cos\frac{\pi}{7} + i\sin\frac{\pi}{7}\right); \quad n = 5$

17. $-1; \quad n = 5$ **18.** $1; \quad n = 7$

19. $i; \quad n = 5$ **20.** $-i; \quad n = 6$

21. $1 + i; \quad n = 2$ **22.** $1 - \sqrt{3}i; \quad n = 3$

In Exercises 23–30, solve the given equation in the complex number system.

23. $x^6 = -1$ **24.** $x^6 + 64 = 0$ **25.** $x^3 = i$

26. $x^4 = i$ **27.** $x^3 + 27i = 0$ **28.** $x^6 + 729 = 0$

29. $x^4 = -1 + \sqrt{3}i$ **30.** $x^4 = -8 - 8\sqrt{3}i$

In Exercises 31–35, represent the roots of unity graphically. Then use the trace feature to obtain approximations of the form $a + bi$ for each root (round to four places).

31. Seventh roots of unity **32.** Fifth roots of unity

33. Eighth roots of unity **34.** Twelfth roots of unity

35. Ninth roots of unity

36. Solve the equation $x^3 + x^2 + x + 1 = 0$. [*Hint:* First find the quotient when $x^4 - 1$ is divided by $x - 1$ and then consider solutions of $x^4 - 1 = 0$.]

37. Solve $x^5 + x^4 + x^3 + x^2 + x + 1 = 0$. [*Hint:* Consider $x^6 - 1$ and $x - 1$ and see Exercise 36.]

38. What do you think are the solutions of $x^{n-1} + x^{n-2} + \cdots + x^3 + x^2 + x + 1 = 0$? (See Exercises 36 and 37.)

Thinkers

39. In the complex plane, identify each point with its complex number label. The unit circle consists of all numbers (points) z such that $|z| = 1$. Suppose v and w are two points (numbers) that move around the unit circle in such a way that $v = w^{12}$ at all times. When w has made one complete trip around the circle, how many trips has v made? [*Hint:* Think polar and DeMoivre.]

40. Suppose u is an nth root of unity. Show that $1/u$ is also an nth root of unity. [*Hint:* Use the definition, *not* polar form.]

41. Let $u_1, u_2, \ldots, u_n$ be the distinct nth roots of unity and suppose v is a nonzero solution of the equation $z^n = r(\cos\theta + i\sin\theta)$. Show that $vu_1, vu_2, \ldots, vu_n$ are n distinct solutions of the equation. [*Remember:* Each u_i is a solution of $x^n = 1$.]

42. Use the formula for nth roots and the identities

$$\cos(x + \pi) = -\cos x \qquad \sin(x + \pi) = -\sin x$$

to show that the nonzero complex number $r(\cos\theta + i\sin\theta)$ has two square roots and that these square roots are negatives of each other.

9.3 Vectors in the Plane

Once a unit of measure has been agreed upon, quantities such as area, length, time, and temperature can be described by a single number. Other quantities, such as an east wind of 10 mph, require two numbers to describe them because they involve both *magnitude* and *direction.* Such quantities are called **vectors** and are represented geometrically by a directed line segment or arrow, as in Figure 9–8.

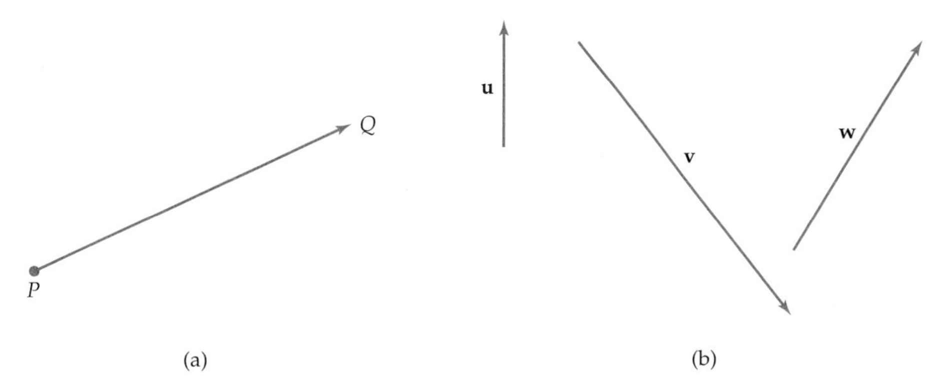

Figure 9–8

When a vector extends from a point P to a point Q, as in Figure 9–8(a), P is called the **initial point** of the vector and Q the **terminal point,** and the vector is written $\overrightarrow{PQ}$. Its **length** is denoted by $\|\overrightarrow{PQ}\|$. When the endpoints are not specified, as in Figure 9–8(b), vectors are denoted by boldface letters such as **u**, **v**, and **w**. The length of a vector **u** is denoted by $\|\mathbf{u}\|$ and is called the **magnitude** of **u**.

If **u** and **v** are vectors with the same magnitude and direction, we say that **u** and **v** are **equivalent** and write u = v. Some examples are shown in Figure 9–9.

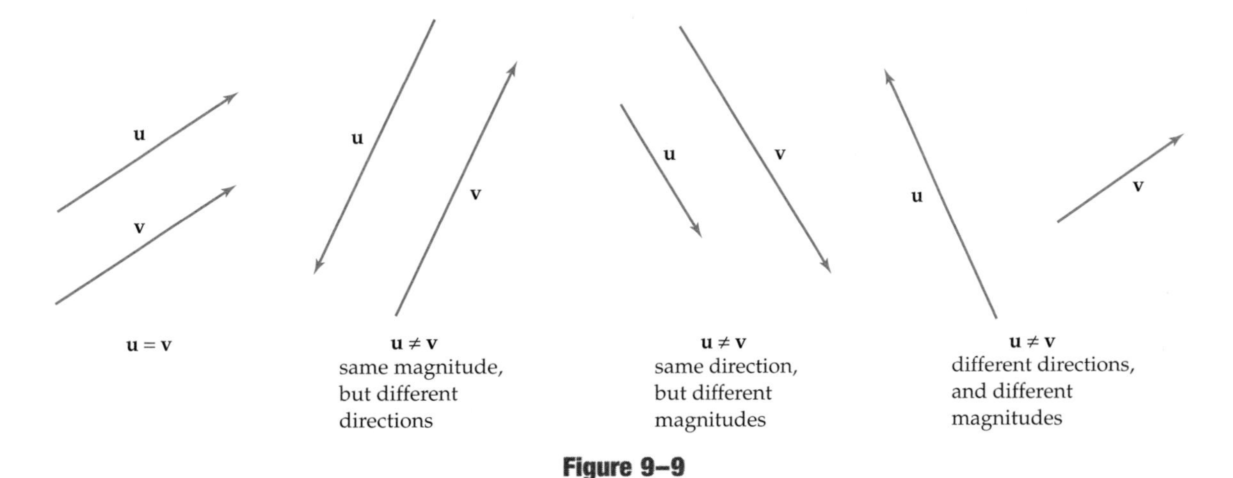

Figure 9–9

Figure 9–10

Example 1 Let $P = (1, 2)$, $Q = (5, 4)$, $O = (0, 0)$, and $R = (4, 2)$, as in Figure 9–10. Show that $\overrightarrow{PQ} = \overrightarrow{OR}$.

Solution The distance formula shows that $\overrightarrow{PQ}$ and $\overrightarrow{OR}$ have the *same length:*

$$\|\overrightarrow{PQ}\| = \sqrt{(5-1)^2 + (4-2)^2} = \sqrt{4^2 + 2^2} = \sqrt{20}.$$

$$\|\overrightarrow{OR}\| = \sqrt{(4-0)^2 + (2-0)^2} = \sqrt{4^2 + 2^2} = \sqrt{20}.$$

Furthermore, the lines through PQ and OR have the same slope:

$$\text{slope } PQ = \frac{4-2}{5-1} = \frac{2}{4} = \frac{1}{2} \qquad \text{slope } OR = \frac{2-0}{4-0} = \frac{2}{4} = \frac{1}{2}.$$

Since $\overrightarrow{PQ}$ and $\overrightarrow{OR}$ both point to the upper right on lines of the same slope, $\overrightarrow{PQ}$ and $\overrightarrow{OR}$ have the *same direction.* Therefore, $\overrightarrow{PQ} = \overrightarrow{OR}$. ■

According to the definition of equivalence, a vector may be moved from one location to another, provided that its magnitude and direction are not changed. Consequently, we have

Equivalent Vectors

Every vector $\overrightarrow{PQ}$ is equivalent to a vector $\overrightarrow{OR}$ with initial point at the origin: If $P = (x_1, y_1)$ and $Q = (x_2, y_2)$, then

$$\overrightarrow{PQ} = \overrightarrow{OR}, \quad \text{where } R = (x_2 - x_1, y_2 - y_1).$$

Proof The proof is similar to the one used in Example 1. It follows from the fact that $\overrightarrow{PQ}$ and $\overrightarrow{OR}$ have the same length:

$$\|\overrightarrow{OR}\| = \sqrt{[(x_2 - x_1) - 0]^2 + [(y_2 - y_1) - 0]^2}$$
$$= \sqrt{(x_2 - x_1)^2 + (y_2 - y_1)^2} = \|\overrightarrow{PQ}\|;$$

and that either the line segments PQ and OR are both vertical or they have the same slope:

$$\text{slope } OR = \frac{(y_2 - y_1) - 0}{(x_2 - x_1) - 0} = \frac{y_2 - y_1}{x_2 - x_1} = \text{slope } PQ$$

as shown in Figure 9–11. ■

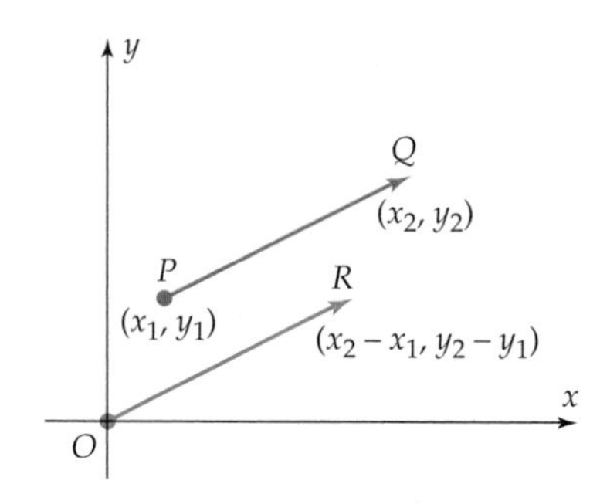

Figure 9–11

Technology Tip

Vectors in component form can be entered on TI-86/89 and HP-38 by using $[a, b]$ in place of $\langle a, b \rangle$.

The magnitude and direction of a vector with the origin as initial point are completely determined by the coordinates of its terminal point. Consequently, we denote the vector with initial point $(0, 0)$ and terminal point (a, b) by $\langle a, b \rangle$. The numbers a and b are called the **components** of the vector $\langle a, b \rangle$.

Since the length of the vector $\langle a, b \rangle$ is the distance from $(0, 0)$ to (a, b), the distance formula shows that

Magnitude

The *magnitude* (or *norm*) of the vector $\mathbf{v} = \langle a, b \rangle$ is

$$\|\mathbf{v}\| = \sqrt{a^2 + b^2}.$$

Example 2 Find the components and the magnitude of the vector with initial point $P = (-2, 6)$ and terminal point $Q = (4, -3)$.

Solution According to the fact in the box on the opposite page (with $x_1 = -2$, $y_1 = 6$, $x_2 = 4$, $y_2 = -3$):

$$\overrightarrow{PQ} = \overrightarrow{OR}, \quad \text{where } R = (4 - (-2), -3 - 6) = (6, -9)$$

that is,

$$\overrightarrow{PQ} = \overrightarrow{OR} = \langle 6, -9 \rangle.$$

Therefore,

$$\|\overrightarrow{PQ}\| = \|\overrightarrow{OR}\| = \sqrt{6^2 + (-9)^2} = \sqrt{36 + 81} = \sqrt{117}. \quad \blacksquare$$

Technology Tip

To find the norm of a vector, use NORM in the MATH submenu of the TI-86 VECTOR menu or in the NORM submenu of the TI-89 MATRIX menu.

Vector Arithmetic

When dealing with vectors, it is customary to refer to ordinary real numbers as **scalars. Scalar multiplication** is an operation in which a scalar k is "multiplied" by a vector $\mathbf{v}$ to produce another *vector* denoted by $k\mathbf{v}$. Here is the formal definition:

Scalar Multiplication

If k is a real number and $\mathbf{v} = \langle a, b \rangle$ is a vector, then

$k\mathbf{v}$ is the vector $\langle ka, kb \rangle$.

The vector $k\mathbf{v}$ is called a *scalar multiple* of $\mathbf{v}$.

Example 3 If $\mathbf{v} = \langle 3, 1 \rangle$, then

$$3\mathbf{v} = 3\langle 3, 1 \rangle = \langle 3 \cdot 3, 3 \cdot 1 \rangle = \langle 9, 3 \rangle$$
$$-2\mathbf{v} = -2\langle 3, 1 \rangle = \langle -2 \cdot 3, -2 \cdot 1 \rangle = \langle -6, -2 \rangle$$

as shown in Figure 9–12:

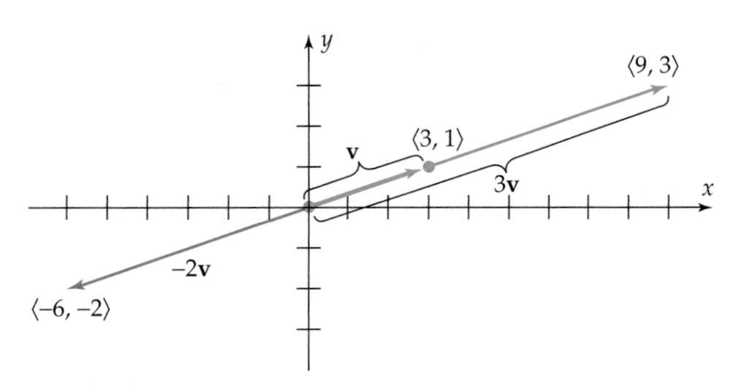

Figure 9–12

Figure 9–12 shows that $3\mathbf{v}$ has the same direction as $\mathbf{v}$, while $-2\mathbf{v}$ has the opposite direction. Also note that

$$\|\mathbf{v}\| = \|\langle 3, 1\rangle\| = \sqrt{3^2 + 1^2} = \sqrt{10}$$
$$\|-2\mathbf{v}\| = \|\langle -6, 2\rangle\| = \sqrt{(-6)^2 + (-2)^2} = \sqrt{40} = 2\sqrt{10}.$$

Therefore,

$$\|-2\mathbf{v}\| = 2\sqrt{10} = 2\|\mathbf{v}\| = |-2| \cdot \|\mathbf{v}\|.$$

Similarly, you can verify that $\|3\mathbf{v}\| = |3| \cdot \|\mathbf{v}\| = 3\|\mathbf{v}\|$. ■

Example 3 is an illustration of the following facts.

Geometric Interpretation of Scalar Multiplication

The *magnitude* of the vector $k\mathbf{v}$ is $|k|$ times the length of $\mathbf{v}$, that is,

$$\|k\mathbf{v}\| = |k| \cdot \|\mathbf{v}\|.$$

The *direction of* $k\mathbf{v}$ is the same as that of $\mathbf{v}$ when k is positive and opposite that of $\mathbf{v}$ when k is negative.

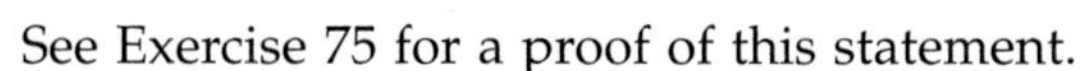

See Exercise 75 for a proof of this statement.

Vector addition is an operation in which two vectors $\mathbf{u}$ and $\mathbf{v}$ are added to produce a new vector denoted $\mathbf{u} + \mathbf{v}$. Formally,

Vector Addition

If $\mathbf{u} = \langle a, b\rangle$ and $\mathbf{v} = \langle c, d\rangle$, then

$$\mathbf{u} + \mathbf{v} = \langle a + c, b + d\rangle.$$

Example 4 If $\mathbf{u} = \langle -5, 2\rangle$ and $\mathbf{v} = \langle 3, 1\rangle$, find $\mathbf{u} + \mathbf{v}$.

Solution

$$\mathbf{u} + \mathbf{v} = \langle -5, 2\rangle + \langle 3, 1\rangle = \langle -5 + 3, 2 + 1\rangle = \langle -2, 3\rangle$$

as shown in Figure 9–13. ■

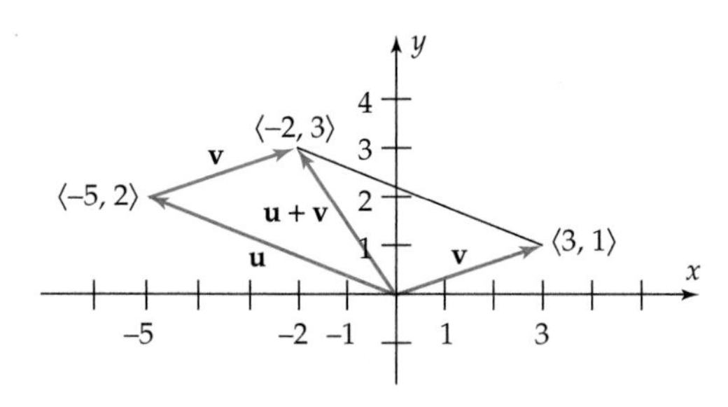

Figure 9–13

Example 4 is an illustration of these facts:

Geometric Interpretations of Vector Addition

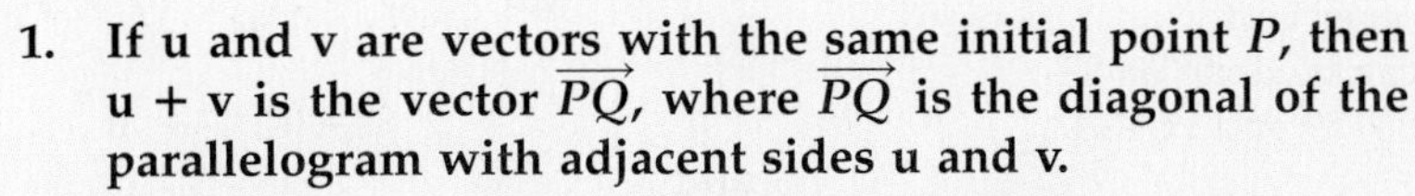

1. **If u and v are vectors with the same initial point P, then $\mathbf{u} + \mathbf{v}$ is the vector $\overrightarrow{PQ}$, where $\overrightarrow{PQ}$ is the diagonal of the parallelogram with adjacent sides u and v.**
2. **If the vector v is moved (without changing its magnitude or direction) so that its initial point lies on the endpoint of the vector u, then $\mathbf{u} + \mathbf{v}$ is the vector with the same initial point P as u and the same terminal point Q as v.**

See Exercise 76 for a proof of these statements.

The **negative** of a vector $\mathbf{v} = \langle c, d \rangle$ is defined to be the vector $(-1)\mathbf{v} = (-1)\langle c, d \rangle = \langle -c, -d \rangle$ and is denoted $-\mathbf{v}$. **Vector subtraction** is then defined as follows.

Vector Subtraction

If $\mathbf{u} = \langle a, b \rangle$ and $\mathbf{v} = \langle c, d \rangle$, then $\mathbf{u} - \mathbf{v}$ is the vector

$$\mathbf{u} + (-\mathbf{v}) = \langle a, b \rangle + \langle -c, -d \rangle$$
$$= \langle a - c, b - d \rangle.$$

A geometric interpretation of vector subtraction is given in Exercise 77.

Example 5 If $\mathbf{u} = \langle 2, 5 \rangle$ and $\mathbf{v} = \langle 6, 1 \rangle$, find $\mathbf{u} - \mathbf{v}$.

Solution

$$\mathbf{u} - \mathbf{v} = \langle 2, 5 \rangle - \langle 6, 1 \rangle = \langle 2 - 6, 5 - 1 \rangle = \langle -4, 4 \rangle$$

as shown in Figure 9–14. ■

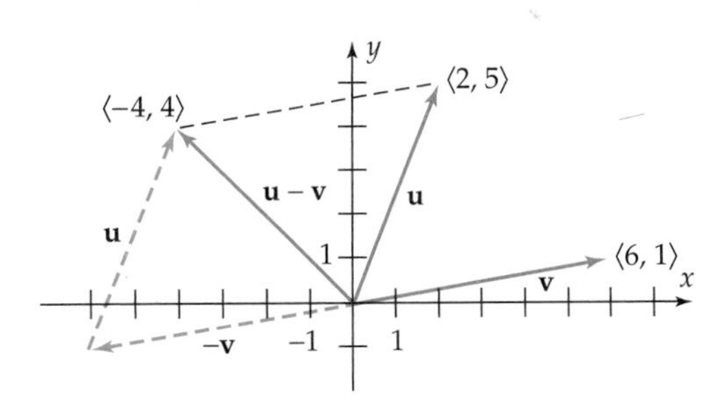

Figure 9–14

The vector $\langle 0, 0 \rangle$ is called the **zero vector** and is denoted **0**.

Example 6 If $\mathbf{u} = \langle -1, 6 \rangle$, $\mathbf{v} = \langle 2/3, -4 \rangle$, and $\mathbf{w} = \langle 2, 5/2 \rangle$, find $2\mathbf{u} + 3\mathbf{v}$ and $4\mathbf{w} - 2\mathbf{u}$.

Technology Tip

Vector arithmetic and other vector operations can be done on TI-86/89 and HP-38.

Solution

$$2\mathbf{u} + 3\mathbf{v} = 2\langle -1, 6\rangle + 3\left\langle \frac{2}{3}, -4\right\rangle = \langle -2, 12\rangle + \langle 2, -12\rangle$$
$$= \langle 0, 0\rangle = \mathbf{0}$$

and

$$4\mathbf{w} - 2\mathbf{u} = 4\left\langle 2, \frac{5}{2}\right\rangle - 2\langle -1, 6\rangle$$
$$= \langle 8, 10\rangle - \langle -2, 12\rangle$$
$$= \langle 8 - (-2), 10 - 12\rangle = \langle 10, -2\rangle. \quad \blacksquare$$

Operations on vectors share many of the same properties as arithmetical operations on numbers.

Properties of Vector Addition and Scalar Multiplication

For any vectors u, v, and w and any scalars *r* and *s*,

1. $\mathbf{u} + (\mathbf{v} + \mathbf{w}) = (\mathbf{u} + \mathbf{v}) + \mathbf{w}$
2. $\mathbf{u} + \mathbf{v} = \mathbf{v} + \mathbf{u}$
3. $\mathbf{v} + \mathbf{0} = \mathbf{v} = \mathbf{0} + \mathbf{v}$
4. $\mathbf{v} + (-\mathbf{v}) = \mathbf{0}$
5. $r(\mathbf{u} + \mathbf{v}) = r\mathbf{u} + r\mathbf{v}$
6. $(r + s)\mathbf{v} = r\mathbf{v} + s\mathbf{v}$
7. $(rs)\mathbf{v} = r(s\mathbf{v}) = s(r\mathbf{v})$
8. $1\mathbf{v} = \mathbf{v}$
9. $0\mathbf{v} = \mathbf{0}$ and $r\mathbf{0} = \mathbf{0}$

Proof If $\mathbf{u} = \langle a, b\rangle$ and $\mathbf{v} = \langle c, d\rangle$, then because addition of real numbers is commutative, we have:

$$\mathbf{u} + \mathbf{v} = \langle a, b\rangle + \langle c, d\rangle = \langle a + c, b + d\rangle$$
$$= \langle c + a, d + b\rangle = \langle c, d\rangle + \langle a, b\rangle = \mathbf{v} + \mathbf{u}.$$

The other properties are proved similarly; see Exercises 53–58. $\blacksquare$

Unit Vectors

A vector with length 1 is called a **unit vector.** For instance, $\langle 3/5, 4/5\rangle$ is a unit vector since

$$\left\|\left\langle \frac{3}{5}, \frac{4}{5}\right\rangle\right\| = \sqrt{\left(\frac{3}{5}\right)^2 + \left(\frac{4}{5}\right)^2} = \sqrt{\frac{9}{25} + \frac{16}{25}} = \sqrt{\frac{25}{25}} = 1.$$

Technology Tip

A unit vector in the same direction as **v** can be found automatically by using UNITV in the MATH submenu of the TI-86 VECTOR menu, or in the VECTOR OPS submenu of the MATRIX submenu of the TI-89 MATH menu.

Example 7 Find a unit vector that has the same direction as the vector $\mathbf{v} = \langle 5, 12\rangle$.

Solution The length of **v** is

$$\|\mathbf{v}\| = \|\langle 5, 12\rangle\| = \sqrt{5^2 + 12^2} = \sqrt{169} = 13.$$

The vector $\mathbf{u} = \frac{1}{13}\mathbf{v} = \left\langle \frac{5}{13}, \frac{12}{13} \right\rangle$ has the same direction as $\mathbf{v}$ (since it is a scalar multiple by a positive number), and $\mathbf{u}$ is a unit vector because

$$\|\mathbf{u}\| = \left\| \frac{1}{13}\mathbf{v} \right\| = \left| \frac{1}{13} \right| \cdot \|\mathbf{v}\| = \frac{1}{13} \cdot 13 = 1. \quad \blacksquare$$

The procedure used in Example 7 (multiplying a vector by the reciprocal of its length) works in the general case:

Unit Vectors

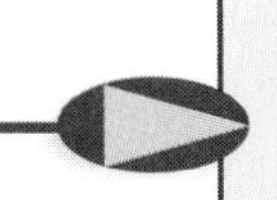

If v is a nonzero vector, then $\frac{1}{\|\mathbf{v}\|}\mathbf{v}$ is a unit vector with the same direction as v.

You can easily verify that the vectors $\mathbf{i} = \langle 1, 0 \rangle$ and $\mathbf{j} = \langle 0, 1 \rangle$ are unit vectors. The vectors **i** and **j** play a special role because they lead to a useful alternate notation for vectors. For example, if $\mathbf{u} = \langle 5, -7 \rangle$, then

$$\mathbf{u} = \langle 5, 0 \rangle + \langle 0, -7 \rangle = 5\langle 1, 0 \rangle - 7\langle 0, 1 \rangle = 5\mathbf{i} - 7\mathbf{j}.$$

Similarly, if $\mathbf{v} = \langle a, b \rangle$ is any vector, then

$$\mathbf{v} = \langle a, b \rangle = \langle a, 0 \rangle + \langle 0, b \rangle = a\langle 1, 0 \rangle + b\langle 0, 1 \rangle = a\mathbf{i} + b\mathbf{j}.$$

The vector **v** is said to be a **linear combination** of **i** and **j**. When vectors are written as linear combinations of **i** and **j**, then the properties in the box on page 584 can be used to write the rules for vector addition and scalar multiplication in this form:

$$(a\mathbf{i} + b\mathbf{j}) + (c\mathbf{i} + d\mathbf{j}) = (a + c)\mathbf{i} + (b + d)\mathbf{j}$$

and

$$c(a\mathbf{i} + b\mathbf{j}) = ca\mathbf{i} + cb\mathbf{j}.$$

Example 8 If $\mathbf{u} = 2\mathbf{i} - 6\mathbf{j}$ and $\mathbf{v} = -5\mathbf{i} + 2\mathbf{j}$, find $3\mathbf{u} - 2\mathbf{v}$.

Solution

$$\begin{aligned} 3\mathbf{u} - 2\mathbf{v} = 3(2\mathbf{i} - 6\mathbf{j}) - 2(-5\mathbf{i} + 2\mathbf{j}) &= 6\mathbf{i} - 18\mathbf{j} + 10\mathbf{i} - 4\mathbf{j} \\ &= 16\mathbf{i} - 22\mathbf{j}. \quad \blacksquare \end{aligned}$$

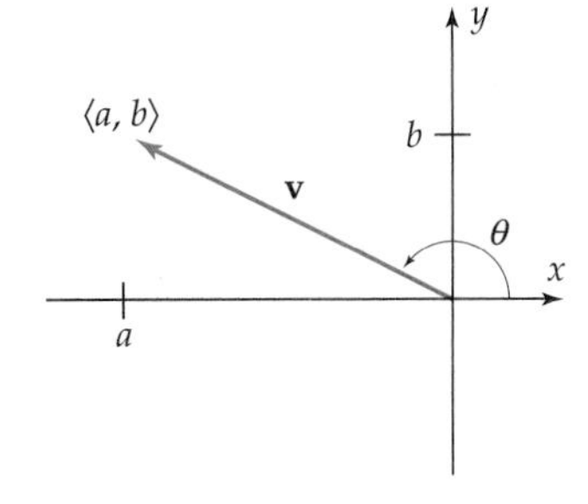

Figure 9–15

Direction Angles

If $\mathbf{v} = \langle a, b \rangle = a\mathbf{i} + b\mathbf{j}$ is a vector, then the direction of **v** is completely determined by the standard position angle θ between 0° and 360°, whose terminal side is **v**, as shown in Figure 9–15. The angle θ is called the

direction angle of the vector **v**. According to the point-in-the-plane description of the trigonometric functions,

$$\cos\theta = \frac{a}{\|\mathbf{v}\|} \qquad \text{and} \qquad \sin\theta = \frac{b}{\|\mathbf{v}\|}.$$

Rewriting each of these equations show that

Components of the Direction Angle

If $\mathbf{v} = \langle a, b\rangle = a\mathbf{i} + b\mathbf{j}$, then

$$a = \|\mathbf{v}\|\cos\theta \qquad \text{and} \qquad b = \|\mathbf{v}\|\sin\theta$$

where θ is the direction angle of v.

Example 9 Find the component form of the vector that represents the velocity of an airplane at the instant its wheels leave the ground, if the plane is going 60 mph and the body of the plane makes a 7° angle with the horizontal.

Solution The velocity vector $\mathbf{v} = a\mathbf{i} + b\mathbf{j}$ has magnitude 60 and direction angle $\theta = 7°$, as shown in Figure 9–16. Hence,

$$\begin{aligned} \mathbf{v} &= (\|\mathbf{v}\|\cos\theta)\mathbf{i} + (\|\mathbf{v}\|\sin\theta)\mathbf{j} \\ &= (60\cos 7°)\mathbf{i} + (60\sin 7°)\mathbf{j} \\ &\approx (60\cdot .9925)\mathbf{i} + (60\cdot .1219)\mathbf{j} \\ &\approx 59.55\mathbf{i} + 7.31\mathbf{j} = \langle 59.55, 7.31\rangle. \quad ■ \end{aligned}$$

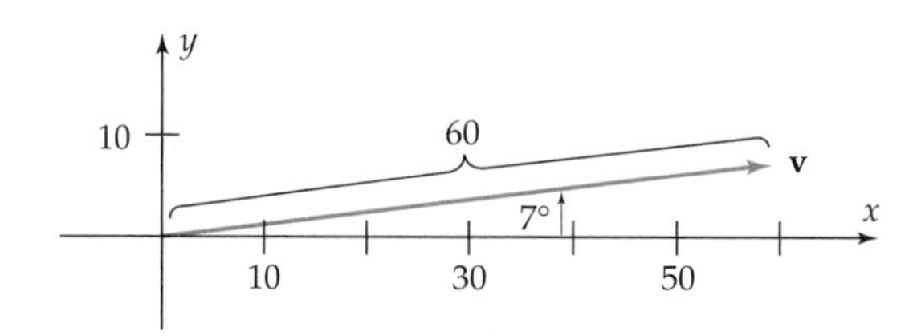

Figure 9–16

If $\mathbf{v} = a\mathbf{i} + b\mathbf{j}$ is a nonzero vector with direction angle θ, then

$$\tan\theta = \frac{\sin\theta}{\cos\theta} = \frac{b/\|\mathbf{v}\|}{a/\|\mathbf{v}\|} = \frac{b}{a}.$$

This fact provides a convenient way to find the direction angle of a vector.

Example 10 Find the direction angle of

(a) $\mathbf{u} = 5\mathbf{i} + 13\mathbf{j}$ (b) $\mathbf{v} = -10\mathbf{i} + 7\mathbf{j}$.

Solution

(a) The direction angle θ of $\mathbf{u}$ satisfies $\tan\theta = b/a = 13/5 = 2.6$. Using the TAN^{-1} key on a calculator we find that $\theta \approx 68.96°$, as shown in Figure 9–17(a).

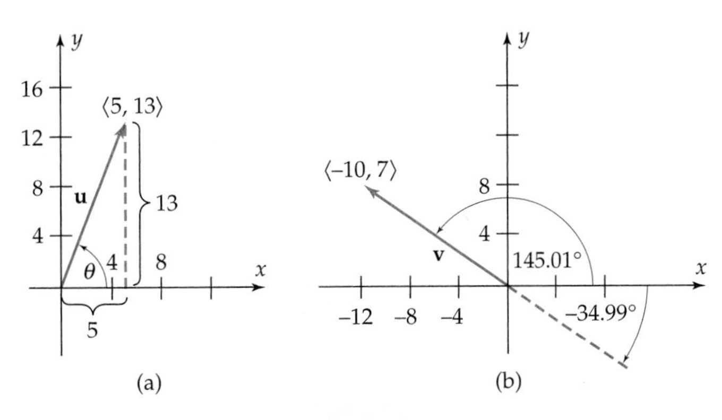

Figure 9–17

(b) The direction angle of $\mathbf{v}$ satisfies $\tan\theta = -7/10 = -.7$. Since $\mathbf{v}$ lies in the second quadrant, θ must be between 90° and 180°. A calculator shows that $-34.99°$ is an angle with tangent (approximately) $-.7$. Since tangent has period $\pi\ (= 180°)$, we know that $\tan t = \tan(t + 180°)$ for every t. Therefore, $\theta = -34.99° + 180° = 145.01°$ is the angle between 90° and 180° such that $\tan\theta \approx -.7$. See Figure 9–17(b). ■

Example 11 An object at the origin is acted upon by two forces. A 150-pound force makes an angle of 20° with the positive x-axis, and the other force of 100 pounds makes an angle of 70°, as shown in Figure 9–18. Find the direction and magnitude of the resultant force.

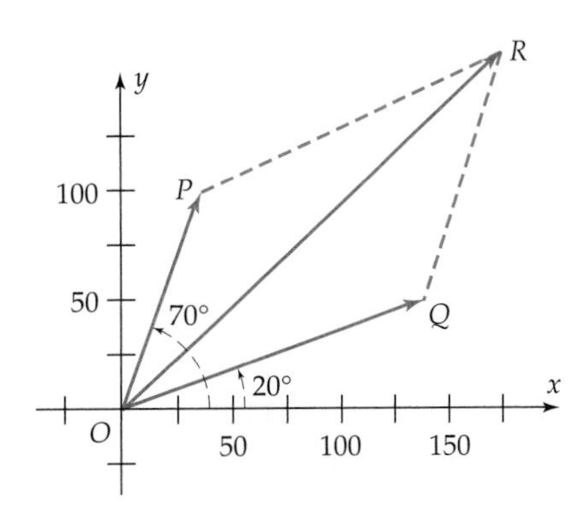

Figure 9–18

Solution The forces acting on the object are

$$\overrightarrow{OP} = (100 \cos 70°)\mathbf{i} + (100 \sin 70°)\mathbf{j}$$

$$\overrightarrow{OQ} = (150 \cos 20°)\mathbf{i} + (150 \sin 20°)\mathbf{j}.$$

The resultant force $\overrightarrow{OR}$ is the sum of $\overrightarrow{OP}$ and $\overrightarrow{OQ}$. Hence,

$$\overrightarrow{OR} = (100 \cos 70° + 150° \cos 20°)\mathbf{i} + (100 \sin 70° + 150 \sin 20°)\mathbf{j}$$

$$\approx 175.16\mathbf{i} + 145.27\mathbf{j}.$$

Therefore, the magnitude of the resultant force is

$$\|\overrightarrow{OR}\| \approx \sqrt{(175.16)^2 + (145.27)^2} \approx 227.56.$$

The direction angle θ of the resultant force satisfies

$$\tan \theta \approx 145.27/175.16 \approx .8294.$$

A calculator shows that $\theta \approx 39.67°$. ■

Applications

Example 12 A 200-pound box lies on a ramp that makes an angle of 24° with the horizontal. A rope is tied to the box from a post at the top of the ramp to keep it in position. Ignoring friction, how much force is being exerted on the rope by the box?

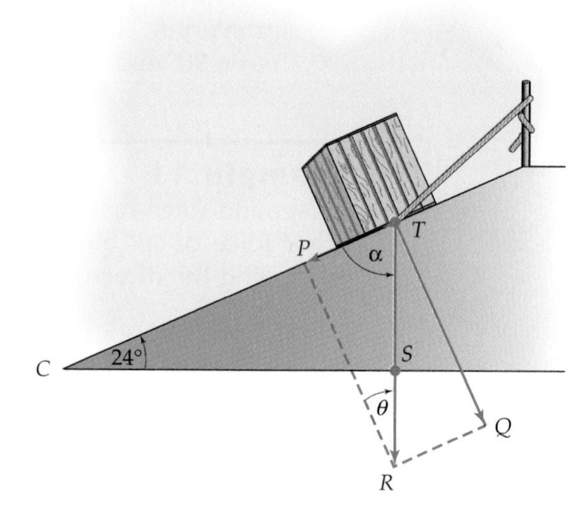

Figure 9–19

Solution Because of gravity the box exerts a 200-pound weight straight down (vector $\overrightarrow{TR}$). As Figure 9–19 shows, $\overrightarrow{TR}$ is the sum of $\overrightarrow{TP}$ and $\overrightarrow{TQ}$. The force on the rope is exerted by $\overrightarrow{TP}$, the vector of the force pulling the box down the ramp; so we must find $\|\overrightarrow{TP}\|$. In right triangle

TSC, $\alpha + 24° = 90°$ and in right triangle TPR, $\alpha + \theta = 90°$. Hence,

$$\alpha + \theta = \alpha + 24°, \qquad \text{so that} \qquad \theta = 24°.$$

Therefore,

$$\frac{\|\overrightarrow{TP}\|}{\|\overrightarrow{TR}\|} = \sin\theta$$

$$\frac{\|\overrightarrow{TP}\|}{200} = \sin 24°$$

$$\|\overrightarrow{TP}\| = 200 \sin 24° \approx 81.35.$$

So, the force on the rope is 81.35 pounds. ■

In aerial navigation, directions are given in terms of the angle measured in degrees clockwise from true north. Thus north is 0°, east is 90°, and so on.

Example 13 An airplane is traveling in the direction 50° with an air speed of 300 mph, and there is a 35-mph wind from the direction 120°, as represented by the vectors **p** and **w** in Figure 9–20. Find the *course* and *ground speed* of the plane (that is, its direction and speed relative to the ground).

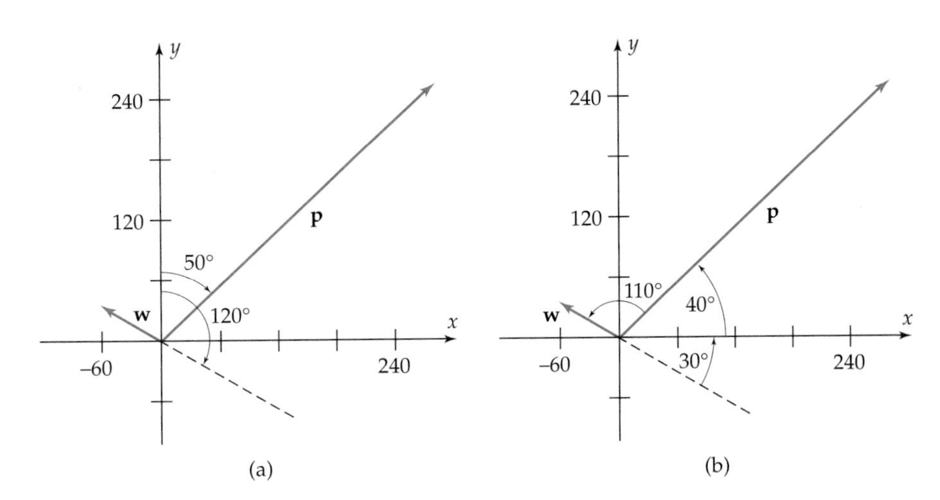

Figure 9–20

Solution The course of the plane is the direction of the vector $\mathbf{p} + \mathbf{w}$, and its ground speed is the magnitude of $\mathbf{p} + \mathbf{w}$. Figure 9–20(b) shows that the direction angle of $\mathbf{p}$ (the angle it makes with the positive x-axis) is 40° and that the direction angle of $\mathbf{w}$ is 150°. Therefore,

$$\begin{aligned}\mathbf{p} + \mathbf{w} &= [(300\cos 40°)\mathbf{i} + (300 \sin 40°)\mathbf{j}] + [(35\cos 150°)\mathbf{i} + (35 \sin 150°)\mathbf{j}] \\ &= (300\cos 40° + 35\cos 150°)\mathbf{i} + (300\sin 40° + 35 \sin 150°)\mathbf{j} \\ &\approx 199.50\mathbf{i} + 210.34\mathbf{j}\end{aligned}$$

The direction angle of $\mathbf{p} + \mathbf{w}$ satisfies $\tan\theta = 210.34/199.50 \approx 1.0543$, and a calculator shows that $\theta \approx 46.5°$. This is the angle $\mathbf{p} + \mathbf{w}$ makes with the positive x-axis; hence, the course of the plane (the angle between true north and $\mathbf{p} + \mathbf{w}$) is $90° - 46.5° = 43.5°$. The ground speed of the plane is

$$\|\mathbf{p} + \mathbf{w}\| \approx \sqrt{(199.5)^2 + (210.34)^2} \approx 289.9 \text{ mph.} \quad ■$$

Exercises 9.3

In Exercises 1–4, find the magnitude of the vector $\overrightarrow{PQ}$.

1. $P = (2, 3)$, $Q = (5, 9)$

2. $P = (-3, 5)$, $Q = (7, -11)$

3. $P = (-7, 0)$, $Q = (-4, -5)$

4. $P = (30, 12)$, $Q = (25, 5)$

In Exercises 5–10, find a vector with the origin as initial point that is equivalent to the vector $\overrightarrow{PQ}$.

5. $P = (1, 5)$, $Q = (7, 11)$ **6.** $P = (2, 7)$, $Q = (-2, 9)$

7. $P = (-4, -8)$, $Q = (-10, 2)$

8. $P = (-5, 6)$, $Q = (-7, -9)$

9. $P = \left(\frac{4}{5}, -2\right)$, $Q = \left(\frac{17}{5}, -\frac{12}{5}\right)$

10. $P = (\sqrt{2}, 4)$, $Q = (\sqrt{3}, -1)$

In Exercises 11–20, find $\mathbf{u} + \mathbf{v}$, $\mathbf{u} - \mathbf{v}$, and $3\mathbf{u} - 2\mathbf{v}$.

11. $\mathbf{u} = \langle -2, 4\rangle$, $\mathbf{v} = \langle 6, 1\rangle$

12. $\mathbf{u} = \langle 4, 0\rangle$, $\mathbf{v} = \langle 1, -3\rangle$

13. $\mathbf{u} = \langle 3, 3\sqrt{2}\rangle$, $\mathbf{v} = \langle 4\sqrt{2}, 1\rangle$

14. $\mathbf{u} = \left\langle \frac{2}{3}, 4\right\rangle$, $\mathbf{v} = \left\langle -7, \frac{19}{3}\right\rangle$

15. $\mathbf{u} = 2\langle -2, 5\rangle$, $\mathbf{v} = \frac{1}{4}\langle -7, 12\rangle$

16. $\mathbf{u} = \mathbf{i} - \mathbf{j}$, $\mathbf{v} = 2\mathbf{i} + \mathbf{j}$ **17.** $\mathbf{u} = 8\mathbf{i}$, $\mathbf{v} = 2(3\mathbf{i} - 2\mathbf{j})$

18. $\mathbf{u} = -4(-\mathbf{i} + \mathbf{j})$, $\mathbf{v} = -3\mathbf{i}$

19. $\mathbf{u} = -\left(2\mathbf{i} + \frac{3}{2}\mathbf{j}\right)$, $\mathbf{v} = \frac{3}{4}\mathbf{i}$

20. $\mathbf{u} = \sqrt{2}\mathbf{j}$, $\mathbf{v} = \sqrt{3}\mathbf{i}$

In Exercises 21–26, find the components of the given vector, where $\mathbf{u} = \mathbf{i} - 2\mathbf{j}$, $\mathbf{v} = 3\mathbf{i} + \mathbf{j}$, $\mathbf{w} = -4\mathbf{i} + \mathbf{j}$.

21. $\mathbf{u} + 2\mathbf{w}$ **22.** $\frac{1}{2}(3\mathbf{v} + \mathbf{w})$

23. $\frac{1}{2}\mathbf{w}$ **24.** $-2\mathbf{u} + 3\mathbf{v}$

25. $\frac{1}{4}(8\mathbf{u} + 4\mathbf{v} - \mathbf{w})$ **26.** $3(\mathbf{u} - 2\mathbf{v}) - 6\mathbf{w}$

In Exercises 27–34, find the component form of the vector $\mathbf{v}$ whose magnitude and direction angle θ are given.

27. $\|\mathbf{v}\| = 4$, $\theta = 0°$ **28.** $\|\mathbf{v}\| = 5$, $\theta = 30°$

29. $\|\mathbf{v}\| = 10$, $\theta = 225°$ **30.** $\|\mathbf{v}\| = 20$, $\theta = 120°$

31. $\|\mathbf{v}\| = 6$, $\theta = 40°$ **32.** $\|\mathbf{v}\| = 8$, $\theta = 160°$

33. $\|\mathbf{v}\| = 1/2$, $\theta = 250°$ **34.** $\|\mathbf{v}\| = 3$, $\theta = 310°$

In Exercises 35–42, find the magnitude and direction angle of the vector $\mathbf{v}$.

35. $\mathbf{v} = \langle 4, 4\rangle$ **36.** $\mathbf{v} = \langle 5, 5\sqrt{3}\rangle$

37. $\mathbf{v} = \langle -8, 0\rangle$ **38.** $\mathbf{v} = \langle 4, 5\rangle$

39. $\mathbf{v} = 6\mathbf{j}$ **40.** $\mathbf{v} = 4\mathbf{i} - 8\mathbf{j}$

41. $-2\mathbf{i} + 8\mathbf{j}$ **42.** $\mathbf{v} = -15\mathbf{i} - 10\mathbf{j}$

In Exercises 43–46, find a unit vector that has the same direction as $\mathbf{v}$.

43. $\langle 4, -5\rangle$ **44.** $-7\mathbf{i} + 8\mathbf{j}$

45. $5\mathbf{i} + 10\mathbf{j}$ **46.** $-3\mathbf{i} - 9\mathbf{j}$

In Exercises 47–50, an object at the origin is acted upon by two forces, $\mathbf{u}$ and $\mathbf{v}$, with direction angle θ_u and θ_v, respectively. Find the direction and magnitude of the resultant force.

47. $\mathbf{u} = 30$ pounds, $\theta_{\mathbf{u}} = 0°$; $\mathbf{v} = 90$ pounds, $\theta_{\mathbf{v}} = 60°$

48. $\mathbf{u} = 6$ pounds, $\theta_{\mathbf{u}} = 45°$; $\mathbf{v} = 6$ pounds, $\theta_{\mathbf{v}} = 120°$

49. $\mathbf{u} = 12$ kilograms, $\theta_{\mathbf{u}} = 130°$; $\mathbf{v} = 20$ kilograms, $\theta_{\mathbf{v}} = 250°$

50. $\mathbf{u} = 30$ kilograms, $\theta_{\mathbf{u}} = 300°$; $\mathbf{v} = 80$ kilograms, $\theta_{\mathbf{v}} = 40°$

If forces $\mathbf{u}_1, \mathbf{u}_2, \ldots, \mathbf{u}_k$ act on an object at the origin, the resultant force is the sum $\mathbf{u}_1 + \mathbf{u}_2 + \cdots + \mathbf{u}_k$. The forces are *said to be in* equilibrium *if their resultant force is* **0**. *In Exercises 51 and 52, find the resultant force and find an additional force v, which, if added to the system, produces equilibrium.*

51. $\mathbf{u}_1 = \langle 2, 5 \rangle$, $\mathbf{u}_2 = \langle -6, 1 \rangle$, $\mathbf{u}_3 = \langle -4, -8 \rangle$

52. $\mathbf{u}_1 = \langle 3, 7 \rangle$, $\mathbf{u}_2 = \langle 8, -2 \rangle$, $\mathbf{u}_3 = \langle -9, 0 \rangle$, $\mathbf{u}_4 = \langle -5, 4 \rangle$

In Exercises 53–58, let $u = \langle a, b \rangle$, $v = \langle c, d \rangle$, and $w = \langle e, f \rangle$, and let r and s be scalars. Prove that the stated property holds by calculating the vector on each side of the equal sign.

53. $\mathbf{v} + \mathbf{0} = \mathbf{v} = \mathbf{0} + \mathbf{v}$

54. $\mathbf{v} + (-\mathbf{v}) = \mathbf{0}$

55. $r(\mathbf{u} + \mathbf{v}) = r\mathbf{u} + r\mathbf{v}$

56. $(r + s)\mathbf{v} = r\mathbf{v} + s\mathbf{v}$

57. $(rs)\mathbf{v} = r(s\mathbf{v}) = s(r\mathbf{v})$

58. $1\mathbf{v} = \mathbf{v}$ and $0\mathbf{v} = \mathbf{0}$

59. Two ropes are tied to a wagon. A child pulls one with a force of 20 pounds, while another child pulls the other with a force of 30 pounds (see figure). If the angle between the two ropes is 28°, how much force must be exerted by a third child, standing behind the wagon, to keep the wagon from moving? [*Hint:* Assume the wagon is at the origin and one rope runs along the positive x-axis. Proceed as in Example 11 to find the resultant force on the wagon from the ropes. The third child must use the same amount in the opposite direction.]

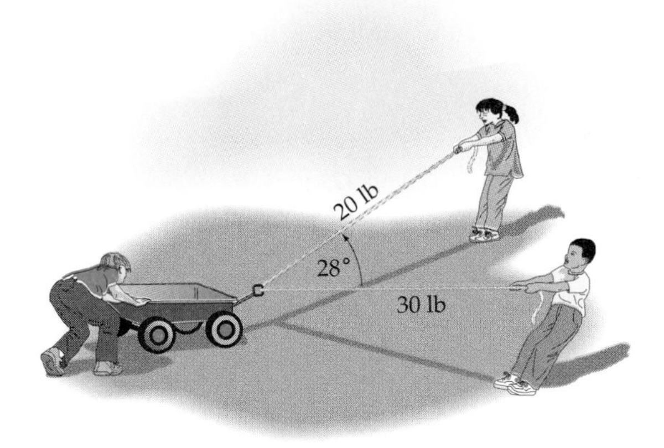

60. Two circus elephants, Bessie and Maybelle, are dragging a large wagon, as shown in the figure. If Bessie pulls with a force of 2200 pounds and Maybelle with a force of 1500 pounds and the wagon moves along the dashed line, what is angle θ?

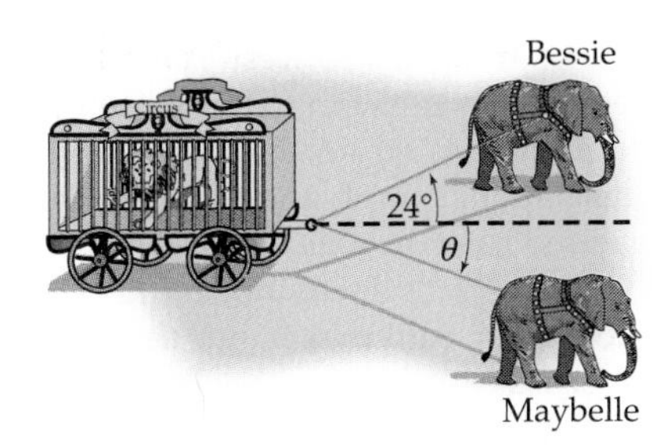

Exercises 61–64 deal with an object on an inclined plane. The situation is similar to that in Figure 9–19 of Example 12, where $\|\overrightarrow{TP}\|$ is the component of the weight of the object parallel to the plane and $\|\overrightarrow{TQ}\|$ is the component of the weight perpendicular to the plane.

61. An object weighing 50 pounds lies on an inclined plane that makes a 40° angle with the horizontal. Find the components of the weight parallel and perpendicular to the plane. [*Hint:* Solve an appropriate triangle.]

62. Do Exercise 61 when the object weighs 200 pounds and the inclined plane makes a 20° angle with the horizontal.

63. If an object on an inclined plane weighs 150 pounds and the component of the weight perpendicular to the plane is 60 pounds, what angle does the plane make with the horizontal?

64. A force of 500 pounds is needed to pull a cart up a ramp that makes a 15° angle with the ground. Assume no friction is involved and find the weight of the cart. [*Hint:* Draw a picture similar to Figure 9–19; the 500-pound force is parallel to the ramp.]

In Exercises 65–70, find the course and ground speed of the plane under the given conditions. (See Example 13.)

65. Air speed 250 mph in the direction 60°; wind speed 40 mph from the direction 330°.

66. Air speed 400 mph in the direction 150°; wind speed 30 mph from the direction 60°.

67. Air speed 300 mph in the direction 300°; wind speed 50 mph in (*not* from) the direction 30°.

68. Air speed 500 mph in the direction 180°; wind speed 70 mph in the direction 40°.

69. The course and ground speed of a plane are 70° and 400 mph respectively. There is a 60-mph wind blowing south. Find the (approximate) direction and air speed of the plane.

70. A plane is flying in the direction 200° with an air speed of 500 mph. Its course and ground speed are 210° and 450 mph, respectively. What is the direction and speed of the wind?

71. A river flows from east to west. A swimmer on the south bank wants to swim to a point on the opposite shore directly north of her starting point. She can swim at 2.8 mph and there is a 1-mph current in the river. In what direction should she head so as to travel directly north (that is, what angle should her path make with the south bank of the river)?

72. A river flows from west to east. A swimmer on the north bank swims at 3.1 mph along a straight course that makes a 75° angle with the north bank of the river and reaches the south bank at a point directly south of his starting point. How fast is the current in the river?

73. A 400-pound weight is suspended by two cables (see figure). What is the force (tension) on each cable? [*Hint:* Imagine that the weight is at the origin and that the dashed line is the x-axis. Then cable **v** is represented by the vector $(c \cos 65°)\mathbf{i} + (c \sin 65°)\mathbf{j}$, which has magnitude c. (Why?) Represent cable **u** similarly, denoting its magnitude by d. Use the fact that $\mathbf{u} + \mathbf{v} = 0\mathbf{i} + 400\mathbf{j}$ (why?) to set up a system of two equations in the unknowns c and d.]

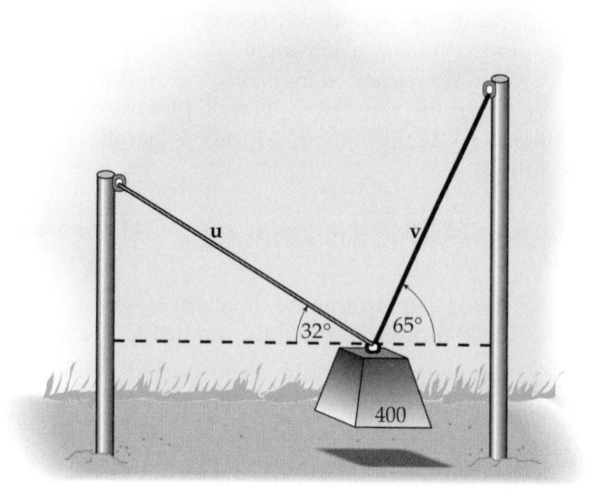

74. A 175-pound high wire artist stands balanced on a tightrope, which sags slightly at the point where he is standing. The rope in front of him makes a 6° angle with the horizontal and the rope behind him makes a 4° angle with the horizontal. Find the force on each end of the rope. [*Hint:* Use a picture and procedure similar to that in Exercise 73.]

75. Let **v** be the vector with initial point (x_1, y_1) and terminal point (x_2, y_2) and let k be any real number.

(a) Find the component form of **v** and $k\mathbf{v}$.

(b) Calculate $\|\mathbf{v}\|$ and $\|k\mathbf{v}\|$.

(c) Use the fact that $\sqrt{k^2} = |k|$ to verify that $\|k\mathbf{v}\| = |k| \cdot \|\mathbf{v}\|$.

(d) Show that $\tan \theta = \tan \beta$, where θ is the direction angle of **v** and β is the direction angle of $k\mathbf{v}$. Use the fact that $\tan t = \tan(t + 180°)$ to conclude that **v** and $k\mathbf{v}$ have either the same or opposite directions.

(e) Use the fact that (c, d) and $(-c, -d)$ lie on the same straight line on opposite sides of the origin (Exercise 77 in Section 1.3) to verify that **v** and $k\mathbf{v}$ have the same direction if $k > 0$ and opposite directions if $k < 0$.

76. Let $\mathbf{u} = \langle a, b \rangle$, $\mathbf{v} = \langle c, d \rangle$. Verify the accuracy of the two geometric interpretations of vector addition given on page 583 as follows.

(a) Show that the distance from (a, b) to $(a + c, b + d)$ is the same as $\|\mathbf{v}\|$.

(b) Show that the distance from (c, d) to $(a + c, b + d)$ is the same as $\|\mathbf{u}\|$.

(c) Show that the line through (a, b) and $(a + c, b + d)$ is parallel to **v** by showing they have the same slope.

(d) Show that the line through (c, d) and $(a + c, b + d)$ is parallel to **u**.

77. Let $\mathbf{u} = \langle a, b \rangle$ and $\mathbf{v} = \langle c, d \rangle$. Show that $\mathbf{u} - \mathbf{v}$ is equivalent to the vector **w** with initial point (c, d) and terminal point (a, b) as follows. (See figure.)

(a) Show that $\|\mathbf{u} - \mathbf{v}\| = \|\mathbf{w}\|$.

(b) Show that $\mathbf{u} - \mathbf{v}$ and **w** have the same direction.

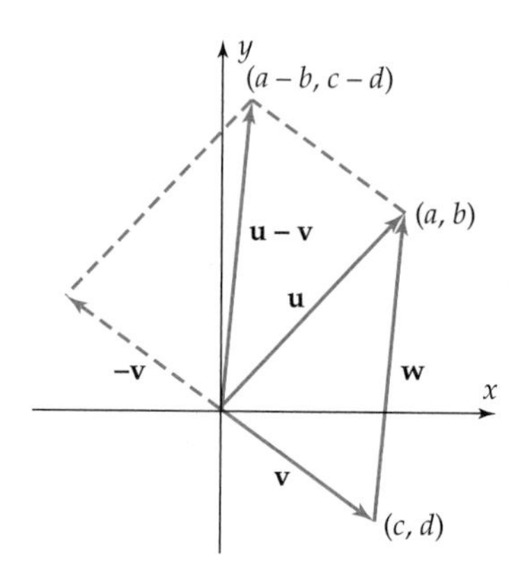

9.4 The Dot Product

We now define a product for vectors and consider some of its many applications. When two vectors are added, their sum is another vector, but the situation is different with products. The dot product of two *vectors* is the *real number* defined as follows.

Dot Product

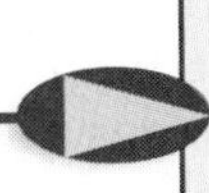

The *dot product* of vectors $\mathbf{u} = \langle a, b \rangle = a\mathbf{i} + b\mathbf{j}$ and $\mathbf{v} = \langle c, d \rangle = c\mathbf{i} + d\mathbf{j}$ is denoted $\mathbf{u} \cdot \mathbf{v}$ and is defined to be the real number $ac + bd$. Thus,

$$\mathbf{u} \cdot \mathbf{v} = ac + bd.$$

Technology Tip

Dot products can be computed automatically by using DOT or DOTP in the MATH submenu of the TI-86 VECTOR menu, or in the VECTOR OPS submenu of the MATRIX submenu of the TI-89 MATH menu, or in the MATRIX submenu of the HP-38 MATH menu.

Example 1

(a) If $\mathbf{u} = \langle 5, 3 \rangle$ and $\mathbf{v} = \langle -2, 6 \rangle$, then

$$\mathbf{u} \cdot \mathbf{v} = 5(-2) + 3 \cdot 6 = 8.$$

(b) If $\mathbf{u} = 4\mathbf{i} - 2\mathbf{j}$ and $\mathbf{v} = 3\mathbf{i} - \mathbf{j}$, then

$$\mathbf{u} \cdot \mathbf{v} = 4 \cdot 3 + (-2)(-1) = 14.$$

(c) $\langle 2, -4 \rangle \cdot \langle 6, 3 \rangle = 2 \cdot 6 + (-4)3 = 0.$ ■

The dot product has a number of useful properties:

Properties of the Dot Product

If u, v, w are vectors and k is a real number, then

1. $\mathbf{u} \cdot \mathbf{u} = \|\mathbf{u}\|^2$.
2. $\mathbf{u} \cdot \mathbf{v} = \mathbf{v} \cdot \mathbf{u}$.
3. $\mathbf{u} \cdot (\mathbf{v} + \mathbf{w}) = \mathbf{u} \cdot \mathbf{v} + \mathbf{u} \cdot \mathbf{w}$.
4. $k\mathbf{u} \cdot \mathbf{v} = k(\mathbf{u} \cdot \mathbf{v}) = \mathbf{u} \cdot k\mathbf{v}$.
5. $\mathbf{0} \cdot \mathbf{u} = 0$.

Proof

1. If $\mathbf{u} = \langle a, b \rangle$, then

$$\|\mathbf{u}\| = \sqrt{a^2 + b^2}.$$

Hence,

$$\mathbf{u} \cdot \mathbf{u} = \langle a, b \rangle \cdot \langle a, b \rangle = a \cdot a + b \cdot b = a^2 + b^2 = \left(\sqrt{a^2 + b^2}\right)^2 = \|\mathbf{u}\|^2.$$

2. If $\mathbf{u} = \langle a, b\rangle$ and $\mathbf{v} = \langle c, d\rangle$, then

$$\mathbf{u}\cdot\mathbf{v} = \langle a, b\rangle\cdot\langle c, d\rangle = ac + bd = ca + db = \langle c, d\rangle\cdot\langle a, b\rangle = \mathbf{v}\cdot\mathbf{u}.$$

The last three statements are proved similarly (Exercises 37–39). ■

Angles

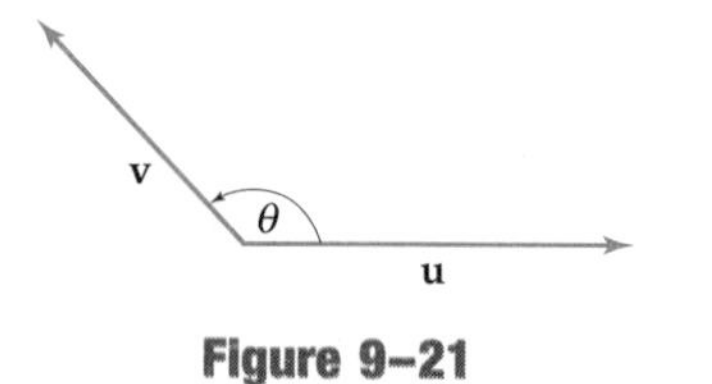

Figure 9–21

If $\mathbf{u} = \langle a, b\rangle$ and $\mathbf{v} = \langle c, d\rangle$ are nonzero vectors, then the **angle between u and v** is the smallest angle θ formed by these two line segments, as shown in Figure 9–21. We ignore clockwise or counterclockwise rotation and consider the angle between **v** and **u** to be the same as the angle between **u** and **v**. Thus, the radian measure of θ ranges from 0 to π.

Nonzero vectors **u** and **v** are said to be **parallel** if the angle between them is either 0 or π radians (that is, **u** and **v** lie on the same straight line through the origin and have either the same or opposite directions). The zero vector **0** is considered to be parallel to every vector.

Any scalar multiple of **u** is parallel to **u** since it lies on the same straight line as **u** (see Example 3 in Section 9.3). Conversely, if **v** is parallel to **u**, it is easy to show that **v** must be a scalar multiple of **u** (Exercise 40). Hence,

Parallel Vectors

Vectors u and v are parallel exactly when

$$\mathbf{v} = k\mathbf{u} \text{ for some real number } k.$$

Example 2 The vectors $\langle 2, 3\rangle$ and $\langle 8, 12\rangle$ are parallel because $\langle 8, 12\rangle = 4\langle 2, 3\rangle$. ■

The angle between nonzero vectors **u** and **v** is closely related to their dot product:

Angle Theorem

If θ is the angle between the nonzero vectors u and v, then

$$\mathbf{u}\cdot\mathbf{v} = \|u\|\,\|v\|\cos\theta,$$

or equivalently,

$$\cos\theta = \frac{\mathbf{u}\cdot\mathbf{v}}{\|\mathbf{u}\|\,\|\mathbf{v}\|}.$$

Proof If $\mathbf{u} = \langle a, b\rangle$, $\mathbf{v} = \langle c, d\rangle$, and the angle θ is not 0 or π, then **u** and **v** form two sides of a triangle, as shown in Figure 9–22.

The lengths of two sides of the triangle are $\|\mathbf{u}\| = \sqrt{a^2 + b^2}$ and $\|\mathbf{v}\| = \sqrt{c^2 + d^2}$. The distance formula shows that the length of the third

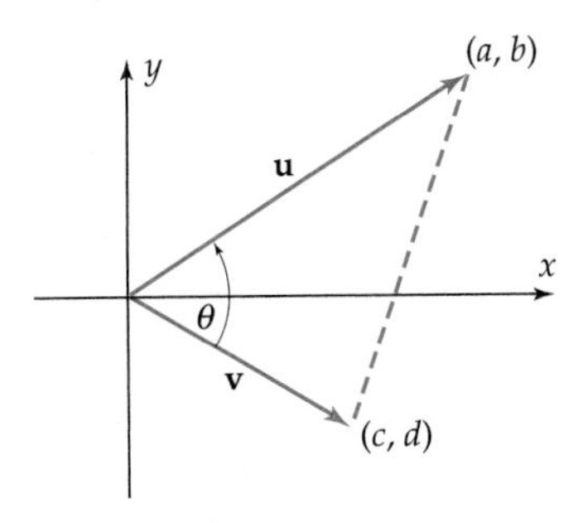

Figure 9–22

side (opposite angle θ) is $\sqrt{(a-c)^2+(b-d)^2}$. Therefore, by the Law of Cosines,

$$\left[\sqrt{(a-c)^2+(b-d)^2}\right]^2 = \|\mathbf{u}\|^2 + \|\mathbf{v}\|^2 - 2\|\mathbf{u}\|\,\|\mathbf{v}\|\cos\theta$$

$$(a-c)^2+(b-d)^2 = (a^2+b^2)+(c^2+d^2) - 2\|\mathbf{u}\|\,\|\mathbf{v}\|\cos\theta$$

$$a^2-2ac+c^2+b^2-2bd+d^2 = (a^2+c^2)+(b^2+d^2) - 2\|\mathbf{u}\|\,\|\mathbf{v}\|\cos\theta$$

$$-2ac-2bd = -2\|\mathbf{u}\|\,\|\mathbf{v}\|\cos\theta.$$

Dividing both sides by -2 shows that

$$ac+bd = \|\mathbf{u}\|\,\|\mathbf{v}\|\cos\theta.$$

Since the left side of this equation is precisely $\mathbf{u}\cdot\mathbf{v}$, the proof is complete in this case. The proof when θ is 0 or π is left to the reader (Exercise 41). ■

Example 3 Find the angle θ between the vectors $\langle -3, 1\rangle$ and $\langle 5, 2\rangle$ shown in Figure 9–23.

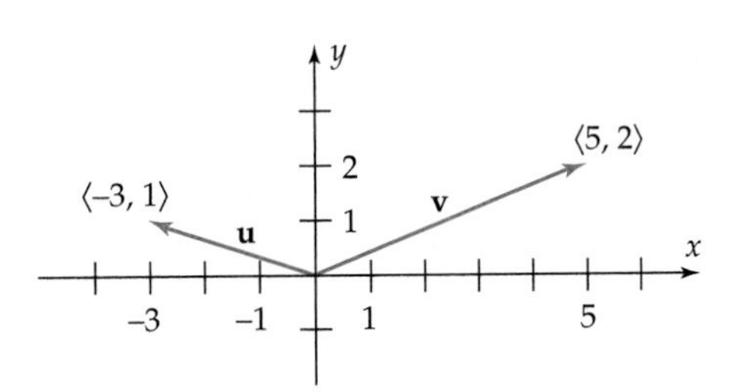

Figure 9–23

Solution Apply the formula in the box above with $\mathbf{u} = \langle -3, 1\rangle$ and $\mathbf{v} = \langle 5, 2\rangle$:

$$\cos\theta = \frac{\mathbf{u}\cdot\mathbf{v}}{\|\mathbf{u}\|\,\|\mathbf{v}\|} = \frac{(-3)5+1\cdot 2}{\sqrt{(-3)^2+1^2}\sqrt{5^2+2^2}} = \frac{-13}{\sqrt{10}\sqrt{29}} = \frac{-13}{\sqrt{290}}.$$

Using the COS^{-1} key, we see that

$$\theta \approx 2.4393 \text{ radians } (\approx 139.76°).$$ ■

The Angle Theorem has several useful consequences. For instance, by taking absolute values on both sides of $\mathbf{u}\cdot\mathbf{v} = \|\mathbf{u}\|\,\|\mathbf{v}\|\cos\theta$, and using the fact that $\big|\|\mathbf{u}\|\,\|\mathbf{v}\|\big| = \|\mathbf{u}\|\,\|\mathbf{v}\|$ (because $\|\mathbf{u}\|\,\|\mathbf{v}\| \ge 0$), we see that

$$|\mathbf{u}\cdot\mathbf{v}| = \big|\|\mathbf{u}\|\,\|\mathbf{v}\|\cos\theta\big| = \big|\|\mathbf{u}\|\,\|\mathbf{v}\|\big|\,|\cos\theta| = \|\mathbf{u}\|\,\|\mathbf{v}\|\,|\cos\theta|.$$

But for any angle θ, $|\cos\theta| \le 1$ so that

$$|\mathbf{u}\cdot\mathbf{v}| = \|\mathbf{u}\|\,\|\mathbf{v}\|\,|\cos\theta| \le \|\mathbf{u}\|\,\|\mathbf{v}\|.$$

This proves the Schwarz inequality:

Schwarz Inequality

For any vectors u and v,

$$|\mathbf{u}\cdot\mathbf{v}| \leq \|\mathbf{u}\|\,\|\mathbf{v}\|.$$

Vectors **u** and **v** are said to be **orthogonal** or (**perpendicular**) if the angle between them is $\pi/2$ radians (90°), or if at least one of them is **0**. Here is the key fact about orthogonal vectors:

Orthogonal Vectors

Let u and v be vectors. Then

u and v are orthogonal exactly when $\mathbf{u}\cdot\mathbf{v} = 0$.

Proof If **u** or **v** is **0**, then $\mathbf{u}\cdot\mathbf{v} = 0$, and if **u** and **v** are nonzero orthogonal vectors, then by the Angle Theorem

$$\mathbf{u}\cdot\mathbf{v} = \|\mathbf{u}\|\,\|\mathbf{v}\|\cos\theta = \|\mathbf{u}\|\,\|\mathbf{v}\|\cos(\pi/2) = \|\mathbf{u}\|\,\|\mathbf{v}\|\,(0) = 0.$$

Conversely, if **u** and **v** are vectors such that $\mathbf{u}\cdot\mathbf{v} = 0$, then Exercise 42 shows that **u** and **v** are orthogonal. ■

Example 4

(a) The vectors $\mathbf{u} = \langle 2, -6\rangle$ and $\mathbf{v} = \langle 9, 3\rangle$ are orthogonal because

$$\mathbf{u}\cdot\mathbf{v} = \langle 2, -6\rangle\cdot\langle 9, 3\rangle = 2\cdot 9 + (-6)3 = 18 - 18 = 0.$$

(b) The vectors $\frac{1}{2}\mathbf{i} + 5\mathbf{j}$ and $10\mathbf{i} - \mathbf{j}$ are orthogonal since

$$\left(\frac{1}{2}\mathbf{i} + 5\mathbf{j}\right)\cdot(10\mathbf{i} - \mathbf{j}) = \frac{1}{2}(10) + 5(-1) = 5 - 5 = 0. \quad ■$$

Projections and Components

If **u** and **v** are nonzero vectors and θ is the angle between them, construct the perpendicular line segment from the terminal point P of **u** to the straight line on which **v** lies. This perpendicular segment intersects the line at a point Q, as shown in Figure 9–24.

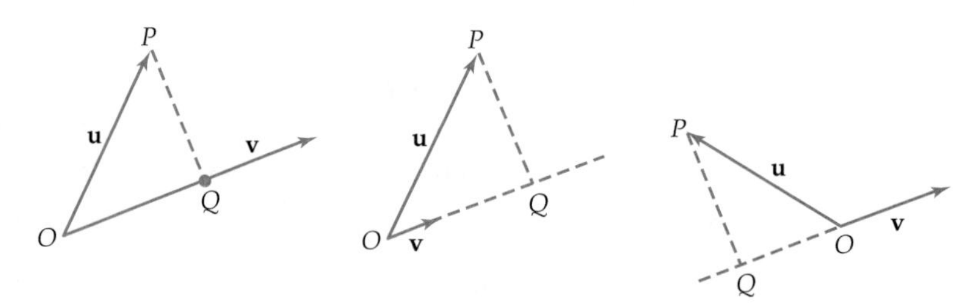

Figure 9–24

The vector $\overrightarrow{OQ}$ is called the **projection of u on v** and is denoted $\text{proj}_{\mathbf{v}}\mathbf{u}$. Here is a useful description of $\text{proj}_{\mathbf{v}}\mathbf{u}$.

Projection of u on v

.If u and v are nonzero vectors, then the projection of u on v is the vector

$$\text{proj}_{\mathbf{v}}\mathbf{u} = \left(\frac{\mathbf{u}\cdot\mathbf{v}}{\|\mathbf{v}\|^2}\right)\mathbf{v}.$$

Proof Since $\text{proj}_{\mathbf{v}}\mathbf{u}$ and $\mathbf{v}$ lie on the same straight line, they are parallel, and hence $\text{proj}_{\mathbf{v}}\mathbf{u} = k\mathbf{v}$ for some real number k. Let $\mathbf{w}$ be the vector with initial point at the origin and the same length and direction as $\overrightarrow{QP}$, as in the two cases shown in Figure 9–25:

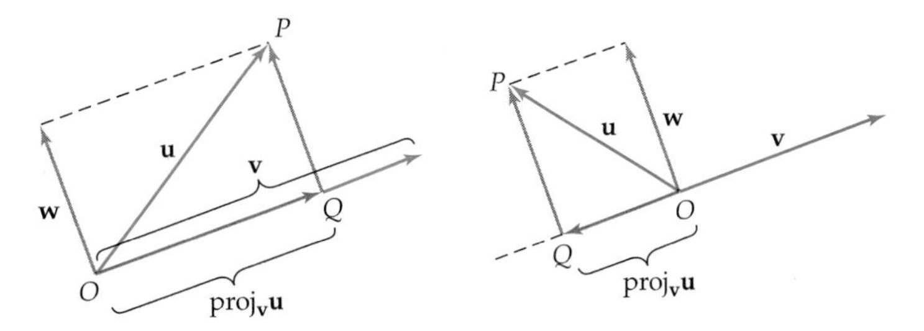

Figure 9–25

Note that $\mathbf{w}$ is parallel to $\overrightarrow{QP}$ and hence is orthogonal to $\mathbf{v}$. As shown in Figure 9–25, $\mathbf{u} = \text{proj}_{\mathbf{v}}\mathbf{u} + \mathbf{w} = k\mathbf{v} + \mathbf{w}$. Consequently, by the properties of the dot project

$$\begin{aligned}\mathbf{u}\cdot\mathbf{v} &= (k\mathbf{v} + \mathbf{w})\cdot\mathbf{v} = (k\mathbf{v})\cdot\mathbf{v} + \mathbf{w}\cdot\mathbf{v}\\ &= k(\mathbf{v}\cdot\mathbf{v}) + \mathbf{w}\cdot\mathbf{v} = k\|\mathbf{v}\|^2 + \mathbf{w}\cdot\mathbf{v}.\end{aligned}$$

But $\mathbf{w}\cdot\mathbf{v} = 0$ because $\mathbf{w}$ and $\mathbf{v}$ are orthogonal. Hence,

$$\mathbf{u}\cdot\mathbf{v} = k\|\mathbf{v}\|^2, \qquad \text{or equivalently,} \qquad k = \frac{\mathbf{u}\cdot\mathbf{v}}{\|\mathbf{v}\|^2}.$$

Therefore,

$$\text{proj}_{\mathbf{v}}\mathbf{u} = k\mathbf{v} = \left(\frac{\mathbf{u}\cdot\mathbf{v}}{\|\mathbf{v}\|^2}\right)\mathbf{v}$$

and the proof is complete. ■

Example 5 If $\mathbf{u} = 8\mathbf{i} + 3\mathbf{j}$ and $\mathbf{v} = 4\mathbf{i} - 2\mathbf{j}$, find $\text{proj}_{\mathbf{v}}\mathbf{u}$ and $\text{proj}_{\mathbf{u}}\mathbf{v}$.

Solution

$$\mathbf{u}\cdot\mathbf{v} = 8\cdot 4 + 3(-2) = 26 \qquad \text{and} \qquad \|\mathbf{v}\|^2 = \mathbf{v}\cdot\mathbf{v} = 4^2 + (-2)^2 = 20.$$

Therefore,

$$\text{proj}_{\mathbf{v}}\mathbf{u} = \left(\frac{\mathbf{u}\cdot\mathbf{v}}{\|\mathbf{v}\|^2}\right)\mathbf{v} = \frac{26}{20}(4\mathbf{i} - 2\mathbf{j}) = \frac{26}{5}\mathbf{i} - \frac{13}{5}\mathbf{j}.$$

as shown in Figure 9–26. We can find the projection of **v** on **u** by noting that $\|\mathbf{u}\|^2 = \mathbf{u}\cdot\mathbf{u} = 8^2 + 3^2 = 73$, and hence

$$\text{proj}_{\mathbf{u}}\mathbf{v} = \left(\frac{\mathbf{v}\cdot\mathbf{u}}{\|\mathbf{u}\|^2}\right)\mathbf{u} = \frac{26}{73}(8\mathbf{i} + 3\mathbf{j}) = \frac{208}{73}\mathbf{i} + \frac{78}{73}\mathbf{j}. \quad \blacksquare$$

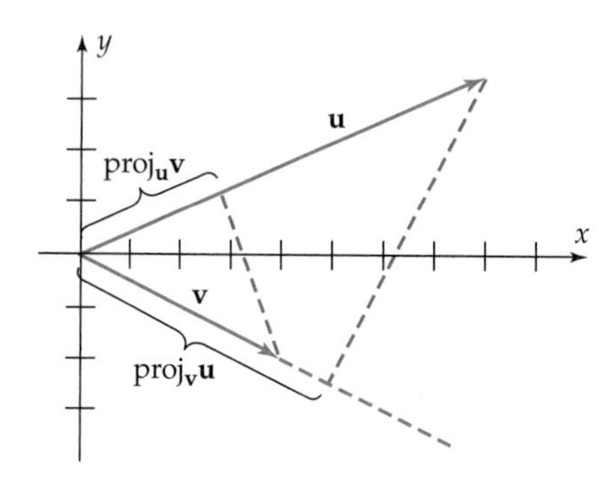

Figure 9–26

Recall that $\frac{1}{\|\mathbf{v}\|}\mathbf{v}$ is a unit vector in the direction of **v** (see page 585). We can express $\text{proj}_{\mathbf{v}}\mathbf{u}$ as a scalar multiple of this unit vector as follows:

$$\text{proj}_{\mathbf{v}}\mathbf{u} = \left(\frac{\mathbf{u}\cdot\mathbf{v}}{\|\mathbf{v}\|^2}\right)\mathbf{v} = \left(\frac{\mathbf{u}\cdot\mathbf{v}}{\|\mathbf{v}\|}\right)\left(\frac{1}{\|\mathbf{v}\|}\mathbf{v}\right).$$

The scalar $\frac{\mathbf{u}\cdot\mathbf{v}}{\|\mathbf{v}\|}$ is called the **component of u along v** and is denoted $\text{comp}_{\mathbf{v}}\mathbf{u}$. Thus,

$$\text{proj}_{\mathbf{v}}\mathbf{u} = \left(\frac{\mathbf{u}\cdot\mathbf{v}}{\|\mathbf{v}\|}\right)\left(\frac{1}{\|\mathbf{v}\|}\mathbf{v}\right) = \text{comp}_{\mathbf{v}}\mathbf{u}\left(\frac{1}{\|\mathbf{v}\|}\mathbf{v}\right).$$

Since $\frac{1}{\|\mathbf{v}\|}\mathbf{v}$ is a unit vector, the length of $\text{proj}_{\mathbf{v}}\mathbf{u}$ is

$$\|\text{proj}_{\mathbf{v}}\mathbf{u}\| = \left\|\text{comp}_{\mathbf{v}}\mathbf{u}\left(\frac{1}{\|\mathbf{v}\|}\mathbf{v}\right)\right\| = |\text{comp}_{\mathbf{v}}\mathbf{u}|\left\|\frac{1}{\|\mathbf{v}\|}\mathbf{v}\right\| = |\text{comp}_{\mathbf{v}}\mathbf{u}|.$$

Furthermore, since $\mathbf{u}\cdot\mathbf{v} = \|\mathbf{u}\|\|\mathbf{v}\|\cos\theta$, where θ is the angle between **u** and **v**, we have

$$\text{comp}_{\mathbf{v}}\mathbf{u} = \frac{\mathbf{u}\cdot\mathbf{v}}{\|\mathbf{v}\|} = \frac{\|\mathbf{u}\|\,\|\mathbf{v}\|\cos\theta}{\|\mathbf{v}\|}.$$

Cancelling $\|\mathbf{v}\|$ on the right side produces this result:

Projections and Components

If u and v are nonzero vectors and θ is the angle between them, then

$$\text{comp}_{\mathbf{v}}\mathbf{u} = \frac{\mathbf{u}\cdot\mathbf{v}}{\|\mathbf{v}\|} = \|\mathbf{u}\|\cos\theta$$

and

$$\|\text{proj}_{\mathbf{v}}\mathbf{u}\| = |\text{comp}_{\mathbf{v}}\mathbf{u}|.$$

Example 6 If $\mathbf{u} = 2\mathbf{i} + 3\mathbf{j}$ and $\mathbf{v} = -5\mathbf{i} + 2\mathbf{j}$, find $\text{comp}_{\mathbf{v}}\mathbf{u}$ and $\text{comp}_{\mathbf{u}}\mathbf{v}$.

Solution

$$\text{comp}_{\mathbf{v}}\mathbf{u} = \frac{\mathbf{u}\cdot\mathbf{v}}{\|\mathbf{v}\|} = \frac{2(-5) + 3\cdot 2}{\sqrt{(-5)^2 + 2^2}} = \frac{-4}{\sqrt{29}}.$$

$$\text{comp}_{\mathbf{u}}\mathbf{v} = \frac{\mathbf{v}\cdot\mathbf{u}}{\|\mathbf{u}\|} = \frac{-4}{\sqrt{2^2 + 3^2}} = \frac{-4}{\sqrt{13}}. \quad \blacksquare$$

Applications

Vectors and the dot product can be used to solve a variety of physical problems.

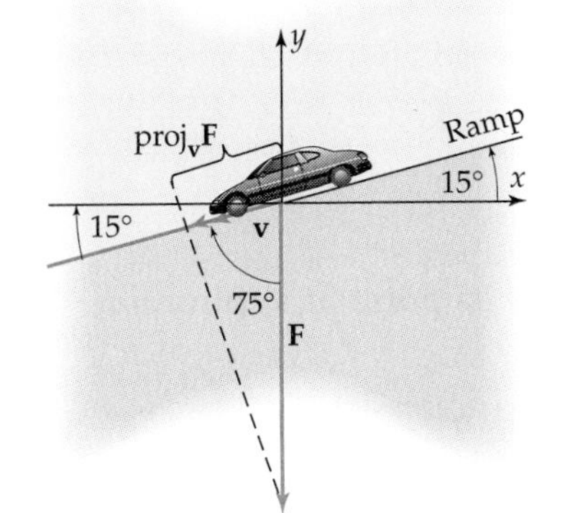

Figure 9–27

Example 7 A 4000-pound automobile is on an inclined ramp that makes a 15° angle with the horizontal. Find the force required to keep it from rolling down the ramp, assuming that the only force that must be overcome is that due to gravity.

Solution The situation is shown in Figure 9–27, where the coordinate system is chosen so that the car is at the origin, the vector **F** representing the downward force of gravity is on the y-axis, and **v** is a unit vector from the origin down the ramp. Since the car weighs 4000 pounds, $\mathbf{F} = -4000\mathbf{j}$. Figure 9–27 shows that the angle between **v** and **F** is 75°. The vector $\text{proj}_{\mathbf{v}}\mathbf{F}$ is the force pulling the car down the ramp, so a force of the same magnitude in the opposite direction is needed to keep the car motionless. As we saw in the preceding box,

$$\|\text{proj}_{\mathbf{v}}\mathbf{F}\| = |\text{comp}_{\mathbf{v}}\mathbf{F}| = \big|\|\mathbf{F}\|\cos 75°\big|$$
$$\approx 4000(.25882) \approx 1035.3.$$

Therefore, a force of 1035.3 pounds is required to hold the car in place. $\blacksquare$

If a constant force **F** is applied to an object, pushing or pulling it a distance d in the direction of the force as shown in Figure 9–28, the amount of **work** done by the force is defined to be the product

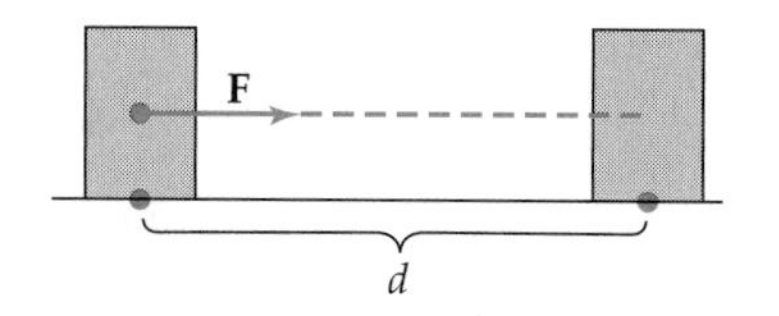

Figure 9–28

$$W = (\text{magnitude of force})(\text{distance}) = \|\mathbf{F}\| \cdot d.$$

If the magnitude of **F** is measured in pounds and d in feet, then the units for W are foot-pounds. For example, if you push a car for 35 feet along a level driveway by exerting a constant force of 110 pounds, the amount of work done is $110 \cdot 35 = 3850$ foot-pounds.

When a force **F** moves an object in the direction of a vector **d** rather than in the direction of **F**, as shown in Figure 9–29, then the motion of the object can be considered as the result of the vector $\text{proj}_{\mathbf{d}}\mathbf{F}$, which is a force in the same direction as **d**.

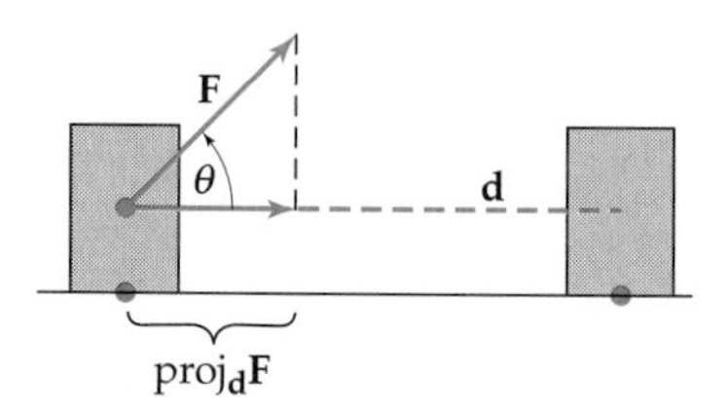

Figure 9–29

Therefore, the amount of work done by **F** is the same as the amount of work done by $\text{proj}_{\mathbf{d}}\mathbf{F}$, namely,

$$W = (\text{magnitude of proj}_{\mathbf{d}}\mathbf{F})(\text{length of } \mathbf{d}) = \|\text{proj}_{\mathbf{d}}\mathbf{F}\| \cdot \|\mathbf{d}\|.$$

The box on page 599 and the Angle Theorem (page 594) show that

$$\begin{aligned} W = \|\text{proj}_{\mathbf{d}}\mathbf{F}\| \cdot \|\mathbf{d}\| &= |\text{comp}_{\mathbf{d}}\mathbf{F}| \cdot \|\mathbf{d}\| \\ &= \|\mathbf{F}\|(\cos\theta)\|\mathbf{d}\| \\ &= \mathbf{F} \cdot \mathbf{d}.^* \end{aligned}$$

Consequently, we have these descriptions of work:

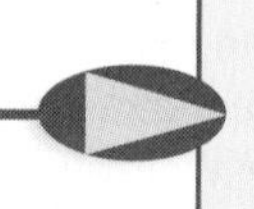

The work W done by a constant force F as its point of application moves along the vector d is

$$W = |\text{comp}_{\mathbf{u}}\mathbf{F}| \cdot \|\mathbf{d}\|, \quad \text{or equivalently,} \quad W = \mathbf{F} \cdot \mathbf{d}.$$

Example 8 How much work is done by a child who pulls a sled 100 feet over level ground by exerting a constant 20-pound force on a rope that makes a 45° angle with the ground?

*This formula reduces to the previous one when **F** and **d** have the same direction because in that case $\cos\theta = \cos 0 = 1$, so that $W = \|\mathbf{F}\| \cdot \|\mathbf{d}\| = (\text{magnitude of force})(\text{distance moved})$.

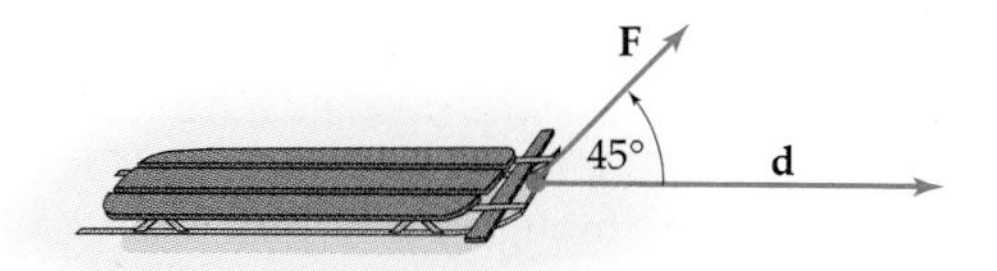

Figure 9–30

Solution The situation is shown in Figure 9–30, where the force **F** on the rope has magnitude 20 and the sled moves along vector **d** of length 100. The work done is

$$W = \mathbf{F} \cdot \mathbf{d} = \|\mathbf{F}\|\,\|\mathbf{d}\| \cdot \cos\theta = 20 \cdot 100 \cdot \frac{\sqrt{2}}{2}$$
$$= 1000\sqrt{2} \approx 1414.2 \text{ foot-pounds.} \quad \blacksquare$$

Exercises 9.4

In Exercises 1–6, find $\mathbf{u} \cdot \mathbf{v}$, $\mathbf{u} \cdot \mathbf{u}$, and $\mathbf{v} \cdot \mathbf{v}$.

1. $\mathbf{u} = \langle 3, 4 \rangle$, $\mathbf{v} = \langle -5, 2 \rangle$
2. $\mathbf{u} = \langle -1, 6 \rangle$, $\mathbf{v} = \langle -4, 1/3 \rangle$
3. $\mathbf{u} = 2\mathbf{i} + \mathbf{j}$, $\mathbf{v} = 3\mathbf{i}$
4. $\mathbf{u} = \mathbf{i} - \mathbf{j}$, $\mathbf{v} = 5\mathbf{j}$
5. $\mathbf{u} = 3\mathbf{i} + 2\mathbf{j}$, $\mathbf{v} = 2\mathbf{i} + 3\mathbf{j}$
6. $\mathbf{u} = 4\mathbf{i} - \mathbf{j}$, $\mathbf{v} = -\mathbf{i} + 2\mathbf{j}$

In Exercises 7–12, find the dot product when $\mathbf{u} = \langle 2, 5 \rangle$, $\mathbf{v} = \langle -4, 3 \rangle$, $\mathbf{w} = \langle 2, -1 \rangle$.

7. $\mathbf{u} \cdot (\mathbf{v} + \mathbf{w})$
8. $\mathbf{u} \cdot (\mathbf{v} - \mathbf{w})$
9. $(\mathbf{u} + \mathbf{v}) \cdot (\mathbf{v} + \mathbf{w})$
10. $(\mathbf{u} + \mathbf{v}) \cdot (\mathbf{u} - \mathbf{v})$
11. $(3\mathbf{u} + \mathbf{v}) \cdot (2\mathbf{w})$
12. $(\mathbf{u} + 4\mathbf{v}) \cdot (2\mathbf{u} + \mathbf{w})$

In Exercises 13–18, find the angle between the two vectors.

13. $\langle 4, -3 \rangle$, $\langle 1, 2 \rangle$
14. $\langle 2, 4 \rangle$, $\langle 0, -5 \rangle$
15. $2\mathbf{i} - 3\mathbf{j}$, $-\mathbf{i}$
16. $2\mathbf{j}$, $4\mathbf{i} + \mathbf{j}$
17. $\sqrt{2}\mathbf{i} + \sqrt{2}\mathbf{j}$, $\mathbf{i} - \mathbf{j}$
18. $3\mathbf{i} - 5\mathbf{j}$, $-2\mathbf{i} + 3\mathbf{j}$

In Exercises 19–24, determine whether the given vectors are parallel, orthogonal, or neither.

19. $\langle 2, 6 \rangle$, $\langle 3, -1 \rangle$
20. $\langle -5, 3 \rangle$, $\langle 2, 6 \rangle$
21. $\langle 9, -6 \rangle$, $\langle -6, 4 \rangle$
22. $-\mathbf{i} + 2\mathbf{j}$, $2\mathbf{i} - 4\mathbf{j}$
23. $2\mathbf{i} - 2\mathbf{j}$, $5\mathbf{i} + 8\mathbf{j}$
24. $6\mathbf{i} - 4\mathbf{j}$, $2\mathbf{i} + 3\mathbf{j}$

In Exercises 25–28, find a real number k such that the two vectors are orthogonal.

25. $2\mathbf{i} + 3\mathbf{j}$, $3\mathbf{i} - k\mathbf{j}$
26. $-3\mathbf{i} + \mathbf{j}$, $2k\mathbf{i} - 4\mathbf{j}$
27. $\mathbf{i} - \mathbf{j}$, $k\mathbf{i} + \sqrt{2}\mathbf{j}$
28. $-4\mathbf{i} + 5\mathbf{j}$, $2\mathbf{i} + 2k\mathbf{j}$

In Exercises 29–32, find $\text{proj}_{\mathbf{u}}\mathbf{v}$ and $\text{proj}_{\mathbf{v}}\mathbf{u}$.

29. $\mathbf{u} = 3\mathbf{i} - 5\mathbf{j}$, $\mathbf{v} = 6\mathbf{i} + 2\mathbf{j}$
30. $\mathbf{u} = 2\mathbf{i} - 3\mathbf{j}$, $\mathbf{v} = \mathbf{i} + 2\mathbf{j}$
31. $\mathbf{u} = \mathbf{i} + \mathbf{j}$, $\mathbf{v} = \mathbf{i} - \mathbf{j}$
32. $\mathbf{u} = 5\mathbf{i} + \mathbf{j}$, $\mathbf{v} = -2\mathbf{i} + 3\mathbf{j}$

In Exercises 33–36, find $\text{comp}_{\mathbf{v}}\mathbf{u}$.

33. $\mathbf{u} = 10\mathbf{i} + 4\mathbf{j}$, $\mathbf{v} = 3\mathbf{i} - 2\mathbf{j}$
34. $\mathbf{u} = \mathbf{i} - 2\mathbf{j}$, $\mathbf{v} = 3\mathbf{i} + \mathbf{j}$
35. $\mathbf{u} = 3\mathbf{i} + 2\mathbf{j}$, $\mathbf{v} = -\mathbf{i} + 3\mathbf{j}$
36. $\mathbf{u} = \mathbf{i} + \mathbf{j}$, $\mathbf{v} = -3\mathbf{i} - 2\mathbf{j}$

In Exercises 37–39, let $\mathbf{u} = \langle a, b \rangle$, $\mathbf{v} = \langle c, d \rangle$, and $\mathbf{w} = \langle r, s \rangle$. Verify that the given property of dot products is valid by calculating the quantities on each side of the equal sign.

37. $\mathbf{u} \cdot (\mathbf{v} + \mathbf{w}) = \mathbf{u} \cdot \mathbf{v} + \mathbf{u} \cdot \mathbf{w}$
38. $k\mathbf{u} \cdot \mathbf{v} = k(\mathbf{u} \cdot \mathbf{v}) = \mathbf{u} \cdot k\mathbf{v}$
39. $\mathbf{0} \cdot \mathbf{u} = 0$

40. Suppose $\mathbf{u} = \langle a, b\rangle$ and $\mathbf{v} = \langle c, d\rangle$ are nonzero parallel vectors.

(a) If $c \neq 0$, show that $\mathbf{u}$ and $\mathbf{v}$ lie on the same nonvertical straight line through the origin.

(b) If $c \neq 0$, show that $\mathbf{v} = \frac{a}{c}\mathbf{u}$ (that is, $\mathbf{v}$ is a scalar multiple of $\mathbf{u}$). [*Hint:* The equation of the line on which $\mathbf{u}$ and $\mathbf{v}$ lie is $y = mx$ for some constant m (why?), which implies that $b = ma$ and $d = mc$.]

(c) If $c = 0$, show that $\mathbf{v}$ is a scalar multiple of $\mathbf{u}$. [*Hint:* If $c = 0$, then $a = 0$ (why?) and hence $b \neq 0$ (otherwise $\mathbf{u} = \mathbf{0}$).]

41. Prove the Angle Theorem in the case when θ is 0 or π.

42. If $\mathbf{u}$ and $\mathbf{v}$ are nonzero vectors such that $\mathbf{u}\cdot\mathbf{v} = 0$, show that $\mathbf{u}$ and $\mathbf{v}$ are orthogonal. [*Hint:* If θ is the angle between $\mathbf{u}$ and $\mathbf{v}$, what is $\cos\theta$ and what does this say about θ?]

43. Show that (1, 2), (3, 4), (5, 2) are the vertices of a right triangle by considering the sides of the triangle as vectors.

44. Find a number x such that the angle between the vectors $\langle 1, 1\rangle$ and $\langle x, 1\rangle$ is $\pi/4$ radians.

45. Find nonzero vectors $\mathbf{u}$, $\mathbf{v}$, and $\mathbf{w}$ such that $\mathbf{u}\cdot\mathbf{v} = \mathbf{u}\cdot\mathbf{w}$ and $\mathbf{v} \neq \mathbf{w}$ and neither $\mathbf{v}$ nor $\mathbf{w}$ is orthogonal to $\mathbf{u}$.

46. If $\mathbf{u}$ and $\mathbf{v}$ are nonzero vectors, show that the vectors $\|\mathbf{u}\|\mathbf{v} + \|\mathbf{v}\|\mathbf{u}$ and $\|\mathbf{u}\|\mathbf{v} - \|\mathbf{v}\|\mathbf{u}$ are orthogonal.

47. A 600-pound trailer is on an inclined ramp that makes a 30° angle with the horizontal. Find the force required to keep it from rolling down the ramp, assuming that the only force that must be overcome is that due to gravity.

48. In Example 7, find the vector that represents the force necessary to keep the car motionless.

In Exercises 49–52, find the work done by a constant force $\mathbf{F}$ *as the point of application of* $\mathbf{F}$ *moves along the vector* $\overrightarrow{PQ}$.

49. $\mathbf{F} = 2\mathbf{i} + 5\mathbf{j}$, $P = (0, 0)$, $Q = (4, 1)$

50. $\mathbf{F} = \mathbf{i} - 2\mathbf{j}$, $P = (0, 0)$, $Q = (-5, 2)$

51. $\mathbf{F} = 2\mathbf{i} + 3\mathbf{j}$, $P = (2, 3)$, $Q = (5, 9)$ [*Hint:* Find the component form of $\overrightarrow{PQ}$.]

52. $\mathbf{F} = 5\mathbf{i} + \mathbf{j}$, $P = (-1, 2)$, $Q = (4, -3)$

53. A lawn mower handle makes an angle of 60° with the ground. A woman pushes on the handle with a force of 30 pounds. How much work is done in moving the lawn mower a distance of 75 feet on level ground?

54. A child pulls a wagon along a level sidewalk by exerting a force of 18 pounds on the wagon handle, which makes an angle of 25° with the horizontal. How much work is done in pulling the wagon 200 feet?

55. A 40-pound cart is pushed 100 feet up a ramp that makes a 20° angle with the horizontal (see figure). How much work is done against gravity? [*Hint:* The amount of work done against gravity is the negative of the amount of work done *by* gravity. Coordinatize the situation so that the cart is at the origin. Then the cart moves along vector $\mathbf{d} = (100\cos 20°)\mathbf{i} + (100\sin 20°)\mathbf{j}$ and the downward force of gravity is $\mathbf{F} = 0\mathbf{i} - 40\mathbf{j}$.]

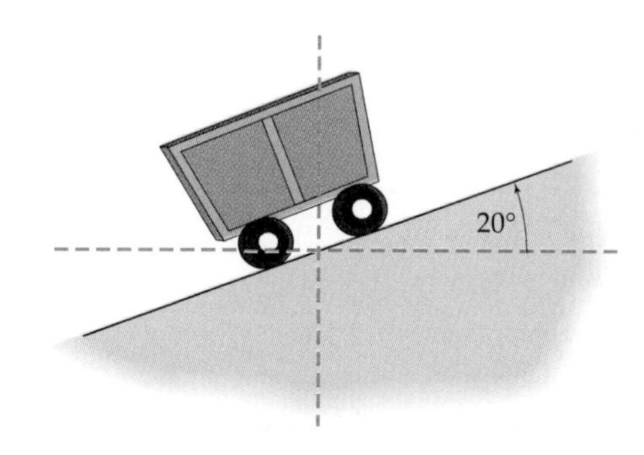

56. Suppose the child in Exercise 54 is pulling the wagon up a hill that makes an angle of 20° with the horizontal and all other facts remain the same. How much work is done in pulling the wagon 150 feet?

Chapter 9 Review

Important Concepts

Important Facts and Formulas

- $|a + bi| = \sqrt{a^2 + b^2}$
- $a + bi = r(\cos\theta + i\sin\theta)$, where
$$r = \sqrt{a^2 + b^2}, \qquad a = r\cos\theta, \qquad b = r\sin\theta$$
- $r_1(\cos\theta_1 + i\sin\theta_1)\cdot r_2(\cos\theta_2 + i\sin\theta_2) = r_1r_2[\cos(\theta_1 + \theta_2) + i\sin(\theta_1 + \theta_2)]$
- $\dfrac{r_1(\cos\theta_1 + i\sin\theta_1)}{r_2(\cos\theta_2 + i\sin\theta_2)} = \dfrac{r_1}{r_2}[\cos(\theta_1 - \theta_2) + i\sin(\theta_1 - \theta_2)]$
- *DeMoivre's Theorem:*
$$[r(\cos\theta + i\sin\theta)]^n = r^n[\cos(n\theta) + i\sin(n\theta)]$$
- The distinct nth roots of $r(\cos\theta + i\sin\theta)$ are
$$\sqrt[n]{r}\left[\cos\left(\frac{\theta + 2k\pi}{n}\right) + i\sin\left(\frac{\theta + 2k\pi}{n}\right)\right] \quad (k = 0, 1, 2, \ldots, n - 1)$$
- The distinct nth roots of unity are
$$\cos\frac{2k\pi}{n} + i\sin\frac{2k\pi}{n} \quad (k = 0, 1, 2, \ldots, n - 1)$$
- If $P = (x_1, y_1)$ and $Q = (x_2, y_2)$, then $\overrightarrow{PQ} = \langle x_2 - x_1, y_2 - y_1\rangle$.

- $\|\langle a, b\rangle\| = \sqrt{a^2 + b^2}$
- If $\mathbf{u} = \langle a, b\rangle$ and k is a scalar, then $k\mathbf{u} = \langle ka, kb\rangle$.
- If $\mathbf{u} = \langle a, b\rangle$ and $\mathbf{v} = \langle c, d\rangle$, then

$$\mathbf{u} + \mathbf{v} = \langle a + c, b + d\rangle \qquad \text{and} \qquad \mathbf{u} - \mathbf{v} = \langle a - c, b - d\rangle$$

- *Properties of Vector Addition and Scalar Multiplication:* For any vectors **u**, **v**, and **w** and any scalars r and s,

1. $\mathbf{u} + (\mathbf{v} + \mathbf{w}) = (\mathbf{u} + \mathbf{v}) + \mathbf{w}$
2. $\mathbf{u} + \mathbf{v} = \mathbf{v} + \mathbf{u}$
3. $\mathbf{v} + \mathbf{0} = \mathbf{v} = \mathbf{0} + \mathbf{v}$
4. $\mathbf{v} + (-\mathbf{v}) = \mathbf{0}$
5. $r(\mathbf{u} + \mathbf{v}) = r\mathbf{u} + r\mathbf{v}$
6. $(r + s)\mathbf{v} = r\mathbf{v} + s\mathbf{v}$
7. $(rs)\mathbf{v} = r(s\mathbf{v}) = s(r\mathbf{v})$
8. $1\mathbf{v} = \mathbf{v}$
9. $0\mathbf{v} = \mathbf{0} = r\mathbf{0}$

- If $\mathbf{v} = \langle a, b\rangle = a\mathbf{i} + b\mathbf{j}$, then

$$a = \|\mathbf{v}\|\cos\theta \qquad \text{and} \qquad b = \|\mathbf{v}\|\sin\theta$$

- where θ is the direction angle of **v**.
- If $\mathbf{u} = \langle a, b\rangle = a\mathbf{i} + b\mathbf{j}$ and $\mathbf{v} = \langle c, d\rangle = c\mathbf{i} + d\mathbf{j}$, then

$$\mathbf{u}\cdot\mathbf{v} = ac + bd.$$

- If θ is the angle between nonzero vectors **u** and **v**, then

$$\mathbf{u}\cdot\mathbf{v} = \|\mathbf{u}\|\,\|\mathbf{v}\|\cos\theta$$

- *Schwarz Inequality:* $|\mathbf{u}\cdot\mathbf{v}| \leq \|\mathbf{u}\|\,\|\mathbf{v}\|$.
- Vectors **u** and **v** are orthogonal exactly when $\mathbf{u}\cdot\mathbf{v} = 0$.
- $\text{proj}_{\mathbf{v}}\mathbf{u} = \left(\dfrac{\mathbf{u}\cdot\mathbf{v}}{\|\mathbf{v}\|^2}\right)\mathbf{v}$
- $\text{comp}_{\mathbf{v}}\mathbf{u} = \dfrac{\mathbf{u}\cdot\mathbf{v}}{\|\mathbf{v}\|} = \|\mathbf{u}\|\cos\theta$, where θ is the angle between **u** and **v**.

Review Questions

1. Simplify: $|i(4 + 2i)| + |3 - i|$.
2. Simplify: $|3 + 2i| - |1 - 2i|$.
3. Graph the equation $|z| = 2$ in the complex plane.
4. Graph the equation $|z - 3| = 1$ in the complex plane.
5. Express in polar form: $1 + \sqrt{3}i$.
6. Express in polar form: $4 - 5i$.

In Questions 7–11, express the given number in the form $a + bi$.

7. $2\left(\cos\dfrac{\pi}{12} + i\sin\dfrac{\pi}{12}\right)\cdot 4\left(\cos\dfrac{\pi}{6} + i\sin\dfrac{\pi}{6}\right)$
8. $3\left(\cos\dfrac{\pi}{8} + i\sin\dfrac{\pi}{8}\right)\cdot 2\left(\cos\dfrac{3\pi}{8} + i\sin\dfrac{3\pi}{8}\right)$

9. $\dfrac{12\left(\cos\frac{7\pi}{12} + i\sin\frac{7\pi}{12}\right)}{3\left(\cos\frac{5\pi}{12} + i\sin\frac{5\pi}{12}\right)}$

10. $\left(\cos\frac{\pi}{12} + i\sin\frac{\pi}{12}\right)^{18}$

11. $\left[\sqrt[3]{3}\left(\cos\frac{5\pi}{36} + i\sin\frac{5\pi}{36}\right)\right]^{12}$

In Questions 12–16, solve the given equation in the complex number system and express your answers in polar form.

12. $x^3 = i$ **13.** $x^6 = 1$ **14.** $x^8 = -\sqrt{3} - 3i$

15. $x^4 = i$ **16.** $x^3 = 1 - i$

In Questions 17–20, let $u = \langle 3, -2\rangle$ and $v = \langle 8, 1\rangle$. Find

17. $\mathbf{u} + \mathbf{v}$ **18.** $\|-3\mathbf{v}\|$ **19.** $\|2\mathbf{v} - 4\mathbf{u}\|$ **20.** $3\mathbf{u} - \frac{1}{2}\mathbf{v}$

In Questions 21–24, let $\mathbf{u} = -2\mathbf{i} + \mathbf{j}$ and $\mathbf{v} = 3\mathbf{i} - 4\mathbf{j}$. Find

21. $4\mathbf{u} - \mathbf{v}$ **22.** $\mathbf{u} + 2\mathbf{v}$ **23.** $\|\mathbf{u} + \mathbf{v}\|$ **24.** $\|\mathbf{u}\| + \|\mathbf{v}\|$

25. Find the components of the vector $\mathbf{v}$ such that $\|\mathbf{v}\| = 5$ and the direction angle of $\mathbf{v}$ is 45°.

26. Find the magnitude and direction angle of $3\mathbf{i} + 4\mathbf{j}$.

27. Find a unit vector whose direction is *opposite* the direction of $3\mathbf{i} - 6\mathbf{j}$.

28. An object at the origin is acted upon by a 10-pound force with direction angle 90° and a 20-pound force with direction angle 30°. Find the magnitude and direction of the resultant force.

29. A plane flies in the direction 120°, with an air speed of 300 mph. The wind is blowing from north to south at 40 mph. Find the course and ground speed of the plane.

30. An object weighing 40 pounds lies on an inclined plane that makes a 30° angle with the horizontal. Find the components of the weight parallel and perpendicular to the plane.

In Questions 31–34, $u = \langle 3, -4\rangle$, $v = \langle -2, 5\rangle$, and $w = \langle 0, 3\rangle$. Find

31. $\mathbf{u}\cdot\mathbf{v}$

32. $\mathbf{u}\cdot\mathbf{u} - \mathbf{v}\cdot\mathbf{v}$

33. $(\mathbf{u} + \mathbf{v})\cdot\mathbf{w}$

34. $(\mathbf{u} + \mathbf{w})\cdot(\mathbf{w} - 3\mathbf{v})$

35. What is the angle between the vectors $5\mathbf{i} - 2\mathbf{j}$ and $3\mathbf{i} + \mathbf{j}$?

36. Is $3\mathbf{i} - 2\mathbf{j}$ orthogonal to $4\mathbf{i} + 6\mathbf{j}$?

In Questions 37 and 38, $u = 4i - 3j$ and $v = 2i + j$. Find

37. $\text{proj}_{\mathbf{v}}\mathbf{u}$ **38.** $\text{comp}_{\mathbf{u}}\mathbf{v}$

39. If $\mathbf{u}$ and $\mathbf{v}$ have the same magnitude, show that $\mathbf{u} + \mathbf{v}$ and $\mathbf{u} - \mathbf{v}$ are orthogonal.

40. If **u** and **v** are nonzero vectors, show that the vector $\mathbf{u} - k\mathbf{v}$ is orthogonal to **v**, where $k = \dfrac{\mathbf{u} \cdot \mathbf{v}}{\|\mathbf{v}\|^2}$.

41. A 3500-pound automobile is on an inclined ramp that makes a 30° angle with the horizontal. Find the force required to keep it from rolling down the ramp, assuming that the only force that must be overcome is that due to gravity.

42. A sled is pulled along level ground by a rope that makes a 50° angle with the horizontal. If a force of 40 pounds is used to pull the sled, how much work is done in pulling it 100 feet?

Discovery Project 9

Surveying

You might ask, how far apart are two points A and B? Under ordinary circumstances, the easiest thing to do would be to measure the distance. However, sometimes it is impractical to make such a direct measurement because of intervening obstacles. Two historical methods for measuring the distance between the two mutually invisible points A and B are given below.

1. Find the distance from A to B, using the classic surveyor's method shown in the figure below. A transit is used to measure the angle between two distant objects, and a reasonably short baseline is measured. Angles are reported in degrees. Triangles are drawn and the law of sines is used to calculate the length of unknown edges. In order to find the distance from A to B, draw an additional triangle which has AB as one of the sides. Caution: the picture is not drawn to scale.

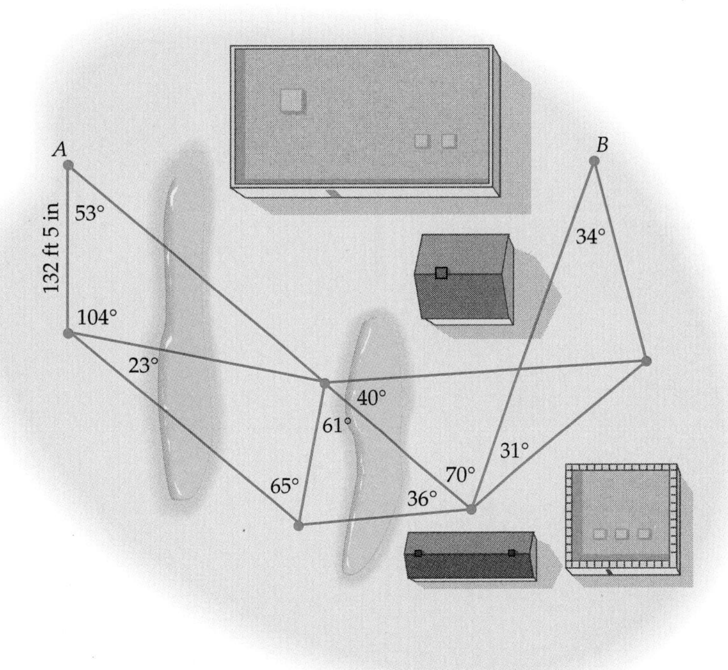

(a)

2. Find the distance from A to B, using a modern laser range finder. Laser technology removes the necessity of drawing triangles; the tool provides a vector (angle and distance) from point to point, so

the distance from A to B can be found using vector arithmetic in place of repeated application of the law of sines. Angles are taken from the internal compass and are reported in degrees measured clockwise from North (0°), so you have to convert compass direction angles to angles in standard position. For example, the angle from point A (straight South) is reported as 180°, but in standard position you would list the angle as 270°.

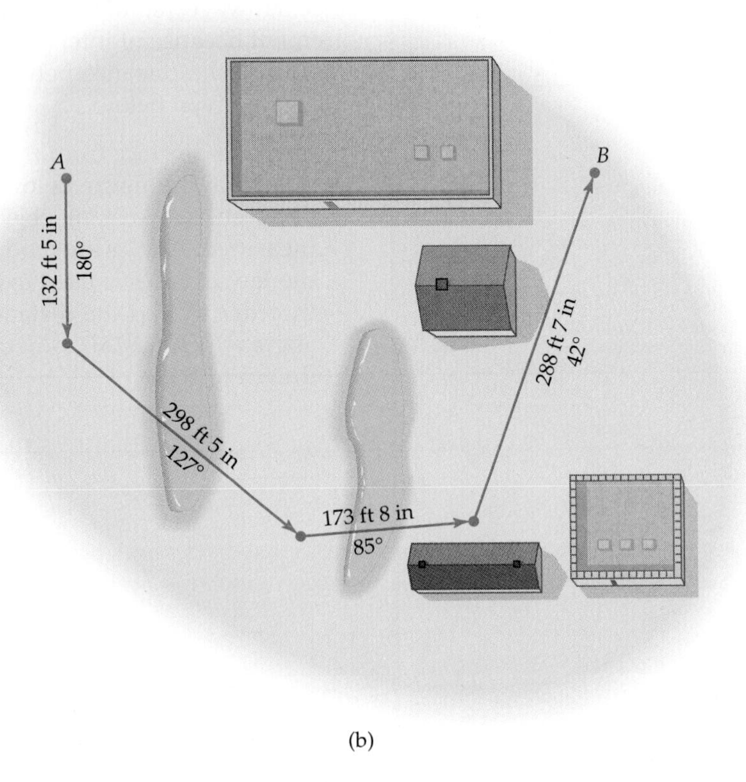

(b)

Chapter

10 Analytic Geometry

Calling all ships!

The planets travel in elliptical orbits around the sun and satellites in elliptical orbits around the earth. Parabolic reflectors are used in spotlights, radar antennas, and satellite dishes. The long-range navigation system (LORAN) uses hyperbolas to enable a ship to determine its exact location. See Exercise 62 on page 649.

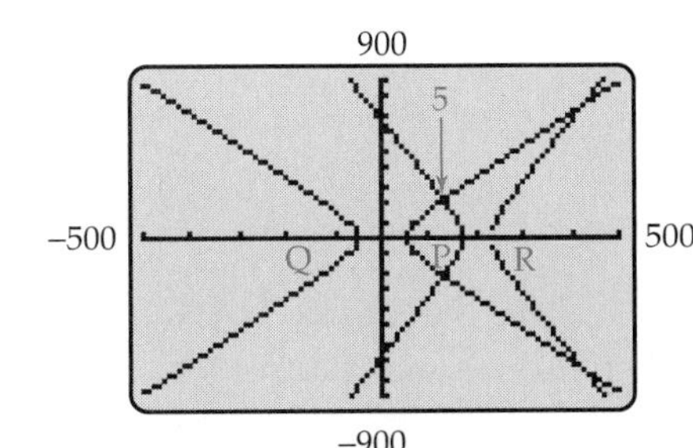

Chapter Outline

Interdependence of Sections

Chapters 10–12 are independent of one another and may be read in any order.

Sections 3.1–3.4 are prerequisites for this chapter. Except for the discussion of standard equations for conics in Sections 10.2 and 10.3, Chapter 6 and Section 7.1 (Trigonometry) are also prerequisites.

The interdependence of sections in this chapter is:

10.1

10.2 → 10.3

10.2 → 10.5

10.4 → 10.5

The discussion of analytic geometry that was begun in Section 1.3 is continued here. A variety of interrelated topics are presented: a more thorough treatment of parametric graphing, an examination of conic sections (which have played a significant role in mathematics since ancient times), and an alternative method of coordinatizing the plane and graphing functions.

10.1 Plane Curves and Parametric Equations

There are many curves in the plane that cannot be represented as the graph of a function $y = f(x)$. Parametric graphing enables us to represent such curves in terms of functions and also provides a formal definition of a curve in the plane.

Consider, for example, an object moving in the plane during a particular time interval. In order to describe both the path of the object and its location at any particular time, three variables are needed: the time t and the coordinates (x, y) of the object at time t. For instance, the coordinates x and y might be given by:

$$x = 4\cos t + 5\cos(3t) \qquad \text{and} \qquad y = \sin(3t) + t.$$

From $t = 0$ to $t = 12.5$, the object traces out the curve shown in Figure 10–1. The points marked on the graph show the location of the object at various times. Note that the object may be at the same location at different times (the points where the graph crosses itself).

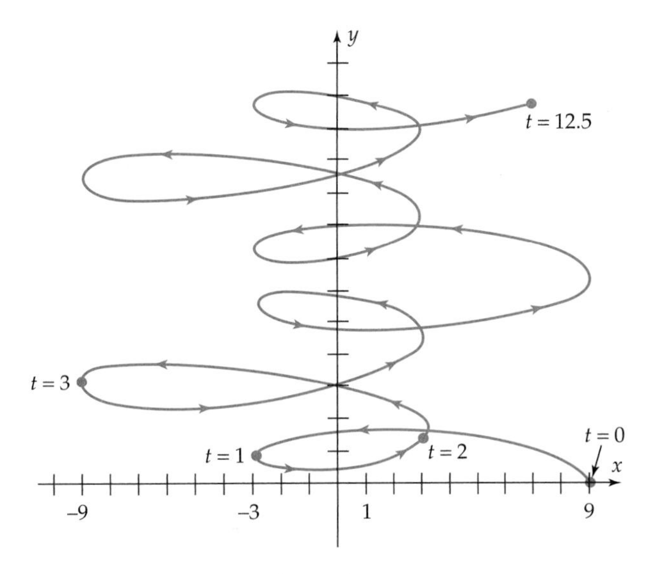

Figure 10–1

In the preceding example, both x and y were determined by continuous functions of t, with t taking values in the interval $[0, 12.5]$.* The example suggests the following definition.

Definition of Plane Curve

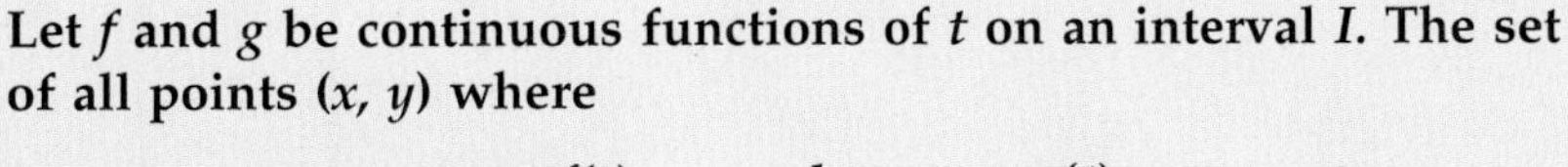

Let f and g be continuous functions of t on an interval I. The set of all points (x, y) where

$$x = f(t) \qquad \text{and} \qquad y = g(t)$$

is called a *plane curve*. The variable t is called a *parameter* and the equations defining x and y are *parametric equations*.

In this general definition of "curve," the variable t need not represent time. As the following examples illustrate, different pairs of parametric equations may produce the same curve. Each such pair of parametric equations is called a **parameterization** of the curve.

A curve given by parametric equations can be graphed by hand [choose values of t, plot the corresponding points (x, y), and make an educated guess about the shape of the curve]. A calculator in parametric mode is more likely to produce an accurate graph, however. When choosing a viewing window in parametric graphing mode, you must specify minimum and maximum values not only for x and y, but also for t.

Technology Tip

For parametric graphing mode, choose PAR (or PARAM or PARAMETRIC or PARM) in the TI MODE menu, or the HP-38 LIB menu, or the COORD submenu of the Sharp 9600 SETUP menu, or in the TYPE submenu of the Casio 9850 GRAPH menu (on the main menu).

*Intuitively, "continuous" means that the graph of the function that determines x, namely, $f(t) = 4\cos t + 5\cos 3t$, is a connected curve with no gaps or holes, and similarly for the function that determines y. Continuous functions are defined more precisely in Chapter 13, which is available as an optional supplement to this book.

Failure to choose an appropriate range for t may lead to an incomplete graph. You must also set the t-step (or t-pitch), which determines how much t changes each time a point is plotted. With a small t-step many points are plotted; with a large one, few points are plotted. A t-step between .05 and .15 usually produces a relatively smooth graph in a reasonable amount of time.

GRAPHING EXPLORATION

Graph the curve in Figure 10–1 in a window with

$$-10 \le x \le 10, \qquad -5 \le y \le 15, \qquad 0 \le t \le 12.5,$$

and t-step = .1. Does the graph look like the one in Figure 10–1? Now change the t-step to 1.5 and graph again. Now how does the graph look? Experiment with various t-steps to see how they affect the graph.

Example 1 A calculator produced the graph in Figure 10–2 of the curve given by

$$x = t^3 - 10t^2 + 29t - 10 \quad \text{and} \quad y = t^2 - 7t + 14 \quad (0 \le t \le 6.5).$$

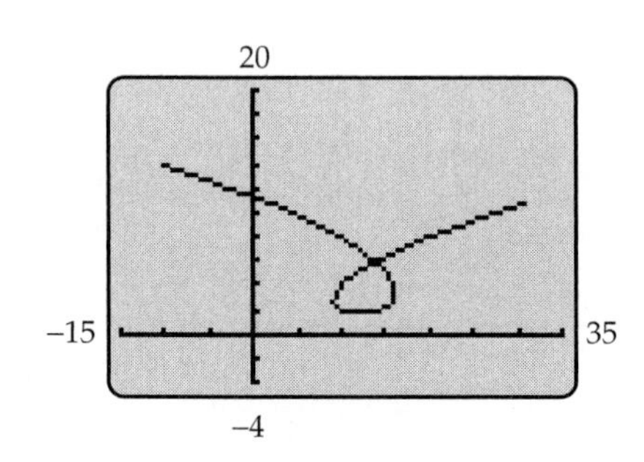

Figure 10–2

GRAPHING EXPLORATION

Set the same viewing window as in Figure 10–2 and graph the curve three separate times, using these ranges for t:

$$0 \le t \le 5; \qquad 1 \le t \le 6; \qquad -1 \le t \le 7.$$

What effect does changing the range of t have on the curve that is graphed? ■

Example 2 The graph of the curve given by

$$x = -2t \quad \text{and} \quad y = 4t^2 - 4 \quad (-1 \le t \le 2)$$

is shown in Figure 10–3.

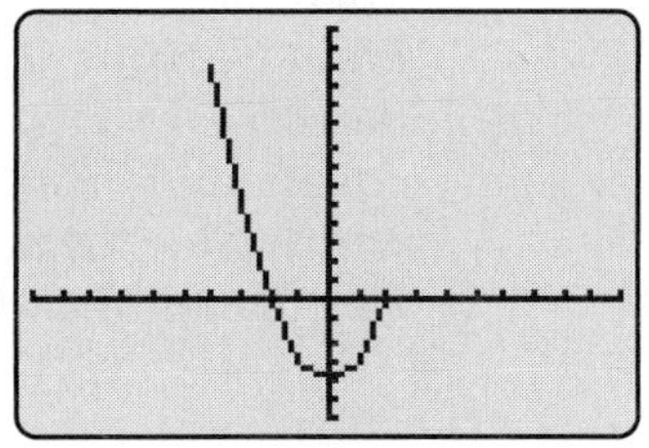

Figure 10–3

GRAPHING EXPLORATION

Graph this curve on your calculator, using the same viewing window and range of t values as in Figure 10–3. In what direction is the curve traced out (that is, what is the first point graphed (corresponding to $t = -1$) and what is the last point graphed (corresponding to $t = 2$)? ■

Example 3 The curve given by

$$x = t + 5\cos t \qquad \text{and} \qquad y = 1 - 3\sin t$$

spirals along the x-axis, as you will see for yourself.

GRAPHING EXPLORATION

Graph this curve in the viewing window with

$$-25 \le x \le 25, \quad -5 \le y \le 5 \qquad \text{and} \qquad -20 \le t \le 20.$$ ■

Some curves given by parametric equations can also be expressed as (part of) the graph of an equation in x and y. The process for doing this, called **eliminating the parameter,** is as follows:

Solve one of the parametric equations for t and substitute this result in the other parametric equation.

Example 4 Consider the curve of Example 2, which was given by

$$x = -2t \qquad \text{and} \qquad y = 4t^2 - 4 \quad (-1 \le t \le 2).$$

Solving $x = -2t$ for t shows that $t = -x/2$. Substituting this in the second equation, we have

$$y = 4t^2 - 4 = 4\left(-\frac{x}{2}\right)^2 - 4 = 4\left(\frac{x^2}{4}\right) - 4 = x^2 - 4.$$

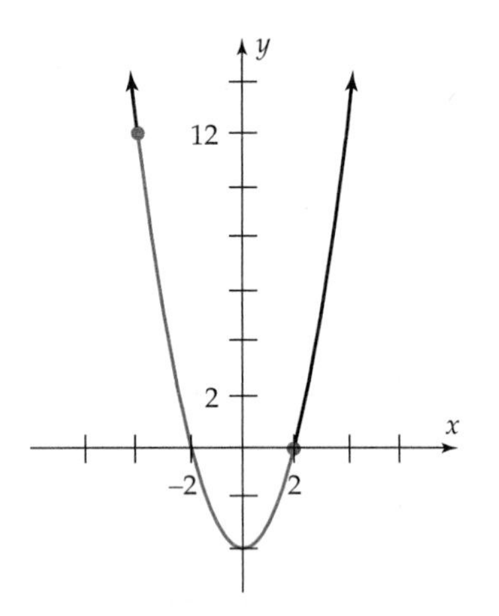

Figure 10–4

Therefore, every point on the curve is also on the graph of $y = x^2 - 4$. From Sections 3.4 or 4.1 we know that the graph of $y = x^2 - 4$ is the parabola in Figure 10–4. However, the curve given by the parametric equations is *not* the entire parabola, but only the part shown in color, which joins the points (2, 0) and (−4, 12) that correspond to the minimum and maximum values of t, namely, $t = -1$ and $t = 2$. ■

Having seen how to graph a curve given by parametric equations, we now consider the reverse problem: finding a parametric respresentation for the graph of an equation in x and y. This is easy when the equation defines y as a function of x, such as

$$y = x^3 + 5x^2 - 3x + 4.$$

A parametric description of this function can be obtained by changing the variable:

$$x = t \qquad \text{and} \qquad y = t^3 + 5t^2 - 3t + 4.$$

The same thing is true for equations that define x as a function of y, such as $x = y^2 + 3$, which can be parameterized by letting

$$x = t^2 + 3 \qquad \text{and} \qquad y = t.$$

In other cases, different techniques may be needed.

Example 5 The graph of $(x - 4)^2 + (y - 1)^2 = 9$ is a circle with center (4, 1) and radius 3 (see page 32), so this equation does not define x as a function of y or y as a function of x. However, the graph of the equation is given by this parameterization:

$$(*) \qquad x = 3\cos t + 4 \qquad \text{and} \qquad y = 3\sin t + 1 \quad (0 \le t \le 2\pi)$$

because by the Pythagorean identity

$$\begin{aligned}(x - 4)^2 + (y - 1)^2 &= (3\cos t + 4 - 4)^2 + (3\sin t + 1 - 1)^2 \\ &= (3\cos t)^2 + (3\sin t)^2 = 9\cos^2 t + 9\sin^2 t \\ &= 9(\cos^2 t + \sin^2 t) = 9 \cdot 1 = 9.\end{aligned}$$

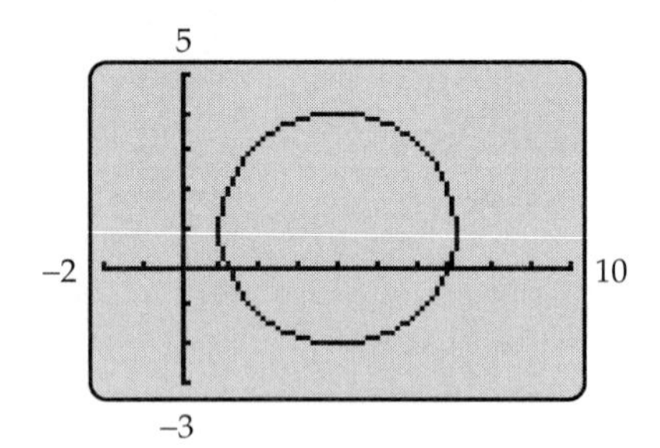

Figure 10–5

With this parameterization the circle is traced out in a counterclockwise direction from the point (7, 1), as shown in Figure 10–5. Another parameterization is given by

$$x = 3\cos 2t + 4 \qquad \text{and} \qquad y = -3\sin 2t + 1 \quad (0 \le t \le \pi).$$

GRAPHING EXPLORATION

Verify that this parameterization traces out the circle in a clockwise direction, twice as fast as the parameterization given by (∗), since t runs from 0 to π, rather than 2π. ■

The procedure in Example 5 works in the general case, as is proved in Exercise 28:

Parametric Equations of a Circle

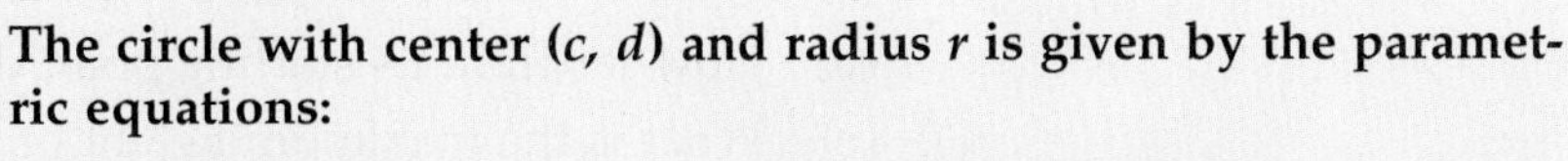

The circle with center (c, d) and radius r is given by the parametric equations:

$$x = r\cos t + c \qquad \text{and} \qquad y = r\sin t + d \quad (0 \le t \le 2\pi).$$

Example 6 Find three parameterizations of the straight line through $(1, -3)$ with slope -2.

Solution The point-slope form of the equation of this line is

$$y - (-3) = -2(x - 1), \qquad \text{or equivalently,} \qquad y = -2x - 1.$$

Its graph is shown in Figure 10–6. Since this equation defines y as a function of x, one parameterization is

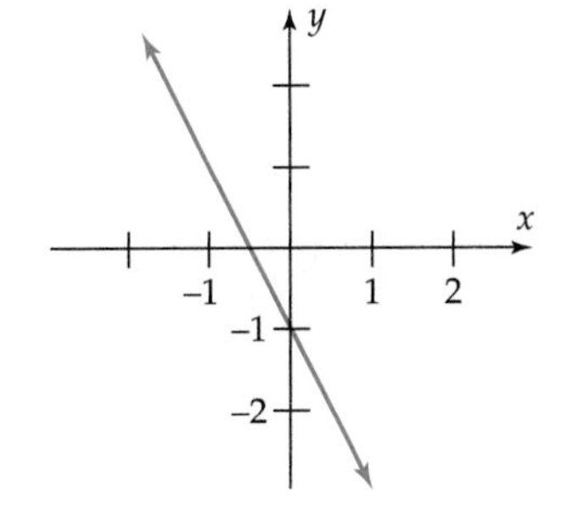

Figure 10–6

$$x = t \qquad \text{and} \qquad y = -2t - 1 \quad (t \text{ any real number}).$$

A second parameterization is given by letting $x = t + 1$; then

$$y = -2x - 1 = -2(t + 1) - 1 = -2t - 3 \quad (t \text{ any real number}).$$

A third parameterization can be obtained by letting

$$x = \tan t \qquad \text{and} \qquad y = -2x - 1 = -2\tan t - 1 \quad (-\pi/2 < t < \pi/2).$$

When t runs from $-\pi/2$ to $\pi/2$, then $x = \tan t$ takes all possible real number values, and hence so does y. ■

CAUTION

Some substitutions in an equation $y = f(x)$ do *not* lead to a parameterization of the entire graph. For instance, in Example 6, letting $x = t^2$ and substituting in the equation $y = -2x - 1$ leads to:

$$x = t^2 \qquad \text{and} \qquad y = -2t^2 - 1 \quad (\text{any real number } t).$$

Thus, x is always nonnegative and y is always negative. So the parameterization produces only the half of the line $y = -2x - 1$ to the right of the y-axis in Figure 10–6.

Applications

In the following applications we ignore air resistance and assume some facts about gravity that are proved in physics.

Example 7 A golfer hits the ball with an initial velocity of 140 feet/second so that its path as it leaves the ground makes an angle of 31° with the horizontal.

(a) When does the ball hit the ground?

(b) How far from its starting point does it land?

(c) What is the maximum height of the ball during its flight?

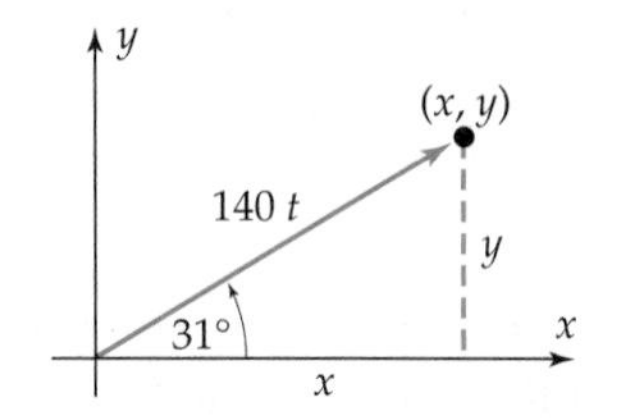

Figure 10–7

Solution Imagine that the golf ball starts at the origin and travels in the direction of the positive x-axis. If there were no gravity, the distance traveled by the ball in t seconds would be $140t$ feet. As shown in Figure 10–7, the coordinates (x, y) of the ball would satisfy

$$\frac{x}{140t} = \cos 31^\circ \qquad\qquad \frac{y}{140t} = \sin 31^\circ$$

$$x = (140 \cos 31^\circ)t \qquad\qquad y = (140 \sin 31^\circ)t.$$

However, there *is* gravity and at time t it exerts a force of $16t^2$ downward (that is, in the negative direction on the y-coordinate). Consequently, the coordinates of the golf ball at time t are

$$x = (140 \cos 31^\circ)t \qquad \text{and} \qquad y = (140 \sin 31^\circ)t - 16t^2.$$

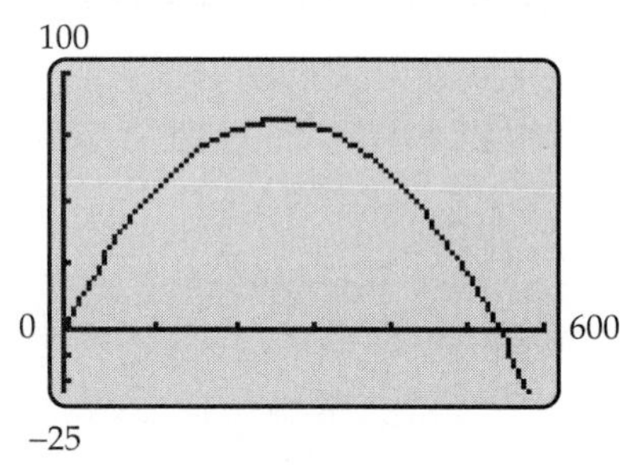

Figure 10–8

The path given by these parametric equations is graphed in Figure 10–8.*

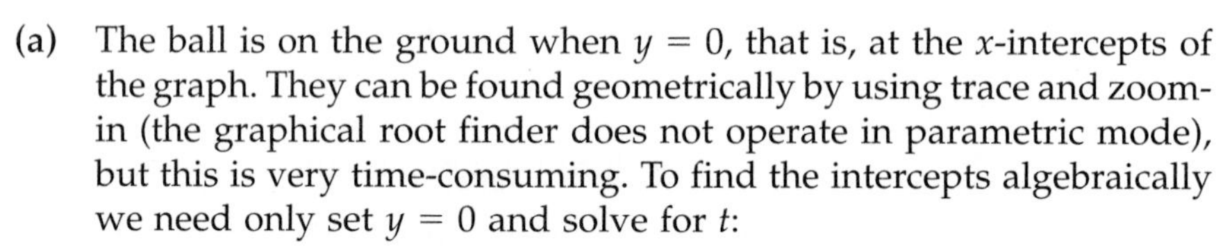

(a) The ball is on the ground when $y = 0$, that is, at the x-intercepts of the graph. They can be found geometrically by using trace and zoom-in (the graphical root finder does not operate in parametric mode), but this is very time-consuming. To find the intercepts algebraically we need only set $y = 0$ and solve for t:

$$(140 \sin 31^\circ)t - 16t^2 = 0$$

$$t(140 \sin 31^\circ - 16t) = 0$$

$$t = 0 \qquad \text{or} \qquad 140 \sin 31^\circ - 16t = 0$$

$$t = \frac{140 \sin 31^\circ}{16} \approx 4.5066.$$

Thus, the ball hits the ground after approximately 4.5066 seconds.

(b) The horizontal distance traveled by the ball is given by the x-coordinate of the intercept. The x-coordinate when $t \approx 4.5066$ is

$$x = (140 \cos 31^\circ)(4.5066) \approx 540.81 \text{ feet.}$$

(c) The graph in Figure 10–8 looks like a parabola and it is, as you can verify by eliminating the parameter t (Exercise 40). The y-coordinate of the vertex is the maximum height of the ball. It can be found geometrically by using trace and zoom-in (the maximum finder doesn't work in parametric mode) or algebraically as follows. The vertex oc-

*Only the part of the graph on or above the x-axis represents the ball's path, since the ball does not go underground after it lands.

curs halfway between its two x-intercepts ($x = 0$ and $x \approx 540.81$), that is, when $x \approx 270.405$. Hence,

$$(140 \cos 31°)t = x = 270.405$$

so that

$$t = \frac{270.45}{140 \cos 31°} \approx 2.2533.$$

Therefore, the y-coordinate of the vertex (the maximum height of the ball) is

$$y = (140 \sin 31°)(2.2533) - 16(2.2533)^2 \approx 81.237 \text{ feet.} \quad ■$$

Example 8 A batter hits a ball that is 3 feet above the ground and leaves the bat with an initial velocity of 138 feet/second, making an angle of 26° with the horizontal and heading toward a 25-foot-high fence that is 400 feet away. Will the ball go over the fence?

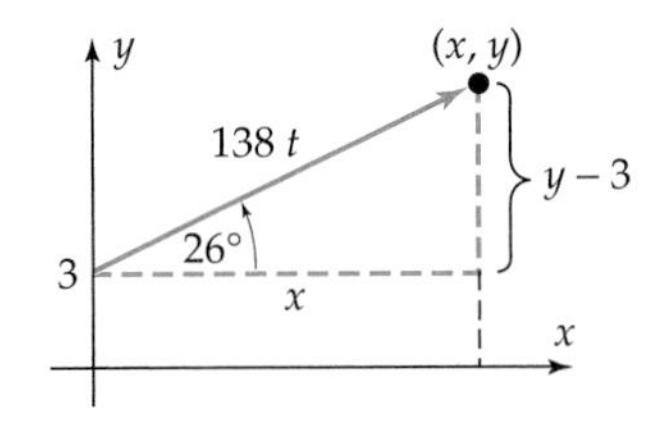

Figure 10–9

Solution Imagine that home plate is at the origin and the ball travels in the direction of the positive x-axis. Using Figure 10–9 we see that without gravity

$$\frac{x}{138t} = \cos 26° \qquad \frac{y-3}{138t} = \sin 26°$$

$$x = (138 \cos 26°)t \qquad y = (138 \sin 26°)t + 3.$$

Allowing for the effect of gravity on the y-coordinate, we find that the ball's path is given by the parametric equations

$$x = (138 \cos 26°)t \quad \text{and} \quad y = (138 \sin 26°)t - 16t^2 + 3.$$

The graph of the ball's path in Figure 10–10 was made with the grid-on feature and vertical tic marks 25 units apart. It shows that the y-coordinate of the ball is greater than 25 when its x-coordinate is 400. So, the ball goes over the fence. ■

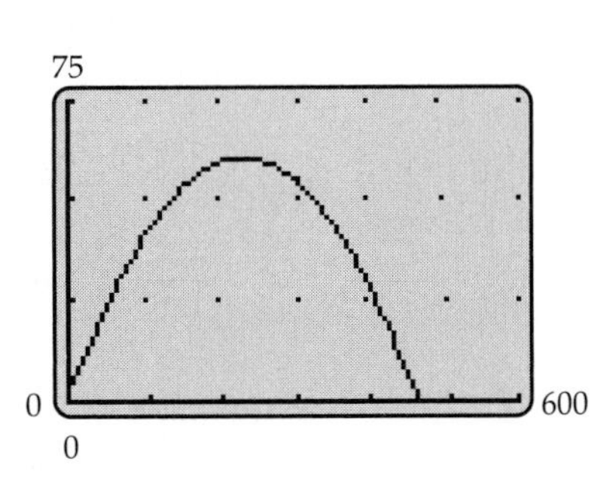

Figure 10–10

The procedure used in Example 8 applies to the general case. Replacing 3 by k, 26° by θ, and 138 by v leads to this conclusion:

Projectile Motion

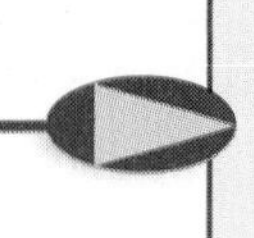

When a projectile is fired from the position (0, k) on the positive y-axis at an angle θ with the horizontal, in the direction of the positive x-axis, with initial velocity v feet/second, with negligible air resistance, then its position at time t seconds is given by the parametric equations:

$$x = (v \cos \theta)t \qquad \text{and} \qquad y = (v \sin \theta)t - 16t^2 + k.$$

Technology Tip

In parametric graphing zoom-in can be very time-consuming. It's often more effective to limit the t range to the values near the points you are interested in and set the t step very small. The picture may be hard to read, but trace can be used to determine coordinates.

GRAPHING EXPLORATION

Will the ball in Example 8 go over the fence if its initial velocity is 135 feet/second? Use degree mode and the viewing window of Figure 10–10 (with $0 \le t \le 4$ and t step = .1) to graph the ball's path. You may need to use trace if the graph is hard to read. If the answer still isn't clear, try changing the t step to .02.

Our final example is a curve that has several interesting applications.

Example 9 Choose a point P on a circle of radius 3 and find a parametric description of the curve that is traced out by P as the circle rolls along the x-axis, as shown in Figure 10–11.

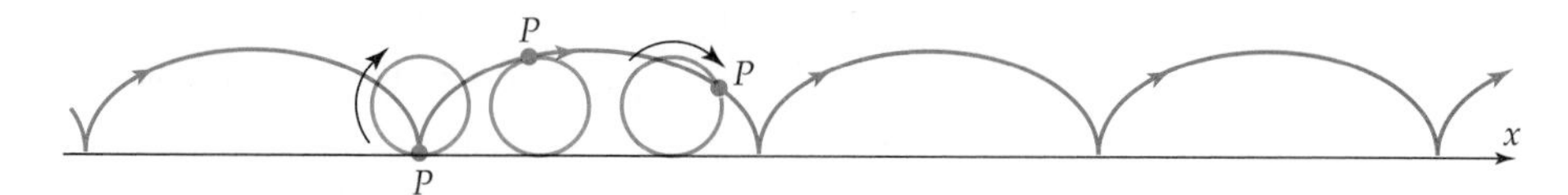

Figure 10–11

Solution This curve is called a **cycloid.** Begin with P at the origin and the center C of the circle at (0, 3). As the circle rolls along the x-axis, the line segment CP moves from vertical through an angle of t radians, as shown in Figure 10–12.

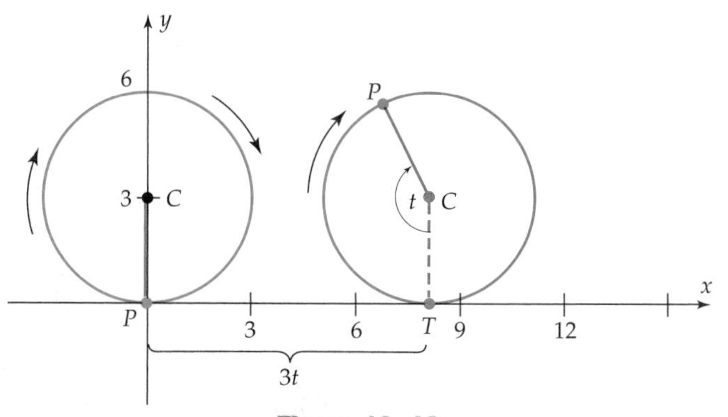

Figure 10–12

The distance from point T to the origin is the length of arc of the circle from T to P. As shown on page 397, this arc has length $3t$. Therefore the center C has coordinates $(3t, 3)$. When $0 < t < \pi/2$, the situation looks like Figure 10–13.

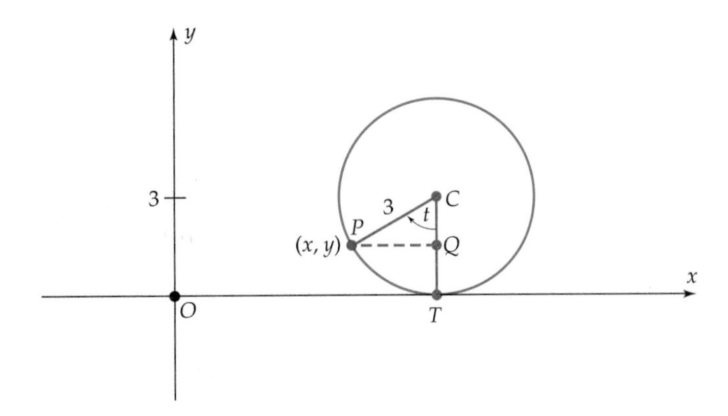

Figure 10–13

Right triangle PQC shows that

$$\sin t = \frac{PQ}{3}, \qquad \text{or equivalently,} \qquad PQ = 3 \sin t$$

and

$$\cos t = \frac{CQ}{3}, \qquad \text{or equivalently,} \qquad CQ = 3 \cos t.$$

In Figure 10–13, P has coordinates (x, y) and we have

$$x = OT - PQ = 3t - 3 \sin t = 3(t - \sin t)$$
$$y = CT - CQ = 3 - 3 \cos t = 3(1 - \cos t).$$

A similar analysis for other values of t (Exercises 47–51) shows that these equations are valid for every t. Therefore, the parametric equations of this cycloid are

$$x = 3(t - \sin t) \qquad \text{and} \qquad y = 3(1 - \cos t) \quad (t \text{ any real number}).$$ ■

If a cycloid is traced out by a circle of radius r, then the argument given in Example 9, with r in place of 3, shows that the parametric equations of the cycloid are

$$x = r(t - \sin t) \qquad \text{and} \qquad y = r(1 - \cos t) \quad (t \text{ any real number}).$$

Cycloids have a number of interesting applications. For example, among all the possible paths joining points P and Q in Figure 10–14 (on the next page), an arch of an inverted cycloid (shown in red) is the curve along which a particle (subject only to gravity) will slide from P to Q in the shortest possible time. This fact was first proved by J. Bernoulli in 1696.

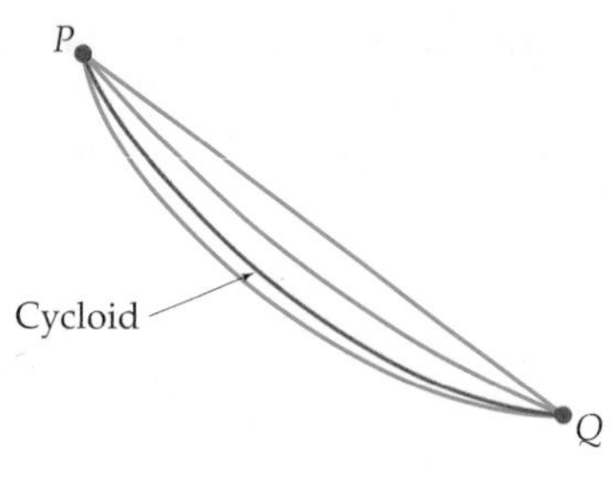

Figure 10–14

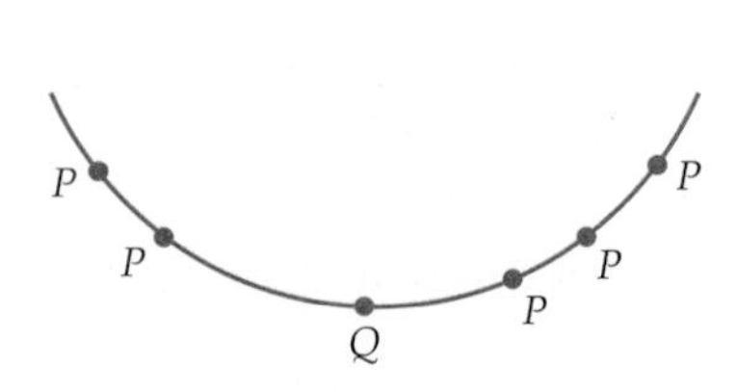

Figure 10–15

The Dutch physicist Christiaan Huygens (who invented the pendulum clock) proved that a particle takes the same time to slide to the bottom point Q of an inverted cycloid arch (as in Figure 10–15) from *any* point P on the curve.

Exercises 10.1

In Exercises 1–14, find a viewing window that shows a complete graph of the curve.

1. $x = t^2 - 4, \quad y = t/2, \quad -2 \le t \le 3$

2. $x = 3t^2, \quad y = 2 + 5t, \quad 0 \le t \le 2$

3. $x = 2t, \quad y = t^2 - 1, \quad -1 \le t \le 2$

4. $x = t - 1, \quad y = \dfrac{t+1}{t-1}, \quad t \ge 1$

5. $x = 4\sin 2t + 9, \quad y = 6\cos t - 8, \quad 0 \le t \le 2\pi$

6. $x = t^3 - 3t - 8, \quad y = 3t^2 - 15, \quad -4 \le t \le 4$

7. $x = 6\cos t + 12\cos^2 t, \quad y = 8\sin t + 8\sin t\cos t,$
$0 \le t \le 2\pi$

8. $x = 12\cos t, \quad y = 12\sin 2t, \quad 0 \le t \le 2\pi$

9. $x = 6\cos t + 5\cos 3t, \quad y = 6\sin t - 5\sin 3t,$
$0 \le t \le 2\pi$

10. $x = 3t^2 + 10, \quad y = 4t^3,$ any real number t

11. $x = 12\cos 3t\cos t + 6, \quad y = 12\cos 3t\sin t - 7,$
$0 \le t \le 2\pi$

12. $x = 2\cos 3t - 6, \quad y = 2\cos 3t\sin t + 7,$
$0 \le t \le 2\pi$

13. $x = t\sin t, \quad y = t\cos t, \quad 0 \le t \le 8\pi$

14. $x = 9\sin t, \quad y = 9t\cos t, \quad 0 \le t \le 20$

In Exercises 15–24, the given curve is part of the graph of an equation in x and y. Find the equation by eliminating the parameter.

15. $x = t - 3, \quad y = 2t + 1, \quad t \ge 0$

16. $x = t + 5, \quad y = \sqrt{t}, \quad t \ge 0$

17. $x = -2 + t^2, \quad y = 1 + 2t^2,$ any real number t

18. $x = t^2 + 1, \quad y = t^2 - 1,$ any real number t

19. $x = e^t, \quad y = t,$ any real number t

20. $x = 2e^t, \quad y = 1 - e^t, \quad t \ge 0$

21. $x = 3\cos t, \quad y = 3\sin t, \quad 0 \le t \le 2\pi$

22. $x = 4\sin 2t, \quad y = 2\cos 2t, \quad 0 \le t \le 2\pi$

23. $x = 3\cos t, \quad y = 4\sin t, \quad 0 \le t \le 2\pi$

24. $x = 2\sin t - 3, \quad y = 2\cos t + 1, \quad 0 \le t \le \pi$

In Exercises 25 and 26, sketch the graphs of the given curves and compare them. Do they differ and if so, how?

25. (a) $x = -4 + 6t, \quad y = 7 - 12t, \quad 0 \le t \le 1$
(b) $x = 2 - 6t, \quad y = -5 + 12t, \quad 0 \le t \le 1$

26. (a) $x = t, \quad y = t^2$
(b) $x = \sqrt{t}, \quad y = t$
(c) $x = e^t, \quad y = e^{2t}$

27. By eliminating the parameter, show that the curve with parametric equations

$$x = a + (c - a)t, \quad y = b + (d - b)t, \quad \text{any real number } t$$

is a straight line.

28. By proceeding as in Example 5, show that the curve with parametric equations

$$x = r\cos t + c, \quad y = r\sin t + d, \quad 0 \le t \le 2$$

is a circle with center (c, d) and radius r.

In Exercises 29–35, find a parameterization of the given curve. Confirm your answer by graphing.

29. line segment from $(-6, 12)$ to $(12, -10)$ [*Hint:* Exercise 27.]

30. line segment from $(14, -5)$ to $(5, -14)$

31. line segment from $(18, 4)$ to $(-16, 14)$

32. circle with center $(7, -4)$ and radius 6

33. circle with center $(9, 12)$ and radius 5

34. $x^2 + y^2 - 14x + 8y + 29 = 0$ [*Hint:* Example 8 in Section 1.3.]

35. $x^2 + y^2 - 4x - 6y + 9 = 0$

In Exercises 36–39, locate all local maxima and minima (other than endpoints) of the curve.

36. $x = 4t^3 - \cos t - 5, \quad y = 3t^2 - 8, \quad -2 \le t \le 2$

37. $x = 4t - 6, \quad y = 3t^2 + 2, \quad -10 \le t \le 10$

38. $x = t^3 + \sin t + 4, \quad y = \cos t, \quad -1.5 \le t \le 2$

39. $x = 4t^3 - t + 4, \quad y = -3t^2 + 5, \quad -2 \le t \le 2$

40. Show that the ball's path in Example 7 is a parabola by eliminating the parameter in the parametric equations

$$x = (140 \cos 31°)t \quad \text{and} \quad y = (140 \sin 31°)t - 16t^2.$$

[*Hint:* Solve the first equation for t and substitute the result in the second equation.]

In Exercises 41–46, use a calculator in degree mode and assume that air resistance is negligible.

41. A skeet is fired from the ground with an initial velocity of 110 feet/second at an angle of 28°.
- **(a)** Graph the skeet's path.
- **(b)** How long is the skeet in the air?
- **(c)** How high does it go?

42. A ball is thrown from a height of 5 feet above the ground with an initial velocity of 60 feet/second at an angle of 50° with the horizontal.
- **(a)** Graph the ball's path.
- **(b)** When and where does the ball hit ground?

43. A medieval bowman shoots an arrow which leaves the bow 4 feet above the ground with an initial velocity of 88 feet/second at an angle of 48° with the horizontal.
- **(a)** Graph the arrow's path.
- **(b)** Will the arrow go over the 40-foot-high castle wall that is 200 feet from the archer?

44. A golfer at a driving range stands on a platform 2 feet above the ground and hits the ball with an initial velocity of 120 feet/second at an angle of 39° with the horizontal. There is a 32-foot-high fence 400 feet away. Will the ball fall short, hit the fence, or go over it?

45. A golf ball is hit off the tee at an angle of 30° and lands 300 feet away. What was its initial velocity? [*Hint:* The ball lands when $x = 300$ and $y = 0$. Use this fact and the parametric equations for the ball's path to find two equations in the variables t and v. Solve for v.]

46. A football kicked from the ground has an initial velocity of 75 feet/second.
- **(a)** Set up the parametric equations that describe the ball's path. Experiment graphically with different angles to find the smallest angle (within one degree) needed so that the ball travels at least 150 feet.
- **(b)** Use algebra and trigonometry to find the angle needed for the ball to travel exactly 150 feet. [*Hint:* The ball lands when $x = 150$ and $y = 0$. Use this fact and the parametric equations for the ball's path to find two equations in the variables t and θ. Solve the "x equation" for t and substitute this result in the other one; then solve for θ. The double-angle identity may be helpful for putting this equation in a form that is easy to solve.]

47. A golf ball is hit off the ground at an angle of θ degrees with an initial velocity of 100 feet/second.
- **(a)** Graph the path of the ball when $\theta = 20°$, $\theta = 40°$, $\theta = 60°$, and $\theta = 80°$.
- **(b)** For what angle in part (a) does the ball land farthest from where it started?
- **(c)** Experiment with different angles, as in parts (a) and (b), and make a conjecture as to which angle results in the ball landing farthest from its starting point.

48. A golf ball is hit off the ground at an angle of θ degrees with an initial velocity of 100 feet/second.
- **(a)** Graph the path of the ball when $\theta = 30°$ and when $\theta = 60°$. In which case does the ball land farthest away?
- **(b)** Do part (a) when $\theta = 25°$ and $\theta = 65°$.
- **(c)** Experiment further and make a conjecture as to the results when the sum of the two angles is 90°.
- **(d)** Prove your conjecture algebraically. [*Hint:* Find the value of t at which a ball hit at angle θ hits the ground (which occurs when $y = 0$); this value of t will be an expression involving θ. Find the corresponding value of x (which is the distance of the ball from the starting point). Then do the same for an angle of $90° - \theta$ and use the cofunction identities (in degrees) to show that you get the same value of x.]

In Exercises 49–51, complete the derivation of the parametric equations of the cycloid in Example 9.

49. (a) If $\pi/2 < t < \pi$, verify that angle θ in the figure has measure $t - \pi/2$ and that

$$x = OT - CQ = 3t - 3\cos\left(t - \frac{\pi}{2}\right)$$

$$y = CT + PQ = 3 + 3\sin\left(t - \frac{\pi}{2}\right).$$

(b) Use the addition and subtraction identities for sine and cosine to show that in this case

$x = 3(t - \sin t)$ and $y = 3(1 - \cos t)$.

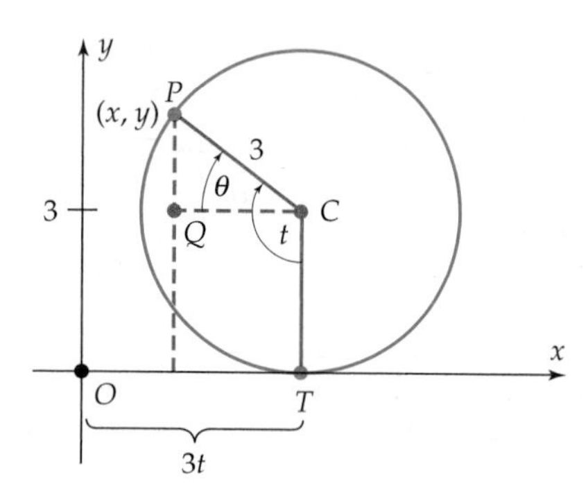

50. (a) If $\pi < t < 3\pi/2$, verify that angle θ in the figure measures $3\pi/2 - t$ and that

$$x = OT + CQ = 3t + 3\cos\left(\frac{3\pi}{2} - t\right)$$

$$y = CT + PQ = 3 + 3\sin\left(\frac{3\pi}{2} - t\right).$$

(b) Use the addition and subtraction identities for sine and cosine to show that in this case

$x = 3(t - \sin t)$ and $y = 3(1 - \cos t)$.

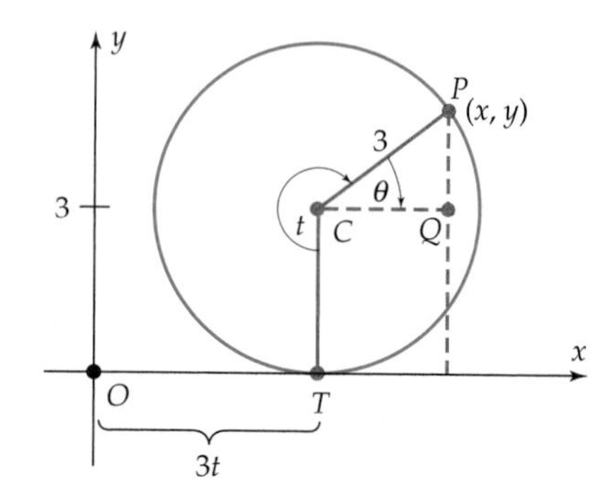

51. (a) If $3\pi/2 < t < 2\pi$, verify that angle θ in the figure has measure $t - 3\pi/2$ and that

$$x = OT + CQ = 3t + 3\cos\left(t - \frac{3\pi}{2}\right)$$

$$y = CT - PQ = 3 - 3\sin\left(t - \frac{3\pi}{2}\right).$$

(b) Use the addition and subtraction identities for sine and cosine to show that in this case

$x = 3(t - \sin t)$ and $y = 3(1 - \cos t)$

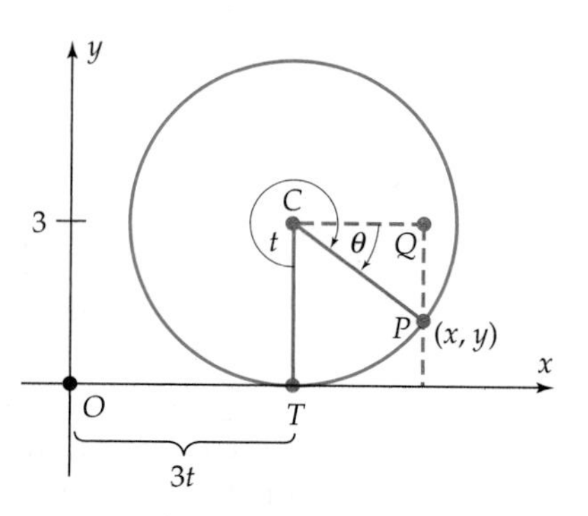

Thinkers

52. A particle moves on the horizontal line $y = 3$. Its x-coordinate at time t seconds is given by $x = 2t^3 - 13t^2 + 23t - 8$. This exercise explores the motion of the particle.

(a) Graph the path of the particle in the viewing window with $-10 \le x \le 10$, $-2 \le y \le 4$, $0 \le t \le 4.3$, and t step $= .05$. Note that the calculator seems to pause before completing the graph.

(b) Use trace (starting with $t = 0$) and watch the path of the particle as you press the right arrow key at regular intervals. How many times does it change direction? When does it appear to be moving the fastest?

(c) At what times t does the particle change direction? What are its x-coordinates at these times?

53. Set your calculator for radian mode and for simultaneous graphing mode [check your instruction manual for how to do this]. Particles A, B, and C are moving in the plane, with their positions at time t seconds given by:

A: $x = 8\cos t$ and $y = 5\sin t$

B: $x = 3t$ and $y = 5t$

C: $x = 3t$ and $y = 4t$.

(a) Graph the paths of A and B in the window with $0 \le x \le 12$, $0 \le y \le 6$, and $0 \le t \le 2$. The paths intersect, but do the particles actually collide? That is, are they at the same point at the same time? [For slow motion, choose a very small t step, such as .01.]

(b) Set t step = .05 and use trace to estimate the time at which A and B are closest to each other.

(c) Graph the paths of A and C and determine geometrically [as in part (b)] whether they collide. Approximately when are they closest?

(d) Confirm your answers in part (c) as follows. Explain why the distance between particles A and C at time t is given by

$$d = \sqrt{(8\cos t - 3t)^2 + (5\sin t - 4t)^2}.$$

A and C will collide if $d = 0$ at some time. Using function graphing mode, graph this distance function when $0 \le t \le 2$ and zoom-in if necessary, show that d is always positive. Find the value of t for which d is smallest.

54. Let P be a point at distance k from the center of a circle of radius r. As the circle rolls along the x-axis, P traces out a curve called a **trochoid.** [When $k \le d$, it may help to think of the circle as a bicycle wheel and P as a point on one of the spokes.]

(a) Assume that P is on the y-axis as close as possible to the x-axis when $t = 0$ and show that the parametric equations of the trochoid are

$$x = rt - k\sin t \qquad \text{and} \qquad y = r - k\cos t.$$

Note that when $k = r$, these are the equations of a cycloid.

(b) Sketch the graph of the trochoid with $r = 3$ and $k = 2$.

(c) Sketch the graph of the trochoid with $r = 3$ and $k = 4$.

55. A circle of radius b rolls along the inside of a larger circle of radius a. The curve traced out by a fixed point P on the smaller circle is called a **hypocycloid.**

(a) Assume that the larger circle has center at the origin and that the smaller circle starts with P located at $(a, 0)$. Use the figure to show that the parametric equations of the hypocycloid are

$$x = (a - b)\cos t + b\cos\left(\frac{a - b}{b}t\right)$$

$$y = (a - b)\sin t - b\sin\left(\frac{a - b}{b}t\right).$$

(b) Sketch the graph of the hypocycloid with $a = 5$, $b = 1$, and $0 \le t \le 2\pi$.

(c) Sketch the graph of the hypocycloid with $a = 5$, $b = 2$, and $0 \le t \le 4\pi$.

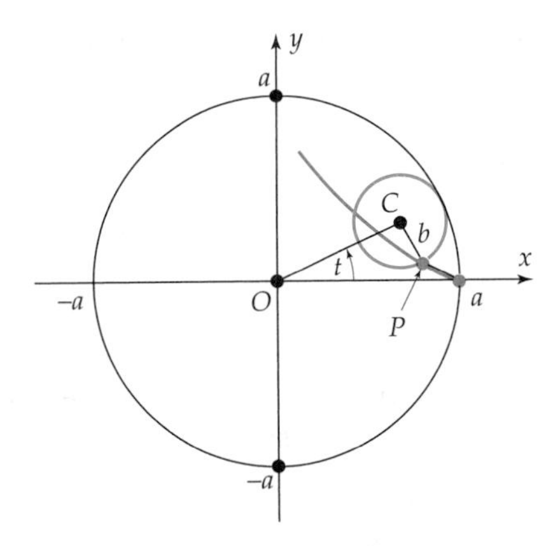

56. Jill is on a ferris wheel of radius 20 feet whose bottom just grazes the ground. The wheel is rotating counterclockwise at the rate of .7 radians per second. Jack is standing 100 feet from the bottom of the ferris wheel. When Jill is at point P, he throws a ball toward the wheel (see figure). The ball leaves Jack's hand 5 feet above the ground with an initial velocity of 62 feet/second at an angle of 1.2 radians with the horizontal. Will Jill be able to catch the ball? Follow the steps below to answer the question.

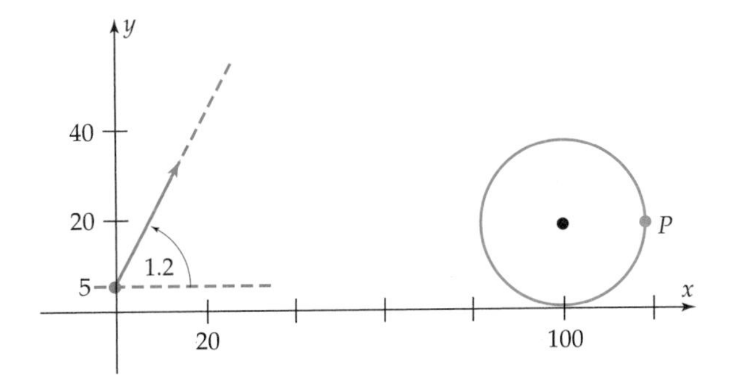

(a) Imagine that Jack is at the origin and the bottom of the ferris wheel is at (100, 0). Then the ball leaves his hand from the point (0, 5) and the wheel is a circle with center (100, 20) and radius 20 (see figure). Therefore, the wheel is the graph of the equation

$$(x - 100)^2 + (y - 20)^2 = 20^2.$$

Show that Jill's movement around the wheel is given by the parametric equations

$$x = 20\cos(.7t) + 100$$
$$y = 20\sin(.7t) + 20$$

by verifying that these equations give a parameterization of the circle [as in Example 5].

(b) Find the parametric equations that describe the position of the ball at time t.

(c) Set your calculator for parametric mode, radian mode, and simultaneous graphing mode [check your instruction manual for how to do this]. Using a square viewing window with $0 \le x \le 130$, $0 \le t \le 9$, and t step = .1 to graph both sets of parametric equations (Jill's motion and the ball's) simultaneously. [For slow motion, make the t step smaller.] Assuming that Jill can reach 2 feet in any direction, can she catch the ball? If not, use the trace to estimate the time at which Jill is closest to the ball.

(d) Experiment by changing the angle or initial velocity (or both) of the ball to find values that will allow Jill to catch the ball.

10.2 Conic Sections

When a right circular cone is cut by a plane, the intersection is a curve called a **conic section,** as shown in Figure 10–16.* Conic sections were studied by the ancient Greeks and are still of interest. For instance, planets travel in elliptical orbits, parabolic mirrors are used in telescopes, and certain atomic particles follow hyperbolic paths.

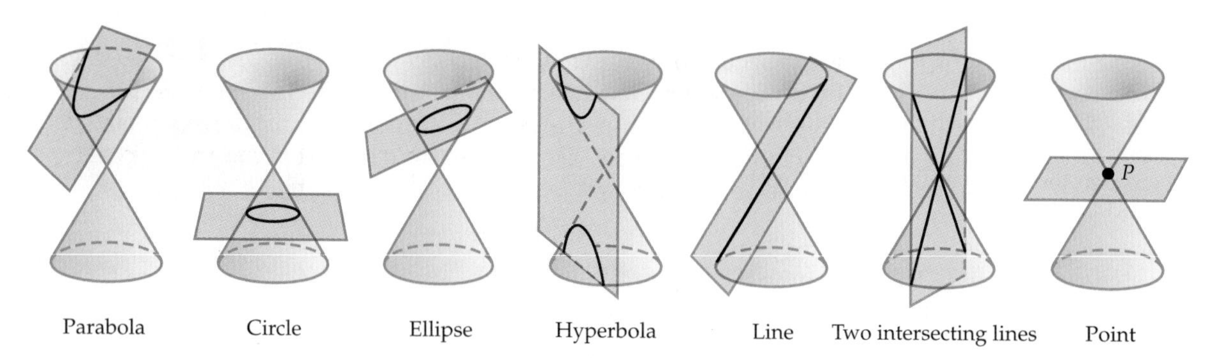

Figure 10–16

Although the Greeks studied conic sections from a purely geometric point of view, the modern approach is to describe them in terms of the coordinate plane and distance, or as the graphs of certain types of equations. This was done for circles in Section 1.3 and will be done here for ellipses, hyperbolas, and parabolas.

In each case the conic is defined in terms of points and distances and its equation determined. The standard form of the equation of a conic includes the key information necessary for a rough sketch of its graph, just

*A point, a line, or two intersecting lines are sometimes called **degenerate conic sections.**

as the standard form of the equation of a circle tells you its center and radius. Techniques for graphing conic sections with a calculator are discussed at the end of the section.

Ellipses

Definition. Let P and Q be points in the plane and r a number greater than the distance from P to Q. The **ellipse** with **foci*** P and Q is the set of all points X such that

$$(\text{distance from } X \text{ to } P) + (\text{distance from } X \text{ to } Q) = r.$$

To draw this ellipse, take a piece of string of length r and pin its ends on P and Q. Put your pencil point against the string and move it, keeping the string taut. You will trace out the ellipse, as shown in Figure 10–17.[†]

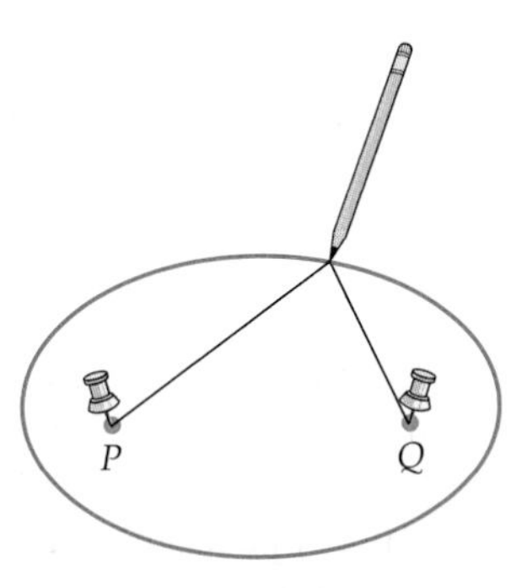

Figure 10–17

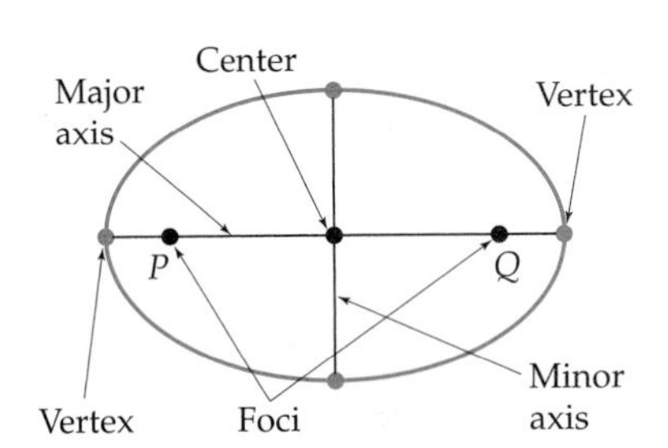

Figure 10–18

The midpoint of the line segment from P to Q is the **center** of the ellipse. The points where the straight line through the foci intersects the ellipse are its **vertices.** The **major axis** of the ellipse is the line segment joining the vertices; its **minor axis** is the line segment through the center, perpendicular to the major axis, as shown in Figure 10–18.

Equation. Suppose that foci are $P = (-c, 0)$ and $Q = (c, 0)$ for some $c > 0$. Let $a = r/2$, so that $r = 2a$. Then (x, y) is on the ellipse exactly when

$$[\text{distance from } (x, y) \text{ to } P] + [\text{distance from } (x, y) \text{ to } Q] = r$$

$$\sqrt{(x + c)^2 + (y - 0)^2} + \sqrt{(x - c)^2 + (y - 0)^2} = 2a$$

$$\sqrt{(x + c)^2 + y^2} = 2a - \sqrt{(x - c)^2 + y^2}$$

Squaring both sides and simplifying (Exercise 54) we obtain

$$a\sqrt{(x - c)^2 + y^2} = a^2 - cx.$$

Again squaring both sides and simplifying, we have

$$(a^2 - c^2)x^2 + a^2y^2 = a^2(a^2 - c^2).$$

*"Foci" is the plural of "focus."

[†]If $P = Q$, you will trace out a circle of radius $r/2$. So a circle is just a special case of an ellipse.

To simplify the form of this equation, let $b = \sqrt{a^2 - c^2}$* so that $b^2 = a^2 - c^2$ and the equation becomes

$$b^2x^2 + a^2y^2 = a^2b^2.$$

Dividing both sides by a^2b^2 shows that the coordinates of every point on the ellipse satisfy the equation

$$\frac{x^2}{a^2} + \frac{y^2}{b^2} = 1.$$

Conversely, it can be shown that every point whose coordinates satisfy this equation is on the ellipse. When the equation is in this form the x- and y-intercepts of the graph are easily found. For instance, to find the x-intercepts, we set $y = 0$ and solve:

$$\frac{x^2}{a^2} + \frac{0^2}{b^2} = 1$$

$$x^2 = a^2$$

$$x = \pm a.$$

Similarly, the y-intercepts are $\pm b$.

A similar argument applies when the foci are on the y-axis and leads to this conclusion:

Standard Equations of Ellipses Centered at the Origin

Let a and b be real numbers with $a > b > 0$. Then the graph of each of the following equations is an ellipse centered at the origin:

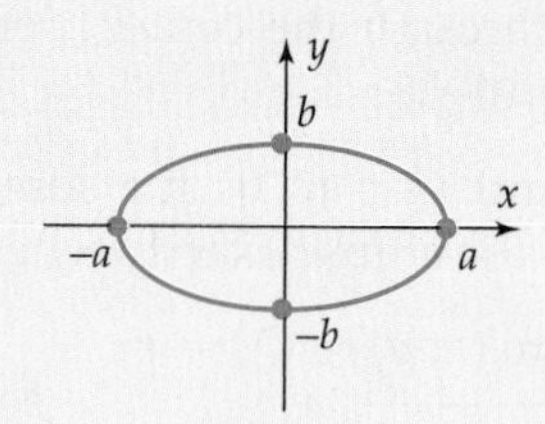

$$\frac{x^2}{a^2} + \frac{y^2}{b^2} = 1$$

- **x-intercepts: $\pm a$ y-intercepts: $\pm b$**
- **major axis on the x-axis, with vertices $(a, 0)$ and $(-a, 0)$**
- **foci: $(c, 0)$ and $(-c, 0)$, where $c = \sqrt{a^2 - b^2}$.**

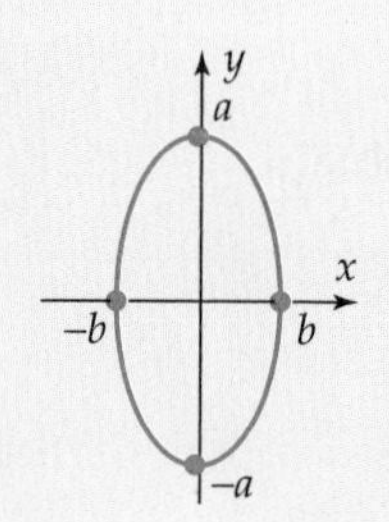

$$\frac{x^2}{b^2} + \frac{y^2}{a^2} = 1$$

- **x-intercepts: $\pm b$ y-intercepts: $\pm a$**
- **major axis on the y-axis, with vertices $(0, a)$ and $(0, -a)$**
- **foci: $(0, c)$ and $(0, -c)$, where $c = \sqrt{a^2 - b^2}$.**

*The distance between the foci is $2c$. Since $r = 2a$ and $r > 2c$ by definition, we have $2a > 2c$ and hence $a > c$. Therefore, $a^2 - c^2$ is a positive number and has a real square root.

In the preceding box $a > b$, but don't let all the letters confuse you: When the equation is in standard form, the denominator of the x term tells you the x-intercepts, the denominator of the y term tells you the y-intercepts, and the major axis is the longer one, as illustrated in the following examples.

Example 1 Identify the graph of $4x^2 + 9y^2 = 36$.

Solution We put the equation in standard form by dividing both sides by 36:

$$\frac{4x^2}{36} + \frac{9y^2}{36} = \frac{36}{36}$$

$$\frac{x^2}{9} + \frac{y^2}{4} = 1$$

$$\frac{x^2}{3^2} + \frac{y^2}{2^2} = 1.$$

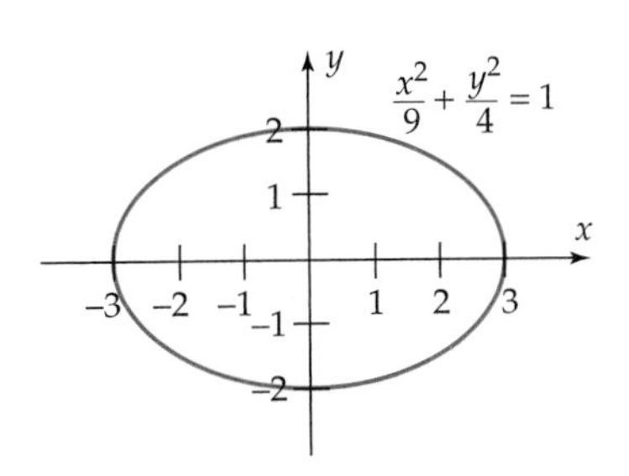

Figure 10–19

The graph now has the form of the first equation in the previous box, with $a = 3$ and $b = 2$. So its graph is an ellipse with x-intercepts ± 3 and y-intercepts ± 2, as shown in Figure 10–19. Its major axis and foci lie on the x-axis, as do its vertices $(\pm 3, 0)$. ■

Example 2 Find the equation of the ellipse with vertices $(0, \pm 6)$ and foci $(0, \pm 2\sqrt{6})$ and sketch its graph.

Solution Since the foci are $(0, 2\sqrt{6})$ and $(0, -2\sqrt{6})$, the center of the ellipse is $(0, 0)$ and its major axis lies on the y-axis. Hence, its equation is of the form

$$\frac{x^2}{b^2} + \frac{y^2}{a^2} = 1.$$

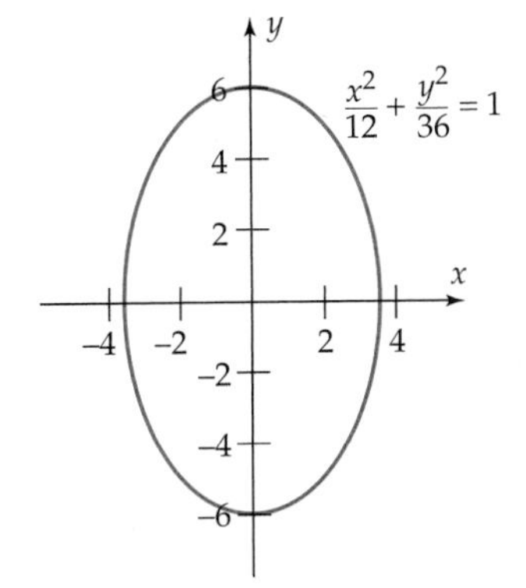

Figure 10–20

From the box on the preceding page, we see that $a = 6$ and $c = 2\sqrt{6}$. Since $c = \sqrt{a^2 - b^2}$, we have $c^2 = a^2 - b^2$, so that

$$b^2 = a^2 - c^2 = 6^2 - (2\sqrt{6})^2 = 36 - 4 \cdot 6 = 12.$$

Hence, $b = \sqrt{12}$ and the equation of the ellipse is

$$\frac{x^2}{(\sqrt{12})^2} + \frac{y^2}{6^2} = 1, \qquad \text{or equivalently,} \qquad \frac{x^2}{12} + \frac{y^2}{36} = 1.$$

The graph has x-intercepts $\pm\sqrt{12} \approx \pm 3.46$ and y-intercepts ± 6, as sketched in Figure 10–20. ■

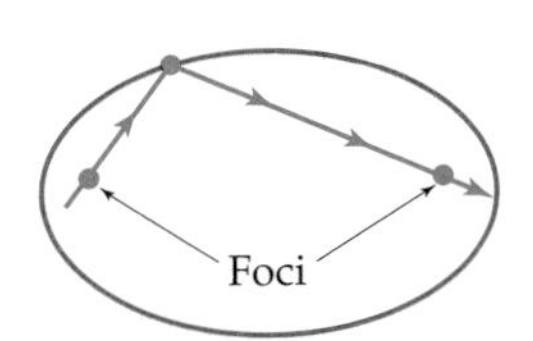

Figure 10–21

Applications. Elliptical surfaces have interesting reflective properties. If a sound or light ray passes through one focus and reflects off an ellipse, the ray will pass through the other focus, as shown in Figure 10–21. Exactly this situation occurs under the elliptical dome of the United States Capitol. A person who stands at one focus and whispers can be clearly heard by anyone at the other focus. Before this fact was widely known, when Congress used to sit under the dome, several political secrets were inadvertently revealed by congressmen to members of the other party.

The planets and many comets have elliptical orbits, with the sun as one focus. The moon travels in an elliptical orbit with the earth as one focus. Satellites are usually put into elliptical orbits around the earth.

Example 3 The earth's orbit around the sun is an ellipse that is almost a circle. The sun is one focus and the major and minor axes have lengths 186,000,000 miles and 185,974,062 miles, respectively. What are the minimum and maximum distances from the earth to the sun?

Solution The orbit is shown in Figure 10–22. If we use a coordinate system with the major axis on the x-axis and the sun having coordinates $(c, 0)$, then we obtain Figure 10–23.

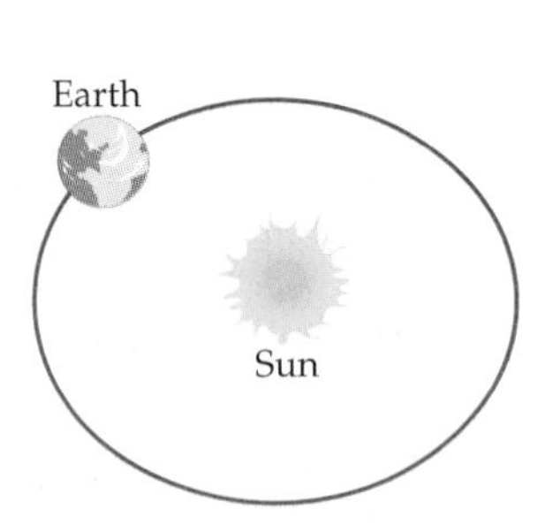

Figure 10–22

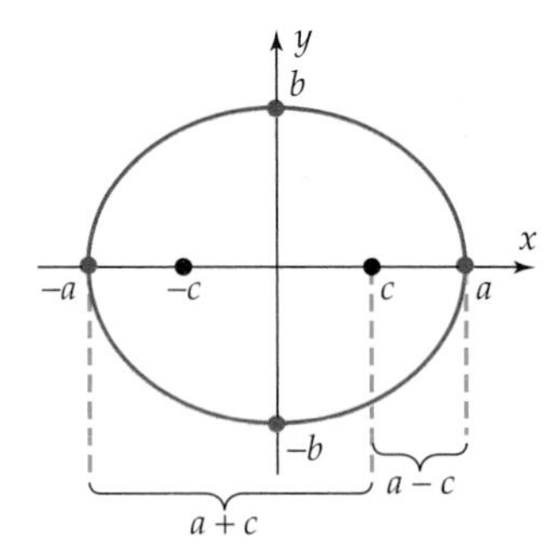

Figure 10–23

The length of the major axis is $2a = 186{,}000{,}000$, so that $a = 93{,}000{,}000$. Similarly, $2b = 185{,}974{,}062$, so that $b = 92{,}987{,}031$. As shown above, the equation of the orbit is $\dfrac{x^2}{a^2} + \dfrac{y^2}{b^2} = 1$, where

$$c = \sqrt{a^2 - b^2} = \sqrt{(93{,}000{,}000)^2 - (92{,}987{,}031)^2} \approx 1{,}553{,}083.$$

Figure 10–23 suggests a fact that can also be proven algebraically: The mininum and maximum distances from a point on the ellipse to the focus $(c, 0)$ occurs at the endpoints of the major axis:

$$\text{minimum distance} = a - c \approx 93{,}000{,}000 - 1{,}553{,}083 = 91{,}446{,}917 \text{ miles}$$

$$\text{maximum distance} = a + c \approx 93{,}000{,}000 + 1{,}553{,}083 = 94{,}553{,}083 \text{ miles.}$$

■

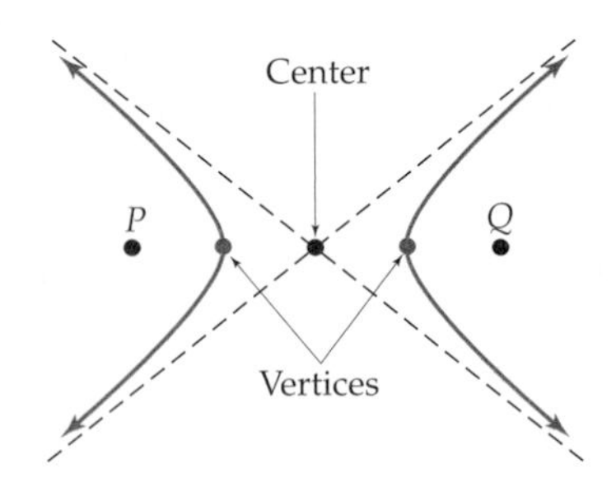

Figure 10–24

Hyperbolas

Definition. Let P and Q be points in the plane and r a positive number. The set of all points X such that

$$|(\text{distance from } P \text{ to } X) - (\text{distance from } Q \text{ to } X)| = r$$

is the **hyperbola** with **foci** P and Q; r will be called the **distance difference.** Every hyperbola has the general shape shown by the red curve in Figure 10–24. The dotted straight lines are the **asymptotes** of the hyperbola; it gets closer and closer to the asymptotes, but never touches them. The asymptotes intersect at the midpoint of the line segment from P to Q; this point is called the **center** of the hyperbola. The **vertices** of the hyperbola are the points where it intersects the line segment from P to Q. The line through P and Q is called the **focal axis.**

Equation. Another complicated exercise in the use of the distance formula, which will be omitted here, leads to the following algebraic description:

Standard Equations of Hyperbolas Centered at the Origin

Let a and b be positive real numbers. Then the graph of each of the following equations is a hyperbola centered at the origin:

$$\frac{x^2}{a^2} - \frac{y^2}{b^2} = 1$$

- **x-intercepts: $\pm a$ $\quad$ y-intercepts: none**
- **focal axis on the x-axis, with vertices $(a, 0)$ and $(-a, 0)$**
- **foci: $(c, 0)$ and $(-c, 0)$, where $c = \sqrt{a^2 + b^2}$.**
- **asymptotes: $y = \frac{b}{a}x$ and $y = -\frac{b}{a}x$**

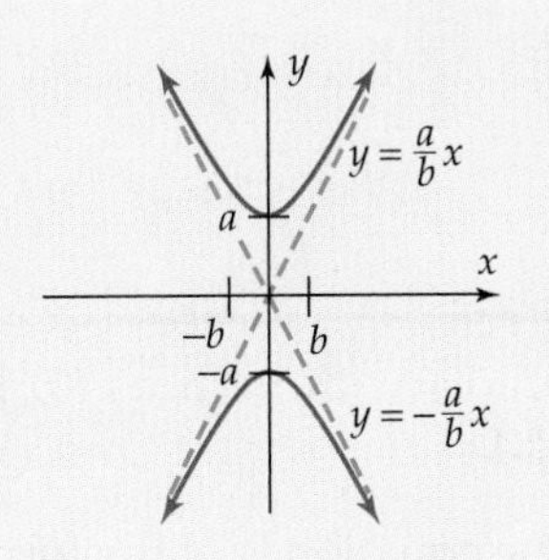

$$\frac{y^2}{a^2} - \frac{x^2}{b^2} = 1$$

- **x-intercepts: none $\quad$ y-intercepts: $\pm a$**
- **focal axis on the y-axis, with vertices $(0, a)$ and $(0, -a)$**
- **foci: $(0, c)$ and $(0, -c)$, where $c = \sqrt{a^2 + b^2}$.**
- **asymptotes: $y = \frac{a}{b}x$ and $y = -\frac{a}{b}x$**

Once again don't worry about all the letters in the box. When the equation is in standard form with the x term positive and y term negative, the hyperbola intersects the x-axis and opens from side to side. When the x term is negative and the y term positive, the hyperbola intersects the y-axis and opens up and down.

Example 4 To graph $9x^2 - 4y^2 = 36$ we first rewrite the equation:

$$\frac{9x^2}{36} - \frac{4y^2}{36} = \frac{36}{36}$$

$$\frac{x^2}{4} - \frac{y^2}{9} = 1$$

$$\frac{x^2}{2^2} - \frac{y^2}{3^2} = 1.$$

Applying the fact in the box with $a = 2$ and $b = 3$ shows that the graph is a hyperbola with vertices (2, 0) and (−2, 0) and asymptotes $y = \frac{3}{2}x$ and $y = -\frac{3}{2}x$. We first plot the vertices and sketch the rectangle determined by the vertical lines $x = \pm 2$ and the horizontal lines $y = \pm 3$. The asymptotes go through the origin and the corners of this rectangle, as shown on the left in Figure 10–25. It is then easy to sketch the hyperbola. ■

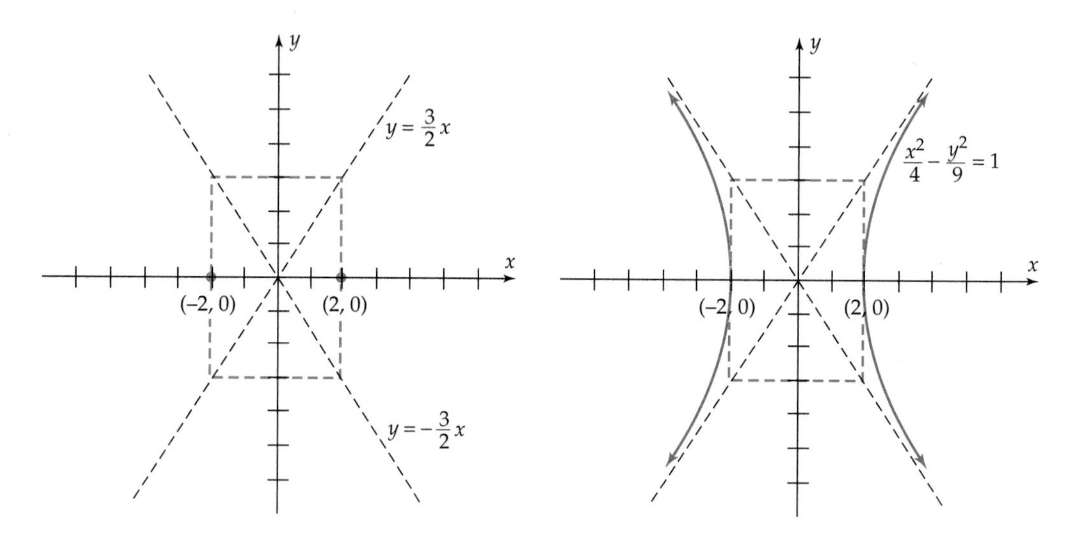

Figure 10–25

Example 5 Find the equation of the hyperbola with vertices (0, 1) and (0, −1) that passes through the point $(3, \sqrt{2}.)$.

Solution The vertices are on the y-axis and the equation is of the form

$$\frac{y^2}{a^2} - \frac{x^2}{b^2} = 1$$

with $a = 1$. Since $\left(3, \sqrt{2}\right)$ is on the graph, we have

$$\frac{(\sqrt{2})^2}{1^2} - \frac{3^2}{b^2} = 1$$

$$2 - \frac{9}{b^2} = 1$$

$$b^2 = 9.$$

Therefore, $b = 3$ and the equation is

$$\frac{y^2}{1^2} - \frac{x^2}{3^2} = 1, \qquad \text{or equivalently,} \qquad y^2 - \frac{x^2}{9} = 1.$$

The asymptotes of the hyperbola are the lines $y = \pm\frac{1}{3}x$. ■

Applications. The reflective properties of hyperbolas are used in the design of camera and telescope lenses. If a light ray passes through one focus of a hyperbola and reflects off the hyperbola at a point P, then the reflected ray moves along the straight line determined by P and the other focus, as shown in Figure 10–26. Other applications of hyperbolas are discussed in Section 10.3.

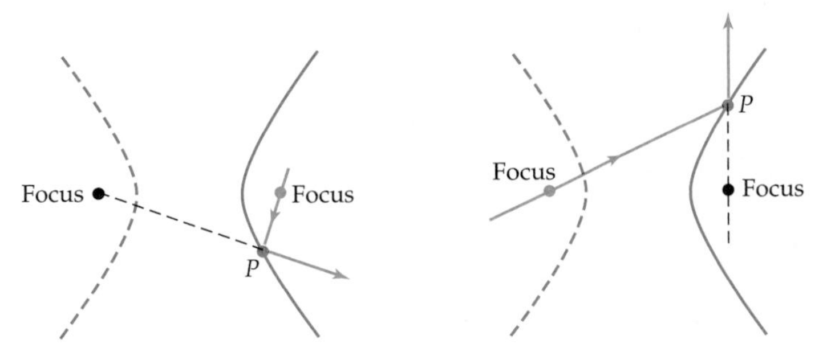

Figure 10–26

Parabolas

Definition. Parabolas appeared in Section 4.1 as the graphs of quadratic functions. Parabolas of this kind are a special case of the following more general definition. Let L be a line in the plane and P a point not on L. If X is any point not on L, the distance from X to L is defined to be the length of the perpendicular line segment from X to L. The **parabola** with **focus** P and **directrix** L is the set of all points X such that

$$\text{distance from } X \text{ to } P = \text{distance from } X \text{ to } L$$

as shown in Figure 10–27.

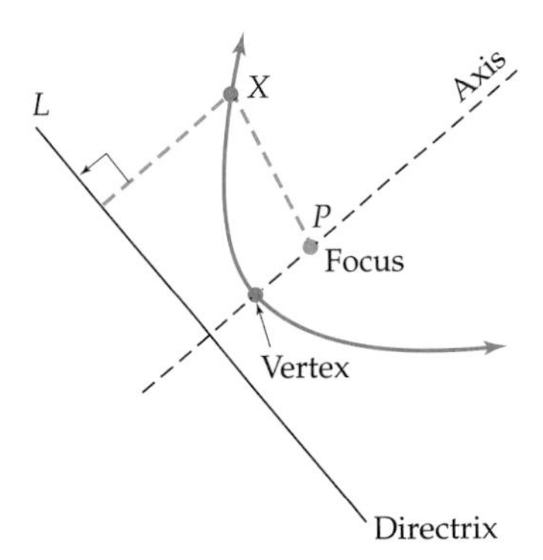

Figure 10–27

The line through P perpendicular to L is called the **axis.** The intersection of the axis with the parabola (the midpoint of the segment of the axis from P to L) is the **vertex** of the parabola, as illustrated in Figure 10–27. The parabola is symmetric with respect to its axis.

Equation. Suppose that the focus is on the y-axis at the point $(0, p)$, where p is a nonzero constant and that the directrix is the horizontal line $y = -p$.

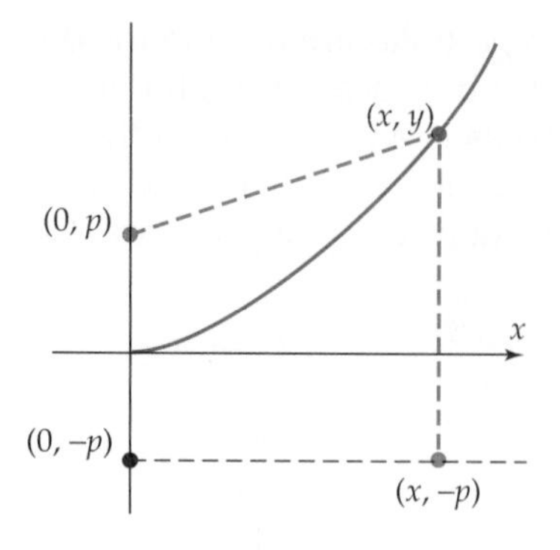

Figure 10–28

If (x, y) is any point on the parabola, then the distance from (x, y) to the horizontal line $y = -p$ is the length of the vertical segment from (x, y) to $(x, -p)$, as shown in Figure 10–28.

By the definition of the parabola,

$$\text{distance from } (x, y) \text{ to } (0, p) = \text{distance from } (x, y) \text{ to } y = -p$$

$$\text{distance from } (x, y) \text{ to } (0, p) = \text{distance from } (x, y) \text{ to } (x, -p)$$

$$\sqrt{(x - 0)^2 + (y - p)^2} = \sqrt{(x - x)^2 + [y - (-p)]^2}.$$

Squaring both sides and simplifying, we have

$$(x - 0)^2 + (y - p)^2 = (x - x)^2 + (y + p)^2$$

$$x^2 + y^2 - 2py + p^2 = 0^2 + y^2 + 2py + p^2$$

$$x^2 = 4py.$$

Conversely, it can be shown that every point whose coordinates satisfy this equation is on the parabola.

A similar argument works for the parabola with focus $(p, 0)$ on the x-axis and directrix the vertical line $x = -p$, and leads to this conclusion:

Standard Equations of Parabolas with Vertex at the Origin

Let p be a nonzero real number. Then the graph of each of the following equations is a parabola with vertex at the origin.

$$x^2 = 4py \quad \begin{cases} \textbf{focus: } (0, p) \\ \textbf{directrix: } y = -p \\ \textbf{axis: } y\textbf{-axis} \\ \textbf{opens upward if } p > 0\textbf{, downward if } p < 0 \end{cases}$$

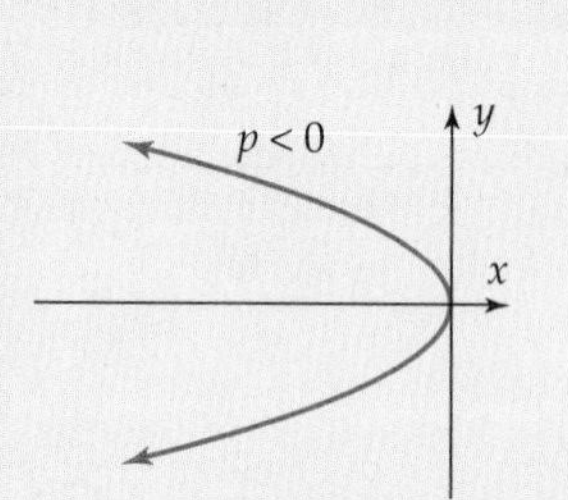

$$y^2 = 4px \quad \begin{cases} \textbf{focus: } (p, 0) \\ \textbf{directrix: } x = -p \\ \textbf{axis: } x\textbf{-axis} \\ \textbf{opens to right if } p > 0\textbf{, to left if } p < 0 \end{cases}$$

Example 6 Show that the graph of $y = -x^2/8$ is a parabola and find its focus and directrix.

Solution The equation $y = -x^2/8$ can be rewritten as $x^2 = -8y$. This equation is of the form $x^2 = 4py$, with $4p = -8$, so that $p = -2$. Hence, the graph is a downward opening parabola with focus $(0, p) = (0, -2)$ and directrix $y = -p = -(-2) = 2$, as shown in Figure 10–29. ■

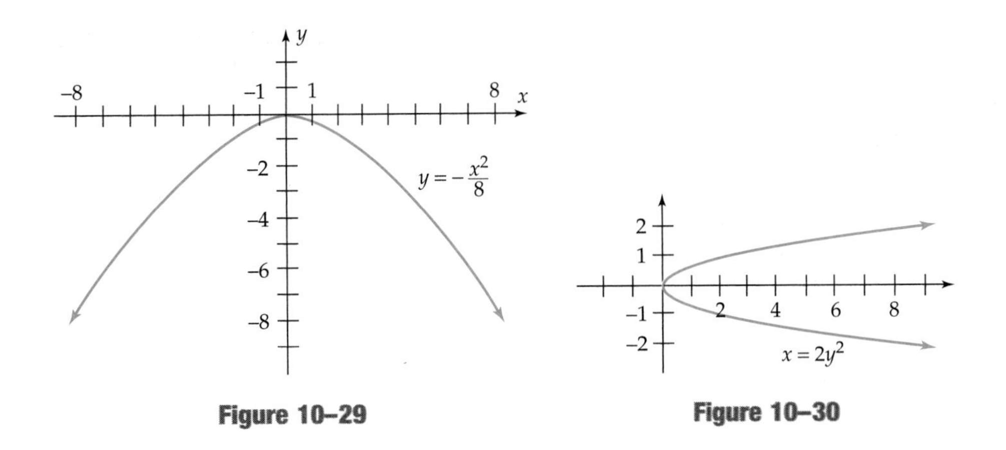

Figure 10–29 **Figure 10–30**

Example 7 Find the focus, directrix, and equation of the parabola that passes through the point (8, 2), has vertex (0, 0), and focus on the x-axis.

Solution The equation is of the form $y^2 = 4px$. Since (8, 2) is on the graph, we have $2^2 = 4p \cdot 8$, so that $p = \frac{1}{8}$. Therefore, the focus is $\left(\frac{1}{8}, 0\right)$ and the directrix is the vertical line $x = -\frac{1}{8}$. The equation is $y^2 = 4\left(\frac{1}{8}\right)x = \frac{1}{2}x$, or equivalently, $x = 2y^2$. Its graph is sketched in Figure 10–30. ■

Applications. Certain laws of physics show that sound waves or light rays from a source at the focus of a parabola will reflect off the parabola in rays parallel to the axis of the parabola, as shown in Figure 10–31. This is the reason that parabolic reflectors are used in automobile headlights and searchlights.

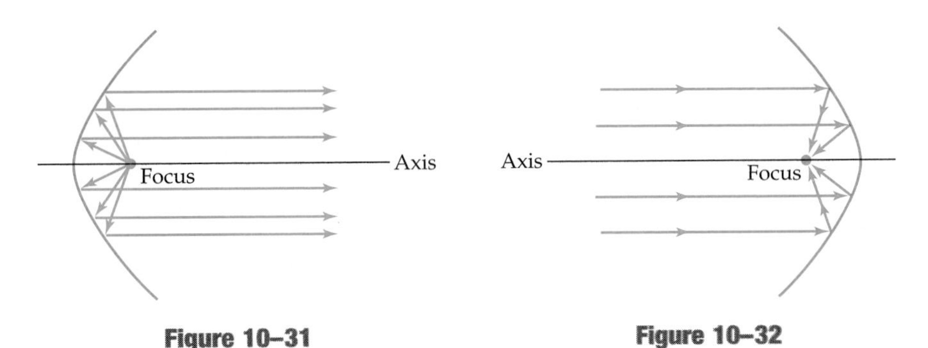

Figure 10–31 **Figure 10–32**

Conversely, a light ray coming toward a parabola will be reflected into the focus, as shown in Figure 10–32. This fact is used in the design of radar antennas, satellite dishes, and field microphones used at outdoor sporting events to pick up conversation on the field.

Projectiles follow a parabolic curve, a fact that is used in the design of water slides in which the rider slides down a sharp incline, then up and over a hill, before plunging downward into a pool. At the peak of the hill, the rider shoots up along a parabolic arc several inches above the slide, experiencing a sensation of weightlessness.

Graphing Conic Sections

Upward- and downward-opening parabolas are the graphs of functions of the form $y = f(x)$, so they can be easily graphed on a calculator. Other conic sections, however, are not the graphs of such functions since they fail the Vertical Line Test (page 139). One method of graphing this kind of conic section is to

> Write its equation in the form $y^2 = f(x)$ and graph both of the functions $y = \sqrt{f(x)}$ and $y = -\sqrt{f(x)}$ on the same screen.

Example 8 The graph of $4x^2 + 25y^2 = 100$ is an ellipse (divide both sides by 100 to put the equation in standard form). Solving this equation for y shows that

$$25y^2 = 100 - 4x^2$$

$$y^2 = \frac{100 - 4x^2}{25}$$

$$y = \sqrt{\frac{100 - 4x^2}{25}} \qquad \text{or} \qquad y = -\sqrt{\frac{100 - 4x^2}{25}}$$

Each of these last equations defines y as a function of x and can be entered in a calculator.

Technology Tip

On Sharp 9600 and TI calculators you can graph both functions in the exploration at the same time by keying in $y =$

$$\{1, -1\}\sqrt{\frac{100 - 4x^2}{25}}.$$

GRAPHING EXPLORATION

Using the standard viewing window in "function" (or "rectangular") graphing mode, graph both of these functions on the same screen and verify that the graph is an ellipse centered at the origin with x-intercepts ± 5 and y-intercepts ± 2. ■

Parametric Equations for Conics

Conic sections can be conveniently graphed in parametric graphing mode. As was the case with circles in Section 10.1, parameterizations of the other conic sections are a consequence of various Pythagorean identities in trigonometry. For example, the hyperbola with equation

$$\frac{x^2}{a^2} - \frac{y^2}{b^2} = 1$$

can be obtained from this parameterization:

$$x = a \sec t \qquad \text{and} \qquad y = b \tan t \quad (0 \le t \le 2\pi)$$

because

$$\frac{x^2}{a^2} - \frac{y^2}{b^2} = \frac{(a \sec t)^2}{a^2} - \frac{(b \tan t)^2}{b^2}$$

$$= \frac{a^2 \sec^2 t}{a^2} - \frac{b^2 \tan^2 t}{b^2} = \sec^2 t - \tan^2 t = 1.$$

Similar arguments (Exercises 56 and 57) lead to the following table. The parameterizations listed in it are not the only possible ones (Exercise 58).

Parametric Equations for Conic Sections

Conic Section	Parameterization
Ellipse: $\frac{x^2}{a^2} + \frac{y^2}{b^2} = 1$	$x = a \cos t, \quad y = b \sin t$ $(0 \le t \le 2\pi)$
Hyperbola: $\frac{x^2}{a^2} - \frac{y^2}{b^2} = 1$	$x = a \sec t = \frac{a}{\cos t}, \quad y = b \tan t$ $(0 \le t \le 2\pi)$
$\frac{y^2}{a^2} - \frac{x^2}{b^2} = 1$	$x = b \tan t, \quad y = a \sec t = \frac{a}{\cos t}$ $(0 \le t \le 2\pi)$
Parabola: $y^2 = 4px$	$x = t^2/4p, \quad y = t$ (any real number t)

Example 9 Find parametric equations for each of these conic sections:

(a) $y^2 = 6x$ (b) $\frac{y^2}{9} - \frac{x^2}{16} = 1.$

Solution

(a) As in the last line of the preceding box, this is the equation of a parabola, with $4p = 6$. Let $y = t$ and $x = t^2/6$, where t is any real number.

GRAPHING EXPLORATION

With your calculator in parametric graphing mode, graph this parabola.

(b) This equation has the form of the second hyperbola equation in the box above, with $a = 3$ and $b = 4$. Let $x = 4 \tan t$ and $y = 3/\cos t$.

GRAPHING EXPLORATION

Set $0 \le t \le 2\pi$ and graph the hyperbola. How is it traced out? Now change the t range so that $-\pi/2 \le t \le 3\pi/2$. Now how is the graph traced out? ■

Example 10 Use parametric equations to graph $4x^2 + 25y^2 = 100$.

Solution Put the equation in standard form by dividing both sides by 100:

$$\frac{x^2}{25} + \frac{y^2}{4} = 1, \qquad \text{or equivalently,} \qquad \frac{x^2}{5^2} + \frac{y^2}{2^2} = 1.$$

Now parameterize by letting $x = 5\cos t$ and $y = 2\sin t$ with $0 \le t \le 2\pi$.

GRAPHING EXPLORATION

Using the standard viewing window in parametric graphing mode, graph the curve given by these parametric equations. Verify that the ellipse is traced out in a counterclockwise direction beginning at the point (5, 0). ■

Exercises 10.2

In Exercises 1–14, find the equation of the conic section that satisfies the given conditions.

1. Ellipse with center (0, 0); foci on x-axis; x-intercepts ± 7; y-intercepts ± 2.
2. Ellipse with center (0, 0); foci on y-axis; x-intercepts ± 1; y-intercepts ± 8.
3. Ellipse with center (0, 0); foci on x-axis; major axis of length 12; minor axis of length 8.
4. Ellipse with center (0, 0); foci on y-axis; major axis of length 20; minor axis of length 18.
5. Hyperbola with center (0, 0); x-intercepts ± 3; asymptote $y = 2x$.
6. Hyperbola with center (0, 0); y-intercepts ± 12; asymptote $y = 3x/2$.
7. Hyperbola with center (0, 0); vertex (2, 0); passing through $(4, \sqrt{3})$.
8. Hyperbola with center (0, 0); vertex $(0, \sqrt{12})$; passing through $(2\sqrt{3}, 6)$.
9. Parabola with vertex (0, 0); axis $x = 0$; passing through (2, 12).
10. Parabola with vertex (0, 0); axis $y = 0$; passing through (2, 12).
11. Parabola with vertex (0, 0) and focus (5, 0).
12. Parabola with vertex (0, 0) and focus (0, 3.5).
13. Ellipse with center (0, 0); endpoints of major and minor axes: (0, −7), (0, 7), (−3, 0), (3, 0).
14. Ellipse with center (0, 0); vertices (8, 0) and (−8, 0); minor axis of length 8.

In Exercises 15–20, determine which of the following equations could possibly have the given graph.

$$y = x^2/4, \quad x^2 = -8y, \quad 6x = y^2, \quad y^2 = -4x,$$
$$2x^2 + y^2 = 12, \quad x^2 + 6y^2 = 18,$$
$$6y^2 - x^2 = 6, \quad 2x^2 - y^2 = 8$$

15.

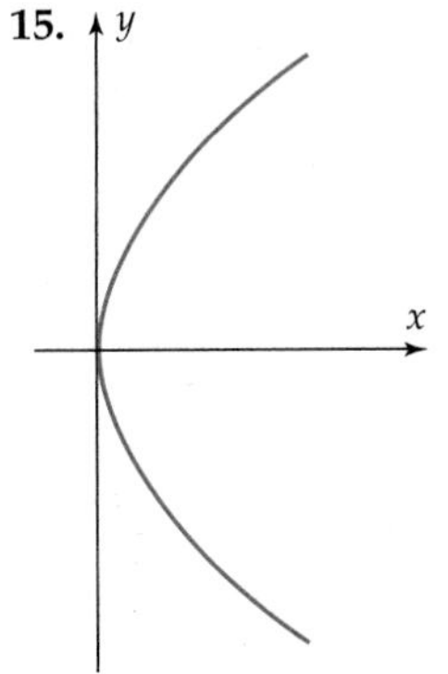

16.

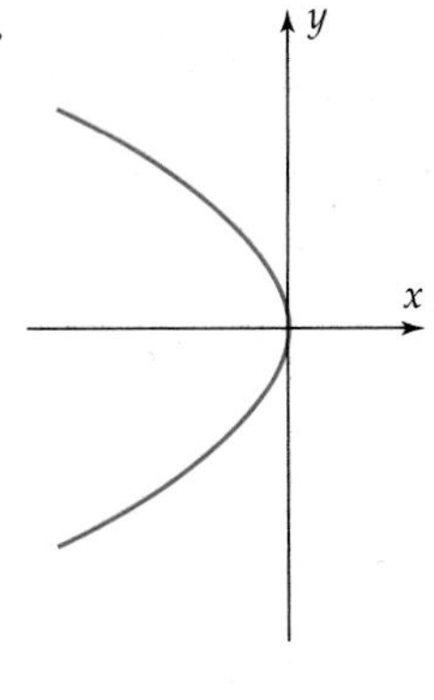

17.

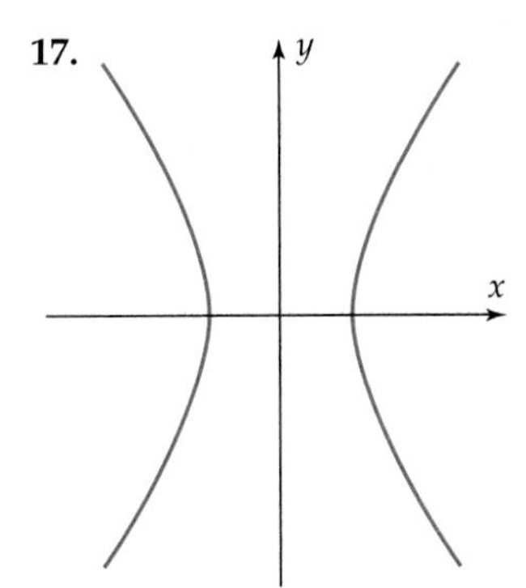

18.

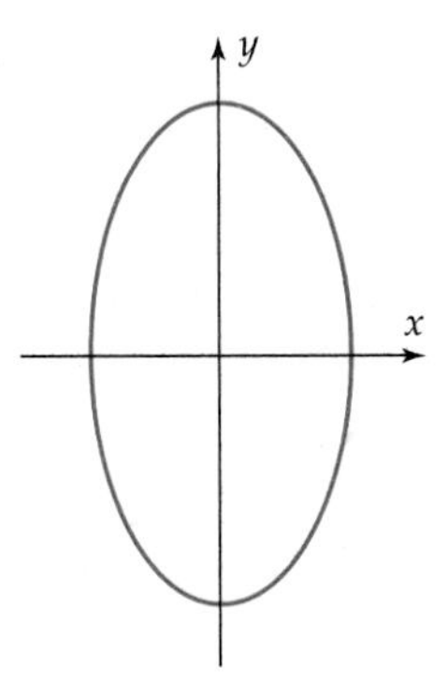

19.

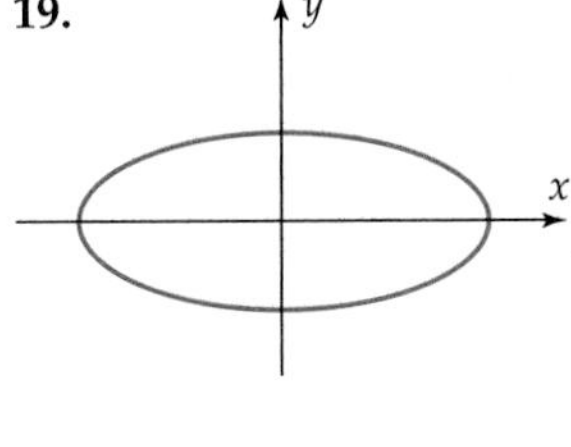

20.

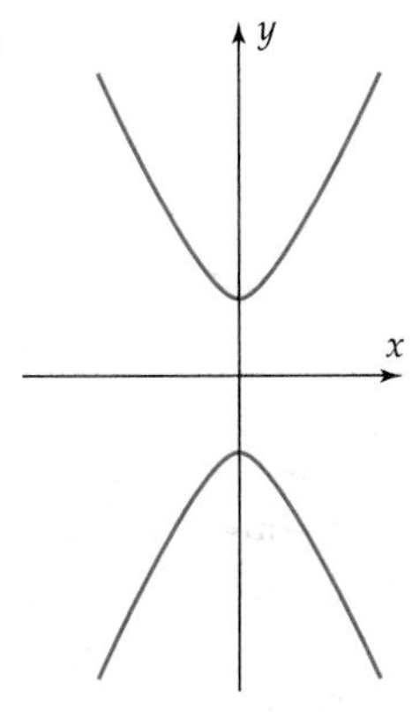

In Exercises 21–32, identify the conic section whose equation is given and find a complete graph of the equation, using the method of Example 8.

21. $\dfrac{x^2}{25} + \dfrac{y^2}{4} = 1$

22. $\dfrac{x^2}{6} + \dfrac{y^2}{16} = 1$

23. $4x^2 + 3y^2 = 12$

24. $9x^2 + 4y^2 = 72$

25. $\dfrac{x^2}{6} - \dfrac{y^2}{16} = 1$

26. $\dfrac{x^2}{4} - y^2 = 1$

27. $4x^2 - y^2 = 16$

28. $3y^2 - 5x^2 = 15$

29. $x = -6y^2$

30. $.5y^2 = 2x$

31. $18y^2 - 8x^2 - 2 = 0$

32. $x^2 - 2y^2 = -1$

In Exercises 33–42, find parametric equations for the curve whose equation is given and use these parametric equations to find a complete graph of the curve.

33. $\dfrac{x^2}{10} - 1 = \dfrac{-y^2}{36}$

34. $\dfrac{y^2}{49} + \dfrac{x^2}{81} = 1$

35. $4x^2 + 4y^2 = 1$

36. $x^2 + 4y^2 = 1$

37. $\dfrac{x^2}{10} - \dfrac{y^2}{36} = 1$

38. $\dfrac{y^2}{9} - \dfrac{x^2}{16} = 1$

39. $x^2 - 4y^2 = 1$

40. $2x^2 - y^2 = 4$

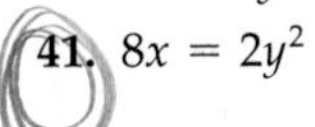

41. $8x = 2y^2$

42. $4y = x^2$

Calculus can be used to show that the area of the ellipse with equation $\dfrac{x^2}{a^2} + \dfrac{y^2}{b^2} = 1$ *is* πab. *Use this fact to find the area of each ellipse in Exercises 43–48.*

43. $\dfrac{x^2}{16} + \dfrac{y^2}{4} = 1$

44. $\dfrac{x^2}{9} + \dfrac{y^2}{5} = 1$

45. $3x^2 + 4y^2 = 12$

46. $7x^2 + 5y^2 = 35$

47. $6x^2 + 2y^2 = 14$

48. $5x^2 + y^2 = 5$

In Exercises 49–52, find the focus and directrix of the parabola.

49. $y = 3x^2$

50. $x = .5y^2$

51. $y = .25x^2$

52. $x = -6y^2$

53. Consider the ellipse whose equation is $\dfrac{x^2}{a^2} + \dfrac{y^2}{b^2} = 1$. Show that if $a = b$, then the graph is actually a circle.

54. Complete the derivation of the equation of the ellipse on page 625 as follows.

(a) By squaring both sides, show that the equation

$$\sqrt{(x + c)^2 + y^2} = 2a - \sqrt{(x - c)^2 + y^2}$$

may be simplified as

$$a\sqrt{(x - c)^2 + y^2} = a^2 - cx.$$

(b) Show that the last equation in part (a) may be further simplified as

$$(a^2 - c^2)x^2 + a^2y^2 = a^2(a^2 - c^2).$$

55. Sketch the graph of $\dfrac{y^2}{4} - \dfrac{x^2}{b^2} = 1$ for $b = 2$, $b = 4$, $b = 8$, $b = 12$, and $b = 20$. What happens to the hyperbola as b takes larger and larger values? Could the graph ever degenerate into a pair of horizontal lines?

56. Show that the curve defined by the parametric equations $x = a\cos t$ and $y = b\sin t$ $(0 \le t \le 2\pi)$ is the ellipse whose equation is $\frac{x^2}{a^2} + \frac{y^2}{b^2} = 1$.

57. Let p be a nonzero constant. Show that the curve defined by the parametric equations $x = t^2/4p$ and $y = t$ (t any real number) is the parabola whose equation is $y^2 = 4px$.

58. Find parameterizations for the ellipse and hyperbola different from those given in the box on page 635. [*Hint:* Example 5 in Section 10.1.]

59. The orbit of the moon around the earth is an ellipse with the earth as one focus. If the length of the major axis of the orbit is 477,736 miles and the length of the minor axis is 477,078 miles, find the minimum and maximum distances from the earth to the moon.

60. Halley's Comet has an elliptical orbit with the sun as one focus and a major axis that is 1,636,484,848 miles long. The closest the comet comes to the sun is 54,004,000 miles. What is the maximum distance from the comet to the sun?

Thinkers

61. The punch bowl and a table holding the punch cups are placed 50 feet apart at a garden party. A portable fence is then set up so that any guest inside the fence can walk straight to the table, then to the punch bowl, and then return to his or her starting point without traveling more than 150 feet. Describe the longest possible such fence that encloses the largest possible area.

62. An arched footbridge over a 100-foot-wide river is shaped like half an ellipse. The maximum height of the bridge over the river is 20 feet. Find the height of the bridge over a point in the river, exactly 25 feet from the center of the river.

10.3 Translations and Rotations of Conics

Now that you are familiar with conic sections centered at the origin, we expand the discussion to include both conics centered at other points in the plane and ones with axes that may not be parallel to the coordinate axes. We begin with a basic fact about graphing that plays a key role in the discussion.

In Section 3.4 we saw that replacing the variable x by $x - 5$ in the rule of the function $y = f(x)$ shifts the graph of the function horizontally 5 units to the right, whereas replacing x by $x + 5$ [that is, $x - (-5)$] shifts the graph horizontally 5 units to the left (see the box on page 152). Similarly, if the rule of a function is given by $y = f(x)$, then replacing y by $y - 4$ shifts the graph 4 units vertically upward [because $y - 4 = f(x)$ is equivalent to $y = f(x) + 4$; see the box on page 151]. For arbitrary equations we have a similar result:

Vertical and Horizontal Shifts

Let h and k be constants. Replacing x by $x - h$ and y by $y - k$ in an equation shifts the graph of the equation

$|h|$ units horizontally (right for positive h, left for negative h) and

$|k|$ units vertically (upward for positive k, downward for negative k).

Example 1 Identify and sketch the graph of

$$\frac{(x-5)^2}{9}+\frac{(y+4)^2}{36}=1.$$

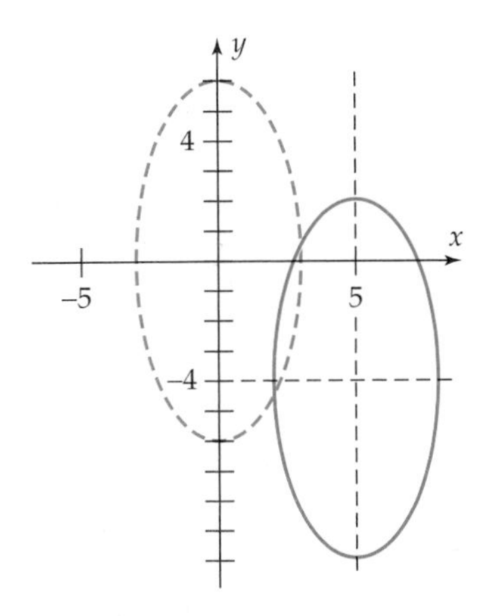

Figure 10–33

Solution This equation can be obtained from the equation $\frac{x^2}{9}+\frac{y^2}{36}=1$ (whose graph is known to be an ellipse) as follows:

replace x by $x-5$ and replace y by $y-(-4)=y+4$.

This is the situation described in the previous box with $h=5$ and $k=-4$. Therefore, the graph is the ellipse $\frac{x^2}{9}+\frac{y^2}{36}=1$ shifted horizontally 5 units to the right and vertically 4 units downward, as shown in Figure 10–33. The center of the ellipse is at $(5,-4)$. Its major axis (the longer one) lies on the vertical line $x=5$, as do its foci. The minor axis is on the horizontal line $y=-4$. ■

Example 2 Identify and sketch the graph of

$$4x^2+9y^2-32x-90y+253=0.$$

Solution We first rewrite the equation:

$$(4x^2-32x)+(9y^2-90y)=-253$$
$$4(x^2-8x)+9(y^2-10y)=-253.$$

Now complete the square in x^2-8x and y^2-10y:

$$4(x^2-8x+16)+9(y^2-10y+25)=-253+?+?$$

Be careful here: On the left side we haven't just added 16 and 25. When the left side is multiplied out we have actually added in $4\cdot16=64$ and $9\cdot25=225$. Therefore, to leave the original equation unchanged, we must add these numbers on the right:

$$4(x^2-8x+16)+9(y^2-10y+25)=-253+64+225$$
$$4(x-4)^2+9(y-5)^2=36$$
$$\frac{4(x-4)^2}{36}+\frac{9(y-5)^2}{36}=\frac{36}{36}$$
$$\frac{(x-4)^2}{9}+\frac{(y-5)^2}{4}=1.$$

The graph of this equation is the ellipse $\frac{x^2}{9}+\frac{y^2}{4}=1$ shifted horizontally 4 units to the right and vertically 5 units upward. Its center is at $(4,5)$. Its major axis lies on the horizontal line $y=5$ and its minor axis on the vertical line $x=4$, as shown in Figure 10–34 on the next page. ■

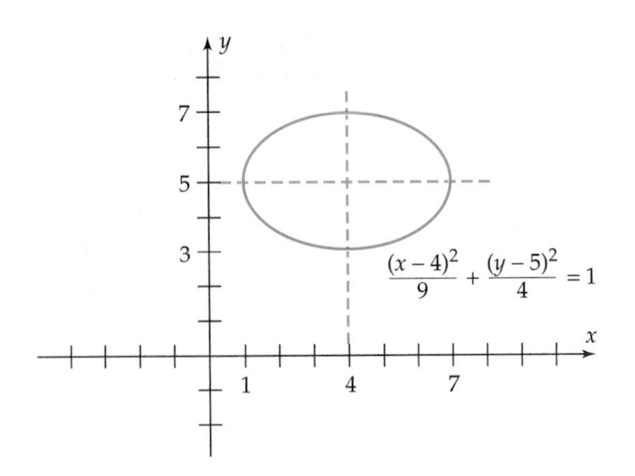

Figure 10–34

Example 3 Identify and sketch the graph of $x = 2y^2 + 12y + 14$.

Solution We rewrite the equation and complete the square in y, being careful to add the appropriate amounts to both sides of the equation:

$$2y^2 + 12y = x - 14$$

$$2(y^2 + 6y) = x - 14$$

$$2(y^2 + 6y + 9) = x - 14 + 2\cdot 9$$

$$2(y + 3)^2 = x + 4$$

$$[y - (-3)]^2 = \frac{1}{2}[x - (-4)].$$

Thus, the graph is the graph of the parabola $y^2 = \frac{1}{2}x$ shifted 4 units horizontally to the left and 3 units vertically downward, as shown in Figure 10–35. The parabola $y^2 = \frac{1}{2}x$ has its vertex at (0, 0) and the x-axis as its axis. When the graph is shifted, the parabola will have its vertex at $(-4, -3)$ and the horizontal line $y = -3$ as its axis. ■

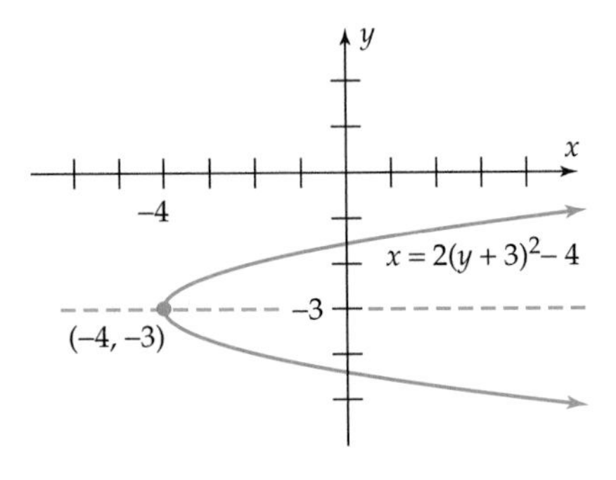

Figure 10–35

By translating the information about conic sections centered at the origin in the boxes on pages 626, 629, and 632, we obtain this summary of the standard equations of conic sections whose axes are parallel to the coordinate axes:

Standard Equations of Conic Sections

Let (h, k) be any point in the plane. If a and b are real numbers with $a > b > 0$, then the graph of each of the following equations is an ellipse with center (h, k).

$$\frac{(x-h)^2}{a^2} + \frac{(y-k)^2}{b^2} = 1 \quad \begin{cases} \text{major axis on the horizontal line } y = k \\ \text{minor axis on the vertical line } x = h \\ \text{foci: } (h-c, k) \text{ and } (h+c, k), \text{ where} \\ \quad c = \sqrt{a^2 - b^2} \end{cases}$$

$$\frac{(x-h)^2}{b^2} + \frac{(y-k)^2}{a^2} = 1 \quad \begin{cases} \text{major axis on the vertical line } x = h \\ \text{minor axis on the horizontal line } y = k \\ \text{foci: } (h, k-c) \text{ and } (h, k+c), \text{ where} \\ \quad c = \sqrt{a^2 - b^2} \end{cases}$$

If a and b are positive real numbers, then the graph of each of the following equations is a hyperbola with center (h, k).

$$\frac{(x-h)^2}{a^2} - \frac{(y-k)^2}{b^2} = 1 \quad \begin{cases} \text{focal axis on the horizontal line } y = k \\ \text{foci: } (h-c, k) \text{ and } (h+c, k), \text{ where} \\ \quad c = \sqrt{a^2 + b^2} \\ \text{vertices: } (h-a, k) \text{ and } (h+a, k) \\ \text{asymptotes: } y = \pm\frac{b}{a}(x-h) + k \end{cases}$$

$$\frac{(y-k)^2}{a^2} - \frac{(x-h)^2}{b^2} = 1 \quad \begin{cases} \text{focal axis on the vertical line } x = h \\ \text{foci: } (h, k-c) \text{ and } (h, k+c), \text{ where} \\ \quad c = \sqrt{a^2 + b^2} \\ \text{vertices: } (h, k-a) \text{ and } (h, k+a) \\ \text{asymptotes: } y = \pm\frac{a}{b}(x-h) + k \end{cases}$$

If p is a nonzero real number, then the graph of each of the following equations is a parabola with vertex (h, k).

$$(x-h)^2 = 4p(y-k) \quad \begin{cases} \text{focus: } (h, k+p) \\ \text{directrix: the horizontal line } y = k - p \\ \text{axis: the vertical line } x = h \\ \text{opens upward if } p > 0, \text{ downward if } p < 0 \end{cases}$$

$$(y-k)^2 = 4p(x-h) \quad \begin{cases} \text{focus: } (h+p, k) \\ \text{directrix: the vertical line } x = h - p \\ \text{axis: the horizontal line } y = k \\ \text{opens to right if } p > 0, \text{ to left if } p < 0 \end{cases}$$

Graphing Techniques

When the equation of a conic section is in standard form, the techniques of Section 10.2 can be used to obtain its graph on a calculator.

Technology Tip

The Casio 9850 has a conic section grapher (on the main menu) that produces the graphs of equations in standard form when the various coefficients are entered.

Example 4 Graph the equation $\dfrac{(y+1)^2}{2} - \dfrac{(x-3)^2}{4} = 1$.

Solution The graph is a hyperbola centered at $(3, -1)$ which can be found in several ways.

Method 1: Solve the equation for y:

$$\frac{(y+1)^2}{2} = 1 + \frac{(x-3)^2}{4}$$

$$(y+1)^2 = 2\left[1 + \frac{(x-3)^2}{4}\right] = 2 + \frac{(x-3)^2}{2}$$

$$y + 1 = \pm\sqrt{2 + \frac{(x-3)^2}{2}}$$

$$y = \sqrt{2 + \frac{(x-3)^2}{2}} - 1 \quad \text{or} \quad y = -\sqrt{2 + \frac{(x-3)^2}{2}} - 1$$

Now graph the last two equations on the same screen. The graph of the first is the top half and the graph of the second the bottom half of the graph of the original equation.

GRAPHING EXPLORATION

Find a viewing window that shows a complete graph of the original equation.

Method 2: Use parametric equations. On page 635 we saw that the hyperbola $\dfrac{y^2}{a^2} - \dfrac{x^2}{b^2} = 1$ can be parameterized by letting $x = b\tan t$ and $y = a/\cos t\ (0 \le t \le 2\pi)$. In this case we have $x - 3$ in place of x and $y + 1$ in place of y, with $a = \sqrt{2}$ and $b = 2$. So, we use this parameterization:

$$x - 3 = 2\tan t \qquad \text{and} \qquad y + 1 = \frac{\sqrt{2}}{\cos t}$$

$$x = 2\tan t + 3 \qquad\qquad y = \frac{\sqrt{2}}{\cos t} - 1 \quad (0 \le t \le 2\pi).$$

GRAPHING EXPLORATION

Use this parameterization to graph the equation. ■

When a second-degree equation is not in standard form, the fastest way to graph it is to use the first method in Example 4, modified as in the next example.

Example 5 Graph the equation $x^2 + 8y^2 + 6x + 9y + 4 = 0$ without first putting it in standard form.

Solution Rewrite it like this:

$$8y^2 + 9y + (x^2 + 6x + 4) = 0.$$

This is a quadratic equation of the form $ay^2 + by + c = 0$, with

$$a = 8, \qquad b = 9, \qquad c = x^2 + 6x + 4$$

and hence can be solved by using the quadratic formula:

$$y = \frac{-b \pm \sqrt{b^2 - 4ac}}{2a}$$

$$y = \frac{-9 \pm \sqrt{9^2 - 4 \cdot 8 \cdot (x^2 + 6x + 4)}}{2 \cdot 8}$$

$$= \frac{-9 \pm \sqrt{81 - 32(x^2 + 6x + 4)}}{16}$$

GRAPHING EXPLORATION

Find a complete graph of the equation by graphing both of the following functions on the same screen and identify the conic:

$$y = \frac{-9 + \sqrt{81 - 32(x^2 + 6x + 4)}}{16}$$

$$y = \frac{-9 - \sqrt{81 - 32(x^2 + 6x + 4)}}{16}.$$

■

Technology Tip

TI-83/86 and Sharp 9600 users can save keystrokes by entering the first function in the exploration as y_1 and then using the RCL key to copy the text of y_1 to y_2. On TI-89, use COPY and PASTE in place of RCL. Then only one sign needs to be changed to make y_2 into the second function of the exploration.

Rotations and Second-Degree Equations

A second-degree equation in x and y is one that can be written in the form

(*) $$Ax^2 + Bxy + Cy^2 + Dx + Ey + F = 0$$

for some constants A, B, C, D, E, F, with at least one of A, B, C nonzero. Every conic section is the graph of a second-degree equation. For instance, the ellipse equation $\frac{x^2}{4} + \frac{(y-3)^2}{6} = 1$ can be rewritten as

$$12\left(\frac{x^2}{4}\right) + 12\left[\frac{(y-3)^2}{6}\right] = 12$$

$$3x^2 + 2(y-3)^2 = 12$$

$$3x^2 + 2(y^2 - 6y + 9) = 12$$

$$3x^2 + 2y^2 - 12y + 18 = 12$$

$$3x^2 + 2y^2 - 12y + 6 = 0.$$

This last equation is equation (*) with $A = 3$, $B = 0$, $C = 2$, $D = 0$, $E = -12$, and $F = 6$. Conversely, it can be shown that the graph of every second-degree equation is a conic section (possibly degenerate). When the equation has an xy term, the conic may be rotated from standard position, so that its axis or axes are not prallel to the coordinate axes.

Example 6 Graph the equation

$$3x^2 + 6xy + y^2 + x - 2y + 7 = 0.$$

Solution We first rewrite it as:

$$y^2 + 6xy - 2y + 3x^2 + x + 7 = 0$$

$$y^2 + (6x - 2)y + (3x^2 + x + 7) = 0.$$

This equation has the form $ay^2 + by + c = 0$, with $a = 1$, $b = 6x - 2$, and $c = 3x^2 + x + 7$. It can be solved with the quadratic formula:

$$y = \frac{-b \pm \sqrt{b^2 - 4ac}}{2a} = \frac{-(6x-2) \pm \sqrt{(6x-2)^2 - 4 \cdot 1 \cdot (3x^2 + x + 7)}}{2 \cdot 1}.$$

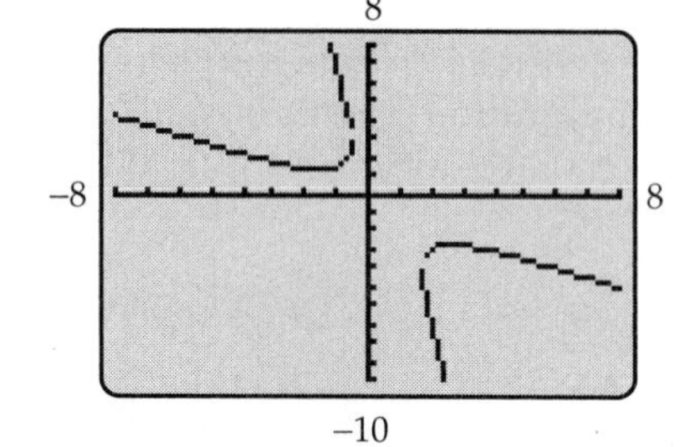

Figure 10–36

The top half of the graph is obtained by graphing

$$y = \frac{-6x + 2 + \sqrt{(6x-2)^2 - 4(3x^2 + x + 7)}}{2}$$

and the bottom half by graphing

$$y = \frac{-6x + 2 - \sqrt{(6x-2)^2 - 4(3x^2 + x + 7)}}{2}.$$

The graph is a hyperbola whose focal axis tilts upward to the left, as shown in Figure 10–36. ■

The Discriminant

The following fact, whose proof is outlined in Exercise 15 of Excursion 10.3.A, makes it easy to identify the graphs of second-degree equations without graphing them.

Graphs of Second-Degree Equations

If the equation

$$Ax^2 + Bxy + Cy^2 + Dx + Ey + F = 0 \quad (A, B, C \text{ not all } 0)$$

has a graph, then that graph is

a circle or an ellipse (or a point), if $B^2 - 4AC < 0$;

a parabola (or a line or two parallel lines), if $B^2 - 4AC = 0$;

a hyperbola (or two intersecting lines), if $B^2 - 4AC > 0$.

The expression $B^2 - 4AC$ is called the **discriminant** of the equation.

Example 7 Identify the graph of

$$2x^2 - 4xy + 3y^2 + 5x + 6y - 8 = 0.$$

Solution We compute the discriminant with $A = 2$, $B = -4$, and $C = 3$:

$$B^2 - 4AC = (-4)^2 - 4 \cdot 2 \cdot 3 = 16 - 24 = -8.$$

Hence, the graph is an ellipse (possibly a circle or a single point). The graph can be found as above by using the quadratic formula to solve

$$3y^2 + (-4x + 6)y + (2x^2 + 5x - 8) = 0$$

and graphing both solutions on the same screen:

$$\begin{aligned} y &= \frac{-b \pm \sqrt{b^2 - 4ac}}{2a} \\ &= \frac{-(-4x + 6) \pm \sqrt{(-4x + 6)^2 - 4 \cdot 3 \cdot (2x^2 + 5x - 8)}}{2 \cdot 3}. \end{aligned}$$

GRAPHING EXPLORATION

Find a viewing window that shows a complete graph of the equation. In what direction does the major axis run? ■

Example 8 The discriminant of

$$3x^2 + 6xy + 3y^2 + 13x + 9y + 53 = 0$$

is $B^2 - 4AC = 6^2 - 4 \cdot 3 \cdot 3 = 0$. Hence, the graph is a parabola (or a line or parallel lines in the degenerate case).

GRAPHING EXPLORATION

Find a viewing window that shows a complete graph of the equation. ■

Applications

The long-range navigation system (LORAN) uses hyperbolas to enable a ship to determine its exact location by radio, as illustrated in the following example.

Example 9 Three LORAN radio transmitters Q, P, and R are located 200 miles apart along a straight line and simultaneously transmit signals at regular intervals. These signals travel at a speed of 980 feet/microsecond.

A ship S receives a signal from P and 305 microseconds later a signal from R. It also receives a signal from Q 528 microseconds after the one from P. Determine the ship's location.

Solution Take the line through the LORAN stations as the x-axis, with the origin located midway between Q and P, so that the situation looks like Figure 10–37.

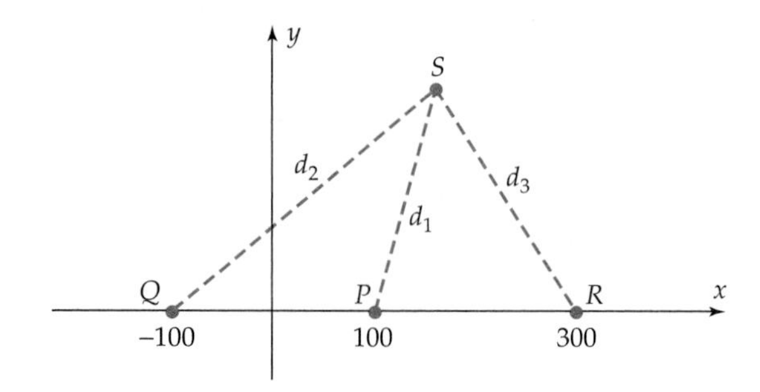

Figure 10–37

If the signal takes t microseconds to go from P to S, then

$$d_1 = 980t \qquad \text{and} \qquad d_2 = 980(t + 528)$$

so that

$$|d_1 - d_2| = |980t - 980(t + 528)| = 980 \cdot 528 = 517{,}440 \text{ feet.}$$

Since one mile is 5280 feet, this means that

$$|d_1 - d_2| = 517{,}440/5{,}280 \text{ miles} = 98 \text{ miles.}$$

In other words,

$$|(\text{distance from } P \text{ to } S) - (\text{distance from } Q \text{ to } S)| = |d_1 - d_2| = 98.$$

This is precisely the situation described in the definition of "hyperbola" on page 628: S is on the hyperbola with foci $P = (100, 0)$, $Q = (-100, 0)$ and distance difference $r = 98$. This hyperbola has an equation of the form

$$\frac{x^2}{a^2} - \frac{y^2}{b^2} = 1,$$

where $(\pm a, 0)$ are the vertices, $(\pm c, 0) = (\pm 100, 0)$ are the foci and $c^2 = a^2 + b^2$. Figure 10–38 and the fact that the vertex $(a, 0)$ is on the hyperbola show that

$$\begin{aligned}
|[\text{distance from } P \text{ to } (a, 0)] - [\text{distance } Q \text{ to } (a, 0)]| &= r = 98 \\
|(100 - a) - (100 + a)| &= 98 \\
|-2a| &= 98 \\
|a| &= 49.
\end{aligned}$$

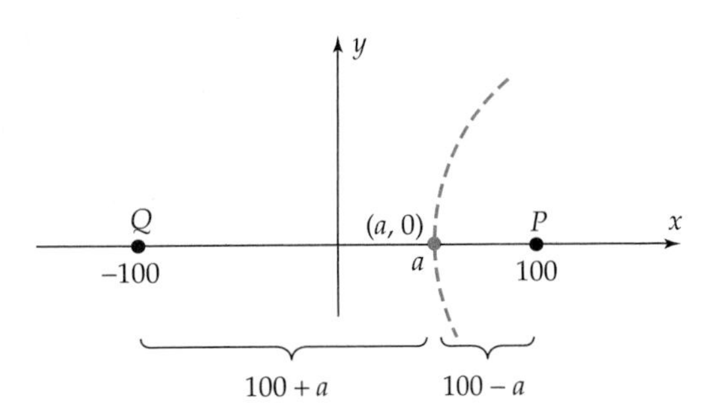

Figure 10–38

Consequently, $a^2 = 49^2 = 2401$ and hence $b^2 = c^2 - a^2 = 100^2 - 49^2 = 7599$. Thus the ship lies on the hyperbola

$$(*) \qquad \frac{x^2}{2401} - \frac{y^2}{7599} = 1.$$

A similar argument using P and R as foci shows that the ship also lies on the hyperbola with foci $P = (100, 0)$ and $R = (300, 0)$ and center $(200, 0)$, whose distance difference r is

$$|d_1 - d_3| = 980 \cdot 305 = 298{,}900 \text{ feet} \approx 56.61 \text{ miles.}$$

As before you can verify that $a = 56.61/2 = 28.305$ and hence $a^2 = 28.305^2 = 801.17$. This hyperbola has center (200, 0) and its foci are $(200 - c, k) = (100, 0)$ and $(200 + c, k) = (300, 0)$, which implies that $c = 100$. Hence, $b^2 = c^2 - a^2 = 100^2 - 801.17 = 9198.83$ and the ship also lies on the hyperbola

$$(**) \qquad \frac{(x - 200)^2}{801.17} - \frac{y^2}{9198.83} = 1.$$

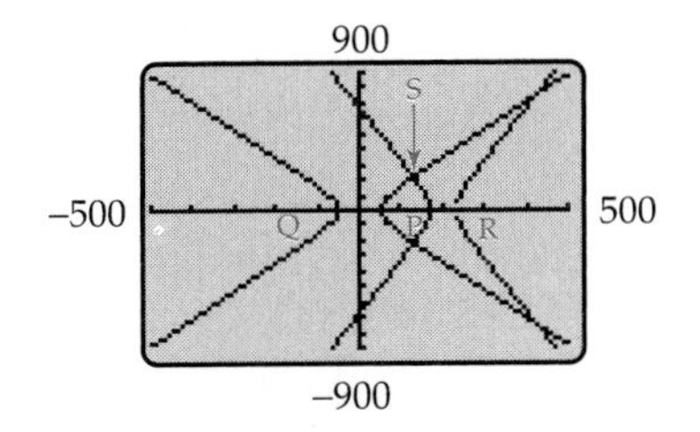

Figure 10–39

Since the ship lies on both hyperbolas, its coordinates are solutions of both the equations (*) and (**). They can be found algebraically by solving each of the equations for y^2, setting the results equal, and solving for x. They can be found geometrically by graphing both hyperbolas and finding the intersection point. As shown in Figure 10–39, there are actually four points of intersection. However, the two below the x-axis represent points on land in our situation. Furthermore, since the signal from P was received first, the ship is closest to P. So it is located at the point S in Figure 10–39. A graphical intersection finder shows that this point is approximately (130.48, 215.14), where the coordinates are in miles from the origin. ■

Exercises 10.3

In Exercises 1–16, find the equation of the conic sections satisfying the given conditions.

1. Ellipse with center (2, 3); endpoints of major and minor axes: (2, −1), (0, 3), (2, 7), (4, 3).
2. Ellipse with center (−5, 2); endpoints of major and minor axes: (0, 2), (−5, 17), (−10, 2), (−5, −13).
3. Ellipse with center (7, −4); foci on the line $x = 7$; major axis of length 12; minor axis of length 5.
4. Ellipse with center (−3, −9); foci on the line $y = -9$; major axis of length 15; minor axis of length 7.
5. Hyperbola with center (−2, 3); vertex (−2, 1); passing through $(-2 + 3\sqrt{10}, 11)$.
6. Hyperbola with center (−5, 1); vertex (−3, 1); passing through $(-1, 1 - 4\sqrt{3})$.
7. Hyperbola with center (4, 2); vertex (7, 2); asymptote $3y = 4x - 10$.
8. Hyperbola with center (−3, −5); vertex (−3, 0); asymptote $6y = 5x - 15$.
9. Parabola with vertex (1, 0); axis $x = 1$; (2, 13) on graph.
10. Parabola with vertex (−3, 0); axis $y = 0$; (−1, 1) on graph.
11. Parabola with vertex (2, 1); axis $y = 1$; (5, 0) on graph.
12. Parabola with vertex (1, −3); axis $y = -3$; (−1, −4) on graph.
13. Ellipse with center (3, −2); passing through (3, −6) and (9, −2).
14. Ellipse with center (2, 5); passing through (2, 4) and (−3, 5).
15. Parabola with vertex (−3, −2) and focus (−47/16, −2).
16. Parabola with vertex (−5, −5) and focus (−5, −99/20).

In Exercises 17–22, assume that the graph of the equation is a nondegenerate conic section. Without graphing, determine whether the graph is a circle, ellipse, hyperbola, or parabola.

17. $x^2 - 2xy + 3y^2 - 1 = 0$
18. $xy - 1 = 0$
19. $x^2 + 2xy + y^2 + 2\sqrt{2}x - 2\sqrt{2}y = 0$
20. $2x^2 - 4xy + 5y^2 - 6 = 0$
21. $17x^2 - 48xy + 31y^2 + 50 = 0$
22. $2x^2 - 4xy - 2y^2 + 3x + 5y - 10 = 0$

In Exercises 23–34, find parametric equations for the conic section whose equation is given and use these parametric equations to find a complete graph.

23. $\dfrac{(x-1)^2}{4} + \dfrac{(y-5)^2}{9} = 1$
24. $\dfrac{(x-2)^2}{16} + \dfrac{(y+3)^2}{12} = 1$
25. $\dfrac{(x+1)^2}{16} + \dfrac{(y-4)^2}{8} = 1$
26. $\dfrac{(x+5)^2}{4} + \dfrac{(y+2)^2}{12} = 1$
27. $y = 4(x-1)^2 + 2$
28. $y = 3(x-2)^2 - 3$
29. $x = 2(y-2)^2$
30. $x = -3(y-1)^2 - 2$
31. $\dfrac{(y+3)^2}{25} - \dfrac{(x+1)^2}{16} = 1$
32. $\dfrac{(y+1)^2}{9} - \dfrac{(x-1)^2}{25} = 1$
33. $\dfrac{(x+3)^2}{1} - \dfrac{(y-2)^2}{4} = 1$
34. $\dfrac{(y+5)^2}{9} - \dfrac{(x-2)^2}{1} = 1$

In Exercises 35–52, use the discriminant to identify the conic section whose equation is given and find a viewing window that shows a complete graph.

35. $9x^2 + 4y^2 + 54x - 8y + 49 = 0$
36. $4x^2 + 5y^2 - 8x + 30y + 29 = 0$
37. $4y^2 - x^2 + 6x - 24y + 11 = 0$
38. $x^2 - 16y^2 = 0$
39. $3y^2 - x - 2y + 1 = 0$
40. $x^2 - 6x + y + 5 = 0$
41. $41x^2 - 24xy + 34y^2 - 25 = 0$
42. $x^2 + 2\sqrt{3}xy + 3y^2 + 8\sqrt{3}x - 8y + 32 = 0$
43. $17x^2 - 48xy + 31y^2 + 49 = 0$
44. $52x^2 - 72xy + 73y^2 = 200$
45. $9x^2 + 24xy + 16y^2 + 90x - 130y = 0$
46. $x^2 + 10xy + y^2 + 1 = 0$
47. $23x^2 + 26\sqrt{3}xy - 3y^2 - 16x + 16\sqrt{3}y + 128 = 0$
48. $x^2 + 2xy + y^2 + 12\sqrt{2}x - 12\sqrt{2}y = 0$
49. $17x^2 - 12xy + 8y^2 - 80 = 0$
50. $11x^2 - 24xy + 4y^2 + 30x + 40y - 45 = 0$

51. $3x^2 + 2\sqrt{3}xy + y^2 + 4x - 4\sqrt{3}y - 16 = 0$

52. $3x^2 + 2\sqrt{2}xy + 2y^2 - 12 = 0$

In Exercises 53 and 54, find the equations of two distinct ellipses satisfying the given conditions.

53. Center at $(-5, 3)$; major axis of length 14; minor axis of length 8.

54. Center at $(2, -6)$; major axis of length 15; minor axis of length 6.

55. Show that the asymptotes of the hyperbola $\frac{x^2}{a^2} - \frac{y^2}{a^2} = 1$ are perpendicular to each other.

56. Find a number k such that $(-2, 1)$ is on the graph of $3x^2 + ky^2 = 4$. Then graph the equation.

57. Find the number b such that the vertex of the parabola $y = x^2 + bx + c$ lies on the y-axis.

58. Find the number d such that the parabola $(y + 1)^2 = dx + 4$ passes through $(-6, 3)$.

59. Find the points of intersection of the parabola $4y^2 + 4y = 5x - 12$ and the line $x = 9$.

60. Find the points of intersection of the parabola $4x^2 - 8x = 2y + 5$ and the line $y = 15$.

61. Two listening stations 1 mile apart record an explosion. One microphone receives the sound 2 seconds after the other does. Use the line through the microphones as the x-axis, with the origin midway between the microphones, and the fact that sound travels at 1100 feet/second to find the equation of the hyperbola on which the explosion is located. Can you determine the exact location of the explosion?

62. Two transmission stations P and Q are located 200 miles apart on a straight shore line. A ship 50 miles from shore is moving parallel to the shore line. A signal from Q reaches the ship 400 microseconds after a signal from P. If the signals travel at 980 feet/microsecond, find the location of the ship (in terms of miles) in the coordinate system with x-axis through P and Q, and origin midway between them.

10.3.A EXCURSION Rotation of Axes

The graph of an equation of the form

$$Ax^2 + Bxy + Cy^2 + Dx + Ey + F = 0,$$

with $B \neq 0$, is a conic section that is rotated so that its axes are not parallel to the coordinate axes, as in Figure 10–40. Although the graph is readily obtained with a calculator (as in Examples 6–8 of Section 10.3), we cannot read off useful information about the center, vertices, etc., from the equation, as we can with an equation in standard form. However, if we replace the xy coordinate system by a new coordinate system, as indicated by the colored uv axes in Figure 10–40, then the conic is not rotated in the new system and has a uv equation in standard form that will provide the desired information.

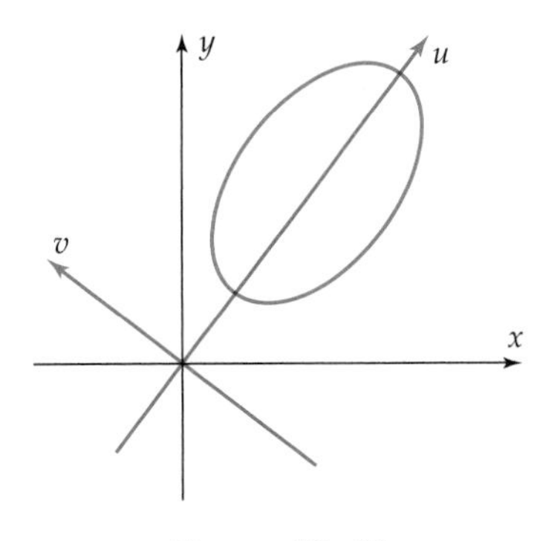

Figure 10–40

In order to use this approach, we must first determine the relationship between the xy coordinates of a point and its coordinates in the uv system. Suppose the uv coordinate system is obtained by rotating the xy axes about the origin, counterclockwise through an angle θ.* If a point P

*All rotations in this section are counterclockwise about the origin, with $0° < \theta < 90°$.

has coordinates (x, y) in the xy system, we can find its coordinates (u, v) in the rotated coordinate system by using Figure 10–41:

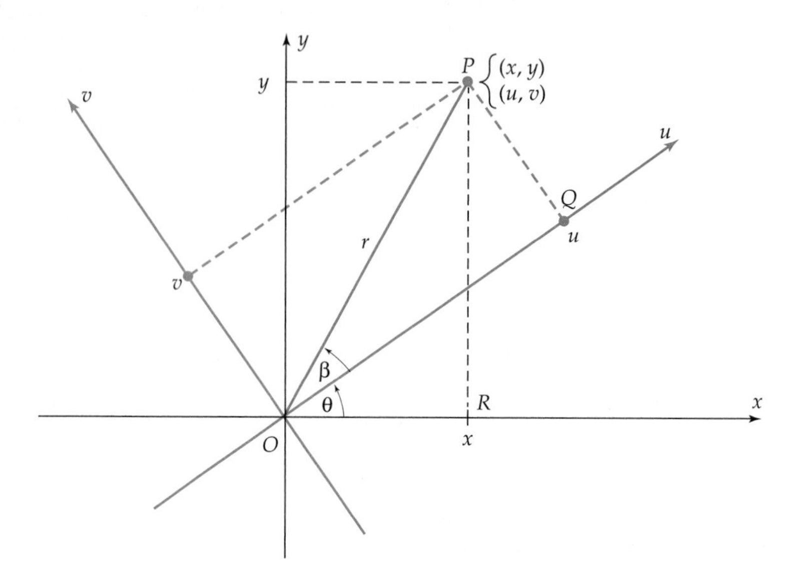

Figure 10–41

Triangle OPQ shows that

$$\cos\beta = \frac{OQ}{OP} = \frac{u}{r} \quad \text{and} \quad \sin\beta = \frac{PQ}{OP} = \frac{v}{r}$$

Therefore

$$u = r\cos\beta \quad \text{and} \quad v = r\sin\beta$$

Similarly, triangle OPR shows that

$$\cos(\theta + \beta) = \frac{OR}{OP} = \frac{x}{r} \quad \text{and} \quad \sin(\theta + \beta) = \frac{PR}{OP} = \frac{y}{r}$$

so that

$$x = r\cos(\theta + \beta) \quad \text{and} \quad y = r\sin(\theta + \beta)$$

Applying the addition identity for cosine (page 518) shows that

$$\begin{aligned} x &= r\cos(\theta + \beta) \\ &= r(\cos\theta\cos\beta - \sin\theta\sin\beta) \\ &= (r\cos\beta)\cos\theta - (r\sin\beta)\sin\theta \\ &= u\cos\theta - v\sin\theta \end{aligned}$$

A similar argument with $y = r\sin(\theta + \beta)$ and the addition identity for sine leads to this result:

The Rotation Equations

If the xy coordinate axes are rotated through an angle θ to produce the uv coordinate axes, then the coordinates (x, y) and (u, v) of a point are related by these equations:

$$x = u\cos\theta - v\sin\theta$$

$$y = u\sin\theta + v\cos\theta$$

Example 1 If the xy axes are rotated 30°, find the equation relative to the uv axes of the graph of

$$3x^2 + 2\sqrt{3}xy + y^2 + x - \sqrt{3}y = 0$$

and graph the equation.

Solution Since $\sin 30° = 1/2$ and $\cos 30° = \sqrt{3}/2$, the rotation equations are

$$x = u\cos 30° - v\sin 30° = \frac{\sqrt{3}}{2}u - \frac{1}{2}v$$

$$y = u\sin 30° + v\cos 30° = \frac{1}{2}u + \frac{\sqrt{3}}{2}v$$

Substitute these expressions in the original equation:

$$3x^2 + 2\sqrt{3}xy + y^2 + x - \sqrt{3}y = 0$$

$$3\left(\frac{\sqrt{3}}{2}u - \frac{1}{2}v\right)^2 + 2\sqrt{3}\left(\frac{\sqrt{3}}{2}u - \frac{1}{2}v\right)\left(\frac{1}{2}u + \frac{\sqrt{3}}{2}v\right) + \left(\frac{1}{2}u + \frac{\sqrt{3}}{2}v\right)^2 + \left(\frac{\sqrt{3}}{2}u - \frac{1}{2}v\right) - \sqrt{3}\left(\frac{1}{2}u + \frac{\sqrt{3}}{2}v\right) = 0$$

$$3\left(\frac{3}{4}u^2 - \frac{\sqrt{3}}{2}uv + \frac{1}{4}v^2\right) + 2\sqrt{3}\left(\frac{\sqrt{3}}{4}u^2 + \frac{1}{2}uv - \frac{\sqrt{3}}{4}v^2\right) + \left(\frac{1}{4}u^2 + \frac{\sqrt{3}}{2}uv + \frac{3}{4}v^2\right) + \left(\frac{\sqrt{3}}{2}u - \frac{1}{2}v\right) - \sqrt{3}\left(\frac{1}{2}u + \frac{\sqrt{3}}{2}v\right) = 0$$

Verify that the last equation simplifies as

$$4u^2 - 2v = 0, \qquad \text{or equivalently,} \qquad v = 2u^2$$

In the uv system, $v = 2u^2$ is the equation of an upward-opening parabola with vertex at $(0, 0)$ as shown in Figure 10–42 on the next page. ■

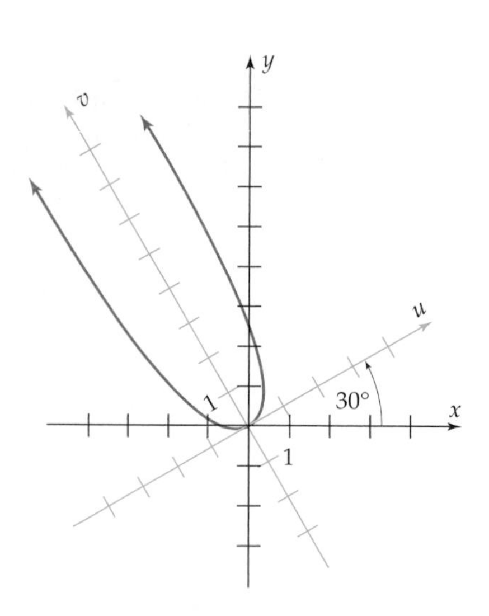

Figure 10–42

Rotating the axes in the preceding example changed the original equation, which included an xy term, to an equation that had no uv term. That enabled us to identify the graph of the new equation as a conic section. This can be done for any second-degree equation by choosing an angle of rotation that will eliminate the xy term. The choice of the angle is determined by this fact, which is proved in Exercise 13:

Rotation Angle

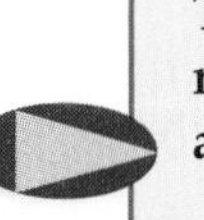

The equation $Ax^2 + Bxy + Cy^2 + Dx + Ey + F = 0$ ($B \neq 0$) can be rewritten as $A'u^2 + C'v^2 + D'u + E'v + F' = 0$ by rotating the xy axes through an angle θ such that

$$\cot 2\theta = \frac{A - C}{B} \qquad (0^\circ < \theta < 90^\circ)$$

and using the rotation equations.

Example 2 What angle of rotation will eliminate the xy term in the equation $153x^2 + 192xy + 97y^2 - 1710x - 1470y + 5625 = 0$, and what are the rotation equations in this case?

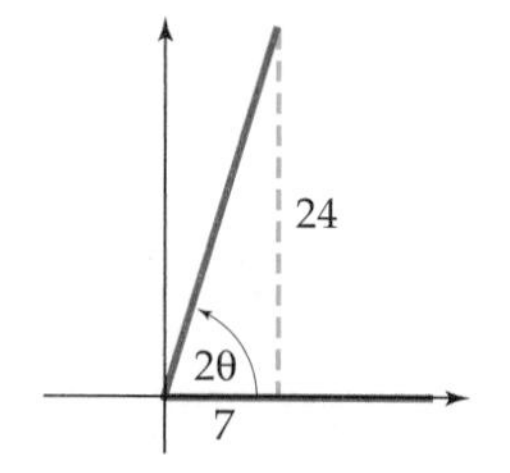

Figure 10–43

Solution According to the fact in the box with $A = 153$, $B = 192$, and $C = 97$, we should rotate through an angle of θ, where $\cot 2\theta = \frac{153 - 197}{192} = \frac{56}{192} = \frac{7}{24}$. Since $0^\circ < 2\theta < 180^\circ$ and $\cot 2\theta$ is positive, the terminal side of the angle 2θ lies in the first quadrant, as shown in Figure 10–43. The hypotenuse of this triangle has length $\sqrt{7^2 + 24^2} =$

$\sqrt{625} = 25$. Hence $\cos 2\theta = 7/25$. The half-angle identities (page 533) show that

$$\sin\theta = \sqrt{\frac{1 - \cos 2\theta}{2}} = \sqrt{\frac{1 - 7/25}{2}} = \sqrt{\frac{9}{25}} = \frac{3}{5}$$

$$\cos\theta = \sqrt{\frac{1 + \cos 2\theta}{2}} = \sqrt{\frac{1 + 7/25}{2}} = \sqrt{\frac{16}{25}} = \frac{4}{5}$$

Using $\sin\theta = 3/5$ and the SIN^{-1} key on a calculator, we find that the angle θ of rotation is approximately 36.87°. The rotation equations are

$$x = u\cos\theta - v\sin\theta = \frac{4}{5}u - \frac{3}{5}v$$

$$y = u\sin\theta + v\cos\theta = \frac{3}{5}u + \frac{4}{5}v. \quad ■$$

Example 3 Graph the equation without using a calculator:

$$153x^2 + 192xy + 97y^2 - 1710x - 1470y + 5625 = 0.$$

Solution The angle θ and the rotation equations for eliminating the xy term were found in the preceding example. Substitute the rotation equations in the given equation:

$$153x^2 + 192xy + 97y^2 - 1710x - 1470y + 5625 = 0$$

$$153\left(\frac{4}{5}u - \frac{3}{5}v\right)^2 + 192\left(\frac{4}{5}u - \frac{3}{5}v\right)\left(\frac{3}{5}u + \frac{4}{5}v\right)$$
$$+ 97\left(\frac{3}{5}u + \frac{4}{5}v\right)^2 - 1710\left(\frac{4}{5}u - \frac{3}{5}v\right) - 1470\left(\frac{3}{5}u + \frac{4}{5}v\right) + 5625 = 0$$

$$153\left(\frac{16}{25}u^2 - \frac{24}{25}uv + \frac{9}{25}v^2\right) + 192\left(\frac{12}{25}u^2 + \frac{7}{25}uv - \frac{12}{25}v^2\right)$$
$$+ 97\left(\frac{9}{25}u^2 + \frac{24}{25}uv + \frac{16}{25}v^2\right) - 2250u - 150v + 5625 = 0$$

$$225u^2 + 25v^2 - 2250u - 150v + 5625 = 0$$

$$9u^2 + v^2 - 90u - 6v + 225 = 0$$

$$9(u^2 - 10u) + (v^2 - 6v) = -225$$

Finally, complete the square in u and v (adding the appropriate amounts to the right side so as not to change the equation):

$$9(u^2 - 10u + 25) + (v^2 - 6v + 9) = -225 + 9\cdot 25 + 9$$

$$9(u - 5)^2 + (v - 3)^2 = 9$$

$$\frac{(u - 5)^2}{1} + \frac{(v - 3)^2}{9} = 1.$$

Therefore the graph is an ellipse centered at (5, 3) in the uv coordinate system, as shown in Figure 10–44. ■

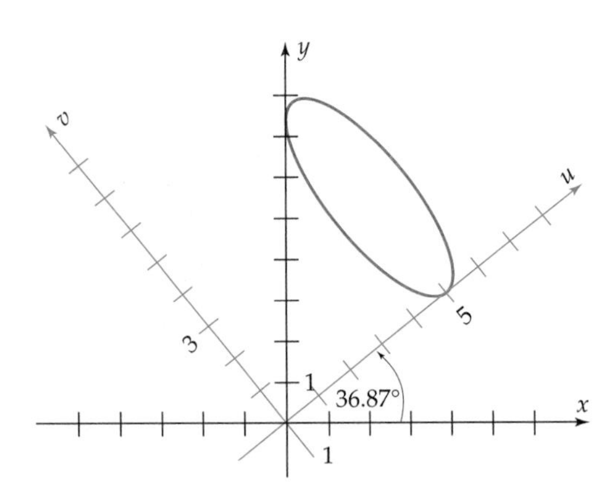

Figure 10–44

Exercises 10.3.A

In Exercises 1–4, find the new coordinates of the point when the coordinate axes are rotated through the given angle.

1. $(3, 2)$; $\theta = 45°$

2. $(-2, 4)$; $\theta = 60°$

3. $(1, 0)$; $\theta = 30°$

4. $(3, 3)$; $\sin\theta = 5/13$

In Exercises 5–8, rotate the axes through the given angle to form the uv coordinate system. Express the given equation in terms of the uv coordinate system.

5. $\theta = 45°$; $xy = 1$

6. $\theta = 45°$; $13x^2 + 10xy + 13y^2 = 72$

7. $\theta = 30°$; $7x^2 - 6\sqrt{3}xy + 13y^2 - 16 = 0$

8. $\sin\theta = 1/\sqrt{5}$; $x^2 - 4xy + 4y^2 + 5\sqrt{5}y + 1 = 0$

In Exercises 9–12 find the angle of rotation that will eliminate the xy term of the equation and list the rotation equations in this case.

9. $41x^2 - 24xy + 34y^2 - 25 = 0$

10. $x^2 + 2\sqrt{3}xy + 3y^2 + 8\sqrt{3}x - 8y + 32 = 0$

11. $17x^2 - 48xy + 31y^2 + 49 = 0$

12. $52x^2 - 72xy + 73y^2 = 200$

Thinkers

13. (a) Given an equation $Ax^2 + Bxy + Cy^2 + Dx + Ey + F = 0$, with $B \neq 0$, and an angle θ, use the rotation equations in the box on page 651 to rewrite the equation in the form $A'u^2 + B'uv + C'v^2 + D'u + E'v + F' = 0$, where $A', \ldots, F'$ are expressions involving $\sin\theta$, $\cos\theta$, and the constants $A, \ldots, F$.

(b) Verify that $B' = 2(C - A)\sin\theta\cos\theta + B(\cos^2\theta - \sin^2\theta)$.

(c) Use the double-angle identities to show that $B' = (C - A)\sin 2\theta + B\cos 2\theta$.

(d) If θ is chosen so that $\cos 2\theta = (A - C)/B$, show that $B' = 0$. This proves the statement in the box on page 652.

14. Assume that the graph of $A'u^2 + C'v^2 + D'u + E'v + F' = 0$ (with at least one of A', C' nonzero) in the uv coordinate system is a nondegenerate conic. Show that its graph is an ellipse if $A'C' > 0$ (A' and C' have the same sign), a hyperbola if $A'C' < 0$ (A' and C' have opposite signs), or a parabola if $A'C' = 0$.

15. Assume the graph of $Ax^2 + Bxy + Cy^2 + Dx + Ey + F = 0$ is a nondegenerate conic section. Prove the statement in the box on page 644 as follows.

(a) In Exercise 13(a) show that $(B')^2 - 4A'C' = B^2 - 4AC$.

(b) Assume θ has been chosen so that $B' = 0$. Use Exercise 14 to show that the graph of the original equation is an ellipse if $B^2 - 4AC < 0$, a parabola if $B^2 - 4AC = 0$, and a hyperbola if $B^2 - 4AC > 0$.

10.4 Polar Coordinates

In the past we have used a rectangular coordinate system in the plane, based on two perpendicular coordinate axes. Now we introduce another coordinate system for the plane, based on angles.

Choose a point O in the plane (called the **origin** or **pole**) and a half-line extending from O (called the **polar axis**). Choose a unit of length. A point P in the plane is given the coordinates (r, θ), where r is the length of segment OP and θ is the angle with the polar axis as initial side, vertex O, and terminal side OP, as shown in Figure 10–45.

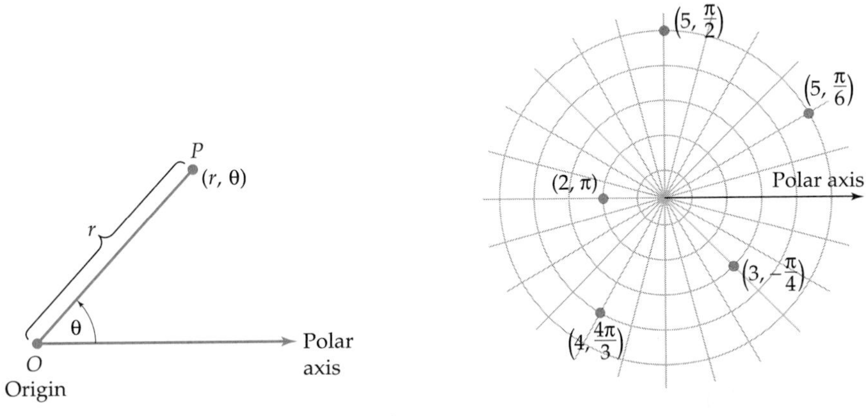

Figure 10–45 **Figure 10–46**

We shall usually measure the angle θ in radians; it may be either positive or negative, depending on whether it is generated by a clockwise or counterclockwise rotation. Some typical points are shown in Figure 10–46, which also illustrates the "circular grid" that a polar coordinate system imposes on the plane.

The polar coordinates of a point P are *not* unique. For instance, angles of radian measure $\pi/3$, $7\pi/3$, and $-5\pi/3$ all have the same terminal side,* so $(2, \pi/3)$, $(2, 7\pi/3)$, and $(2, -5\pi/3)$ represent the same point (Figure 10–47). Furthermore, we shall consider the coordinates of the origin to be $(0, \theta)$, where θ is *any* angle.

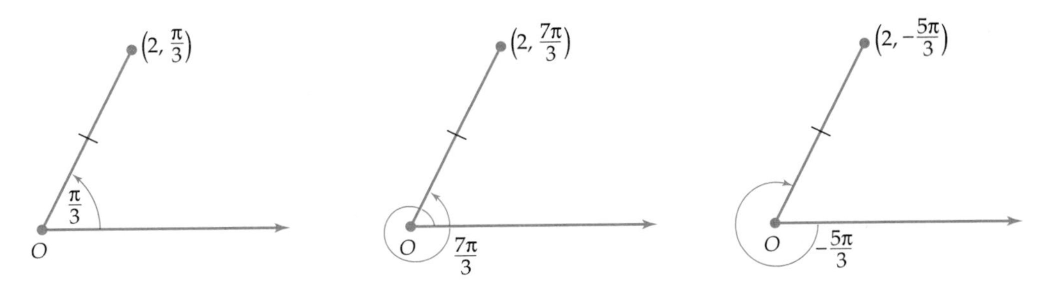

Figure 10–47

*Because $\frac{7\pi}{3} = \frac{\pi}{3} + 2\pi$ and $\frac{-5\pi}{3} = \frac{\pi}{3} - 2\pi$.

Negative values for the first coordinate will be allowed according to this convention: For each positive r, the point $(-r, \theta)$ lies on the straight line containing the terminal side of θ, at distance r from the origin, on the *opposite* side of the origin from the point (r, θ), as shown in Figure 10–48.

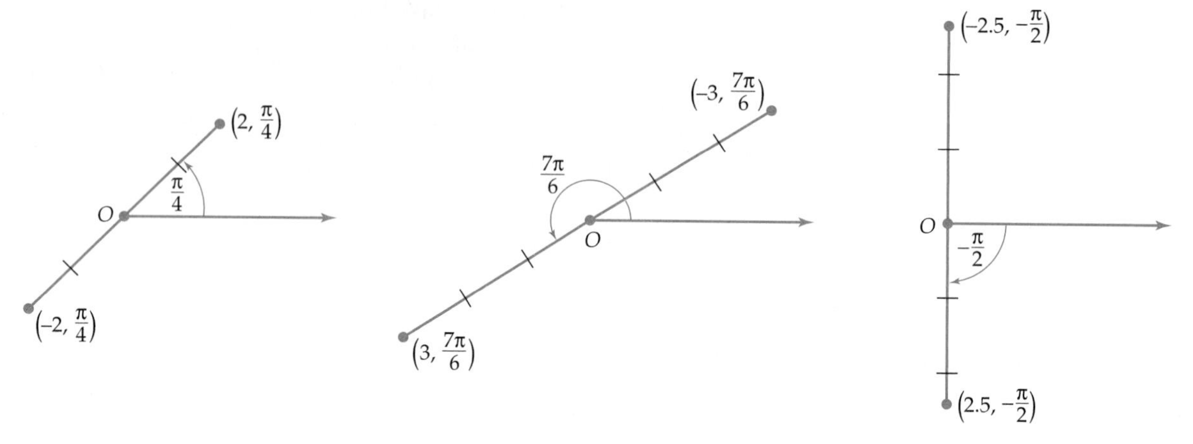

Figure 10–48

It is sometimes convenient to use both a rectangular and a polar coordinate system in the plane, with the polar axis coinciding with the positive x-axis. Then the y-axis is the polar line $\theta = \pi/2$. Suppose P has rectangular coordinates (x, y) and polar coordinates (r, θ), with $r > 0$, as in Figure 10–49.

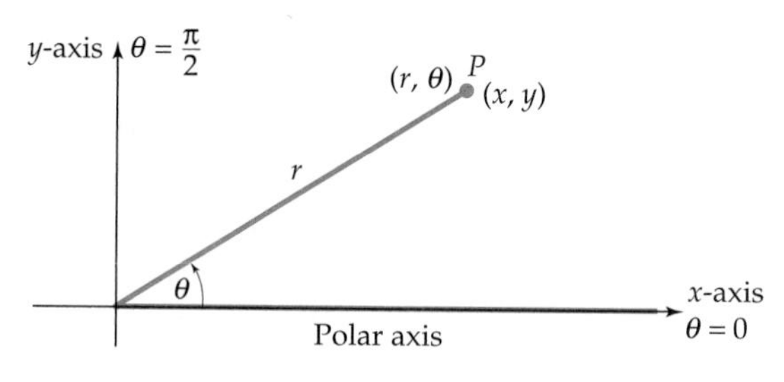

Figure 10–49

Since r is the distance from (x, y) to $(0, 0)$, the distance formula shows that $r = \sqrt{x^2 + y^2}$, and hence $r^2 = x^2 + y^2$. The point-in-the-plane description of the trigonometric functions shows that

$$\cos\theta = \frac{x}{r} \qquad \sin\theta = \frac{y}{r} \qquad \tan\theta = \frac{y}{x}.$$

Solving the first two equations for x and y, we obtain the relationship between polar and rectangular coordinates:*

*The conclusions in the next box are also true when $r < 0$ (Exercise 50).

Coordinate Conversion Formulas

If a point has rectangular coordinates (x, y) and polar coordinates (r, θ), then

$$x = r\cos\theta \quad \text{and} \quad y = r\sin\theta$$

$$r^2 = x^2 + y^2 \quad \text{and} \quad \tan\theta = \frac{y}{x}.$$

Example 1 Convert each of the following points in polar coordinates to rectangular coordinates:

(a) $(2, \pi/6)$ (b) $(3, 4)$.

Solution

Technology Tip

Keys to convert from rectangular to polar coordinates, or vice versa, are in the TI-83 ANGLE menu, in the OPS submenu of the TI-86 VECTOR menu, in the ANGLE submenu of the TI-89 MATH menu, in the CONV submenu of the Sharp 9600 MATH menu, and in the ANGLE submenu of the Casio 9850 OPTN menu. Conversion programs for HP-38 are in the Program Appendix.

(a) Apply the first set of equations in the box with $r = 2$ and $\theta = \pi/6$:

$$x = 2\cos\frac{\pi}{6} = 2\cdot\frac{\sqrt{3}}{2} = \sqrt{3} \quad \text{and} \quad y = 2\sin\frac{\pi}{6} = 2\cdot\frac{1}{2} = 1.$$

So the rectangular coordinates are $(\sqrt{3}, 1)$.

(b) The point with polar coordinates $(3, 4)$ has $r = 3$ and $\theta = 4$ radians. Therefore, its rectangular coordinates are

$$(3\cos 4, 3\sin 4) \approx (-1.9609, -2.2704). \quad \blacksquare$$

Example 2 Convert each of the following points in rectangular coordinates to polar coordinates:

(a) $(2, -2)$ (b) $(3, 5)$ (c) $(-2, 4)$.

Solution

(a) The second set of equations in the box, with $x = 2$, $y = -2$, shows that

$$r = \sqrt{2^2 + (-2)^2} = \sqrt{8} = 2\sqrt{2} \quad \text{and} \quad \tan\theta = -2/2 = -1.$$

We must find an angle θ whose terminal side passes through $(2, -2)$ and whose tangent is -1. Two of the many possibilities are $\theta = -\pi/4$ and $\theta = 7\pi/4$. So $(2\sqrt{2}, -\pi/4)$ is one pair of polar coordinates and $(2\sqrt{2}, 7\pi/4)$ is another.

(b) Applying the second set of equations in the box, with $x = 3$, $y = 5$, we have

$$r = \sqrt{3^2 + 5^2} = \sqrt{34} \quad \text{and} \quad \tan\theta = 5/3.$$

The TAN^{-1} key on a calculator shows that $\theta \approx 1.0304$ radians is an angle between 0 and $\pi/2$ with tangent $5/3$. Since $(3, 5)$ is in the first quadrant, one pair of (approximate) polar coordinates is $(\sqrt{34}, 1.0304)$.

(c) In this case $r = \sqrt{(-2)^2 + 4^2} = \sqrt{20} = 2\sqrt{5}$ and $\tan\theta = 4/(-2) = -2$. The TAN^{-1} key shows that $\theta \approx -1.1071$ is an angle between

$-\pi/2$ and 0 with tangent -2. However, we want an angle between $\pi/2$ and π with tangent -2 because $(-2, 4)$ is in the second quadrant. The tangent function has period π; hence,

$$-2 = \tan(-1.1071) = \tan(-1.1071 + \pi),$$

with $-1.1071 + \pi \approx 2.0344$ an angle between $\pi/2$ and π. Therefore, one pair of polar coordinates is $\left(2\sqrt{5}, 2.0344\right)$. ■

Polar Graphs

The graphs of a few polar coordinate equations can be easily determined from the appropriate definitions.

Example 3 The graph of the equation $r = 3$* consists of all points (r, θ) with first coordinate 3, that is, all points whose distance from the origin is 3. So the graph is a circle with center O and radius 3 (Figure 10–50). ■

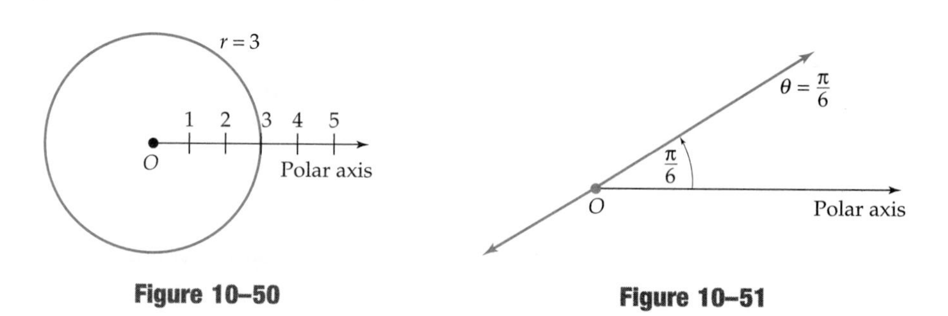

Figure 10–50

Figure 10–51

Example 4 The graph of $\theta = \pi/6$ consists of all points $(r, \pi/6)$. If $r \geq 0$, then $(r, \pi/6)$ lies on the terminal side of an angle of $\pi/6$ radians, whose initial side is the polar axis. A point $(r, \pi/6)$, with $r < 0$, lies on the extension of this terminal side across the origin. So, the graph is the straight line in Figure 10–51. ■

Some polar graphs can be sketched by hand by using basic facts about trigonometric functions.

Example 5 Graph $r = 1 + \sin\theta$.

Solution Remember the behavior of $\sin\theta$ between 0 and 2π:

*Every equation here is understood to involve two variables, but one may have coefficient 0; in this case, $r = 3 + 0\cdot\theta$. An analogous situation occurs in rectangular coordinates with equations such as $y = 3$ or $x = -2$.

As θ increases from 0 to $\pi/2$, $\sin\theta$ increases from 0 to 1. So $r = 1 + \sin\theta$ increases from 1 to 2.

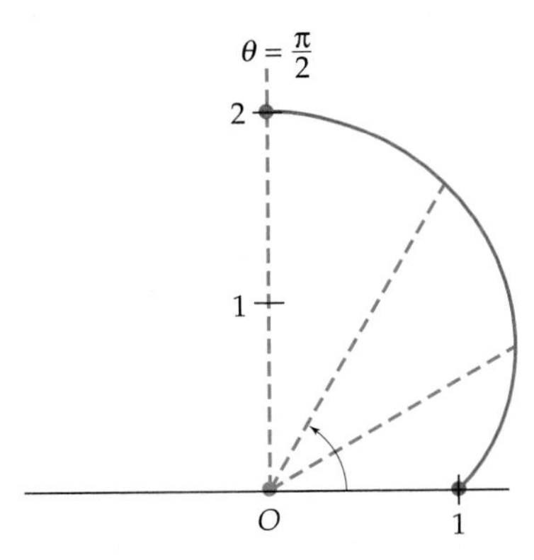

As θ increases from $\pi/2$ to π, $\sin\theta$ decreases from 1 to 0. So $r = 1 + \sin\theta$ decreases from 2 to 1.

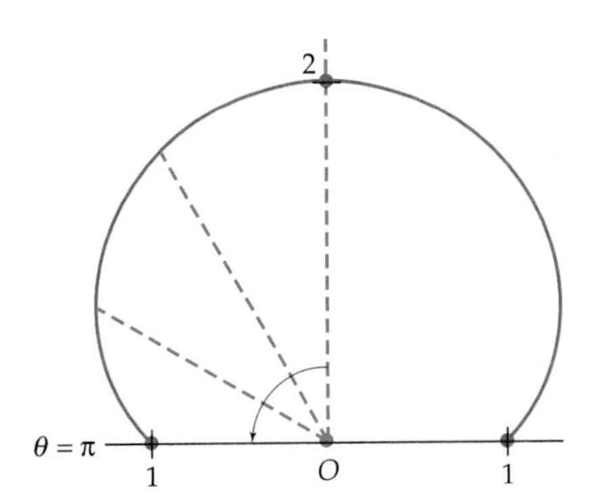

As θ increases from π to $3\pi/2$, $\sin\theta$ decreases from 0 to −1. So $r = 1 + \sin\theta$ decreases from 1 to 0.

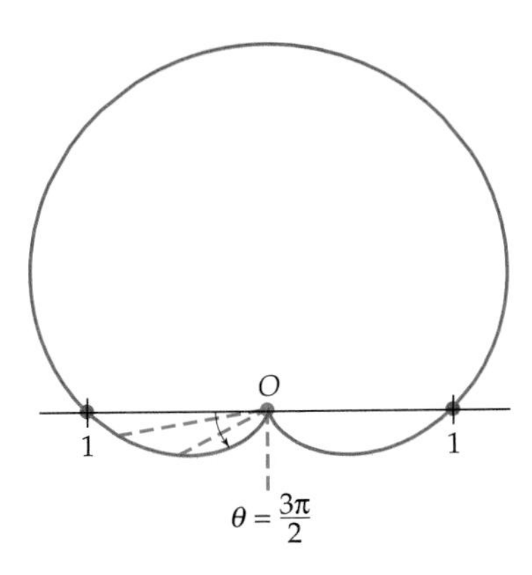

As θ increases from $3\pi/2$ to 2π, $\sin\theta$ increases from −1 to 0. So $r = 1 + \sin\theta$ increases from 0 to 1.

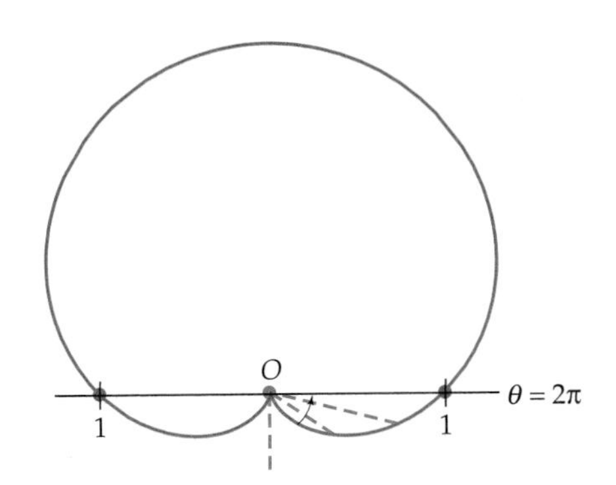

Figure 10–52

As θ takes values larger than 2π, $\sin\theta$ repeats the same pattern, and hence so does $r = 1 + \sin\theta$. The same is true for negative values of θ. The full graph (called a **cardioid**) is at the lower right in Figure 10–52. ■

The easiest way to graph polar equation $r = f(\theta)$ is to use a calculator in polar graphing mode. A second way is to use parametric graphing mode, with the coordinate converison formulas as a parameterization:

$$x = r\cos\theta = f(\theta)\cos\theta$$

$$y = r\sin\theta = f(\theta)\sin\theta.$$

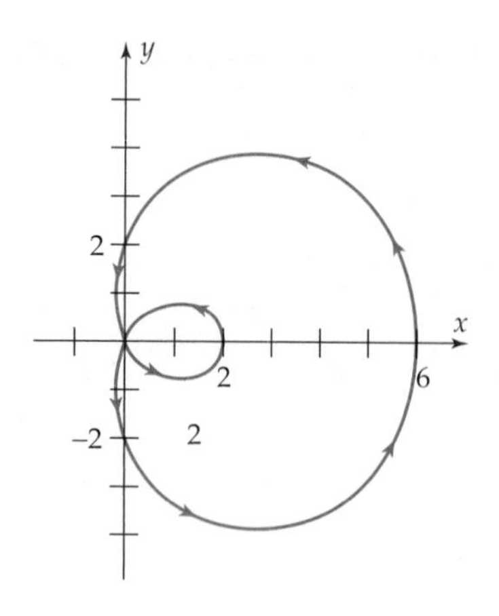

Figure 10–53

Example 6 Graph $r = 2 + 4\cos\theta$.

Solution *Polar Method:* Put your calculator in polar graphing mode and enter $r = 2 + 4\cos\theta$ in the function memory. Set the viewing window by entering minimum and maximum values for x, y, and θ. Since sine has period 2π, a complete graph can be obtained by taking $0 \le \theta \le 2\pi$. You must also set the θ step (or θ pitch), which determines how many values of θ the calculator uses to plot the graph. With an appropriate θ step, the graph should look like Figure 10–53.

GRAPHING EXPLORATION

Graph $r = 2 + 4\cos\theta$, using the viewing window with $-4 \le x \le 8$, $-4 \le y \le 4$, $0 \le \theta \le 6.3$, and θ step $= 1$. If the graph does not resemble Figure 10–53, try a smaller θ step until you find one that produces a graph like Figure 10–53.

To understand what the θ step does, set your calculator to graph in "dot" mode rather than "connected" mode, so you can see the points it actually plots. Now graph again, beginning with θ step $= 1$ and then using smaller values.

Parametric Method: Put your calculator in parametric graphing mode. The parametric equations for $r = 2 + 4\cos\theta$ are as follows (using t as the variable instead of θ with $0 \le t \le 2\pi$):

$$x = r\cos t = (2 + 4\cos t)\cos t = 2\cos t + 4\cos^2 t$$

$$y = r\sin t = (2 + 4\cos t)\sin t = 2\sin t + 4\sin t\cos t.$$

They also produce the graph in Figure 10–53. ■

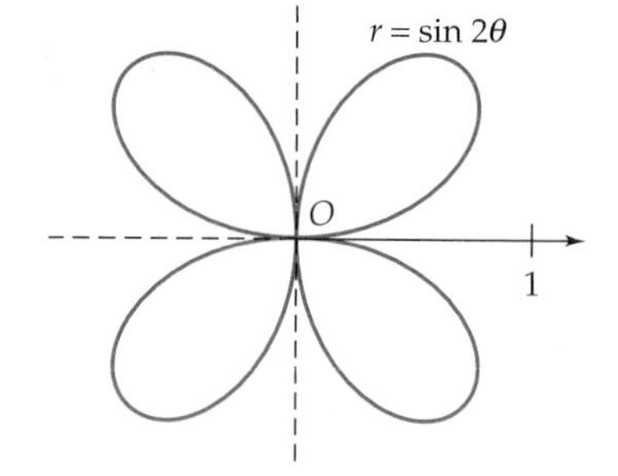

Figure 10–54

Example 7 The graph of $r = \sin 2\theta$ in Figure 10–54 can be obtained either by graphing directly in polar mode or by using parametric mode and the equations:

$$x = r\cos t = \sin 2t\cos t \quad \text{and} \quad y = r\sin t = \sin 2t\sin t \quad (0 \le t \le 2\pi).$$ ■

Here is a summary of commonly encountered polar graphs (in each case, a and b are constants):

Equation	Name of Graph	Shape of Graph*	
$r = a\theta\ (\theta \geq 0)$ $r = a\theta\ (\theta \leq 0)$	Archimedean spiral	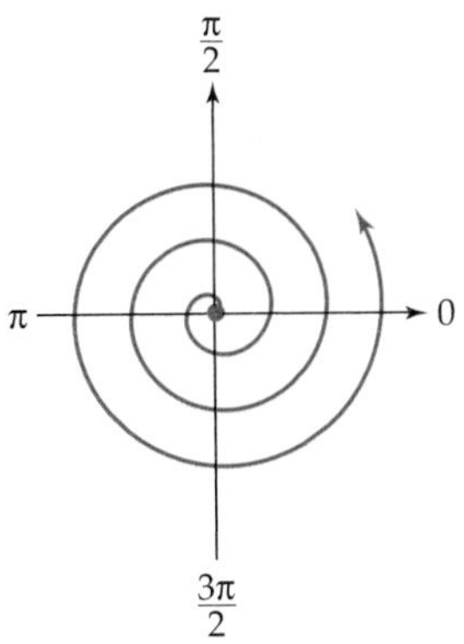 $r = a\theta\ (\theta \geq 0)$	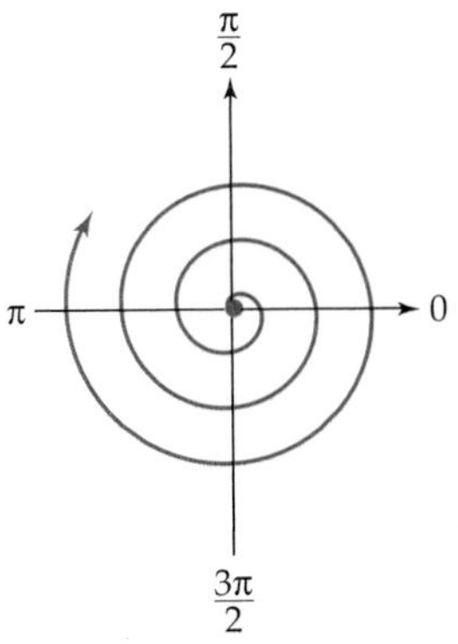 $r = a\theta\ (\theta \leq 0)$
$r = a(1 \pm \sin\theta)$ $r = a(1 \pm \cos\theta)$	cardioid	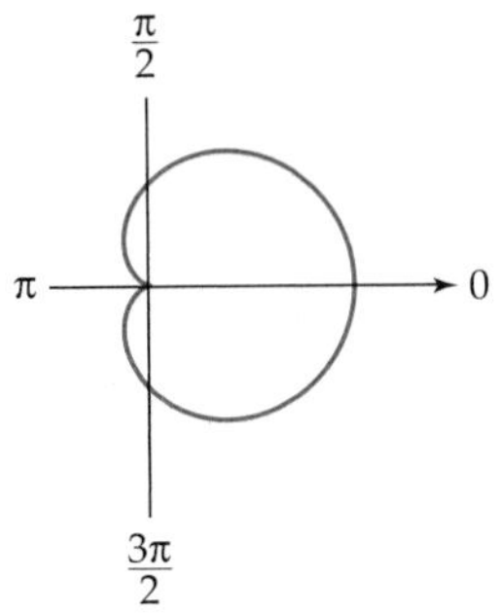 $r = a(1 + \cos\theta)$	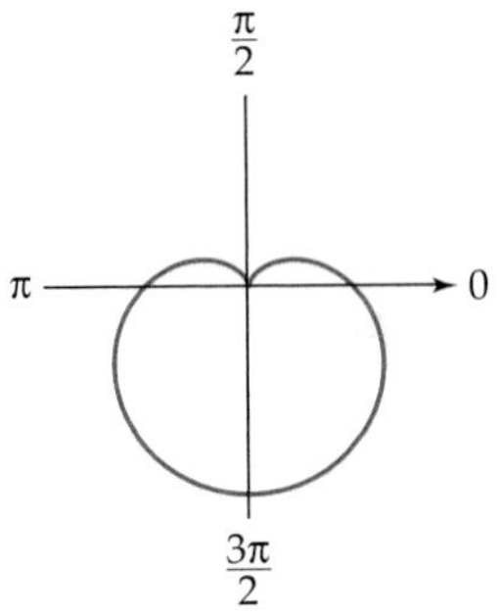 $r = a(1 - \sin\theta)$
$r = a \sin n\theta$ $r = a \cos n\theta$ $(n \geq 2)$	rose (There are n petals when n is odd and $2n$ petals when n is even.)	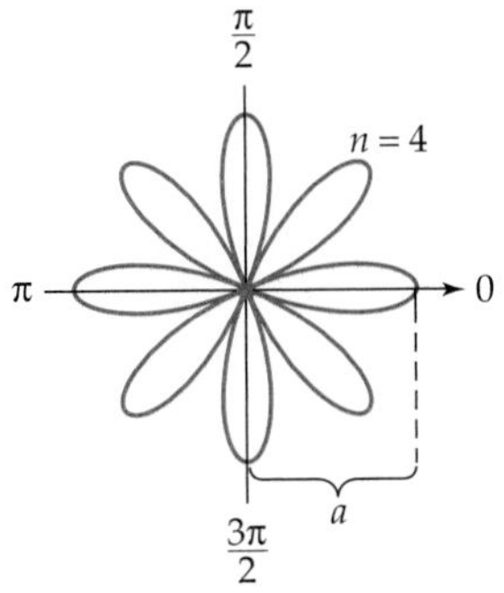 $n = 4$ $r = a \cos n\theta$	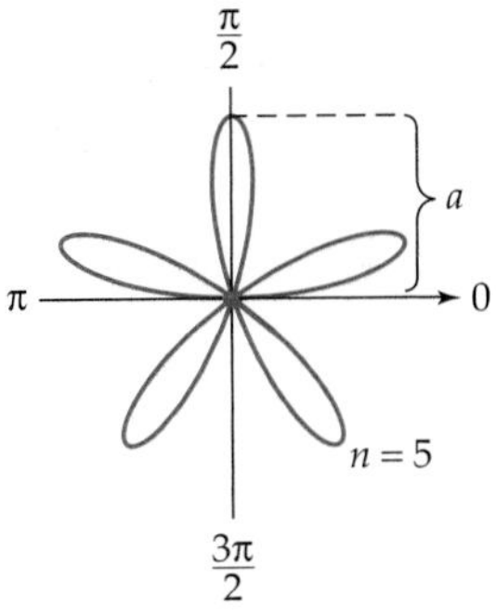 $n = 5$ $r = a \sin n\theta$

*Depending on the plus or minus sign and whether sine or cosine is involved, the basic shape of a specific graph may differ from those shown by a rotation, reversal, or horizontal or vertical shift.

Equation	Name of Graph	Shape of Graph*
$r = a \sin \theta$ $r = a \cos \theta$	circle	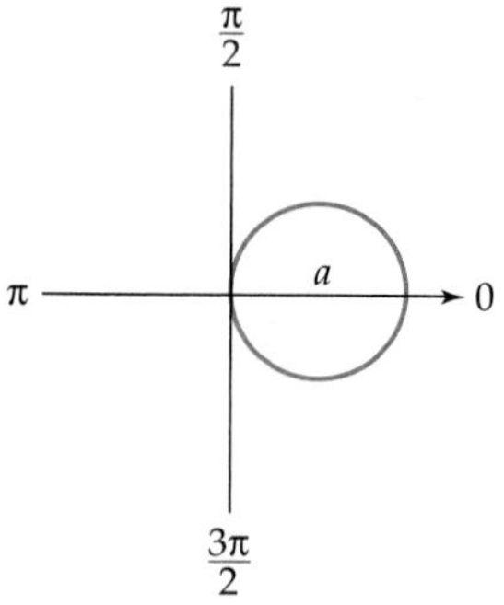 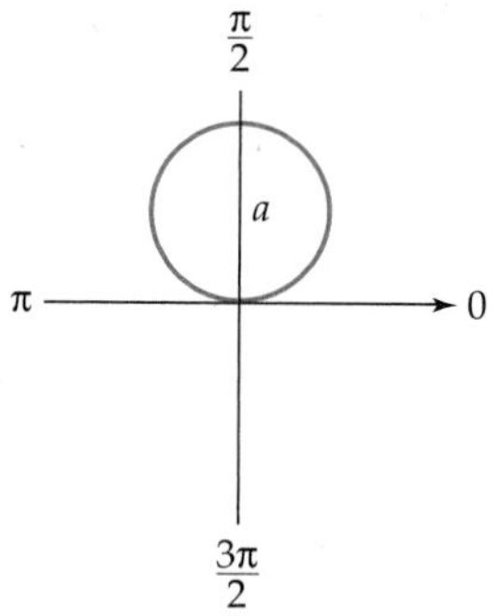 $r = a \cos \theta$ $r = a \sin \theta$
$r^2 = \pm a^2 \sin 2\theta$ $r^2 = \pm a^2 \cos 2\theta$	lemniscate	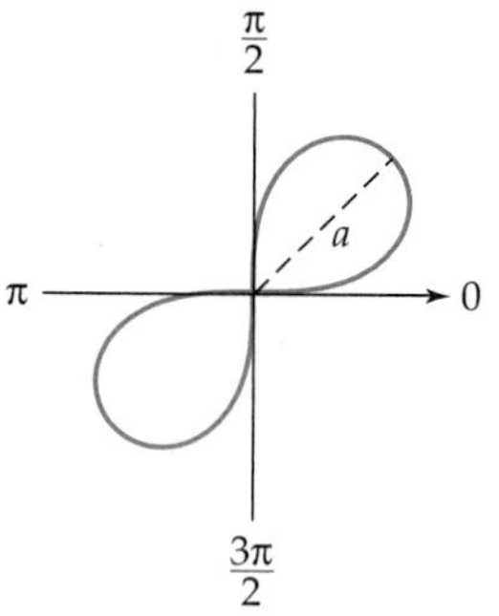 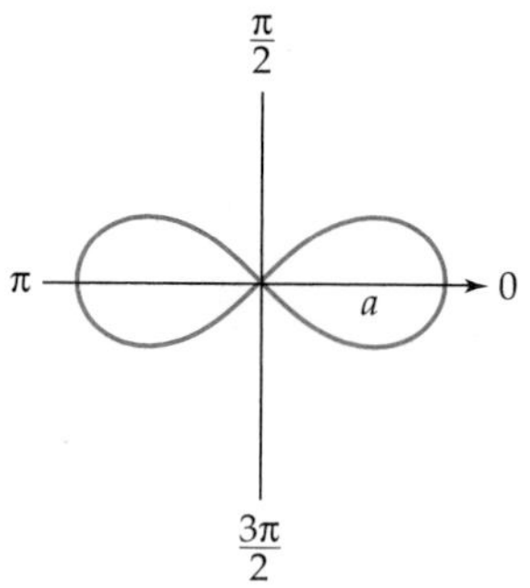 $r^2 = a^2 \sin 2\theta$ $r^2 = a^2 \cos 2\theta$
$r = a + b \sin \theta$ $r = a + b \cos \theta$ $(a, b > 0; a \neq b)$	limaçon	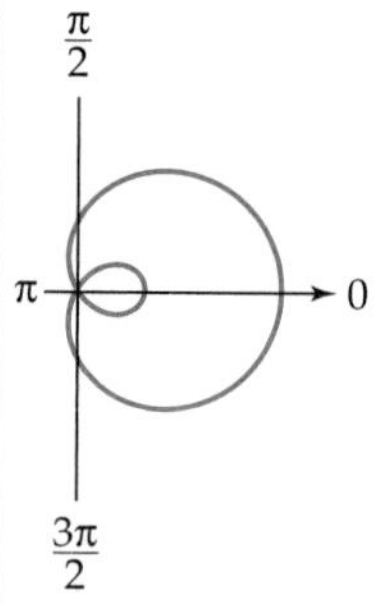 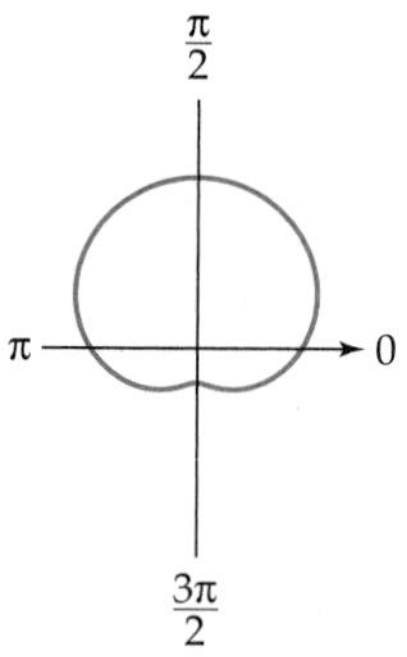 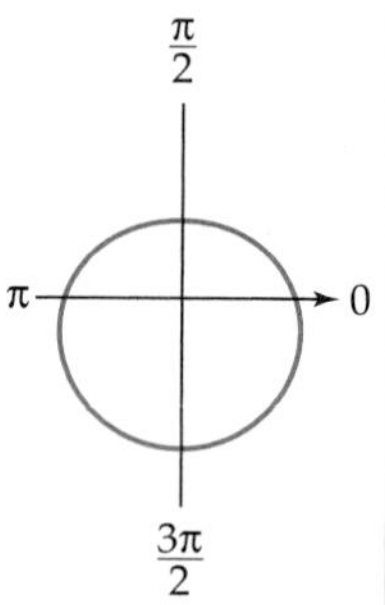 $a < b$ $r = a + b \cos \theta$ $b < a < 2b$ $r = a + b \sin \theta$ $a \geq 2b$ $r = a - b \sin \theta$

Exercises 10.4

1. What are the polar coordinates of the points P, Q, R, S, T, U, V in the figure?

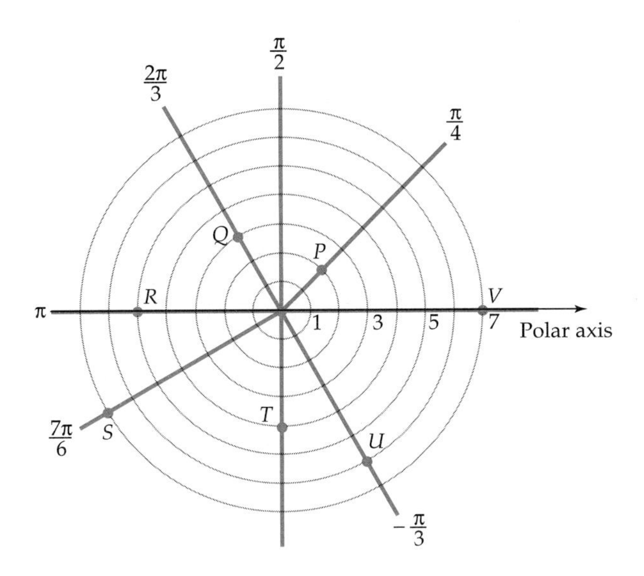

In Exercises 2–6, list four other pairs of polar coordinates for the given point, each with a different combination of signs (that is, $r > 0, \theta > 0$; $r > 0, \theta < 0$; $r < 0, \theta > 0$; $r < 0, \theta < 0$).

2. $(3, \pi/3)$ **3.** $(-5, \pi)$ **4.** $(2, -2\pi/3)$

5. $(-1, -\pi/6)$ **6.** $(\sqrt{3}, 3\pi/4)$

In Exercises 7–10, convert the polar coordinates to rectangular coordinates.

7. $(3, \pi/3)$ **8.** $(-2, \pi/4)$ **9.** $(-1, 5\pi/6)$

10. $(2, 0)$

In Exercises 11–16, convert the rectangular coordinates to polar coordinates.

11. $(3\sqrt{3}, -3)$ **12.** $(2\sqrt{3}, -2)$ **13.** $(2, 4)$

14. $(3, -2)$ **15.** $(-5, 2.5)$ **16.** $(-6.2, -3)$

In Exercises 17–22, sketch the graph of the equation without using a calculator.

17. $r = 4$ **18.** $r = -1$ **19.** $\theta = -\pi/3$

20. $\theta = 5\pi/6$ **21.** $\theta = 1$ **22.** $\theta = -4$

In Exercises 23–46, sketch the graph of the equation.

23. $r = \theta \quad (\theta \le 0)$ **24.** $r = 3\theta \quad (\theta \ge 0)$

25. $r = 1 - \sin\theta$ **26.** $r = 3 - 3\cos\theta$

27. $r = -2\cos\theta$ **28.** $r = -6\sin\theta$

29. $r = \cos 2\theta$ **30.** $r = \cos 3\theta$

31. $r = \sin 3\theta$ **32.** $r = \sin 4\theta$

33. $r^2 = 4\cos 2\theta$ **34.** $r^2 = \sin 2\theta$

35. $r = 2 + 4\cos\theta$ **36.** $r = 1 + 2\cos\theta$

37. $r = \sin\theta + \cos\theta$ **38.** $r = 4\cos\theta + 4\sin\theta$

39. $r = \sin(\theta/2)$ **40.** $r = 4\tan\theta$

41. $r = \sin\theta\tan\theta$ (cissoid)

42. $r = 4 + 2\sec\theta$ (conchoid)

43. $r = e^{\theta}$ (logarithmic spiral)

44. $r^2 = 1/\theta$ **45.** $r = 1/\theta \quad (\theta > 0)$ **46.** $r^2 = \theta$

47. **(a)** Find a complete graph of $r = 1 - 2\sin 3\theta$.
(b) Predict what the graph of $r = 1 - 2\sin 4\theta$ will look like. Then check your prediction with a calculator.
(c) Predict what the graph of $r = 1 - 2\sin 5\theta$ will look like. Then check your prediction with a calculator.

48. **(a)** Find a complete graph of $r = 1 - 3\sin 2\theta$.
(b) Predict what the graph of $r = 1 - 3\sin 3\theta$ will look like. Then check your prediction with a calculator.
(c) Predict what the graph of $r = 1 - 3\sin 4\theta$ will look like. Then check your prediction with a calculator.

49. If a, b are constants such that $ab \ne 0$, show that the graph of $r = a\sin\theta + b\cos\theta$ is a circle. [*Hint:* Multiply both sides by r and convert to rectangular coordinates.]

50. Prove that the coordinate conversion formulas are valid when $r < 0$. [*Hint:* If P has coordinates (x, y) and (r, θ), with $r < 0$, verify that the point Q with rectangular coordinates $(-x, -y)$ has polar coordinates $(-r, \theta)$. Since $r < 0$, $-r$ is positive and the conversion formulas proved in the text apply to Q. For instance, $-x = -r\cos\theta$, which implies that $x = r\cos\theta$.]

51. *Distance Formula for Polar Coordinates:* Prove that the distance from (r, θ) to (s, β) is $\sqrt{r^2 + s^2 - 2rs\cos(\theta - \beta)}$. [*Hint:* If $r > 0, s > 0$, and $\theta > \beta$, then the triangle with vertices (r, θ), (s, β), $(0, 0)$ has an angle of $\theta - \beta$, whose sides have lengths r and s. Use the Law of Cosines.]

52. Explain why the following symmetry tests for the graphs of polar equations are valid.

(a) If replacing θ by $-\theta$ produces an equivalent equation, then the graph is symmetric with respect to the line $\theta = 0$ (the x-axis).

(b) If replacing θ by $\pi - \theta$ produces an equivalent equation, then the graph is symmetric with respect to the line $\theta = \pi/2$ (the y-axis).

(c) If replacing r by $-r$ produces an equivalent equation, then the graph is symmetric with respect to the origin.

10.5 Polar Equations of Conics

In a rectangular coordinate system each type of conic section has a different definition. By using polar coordinates it is possible to give a unified treatment of conics and their equations. Before doing this we must first introduce a concept that will play a key role in the development.

Recall that both ellipses and hyperbolas are defined in terms of two foci and both have two vertices that lie on the line through the foci (see pages 625 and 628). The **eccentricity** of an ellipse or a hyperbola is denoted e and is defined to be the ratio

$$e = \frac{\text{distance between the foci}}{\text{distance between the vertices}}.$$

For conics centered at the origin, with foci on the x-axis, the situation is as follows:

Ellipse

$$\frac{x^2}{a^2} + \frac{y^2}{b^2} = 1 \quad (a > b)$$

foci: $(\pm c, 0)$ vertices: $(\pm a, 0)$

$$c = \sqrt{a^2 - b^2}$$

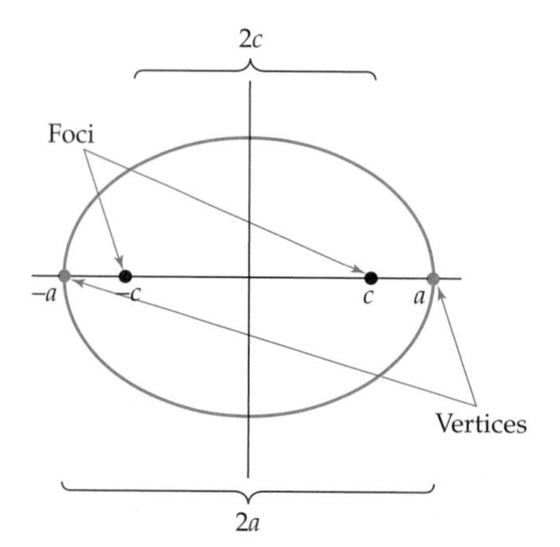

$$e = \frac{2c}{2a} = \frac{c}{a} = \frac{\sqrt{a^2 - b^2}}{a}$$

Hyperbola

$$\frac{x^2}{a^2} - \frac{y^2}{b^2} = 1$$

foci: $(\pm c, 0)$ vertices: $(\pm a, 0)$

$$c = \sqrt{a^2 + b^2}$$

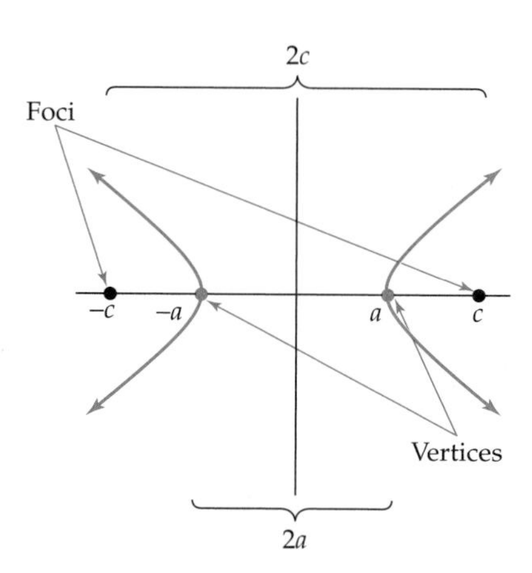

$$e = \frac{2c}{2a} = \frac{c}{a} = \frac{\sqrt{a^2 + b^2}}{a}$$

A similar analysis shows that the formulas for e are also valid for conics whose equations are of the form

$$\frac{x^2}{b^2} + \frac{y^2}{a^2} = 1 \quad (a > b) \qquad \text{or} \qquad \frac{y^2}{a^2} - \frac{x^2}{b^2} = 1.$$

These formulas can be used to compute the eccentricity of any ellipse or hyperbola whose equation can be put in standard form.

Example 1 Find the eccentricity of the conic with equation

(a) $\dfrac{y^2}{4} - \dfrac{x^2}{21} = 1$ (b) $4x^2 + 9y^2 - 32x - 90y + 253 = 0.$

Solution

(a) In this case $a^2 = 4$ (so that $a = 2$) and $b^2 = 21$. Hence, the eccentricity is

$$e = \frac{\sqrt{a^2 + b^2}}{a} = \frac{\sqrt{4 + 21}}{2} = \frac{\sqrt{25}}{2} = \frac{5}{2} = 2.5.$$

(b) In Example 2 of Section 10.3 we saw that the equation can be put into this standard form: $\dfrac{(x-4)^2}{9} + \dfrac{(y-5)^2}{4} = 1$. Hence, its graph is just the ellipse $\dfrac{x^2}{9} + \dfrac{y^2}{4} = 1$ shifted vertically and horizontally. Since the shifting does not change the distances between foci or vertices, both ellipses have the same eccentricity, which can be computed using $a^2 = 9$ and $b^2 = 4$:

$$e = \frac{\sqrt{a^2 - b^2}}{a} = \frac{\sqrt{9 - 4}}{3} = \frac{\sqrt{5}}{3} \approx .745. \quad ■$$

Example 1 and the preceding pictures illustrate the following fact. For ellipses the distance between the foci (numerator of e) is less than the distance between the vertices (denominator), so $e < 1$. For hyperbolas, however, $e > 1$ because the distance between the foci is greater than that between the vertices.

The eccentricity of an ellipse measures its "roundness." An ellipse whose eccentricity is close to 0 is almost circular (Exercise 19). The eccentricity of a hyperbola measures how "flat" its branches are. The branches of a hyperbola with large eccentricity look almost like parallel lines (Exercise 20).

Conics and Polar Equations

The polar analogues of the standard equations of ellipses, parabolas, and hyperbolas are given in the following chart. The proof of these statements is given at the end of the section. In the chart e and d are constants, with $e > 0$. Remember that in a rectangular coordinate system whose positive x-axis coincides with the polar axis, a point with polar coordinates (r, θ) is on the x-axis when $\theta = 0$ or π and on the y-axis when $\theta = \pi/2$ or $3\pi/2$.

Polar Equations for Conic Sections

Equation		Graph
$r = \dfrac{ed}{1 + e\cos\theta}$ or $r = \dfrac{ed}{1 - e\cos\theta}$	$0 < e < 1$	*Ellipse* with eccentricity e One of the foci: (0, 0) Vertices at $\theta = 0$ and $\theta = \pi$
	$e = 1$	*Parabola* with focus (0, 0) Vertex at $\theta = 0$ or $\theta = \pi$; (r is not defined for the other value of θ)
	$e > 1$	*Hyperbola* with eccentricity e One of the foci: (0, 0) Vertices at $\theta = 0$ and $\theta = \pi$
$r = \dfrac{ed}{1 + e\sin\theta}$ or $r = \dfrac{ed}{1 - e\sin\theta}$	$0 < e < 1$	*Ellipse* with eccentricity e One of the foci: (0, 0) Vertices at $\theta = \pi/2$ and $\theta = 3\pi/2$
	$e = 1$	*Parabola* with focus (0, 0) Vertex at $\theta = \pi/2$ or $\theta = 3\pi/2$; (r is not defined for the other value of θ)
	$e > 1$	*Hyperbola* with eccentricity e One of the foci: (0, 0) Vertices at $\theta = \pi/2$ and $\theta = 3\pi/2$

Example 2 Find a complete graph of $r = \dfrac{3e}{1 + e\cos\theta}$ when

(a) $e = .7$ (b) $e = 1$ (c) $e = 2$.

Solution From the first equation in the preceding chart (with $d = 3$) we know that the graphs are an ellipse, parabola, and hyperbola, respectively, as shown in Figure 10–55.

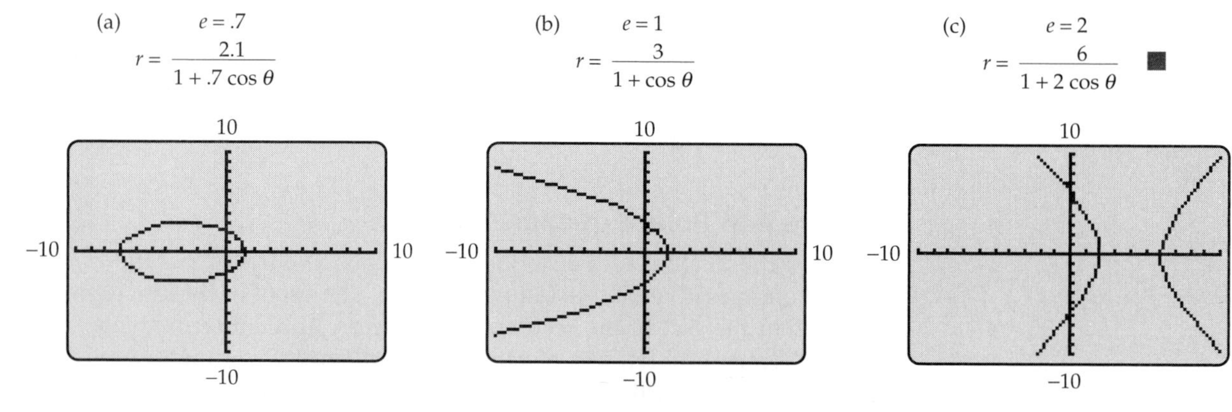

Figure 10–55

Example 3 Identify the conic section that is the graph of

$$r = \frac{20}{4 - 10\sin\theta}$$

and find its vertices and eccentricity.

Solution First, rewrite the equation in one of the forms listed in the preceding box:

$$r = \frac{20}{4 - 10\sin\theta} = \frac{20}{4\left(1 - \frac{10}{4}\sin\theta\right)} = \frac{5}{1 - 2.5\sin\theta}.$$

This is one such form, with $e = 2.5$ and $ed = 5$ (so that $d = 2$). Consequently, the graph is a hyperbola with eccentricity $e = 2.5$ whose vertices are at

$$\theta = \frac{\pi}{2}, \qquad r = \frac{20}{4 - 10\sin\frac{\pi}{2}} = \frac{20}{4 - 10\cdot 1} = -\frac{20}{6} = -\frac{10}{3}$$

and

$$\theta = \frac{3\pi}{2}, \qquad r = \frac{20}{4 - 10\sin\frac{3\pi}{2}} = \frac{20}{4 - 10(-1)} = \frac{20}{14} = \frac{10}{7}.$$

GRAPHING EXPLORATION

Find a viewing window that shows a complete graph of this hyperbola. ■

Example 4 Find a polar equation of the ellipse with (0, 0) as a focus and vertices (3, 0) and $(6, \pi)$.

Solution Because of the location of the vertices, the polar equation is of the form $r = ed/(1 \pm e\cos\theta)$. We first consider the equation

$$r = \frac{ed}{1 + e\cos\theta}.$$

Since the coordinates of the vertices satisfy the equation, we must have:

$$3 = \frac{ed}{1 + e\cos 0} = \frac{ed}{1 + e} \qquad \text{and} \qquad 6 = \frac{ed}{1 + e\cos\pi} = \frac{ed}{1 - e}$$

which imply that

$$3(1 + e) = ed \qquad \text{and} \qquad 6(1 - e) = ed.$$

Therefore,

$$3(1 + e) = 6(1 - e)$$
$$3 + 3e = 6 - 6e$$
$$9e = 3$$
$$e = 1/3.$$

Substituting $e = 1/3$ in either of the original equations shows that $d = 12$. So an equation of the ellipse is

$$r = \frac{ed}{1 + e\cos\theta} = \frac{\frac{1}{3}\cdot 12}{1 + \frac{1}{3}\cos\theta} = \frac{12}{3 + \cos\theta}.$$

If we had started instead with the equation $r = ed/(1 - e\cos\theta)$ and solved for e as above, we would have obtained $e = -1/3$, which is impossible since $e > 0$.

Alternate Solution Verify that the vertex (3, 0) also has polar coordinates $(-3, \pi)$. Similarly, $(6, \pi)$ also has polar coordinates $(-6, 0)$. If you begin with the equation $r = ed/(1 - e\cos\theta)$ and the vertices $(-3, \pi)$ and $(-6, 0)$ and proceed as before to find e and d, you obtain the equation

$$r = \frac{-12}{3 - \cos\theta}. \quad ■$$

Alternate Definition of Conics

The theorem stated in the following box is sometimes used as a definition of the conic sections because it provides a unified approach instead of the variety of descriptions given in Section 10.2. Its proof also provides a proof of the statements in the box on page 666.

The basic idea is to describe every conic in terms of a straight line L (the **directrix**) and a point P not on L (the **focus**), in much the same way that parabolas were defined in Section 10.2. The number e in the theorem turns out to be the eccentricity of the conic.

Conic Section Theorem

Let L be a straight line, P a point not on L, and e a positive constant. The set of all points X in the plane such that

$$\frac{\text{distance from } X \text{ to } P}{\text{distance from } X \text{ to } L} = e$$

is a conic section with P as one of the foci.* The conic is an ellipse if $0 < e < 1$, a parabola if $e = 1$,† and a hyperbola if $e > 1$.

*The distance from X to L is measured along the line through X that is perpendicular to L.

†When $e = 1$, the given condition is equivalent to

distance from X to P = distance from X to L

which is the definition of a parabola given in Section 10.2.

Proof Coordinatize the plane so that the pole is the point P, the polar axis is horizontal, and the directrix L is a vertical line to the left of the pole, as in Figure 10–56. Let d be the distance from P to L and (r, θ) the polar coordinates of X.

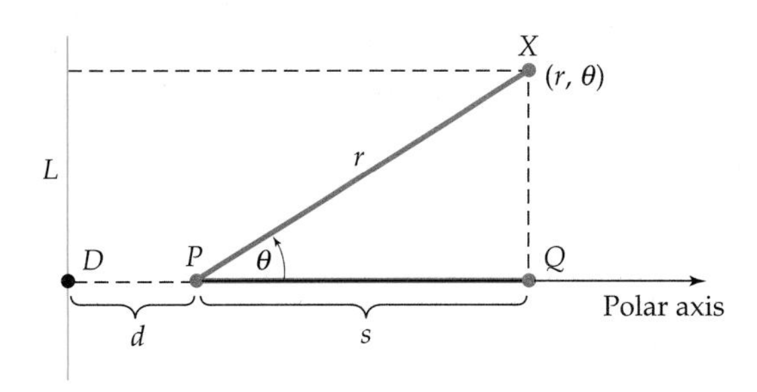

Figure 10–56

If X satisfies the condition

$$\frac{\text{distance from } X \text{ to } P}{\text{distance from } X \text{ to } L} = e$$

then

$$(*) \qquad \text{distance from } X \text{ to } P = e(\text{distance from } X \text{ to } L).$$

Figure 10–56 shows that r is the distance from P to X and that the distance from X to L is the same as that from D to Q, namely, $d + s$. Furthermore, $\cos\theta = s/r$ so that $s = r\cos\theta$. Consequently, equation $(*)$ can be written in polar coordinates as follows:

$$\begin{aligned}
\text{distance from } X \text{ to } P &= e(\text{distance from } X \text{ to } L)\\
r &= e(d + s)\\
r &= e(d + r\cos\theta)\\
r - er\cos\theta &= ed\\
r(1 - e\cos\theta) &= ed\\
r &= \frac{ed}{1 - e\cos\theta}.
\end{aligned}$$

To show that this is actually the equation of a conic, we translate it into rectangular coordinates using the conversion formulas from Section 10.4:

$$r^2 = x^2 + y^2 \qquad \text{and} \qquad \cos\theta = \frac{x}{r} = \frac{x}{\pm\sqrt{x^2 + y^2}}.$$

Then the polar coordinate equation becomes

$$\pm\sqrt{x^2 + y^2} = \frac{ed}{1 - e\left(\dfrac{x}{\pm\sqrt{x^2 + y^2}}\right)}$$

$$\pm\sqrt{x^2 + y^2}\left(1 - \frac{ex}{\pm\sqrt{x^2 + y^2}}\right) = ed$$

$$\pm\sqrt{x^2 + y^2} - ex = ed$$

$$\pm\sqrt{x^2 + y^2} = ed + ex.$$

Squaring both sides and rearranging terms, we have

$$x^2 + y^2 = e^2d^2 + 2de^2x + e^2x^2$$

$$(**) \qquad (1 - e^2)x^2 - 2de^2x + y^2 = e^2d^2.$$

Now we consider the two possibilities, $e = 1$ and $e \neq 1$.

Case 1. If $e = 1$, then equation (**) becomes

$$-2dx + y^2 = d^2$$

$$y^2 = 2dx + d^2$$

$$(y - 0)^2 = 2d\left(x + \frac{d}{2}\right)$$

$$(y - 0)^2 = 4\left(\frac{d}{2}\right)\left(x - \left(-\frac{d}{2}\right)\right).$$

The box on page 641 (with $k = 0$, $p = d/2$, $h = -d/2$) shows that this is the standard equation of a parabola with

$$\text{vertex}\left(-\frac{d}{2}, 0\right), \qquad \text{focus}(0, 0), \qquad \text{directrix } x = -d.$$

Case 2. If $e \neq 1$, then we can divide both sides of equation (**) by the nonzero number $1 - e^2$:

$$\left(x^2 - \frac{2de^2}{1 - e^2}x\right) + \frac{y^2}{1 - e^2} = \frac{e^2d^2}{1 - e^2}.$$

Next we complete the square on the expression in parentheses by adding the square of half of the coefficient of x to both sides of the equation and simplify the result:

$$\left[x^2 - \frac{2de^2}{1 - e^2}x + \left(\frac{de^2}{1 - e^2}\right)^2\right] + \frac{y^2}{1 - e^2} = \frac{e^2d^2}{1 - e^2} + \left(\frac{de^2}{1 - e^2}\right)^2$$

$$\left(x - \frac{de^2}{1 - e^2}\right)^2 + \frac{y^2}{1 - e^2} = \frac{(1 - e^2)e^2d^2 + (de^2)^2}{(1 - e^2)^2}$$

$$\left(x - \frac{de^2}{1 - e^2}\right)^2 + \frac{y^2}{1 - e^2} = \frac{e^2d^2}{(1 - e^2)^2}.$$

Dividing both sides of the last equation by $e^2d^2/(1-e^2)^2$ produces the equation

$$(***)\qquad \frac{\left(x-\dfrac{de^2}{1-e^2}\right)^2}{\dfrac{e^2d^2}{(1-e^2)^2}}+\frac{y^2}{\dfrac{e^2d^2}{1-e^2}}=1.$$

Now we consider the two possibilities, $e<1$ and $e>1$.

Case 2A. If $e<1$, then $1-e^2>0$ and the constants in the denominators on the left side of equation (***) are positive. Therefore, equation (***) is of the form

$$\frac{(x-h)^2}{a^2}+\frac{(y-k)^2}{b^2}=1$$

with $h=de^2/(1-e^2)$, $k=0$, and a and b positive numbers such that

$$a^2=\frac{e^2d^2}{(1-e^2)^2}\qquad\text{and}\qquad b^2=\frac{e^2d^2}{1-e^2}.$$

In this case $a>b$ by Exercise 47. According to the box on page 641 this is the standard equation of an ellipse with center $(h,0)$ and foci $(h-c,0)$ and $(h+c,0)$, where $c^2=a^2-b^2$. Its eccentricity is the number

$$\frac{c}{a}=\sqrt{\frac{c^2}{a^2}}=\sqrt{\frac{a^2-b^2}{a^2}}=\sqrt{1-b^2\cdot\frac{1}{a^2}}$$

$$=\sqrt{1-\frac{e^2d^2}{1-e^2}\cdot\frac{(1-e^2)^2}{e^2d^2}}=\sqrt{e^2}=e.$$

Case 2B. If $e>1$, then $1-e^2<0$. Therefore,

$$e^2-1=-(1-e^2)>0$$

so that equation (***) may be written as

$$\frac{\left(x-\dfrac{de^2}{1-e^2}\right)^2}{\dfrac{e^2d^2}{(1-e^2)^2}}-\frac{y^2}{\dfrac{e^2d^2}{e^2-1}}=1.$$

This is an equation of the form

$$\frac{(x-h)^2}{a^2}-\frac{(y-k)^2}{b^2}=1$$

with a and b positive. The box on page 641 shows that it is the standard equation of a hyperbola with foci $(h-c,0)$ and $(h+c,0)$, where $c^2=a^2+b^2$. Exercise 48 shows that its eccentricity is e.

The preceding argument depends on coordinatizing the plane in a certain way and taking d to be the distance from the pole to L. Similar arguments, in which d is the distance from the pole to L and the plane is coordinatized so that L is to the right of the pole or parallel to the polar axis, lead to the other polar equations shown in Figure 10–57.

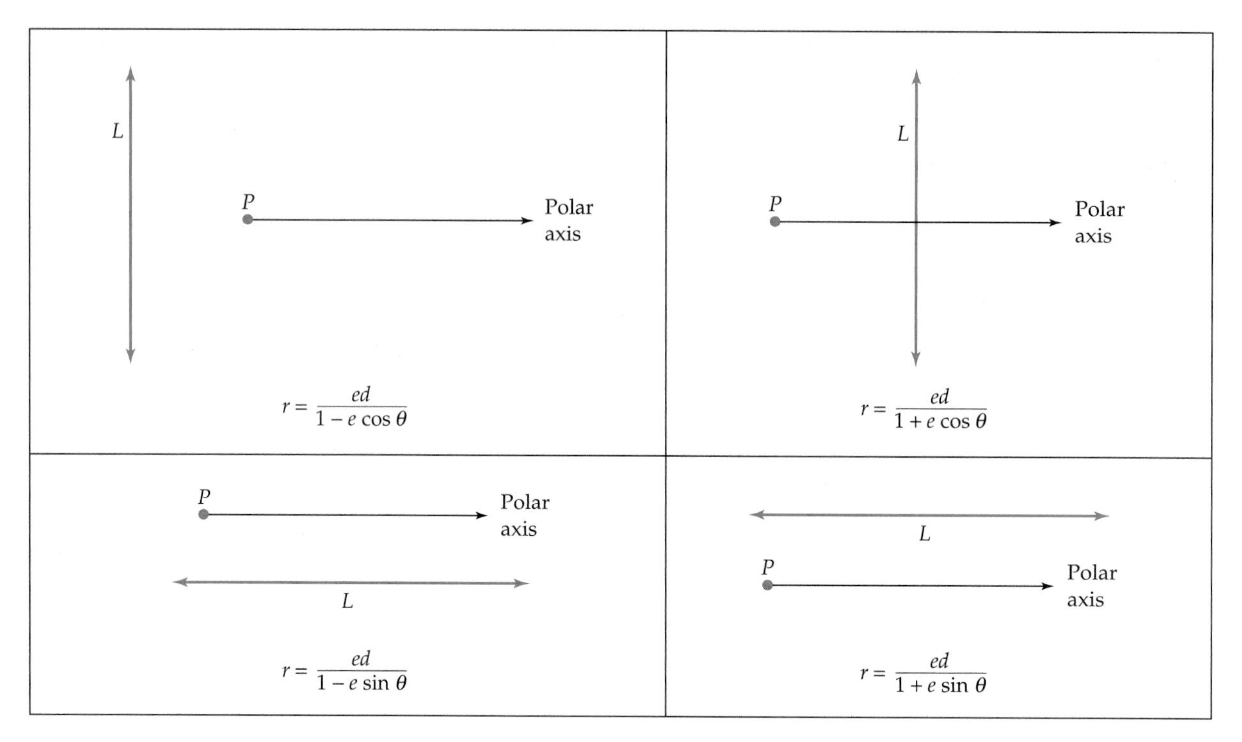

Figure 10–57

Analogous arguments when $d < 0$ (using $-d$ as the distance from the pole to L) then complete the proof. ■

Example 5 Find the polar equation of the hyperbola with focus at the pole, directrix $r = -4\csc\theta$, and eccentricity 3.

Solution The equation of the directrix can be written as

$$r = \frac{-4}{\sin\theta}, \qquad \text{or equivalently,} \qquad r\sin\theta = -4.$$

With the conversion formulas for a rectangular coordinate system whose positive x-axis coincides with the polar axis, this equation becomes $y = -4$. So the directrix is a line parallel to the polar axis and 4 units below it. Using Figure 10–57, we see that $d = 4$ and the equation is

$$r = \frac{ed}{1 - e\sin\theta} = \frac{3 \cdot 4}{1 - 3\sin\theta} = \frac{12}{1 - 3\sin\theta}. \quad ■$$

Exercises 10.5

In Exercises 1–6, which of the graphs (a)–(f) at the bottom of the page could possibly be the graph of the equation?

1. $r = \dfrac{3}{1 - \cos\theta}$

2. $r = \dfrac{6}{2 + \cos\theta}$

3. $r = \dfrac{6}{2 - 4\sin\theta}$

4. $r = \dfrac{15}{1 + 4\cos\theta}$

5. $r = \dfrac{6}{3 - 2\sin\theta}$

6. $r = \dfrac{6}{\frac{3}{2} + \frac{3}{2}\sin\theta}$

In Exercises 7–12, identify the conic section whose equation is given; if it is an ellipse or hyperbola, state its eccentricity.

7. $r = \dfrac{12}{3 + 4\sin\theta}$

8. $r = \dfrac{-10}{2 + 3\cos\theta}$

9. $r = \dfrac{8}{3 + 3\sin\theta}$

10. $r = \dfrac{20}{5 - 10\sin\theta}$

11. $r = \dfrac{2}{6 - 4\cos\theta}$

12. $r = \dfrac{-6}{5 + 2\cos\theta}$

In Exercises 13–18, find the eccentricity of the conic whose equation is given.

13. $\dfrac{x^2}{100} + \dfrac{y^2}{99} = 1$

14. $\dfrac{(x-4)^2}{18} + \dfrac{(y+5)^2}{25} = 1$

15. $\dfrac{(x-6)^2}{10} - \dfrac{y^2}{40} = 1$

16. $4x^2 + 9y^2 - 24x + 36y + 36 = 0$

17. $16x^2 - 9y^2 - 32x + 36y + 124 = 0$

18. $4x^2 - 5y^2 - 16x - 50y + 71 = 0$

19. **(a)** Using a square viewing window (so that circles look like circles), graph these ellipses (on the same screen if possible):

$$\frac{x^2}{16} + \frac{y^2}{1} = 1 \qquad \frac{x^2}{16} + \frac{y^2}{6} = 1 \qquad \frac{x^2}{16} + \frac{y^2}{14} = 1$$

(b) Compute the eccentricity of each ellipse in part (a).

(c) Based on parts (a) and (b), how is the shape of an ellipse related to its eccentricity?

20. **(a)** Graph these hyperbolas (on the same screen if possible):

$$\frac{y^2}{4} - \frac{x^2}{1} = 1 \qquad \frac{y^2}{4} - \frac{x^2}{12} = 1 \qquad \frac{y^2}{4} - \frac{x^2}{96} = 1$$

(b) Compute the eccentricity of each hyperbola in part (a).

(c) Based on parts (a) and (b), how is the shape of a hyperbola related to its eccentricity?

In Exercises 21–32, sketch the graph of the equation and label the vertices.

21. $r = \dfrac{8}{1 - \cos\theta}$

22. $r = \dfrac{5}{3 + 2\sin\theta}$

23. $r = \dfrac{4}{2 - 4\cos\theta}$

24. $r = \dfrac{5}{1 + \cos\theta}$

25. $r = \dfrac{10}{4 - 3\sin\theta}$

26. $r = \dfrac{12}{3 + 4\sin\theta}$

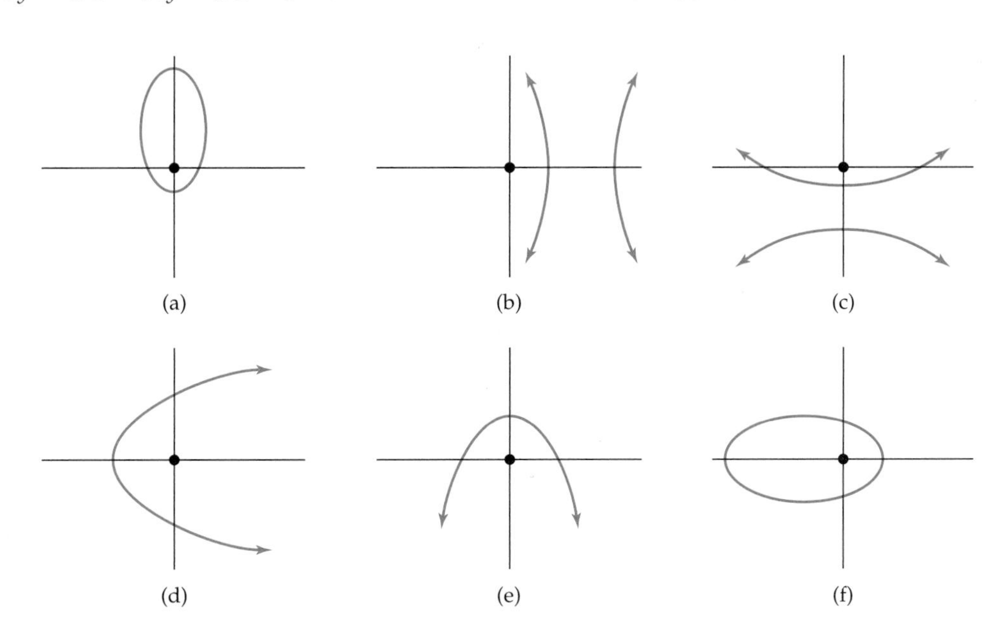

27. $r = \dfrac{15}{3 - 2\cos\theta}$

28. $r = \dfrac{32}{3 + 5\sin\theta}$

29. $r = \dfrac{3}{1 + \sin\theta}$

30. $r = \dfrac{10}{3 + 2\cos\theta}$

31. $r = \dfrac{10}{2 + 3\sin\theta}$

32. $r = \dfrac{15}{4 - 4\cos\theta}$

In Exercises 33–46, find the polar equation of the conic section that has focus (0, 0) and satisfies the given conditions.

33. Parabola; vertex $(3, \pi)$

34. Parabola; vertex $(2, \pi/2)$

35. Ellipse; vertices $(2, \pi/2)$ and $(8, 3\pi/2)$

36. Ellipse; vertices $(2, 0)$ and $(4, \pi)$

37. Hyperbola; vertices $(1, 0)$ and $(-3, \pi)$

38. Hyperbola; vertices $(-2, \pi/2)$ and $(4, 3\pi/2)$

39. Eccentricity 4; directrix; $r = -2\sec\theta$

40. Eccentricity 2; directrix: $r = 4\csc\theta$

41. Eccentricity 1; directrix: $r = -3\csc\theta$

42. Eccentricity 1; directrix: $r = 5\sec\theta$

43. Eccentricity 1/2; directrix: $r = 2\sec\theta$

44. Eccentricity 4/5; directrix: $r = 3\csc\theta$

45. Hyperbola; vertical directrix to the left of the pole; eccentricity 2; $(1, 2\pi/3)$ is on the graph.

46. Hyperbola; horizontal directrix above the pole; eccentricity 2; $(1, 2\pi/3)$ is on the graph.

47. In Case 2A of the proof of the Conic Section Theorem, show that $a > b$.

48. In Case 2B of the proof of the Conic Section Theorem, show that the hyperbola has eccentricity e.

49. A comet travels in a parabolic orbit with the sun as focus. When the comet is 60 million miles from the sun, the line segment from the sun to the comet makes an angle of $\pi/3$ radians with the axis of the parabolic orbit. Using the sun as the pole and assuming the axis of the orbit lies along the polar axis, find a polar equation for the orbit.

50. Halley's Comet has an elliptical orbit, with eccentricity .97 and the sun as a focus. The length of the major axis of the orbit is 3364.74 million miles. Using the sun as the pole and assuming the major axis of the orbit is perpendicular to the polar axis, find a polar equation for the orbit.

Chapter 10 Review

Important Concepts

Section 10.1	Plane curve 611 Parameter 611 Parametric equations 611 Eliminating the parameter 613 Finding parameterizations 614–615 Cycloid 618
Section 10.2	Ellipse: foci, center, vertices, major and minor axes 625 Equations of ellipses centered at the origin 626 Hyperbola: foci, center, vertices, focal axis 628–629 Equations of hyperbolas centered at the origin 629 Parabola: focus, directrix, vertex, axis 631 Equations of parabolas with vertex at the origin 632 Parametric equations for conics 635
Section 10.3	Vertical and horizontal shifts 638 Standard equations of conic sections 641 Graphs of rotated conics 643 Discriminant 644
Excursion 10.3.A	Rotation equations 651 Eliminating the xy term 652
Section 10.4	Polar coordinates 655 Pole 655 Polar axis 655 Polar/rectangular coordinate conversion 657 Polar graphs 658–662
Section 10.5	Eccentricity 664 Polar equations for conics 666 Focus 668 Directrix 668

Important Facts and Formulas

- Equation of ellipse with center (h, k) and axes on the lines $x = h$, $y = k$:

$$\frac{(x-h)^2}{a^2} + \frac{(y-k)^2}{b^2} = 1$$

- Equation of hyperbola with center (h, k) and vertices on the line $y = k$:

$$\frac{(x-h)^2}{a^2} - \frac{(y-k)^2}{b^2} = 1$$

- Equation of hyperbola with center (h, k) and vertices on the line $x = h$:

$$\frac{(y-k)^2}{a^2} - \frac{(x-h)^2}{b^2} = 1$$

- Equation of a parabola with vertex (h, k) and axis $x = h$:

$$(x-h)^2 = 4p(y-k)$$

- Equation of a parabola with vertex (h, k) and axis $y = k$:

$$(y-k)^2 = 4p(x-h)$$

- Rotation equations:

$$x = u\cos\theta - v\sin\theta$$
$$y = u\sin\theta + v\cos\theta$$

- To eliminate the xy term in $Ax^2 + Bxy + Cy^2 + Dx + Ey + F = 0$, rotate the axes through an angle θ such that $\cot 2\theta = \dfrac{A - C}{B}$.
- The rectangular and polar coordinates of a point are related by:

$$x = r\cos\theta \quad \text{and} \quad y = r\sin\theta;$$
$$r^2 = x^2 + y^2 \quad \text{and} \quad \tan\theta = \frac{y}{x}$$

- If e and d are constants with $e > 0$, then the graph of a polar equation of the form

$$r = \frac{ed}{1 \pm e\cos\theta} \quad \text{or} \quad r = \frac{ed}{1 \pm e\sin\theta}$$

is an ellipse if $0 < e < 1$, a parabola if $e = 1$, and a hyperbola if $e > 1$.

Review Questions

In Questions 1–4, find a viewing window that shows a complete graph of the curve whose parametric equations are given.

1. $x = 8\cos t + \cos 8t$ and $y = 8\sin t - \sin 8t$ $(0 \le t \le 2\pi)$

2. $x = [64\cos(\pi/6)]t$ and $y = -16t^2 + [64\sin(\pi/6)]t$ $(0 \le t \le \pi)$

3. $x = t^3 + t + 1$ and $y = t^2 + 2t$ $(-3 \le t \le 3)$

4. $x = t^2 - t + 3$ and $y = t^3 - 5t$ $(-3 \le t \le 3)$

In Questions 5–8, sketch the graph of the curve whose parametric equations are given and find an equation in x and y whose graph contains the given curve.

5. $x = 2t - 1,\quad y = 2 - t,\quad -3 \le t \le 3$

6. $x = 3\cos t,\quad y = 5\sin t,\quad 0 \le t \le 2\pi$

7. $x = \cos t,\quad y = 2\sin^2 t,\quad 0 \le t \le 2\pi$

8. $x = e^t,\quad y = \sqrt{t + 1},\quad t \ge 1$

9. Which of the following is *not* a parameterization of the curve $x = y^2 + 1$?
(a) $x = t^2 + 1,\quad y = t,$ any real number t
(b) $x = \sin^2 t + 1,\quad y = \sin t,$ any real number t
(c) $x = t^4 + 1,\quad y = t^2,$ any real number t
(d) $x = t^6 + 1,\quad y = t^3,$ any real number t

10. Which of the curves in Questions 1–4 appear to be the graphs of functions of the form $y = f(x)$?

In Questions 11–14, find the foci and vertices of the conic and state whether it is an ellipse or a hyperbola.

11. $\dfrac{x^2}{16} + \dfrac{y^2}{20} = 1$

12. $\dfrac{x^2}{9} - \dfrac{y^2}{16} = 1$

13. $\dfrac{(x - 1)^2}{7} + \dfrac{(y - 3)^2}{16} = 1$

14. $3x^2 = 1 + 2y^2$

15. Find the focus and directrix of the parabola $10y = 7x^2$.

16. Find the focus and directrix of the parabola

$$3y^2 - x - 4y + 4 = 0.$$

In Questions 17–28, sketch the graph of the equation. If there are asymptotes, give their equations.

17. $\dfrac{x^2}{4} + \dfrac{y^2}{25} = 1$

18. $25x^2 + 4y^2 = 100$

19. $\dfrac{(x-3)^2}{9} + \dfrac{(y+5)^2}{4} = 1$

20. $\dfrac{x^2}{9} - \dfrac{y^2}{16} = 1$

21. $\dfrac{(y+4)^2}{25} - \dfrac{(x-1)^2}{4} = 1$

22. $4x^2 - 9y^2 = 144$

23. $x^2 + 4y^2 - 10x + 9 = 0$

24. $9x^2 - 4y^2 - 36x + 24y - 36 = 0$

25. $2y = 4(x-3)^2 + 6$

26. $3y = 6(x+1)^2 - 9$

27. $x = y^2 + 2y + 2$

28. $y = x^2 - 2x + 3$

29. What is the center of the ellipse $4x^2 + 3y^2 - 32x + 36y + 124 = 0$?

30. Find the equation of the ellipse with center at the origin, one vertex at $(0, 4)$, passing through $(\sqrt{3}, 2\sqrt{3})$.

31. Find the equation of the ellipse with center at $(3, 1)$, one vertex at $(1, 1)$, passing through $(2, 1 + \sqrt{3/2})$.

32. Find the equation of the hyperbola with center at the origin, one vertex at $(0, 5)$, passing through $(1, 3\sqrt{5})$.

33. Find the equation of the hyperbola with center at $(3, 0)$, one vertex at $(3, 2)$, passing through $(1, \sqrt{5})$.

34. Find the equation of the parabola with vertex $(2, 5)$, axis $x = 2$, and passing through $(3, 12)$.

35. Find the equation of the parabola with vertex $(3/2, -1/2)$, axis $y = -1/2$, and passing through $(-3, 1)$.

36. Find the equation of the parabola with vertex $(5, 2)$ that passes through the points $(7, 3)$ and $(9, 6)$.

In Questions 37–40, assume that the graph of the equation is a nondegenerate conic. Use the discriminant to identify the graph.

37. $3x^2 + 2\sqrt{2}xy + 2y^2 - 12 = 0$

38. $x^2 + y^2 - xy - 4y = 0$

39. $4xy - 3x^2 - 20 = 0$

40. $4x^2 - 4xy + y^2 - \sqrt{5}x - 2\sqrt{5}y = 0$

In Questions 41–46, find a viewing window that shows a complete graph of the equation.

41. $x^2 - xy + y^2 - 6 = 0$

42. $x^2 + xy + y^2 - 3y - 6 = 0$

43. $x^2 + xy - 2 = 0$

44. $x^2 - 4xy + y^2 + 5 = 0$

45. $x^2 + 3xy + y^2 - 2\sqrt{2}x + 2\sqrt{2}y = 0$

46. $x^2 + 2xy + y^2 - 4\sqrt{2}y = 0$

In Questions 47–48, find the rotation equations when the xy axes are rotated through the given angle.

47. $60°$

48. $45°$

In Questions 49–50, find the angle through which the x-y axes should be rotated to eliminate the xy term in the equation.

49. $x^2 + xy + y^2 - 3y - 6 = 0$

50. $x^2 - 4xy + y^2 + 5 = 0$

51. Plot the points $(2, 3\pi/4)$ and $(-3, -2\pi/3)$ on a polar coordinate graph.

52. List four other pairs of polar coordinates for the point $(-2, \pi/4)$.

In Questions 53–62, sketch the graph of the equation in a polar coordinate system.

53. $r = 5$

54. $r = -2$

55. $\theta = 2\pi/3$

56. $\theta = -5\pi/6$

57. $r = 2\theta \quad (\theta \leq 0)$

58. $r = 4 \cos \theta$

59. $r = 2 - 2 \sin \theta$

60. $r = \cos 3\theta$

61. $r^2 = \cos 2\theta$

62. $r = 1 + 2 \sin \theta$

63. Convert $(3, -2\pi/3)$ from polar to rectangular coordinates.

64. Convert $(3, \sqrt{3})$ from rectangular to polar coordinates.

65. What is the eccentricity of the ellipse $3x^2 + y^2 = 84$?

66. What is the eccentricity of the ellipse $24x^2 + 30y^2 = 120$?

In Questions 67–70, sketch the graph of the equation, labeling the vertices and identifying the conic.

67. $r = \dfrac{12}{2 - \sin \theta}$

68. $r = \dfrac{14}{7 + 7 \cos \theta}$

69. $r = \dfrac{-24}{3 - 9 \cos \theta}$

70. $r = \dfrac{10}{3 + 4 \sin \theta}$

In Questions 71–74, find a polar equation of the conic that has focus (0, 0) and satisfies the given conditions.

71. Ellipse; vertices $(4, 0)$ and $(6, \pi)$

72. Hyperbola; vertices $(5, \pi/2)$ and $(-3, 3\pi/2)$

73. Eccentricity 1; directrix $r = 2 \sec \theta$

74. Eccentricity .75; directrix $r = -3 \csc \theta$

Discovery Project 10

Designing Machines to Make Designs

Parametric equations are extremely useful for modeling the behavior of moving parts, particularly when the action can be decomposed into two or more discrete movements. You saw this in the first section of the chapter in the form of the cycloid curves—the rolling of a circle was decomposed into the spinning of the circle and the linear motion of the center of the circle. Similarly, the motion of machine parts can be examined from the reverse point of view. That is, you can look at how two or more discrete parts act in concert to create a desirable pattern. In the case of the cycloid, the parametric equations $x = r(t - \sin t)$ and $y = r(1 - \cos t)$ model the motion of a point on the rim of a wheel of radius r which is rolling across the x-axis. The first component of each formula, the t and the 1, show the linear movement of the axle or center of the circle, while the $\sin t$ and $\cos t$ take care of the rotation of the rim around the axle. What happens if the motion of the center and rim are no longer synchronized?

1. Graph the parametric equations $x = t + \cos(kt)$ and $y = 5 + \sin(kt)$ for various values of k greater than one. What does it mean physically when k is larger than one?

2. Graph the parametric equations $x = t + \cos(kt)$ and $y = 5 + 5\sin(kt)$ for various values of k less than one. What does it mean physically when k is less than one?

3. The formulas from the previous questions can also be used to design a machine where objects move at a constant speed below a device which spins in a circle above the objects. This is a popular method for decorating mass-produced pastries. Find a value of k so that the graph of the parametric system looks like the picture to the right. Interpret what the value of k means in this instance.

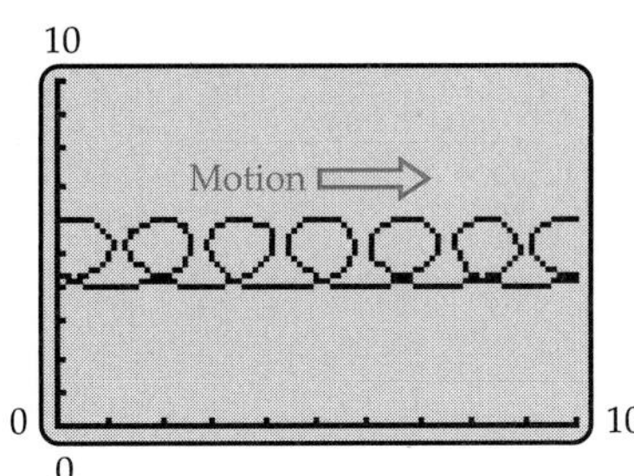

4. Based on your answer to the previous question, modify the parametric equations so that the pastries will have a strip of decorative icing approximately 1 cm wide, measured perpendicularly to the linear motion. Be sure to give time and length units in your answer.

5. Sewing machines use similar devices, called feed dogs, to move the cloth beneath the presser foot. The motion of the feed dogs can be simulated using functions of the type $\frac{|\sin kt|}{\sin kt}$ in place of sines and cosines. Use this to design a set of parametric equations which would produce stitches like the ones shown to the right.

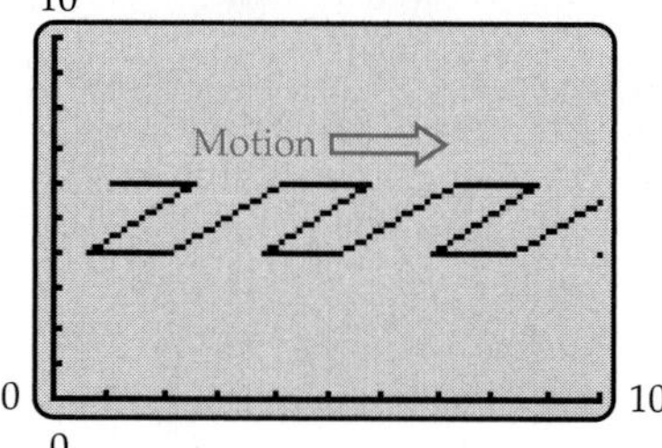

6. How would you change your answer to question 5 if the width of the stitches, measured perpendicularly to the direction of motion, needs to be 5 times as wide but the length when measured from peak to peak in the direction of motion, remains the same?

Systems of Equations

Is this a diamond in the rough?

```
rref C
      [[1 0 0 3 ]
       [0 1 0 -1]
       [0 0 1 4 ]]
```

The structure of certain crystals can be described by a large system of linear equations (more than a hundred equations and variables). A variety of resource allocation problems involving many variables can be handled by solving an appropriate system of equations. The fastest solution methods involve matrices and are easily implemented on a computer or calculator. See Exercise 53 on page 707.

Interdependence of Sections

Chapters 11 and 12 are independent of each other and may be read in any order.

In this chapter, Section 11.4 is independent of the rest of the chapter and may be read at any time. So, the interdependence of sections is

11.1 ⟶ 11.2 ⟶ 11.3

11.4

Section 11.1 may be omitted by readers who are familiar with solving systems of two linear equations.

Chapter Outline

This chapter deals with *systems of equations,* such as

$$\begin{aligned} 2x - 5y + 3z &= 1 \\ x + 2y - z &= 2 \\ 3x + y + 2z &= 11 \end{aligned}$$

Three equations in three variables

$$\begin{aligned} 2x + 5y + z + w &= 0 \\ 2y - 4z + 41w &= 5 \\ 3x + 7y + 5z - 8w &= -6 \end{aligned}$$

Three equations in four variables

$$\begin{aligned} x^2 + y^2 &= 25 \\ x^2 - y &= 7 \end{aligned}$$

Two equations in two variables

Sections 11.1–11.3 deal with systems of linear equations (such as the first two shown above). Systems involving nonlinear equations are considered in Section 11.4.

A *solution of a system* is a solution that satisfies *all* the equations in the system. For instance, in the first system of equations above, $x = 1, y = 2, z = 3$ is a solution of all three equations (check it) and hence is a solution of the system. On the other hand, $x = 0, y = 7, z = 12$ is a solution of the first two equations, but not of the third (check it). So $x = 0, y = 7, z = 12$ is not a solution of the system.

11.1 Systems of Linear Equations in Two Variables

Systems of linear equations in two variables may be solved geometrically or algebraically. The geometric method is similar to what we have done previously.

Example 1 Solve this system geometrically:

$$\begin{aligned} 2x - y &= 1 \\ 3x + 2y &= 4. \end{aligned}$$

Solution First, we solve each equation for y:

$$\begin{aligned} 2x - y &= 1 \\ -y &= -2x + 1 \\ y &= 2x - 1 \end{aligned} \qquad \begin{aligned} 3x + 2y &= 4 \\ 2y &= -3x + 4 \\ y &= \frac{-3x + 4}{2}. \end{aligned}$$

Next, we graph both equations on the same screen (Figure 11–1). As we saw in Section 1.4, each graph is a straight line and every point on the graph represents a solution of the equation. Therefore, the solution of the system is given by the coordinates of the point that lies on both lines. An intersection finder (Figure 11–2) shows that the approximate coordinates of this point are

$$x \approx .85714286 \qquad \text{and} \qquad y \approx .71428571. \blacksquare$$

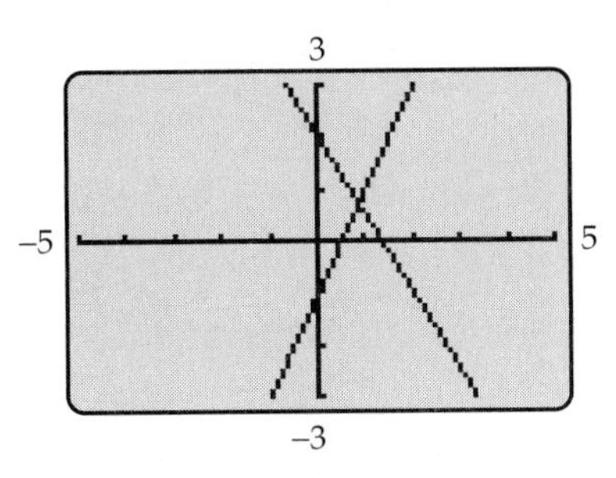

Figure 11–1

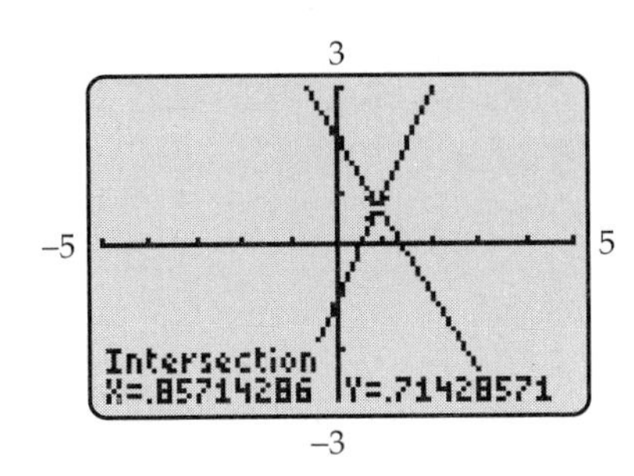

Figure 11–2

As shown in Example 1, the solutions of a system of linear equations are determined by the points where their graphs intersect. There are exactly three geometric possibilities for two lines in the plane: They are parallel, they intersect at a single point, or they coincide, as illustrated in Figure 11–3. Each of these possibilities leads to a different number of solutions for the system.

Number of Solutions of a System

A system of two linear equations in two variables must have

No solutions ***or***

Exactly one solution ***or***

An infinite number of solutions.

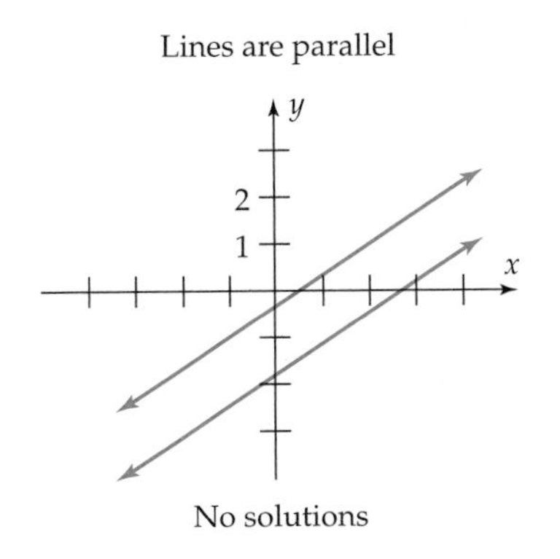

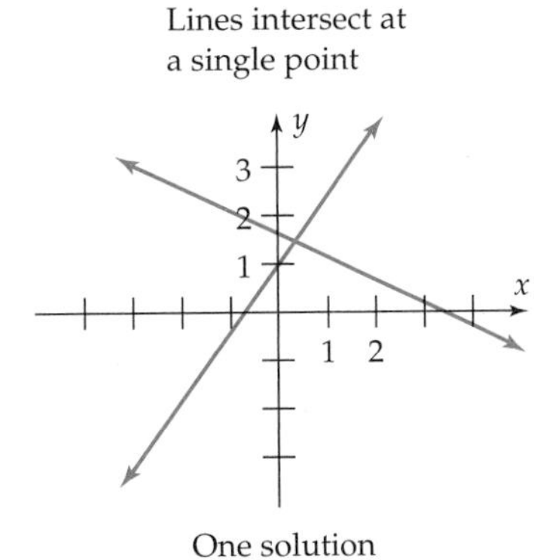

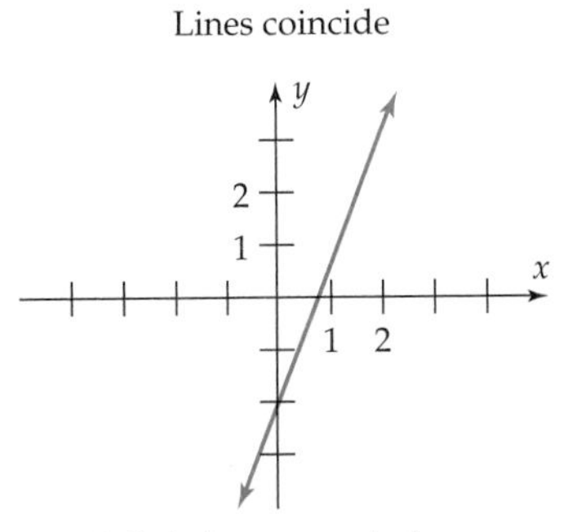

Figure 11–3

Algebraic Methods

When you use a calculator to solve systems geometrically, you may have to settle for an approximate solution, as we did in Example 1. Algebraic methods, however, produce exact solutions. Furthermore, algebraic methods are often as easy to implement as geometric ones, so we shall use them, whenever practical, to obtain exact solutions. One algebraic method is **substitution,** which is explained in the next example.

Example 2 Solve the system from Example 1 by substitution:

$$2x - y = 1$$
$$3x + 2y = 4.$$

Solution Any solution of this system must satisfy the first equation: $2x - y = 1$. Solving this equation for y, as in Example 1, shows that

$$y = 2x - 1.$$

Substituting this expression for y in the second equation, we have

$$\begin{aligned} 3x + 2y &= 4 \\ 3x + 2(2x - 1) &= 4 \\ 3x + 4x - 2 &= 4 \\ 7x &= 6 \\ x &= 6/7. \end{aligned}$$

Therefore, every solution of the original system must have $x = 6/7$. But when $x = 6/7$, we see from the first equation that

$$\begin{aligned} 2x - y &= 1 \\ 2\left(\frac{6}{7}\right) - y &= 1 \\ \frac{12}{7} - y &= 1 \\ -y &= -\frac{12}{7} + 1 \\ y &= \frac{12}{7} - 1 = \frac{5}{7}. \end{aligned}$$

(We would also have found that $y = 5/7$ if we had substituted $x = 6/7$ in the second equation.) Consequently, the exact solution of the original system is $x = 6/7$, $y = 5/7$. ■

CAUTION

In order to guard against arithmetic mistakes, you should always *check your answers* by substituting them into *all* the equations of the original system. We have in fact checked the answers in all the examples, but these checks are omitted to save space.

When using the substitution method, you may solve either of the given equations for either one of the variables and then substitute that result in the other equation. In Example 2, we solved for y in the first equation because that avoided the fractional expression that would have occurred if we had solved for x or had solved the second equation for x or y.

The Elimination Method

The **elimination method** of solving systems of linear equations is often more convenient than substitution. It depends on this fact:

> **Multiplying both sides of an equation by a nonzero constant does not change the solutions of the equation.**

For example, the equation $x + 3 = 5$ has the same solution as $2x + 6 = 10$ (the first equation multiplied by 2). The elimination method also uses this fact from basic algebra:

If $A = B$ and $C = D$, then $A + C = B + D$ and $A - C = B - D$.

Example 3 To solve this system

$$\begin{aligned} x - 3y &= 4 \\ 2x + y &= 1 \end{aligned}$$

we replace the first equation by an equivalent one (that is, one with the same solutions):

$$\begin{aligned} -2x + 6y &= -8 \qquad \textit{[First equation multiplied by } -2] \\ 2x + y &= 1. \end{aligned}$$

The multiplier -2 was chosen so that the coefficients of x in the two equations would be negatives of each other. Any solution of this last system must also be a solution of the sum of the two equations:

$$\begin{aligned} -2x + 6y &= -8 \\ 2x + y &= 1 \\ \hline 7y &= -7. \qquad \textit{[The first variable has been eliminated]} \end{aligned}$$

Solving this last equation we see that $y = -1$. Substituting this value in the first of the original equations shows that

$$\begin{aligned} x - 3(-1) &= 4 \\ x &= 1. \end{aligned}$$

Therefore, $x = 1, y = -1$ is the solution of the original system. ■

Example 4 Any solution of the system

$$\begin{aligned} 5x - 3y &= 3 \\ 3x - 2y &= 1 \end{aligned}$$

must also be a solution of this system:

$$\begin{aligned} 10x - 6y &= 6 \qquad \textit{[First equation multiplied by 2]} \\ -9x + 6y &= -3. \qquad \textit{[Second equation multiplied by } -3] \end{aligned}$$

The multipliers 2 and -3 were chosen so that the coefficients of y in the new equations would be negatives of each other. Any solution of this last system must also be a solution of the equation obtained by adding these two equations:

$$\begin{aligned} 10x - 6y &= 6 \\ -9x + 6y &= -3 \\ \hline x &= 3. \end{aligned} \qquad \text{[The second variable has been eliminated]}$$

Substituting $x = 3$ in the first of the original equations shows that

$$\begin{aligned} 5(3) - 3y &= 3 \\ -3y &= -12 \\ y &= 4. \end{aligned}$$

Therefore the solution of the original system is $x = 3$, $y = 4$. ■

Example 5 To solve the system

$$\begin{aligned} 2x - 3y &= 5 \\ 4x - 6y &= 1 \end{aligned}$$

we multiply the first equation by -2 and add:

$$\begin{aligned} -4x + 6y &= -10 \\ 4x - 6y &= 1 \\ \hline 0 \qquad &= -9. \end{aligned}$$

Since $0 = -9$ is always false, the original system cannot possibly have any solutions. A system with no solution is said to be **inconsistent.** ■

GRAPHING EXPLORATION

Confirm the result of Example 5 geometrically by graphing the two equations in the system. Do these lines intersect or are they parallel?

Example 6 To solve the system

$$\begin{aligned} 3x - y &= 2 \\ 6x - 2y &= 4 \end{aligned}$$

we multiply the first equation by 2 to obtain the system:

$$\begin{aligned} 6x - 2y &= 4 \\ 6x - 2y &= 4. \end{aligned}$$

The two equations are identical. So the solutions of this system are the same as the solutions of the single equation $6x - 2y = 4$, which can be rewritten as

$$2y = 6x - 4$$
$$y = 3x - 2.$$

This equation, and hence the original system, has infinitely many solutions. They can be described as follows: Choose any real number for x, say $x = b$. Then $y = 3x - 2 = 3b - 2$. So the solutions of the system are all pairs of numbers of the form

$$x = b, \qquad y = 3b - 2 \quad \text{where } b \text{ is any real number.}$$

A system such as this is said to be **dependent.** ■

Some nonlinear systems can be solved by replacing them with equivalent linear systems.

Example 7 To solve the system

$$\frac{1}{x} + \frac{3}{y} = -1$$
$$\frac{2}{x} - \frac{1}{y} = 5$$

we let $u = 1/x$ and $v = 1/y$ so that the system becomes:

$$u + 3v = -1$$
$$2u - v = 5.$$

We can solve this system by multiplying the first equation by -2 and adding it to the second equation:

$$\begin{aligned} -2u - 6v &= 2 \\ 2u - v &= 5 \\ \hline -7v &= 7 \\ v &= -1. \end{aligned}$$

Substituting $v = -1$ in the equation $u + 3v = -1$, we see that $u = -3(-1) - 1 = 2$. Consequently, the possible solution of the original system is

$$x = \frac{1}{u} = \frac{1}{2} \qquad \text{and} \qquad y = \frac{1}{v} = \frac{1}{(-1)} = -1.$$

You should substitute this possible solution in both equations of the original system to check that it is actually a solution of the system. ■

Applications

Example 8 575 people attend a ball game, and total ticket sales are \$2575. If adult tickets cost \$5 and children's tickets \$3, how many adults attended the game? How many children?

Solution Let x be the number of adults and y the number of children. Then,

Number of adults + Number of children = Total attendance

$$x + y = 575.$$

We can obtain a second equation by using the information about ticket sales:

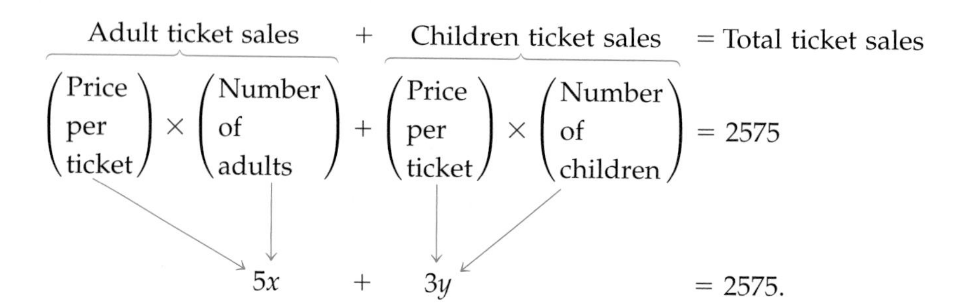

In order to find x and y we need only solve this system of equations:

$$\begin{aligned} x + y &= 575 \\ 5x + 3y &= 2575 \end{aligned}$$

Multiplying the first equation by -3 and adding we have:

$$\begin{aligned} -3x - 3y &= -1725 \\ 5x + 3y &= 2575 \\ \hline 2x &= 850 \\ x &= 425 \end{aligned}$$

So, 425 adults attended the game. The number of children was $y = 575 - x = 575 - 425 = 150$. ■

Example 9 A plane flies 3000 miles from San Francisco to Boston in 5 hours, with a tailwind all the way. The return trip on the same route, now with a headwind, takes 6 hours. Assuming both remain constant, find the speed of the plane and the speed of the wind.

Solution Let x be the plane's speed and y the wind speed (both in miles per hour). Then on the trip to Boston with a tailwind,

$x + y =$ actual speed of the plane (wind and plane go in same direction).

On the return trip against a headwind,

$x - y =$ actual speed of plane (wind and plane go in opposite directions).

Using the basic rate/distance equation we have:

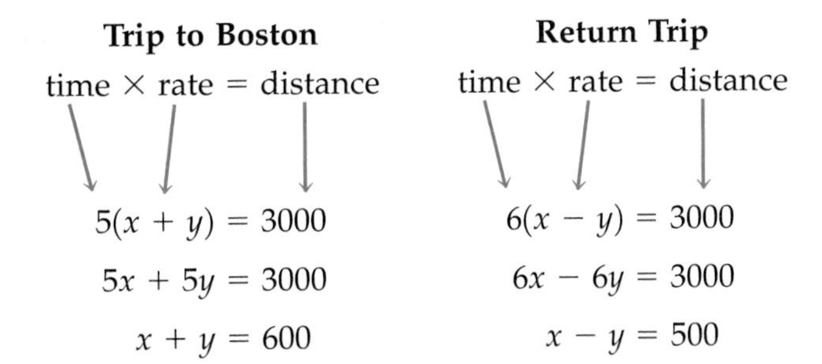

$$5(x + y) = 3000 \qquad 6(x - y) = 3000$$
$$5x + 5y = 3000 \qquad 6x - 6y = 3000$$
$$x + y = 600 \qquad x - y = 500$$

Thus, we need only solve this system of equations:

$$x + y = 600$$
$$x - y = 500.$$

Adding the two equations shows that

$$2x = 1100$$
$$x = 550$$

Substituting this result in the first equation, we have

$$550 + y = 600$$
$$y = 50$$

Thus, the plane's speed is 550 mph and the wind speed is 50 mph. ■

Example 10 How many pounds of tin and how many pounds of copper should be added to 1000 pounds of an alloy that is 10% tin and 30% copper in order to produce a new alloy that is 27.5% tin and 35% copper?

Solution Let x be the number of pounds of tin and y the number of pounds of copper to be added to the 1000 pounds of the old alloy. Then there will be $1000 + x + y$ pounds of the new alloy. We first find the *amounts* of tin and copper in the new alloy:

	Pounds in old alloy	+	Pounds added	=	Pounds in new alloy
Tin	10% of 1000	+	x	=	$100 + x$
Copper	30% of 1000	+	y	=	$300 + y$

Now consider the *percentages* of tin and copper in the new alloy.

	Percentage in new alloy	×	Total weight of new alloy	=	Pounds in new alloy
Tin	27.5%	of	$1000 + x + y$	=	$.275(1000 + x + y)$
Copper	35%	of	$1000 + x + y$	=	$.35(1000 + x + y)$

The two ways of computing the weight of each metal in the alloy must produce the same result, that is,

$$100 + x = .275(1000 + x + y) \qquad \textit{[pounds of tin]}$$

$$300 + y = .35(1000 + x + y). \qquad \textit{[pounds of copper]}$$

Multiplying out the right sides and rearranging terms produces this system of equations:

$$.725x - .275y = 175$$
$$-.35x + .65y = 50.$$

Multiplying the first equation by .65 and the second by .275 and adding the results, we have:

$$\begin{aligned} .47125x - .17875y &= 113.75 \\ -.09625x + .17875y &= 13.75 \\ \hline .37500x \qquad\qquad &= 127.50 \\ x &= 340. \end{aligned}$$

Substituting this in the first equation and solving for y shows that $y = 260$. Therefore, 340 pounds of tin and 260 pounds of copper should be added. ■

Exercises 11.1

In Exercises 1–6, determine whether the given values of x, y, and z are a solution of the system of equations.

1. $x = -1, y = 3$
$2x + y = 1$
$-3x + 2y = 9$

2. $x = 3, y = 4$
$2x + 6y = 30$
$x + 2y = 11$

3. $x = 2, y = -1$
$\frac{1}{3}x + \frac{1}{2}y = \frac{1}{6}$
$\frac{1}{2}x + \frac{1}{3}y = \frac{2}{3}$

4. $x = .4, \quad y = .7$
$3.1x - 2y = -.16$
$5x - 3.5y = -.48$

5. $x = \frac{1}{2}, y = 3, z = -1$
$2x - y + 4z = -6$
$3y + 3z = 6$
$2z = 2$

6. $x = 2, y = \frac{3}{2}, z = -\frac{1}{2}$
$3x + 4y - 2z = 13$
$\frac{1}{2}x + 8z = -3$
$x - 3y + 5z = -5$

In Exercises 7–14, use substitution to solve the system.

7. $x - 2y = 5$
$2x + y = 3$

8. $3x - y = 1$
$-x + 2y = 4$

9. $3x - 2y = 4$
$2x + y = -1$

10. $5x - 3y = -2$
$-x + 2y = 3$

11. $r + s = 0$
$r - s = 5$

12. $t = 3u + 5$
$t = u + 5$

13. $x + y = c + d$ (where c, d are constants)
$x - y = 2c - d$

14. $x + 3y = c - d$ (where c, d are constants)
$2x - y = c + d$

In Exercises 15–40, use the elimination method to solve the system.

15. $2x - 2y = 12$
$-2x + 3y = 10$

16. $3x + 2y = -4$
$4x - 2y = -10$

17. $x + 3y = -1$
$2x - y = 5$

18. $4x - 3y = -1$
$x + 2y = 19$

19. $2x + 3y = 15$
$8x + 12y = 40$

20. $2x + 5y = 8$
$6x + 15y = 18$

21. $3x - 2y = 4$
$6x - 4y = 8$

22. $2x - 8y = 2$
$3x - 12y = 3$

23. $12x - 16y = 8$
$42x - 56y = 28$

24. $\frac{1}{3}x + \frac{2}{5}y = \frac{1}{6}$
$20x + 24y = 10$

25. $9x - 3y = 1$
$6x - 2y = -5$

26. $8x + 4y = 3$
$10x + 5y = 1$

27. $\frac{x}{3} - \frac{y}{2} = -3$
$\frac{2x}{5} + \frac{y}{5} = -2$

28. $\frac{x}{3} + \frac{3y}{5} = 4$
$\frac{x}{6} - \frac{y}{2} = -3$

29. $\frac{x+y}{4} - \frac{x-y}{3} = 1$
$\frac{x+y}{4} + \frac{x-y}{2} = 9$

30. $\frac{x-y}{4} + \frac{x+y}{3} = 1$
$\frac{x+2y}{3} + \frac{3x-y}{2} = -2$

31. $\frac{1}{x} - \frac{3}{y} = 2$
$\frac{2}{x} + \frac{1}{y} = 3$

32. $\frac{5}{x} + \frac{2}{y} = 0$
$\frac{6}{x} + \frac{4}{y} = 3$

33. $\frac{2}{x} + \frac{3}{y} = 8$
$\frac{3}{x} - \frac{1}{y} = 1$

34. $\frac{3}{x^2} + \frac{2}{y^2} = 11$ $\left[\textit{Hint}: \text{Let } u = \frac{1}{x^2} \text{ and } v = \frac{1}{y^2}.\right]$
$\frac{1}{x^2} - \frac{3}{y^2} = -11$

35. $\frac{3}{x+1} - \frac{4}{y-2} = 2$
$\frac{1}{x+1} + \frac{4}{y-2} = 5$
$\left[\textit{Hint}: \text{Let } u = \frac{1}{x+1} \text{ and } v = \frac{1}{y-2}.\right]$

36. $\frac{-5}{x^2+3} - \frac{2}{y^2-2} = -12$
$\frac{3}{x^2+3} + \frac{1}{y^2-2} = 5$

37. $3.5x - 2.18y = 2.00782$
$1.92x + 6.77y = -3.86928$

38. $463x - 80y = -13781.6$
$.0375x + .912y = 50.79624$

39. $ax + by = r$ (where a, b, c, d, r, s are
$cx + dy = s$ constants and $ad - bc \neq 0$)

40. $ax + by = ab$ (where a, b are nonzero
$bx - ay = ab$ constants)

41. Let c be any real number. Show that this system has exactly one solution:

$$x + 2y = c$$
$$6x - 3y = 4$$

42. (a) Find the values of c for which this system has an infinite number of solutions.

$$2x - 4y = 6$$
$$-3x + 6y = c$$

(b) Find the values of c for which the system in part (a) has no solutions.

In Exercises 43 and 44, find the values of c and d for which both given points lie on the given straight line.

43. $cx + dy = 2$; $(0, 4)$ and $(2, 16)$

44. $cx + dy = -6$; $(1,3)$ and $(-2, 12)$

45. Bill and Ann plan to install a heating system for their swimming pool. Since gas is not available, they have a choice of electric or solar heat. They have gathered the following cost information.

System	Installation Costs	Monthly Operational Cost
Electric	\$2,000	\$80
Solar	\$14,000	\$ 9.50

(a) Ignoring changes in fuel prices, write a linear equation for each heating system that expresses its total cost y in terms of the number of *years* x of operation.
(b) What is the five-year total cost of electric heat? Of solar heat?
(c) In what year will the total cost of the two heating systems be the same? Which is the cheapest system before that time? After that time?

46. One parcel of land is worth \$100,000 now and is increasing in value at the rate of \$3000 per year. A second parcel is now worth \$60,000 and is increasing in value at the rate of \$7500 per year.
(a) For each parcel of land write an equation that expresses the value y of the land and in year x.
(b) Graph the equations in part (a).
(c) Where do the lines intersect? What is the significance of this point?
(d) Which parcel will be worth more in five years? In 15 years?

47. A toy company makes Boomie Babies, as well as collector cases for each Boomie Baby. To make x cases costs the company \$5000 in fixed overhead, plus \$7.50 per case. An outside supplier has offered to produce any desired volume of cases for \$8.20 per case.

(a) Write an equation that expresses the company's cost to make x cases itself.
(b) Write an equation that expresses the cost of buying x cases from the outside supplier.
(c) Graph both equations on the same axes and determine when the two costs are the same.
(d) When should the company make the cases themselves and when should they buy them from the outside supplier?

48. The sum of two numbers is 40. The difference between twice the first number and the second is 11. What are the numbers?

49. A 200-seat theater charges $3 for adults and $1.50 for children. If all seats were filled and the total ticket income was $510, how many adults and how many children were in the audience?

50. A theater charges $4 for main floor seats and $2.50 for balcony seats. If all seats are sold, the ticket income is $2100. At one show, 25% of the main floor seats and 40% of the balcony seats were sold and ticket income was $600. How many seats are on the main floor and how many in the balcony?

51. An investor has part of her money in an account that pays 9% annual interest, and the rest in an account that pays 11% annual interest. If she has $8000 less in the higher paying account than in the lower paying one and her total annual interest income is $2010, how much does she have invested in each account?

52. Joyce has money in two investment funds. Last year the first fund paid a dividend of 8% and the second a dividend of 2% and Joyce received a total of $780. This year the first fund paid a 10% dividend and the second only 1% and Joyce received $810. How much money does she have invested in each fund?

53. At a certain store, cashews cost $4.40/pound and peanuts $1.20/pound. If you want to buy exactly 3 pounds of nuts for $6.00, how many pounds of each kind of nuts should you buy? [*Hint:* If you buy x pounds of cashews and y pounds of peanuts, then $x + y = 3$. Find a second equation by considering cost and solve the resulting system.]

54. A store sells deluxe tape recorders for $150. The regular model costs $120. The total tape recorder inventory would sell for $43,800. But during a recent month the store actually sold half of its deluxe models and two-thirds of the regular models and took in a total of $26,700. How many of each kind of recorder did they have at the beginning of the month?

55. A boat made a 4-mile trip upstream against a constant current in 15 minutes. The return trip at the same constant speed with the same current took 12 minutes. What is the speed of the boat and of the current?

56. A plane flying into a headwind travels 2000 miles in 4 hours and 24 minutes. The return flight along the same route with a tailwind takes 4 hours. Find the wind speed and the plane's speed (assuming both are constant).

57. A winemaker has two large casks of wine. One wine is 8% alcohol and the other 18% alcohol. How many liters of each wine should be mixed to produce 30 liters of wine that is 12% alcohol?

58. How many cubic centimeters (cm^3) of a solution that is 20% acid and of another solution that is 45% acid should be mixed to produce 100 cm^3 of a solution that is 30% acid?

59. How many grams of a 50%-silver alloy should be mixed with a 75%-silver alloy to obtain 40 grams of a 60%-silver alloy?

60. A machine in a pottery factory takes 3 minutes to form a bowl and 2 minutes to form a plate. The material for a bowl costs $.25 and the material for a plate costs $.20. If the machine runs for 8 hours straight and exactly $44 is spent for material, how many bowls and plates can be produced?

61. Because Chevrolet and Saturn produce cars in the same price range, Chevrolet's sales are not only a function of Chevy prices (x), but of Saturn prices (y) as well. Saturn prices are related similarly to both Saturn and Chevy prices. Suppose General Motors forecasts the demand z_1 for Chevrolets and the demand z_2 for Saturns to be given by

$$z_1 = 68{,}000 - 6x + 4y \quad \text{and} \quad z_2 = 42{,}000 + 3x - 3y.$$

Solve this system of equations and express
(a) the price x of Chevrolets as a function of z_1 and z_2;
(b) the price y of Saturns as a function of z_1 and z_2.

11.2 Large Systems of Linear Equations

Systems of linear equations in three variables can be interpreted geometrically as the intersections of planes. However, algebraic methods are the only practical means to solve systems of linear equations with three or more variables. We begin with **Gaussian elimination,*** which is an extension of the elimination method used in Section 11.1. In order to understand why Gaussian elimination works, it is helpful to examine a system of two equations from a different viewpoint.

Example 1 In Example 3 of Section 11.1, we solved the system

$$\begin{aligned} x - 3y &= 4 \\ 2x + y &= 1 \end{aligned}$$

by multiplying the first equation by -2 and adding it to the second in order to eliminate the variable x:

$$\begin{aligned} -2x + 6y &= -8 && [-2 \textit{ times first equation}] \\ 2x + y &= 1 && [\textit{Second equation}] \\ \hline 7y &= -7. && [\textit{Sum of second equation and } -2 \textit{ times first equation}] \end{aligned}$$

We then solved this last equation for y and substituted the answer, $y = -1$, in the original first equation to find that $x = 1$. What we did, in effect, was

Replace the original system by the following system, in which x has been eliminated from the second equation, then solve this new system.

$$(*) \quad \begin{aligned} x - 3y &= 4 && [\textit{First equation}] \\ 7y &= -7. && [\textit{Sum of second equation and } -2 \textit{ times first equation}] \end{aligned}$$

As we saw on page 685, any solution of the original system must be a solution of the first equation and of the sum equation $7y = -7$, and hence of system **(*)**. Conversely, it is easy to check that any solution of system **(*)** is also a solution of the original system. *Note:* We are not claiming that the second equations in the two systems have the same solutions—they don't—but only that the two *systems* have the same solution, namely, $x = 1, y = -1$. ■

Two systems of equations are said to be **equivalent** if they have the same solutions, that is, every solution of one system is a solution of the other system and vice versa. The basic technique in Gaussian elimination is to do what was done in Example 1: Replace the given system by an equivalent one (perhaps several times) until you obtain an equivalent

*Named after the great German mathematician K. F. Gauss (1777–1855).

system in which enough variables have been eliminated to make the system easy to solve. There are several ways to produce an equivalent system from a given one.

Elementary Operations

Performing any of the following operations on a system of equations produces an equivalent system:

1. **Interchange any two equations in the system.**
2. **Replace an equation in the system by a nonzero constant multiple of itself.**
3. **Replace an equation in the system by the sum of itself and a constant multiple of another equation in the system.**

The third elementary operation was illustrated in Example 1. The reason that the first elementary operation produces an equivalent system is that rearranging the order of the equations certainly doesn't affect their solutions, and hence doesn't affect the solutions of the system. The second elementary operation produces an equivalent system because multiplying a single equation by a constant does not change the solutions of that equation, and hence does not change the solutions of any system including that equation.

Example 2 Solve the system

$$\begin{aligned} x + 4y - 3z &= 1 && \text{[Equation A]} \\ -3x - 6y + z &= 3 && \text{[Equation B]} \\ 2x + 11y - 5z &= 0 && \text{[Equation C]} \end{aligned}$$

Solution We first use elementary operations to produce an equivalent system in which the variable x has been eliminated from the second and third equations.

To eliminate x from equation B, replace equation B by the sum of itself and 3 times equation A:

$$\begin{aligned} \text{[3 times A]} \quad 3x + 12y - 9z &= 3 \\ \text{[B]} \quad -3x - 6y + z &= 3 \\ \hline 6y - 8z &= 6 \end{aligned}$$

$$\begin{aligned} x + 4y - 3z &= 1 && \text{[A]} \\ 6y - 8z &= 6 && \text{[Sum of B and 3 times A]} \\ 2x + 11y - 5z &= 0 && \text{[C]} \end{aligned}$$

To eliminate x from equation C we replace equation C by the sum of itself and -2 times equation A:

$$\begin{array}{lrcl} [-2 \text{ times } A] & -2x - 8y + 6z &=& -2 \\ [C] & 2x + 11y - 5z &=& 0 \\ \hline & 3y + z &=& -2 \end{array}$$

$$\begin{aligned} x + 4y - 3z &= 1 \\ 6y - 8z &= 6 \\ 3y + z &= -2 \quad \text{[Sum of } C \text{ and } -2 \text{ times } A] \end{aligned}$$

The next step is to eliminate the y term in one of the last two equations. This can be done by replacing the second equation by the sum of itself and -2 times the third equation:

$$\begin{aligned} x + 4y - 3z &= 1 \\ -10z &= 10 \quad \text{[Sum of second equation and } -2 \text{ times third equation]} \\ 3y + z &= -2 \end{aligned}$$

Finally, interchange the last two equations:

$$(*) \qquad \begin{aligned} x + 4y - 3z &= 1 \\ 3y + z &= -2 \\ -10z &= 10. \end{aligned}$$

This last system, which is equivalent to the original one, is easily solved. The last equation shows that

$$-10z = 10, \qquad \text{or equivalently,} \qquad z = -1.$$

Substituting $z = -1$ in the second equation shows that

$$\begin{aligned} 3y + z &= -2 \\ 3y + (-1) &= -2 \\ 3y &= -1 \\ y &= -\frac{1}{3}. \end{aligned}$$

Substituting $y = -1/3$ and $z = -1$ in the first equation yields:

$$\begin{aligned} x + 4y - 3z &= 1 \\ x + 4\left(-\frac{1}{3}\right) - 3(-1) &= 1 \\ x = 1 + \frac{4}{3} - 3 &= -\frac{2}{3}. \end{aligned}$$

Therefore, the original system has just one solution: $x = -2/3$, $y = -1/3$, $z = -1$. ■

The process used to solve the final system **(*)** in Example 2 is called **back substitution** because you begin with the last equation and work back to the first. It works because system **(*)** is in **triangular form:** The first variable in the first equation, x, does not appear in any subsequent equation; the first variable in the second equation, y, does not appear in any subsequent equation, and so on. It can be shown that the procedure in Example 2 works in every case:

Gaussian Elimination

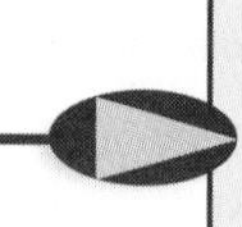

Any system of linear equations can be transformed into an equivalent system in triangular form by using a finite number of elementary operations. If the system has solutions, they can then be found by back substitution in the triangular form system.

Technology Tip

Some (but not all) systems with the same number of linear equations as variables can be solved directly by using SIMULT on TI-86, SYSTEM in the TOOL menu of Sharp 9600, or SIMULTANEOUS in the Casio 9850 EQUATION menu (on the main menu).

In terms of convenience, most people prefer to use a calculator or computer to solve large systems of equations. However, the system solvers on calculators (see the Tip in the margin) only work when the system has the same number of equations as variables *and* has a unique solution. When the solver gives an error message, you can't tell whether the system has many solutions or none at all. So we now develop a version of Gaussian elimination that also handles these situations and can be implemented on a calculator.

Matrix Methods

When solving systems by hand, a lot of time is wasted copying the x's, y's, z's, and so on. This fact suggests a shorthand system for representing a system of equations. For example, the system

$$(*)\qquad \begin{aligned} x + 2y + 3z &= -2 \\ 2x + 6y + z &= 2 \\ 3x + 3y + 10z &= -2 \end{aligned}$$

can be represented by the following rectangular array of numbers, consisting of the coefficients of the variables and the constants on the right of the equal sign, arranged in the same order they appear in the system:

$$\left(\begin{array}{ccc|c} 1 & 2 & 3 & -2 \\ 2 & 6 & 1 & 2 \\ 3 & 3 & 10 & -2 \end{array}\right)$$

This array is called the **augmented matrix** of the system. It has three horizontal **rows** and four vertical **columns.**

Example 3 Use the matrix form of the preceding system (*) to solve the system.

Solution To solve the system in its original equation form, we would use elementary operations to eliminate the x terms from the last two equations, and then eliminate the y term from the last equation. With matrices we do essentially the same thing, with the elementary operations on equations being replaced by the corresponding **row operations** on the augmented matrix in order to make certain entries in the first and second columns 0. Here is a side-by-side development of the two solution methods.

Technology Tip

Check your instruction manual to learn how to enter and store matrices in the matrix memory and to perform row operations on them.

Equation Method

Replace the second equation by the sum of itself and -2 times the first equation:

$$\begin{aligned} x + 2y + 3z &= -2 \\ 2y - 5z &= 6 \\ 3x + 3y + 10z &= -2 \end{aligned}$$

Replace the third equation by the sum of itself and -3 times the first equation:

$$\begin{aligned} x + 2y + 3z &= -2 \\ 2y - 5z &= 6 \\ -3y + z &= 4 \end{aligned}$$

Multiply the second equation by $1/2$ (so that y has coefficient 1):

$$\begin{aligned} x + 2y + 3z &= -2 \\ y - \frac{5}{2}z &= 3 \\ -3y + z &= 4 \end{aligned}$$

Replace the third equation by the sum of itself and 3 times the second equation:

$$\begin{aligned} x + 2y + 3z &= -2 \\ y - \frac{5}{2}z &= 3 \\ -\frac{13}{2}z &= 13 \end{aligned}$$

Finally, multiply the last equation by $-2/13$:[†]

$$(**)\qquad \begin{aligned} x + 2y + 3z &= -2 \\ y - \frac{5}{2}z &= 3 \\ z &= -2 \end{aligned}$$

Matrix Method

Replace the second row by the sum of itself and -2 times the first row:

$$\left(\begin{array}{ccc|c} 1 & 2 & 3 & -2 \\ 0 & 2 & -5 & 6 \\ 3 & 3 & 10 & -2 \end{array}\right)$$

Replace the third row by the sum of itself and -3 times the first row:

$$\left(\begin{array}{ccc|c} 1 & 2 & 3 & -2 \\ 0 & 2 & -5 & 6 \\ 0 & -3 & 1 & 4 \end{array}\right)$$

Multiply the second row by $1/2$:

$$\left(\begin{array}{ccc|c} 1 & 2 & 3 & -2 \\ 0 & 1 & -\frac{5}{2} & 3 \\ 0 & -3 & 1 & 4 \end{array}\right)$$

Replace the third row by the sum of itself and 3 times the second row:

$$\left(\begin{array}{ccc|c} 1 & 2 & 3 & -2 \\ 0 & 1 & -\frac{5}{2} & 3 \\ 0 & 0 & -\frac{13}{2} & 13 \end{array}\right)$$

Finally, multiply the last row by $-2/13$:

$$\left(\begin{array}{ccc|c} 1 & 2 & 3 & -2 \\ 0 & 1 & -\frac{5}{2} & 3 \\ 0 & 0 & 1 & -2 \end{array}\right)$$

[†]This step isn't necessary, but it is often convenient to have 1 as the coefficient of the first variable in each equation.

System (**) is easily solved. The third equation shows that $z = -2$ and substituting this in the second equation shows that

$$y - \frac{5}{2}(-2) = 3$$

$$y = 3 - 5 = -2.$$

Substituting $y = -2$ and $z = -2$ in the first equation yields

$$x + 2(-2) + 3(-2) = -2$$

$$x = -2 + 4 + 6 = 8.$$

Therefore, the only solution of the original system is $x = 8$, $y = -2$, $z = -2$. ■

When using matrix notation, row operations replace elementary operations on equations, as shown in Example 3. The solution process ends when you reach a matrix, such as the last one in Example 3, that satisfies these conditions:

All rows consisting entirely of zeros (if any) are at the bottom.

The first nonzero entry in each nonzero row is a 1 (called a **leading 1**).

Each leading 1 appears to the right of leading 1's in any preceding rows.

Such a matrix is said to be in **row echelon form.**

Technology Tip

To put a matrix in row echelon form, use REF in the MATH or OPS submenu of the TI-83/86 MATRIX menu; or in the MATRIX submenu of the TI-89 MATH menu; or use rowEF in the MATH submenu of the Sharp 9600 MATRIX menu.

Most calculators have a key that uses row operations to put a given matrix into row echelon form (see the tip in the margin). For example, using the TI-83 REF key on the first matrix in Example 3 produced this row echelon matrix and corresponding system of equations:

$$\begin{pmatrix} 1 & 1 & \frac{10}{3} & -\frac{2}{3} \\ 0 & 1 & -\frac{17}{12} & \frac{5}{6} \\ 0 & 0 & 1 & -2 \end{pmatrix} \qquad \begin{aligned} x + y + \frac{10}{3}z &= -\frac{2}{3} \\ y - \frac{17}{12}z &= \frac{5}{6} \\ z &= -2. \end{aligned}$$

Because the calculator used a different sequence of row operations, it produced a different row echelon matrix (and corresponding equations) than the one in (**) of Example 3. You can easily verify, however, that the preceding system has the same solutions as (**) in Example 3: $x - 8$, $y = -2$, $z = -2$.

Example 4 Solve the system

$$\begin{aligned} x + y + 2z &= 1 \\ 2x + 4y + 5z &= 2 \\ 3x + 5y + 7z &= 4 \end{aligned}$$

Solution If you try to use a systems equation solver on a calculator, you will get an error message. So we form the augmented matrix and reduce it to row echelon form via row operations. A TI-86 produced the row echelon matrix in Figure 11–4 and working by hand produces the following:

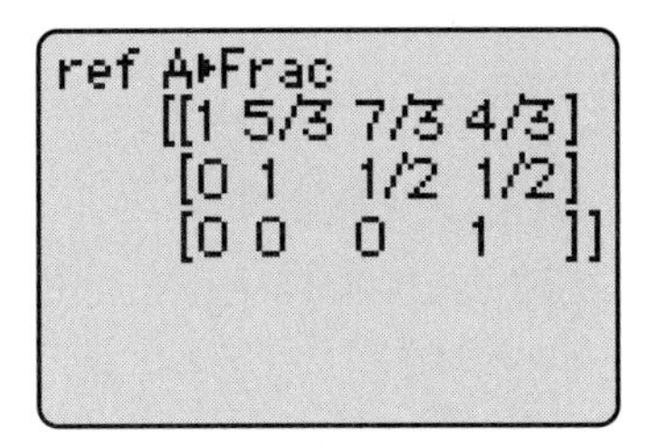

Figure 11–4

Technology Tip

On TI calculators, using the FRAC key in conjunction with the REF key (as in Figure 11–4) usually eliminates long decimals and makes the matrix easier to read.

$$\begin{pmatrix} 1 & 1 & 2 & 1 \\ 2 & 4 & 5 & 2 \\ 3 & 5 & 7 & 4 \end{pmatrix}$$

Replace second row by the sum of itself and −2 times the first row:

$$\begin{pmatrix} 1 & 1 & 2 & 1 \\ 0 & 2 & 1 & 0 \\ 3 & 5 & 7 & 4 \end{pmatrix}$$

Replace third row by the sum of itself and −3 times the first row:

$$\begin{pmatrix} 1 & 1 & 2 & 1 \\ 0 & 2 & 1 & 0 \\ 0 & 2 & 1 & 1 \end{pmatrix}$$

Multiply second row by 1/2:

$$\begin{pmatrix} 1 & 1 & 2 & 1 \\ 0 & 1 & \frac{1}{2} & 0 \\ 0 & 2 & 1 & 1 \end{pmatrix}$$

Replace third row by the sum of itself and −2 times the second row:

$$\begin{pmatrix} 1 & 1 & 2 & 1 \\ 0 & 1 & \frac{1}{2} & 0 \\ 0 & 0 & 0 & 1 \end{pmatrix}$$

Look at the last row in the matrix above or the one in Figure 11–4; it represents the equation

$$0x + 0y + 0z = 1.$$

Since this equation has no solutions (the left side is always 0 and the right side 1), neither does the original system. Such a system is said to be **inconsistent.** ■

The Gauss-Jordan Method

Gaussian elimination on a calculator is an efficient method of solving systems of equations, but may involve some messy calculations when you solve the final triangular form system by hand. Most hand computations can be eliminated by using a slight variation, known as the **Gauss-Jordan method.*** In order to see how this method works and its relationship to Gaussian elimination, we work the first example by hand.

*This method was developed by the German engineer Wilhelm Jordan (1842–1899).

Example 5 Use the Gauss-Jordan method to solve this system:

$$\begin{aligned} x - y + 5z &= -6 \\ 3x + 3y - z &= 10 \\ x - 5y + 8z &= -17 \\ x + 3y + 2z &= 5 \end{aligned}$$

Solution We form the augmented matrix and use Gaussian elimination to reduce it to row echelon form.

$$\begin{pmatrix} 1 & -1 & 5 & -6 \\ 3 & 3 & -1 & 10 \\ 1 & -5 & 8 & -17 \\ 1 & 3 & 2 & 5 \end{pmatrix}$$

Replace second row by the sum of itself and −3 times the first row:

$$\begin{pmatrix} 1 & -1 & 5 & -6 \\ 0 & 6 & -16 & 28 \\ 1 & -5 & 8 & -17 \\ 1 & 3 & 2 & 5 \end{pmatrix}$$

Replace third row by the sum of itself and −1 times the first row:

$$\begin{pmatrix} 1 & -1 & 5 & -6 \\ 0 & 6 & -16 & 28 \\ 0 & -4 & 3 & -11 \\ 1 & 3 & 2 & 5 \end{pmatrix}$$

Replace fourth row by the sum of itself and −1 times the first row:

$$\begin{pmatrix} 1 & -1 & 5 & -6 \\ 0 & 6 & -16 & 28 \\ 0 & -4 & 3 & -11 \\ 0 & 4 & -3 & 11 \end{pmatrix}$$

Since the last two rows are negatives of each other we can simplify the situation by replacing the fourth row by the sum of itself and (1 times) the third row:

$$\begin{pmatrix} 1 & -1 & 5 & -6 \\ 0 & 6 & -16 & 28 \\ 0 & -4 & 3 & -11 \\ 0 & 0 & 0 & 0 \end{pmatrix}$$

Multiply second row by 1/6:

$$\begin{pmatrix} 1 & -1 & 5 & -6 \\ 0 & 1 & -\frac{8}{3} & \frac{14}{3} \\ 0 & -4 & 3 & -11 \\ 0 & 0 & 0 & 0 \end{pmatrix}$$

Replace third row by the sum of itself and 4 times the second row:

$$\begin{pmatrix} 1 & -1 & 5 & -6 \\ 0 & 1 & -\frac{8}{3} & \frac{14}{3} \\ 0 & 0 & -\frac{23}{3} & \frac{23}{3} \\ 0 & 0 & 0 & 0 \end{pmatrix}$$

Multiply third row by −3/23:

$$\begin{pmatrix} 1 & -1 & 5 & -6 \\ 0 & 1 & -\frac{8}{3} & \frac{14}{3} \\ 0 & 0 & 1 & -1 \\ 0 & 0 & 0 & 0 \end{pmatrix}$$

At this point in Gaussian elimination we could use back substitution to solve the triangular form system represented by the last matrix. In the Gauss-Jordan method, however, additional elimination of variables replaces back substitution, as follows:

Replace second row by the sum of itself and 8/3 times the third row:

$$\begin{pmatrix} 1 & -1 & 5 & -6 \\ 0 & 1 & 0 & 2 \\ 0 & 0 & 1 & -1 \\ 0 & 0 & 0 & 0 \end{pmatrix}$$

Replace first row by the sum of itself and −5 times the third row:

$$\begin{pmatrix} 1 & -1 & 0 & -1 \\ 0 & 1 & 0 & 2 \\ 0 & 0 & 1 & -1 \\ 0 & 0 & 0 & 0 \end{pmatrix}$$

Replace first row by the sum of itself and 1 times the second row:

$$\begin{pmatrix} 1 & 0 & 0 & 1 \\ 0 & 1 & 0 & 2 \\ 0 & 0 & 1 & -1 \\ 0 & 0 & 0 & 0 \end{pmatrix}$$

The last matrix represents the following system, whose solution is obvious:

$$\begin{aligned} x \qquad &= 1 \\ y \quad &= 2 \\ z &= -1 \end{aligned} \quad ■$$

Technology Tip

To put a matrix in reduced row echelon form, use RREF in the MATH or OPS submenu of the TI-83/86 MATRIX menu, rrowEF in the MATH submenu of the Sharp 9600 MATRIX menu, or RREF in the MATRIX submenu of the HP-38 or TI-89 MATH menu.

In the Gauss-Jordan method, row operations can be performed in any order. For instance, instead of first setting up the system for back substitution (as in Example 5), you can work column by column to obtain columns with one entry 1 and the rest 0. All that matters is that you finish with a matrix in row echelon form in which any column containing a leading 1 has zeros in all other positions, such as the last matrix in Example 5. Such a matrix is said to be in **reduced row echelon form.**

As a general rule, Gaussian elimination (matrix version) is the method of choice when working by hand (the additional row operations needed to put a matrix in reduced row echelon form are usually more time-consuming—and error-prone—than back substitution). With a calculator or computer, however, it's better to find a reduced row echelon matrix for the system.

Example 6 Solve this system:

$$\begin{aligned} 2x + 5y + z + 3w &= 0 \\ 2y - 4z + 6w &= 0 \\ 2x + 17y - 23z + 40w &= 0 \end{aligned}$$

Solution A system such as this, in which all the constants on the right side are zero, is called a **homogeneous system.** Every homogeneous system has at least one solution, namely, $x = 0$, $y = 0$, $z = 0$, $w = 0$, which is called the **trivial solution.** The issue with homogeneous systems is whether or not they have any nonzero solutions. The augmented matrix of the system is shown in Figure 11–5 and an equivalent reduced row echelon form matrix in Figure 11–6.*

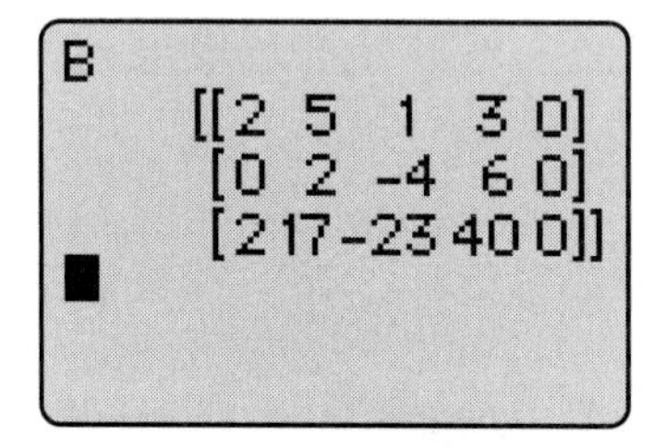

Figure 11–5

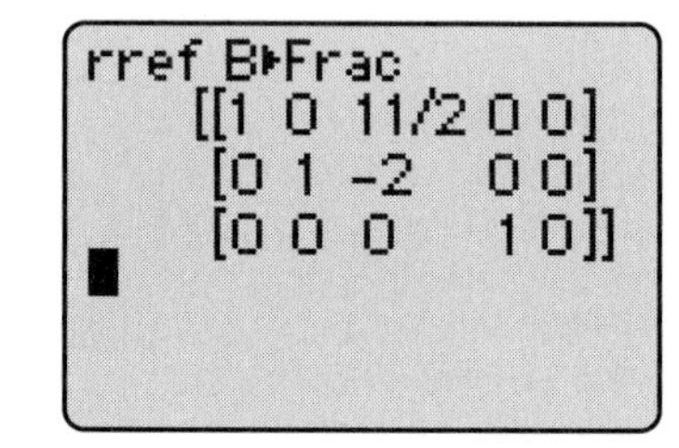

Figure 11–6

The system corresponding to the reduced echelon form matrix in Figure 11–6 is

$$\begin{aligned} x + \frac{11}{2}z &= 0 \\ y - 2z &= 0 \\ w &= 0. \end{aligned}$$

The second equation shows that

$$y = 2z.$$

This equation has an infinite number of solutions, for instance,

$$z = 1, y = 2 \qquad \text{or} \qquad z = 3, y = 6 \qquad \text{or} \qquad z = -2.5, y = -5.$$

*When dealing with homogeneous systems, it's not really necessary to include the last column of zeros, as is done here, because row operations do not change this column.

In fact, for each real number t, there is a solution: $z = t$, $y = 2t$. Substituting $z = t$ into the first equation shows that

$$x + \frac{11}{2}t = 0$$

$$x = -\frac{11}{2}t.$$

Therefore, this system, and hence the original one, has an infinite number of solutions, one for each real number t:

$$x = -\frac{11}{2}t, \qquad y = 2t, \qquad z = t, \qquad w = 0.$$

A system with infinitely many solutions, such as this one, is said to be **dependent.** ■

NOTE Every system that has more variables than equations (as in Example 6) is dependent (or inconsistent), but other systems may be dependent as well.

Applications

In calculus it is sometimes necessary to write a complicated rational expression as the sum of simpler ones. One technique for doing this involves systems of equations.

Example 7 Find constants A, B, and C such that

$$\frac{2x^2 + 15x + 10}{(x-1)(x+2)^2} = \frac{A}{x-1} + \frac{B}{x+2} + \frac{C}{(x+2)^2}.$$

Solution Multiply both sides of the equation by the common denominator $(x - 1)(x + 2)^2$ and collect like terms on the right side:

$$\begin{aligned}
2x^2 + 15x + 10 &= A(x+2)^2 + B(x-1)(x+2) + C(x-1)\\
&= A(x^2 + 4x + 4) + B(x^2 + x - 2) + C(x - 1)\\
&= Ax^2 + 4Ax + 4A + Bx^2 + Bx - 2B + Cx - C\\
&= (A + B)x^2 + (4A + B + C)x + (4A - 2B - C).
\end{aligned}$$

Since the polynomials on the left and right sides of the last equation are equal, their coefficients must be equal term by term, that is,

$$\begin{aligned}
A + B &= 2 \qquad &&\text{[Coefficients of } x^2\text{]}\\
4A + B + C &= 15 \qquad &&\text{[Coefficients of } x\text{]}\\
4A - 2B - C &= 10 \qquad &&\text{[Constant terms]}
\end{aligned}$$

We can consider this as a system of equations with unknowns A, B, C. The augmented matrix of the system is shown in Figure 11–7 and an equivalent reduced row echelon form matrix in Figure 11–8 on the next page.

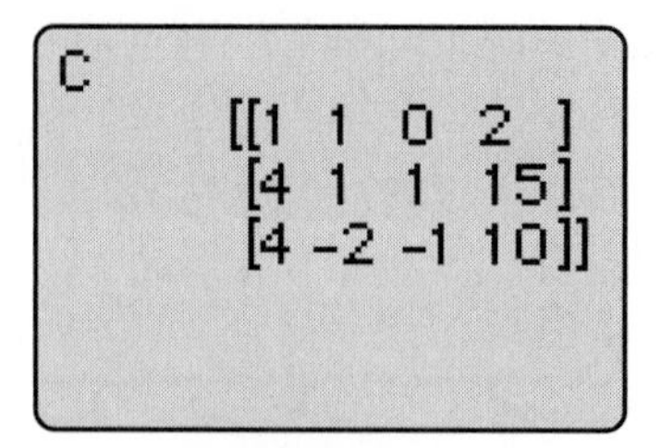

Figure 11–7

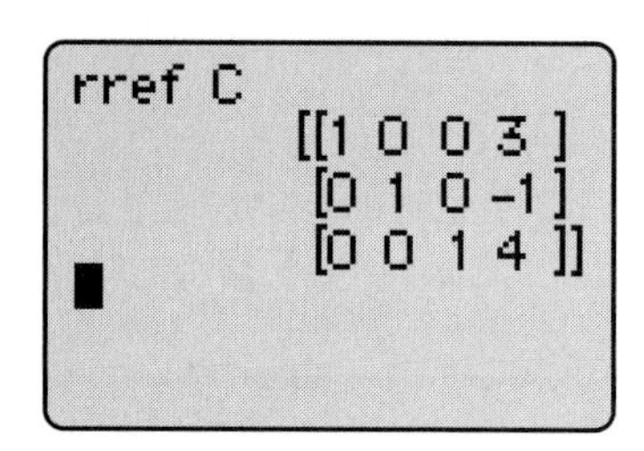

Figure 11–8

The solutions of the system can be read from the reduced row echelon form matrix in Figure 11–8:

$$A = 3, \qquad B = -1, \qquad C = 4.$$

Therefore,

$$\frac{2x^2 + 15x + 10}{(x-1)(x+2)^2} = \frac{3}{x-1} + \frac{-1}{x+2} + \frac{4}{(x+2)^2}.$$

The right side of this equation is called the **partial fraction decomposition** of the fraction on the left side. ■

Example 8 Charlie is starting a small business and borrows \$10,000 on three different credit cards, with annual interest rates of 18%, 15%, and 9%, respectively. He borrows three times as much on the 15% card as on the 18% card, and his total annual interest on all three cards is \$1244.25. How much did he borrow on each credit card?

Solution Let x be the amount on the 18% card, y the amount on the 15% card, and z the amount on the 9% card. Then, $x + y + z = 10{,}000$. Furthermore,

Interest on 18% card	+	Interest on 15% card	+	Interest on 9% card	=	Total interest
↓		↓		↓		↓
$.18x$	+	$.15y$	+	$.09z$	=	1244.25

Finally, we have

Amount on 15% card	=	3 times amount on 18% card
y	=	$3x,$

which is equivalent to $3x - y = 0$. Therefore, we must solve this system of equations:

$$\begin{aligned} x + y + z &= 10{,}000 \\ .18x + .15y + .09z &= 1{,}244.25 \\ 3x - y &= 0 \end{aligned}$$

whose augmented matrix is

$$\begin{pmatrix} 1 & 1 & 1 & 10{,}000 \\ .18 & .15 & .09 & 1{,}244.25 \\ 3 & -1 & 0 & 0 \end{pmatrix}.$$

CALCULATOR EXPLORATION

Enter this matrix in your calculator. Use row operations or the reduced row echelon form key to put it in reduced row echelon form. Read the solutions of the system from this last matrix.

The Exploration shows that Charlie borrowed \$1275 on the 18% card, \$3825 on the 15% card, and \$4900 on the 9% card. ■

The preceding examples illustrate the following fact, whose proof is omitted.

Number of Solutions of a System

Any system of linear equations must have

No solutions (an inconsistent system) ***or***

Exactly one solution ***or***

An infinite number of solutions (a dependent system).

Exercises 11.2

In Exercises 1–4, write the augmented matrix of the system.

1.
$$\begin{aligned} 2x - 3y + 4z &= 1 \\ x + 2y - 6z &= 0 \\ 3x - 7y + 4z &= -3 \end{aligned}$$

2.
$$\begin{aligned} x + 2y - 3w + 7z &= -5 \\ 2x - y \qquad + 2z &= 4 \\ 3x \qquad + 7w - 6z &= 0 \end{aligned}$$

3.
$$\begin{aligned} x - \frac{1}{2}y + \frac{7}{4}z &= 0 \\ 2x - \frac{3}{2}y + 5z &= 0 \\ -2y + \frac{1}{3}z &= 0 \end{aligned}$$

4.
$$\begin{aligned} 2x - \frac{1}{2}y + \frac{7}{2}w - 6z &= 1 \\ \frac{1}{4}x - 6y + 2w - z &= 2 \\ 4y - \frac{1}{2}w + z &= 3 \\ 2x + 3y \qquad + \frac{1}{2}z &= 4 \end{aligned}$$

In Exercises 5–8, the augmented matrix of a system of equations is given. Express the system in equation notation.

5. $\begin{pmatrix} 2 & -3 & 1 \\ 4 & 7 & 2 \end{pmatrix}$

6. $\begin{pmatrix} 2 & 3 & 5 & 2 \\ 1 & 6 & 9 & 0 \end{pmatrix}$

7. $\begin{pmatrix} 1 & 0 & 1 & 0 & 1 \\ 1 & -1 & 4 & -2 & 3 \\ 4 & 2 & 5 & 0 & 2 \end{pmatrix}$

8. $\begin{pmatrix} 1 & 7 & 0 & 4 \\ 2 & 3 & 1 & 6 \\ -1 & 0 & 2 & 3 \end{pmatrix}$

In Exercises 9–12, the reduced row echelon form of the augmented matrix of a system of equations is given. Find the solutions of the system.

9. $\begin{pmatrix} 1 & 0 & 0 & 0 & 3/2 \\ 0 & 1 & 0 & 0 & 5 \\ 0 & 0 & 1 & 0 & -2 \\ 0 & 0 & 0 & 1 & 0 \end{pmatrix}$

10. $\begin{pmatrix} 1 & 0 & 0 & 0 & 0 & 5 \\ 0 & 1 & 0 & 0 & 0 & 4 \\ 0 & 0 & 1 & 0 & 0 & 3 \\ 0 & 0 & 0 & 0 & 1 & 2 \\ 0 & 0 & 0 & 0 & 0 & 1 \end{pmatrix}$

11. $\begin{pmatrix} 1 & 0 & 0 & 1 & 2 \\ 0 & 1 & 0 & 2 & -3 \\ 0 & 0 & 1 & 0 & 4 \\ 0 & 0 & 0 & 0 & 0 \end{pmatrix}$

12. $\begin{pmatrix} 1 & 0 & 0 & 0 & 7 \\ 0 & 1 & 0 & 0 & 1 \\ 0 & 0 & 1 & 0 & -5 \\ 0 & 0 & 0 & 1 & 4 \\ 0 & 0 & 0 & 0 & 0 \\ 0 & 0 & 0 & 0 & 0 \end{pmatrix}$

In Exercises 13–16, use Gaussian elimination to solve the system.

13. $\begin{aligned} -x + 3y + 2z &= 0 \\ 2x - y - z &= 3 \\ x + 2y + 3z &= 0 \end{aligned}$

14. $\begin{aligned} 3x + 7y + 9z &= 0 \\ x + 2y + 3z &= 2 \\ x + 4y + z &= 2 \end{aligned}$

15. $\begin{aligned} x + y + z &= 1 \\ x - 2y + 2z &= 4 \\ 2x - y + 3z &= 5 \end{aligned}$

16. $\begin{aligned} 2x - y + z &= 1 \\ 3x + y + z &= 0 \\ 7x - y + 3z &= 2 \end{aligned}$

In Exercises 17–20, use the Gauss-Jordan method to solve the system.

17. $\begin{aligned} x - 2y + 4z &= 6 \\ x + y + 13z &= 6 \\ -2x + 6y - z &= -10 \end{aligned}$

18. $\begin{aligned} x - y + 5z &= -6 \\ 3x + 3y - z &= 10 \\ x + 3y + 2z &= 5 \end{aligned}$

19. $\begin{aligned} x + y + z &= 200 \\ x - 2y &= 0 \\ 2x + 3y + 5z &= 600 \\ 2x - y + z &= 200 \end{aligned}$

20. $\begin{aligned} 3x - y + z &= 6 \\ x + 2y - z &= 0 \end{aligned}$

In Exercises 21–36, solve the system.

21. $\begin{aligned} 11x + 10y + 9z &= 5 \\ x + 2y + 3z &= 1 \\ 3x + 2y + z &= 1 \end{aligned}$

22. $\begin{aligned} -x + 2y - 3z + 4w &= 8 \\ 2x - 4y + z + 2w &= -3 \\ 5x - 4y + z + 2w &= -3 \end{aligned}$

23. $\begin{aligned} x + y &= 3 \\ 5x - y &= 3 \\ 9x - 4y &= 1 \end{aligned}$

24. $\begin{aligned} 2x - y + 2z &= 3 \\ -x + 2y - z &= 0 \\ x + y - z &= 1 \end{aligned}$

25. $\begin{aligned} x - 4y - 13z &= 4 \\ x - 2y - 3z &= 2 \\ -3x + 5y + 4z &= 2 \end{aligned}$

26. $\begin{aligned} 2x - 4y + z &= 3 \\ x + 3y - 7z &= 1 \\ -2x + 4y - z &= 10 \end{aligned}$

27. $\begin{aligned} 4x + y + 3z &= 7 \\ x - y + 2z &= 3 \\ 3x + 2y + z &= 4 \end{aligned}$

28. $\begin{aligned} x + 4y + z &= 3 \\ -x + 2y + 2z &= 0 \\ 2x + 2y - z &= 3 \end{aligned}$

29. $\begin{aligned} x + y + z &= 0 \\ 3x - y + z &= 0 \\ -5x - y + z &= 0 \end{aligned}$

30. $\begin{aligned} x + y + z &= 0 \\ x - y - z &= 0 \\ x - y + z &= 0 \end{aligned}$

31. $\begin{aligned} 2x + y + 3z - 2w &= -6 \\ 4x + 3y + z - w &= -2 \\ x + y + z + w &= -5 \\ -2x - 2y + 2z + 2w &= -10 \end{aligned}$

32. $\begin{aligned} x + y + z + w &= -1 \\ -x + 4y + z - w &= 0 \\ x - 2y + z - 2w &= 11 \\ -x - 2y + z + 2w &= -3 \end{aligned}$

33. $\begin{aligned} x - 2y - z - 3w &= -3 \\ -x + y + z &= 0 \\ 4y + 3z - 2w &= -1 \\ 2x - 2y + w &= 1 \end{aligned}$

34. $\begin{aligned} 3x - y + 2z &= 0 \\ -x + 3y + 2z + 5w &= 0 \\ x + 2y + 5z - 4w &= 0 \\ 2x - y + 3w &= 0 \end{aligned}$

35. $\begin{aligned} \frac{3}{x} - \frac{1}{y} + \frac{4}{z} &= -13 \\ \frac{1}{x} + \frac{2}{y} - \frac{1}{z} &= 12 \\ \frac{4}{x} - \frac{1}{y} + \frac{3}{z} &= -7 \end{aligned}$

[*Hint:* Let $u = 1/x$, $v = 1/y$, $w = 1/z$.]

36. $\begin{aligned} \frac{1}{x+1} - \frac{2}{y-3} + \frac{3}{z-2} &= 4 \\ \frac{5}{y-3} - \frac{10}{z-2} &= -5 \\ \frac{-3}{x+1} + \frac{4}{y-3} - \frac{1}{z-2} &= -2 \end{aligned}$

[*Hint:* Let $u = 1/(x+1)$, $v = 1/(y-3)$, $w = 1/(z-2)$.]

In Exercises 37–42, find the constants A, B, C.

37. $\dfrac{x}{(x+1)(x+2)} = \dfrac{A}{x+1} + \dfrac{B}{x+2}$

38. $\dfrac{1}{(x+1)(x-1)} = \dfrac{A}{x+1} + \dfrac{B}{x-1}$

39. $\dfrac{2x+1}{(x+2)(x-3)^2} = \dfrac{A}{x+2} + \dfrac{B}{x-3} + \dfrac{C}{(x-3)^2}$

40. $\dfrac{x^2-x-21}{(2x-1)(x^2+4)} = \dfrac{A}{2x-1} + \dfrac{Bx+C}{x^2+4}$

41. $\dfrac{5x^2+1}{(x+1)(x^2-x+1)} = \dfrac{A}{x+1} + \dfrac{Bx+C}{x^2-x+1}$

42. $\dfrac{x-2}{(x+4)(x^2+2x+2)} = \dfrac{A}{x+4} + \dfrac{Bx+C}{x^2+2x+2}$

Exercises 43–46 deal with systems such as this one:

$$\begin{aligned} x + y + 4z - w &= 1 \\ y - 2z + 3w &= 0 \end{aligned}$$

Verify that for each pair of real numbers, s and t, the system has a solution:

$$w = s, \quad z = t, \quad y = 2t - 3s,$$
$$x = 1 - (2t - 3s) - 4t + s = 1 - 6t + 4s$$

With this method in mind, solve these dependent systems by the elimination method.

43. $\begin{aligned} x - y + 2z + 3w &= 0 \\ x + z + w &= 0 \\ 3x - 2y + 5z + 7w &= 0 \end{aligned}$

44. $\begin{aligned} x + 2y + z + 4w &= 1 \\ y + 3z - w &= 2 \\ x + 4y + 7z - 2w &= 5 \\ 3x + 7y + 6z + 11w &= 5 \end{aligned}$

45. $\begin{aligned} x + y + z - w &= 0 \\ 2x - 4y - 4z + w &= 0 \\ 4x - 2y + 2z - 3w &= 0 \\ 7x - y - z - 3w &= 0 \end{aligned}$

46. $\begin{aligned} x + 2y + 3z + 4v &= 0 \\ 2x + 4y + 6z + w + 9v &= 0 \\ x + 2y + 3z + w + 5v &= 0 \end{aligned}$

47. A collection of nickels, dimes, and quarters totals \$6.00. If there are 52 coins altogether and twice as many dimes as nickels, how many of each kind of coin are there?

48. A collection of nickels, dimes, and quarters totals \$8.20. The number of nickels and dimes together is twice the number of quarters. The value of the nickels is one-third of the value of the dimes. How many of each kind of coin are there?

49. Lillian borrows \$10,000. She borrows some from her friend at 8% annual interest, twice as much as that from her bank at 9%, and the remainder from her insurance company at 5%. She pays a total of \$830 in interest for the first year. How much did she borrow from each source?

50. An investor puts a total of \$25,000 into three very speculative stocks. She invests some of it in Crystalcomp and \$2000 more than one-half that amount in Flyboys. The remainder is invested in Zumcorp. Crystalcomp rises 16% in value, Flyboys 20%, and Zumcorp 18%. Her investment in the three stocks is now worth \$29,440. How much was originally invested in each stock?

51. An investor has \$70,000 invested in a mutual fund, bonds, and a fast food franchise. She has twice as much invested in bonds as in the mutual fund. Last year the mutual fund paid a 2% dividend, the bonds 10%, and the fast food franchise 6%; her dividend income was \$4800. How much is invested in each of the three investments?

52. Tickets to a band concert cost \$2 for children, \$3 for teenagers, and \$5 for adults. 570 people attended the concert and total ticket receipts were \$1950. Three-fourths as many teenagers as children attended. How many children, adults, and teenagers attended?

53. Comfort Systems, Inc., sells three models of humidifiers: the bedroom model weighs 10 pounds and comes in an 8-cubic-foot box; the living room model weighs 20 pounds and comes in an 8-cubic-foot box; the whole-house model weighs 60 pounds and comes in a 28-cubic-foot box. Each of their delivery vans has 248 cubic feet of space and can hold a maximum of 440 pounds. In order for a van to be as fully loaded as possible, how many of each model should it carry?

54. Peanuts cost \$3 per pound, almonds \$4 per pound, and cashews \$8 per pound. How many pounds of each should be used to produce 140 pounds of a mixture costing \$6 per pound, in which there are twice as many peanuts as almonds?

55. If Tom, Dick, and Harry work together, they can paint a large room in 4 hours. When only Dick and Harry work together, it takes 8 hours to paint the room. Tom and Dick, working together, take 6 hours to paint the room. How long would it take each of them to paint the room alone? [*Hint:* If x is the amount of the room painted in 1 hour by Tom, y is the amount painted by Dick, and z the amount painted by Harry, then $x + y + z = 1/4$.]

56. Pipes R, S, T are connected to the same tank. When all three pipes are running, they can fill the tank in 2 hours. When only pipes S and T are running, they can fill the tank in 4 hours. When only R and T are running, they can fill the tank in 2.4 hours. How long would it take each pipe running alone to fill the tank?

57. A furniture manufacturer has 1950 machine hours available each week in the cutting department, 1490 hours in the assembly department, and 2160 in the finishing department. Manufacturing a chair requires .2 hours of cutting, .3 hours of assembly, and .1 hours of finishing. A chest requires .5 hours of cutting, .4 hours of assembly, and .6 hours of finishing. A table requires .3 hours of cutting, .1 hours of assembly, and .4 hours of finishing. How many chairs, chests, and tables should be produced in order to use all the available production capacity?

58. A stereo equipment manufacturer produces three models of speakers, R, S, and T, and has three kinds of delivery vehicles: trucks, vans, and station wagons. A truck holds 2 boxes of model R, 1 of model S, and 3 of model T. A van holds 1 box of model R, 3 of model S, and 2 of model T. A station wagon holds 1 box of model R, 3 of model S, and 1 of model T. If 15 boxes of model R, 20 boxes of model S, and 22 boxes of model T are to be delivered, how many vehicles of each type should be used so that all operate at full capacity?

59. A company produces three camera models: A, B, and C. Each model A requires 3 hours of lens polishing, 2 hours of assembly time, and 2 hours of finishing time. Each model B requires 2 hours of lens polishing, 2 hours of assembly time, and 1 hour of finishing time. Each model C requires 1, 3, and 1 hours of lens polishing, assembly, and finishing time. There are 100 hours available for lens polishing, 100 hours for assembly, and 65 hours for finishing each week. How many of each model should be produced if all available time is to be used?

11.3 Matrix Methods for Square Systems

Matrices were used in Section 11.2 as a convenient shorthand for solving systems of linear equations. We now consider matrices in a more general setting and show how the algebra of matrices provides an alternative method for solving systems of equations that are not dependent and have the same number of equations as variables.

Let m and n be positive integers. An $\boldsymbol{m \times n}$ **matrix** (read "m by n matrix") is a rectangular array of numbers, with m horizontal **rows** (numbered from top to bottom) and n vertical **columns** (numbered from left to right). For example,

$$\begin{pmatrix} 3 & 2 & -5 \\ 6 & 1 & 7 \\ -2 & 0 & 5 \end{pmatrix} \qquad \begin{pmatrix} 3/2 \\ -5 \\ 0 \\ 12 \end{pmatrix}$$

3 × 3 matrix *4 × 1 matrix*

Each entry in a matrix can be located by stating the row and column in which it appears. For instance, in the 3×3 matrix above, -5 is the entry in row 1 column 3. When you enter a matrix on a calculator, the words "row" and "column" won't be displayed, but the row number will always be listed before the column number. Thus a display such as "A[3, 2]," or simply "3, 2," indicates the entry in row 3, column 2.

Two matrices are said to be **equal** if they have the same size (same number of rows and columns) and the corresponding entries are equal. For example,

$$\begin{pmatrix} 3 & (-1)^2 \\ 6 & 12 \end{pmatrix} = \begin{pmatrix} 3 & 1 \\ \sqrt{36} & 12 \end{pmatrix}, \quad \text{but} \quad \begin{pmatrix} 6 & 4 \\ 5 & 1 \end{pmatrix} \neq \begin{pmatrix} 6 & 5 \\ 4 & 1 \end{pmatrix}.$$

Matrix Multiplication

Although there is an extensive arithmetic of matrices, we shall need only matrix multiplication. The simplest case is the product of a matrix with a single row and a matrix with a single column, where the row and column have the same number of entries. This is done by multiplying corresponding entries (first by first, second by second, and so on) and then adding the results. An example is shown in Figure 11–9.

$$\begin{pmatrix} 3 & 1 & 2 \end{pmatrix} \begin{pmatrix} 2 \\ 0 \\ 1 \end{pmatrix} = \underset{\substack{\uparrow \\ \textit{First} \\ \textit{Terms}}}{3 \cdot 2} + \underset{\substack{\uparrow \\ \textit{Second} \\ \textit{Terms}}}{1 \cdot 0} + \underset{\substack{\uparrow \\ \textit{Third} \\ \textit{Terms}}}{2 \cdot 1} = 8$$

Figure 11–9

Note that the product of a row and a column is a single number.

Now let A be an $m \times n$ matrix and B an $n \times p$ matrix, so that the number of columns of A is the same as the number of rows of B (namely, n). The product matrix AB is defined to be an $m \times p$ matrix (same number of rows as A and same number of columns as B). The entry in row i, column j, of AB is this number:

the product of row i of A and column j of B.

Example 1 If it is defined, find the product AB, where

$$A = \begin{pmatrix} 3 & 1 & 2 \\ -1 & 0 & 4 \end{pmatrix} \quad \text{and} \quad B = \begin{pmatrix} 2 & -3 & 0 & 1 \\ 0 & 5 & 2 & 7 \\ 1 & 8 & -4 & 1 \end{pmatrix}.$$

Solution A has 3 columns and B has 3 rows. So the product matrix AB is defined. AB has 2 rows (same as A) and 4 columns (same as B). Its entries are calculated as follows. The entry in row 1, column 1 of AB is the product of row 1 of A and column 1 of B, which is the number 8, as shown in Figure 11–9 and indicated at the right here:

row 1, column 1 $\begin{pmatrix} 3 & 1 & 2 \\ -1 & 0 & 4 \end{pmatrix}\begin{pmatrix} 2 & -3 & 0 & 1 \\ 0 & 5 & 2 & 7 \\ 1 & 8 & -4 & 1 \end{pmatrix} = \begin{pmatrix} 8 & & & \\ & & & \end{pmatrix}$ $3 \cdot 2 + 1 \cdot 0 + 2 \cdot 1 = 8$

The other entries in AB are obtained similarly:

row 1, column 2 $\begin{pmatrix} 3 & 1 & 2 \\ -1 & 0 & 4 \end{pmatrix}\begin{pmatrix} 2 & -3 & 0 & 1 \\ 0 & 5 & 2 & 7 \\ 1 & 8 & -4 & 1 \end{pmatrix} = \begin{pmatrix} 8 & 12 & & \\ & & & \end{pmatrix}$ $3(-3) + 1 \cdot 5 + 2 \cdot 8 = 12$

row 1, column 3 $\begin{pmatrix} 3 & 1 & 2 \\ -1 & 0 & 4 \end{pmatrix}\begin{pmatrix} 2 & -3 & 0 & 1 \\ 0 & 5 & 2 & 7 \\ 1 & 8 & -4 & 1 \end{pmatrix} = \begin{pmatrix} 8 & 12 & -6 & \\ & & & \end{pmatrix}$ $3 \cdot 0 + 1 \cdot 2 + 2(-4) = -6$

row 1, column 4 $\begin{pmatrix} 3 & 1 & 2 \\ -1 & 0 & 4 \end{pmatrix}\begin{pmatrix} 2 & -3 & 0 & 1 \\ 0 & 5 & 2 & 7 \\ 1 & 8 & -4 & 1 \end{pmatrix} = \begin{pmatrix} 8 & 12 & -6 & 12 \\ & & & \end{pmatrix}$ $3 \cdot 1 + 1 \cdot 7 + 2 \cdot 1 = 12$

row 2, column 1 $\begin{pmatrix} 3 & 1 & 2 \\ -1 & 0 & 4 \end{pmatrix}\begin{pmatrix} 2 & -3 & 0 & 1 \\ 0 & 5 & 2 & 7 \\ 1 & 8 & -4 & 1 \end{pmatrix} = \begin{pmatrix} 8 & 12 & -6 & 12 \\ 2 & & & \end{pmatrix}$ $(-1)2 + 0 \cdot 0 + 4 \cdot 1 = 2$

row 2, column 2 $\begin{pmatrix} 3 & 1 & 2 \\ -1 & 0 & 4 \end{pmatrix}\begin{pmatrix} 2 & -3 & 0 & 1 \\ 0 & 5 & 2 & 7 \\ 1 & 8 & -4 & 1 \end{pmatrix} = \begin{pmatrix} 8 & 12 & -6 & 12 \\ 2 & 35 & & \end{pmatrix}$ $(-1)(-3) + 0 \cdot 5 + 4 \cdot 8 = 35$

row 2, column 3 $\begin{pmatrix} 3 & 1 & 2 \\ -1 & 0 & 4 \end{pmatrix}\begin{pmatrix} 2 & -3 & 0 & 1 \\ 0 & 5 & 2 & 7 \\ 1 & 8 & -4 & 1 \end{pmatrix} = \begin{pmatrix} 8 & 12 & -6 & 12 \\ 2 & 35 & -16 & \end{pmatrix}$ $(-1)0 + 0 \cdot 2 + 4(-4) = -16$

row 2, column 4 $\begin{pmatrix} 3 & 1 & 2 \\ -1 & 0 & 4 \end{pmatrix}\begin{pmatrix} 2 & -3 & 0 & 1 \\ 0 & 5 & 2 & 7 \\ 1 & 8 & -4 & 1 \end{pmatrix} = \begin{pmatrix} 8 & 12 & -6 & 12 \\ 2 & 35 & -16 & 3 \end{pmatrix}$ $(-1)1 + 0 \cdot 7 + 4 \cdot 1 = 3$

The last matrix on the right is the product AB. ■

Example 2 Use matrix multiplication to express this system of equations in matrix form:

$$\begin{aligned} x + y + z &= 2 \\ 2x + 3y \quad &= 5 \\ x + 2y + z &= -1. \end{aligned}$$

Solution Let A be the 3×3 matrix of coefficients on the left side of the equations, B the column matrix of constants on the right side, and X the column matrix of unknowns:

$$A = \begin{pmatrix} 1 & 1 & 1 \\ 2 & 3 & 0 \\ 1 & 2 & 1 \end{pmatrix}, \quad X = \begin{pmatrix} x \\ y \\ z \end{pmatrix} \quad B = \begin{pmatrix} 2 \\ 5 \\ -1 \end{pmatrix}.$$

Technology Tip

Learn how to enter matrices in the matrix memory of your calculator. All calculators will perform matrix multiplication, using the ordinary multiplication key on the keyboard.

Then AX is a matrix with 3 rows and 1 column, as is B:

$$AX = \begin{pmatrix} 1 & 1 & 1 \\ 2 & 3 & 0 \\ 1 & 2 & 1 \end{pmatrix}\begin{pmatrix} x \\ y \\ z \end{pmatrix} = \begin{pmatrix} x + & y + & z \\ 2x + & 3y + & 0z \\ x + & 2y + & z \end{pmatrix} \quad \text{and} \quad B = \begin{pmatrix} 2 \\ 5 \\ -1 \end{pmatrix}.$$

The entries in AX are just the left sides of the equations of the system and the entries in B are the constants on the right sides. Therefore, the system can be expressed as the matrix equation $AX = B$. ■

It can be shown that matrix multiplication is associative, that is, $A(BC) = (AB)C$ whenever all the products are defined. However, matrix multiplication is not commutative, that is, AB may not be equal to BA, even when both products are defined (Exercises 13–16).

Technology Tip

To display an $n \times n$ identity matrix, use IDENT(ITY) n in the OPS or MATH submenu of the TI MATRIX menu, in the OPE submenu of the Sharp 9600 MATRIX menu, or in the MAT submenu of the Casio 9850 OPTN menu. On HP-38 and TI-89, use IDENMAT n or IDENTITY in the MATRIX submenu of the MATH menu.

Identity Matrices and Inverses

The $n \times n$ **identity matrix** I_n is the matrix with 1's on the diagonal from upper left to lower right and 0's everywhere else; for example,

$$I_2 = \begin{pmatrix} 1 & 0 \\ 0 & 1 \end{pmatrix} \quad I_3 = \begin{pmatrix} 1 & 0 & 0 \\ 0 & 1 & 0 \\ 0 & 0 & 1 \end{pmatrix} \quad I_4 = \begin{pmatrix} 1 & 0 & 0 & 0 \\ 0 & 1 & 0 & 0 \\ 0 & 0 & 1 & 0 \\ 0 & 0 & 0 & 1 \end{pmatrix}.$$

The number 1 is the multiplicative identity of the number system because $a \cdot 1 = a = 1 \cdot a$ for every number a. The identity matrix I_n is the multiplicative identity for $n \times n$ matrices:

Identity Matrix

For any $n \times n$ matrix A,

$$AI_n = A = I_nA.$$

For example, in the 2×2 case

$$\begin{pmatrix} a & b \\ c & d \end{pmatrix}\begin{pmatrix} 1 & 0 \\ 0 & 1 \end{pmatrix} = \begin{pmatrix} a \cdot 1 + b \cdot 0 & a \cdot 0 + b \cdot 1 \\ c \cdot 1 + d \cdot 0 & c \cdot 0 + d \cdot 1 \end{pmatrix} = \begin{pmatrix} a & b \\ c & d \end{pmatrix}.$$

Verify that the same answer results if you reverse the order of multiplication.

Every nonzero number c has a multiplicative inverse $c^{-1} = 1/c$ with the property that $cc^{-1} = 1$. The analogous statement for matrix multiplication does not always hold, and special terminology is used when it does. An $n \times n$ matrix A is said to be **invertible** (or **nonsingular**) if there is an $n \times n$ matrix B such that $AB = I_n$. In this case it can be proved that $BA = I_n$ also. The matrix B is called the **inverse** of A and is sometimes denoted A^{-1}.

Example 3 You can readily verify that

$$\begin{pmatrix} 2 & 1 \\ 3 & 1 \end{pmatrix}\begin{pmatrix} -1 & 1 \\ 3 & -2 \end{pmatrix} = \begin{pmatrix} 1 & 0 \\ 0 & 1 \end{pmatrix} = \begin{pmatrix} -1 & 1 \\ 3 & -2 \end{pmatrix}\begin{pmatrix} 2 & 1 \\ 3 & 1 \end{pmatrix}.$$

Therefore, $A = \begin{pmatrix} 2 & 1 \\ 3 & 1 \end{pmatrix}$ is an invertible matrix with inverse $A^{-1} = \begin{pmatrix} -1 & 1 \\ 3 & -2 \end{pmatrix}$. ■

Example 4 Find the inverse of the matrix $\begin{pmatrix} 2 & 6 \\ 1 & 4 \end{pmatrix}$.

Solution We must find numbers x, y, u, v such that

$$\begin{pmatrix} 2 & 6 \\ 1 & 4 \end{pmatrix}\begin{pmatrix} x & u \\ y & v \end{pmatrix} = \begin{pmatrix} 1 & 0 \\ 0 & 1 \end{pmatrix}$$

which is the same as

$$\begin{pmatrix} 2x + 6y & 2u + 6v \\ x + 4y & u + 4v \end{pmatrix} = \begin{pmatrix} 1 & 0 \\ 0 & 1 \end{pmatrix}.$$

Since corresponding entries in these last two matrices are equal, finding x, y, u, v amounts to solving these systems of equations:

$$\begin{aligned} 2x + 6y &= 1 \\ x + 4y &= 0 \end{aligned} \qquad \text{and} \qquad \begin{aligned} 2u + 6v &= 0 \\ u + 4v &= 1. \end{aligned}$$

We shall solve the systems by the Gauss-Jordan method of Section 11.2. Since the coefficient matrices of the two systems are the same, we shall use the following matrix, whose first three columns form the augmented matrix of the first system and whose first two and last columns form the augmented matrix of the second system:

$$\left(\begin{array}{cc|cc} 2 & 6 & 1 & 0 \\ 1 & 4 & 0 & 1 \end{array}\right).$$

Performing row operations on this matrix amounts to simultaneously doing the operations on the two augmented matrices:

Multiply row 1 by 1/2:

$$\left(\begin{array}{cc|cc} 1 & 3 & 1/2 & 0 \\ 1 & 4 & 0 & 1 \end{array}\right)$$

Replace row 2 by the sum of itself and −1 times row 1:

$$\left(\begin{array}{cc|cc} 1 & 3 & 1/2 & 0 \\ 0 & 1 & -1/2 & 1 \end{array}\right)$$

Replace row 1 by the sum of itself and −3 times row 2:

$$\left(\begin{array}{cc|cc} 1 & 0 & 2 & -3 \\ 0 & 1 & -1/2 & 1 \end{array}\right)$$

The first three columns of the last matrix show that $x = 2$ and $y = -1/2$. Similarly, the first two and last columns show that $u = -3$ and $v = 1$. Therefore,

$$A^{-1} = \begin{pmatrix} 2 & -3 \\ -1/2 & 1 \end{pmatrix}.$$

Observe that A^{-1} is just the right half of the final form of the preceding augmented matrix and that the left half is the identity matrix I_2. ■

Technology Tip

On calculators other than TI-89 you can find the inverse A^{-1} of matrix A by keying in A (or Mat A) and using the x^{-1} key. Using the ^ key and -1 produces A^{-1} on TI-89 and HP-38, but leads to an error message on other calculators.

Although the technique in Example 4 can be used to find the inverse of any matrix that has one, it's quicker to use a calculator. Any calculator with matrix capabilities can find the inverse of an invertible matrix (see the tip in the margin).

CAUTION

A calculator should produce an error message when asked for the inverse of a matrix A that does not have one. However, because of round-off errors, it may sometimes display a matrix which it says is A^{-1}. As an accuracy check when finding inverses, multiply A by A^{-1} to see if the product is the identity matrix. If it isn't, A does not have an inverse.

Inverse Matrices and Systems of Equations

Suppose a system of equations is written in matrix form $AX = B$ as in Example 2 and that the matrix A has an inverse. Then we can solve $AX = B$ by multiplying both sides by A^{-1}:

$$\begin{aligned} A^{-1}(AX) &= A^{-1}B \\ (A^{-1}A)X &= A^{-1}B && \text{[Matrix multiplication is associative]} \\ I_nX &= A^{-1}B && [A^{-1}A \text{ is the identity matrix}] \\ X &= A^{-1}B && \text{[Product of identity matrix and } X \text{ is } X] \end{aligned}$$

The next example shows how this works in practice.

Example 5 Solve the system

$$\begin{aligned} x + y + z &= 2 \\ 2x + 3y &= 5 \\ x + 2y + z &= -1. \end{aligned}$$

Solution As we saw in Example 2, this system is equivalent to the matrix equation

$$AX = B$$

$$\begin{pmatrix} 1 & 1 & 1 \\ 2 & 3 & 0 \\ 1 & 2 & 1 \end{pmatrix}\begin{pmatrix} x \\ y \\ z \end{pmatrix} = \begin{pmatrix} 2 \\ 5 \\ -1 \end{pmatrix}.$$

Use a calculator to find the inverse of the coefficient matrix A and multiply both sides of the equation by A^{-1}:

$$A^{-1}AX = A^{-1}B$$

$$\begin{pmatrix} 1.5 & .5 & -1.5 \\ -1 & 0 & 1 \\ .5 & -.5 & .5 \end{pmatrix}\begin{pmatrix} 1 & 1 & 1 \\ 2 & 3 & 0 \\ 1 & 2 & 1 \end{pmatrix}\begin{pmatrix} x \\ y \\ z \end{pmatrix} = \begin{pmatrix} 1.5 & .5 & -1.5 \\ -1 & 0 & 1 \\ .5 & -.5 & .5 \end{pmatrix}\begin{pmatrix} 2 \\ 5 \\ -1 \end{pmatrix}$$

$$\begin{pmatrix} 1 & 0 & 0 \\ 0 & 1 & 0 \\ 0 & 0 & 1 \end{pmatrix}\begin{pmatrix} x \\ y \\ z \end{pmatrix} = \begin{pmatrix} 1.5 & .5 & -1.5 \\ -1 & 0 & 1 \\ .5 & -.5 & .5 \end{pmatrix}\begin{pmatrix} 2 \\ 5 \\ -1 \end{pmatrix} \qquad [\text{since } A^{-1}A = I_3]$$

$$\begin{pmatrix} x \\ y \\ z \end{pmatrix} = \begin{pmatrix} 7 \\ -3 \\ -2 \end{pmatrix} \qquad [\text{since } I_3X = X]$$

Therefore, the solution of the original system is $x = 7, y = -3, z = -2$. ■

Only a matrix with the same number of rows as columns can possibly have an inverse. Consequently, the method of Example 5 can be tried only when the system has the same number of equations as unknowns (so that the coefficient matrix has the same number of rows as columns). In this case, you should use your calculator to verify that the coefficient matrix actually has an inverse (see the caution on page 713). If it does not, other methods must be used. Here is a summary of the possibilities:

Matrix Solution of a System of Equations

Suppose the system of equations is written in matrix form as $AX = B$.

If the matrix A has an inverse, then the unique solution of the system is

$$X = A^{-1}B.$$

If A does not have an inverse (which is always the case when the number of equations differs from the number of unknowns), then the system either has no solutions or has infinitely many solutions. Its solutions (if any) may be found using Gaussian elimination (Section 11.2).

Example 6 Solve the system

$$\begin{aligned} 2x + y - z &= 2 \\ x + 3y + 2z &= 1 \\ x + y + z &= 2. \end{aligned}$$

Solution Since there are the same number of equations as unknowns, we can try the method of Example 5. In this case we have

$$A = \begin{pmatrix} 2 & 1 & -1 \\ 1 & 3 & 2 \\ 1 & 1 & 1 \end{pmatrix} \quad \text{and} \quad B = \begin{pmatrix} 2 \\ 1 \\ 2 \end{pmatrix}.$$

CALCULATOR EXPLORATION

Verify that the matrix A does have an inverse. Show that the solutions of the system are $x = 2$, $y = -1$, $z = 1$ by computing $A^{-1}B$. ■

Applications

Just as two points determine a unique line, three points (that aren't on the same line) determine a unique parabola, as the next example demonstrates.

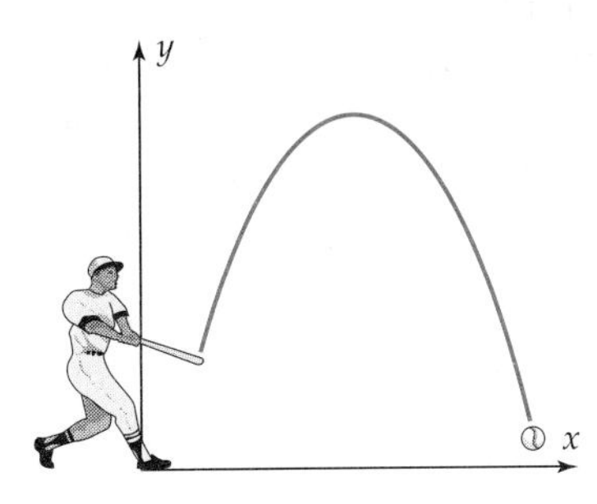

Figure 11–10

Example 7 A batter hits a baseball and special measuring devices locate its position at various times during its flight. If the path of the ball is drawn on a coordinate plane, with the batter at the origin, it looks like Figure 11–10. According to the measuring devices, the ball passes through the points (7, 9), (47, 38), and (136, 70).

(a) What is the equation of the path of the ball?

(b) How far from the batter does the ball hit the ground?

Solution

(a) The path of the ball appears to be part of a parabola (a fact that can be proved by eliminating the parameter in the equations for projectile motion on page 618) and hence has an equation of the form

$$y = ax^2 + bx + c$$

for some constants a, b, c. Since (7, 9) is on the graph, we know that

$$\text{when } x = 7, \text{ then } y = 9,$$

so that

$$9 = a(7^2) + b(7) + c, \quad \text{or equivalently,} \quad 49a + 7b + c = 9.$$

Similarly, since (47, 38) and (136, 70) are on the graph, we have

$$38 = a(47^2) + b(47) + c, \quad \text{or equivalently,} \quad 2209a + 47b + c = 38$$

and

$$70 = a(136^2) + b(136) + c, \quad \text{or equivalently,} \quad 18{,}496a + 136b + c = 70.$$

We can determine a, b, and c by solving this system of equations:

$$\begin{aligned} 49a + 7b + c &= 9 \\ 2209a + 47b + c &= 38 \\ 18{,}496a + 136b + c &= 70, \end{aligned}$$

or in matrix form,

$$\begin{pmatrix} 49 & 7 & 1 \\ 2209 & 47 & 1 \\ 18{,}496 & 136 & 1 \end{pmatrix}\begin{pmatrix} a \\ b \\ c \end{pmatrix} = \begin{pmatrix} 9 \\ 38 \\ 70 \end{pmatrix}.$$

A calculator shows that the solution is

$$\begin{pmatrix} a \\ b \\ c \end{pmatrix} = \begin{pmatrix} 49 & 7 & 1 \\ 2209 & 47 & 1 \\ 18{,}496 & 136 & 1 \end{pmatrix}^{-1}\begin{pmatrix} 9 \\ 38 \\ 70 \end{pmatrix} \approx \begin{pmatrix} -.00283 \\ .87798 \\ 2.99296 \end{pmatrix}.$$

Therefore the approximate equation for the ball's path is

$$y = -.00283x^2 + .87798x + 2.99296.$$

(b) The ball hits the ground at a point where its height y is 0, that is, when

$$-.00283x^2 + .87798x + 2.99296 = 0.$$

Use the quadratic formula or an equation solver to verify that the solutions of this equation are $x \approx -3.37$ (which is not applicable here) and $x \approx 313.61$. Therefore, the ball hits the ground approximately 314 feet from the batter. ■

Given any three points not on a straight line, the method in Example 7 can be used to find the unique parabola that passes through these points. This parabola can also be found by quadratic regression.*

CALCULATOR EXPLORATION

Use quadratic regression on the three points in Example 7 and verify that the equation obtained is the same as the one in Example 7.

To see why regression produces the parabola that actually passes through the three points, recall that the error in a quadratic model is measured by taking the difference between the y-coordinate of each data point and the y-coordinate of the corresponding point on the model and summing the squares of these errors. If the data points actually lie on a parabola (which is always the case with three points not on a line), then the error for that parabola will be 0 (smallest possible error). Hence, it will be the parabola produced by the least squares quadratic regression procedure.

*If you have not read Section 2.5 and Excursion 4.4.A, you may skip this discussion.

Exercises 11.3

In Exercises 1–6, determine if the product AB or BA is defined. If a product is defined, state its size (number of rows and columns). Do not calculate any products.

1. $A = \begin{pmatrix} 3 & 6 & 7 \\ 8 & 0 & 1 \end{pmatrix}$, $B = \begin{pmatrix} 2 & 5 & 9 & 1 \\ 7 & 0 & 0 & 6 \\ -1 & 3 & 8 & 7 \end{pmatrix}$

2. $A = \begin{pmatrix} -1 & -2 & -5 \\ 9 & 2 & -1 \\ 10 & 34 & 5 \end{pmatrix}$, $B = \begin{pmatrix} 17 & -9 \\ -6 & 12 \\ 3 & 5 \end{pmatrix}$

3. $A = \begin{pmatrix} 1 & 0 \\ 1 & 1 \\ 0 & 1 \end{pmatrix}$, $B = \begin{pmatrix} 5 & 6 & 11 \\ 7 & 8 & 15 \end{pmatrix}$

4. $A = \begin{pmatrix} 1 & -5 & 7 \\ 2 & 4 & 8 \\ 1 & -1 & 2 \end{pmatrix}$, $B = \begin{pmatrix} -2 & 4 & 9 \\ 13 & -2 & 1 \\ 5 & 25 & 0 \end{pmatrix}$

5. $A = \begin{pmatrix} -4 & 15 \\ 3 & -7 \\ 2 & 10 \end{pmatrix}$, $B = \begin{pmatrix} 1 & 2 \\ 3 & 4 \end{pmatrix}$

6. $A = \begin{pmatrix} 10 & 12 \\ -6 & 0 \\ 1 & 23 \\ -4 & 3 \end{pmatrix}$, $B = \begin{pmatrix} 1 & 2 & 3 \\ 3 & 2 & 1 \end{pmatrix}$

In Exercises 7–12, find AB.

7. $A = \begin{pmatrix} 3 & 2 \\ 2 & 4 \end{pmatrix}$, $B = \begin{pmatrix} 1 & -2 & 3 \\ 0 & 3 & 1 \end{pmatrix}$

8. $A = \begin{pmatrix} -1 & 2 & 3 \\ 0 & -1 & 2 \\ 1 & 2 & 0 \end{pmatrix}$, $B = \begin{pmatrix} 3 & -2 & -1 \\ 1 & 0 & 5 \\ 1 & -1 & -1 \end{pmatrix}$

9. $A = \begin{pmatrix} 1 & 0 & -4 \\ 0 & 2 & -1 \\ 2 & 3 & 4 \end{pmatrix}$, $B = \begin{pmatrix} 1 & 1 \\ 1 & 0 \\ 0 & 1 \end{pmatrix}$

10. $A = \begin{pmatrix} 1 & -2 \\ 3 & 0 \\ 0 & -1 \\ 2 & 1 \end{pmatrix}$, $B = \begin{pmatrix} -1 & 3 & -2 & 0 \\ 6 & 1 & 0 & -2 \end{pmatrix}$

11. $A = \begin{pmatrix} 2 & 0 & -1 \\ 1 & 1 & 2 \\ 0 & 2 & -3 \\ 2 & 3 & 0 \end{pmatrix}$, $B = \begin{pmatrix} 1 & 0 & 1 & 1 \\ 1 & 1 & 0 & 1 \\ 1 & 1 & 1 & 0 \end{pmatrix}$

12. $A = \begin{pmatrix} 10 & 0 & 1 & 0 \\ -1 & 1 & 0 & 1 \end{pmatrix}$, $B = \begin{pmatrix} 2 & -1 & 0 & 1 \\ -2 & 3 & 1 & -4 \\ 3 & 5 & 2 & -5 \end{pmatrix}$

In Exercises 13–16, show that AB is not equal to BA by computing both products.

13. $A = \begin{pmatrix} 3 & 2 \\ 5 & 1 \end{pmatrix}$, $B = \begin{pmatrix} 7 & -5 \\ -2 & 6 \end{pmatrix}$

14. $A = \begin{pmatrix} 3/2 & 2 \\ 4 & 7/2 \end{pmatrix}$, $B = \begin{pmatrix} 1 & -3/2 \\ 5/2 & 1 \end{pmatrix}$

15. $A = \begin{pmatrix} 4 & 2 & -1 \\ 0 & 1 & 2 \\ -3 & 0 & 1 \end{pmatrix}$, $B = \begin{pmatrix} 1 & 7 & -5 \\ 2 & -2 & 6 \\ 0 & 0 & 0 \end{pmatrix}$

16. $A = \begin{pmatrix} 1 & 1 & -1 & 1 \\ 2 & 0 & 3 & 2 \\ -3 & 0 & 0 & 1 \\ 1 & -1 & 1 & 2 \end{pmatrix}$, $B = \begin{pmatrix} 0 & 1 & 7 & 7 \\ 2 & 3 & -2 & 1 \\ 5 & 0 & 1 & 0 \\ -1 & 0 & 1 & 0 \end{pmatrix}$

In Exercises 17–24, find the inverse of the matrix, if it exists.

17. $\begin{pmatrix} 1 & 2 \\ 3 & 4 \end{pmatrix}$

18. $\begin{pmatrix} 3 & 5 \\ 1 & 4 \end{pmatrix}$

19. $\begin{pmatrix} 3 & -1 \\ -6 & 2 \end{pmatrix}$

20. $\begin{pmatrix} 1 & -1 & 0 \\ 1 & 0 & -1 \\ 6 & -2 & -3 \end{pmatrix}$

21. $\begin{pmatrix} 1 & 2 & 0 \\ 3 & -1 & 2 \\ -2 & 3 & -2 \end{pmatrix}$

22. $\begin{pmatrix} 1 & -3 & 4 \\ 2 & -5 & 7 \\ 0 & -1 & 1 \end{pmatrix}$

23. $\begin{pmatrix} 5 & 0 & 2 \\ 2 & 2 & 1 \\ -3 & 1 & -1 \end{pmatrix}$

24. $\begin{pmatrix} -1 & 3 & 1 \\ 2 & 5 & 0 \\ 3 & 1 & -2 \end{pmatrix}$

In Exercises 25–28, solve the system of equations by using the method of Example 5.

25. $\begin{aligned} -x + y &= 1 \\ -x + z &= -2 \\ 6x - 2y - 3z &= 3 \end{aligned}$

26. $\begin{aligned} x + 2y + 3z &= 1 \\ 2x + 5y + 3z &= 0 \\ x + 8z &= -1 \end{aligned}$

27. $\begin{aligned} 2x + y &= 0 \\ -4x - y - 3z &= 1 \\ 3x + y + 2z &= 2 \end{aligned}$

28. $\begin{aligned} -3x - 3y - 4z &= 2 \\ y + z &= 1 \\ 4x + 3y + 4z &= 3 \end{aligned}$

In Exercises 29–39, solve the system by any method.

29. $\begin{aligned} x + y + 2w &= 3 \\ 2x - y + z - w &= 5 \\ 3x + 3y + 2z - 2w &= 0 \\ x + 2y + z &= 2 \end{aligned}$

30.
$$\begin{aligned} x - 2y + 3z \quad\quad &= 1 \\ y - z + w &= -2 \\ -2x + 2y - 2z + 4w &= 5 \\ 2y - 3z + w &= 8 \end{aligned}$$

31.
$$\begin{aligned} x + y + 6z + 2v \quad\quad &= 1.5 \\ x + 5z + 2v - 3w &= 2 \\ 3x + 2y + 17z + 6v - 4w &= 2.5 \\ 4x + 3y + 21z + 7v - 2w &= 3 \\ -6x - 5y - 36z - 12v + 3w &= 3.5 \end{aligned}$$

32.
$$\begin{aligned} x - 1.5y + 1.5v - 4w &= 3 \\ -1.5y + .5w &= 0 \\ -x + 2y + .5z - 2v + 4.5w &= 2 \\ -.5y - 2.5z + .5v - .75w &= 8 \\ y - .5z + .5w &= 4 \end{aligned}$$

33.
$$\begin{aligned} x + 2y + 2z - 2w &= -23 \\ 4x + 4y - z + 5w &= 7 \\ -2x + 5y + 6z + 4w &= 0 \\ 5x + 13y + 7z + 12w &= -7 \end{aligned}$$

34.
$$\begin{aligned} x + 4y + 5z + 2w &= 0 \\ 2x + y + 4z - 2w &= 0 \\ -x + 7y + 10z + 5w &= 0 \\ -4x + 2y + z + 5w &= 0 \end{aligned}$$

35.
$$\begin{aligned} x + 2y + 5z - 2v + 4w &= 0 \\ 2x - 4y + 6z + v + 4w &= 0 \\ 5x + 2y - 3z + 2v + 3w &= 0 \\ 6x - 5y - 2z + 5v + 3w &= 0 \\ x + 2y - z - 2v + 4w &= 0 \end{aligned}$$

36.
$$\begin{aligned} 4x + 2y + 3z + 3v + 2w &= 1 \\ 2x + 2z + 2u + v - w &= 2 \\ 10x + 2y + 10z + 10u + 3v + 4w &= 5 \\ 16x + 4y + 16z + 18u + 7v - 2w &= -3 \\ x + 2y + 4z - 6u + 2v + w &= -2 \\ 6x + 2y + 6z + 6u + 3v \quad\quad &= 1 \end{aligned}$$

37.
$$\begin{aligned} x + 2y + 3z \quad\quad &= 1 \\ 3x + 2y + 4z \quad\quad &= -1 \\ 2x + 6y + 8z + w &= 3 \\ 2x + 2z - 2w &= 3 \end{aligned}$$

38.
$$\begin{aligned} x + 2y + 4z &= 6 \\ y + z &= 1 \\ x + 3y + 5z &= 10 \end{aligned}$$

39.
$$\begin{aligned} x \quad\quad + 3w &= -2 \\ x - 4y - z + 3w &= -7 \\ 4y + z \quad\quad &= 5 \\ -x + 12y + 3z - 3w &= 17 \end{aligned}$$

In Exercises 40–46, find constants a, b, c such that the three given points lie on the parabola $y = ax^2 + bx + c$. See Example 7.

40. $(-3, 2), (1, 1), (2, -1)$

41. $(-3, 15), (1, -7), (5, 111)$

42. $(1, -2), (3, 1), (4, -1)$

43. $(1, 0), (-1, 6), (2, 3)$

44. $(1, 1), (0, 0), (-1, 2)$

45. $(-1, 6), (-2, 16), (1, 4)$

46. $(-1, -6), (2, -3), (4, -25)$

47. Find constants a, b, c such that the points $(0, -2)$, $(\ln 2, 1)$, and $(\ln 4, 4)$ lie on the graph of $f(x) = ae^x + be^{-x} + c$. [*Hint:* Proceed as in Example 7.]

48. Find constants a, b, c such that the points $(0, -1)$, $(\ln 2, 4)$, and $(\ln 3, 7)$ lie on the graph of $f(x) = ae^x + be^{-x} + c$.

49. A candy company produces three types of gift boxes: *A*, *B*, and *C*. A box of variety *A* contains .6 pound of chocolates and .4 pound of mints. A box of variety *B* contains .3 pound of chocolates, .4 pound of mints, and .3 pound of caramels. A box of variety *C* contains .5 pound of chocolates, .3 pound of mints, and .2 pound of caramels. The company has 41,400 pounds of chocolates, 29,400 pounds of mints, and 16,200 pounds of caramels in stock. How many boxes of each kind should be made in order to use up all their stock?

50. Certain circus animals are fed the same three food mixes: *R*, *S*, and *T*. Lions receive 1.1 units of *R*, 2.4 units of *S*, and 3.7 units of *T* each day. Horses receive 8.1 units of *R*, 2.9 units of *S*, and 5.1 units of *T* each day. Bears receive 1.3 units of *R*, 1.3 units of *S*, and 2.3 units of *T* each day. If 16,000 units of *R*, 28,000 units of *S*, and 44,000 units of *T* are available each day, how many of each type of animal can be supported?

11.4 Systems of Nonlinear Equations

Some systems that include nonlinear equations can be solved algebraically.

Example 1 Solve the system

$$-2x + y = -1$$
$$xy = 3.$$

Solution Solve the first equation for y:

$$y = 2x - 1$$

and substitute this into the second equation:

$$\begin{aligned} xy &= 3 \\ x(2x - 1) &= 3 \\ 2x^2 - x &= 3 \\ 2x^2 - x - 3 &= 0 \\ (2x - 3)(x + 1) &= 0 \end{aligned}$$

$$2x - 3 = 0 \quad \text{or} \quad x + 1 = 0$$
$$x = 3/2 \qquad\qquad x = -1$$

Using the equation $y = 2x - 1$ to find the corresponding values of y, we see that:

$$\text{If } x = \frac{3}{2}, \quad \text{then } y = 2\left(\frac{3}{2}\right) - 1 = 2.$$

$$\text{If } x = -1, \quad \text{then } y = 2(-1) - 1 = -3.$$

Therefore, the solutions of the system are $x = 3/2$, $y = 2$, and $x = -1$, $y = -3$. ■

Example 2 Solve the system

$$x^2 + y^2 = 8$$
$$x^2 - y = 6.$$

Solution Solve the second equation for y, obtaining $y = x^2 - 6$, and substitute this into the first equation:

$$x^2 + y^2 = 8$$
$$x^2 + (x^2 - 6)^2 = 8$$
$$x^2 + x^4 - 12x^2 + 36 = 8$$
$$x^4 - 11x^2 + 28 = 0$$
$$(x^2 - 4)(x^2 - 7) = 0$$
$$x^2 - 4 = 0 \qquad \text{or} \qquad x^2 - 7 = 0$$
$$x^2 = 4 \qquad\qquad x^2 = 7$$
$$x = \pm 2 \qquad\qquad x = \pm\sqrt{7}$$

Using the equation $y = x^2 - 6$ to find the corresponding values of y, we find that the solutions of the system are

$$x = 2, y = -2; \qquad x = -2, y = -2; \qquad x = \sqrt{7}, y = 1;$$
$$x = -\sqrt{7}, y = 1. \quad ■$$

Algebraic techniques were successful in Examples 1 and 2 because substitution led to equations whose solutions could be found exactly. When this is not the case, graphical methods are needed. Recall that the graph of an equation in two variables consists of points whose coordinates are solutions of the equation. So the solutions of a system of such equations are the points that are on the graphs of all the equations in the system. They can be approximated with a graphical intersection finder.

Example 3 Solve the system

$$y = x^4 - 4x^3 + 9x - 1$$
$$y = 3x^2 - 3x - 7.$$

Solution If you try substitution on this system, say by substituting the expression for y from the first equation into the second, you obtain

$$x^4 - 4x^3 + 9x - 1 = 3x^2 - 3x - 7$$
$$x^4 - 4x^3 - 3x^2 + 12x + 6 = 0.$$

This fourth-degree equation cannot be readily solved algebraically, so a graphical approach is appropriate.

Graph both equations of the original system on the same screen. In the viewing window of Figure 11–11, the graphs intersect at three points. However, the graphs seem to be getting closer together as they run off the screen at the top right, which suggests that there may be another intersection point.

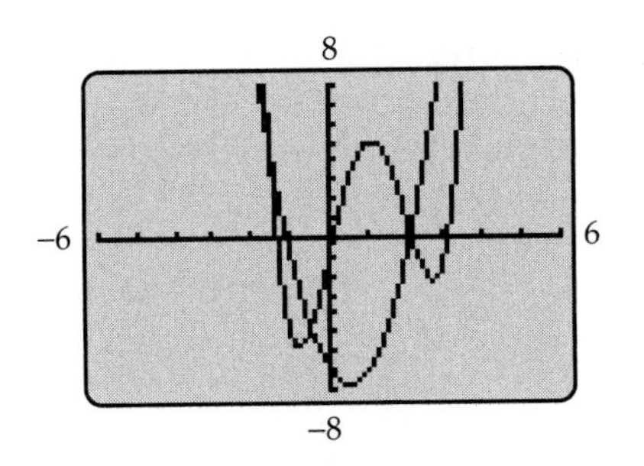

Figure 11–11

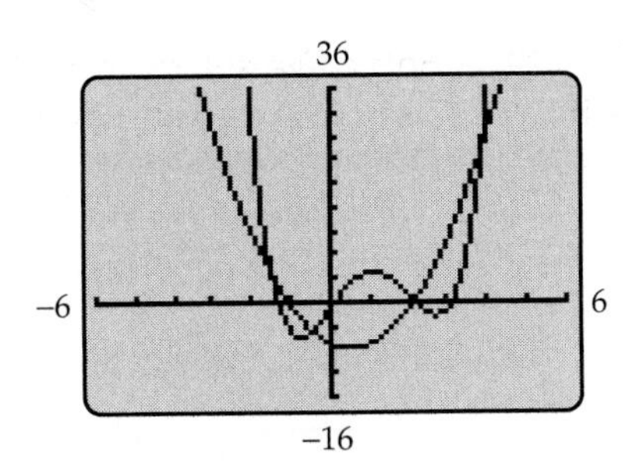

Figure 11–12

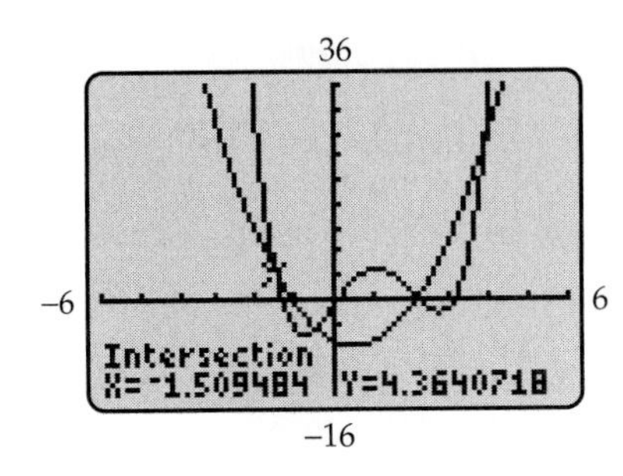

Figure 11–13

The larger window in Figure 11–12 shows four intersection points. There cannot be any more because, as we saw in the previous paragraph, the intersection points (solutions of the system) correspond to the solutions of a fourth-degree polynomial equation, which has a maximum of four solutions. An intersection finder (Figure 11–13) shows that one of the approximate solutions of the system is

$$x = -1.509484, \qquad y = 4.3640718.$$

GRAPHING EXPLORATION

Graph the two equations in the viewing window of Figure 11–12 and use your intersection finder to approximate the other three solutions of this system. ■

Systems with Second-Degree Equations

Solving systems of equations graphically depends on our ability to graph each equation in the system. With some equations of higher degree, this may require special techniques.

Example 4 Solve this system graphically:

$$x^2 - 4x - y + 1 = 0$$
$$10x^2 + 25y^2 = 100.$$

Solution It's easy to graph the first equation since it can be rewritten as $y = x^2 - 4x + 1$. To graph the second equation, we must first solve for y:

$$10x^2 + 25y^2 = 100$$
$$25y^2 = 100 - 10x^2$$
$$y^2 = \frac{100 - 10x^2}{25}$$
$$y = \sqrt{\frac{100 - 10x^2}{25}} \quad \text{or} \quad y = -\sqrt{\frac{100 - 10x^2}{25}}$$

> **Technology Tip**
>
> On Sharp 9600 and TI calculators, you can graph both of these functions at once by keying in
>
> $$y = \{1, -1\}\sqrt{\frac{100 - 10x^2}{25}}.$$

Each of these last equations does define y as a function of x and hence can be graphed on a calculator (see the Tip). By graphing both these functions on the same screen, we obtain the complete graph of the equation $10x^2 + 25y^2 = 100$, as shown in Figure 11–14. The graphs of both equations in the system are shown in Figure 11–15.

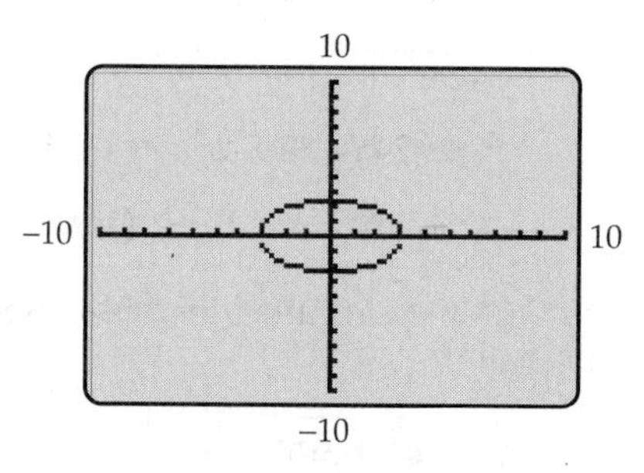

Figure 11–14

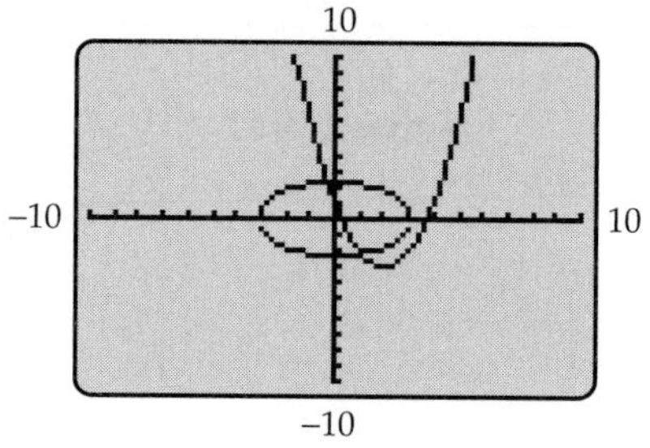

Figure 11–15

The two intersection points (solutions of the system) in Figure 11–15 can now be determined by an intersection finder:

$$x = -.2348, y = 1.9945 \qquad \text{and} \qquad x = .9544, y = -1.9067. \quad \blacksquare$$

Example 5 In order to solve the system

$$-4x^2 + 24xy + 3y^2 - 48 = 0$$
$$16x^2 + 24xy + 9y^2 + 100x + 50y + 100 = 0$$

we must express each equation in terms of functions in order to graph it. The first equation may be rewritten as

$$3y^2 + (24x)y + (-4x^2 - 48) = 0.$$

This is a quadratic equation of the form $ay^2 + by + c = 0$, with

$$a = 3, \qquad b = 24x, \qquad c = -4x^2 - 48$$

and hence can be solved by using the quadratic formula:

$$y = \frac{-b \pm \sqrt{b^2 - 4ac}}{2a}$$

$$y = \frac{-24x \pm \sqrt{(24x)^2 - 4\cdot 3\cdot(-4x^2 - 48)}}{2\cdot 3}$$

$$= \frac{-24x \pm \sqrt{(24x)^2 - 12(-4x^2 - 48)}}{6}.$$

Consequently, the graph of the first equation can be obtained by graphing both of these functions on the same screen (Figure 11–16):

Technology Tip

TI-83/86 and Sharp 9600 users can save keystrokes by entering the first equation of (*) as y_1 and then using the RCL key to copy the text of y_1 to y_2. On TI-89, use COPY and PASTE in place of RCL. Then only one sign needs to be changed to make y_2 into the second equation to be graphed.

(*)
$$y = \frac{-24x + \sqrt{(24x)^2 - 12(-4x^2 - 48)}}{6}.$$
$$y = \frac{-24x - \sqrt{(24x)^2 - 12(-4x^2 - 48)}}{6}.$$

The second equation can also be solved for y by rewriting it as follows:

$$16x^2 + 24xy + 9y^2 + 100x + 50y + 100 = 0$$
$$9y^2 + (24xy + 50y) + (16x^2 + 100x + 100) = 0$$
$$9y^2 + (24x + 50)y + (16x^2 + 100x + 100) = 0.$$

Now apply the quadratic formula with $a = 9$, $b = 24x + 50$, and $c = 16x^2 + 100x + 100$:

$$y = \frac{-b \pm \sqrt{b^2 - 4ac}}{2a}$$
$$= \frac{-(24x + 50) \pm \sqrt{(24x + 50)^2 - 4\cdot 9(16x^2 + 100x + 100)}}{2\cdot 9}.$$

Thus, the graph of the second equation (Figure 11–17) consists of the graphs of these two functions:

(**)
$$y = \frac{-(24x + 50) + \sqrt{(24x + 50)^2 - 36(16x^2 + 100x + 100)}}{18}$$
$$y = \frac{-(24x + 50) - \sqrt{(24x + 50)^2 - 36(16x^2 + 100x + 100)}}{18}.$$

By graphing both equations (that is, all four functions shown in (*) and (**)), we obtain Figure 11–18. Then an intersection finder shows that the solutions of the system (points of intersection) are

$$x = -3.623, y = -1.113 \quad \text{and} \quad x = -.943, y = -1.833. \blacksquare$$

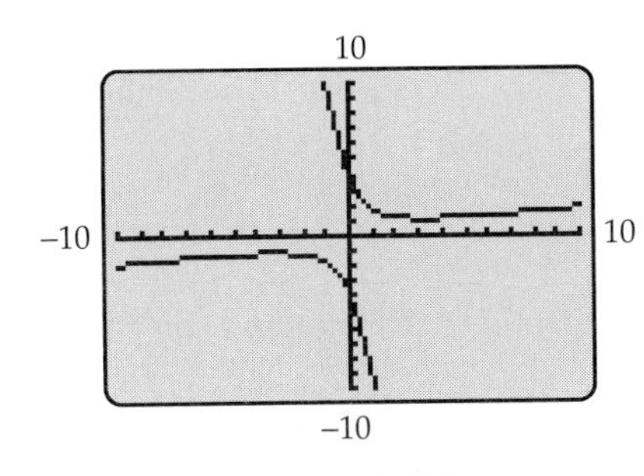

Figure 11–16

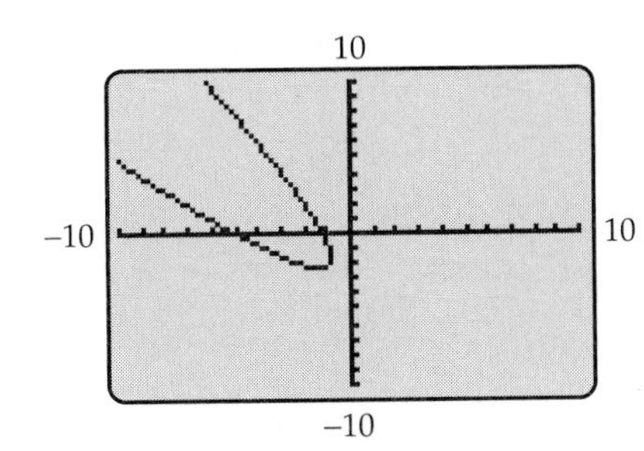

Figure 11–17

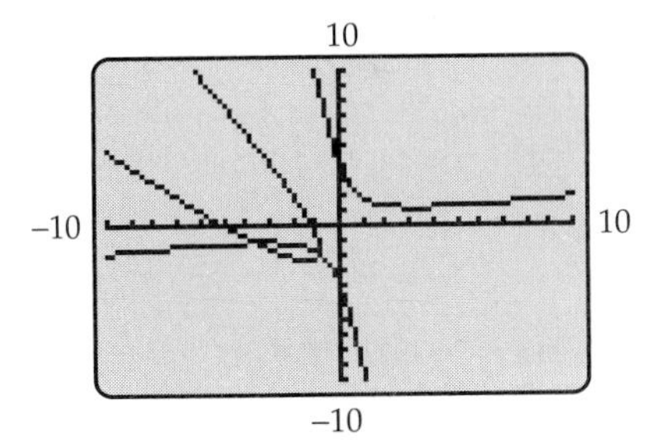

Figure 11–18

Example 6 A 52-foot-long piece of wire is to be cut into three pieces, two of which are the same length. The two equal pieces are to be bent into squares and the third piece into a circle. What should the length of each piece be if the total area enclosed by the two squares and the circle is 100 square feet?

Solution Let x be the length of each piece of wire that is to be bent into a square and y the length of the piece that is to be bent into a circle. Since the original wire is 52 feet long,

$$x + x + y = 52, \qquad \text{or equivalently,} \qquad y = 52 - 2x.$$

If a piece of wire of length x is bent into a square, the side of the square will have length $x/4$ and hence the area of the square will be $(x/4)^2 = x^2/16$. The remaining piece of wire will be made into a circle of circumference (length) y. Since the circumference is 2π times the radius (that is, $y = 2\pi r$), the circle has radius $r = y/2\pi$. Therefore, the area of the circle is

$$\pi r^2 = \pi\left(\frac{y}{2\pi}\right)^2 = \frac{\pi y^2}{4\pi^2} = \frac{y^2}{4\pi}.$$

The sum of the areas of the two squares and the circle is 100, that is,

$$\frac{x^2}{16} + \frac{x^2}{16} + \frac{y^2}{4\pi} = 100$$

$$\frac{y^2}{4\pi} = 100 - \frac{2x^2}{16}$$

$$y^2 = 4\pi\left(100 - \frac{x^2}{8}\right).$$

Therefore, the lengths x and y are solutions of this system of equations:

$$y = 52 - 2x$$

$$y^2 = 4\pi\left(100 - \frac{x^2}{8}\right).$$

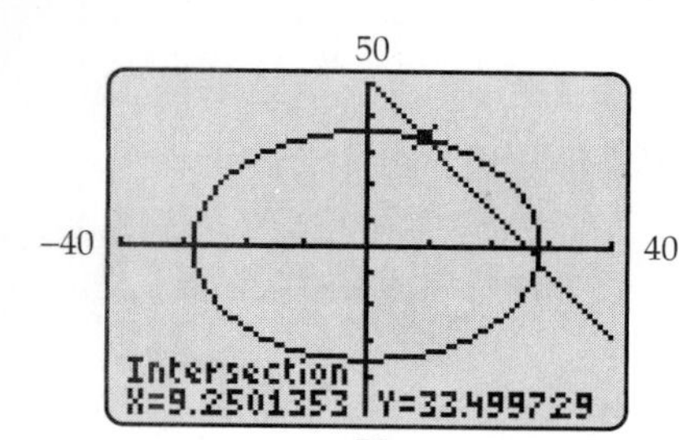

Figure 11–19

The system may be solved either algebraically or graphically, using Figure 11–19. Since x and y are lengths, both must be positive. Consequently, we need only consider the intersection point in the first quadrant. An intersection finder shows that its coordinates are $x \approx 9.25$, $y \approx 33.50$. Therefore, the wire should be cut into two 9.25-foot pieces and one 33.5-foot piece. ■

Exercises 11.4

In Exercises 1–12, solve the system algebraically.

1. $x^2 - y = 0$
   $-2x + y = 3$

2. $x^2 - y = 0$
   $-3x + y = -2$

3. $x^2 - y = 0$
   $x + 3y = 6$

4. $x^2 - y = 0$
   $x + 4y = 4$

5. $x + y = 10$
   $xy = 21$

6. $2x + y = 4$
   $xy = 2$

7. $xy + 2y^2 = 8$
   $x - 2y = 4$

8. $xy + 4x^2 = 3$
   $3x + y = 2$

9. $x^2 + y^2 - 4x - 4y = -4$
   $x - y = 2$

10. $x^2 + y^2 - 4x - 2y = -1$
$x + 2y = 2$

11. $x^2 + y^2 = 25$
$x^2 + y = 19$

12. $x^2 + y^2 = 1$
$x^2 - y = 5$

In Exercises 13–28, solve the system by any means.

13. $y = x^3 - 3x^2 + 4$
$y = -.5x^2 + 3x - 2$

14. $y = -x^3 + 3x^2 + x - 3$
$y = -2x^2 + 5$

15. $y = x^3 - 3x + 2$
$y = \dfrac{3}{x^2 + 3}$

16. $y = .25x^4 - 2x^2 + 4$
$y = x^3 - x^2 - 2x + 1$

17. $y = x^3 + x + 1$
$y = \sin x$

18. $y = x^2 - 4$
$y = \cos x$

19. $25x^2 - 16y^2 = 400$
$-9x^2 + 4y^2 = -36$

20. $9x^2 + 16y^2 = 140$
$-x^2 + 4y^2 = -4$

21. $5x^2 + 3y^2 - 20x + 6y = -8$
$x - y = 2$

22. $4x^2 + 9y^2 = 36$
$2x - y = -1$

23. $x^2 + 4xy + 4y^2 - 30x - 90y + 450 = 0$
$x^2 + x - y + 1 = 0$

24. $3x^2 + 4xy + 3y^2 - 12x + 2y + 7 = 0$
$x^2 - 10x - y + 21 = 0$

25. $4x^2 - 6xy + 2y^2 - 3x + 10y = 6$
$4x^2 + y^2 = 64$

26. $5x^2 + xy + 6y^2 - 79x - 73y + 196 = 0$
$x^2 - 2xy + y^2 - 8x - 8y + 48 = 0$

27. $x^2 + 3xy + y^2 = 2$
$3x^2 - 5xy + 3y^2 = 7$

28. $2x^2 - 8xy + 8y^2 + 2x - 5 = 0$
$16x^2 - 24xy + 9y^2 + 100x - 200y + 100 = 0$

29. What would be the answer in Example 7 if the original piece of wire were 70 feet long? Explain.

30. A 52-foot-long piece of wire is to be cut into three pieces, two of which are the same length. The two equal pieces are to be bent into circles and the third piece into a square. What should the length of each piece be if the total area enclosed by the two circles and the square is 100 square feet?

31. A rectangular box (including top) with square ends and a volume of 16 cubic meters is to be constructed from 40 square meters of cardboard. What should its dimensions be?

32. A rectangular sheet of metal is to be rolled into a circular tube. If the tube is to have a surface area (excluding ends) of 210 square inches and a volume of 252 cubic inches, what size sheet of metal should be used?

33. Find two real numbers whose sum is -16 and whose product is 48.

34. Find two real numbers whose sum is 34.5 and whose product is 297.

35. Find two positive real numbers whose difference is 1 and whose product is 4.16.

36. Find two real numbers whose difference is 25.75 and whose product is 127.5.

37. Find two real numbers whose sum is 3 such that the sum of their squares is 369.

38. Find two real numbers whose sum is 2 such that the difference of their squares is 60.

39. Find the dimensions of a rectangular room whose perimeter is 58 feet and whose area is 204 square feet.

40. Find the dimensions of a rectangular room whose perimeter is 53 feet and whose area is 165 square feet.

41. A rectangle has area 120 square inches and a diagonal of length 17 inches. What are its dimensions?

42. A right triangle has area 225 square centimeters and a hypotenuse of length 35 centimeters. To the nearest tenth of a centimeter, how long are the legs of the triangle?

43. Find the equation of the straight line that intersects the parabola $y = x^2$ *only* at the point (3, 9). [*Hint:* What condition on the discriminant guarantees that a quadratic equation has exactly one real solution?]

Chapter 11 Review

Important Concepts

Review Questions

In Questions 1–10, solve the system of linear equations by any means you want.

1. $$\begin{aligned} -5x + 3y &= 4 \\ 2x - y &= -3 \end{aligned}$$

2. $$\begin{aligned} 3x - y &= 6 \\ 2x + 3y &= 7 \end{aligned}$$

3. $$\begin{aligned} 3x - 5y &= 10 \\ 4x - 3y &= 6 \end{aligned}$$

4. $$\begin{aligned} \frac{1}{4}x - \frac{1}{3}y &= -\frac{1}{4} \\ \frac{1}{10}x + \frac{2}{5}y &= \frac{2}{5} \end{aligned}$$

5. $$\begin{aligned} 3x + y - z &= 13 \\ x + 2z &= 9 \\ -3x - y + 2z &= 9 \end{aligned}$$

6. $$\begin{aligned} x + 2y + 3z &= 1 \\ 4x + 4y + 4z &= 2 \\ 10x + 8y + 6z &= 4 \end{aligned}$$

7. $$\begin{aligned} 4x + 3y - 3z &= 2 \\ 5x - 3y + 2z &= 10 \\ 2x - 2y + 3z &= 14 \end{aligned}$$

8. $$\begin{aligned} x + y - 4z &= 0 \\ 2x + y - 3z &= 2 \\ -3x - y + 2z &= -4 \end{aligned}$$

9. $$\begin{aligned} x - 2y - 3z &= 1 \\ 5y + 10z &= 0 \\ 8x - 6y - 4z &= 8 \end{aligned}$$

10. $$\begin{aligned} 4x - y - 2z &= 4 \\ x - y - \tfrac{1}{2}z &= 1 \\ 2x - y - z &= 8 \end{aligned}$$

11. The sum of one number and three times a second number is -20. The sum of the second number and two times the first number is 55. Find the two numbers.

12. You are given \$144 in \$1, \$5, and \$10 bills. There are 35 bills. There are two more \$10 bills than \$5 bills. How many bills of each type do you have?

13. Let L be the line with equation $4x - 2y = 6$ and M the line with equation $-10x + 5y = -15$. Which of the following statements is true?

(a) L and M do not intersect.
(b) L and M intersect at a single point.
(c) L and M are the same line.
(d) All of the above are true.
(e) None of the above are true.

14. Which of the following statements about this system of equations are *false*?

$$\begin{aligned} x \quad + z &= 2 \\ 6x + 4y + 14z &= 24 \\ 2x + y + 4z &= 7 \end{aligned}$$

(a) $x = 2, y = 3, z = 0$ is a solution.
(b) $x = 1, y = 1, z = 1$ is a solution.
(c) $x = 1, y = -3, z = 3$ is a solution.
(d) The system has an infinite number of solutions.
(e) $x = 2, y = 5, z = -1$ is not a solution.

15. Tickets to a lecture cost \$1 for students, \$1.50 for faculty, and \$2 for others. Total attendance at the lecture was 460, and the total income from tickets was \$570. Three times as many students as faculty attended. How many faculty members attended the lecture?

16. An alloy containing 40% gold and an alloy containing 70% gold are to be mixed to produce 50 pounds of an alloy containing 60% gold. How much of each alloy is needed?

17. Write the augmented matrix of the system

$$\begin{aligned} x - 2y + 3z &= 4 \\ 2x + y - 4z &= 3 \\ -3x + 4y - z &= -2 \end{aligned}$$

18. Use matrix methods to solve the system in Question 17.

19. Write the coefficient matrix of the system

$$\begin{aligned} 2x - y - 2z + 2u &= 0 \\ x + 3y - 2z + u &= 0 \\ -x + 4y + 2z - 3u &= 0 \end{aligned}$$

20. Use matrix methods to solve the system in Question 19.

In Questions 21 and 22, find the constants A, B, C that make the statement true.

21. $$\frac{4x - 7}{x^2 - x - 6} = \frac{A}{x - 3} + \frac{B}{x + 2}$$

22. $$\frac{6x^2 + 6x - 6}{(x^2 - 1)(x + 2)} = \frac{A}{x + 1} + \frac{B}{x - 1} + \frac{C}{x + 2}$$

In Questions 23–26, perform the indicated matrix multiplication or state that the product is not defined. Use these matrices:

$$A = \begin{pmatrix} -1 & 0 \\ 0 & -1 \end{pmatrix}, \quad B = \begin{pmatrix} 2 & -3 \\ 4 & 1 \end{pmatrix}, \quad C = \begin{pmatrix} 3 & 2 \\ 2 & 4 \end{pmatrix},$$

$$D = \begin{pmatrix} -3 & 1 & 2 \\ 1 & 0 & 4 \end{pmatrix}, \quad E = \begin{pmatrix} 1 & 2 \\ -3 & 4 \\ 0 & 5 \end{pmatrix}, \quad F = \begin{pmatrix} 2 & 3 \\ 6 & 3 \\ 6 & 1 \end{pmatrix}$$

23. AB **24.** CD **25.** AE **26.** DF

In Questions 27–30, find the inverse of the matrix, if it exists.

27. $\begin{pmatrix} 3 & -7 \\ 4 & -9 \end{pmatrix}$

28. $\begin{pmatrix} 2 & 6 \\ 1 & 3 \end{pmatrix}$

29. $\begin{pmatrix} 3 & 2 & 6 \\ 1 & 1 & 2 \\ 2 & 2 & 5 \end{pmatrix}$

30. $\begin{pmatrix} 1 & -1 & 1 \\ 2 & -3 & 2 \\ -4 & 6 & 1 \end{pmatrix}$

In Questions 31 and 32, use matrix inverses to solve the system.

31.
$$\begin{aligned} x + \phantom{4y -{}} 2z + 6w &= 2 \\ 3x + 4y - 2z - w &= 0 \\ 5x + \phantom{4y -{}} 2z - 5w &= -4 \\ 4x - 4y + 2z + 3w &= 1 \end{aligned}$$

32.
$$\begin{aligned} 2x + y + 2z \phantom{{}- 2u} + u &= 2 \\ x + 3y - 4z - 2u + 2v &= -2 \\ 2x + 3y + 5z - 4u + v &= 1 \\ x \phantom{{}+ 3y} - 2z \phantom{{}- 2u} + 4v &= 4 \\ 2x \phantom{{}+ 3y} + 6z \phantom{{}- 2u} - 5v &= 0 \end{aligned}$$

In Questions 33 and 34, find the equation of the parabola passing through the given points.

33. $(-3, 52), (2, 17), (8, 305)$

34. $(-2, -18), (2, 6), (4, -12)$

In Questions 35–40, solve the system.

35. $x^2 - y = 0$
$y - 2x = 3$

36. $x^2 + y^2 = 25$
$x^2 + y = 19$

37. $x^2 + y^2 = 16$
$x + y = 2$

38. $6x^2 + 4xy + 3y^2 = 36$
$x^2 - xy + y^2 = 9$

39. $x^3 + y^3 = 26$
$x^2 + y = 6$

40. $x^2 - 3xy + 2y^2 - y + x = 0$
$5x^2 - 10xy + 5y^2 = 8$

41. An animal feed is to be made from corn, soybeans, and meat by-products. One bag is to supply 1800 units of fiber, 2800 units of fat, and 2200 units of protein. Each pound of corn has 10 units of fiber, 30 units of fat, and 20 units of protein. Each pound of soybeans has 20 units of fiber, 20 units of fat, and 40 units of protein. Each pound of by-products has 30 units of fiber, 40 units of fat, and 25 units of protein. How many pounds of corn, soybeans, and by-products should each bag contain?

42. A home supply store sells three models of dehumidifiers. The Standard model weighs 10 pounds and comes in a 10-cubic-foot box. The Sleek model weighs 20 pounds and comes in an 8-cubic-foot box. The Super model weighs 60 pounds and comes in a 28-cubic-foot box. The store's delivery van has 248 cubic feet of space and can hold a maximum of 440 pounds. In order for the van to be fully loaded, how many boxes of each model should it carry? [There are several correct answers.]

Discovery Project 11

Input-Output Analysis

Wassily Leontief won the Nobel Prize in economics in 1973 for his method of input-output analysis of the economies of industrialized nations. This method has become a permanent part of production planning and forecasting by both national governments and private corporations. During the Arab oil boycott in 1973, for example, General Electric used input-output analysis on 184 sectors of the economy (such as energy, agriculture, and transportation) to predict the effect of the energy crisis on public demand for its products. The key to Leontief's method is knowing how much each sector of the economy needs from other sectors in order to do its job.

A simple economic model will illustrate the basic ideas behind input-output analysis. Suppose the country Hypothetica has just two sectors in its economy: agriculture and manufacturing. The production of a ton of agricultural products requires the use of .1 ton of agricultural products and .1 ton of manufactured products. Similarly, the production of a ton of manufactured products consumes .1 ton of agricultural products and .3 ton of manufactured products. The key economic question is: How many tons of each sector must Hypothetica produce in order to have enough surplus to export 10,000 tons of agricultural products and 10,000 tons of manufactured goods?

Let A be the total amount of agricultural goods and M the total amount of manufactured goods produced. The total agricultural production A is the sum of the amount needed for producing the agricultural products, plus the amount needed for producing manufactured products, plus the amount targeted for export. In order words,

agricultural needs	+	manufacturing needs	+	export needs	=	total agricultural production
$.1A$	+	$.1M$	+	100,000	=	A.

1. Write a similar equation for manufactured goods.
2. Solve the system of equations given by the agricultural equation above and the manufacturing equation of Problem 1. What level of production of agricultural and manufactured goods should Hypothetica work toward in order to reach its export goals?
3. The same kind of analysis can be used to determine the surplus available for exports in an economy. Suppose that Hypothetica can produce a total of 120,000 tons of agricultural products and 28,000 tons of manufactured goods. How much of each is available for export?
4. Input-output analysis can also be used for allocating resources in smaller-scale situations. Suppose a horse outfitter is hired by the State of Washington Department of Fish and Wildlife to haul 20 salt blocks into a game area for winter feeding. Each horse can carry a 200-pound load. A single person can handle three horses. Each person requires one horse to ride and one-half horse to carry his or her personal gear. A single horse can carry the feed and equipment for five horses. How many people and horses must go on the trip? [*Note:* Each horse load of salt must consist of an even number of blocks. Each salt block weighs 50 pounds. Also, fractional numbers of horses and people are not allowed. Make sure to adjust your answer to integer values and be sure that the answer still provides sufficient carrying capacity for the salt.]

Chapter

12 Discrete Algebra

What's next?

CD and CD-ROM players, fax machines, cameras, and other devices incorporate digital technology, which uses sequences of 0's and 1's to send signals. Determining the monthly payment on a car loan involves the sum of a geometric sequence. Problems involving the action of bouncing balls, vacuum pumps, and other devices can sometimes be solved by using sequences. See Exercises 37, 41 and 42 on page 751.

Interdependence of Sections

Sections 12.1, 12.4, and 12.5 are independent of one another and may be read in any order. The interdependence of sections is:

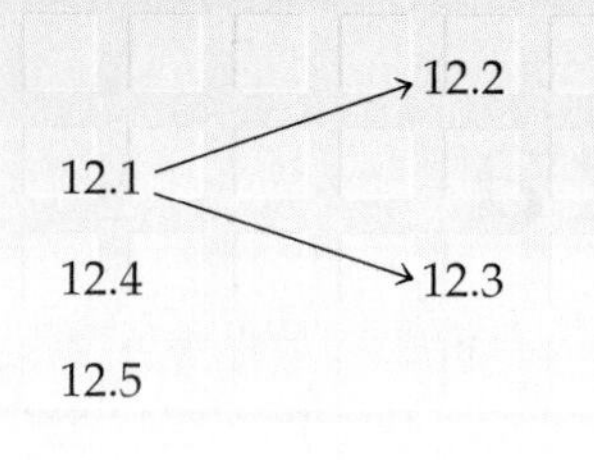

Chapter Outline

This chapter deals with a variety of subjects involving counting processes and the nonnegative integers 0, 1, 2, 3, ...

12.1 Sequences and Sums

A **sequence** is an ordered list of numbers. We usually write them horizontally, with the ordering understood to be from left to right. The same number may appear several times on the list. Each number on the list is called a **term** of the sequence. We are primarily interested in infinite sequences, such as

$$2, 4, 6, 8, 10, 12, \ldots$$

$$1, -3, 5, -7, 9, -11, 13, \ldots$$

$$2, 1, \frac{2}{3}, \frac{2}{4}, \frac{2}{5}, \frac{2}{6}, \frac{2}{7}, \ldots$$

where the dots indicate that the same pattern continues forever.*

When the pattern in an ordered list of numbers isn't obvious, the sequence is usually described as follows: The first term is denoted a_1, the second term a_2, and so on. Then a formula is given for the nth term a_n.

Example 1 Consider the sequence $a_1, a_2, a_3, \ldots, a_n, \ldots$ where a_n is given by the formula

$$a_n = \frac{n^2 - 3n + 1}{2n + 5}.$$

*Such a list defines a function f whose domain is the set of positive integers. The rule is $f(1)$ = first number on the list, $f(2)$ = second number on the list, and so on. Conversely, any function g whose domain is the set of positive integers leads to an ordered list of numbers, namely, $g(1), g(2), g(3), \ldots$. So a sequence is formally defined to be a function whose domain is the set of positive integers.

To find a_1 we substitute $n = 1$ in the formula for a_n; to find a_2 we substitute $n = 2$ in the formula; and so on:

$$a_1 = \frac{1^2 - 3 \cdot 1 + 1}{2 \cdot 1 + 5} = -\frac{1}{7}$$

$$a_2 = \frac{2^2 - 3 \cdot 2 + 1}{2 \cdot 2 + 5} = -\frac{1}{9}$$

$$a_3 = \frac{3^2 - 3 \cdot 3 + 1}{2 \cdot 3 + 5} = \frac{1}{11}.$$

Thus, the sequence begins $-1/7, -1/9, 1/11, \ldots$. The 39th term is

$$a_{39} = \frac{39^2 - 3 \cdot 39 + 1}{2 \cdot 39 + 5} = \frac{1405}{83}. \quad \blacksquare$$

Example 2 It is easy to list the first few terms of the sequence

$$a_1, a_2, a_3, \ldots \qquad \text{where} \qquad a_n = \frac{(-1)^n}{n + 2}.$$

Substituting $n = 1$, $n = 2$, and so on, in the formula for a_n shows that:

$$a_1 = \frac{(-1)^1}{1 + 2} = -\frac{1}{3}, \qquad a_2 = \frac{(-1)^2}{2 + 2} = \frac{1}{4}, \qquad a_3 = \frac{(-1)^3}{3 + 2} = -\frac{1}{5}.$$

Similarly,

$$a_{41} = \frac{(-1)^{41}}{41 + 2} = -\frac{1}{43} \qquad \text{and} \qquad a_{206} = \frac{(-1)^{206}}{206 + 2} = \frac{1}{208}. \quad \blacksquare$$

Example 3 Here are some other sequences whose nth term can be described by a formula:

Sequence	nth Term	First 5 Terms
$a_1, a_2, a_3, \ldots$	$a_n = n^2 + 1$	$2, 5, 10, 17, 26$
$b_1, b_2, b_3, \ldots$	$b_n = \frac{1}{n}$	$1, \frac{1}{2}, \frac{1}{3}, \frac{1}{4}, \frac{1}{5}$
$c_1, c_2, c_3, \ldots$	$c_n = \frac{(-1)^{n+1}2n}{(n + 1)(n + 2)}$	$\frac{1}{3}, -\frac{1}{3}, \frac{3}{10}, -\frac{4}{15}, \frac{5}{21}$
$x_1, x_2, x_3, \ldots$	$x_n = 3 + \frac{1}{10^n}$	$3.1, 3.01, 3.001, 3.0001, 3.00001.$ ■

A **constant sequence** is a sequence in which every term is the same number, such as the sequence $7, 7, 7, 7, \ldots$ or the sequence $a_1, a_2, a_3, \ldots$ where $a_n = -18$ for every $n \geq 1$.

The subscript notation for sequences is sometimes abbreviated by writing $\{a_n\}$ in place of $a_1, a_2, a_3, \ldots$.

Example 4 $\{1/2^n\}$ denotes the sequence whose first four terms are

$$a_1 = \frac{1}{2^1}, \qquad a_2 = \frac{1}{2^2} = \frac{1}{4}, \qquad a_3 = \frac{1}{2^3} = \frac{1}{8}, \qquad a_4 = \frac{1}{2^4} = \frac{1}{16}.$$

Similarly, $\{(-1)^n n^2\}$ denotes the sequence with first three terms

$$a_1 = (-1)^1 \cdot 1^2 = -1, \qquad a_2 = (-1)^2 \cdot 2^2 = 4, \qquad a_3 = (-1)^3 \cdot 3^2 = -9$$

and 23rd term $a_{23} = (-1)^{23} \cdot 23^2 = -529$. ■

There are several ways to display the terms of a sequence on a calculator, as we now see.

Example 5

(a) Display the first five terms of the sequence $\{a_n = n^2 - n - 3\}$ on your calculator screen.

(b) Display the first, fifth, ninth, and thirteenth terms of this sequence.

Solution

Method 1: Enter the sequence in the function memory as $y_1 = x^2 - x - 3$. In the table set-up screen, begin the table at $x = 1$ and set the increment at 1 and display a table of values (Figure 12–1). To display the first, fifth, ninth, and thirteenth terms, set the table increment at 4 (Figure 12–2).

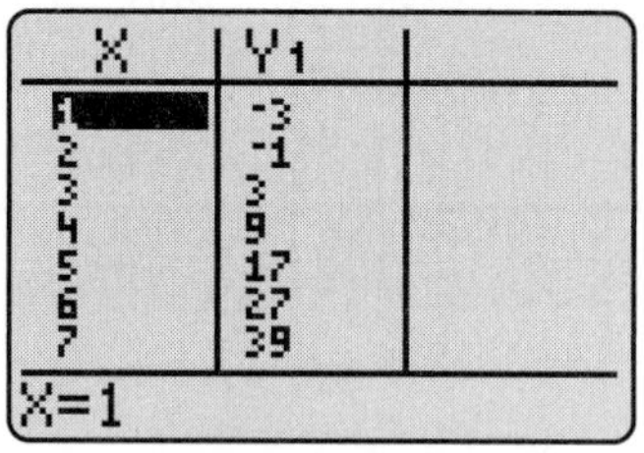

X	Y1
1	-3
2	-1
3	3
4	9
5	17
6	27
7	39

X=1

Figure 12–1

X	Y1
1	-3
5	17
9	69
13	153
17	269
21	417
25	597

X=1

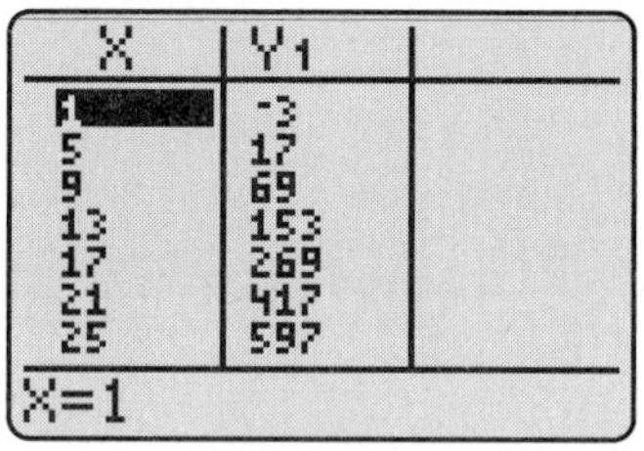

Figure 12–2

Technology Tip

SEQ is in the OPS submenu of the TI-83/86 LIST menu, in the LIST submenu of the TI-89 MATH menu, in the OPE submenu of the Sharp 9600 LIST menu, in the LIST submenu of the Casio 9850 OPTN menu. MAKELIST is in the LIST submenu of the HP-38 MATH menu.

Method 2: Using the tip in the margin, enter the following, which produces Figure 12–3.

$$\text{SEQ}(x^2 - x - 3, x, 1, 5, 1).^*$$

To display the first, fifth, ninth and thirteenth terms, enter

$$\text{SEQ}(x^2 - x - 3, x, 1, 13, 4),^*$$

which tells the calculator to look at every fourth term from 1 to 13, and produces Figure 12–4. ■

*The final 1 is optional on TI and Sharp 9600; omit "x" on Sharp 9600 and use MAKELIST in place of SEQ on HP-38.

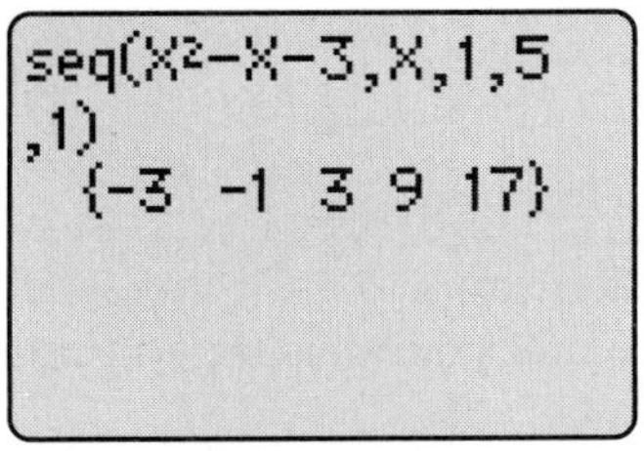

```
seq(X²-X-3,X,1,5
,1)
 {-3 -1 3 9 17}
```

Figure 12–3

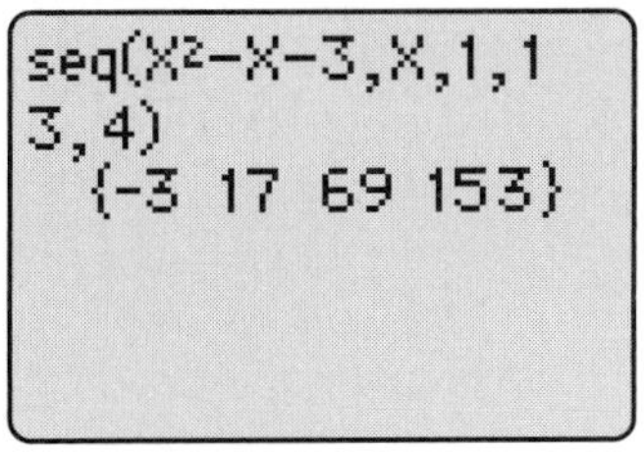

```
seq(X²-X-3,X,1,1
3,4)
 {-3 17 69 153}
```

Figure 12–4

Example 6 Display the first ten terms of the sequence $\left\{a_n = \dfrac{n}{n+1}\right\}$ on your calculator screen in fractional form, if possible.

Solution Creating a table always produces decimal approximations, as you can easily verify. The same is usually true of the SEQ key, unless you take special steps. On HP-38, change the number format mode to "fraction" (MODE menu). On TI calculators (other than TI-89), either use the FRAC key after obtaining the sequence (Figure 12–5) or use parentheses and the FRAC key as part of the function: Entering

$$\text{SEQ}(x/(x+1)\ \blacktriangleright\text{FRAC}, x, 1, 10, 1)$$

produces Figure 12–6. In each figure, you must use the arrow key to scroll to the right in order to see all the terms. ■

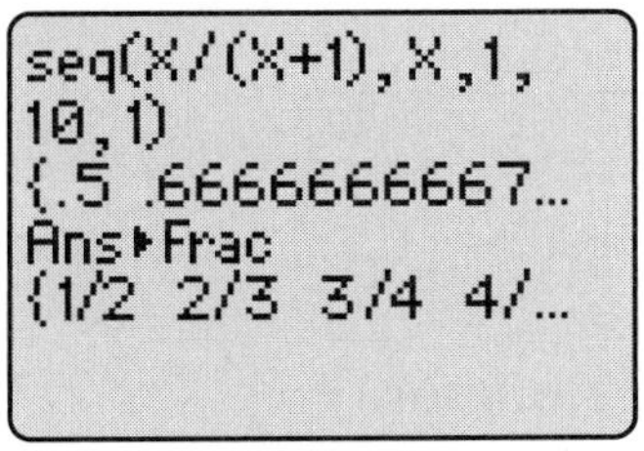

```
seq(X/(X+1),X,1,
10,1)
{.5 .6666666667…
Ans▸Frac
{1/2 2/3 3/4 4/…
```

Figure 12–5

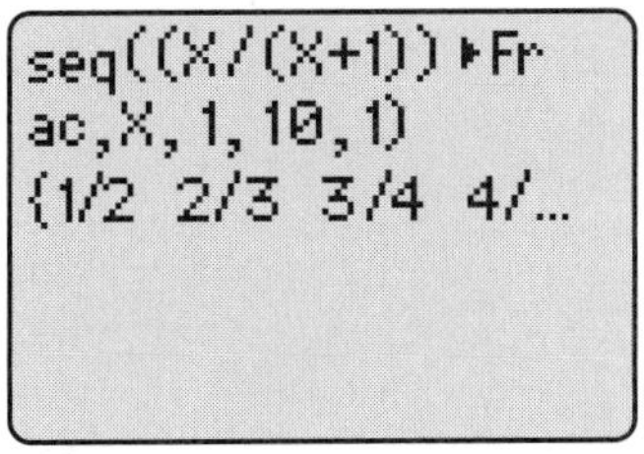

```
seq((X/(X+1))▸Fr
ac,X,1,10,1)
{1/2 2/3 3/4 4/…
```

Figure 12–6

A sequence is said to be defined **recursively** (or **inductively**) if the first term is given (or the first several terms) and there is a method of determining the nth term by using the terms that precede it.

Example 7 Consider the sequence whose first two terms are

$$a_1 = 1 \qquad \text{and} \qquad a_2 = 1$$

and whose nth term (for $n \geq 3$) is the sum of the two preceding terms:

$$a_3 = a_2 + a_1 = 1 + 1 = 2$$

$$a_4 = a_3 + a_2 = 2 + 1 = 3$$

$$a_5 = a_4 + a_3 = 3 + 2 = 5.$$

Technology Tip

Some calculators have a sequence graphing mode. It can be chosen in the TI-83/89 MODE menu, the HP-38 LIB menu, the COORD submenu of the Sharp 9600 SETUP menu, or the RECUR submenu of the Casio 9850 main menu. On such calculators, recursively defined functions may be entered into the sequence memory (Y = list). Check your instruction manual for the correct syntax.

For each integer n, the two preceding integers are $n - 1$ and $n - 2$. So

$$a_n = a_{n-1} + a_{n-2} \quad (n \geq 3).$$

This sequence 1, 1, 2, 3, 5, 8, 13, . . . is called the **Fibonacci sequence,** and the numbers that appear in it are called **Fibonacci numbers.** Fibonacci numbers have many surprising and interesting properties. See Exercises 58–64 for details. ■

Example 8 The sequence given by

$$a_1 = -7 \qquad \text{and} \qquad a_n = a_{n-1} + 3 \quad \text{for } n \geq 2$$

is defined recursively. Its first three terms are

$$a_1 = -7, \qquad a_2 = a_1 + 3 = -7 + 3 = -4,$$
$$a_3 = a_2 + 3 = -4 + 3 = -1. \quad ■$$

Sometimes it is convenient or more natural to begin numbering the terms of a sequence with a number other than 1. So we may consider sequences such as

$$b_4, b_5, b_6, \ldots \qquad \text{or} \qquad c_0, c_1, c_2, \ldots.$$

Example 9 The sequence 4, 5, 6, 7, . . . can be conveniently described by saying $b_n = n$, with $n \geq 4$. In the brackets notation, we write $\{n\}_{n\geq4}$. Similarly, the sequence

$$2^0, 2^1, 2^2, 2^3, \ldots$$

may be described as $\{2^n\}_{n\geq0}$ or by saying $c_n = 2^n$, with $n \geq 0$. ■

Summation Notation

It is sometimes necessary to find the sum of various terms in a sequence. For instance, we might want to find the sum of the first nine terms of the sequence $\{a_n\}$. Mathematicians often use the Greek letter sigma (Σ) to abbreviate such a sum:*

$$\sum_{k=1}^{9} a_k = a_1 + a_2 + a_3 + a_4 + a_5 + a_6 + a_7 + a_8 + a_9.$$

Similarly, for any positive integer m and numbers $c_1, c_2, \ldots, c_m$

Summation Notation

$$\sum_{k=1}^{m} c_k \quad \textbf{means} \quad c_1 + c_2 + c_3 + \cdots + c_m.$$

*Σ is the letter S in the Greek alphabet, the first letter in *Sum*.

Example 10 Compute each of these sums:

(a) $\sum_{k=1}^{5} k^2$ (b) $\sum_{k=1}^{4} k^2(k-2)$ (c) $\sum_{k=1}^{6} (-1)^k k$.

Solution

(a) We successively substitute 1, 2, 3, 4, 5 for k in the expression k^2 and add up the results:

$$\sum_{k=1}^{5} k^2 = 1^2 + 2^2 + 3^2 + 4^2 + 5^2 = 55.$$

(b) Successively substituting 1, 2, 3, 4 for k in $k^2(k-2)$ and adding the results, we have:

$$\sum_{k=1}^{4} k^2(k-2) = 1^2(1-2) + 2^2(2-2) + 3^2(3-2) + 4^2(4-2)$$

$$= 1(-1) + 4(0) + 9(1) + 16(2) = 40.$$

(c) $\sum_{k=1}^{6} (-1)^k k =$

$$(-1)^1 \cdot 1 + (-1)^2 \cdot 2 + (-1)^3 \cdot 3 + (-1)^4 \cdot 4 + (-1)^5 \cdot 5 + (-1)^6 \cdot 6$$

$$= -1 + 2 - 3 + 4 - 5 + 6 = 3. \quad \blacksquare$$

In sums such as $\sum_{k=1}^{5} k^2$ and $\sum_{k=1}^{6} (-1)^k k$, the letter k is called the **summation index.** Any letter may be used for the summation index, just as the rule of a function f may be denoted by $f(x)$ or $f(t)$ or $f(k)$, etc. For example, $\sum_{n=1}^{5} n^2$ means: Take the sum of the terms n^2 as n takes values from 1 to 5. In other words, $\sum_{n=1}^{5} n^2 = \sum_{k=1}^{5} k^2$. Similarly,

$$\sum_{k=1}^{4} k^2(k-2) = \sum_{j=1}^{4} j^2(j-2) = \sum_{n=1}^{4} n^2(n-2).$$

The Σ notation for sums can also be used for sums that don't begin with $k = 1$. For instance,

$$\sum_{k=4}^{10} k^2 = 4^2 + 5^2 + 6^2 + 7^2 + 8^2 + 9^2 + 10^2 = 371$$

$$\sum_{j=0}^{3} j^2(2j+5) = 0^2(2 \cdot 0 + 5) + 1^2(2 \cdot 1 + 5) + 2^2(2 \cdot 2 + 5) + 3^2(2 \cdot 3 + 5).$$

Example 11 Use a calculator to compute these sums:

(a) $\sum_{k=1}^{50} k^2$ (b) $\sum_{k=38}^{75} k^2$.

Technology Tip

On TI-86/89, Casio 9850, and HP-38, SUM (or ΣLIST) is in the same submenu as SEQ (or MAKELIST). On TI-83 and Sharp 9600, SUM is in the MATH submenu of the LIST menu.

Solution In each case, use SUM together with SEQ (or ΣLIST and MAKELIST on HP-38), with the same syntax for SEQ as in Example 5:

$$\sum_{k=1}^{50} k^2 = \text{SUM SEQ}(x^2, x, 1, 50, 1) = 42{,}925$$

and

$$\sum_{k=38}^{75} k^2 = \text{SUM SEQ}(x^2, x, 38, 75, 1) = 125{,}875$$

as shown in Figure 12–7. ■

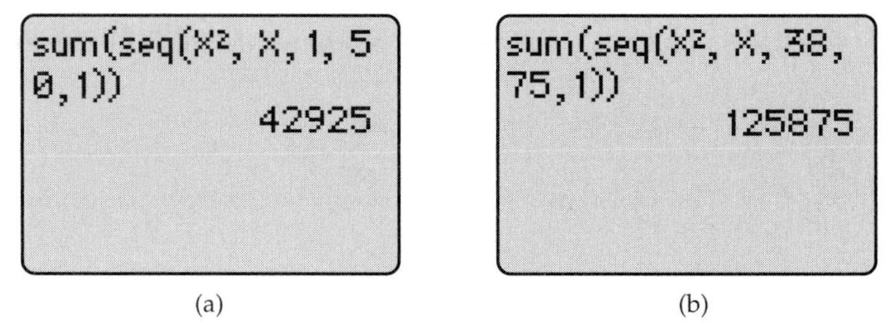

(a) (b)

Figure 12–7

Partial Sums

Suppose $\{a_n\}$ is a sequence and k is a positive integer. The sum of the first k terms of the sequence is called the ***k*th partial sum** of the sequence. Thus,

Partial Sums

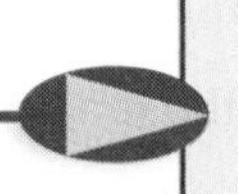

The *k*th sum of $\{a_n\} = \sum_{n=1}^{k} a_n = a_1 + a_2 + a_3 + \cdots + a_k.$

Example 12 Here are some partial sums of the sequence $\{n^3\}$:

First partial sum: $\sum_{n=1}^{1} n^3 = 1^3 = 1$

Second partial sum: $\sum_{n=1}^{2} n^3 = 1^3 + 2^3 = 9$

Sixth partial sum: $\sum_{n=1}^{6} n^3 = 1^3 + 2^3 + 3^3 + 4^3 + 5^3 + 6^3 = 441.$ ■

Example 13 The sequence $\{2^n\}_{n \geq 0}$ begins with the 0th term, so the fourth partial sum (the sum of the first four terms) is

$$2^0 + 2^1 + 2^2 + 2^3 = \sum_{n=0}^{3} 2^n.$$

Similarly, the fifth partial sum of the sequence $\left\{\dfrac{1}{n(n-2)}\right\}_{n\geq 3}$ is the sum of the first five terms:

$$\frac{1}{3(3-2)}+\frac{1}{4(4-2)}+\frac{1}{5(5-2)}+\frac{1}{6(6-2)}+\frac{1}{7(7-2)}=\sum_{n=3}^{7}\frac{1}{n(n-2)}. \quad \blacksquare$$

Certain calculations can be written very compactly in summation notation. For example, the distributive law shows that

$$ca_1 + ca_2 + ca_3 + \cdots + ca_r = c(a_1 + a_2 + a_3 + \cdots + a_r).$$

In summation notation this becomes

$$\sum_{n=1}^{r} ca_n = c\left(\sum_{n=1}^{r} a_n\right).$$

This proves the first of the following statements.

Properties of Sums

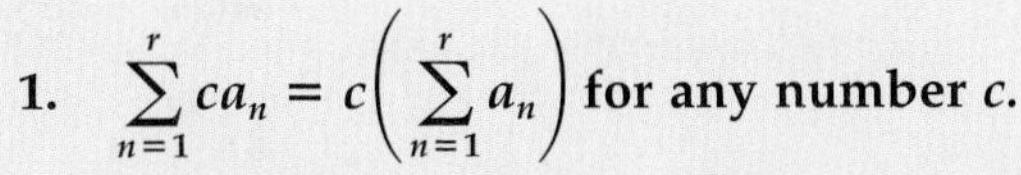

1. $\displaystyle\sum_{n=1}^{r} ca_n = c\left(\sum_{n=1}^{r} a_n\right)$ **for any number** c.
2. $\displaystyle\sum_{n=1}^{r} (a_n + b_n) = \sum_{n=1}^{r} a_n + \sum_{n=1}^{r} b_n$
3. $\displaystyle\sum_{n=1}^{r} (a_n - b_n) = \sum_{n=1}^{r} a_n - \sum_{n=1}^{r} b_n$

To prove statement 2, use the commutative and associative laws repeatedly to show that:

$$(a_1 + b_1) + (a_2 + b_2) + (a_3 + b_3) + \cdots + (a_r + b_r)$$
$$= (a_1 + a_2 + a_3 + \cdots + a_r) + (b_1 + b_2 + b_3 + \cdots + b_r)$$

which can be written in summation notation as

$$\sum_{n=1}^{r} (a_n + b_n) = \sum_{n=1}^{r} a_n + \sum_{n=1}^{r} b_n.$$

The last statement is proved similarly.

Exercises 12.1

In Exercises 1–10, find the first five terms of the sequence $\{a_n\}$.

1. $a_n = 2n + 6$
2. $a_n = 2^n - 7$
3. $a_n = \frac{1}{n^3}$
4. $a_n = \frac{1}{(n+3)(n+1)}$
5. $a_n = (-1)^n\sqrt{n+2}$
6. $a_n = (-1)^{n+1}n(n-1)$
7. $a_n = 4 + (-.1)^n$
8. $a_n = 5 - (.1)^n$
9. $a_n = (-1)^n + 3n$
10. $a_n = (-1)^{n+2} - (n+1)$

In Exercises 11–14, express the sum in Σ *notation.*

11. $1 + 2 + 3 + 4 + 5 + 6 + 7 + 8 + 9 + 10 + 11$
12. $1^1 + 2^2 + 3^3 + 4^4 + 5^5$
13. $\frac{1}{2^7} + \frac{1}{2^8} + \frac{1}{2^9} + \frac{1}{2^{10}} + \frac{1}{2^{11}} + \frac{1}{2^{12}} + \frac{1}{2^{13}}$
14. $(-6)^{11} + (-6)^{12} + (-6)^{13} + (-6)^{14} + (-6)^{15}$

In Exercises 15–20, find the sum.

15. $\sum_{i=1}^{5} 3i$
16. $\sum_{i=1}^{4} \frac{1}{2^i}$
17. $\sum_{n=1}^{16} (2n - 3)$
18. $\sum_{n=1}^{75} (-1)^n(3n + 1)$
19. $\sum_{n=15}^{36} (n^2 - 8)$
20. $\sum_{k=0}^{25} (2k^2 - 5k + 1)$

In Exercises 21–26, find a formula for the nth term of the sequence whose first few terms are given.

21. $-1, 1, -1, 1, -1, 1, \ldots$
22. $2, -2, 2, -2, 2, -2, \ldots$
23. $\frac{1}{2}, \frac{2}{3}, \frac{3}{4}, \frac{4}{5}, \frac{5}{6}, \ldots$
24. $\frac{1}{2\cdot 3}, \frac{1}{3\cdot 4}, \frac{1}{4\cdot 5}, \frac{1}{5\cdot 6}, \frac{1}{6\cdot 7}, \ldots$
25. $2, 7, 12, 17, 22, 27, \ldots$
26. $8, -5, 2, -11, -4, -17, -10, \ldots$

In Exercises 27–34, find the first five terms of the given sequence.

27. $a_1 = 4$ and $a_n = 2a_{n-1} + 3$ for $n \geq 2$
28. $a_1 = -3$ and $a_n = (-1)^n 4a_{n-1} - 5$ for $n \geq 2$
29. $a_1 = 1, a_2 = -2, a_3 = 3,$ and $a_n = a_{n-1} + a_{n-2} + a_{n-3}$ for $n \geq 4$
30. $a_1 = 1, a_2 = 3,$ and $a_n = 2a_{n-1} + 3a_{n-2}$ for $n \geq 3$
31. $a_0 = 2, a_1 = 3,$ and $a_n = (a_{n-1})\left(\frac{1}{2}a_{n-2}\right)$ for $n \geq 2$
32. $a_0 = 1, a_1 = 1,$ and $a_n = na_{n-1}$ for $n \geq 2$
33. a_n is the nth digit in the decimal expansion of π.
34. a_n is the nth digit in the decimal expansion of $1/13$.

In Exercises 35–38, find the third and the sixth partial sums of the sequence.

35. $\{n^2 - 5n + 2\}$
36. $\{(2n - 3n^2)^2\}$
37. $\{(-1)^{n+1}5\}$
38. $\{2^n(2 - n^2)\}_{n\geq 0}$

In Exercises 39–42, express the given sum in Σ *notation.*

39. $\frac{1}{3} + \frac{1}{5} + \frac{1}{7} + \frac{1}{9} + \frac{1}{11} + \frac{1}{13}$
40. $2 + 1 + \frac{4}{5} + \frac{5}{7} + \frac{2}{3} + \frac{7}{11} + \frac{8}{13}$
41. $\frac{1}{8} - \frac{2}{9} + \frac{3}{10} - \frac{4}{11} + \frac{5}{12}$
42. $\frac{2}{3\cdot 5} + \frac{4}{5\cdot 7} + \frac{8}{7\cdot 9} + \frac{16}{9\cdot 11} + \frac{32}{11\cdot 13} + \frac{64}{13\cdot 15} + \frac{128}{15\cdot 17}$

In Exercises 43–48, use a calculator to approximate the required term or sum.

43. a_{12} where $a_n = \left(1 + \frac{1}{n}\right)^n$
44. a_{50} where $a_n = \frac{\ln n}{n^2}$
45. a_{102} where $a_n = \frac{n^3 - n^2 + 5n}{3n^2 + 2n - 1}$
46. a_{125} where $a_n = \sqrt[n]{n}$
47. $\sum_{k=1}^{14} \frac{1}{k^2}$
48. $\sum_{n=8}^{22} \frac{1}{n}$
49. From 1991 through 1998, the annual dividends per share of Coca-Cola stock were approximated by the sequence $\{a_n\}$, where $n = 1$ corresponds to 1991 and $a_n = .175 + .055n$.
    (a) What were the approximate dividends per share in 1993, 1995, and 1998?
    (b) If you owned one share from 1991 through 1998, how much would you have collected in dividends?
50. Calcuplay Company offers you a job, with salary b_n in year n, where $\{b_n\}$ is the sequence defined by $b_n = (1.06)^{n-1}30{,}000$.

(a) What is your starting salary ($n = 0$)?
(b) What is your salary two years from now? Ten years from now?
(c) If you work for Calcuplay for 25 years, how much will you earn?

51. Book sales in the United States (in billions of dollars) in year n are projected to be given by the sequence $\{a_n\}$, where $n = 0$ corresponds to 1990 and $a_n = .6n + 15.2$.
(a) What were book sales in 1993? In 2000?
(b) Find the projected total book sales from 1995 through 2003.

52. Annual revenue (in billions of dollars) from home video sales and rentals is approximated by the sequence $\{b_n\}$, where $n = 1$ corresponds to 1980 and $b_n = .972n - .768$. What will be the total revenue for the period from 1995 through 2005?

Thinkers

Exercises 53–57 deal with prime numbers. A positive integer greater than 1 is prime *if its only positive integer factors are itself and 1. For example, 7 is prime because its only factors are 7 and 1, but 15 is not prime because it has factors other than 15 and 1 (namely, 3 and 5).*

53. (a) Let $\{a_n\}$ be the sequence of prime integers in their usual ordering. Verify that the first ten terms are 2, 3, 5, 7, 11, 13, 17, 19, 23, 29.
(b) Find $a_{17}, a_{18}, a_{19}, a_{20}$.

In Exercises 54–57, find the first five terms of the sequence.

54. a_n is the nth prime integer larger than 10. [*Hint:* $a_1 = 11$.]

55. a_n is the square of the nth prime integer.

56. a_n is the number of prime integers less than n.

57. a_n is the largest prime integer less than $5n$.

Exercises 58–64 deal with the Fibonacci sequence $\{a_n\}$ that was discussed in Example 7.

58. Leonardo Fibonacci discovered the sequence in the 13th century in connection with this problem: A rabbit colony begins with one pair of adult rabbits (one male, one female). Each adult pair produces one pair of babies (one male, one female) every month. Each pair of baby rabbits becomes adult and produces the first offspring at age two months. Assuming that no rabbits die, how many adult pairs of rabbits are in the colony at the end of n months ($n = 1, 2, 3. \ldots$)? [*Hint:* It may be helpful to make up a chart listing for each month the number of adult pairs, the number of one-month-old pairs, and the number of baby pairs.]

59. (a) List the first ten terms of the Fibonacci sequence.
(b) List the first ten partial sums of the sequence.
(c) Do the partial sums follow an identifiable pattern?

60. Verify that every positive integer less than or equal to 15 can be written as a sum of Fibonacci numbers, with none used more than once.

61. Verify that $5(a_n)^2 + 4(-1)^n$ is always a perfect square for $n = 1, 2, \ldots, 10$.

62. Verify that $(a_n)^2 = a_{n+1}a_{n-1} + (-1)^{n-1}$ for $n = 2, \ldots, 10$.

63. Show that $\sum_{n=1}^{k} a_n = a_{k+2} - 1$. [*Hint:* $a_1 = a_3 - a_2$; $a_2 = a_4 - a_3$; etc.]

64. Show that $\sum_{n=1}^{k} a_{2n-1} = a_{2k}$, that is, the sum of the first k odd-numbered terms is the kth even-numbered term. [*Hint:* $a_3 = a_4 - a_2$; $a_5 = a_6 - a_4$; etc.]

12.2 Arithmetic Sequences

An **arithmetic sequence** (sometimes called an **arithmetic progression**) is a sequence in which the difference between each term and the preceding one is always the same constant.

Example 1 In the sequence 3, 8, 13, 18, 23, 28, . . . the difference between each term and the preceding one is always 5. So this is an arithmetic sequence. ■

Example 2 In the sequence 14, 10, 6, 2, −2, −6, −10, −14, . . . the difference between each term and the preceding one is −4 (for example, $10 - 14 = -4$ and $-6 - (-2) = -4$). Hence, this sequence is arithmetic. ■

If $\{a_n\}$ is an arithmetic sequence, then for each $n \geq 2$, the term preceding a_n is a_{n-1} and the difference $a_n - a_{n-1}$ is some constant—call it d. Therefore, $a_n - a_{n-1} = d$, or equivalently,

Arithmetic Sequences

In an arithmetic sequence $\{a_n\}$

$$a_n = a_{n-1} + d$$

for some constant d and all $n \geq 2$.

The number d is called the **common difference** of the arithmetic sequence.

Example 3 If $\{a_n\}$ is an arithmetic sequence with $a_1 = 3$ and $a_2 = 4.5$, then the common difference is $d = a_2 - a_1 = 4.5 - 3 = 1.5$. So the sequence begins with 3, 4.5, 6, 7.5, 9, 10.5, 12, 13.5, ... ■

Example 4 The sequence $\{-7 + 4n\}$ is an arithmetic sequence because for each $n \geq 2$,

$$\begin{aligned} a_n - a_{n-1} &= (-7 + 4n) - [-7 + 4(n - 1)] \\ &= (-7 + 4n) - (-7 + 4n - 4) = 4. \end{aligned}$$

Therefore, the common difference is $d = 4$. ■

If $\{a_n\}$ is an arithmetic sequence with common difference d, then for each $n \geq 2$ we know that $a_n = a_{n-1} + d$. Applying this fact repeatedly shows that

$$a_2 = a_1 + d$$
$$a_3 = a_2 + d = (a_1 + d) + d = a_1 + 2d$$
$$a_4 = a_3 + d = (a_1 + 2d) + d = a_1 + 3d$$
$$a_5 = a_4 + d = (a_1 + 3d) + d = a_1 + 4d$$

and in general,

nth Term of an Arithmetic Sequence

In an arithmetic sequence $\{a_n\}$ with common difference d

$$a_n = a_1 + (n-1)d \quad \text{for every } n \geq 1.$$

Example 5 Find the nth term of the arithmetic sequence with first term -5 and common difference 3. Since $a_1 = -5$ and $d = 3$, the formula in the box shows that

$$a_n = a_1 + (n-1)d = -5 + (n-1)3 = 3n - 8. \quad ■$$

Example 6 What is the 45th term of the arithmetic sequence whose first three terms are 5, 9, 13?

Solution The first three terms show that $a_1 = 5$ and that the common difference d is 4. Applying the formula in the box with $n = 45$, we have

$$a_{45} = a_1 + (45-1)d = 5 + (44)4 = 181. \quad ■$$

Example 7 If $\{a_n\}$ is an arithmetic sequence with $a_6 = 57$ and $a_{10} = 93$, find a_1 and a formula for a_n.

Solution Apply the formula $a_n = a_1 + (n-1)d$ with $n = 6$ and $n = 10$:

$$a_6 = a_1 + (6-1)d \qquad \text{and} \qquad a_{10} = a_1 + (10-1)d$$
$$57 = a_1 + 5d \qquad\qquad\qquad 93 = a_1 + 9d$$

We can find a_1 and d by solving this system:

$$a_1 + 9d = 93$$
$$a_1 + 5d = 57.$$

Subtracting the second equation from the first shows that $4d = 36$, and hence $d = 9$. Substituting $d = 9$ in the second equation shows that $a_1 = 12$. So the formula for a_n is

$$a_n = a_1 + (n-1)d = 12 + (n-1)9 = 9n + 3. \quad \blacksquare$$

Partial Sums

It's easy to compute partial sums of arithmetic sequences by using the following formulas.

Partial Sums of an Arithmetic Sequence

If $\{a_n\}$ is an arithmetic sequence with common difference d, then for each positive integer k the kth partial sum can be found by using *either* of these formulas:

1. $\displaystyle\sum_{n=1}^{k} a_n = \frac{k}{2}(a_1 + a_k)$ **or**

2. $\displaystyle\sum_{n=1}^{k} a_n = ka_1 + \frac{k(k-1)}{2}d$

Proof Let S denote the kth partial sum $a_1 + a_2 + \cdots + a_k$. For reasons that will become apparent later we shall calculate the number $2S$:

$$2S = S + S = (a_1 + a_2 + \cdots + a_k) + (a_1 + a_2 + \cdots + a_k).$$

Now we rearrange the terms on the right by grouping the first and last terms together, then the first and last of the remaining terms, and so on:

$$2S = (a_1 + a_k) + (a_2 + a_{k-1}) + (a_3 + a_{k-2}) + \cdots + (a_k + a_1).$$

Since adjacent terms of the sequence differ by d we have:

$$a_2 + a_{k-1} = (a_1 + d) + (a_k - d) = a_1 + a_k.$$

Using this fact,

$$a_3 + a_{k-2} = (a_2 + d) + (a_{k-1} - d) = a_2 + a_{k-1} = a_1 + a_k.$$

Continuing in this manner we see that every pair in the sum for $2S$ is equal to $a_1 + a_k$. Therefore,

$$\begin{aligned} 2S &= (a_1 + a_k) + (a_2 + a_{k-1}) + (a_3 + a_{k-2}) + \cdots + (a_k + a_1) \\ &= (a_1 + a_k) + (a_1 + a_k) + (a_1 + a_k) + \cdots + (a_1 + a_k) \quad (k \text{ terms}) \\ &= k(a_1 + a_k). \end{aligned}$$

Dividing both sides of this last equation by 2 shows that $S = \frac{k}{2}(a_1 + a_k)$.This proves the first formula. To obtain the second one, note that

$$a_1 + a_k = a_1 + [a_1 + (k - 1)d] = 2a_1 + (k - 1)d.$$

Substituting the right side of this equation in the first formula for S shows that

$$S = \frac{k}{2}(a_1 + a_k) = \frac{k}{2}[2a_1 + (k - 1)d] = ka_1 + \frac{k(k - 1)}{2}d.$$

This proves the second formula. ■

Example 8 Find the 12th partial sum of the arithmetic sequence that begins $-8, -3, 2, 7, \ldots$.

Solution We first note that the common difference d is 5. Since $a_1 = -8$ and $d = 5$, the second formula in the box with $k = 12$ shows that

$$\sum_{n=1}^{12} a_n = 12(-8) + \frac{12(11)}{2}5 = -96 + 330 = 234. \quad ■$$

Example 9 Find the sum of all multiples of 3 from 3 to 333.

Solution Note that this sum is just a partial sum of the arithmetic sequence $3, 6, 9, 12, \ldots$. Since this sequence can be written in the form

$$3 \cdot 1, 3 \cdot 2, 3 \cdot 3, 3 \cdot 4, 3 \cdot 5, 3 \cdot 6, \ldots$$

we see that $333 = 3 \cdot 111$ is the 111th term. The 111th partial sum of this sequence can be found by using the first formula in the box with $k = 111$, $a_1 = 3$, and $a_{111} = 333$:

$$\sum_{n=1}^{111} a_n = \frac{111}{2}(3 + 333) = \frac{111}{2}(336) = 18{,}648. \quad ■$$

Example 10 If the starting salary for a job is \$20,000 and you get a \$2000 raise at the beginning of each subsequent year, what will your salary be during the tenth year? How much will you earn during the first ten years?

Solution Your yearly salary rates form a sequence: 20,000, 22,000, 24,000, 26,000, and so on. It is an arithmetic sequence with $a_1 = 20{,}000$ and $d = 2000$. Your tenth-year salary is

$$a_{10} = a_1 + (10 - 1)d = 20{,}000 + 9 \cdot 2000 = \$38{,}000.$$

Your ten-year total earnings are the tenth partial sum of the sequence:

$$\frac{10}{2}(a_1 + a_{10}) + \frac{10}{2}(20{,}000 + 38{,}000) = 5(58{,}000) = \$290{,}000. \quad ■$$

Exercises 12.2

In Exercises 1–6, the first term a_1 and the common difference d of an arithmetic sequence are given. Find the fifth term and the formula for the nth term.

1. $a_1 = 5, d = 2$

2. $a_1 = -4, d = 5$

3. $a_1 = 4, d = \frac{1}{4}$

4. $a_1 = -6, d = \frac{2}{3}$

5. $a_1 = 10, d = -\frac{1}{2}$

6. $a_1 = \pi, d = \frac{1}{5}$

In Exercises 7–12, find the kth partial sum of the arithmetic sequence $\{a_n\}$ with common difference d.

7. $k = 6, a_1 = 2, d = 5$

8. $k = 8, a_1 = \frac{2}{3}, d = -\frac{4}{3}$

9. $k = 7, a_1 = \frac{3}{4}, d = -\frac{1}{2}$

10. $k = 9, a_1 = 6, a_9 = -24$

11. $k = 6, a_1 = -4, a_6 = 14$

12. $k = 10, a_1 = 0, a_{10} = 30$

In Exercises 13–18, show that the sequence is arithmetic and find its common difference.

13. $\{3 - 2n\}$

14. $\left\{4 + \frac{n}{3}\right\}$

15. $\left\{\frac{5 + 3n}{2}\right\}$

16. $\left\{\frac{\pi - n}{2}\right\}$

17. $\{c + 2n\}$ (c constant)

18. $\{2b + 3nc\}$ (b, c constants)

In Exercises 19–24, use the given information about the arithmetic sequence with common difference d to find a_5 and a formula for a_n.

19. $a_4 = 12, d = 2$

20. $a_7 = -8, d = 3$

21. $a_2 = 4, a_6 = 32$

22. $a_7 = 6, a_{12} = -4$

23. $a_5 = 0, a_9 = 6$

24. $a_5 = -3, a_9 = -18$

In Exercises 25–28, find the sum.

25. $\sum_{n=1}^{20} (3n + 4)$

26. $\sum_{n=1}^{25} \left(\frac{n}{4} + 5\right)$

27. $\sum_{n=1}^{40} \frac{n + 3}{6}$

28. $\sum_{n=1}^{30} \frac{4 - 6n}{3}$

29. Find the sum of all the even integers from 2 to 100.

30. Find the sum of all the integer multiples of 7 from 7 to 700.

31. Find the sum of the first 200 positive integers.

32. Find the sum of the positive integers from 101 to 200 (inclusive). [*Hint:* What's the sum from 1 to 100? Use it and Exercise 31.]

33. A business makes a $10,000 profit during its first year. If the yearly profit increases by $7500 in each subsequent year, what will the profit be in the tenth year and what will the total profit for the first ten years be?

34. If a man's starting salary is $15,000 and he receives a $1000 increase every six months, what will his salary be during the last six months of the sixth year? How much will he earn during the first six years?

35. A lecture hall has 6 seats in the first row, 8 in the second, 10 in the third, and so on, through row 12. Rows 12 through 20 (the last row) all have the same number of seats. Find the number of seats in the lecture hall.

36. A monument is constructed by laying a row of 60 bricks at ground level. A second row, with two fewer bricks, is centered on that; a third row, with two fewer bricks, is centered on the second; and so on. The top row contains ten bricks. How many bricks are there in the monument?

37. A ladder with nine rungs is to be built, with the bottom rung 24 inches wide and the top rung 18 inches wide. If the lengths of the rungs decrease uniformly from bottom to top, how long should each of the seven intermediate rungs be?

38. Find the first eight numbers in an arithmetic sequence in which the sum of the first and seventh term is 40 and the product of the first and fourth terms is 160.

12.3 Geometric Sequences

A **geometric sequence** (sometimes called a **geometric progression**) is a sequence in which the quotient of each term and the preceding one is the same constant r. This constant r is called the **common ratio** of the geometric sequence.

Example 1 The sequence $3, 9, 27, \ldots 3^n, \ldots$ is geometric with common ratio 3. For instance, $a_2/a_1 = 9/3 = 3$ and $a_3/a_2 = 27/9 = 3$. If 3^n is any term ($n \geq 2$), then the preceding term is 3^{n-1} and

$$\frac{3^n}{3^{n-1}} = \frac{3 \cdot 3^{n-1}}{3^{n-1}} = 3. \quad \blacksquare$$

Example 2 The sequence $\{5/2^n\}$ which begins $5/2, 5/4, 5/8, 5/16, \ldots$ is geometric with common ratio $r = 1/2$ because for each $n \geq 1$

$$\frac{5/2^n}{5/2^{n-1}} = \frac{5}{2^n} \cdot \frac{2^{n-1}}{5} = \frac{2^{n-1}}{2^n} = \frac{2^{n-1}}{2^{n-1} \cdot 2} = \frac{1}{2}. \quad \blacksquare$$

If $\{a_n\}$ is a geometric sequence with common ratio r, then for each $n \geq 2$ the term preceding a_n is a_{n-1} and

$$\frac{a_n}{a_{n-1}} = r, \qquad \text{or equivalently,} \qquad a_n = ra_{n-1}.$$

Applying this last formula for $n = 2, 3, 4, \ldots$ we have

$$a_2 = ra_1$$

$$a_3 = ra_2 = r(ra_1) = r^2a_1$$

$$a_4 = ra_3 = r(r^2a_1) = r^3a_1$$

$$a_5 = ra_4 = r(r^3a_1) = r^4a_1$$

and in general

nth Term of a Geometric Sequence

If $\{a_n\}$ is a geometric sequence with common ratio r, then for all $n \geq 1$,

$$a_n = r^{n-1}a_1.$$

Example 3 To find a formula for the nth term of the geometric sequence $\{a_n\}$ where $a_1 = 7$ and $r = 2$, we use the equation in the box:

$$a_n = r^{n-1}a_1 = 2^{n-1} \cdot 7.$$

So the sequence is $\{7 \cdot 2^{n-1}\}$. $\blacksquare$

Example 4 If the first two terms of a geometric sequence are 2 and $-2/5$, then the common ratio must be

$$r = \frac{a_2}{a_1} = \frac{-2/5}{2} = \frac{-2}{5} \cdot \frac{1}{2} = -\frac{1}{5}.$$

Using the equation in the box, we now see that the formula for the nth term is

$$a_n = r^{n-1}a_1 = \left(-\frac{1}{5}\right)^{n-1}(2) = \frac{(1)^{n-1}}{(-5)^{n-1}}(2) = \frac{2}{(-5)^{n-1}}.$$

So, the sequence begins $2, -2/5, 2/5^2, -2/5^3, 2/5^4, \ldots$. ■

Example 5 If $\{a_n\}$ is a geometric sequence with $a_2 = 20/9$ and $a_5 = 160/243$, then by the equation in the preceding box

$$\frac{160/243}{20/9} = \frac{a_5}{a_2} = \frac{r^4 a_1}{r a_1} = r^3.$$

Consequently,

$$r = \sqrt[3]{\frac{160/243}{20/9}} = \sqrt[3]{\frac{160}{243} \cdot \frac{9}{20}} = \sqrt[3]{\frac{8 \cdot 9}{243}} = \sqrt[3]{\frac{8}{27}} = \frac{2}{3}.$$

Since $a_2 = ra_1$ we see that

$$a_1 = \frac{a_2}{r} = \frac{20/9}{2/3} = \frac{20}{9} \cdot \frac{3}{2} = \frac{10}{3}.$$

Therefore,

$$a_n = r^{n-1}a_1 = \left(\frac{2}{3}\right)^{n-1} \cdot \frac{10}{3} = \frac{2^{n-1} \cdot 2 \cdot 5}{3^{n-1} \cdot 3} = \frac{2^n \cdot 5}{3^n} = 5\left(\frac{2}{3}\right)^n. \quad ■$$

Partial Sums

If the common ratio r of a geometric sequence is the number 1, then we have

$$a_n = 1^{n-1}a_1 \quad \text{for every } n \geq 1.$$

Therefore, the sequence is just the constant sequence $a_1, a_1, a_1, \ldots$. For any positive integer k, the kth partial sum of this constant sequence is

$$\underbrace{a_1 + a_1 + \cdots + a_1}_{k \text{ terms}} = ka_1.$$

In other words, the kth partial sum of a constant sequence is just k times the constant. If a geometric sequence is not constant (that is, $r \neq 1$), then its partial sums are given by the following formula.

Partial Sums of a Geometric Sequence

The kth partial sum of the geometric sequence $\{a_n\}$ with common ratio $r \neq 1$ is

$$\sum_{n=1}^{k} a_n = a_1\left(\frac{1 - r^k}{1 - r}\right).$$

Proof If S denotes the kth partial sum, then the formula for the nth term of a geometric sequence shows that

$$S = a_1 + a_2 + \cdots + a_k = a_1 + a_1r + a_1r^2 + a_1r^3 + \cdots + a_1r^{k-1}.$$

Use this equation to compute $S - rS$:

$$\begin{aligned} S &= a_1 + a_1r + a_1r^2 + a_1r^3 + \cdots + a_1r^{k-1} \\ rS &= \phantom{a_1 + {}} a_1r + a_1r^2 + a_1r^3 + \cdots + a_1r^{k-1} + a_1r^k \\ \hline S - rS &= a_1 \qquad\qquad\qquad\qquad\qquad\qquad - a_1r^k \\ (1 - r)S &= a_1(1 - r^k) \end{aligned}$$

Since $r \neq 1$, we can divide both sides of this last equation by $1 - r$ to complete the proof:

$$S = \frac{a_1(1 - r^k)}{1 - r} = a_1\left(\frac{1 - r^k}{1 - r}\right). \quad \blacksquare$$

Example 6 Find the sum

$$-\frac{3}{2} + \frac{3}{4} - \frac{3}{8} + \frac{3}{16} - \frac{3}{32} + \frac{3}{64} - \frac{3}{128} + \frac{3}{256} - \frac{3}{512}.$$

Solution Note that this is the ninth partial sum of the geometric sequence $\left\{3\left(\frac{-1}{2}\right)^n\right\}$. The common ratio is $r = -1/2$. The formula in the box shows that

$$\begin{aligned} \sum_{n=1}^{9} 3\left(\frac{-1}{2}\right)^n &= a_1\left(\frac{1 - r^9}{1 - r}\right) = \left(\frac{-3}{2}\right)\left[\frac{1 - (-1/2)^9}{1 - (-1/2)}\right] \\ &= \left(\frac{-3}{2}\right)\left(\frac{1 + 1/2^9}{3/2}\right) = \left(\frac{-3}{2}\right)\left(\frac{2}{3}\right)\left(1 + \frac{1}{2^9}\right) \\ &= -1 - \frac{1}{2^9} = -1 - \frac{1}{512} = -\frac{513}{512}. \end{aligned} \quad \blacksquare$$

Example 7 A superball is dropped from a height of 9 feet. It hits the ground and bounces to a height of 6 feet. It continues to bounce up and down. On each bounce it rises to 2/3 of the height of the previous bounce.

How far has the ball traveled (both up and down) when it hits the ground for the seventh time?

Solution We first consider how far the ball travels on each bounce. On the first bounce it rises 6 feet and falls 6 feet for a total of 12 feet. On the second bounce it rises and falls 2/3 of the previous height, and hence travels 2/3 of 12 feet. If a_n denotes the distance traveled on the nth bounce, then

$$a_1 = 12 \qquad a_2 = \left(\frac{2}{3}\right)a_1 \qquad a_3 = \left(\frac{2}{3}\right)a_2 = \left(\frac{2}{3}\right)^2 a_1$$

and in general

$$a_n = \left(\frac{2}{3}\right)a_{n-1} = \left(\frac{2}{3}\right)^{n-1} a_1.$$

So $\{a_n\}$ is a geometric sequence with common ratio $r = 2/3$. When the ball hits the ground for the seventh time it has completed six bounces. Therefore, the total distance it has traveled is the distance it was originally dropped (9 feet) plus the distance traveled in six bounces, namely,

$$9 + a_1 + a_2 + a_3 + a_4 + a_5 + a_6 = 9 + \sum_{n=1}^{6} a_n = 9 + a_1\left(\frac{1 - r^6}{1 - r}\right)$$

$$= 9 + 12\left[\frac{1 - (2/3)^6}{1 - (2/3)}\right] \approx 41.84 \text{ feet} \quad ■$$

Exercises 12.3

In Exercises 1–8, determine whether the sequence is arithmetic, geometric, or neither.

1. 2, 7, 12, 17, 22, . . .

2. 2, 6, 18, 54, 162, . . .

3. 13, 13/2, 13/4, 13/8, . . .

4. $-1, -\frac{1}{2}, 0, \frac{1}{2}, \ldots$

5. 50, 48, 46, 44, . . .

6. $2, -3, 9/2, -27/4, -81/8, \ldots$

7. $3, -3/2, 3/4, -3/8, 3/16, \ldots$

8. $-6, -3.7, -1.4, 9, 3.2, \ldots$

In Exercises 9–14, the first term a_1 and the common ratio r of a geometric sequence are given. Find the sixth term and a formula for the nth term.

9. $a_1 = 5, r = 2$

10. $a_1 = 1, r = -2$

11. $a_1 = 4, r = \frac{1}{4}$

12. $a_1 = -6, r = \frac{2}{3}$

13. $a_1 = 10, r = -\frac{1}{2}$

14. $a_1 = \pi, r = \frac{1}{5}$

In Exercises 15–18, find the kth partial sum of the geometric sequence $\{a_n\}$ with common ratio r.

15. $k = 6, a_1 = 5, r = \frac{1}{2}$

16. $k = 8, a_1 = 9, r = \frac{1}{3}$

17. $k = 7, a_2 = 6, r = 2$

18. $k = 9, a_2 = 6, r = \frac{1}{4}$

In Exercises 19–22, show that the given sequence is geometric and find the common ratio.

19. $\left\{\left(-\frac{1}{2}\right)^n\right\}$

20. $\{2^{3n}\}$

21. $\{5^{n+2}\}$

22. $\{3^{n/2}\}$

In Exercises 23–28, use the given information about the geometric sequence $\{a_n\}$ to find a_5 and a formula for a_n.

23. $a_1 = 256, a_2 = -64$

24. $a_1 = 1/6, a_2 = -1/18$

25. $a_2 = 4, a_5 = 1/16$

26. $a_3 = 4, a_6 = -32$

27. $a_4 = -4/5, r = 2/5$

28. $a_2 = 6, a_7 = 192$

In Exercises 29–34, find the sum.

29. $\sum_{n=1}^{7} 2^n$

30. $\sum_{k=1}^{6} 3\left(\frac{1}{2}\right)^k$

31. $\sum_{n=1}^{9} \left(-\frac{1}{3}\right)^n$

32. $\sum_{n=1}^{5} 5 \cdot 3^{n-1}$

33. $\sum_{j=1}^{6} 4\left(\frac{3}{2}\right)^{j-1}$

34. $\sum_{t=1}^{8} 6(.9)^{t-1}$

35. For 1987–1998, the annual revenues per share in year n of Walt Disney stock are approximated by $a_n = 1.71(1.191)^n$, where $n = 7$ represents 1987.
(a) Show that the sequence $\{a_n\}$ is a geometric sequence.
(b) Approximate the total revenues per share for the period 1987–1998.

36. The annual dividends per share of Walt Disney stock from 1989 through 1998 are approximated by the sequence $\{b_n\}$, where $n = 9$ corresponds to 1989 and $b_n = .0228(1.1999)^n$.
(a) Show that the sequence $\{b_n\}$ is a geometric sequence.
(b) Approximate the total dividends per share for the period 1989–1998.

37. A ball is dropped from a height of 8 feet. On each bounce it rises to half its previous height. When the ball hits the ground for the seventh time, how far has it traveled?

38. A ball is dropped from a height of 10 feet. On each bounce it rises to 45% of its previous height. When it hits the ground for the tenth time, how far has it traveled?

39. If you are paid a salary of 1¢ on the first day of March, 2¢ on the second day, and your salary continues to double each day, how much will you earn in the month of March?

40. Starting with your parents, how many ancestors do you have for the preceding ten generations?

41. A car that sold for $8000 depreciates in value 25% each year. What is it worth after five years?

42. A vacuum pump removes 60% of the air in a container at each stroke. What percentage of the original amount of air remains after six strokes?

Thinkers

43. Suppose $\{a_n\}$ is a geometric sequence with common ratio $r > 0$ and each $a_n > 0$. Show that the sequence $\{\log a_n\}$ is an arithmetic sequence with common difference $\log r$.

44. Suppose $\{a_n\}$ is an arithmetic sequence with common difference d. Let C be any positive number. Show that the sequence $\{C^{a_n}\}$ is a geometric sequence with common ratio C^d.

45. In the geometric sequence 1, 2, 4, 8, 16, ..., show that each term is 1 plus the sum of all preceding terms.

46. In the geometric sequence 2, 6, 18, 54, ..., show that each term is twice the sum of 1 and all preceding terms.

47. The minimum monthly payment for a certain bank credit card is the larger of 1/25 of the outstanding balance or $5. If the balance is less than $5, the entire balance is due. If you make only the minimum payment each month, how long will it take to pay off a balance of $200 (excluding any interest that might be due)?

12.3.A EXCURSION Infinite Series

We now introduce a topic that is closely related to infinite sequences and has some very useful applications. We can give only a few highlights here; complete coverage requires calculus.

Consider the sequence $\{3/10^n\}$ and let S_k denote its kth partial sum; then

$$S_1 = \frac{3}{10}$$

$$S_2 = \frac{3}{10} + \frac{3}{10^2} = \frac{33}{100}$$

$$S_3 = \frac{3}{10} + \frac{3}{10^2} + \frac{3}{10^3} = \frac{333}{1000}$$

$$S_4 = \frac{3}{10} + \frac{3}{10^2} + \frac{3}{10^3} + \frac{3}{10^4} = \frac{3333}{10{,}000}.$$

These partial sums $S_1, S_2, S_3, S_4, \ldots$ themselves form a sequence:

$$\frac{3}{10}, \frac{33}{100}, \frac{333}{1000}, \frac{3333}{10{,}000}, \ldots.$$

The terms in the sequence of partial sums appear to be getting closer and closer to 1/3. In other words, as k gets larger and larger, the corresponding partial sum S_k gets closer and closer to 1/3. Consequently, we write

$$\frac{3}{10} + \frac{3}{10^2} + \frac{3}{10^3} + \frac{3}{10^4} + \cdots = \frac{1}{3}$$

and say that 1/3 is the *sum* of the *infinite series*

$$\frac{3}{10} + \frac{3}{10^2} + \frac{3}{10^3} + \frac{3}{10^4} + \cdots.$$

In the general case, an **infinite series** (or simply **series**) is defined to be an expression of the form

$$a_1 + a_2 + a_3 + a_4 + a_5 + \cdots$$

in which each a_n is a real number. This series is also denoted by the symbol $\sum_{n=1}^{\infty} a_n$.

Example 1

(a) $\sum_{n=1}^{\infty} 2(.6)^n$ denotes the series

$$2(.6) + 2(.6)^2 + 2(.6)^3 + 2(.6)^4 + \cdots.$$

(b) $\sum_{n=1}^{\infty}\left(\frac{-1}{2}\right)^n$ denotes the series

$$\frac{-1}{2}+\left(\frac{-1}{2}\right)^2+\left(\frac{-1}{2}\right)^3+\left(\frac{-1}{2}\right)^4+\cdots=-\frac{1}{2}+\frac{1}{4}-\frac{1}{8}+\frac{1}{16}+\cdots. \quad \blacksquare$$

The **partial sums** of the series $a_1 + a_2 + a_3 + a_4 + \cdots$ are

$$S_1 = a_1$$
$$S_2 = a_1 + a_2$$
$$S_3 = a_1 + a_2 + a_3$$

and in general, for any $k \geq 1$

$$S_k = a_1 + a_2 + a_3 + a_4 + \cdots + a_k.$$

If it happens that the terms $S_1, S_2, S_3, S_4, \ldots$ of the *sequence* of partial sums get closer and closer to a particular real number S in such a way that the partial sum S_k is arbitrarily close to S when k is large enough, then we say that the series **converges** and that S is the **sum of the convergent series.** For example, we just saw that the series

$$\frac{3}{10}+\frac{3}{10^2}+\frac{3}{10^3}+\frac{3}{10^4}+\cdots$$

converges and that its sum is $1/3$.

A sequence is a *list* of numbers $a_1, a_2, a_3, \ldots$. Intuitively, you can think of a convergent series $a_1 + a_2 + a_3 + \cdots$ as an "infinite sum" of numbers. But be careful: Not every series has a sum. For instance, the partial sums of the series

$$1 + 2 + 3 + 4 + \cdots$$

get larger and larger (compute some) and do not get closer and closer to a single real number. So this series is not convergent.

Example 2 Although no proof will be given here, it is intuitively clear that every infinite decimal may be thought of as the sum of a convergent series. For instance,

$$\pi = 3.1415926\cdots = 3 + .1 + .04 + .001 + .0005 + .00009 + \cdots.$$

Note that the third partial sum is $3 + .1 + .04 = 3.14$, which is π to two decimal places. Similarly, the kth partial sum of this series is just π to $k - 1$ decimal places. $\blacksquare$

Infinite Geometric Series

If $\{a_n\}$ is a geometric sequence with common ratio r, then the corresponding infinite series

$$a_1 + a_2 + a_3 + a_4 + a_5 + \cdots$$

is called an **infinite geometric series.** By using the formula for the nth term of a geometric sequence, we can also express the corresponding geometric series in the form

$$a_1 + ra_1 + r^2a_1 + r^3a_1 + r^4a_1 + \cdots.$$

Under certain circumstances, an infinite geometric series is convergent and has a sum:

Sum of an Infinite Geometric Series

If $|r| < 1$, then the infinite geometric series

$$a_1 + ra_1 + r^2a_1 + r^3a_1 + r^4a_1 + \cdots$$

converges and its sum is

$$\frac{a_1}{1 - r}.$$

Although we cannot prove this fact rigorously here, we can make it highly plausible both geometrically and algebraically.

Example 3 $\sum_{n=1}^{\infty} \frac{8}{5^n} = \frac{8}{5} + \frac{8}{5^2} + \frac{8}{5^3} + \cdots$ is an infinite geometric series with $a_1 = 8/5$ and $r = 1/5$. The kth partial sum of this series is the same as the kth partial sum of the sequence $\{8/5^n\}$ and hence from the box on page 749 we know that

$$S_k = a_1\left(\frac{1 - r^k}{1 - r}\right) = \frac{8}{5}\left[\frac{1 - \left(\frac{1}{5}\right)^k}{1 - \frac{1}{5}}\right]$$

$$= \frac{8}{5}\left(\frac{1 - \frac{1}{5^k}}{\frac{4}{5}}\right) = \frac{8}{5}\cdot\frac{5}{4}\left(1 - \frac{1}{5^k}\right)$$

$$= 2\left(1 - \frac{1}{5^k}\right) = 2 - \frac{2}{5^k}.$$

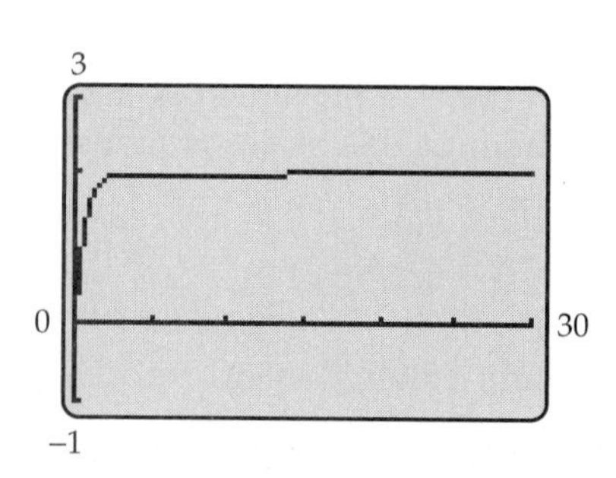

Figure 12–8

The function $f(x) = 2 - 2/5^x$ is defined for all real numbers. When $x = k$ is a positive integer, then $f(k)$ is S_k, the kth partial sum of the series. Using a calculator we obtain the graph of $f(x)$ in Figure 12–8.

GRAPHING EXPLORATION

Graph $f(x)$ in the same viewing window as in Figure 12–8. Use the trace feature to move the cursor along the graph. As x gets larger, what is the apparent value of $f(x)$ (that is, the value of the partial sum)?

Your calculator will probably tell you that every partial sum is 2, once you move beyond approximately $x = 15$. Actually, the partial sums are slightly smaller than 2, but are rounded to 2 by the calculator. In any case, the horizontal line through 2 is a horizontal asymptote of the graph (meaning that the graph gets very close to the line as x gets larger), so it is very plausible that the sequence converges to the number 2. But 2 is exactly what the preceding box says the sum should be:

$$\frac{a_1}{1-r} = \frac{\frac{8}{5}}{1-\frac{1}{5}} = \frac{\frac{8}{5}}{\frac{4}{5}} = \frac{8}{4} = 2. \quad ■$$

Example 3 is typical of the general case, as can be seen algebraically. Consider the geometric series $a_1 + a_2 + a_3 + \cdots$ with common ratio r such that $|r| < 1$. The kth partial sum S_k is the same as the kth partial sum of the geometric sequence $\{a_n\}$ and hence

$$S_k = a_1\left(\frac{1-r^k}{1-r}\right).$$

As k gets larger and larger, the number r^k gets very close to 0 because $|r| < 1$ (for instance, $(-.6)^{20} \approx .0000366$ and $.2^9 \approx .000000512$). Consequently, when k is very large, $1 - r^k$ is very close to $1 - 0$ so that

$$S_k = a_1\left(\frac{1-r^k}{1-r}\right) \quad \text{is very close to} \quad a_1\left(\frac{1-0}{1-r}\right) = \frac{a_1}{1-r}.$$

Example 4 $\displaystyle\sum_{n=1}^{\infty}\left(\frac{-1}{2}\right)^n = -\frac{1}{2} + \frac{1}{4} - \frac{1}{8} + \frac{1}{16} + \cdots$ is an infinite geometric series with $a_1 = -1/2$ and $r = -1/2$. Since $|r| < 1$, this series converges and its sum is

$$\frac{a_1}{1-r} = \frac{-\frac{1}{2}}{1-\left(-\frac{1}{2}\right)} = \frac{-\frac{1}{2}}{\frac{3}{2}} = -\frac{1}{3}. \quad ■$$

Infinite geometric series provide another way of writing an infinite repeating decimal as a rational number.

Example 5 To express 6.8573573573⋯ as a rational number, we first write it as 6.8 + 0.573573573⋯. Consider 0.573573573⋯ as an infinite series:

$$.0573 + .0000573 + .0000000573 + .0000000000573 + \cdots$$

which is the same as

$$.0573 + (.001)(.0573) + (.001)^2(.0573) + (.001)^3(.0573) + \cdots.$$

This is a convergent geometric series with $a_1 = .0573$ and $r = .001$. Its sum is

$$\frac{a_1}{1-r} = \frac{.0573}{1-.001} = \frac{.0573}{.999} = \frac{573}{9990}.$$

Therefore,

$$\begin{aligned}6.8573573573\cdots &= 6.8 + [.0573 + .0000573 + \cdots]\\ &= 6.8 + \frac{573}{9990}\\ &= \frac{68}{10} + \frac{573}{9990}\\ &= \frac{68{,}505}{9990} = \frac{4567}{666}.\end{aligned}$$ ■

Exercises 12.3.A

In Exercises 1–8, find the sum of the infinite series, if it has one.

1. $\sum_{n=1}^{\infty} \frac{1}{2^n}$ **2.** $\sum_{n=1}^{\infty} \left(-\frac{3}{4}\right)^n$ **3.** $\sum_{n=1}^{\infty} (.06)^n$

4. $1 - .5 + .25 - .125 + .0625 - \cdots$

5. $500 + 200 + 80 + 32 + \cdots$

6. $9 - 3\sqrt{3} + 3 - \sqrt{3} + 1 - \frac{1}{\sqrt{3}} + \cdots$

7. $2 + \sqrt{2} + 1 + \frac{1}{\sqrt{2}} + \frac{1}{2} + \cdots$

8. $\sum_{n=1}^{\infty} \left(\frac{1}{2^n} - \frac{1}{3^n}\right)$

In Exercises 9–15, express the repeating decimal as a rational number.

9. .22222⋯ **10.** .37373737⋯

11. 5.4272727⋯ **12.** 85.131313⋯

13. 2.1425425425⋯ **14.** 3.7165165165⋯

15. 1.74241241241⋯

16. If $\{a_n\}$ is an arithmetic sequence with common difference $d > 0$ and each $a_i > 0$, explain why the infinite series $a_1 + a_2 + a_3 + a_4 + \cdots$ is not convergent.

17. (a) Verify that $\sum_{n=1}^{\infty} 2(1.5)^n$ is a geometric series with $a_1 = 3$ and $r = 1.5$.

(b) Find the kth partial sum of the series and use this expression to define a function f, as in Example 3.

(c) Graph the function f in a viewing window with $0 \le x \le 30$. As x gets very large, what happens to the corresponding value of $f(x)$? Does the graph get closer and closer to some horizontal line, as in Example 3? What does this say about the convergence of the series?

18. Use the graphical approach illustrated in Example 3 to find the sum of the series in Example 4. Does the graph get very close to the horizontal line through $-1/3$? What's going on?

12.4 The Binomial Theorem

The Binomial Theorem provides a formula for calculating the product $(x + y)^n$ for any positive integer n. Before we state the theorem, some preliminaries are needed.

Let n be a positive integer. The symbol $n!$ (read **n factorial**) denotes the product of all the integers from 1 to n. For example,

$$2! = 1 \cdot 2 = 2, \qquad 3! = 1 \cdot 2 \cdot 3 = 6, \qquad 4! = 1 \cdot 2 \cdot 3 \cdot 4 = 24,$$

$$5! = 1 \cdot 2 \cdot 3 \cdot 4 \cdot 5 = 120,$$

$$10! = 1 \cdot 2 \cdot 3 \cdot 4 \cdot 5 \cdot 6 \cdot 7 \cdot 8 \cdot 9 \cdot 10 = 3{,}628{,}800.$$

In general, we have this result:

***n* Factorial**

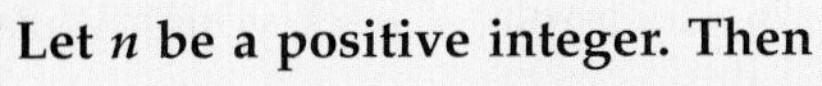

Let n be a positive integer. Then

$$n! = 1 \cdot 2 \cdot 3 \cdot 4 \cdots (n-2)(n-1)n.$$

0! is defined to be the number 1.

Learn to use your calculator to compute factorials. You will find ! in the PROB (or PRB) submenu of the MATH or OPTN menu.

CALCULATOR EXPLORATION

15! is such a large number your calculator will switch to scientific notation to express it. What is this approximation? Many calculators cannot compute factorials larger than 69! If yours does compute larger ones, how large a one can you compute without getting an error message [or on HP-38, getting the number $9.9999 \cdots$ E499]?

If r and n are integers with $0 \le r \le n$, then

Binomial Coefficients

Either of the symbols $\binom{n}{r}$ or ${}_nC_r$ denotes the number $\dfrac{n!}{r!(n-r)!}$.

$\binom{n}{r}$ is called a *binomial coefficient*.

For example,

$$_5C_3 = \binom{5}{3} = \frac{5!}{3!(5-3)!} = \frac{5}{3!2!} = \frac{1\cdot2\cdot3\cdot4\cdot5}{(1\cdot2\cdot3)(1\cdot2)} = \frac{4\cdot5}{2} = 10$$

$$_4C_2 = \binom{4}{2} = \frac{4!}{2!(4-2)!} = \frac{4!}{2!2!} = \frac{1\cdot2\cdot3\cdot4}{(1\cdot2)(1\cdot2)} = \frac{3\cdot4}{2} = 6.$$

Binomial coefficients can be computed on a calculator by using $_nC_r$ or Comb in the PROB (or PRB) submenu of the MATH or OPTN menu.

CALCULATOR EXPLORATION

Compute $_{56}C_{47} = \binom{56}{47}$. Although calculators cannot compute 475!, they can compute many binomial coefficients, such as $\binom{475}{400}$, because most of the factors cancel out (as in the previous example). Check yours. Will it also compute $\binom{475}{50}$?

The preceding examples illustrate a fact whose proof will be omitted: *Every binomial coefficient is an integer.* Furthermore, for every nonnegative integer n,

$$\binom{n}{0} = 1 \quad \text{and} \quad \binom{n}{n} = 1$$

because

$$\binom{n}{0} = \frac{n!}{0!(n-0)!} + \frac{n!}{0!n!} = \frac{n!}{n!} = 1 \quad \text{and}$$

$$\binom{n}{n} = \frac{n!}{n!(n-n)!} = \frac{n!}{n!0!} = \frac{n!}{n!} = 1.$$

If we list the binomial coefficients for each value of n in this manner:

$$n = 0 \qquad \binom{0}{0}$$

$$n = 1 \qquad \binom{1}{0} \quad \binom{1}{1}$$

$$n = 2 \qquad \binom{2}{0} \quad \binom{2}{1} \quad \binom{2}{2}$$

$$n = 3 \qquad \binom{3}{0} \quad \binom{3}{1} \quad \binom{3}{2} \quad \binom{3}{3}$$

$$n = 4 \qquad \binom{4}{0} \quad \binom{4}{1} \quad \binom{4}{2} \quad \binom{4}{3} \quad \binom{4}{4}$$

$$\vdots \quad \therefore \qquad \qquad \ddots$$

and then calculate each of them, we obtain the following array of numbers:

$$\begin{array}{lccccccccc}
\text{row } 0 & & & & & 1 & & & & \\
\text{row } 1 & & & & 1 & & 1 & & & \\
\text{row } 2 & & & 1 & & 2 & & 1 & & \\
\text{row } 3 & & 1 & & 3 & & 3 & & 1 & \\
\text{row } 4 & 1 & & 4 & & 6 & & 4 & & 1 \\
\vdots & \cdot^{\cdot^{\cdot}} & & & & & & & & \ddots
\end{array}$$

This array is called **Pascal's triangle.** Its pattern is easy to remember: Each entry (except the 1's at the beginning or end of a row) is the sum of the two closest entries in the row above it. In the fourth row, for instance, 6 is the sum of the two 3's above it, and each 4 is the sum of the 1 and 3 above it. See Exercise 47 for a proof.

In order to develop a formula for calculating $(x + y)^n$, we first calculate these products for small values of n to see if we can find some kind of pattern:

$$\begin{array}{llrl}
 & n = 0 & (x + y)^0 = & 1 \\
 & n = 1 & (x + y)^1 = & 1x + 1y \\
(*) & n = 2 & (x + y)^2 = & 1x^2 + 2xy + 1y^2 \\
 & n = 3 & (x + y)^3 = & 1x^3 + 3x^2y + 3xy^2 + 1y^3 \\
 & n = 4 & (x + y)^4 = & 1x^4 + 4x^3y + 6x^2y^2 + 4xy^3 + 1y^4
\end{array}$$

One pattern is immediately obvious: the coefficients here (shown in color) are the top part of Pascal's triangle! In the case $n = 4$, for example, this means that the coefficients are the numbers

$$\begin{array}{ccccc}
1 & 4 & 6 & 4 & 1 \\
\binom{4}{0}, & \binom{4}{1}, & \binom{4}{2}, & \binom{4}{3}, & \binom{4}{4}.
\end{array}$$

If this pattern holds for larger n, then the coefficients in the expansion of $(x + y)^n$ are

$$\binom{n}{0}, \binom{n}{1}, \binom{n}{2}, \binom{n}{3}, \ldots, \binom{n}{n-1}, \binom{n}{n}.$$

As for the xy-terms associated with each of these coefficients, look at the pattern in (*) above: the exponent of x goes down by 1 and the exponent of y goes up by 1 as you go from term to term, which suggests that the terms of the expansion of $(x + y)^n$ (without the coefficients) are:

$$x^n, \quad x^{n-1}y, \quad x^{n-2}y^2, \quad x^{n-3}y^3, \ldots, xy^{n-1}, \quad y^n.$$

Combining the patterns of coefficients and xy-terms and using the fact that $\binom{n}{0} = 1$ and $\binom{n}{n} = 1$ suggests that the following result is true about the expansion of $(x + y)^n$.

The Binomial Theorem

For each positive integer n,

$$(x + y)^n = x^n + \binom{n}{1}x^{n-1}y + \binom{n}{2}x^{n-2}y^2 + \binom{n}{3}x^{n-3}y^3 + \cdots + \binom{n}{n-1}xy^{n-1} + y^n.$$

Using summation notation and the fact that $\binom{n}{0} = 1 = \binom{n}{n}$, we can write the Binomial Theorem compactly as

$$(x + y)^n = \sum_{j=0}^{n}\binom{n}{j}x^{n-j}y^j.$$

The Binomial Theorem will be proved in Section 12.5 by means of mathematical induction. We shall assume its truth for now and illustrate some of its uses.

Example 1 Expand $(x + y)^8$.

Solution We apply the Binomial Theorem in the case $n = 8$:

$$(x + y)^8 = x^8 + \binom{8}{1}x^7y + \binom{8}{2}x^6y^2 + \binom{8}{3}x^5y^3 + \binom{8}{4}x^4y^4 + \binom{8}{5}x^3y^5 + \binom{8}{6}x^2y^6 + \binom{8}{7}xy^7 + y^8.$$

The coefficients can be computed individually by hand or by using ${}_nC_r$ (or COMB) on a calculator; for instance,

$${}_8C_2 = \binom{8}{2} = \frac{8!}{2!6!} = 28 \quad \text{or} \quad {}_8C_3 = \binom{8}{3} = \frac{8!}{3!5!} = 56.$$

Alternatively, you can display all the coefficients at once by making a table of values for the function $f(x) = {}_8C_x$, as shown in Figure 12–9.*

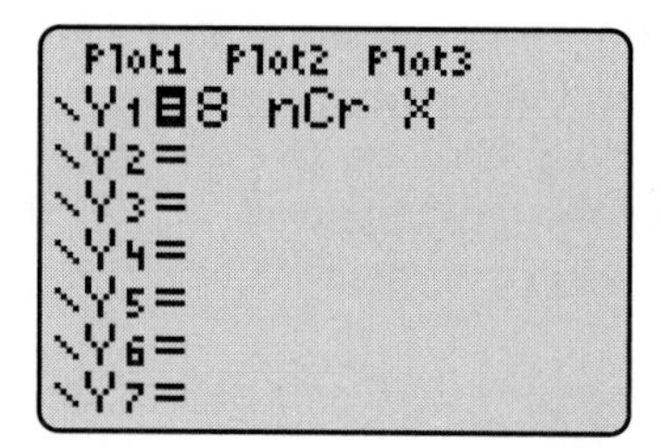

X	Y1
1	8
2	28
3	56
4	70
5	56
6	28
7	8

X=7

Figure 12–9

*Thanks to Nick Goodbody for suggesting this.

Substituting these values in the preceding expansion, we have

$$(x + y)^8 = x^8 + 8x^7y + 28x^6y^2 + 56x^5y^3 + 70x^4y^4 + 56x^3y^5 + 28x^2y^6 + 8xy^7 + y^8. \blacksquare$$

Example 2 Expand $(1 - z)^6$.

Solution Note that $1 - z = 1 + (-z)$ and apply the Binomial Theorem with $x = 1$, $y = -z$, and $n = 6$:

$$\begin{aligned}(1 - z)^6 &= 1^6 + \binom{6}{1}1^5(-z) + \binom{6}{2}1^4(-z)^2 + \binom{6}{3}1^3(-z)^3 \\ &\quad + \binom{6}{4}1^2(-z)^4 + \binom{6}{5}1(-z)^5 + (-z)^6 \\ &= 1 - \binom{6}{1}z + \binom{6}{3}z^2 - \binom{6}{3}z^3 + \binom{6}{4}z^4 - \binom{6}{5}z^5 + z^6 \\ &= 1 - 6z + 15z^2 - 20z^3 + 15z^4 - 6z^5 + z^6. \blacksquare\end{aligned}$$

Example 3 Expand $(x^2 + x^{-1})^4$.

Solution Use the Binomial Theorem with x^2 in place of x and x^{-1} in place of y:

Technology Tip

Binomial expansions, such as those in Examples 1–3, can be done on TI-89 by using EXPAND in the ALGEBRA menu.

$$\begin{aligned}(x^2 + x^{-1})^4 &= (x^2)^4 + \binom{4}{1}(x^2)^3(x^{-1}) + \binom{4}{2}(x^2)^2(x^{-1})^2 \\ &\quad + \binom{4}{3}(x^2)(x^{-1})^3 + (x^{-1})^4 \\ &= x^8 + 4x^6x^{-1} + 6x^4x^{-2} + 4x^2x^{-3} + x^{-4} \\ &= x^8 + 4x^5 + 6x^2 + 4x^{-1} + x^{-4}. \blacksquare\end{aligned}$$

Example 4 Show that $(1.001)^{1000} > 2$ without using a calculator.

Solution We write 1.001 as $1 + .001$ and apply the Binomial Theorem with $x = 1$, $y = .001$, and $n = 1000$:

$$\begin{aligned}(1.001)^{1000} &= (1 + .001)^{1000} \\ &= 1^{1000} + \binom{1000}{1}1^{999}(.001) + \text{other positive terms} \\ &= 1 + \binom{1000}{1}(.001) + \text{other positive terms.}\end{aligned}$$

But $\binom{1000}{1} = \frac{1000!}{1!999!} = \frac{1000 \cdot 999!}{999!} = 1000$. Therefore, $\binom{1000}{1}(.001) = 1{,}000(.001) = 1$ and

$$(1.001)^{1000} = 1 + 1 + \text{other positive terms} = 2 + \text{other positive terms}.$$

Hence, $(1.001)^{1000} > 2$. ■

Sometimes we need to know only one term in the expansion of $(x + y)^n$. If you examine the expansion given by the Binomial Theorem, you will see that in the second term y has exponent 1, in the third term y has exponent 2, and so on. Thus,

Properties of the Binomial Expansion

In the binomial expansion of $(x + y)^n$,

The exponent of y is always one less than the number of the term.

Furthermore, in each of the middle terms of the expansion,

The coefficient of the term containing y^r is $\binom{n}{r}$.

The sum of the x exponent and the y exponent is n.

For instance, in the *ninth* term of the expansion of $(x + y)^{13}$, y has exponent 8, the coefficient is $\binom{13}{8}$, and x must have exponent 5 (since $8 + 5 = 13$). Thus, the ninth term is $\binom{13}{8}x^5y^8$.

Example 5 Find the ninth term of the expansion of $\left(2x^2 + \frac{\sqrt[4]{y}}{\sqrt{6}}\right)^{13}$.

Solution We shall use the Binomial Theorem with $n = 13$ and with $2x^2$ in place of x and $\sqrt[4]{y}/\sqrt{6}$ in place of y. The remarks above show that the ninth term is

$$\binom{13}{8}(2x^2)^5\left(\frac{\sqrt[4]{y}}{\sqrt{6}}\right)^8.$$

Since $\sqrt[4]{y} = y^{1/4}$ and $\sqrt{6} = \sqrt{3}\sqrt{2} = 3^{1/2}2^{1/2}$, we can simplify as follows:

$$\binom{13}{8}(2x^2)^5\left(\frac{\sqrt[4]{y}}{\sqrt{6}}\right)^8 = \binom{13}{8}2^5(x^2)^5\frac{(y^{1/4})^8}{(3^{1/2})^8(2^{1/2})^8} = \binom{13}{8}2^5x^{10}\frac{y^2}{3^4 \cdot 2^4}$$

$$= \binom{13}{8}\frac{2}{3^4}x^{10}y^2 = \frac{13 \cdot 12 \cdot 11 \cdot 10 \cdot 9}{5 \cdot 4 \cdot 3 \cdot 2} \cdot \frac{2}{3^4}x^{10}y^2$$

$$= \frac{286}{9}x^{10}y^2. \quad ■$$

Exercises 12.4

In Exercises 1–10, evaluate the expression.

1. $6!$ **2.** $\frac{11!}{8!}$ **3.** $\frac{12!}{9!3!}$ **4.** $\frac{9! - 8!}{7!}$

5. $\binom{5}{3} + \binom{5}{2} - \binom{6}{3}$ **6.** $\binom{12}{11} - \binom{11}{10} + \binom{7}{0}$

7. $\binom{6}{0} + \binom{6}{1} + \binom{6}{2} + \binom{6}{3} + \binom{6}{4} + \binom{6}{5} + \binom{6}{6}$

8. $\binom{6}{0} - \binom{6}{1} + \binom{6}{2} - \binom{6}{3} + \binom{6}{4} - \binom{6}{5} + \binom{6}{6}$

9. $\binom{100}{96}$ **10.** $\binom{75}{72}$

In Exercises 11–16, expand the expression.

11. $(x + y)^5$ **12.** $(a + b)^7$ **13.** $(a - b)^5$
14. $(c - d)^8$ **15.** $(2x + y^2)^5$ **16.** $(3u - v^3)^6$

In Exercises 17–26, use the Binomial Theorem to expand and (where possible) simplify the expression.

17. $(\sqrt{x} + 1)^6$ **18.** $(2 - \sqrt{y})^5$

19. $(1 - c)^{10}$ **20.** $\left(\sqrt{c} + \frac{1}{\sqrt{c}}\right)^7$

21. $(x^{-3} + x)^4$ **22.** $(3x^{-2} - x^2)^6$

23. $(1 + \sqrt{3})^4 + (1 - \sqrt{3})^4$

24. $(\sqrt{3} + 1)^6 - (\sqrt{3} - 1)^6$

25. $(1 + i)^6$, where $i^2 = -1$

26. $(\sqrt{2} - i)^4$, where $i^2 = -1$

In Exercises 27–32, find the indicated term of the expansion of the given expression.

27. third, $(x + y)^5$ **28.** fourth, $(a + b)^6$

29. fifth, $(c - d)^7$ **30.** third, $(a + 2)^8$

31. fourth, $\left(u^{-2} + \frac{u}{2}\right)^7$ **32.** fifth, $(\sqrt{x} - \sqrt{2})^7$

33. Find the coefficient of x^5y^8 in the expansion of $(2x - y^2)^9$.

34. Find the coefficient of $x^{12}y^6$ in the expansion of $(x^3 - 3y)^{10}$.

35. Find the coefficient of $1/x^3$ in the expansion of $\left(2x + \frac{1}{x^2}\right)^6$.

36. Find the constant term in the expansion of $\left(y - \frac{1}{2y}\right)^{10}$.

37. (a) Verify that $\binom{9}{1} = 9$ and $\binom{9}{8} = 9$.

(b) Prove that for each positive integer n, $\binom{n}{1} = n$ and $\binom{n}{n-1} = n$. [*Note:* Part (a) is just the case when $n = 9$ and $n - 1 = 8$.]

38. (a) Verify that $\binom{7}{2} = \binom{7}{5}$.

(b) Let r and n be integers with $0 \le r \le n$. Prove that $\binom{n}{r} = \binom{n}{n-r}$. [*Note:* Part (a) is just the case when $n = 7$ and $r = 2$.]

39. Prove that for any positive integer n,

$$2^n = \binom{n}{0} + \binom{n}{1} + \binom{n}{2} + \cdots + \binom{n}{n}.$$

[*Hint:* $2 = 1 + 1$.]

40. Prove that for any positive integer n,

$$\binom{n}{0} - \binom{n}{1} + \binom{n}{2} - \binom{n}{3} + \binom{n}{4} - \cdots$$
$$+ (-1)^k\binom{n}{k} + \cdots + (-1)^n\binom{n}{n} = 0.$$

41. Use the Binomial Theorem with $x = \sin\theta$ and $y = \cos\theta$ to find $(\cos\theta + i\sin\theta)^4$ where $i^2 = -1$.

42. (a) Use DeMoivre's Theorem to find

$$(\cos\theta + i\sin\theta)^4.$$

(b) Use the fact that the two expressions obtained in part (a) and in Exercise 41 must be equal to express $\cos 4\theta$ and $\sin 4\theta$ in terms of $\sin\theta$ and $\cos\theta$.

43. (a) Let f be the function given by $f(x) = x^5$. Let h be a nonzero number and compute $f(x + h) - f(x)$ (but leave all binomial coefficients in the form $\binom{5}{r}$ here and below).

(b) Use part (a) to show that h is a factor of $f(x + h) - f(x)$ and find $\frac{f(x + h) - f(x)}{h}$.

(c) If h is *very* close to 0, find a simple approximation of the quantity $\frac{f(x + h) - f(x)}{h}$. [See part (b).]

44. Do Exercise 43 with $f(x) = x^8$ in place of $f(x) = x^5$.

45. Do Exercise 43 with $f(x) = x^{12}$ in place of $f(x) = x^5$.

46. Let n be a fixed positive integer. Do Exercise 43 with $f(x) = x^n$ in place of $f(x) = x^5$.

Thinkers

47. Let r and n be integers such that $0 \le r \le n$.

(a) Verify that $(n - r)! = (n - r)[n - (r + 1)]!$

(b) Verify that $(n - r)! = [(n + 1) - (r + 1)]!$

(c) Prove that $\binom{n}{r+1} + \binom{n}{r} = \binom{n+1}{r+1}$ for any $r \le n - 1$. [*Hint:* Write out the terms on the left side and use parts (a) and (b) to express each of them as a fraction with denominator $(r + 1)!(n - r)!$. Then add these two fractions, simplify the numerator, and compare the result with $\binom{n+1}{r+1}$.]

(d) Use part (c) to explain why each entry in Pascal's triangle (except the 1's at the beginning or end of a row) is the sum of the two closest entries in the row above it.

48. (a) Find these numbers and write them one *below* the next: $11^0, 11^1, 11^2, 11^3, 11^4$.

(b) Compare the list in part (a) with rows 0 to 4 of Pascal's triangle. What's the explanation?

(c) What can be said about 11^5 and row 5 of Pascal's triangle?

(d) Calculate all integer powers of 101 from 101^0 to 101^8, list the results one under the other, and compare the list with rows 0 to 8 of Pascal's triangle. What's the explanation? What happens with 101^9?

12.5 Mathematical Induction

Mathematical induction is a method of proof that can be used to prove a wide variety of mathematical facts, including the Binomial Theorem, DeMoivre's Theorem, and statements such as:

The sum of the first n positive integers is the number $\frac{n(n+1)}{2}$.

$2^n > n$ for every positive integer n.

For each positive integer n, 4 is a factor of $7^n - 3^n$.

All of the preceding statements have a common property. For example, a statement such as

The sum of the first n positive integers is the number $\frac{n(n+1)}{2}$

or, in symbols,

$$1 + 2 + 3 + \cdots + n = \frac{n(n+1)}{2}$$

is really an infinite sequence of statements, one for each possible value of n:

$$n = 1: \qquad 1 = \frac{1(2)}{2}$$

$$n = 2: \qquad 1 + 2 = \frac{2(3)}{2}$$

$$n = 3: \qquad 1 + 2 + 3 = \frac{3(4)}{2}$$

and so on. Obviously, there isn't time enough to verify every one of the statements on this list, one at a time. But we can find a workable method of proof by examining how each statement on the list is *related* to the *next* statement on the list.

For example, for $n = 50$, the statement is

$$1 + 2 + 3 + \cdots + 50 = \frac{50(51)}{2}.$$

At the moment, we don't know whether or not this statement is true. But just *suppose* that it were true. What could then be said about the next statement, the one for $n = 51$:

$$1 + 2 + 3 + \cdots + 50 + 51 = \frac{51(52)}{2}?$$

Well, *if* it is true that

$$1 + 2 + 3 + \cdots + 50 = \frac{50(51)}{2}$$

then adding 51 to both sides and simplifying the right side would yield these equalities:

$$1 + 2 + 3 + \cdots + 50 + 51 = \frac{50(51)}{2} + 51$$

$$1 + 2 + 3 + \cdots + 50 + 51 = \frac{50(51)}{2} + \frac{2(51)}{2} = \frac{50(51) + 2(51)}{2}$$

$$1 + 2 + 3 + \cdots + 50 + 51 = \frac{(50 + 2)51}{2}$$

$$1 + 2 + 3 + \cdots + 50 + 51 = \frac{51(52)}{2}.$$

Since this last equality is just the original statement for $n = 51$, we conclude that

If the statement is true for $n = 50$, *then* it is also true for $n = 51$.

We have *not* proved that the statement actually *is* true for $n = 50$, but only that *if* it is, then it is also true for $n = 51$.

We claim that this same conditional relationship holds for any two consecutive values of n. In other words, we claim that for any positive integer k,

① ***If* the statement is true for $n = k$, *then* it is also true for $n = k + 1$.**

The proof of this claim is the same argument used earlier (with k and $k + 1$ in place of 50 and 51): *If* it is true that

$$1 + 2 + 3 + \cdots + k = \frac{k(k + 1)}{2} \qquad \text{[Original statement for } n = k\text{]}$$

then adding $k + 1$ to both sides and simplifying the right side produces these equalities:

$$1 + 2 + 3 + \cdots + k + (k + 1) = \frac{k(k + 1)}{2} + (k + 1)$$

$$1 + 2 + 3 + \cdots + k + (k + 1) = \frac{k(k + 1)}{2} + \frac{2(k + 1)}{2} = \frac{k(k + 1) + 2(k + 1)}{2}$$

$$1 + 2 + 3 + \cdots + k + (k + 1) = \frac{(k + 2)(k + 1)}{2}$$

$$1 + 2 + 3 + \cdots + k + (k + 1) = \frac{(k + 1)[(k + 1) + 1]}{2}.$$

[*Original statement for* $n = k + 1$]

We have proved that claim ① is valid for each positive integer k. We have *not* proved that the original statement is true for any value of n, but only that *if* it is true for $n = k$, then it is also true for $n = k + 1$. Applying this fact when $k = 1, 2, 3, \ldots$, we see that

② {

If the statement is true for $n = 1$, *then* it is also true for $n = 1 + 1 = 2$;

If the statement is true for $n = 2$, *then* it is also true for $n = 2 + 1 = 3$;

If the statement is true for $n = 3$, *then* it is also true for $n = 3 + 1 = 4$;

$\vdots$

If the statement is true for $n = 50$, *then* it is also true for $n = 50 + 1 = 51$;

If the statement is true for $n = 51$, *then* it is also true for $n = 51 + 1 = 52$;

$\vdots$

and so on.

We are finally in a position to *prove* the original statement: $1 + 2 + 3 + \cdots + n = n(n + 1)/2$. Obviously, it *is true* for $n = 1$ since $1 = 1(2)/2$. Now apply in turn each of the propositions on list ②. Since the statement *is* true for $n = 1$, it must also be true for $n = 2$, and hence for $n = 3$, and hence for $n = 4$, and so on, for every value of n. Therefore, the original statement is true for *every* positive integer n.

The preceding proof is an illustration of the following principle:

Principle of Mathematical Induction

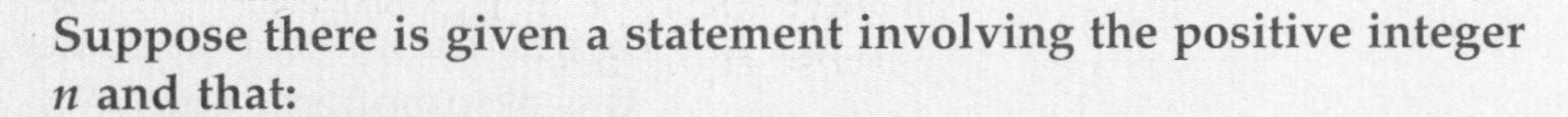

Suppose there is given a statement involving the positive integer n and that:

> **(i) The statement is true for $n = 1$.**
>
> **(ii) If the statement is true for $n = k$ (where k is any positive integer), then the statement is also true for $n = k + 1$.**

Then the statement is true for every positive integer n.

Property (i) is simply a statement of fact. To verify that it holds, you must prove the given statement is true for $n = 1$. This is usually easy, as in the preceding example.

Property (ii) is a *conditional* property. It does not assert that the given statement *is* true for $n = k$, but only that *if* it is true for $n = k$, then it is also true for $n = k + 1$. So to verify that property (ii) holds, you need only prove this conditional proposition:

If the statement is true for $n = k$, *then* it is also true for $n = k + 1$.

In order to prove this, or any conditional proposition, you must proceed as in the previous example: Assume the "if" part and use this assumption to prove the "then" part. As we saw earlier, the same argument will usually work for any possible k. Once this conditional proposition has been proved, you can use it *together* with property (i) to conclude that the given statement is necessarily true for every n, just as in the preceding example.

Thus proof by mathematical induction reduces to two steps:

Step 1 Prove that the given statement is true for $n = 1$.

Step 2 Let k be a positive integer. Assume that the given statement is true for $n = k$. Use this assumption to prove that the statement is true for $n = k + 1$.

Step 2 may be performed before step 1 if you wish. Step 2 is sometimes referred to as the **inductive step.** The assumption that the given statement is true for $n = k$ in this inductive step is called the **induction hypothesis.**

Example 1 Prove that $2^n > n$ for every positive integer n.

Solution Here the statement involving n is $2^n > n$.

Step 1 When $n = 1$, we have the statement $2^1 > 1$. This is obviously true.

Step 2 Let k be any positive integer. We assume that the statement is true for $n = k$, that is, we assume that $2^k > k$. We shall use this assumption to prove that the statement is true for $n = k + 1$, that is, that $2^{k+1} > k + 1$.

We begin with the induction hypothesis:* $2^k > k$. Multiplying both sides of this inequality by 2 yields:

$$2 \cdot 2^k > 2k$$

$$2^{k+1} > 2k. \qquad (3)$$

Since k is a positive integer, we know that $k \geq 1$. Adding k to each side of the inequality $k \geq 1$, we have

$$k + k \geq k + 1$$

$$2k \geq k + 1.$$

Combining this result with inequality (3), we see that

$$2^{k+1} > 2k \geq k + 1.$$

The first and last terms of this inequality show that $2^{k+1} > k + 1$. Therefore, the statement is true for $n = k + 1$. This argument works for any positive integer k. Thus, we have completed the inductive step. By the Principle of Mathematical Induction, we conclude that $2^n > n$ for every positive integer n. ■

Example 2 Simple arithmetic shows that

$$7^2 - 3^2 = 49 - 9 = 40 = 4 \cdot 10$$

and

$$7^3 - 3^3 = 343 - 27 = 316 = 4 \cdot 79.$$

In each case, 4 is a factor. These examples suggest that

For each positive integer n, 4 is a factor of $7^n - 3^n$.

This conjecture can be proved by induction as follows.

Step 1 When $n = 1$, the statement is "4 is a factor of $7^1 - 3^1$." Since $7^1 - 3^1 = 4 = 4 \cdot 1$, the statement is true for $n = 1$.

Step 2 Let k be a positive integer and assume that the statement is true for $n = k$, that is, that 4 is a factor of $7^k - 3^k$. Let us denote the other factor by D, so that the induction hypothesis is: $7^k - 3^k = 4D$. We must use this assumption to prove that the statement is true for $n = k + 1$, that is, that 4 is a factor of $7^{k+1} - 3^{k+1}$. Here is the proof:

*This is the point at which you usually must do some work. Remember that what follows is the "finished proof." It does not include all the thought, scratch work, false starts, and so on that were done before this proof was actually found.

$$
\begin{aligned}
7^{k+1} - 3^{k+1} &= 7^{k+1} - 7 \cdot 3^k + 7 \cdot 3^k - 3^{k+1} && [\textit{Since } -7 \cdot 3^k + 7 \cdot 3^k = 0] \\
&= 7(7^k - 3^k) + (7 - 3)3^k && [\textit{Factor}] \\
&= 7(4D) + (7 - 3)3^k && [\textit{Induction hypothesis}] \\
&= 7(4D) + 4 \cdot 3^k && [7 - 3 = 4] \\
&= 4(7D + 3^k). && [\textit{Factor out } 4]
\end{aligned}
$$

From this last line, we see that 4 is a factor of $7^{k+1} - 3^{k+1}$. Thus, the statement is true for $n = k + 1$, and the inductive step is complete. Therefore, by the Principle of Mathematical Induction the conjecture is actually true for every positive integer n. ■

Another example of mathematical induction, the proof of the Binomial Theorem, is given at the end of this section.

Sometimes a statement involving the integer n may be false for $n = 1$ and (possibly) other small values of n, but true for all values of n beyond a particular number. For instance, the statement $2^n > n^2$ is false for $n = 1$, 2, 3, 4. But it is true for $n = 5$ and all larger values of n. A variation on the Principle of Mathematical Induction can be used to prove this fact and similar statements. See Exercise 28 for details.

A Common Mistake with Induction

It is sometimes tempting to omit step 2 of an inductive proof when the given statement can easily be verified for small values of n, especially if a clear pattern seems to be developing. As the next example shows, however, *omitting step 2 may lead to error.*

Example 3 An integer (>1) is said to be *prime* if its only positive integer factors are itself and 1. For instance, 11 is prime since its only positive integer factors are 11 and 1. But 15 is not prime because it has factors other than 15 and 1 (namely, 3 and 5). For each positive integer n, consider the number

$$f(n) = n^2 - n + 11.$$

You can readily verify that

$$f(1) = 11, \quad f(2) = 13, \quad f(3) = 17, \quad f(4) = 23, \quad f(5) = 31$$

and that *each of these numbers is prime.* Furthermore, there is a clear pattern: The first two numbers (11 and 13) differ by 2; the next two (13 and 17) differ by 4; the next two (17 and 23) differ by 6; and so on. On the basis of this evidence, we might conjecture:

For each positive integer n, the number $f(n) = n^2 - n + 11$ is prime.

We have seen that this conjecture is true for $n = 1, 2, 3, 4, 5$. Unfortunately, however, it is *false* for some values of n. For instance, when $n = 11$,

$$f(11) = 11^2 - 11 + 11 = 11^2 = 121.$$

But 121 is obviously *not* prime since it has a factor other than 121 and 1, namely, 11. You can verify that the statement is also false for $n = 12$ but true for $n = 13$. ■

In the preceding example, the proposition

If the statement is true for $n = k$, then it is true for $n = k + 1$

is false when $k = 10$ and $k + 1 = 11$. If you were not aware of this and tried to complete step 2 of an inductive proof, you would not have been able to find a valid proof for it. Of course, the fact that you can't find a proof of a proposition doesn't always mean that no proof exists. But when you are unable to complete step 2, you are warned that there is a possibility that the given statement may be false for some values of n. This warning should prevent you from drawing any wrong conclusions.

Proof of the Binomial Theorem

We shall use induction to prove that for every positive integer n,

$$(x + y)^n = x^n + \binom{n}{1}x^{n-1}y + \binom{n}{2}x^{n-2}y^2 + \binom{n}{3}x^{n-3}y^3 + \cdots + \binom{n}{n-1}xy^{n-1} + y^n.$$

This theorem was discussed and its notation explained in Section 12.4.

Step 1 When $n = 1$, there are only two terms on the right side of the preceding equation, and the statement reads $(x + y)^1 = x^1 + y^1$. This is certainly true.

Step 2 Let k be any positive integer and assume that the theorem is true for $n = k$, that is, that

$$(x + y)^k = x^k + \binom{k}{1}x^{k-1}y + \binom{k}{2}x^{k-2}y^2 + \cdots + \binom{k}{r}x^{k-r}y^r + \cdots + \binom{k}{k-1}xy^{k-1} + y^k.$$

[On the right side of this equation, we have included a typical middle term $\binom{k}{r}x^{k-r}y^r$. The sum of the exponents is k, and the bottom part of the binomial coefficient is the same as the y exponent.] We shall use this assumption to prove that the theorem is true for $n = k + 1$, that is, that

$$(x + y)^{k+1} = x^{k+1} + \binom{k+1}{1}x^k y + \binom{k+1}{2}x^{k-1}y^2 + \cdots + \binom{k+1}{r+1}x^{k-r}y^{r+1} + \cdots + \binom{k+1}{k}xy^k + y^{k+1}.$$

We have simplified some of the terms on the right side; for instance, $(k + 1) - 1 = k$ and $(k + 1) - (r + 1) = k - r$. But this is the correct statement for $n = k + 1$: The coefficients of the middle terms are $\binom{k+1}{1}$, $\binom{k+1}{2}$, $\binom{k+1}{3}$, and so on; the sum of the exponents of each middle term is $k + 1$, and the bottom part of each binomial coefficient is the same as the y exponent.

In order to prove the theorem for $n = k + 1$, we shall need this fact about binomial coefficients: For any integers r and k with $0 \leq r < k$,

④ $$\binom{k}{r+1} + \binom{k}{r} = \binom{k+1}{r+1}.$$

A proof of this fact is outlined in Exercise 47 on page 764.

To prove the theorem for $n = k + 1$, we first note that

$$(x + y)^{k+1} = (x + y)(x + y)^k.$$

Applying the induction hypothesis to $(x + y)^k$, we see that

$$\begin{aligned}(x + y)^{k+1} &= (x + y)\left[x^k + \binom{k}{1}x^{k-1}y + \binom{k}{2}x^{k-2}y^2 + \cdots + \binom{k}{r}x^{k-r}y^r \right.\\ &\qquad \left. + \binom{k}{r+1}x^{k-(r+1)}y^{r+1} + \cdots + \binom{k}{k-1}xy^{k-1} + y^k\right]\\ &= x\left[x^k + \binom{k}{1}x^{k-1}y + \cdots + y^k\right] + y\left[x^k + \binom{k}{1}x^{k-1}y + \cdots + y^k\right].\end{aligned}$$

Next we multiply out the right-hand side. Remember that multiplying by x increases the x exponent by 1 and multiplying by y increases the y exponent by 1.

$$\begin{aligned}(x + y)^{k+1} &= \left[x^{k+1} + \binom{k}{1}x^k y + \binom{k}{2}x^{k-1}y^2 + \cdots + \binom{k}{r}x^{k-r+1}y^r \right.\\ &\qquad \left. + \binom{k}{r+1}x^{k-r}y^{r+1} + \cdots + \binom{k}{k-1}x^2y^{k-1} + xy^k\right]\\ &\quad + \left[x^k y + \binom{k}{1}x^{k-1}y^2 + \binom{k}{2}x^{k-2}y^3 + \cdots + \binom{k}{r}x^{k-r}y^{r+1} \right.\\ &\qquad \left. + \binom{k}{r+1}x^{k-(r+1)}y^{r+2} + \cdots + \binom{k}{k-1}xy^k + y^{k+1}\right]\\ &= x^{k+1} + \left[\binom{k}{1} + 1\right]x^k y + \left[\binom{k}{2} + \binom{k}{1}\right]x^{k-1}y^2 + \cdots\\ &\quad + \left[\binom{k}{r+1} + \binom{k}{r}\right]x^{k-r}y^{r+1} + \cdots + \left[1 + \binom{k}{k-1}\right]xy^k + y^{k+1}.\end{aligned}$$

Now apply statement ④ to each of the coefficients of the middle terms.

For instance, with $r = 1$, statement ④ shows that $\binom{k}{2} + \binom{k}{1} = \binom{k+1}{2}$.

Similarly, with $r = 0$, $\binom{k}{1} + 1 = \binom{k}{1} + \binom{k}{0} = \binom{k+1}{1}$, and so on. Then the expression above for $(x + y)^{k+1}$ becomes

$$(x + y)^{k+1} = x^{k+1} + \binom{k+1}{1}x^k y + \binom{k+1}{2}x^{k-1}y^2 + \cdots$$
$$+ \binom{k+1}{r+1}x^{k-r}y^{r+1} + \cdots + \binom{k+1}{k}xy^k + y^{k+1}.$$

Since this last statement says the theorem is true for $n = k + 1$, the inductive step is complete. By the Principle of Mathematical Induction the theorem is true for every positive integer n.

Exercises 12.5

In Exercises 1–18, use mathematical induction to prove that each of the given statements is true for every positive integer n.

1. $1 + 2 + 2^2 + 2^3 + 2^4 + \cdots + 2^{n-1} = 2^n - 1$

2. $1 + 3 + 3^2 + 3^3 + 3^4 + \cdots + 3^{n-1} = \frac{3^n - 1}{2}$

3. $1 + 3 + 5 + 7 + \cdots + (2n - 1) = n^2$

4. $2 + 4 + 6 + 8 + \cdots + 2n = n^2 + n$

5. $1^2 + 2^2 + 3^2 + \cdots + n^2 = \frac{n(n+1)(2n+1)}{6}$

6. $\frac{1}{2} + \frac{1}{4} + \frac{1}{8} + \cdots + \frac{1}{2^n} = 1 - \frac{1}{2^n}$

7. $\frac{1}{1\cdot 2} + \frac{1}{2\cdot 3} + \frac{1}{3\cdot 4} + \cdots + \frac{1}{n(n+1)} = \frac{n}{n+1}$

8. $\left(1 + \frac{1}{1}\right)\left(1 + \frac{1}{2}\right)\left(1 + \frac{1}{3}\right)\cdots\left(1 + \frac{1}{n}\right) = n + 1$

9. $n + 2 > n$ **10.** $2n + 2 > n$

11. $3^n \geq 3n$ **12.** $3^n \geq 1 + 2n$

13. $3n > n + 1$ **14.** $\left(\frac{3}{2}\right)^n > n$

15. 3 is a factor of $2^{2n+1} + 1$

16. 5 is a factor of $2^{4n-2} + 1$

17. 64 is a factor of $3^{2n+2} - 8n - 9$

18. 64 is a factor of $9^n - 8n - 1$

19. Let c and d be fixed real numbers. Prove that

$$c + (c + d) + (c + 2d) + (c + 3d) + \cdots + [c + (n-1)d] = \frac{n[2c + (n-1)d]}{2}$$

20. Let r be a fixed real number with $r \neq 1$. Prove that

$$1 + r + r^2 + r^3 + \cdots + r^{n-1} = \frac{r^n - 1}{r - 1}.$$

[*Remember:* $1 = r^0$; so when $n = 1$ the left side reduces to $r^0 = 1$.]

21. (a) Write *each* of $x^2 - y^2$, $x^3 - y^3$, and $x^4 - y^4$ as a product of $x - y$ and another factor.

(b) Make a conjecture as to how $x^n - y^n$ can be written as a product of $x - y$ and another factor. Use induction to prove your conjecture.

22. Let $x_1 = \sqrt{2}$; $x_2 = \sqrt{2 + \sqrt{2}}$; $x_3 = \sqrt{2 + \sqrt{2 + \sqrt{2}}}$; and so on. Prove that $x_n < 2$ for every positive integer n.

In Exercises 23–27, if the given statement is true, prove it. If it is false, give a counterexample.

23. Every odd positive integer is prime.

24. The number $n^2 + n + 17$ is prime for every positive integer n.

25. $(n + 1)^2 > n^2 + 1$ for every positive integer n.

26. 3 is a factor of the number $n^3 - n + 3$ for every positive integer n.

27. 4 is a factor of the number $n^4 - n + 4$ for every positive integer n.

28. Let q be a *fixed* integer. Suppose a statement involving the integer n has these two properties:

(i) The statement is true for $n = q$.

(ii) *If* the statement is true for $n = k$ (where k is any integer with $k \geq q$), then the statement is also true for $n = k + 1$.

Then we claim that the statement is true for every integer n greater than or equal to q.

(a) Give an informal explanation that shows why this claim should be valid. Note that when $q = 1$, this claim is precisely the Principle of Mathematical Induction.

(b) The claim made before part (a) will be called the *Extended Principle of Mathematical Induction.* State the two steps necessary to use this principle to prove that a given statement is true for all $n \geq q$. (See discussion on page 767.)

In Exercises 29–34, use the Extended Principle of Mathematical Induction (Exercise 28) to prove the given statement.

29. $2n - 4 > n$ for every $n \geq 5$. (Use 5 for q here.)

30. Let r be a fixed real number with $r > 1$. Then $(1 + r)^n > 1 + nr$ for every integer $n \geq 2$. (Use 2 for q here.)

31. $n^2 > n$ for all $n \geq 2$

32. $2^n > n^2$ for all $n \geq 5$

33. $3^n > 2^n + 10n$ for all $n \geq 4$

34. $2n < n!$ for all $n \geq 4$

Thinkers

35. Let n be a positive integer. Suppose that there are three pegs and on one of them n rings are stacked, with each ring being smaller in diameter than the one below it (see the figure). We want to transfer the stack of rings to another peg according to these rules: (i) Only one ring may be moved at a time; (ii) a ring can be moved to any peg, provided it is never placed on top of a smaller ring; (iii) the final order of the rings on the new peg must be the same as the original order on the first peg.

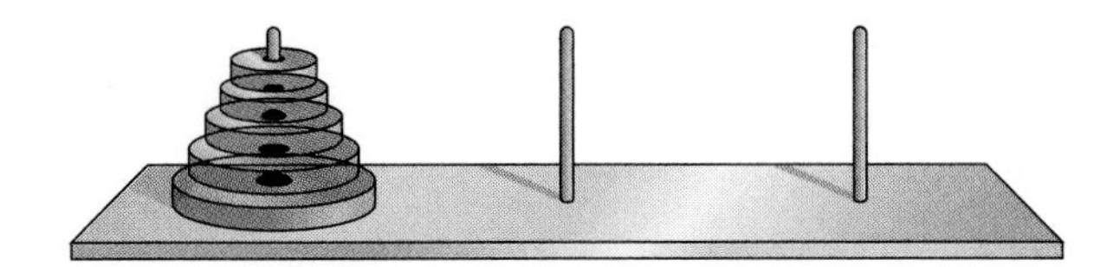

(a) What is the smallest possible number of moves when $n = 2$? $n = 3$? $n = 4$?

(b) Make a conjecture as to the smallest possible number of moves required for any n. Prove your conjecture by induction.

36. The basic formula for compound interest $T(x) = P(1 + r)^x$ was discussed on page 334. Prove by induction that the formula is valid whenever x is a positive integer. [*Note:* P and r are assumed to be constant.]

37. Use induction to prove DeMoivre's Theorem: For any complex number $z = r(\cos\theta + i\sin\theta)$ and any positive integer n,

$$z^n = r^n[\cos(n\theta) + i\sin(n\theta)].$$

Chapter 12 Review

Important Concepts

Important Facts and Formulas

- In an arithmetic sequence $\{a_n\}$ with common difference d:

$$a_n = a_1 + (n-1)d \qquad \sum_{n=1}^{k} a_n = \frac{k}{2}(a_1 + a_k)$$

$$\sum_{n=1}^{k} a_n = ka_1 + \frac{k(k-1)}{2}d$$

- In a geometric sequence $\{a_n\}$ with common ratio $r \neq 1$:

$$a_n = r^{n-1}a_1 \qquad \sum_{n=1}^{k} a_n = a_1\left(\frac{1-r^k}{1-r}\right)$$

- $n! = 1 \cdot 2 \cdot 3 \cdots (n-2)(n-1)n$
- $\binom{n}{r} = \dfrac{n!}{r!(n-r)!} = {}_nC_r$
- *The Binomial Theorem:*

$$(x+y)^n = x^n + \binom{n}{1}x^{n-1}y + \binom{n}{2}x^{n-2}y^2 + \binom{n}{3}x^{n-3}y^3 + \cdots + \binom{n}{n-1}xy^{n-1} + y^n$$

$$= \sum_{j=0}^{n}\binom{n}{j}x^{n-j}y^j$$

Review Questions

In Questions 1–4, find the first four terms of the sequence $\{a_n\}$.

1. $a_n = 2n - 5$
2. $a_n = 3^n - 27$
3. $a_n = \left(\frac{-1}{n}\right)^2$
4. $a_n = (-1)^{n+1}(n - 1)$
5. Find the fifth partial sum of the sequence $\{a_n\}$, where $a_1 = -4$ and $a_n = 3a_{n-1} + 2$.
6. Find the fourth partial sum of the sequence $\{a_n\}$, where $a_1 = 1/9$ and $a_n = 3a_{n-1}$.
7. $\sum_{n=0}^{4} 2^n(n + 1) = ?$
8. $\sum_{n=2}^{4} (3n^2 - n + 1) = ?$

In Questions 9–12, find a formula for a_n; assume that the sequence is arithmetic.

9. $a_1 = 3$ and the common difference is -6.
10. $a_2 = 4$ and the common difference is 3.
11. $a_1 = -5$ and $a_3 = 7$.
12. $a_3 = 2$ and $a_7 = -1$.

In Questions 13–16, find a formula for a_n; assume that the sequence is geometric.

13. $a_1 = 2$ and the common ratio is 3.
14. $a_1 = 5$ and the common ratio is $-1/2$.
15. $a_2 = 192$ and $a_7 = 6$.
16. $a_3 = 9/2$ and $a_6 = -243/16$.
17. Find the 11th partial sum of the arithmetic sequence with $a_1 = 5$ and common difference -2.
18. Find the 12th partial sum of the arithmetic sequence with $a_1 = -3$ and $a_{12} = 16$.
19. Find the fifth partial sum of the geometric sequence with $a_1 = 1/4$ and common ratio 3.
20. Find the sixth partial sum of the geometric sequence with $a_1 = 5$ and common ratio $1/2$.
21. Find numbers b, c, d such that 4, b, c, d, 23 are the first five terms of an arithmetic sequence.
22. Find numbers c and d such that 8, c, d, 27 are the first four terms of a geometric sequence.
23. Is it better to be paid \$5 per day for 100 days or to be paid 5¢ the first day, 10¢ the second day, 20¢ the third day, and have your salary increase in this fashion every day for 100 days?
24. Tuition at Bigstate University is now \$3000 per year and will increase \$150 per year in succeeding years. If a student starts school now, spends four years as an undergraduate, three years in law school, and five years getting a Ph.D., how much tuition will she have paid?

Find the following sums, if they exist.

25. $\sum_{n=1}^{\infty} \frac{1}{2^{n-1}}$
26. $\sum_{n=1}^{\infty} \left(\frac{-1}{4^n}\right)$
27. Use the Binomial Theorem to show that $(1.02)^{51} > 2.5$.
28. What is the coefficient of u^3v^2 in the expansion of $(u + 5v)^5$?

29. $\binom{15}{12} = ?$ **30.** $\binom{18}{3} = ?$

31. Let n be a positive integer. Simplify $\binom{n+1}{n}$.

32. Use the Binomial Theorem to expand $(\sqrt{x} + 1)^5$. Simplify your answer.

33. $\dfrac{20!5!}{6!17!} = ?$ **34.** $\dfrac{7! - 5!}{4!} = ?$

35. Find the coefficient of x^2y^4 in the expansion of $(2y + x^2)^5$.

36. Prove that for every positive integer n,

$$1^3 + 2^3 + 3^3 + \cdots + n^3 = \frac{n^2(n+1)^2}{4}.$$

37. Prove that for every positive integer n,

$$1 + 5 + 5^2 + 5^3 + \cdots + 5^{n-1} = \frac{5^n - 1}{4}.$$

38. Prove that $2^n \geq 2n$ for every positive integer n.

39. If x is a real number with $|x| < 1$, then prove that $|x^n| < 1$ for all $n \geq 1$.

40. Prove that for any positive integer n, $1 + 5 + 9 + \cdots + (4n - 3) = n(2n - 1)$.

41. Prove that for any positive integer n, $1 + 4 + 4^2 + 4^3 + \cdots + 4^{n-1} = \frac{1}{3}(4^n - 1)$.

42. Prove that $3n < n!$ for every $n \geq 4$.

43. Prove that for every positive integer n, 8 is a factor of $9^n - 8n - 1$.

Discovery Project 12

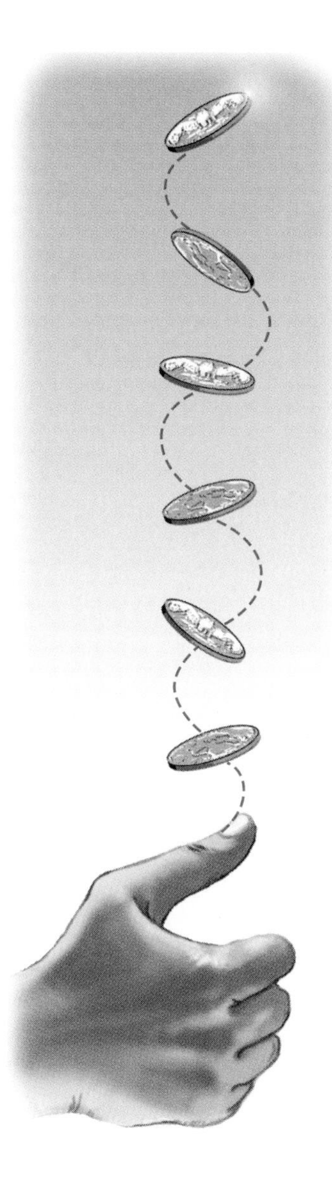

Taking Your Chances

Games of chance have been played in many societies throughout history, so it should not be surprising that as mathematics becomes more sophisticated its principles are applied to the study of games of chance. Many of the common games played can be analyzed using the techniques of discrete mathematics, particularly the Binomial Theorem. The Binomial Theorem comes into play because of the nature of gaming; each play is either a win or a loss, one of two choices. Consider the simple game of flipping a coin. What does a sequence of two games look like?

The tree diagram indicates four paths through the sequence of two games: head-head, head-tail, tail-head, and tail-tail.

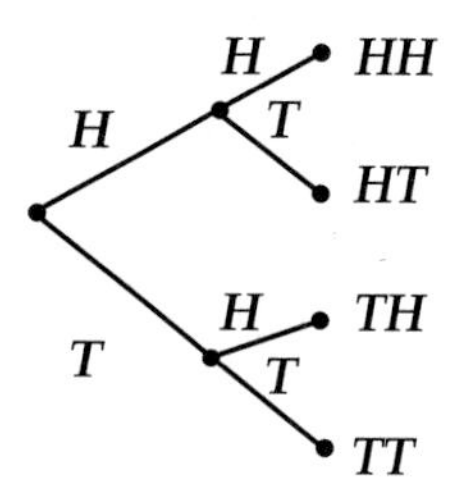

If you think of heads and tails as the variables H and T, then you have three final results: H^2, T^2, and two HT's if you consider $HT = TH$. Indeed, you can think of the last column of the tree as the polynomial of two variables

$$H^2 + 2HT + T^2.$$

The coefficients of the terms are the binomial coefficients you learned about in the Binomial Theorem. You can use the theorem to help you describe many such sequential games.

1. Write a polynomial which describes the results of a game where you flip a coin eight times in a row.

The variables H and T can also be used to represent the probability that you will get heads or tails, respectively, when you flip a coin. If the coin you are using is a fair one, then the probability of heads or tails is equally likely; that is, you would expect to see heads half the time and tails half the time. These values of $\frac{1}{2}$ are the probability that heads or tails will appear on a given flip.

2. What is the probability that you will get exactly four heads and four tails when you flip a coin eight times? Answer this by evaluating the H^4T^4 term of the probability polynomial using the values $H = \frac{1}{2}$ and $T = \frac{1}{2}$.

3. Show that the probability polynomial for flipping eight coins adds to exactly 1.

There is no magic to the values of $\frac{1}{2}$ for winning or losing a game. The only constraint is that the probability values for winning and losing must add to one.

4. Suppose that you play a game where the probability that you win is 0.47 and that you lose is 0.53. What is the probability that you win exactly four out of seven games? Evaluate the W^4L^3 term of the probability polynomial $(W + L)^7$.

Algebra Review

This appendix reviews the fundamental algebraic facts that are used frequently in this book. You must be able to handle these algebraic manipulations in order to succeed in this course and in calculus.

1.A Integral Exponents

Exponents provide a convenient shorthand for certain products. If c is a real number, then c^2 denotes cc and c^3 denotes ccc. More generally, for any positive integer n

$$c^n \textbf{ denotes the product } ccc\cdots c \quad (n \textbf{ factors}).$$

In this notation c^1 is just c, so we usually omit the exponent 1.

Example 1 $3^4 = 3 \cdot 3 \cdot 3 \cdot 3 = 81$ and

$$(-2)^5 = (-2)(-2)(-2)(-2)(-2) = -32.$$

For every positive integer n, $0^n = 0 \cdots 0 = 0$. ■

Example 2 To find $(2.4)^9$ use the $\wedge$ (or a^b or x^y) key on your calculator:*

$$2.4 \wedge 9 \text{ ENTER*}$$

which produces the (approximate) answer 2641.80754. ■

*The ENTER key is labeled EXE on Casio calculators.

Because exponents are just shorthand for multiplication, it is easy to determine the rules they obey. For instance,

$$c^3c^5 = (ccc)(ccccc) = c^8, \qquad \text{that is,} \qquad c^3c^5 = c^{3+5}.$$

$$\frac{c^7}{c^4} = \frac{ccccccc}{cccc} = \frac{\cancel{cccc}ccc}{\cancel{cccc}} = ccc = c^3, \qquad \text{that is,} \qquad \frac{c^7}{c^4} = c^{7-4}.$$

Similar arguments work in the general case:

To multiply c^m by c^n, add the exponents: $c^mc^n = c^{m+n}$.
To divide c^m by c^n, subtract the exponents: $c^m/c^n = c^{m-n}$.

Example 3 $4^2 \cdot 4^7 = 4^{2+7} = 4^9$ and $2^8/2^3 = 2^{8-3} = 2^5$. ■

The notation c^n can be extended to the cases when n is zero or negative as follows:

If $c \neq 0$, then c^0 is defined to be the number 1.
If $c \neq 0$ and n is a positive integer, then

$$c^{-n} \text{ is defined to be the number } \frac{1}{c^n}.$$

Note that 0^0 and negative powers of 0 are *not* defined (negative powers of 0 would involve division by 0). The reason for choosing these definitions of c^{-n} for nonzero c is that the multiplication and division rules for exponents remain valid. For instance,

$$c^5 \cdot c^0 = c^5 \cdot 1 = c^5, \qquad \text{so that} \qquad c^5c^0 = c^{5+0}.$$
$$c^7c^{-7} = c^7(1/c^7) = 1 = c^0, \qquad \text{so that} \qquad c^7c^{-7} = c^{7-7}.$$

Example 4 $6^{-3} = 1/6^3 = 1/216$ and $(-2)^{-5} = 1/(-2)^5 = -1/32$. A calculator shows that $(.287)^{-12} \approx 3{,}201{,}969.857$.* ■

If c and d are nonzero real numbers and m and n are integers (positive, negative, or zero), then we have these

* $\approx$ means "approximately equal to."

Exponent Laws

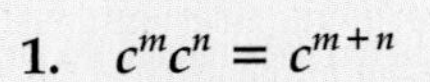

1. $c^m c^n = c^{m+n}$
2. $\dfrac{c^m}{c^n} = c^{m-n}$
3. $(c^m)^n = c^{mn}$
4. $(cd)^n = c^n d^n$
5. $\left(\dfrac{c}{d}\right)^n = \dfrac{c^n}{d^n}$
6. $\dfrac{1}{c^{-n}} = c^n$

Example 5 Here are examples of each of the six exponent laws.

1. $\pi^{-5}\pi^2 = \pi^{-5+2} = \pi^{-3} = 1/\pi^3.$
2. $x^9/x^4 = x^{9-4} = x^5.$
3. $(5^{-3})^2 = 5^{(-3)2} = 5^{-6}.$
4. $(2x)^5 = 2^5x^5 = 32x^5.$ ■
5. $\left(\dfrac{7}{3}\right)^{10} = \dfrac{7^{10}}{3^{10}}.$
6. $1/x^{-5} = \dfrac{1}{\left(\dfrac{1}{x^5}\right)} = x^5.$

> **CAUTION**
> $(2x)^5$ is *not* the same as $2x^5$. Part 4 of Example 5 shows that $(2x)^5 = 32x^5$ and *not* $2x^5$.

The exponent laws can often be used to simplify complicated expressions.

Example 6

(a) $(2x^2y^3z)^4 = 2^4(x^2)^4(y^3)^4z^4 = 16x^8y^{12}z^4.$

(first step: Law (4); second step: Law (3))

(b) $(r^{-3}s^2)^{-2} = (r^{-3})^{-2}(s^2)^{-2} = r^6s^{-4} = r^6/s^4.$

(first step: Law (4); second step: Law (3))

(c) $\dfrac{x^5(y^2)^3}{(x^2y)^2} = \dfrac{x^5y^6}{(x^2y)^2} = \dfrac{x^5y^6}{(x^2)^2y^2} = \dfrac{x^5y^6}{x^4y^2} = x^{5-4}y^{6-2} = xy^4.$ ■

(steps: Law (3), Law (4), Law (3), Law (2))

It is usually more efficient to use the exponent laws with the negative exponents rather than first converting to positive exponents. If positive exponents are required, the conversion can be made in the last step.

Example 7 Simplify and express without negative exponents

$$\frac{a^{-2}(b^2c^3)^{-2}}{(a^{-3}b^{-5})^2c}.$$

Solution

$$\frac{a^{-2}(b^2c^3)^{-2}}{(a^{-3}b^{-5})^2c} \underset{\text{Law (4)}}{=} \frac{a^{-2}(b^2)^{-2}(c^3)^{-2}}{(a^{-3})^2(b^{-5})^2c} \underset{\text{Law (3)}}{=} \frac{a^{-2}b^{-4}c^{-6}}{a^{-6}b^{-10}c}$$

$$\underset{\text{Law (2)}}{=} a^{-2-(-6)}b^{-4-(-10)}c^{-6-1} = a^4b^6c^{-7} = \frac{a^4b^6}{c^7}. \quad \blacksquare$$

Since $(-1)(-1) = +1$, any even power of -1, such as $(-1)^4$ or $(-1)^{12}$, will be equal to 1. Every odd power of -1 is equal to -1; for instance $(-1)^5 = (-1)^4(-1) = 1(-1) = -1$. Consequently, for every positive number c

$$(-c)^n = [(-1)c]^n = (-1)^n c^n = \begin{cases} c^n & \text{if } n \text{ is even} \\ -c^n & \text{if } n \text{ is odd} \end{cases}.$$

Example 8 $(-3)^4 = 3^4 = 81$ and $(-5)^3 = -5^3 = -125$. $\blacksquare$

CAUTION

Be careful with negative bases. For instance, if you want to compute $(-12)^4$, which is a positive number, but you key in (−) 12 ∧ 4 ENTER the calculator will interpret this as $-(12^4)$ and produce a negative answer. To get the correct answer, you must key in the parentheses:

((−) 12) ∧ 4 ENTER.

Exercises 1.A

In Exercises 1–18, evaluate the expression.

1. $(-6)^2$

2. -6^2

3. $5 + 4(3^2 + 2^3)$

4. $(-3)2^2 + 4^2 - 1$

5. $\dfrac{(-3)^2 + (-2)^4}{-2^2 - 1}$

6. $\dfrac{(-4)^2 + 2}{(-4)^2 - 7} + 1$

7. $\left(\dfrac{-5}{4}\right)^3$

8. $-\left(\dfrac{7}{4} + \dfrac{3}{4}\right)^2$

9. $\left(\dfrac{1}{3}\right)^3 + \left(\dfrac{2}{3}\right)^3$

10. $\left(\dfrac{5}{7}\right)^2 + \left(\dfrac{2}{7}\right)^2$

11. $2^4 - 2^7$

12. $3^3 - 3^{-7}$

13. $(2^{-2} + 2)^2$

14. $(3^{-1} - 3^3)^2$

15. $2^2 \cdot 3^{-3} - 3^2 \cdot 2^{-3}$

16. $4^3 \cdot 5^{-2} + 4^2 \cdot 5^{-1}$

17. $\dfrac{1}{2^3} + \dfrac{1}{2^{-4}}$

18. $3^2\left(\dfrac{1}{3} + \dfrac{1}{3^{-2}}\right)$

In Exercises 19–38, simplify the expression. Each letter represents a nonzero real number and should appear at most once in your answer.

19. $x^2 \cdot x^3 \cdot x^5$
20. $y \cdot y^4 \cdot y^6$
21. $(.03)y^2 \cdot y^7$
22. $(1.3)z^3 \cdot z^5$
23. $(2x^2)^3 3x$
24. $(3y^3)^4 5y^2$
25. $(3x^2y)^2$
26. $(2xy^3)^3$
27. $(a^2)(7a)(-3a^3)$
28. $(b^3)(-b^2)(3b)$
29. $(2w)^3(3w)(4w)^2$
30. $(3d)^4(2d)^2(5d)$
31. $a^{-2}b^3a^3$
32. $c^4d^5c^{-3}$
33. $(2x)^{-2}(2y)^3(4x)$
34. $(3x)^{-3}(2y)^{-2}(2x)$
35. $(-3a^4)^2(9x^3)^{-1}$
36. $(2y^3)^3(3y^2)^{-2}$
37. $(2x^2y)^0(3xy)$
38. $(3x^2y^4)^0$

In Exercises 39–42, express the given number as a power of 2.

39. $(64)^2$
40. $(1/8)^3$
41. $(2^4 \cdot 16^{-2})^3$
42. $(1/2)^{-8}(1/4)^4(1/16)^{-3}$

In Exercises 43–60, simplify and write the given expression without negative exponents. All letters represent nonzero real numbers.

43. $\dfrac{x^4(x^2)^3}{x^3}$
44. $\left(\dfrac{z^2}{t^3}\right)^4 \cdot \left(\dfrac{z^3}{t}\right)^5$
45. $\left(\dfrac{e^6}{c^4}\right)^2 \cdot \left(\dfrac{c^3}{e}\right)^3$
46. $\left(\dfrac{x^7}{y^6}\right)^2 \cdot \left(\dfrac{y^2}{x}\right)^4$
47. $\left(\dfrac{ab^2c^3d^4}{abc^2d}\right)^2$
48. $\dfrac{(3x)^2(y^2)^3x^2}{(2xy^2)^3}$
49. $\left(\dfrac{a^6}{b^{-4}}\right)^2$
50. $\left(\dfrac{x^{-2}}{y^{-2}}\right)^2$
51. $\left(\dfrac{c^5}{d^{-3}}\right)^{-2}$
52. $\left(\dfrac{x^{-1}}{2y^{-1}}\right)\left(\dfrac{2y}{x}\right)^{-2}$
53. $\left(\dfrac{3x}{y^2}\right)^{-3}\left(\dfrac{-x}{2y^3}\right)^2$
54. $\left(\dfrac{5u^2v}{2uv^2}\right)^2\left(\dfrac{-3uv}{2u^2v}\right)^{-3}$
55. $\dfrac{(a^{-3}b^2c)^{-2}}{(ab^{-2}c^3)^{-1}}$
56. $\dfrac{(-2cd^2e^{-1})^3}{(5c^{-3}de)^{-2}}$
57. $(c^{-1}d^{-2})^{-3}$
58. $[(x^2y^{-1})^2]^{-3}$
59. $a^2(a^{-1} + a^{-3})$
60. $\dfrac{a^{-2}}{b^{-2}} + \dfrac{b^2}{a^2}$

In Exercises 61–66, determine the sign of the given number without calculating the product.

61. $(-2.6)^3(-4.3)^{-2}$
62. $(4.1)^{-2}(2.5)^{-3}$
63. $(-1)^9(6.7)^5$
64. $(-4)^{12}6^9$
65. $(-3.1)^{-3}(4.6)^{-6}(7.2)^7$
66. $(45.8)^{-7}(-7.9)^{-9}(-8.5)^{-4}$

In Exercises 67–72, r, s, and t are positive integers and a, b, and c are nonzero real numbers. Simplify and write the given expression without negative exponents.

67. $\dfrac{3^{-r}}{3^{-s-r}}$
68. $\dfrac{4^{-(t+1)}}{4^{2-t}}$
69. $\left(\dfrac{a^6}{b^{-4}}\right)^t$
70. $\dfrac{c^{-t}}{(6b)^{-s}}$
71. $\dfrac{(c^{-r}b^s)^t}{(c^tb^{-s})^r}$
72. $\dfrac{(a^rb^{-s})^{-t}}{(b^tc^r)^{-s}}$

Errors to Avoid

In Exercises 73–80, give an example to show that the statement may be false for some numbers.

73. $a^r + b^r = (a + b)^r$
74. $a^ra^s = a^{rs}$
75. $a^rb^s = (ab)^{r+s}$
76. $c^{-r} = -c^r$
77. $\dfrac{c^r}{c^s} = c^{r/s}$
78. $(a + 1)(b + 1) = ab + 1$
79. $(-a)^2 = -a^2$
80. $(-a)(-b) = -ab$

1.B Arithmetic of Algebraic Expressions

Expressions such as

$$b + 3c^2, \qquad 3x^2 - 5x + 4, \qquad \sqrt{x^3 + z}, \qquad \frac{x^3 + 4xy - \pi}{x^2 + xy}$$

are called **algebraic expressions.** Each expression represents a number that is obtained by performing various algebraic operations (such as addition or taking roots) on one or more numbers, some of which may be denoted by letters.

A letter that denotes a particular real number is called a **constant;** its value remains unchanged throughout the discussion. For example, the Greek letter π has long been used to denote the number $3.14159\cdots$. Sometimes a constant is a fixed but unspecified real number, as in "an angle of k degrees" or "a triangle with base of length b."

A letter that can represent *any* real number is called a **variable.** In the expression $2x + 5$, for example, the variable x can be any real number. If $x = 3$, then $2x + 5 = 2 \cdot 3 + 5 = 11$. If $x = \frac{1}{2}$, then $2x + 5 = 2 \cdot \frac{1}{2} + 5 = 6$, and so on.*

Constants are usually denoted by letters near the beginning of the alphabet and variables by letters near the end of the alphabet. Consequently, in expressions such as $cx + d$ and $cy^2 + dy$, it is understood that c and d are constants and x and y are variables.

The usual rules of arithmetic are valid for algebraic expressions:

Commutative Laws:

$$a + b = b + a \quad \text{and} \quad ab = ba$$

Associative Laws:

$$(a + b) + c = a + (b + c) \quad \text{and} \quad (ab)c = a(bc)$$

Distributive Laws:

$$a(b + c) = ab + ac \quad \text{and} \quad (b + c)a = ba + ca.$$

Example 1 Use the distributive law to *combine like terms;* for instance,

$$3x + 5x + 4x = (3 + 5 + 4)x = 12x.$$

In practice, you do the middle part in your head and simply write $3x + 5x + 4x = 12x$. ■

Example 2 In more complicated expressions, eliminate parentheses, use the commutative law to group like terms together, and then combine them.

*We assume any conditions on the constants and variables necessary to guarantee that an algebraic expression does represent a real number. For instance, in $\sqrt{z}$ we assume $z \geq 0$ and in $1/c$ we assume $c \neq 0$.

$$\left(a^2b - 3\sqrt{c}\right) + \left(5ab + 7\sqrt{c}\right) + 7a^2b = a^2b - 3\sqrt{c} + 5ab + 7\sqrt{c} + 7a^2b$$

Regroup: $= \underbrace{a^2b + 7a^2b} - \underbrace{3\sqrt{c} + 7\sqrt{c}} + 5ab$

Combine like terms: $= 8a^2b + 4\sqrt{c} + 5ab.$ ■

> **CAUTION**
>
> Be careful when parentheses are preceded by a minus sign: $-(b + 3) = -b - 3$ and *not* $-b + 3$. Here's the reason: $-(b + 3)$ means $(-1)(b + 3)$, so that by the distributive law,
>
> $$-(b + 3) = (-1)(b + 3) = (-1)b + (-1)3 = -b - 3.$$
>
> Similarly, $-(7 - y) = -7 - (-y) = -7 + y$.

The examples in the Caution Box illustrate the following.

Rules for Eliminating Parentheses

Parentheses preceded by a plus sign (or no sign) may be deleted.

Parentheses preceded by a minus sign may be deleted *if* the sign of every term within the parentheses is changed.

The usual method of multiplying algebraic expressions is to use the distributive laws repeatedly, as shown in the following examples. The net result is to *multiply every term in the first sum by every term in the second sum.*

Example 3 To compute $(y - 2)(3y^2 - 7y + 4)$, we first apply the distributive law, treating $(3y^2 - 7y + 4)$ as a single number:

$$(y - 2)(3y^2 - 7y + 4) = y(3y^2 - 7y + 4) - 2(3y^2 - 7y + 4)$$

Distributive law: $= 3y^3 - 7y^2 + 4y - 6y^2 + 14y - 8$

Regroup: $= 3y^3 - \underbrace{7y^2 - 6y^2} + \underbrace{4y + 14y} - 8$

Combine like terms: $= 3y^3 - 13y^2 + 18y - 8.$ ■

Example 4 We follow the same procedure with $(2x - 5y)(3x + 4y)$:

$$\begin{aligned}(2x - 5y)(3x + 4y) &= 2x(3x + 4y) - 5y(3x + 4y)\\ &= 2x\cdot 3x + 2x\cdot 4y + (-5y)\cdot 3x + (-5y)\cdot 4y\\ &= 6x^2 + \underbrace{8xy - 15xy} - 20y^2\\ &= 6x^2 - 7xy - 20y^2.\end{aligned}$$ ■

Observe the pattern in the second line of Example 4 and its relationship to the terms being multiplied:

$$(2x - 5y)(3x + 4y) = 2x \cdot 3x + 2x \cdot 4y + (-5y) \cdot 3x + (-5y) \cdot 4y$$

(2x − 5y)(3x + 4y) — *First terms*: $2x \cdot 3x$

(2x − 5y)(3x + 4y) — *Outside terms*: $2x \cdot 4y$

(2x − 5y)(3x + 4y) — *Inside terms*: $(-5y) \cdot 3x$

(2x − 5y)(3x + 4y) — *Last terms*: $(-5y) \cdot 4y$

> **CAUTION**
> The FOIL method can be used only when multiplying two expressions that each have two terms.

This pattern is easy to remember by using the acronym FOIL (**F**irst, **O**utside, **I**nside, **L**ast). The FOIL method makes it easy to find products such as this one mentally, without the necessity of writing out the intermediate steps.

Example 5

$$(3x + 2)(x + 5) = \underset{\text{First}}{3x^2} + \underset{\text{Outside}}{15x} + \underset{\text{Inside}}{2x} + \underset{\text{Last}}{10} = 3x^2 + 17x + 10. \quad \blacksquare$$

Exercises 1.B

In Exercises 1–54, perform the indicated operations and simplify your answer.

1. $x + 7x$

2. $5w + 7w - 3w$

3. $6a^2b + (-8b)a^2$

4. $-6x^3\sqrt{t} + 7x^3\sqrt{t} + 15x^3\sqrt{t}$

5. $(x^2 + 2x + 1) - (x^3 - 3x^2 + 4)$

6. $\left[u^4 - (-3)u^3 + \frac{u}{2} + 1\right] + \left(u^4 - 2u^3 + 5 - \frac{u}{2}\right)$

7. $\left[u^4 - (-3)u^3 + \frac{u}{2} + 1\right] - \left(u^4 - 2u^3 + 5 - \frac{u}{2}\right)$

8. $\left(6a^2b + 3a\sqrt{c} - 5ab\sqrt{c}\right) + \left(-6ab^2 - 3ab + 6ab\sqrt{c}\right)$

9. $[4z - 6z^2w - (-2)z^3w^2] + (8 - 6z^2w - zw^3 + 4z^3w^2)$

10. $(x^5y - 2x + 3xy^3) - (-2x - x^5y + 2xy^3)$

11. $(9x - x^3 + 1) - [2x^3 + (-6)x + (-7)]$

12. $\left(x - \sqrt{y} - z\right) - \left(x + \sqrt{y} - z\right) - \left(\sqrt{y} + z - x\right)$

13. $(x^2 - 3xy) - (x + xy) - (x^2 + xy)$

14. $2x(x^2 + 2)$

15. $(-5y)(-3y^2 + 1)$

16. $x^2y(xy - 6xy^2)$

17. $3ax(4ax - 2a^2y + 2ay)$

18. $2x(x^2 - 3xy + 2y^2)$

19. $6z^3(2z + 5)$

20. $-3x^2(12x^6 - 7x^5)$

21. $3ab(4a - 6b + 2a^2b)$

22. $(-3ay)(4ay - 5y)$

23. $(x + 1)(x - 2)$

24. $(x + 2)(2x - 5)$

25. $(-2x + 4)(-x - 3)$

26. $(y - 6)(2y + 2)$

27. $(y + 3)(y + 4)$

28. $(w - 2)(3w + 1)$

29. $(3x + 7)(-2x + 5)$

30. $(ab + 1)(a - 2)$

31. $(y - 3)(3y^2 + 4)$

32. $(y + 8)(y - 8)$

33. $(x + 4)(x - 4)$

34. $(3x - y)(3x + y)$

35. $(4a + 5b)(4a - 5b)$

36. $(x + 6)^2$

37. $(y - 11)^2$

38. $(2x + 3y)^2$

39. $(5x - b)^2$

40. $(2s^2 - 9y)(2s^2 + 9y)$

41. $(4x^3 - y^4)^2$

42. $(4x^3 - 5y^2)(4x^3 + 5y^2)$

43. $(-3x^2 + 2y^4)^2$

44. $(c - 2)(2c^2 - 3c + 1)$

45. $(2y + 3)(y^2 + 3y - 1)$

46. $(x + 2y)(2x^2 - xy + y^2)$

47. $(5w + 6)(-3w^2 + 4w - 3)$

48. $(5x - 2y)(x^2 - 2xy + 3y^2)$

49. $2x(3x + 1)(4x - 2)$

50. $3y(-y + 2)(3y + 1)$

51. $(x - 1)(x - 2)(x - 3)$

52. $(y - 2)(3y + 2)(y + 2)$

53. $(x + 4y)(2y - x)(3x - y)$

54. $(2x - y)(3x + 2y)(y - x)$

In Exercises 55–64, find the coefficient of x^2 in the given product. Avoid doing any more multiplying than necessary.

55. $(x^2 + 3x + 1)(2x - 3)$

56. $(x^2 - 1)(x + 1)$

57. $(x^3 + 2x - 6)(x^2 + 1)$

58. $(\sqrt{3} + x)(\sqrt{3} - x)$

59. $(x + 2)^3$

60. $(x^2 + x + 1)(x - 1)$

61. $(x^2 + x + 1)(x^2 - x + 1)$

62. $(2x^2 + 1)(2x^2 - 1)$

63. $(2x - 1)(x^2 + 3x + 2)$

64. $(1 - 2x)(4x^2 + x - 1)$

In Exercises 65–70, perform the indicated multiplication and simplify your answer if possible.

65. $(\sqrt{x} + 5)(\sqrt{x} - 5)$

66. $(2\sqrt{x} + \sqrt{2y})(2\sqrt{x} - \sqrt{2y})$

67. $(3 + \sqrt{y})^2$

68. $(7w - \sqrt{2x})^2$

69. $(1 + \sqrt{3}x)(x + \sqrt{3})$

70. $(2y + \sqrt{3})(\sqrt{5}y - 1)$

In Exercises 71–76, compute the product and arrange the terms of your answer according to decreasing powers of x, with each power of x appearing at most once.

Example: $(ax + b)(4x - c) = 4ax^2 + (4b - ac)x - bc$.

71. $(ax + b)(3x + 2)$

72. $(4x - c)(dx + c)$

73. $(ax + b)(bx + a)$

74. $rx(3rx + 1)(4x - r)$

75. $(x - a)(x - b)(x - c)$

76. $(2dx - c)(3cx + d)$

In Exercises 77–82, assume that all exponents are nonnegative integers and find the product.

Example: $2x^k(3x + x^{n+1}) = (2x^k)(3x) + (2x^k)(x^{n+1})$
$= 6x^{k+1} + 2x^{k+n+1}$.

77. $3^r 3^4 3^t$

78. $(2x^n)(8x^k)$

79. $(x^m + 2)(x^n - 3)$

80. $(y^r + 1)(y^s - 4)$

81. $(2x^n - 5)(x^{3n} + 4x^n + 1)$

82. $(3y^{2k} + y^k + 1)(y^k - 3)$

Errors to Avoid

In Exercises 83–92, find a numerical example to show that the given statement is false. Then find the mistake in the statement and correct it.

Example: *The statement* $-(b + 2) = -b + 2$ *is false when* $b = 5$, *since* $-(5 + 2) = -7$ *but* $-5 + 2 = -3$. *The mistake is the sign on the* 2. *The correct statement is* $-(b + 2) = -b - 2$.

83. $3(y + 2) = 3y + 2$

84. $x - (3y + 4) = x - 3y + 4$

85. $(x + y)^2 = x + y^2$

86. $(2x)^3 = 2x^3$

87. $(7x)(7y) = 7xy$

88. $(x + y)^2 = x^2 + y^2$

89. $y + y + y = y^3$

90. $(a - b)^2 = a^2 - b^2$

91. $(x - 3)(x - 2) = x^2 - 5x - 6$

92. $(a + b)(a^2 + b^2) = a^3 + b^3$

Thinkers

In Exercises 93 and 94, explain algebraically why each of these parlor tricks always works.

93. Write down a nonzero number. Add 1 to it and square the result. Subtract 1 from the original number and square the result. Subtract this second square from the first one. Divide by the number with which you started. The answer is 4.

94. Write down a positive number. Add 4 to it. Multiply the result by the original number. Add 4 to this result and then take the square root. Subtract the number with which you started. The answer is 2.

95. Invent a similar parlor trick in which the answer is always the number with which you started.

1.C Factoring

Factoring is the reverse of multiplication: We begin with a product and find the factors that multiply together to produce this product. Factoring skills are necessary to simplify expressions, to do arithmetic with fractional expressions, and to solve equations and inequalities.

The first general rule for factoring is

Common Factors

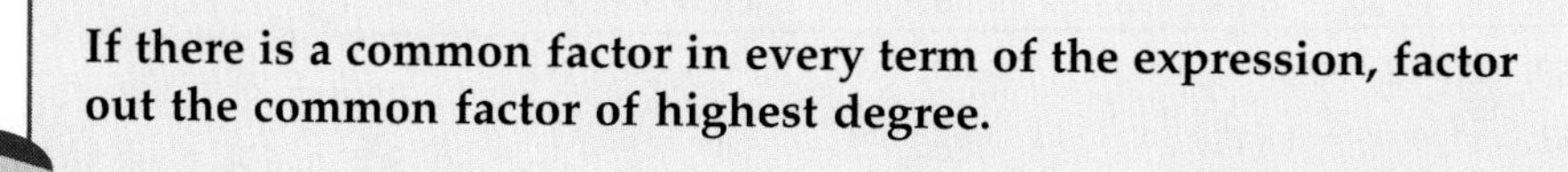

If there is a common factor in every term of the expression, factor out the common factor of highest degree.

Example 1 In $4x^6 - 8x$, for example, each term contains a factor of $4x$, so that $4x^6 - 8x = 4x(x^5 - 2)$. Similarly, the common factor of highest degree in $x^3y^2 + 2xy^3 - 3x^2y^4$ is xy^2 and

$$x^3y^2 + 2xy^3 - 3x^2y^4 = xy^2(x^2 + 2y - 3xy^2). \quad ■$$

You can greatly increase your factoring proficiency by learning to recognize multiplication patterns that appear frequently. Here are the most common ones.

Quadratic Factoring Patterns

Difference of Squares	$u^2 - v^2 = (u + v)(u - v)$
Perfect Squares	$u^2 + 2uv + v^2 = (u + v)^2$
	$u^2 - 2uv + v^2 = (u - v)^2$

Example 2

(a) $x^2 - 9y^2$ can be written $x^2 - (3y)^2$, a difference of squares. Therefore, $x^2 - 9y^2 = (x + 3y)(x - 3y)$.

(b) $y^2 - 7 = y^2 - (\sqrt{7})^2 = (y + \sqrt{7})(y - \sqrt{7})$.*

(c) $36r^2 - 64s^2 = (6r)^2 - (8s)^2 = (6r + 8s)(6r - 8s)$

$$= 2(3r + 4s)2(3r - 4s) = 4(3r + 4s)(3r - 4s). \quad ■$$

*When a polynomial has integer coefficients, we normally look only for factors with integer coefficients. But when it is easy to find other factors, as here, we shall do so.

Example 3 Since the first and last terms of $4x^2 - 36x + 81$ are perfect squares, we try to use the perfect square pattern with $u = 2x$ and $v = 9$:

$$\begin{aligned} 4x^2 - 36x + 81 &= (2x)^2 - 36x + 9^2 \\ &= (2x)^2 - 2\cdot 2x \cdot 9 + 9^2 = (2x - 9)^2. \quad \blacksquare \end{aligned}$$

Cubic Factoring Patterns

Difference of Cubes	$u^3 - v^3 = (u - v)(u^2 + uv + v^2)$
Sum of Cubes	$u^3 + v^3 = (u + v)(u^2 - uv + v^2)$
Perfect Cubes	$u^3 + 3u^2v + 3uv^2 + v^3 = (u + v)^3$
	$u^3 - 3u^2v + 3uv^2 - v^3 = (u - v)^3$

Example 4

(a) $$\begin{aligned} x^3 - 125 = x^3 - 5^3 &= (x - 5)(x^2 + 5x + 5^2) \\ &= (x - 5)(x^2 + 5x + 25). \end{aligned}$$

(b) $$\begin{aligned} x^3 + 8y^3 = x^3 + (2y)^3 &= (x + 2y)[x^2 - x\cdot 2y + (2y)^2] \\ &= (x + 2y)(x^2 - 2xy + 4y^2). \end{aligned}$$

(c) $$\begin{aligned} x^3 - 12x^2 + 48x - 64 &= x^3 - 12x^2 + 48x - 4^3 \\ &= x^3 - 3x^2\cdot 4 + 3x\cdot 4^2 - 4^3 \\ &= (x - 4)^3. \quad \blacksquare \end{aligned}$$

When none of the multiplication patterns applies, use trial and error to factor quadratic polynomials. If a quadratic has two first-degree factors, then the factors must be of the form $ax + b$ and $cx + d$ for some constants a, b, c, d. The product of such factors is

$$\begin{aligned} (ax + b)(cx + d) &= acx^2 + adx + bcx + bd \\ &= acx^2 + (ad + bc)x + bd. \end{aligned}$$

Note that *ac is the coefficient of x^2* and *bd is the constant term* of the product polynomial. This pattern can be used to factor quadratics by reversing the FOIL process.

Example 5 If $x^2 + 9x + 18$ factors as $(ax + b)(cx + d)$, then we must have $ac = 1$ (coefficient of x^2) and $bd = 18$ (constant term). Thus, $a = \pm 1$ and $c = \pm 1$ (the only integer factors of 1). The only possibilities for b and d are

$$\pm 1, \pm 18 \qquad \text{or} \qquad \pm 2, \pm 9 \qquad \text{or} \qquad \pm 3, \pm 6.$$

We mentally try the various possibilities, using FOIL as our guide. For example, we try $b = 2$, $d = 9$ and check this factorization: $(x + 2)(x + 9)$. The sum of the outside and inside terms is $9x + 2x = 11x$, so this product can't be $x^2 + 9x + 18$. By trying other possibilities we find that $b = 3$, $d = 6$ leads to the correct factorization: $x^2 + 9x + 18 = (x + 3)(x + 6)$. ■

Example 6 To factor $6x^2 + 11x + 4$ as $(ax + b)(cx + d)$, we must find numbers a and c whose product is 6, the coefficient of x^2, and numbers b and d whose product is the constant term 4. Some possibilities are

$ac = 6$

a	± 1	± 2	± 3	± 6
c	± 6	± 3	± 2	± 1

$bd = 4$

b	± 1	± 2	± 4
d	± 4	± 2	± 1

Trial and error shows that $(2x + 1)(3x + 4) = 6x^2 + 11x + 4$. ■

Occasionally the patterns above can be used to factor expressions involving larger exponents than 2.

Example 7

(a) $$x^6 - y^6 = (x^3)^2 - (y^3)^2 = (x^3 + y^3)(x^3 - y^3)$$
$$= (x + y)(x^2 - xy + y^2)(x - y)(x^2 + xy + y^2).$$

(b) $$x^8 - 1 = (x^4)^2 - 1 = (x^4 + 1)(x^4 - 1)$$
$$= (x^4 + 1)(x^2 + 1)(x^2 - 1)$$
$$= (x^4 + 1)(x^2 + 1)(x + 1)(x - 1).$$ ■

Example 8 To factor $x^4 - 2x^2 - 3$, let $u = x^2$. Then,

$$x^4 - 2x^2 - 3 = (x^2)^2 - 2x^2 - 3$$
$$= u^2 - 2u - 3 = (u + 1)(u - 3)$$
$$= (x^2 + 1)(x^2 - 3)$$
$$= (x^2 + 1)(x + \sqrt{3})(x - \sqrt{3}).$$ ■

Example 9 $3x^3 + 3x^2 + 2x + 2$ can be factored by regrouping and using the distributive law to factor out a common factor:

$$(3x^3 + 3x^2) + (2x + 2) = 3x^2(x + 1) + 2(x + 1)$$
$$= (3x^2 + 2)(x + 1).$$ ■

Exercises 1.C

In Exercises 1–58, factor the expression.

1. $x^2 - 4$

2. $x^2 + 6x + 9$

3. $9y^2 - 25$

4. $y^2 - 4y + 4$

5. $81x^2 + 36x + 4$

6. $4x^2 - 12x + 9$

7. $5 - x^2$

8. $1 - 36u^2$

9. $49 + 28z + 4z^2$

10. $25u^2 - 20uv + 4v^2$

11. $x^4 - y^4$

12. $x^2 - 1/9$

13. $x^2 + x - 6$

14. $y^2 + 11y + 30$

15. $z^2 + 4z + 3$

16. $x^2 - 8x + 15$

17. $y^2 + 5y - 36$

18. $z^2 - 9z + 14$

19. $x^2 - 6x + 9$

20. $4y^2 - 81$

21. $x^2 + 7x + 10$

22. $w^2 - 6w - 16$

23. $x^2 + 11x + 18$

24. $x^2 + 3xy - 28y^2$

25. $3x^2 + 4x + 1$

26. $4y^2 + 4y + 1$

27. $2z^2 + 11z + 12$

28. $10x^2 - 17x + 3$

29. $9x^2 - 72x$

30. $4x^2 - 4x - 3$

31. $10x^2 - 8x - 2$

32. $7z^2 + 23z + 6$

33. $8u^2 + 6u - 9$

34. $2y^2 - 4y + 2$

35. $4x^2 + 20xy + 25y^2$

36. $63u^2 - 46uv + 8v^2$

37. $x^3 - 125$

38. $y^3 + 64$

39. $x^3 + 6x^2 + 12x + 8$

40. $y^3 - 3y^2 + 3y - 1$

41. $8 + x^3$

42. $z^3 - 9z^2 + 27z - 27$

43. $-x^3 + 15x^2 - 75x + 125$

44. $27 - t^3$

45. $x^3 + 1$

46. $x^3 - 1$

47. $8x^3 - y^3$

48. $(x - 1)^3 + 1$

49. $x^6 - 64$

50. $x^5 - 8x^2$

51. $y^4 + 7y^2 + 10$

52. $z^4 - 5z^2 + 6$

53. $81 - y^4$

54. $x^6 + 16x^3 + 64$

55. $z^6 - 1$

56. $y^6 + 26y^3 - 27$

57. $x^4 + 2x^2y - 3y^2$

58. $x^8 - 17x^4 + 16$

In Exercises 59–64, factor by regrouping and using the distributive law (as in Example 9).

59. $x^2 - yz + xz - xy$

60. $x^6 - 2x^4 - 8x^2 + 16$

61. $a^3 - 2b^2 + 2a^2b - ab$

62. $u^2v - 2w^2 - 2uvw + uw$

63. $x^3 + 4x^2 - 8x - 32$

64. $z^8 - 5z^7 + 2z - 10$

Thinker

65. Show that there do *not* exist real numbers c and d such that $x^2 + 1 = (x + c)(x + d)$.

1.D Fractional Expressions

Quotients of algebraic expressions are called **fractional expressions.** A quotient of two polynomials is sometimes called a **rational expression.** The basic rules for dealing with fractional expressions are essentially the same as those for ordinary numerical fractions. For instance, $\frac{2}{4} = \frac{3}{6}$ and the "cross products" are equal: $2 \cdot 6 = 4 \cdot 3$. In the general case we have

Properties of Fractions

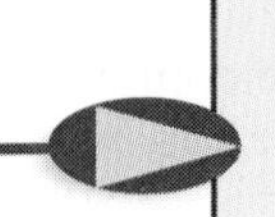

1. *Equality rule:* $\dfrac{a}{b} = \dfrac{c}{d}$ **exactly when** $ad = bc$.*

2. *Cancellation property:* **If** $k \neq 0$, **then** $\dfrac{ka}{kb} = \dfrac{a}{b}$.

*Throughout this section we assume that all denominators are nonzero.

The cancellation property follows directly from the equality rule because $(ka)b = (kb)a$.

Example 1 Here are examples of the two properties:

1. $\dfrac{x^2 + 2x}{x^2 + x - 2} = \dfrac{x}{x - 1}$ because the cross products are equal:

$$(x^2 + 2x)(x - 1) = x^3 + x^2 - 2x = (x^2 + x - 2)x.$$

2. $\dfrac{x^4 - 1}{x^2 + 1} = \dfrac{(x^2 + 1)(x^2 - 1)}{(x^2 + 1)} = \dfrac{x^2 - 1}{1} = x^2 - 1.$ ■

A fraction is in **lowest terms** if its **numerator** (top) and **denominator** (bottom) have no common factors except ± 1. To express a fraction in lowest terms, factor numerator and denominator and cancel common factors.

Example 2 $\dfrac{x^2 + x - 6}{x^2 - 3x + 2} = \dfrac{(x - 2)(x + 3)}{(x - 2)(x - 1)} = \dfrac{x + 3}{x - 1}.$ ■

To add two fractions with the same denominator, simply add the numerators as in ordinary arithmetic: $\dfrac{a}{b} + \dfrac{c}{b} = \dfrac{a + c}{b}$. Subtraction is done similarly.

Example 3

$$\frac{7x^2 + 2}{x^2 + 3} - \frac{4x^2 + 2x - 5}{x^2 + 3} = \frac{(7x^2 + 2) - (4x^2 + 2x - 5)}{x^2 + 3}$$

$$= \frac{7x^2 + 2 - 4x^2 - 2x + 5}{x^2 + 3}$$

$$= \frac{3x^2 - 2x + 7}{x^2 + 3}. \quad ■$$

To add or subtract fractions with different denominators, you must first find a common denominator. One common denominator for a/b and c/d is the product of the two denominators bd because both fractions can be expressed with this denominator:

$$\frac{a}{b} = \frac{ad}{bd} \quad \text{and} \quad \frac{c}{d} = \frac{bc}{bd}.$$

Consequently,

$$\frac{a}{b} + \frac{c}{d} = \frac{ad}{bd} + \frac{bc}{bd} = \frac{ad + bc}{bd} \quad \text{and} \quad \frac{a}{b} - \frac{c}{d} = \frac{ad}{bd} - \frac{bc}{bd} = \frac{ad - bc}{bd}.$$

Example 4

$$\begin{aligned} \frac{2x+1}{3x} - \frac{x^2-2}{x-1} &= \frac{(2x+1)(x-1)}{3x(x-1)} - \frac{3x(x^2-2)}{3x(x-1)} \\ &= \frac{(2x+1)(x-1) - 3x(x^2-2)}{3x(x-1)} \\ &= \frac{2x^2 - x - 1 - 3x^3 + 6x}{3x^2 - 3x} \\ &= \frac{-3x^3 + 2x^2 + 5x - 1}{3x^2 - 3x}. \quad \blacksquare \end{aligned}$$

Although the product of the denominators can always be used as a common denominator, it's often more efficient to use the *least common denominator.* The least common denominator can be found by factoring each denominator completely (with integer coefficients) and then taking the product of the highest power of each of the distinct factors.

Example 5 In the sum $\frac{1}{100} + \frac{1}{120}$, the denominators are $100 = 2^2 \cdot 5^2$ and $120 = 2^3 \cdot 3 \cdot 5$. The distinct factors are 2, 3, 5. The highest exponent of 2 in either denominator is 3, the highest of 3 is 1, and the highest of 5 is 2. So the least common denominator is $2^3 \cdot 3 \cdot 5^2 = 600$ and

$$\frac{1}{100} + \frac{1}{120} = \frac{6}{600} + \frac{5}{600} = \frac{11}{600}. \quad \blacksquare$$

Example 6 To find the least common denominator of $\frac{1}{x^2 + 2x + 1}$, $\frac{5x}{x^2 - x}$, and $\frac{3x - 7}{x^4 + x^3}$, factor each of the denominators completely:

$$(x + 1)^2, \qquad x(x - 1), \qquad x^3(x + 1),$$

The distinct factors are x, $x + 1$, and $x - 1$. The least common denominator is determined by the highest power of each factor:

$$x^3(x + 1)^2(x - 1). \quad \blacksquare$$

To express one of several fractions in terms of the least common denominator, multiply its numerator and denominator by those factors in the common denominator that *don't* appear in the denominator of the fraction.

Example 7 The preceding example shows that the least common denominator of $\frac{1}{(x+1)^2}$, $\frac{5x}{x(x-1)}$, and $\frac{3x-7}{x^3(x+1)}$ is $x^3(x+1)^2(x-1)$. Therefore,

$$\frac{1}{(x+1)^2} = \frac{1}{(x+1)^2}\cdot\frac{x^3(x-1)}{x^3(x-1)} = \frac{x^3(x-1)}{x^3(x+1)^2(x-1)}$$

$$\frac{5x}{x(x-1)} = \frac{5x}{x(x-1)}\cdot\frac{x^2(x+1)^2}{x^2(x+1)^2} = \frac{5x^3(x+1)^2}{x^3(x+1)^2(x-1)}$$

$$\frac{3x-7}{x^3(x+1)} = \frac{3x-7}{x^3(x+1)}\cdot\frac{(x+1)(x-1)}{(x+1)(x-1)} = \frac{(3x-7)(x+1)(x-1)}{x^3(x+1)^2(x-1)}. \quad \blacksquare$$

Example 8 To find $\frac{1}{z} + \frac{3z}{z+1} - \frac{z^2}{(z+1)^2}$ we use the least common denominator $z(z+1)^2$:

$$\begin{aligned}\frac{1}{z} + \frac{3z}{z+1} - \frac{z^2}{(z+1)^2} &= \frac{(z+1)^2}{z(z+1)^2} + \frac{3z^2(z+1)}{z(z+1)^2} - \frac{z^3}{z(z+1)^2}\\ &= \frac{(z+1)^2 + 3z^2(z+1) - z^3}{z(z+1)^2}\\ &= \frac{z^2 + 2z + 1 + 3z^3 + 3z^2 - z^3}{z(z+1)^2}\\ &= \frac{2z^3 + 4z^2 + 2z + 1}{z(z+1)^2}. \quad \blacksquare\end{aligned}$$

Multiplication of fractions is easy: Multiply corresponding numerators and denominators, then simplify your answer.

Example 9

$$\begin{aligned}\frac{x^2-1}{x^2+2}\cdot\frac{3x-4}{x+1} &= \frac{(x^2-1)(3x-4)}{(x^2+2)(x+1)}\\ &= \frac{(x-1)(x+1)(3x-4)}{(x^2+2)(x+1)} = \frac{(x-1)(3x-4)}{x^2+2}. \quad \blacksquare\end{aligned}$$

Division of fractions is given by the rule:

Invert the divisor and multiply: $\frac{a}{b} \div \frac{c}{d} = \frac{a}{b} \cdot \frac{d}{c} = \frac{ad}{bc}.$

Example 10

$$\frac{x^2 + x - 2}{x^2 - 6x + 9} \div \frac{x^2 - 1}{x - 3} = \frac{x^2 + x - 2}{x^2 - 6x + 9} \cdot \frac{x - 3}{x^2 - 1}$$
$$= \frac{(x + 2)(x - 1)}{(x - 3)^2} \cdot \frac{x - 3}{(x - 1)(x + 1)}$$
$$= \frac{x + 2}{(x - 3)(x + 1)}. \quad \blacksquare$$

Division problems can also be written as fractions. For instance, $8/2$ means $8 \div 2 = 4$. Similarly, the compound fraction $\frac{a/b}{c/d}$ means $\frac{a}{b} \div \frac{c}{d}$. So, the basic rule for simplifying compound fractions is: *Invert the denominator and multiply it by the numerator.*

Example 11

(a) $$\frac{16y^2z/8yz^2}{yz/6y^3z^3} = \frac{16y^2z}{8yz^2} \cdot \frac{6y^3z^3}{yz} = \frac{16 \cdot 6 \cdot y^5z^4}{8y^2z^3}$$
$$= 2 \cdot 6 \cdot y^{5-2}z^{4-3} = 12y^3z.$$

(b) $$\frac{\frac{y^2}{y + 2}}{y^3 + y} = \frac{y^2}{y + 2} \cdot \frac{1}{y^3 + y} = \frac{y^2}{(y + 2)(y^3 + y)}$$
$$= \frac{y^2}{(y + 2)y(y^2 + 1)}$$
$$= \frac{y}{(y + 2)(y^2 + 1)}. \quad \blacksquare$$

Exercises 1.D

In Exercises 1–10, express the fraction in lowest terms.

1. $\frac{63}{49}$
2. $\frac{121}{33}$
3. $\frac{13 \cdot 27 \cdot 22 \cdot 10}{6 \cdot 4 \cdot 11 \cdot 12}$
4. $\frac{x^2 - 4}{x + 2}$
5. $\frac{x^2 - x - 2}{x^2 + 2x + 1}$
6. $\frac{z + 1}{z^3 + 1}$
7. $\frac{a^2 - b^2}{a^3 - b^3}$
8. $\frac{x^4 - 3x^2}{x^3}$
9. $\frac{(x + c)(x^2 - cx + c^2)}{x^4 + c^3x}$
10. $\frac{x^4 - y^4}{(x^2 + y^2)(x^2 - xy)}$

In Exercises 11–28, perform the indicated operations.

11. $\frac{3}{7}+\frac{2}{5}$ **12.** $\frac{7}{8}-\frac{5}{6}$ **13.** $\left(\frac{19}{7}+\frac{1}{2}\right)-\frac{1}{3}$

14. $\frac{1}{a}-\frac{2a}{b}$ **15.** $\frac{c}{d}+\frac{3c}{e}$ **16.** $\frac{r}{s}+\frac{s}{t}+\frac{t}{r}$

17. $\frac{b}{c}-\frac{c}{b}$ **18.** $\frac{a}{b}+\frac{2a}{b^2}+\frac{3a}{b^3}$ **19.** $\frac{1}{x+1}-\frac{1}{x}$

20. $\frac{1}{2x+1}+\frac{1}{2x-1}$

21. $\frac{1}{x+4}+\frac{2}{(x+4)^2}-\frac{3}{x^2+8x+16}$

22. $\frac{1}{x}+\frac{1}{xy}+\frac{1}{xy^2}$ **23.** $\frac{1}{x}-\frac{1}{3x-4}$

24. $\frac{3}{x-1}+\frac{4}{x+1}$ **25.** $\frac{1}{x+y}+\frac{x+y}{x^3+y^3}$

26. $\frac{6}{5(x-1)(x-2)^2}+\frac{x}{3(x-1)^2(x-2)}$

27. $\frac{1}{4x(x+1)(x+2)^3}-\frac{6x+2}{4(x+1)^3}$

28. $\frac{x+y}{(x^2-xy)(x-y)^2}-\frac{2}{(x^2-y^2)^2}$

In Exercises 29–42, express in lowest terms.

29. $\frac{3}{4}\cdot\frac{12}{5}\cdot\frac{10}{9}$ **30.** $\frac{10}{45}\cdot\frac{6}{14}\cdot\frac{1}{2}$ **31.** $\frac{3a^2c}{4ac}\cdot\frac{8ac^3}{9a^2c^4}$

32. $\frac{6x^2y}{2x}\cdot\frac{y}{21xy}$ **33.** $\frac{7x}{11y}\cdot\frac{66y^2}{14x^3}$ **34.** $\frac{ab}{c^2}\cdot\frac{cd}{a^2b}\cdot\frac{ad}{bc^2}$

35. $\frac{3x+9}{2x}\cdot\frac{8x^2}{x^2-9}$ **36.** $\frac{4x+16}{3x+15}\cdot\frac{2x+10}{x+4}$

37. $\frac{5y-25}{3}\cdot\frac{y^2}{y^2-25}$ **38.** $\frac{6x-12}{6x}\cdot\frac{8x^2}{x-2}$

39. $\frac{u}{u-1}\cdot\frac{u^2-1}{u^2}$ **40.** $\frac{t^2-t-6}{t^2-6t+9}\cdot\frac{t^2+4t-5}{t^2-25}$

41. $\frac{2u^2+uv-v^2}{4u^2-4uv+v^2}\cdot\frac{8u^2+6uv-9v^2}{4u^2-9v^2}$

42. $\frac{2x^2-3xy-2y^2}{6x^2-5xy-4y^2}\cdot\frac{6x^2+6xy}{x^2-xy-2y^2}$

In Exercises 43–60, compute the quotient and express in lowest terms.

43. $\frac{5}{12}\div\frac{4}{14}$ **44.** $\frac{\frac{100}{52}}{\frac{27}{26}}$ **45.** $\frac{uv}{v^2w}\div\frac{uv}{u^2v}$

46. $\frac{3x^2y}{(xy)^2}\div\frac{3xyz}{x^2y}$ **47.** $\frac{\frac{x+3}{x+4}}{\frac{2x}{x+4}}$ **48.** $\frac{\frac{(x+2)^2}{(x-2)^2}}{\frac{x^2+2x}{x^2-4}}$

49. $\frac{x+y}{x+2y}\div\left(\frac{x+y}{xy}\right)^2$ **50.** $\frac{\frac{u^3+v^3}{u^2-v^2}}{\frac{u^2-uv+v^2}{u+v}}$

51. $\frac{\frac{(c+d)^2}{c^2-d^2}}{cd}$ **52.** $\frac{\frac{1}{x}-\frac{3}{2}}{\frac{2}{x-2}+\frac{5}{x}}$ **53.** $\frac{\frac{1}{x^2}-\frac{1}{y^2}}{\frac{1}{x}+\frac{1}{y}}$

54. $\frac{\frac{x}{x+1}+\frac{1}{x}}{\frac{1}{x}+\frac{1}{x+1}}$ **55.** $\frac{\frac{6}{y}-3}{1-\frac{1}{y-1}}$

56. $\frac{\frac{1}{3x}-\frac{1}{4y}}{\frac{5}{6x^2}+\frac{1}{y}}$ **57.** $\frac{\frac{1}{x+h}-\frac{1}{x}}{h}$

58. $\frac{\frac{1}{(x+h)^2}-\frac{1}{x^2}}{h}$ **59.** $(x^{-1}+y^{-1})^{-1}$

60. $\frac{(x+y)^{-1}}{x^{-1}+y^{-1}}$

Errors to Avoid

In Exercises 61–67, find a numerical example to show that the given statement is false. *Then find the mistake in the statement and correct it.*

61. $\frac{1}{a}+\frac{1}{b}=\frac{1}{a+b}$ **62.** $\frac{x^2}{x^2+x^6}=1+x^3$

63. $\left(\frac{1}{\sqrt{a}+\sqrt{b}}\right)^2=\frac{1}{a+b}$ **64.** $\frac{r+s}{r+t}=1+\frac{s}{t}$

65. $\frac{u}{v}+\frac{v}{u}=1$ **66.** $\frac{\frac{1}{x}}{\frac{1}{y}}=\frac{1}{xy}$

67. $(\sqrt{x}+\sqrt{y})\frac{1}{\sqrt{x}+\sqrt{y}}=x+y$

Appendix 2

Geometry Review

An **angle** consists of two half-lines that begin at the same point P, as in Figure A–1. The point P is called the **vertex** of the angle and the half-lines the **sides** of the angle.

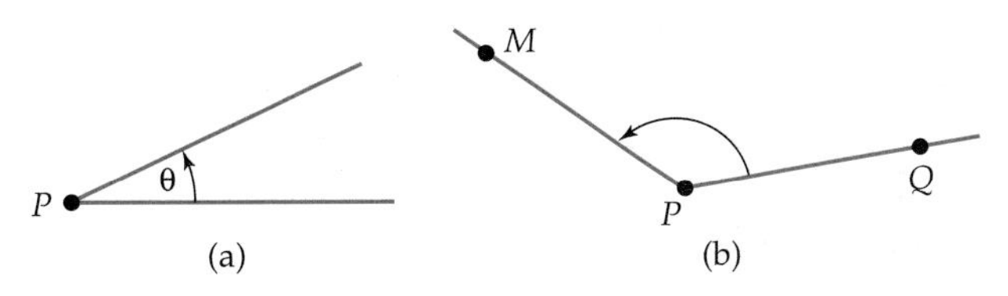

Figure A–1

An angle may be labeled by a Greek letter, such as angle θ in Figure A–1(a), or by listing three points (a point on one side, the vertex, a point on the other side), such as angle QPM in Figure A–1(b).

In order to measure the size of an angle, we must assign a number to each angle. Here is the classical method for doing this:

1. Construct a circle whose center is the vertex of the angle.
2. Divide the circumference of the circle into 360 equal parts (called **degrees**) by marking 360 points on the circumference, beginning with the point where one side of the angle intersects the circle. Label these points 0°, 1°, 2°, 3°, and so on.
3. The label of the point where the second side of the angle intersects the circle is the degree measure of the angle.

For example, Figure A–2 on the next page shows an angle θ of measure 25 degrees (in symbols, 25°) and an angle β of measure 135°.

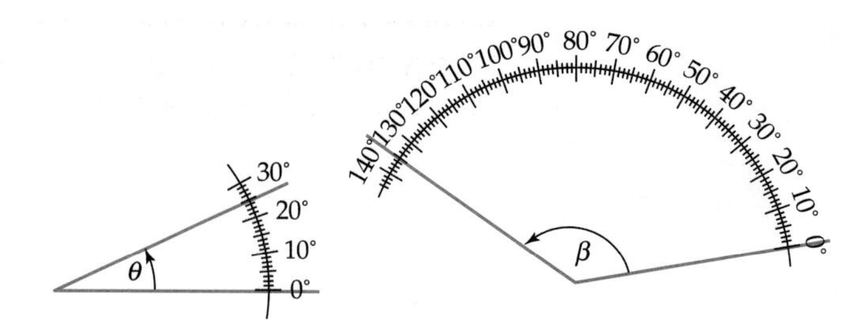

Figure A–2

An **acute angle** is an angle whose measure is strictly between 0° and 90°, such as angle θ in Figure A–2. A **right angle** is an angle that measures 90°. An **obtuse angle** is an angle whose measure is strictly between 90° and 180°, such as angle β in Figure A–2.

A **triangle** has three sides (straight line segments) and three angles, formed at the points where the various sides meet. When angles are measured in degrees,

> **the sum of the measures of all three angles of a triangle is *always* 180°.**

For instance, see Figure A–3.

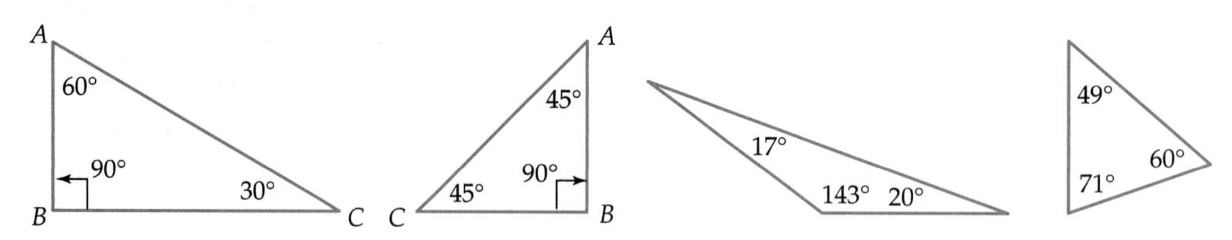

Figure A–3

A **right triangle** is a triangle, one of whose angles is a right angle, such as the first two triangles shown in Figure A–3. The side of a right triangle that lies opposite the right angle is called the **hypotenuse.** In each of the right triangles in Figure A–3, side AC is the hypotenuse.

Pythagorean Theorem

If the sides of a right triangle have lengths a and b and the hypotenuse has length c, then

$$c^2 = a^2 + b^2.$$

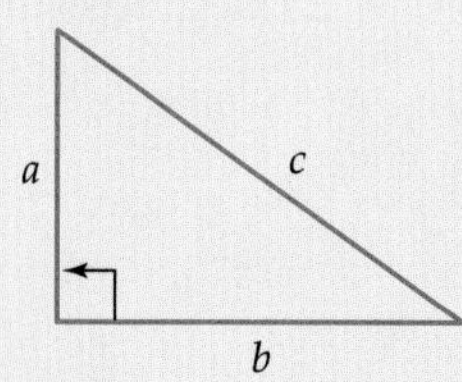

Example 1 Consider the right triangle with sides of lengths 5 and 12, as shown in Figure A–4.

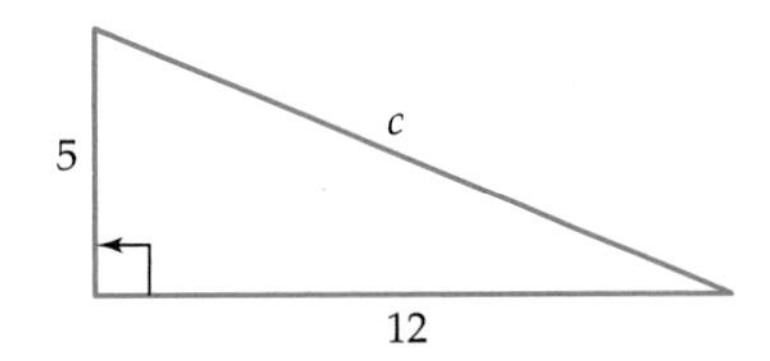

Figure A–4

According to the Pythagorean Theorem the length c of the hypotenuse satisfies the equation: $c^2 = 5^2 + 12^2 = 25 + 144 = 169$. Since $169 = 13^2$, we see that c must be 13. ■

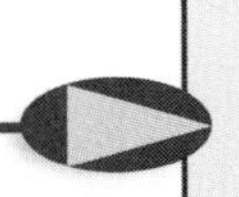

Theorem I

If two angles of a triangle are equal, then the two sides opposite these angles have the same length.

Example 2 Suppose the hypotenuse of the right triangle shown in Figure A–5 has length 1 and that angles B and C measure 45° each.

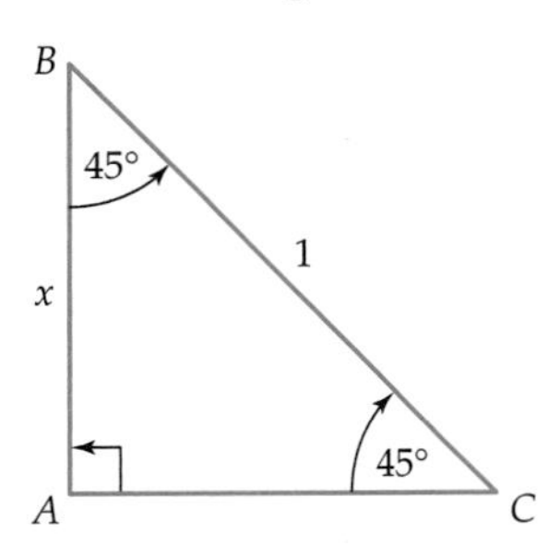

Figure A–5

Then by Theorem I, sides AB and AC have the same length. If x is the length of side AB, then by the Pythagorean Theorem:

$$x^2 + x^2 = 1^2$$
$$2x^2 = 1$$
$$x^2 = \frac{1}{2}$$
$$x = \sqrt{\frac{1}{2}} = \frac{1}{\sqrt{2}} = \frac{\sqrt{2}}{2}.$$

(We ignore the other solution of this equation, namely, $x = -\sqrt{1/2}$, since x represents a length here and thus must be nonnegative.) Therefore, the sides of a 90°–45°–45° triangle with hypotenuse 1 are each of length $\sqrt{2}/2$. ■

Theorem II

In a right triangle that has an angle of 30°, the length of the side opposite the 30° angle is one-half the length of the hypotenuse.

Example 3 Suppose that in the right triangle shown in Figure A–6 angle B is 30° and the length of hypotenuse BC is 2.

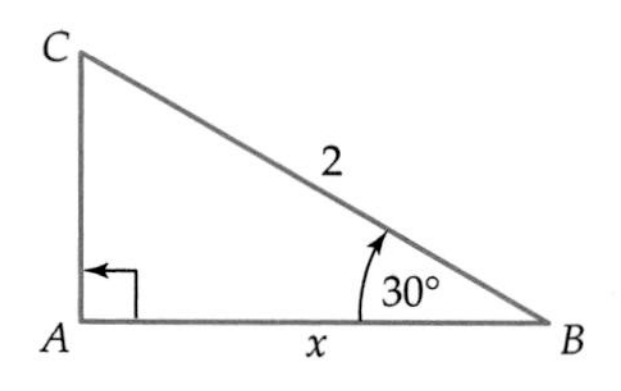

Figure A–6

By Theorem II the side opposite the 30° angle, namely, side AC, has length 1. If x denotes the length of side AB, then by the Pythagorean Theorem:

$$1^2 + x^2 = 2^2$$

$$x^2 = 3$$

$$x = \sqrt{3}. \quad \blacksquare$$

Example 4 The right triangle shown in Figure A–7 has a 30° angle at C, and side AC has length $\sqrt{3}/2$.

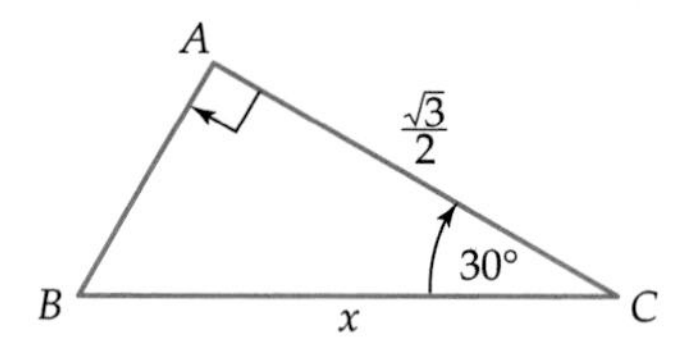

Figure A–7

Let x denote the length of the hypotenuse BC. By Theorem II, side AB has length $\frac{1}{2}x$. By the Pythagorean Theorem:

$$\left(\frac{1}{2}x\right)^2 + \left(\frac{\sqrt{3}}{2}\right)^2 = x^2$$

$$\frac{x^2}{4} + \frac{3}{4} = x^2$$

$$\frac{3}{4} = \frac{3}{4}x^2$$

$$x^2 = 1$$

$$x = 1.$$

Therefore, the triangle has hypotenuse of length 1 and sides of lengths $1/2$ and $\sqrt{3}/2$. ■

Two triangles, as in Figure A–8, are said to be **similar** if their corresponding angles are equal (that is, $\measuredangle A = \measuredangle D$; $\measuredangle B = \measuredangle E$; and $\measuredangle C = \measuredangle F$). Thus, similar triangles have the same *shape* but not necessarily the same *size*.

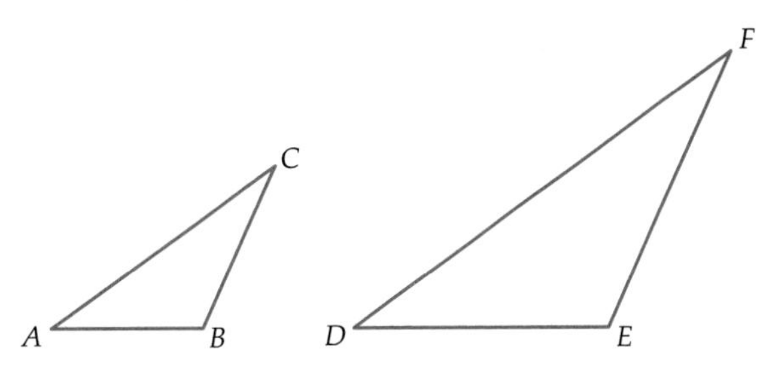

Figure A–8

Theorem III

Suppose triangle *ABC* with sides *a, b, c* is similar to triangle *DEF* with sides *d, e, f* (that is, $\measuredangle A = \measuredangle D$; $\measuredangle B = \measuredangle E$; $\measuredangle C = \measuredangle F$).

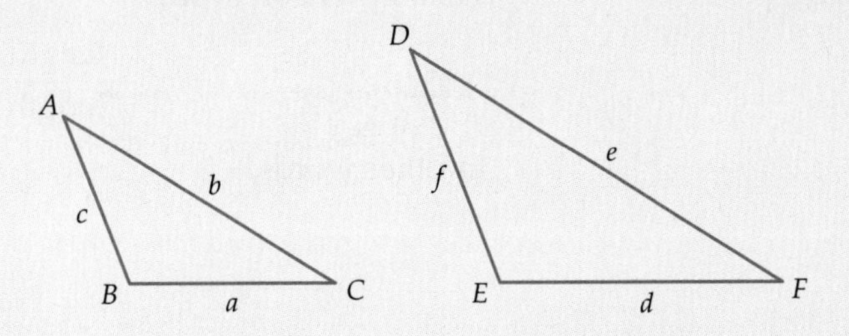

then

$$\frac{a}{d} = \frac{b}{e} = \frac{c}{f}.$$

These equalities are equivalent to:

$$\frac{a}{b} = \frac{d}{e}, \quad \frac{b}{c} = \frac{e}{f}, \quad \frac{a}{c} = \frac{d}{f}.$$

The equivalence of the equalities in the conclusion of the theorem is easily verified. For example, since

$$\frac{a}{d} = \frac{b}{e}$$

we have

$$ae = db$$

Dividing both sides of this equation by *be* yields:

$$\frac{ae}{be} = \frac{db}{be}$$

$$\frac{a}{b} = \frac{d}{e}.$$

The other equivalences are proved similarly.

Example 5 Suppose the triangles in Figure A–9 are similar and that the sides have the lengths indicated.

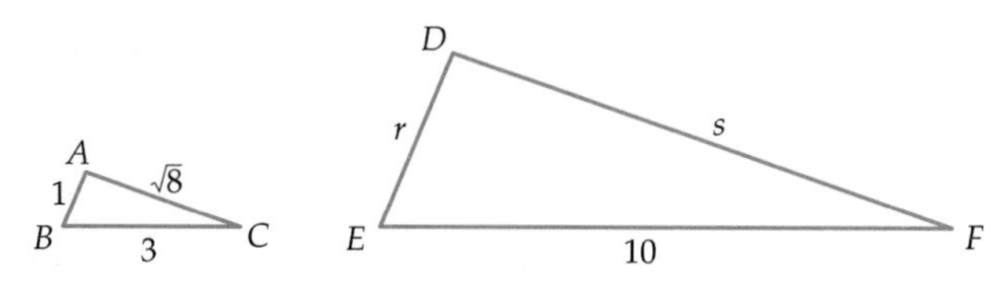

Figure A–9

Then by Theorem III,

$$\frac{\text{length } AC}{\text{length } DF} = \frac{\text{length } BC}{\text{length } EF}$$

In other words,

$$\frac{\sqrt{8}}{s} = \frac{3}{10}$$

so that

$$3s = 10\sqrt{8}$$

$$s = \left(\frac{10}{3}\right)\sqrt{8}.$$

Similarly, by Theorem III,

$$\frac{\text{length } AB}{\text{length } DE} = \frac{\text{length } BC}{\text{length } EF}$$

so that

$$\frac{1}{r} = \frac{3}{10}$$

$$3r = 10$$

$$r = \frac{10}{3}$$

Therefore the sides of triangle *DEF* are of lengths 10, $\frac{10}{3}$, and $\frac{10}{3}\sqrt{8}$. ■

Appendix

3 Programs

The programs listed here are of two types: programs to give older calculators some of the features that are built-in on newer ones (such as a table maker) and programs to do specific tasks discussed in this book (such as synthetic division). Each program is preceded by a *Description,* which describes, in general terms, how the program operates and what it does. Some programs require that certain things be done before the program is run (such as entering a function in the function memory); these requirements are listed as *Preliminaries.* Occasionally, italic remarks appear in brackets after a program step; they are *not* part of the program, but are intended to provide assistance when you are entering the program into your calculator. A remark such as "[*MATH NUM menu*]" means that the symbols or commands needed for that step of the program are in the NUM submenu of the MATH menu.

Fraction Conversion (Built-in on TI-82/83/85/86 and HP-38)

Description: Enter a repeating decimal; the program converts it into a fraction. The denominator is displayed on the last line and the numerator on the line above.

Sharp 9600

Input N
$0 \rightarrow D$
Label 1
$D + 1 \rightarrow D$
If fpart (round ($N \times D$, 7)) $\neq 0$ Goto 1 [*MATH NUM menu*]
int($N \times D + .5$) $\rightarrow N$
Print N
Print D

Casio 9850

Fix 7
"N = "? $\rightarrow N$
$0 \rightarrow D$
Lbl 1
$D + 1 \rightarrow D$
$N \times D$
Rnd [*MATH NUM menu*]
(Frac Ans) $\neq 0 \Rightarrow$ Goto 1 [*MATH NUM menu*]
(Ans + .5) $\rightarrow N$
Norm [*DISP menu*]
(Int N)◢
D

TI-82/83 Quadratic Formula (Built-in on other calculators)

Description: Enter the coefficients of the quadratic equation $ax^2 + bx + c = 0$; the program finds all real solutions.

```
:ClrHome [Optional]
:Disp "AX² + BX + C = 0" [Optional]
:Prompt A
:Prompt B
:Prompt C
:(B² − 4AC) → S
:If S < 0
:Goto 1
:Disp (−B + √S)/(2A)
:Disp (−B − √S)/(2A)
:Stop
:Lbl 1
:Disp "NO REAL ROOTS"
```

TI-85 Table Maker (Built-in on other calculators)

Preliminaries: Enter the function to be evaluated in the function memory as Y_1.

Description: Select a starting point and an increment (the amount by which adjacent x entries differ); the program displays a table of function values. To scroll through the table a page at a time, press "down" or "up." Press "quit" to end the program. [*Note:* An error message will occur if the calculator attempts to evaluate the function at a number for which it is not defined. In this case, change the starting point or increment to avoid the undefined point.]

```
:Lbl SETUP
:ClLCD [Optional]
Disp " "TABLE SETUP"
:Input "TblMin = ", tblmin
:Input "ΔTbl = ", dtbl      [CHAR GREEK menu]
:tblmin → x      [Use the x-var key for x.]
:Lbl CONTD
:ClLCD
:Output(1,1,"x")      [I/O menu]
:Output(1,10,"y1")
:For (cnt,2,7,1)
:Output(cnt,1,x)
Output(cnt,8," ")
:Output(cnt,9,y1)
:x + dtbl → x
:End
:Menu(1, "Down",CONTD, 2,"UP",CONTU, 4,
"Setup," SETUP, 5, "quit", TQUIT)
:Lbl CONTU
:x − 12*dtbl → x
:Goto CONTD
:Lbl TQUIT
:ClLCD
:Stop
```

Synthetic Division (Built-in on TI-89/92)

Preliminaries: Enter the coefficients of the dividend $F(x)$ (in order of decreasing powers of x, putting in zeros for missing coefficients) as list L_1 (or List 1). If the coefficients are 1, 2, 3, for example, key in {1, 2, 3,} and store it in L_1. The symbols { } are on the keyboard or in the LIST menu. The list name L_1 is on the keyboard of TI-82/83 and Sharp 9600. On TI-85/86 and HP-38, type in L1. On Casio 9850, use "List" in the LIST submenu of the OPTN menu to type in List 1.

Description: Write the divisor in the form $x - a$ and enter a. The program displays the degree of the quotient $Q(x)$, the coefficients of $Q(x)$ (in order of decreasing powers of x), and the remainder. If the program pauses before it has displayed all these items, you can use the arrow keys to scroll through the display; then press ENTER (or OK) to continue.

TI-82/83/85/86

```
:ClrHome [ClLCD on TI-85/86]
:Disp "DIVISOR IS X − A"
:Prompt A
:L1 → L2 [See Preliminaries for how to enter list names]
:dim L1 → N [dimL L1 on TI-85/86]     [LIST OPS menu]
:For (K, 2, N)
:(L1(K) + A × L2(K − 1)) → L2(K)
:End
:round(L2(N),9) → R [MATH NUM menu]
:(N − 1) → dim L2
:Disp "DEGREE OF QUOTIENT"
:Disp dim L2 − 1
:Disp "COEFFICIENTS"
:Pause L2
:Disp "REMAINDER"
:Disp R
```

Sharp 9600

```
Clr T
Print "DIVISOR IS X − A"
Input A
L1 → L2     [See Preliminaries for how to enter list names]
dim (L1) → N     [LIST OPE menu]
2 → K
Label 1
(L1(K) + A × L2(K − 1)) → L2(K)
K + 1 → K
If K ≤ N Goto 1
round(L2(N),9) → R     [MATH NUM menu]
(N − 1) → dim L2
Clr T
Print "DEGREE OF QUOTIENT"
Print dim L2 − 1
Print "COEFFICIENTS"
Print L2
Print "REMAINDER"
Print R
```

Casio 9850

```
"DIVISOR IS X − A"
"A = "? → A
List 1 → List 2      [See Preliminaries for how to enter list names]
dim List 1 → N       [OPTN LIST menu]
2 → K
Lbl 1
(List 1[K] + A × List 2 [K − 1]) → List 2 [K]
K + 1 → K
K ≤ N ⇒ Goto 1
Fix 9     [SETUP DISP menu]
List 2 [N]
Rnd     [OPTN MATH NUM menu]
Ans → R
Norm     [SETUP DISP menu]
Seq(List 2 [X], X, 1, N − 1, 1) → List 2
                          [OPTN LIST menu]
"DEGREE OF QUOTIENT"
dim List 2 − 1◢
"COEFFICIENTS"
List 2◢
"REMAINDER"
R
```

HP-38

```
Input A; "SYNDIV", "X − A"; "ENTER A"; 0:
L1 → L2:          [See Preliminaries for how to enter list names]
Size(L2) → N:     [MATH LIST menu]
For C = 1 to (N − 1)    Step 1; L2(C) × A + L2(C + 1) → L2(C + 1) End:
Makelist (L2(J), J, 1, N − 1, 1) → L3:     [MATH LIST menu]
Msgbox "DEGREE OF QUOTIENT" Size(L3) − 1:
Msgbox "COEFFICIENTS" L3:
Msgbox "REMAINDER" L2(N):
```

HP-38 Rectangular/Polar Conversion Program

Description: Enter the rectangular coordinates of a point in the plane; the program displays the polar coordinates of the point.

```
Input X; "RECTANGULAR TO POLAR"; "X = "; "ENTER X"; 0:
Input Y: "RECTANGULAR TO POLAR"; "Y = "; "ENTER Y"; 0:
If X > 0
Then (ATAN(Y/X)) → C
Else (ATAN(X,Y) + π) → C:
End:
MSGBOX "R = " √(X² + Y²) "θ = " C:
```

HP-38 Polar/Rectangular Conversion Program

Description: Enter the polar coordinates of a point in the plane; the program displays the rectangular coordinates of the point.

```
Input R; "POLAR TO RECTANGULAR"; "R = "; "ENTER R"; 0:
Input θ; "POLAR TO RECTANGULAR"; "θ = "; "ENTER θ"; 0: [CHARS menu]
MSGBOX "X = " R(cos θ) "Y = " R(sin θ):
```

Answers to Odd-Numbered Exercises

Chapter 1

Section 1.1, page 12

1.

-7, -5, -1, 0, 2.25, 10

-4.75, $\frac{1}{2}$, $\frac{8}{3}$

3. 2.70312×10^8 **5.** 6.529×10^6 **7.** 2×10^{-9}

9. 150,000,000,000 m

11. .00000000000000000016726

13. $-4 > -8$ **15.** $\pi < 100$ **17.** $y \leq 7.5$

19. $t > 0$ **21.** $c \leq 3$ **23.** $<$

25. $=$ **27.** $>$ **29.** $b + c = a$

31. a lies to the right of b.

33. $a < b$ **35.** 11 **37.** 0 **39.** 169

41. $\pi - \sqrt{2}$ **43.** π **45.** $<$ **47.** $>$ **49.** $<$

51. -2 0 2 4 6 8

53. -3 -2 -1 0 1 2 3

55. -4 -2 0 1

57. $[5, 8]$ **59.** $(-3, 14)$ **61.** $[-8, \infty)$

63. 7 **65.** 14.5 **67.** $\pi - 3$ **69.** $\sqrt{3} - \sqrt{2}$

71. **(a)** 45.05 miles **(b)** 1.65 miles

73. $|p - 25.75| \leq 4$ **75.** t^2

77. $b - 3$ **79.** $-(c - d) = d - c$ **81.** 0

83. $|(c - d)^2| = (c - d)^2 = c^2 - 2cd + d^2$

85. $|x - 5| < 4$ **87.** $|x + 4| \leq 17$ **89.** $|c| < |b|$

91. The distance from x to 3 is less than 2 units.

93. The distance from x to -7 is at most 3 units.

95. **(a)** iii **(b)** i **(c)** ii **(d)** v **(e)** iv

97. $x = 1$ or -1 **99.** $x = 1$ or 3

101. $x = -\pi + 4$ or $-\pi - 4$

103. $-7 < x < 7$ **105.** $3 < x < 7$

107. $x \leq -5$ or $x \geq 1$

109. Since $|a| \geq 0$, $|b| \geq 0$, and $|c| \geq 0$, the sum $|a| + |b| + |c|$ is positive only when one or more of $|a|, |b|, |c|$ is positive. But $|a|$ is positive only when $a \neq 0$; similarly for b, c.

Excursion 1.1.A, page 17

1. $.7777\cdots$ **3.** $1.6428571428571\cdots$

5. $.052631578947368421052\cdots$

7. No; $\frac{2}{3} = .6666\cdots$ **9.** Yes; $\frac{1}{64} = .015625$

11. No **13.** Yes; $\frac{1}{.625} = 1.6$

15. $\frac{37}{99}$ **17.** $\frac{758{,}679}{9900} = \frac{252{,}893}{3300}$

19. $\frac{5}{37}$ **21.** $\frac{517{,}896}{9900} = \frac{14{,}386}{275}$

23. If $d = .74999\cdots$, then $1000d - 100d = (749.999\cdots) - (74.999\cdots) = 675$. Hence $900d = 675$ so that $d = \frac{675}{900} = \frac{3}{4}$. Also $.75000\cdots = .75 = \frac{75}{100} = \frac{3}{4}$.

25. $\frac{6}{17} = .352941176470588235 29\cdots$

27. $\dfrac{1}{29} = .0344827586206896551724137931034 4\cdots$

29. $\dfrac{283}{47} =$
$6.0212765957446808510638297872340425531914893617021 2\cdots$

31. (a) One of many possible ways is to use the nonrepeating decimal expansion of π. For instance, with .75, associate $.7531415926\cdots$; with 6.593 associate $6.59331415926\cdots$, etc. Thus different terminating decimals correspond to different nonrepeating ones.
(b) As suggested in the *Hint*, associate with $.134134134\cdots$ the irrational number $.134013400134000134\cdots$. With $6.17398419841\cdots$ associate $6.17398410984100984100009841\cdots$, etc. Thus, different repeating decimals lead to different nonrepeating ones.

Section 1.2, page 25

1. $x = 3$ or 5 **3.** $x = -2$ or 7

5. $y = \dfrac{1}{2}$ or -3 **7.** $t = -2$ or $-\dfrac{1}{4}$

9. $u = 1$ or $-\dfrac{4}{3}$ **11.** $x = \dfrac{1}{4}$ or $-\dfrac{4}{3}$

13. $x = 1 \pm \sqrt{13}$

15. $x = (1 + \sqrt{5})/2$ or $(1 - \sqrt{5})/2$

17. $x = 2 \pm \sqrt{3}$ **19.** $x = -3 \pm \sqrt{2}$

21. No real number solutions

23. $x = \dfrac{1}{2} \pm \sqrt{2}$ **25.** $x = \dfrac{2 \pm \sqrt{3}}{2}$

27. $u = \dfrac{-4 \pm \sqrt{6}}{5}$ **29.** 2

31. 2 **33.** 1

35. $Y = 800$ **37.** $Y = 1300$

39. $Y = 1500$ **41.** $Y = 3040$

43. $x = -5$ or 8 **45.** $x = \dfrac{-5 \pm \sqrt{57}}{8}$

47. $x = -3$ or -6 **49.** $x = \dfrac{-1 \pm \sqrt{2}}{2}$

51. $x = 5$ or $-\dfrac{3}{2}$

53. No real number solutions

55. No real number solutions

57. $x \approx 1.824$ or $.470$ **59.** $x = 13.79$

61. $y = \pm 1$ or $\pm\sqrt{6}$ **63.** $x = \pm\sqrt{7}$

65. $y = \pm 2$ or $\pm 1/\sqrt{2}$ **67.** $x = \pm 1/\sqrt{5}$

69. $k = 10$ or -10 **71.** $k = 16$

73. Yes **75.** No

77. $k = 4$

Excursion 1.2.A, page 27

1. $x = -6$ or 3 **3.** $x = 3/2$

5. $x = 2$ **7.** $x = 3/2$

9. $x = -5$ or 1 or -3 or -1

11. $x = 1$ or 4 or $\dfrac{5 + \sqrt{33}}{2}$ or $\dfrac{5 - \sqrt{33}}{2}$

13. $\text{CL} = .02 + \sqrt{.000882}$ or $.02 - \sqrt{.000882}$

Section 1.3, page 34

1. $A(-3, 3)$; $B(-1.5, 3)$; $C(-2.3, 0)$; $D(-1.5, -3)$; $E(0, 2)$; $F(0, 0)$; $G(2, 0)$; $H(3, 1)$; $I(3, -1)$

3. $P(-6, 3)$ **5.** $P(4, 2)$

7.

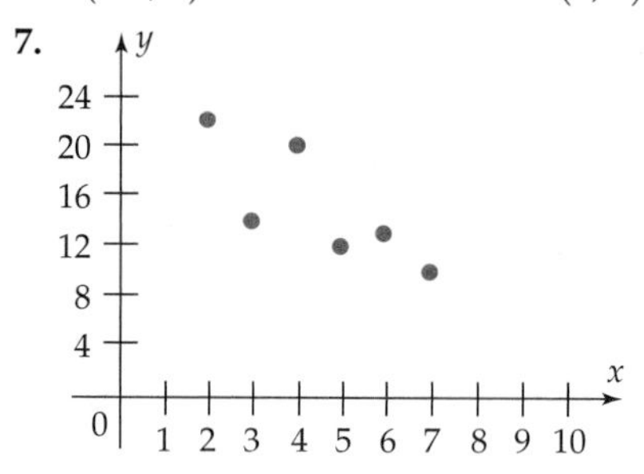

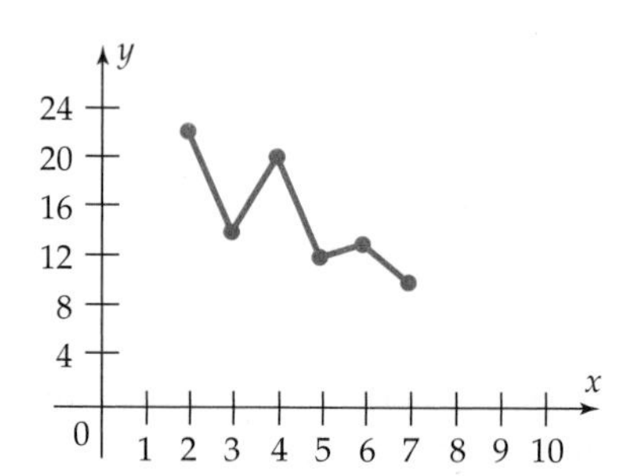

9. (a) About \$0.94 in 1987 and \$1.19 in 1995.
(b) About 27.66%
(c) In the first third of 1985 and from 1989 onward.

11. (a) Quadrant IV
(b) Quadrants III or IV

13. (a)

(−2, 3) (3, 2) (4, −1) (−5, −4)

(b)

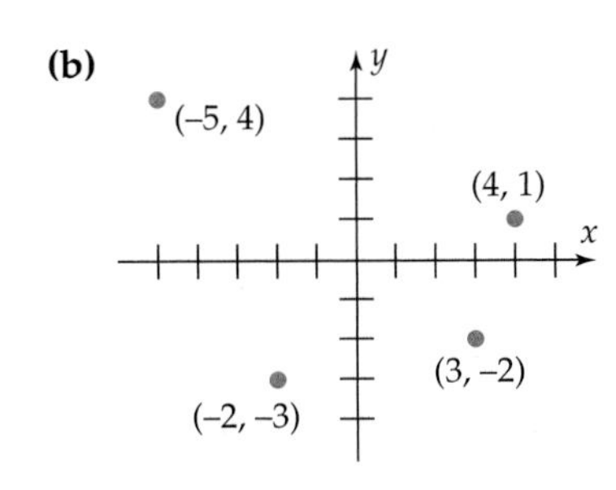

(c) They are mirror images of each other, with the x-axis being the mirror. In other words, they lie on the same vertical line, on opposite sides of the x-axis, the same distance from the axis.

15. $13; \left(-\frac{1}{2}, -1\right)$ **17.** $\sqrt{17}; \left(\frac{3}{2}, -3\right)$

19. $\sqrt{6 - 2\sqrt{6}} \approx 1.05; \left(\frac{\sqrt{2} + \sqrt{3}}{2}, \frac{3}{2}\right)$

21. $\sqrt{2}|a - b|; \left(\frac{a + b}{2}, \frac{a + b}{2}\right)$

23. (a)

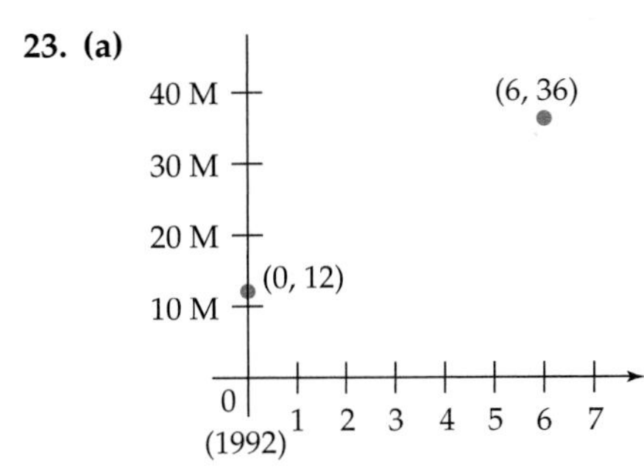

(b)

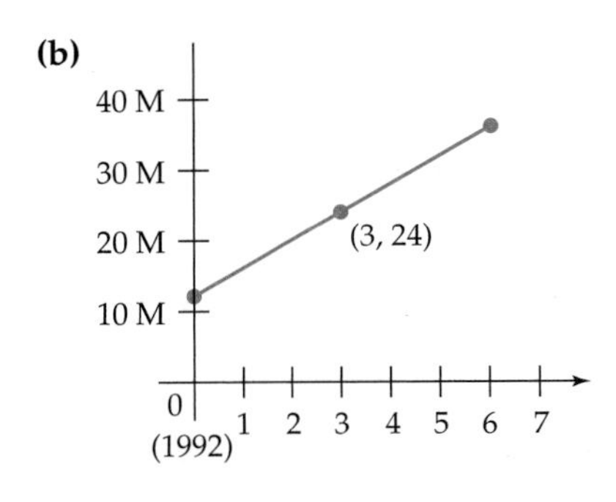

(c) In 1995, about 24 million personal computers were sold. We must assume that sales increased steadily.

25. Yes **27.** Yes **29.** No

31. $(x + 3)^2 + (y - 4)^2 = 4$ **33.** $x^2 + y^2 = 2$

35.

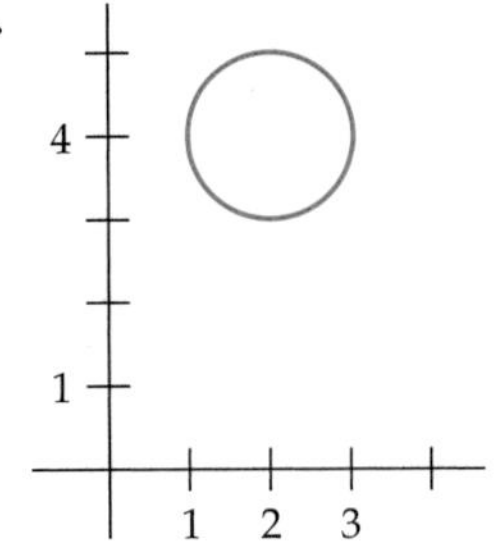

37.

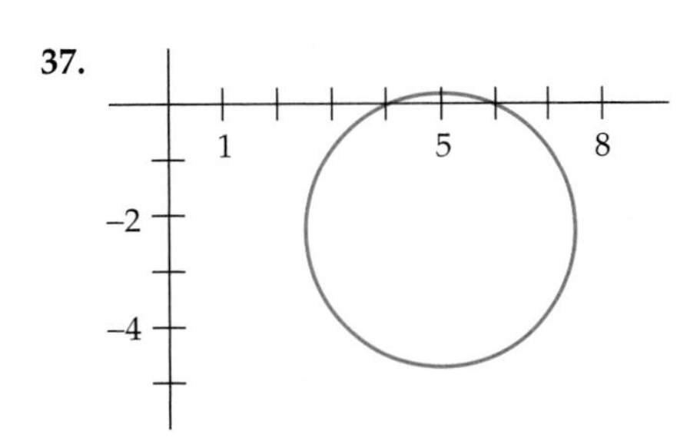

39. Center $(-4, 3)$, radius $2\sqrt{10}$

41. Center $(-3, 2)$, radius $2\sqrt{7}$

43. Center $(-12.5, -5)$, radius $\sqrt{169.25}$

45. Hypotenuse from $(1, 1)$ to $(2, -2)$ has length $\sqrt{10}$; other sides have lengths $\sqrt{2}$ and $\sqrt{8}$. Since $(\sqrt{2})^2 + (\sqrt{8})^2 = (\sqrt{10})^2$, this is a right triangle.

47. Hypotenuse from $(-2, 3)$ to $(3, -2)$ has length $\sqrt{50}$; other sides have lengths $\sqrt{5}$ and $\sqrt{45}$. Since $(\sqrt{5})^2 + (\sqrt{45})^2 = (\sqrt{50})^2$, this is a right triangle.

49. B **51.** C

53. $(x - 2)^2 + (y - 2)^2 = 8$

55. $(x - 1)^2 + (y - 2)^2 = 8$

57. $(x + 5)^2 + (y - 4)^2 = 16$

59. $x^2 + y^2 - 4x - 2y = 0$

61. $(-3, -4)$ and $(2, 1)$

63. Assume $k > d$. The other two vertices of one possible square are $(c + k - d, d), (c + k - d, k)$; those of another square are $(c - (k - d), d)$, $(c - (k - d), k)$; those of a third square are $\left(c + \frac{k - d}{2}, \frac{k + d}{2}\right), \left(c - \frac{k - d}{2}, \frac{k + d}{2}\right)$.

65. $(0, 0), (6, 0)$ **67.** $(3, -5 + \sqrt{11}), (3, -5 - \sqrt{11})$

69. $x = 6$

71. M has coordinates $(s/2, r/2)$ by the midpoint formula. Hence the distance from M to $(0, 0)$ is

$$\sqrt{\left(\frac{s}{2} - 0\right)^2 + \left(\frac{r}{2} - 0\right)^2} = \sqrt{\frac{s^2}{4} + \frac{r^2}{4}},$$

and the distance from M to $(0, r)$ is the same:

$$\sqrt{\left(\frac{s}{2} - 0\right)^2 + \left(\frac{r}{2} - r\right)^2} = \sqrt{\left(\frac{s}{2}\right)^2 + \left(-\frac{r}{2}\right)^2}$$
$$= \sqrt{\frac{s^2}{4} + \frac{r^2}{4}}$$

as is the distance from M to $(s, 0)$:

$$\sqrt{\left(\frac{s}{2} - s\right)^2 + \left(\frac{r}{2} - 0\right)^2} = \sqrt{\left(-\frac{s}{2}\right)^2 + \left(\frac{r}{s}\right)^2}$$
$$= \sqrt{\frac{s^2}{4} + \frac{r^2}{4}}.$$

73. Place one vertex of the rectangle at the origin, with one side on the positive x-axis and another on the positive y-axis. Let $(a, 0)$ be the coordinates of the vertex on the x-axis and $(0, b)$ the coordinates of the vertex on the y-axis. Then the fourth vertex has coordinates (a, b) (draw a picture!). One diagonal has endpoints $(0, b)$ and $(a, 0)$, so that its length is $\sqrt{(0 - a)^2 + (b - 0)^2} = \sqrt{a^2 + b^2}$. The other diagonal has endpoints $(0, 0)$ and (a, b) and hence has the same length: $\sqrt{(0 - a)^2 + (0 - b)^2} = \sqrt{a^2 + b^2}$.

75. The circle $(x - k)^2 + y^2 = k^2$ has center $(k, 0)$ and radius $|k|$ (the distance from $(k, 0)$ to $(0, 0)$). So the family consists of every circle that is tangent to the y-axis *and* has center on the x-axis.

77. The points are on opposite sides of the origin because one first coordinate is positive and one is negative. They are equidistant from the origin because the midpoint on the line segment joining them is

$$\left[\frac{c + (-c)}{2}, \frac{d + (-d)}{2}\right] = (0, 0).$$

Section 1.4, page 46

1. **(a)** C **(b)** B **(c)** B **(d)** D

3. Slope, 2; y-intercept, $x = 5$

5. Slope, $-\frac{3}{7}$; y-intercept, $x = \frac{-11}{7}$

7. Slope, $\frac{5}{2}$

9. Slope, 4

11. $t = 22$

13. $t = \frac{12}{5}$

15.

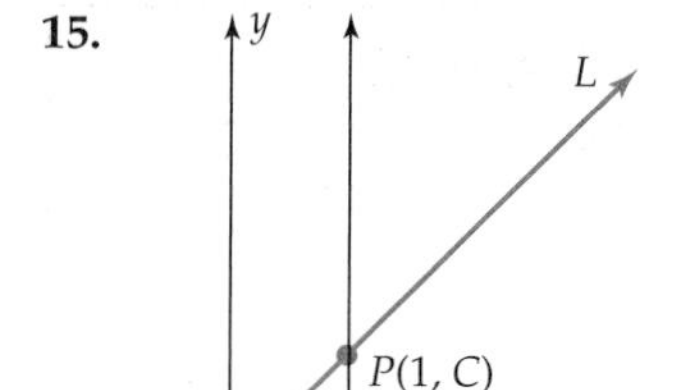

Slope of $L = \frac{C - 0}{1 - 0} = C$

17. $y = x + 2$

19. $y = -x + 8$

21. $y = -x - 5$

23. $y = \frac{-7}{3}x + \frac{34}{9}$

25. Perpendicular

27. Parallel

29. Parallel

31. Perpendicular

33. Yes

35. $y = 3x + 7$

37. $y = \frac{3}{2}x$

39. $y = x - 5$

41. $y = -x + 2$

43. $k = \frac{-11}{3}$

45. $y = \frac{-3}{4}x + \frac{25}{4}$

47. $y = -x/2 + 6$

49. Both have slope $-A/B$

51. **(a)** $y = 2x + 12$ **(b)** $x = 5$; $y = \$220{,}000$

53. **(a)** $y = 310x + 2125$
(b) $x = 5$; $y = 3675$ billion (1991)
$x = 14$; $y = 6465$ billion (2000)

55. \$375,000; \$60,000

57. **(a)** $y = 58.2x + 698$
(b) $x = 3$, $y = 872.6$ billion (1993);
$x = 9$, $y = 1221.8$ billion (1999)

59. **(a)** $c(x) = 25x + 180{,}000$ **(b)** $r(x) = 40x$
(c) $p(x) = 15x - 180{,}000$ **(d)** $x = 12{,}000$

61. **(a)** $y = 8.50x + 50{,}000$
(b) \$11, \$9.50, \$9 per hat

63. **(a)** $x = 10$ **(b)** $x = 30$

65. **(a)**

Women's Target Weight	
***x* inches over 5 ft**	**Weight (lb)**
0	100
1	105
2	110
3	115
6	130
12	160
15	175

Men's Target Weight	
x inches over 5 ft	Weight (lb)
0	106
1	112
2	118
3	124
6	142
12	178
15	196

(b) $y = 5x + 100$ **(c)** 145 lb
(d) $y = 6x + 106$; 160 lb

67. Let $y = mx + b$ and $y = mx + c$ be equations of lines with same slope m, and $b \neq c$. Suppose (x_1, y_1) is an arbitrary point lying on both lines. Then $y_1 = mx_1 + b$ and $y_1 = mx_1 + c$. So, $mx_1 + b = mx_1 + c$ and,
$b = c$, a contradiction.
Thus, the lines share no point in common so must be parallel.

69.

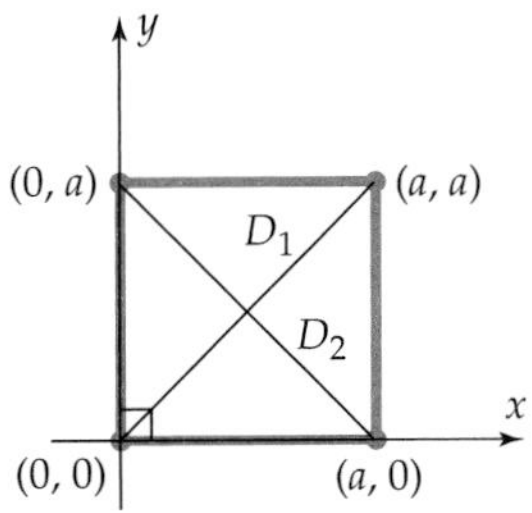

Equation of D_1: $y = x$, with slope 1
Equation of D_2: $y = -x + a$, with slope -1
(slope D_1)(slope D_2) = $(1)(-1) = -1$

Chapter 1 Review, page 52

1. (a) $>$ **(b)** $<$ **(c)** $<$ **(d)** $>$ **(e)** $=$

3. (a) 1.232×10^{16} **(b)** 7.89×10^{-11}

5. (a) $-10 < y < 0$ **(b)** $0 \leq x \leq 10$

7. (a) $|x + 7| < 3$ **(b)** $|y| > |x - 3|$

9. $x = 2$ or 8 **11.** $x = -1/2$ or $-11/2$

13. $-4 \leq x \leq 0$ **15. (a)** $7 - \pi$ **(b)** $\sqrt{23} - \sqrt{3}$

17. (a) $(-8, \infty)$ **(b)** $(-\infty, 5]$

19. 28/99 **21.** $x = 44/7$

23. No real solutions **25.** $z = \dfrac{-3 \pm 2\sqrt{11}}{5}$

27. 2 **29.** $x = 3$ or -3 or $\sqrt{2}$ or $-\sqrt{2}$

31. $x = -1$ or $5/3$

33. $\sqrt{58}$ **35.** $\sqrt{c^2 + d^2}$ **37.** $\left(d, \dfrac{c + 2d}{2}\right)$

39. (a) $\sqrt{17}$ **(b)** $(x - 2)^2 + (y + 3)^2 = 17$

41.

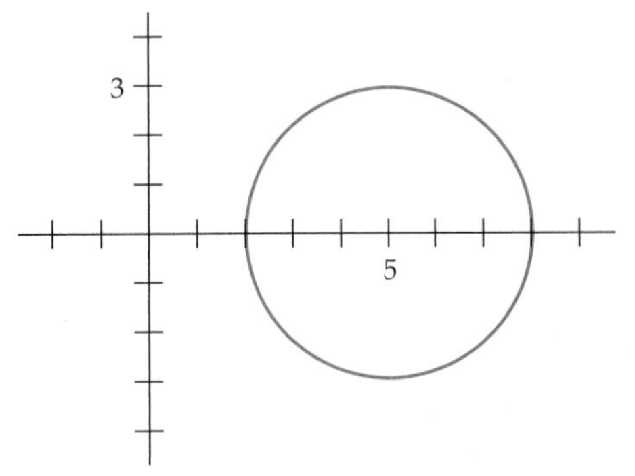

43. (b) and (d) **45.** (c)

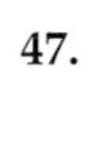

47.

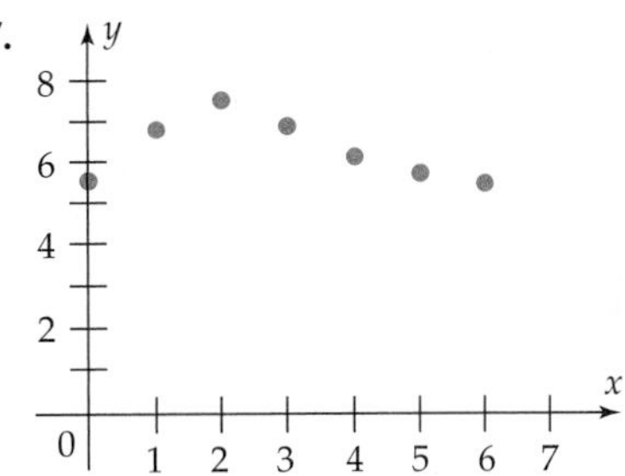

49. (a) 1 **(b)** 4/5

51. $y = 3x - 7$ **53.** $y = -2x + 1$

55. $x - 5y = -29$ **57.** 25,000 ft **59.** False

61. False **63.** False **65.** False **67.** (d)

69. (e) **71.** 5/3

73. (a) $y = .25x + 62.9$ **(b)** $x = 40$, $y = 72.9$ yrs.

75. (c) **77.** (d)

Chapter 2

Section 2.1, page 70

1. $P = (3, 5)$; $Q = (-6, 2)$

3. $P = (-12, 8)$; $Q = (-3, -8.5)$

5–10. Answers vary.

11.

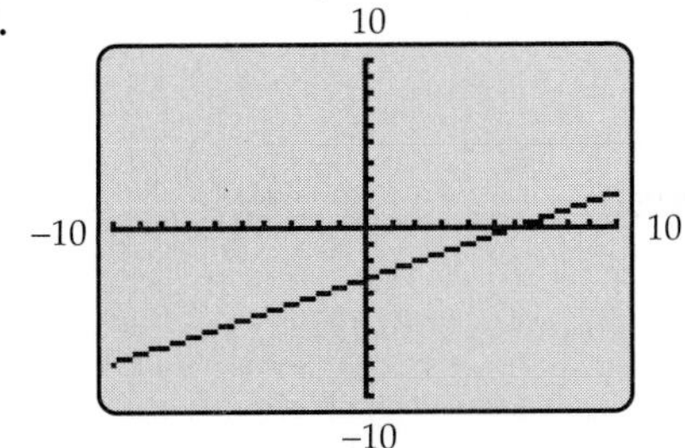

13.

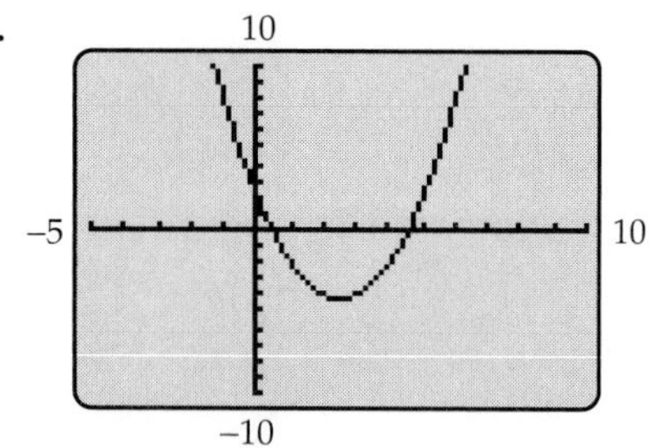

15.

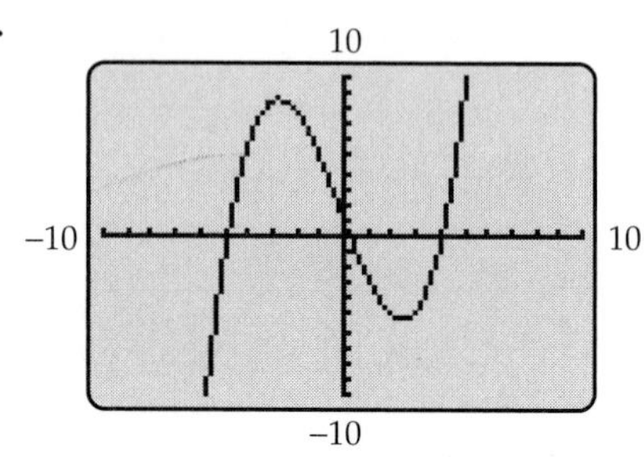

17. c **19.** d **21.** e **23.** (c)

25. $0 \le x \le 42$ and $0 \le y \le 4500$, with x-scl $= 5$, y-scl $= 500$. Both x (time) and y (concentration) must be nonnegative and this window shows the part of the graph where they are.

27. $0 \le x \le 4000$ and $0 \le y \le 100$, with x-scl $= 500$, y-scl $= 10$. Both x and y (numbers of barrels) must be nonnegative and this window shows the part of the graph where they are.

29. (a)

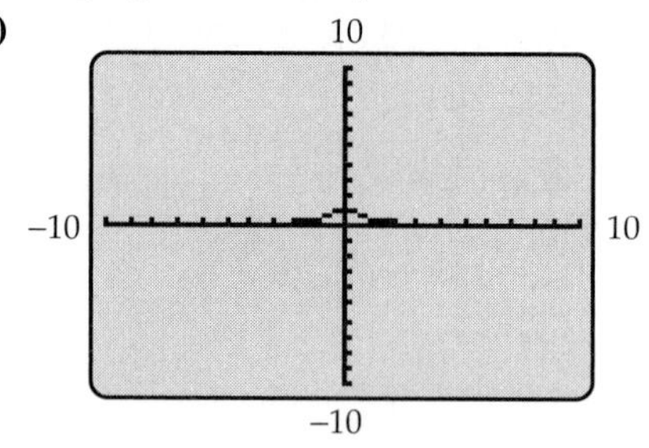

(c) $-10 \le x \le 10$ and $0 \le y \le 1$

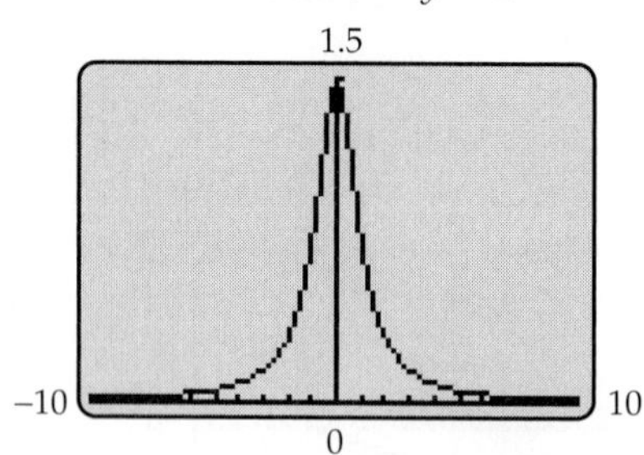

31. Maximum at about (.7922, 4.48490)
Minimum at about (4.2078, −3.4849)

33. 1961

35. Not the same

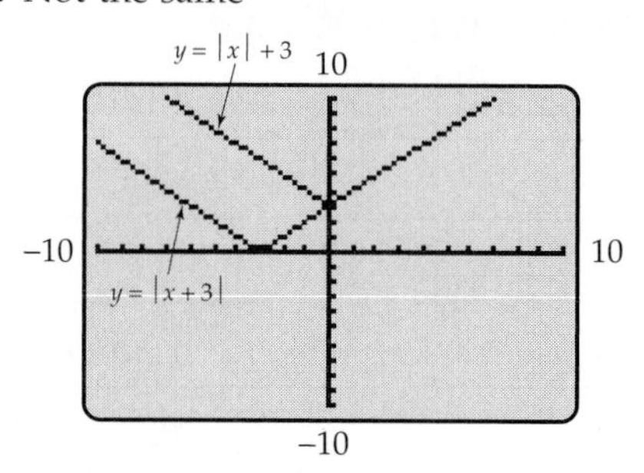

37. Same

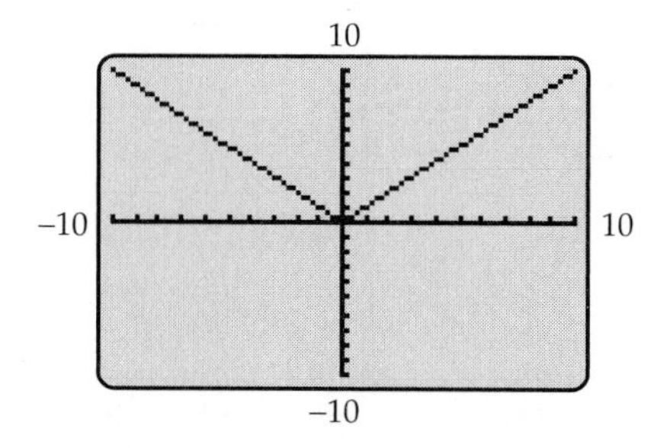

39. Not the same

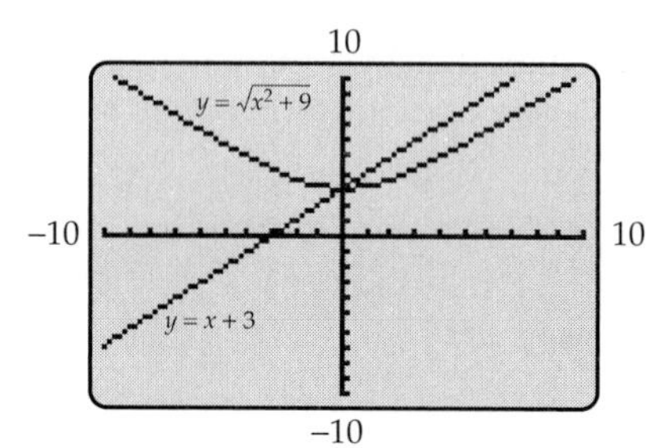

41. (a) Graphs appear identical.

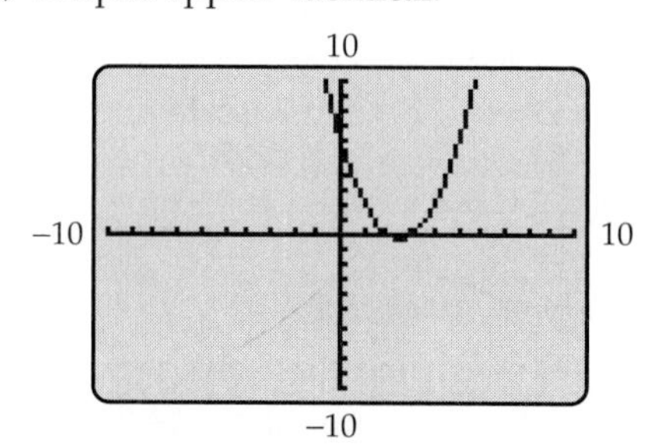

(b) Graphs are different.

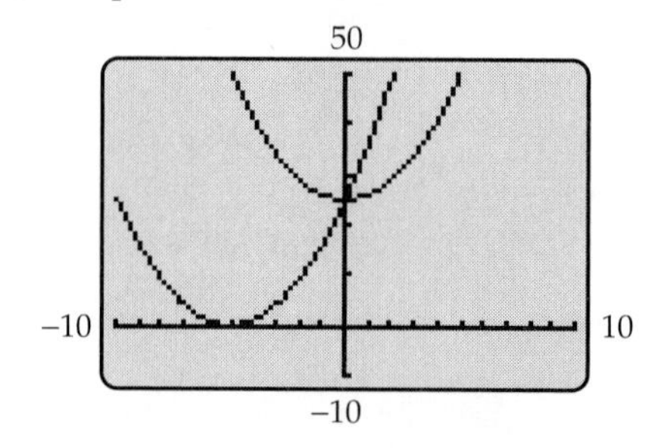

43. Possibly true

In Answers 45–49, the graphs shown were made on a TI-83. For wide-screen calculators such as TI-86/89, Sharp 9600, Casio 9850, and HP-38, the x-axis should be longer than this one to have a square window with the same y-axis as here.

45.

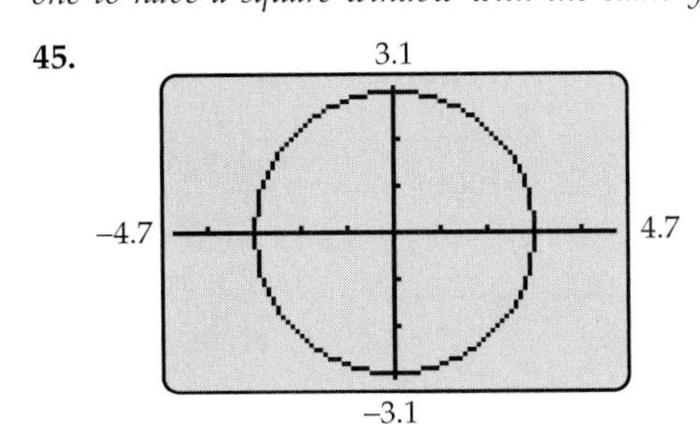

47. The two halves of the graph should be connected, but the ends of the two pieces are almost vertical so the calculator could not plot enough points near them to make the graph appear connected.

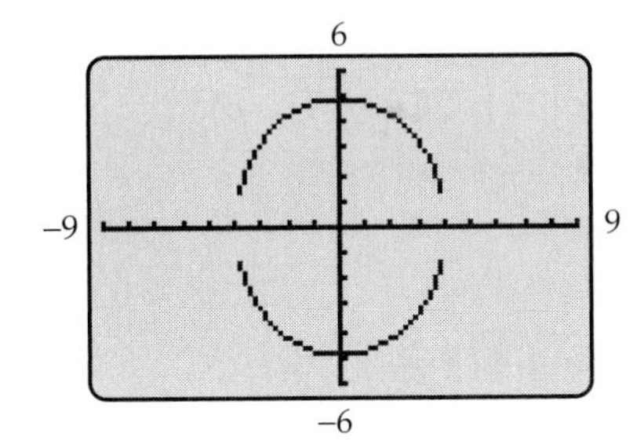

49.

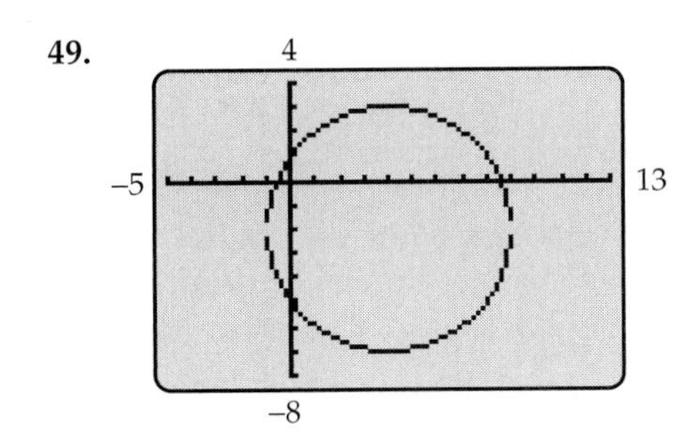

51.

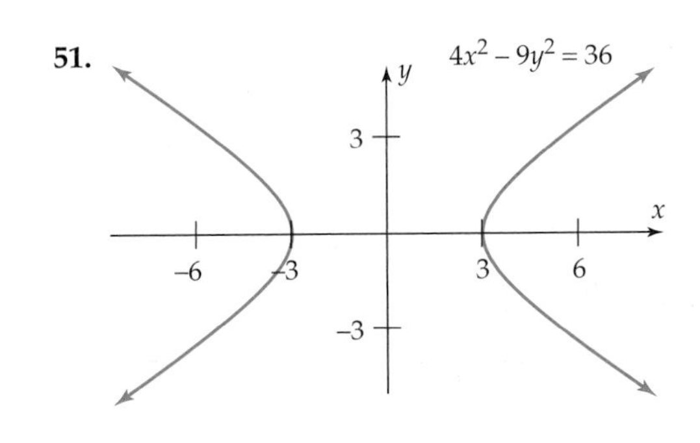

53.

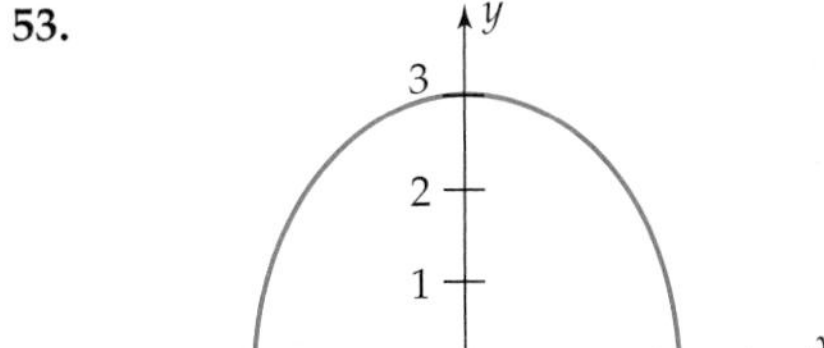
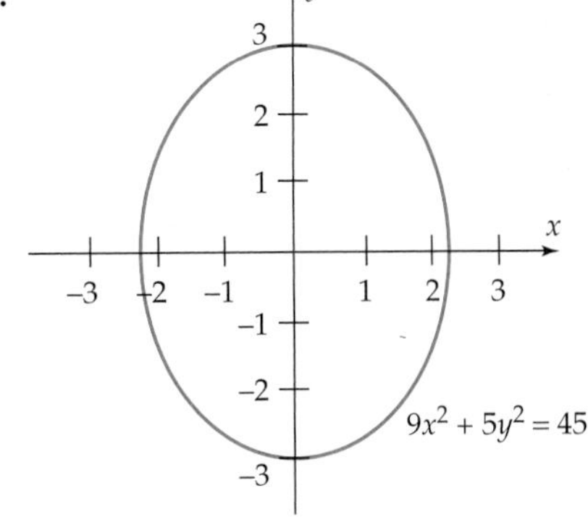

55. $-5 \le x \le 5$ and $-100 \le y \le 100$

57. $-10 \le x \le 10$ and $-2 \le y \le 20$ [Where is the right half of the graph?]

59. $-6 \le x \le 12$ and $-100 \le y \le 250$

61. All four graphs have the same shape. Graph (b) is graph (a) shifted 5 units vertically upward, graph (c) is graph (a) shifted 5 units vertically downward, and graph (d) is graph (a) shifted 2 units vertically downward.

63. Graph (b) is graph (a) stretched by a factor of 2 vertically away from the x-axis; similarly for graph (c), except the stretching factor is 3. Graph (d) is graph (a) shrunk vertically toward the x-axis by a factor of $\frac{1}{2}$.

65. The graphs are "mirror images" of each other, with the straight line $y = x$ being the mirror.

67. Same answer as Exercise 65

Section 2.2, page 79

1. 3 **3.** 3 **5.** 2

7. $x = -2.42645$ **9.** $x = 1.1640$ **11.** $x = 1.23725$

13. $x = 1.1921$ **15.** $x = -1.3794$ **17.** $x = -1.6005$

19. $x = 1.8608059$ **21.** $x = 2.1017$

23. $x = -1.7521$ **25.** $x = .9505$

27. $x = 0$ or 2.2074 **29.** $x = 2.3901454$

31. $x = -.6513878188$ or 1.151387819

33. $x = 7.033393042$ **35.** $x = 2/3$

37. $x = 1/12$ **39.** $x = \sqrt{3}$

41. $x = 1.4528$ **43.** $x = .0499$ or 1.9097

45. 2001 **47.** 1991

Section 2.3, page 88

1. The two numbers: x and y; their sum is 15: $x + y = 15$; the difference of their squares is 5: $x^2 - y^2 = 5$.

3.

English Language	Mathematical Language
Length of rectangle	x
Width of rectangle	y
Perimeter is 45	$x + y + x + y = 45$ or $2x + 2y = 45$
Area is 112.5	$xy = 112.5$

5. Let x be the old salary. Then the raise is 8% of x. Hence,

$$\text{old salary} + \text{raise} = \$1600$$
$$x + (8\% \text{ of } x) = 1600$$
$$x + .08x = 1600.$$

7. The circle has radius $r = 16/2 = 8$, so its area is $\pi r^2 = \pi \cdot 8^2 = 64\pi$. Let x be the amount by which the radius is to be reduced. Then $r = 8 - x$ and the new area is $\pi(8 - x)^2$, which must be 48π less than the original area, that is, $\pi(8 - x)^2 = 64\pi - 48\pi$, or equivalently, $\pi(8 - x)^2 = 16\pi$.

9. \$366.67 at 12% and \$733.33 at 6%

11. $2\frac{2}{3}$ qt

13. 65 mph

15. 38.25 and 44

17. Approximately 1.753 ft

19. Red Riding Hood, 54 mph; wolf, 48 mph

21. **(a)** Approximately 6.3 sec **(b)** Approximately 4.9 sec

23. **(a)** Approximately 4.4 sec **(b)** After 50 sec

25. 23 cm by 24 cm by 25 cm

27. $r = 4.658$

29. $x = 2.234$

31. 2.2 by 4.4 by 4 ft high

Section 2.4, page 94

1. $(-1, 8)$

3. $(1.1428562, -2.0625)$

5. $(-.3409, .0003222)$

7. **(a)** $(-1, 4)$ **(b)** $(-1, 4)$ and $(2, 4)$ **(c)** $(3, 20)$

9. 39.1487 by 39.1487 by 19.5743

11. **(a)** 4.4267 by 4.4267 inches. **(b)** 4 by 4 inches (sides must be no more than 12 inches).

13. **(a)** Approximately 206 **(b)** Approximately 269; approximately \$577

15. **(a)** 600 **(b)** 958

17. $x = 9.306$

19. Approximately (1.871, 1.5)

21. 12 times

Section 2.5, page 104

1. **(a)** $y = \frac{3}{4}x + \frac{5}{4}$ **(b)** Sum of squares $= 2\frac{3}{8}$ Model C still has least error.

3. **(a)** Slope = 1.05640540541 **(b)** $y = 1.05640540541x + 21.0778918918$ **(c)** Line described in (b) predicts a higher number of workers.

5. Positive

7. Very little

9. **(a)** Linear **(b)** Positive

11. **(a)** Nonlinear

13. **(a)** Linear **(b)** Negative

15. **(a)**

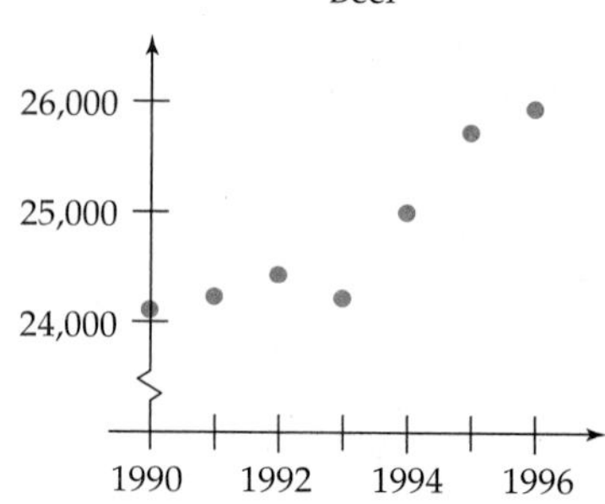

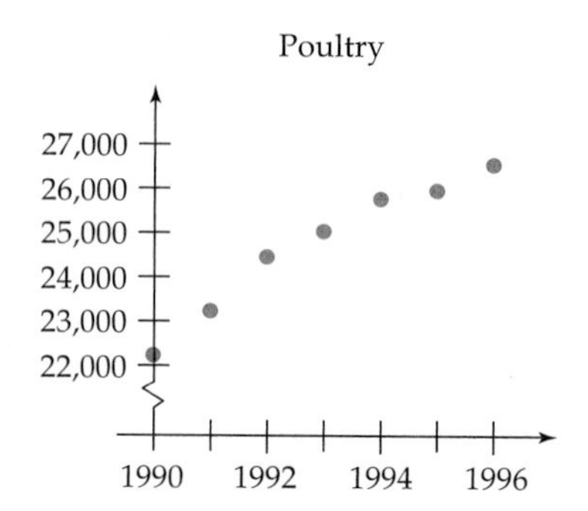

(b) y_1, poultry; y_2, beef

17. **(a)** $y = -3.866929638x + 542.7452878$ **(b)** 2090

19. **(a)** $y = 4x + 18$ **(b)** Slope indicates a positive correlation between time and percentage. y-intercept shows that regression line passes through actual plotted values. **(c)** 100%

21. **(a)**

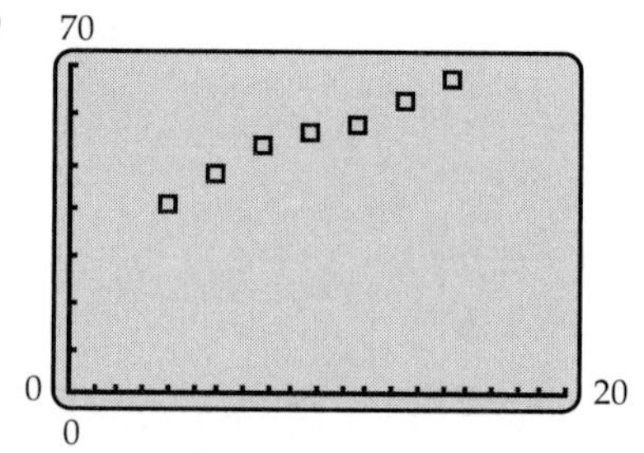

(b) $y = 1.714285714x + 37.42857143$ **(c)** About 60, about 72

23. (a)

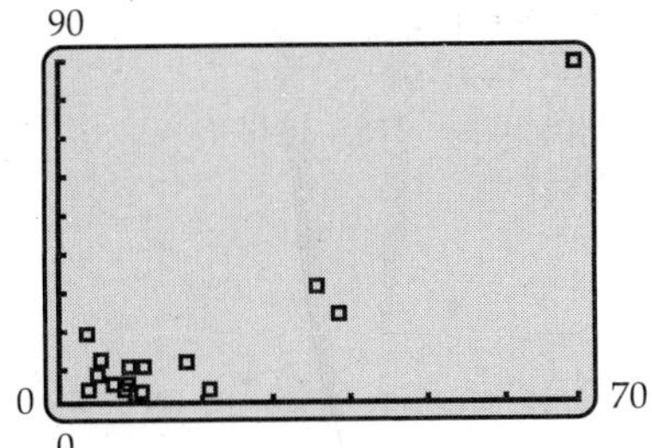

(b) $y = 1.089256088x - 1.413259148$

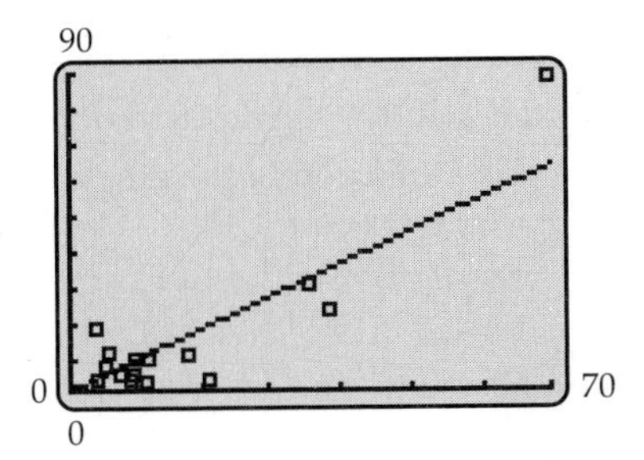

(c) United States, Russia, China
(d) The country uses more energy than it produces.
(e) The country produces more energy than it uses.
(f) United States, Russia, China, Saudi Arabia, Canada, Japan

Chapter 2 Review, page 110

1. (a) a, d
(b) b, c do not show peaks and valleys near the origin; e crowds the graph onto the y-axis.
(c) a or $-4 \le x \le 6$ and $-10 \le y \le 10$

3. (a) None of them
(b) a, b, d do not show any peaks or valleys; c does not show the valleys; e shows only one point on the graph (which can't be distinguished because it's on the y-axis).
(c) $-7 \le x \le 11$ and $-1000 \le y \le 500$

5. (a) b, c
(b) a shows no peaks or valleys; d is too crowded horizontally; e shows only one point.
(c) $-10 \le x \le 10$ and $-150 \le y \le 150$

7. As x moves left or right from the origin, x^2 grows larger and larger, and hence so does $x^2 - 10$. Therefore, the graph always rises as it moves away from the y-axis and is complete.

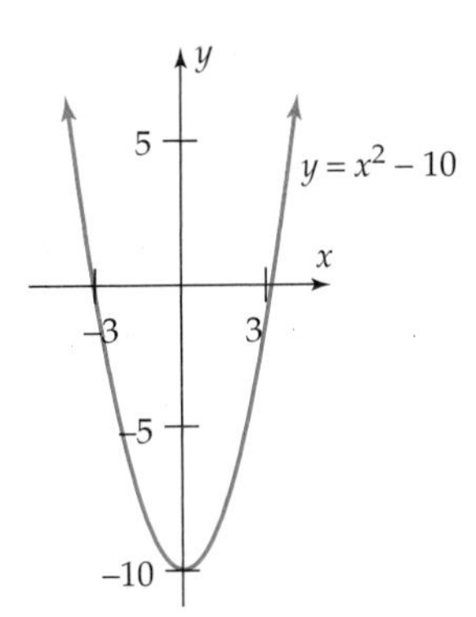

9. Note that y is defined only when $x \ge 5$ (why?). Also, $x - 5$ grows larger as x gets larger (that is, as you move to the right) and hence the same is true of $\sqrt{x-5}$. Therefore, the graph always rises as you move to the right and is complete.

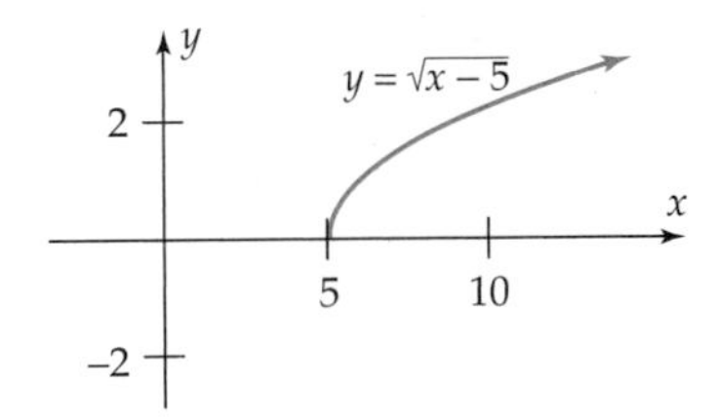

11.

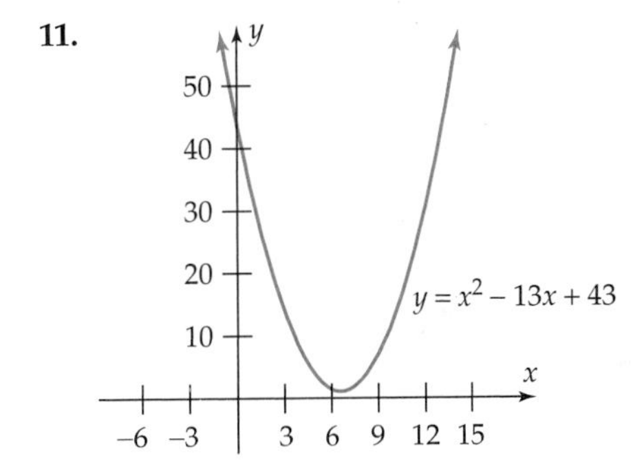

13.

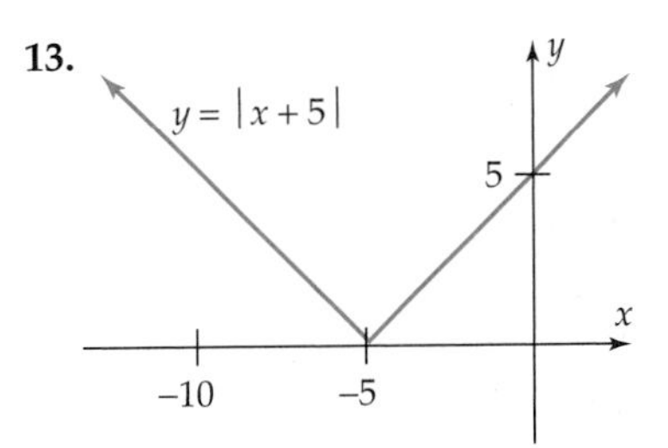

15. $x = 2.7644$ **17.** $x = 3.2678$
19. $x = -3.2843$ **21.** $x = 1.6511$

23. gold, $\frac{3}{11}$ oz; silver, $\frac{8}{11}$ oz

25. $2\frac{2}{9}$ hrs **27.** 9.6 ft

29. 4 ft **31.** 25

33. 20 yd by 30 yd (interior fence is 20 yd)

35. $x = \sqrt{3} \approx 1.732$

37. (a)

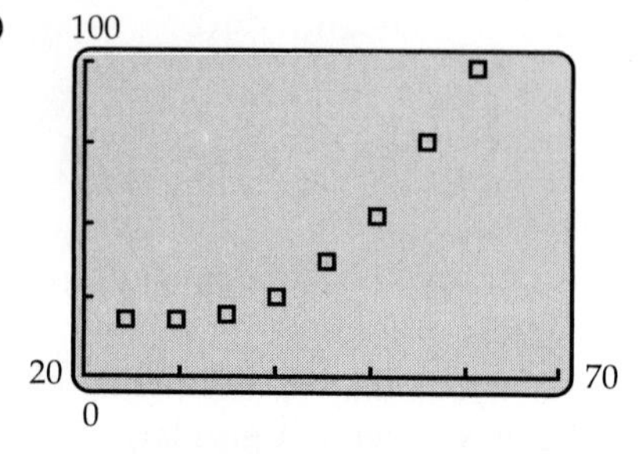

(b) Nonlinear

39. (a)

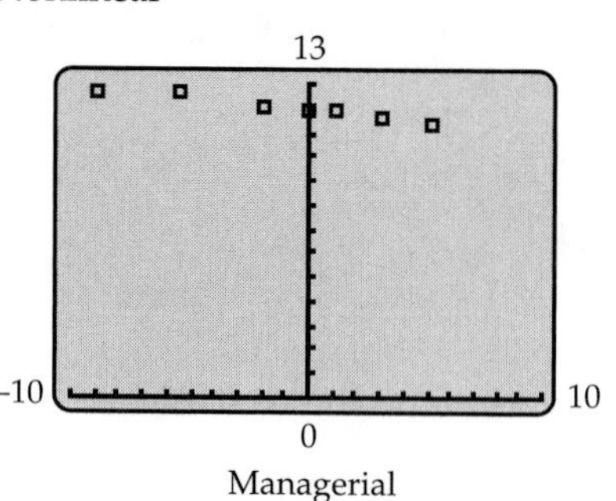

Managerial

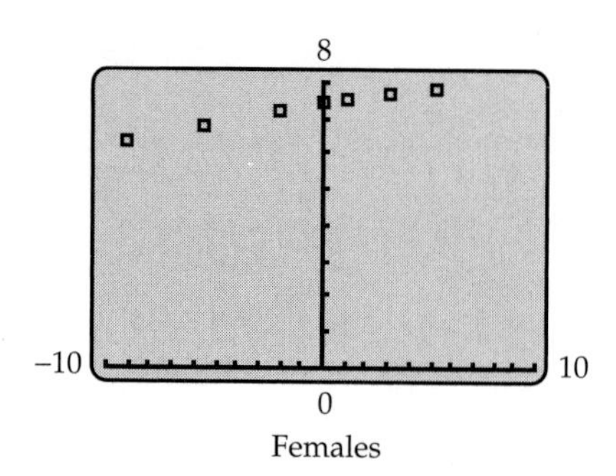

Females

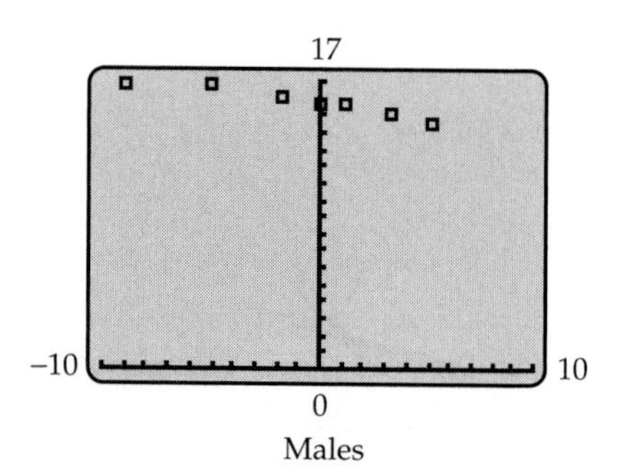

Males

(b) Managerial, y_2 (negative slope, y-intercept at 11.74); Female, y_1 (positive slope, y-intercept at 7.34); Male, y_3 (negative slope, y-intercept at 15.48)

41. (a) $y = 0.3055833333x + 6.912$
(b) \$10.27, \$13.33, \$0.05

43. (a) $y = 6.274090909x + 47.46409091$
(b) 141.58; 185.49

Chapter 3

Section 3.1, page 124

1. Yes. Each input produces only one output.

3. No. The value −5 produces two outputs.

5. 6 **7.** −2 **9.** −17

11. y is a function of x. **13.** x is a function of y.

15. y is a function of x and x is a function of y.

17. Neither x nor y is a function of the other.

19.

X	Y1
-2	-2
-1.5	-3.25
-1	-4
-.5	-4.25
0	-4
.5	-3.25
1	-2

X= -2

X	Y1
1.5	-.25
2	2
2.5	4.75
3	8
3.5	11.75
4	16
4.5	20.75

X=4.5

21.

X	Y1
-2	0
-1.2	1.6
-.04	1.9996
.04	1.9996
1.2	1.6
2	0

X=

23. (500, 0); (1509, 0); (3754, 35.08); (6783, 119.15); (12500, 405); (55342, 2547.10)

25. Each input (income) yields only one output (tax).

27. Postage is a function of weight since each weight determines one and only one postage amount. But weight is *not* a function of postage since a given postage amount may apply to several different weights. For instance, *all* letters under 1 oz use just one first-class stamp.

29. (a) $A = \pi r^2$ **(b)** $A = \frac{1}{4}\pi d^2$

31. $V = 4x^3$

33. $D = 400 - 16t^2$; range $= [0, 400]$

35. Theoretically, neither height nor weight is a function of the other. Each, as input, yields various outputs.

37. Domain: $[-3, 3]$; range: $[-4, 3]$

39. 2 is output of 1/2; 0 of 5/2; and -3 of $-5/2$.

41. -1 is output of -2; 3 of 0; 2 of 1; -1 of 2.5; and 1 of -1.5.

43. 1 is output of -2; -3 of -1; -1 of 0; .5 (approximately) of 1/2; and 1.5 of 1.

45. (a) All positive numbers that can be entered in your calculator
(b) All numbers between -1 and 1 (inclusive) that can be displayed on your calculator

47. (a) The integer part of a positive number c is the integer that is closest to the number and less than or equal to the number, that is, $[c]$.
(b) All negative integers: $-1, -2, -3, \ldots$
(c) All negative numbers that are not integers

Section 3.2, page 133

1. (a) $-4/5$ (b) $-3/4$ (c) $-2/5$ (d) $-1/8$ (e) 0

3. $\sqrt{3} + 1$

5. $\sqrt{\sqrt{2} + 3} - \sqrt{2} + 1$

7. 4

9. 34/3

11. 59/12

13. $(a + k)^2 + \dfrac{1}{a + k} + 2$

15. $(2 - x)^2 + \dfrac{1}{2 - x} + 2 = 6 - 4x + x^2 + \dfrac{1}{2 - x}$

17. 8

19. -1

21. $(s + 1)^2 - 1 = s^2 + 2s$

23. $t^2 - 1$

	$f(r)$	$f(r) - f(x)$	$\dfrac{f(r) - f(x)}{r - x}$
25.	r	$r - x$	1
27.	$3r + 7$	$3(r - x)$	3
29.	$r - r^2$	$r - r^2 - x + x^2$	$1 - r - x$
31.	$\sqrt{r}$	$\sqrt{r} - \sqrt{x}$	$\dfrac{1}{\sqrt{r} + \sqrt{x}}$

33. 1

35. 3

37. $-2x - h + 1$

39. $\dfrac{1}{\sqrt{x + h} + \sqrt{x}}$

41. (a) $[-3, 4]$ (b) $[-2, 3]$ (c) -2 (d) .5 (e) 1 (f) -1

43. (a) $(-\infty, 20]$ (b) 3 (c) -1 (d) 1 (e) 2

45. (iii) or (v)

47. All real numbers

49. All real numbers

51. All nonnegative real numbers

53. All nonzero real numbers

55. All real numbers

57. All real numbers except -2 and 3

59. [6, 12]

61. Many possible answers, including $f(x) = x^2$ and $g(x) = |x|$

63. $f(x) = \sqrt{2x - 5}$

65. $f(x) = \dfrac{x^3 + 6}{5}$

67. (a) $f(x) = 5x + 200$; $g(x) = 10x$
(b) 300 and 200; 375 and 350; 450 and 500
(c) 45

69. $d(t) = \begin{cases} 55t \text{ if } 0 \le t \le 2 \\ 110 + 45(t - 2) \text{ if } t > 2 \end{cases}$, where t = time (in hours) and $d(t)$ = distance (in miles)

71. (a) $p(x) = \begin{cases} 12 \text{ for } 1 \le x \le 10 \\ 12 - .25(x - 10) \text{ for } x > 10 \end{cases}$, where $p(x)$ is the price per copy when x copies are purchased
(b) $T(x) = \begin{cases} 12x \text{ for } 1 \le x \le 10 \\ 14.5x - .25x^2 \text{ for } x \ge 10 \end{cases}$, where $T(x)$ is the total cost of x copies.

73. $C(x) = 5.75x + \dfrac{45{,}000}{x}$

75. (a) $y = 677.4571429x + 8884.942857$
(b) 10,917; 12,272; 13,627
All are within \$60 of the actual figure.
(c) 14,982

Section 3.3, page 145

1.

3.

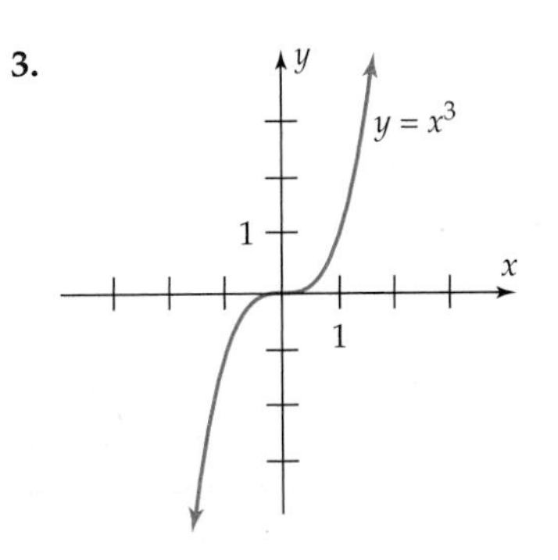

5.

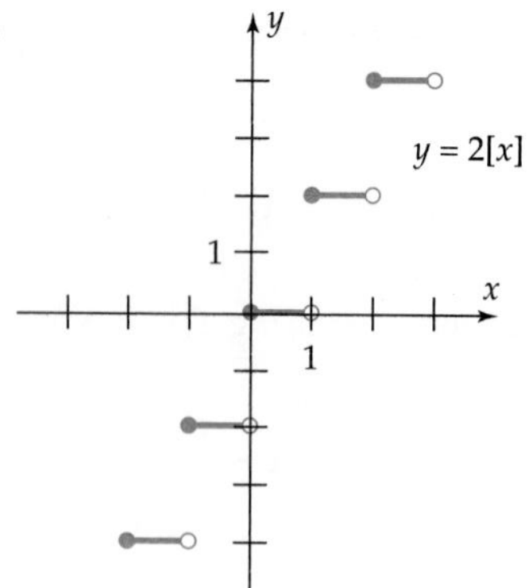

7.

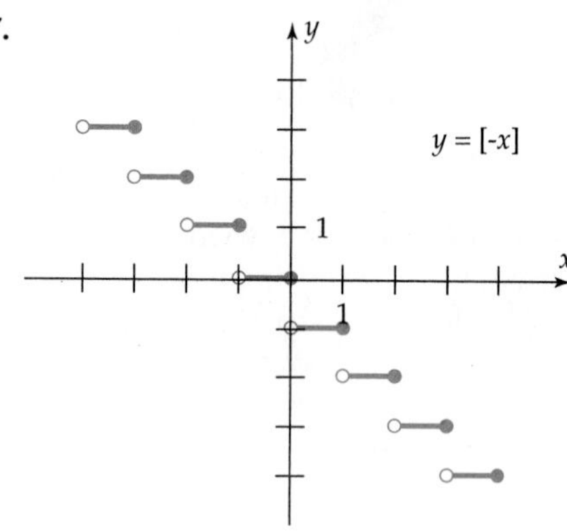

9.

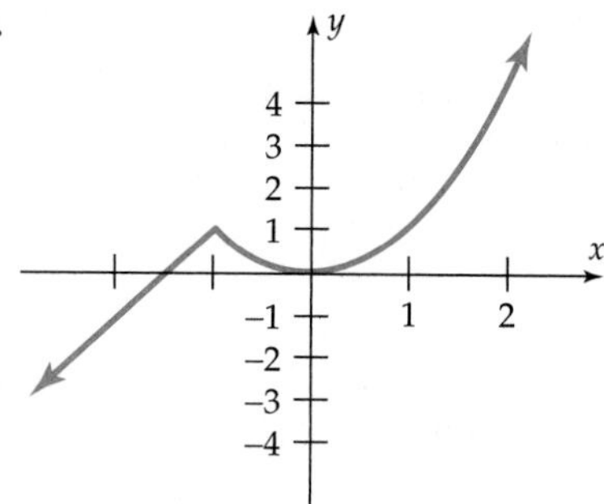

11.

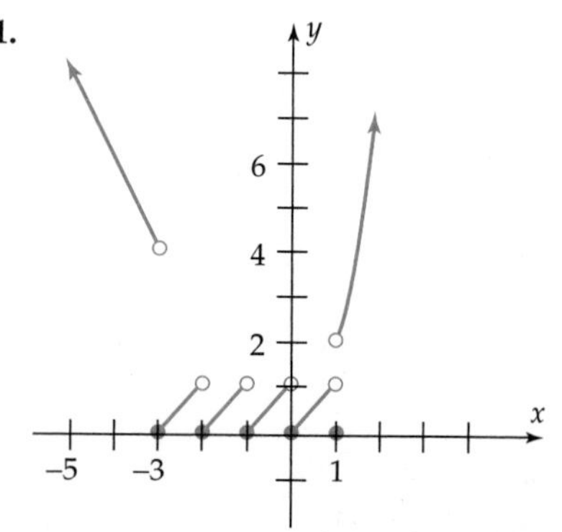

13. **(a)** Several possibilities, including

$$p(x) = \begin{cases} [x] \text{ if } x \text{ is an integer} \\ [x] + 1 \text{ if } x \text{ is not an integer} \end{cases}$$

or $p(x) = -[-x]$, with $x > 0$ in all cases, where x is the weight in ounces

(b) Graph for $0 < x \le 4$:

(c)

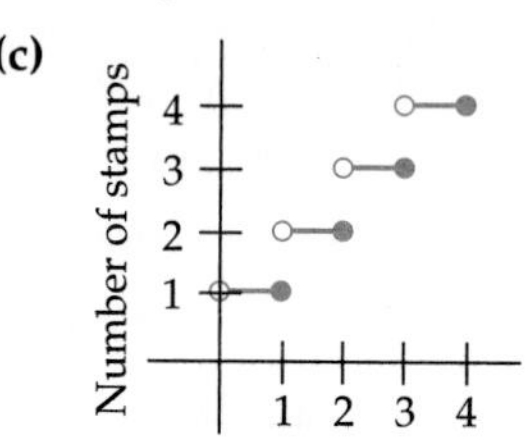

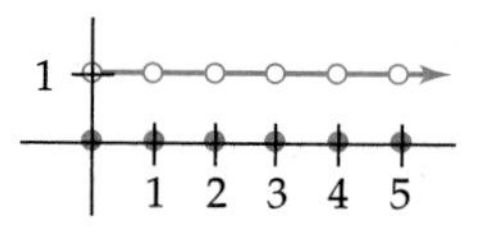

15. **(a)** $f(x) = \begin{cases} x + 2 & x \ge 0 \\ -x + 2 & x < 0 \end{cases}$

(b)

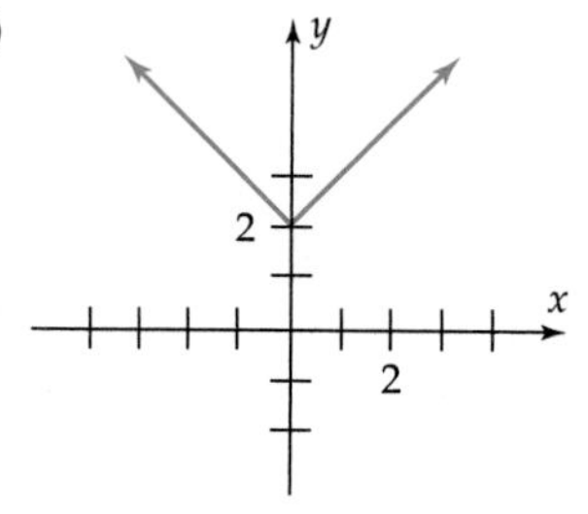

17. **(a)** $h(x) = \begin{cases} \frac{x}{2} - 2 & x \ge 0 \\ -\frac{x}{2} - 2 & x < 0 \end{cases}$

(b)

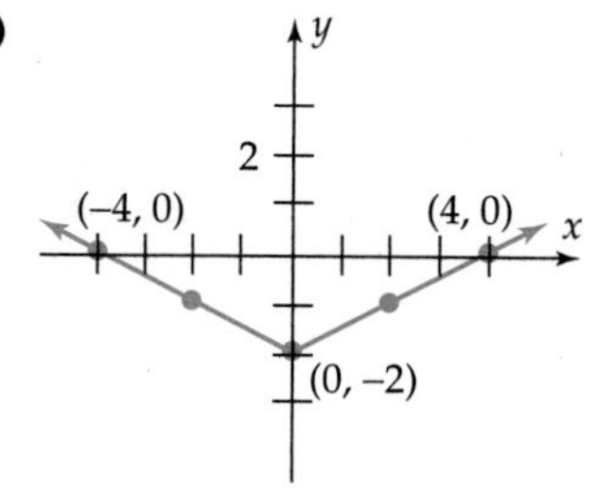

19. **(a)** $f(x) = \begin{cases} x - 5 & x \ge 5 \\ 5 - x & x < 5 \end{cases}$

(b)

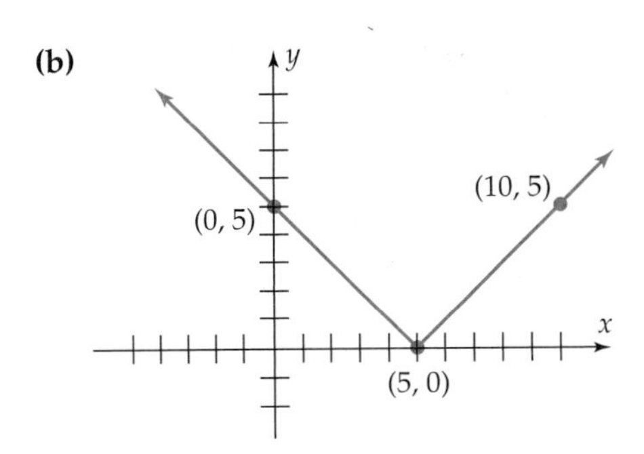

21. Minimum at $x = .57735$; maximum at $x = -.57735$

23. Minimum at $x = -1$; maximum at $x = 1$

25. Minimum at $x = 0.7633$; maximum at $x = 0.4367$

27. **(a)** $A(x) = 50x - x^2$

    (b) To maximize area each side should be 25 in. long.

29. **(a)** $SA(x) = 2x^2 + 3468/x$

    (b) Base is approximately 9.5354 in. × 9.5354 in.; height is same (that is, 9.5354 in.).

31. Increasing on $(-2.5, 0)$ and $(1.7, 4)$; decreasing on $(-6, -2.5)$ and $(0, 1.7)$

33. Constant on $(-\infty, -1]$ and $[1, \infty)$; decreasing on $(-1, 1)$.

35. Increasing on $[-5.7936, .46028]$; decreasing on $(-\infty, -5.7936]$ and $[.46028, \infty)$

37. Increasing on $(0, 0.867)$ and $(2.883, \infty)$; decreasing on $(-\infty, 0)$ and $(0.867, 2.883)$

39. **(a)** iv **(b)** i **(c)** v **(d)** iii **(e)** ii

41. Many correct answers, including:

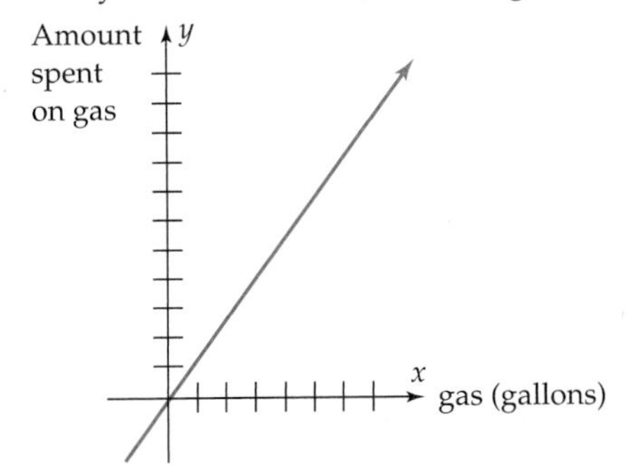

Domain ≈ $[0, 150]$
Range ≈ $[0, 300]$

43. Many correct answers, including

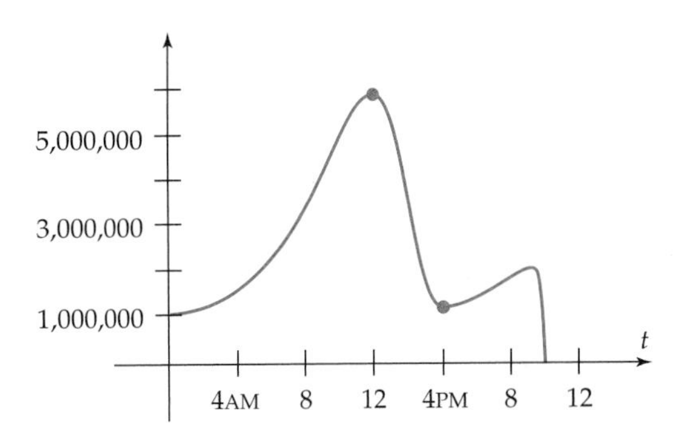

45.

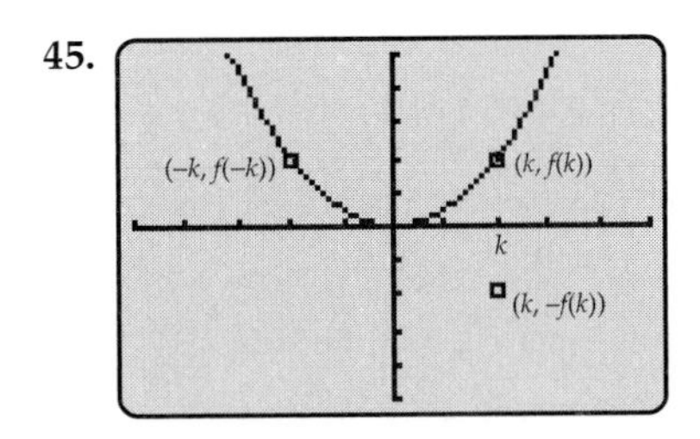

47. Domain: all real numbers; range: all real numbers

49. Domain: all real numbers x such that $x \le -2$ or $x \ge 2$; range: all nonnegative real numbers

51. $[-3, 5]$ 53. 4 55. 3.5

57. 4.5 59. 4 61. 1, 5

63. $[-8, 9]$ 65. $-7, -3, 3, 6.8$ (approximately)

67. 1, 3, 5, and others 69. 1 and -9

71. Domain $f = [-6, 7]$; domain $g = [-8, 9]$

73. Approximately -1.5 and $-.2$ 75. $x = 3$

77. $[-2, -1]$ and $[3, 7]$

79. Approximately −\$13,000 (that is, a loss of \$13,000)

81. **(a)** Approximately 12,300

    (b) Approximately 28,500

83. Many correct answers, including

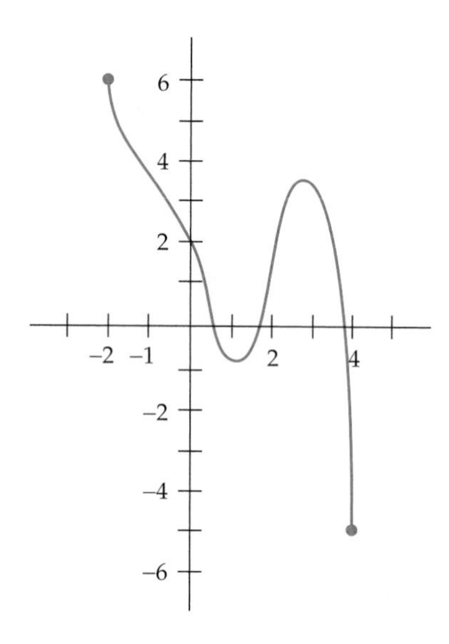

85. $-15 \le x \le 30$ and $-10 \le y \le 10$ $(-10 \le t \le 10)$

87. $-10 \le x \le 6$ and $-7 \le y \le 7$ $(-7 \le t \le 7)$

89. $-15 \le x \le 0$ and $0 \le y \le 4$ $(0 \le t \le 4)$

91. $-5 \le x \le 45$ and $-65 \le y \le 65$

93. Entire graph: $-2 \le x \le 32$ and $-10 \le y \le 75$; near the origin: $-2 \le x \le 5$ and $-10 \le y \le 10$

95. Entire graph: $-16 \le x \le 2$ and $-62 \le y \le 60$; near the origin: $-2 \le x \le 2$ and $-4 \le y \le 2$

97. At 15 min she stopped for 5 min, then continued running steadily for 10 min. At 30 min she turned back and jogged home without stopping, arriving home 55 min after her start.

99. 7 times

101.

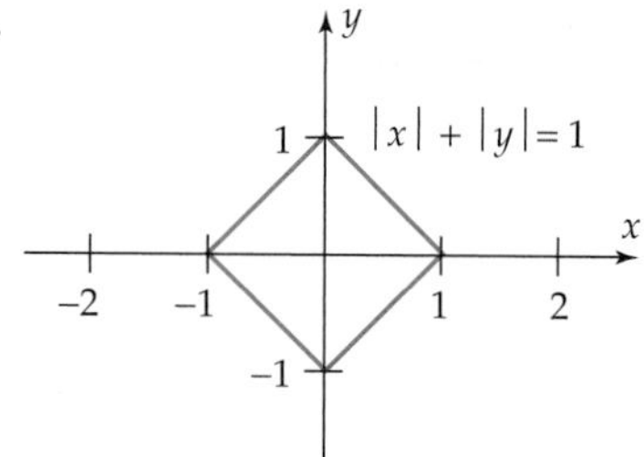

Section 3.4, page 157

1. Viewing window: $-10 \le x \le 10$ and $-36 \le y \le 42$

3. Viewing window: $-13 \le x \le 12$ and $-2 \le y \le 14$

5.

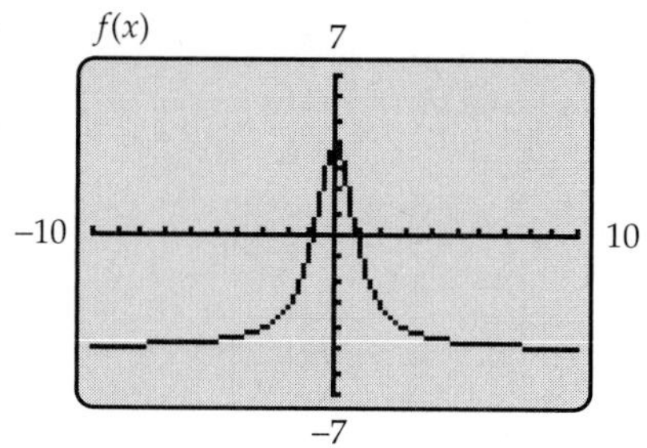

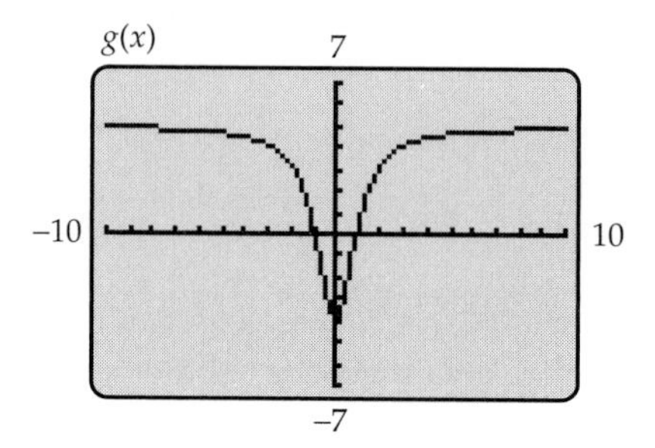

7. Shift the graph of f horizontally 3 units to the right; then shift it 2 units vertically upward.

9. Reflect the graph of f in the x-axis; shrink it toward the x-axis by a factor of $1/2$; shift it vertically 6 units downward.

11. Shift the graph of f horizontally 2 units to the right; stretch it away from the x-axis by a factor of 2; then reflect it in the x-axis.

13. $g(x) = f(x+5) + 4 = (x+5)^2 + 2 + 4 = (x+5)^2 + 6$

15. $g(x) = 2f(x-6) - 3 = 2\sqrt{x-6} - 3$

17. $g(x) = 2f(x-2) + 2$
$= 2[(x-2)^2 + 3(x-2) + 1] + 2$
$= 2(x-2)^2 + 6(x-2) + 4$

19.

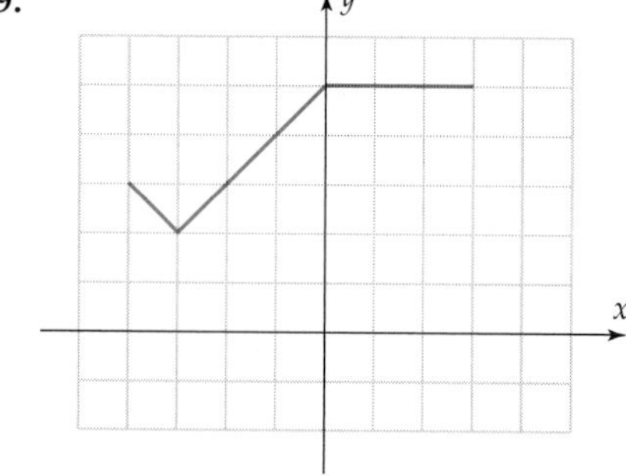

21.

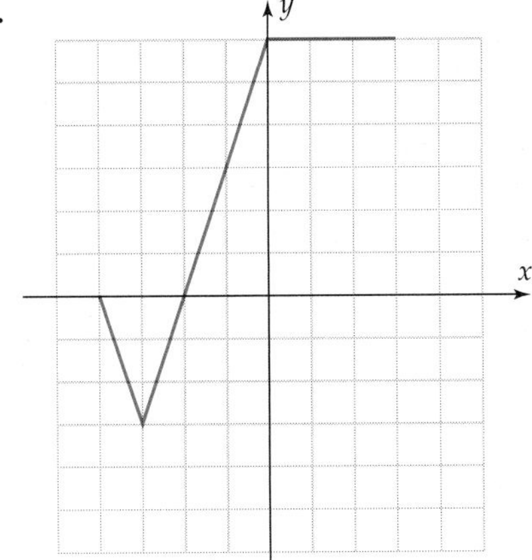

23.

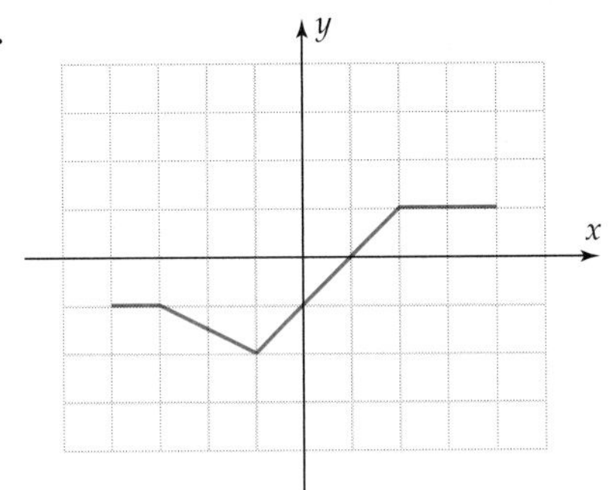

25.

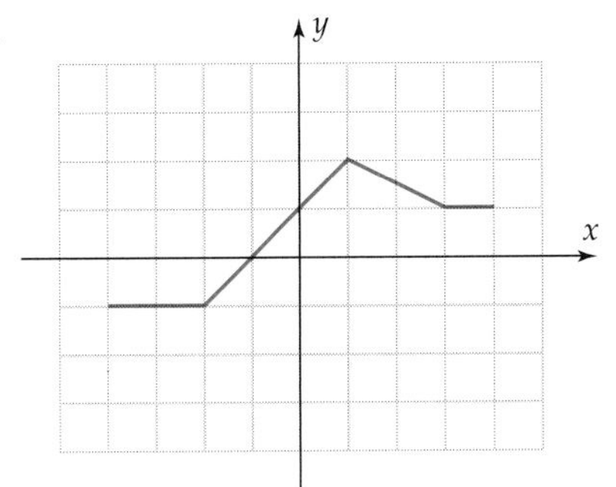

27.

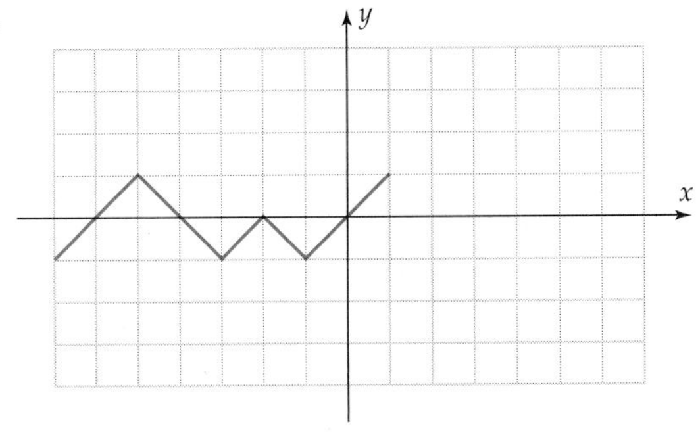

29.

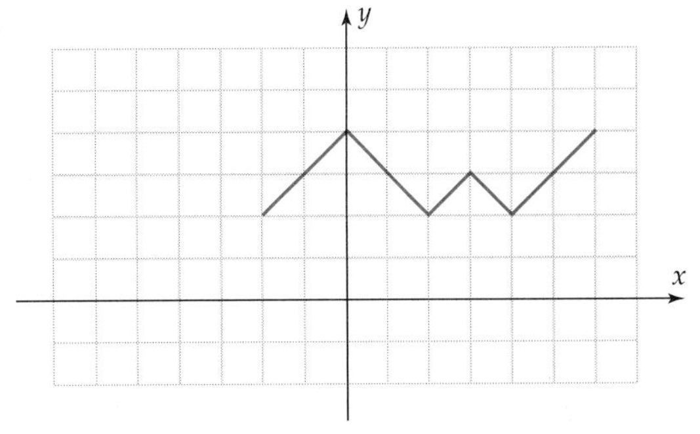

31.

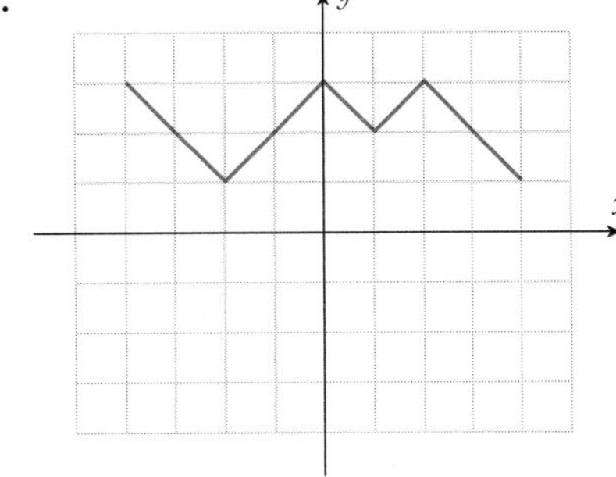

33.

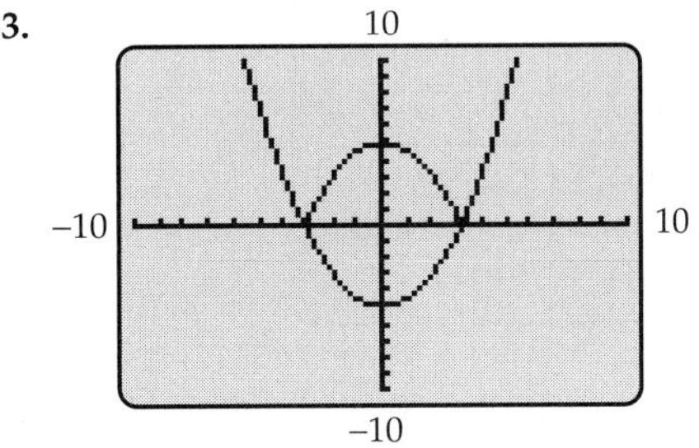

35.

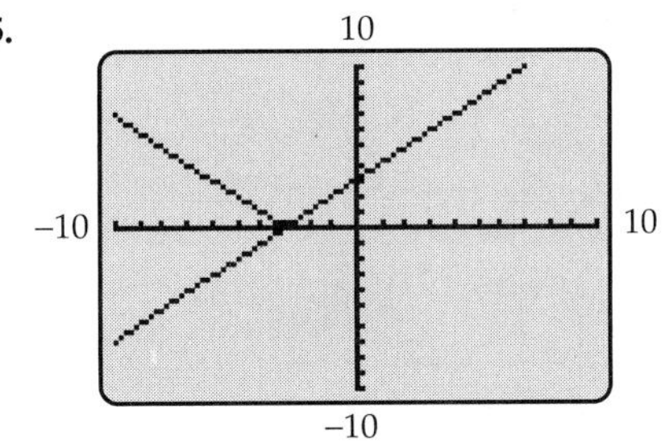

37.

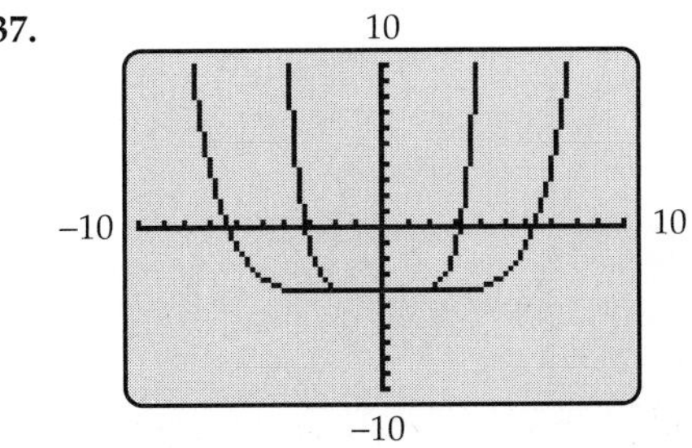

39.

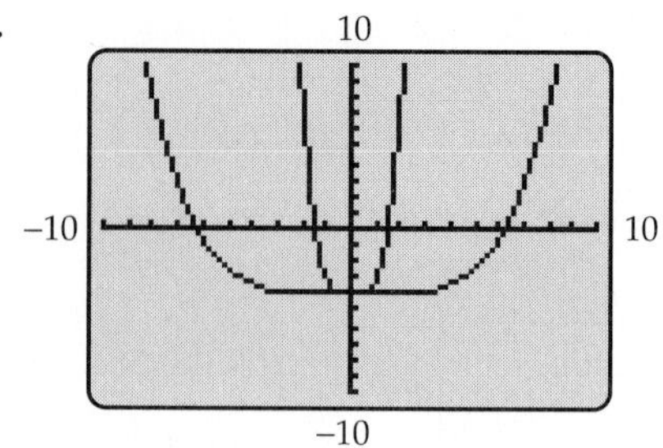

41.

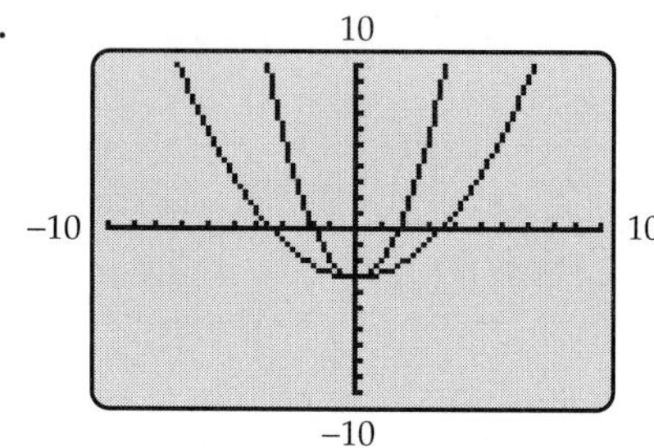

43.

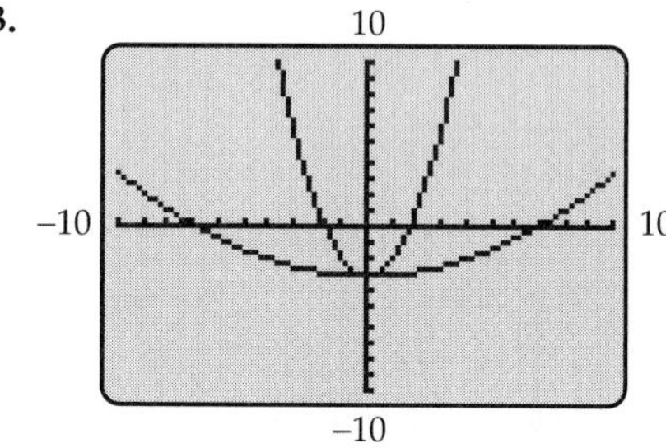

45.

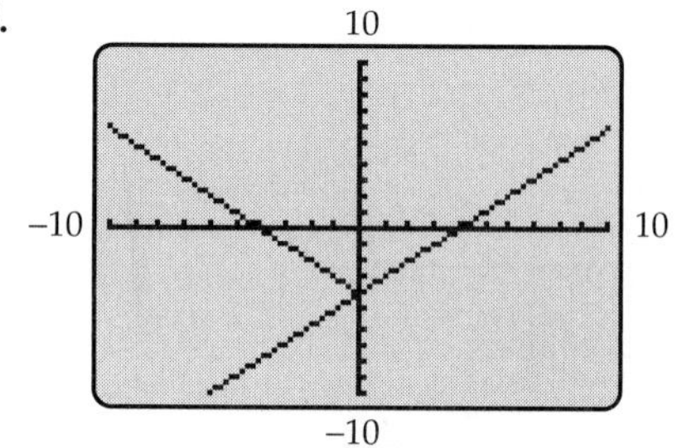

47.

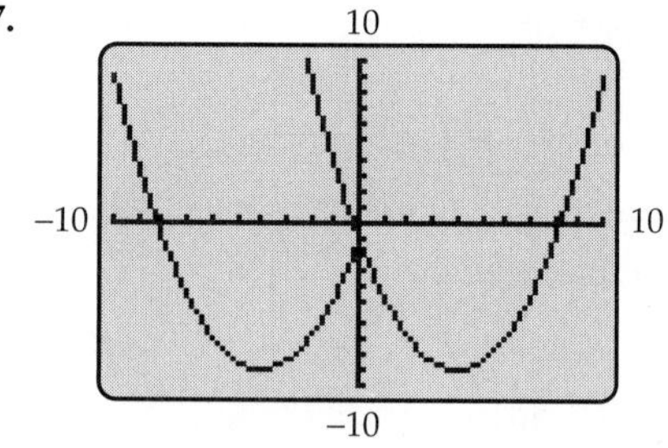

49. (a) Shifts upward by 28,000 units
(b) The slope of the graph will increase by 0.12.

Excursion 3.4.A, page 163

1. Symmetric with respect to the y-axis

3. No axis or origin symmetry

5. Odd **7.** Even **9.** Even **11.** Even

13. Neither **15.** Yes **17.** No **19.** Origin

21. Origin **23.** y-axis

25.

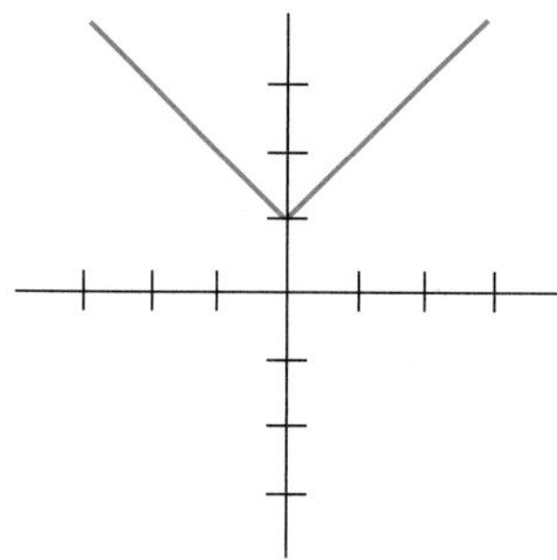

27.

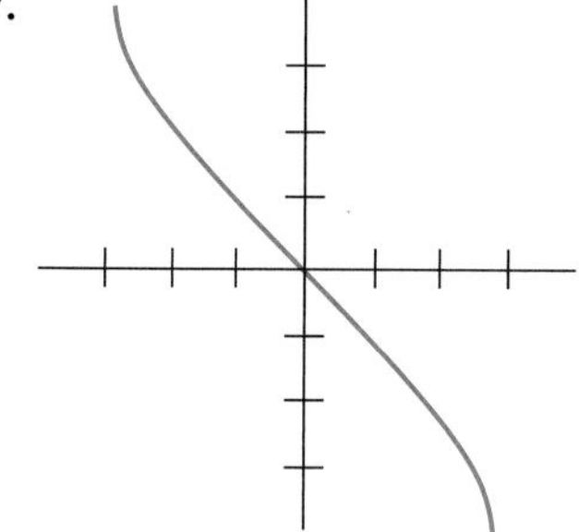

29.

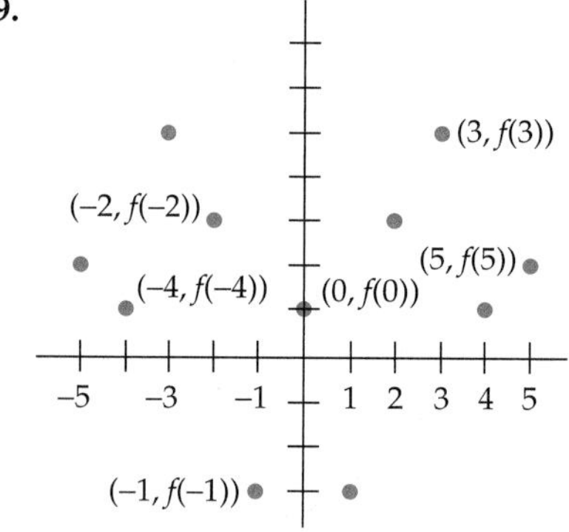

31. Many correct graphs, including the one shown here:

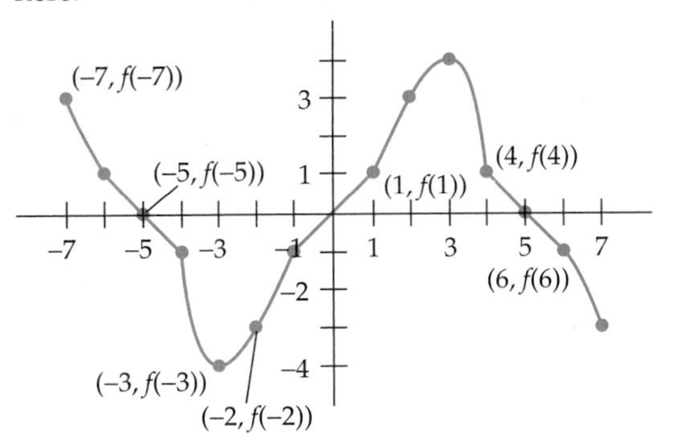

33. Suppose the graph is symmetric to the x-axis and the y-axis. If (x, y) is on the graph, then $(x, -y)$ is on the graph by x-axis symmetry. Hence, $(-x, -y)$ is on the graph by y-axis symmetry. Therefore, (x, y) on the graph implies that $(-x, -y)$ is on the graph, so the graph is symmetric with respect to the origin. Next suppose that the graph is symmetric to the y-axis and the origin. If (x, y) is on the graph, then $(-x, y)$ is on the graph by y-axis symmetry. Hence, $(-(-x), -y) = (x, -y)$ is on the graph by origin symmetry. Therefore, (x, y) on the graph implies that $(x, -y)$ is on the graph, so the graph is symmetric with respect to the x-axis. The proof of the third case is similar to that of the second case.

Section 3.5, page 171

1. $(f+g)(x) = x^3 - 3x + 2$;
$(f-g)(x) = -x^3 - 3x + 2$;
$(g-f)(x) = x^3 + 3x - 2$

3. $(f+g)(x) = \dfrac{1}{x} + x^2 + 2x - 5$; $(f-g)(x) = \dfrac{1}{x} - x^2 - 2x + 5$; $(g-f)(x) = x^2 + 2x - 5 - \dfrac{1}{x}$

5. $(fg)(x) = -3x^4 + 2x^3$;
$\left(\dfrac{f}{g}\right)(x) = \dfrac{-3x+2}{x^3}$; $\left(\dfrac{g}{f}\right)(x) = \dfrac{x^3}{-3x+2}$

7. $(fg)(x) = x^2\sqrt{x-3} - 3\sqrt{x-3}$
$(f/g)(x) = \dfrac{x^2-3}{\sqrt{x-3}}$
$(g/f)(x) = \dfrac{\sqrt{x-3}}{x^2-3}$

9. Domain of fg: all real numbers except 0; domain of f/g: all real numbers except 0

11. Domain of fg: $[-2, 2]$; domain of f/g: $(-2, 2]$

13. 0 **15.** 30 **17.** 49; 1; −8 **19.** −3; −3; 0

21. $(f \circ g)(x) = (x+3)^2$; $(g \circ f)(x) = x^2 + 3$; domain of $f \circ g$ and $g \circ f$ is all real numbers.

23. $(f \circ g)(x) = 1/\sqrt{x}$; $(g \circ f)(x) = 1/\sqrt{x}$; domain of $f \circ g$ and $g \circ f$ is $(0, \infty)$.

25. $(ff)(x) = x^6$; $(f \circ f)(x) = x^9$

27. $(ff)(x) = \dfrac{1}{x^2}$; $(f \circ f)(x) = x$

29. $(f \circ g)(x) = f\left(\dfrac{x-2}{9}\right) = 9\left(\dfrac{x-2}{9}\right) + 2 = x$ and
$(g \circ f)(x) = g(9x+2) = \dfrac{(9x+2)-2}{9} = x$

31. $(f \circ g)(x) = f(x-2)^3 = \sqrt[3]{(x-2)^3} + 2 = x$ and
$(g \circ f)(x) = g(\sqrt[3]{x} + 2) = (\sqrt[3]{x} + 2 - 2)^3 = x$

33.

x	−4	−3	−2	−1	0
$f(x)$	−3	−1	0	1/2	1
$g(x) = f(f(x))$	−1	1/2	1	1.2	1.5

x	1	2	3	4
$f(x)$	1.5	1	−2	−2
$g(x) = f(f(x))$	2	1.5	0	0

35.

x	1	2	3	4	5
$(g \circ f)(x)$	4	2	5	4	4

37.

x	1	2	3	4	5
$(f \circ f)(x)$	1	3	3	5	1

In Answers 39–45, the given function is $B \circ A$, where A and B are the functions listed here. In some cases other correct answers are possible.

39. $A(x) = x^2 + 2$, $B(x) = \sqrt[3]{x}$

41. $A(x) = 7x^3 - 10x + 17$, $B(x) = x^7$

43. $A(x) = 3x^2 + 5x - 7$, $B(x) = \dfrac{1}{x}$

45. Several possible answers, including $h(x) = x^2 + 2x$; $k(t) = t + 1$

47. $(f \circ g)(x) = (\sqrt{x})^3$, domain $[0, \infty)$; $(g \circ f)(x) = \sqrt{x^3}$, domain $[0, \infty)$

49. $(f \circ g)(x) = \sqrt{5x+10}$, domain $[-2, \infty)$;
$(g \circ f)(x) = 5\sqrt{x+10}$, domain $[-10, \infty)$

51. **(a)** $f(x^2) = 2x^6 + 5x^2 - 1$
(b) $(f(x))^2 = (2x^3 + 5x - 1)^2 = 4x^6 + 20x^4 - 4x^3 + 25x^2 - 10x + 1$
(c) No; $f(x^2) \neq (f(x))^2$ in general

53. Not the same

55. (a) One day, .00012246 sq in; one week, .0000025 sq in.; one 31-day month, .00000013 sq in.

(b) No; no. The model is probably reasonable until the puddle is about the size of a period, with a radius of approximately .01. This occurs after approximately 15 hours.

57. (a) $A = \pi\left(\frac{d}{2}\right)^2 = \pi \cdot \frac{d^2}{4} = \frac{\pi}{4}\left(6 - \frac{50}{t^2 + 10}\right)^2$

(b) $\pi/4 \approx .7854$ sq in.; 22.2648 sq in.

(c) In approximately 11.39 weeks

59. $V = 256\pi t^3/3$; 17,157.28 cm^3 **61.** $s = 10t/3$

63. One such function is $f(x) = \frac{x-1}{x}$.

Section 3.6, page 183

1. (a) 14 ft/sec **(b)** 54 ft/sec **(c)** 112 ft/sec **(d)** $93\frac{1}{3}$ ft/sec

3. (a) .709 gal/in. **(b)** 2.036 gal/in.

5. (a) 250 ties/mo **(b)** 438 ties/mo **(c)** 500 ties/mo **(d)** 563 ties/mo **(e)** -188 ties/mo **(f)** -750 ties/mo **(g)** -1500 ties/mo **(h)** -375 ties/mo

7. (a) -55.5 **(b)** -92.5 **(c)** -462.5

9. -2 **11.** -1 **13.** 1.5858

15. 1 **17.** $2x + h$ **19.** $2t + h - 8000$

21. $2\pi r + \pi h$

23. (a) Average rate of change is -7979.9, which means that water is leaving the tank at a rate of 7979.9 gal/min.

(b) -7979.99 **(c)** -7980

25. (a) 6.5π **(b)** 6.2π **(c)** 6.1π **(d)** 6π

(e) It's the same.

27. (a) C, 62.5 ft/sec; D, 75 ft/sec

(b) Approximately $t = 4$ to $t = 9.8$ sec

(c) The average speed of car D from $t = 4$ to $t = 10$ sec is the slope of the secant line joining the (approximate) points (4, 100) and (10, 600), namely, $\frac{600 - 100}{10 - 4} \approx 83.33$ ft/sec. The average speed of car C is the slope of the secant line joining the (approximate) points (4, 475) and (10, 800), namely, $\frac{800 - 475}{10 - 4} \approx 54.17$ ft/sec.

29. (a) From day 0 until any day up to day 94, the average growth rate is positive.

(b) From day 0 to day 95

(c) -28, meaning that the population is decreasing at a rate of 28 chipmunks per day

(d) 20, -20, and 0 chipmunks per day

Section 3.7, page 193

1. No **3.** Yes **5.** Yes **7.** No

9. $g(x) = -x$ **11.** $g(x) = \frac{x+4}{5}$

13. $g(x) = \sqrt[3]{\frac{5-x}{2}}$ **15.** $g(x) = \frac{x^2+7}{4}$, $(x \geq 0)$

17. $g(x) = \frac{1}{x}$ **19.** $g(x) = \frac{1}{2x} - \frac{1}{2}$

21. $g(x) = \sqrt[3]{\frac{5x+1}{1-x}}$

23. $(f \circ g)(x) = f(g(x)) = f(x-1) = (x-1) + 1 = x$ and $(g \circ f)(x) = g(f(x)) = g(x+1) = (x+1) - 1 = x$

25. $(f \circ g)(x) = f\left(\frac{1-x}{x}\right) = \frac{1}{\left(\frac{1-x}{x}\right) + 1} = \frac{1}{\frac{(1-x)+x}{x}} = x$ and $(g \circ f)(x) = g\left(\frac{1}{x+1}\right) = \frac{1 - \frac{1}{x+1}}{\frac{1}{x+1}} = \frac{\frac{(x+1)-1}{x+1}}{\frac{1}{x+1}} = x$

27. $(f \circ g)(x) = f(\sqrt[5]{x}) = (\sqrt[5]{x})^5 = x$ and $(g \circ f)(x) = g(x^5) = \sqrt[5]{x^5} = x$

29. $(f \circ f)(x) = f(f(x)) = \frac{2f(x)+1}{3f(x)-2} = \frac{2\left[\frac{2x+1}{3x-2}\right] + 1}{3\left[\frac{2x+1}{3x-2}\right] - 2} = \frac{\frac{2(2x+1) + (3x-2)}{3x-2}}{\frac{3(2x+1) - 2(3x-2)}{3x-2}} = \frac{\frac{7x}{3x-2}}{\frac{7}{3x-2}} = x$

31.

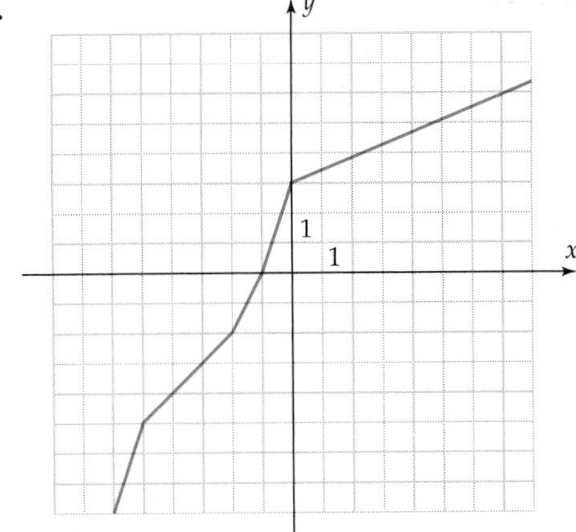

33.

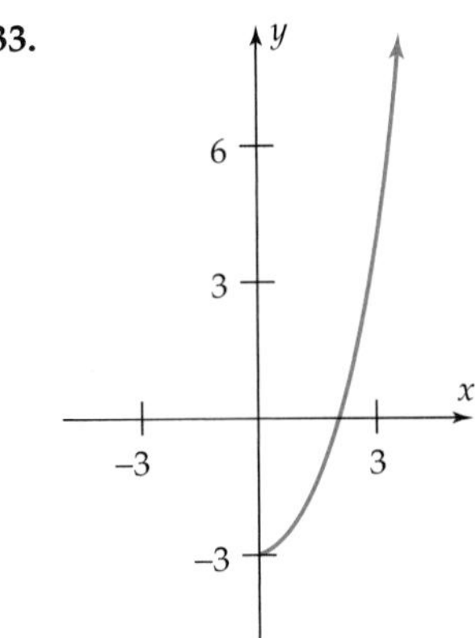

35.

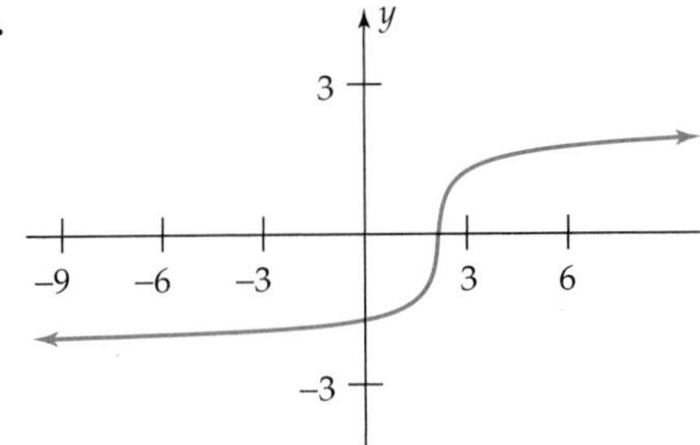

37.

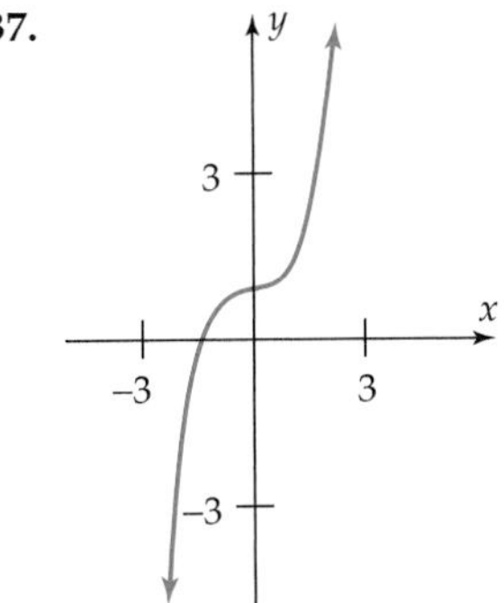

In Exercises 39–45, there are several correct answers for each, including these:

39. One restricted function is $h(x) = |x|$ with $x \geq 0$ (so that $h(x) = x$); inverse function $g(y) = y$ with $y \geq 0$.

41. One restricted function is $h(x) = -x^2$ with $x \leq 0$; inverse function $g(y) = -\sqrt{-y}$ with $y \leq 0$. Another restricted function is $h(x) = -x^2$ with $x \geq 0$; inverse function $g(y) = \sqrt{-y}$ with $y \leq 0$.

43. One restricted function is $h(x) = \dfrac{x^2 + 6}{2}$ with $x \geq 0$; inverse function $g(y) = \sqrt{2y - 6}$ with $y \geq 3$.

45. One restricted function is $f(x) = \dfrac{1}{x^2 + 1}$ with $x \leq 0$; inverse function $g(y) = -\sqrt{\dfrac{1}{y} - 1} = -\sqrt{\dfrac{1 - y}{y}}$ with $0 < y \leq 1$.

47. **(a)** $f^{-1}(x) = \dfrac{x - 2}{3}$ **(b)** $f^{-1}(1) = -1/3$ and $1/f(1) = 1/5$

49. Let $y = f(x) = mx + b$. Since $m \neq 0$, we can solve for x and obtain $x = \dfrac{y - b}{m}$. Hence, the rule of the inverse function g is $g(y) = \dfrac{y - b}{m}$, and we have: $(f \circ g)(y) = f(g(y)) = f\left(\dfrac{y - b}{m}\right) = m\left(\dfrac{y - b}{m}\right) + b = y$ and $(g \circ f)(x) = g(f(x)) = g(mx + b) = \dfrac{(mx + b) - b}{m} = x$.

51. **(a)** Slope $= \dfrac{a - b}{b - a} = \dfrac{-(b - a)}{b - a} = -1$.

(b) The line $y = x$ has slope 1 (why?) and by (a), line PQ has slope -1. Since the product of their slopes is -1, the lines are perpendicular.

(c) Length $PR = \sqrt{(a - c)^2 + (b - c)^2}$
$= \sqrt{a^2 - 2ac + c^2 + b^2 - 2bc + c^2}$
$= \sqrt{a^2 + b^2 + 2c^2 - 2ac - 2bc}$;
Length $RQ = \sqrt{(c - b)^2 + (c - a)^2}$
$= \sqrt{c^2 - 2bc + b^2 + c^2 - 2ac + a^2}$
$= \sqrt{a^2 + b^2 + 2c^2 - 2ac - 2bc}$.
Since the two lengths are the same, $y = x$ is the perpendicular bisector of segment PQ.

53. **(a)** Suppose $a \neq b$. If $f(a) = f(b)$, then $g(f(a)) = g(f(b))$. But $a = g(f(a))$ by (1) and $b = g(f(b))$. Hence, $a = b$, contrary to our hypothesis. Therefore, it cannot happen that $f(a) = f(b)$, that is, $f(a) \neq f(b)$. Hence, f is one-to-one.

(b) If $g(y) = x$, then $f(g(y)) = f(x)$. But $f(g(y)) = y$ by (2). Hence, $y = f(g(y)) = f(x)$.

(c) If $f(x) = y$, then $g(f(x)) = g(y)$. But $g(f(x)) = x$ by (1). Hence, $x = g(f(x)) = g(y)$.

Chapter 3 Review, page 196

1. **(a)** -3 **(b)** 1755 **(c)** 2 **(d)** 14

3.

x	0	1	2	-4	t	k
$f(x)$	7	5	3	15	$7 - 2t$	$7 - 2k$

x	$b - 1$	$1 - b$	$6 - 2u$
$f(x)$	$9 - 2b$	$5 + 2b$	$-5 + 4u$

5. Many possible answers, including:

(a) $f(x) = x^2, a = 2, b = 3; f(a + b) = f(2 + 3) = 5^2 = 25$, but $f(a) + f(b) = f(2) + f(3) = 2^2 + 3^2 = 13$, so the statement is false.

(b) $f(x) = x + 1, a = 0, b = 1; f(ab) = f(0) = 1$, but $f(a)f(b) = f(0)f(1) = 1 \cdot 2 = 2$, so the statement is false.

7. $[4, \infty)$

9. $(t + 2)^2 - 3(t + 2) = t^2 + t - 2$

11. $2\left(\frac{x}{2}\right)^3 + \left(\frac{x}{2}\right) + 1 = \frac{x^3}{4} + \frac{x}{2} + 1$

13. **(a)** $f(t) = 50\sqrt{t}$
(b) $g(t) = 2500\pi t$
(c) radius: 150 meters; area: 70,685.83 square meters
(d) 12.73 hr

15.

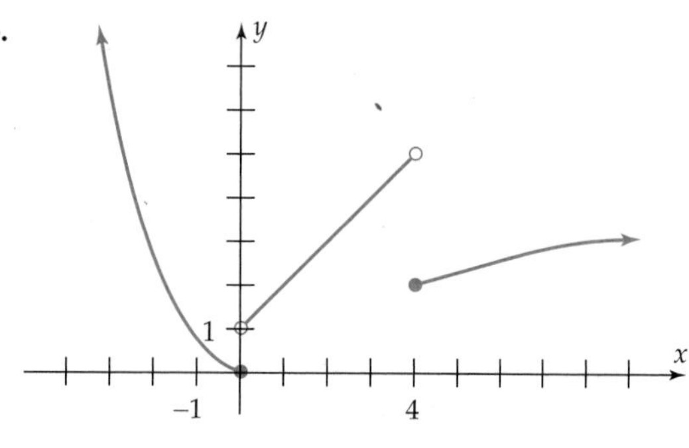

17. (b)

19. No local maxima; minimum at $x = -.5$; increasing on $[-.5, \infty)$; decreasing on $(-\infty, -.5]$

21. Maximum at $x \approx -5.0704$; minimum at $x \approx -.2629$. Increasing on $(-\infty, -5.0704]$ and $[-.2629, \infty)$; decreasing on $[-5.0704, -.2629]$

23.

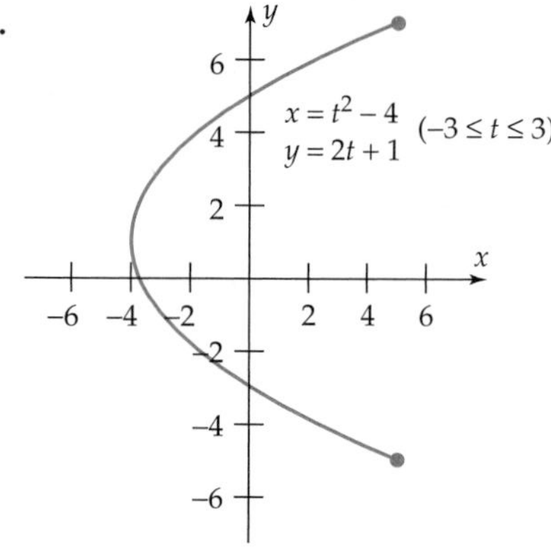

25. Many correct answers, including

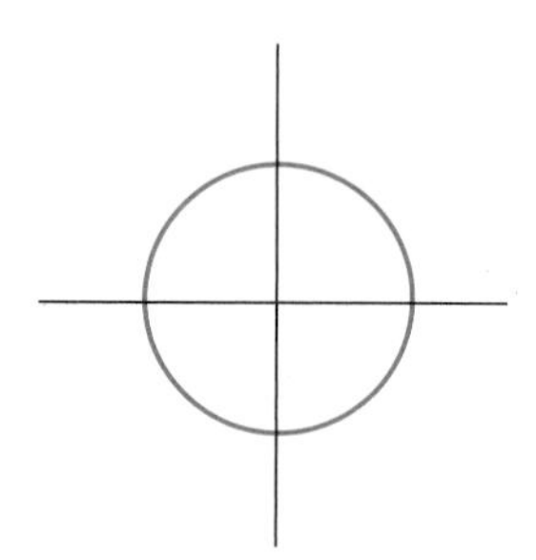

27. x-axis, y-axis, origin

29. Even

31. Odd

33. y-axis

35.

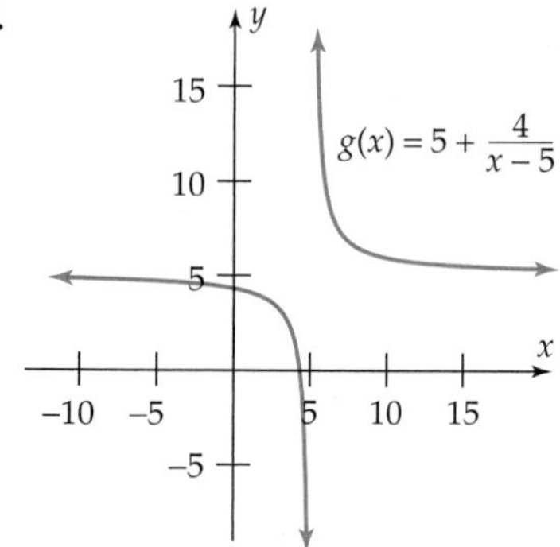

37. Approximately $[-3, 3.8]$

39. Many correct answers, including $x = -2$; all x in the interval (2.5, 3.8); all x in the interval [5, 6]

41. 1 **43.** -3 **45.** True

47. $x = 4$ **49.** $x \leq 3$ **51.** $x < 3$

53. **(a)** King Richard's car
(b) King Richard's
(c) Fireball Bob's

55. Shrink the graph of g toward the x-axis by a factor of .25, then shift the graph vertically 2 units upward.

57. Shift the graph of g horizontally 7 units to the right; then stretch it away from the x-axis by a factor of 3; then reflect it in the x-axis; finally, shift the graph vertically 2 units upward.

59. (e)

61. **(a)** 1/3 **(b)** $(x-1)\sqrt{x^2+5}\quad (x \neq 1)$
(c) $\dfrac{\sqrt{c^2+2c+6}}{c}$

63.

x	-4	-3	-2	-1	0
$g(x)$	1	4	3	1	-1
$h(x)$	-3	-3	-4	-3	1

x	1	2	3	4
$g(x)$	-3	-2	-4	-3
$h(x)$	4	3	1	4

65. $\dfrac{82}{27}$ **67.** $\dfrac{1}{x^3} + 3$ **69.** $\dfrac{1}{4}$

71. $(f \circ g)(x) = f(x^2 - 1) = \dfrac{1}{x^2-1}$; $(g \circ f)(x) = g\left(\dfrac{1}{x}\right) = \dfrac{1}{x^2} - 1$

73. All nonnegative numbers except 1

75. **(a)** $-1/3$ **(b)** 5/8

77. 6 **79.** 3 **81.** $2x + h$

83. **(a)** \$290/ton **(b)** \$230/ton **(c)** \$212/ton

85. **(a)** Approximately day 45 to day 59
(b) Approximately any 10-day interval between day 20 and day 35
(c) Approximately day 30 to day 40

87. **(a)** Whites: $-.08$; blacks: $-.48$; hispanics: $-.1$
(b) Whites: 2045; blacks: 2015; hispanics: 2098

89. $g(x) = 5 - (x-7)^2 = -x^2 + 14x - 44;\quad x \geq 7$

91.

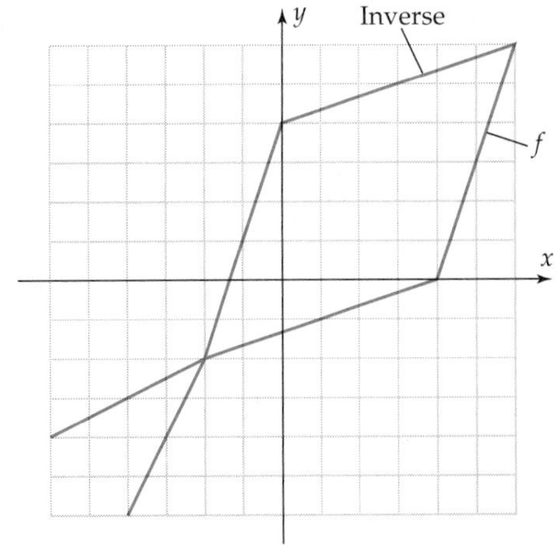

93. The graph of f passes the horizontal line test and hence has an inverse function. It is easy to verify either geometrically [by reflecting the graph of f in the line $y = x$] or algebraically [by calculating $f(f(x))$] that f is its own inverse function.

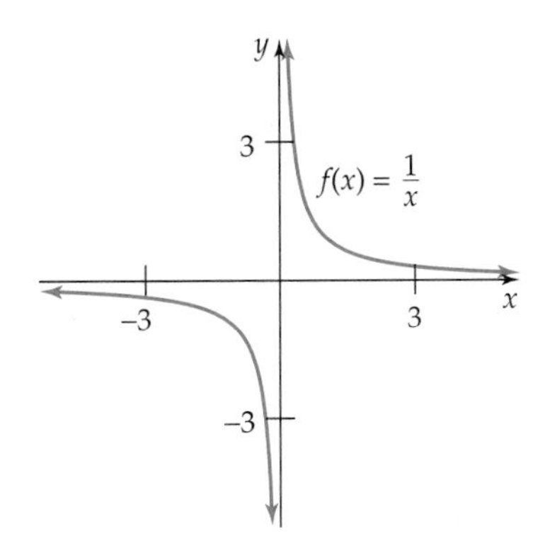

95. There is no inverse function because the graph of f fails the horizontal line test (use the viewing window with $-10 \leq x \leq 20$ and $-200 \leq y \leq 100$).

Chapter 4

Section 4.1, page 211

1. (5, 2), upward **3.** (1, 2), downward

5. (3, -6), upward **7.** ($-3/2$, 15/4), upward

9. (-3, -21), upward **11.** (4, 14), downward

13. (2/3, 19/3), downward **15.** (1/2, 1/4), downward

17. $g(x) = 2x^2 - 5$; vertex $(0, -5)$

19. $h(x) = 2(x-3)^2 + 4$; vertex $(3, 4)$

21. $f(x) = 3x^2$ **23.** $b = 0$ **25.** $b = -4, c = 8$

27. (a) At 30 km/hr, 30 m; at 100 km/hr, 170 m
(b) 50 km/hr

29. Minimum product is -4; numbers are 2 and -2

31. $h = 15, b = 15$

33. Two 50-ft sides and one 100-ft side

35. (a) $f(x) = \frac{1}{1152}x^2 - 2$
(b) Approximately 33.94 in. from center

37. \$3.50

39. \$3.67 (but if tickets must be priced in multiples of .20, then \$3.60 is best)

41. $t = 2.5$ sec, $h = 196$ ft 43. 22 ft

45. $t = \frac{125}{8}$ sec, $h = \frac{125^2}{4} = 3906.25$ ft

Section 4.2, page 219

1. Polynomial of degree 3; leading coefficient 1; constant term 1

3. Polynomial of degree 3; leading coefficient 1; constant term -1

5. Polynomial of degree 2; leading coefficient 1; constant term -3

7. Not a polynomial

9. Quotient $3x^3 - 3x^2 + 5x - 11$; remainder 12

11. Quotient $x^2 + 2x - 6$; remainder $-7x + 7$

13. Quotient $5x^2 + 5x + 5$; remainder 0

15. No 17. Yes 19. 0, 2 21. $2\sqrt{2}, -1$

23. 2 25. 6 27. -30 29. 170,802

31. 5,935,832 33. No 35. No 37. Yes

39. $(x + 4)(2x - 7)(3x - 5)$

41. $(x - 3)(x + 3)(2x + 1)^2$

43. $f(x) = (x + 2)(x + 1)(x - 1)(x - 2)(x - 3)$
$= x^5 - 3x^4 - 5x^3 + 15x^2 + 4x - 12$

45. $f(x) = x(x + 1)(x - 1)(x - 2)(x - 3)$
$= x^5 - 5x^4 + 5x^3 + 5x^2 - 6x$

47. Many correct answers, including $(x - 1)(x - 7)(x + 4)$

49. Many correct answers, including $(x - 1)(x - 2)^2(x - \pi)^3$

51. $f(x) = \frac{17}{100}(x - 5)(x - 8)x$

53. $k = 1$ 55. $k = 1$

57. If $x - c$ were a factor of $x^4 + x^2 + 1$, then c would be a solution of $x^4 + x^2 + 1 = 0$, that is, c would satisfy $c^4 + c^2 = -1$. But $c^4 \geq 0$ and $c^2 \geq 0$, so that is impossible. Hence, $x - c$ is not a factor.

59. (a) Many possible answers, including: if $n = 3$ and $c = 1$, then $x + 1 = x - (-1)$ is not a factor of $x^3 - 1$ since -1 is not a solution of $x^3 - 1 = 0$.
(b) Since n is odd $(-c)^n = -c^n$ and hence $-c$ is a solution of $x^n + c^n = 0$. Thus, $x - (-c) = x + c$ is a factor of $x^n + c^n$ by the Factor Theorem.

61. $k = 5$

Excursion 4.2.A, page 223

1.
$$\begin{array}{r|rrrrr} 2 & 3 & -8 & 0 & 9 & 5 \\ & & 6 & -4 & -8 & 2 \\ \hline & 3 & -2 & -4 & 1 & \boxed{7} \end{array}$$
quotient $3x^3 - 2x^2 - 4x + 1$; remainder 7

3.
$$\begin{array}{r|rrrrr} -3 & 2 & 5 & 0 & -2 & -8 \\ & & -6 & 3 & -9 & 33 \\ \hline & 2 & -1 & 3 & -11 & \boxed{25} \end{array}$$
quotient $2x^3 - x^2 + 3x - 11$; remainder 25

5.
$$\begin{array}{r|rrrrr} 7 & 5 & 0 & -3 & -4 & 6 \\ & & 35 & 245 & 1{,}694 & 11{,}830 \\ \hline & 5 & 35 & 242 & 1{,}690 & \boxed{11{,}836} \end{array}$$
quotient $5x^3 + 35x^2 + 242x + 1690$; remainder 11,836

7.
$$\begin{array}{r|rrrrr} 2 & 1 & -6 & 4 & 2 & -7 \\ & & 2 & -8 & -8 & -12 \\ \hline & 1 & -4 & -4 & -6 & \boxed{-19} \end{array}$$
quotient $x^3 - 4x^2 - 4x - 6$; remainder -19

9. Quotient $3x^3 + \frac{3}{4}x^2 - \frac{29}{16}x - \frac{29}{64}$; remainder $\frac{483}{256}$

11. Quotient $2x^3 - 6x^2 + 2x + 2$; remainder 1

13. $g(x) = (x + 4)(3x^2 - 3x + 1)$

15. $g(x) = \left(x - \frac{1}{2}\right)(2x^4 - 6x^3 + 12x^2 - 10)$

17. Quotient $x^2 - 2.15x + 4$; remainder 2.25

19. $c = -4$

Section 4.3, page 230

1. $x = \pm 1$ or -3 3. $x = \pm 1$ or -5

5. $x = -4, 0, 1,$ or $1/2$ 7. $x = -3$ or 2

9. $x = 2$

11. $x = -5, 2,$ or 3

13. $(x - 2)(2x^2 + 1)$

15. $x^3(x^2 + 3)(x + 2)$

17. $(x - 2)(x - 1)^2(x^2 + 3)$

19. Lower -5; upper 2

21. Lower -7; upper 3

23. $x = 1, 2,$ or $-1/2$

25. $x = 1, 1/2,$ or $1/3$

27. $x = 2$ or $\dfrac{-5 \pm \sqrt{37}}{2}$

29. $x = 1/2$ or $\pm\sqrt{2}$ or $\pm\sqrt{3}$

31. $x = -1, 5,$ or $\pm\sqrt{3}$

33. $x = 1/3$ or -1.8393

35. $x = -2.2470$ or $-.5550$ or $.8019$ or 50

37. **(a)** The only possible rational roots of $f(x) = x^2 - 2$ are ± 1 or ± 2 (why?). But $\sqrt{2}$ is a root of $f(x)$ and $\sqrt{2} \neq \pm 1$ or ± 2. Hence, $\sqrt{2}$ is irrational.
(b) $\sqrt{3}$ is a root of $x^2 - 3$ whose only possible rational roots are ± 1 or ± 3 (why?). But $\sqrt{3} \neq \pm 1$ or ± 3.

39. **(a)** 8.6378 people per 100,000
(b) 1995
(c) 1991

41. 2 by 2 in.

43. **(a)** 6°/day at the beginning; 6.6435°/day at the end
(b) Day 2.0330 and day 10.7069
(c) Day 5.0768 and day 9.6126
(d) Day 7.6813

Section 4.4, page 239

1. Yes

3. Yes

5. No

7. Degree 3, yes; degree 4, no; degree 5, yes

9. No

11. Degree 3, no; degree 4, no; degree 5, yes

13. The graphs have the same *shape* in the window with $-40 \le x \le 40$ and $-1000 \le y \le 5000$ but don't look identical.

15. -2 is a root of odd multiplicity, as are 1 and 3

17. -2 and -1 are roots of odd multiplicity; 2 is a root of even multiplicity.

19. (e)

21. (f)

23. (c)

25. The graph in the standard viewing window does not rise at the far right as does the graph of the highest degree term x^3, so it is not complete.

27. The graph in the standard viewing window does not rise at the far left and far right as does the graph of the highest degree term $.005x^4$, so it is not complete.

29. $-9 \le x \le 3$ and $-20 \le y \le 40$

31. $-6 \le x \le 6$ and $-60 \le y \le 320$

33. $-3 \le x \le 4$ and $-35 \le y \le 20$

35. Left half: $-33 \le x \le -2$ and $-50{,}000 \le y \le 250{,}000$; right half: $-2 \le x \le 3$ and $-20 \le y \le 30$

37. $-90 \le x \le 120$ and $-15{,}000 \le y \le 5000$

39. Overall: $-3 \le x \le 3$ and $-20 \le y \le 20$; near y-axis: $-.1 \le x \le .2$ and $4.997 \le y \le 5.001$

41. **(a)** The graph of a cubic polynomial (degree 3) has at most $3 - 1 = 2$ local extrema. When $|x|$ is large, the graph resembles the graph of ax^3, that is, one end shoots upward and the other end downward. If the graph had only one local extremum, both ends of the graph would go in the same direction (both up or down). Thus, the graph of a cubic polynomial has either two local extrema or none.
(b) These are the only possible shapes for a graph that has 0 or 2 local extrema, 1 point of inflection, and resembles the graph of ax^3 when $|x|$ is large.

43. **(a)** Odd **(b)** Positive **(c)** $-2, 0, 4,$ and 6
(d) 5

45. (d)

47.

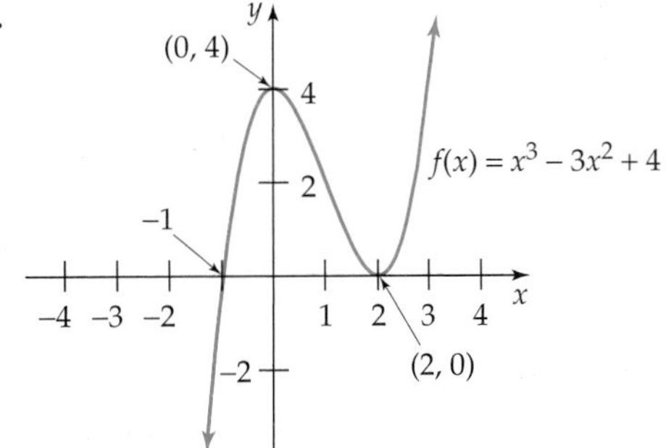

49.

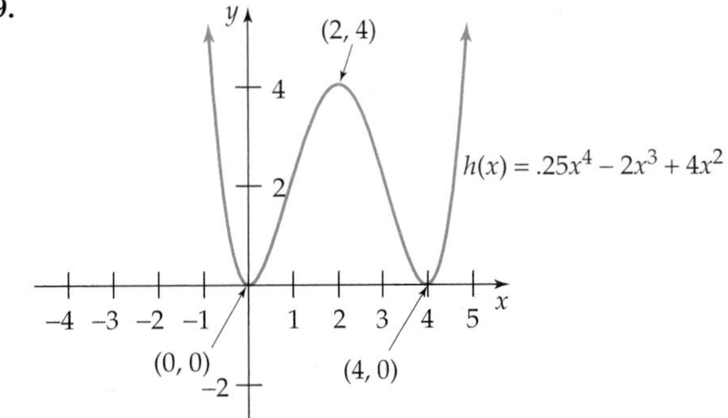

51.

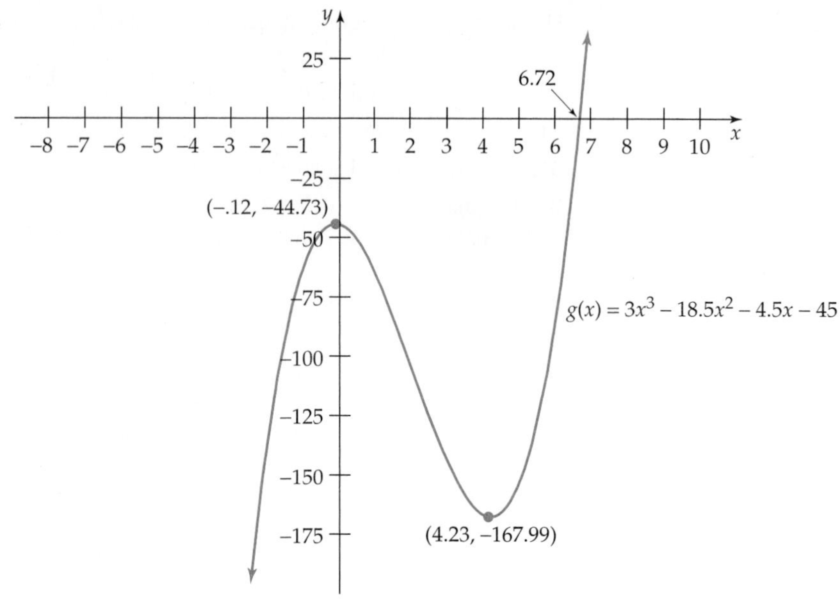

53.

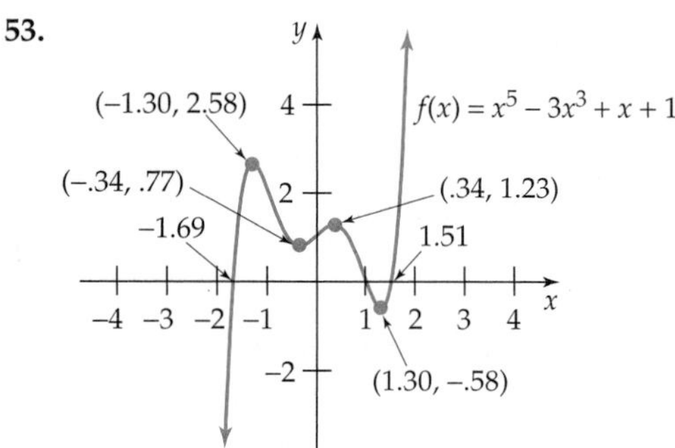

55.

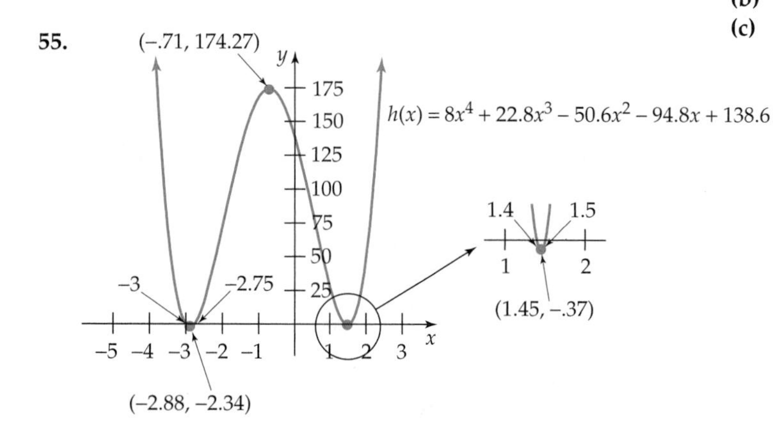

57. **(a)** 82,794; \$1546.39 **(b)** 147,200

59. **(a)** $R(x) = 350x$; $P(x) = -.01x^2 + 375x - 600{,}000$
(b) 1675 **(c)** 35,825 **(d)** 18,750

61. 2.3542 by 2.3542 in.

63. **(a)** The solutions are roots of $g(x) - 4 = .01x^3 - .06x^2 + .12x - .08$. This polynomial has degree 3 and hence has at most 3 roots.
(b) $1 \le x \le 3$ and $3.99 \le y \le 4.01$
(c) Suppose $f(x)$ has degree n. If the graph of $f(x)$ had a horizontal segment lying on the line $y = k$ for some constant k, then the equation $f(x) = k$ would have infinitely many solutions (why?). But the polynomial $f(x) - k$ has degree n (why?) and thus has at most n roots. Hence the equation $f(x) = k$ has at most n solutions, which means the graph cannot have a horizontal segment.

65. **(a)** No, except between $x = 0$ and $x = 2$
(b) Yes
(c) Approximately .0007399

67. (a)

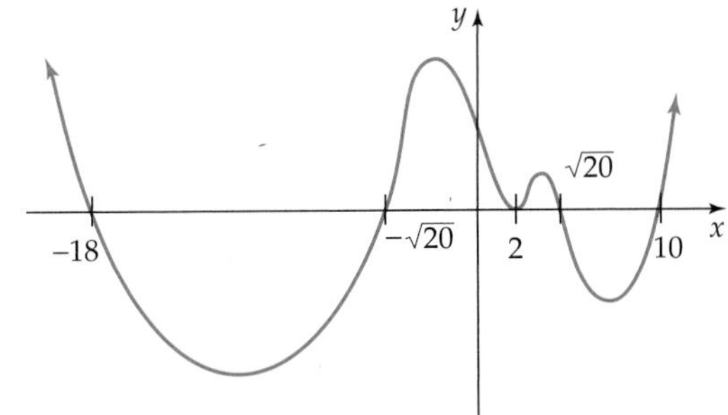

(b)–(d) Depends on the calculator

Excursion 4.4.A, page 246

1. Cubic

3. Quadratic

5. (a) $y = -2.134090909x^3 + 52.00963203x^2 - 359.2162338x + 5512.618182$

(b) 1987: 4814.58 per 100,000; 1995: 4623.99

(c) 3951.94

(d) Answers vary; perhaps through the year 2000.

7. (a)

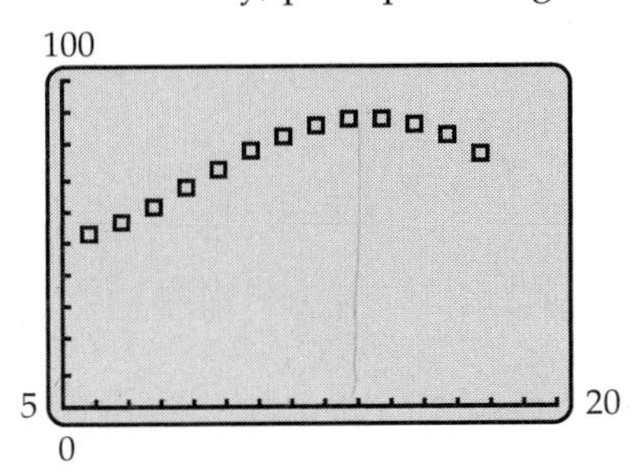

(b) $y = -.5179820180x^2 + 14.65684316x - 20.88711289$

(c) Noon: 80°; 9 A.M.: 69°; 2 P.M.: 83°

9. (a)

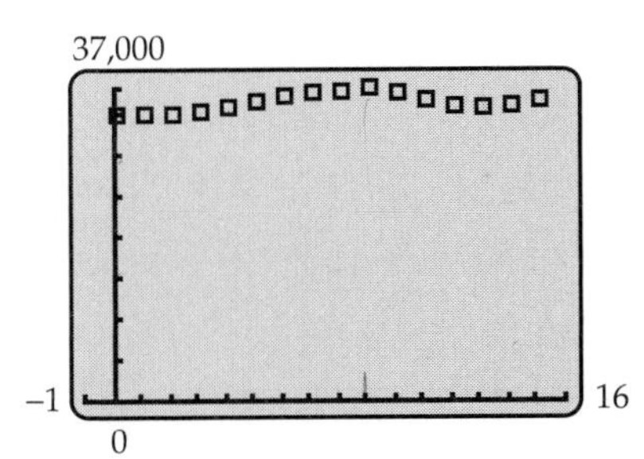

(b) Quartic

(c) $y = 2.397660405x^4 - 73.93709233x^3 + 691.3121614x^2 - 1762.930366x + 33079.5044$

(d) $66,475; no; this is double that of 1994.

(e) 1998

11. (a)

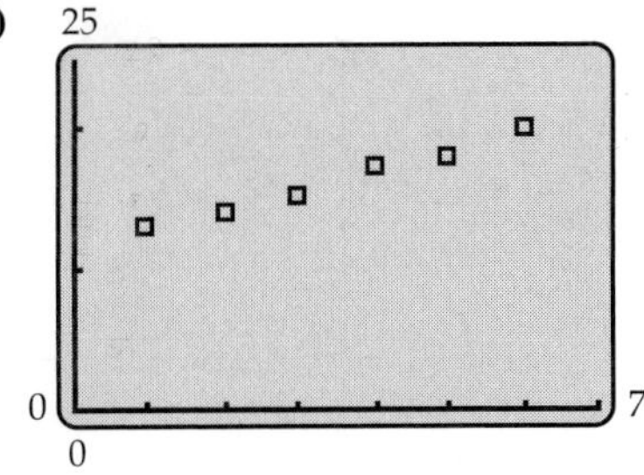

(b) $y = 1.32x + 12.9133$; a linear model, chosen from the scatter plot, although a quadratic or cubic model fits the data points reasonably well. A quartic model fits these data points, but indicates a very large increase after 1996, which seems unlikely.

13. (a) Cubic: $y = -.01897x^3 + .56273x^2 - 5.13900x + 21.94073$;
Quartic:
$y = -.00504x^4 + .17273x^3 - 2.01756x^2 + 9.29074x - 6.03968$

(b) No

Section 4.5, page 260

1. All real numbers except $-5/2$

3. All real numbers except $3 + \sqrt{5}$ and $3 - \sqrt{5}$

5. All real numbers except $-\sqrt{2}$, 1, and $\sqrt{2}$

7. Vertical asymptotes $x = -1$ and $x = 6$

9. Hole at $x = 0$; vertical asymptote $x = -1$

11. Vertical asymptotes: $x = -2$ and $x = 2$

13. $y = 3$; any window with $-115 \le x \le 110$

15. $y = -1$; any window with $-31 \le x \le 35$

17. $y = 5/2$; any window with $-40 \le x \le 42$

19.

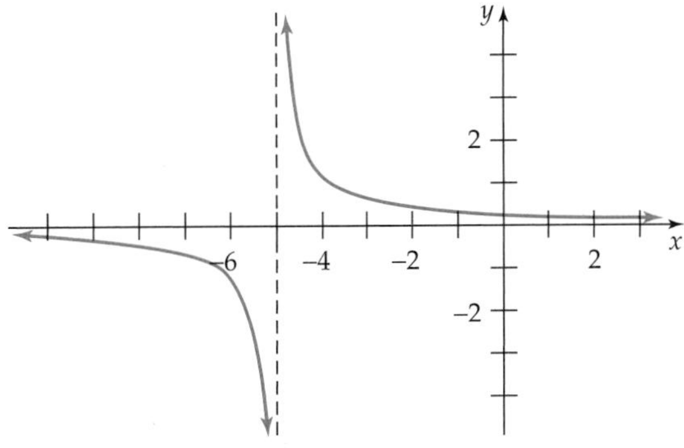

vertical asymptote $x = -5$
horizontal asymptote $y = 0$

21.

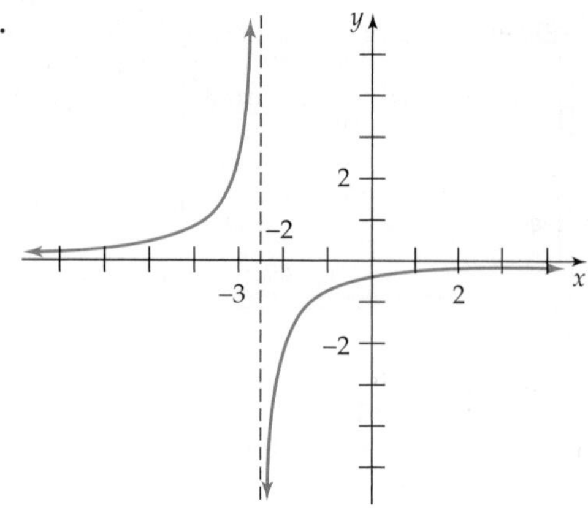

vertical asymptote $x = -2.5$
horizontal asymptote $y = 0$

23.

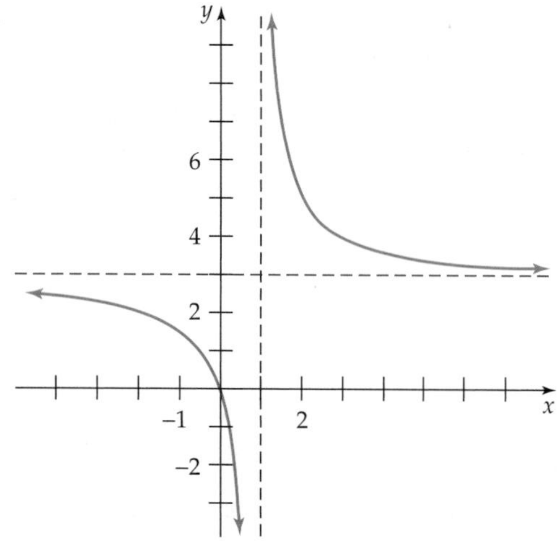

vertical asymptote $x = 1$
horizontal asymptote $y = 3$

25.

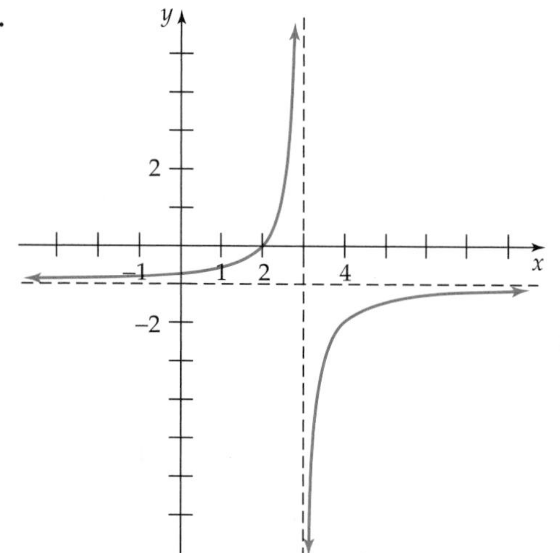

vertical asymptote $x = 3$
horizontal asymptote $y = -1$

27.

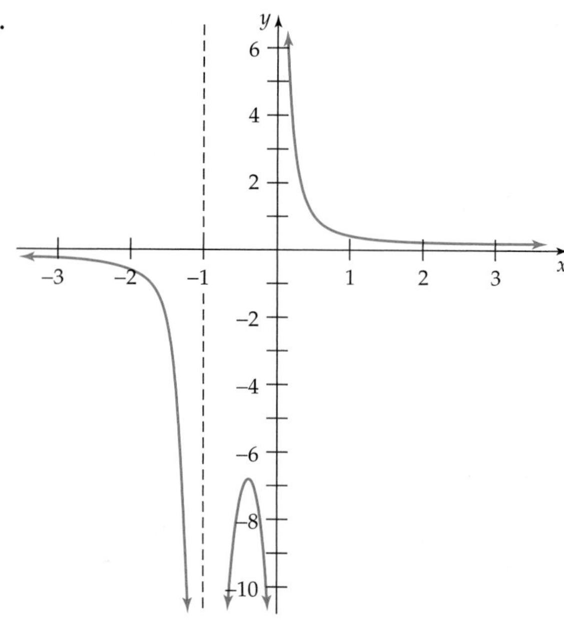

vertical asymptotes $x = -1, x = 0$
horizontal asymptote $y = 0$

29.

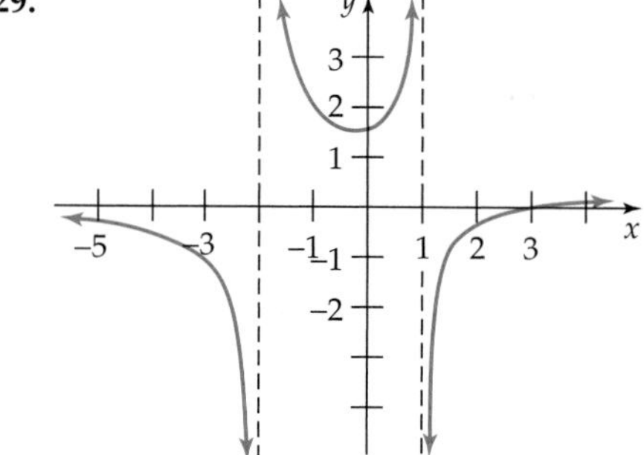

vertical asymptotes $x = -2, x = 1$
horizontal asymptote $y = 0$

31.

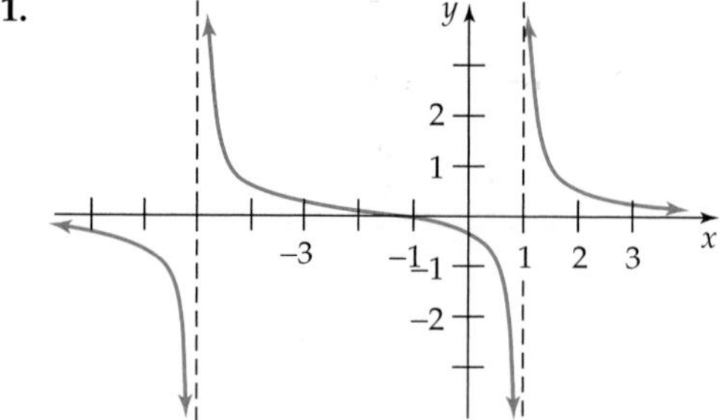

vertical asymptotes $x = -5, x = 1$
horizontal asymptote $y = 0$

33.

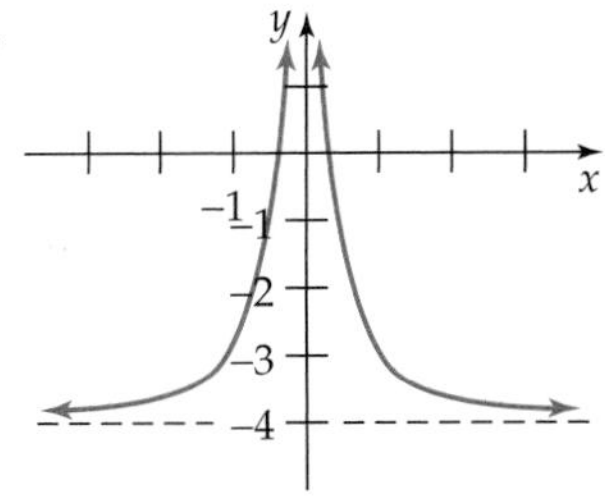

vertical asymptote $x = 0$
horizontal asymptote $y = -4$

35.

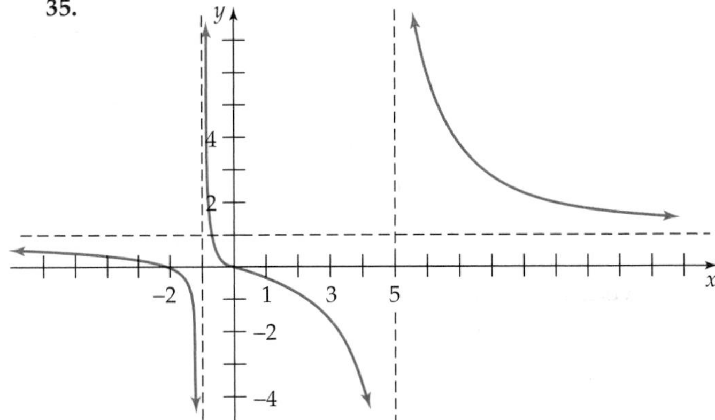

vertical asymptotes $x = -1, x = 5$
horizontal asymptote $y = 1$

37.

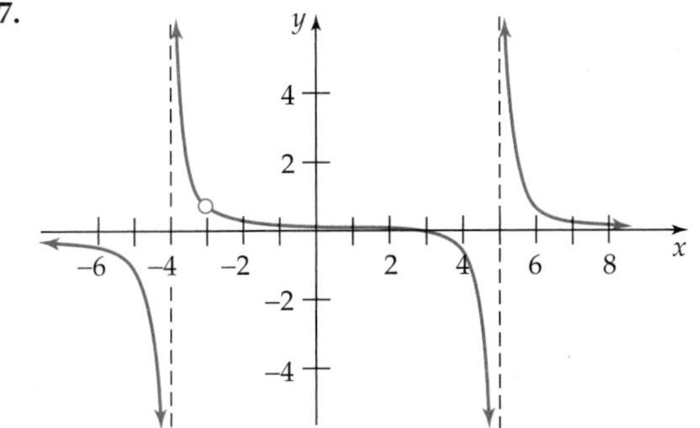

vertical asymptotes $x = -4, x = 5$
hole at $x = -3$
horizontal asymptote $y = 0$

39. Overall: $-5 \le x \le 4.4$ and $-8 \le y \le 4$; hidden area near origin: $-2 \le x \le 2$ and $-.5 \le y \le .5$; hidden area near $x = -5$: $-15 \le x \le -3$ and $-.07 \le y \le .02$

41. $-9.4 \le x \le 9.4$ and $-4 \le y \le 4$; there is a hole at $x = 2$.

43. Overall: $-4.7 \le x \le 4.7$ and $-2 \le y \le 2$; there is a hole at $x = -1$; to see the vertical asymptote, use $.65 \le x \le .75$ and $-3 \le y \le 3$.

45. For vertical asymptotes and x-intercepts: $-4.7 \le x \le 4.7$ and $-8 \le y \le 8$; to see graph get close to the horizontal asymptote: $-40 \le x \le 35$ and $-2 \le y \le 3$

47. Overall: $-4.7 \le x \le 4.7$ and $-3 \le y \le 3$; hidden area near $x = 4$: $3 \le x \le 15$ and $-.02 \le y \le .01$

49. **(b)** Stretch the graph of $f(x)$ away from the x-axis by a factor of 2.
(c) The graph of $h(x)$ is the graph of $f(x)$ shifted vertically 4 units upward; the graph of $k(x)$ is the graph of $f(x)$ shifted horizontally 3 units to the right; the graph of $t(x)$ is the graph of $f(x)$ shifted horizontally 2 units to the left.
(d) Shift the graph of $f(x)$ horizontally 3 units to the right, stretch by a factor of 2, then shift vertically 4 units upward.
(e) $p(x) = \dfrac{4x - 10}{x - 3}$
(f) Shift the graph of $f(x)$ horizontally $|s|$ units (to the left if $s > 0$; to the right if $s < 0$); stretch (or shrink) the graph by a factor of $|r|$ (away from the x-axis if $|r| > 1$, toward the x-axis if $0 < |r| < 1$); also if $r < 0$, reflect the graph in the x-axis; then shift vertically $|t|$ units (upward if $t > 0$; downward if $t < 0$).
(g) $q(x) = \dfrac{tx + (r + ts)}{x + s}$

51. **(a)** $\dfrac{-1}{x(x + h)}$
(b) $-1/4.2 \approx -.2381$; $-1/4.02 \approx -.2488$; $-1/4.002 \approx -.2499$; instantaneous rate of change $-1/4 = -.25$
(c) $-1/9.3 \approx -.1075$; $-1/9.03 \approx -.1107$; $-1/9.003 \approx -.1111$; instantaneous rate of change $-1/9 = -.1111\cdots$
(d) They are the same.

53. **(a)** $y = \dfrac{x - 1}{x - 2}$ **(b)** $y = \dfrac{x - 1}{-x - 2}$
(c) Graph (a) has a vertical asymptote at $x = 2$ and graph (b) has a vertical asymptote at $x = -2$.

55. 8.4343 in. × 8.4343 in. × 14.057 in.

57. **(a)** $c(x) = \dfrac{2800 + 3.5x^2}{x}$ **(b)** $13.91 \le x \le 57.52$
(c) 28.28 mph

59. **(a)** $p(x) = \dfrac{500 + x^2}{x}$ **(b)** $10 \le x \le 50$
(c) $x = 22.36$; 22.36 m by 11.18 m

61. **(a)** $h_1 = h - 2$
(b) $h_1 = \dfrac{150}{\pi r^2} - 2$ (because $\pi r^2 h = 150$)

(c) $V = \pi(r-1)^2\left(\frac{150}{\pi r^2} - 2\right)$

(d) The walls are 1 ft thick.

(e) $r \approx 2.88$ ft; $h \approx 5.76$ ft

63. (a) $g(0) = 9.801 \text{ m/sec}^2$

(b)

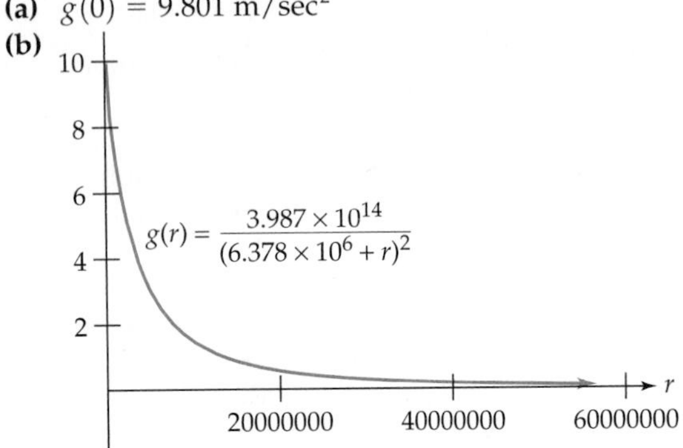

(c) There are no r-intercepts because the numerator is never zero. So you can never completely escape the pull of gravity.

7.

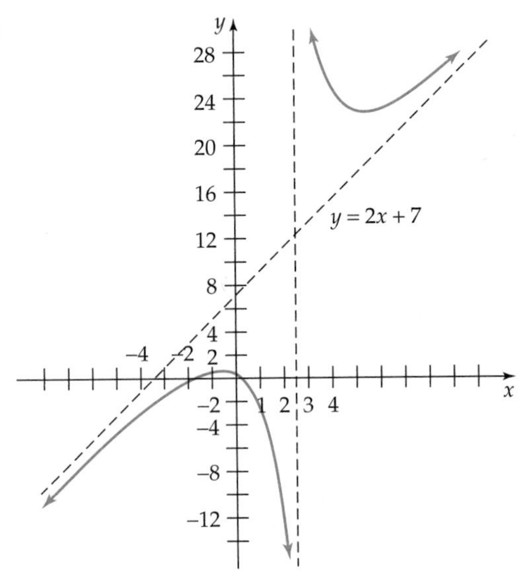

vertical asymptote $x = 5/2$
oblique asymptote $y = 2x + 7$

9.

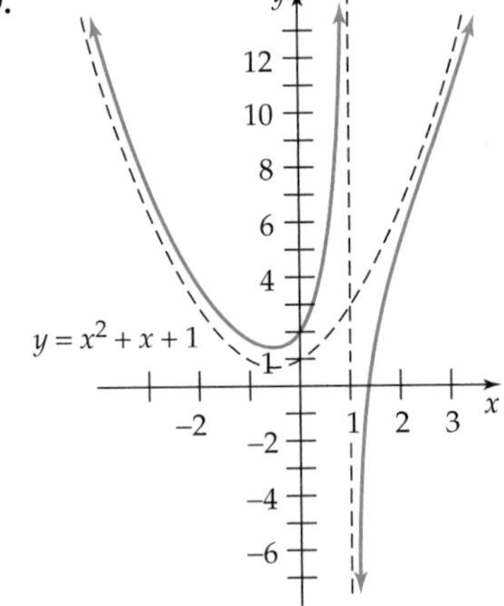

vertical asymptote $x = 1$
parabolic asymptote $y = x^2 + x + 1$

Excursion 4.5.A, page 267

1. Asymptote: $y = x$; window: $-14 \le x \le 14$ and $-15 \le y \le 15$

3. Asymptote: $y = x^2 - x$; window: $-15 \le x \le 6$ and $-40 \le y \le 240$

5.

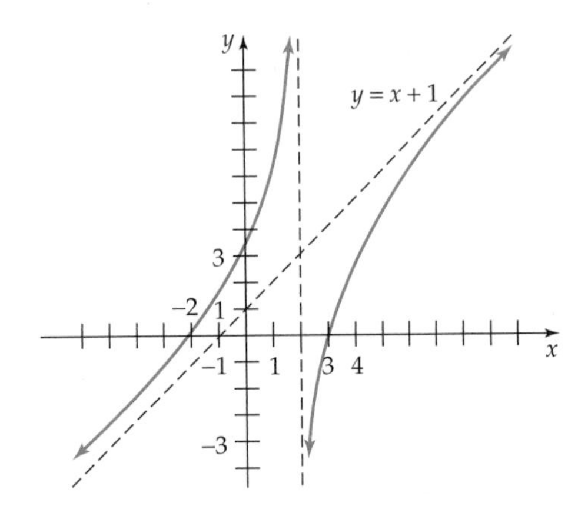

vertical asymptote $x = 2$
oblique asymptote $y = x + 1$

11. 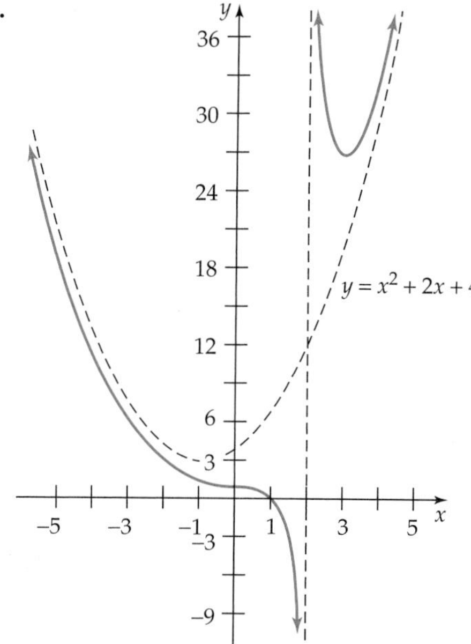

vertical asymptote $x = 2$
parabolic asymptote $y = x^2 + 2x + 4$

13. $-15.5 \le x \le 8.5$ and $-16 \le y \le 8$
15. $-4.7 \le x \le 4.7$ and $-12 \le y \le 8$
17. Overall: $-13 \le x \le 7$ and $-20 \le y \le 20$; hidden area near the origin: $-2.5 \le x \le 1$ and $-.02 \le y \le .02$
19. **(b)** Approximately $.06 \le x \le 2.78$

Section 4.6, page 276

1. $(-\infty, 3/2]$ 3. $(-2, \infty)$ 5. $(-\infty, -8/5]$
7. $(1, \infty)$ 9. $(2, 4)$ 11. $[-3, 5/2)$
13. $(-\infty, 4/7)$ 15. $[-7/17, \infty)$ 17. $[-1, 1/8)$
19. $[5, \infty)$ 21. $x < \dfrac{b+c}{a}$ 23. $c < x < a + c$
25. $1 \le x \le 3$
27. $x \le \dfrac{-9 - \sqrt{21}}{2}$ or $x \ge \dfrac{-9 + \sqrt{21}}{2}$
29. $x \le \dfrac{1 - \sqrt{33}}{2}$ or $x \ge \dfrac{1 + \sqrt{33}}{2}$
31. $-1 \le x \le 0$ or $x \ge 1$
33. $x < -1$ or $0 < x < 3$
35. $-2 < x < -1$ or $1 < x < 2$
37. $-2.26 \le x \le 0.76$ or $x \ge 3.51$
39. $.5 < x < .84$ 41. $x < -1/3$ or $x > 2$
43. $-2 < x < -1$ or $1 < x < 3$ 45. $x > 1$
47. $x \le -9/2$ or $x > -3$ 49. $-3 < x < 1$ or $x \ge 5$
51. $-\sqrt{7} < x < \sqrt{7}$ or $x > 5.34$
53. $x < -3$ or $1/2 < x < 5$
55. $x > -1.43$
57. $x \le -3.79$ or $x \ge .79$
59. Approximately 8.608 cents per kwh
61. More than $12,500 63. Between $4000 and $5400
65. $1 < x < 19$ and $y = 20 - x$ 67. $10 < x < 35$
69. $1 \le t \le 4$ 71. $2 < t < 2.25$
73. **(a)** $x^2 < x$ when $0 < x < 1$ and $x^2 > x$ when $x < 0$ or $x > 1$.
(b) If c is nonzero and $|c| < 1$, then either $0 < c < 1$ or $-1 < c < 0$ (which is equivalent to $1 > -c > 0$). If $0 < c < 1$, then $|c| = c$ and c is a solution of $x^2 < x$ by part (a), so that $c^2 < c = |c|$. If $1 > -c > 0$, then $|c| = -c$, which is a solution of $x^2 < x$ by part (a), so that $c^2 = (-c)^2 < (-c) = |c|$.
(c) If $|c| > 1$, then either $c < -1$ or $c > 1$. In either case, c is a solution of $x^2 > x$ by part (a).

Excursion 4.6.A, page 281

1. $-4/3 \le x \le 0$ 3. $7/6 < x < 11/6$
5. $x < -2$ or $x > -1$
7. $x \le -11/20$ or $x \ge -1/4$
9. $x < -53/40$ or $x > -43/40$
11. $x \le -7/2$ or $x \ge -5/4$
13. $x < -5$ or $-5 < x < -4/3$ or $x > 6$
15. $-1/7 < x < 3$
17. $-\sqrt{3} < x < -1$ or $1 < x < \sqrt{3}$
19. $x < -\sqrt{6}$ or $x > \sqrt{6}$
21. $x \le 2$ or $-1 \le x \le 0$ or $x \ge 1$
23. $0 < x < 2/3$ or $2 < x < 8/3$
25. $-1.43 < x < 1.24$ 27. $x < -.89$ or $x > 1.56$
29. $x \le 2$ or $x \ge 14/3$
31. $-1.13 < x < 1.35$ or $1.35 < x < 1.67$
33. If $|x - 3| < E/5$, then multiplying both sides by 5 shows that $5|x - 3| < E$. But $5|x - 3| = |5| \cdot |x - 3| = |5(x - 3)| = |5x - 15| = |(5x - 4) - 11|$. Thus, $|(5x - 4) - 11| < E$.

Section 4.7, page 287

1. $8 + 2i$ 3. $-2 - 10i$ 5. $-\dfrac{1}{2} - 2i$
7. $\left(\dfrac{\sqrt{2} - \sqrt{3}}{2}\right) + 2i$ 9. $1 + 13i$ 11. $-10 + 11i$

13. $-21 - 20i$ 15. 4 17. $-i$

19. i 21. i

23. $\frac{5}{29} + \frac{2}{29}i$ 25. $-\frac{1}{3}i$ 27. $\frac{12}{41} - \frac{15}{41}i$

29. $\frac{-5}{41} - \frac{4}{41}i$ 31. $\frac{10}{17} - \frac{11}{17}i$ 33. $\frac{7}{10} + \frac{11}{10}i$

35. $-\frac{113}{170} + \frac{41}{170}i$ 37. $6i$ 39. $\sqrt{14}i$

41. $-4i$ 43. $11i$ 45. $(\sqrt{15} - 3\sqrt{2})i$

47. $\frac{2}{3}$ 49. $-41 - i$

51. $(2 + 5\sqrt{2}) + (\sqrt{5} - 2\sqrt{10})i$

53. $\frac{1}{3} - \frac{\sqrt{2}}{3}i$ 55. $x = 2, y = -2$

57. $x = -3/4, y = 3/2$ 59. $x = \frac{1}{3} \pm \frac{\sqrt{14}}{3}i$

61. $x = -\frac{1}{2} \pm \frac{\sqrt{7}}{2}i$ 63. $x = \frac{1}{4} \pm \frac{\sqrt{31}}{4}i$

65. $x = \frac{3 \pm \sqrt{3}}{2}$

67. $x = 2, -1 + \sqrt{3}i, -1 - \sqrt{3}i$

69. $x = 1, -1, i, -i$ 71. -1

73. $z + w = (a + bi) + (c + di) = (a + c) + (b + d)i$ and hence
$$\overline{z + w} = (a + c) - (b + d)i = a + c - bi - di = (a - bi) + (c - di) = \overline{z} + \overline{w}$$

75. We first express z/w in standard form:
$\frac{z}{w} = \frac{a + bi}{c + di} = \frac{a + bi}{c + di} \cdot \frac{c - di}{c - di} = \frac{(ac + bd) + (bc - ad)i}{c^2 + d^2}$. Hence,
$\overline{\left(\frac{z}{w}\right)} = \frac{(ac + bd) - (bc - ad)i}{c^2 + d^2} = \frac{ac + bd - bci + adi}{c^2 + d^2}$. On the other hand,
$\frac{\overline{z}}{\overline{w}} = \frac{a - bi}{c - di} = \frac{a - bi}{c - di} \cdot \frac{c + di}{c + di} = \frac{ac + bd - bci + adi}{c^2 + d^2}$.

77. If $z = a + bi$, with a, b real numbers, then $z - \overline{z} = (a + bi) - (a - bi) = 2bi$. If $z = a + bi$ is real, then $b = 0$ and hence, $z - \overline{z} = 2bi = 0$. Therefore, $z = \overline{z}$. Conversely, if $z = \overline{z}$, then $0 = z - \overline{z} = 2bi$, which implies that $b = 0$. Hence, $z = a$ is real.

79. $\frac{1}{z} = \left(\frac{a}{a^2 + b^2}\right) + \left(\frac{-b}{a^2 + b^2}\right)i$

Section 4.8, page 293

1. 2 3. 6 5. -30

7. $x = 0$ (multiplicity 54); $x = -4/5$ (multiplicity 1)

9. $x = 0$ (multiplicity 15); $x = \pi$ (multiplicity 14); $x = \pi + 1$ (multiplicity 13)

11. $x = 1 + 2i$ or $1 - 2i$; $f(x) = (x - 1 - 2i)(x - 1 + 2i)$

13. $x = -\frac{1}{3} + \frac{2\sqrt{5}}{3}i$ or $-\frac{1}{3} - \frac{2\sqrt{5}}{3}i$;
$$f(x) = \left(x + \frac{1}{3} - \frac{2\sqrt{5}}{3}i\right)\left(x + \frac{1}{3} + \frac{2\sqrt{5}}{3}i\right)$$

15. $x = 3$ or $-\frac{3}{2} + \frac{3\sqrt{3}}{2}i$ or $-\frac{3}{2} - \frac{3\sqrt{3}}{2}i$;
$$f(x) = (x - 3)\left(x + \frac{3}{2} - \frac{3\sqrt{3}}{2}i\right)\left(x + \frac{3}{2} + \frac{3\sqrt{3}}{2}i\right)$$

17. $x = -2$ or $1 + \sqrt{3}i$ or $1 - \sqrt{3}i$; $f(x) = (x + 2)(x - 1 - \sqrt{3}i)(x - 1 + \sqrt{3}i)$

19. $x = 1$ or i or -1 or $-i$; $f(x) = (x - 1)(x - i)(x + 1)(x + i)$

21. $x = \sqrt{5}$ or $-\sqrt{5}$ or $\sqrt{2}i$ or $-\sqrt{2}i$; $f(x) = (x - \sqrt{5})(x + \sqrt{5})(x - \sqrt{2}i)(x + \sqrt{2}i)$

23. Many correct answers, including $(x - 1)(x - 7)(x + 4)$

25. Many correct answers, including $(x - 1)(x - 2)^2(x - \pi)^3$

27. $f(x) = 2x(x - 4)(x + 3)$

In Exercises 29–40, there are many correct answers, including the following.

29. $x^2 - 4x + 5$ 31. $(x - 2)(x^2 - 4x + 5)$

33. $(x + 3)(x^2 - 2x + 2)(x^2 - 2x + 5)$

35. $x^2 - 2x + 5$ 37. $(x - 4)^2(x^2 - 6x + 10)$

39. $(x^4 - 3x^3)(x^2 - 2x + 2)$

41. $3x^2 - 6x + 6$ 43. $-2x^3 + 2x^2 - 2x + 2$

45. Many correct answers, including
$$x^2 - (1 - i)x + (2 + i)$$

47. Many correct answers, including
$$x^3 - 5x^2 + (7 + 2i)x - (3 + 6i)$$

49. $3, -\frac{1}{2} + \frac{\sqrt{3}}{2}i, -\frac{1}{2} - \frac{\sqrt{3}}{2}i$ 51. $i, -i, -1, -2$

53. $1, 2i, -2i$ 55. $i, -i, 2 + i, 2 - i$

57. **(a)** Since $z + w = (a + c) + (b + d)i$, $\overline{z + w} = (a + c) - (b + d)i$. Since $\overline{z} = a - bi$ and $\overline{w} = c - di$, $\overline{z} + \overline{w} = (a - bi) + (c - di) = (a + c) - (b + d)i$. Hence $\overline{z + w} = \overline{z} + \overline{w}$.

(b) Since $zw = (ac - bd) + (ad + bc)i$, $\overline{zw} = (ac - bd) - (ad + bc)i$. Since $\overline{z} = a - bi$ and $\overline{w} = c - di$, $\overline{z}\,\overline{w} = (a - bi)(c - di) = (ac - bd) - (ad + bc)i$. Hence $\overline{zw} = \overline{z}\cdot\overline{w}$.

59. **(a)**

$\overline{f(z)} = \overline{az^3 + bz^2 + cz + d}$	(definition of $f(z)$)
$= \overline{az^3} + \overline{bz^2} + \overline{cz} + \overline{d}$	(Exercise 57(a))
$= \overline{a}\overline{z^3} + \overline{b}\overline{z^2} + \overline{c}\,\overline{z} + \overline{d}$	(Exercise 57(b))
$= a\overline{z^3} + b\overline{z^2} + c\overline{z} + d$	($r = \overline{r}$ for r real)
$= a\overline{z}^3 + b\overline{z}^2 + c\overline{z} + d$	(Exercise 57(b))
$= f(\overline{z})$	(definition of f)

(b) Since $f(z) = 0$, we have $0 = \overline{0} = \overline{f(z)} = f(\overline{z})$. Hence $\overline{z}$ is a root of $f(x)$.

61. If $f(z)$ is a polynomial with real coefficients, then $f(z)$ can be factored as $g_1(z)g_2(z)g_3(z)\cdots g_k(z)$, where each $g_i(z)$ is a polynomial with real coefficients and degree 1 or 2. The rules of polynomial multiplication show that the degree of $f(z)$ is the sum: degree $g_1(z)$ + degree $g_2(z)$ + degree $g_3(z)$ + ⋯ + degree $g_k(z)$. If all of the $g_i(z)$ have degree 2, then this last sum is an even number. But $f(z)$ has odd degree, so this can't occur. Therefore, at least one of the $g_i(z)$ is a first-degree polynomial and hence must have a real root. This root is also a root of $f(z)$.

Chapter 4 Review, page 296

1. (2, 3) **3.** (4, −4) **5.** (1.5, −5.75)

7. **(a)** $y = 260 - x$
(b) $A = -x^2 + 260x - 350$
(c) $x = 130$ ft; $y = 130$ ft

9. $x = 7.5$ ft; $y = 105$ ft

11. (a), (c), (e), (f) **13.** 0

15.

2	1	−5	8	1	−17	16	−4
		2	−6	4	10	−14	4
	1	−3	2	5	−7	2	0

other factor: $x^5 - 3x^4 + 2x^3 + 5x^2 - 7x + 2$

17. Many correct answers, including $f(x) = 5(x - 1)^2(x + 1) = 5x^3 - 5x^2 - 5x + 5$

19. −1 and 5/3 **21.** $\sqrt[3]{2}$ **23.** $0, \pm\sqrt{\dfrac{2 + \sqrt{21}}{2}}$

25. **(a)** $1, -1, 3, -3, \frac{1}{2}, -\frac{1}{2}, \frac{3}{2}, -\frac{3}{2}$ **(b)** 3
(c) $3, \dfrac{1 + \sqrt{3}}{2}, \dfrac{1 - \sqrt{3}}{2}$

27. 1 **29.** $3, -3, \sqrt{2}, -\sqrt{2}$

31. $(x^4 - 4x^3 + 16x - 16) \div (x - 5)$ is $x^3 + x^2 + 5x + 41$ with remainder 189. Since all coefficients and the remainder are positive, 5 is an upper bound for the roots.

33. 1, −1, 1.867, −.867 **35.** $2x + h + 1$

37. Many correct answers

39. (c)

41. $-3 \le x \le 9$ and $-35 \le y \le 15$

43. $-2 \le x \le 18$ and $-500 \le y \le 1100$

45. **(a)**

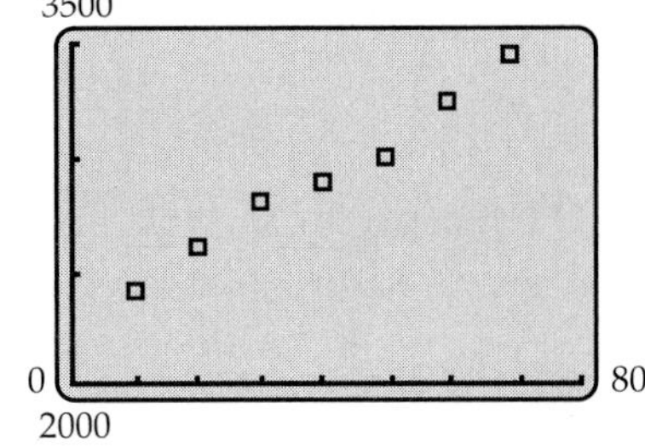

(b) $c(x) \approx 0.008515x^3 - 1.1094x^2 + 56.2583x + 2017.2576$
(c) \$3466.54
(d) Average cost of 35: \$85.49; average cost of 75: \$47.84

47.

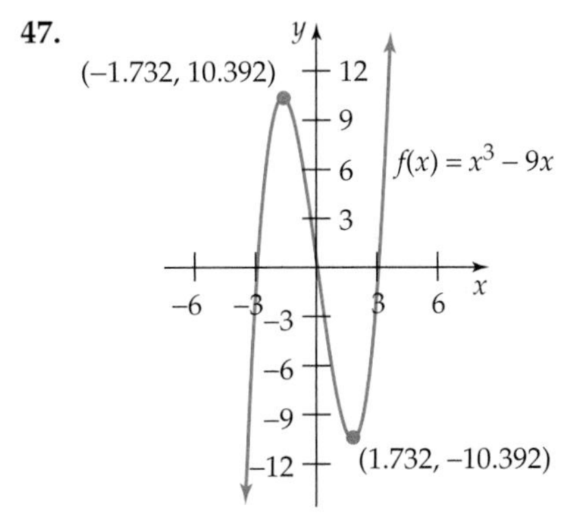

49.

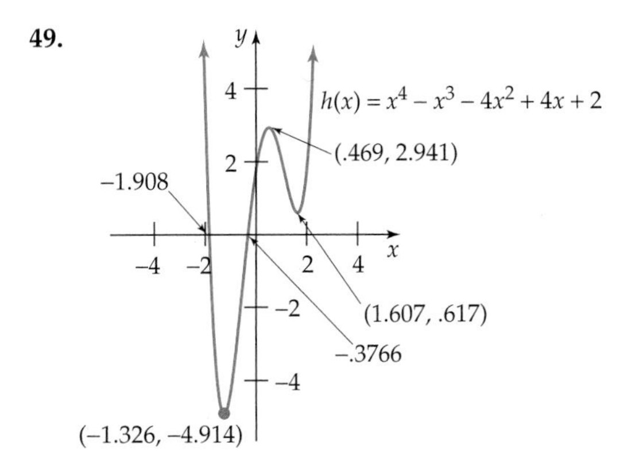

51.

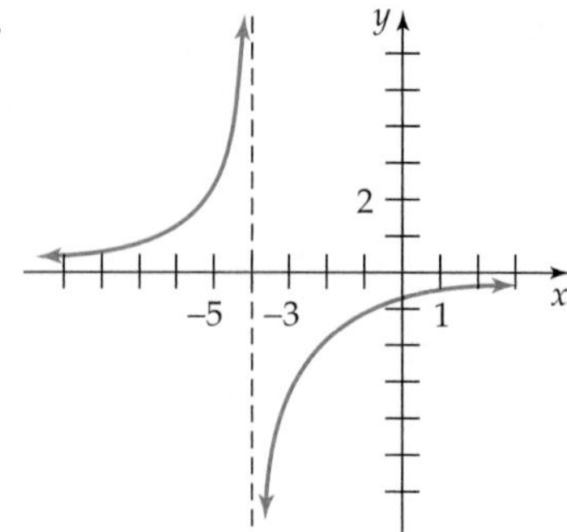

53.

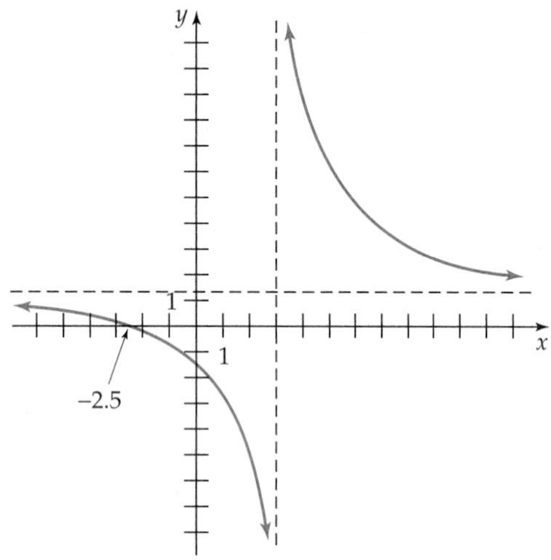

55. Vertical asymptotes $x = -2$ and $x = 3$; horizontal asymptote $y = 0$; hole at $x = 1$

57. Overall: $-4.7 \le x \le 4.7$ and $-5 \le y \le 5$ (Adjust x range to avoid erroneous vertical lines on widescreen calculators.) Near $x = 3$: $1 \le x \le 25$ and $-.5 \le y \le .1$

59. $-19 \le x \le 19$ and $-8 \le y \le 8$

61. At least 400 bags (priced at \$1.75 each) and at most 2500 bags (priced at \$0.70 each)

63. **(a)** $T = \dfrac{40}{x} + \dfrac{110}{x+25}$, where x is the speed of the car and $0 < x < 55$ (speed limit)

(b) At least 44.08 mph

65. $\dfrac{1}{(x+h+1)(x+1)}$ 67. (d)

69. $y \le -17$ or $y \ge 13$

71. $(-\infty, -2)$ and $\left(-\dfrac{1}{3}, \infty\right)$

73. $-\sqrt{3} \le x < -1$ or $-1 < x < 1$ or $1 < x \le \sqrt{3}$

75. $x \le -1$ or $0 \le x \le 1$

77. (e) 79. $x \le -4/3$ or $x \ge 0$

81. $x \le -7$ or $x > -4$

83. $x < -2\sqrt{3}$ or $-3 < x < 2\sqrt{3}$

85. $x < \dfrac{1-\sqrt{13}}{6}$ or $x > \dfrac{1+\sqrt{13}}{6}$

87. $x = \dfrac{-3 \pm \sqrt{31}i}{2}$ 89. $x = \dfrac{3 \pm \sqrt{31}i}{10}$

91. $x = \sqrt{2/3}$ or $-\sqrt{2/3}$ or i or $-i$

93. $x = -2$ or $1 + \sqrt{3}i$ or $1 - \sqrt{3}i$

95. $i, -i, 2, -1$

97. Many correct answers, including $x^4 - 2x^3 + 2x^2$

Chapter 5

Section 5.1, page 312

1. $.09$ 3. $.08^6$ 5. $6\sqrt{2}$

7. $1/2$ 9. $-1 + \sqrt{3}$ 11. $14 + 3\sqrt{3}$

13. $(a^2 + b^2)^{1/3}$ 15. $a^{3/16}$ 17. $4t^{27/10}$

19. $4a^4/b$ 21. $d^5/(2\sqrt{c})$ 23. $15\sqrt{5}$

25. $(4x + 2y)^2$ 27. 1 29. $x^{9/2}$

31. $c^{42/5}d^{10/3}$ 33. $\dfrac{a^{1/2}}{49b^{5/2}}$ 35. $\dfrac{2^{9/2}a^{12/5}}{3^4b^4}$

37. a^x 39. $\dfrac{1}{x^{1/5}y^{2/5}}$ 41. 1

43. $x^{7/6} - x^{11/6}$ 45. $x - y$

47. $x + y - (x + y)^{3/2}$ 49. $(x^{1/3} + 3)(x^{1/3} - 2)$

51. $(x^{1/2} + 3)(x^{1/2} + 1)$

53. $(x^{2/5} + 9)(x^{1/5} + 3)(x^{1/5} - 3)$

55. $3\sqrt{2}/4$ 57. $\dfrac{3\sqrt{3} - 3}{4}$ 59. $\dfrac{2\sqrt{x} - 4}{x - 4}$

61. $\dfrac{1}{\sqrt{x+h+1} + \sqrt{x+1}}$

63. $\dfrac{2x + h}{\sqrt{(x+h)^2 + 1} + \sqrt{x^2 + 1}}$

65. **(a)** The square (or any even power) of a real number is never negative. Graphically these equations lie strictly above or on the x-axis.

(b) $\sqrt[3]{-8} = -2$, whereas $\sqrt[6]{(-8)^2} = 2$

67. 3. $\sqrt[n]{\sqrt[m]{c}} = \sqrt[mn]{c}$ 4. $\sqrt[m]{cd} = \sqrt[m]{c}\sqrt[m]{d}$

5. $\sqrt[m]{\dfrac{c}{d}} = \dfrac{\sqrt[m]{c}}{\sqrt[m]{d}}$

69. (a) Since its graph passes the horizontal line test, f is one-to-one and hence has an inverse.

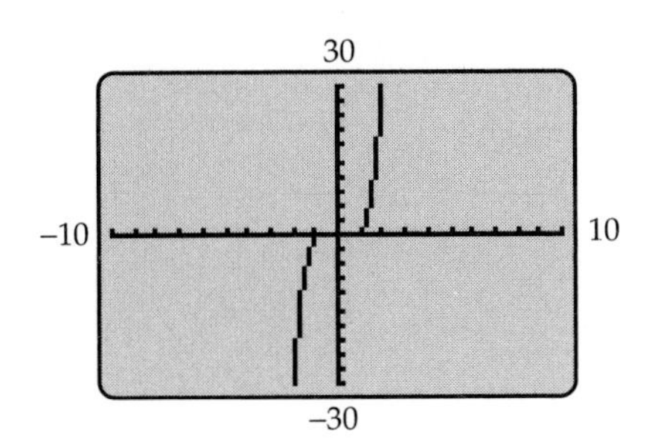

(b) $(g \circ f)(x) = g(f(x)) = (f(x))^{1/5} = (x^5)^{1/5} = x$ and $(f \circ g)(x) = f(g(x)) = (g(x))^5 = (x^{1/5})^5 = x$

71. (a) $x^{1/2} < x^{1/4} < x^{1/6}$ when $0 < x < 1$ because the graph of $y = x^{1/2}$ lies below the graph of $y = x^{1/4}$, which lies below the graph of $y = x^{1/6}$.

(b) $x^{1/2} > x^{1/4} > x^{1/6}$ when $x > 1$ because the graph of $y = x^{1/2}$ lies above the graph of $y = x^{1/4}$, which lies above the graph of $y = x^{1/6}$.

73. (a) The graph of g is the graph of f shifted horizontally 3 units to the left.

(b) The graph of h is the graph of f shifted vertically 2 units downward.

(c) The graph of k is the graph of f shifted horizontally 3 units to the left, then vertically 2 units downward.

Excursion 5.1.A, page 319

1. $x = 7$ **3.** $x = 4$ **5.** $x = -2$

7. $x = \pm 3$ **9.** $x = -1$ or 2 **11.** $x = 9$

13. $x = 1/2$ **15.** $x = 1/2$ or -4 **17.** $x \approx \pm .73$

19. $x \approx -1.17$ or 2.59 or $x = -1$

21. $x \approx 1.658$ **23.** $x = 6$

25. $x = 3$ or 7 **27.** No solutions

29. $x \approx -.457$ or 1.40 **31.** $b = \sqrt{\dfrac{a^2}{A^2 - 1}}$

33. $u = \sqrt{\dfrac{x^2}{1 - K^2}}$ **35.** $x = 4$ **37.** $x = 4$

39. $x = -1$ or -8 **41.** $x = -64$ or 8

43. $x = 16$ **45.** $x = -1/2$ or $1/3$

47. $x \approx 105.236$ **49.** $x \approx -.283$

51. (a) 11.47 ft or 29.91 ft **(b)** 21.00 ft

53. (a) $I = \dfrac{x}{(x^2 + 1024)^{3/2}}$ **(b)** 22.63 ft

Section 5.2, page 330

1.

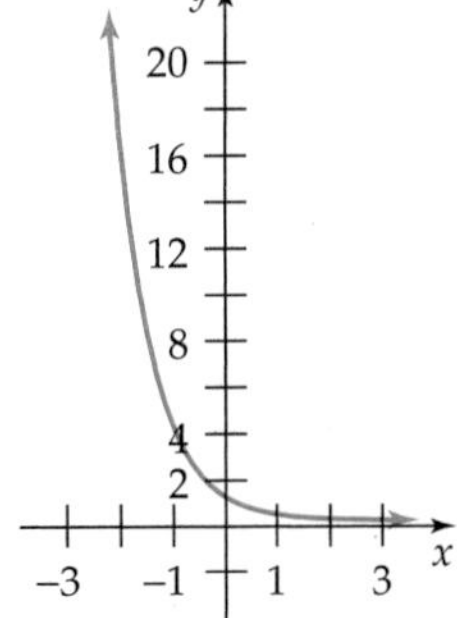

3.

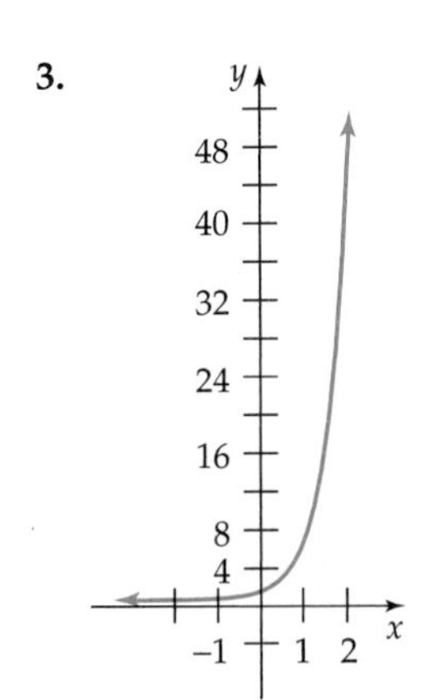

5.

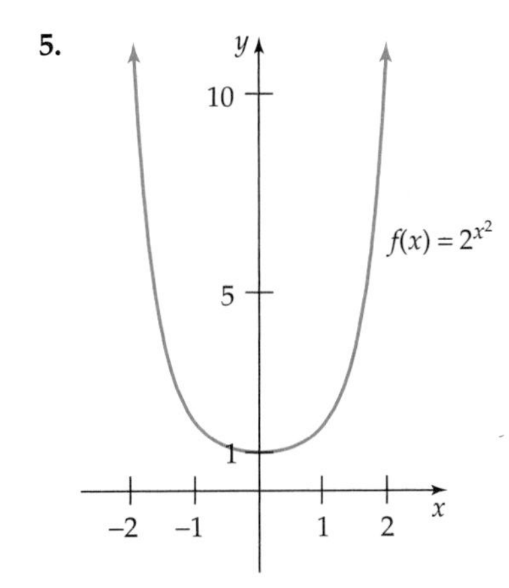

7. Shift the graph of h vertically 5 units downward.

9. Stretch the graph of h away from the x-axis by a factor of 3.

11. Shift the graph of h horizontally 2 units to the left, then vertically 5 units downward.

13. $f(x)$: C; $g(x)$: A; $h(x)$: B

15. Neither

17. Even

19. Even **21.** 11 **23.** .8

25. $\dfrac{10^{x+h} - 10^x}{h}$ **27.** $\dfrac{(2^{x+h} + 2^{-x-h}) - 2^x - 2^{-x}}{h}$

29. $-3 \le x \le 3$ and $0 \le y \le 12$

31. $-4 \le x \le 4$ and $0 \le y \le 10$

33. $-10 \le x \le 10$ and $0 \le y \le 20$

35. $-10 \le x \le 10$ and $0 \le y \le 6$

37. The x-axis is a horizontal asymptote for the left side of the graph; local maximum at $(-1.44, -0.53)$.

39. No asymptotes; local minimum at (0, 1)

41. The x-axis is a horizontal asymptote; local maximum at (0, 1).

43. (a) $t = 0.75, C = 8; t = 1, C = 16$
(b) $C(t) = 2^{4t}$

45. $f(x) = -315(e^{0.00418} + e^{-0.00418x}) + 1260$

47. (a) About 520 in 15 days; about 1559 in 25 days
(b) In 29.3 days

49. (a) 1980: 74.06; 2000: 76.34
(b) 1930

51. (a) $g(x) = 67.4(1.026)^x$ **(b)** 112.62 million

53. (a) $f(x) = 75(1.05)^x$ **(b)** About 164

55. (a) $f(x) = 0.75^x$ **(b)** About 8 ft

57. (a) $f(x) = 6(3^x)$ or $f(x) = 18(3^{x-1})$
(b) 3 **(c)** No; yes

59. (a)

Folds	0	1	2	3	4
Thickness	.002	.004	.008	.016	.032

(b) $f(x) = .002(2^x)$ **(c)** 2097.15 in. = 174.76 ft
(c) 43

61. About 256; about 654

63. (a) $M(x) = 5(.5^{x/5730})$ **(b)** 3.08 g; 1.90 g
(c) After 13,304.65 years

65. (a) 100,000 now; 83,527 in 2 months; 58,275 in 6 months
(b) No. The graph continues to decrease toward zero.

67. (a) The current population is 10, and in 5 years it will be about 149.
(b) After about 9.55 years

69. Many correct answers: $f(x) = a^x$ for any nonnegative constant a

71. (a) The graph of f is the mirror image of the graph of g.
(b) $k(x) = f(x)$; see (a).

73. (a) Not entirely
(b) The graph of $f_8(x)$ appears to coincide with the graph of $g(x)$ on most calculator screens; when $-2.4 \le x \le 2.4$, the maximum error is at most .01.
(c) Not at the right side of the viewing window; $f_{12}(x)$

Excursion 5.2.A, page 339

1. Annually: \$1469.33; quarterly: \$1485.95; monthly: \$1489.85; weekly: \$1491.37

3. \$585.83 **5.** \$610.40 **7.** \$639.76

9. \$563.75 **11.** \$582.02 **13.** \$568.59

15. About 9.9 years **17.** About 5.00%

19. About 5.92%

Section 5.3, page 349

1. 4 **3.** -2.5 **5.** $10^3 = 1000$

7. $10^{2.88} = 750$ **9.** $e^{1.0986} = 3$ **11.** $e^{-4.6052} = .01$

13. $e^{z+w} = x^2 + 2y$ **15.** $\log .01 = -2$

17. $\log 3 = .4771$ **19.** $\ln 25.79 = 3.25$

21. $\ln 5.5527 = 12/7$ **23.** $\ln w = 2/r$ **25.** $\sqrt{43}$

27. 15 **29.** 1/2 **31.** 931 **33.** $x + y$

35. x^2 **37.** $\ln(x^2y^3)$ **39.** $\log(x - 3)$

41. $\ln(x^{-7})$ **43.** $3\ln(e - 1)$ **45.** $\log(20xy)$

47. $2u + 5v$ **49.** $\frac{1}{2}u + 2v$ **51.** $\frac{2}{3}u + \frac{1}{6}v$

53. $(-1, \infty)$ **55.** $(-\infty, 0)$

57. (a) For all $x > 0$
(b) According to the fourth property of natural logarithms on page 344, $e^{\ln x} = x$ for every $x > 0$.

59. False; the right side is not defined when $x < 0$, but the left side is.

61. True by the Power Law

63. False; the graph of the left side differs from the graph of the right side.

65. Stretch the graph of g away from the x-axis by a factor of 2.

67. Shift the graph of g horizontally 4 units to the right.

69. Shift the graph of g horizontally 3 units to the left, then shift it vertically 4 units downward.

71.

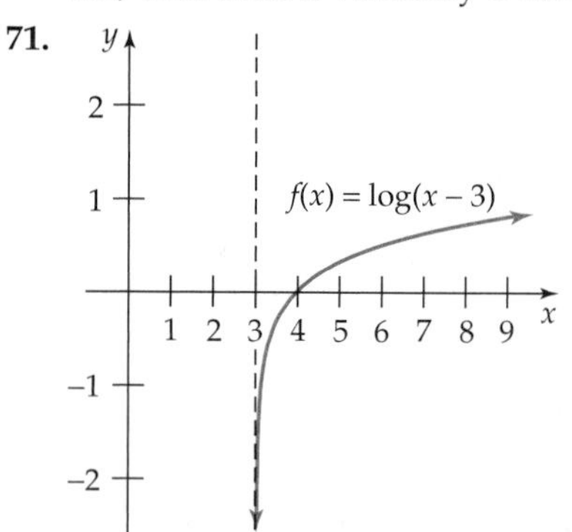

73.

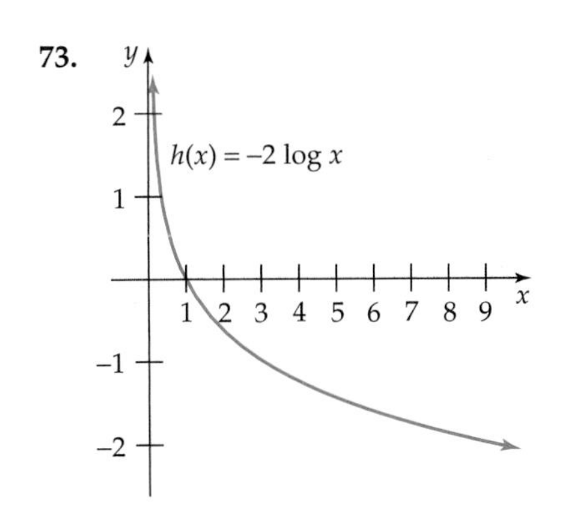

75. $0 \le x \le 9.4$ and $-6 \le y \le 6$ (vertical asymptote at $x = 1$)

77. $-10 \le x \le 10$ and $-3 \le y \le 3$

79. $0 \le x \le 20$ and $-6 \le y \le 3$

81. .5493 **83.** $-.2386$

85. **(a)** $\dfrac{\ln(3+h) - \ln 3}{h}$ **(b)** $h \approx 2.2$

87. **(a)** Both are acid. **(b)** .01 **(c)** Yes

89. $A = -9$; $B = 10$

91. $(f \circ g)(x) = \dfrac{1}{1 + e^{-\ln[x/(1-x)]}} = \dfrac{1}{1 + \dfrac{1}{e^{\ln[x/(1-x)]}}}$

$$= \frac{1}{1 + \dfrac{1}{\dfrac{x}{1-x}}} = \frac{1}{1 + \dfrac{1-x}{x}} = \frac{1}{\dfrac{1}{x}} = x;$$

$$(g \circ f)(x) = \ln\left(\frac{\dfrac{1}{1+e^{-x}}}{1 - \dfrac{1}{1+e^{-x}}}\right)$$

$$= \ln\left(\frac{\dfrac{1}{1+e^{-x}}}{1 - \dfrac{1}{1+e^{-x}}} \cdot \frac{1+e^{-x}}{1+e^{-x}}\right)$$

$$= \ln\left(\frac{1}{1 + e^{-x} - 1}\right) = \ln(e^x) = x$$

93. 2 **95.** Approximately 2.54

97. 20 decibels

99. Approximately 66 decibels

101. About 4392 meters

103. **(a)** 77 **(b)** 66; 59 **(c)** About 14 weeks

105. **(a)** 9.9 days **(b)** About 6986

107. $n = 30$ gives an approximation with a maximum error of .00001 when $-.7 \le x \le .7$.

109. **(a)** No advertising: about 120 bikes; $1000: about 299 bikes; $10,000: about 513 bikes
(b) $1000, yes; $10,000; no
(c) Yes; yes (but not as worthwhile as spending $1000)

Excursion 5.3.A, page 358

1.

x	0	1	2	4
$f(x) = \log_4 x$	Not defined	0	.5	1

3.

x	1/36	1/6	1	216
$h(x) = \log_6 x$	-2	-1	0	3

5.

x	0	1/7	$\sqrt{7}$	49
$f(x) = 2\log_7 x$	Not defined	-2	1	4

7.

x	-2.75	-1	1	29
$h(x) = 3\log_2(x+3)$	-6	3	6	15

9. $\log .01 = -2$ **11.** $\log \sqrt[3]{10} = 1/3$

13. $\log r = 7k$ **15.** $\log_7 5{,}764{,}801 = 8$

17. $\log_3(1/9) = -2$ **19.** $10^4 = 10{,}000$

21. $10^{2.88} \approx 750$ **23.** $5^3 = 125$ **25.** $2^{-2} = \dfrac{1}{4}$

27. $10^{z+w} = x^2 + 2y$ **29.** $\sqrt{43}$ **31.** $\sqrt{x^2 + y^2}$

33. 1/2 **35.** 6 **37.** $b = 3$ **39.** $b = 20$

41. 5 **43.** 3 **45.** 4 **47.** $\log \dfrac{x^2y^3}{z^6}$

49. $\log(x^2 - 3x)$ **51.** $\log_2(5c)$ **53.** $\log_4\left(\dfrac{1}{49c^2}\right)$

55. $\ln\left(\dfrac{(x+1)^2}{x+2}\right)$ **57.** $\log_2(x)$ **59.** $\ln(e^2 - 2e + 1)$

61. 3.3219 **63.** .8271 **65.** 1.1115 **67.** 1.6199

69. True **71.** True **73.** False **75.** 397^{398}

77. $\log_b u = \dfrac{\log_a u}{\log_a b}$ **79.** $\log_{10} u = 2\log_{100} u$

81. $\log_b x = \dfrac{1}{2}\log_b v + 3 = \log_b \sqrt{v} + \log_b b^3 = \log_b(b^3/\sqrt{v})$; hence $x = b^3\sqrt{v}$.

83. $f(x) = g(x)$ only when $x \approx .123$, so the statement is false.

85.

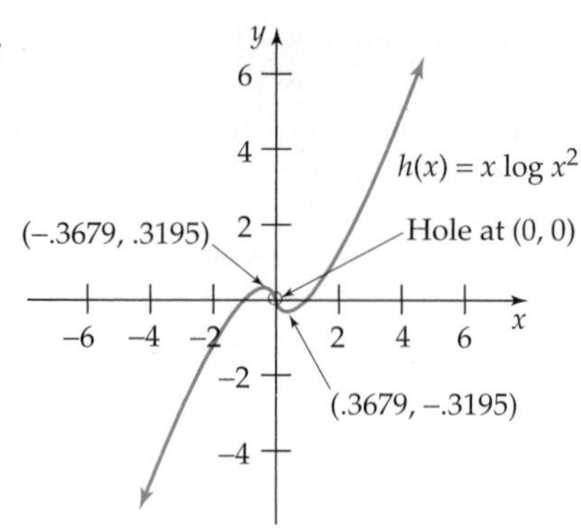

Section 5.4, page 366

1. $x = 4$ **3.** $x = 1/9$ **5.** $x = \dfrac{1}{2}$ or -3

7. $x = -2$ or $-1/2$ **9.** $x = \ln 5/\ln 3 \approx 1.465$

11. $x = \ln 3/\ln 1.5 \approx 2.7095$

13. $x = \dfrac{\ln 3 - 5\ln 5}{\ln 5 + 2\ln 3} \approx -1.825$

15. $x = \dfrac{\ln 2 - \ln 3}{3\ln 2 + \ln 3} \approx -.1276$

17. $x = (\ln 5)/2 \approx .805$

19. $x = (-\ln 3.5)/1.4 \approx -.895$

21. $x = 2\ln(5/2.1)/\ln 3 \approx 1.579$ **23.** $x = 0$ or 1

25. $x = \ln 2 \approx .693$ or $x = \ln 3 \approx 1.099$

27. $x = \ln 3 \approx 1.099$

29. $x = \ln 2/\ln 4 = 1/2$ or $x = \ln 3/\ln 4 \approx .792$

31. $x = \ln(t + \sqrt{t^2 + 1})$

33. If $\ln u = \ln v$, then $e^{\ln u} = e^{\ln v}$, so $u = v$.

35. $x = 9$ **37.** $x = 5$ **39.** $x = 6$ **41.** $x = 3$

43. $x = \dfrac{-5 + \sqrt{37}}{2}$ **45.** $x = 9/(e - 1)$ **47.** $x = 5$

49. $x = \pm\sqrt{10001}$ **51.** $x = \sqrt{\dfrac{e + 1}{e - 1}}$

53. Approximately 3689 years old

55. Approximately 2534 years ago

57. Approximately 444,000,000 years

59. Approximately 10.413 years

61. Approximately 9.853 days

63. Approximately 6.99%

65. **(a)** Approximately 22.5 years
(b) Approximately 22.1 years

67. \$3197.05 **69.** 79.36 years

71. **(a)** About 2.1548% **(b)** In the year 2013

73. **(a)** $k \approx 21.459$ **(b)** $t \approx .182$

75. **(a)** There are 20 bacteria at the beginning and 2500 three hours later.
(b) $\dfrac{\ln 2}{\ln 5} \approx .43$

77. **(a)** At the outbreak: 200 people; after 3 weeks: about 2795 people
(b) In about 6.02 weeks

79. **(a)** $k \approx .229$, $c \approx 83.3$
(b) 12.43 weeks

Section 5.5, page 375

1. Cubic, logistic

3. Exponential, quadratic, cubic

5. Exponential, logarithmic, quadratic, cubic

7. Quadratic, cubic

9. Quadratic, cubic

11. Ratios range from approximately 5.059 to 5.076; exponential is appropriate

13. **(a)** For large values of x the term $54e^{-.0228x}$ is close to zero so the quantity $(1 + 54e^{-.0228x})$ is slightly larger than 1, which means $\dfrac{384.57}{1 + 54e^{-.0228x}}$ is always less than (but very close to) 384.57.

(b)

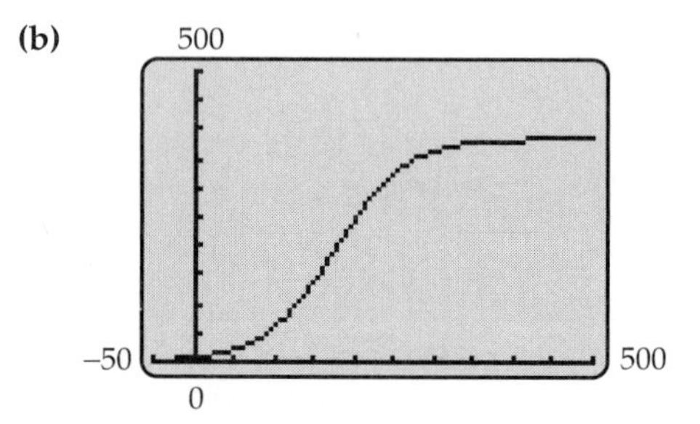

15. **(a)**

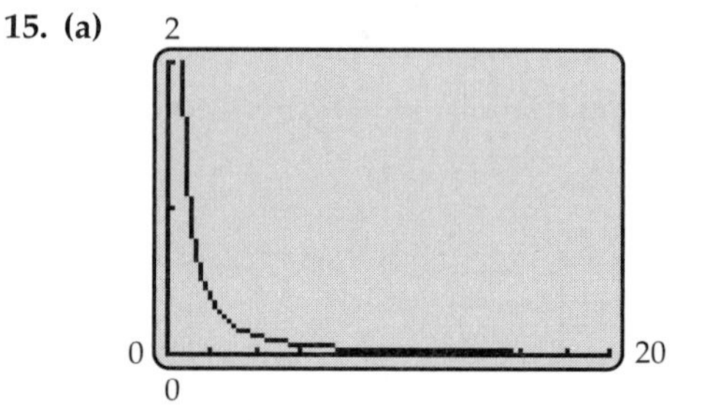

(b)

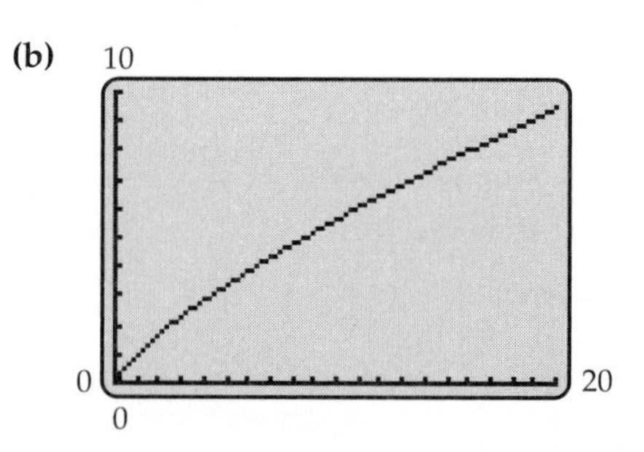

(c)

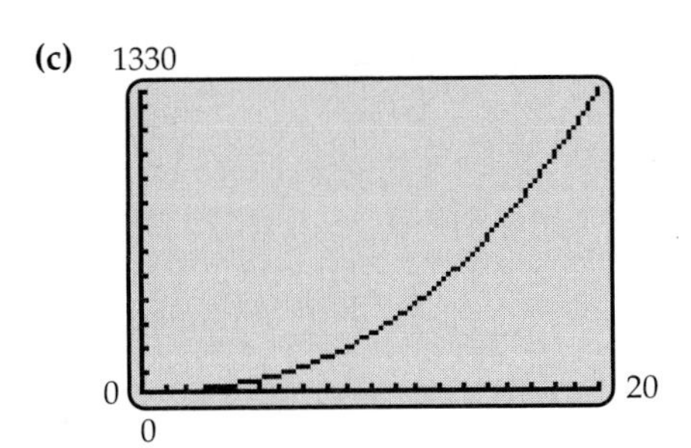

17. Power model

19. Power or logarithmic model

21. (a)

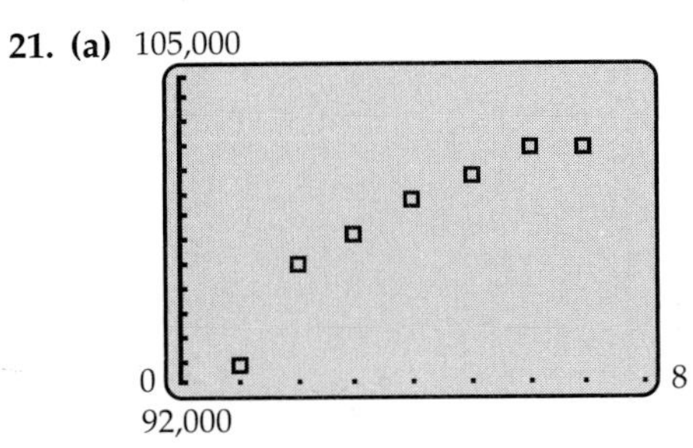

(b)

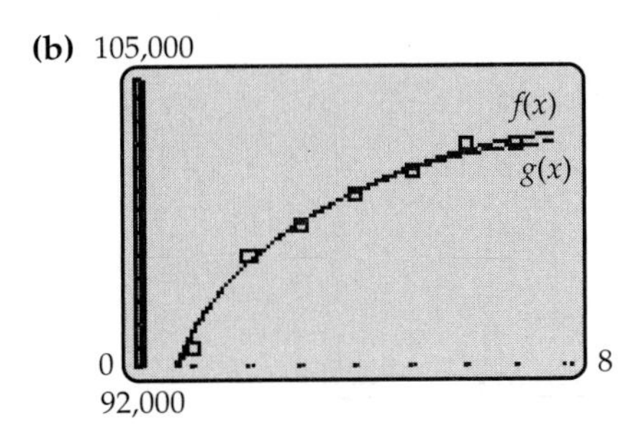

(c)

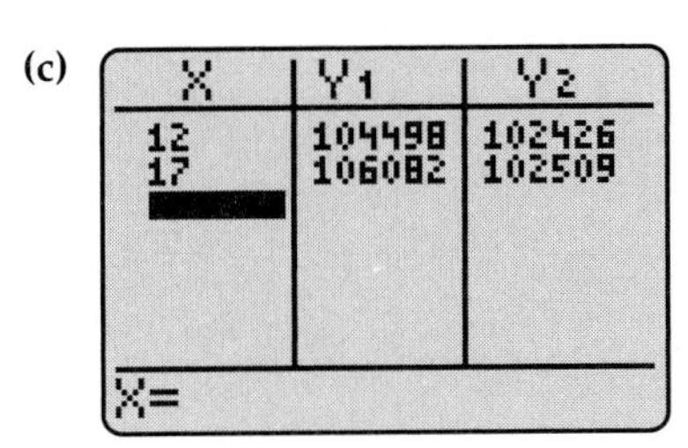

X	Y1	Y2
12	104498	102426
17	106082	102509

X=

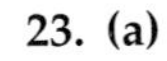

23. (a)

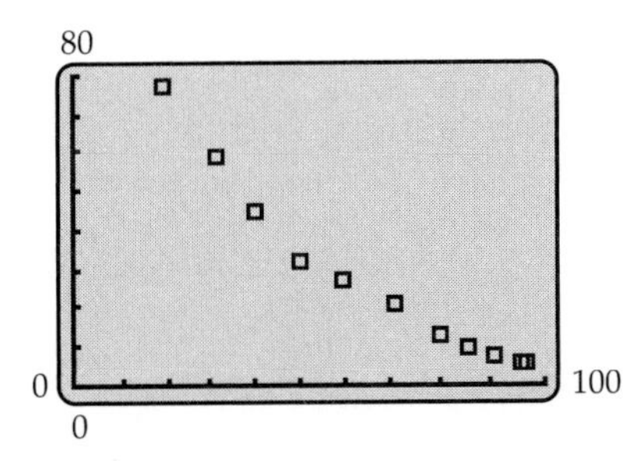

(b)

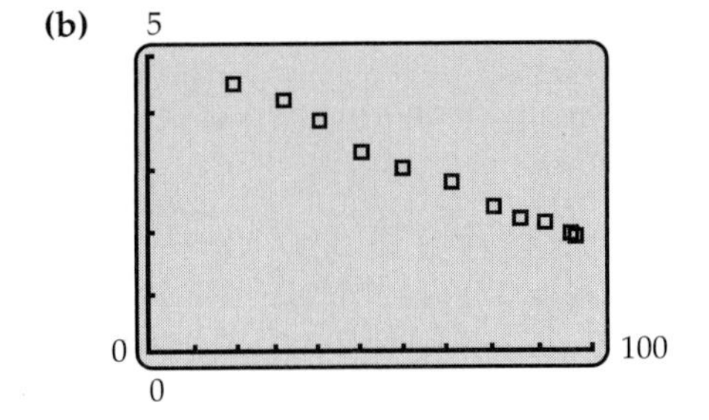

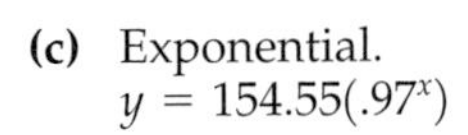

(c) Exponential.
$y = 154.55(.97^x)$

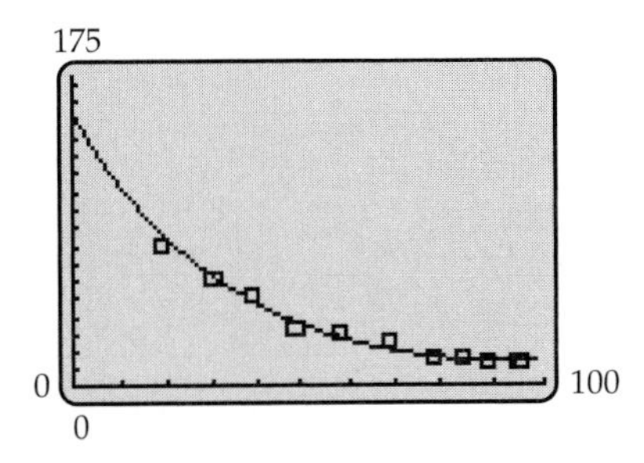

25. (a)

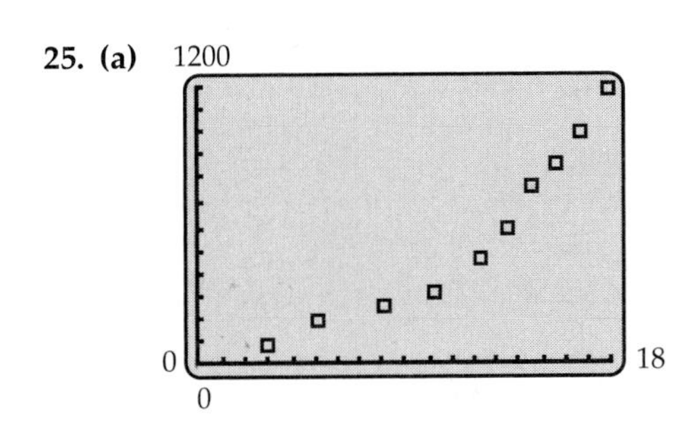

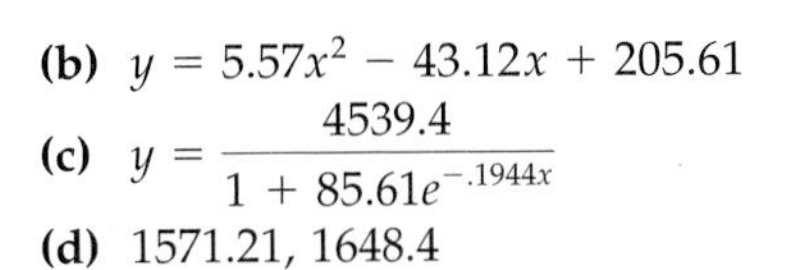

(b) $y = 5.57x^2 - 43.12x + 205.61$

(c) $y = \dfrac{4539.4}{1 + 85.61e^{-.1944x}}$

(d) 1571.21, 1648.4

27. (a)

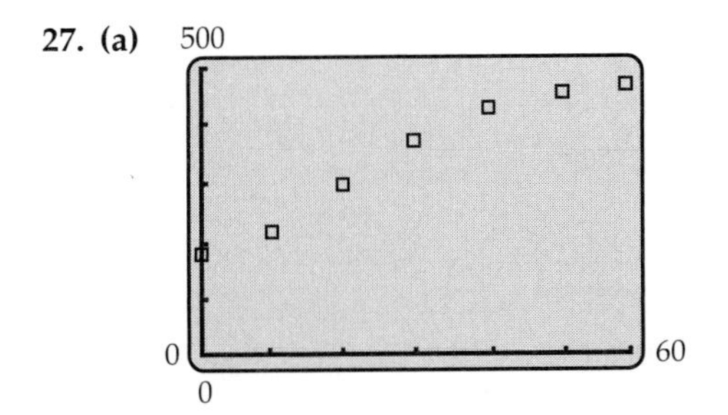

(b) $y = -.0025x^3 + .167x^2 + 3.784x + 170.411$
This model is not reasonable because it predicts smaller and smaller farm sizes after 1997, decreasing to 0 in about 30 years.

(c) $y = \dfrac{519.79}{1 + 2.236e^{-.056x}}$

(d) 482.7 acres

(e) Never; the model levels off at approximately 519.8 acres.

29. (a)

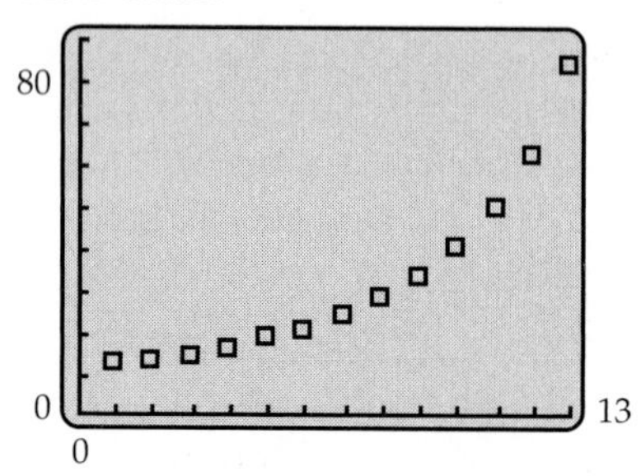

(b) $y = 10.48(1.16^x)$

(c–d)

Year	Worldwide shipments (thousands)	Predicted number shipments (thousands)	Worldwide shipments ratio (current to previous)
1985	14.7	12.2	
1986	15.1	14.1	1.03
1987	16.7	16.4	1.11
1988	18.1	19	1.08
1989	21.3	22	1.18
1990	23.7	25.5	1.11
1991	27	29.6	1.14
1992	32.4	34.4	1.2
1993	38.9	39.9	1.2
1994	47.9	46.2	1.23
1995	60.2	53.6	1.26
1996	70.9	62.2	1.18
1997	84.3	72.2	1.19

An exponential model may not be appropriate.

Chapter 5 Review, page 382

1. c^2 **3.** $a^{10/3}b^{42/5}$ **5.** $u^{1/2} - v^{1/2}$ **7.** $c^2d^4/2$

9. $\dfrac{2}{\sqrt{2x + 2h + 1} + \sqrt{2x + 1}}$

11. $x = \dfrac{5 - \sqrt{5}}{2}$ **13.** No solutions

15. $x = -1.733$ or 5.521

17. $-3 \le x \le 3$ and $0 \le y \le 2$

19.

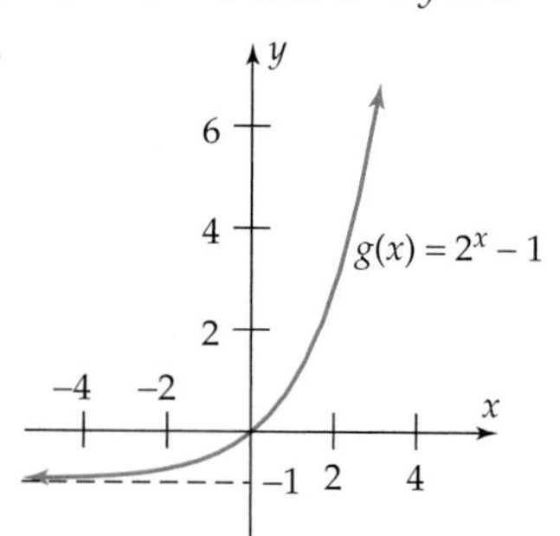

21.

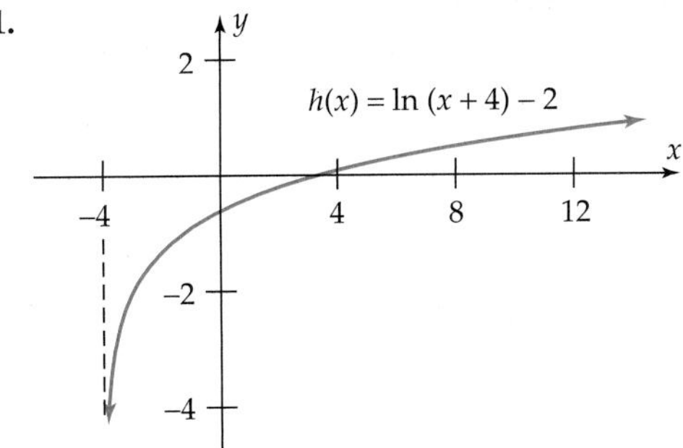

23. (a) About 2.03
(b) About 31.97
(c) Approximately 6 to 10 months
(d) Never; however, at the end of 18 months about 99.6% of the program will be mastered.

25. $\ln 756 = 6.628$ **27.** $\ln(u + v) = r^2 - 1$

29. $\log 756 = 2.8785$ **31.** $e^{7.118} = 1234$

33. $e^t = rs$ **35.** $5^u = cd - k$ **37.** 3 **39.** 3/4

41. $2 \ln x$ **43.** $\ln(9y/x^2)$ **45.** $\ln(1/x^{10})$

47. 2 **49.** (c)

51. (c) **53.** $x = \pm\sqrt{2}$ **55.** $x = \dfrac{3 \pm \sqrt{57}}{4}$

57. $x = -1/2$ **59.** $x = e^{(u-c)/d}$ **61.** $x = 2$

63. $x = 101$ **65.** About 1.64 mg

67. Approximately 12 years **69.** \$452.89 **71.** 7.6

73. (a) 12°F
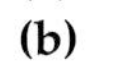
(b)

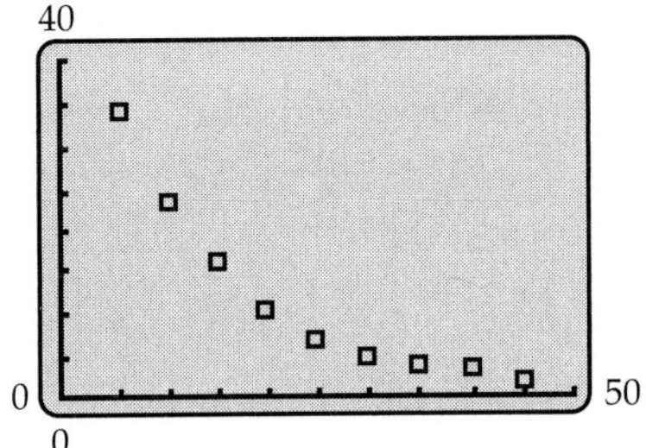

(c) The points $(x, \ln(y))$ are approximately linear.
(d) $y = 41.639(.936^x)$
(e) 9.72°F

Chapter 6

Section 6.1, page 398

1. $2\pi/9$ **3.** $\pi/9$ **5.** $\pi/18$

7. $9\pi/4, 17\pi/4, -7\pi/4, -15\pi/4$

9. $11\pi/6, 23\pi/6, -13\pi/6, -25\pi/6$

11. $5\pi/3$ **13.** $3\pi/4$ **15.** $3\pi/5$ **17.** $7 - 2\pi$

19. $\pi/30$ **21.** $-\pi/15$ **23.** $5\pi/12$ **25.** $3\pi/4$

27. $-5\pi/4$ **29.** $31\pi/6$ **31.** 36° **33.** −18°

35. 135° **37.** 4° **39.** −75° **41.** 972°

43. $4\pi/3$ **45.** $7\pi/6$ **47.** $41\pi/6$ **49.** 8π cm

51. 17/4 **53.** 50/9 **55.** 2000 **57.** 5

59. 8.75 **61.** 3490.66 mi **63.** 942.48 mi

65. 7π **67.** 4π **69.** 42.5π **71.** $2\pi k$

73. 3 radians ($\approx 171.9°$)

75. (a) 400π radians per min
(b) 800π in. per min or $\dfrac{200\pi}{3}$ ft per min

77. (a) 5π radians per sec
(b) 6.69 mph

79. 15.92 ft

Section 6.2, page 407

1. −1 **3.** 0 **5.** 0 **7.** 0 **9.** 0

11. $\sin t = 1/\sqrt{5}, \cos t = -2/\sqrt{5}, \tan t = -1/2$

13. $\sin t = -4/5, \cos t = -3/5, \tan t = 4/3$

15. $\sin\left(\frac{5\pi}{6}\right) = \frac{1}{2}, \cos\left(\frac{5\pi}{6}\right) = \frac{-\sqrt{3}}{2}, \tan\left(\frac{5\pi}{6}\right) = -\frac{\sqrt{3}}{3}$

17. $\sin\left(\frac{7\pi}{3}\right) = \frac{\sqrt{3}}{2}, \cos\left(\frac{7\pi}{3}\right) = \frac{1}{2}, \tan\left(\frac{7\pi}{3}\right) = \sqrt{3}$

19. $\sin\left(\frac{11\pi}{4}\right) = \frac{\sqrt{2}}{2}, \cos\left(\frac{11\pi}{4}\right) = -\frac{\sqrt{2}}{2}, \tan\left(\frac{11\pi}{4}\right) = -1$

21. $\sin\left(-\frac{3\pi}{2}\right) = 1, \cos\left(-\frac{3\pi}{2}\right) = 0,$
$\tan\left(-\frac{3\pi}{2}\right)$ not defined

23. $\sin\left(-\frac{23\pi}{6}\right) = \frac{1}{2}, \cos\left(-\frac{23\pi}{6}\right) = \frac{\sqrt{3}}{2},$
$\tan\left(-\frac{23\pi}{6}\right) = \frac{\sqrt{3}}{3}$

25. $\sin\left(-\frac{19\pi}{3}\right) = -\frac{\sqrt{3}}{2}, \cos\left(-\frac{19\pi}{3}\right) = \frac{1}{2},$
$\tan\left(-\frac{19\pi}{3}\right) = -\sqrt{3}$

27. $\sin\left(-\frac{15\pi}{4}\right) = \frac{\sqrt{2}}{2}, \cos\left(-\frac{15\pi}{4}\right) = \frac{\sqrt{2}}{2},$
$\tan\left(-\frac{15\pi}{4}\right) = 1$

29. $\sin\left(-\frac{17\pi}{2}\right) = -1, \cos\left(-\frac{17\pi}{2}\right) = 0,$
$\tan\left(-\frac{17\pi}{2}\right)$ is not defined

31. $-\sqrt{3}/2$ **33.** $-\sqrt{2}/2$ **35.** $\frac{\sqrt{2}}{4}\left(1 - \sqrt{3}\right)$

37. $\sin t = 7/\sqrt{53}, \cos t = 2/\sqrt{53}, \tan t = 7/2$

39. $\sin t = -6/\sqrt{61}, \cos t = -5/\sqrt{61}, \tan t = 6/5$

41. $\sin t = -10/\sqrt{103}$, $\cos t = \sqrt{3}/\sqrt{103}$, $\tan t = -10/\sqrt{3}$

43. $\sin t = -3/\sqrt{10}$, $\cos t = 1/\sqrt{10}$, $\tan t = -3$

45. $\sin t = -5/\sqrt{34}$, $\cos t = 3/\sqrt{34}$, $\tan t = -5/3$

47. $\sin t = 1/\sqrt{5}$, $\cos t = -2/\sqrt{5}$, $\tan t = -1/2$

49. Quadrant I: $\sin t(+)$, $\cos t(+)$, $\tan t(+)$; Quadrant II: $\sin t(+)$, $\cos t(-)$, $\tan t(-)$; Quadrant III: $\sin t(-)$, $\cos t(-)$, $\tan t(+)$; Quadrant IV: $\sin t(-)$, $\cos t(+)$, $\tan t(-)$

51. Positive since $0 < 1 < \pi/2$

53. Negative since $\pi/2 < 3 < \pi$

55. Positive since $0 < 1.5 < \pi/2$

57. $t = \dfrac{\pi}{2} + 2\pi n$, n any integer

59. $t = \pi n$, n any integer

61. $t = \dfrac{\pi}{2} + \pi n$, n any integer

63. $\sin(\cos 0) = \sin 1$, while $\cos(\sin 0) = \cos 0 = 1$. Since $\sin 1 < 1$ (draw a picture!), $\cos(\sin 0)$ is larger than $\sin(\cos 0)$.

65. **(a)** Each horse moves through an angle of 2π radians in 1 min. The angle between horses A and B is $\pi/4$ radians ($= \frac{1}{8}$ of 2π radians). It takes $\frac{1}{8}$ min for each horse to move through an angle of $\pi/4$ radians. Thus the position occupied by B at time t will be occupied by A $\frac{1}{8}$ min later, that is, at time $t + \frac{1}{8}$. Therefore, $B(t) = A(t + \frac{1}{8})$.
(b) $C(t) = A(t + \frac{1}{3})$
(c) $E(t) = D(t + \frac{1}{8})$; $F(t) = D(t + \frac{1}{3})$
(d) The triangles in Figure B are similar, so that $\dfrac{5}{1} = \dfrac{D(t)}{A(t)}$. Therefore, $D(t) = 5A(t)$.
(e) $E(t) = 5B(t) = 5A(t + \frac{1}{8})$; $F(t) = 5C(t) = 5A(t + \frac{1}{3})$
(f) Since horse A travels through an angle of 2π radians each minute and its starting angle is 0 radians, then at the end of t min horse A will be on the terminal side of an angle of $2\pi t$ radians, at the point where it intersects the unit circle. $A(t)$ is the second coordinate of this point; hence, $A(t) = \sin(2\pi t)$.
(g) $B(t) = A(t + \frac{1}{8}) = \sin[2\pi(t + \frac{1}{8})] = \sin(2\pi t + \pi/4)$; $C(t) = \sin(2\pi t + 2\pi/3)$
(h) $D(t) = 5\sin(2\pi t)$; $E(t) = 5\sin(2\pi t + \pi/4)$; $f(t) = 5\sin(2\pi t + 2\pi/3)$

Section 6.3, page 416

1. $(fg)(t) = 3\sin^2 t + 6\sin t\cos t$

3. $3\sin^3 t + 3\sin^2 t\tan t$ **5.** $(\cos t - 2)(\cos t + 2)$

7. $(\sin t - \cos t)(\sin t + \cos t)$ **9.** $(\tan t + 3)^2$

11. $(3\sin t + 1)(2\sin t - 1)$

13. $(\cos^2 t + 5)(\cos t + 1)(\cos t - 1)$

15. $(f \circ g)(t) = \cos(2t + 4)$, $(g \circ f)(t) = 2\cos t + 4$

17. $(f \circ g)(t) = \tan(t^2 + 2)$, $(g \circ f)(t) = \tan^2(t + 3) - 1$

19. Yes **21.** No **23.** No **25.** $\sin t = -\sqrt{3}/2$

27. $\sin t = \sqrt{3}/2$ **29.** $-3/5$ **31.** $-3/5$

33. $3/4$ **35.** $-3/4$

37. $-\sqrt{21}/5$ **39.** $-2/5$ **41.** $-\sqrt{21}/5$

43. $\dfrac{\sqrt{2 + \sqrt{2}}}{2}$ **45.** $\dfrac{\sqrt{2 - \sqrt{2}}}{2}$ **47.** $\sin^2 t - \cos^2 t$

49. $\sin t$ **51.** $|\sin t \cos t|\sqrt{\sin t}$

53. $1/4$ **55.** $\cos t + 2$ **57.** $\cos t$

59. $f(t + \pi) = \sin 2(t + \pi) = \sin(2t + 2\pi) = \sin 2t = f(t)$

61. $f(t + \pi/2) = \sin 4(t + \pi/2) = \sin(4t + 2\pi) = \sin 4t = f(t)$

63. $f(t + \pi/2) = \tan 2(t + \pi/2) = \tan(2t + \pi) = \tan 2t = f(t)$

65. **(a)** There is no such number k.
(b) If we substitute $t = 0$ in $\cos(t + k) = \cos t$, we get $\cos k = \cos 0 = 1$.
(c) If there were such a number k, then by part (b), $\cos k = 1$, which is impossible by part (a). Therefore, there is no such number k, and the period is 2π.

Section 6.4, page 426

1. $t = \ldots, -2\pi, -\pi, 0, \pi, 2\pi, \ldots$; or $t = \pi k$, where k is any integer

3. $t = \ldots, -7\pi/2, -3\pi/2, \pi/2, 5\pi/2, 9\pi/2, \ldots$; or $t = \pi/2 + 2\pi k$, where k is any integer

5. $t = \ldots, -3\pi, -\pi, \pi, 3\pi, \ldots$; or $t = \pi + 2k\pi$, where k is any integer

7. 11 **9.** 1.4

11. Shift the graph of f vertically 3 units upward.

13. Reflect the graph of f in the horizontal axis.

15. Shift the graph of f vertically 5 units upward.

17. Stretch the graph of f away from the horizontal axis by a factor of 3.

19. Stretch the graph of f away from the horizontal axis by a factor of 3, then shift the resulting graph vertically 2 units upward.

21. Shift the graph of f horizontally 2 units to the right.

23. 2 solutions **25.** 2 solutions

27. 2 solutions **29.** 2 solutions

31. Possibly an identity **33.** Possibly an identity
35. Possibly an identity **37.** Not an identity
39. Possibly an identity **41.** Not an identity

43. (a) Yes if proper value of k is used; no
(b) $0, 2\pi, 4\pi, 6\pi$, etc. So why do the graphs look identical?

45. (a) 80
(b) 14 or 15 on 96-pixel-wide screens; up to 40–50 on wider screens; quite different from part (a). Explain what's going on. [*Hint:* How many points have to be plotted in order to get even a rough approximation of one full wave? How many points is the calculator plotting for the entire graph?]

47. (a) $-\pi \leq t \leq \pi$
(b) $n = 15$; $f_{15}(2)$ and $g(2)$ are identical in the first nine decimal places and differ in the tenth, a very good approximation.

49. $r(t)/s(t)$, where $r(t) = f_{15}(t)$ in Exercise 47 and $s(t) = f_{16}(t)$ in Exercise 48.

51. The y-coordinate of the new point is the same as the x-coordinate of the point on the unit circle. To explain what's going on, look at the definition of the cosine function.

Section 6.5, page 437

1. Amplitude: 3; period: π, phase shift: $+\dfrac{\pi}{2}$

3. Amplitude: 7; period: $\dfrac{2\pi}{7}$, phase shift: $-\dfrac{1}{49}$

5. Amplitude: 1; period: 1; phase shift: 0

7. Amplitude: 6; period: $\dfrac{2}{3}$; phase shift: $-\dfrac{1}{3\pi}$

9. $f(t) = 3\sin\left(8t - \dfrac{8\pi}{5}\right)$, (other answers possible)

11. $f(t) = \dfrac{2}{3}\sin(2\pi t)$

13. $f(t) = 7\sin\left(\dfrac{6\pi}{5}t + \dfrac{3\pi^2}{5}\right)$

15. $f(t) = 2\sin 4t$ **17.** $f(t) = 1.5\cos\dfrac{t}{2}$

19. (a) $f(t) = -12\sin\left(10t + \dfrac{\pi}{2}\right)$
(b) $g(t) = -12\cos 10t$

21. (a) $f(t) = -\sin 2t$ **(b)** $g(t) = -\cos\left(2t - \dfrac{\pi}{2}\right)$

23. (a) $f(t) = \dfrac{1}{2}\sin 8t$ **(b)** $g(t) = \dfrac{1}{2}\cos\left(8t - \dfrac{\pi}{2}\right)$

25.

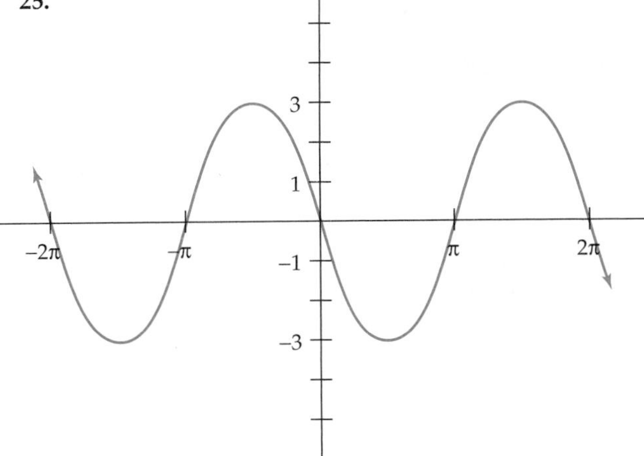

27.

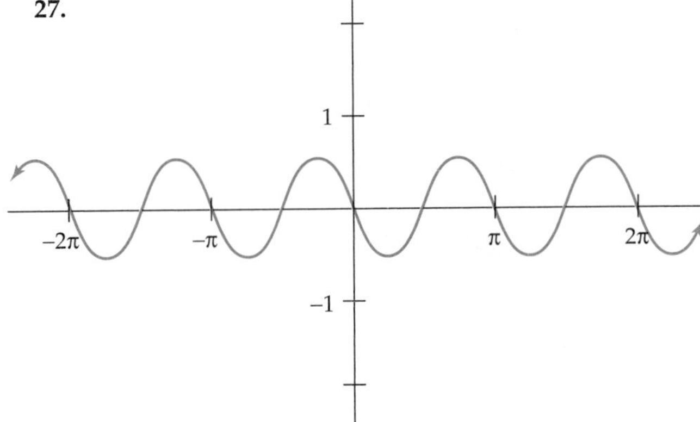

29.

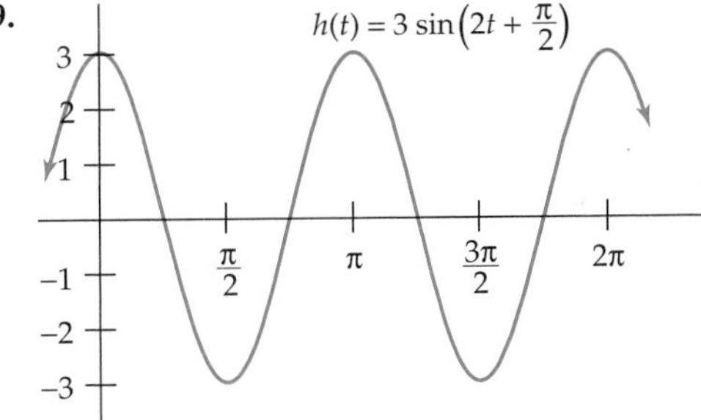

31. Local maximum at $t = 5\pi/6 \approx 2.6180$; local minimum at $t = 11\pi/6 \approx 5.7596$

33. Local maxima at $t = \pi/6 \approx .5236$, $t = 5\pi/6 \approx 2.6180$, $t = 3\pi/2 \approx 4.7124$; local minima at $t = \pi/2 \approx 1.5708$, $t = 7\pi/6 \approx 3.6652$, $t = 11\pi/6 \approx 5.7596$

35. $A \approx 5.3852$, $b = 1$, $c \approx 1.1903$

37. $A \approx 3.8332$, $b = 4$, $c \approx 1.4572$

39. All waves in the graph of g are of equal height, which is not the case with the graph of f.

41. 1/980,000; 980,000 **43.** $f(t) = 125\sin(\pi t/5)$

45. $f(t) = \cos 20\pi t + \sqrt{16 - \sin^2(20\pi t)}$

47. $h(t) = 6\sin(\pi t/2)$ **49.** $h(t) = 6\cos(\pi t/2)$

51. $d(t) = 10\sin(\pi t/2)$

53. **(a)** At least four (starting point, high point, low point, ending point)
(b) At least 301 (4 points for the first wave and 3 for each of the remaining 99 waves because the starting point of one is the ending point of the preceding one)
(c) Answers vary from 95 to 239.

55. **(a)**

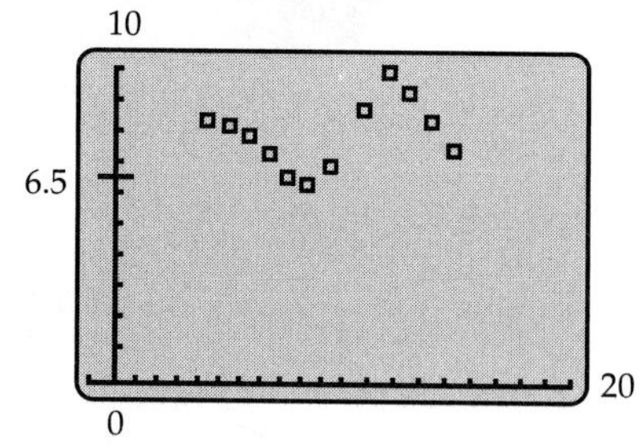

(b) Roughly periodic;
$y = 1.22\sin(.8134x - 2.22) + 7.77$

57. **(a)** $y = 2.515\sin(.414x + 1.785) + 2.262$
(b) About 15.177, which appears reasonable.
(c)

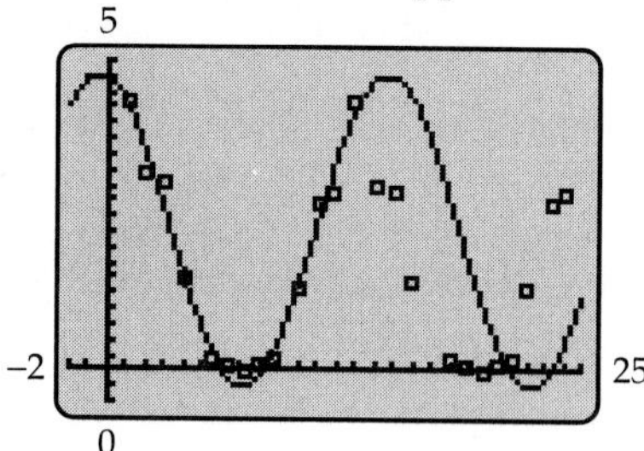

The model is not a good fit in the second year.
(d) $y = 2.048\sin(.522x + 1.049) + 1.67$
(e) About 12.037

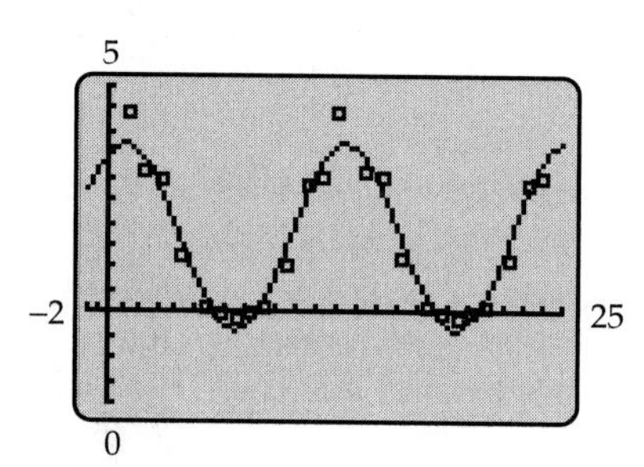

This model provides a much better fit.

59. **(a)** $k = 9.8/\pi^2$
(b) When k is replaced by (k + .01% of k), the value of ω changes and the period of the pendulum becomes approximately 2.000099998 sec, meaning that the clock loses .000099998 sec every 2 sec, for a total of approximately 397.43 sec (6.62 min) during the three months.

Excursion 6.5.A, page 447

1. $A \approx 2.2361, b = 1, c \approx 1.1071$

3. $A \approx 5.3852, b = 4, c \approx -1.1903$

5. $A \approx 5.1164, b = 3, c \approx -.7442$

7. $0 \le t \le 2\pi$ and $-5 \le y \le 5$ (one period)

9. $-10 \le t \le 10$ and $-10 \le y \le 10$

11. $0 \le t \le \pi/50$ and $-2 \le y \le 2$ (one period)

13. $0 \le t \le .04$ and $-7 \le y \le 7$ (one period)

15. $0 \le t \le 10$ and $-6 \le y \le 10$ (one period)

17. To the left of the y-axis, the graph lies above the t-axis, which is a horizontal asymptote of the graph. To the right of the y-axis, the graph makes waves of amplitude 1, of shorter and shorter period.
Window: $-3 \le t \le 3.2$ and $-2 \le y \le 2$

19. The graph is symmetric with respect to the y-axis and consists of waves along the t-axis, whose amplitude slowly increases as you move farther from the origin in either direction.
Window: $-30 \le t \le 30$ and $-6 \le y \le 6$

21. The graph is symmetric with respect to the y-axis and consists of waves along the t-axis whose amplitude rapidly decreases as you move farther from the origin in either direction.
Window: $-30 \le t \le 30$ and $-.3 \le y \le 1$

23. The function is periodic with period π. (Why?) The graph lies on or below the t-axis because the logarithmic function is negative for numbers between 0 and 1 and $|\cos t|$ is always between 0 and 1. The graph has vertical asymptotes when $t = \pm\pi/2, \pm 3\pi/2, \pm 5\pi/2, \pm 7\pi/2, \ldots$ ($\cos t = 0$ at these points and ln 0 is not defined).
Window: $-2\pi \le t \le 2\pi$ and $-3 \le y \le 1$ (two periods)

Section 6.6, page 453

1. Fourth quadrant **3.** Second quadrant

5. Fourth quadrant

7. $\sin t = 4/5, \cos t = 3/5, \tan t = 4/3, \cot t = 3/4, \sec t = 5/3, \csc t = 5/4$

9. $\sin t = 12/13, \cos t = -5/13, \tan t = -12/5, \cot t = -5/12, \sec t = -13/5, \csc t = 13/12$

11. $\sin t = 5/\sqrt{26}$, $\cos t = -1/\sqrt{26}$, $\tan t = -5$, $\cot t = -1/5$, $\sec t = -\sqrt{26}$, $\csc t = \sqrt{26}/5$

13. $\sin t = \sqrt{3}/\sqrt{5}$, $\cos t = \sqrt{2}/\sqrt{5}$, $\tan t = \sqrt{3}/\sqrt{2}$, $\cot t = \sqrt{2}/\sqrt{3}$, $\sec t = \sqrt{5}/\sqrt{2}$, $\csc t = \sqrt{5}/\sqrt{3}$

15. $\sin t = \dfrac{3}{\sqrt{12 + 2\sqrt{2}}}$, $\cos t = \dfrac{1 + \sqrt{2}}{\sqrt{12 + 2\sqrt{2}}}$, $\tan t = \dfrac{3}{1 + \sqrt{2}}$, $\cot t = \dfrac{1 + \sqrt{2}}{3}$, $\sec t = \dfrac{\sqrt{12 + 2\sqrt{2}}}{1 + \sqrt{2}}$, $\csc t = \dfrac{\sqrt{12 + 2\sqrt{2}}}{3}$

17. $\sin\left(\dfrac{4\pi}{3}\right) = -\dfrac{\sqrt{3}}{2}$, $\cos\left(\dfrac{4\pi}{3}\right) = -\dfrac{1}{2}$, $\tan\left(\dfrac{4\pi}{3}\right) = \sqrt{3}$, $\cot\left(\dfrac{4\pi}{3}\right) = \dfrac{1}{\sqrt{3}}$, $\sec\left(\dfrac{4\pi}{3}\right) = -2$, $\csc\left(\dfrac{4\pi}{3}\right) = \dfrac{-2}{\sqrt{3}}$

19. $\sin\left(\dfrac{7\pi}{4}\right) = -\dfrac{\sqrt{2}}{2}$, $\cos\left(\dfrac{7\pi}{4}\right) = \dfrac{\sqrt{2}}{2}$, $\tan\left(\dfrac{7\pi}{4}\right) = -1$, $\cot\left(\dfrac{7\pi}{4}\right) = -1$, $\sec\left(\dfrac{7\pi}{4}\right) = \sqrt{2}$, $\csc\left(\dfrac{7\pi}{4}\right) = -\sqrt{2}$

21. -3.8287

23. (a) 5.6511; 5.7618; 5.7731; 5.7743
(b) $(\sec 2)^2$

25. $\cos t + \sin t$ **27.** $1 - 2\sec t + \sec^2 t$

29. $\cot^3 t - \tan^3 t$

31. $\csc t(\sec t - \csc t)$ **33.** $(-1)(\tan^2 t + \sec^2 t)$

35. $(\cos^2 t + 1 + \sec^2 t)(\cos t - \sec t)$ **37.** $\cot t$

39. $\dfrac{2\tan t + 1}{3\sin t + 1}$ **41.** $4 - \tan t$

43. $1 + \cot^2 t = 1 + \dfrac{\cos^2 t}{\sin^2 t} = \dfrac{\sin^2 t + \cos^2 t}{\sin^2 t} = \dfrac{1}{\sin^2 t} = \csc^2 t$

45. $\sec(-t) = \dfrac{1}{\cos(-t)} = \dfrac{1}{\cos t} = \sec t$

47. $\sin t = \dfrac{\sqrt{3}}{2}$, $\cos t = -\dfrac{1}{2}$, $\tan t = -\sqrt{3}$, $\csc t = \dfrac{2\sqrt{3}}{3}$, $\sec t = -2$, $\cot t = -\dfrac{\sqrt{3}}{3}$

49. $\sin t = 1$, $\cos t = 0$, $\tan t$ is undefined, $\cot t = 0$, $\sec t$ is undefined, $\csc t = 1$

51. $\sin t = \dfrac{12}{13}$, $\cos t = -\dfrac{5}{13}$, $\tan t = -\dfrac{12}{5}$, $\cot t = -\dfrac{5}{12}$, $\sec t = -\dfrac{13}{5}$, $\csc t = \dfrac{13}{12}$

53. Possibly an identity **55.** Not an identity

57. Look at the graph of $y = \sec t$ in Figure 6–62 on page 452. If you draw in the line $y = t$, it will pass through $(-\pi/2, -\pi/2)$ and $(\pi/2, \pi/2)$, and obviously will not intersect the graph of $y = \sec t$ when $-\pi/2 \le t \le \pi/2$. But it will intersect each part of the graph that lies above the horizontal axis, to the right of $t = \pi/2$; it will also intersect those parts that lie below the horizontal axis, to the left of $-\pi/2$. The first coordinate of each of these infinitely many intersection points will be a solution of $\sec t = t$.

59. (a) $\dfrac{\cos\theta\sin\theta}{2}$ **(b)** $\dfrac{\tan\theta}{2}$ **(c)** $\dfrac{\theta}{2}$

Chapter 6 Review, page 457

1. $\dfrac{\pi}{3}$ **3.** 324° **5.** $\dfrac{11\pi}{9}$

7. $-495°$ **9.** $\cos v = -\dfrac{1}{3}$ **11.** 0

13. .809 **15.** $-\dfrac{\sqrt{3}}{2}$ **17.** $\dfrac{\sqrt{3}}{2}$

19.

t	0	$\pi/6$	$\pi/4$	$\pi/3$	$\pi/2$
$\sin t$	0	$1/2$	$\sqrt{2}/2$	$\sqrt{3}/2$	1
$\cos t$	1	$\sqrt{3}/2$	$\sqrt{2}/2$	$1/2$	0

21. $9/4$ **23.** 0 **25.** $-\sqrt{\dfrac{2}{3}}$ **27.** $-\dfrac{3}{5}$

29. $-\dfrac{12}{13}$ **31.** $-\dfrac{1}{2}$ **33.** $\dfrac{\sqrt{3}}{2}$ **35.** (c)

37. $\dfrac{7}{\sqrt{58}}$ **39.** $-\dfrac{7}{3}$ **41.** $y = -\sqrt{3}x$ **43.** $-\dfrac{3}{2}$

45. $\dfrac{2}{\sqrt{13}}$ **47.** See Figure 6–63 on page 453.

49. (d) **51.** $-\dfrac{3}{2}$ **53.** 0

55. (d) **57.** $-\dfrac{1}{\sqrt{3}}$

59. (a) $\dfrac{3}{2}$ **(b)** $t = \dfrac{\pi}{5}$

61.

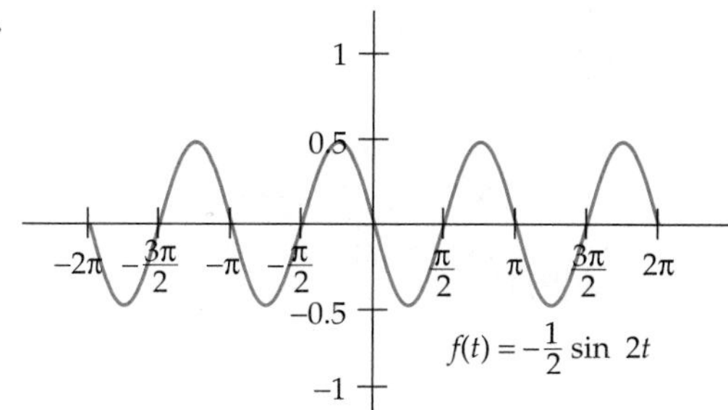

63. Not an identity **65.** Possibly an identity

67. $1/2$ **69.** $2\cos(5t/2)$

71. $f(t) = 8\sin\left(\frac{2\pi t - 28\pi}{5}\right)$ is one possibility.

73. $A \approx 10.5588, b = 4, c = .4581$

75. $0 \leq t \leq \pi/50$ and $-5 \leq y \leq 5$ (one period)

Chapter 7

Section 7.1, page 475

	sin	cos	tan	csc	sec	cot
1.	$\frac{3\sqrt{13}}{13}$	$\frac{2\sqrt{13}}{13}$	$\frac{3}{2}$	$\frac{\sqrt{13}}{3}$	$\frac{\sqrt{13}}{2}$	$\frac{2}{3}$
3.	$\frac{6\sqrt{61}}{61}$	$\frac{-5\sqrt{61}}{61}$	$\frac{-6}{5}$	$\frac{\sqrt{61}}{6}$	$\frac{-\sqrt{61}}{5}$	$\frac{-5}{6}$
5.	$\frac{-\sqrt{22}}{11}$	$\frac{-3\sqrt{11}}{11}$	$\frac{\sqrt{2}}{3}$	$\frac{-\sqrt{22}}{2}$	$\frac{-\sqrt{11}}{3}$	$\frac{3\sqrt{2}}{2}$

7. $\sin\theta = \sqrt{\frac{2}{11}}, \cos\theta = \frac{3}{\sqrt{11}}, \tan\theta = \frac{\sqrt{2}}{3}$

9. $\sin\theta = \sqrt{\frac{3}{7}}, \cos\theta = \frac{2}{\sqrt{7}}, \tan\theta = \frac{\sqrt{3}}{2}$

11. $\sin\theta = \frac{h}{m}, \cos\theta = \frac{d}{m}, \tan\theta = \frac{h}{d}$

13. $c = 36$ **15.** $c = 36$ **17.** $c = 8.4$

19. $h = 25\sqrt{2}/2$ **21.** $h = 300$ **23.** $h = 50\sqrt{3}$

25. $c = 4\sqrt{3}/3$ **27.** $a = 10\sqrt{3}/3$

29. $\measuredangle A = 40°, a = 10\cos 50° = 6.4, c = 10\sin 50° = 7.7$

31. $\measuredangle C = 76°, b = 6/\sin 14° = 24.8,$ $c = 6/\tan 14° = 24.1$

33. $\measuredangle C = 25°, a = 5\tan 65° = 10.7,$ $b = 5/\cos 65° = 11.8$

35. $\measuredangle C = 18°, a = 3.5\sin 72° = 3.3,$ $c = 3.5\cos 72° = 1.1$

37. About 48.59° **39.** About 48.19°

41. $\measuredangle A = 33.7°, \measuredangle C = 56.3°$

43. $\measuredangle A = 44.4°, \measuredangle C = 45.6°$

45. $\measuredangle A = 48.2°, \measuredangle C = 41.8°$

47. $\measuredangle A = 60.8°, \measuredangle C = 29.2°$

49. About 15.32 ft **51.** About 3.06°

53. 30°

55. 460.2 ft **57.** 8598.3 ft

59. No **61.** 27.5 ft; no **63.** 19.25 ft; 53.97°

65. 351.1 m **67.** 10.1 ft **69.** 1.6 mi

71. **(a)** 56.7 ft **(b)** 9.7 ft

73. 205.7 ft **75.** 173.2 mi **77.** 52.5 mph

79. **(a)** $A(t) = 200\cos t\sin t$
(b) $t \approx .7854$; approximately 14.1421 by 7.0711

81. 449.1 ft

Section 7.2, page 485

1. $a = 4.2, \measuredangle B = 125.0°, \measuredangle C = 35.0°$

3. $c = 13.9, \measuredangle A = 22.5°, \measuredangle B = 39.5°$

5. $a = 24.4, \measuredangle B = 18.4°, \measuredangle C = 21.6°$

7. $c = 21.5, \measuredangle A = 33.5°, \measuredangle B = 67.9°$

9. $\measuredangle A = 120°, \measuredangle B = 21.8°, \measuredangle C = 38.2°$

11. $\measuredangle A = 24.1°, \measuredangle B = 30.8°, \measuredangle C = 125.1°$

13. $\measuredangle A = 38.8°, \measuredangle B = 34.5°, \measuredangle C = 106.7°$

15. $\measuredangle A = 34.1°, \measuredangle B = 50.5°, \measuredangle C = 95.4°$

17. 54.2° at vertex (0, 0); 48.4° at vertex (5, −2); 77.4° at vertex (1, −4)

19. 334.9 km **21.** 63.7 ft **23.** 84.9°

25. 8.4 km **27.** 231.9 ft **29.** 154.5 ft

31. 4.7 cm and 9.0 cm **33.** 33.44° **35.** 978.7 mi

37. $AB = 24.27, AC = 21.23, BC = 19.5, \measuredangle A = 50.2°,$ $\measuredangle B = 56.8°, \measuredangle C = 73.0°$

39. 16.99 m

Section 7.3, page 497

1. $\measuredangle C = 110°, b = 2.5, c = 6.3$

3. $\measuredangle B = 14°, b = 2.2, c = 6.8$

5. $\measuredangle A = 88°, a = 17.3, c = 12.8$

7. $\measuredangle C = 41.5°, b = 9.7, c = 10.9$

9. 7.3 **11.** 32.5 **13.** 82.3 **15.** 31.4

17. No solution

19. $\measuredangle A_1 = 55.2°, \measuredangle C_1 = 104.8°, c_1 = 14.1;$ $\measuredangle A_2 = 124.8°, \measuredangle C_2 = 35.2°, c_2 = 8.4$

21. No solution

23. $\measuredangle B_1 = 65.8°, \measuredangle A_1 = 58.2°, a_1 = 10.3;$ $\measuredangle B_2 = 114.2°, \measuredangle A_2 = 9.8°, a_2 = 2.1$

25. $\measuredangle C = 72°, b = 14.7, c = 15.2$

27. $a = 9.8, \measuredangle B = 23.3°, \measuredangle C = 81.7°$

29. $\measuredangle A = 18.6°, \measuredangle B = 39.6°, \measuredangle C = 121.9°$

31. $c = 13.9, \measuredangle A = 60.1°, \measuredangle B = 72.9°$

33. $\measuredangle C = 39.8°, \measuredangle A = 77.7°, a = 18.9$

35. No solution **37.** 6.5 **39.** About 7691

41. 135.5 m **43.** 5.4° **45.** 5 ft

47. 5.3° **49.** 30.1 km **51.** About 9642 ft

53. **(a)** Use the Law of Cosines in triangle ABD to find $\measuredangle ABD$; then $\measuredangle EBA$ is $180° - \measuredangle ABD$. (Why?)

Use the Law of Cosines in triangle ABC to find $\angle CAB$; then $\angle EAB$ is $180° - \angle CAB$. You now have two of the angles in triangle EAB and can easily find the third. Use these angles, side AB, and the Law of Sines to find AE.

(b) 94.24 ft

55. 13.36 m **57.** 5.8 gal **59.** 11.18 sq units

61. No such triangle exists because the sum of the lengths of any two sides of a triangle must be greater than the length of the third side, which is not the case here.

Chapter 7 Review, page 501

1. (d) **3.** (e)

5. $b = \sqrt{313} \approx 17.7, A \approx 42.7°, C \approx 47.3°$

7. $b \approx 14.6, c \approx 8.40, A = 55°$

9. 225.9 ft **11.** 1.5°

13. $A = 52.9°, B = 41.6°, C = 85.5°$

15. $A = 20.6°, b = 21.8, C = 29.4°$

17. Approximately 301 mi

19. $A = 25°, a = 2.9, b = 5.6$

21. $A = 52.03°, B = 65.97°, b = 86.9$

23. $B = 81.8°, C = 38.2°, c = 2.5$ and $B = 98.2°, C = 21.8°, c = 1.5$

25. 147.4 **27.** 13.4 km

29. $a = 41.6; C = 75°, c = 54.1$

31. $A = 35.5°, b = 8.3, C = 68.5°$

33. Joe is 217.9 m from the pole and Alice is 240 m from the pole.

35. **(a)** 3940.65 ft **(b)** 4377.53 ft **(c)** 3967.39 ft

37. 71.89° **39.** 10 **41.** 37.95

Chapter 8

Section 8.1, page 516

1. Possibly an identity

3. Possibly an identity

5. B **7.** E

9. $\tan x \cos x = \left(\dfrac{\sin x}{\cos x}\right)\cos x = \sin x$

11. $\cos x \sec x = \cos x\left(\dfrac{1}{\cos x}\right) = 1$

13. $\tan x \csc x = \left(\dfrac{\sin x}{\cos x}\right)\left(\dfrac{1}{\sin x}\right) = \dfrac{1}{\cos x} = \sec x$

15. $\dfrac{\tan x}{\sec x} = \dfrac{\sin x/\cos x}{1/\cos x} = \sin x$

17. $(1 + \cos x)(1 - \cos x) = 1 - \cos^2 x = \sin^2 x$

19. Not an identity

21. $\dfrac{\sin(-x)}{\cos(-x)} = \dfrac{-\sin x}{\cos x} = -\tan x$

23. $\cot(-x) = \dfrac{\cos(-x)}{\sin(-x)} = \dfrac{\cos x}{-\sin x} = -\cot x$

25. Not an identity

27. $\sec^2 x - \csc^2 x = (1 + \tan^2 x) - (1 + \cot^2 x) = \tan^2 x - \cot^2 x$

29. $\sin^2 x(\cot x + 1)^2 = [\sin x(\cot x + 1)]^2 = (\cos x + \sin x)^2 = [\cos x(1 + \tan x)]^2 = \cos^2 x(\tan x + 1)^2$

31. $\sin^2 x - \tan^2 x = \sin^2 x - \dfrac{\sin^2 x}{\cos^2 x} = \sin^2 x\left(1 - \dfrac{1}{\cos^2 x}\right) = \sin^2 x(1 - \sec^2 x) = \sin^2 x(-\tan^2 x) = -\sin^2 x \tan^2 x$

33. $(\cos^2 x - 1)(\tan^2 x + 1) = (-\sin^2 x)(\sec^2 x) = (-\sin^2 x)\left(\dfrac{1}{\cos^2 x}\right) = -\tan^2 x$

35. $\dfrac{\sec x}{\csc x} = \dfrac{1/\cos x}{1/\sin x} = \dfrac{\sin x}{\cos x} = \tan x$

37. $\cos^4 x - \sin^4 x = (\cos^2 x - \sin^2 x)(\cos^2 x + \sin^2 x) = \cos^2 x - \sin^2 x$

39. Not an identity

41. $\dfrac{\sec x}{\csc x} + \dfrac{\sin x}{\cos x} = \dfrac{1/\cos x}{1/\sin x} + \tan x = \dfrac{\sin x}{\cos x} + \tan x = \tan x + \tan x = 2\tan x$

43. $\dfrac{\sec x + \csc x}{1 + \tan x} = \dfrac{\dfrac{1}{\cos x} + \dfrac{1}{\sin x}}{1 + \dfrac{\sin x}{\cos x}} \cdot \dfrac{\sin x \cos x}{\sin x \cos x} = \dfrac{\sin x + \cos x}{\sin x \cos x + \sin^2 x} = \dfrac{\sin x + \cos x}{\sin x(\cos x + \sin x)} = \dfrac{1}{\sin x} = \csc x$

45. $\dfrac{1}{\csc x - \sin x} = \dfrac{1}{\dfrac{1}{\sin x} - \sin x} \cdot \dfrac{\sin x}{\sin x} = \dfrac{\sin x}{1 - \sin^2 x} = \dfrac{\sin x}{\cos^2 x} = \left(\dfrac{1}{\cos x}\right)\left(\dfrac{\sin x}{\cos x}\right) = \sec x \tan x$

47. Not an identity

49. Conjecture: $\cos x$. Proof: $1 - \dfrac{\sin^2 x}{1 + \cos x} = \dfrac{1 + \cos x - \sin^2 x}{1 + \cos x} = \dfrac{\cos x + (1 - \sin^2 x)}{1 + \cos x} = \dfrac{\cos x + \cos^2 x}{1 + \cos x} = \dfrac{\cos x(1 + \cos x)}{1 + \cos x} = \cos x$

51. Conjecture: $\tan x$. Proof:
$(\sin x + \cos x)(\sec x + \csc x) - \cot x - 2 =$
$\sin x \sec x + \sin x \csc x + \cos x \sec x +$
$\cos x \csc x = \sin x \cdot \dfrac{1}{\cos x} + \sin x \cdot \dfrac{1}{\sin x} +$
$\cos x \cdot \dfrac{1}{\cos x} + \cos x \cdot \dfrac{1}{\sin x} - \cot x - 2 =$
$\dfrac{\sin x}{\cos x} + 1 + 1 + \dfrac{\cos x}{\sin x} - \cot x - 2$
$= \tan x + \cot x - \cot x = \tan x$

53. $\dfrac{1 - \sin x}{\sec x} = \dfrac{1 - \sin x}{\sec x} \cdot \dfrac{(1 + \sin x)}{(1 + \sin x)}$
$= \dfrac{1 - \sin^2 x}{\sec x(1 + \sin x)} = \dfrac{\cos^2 x}{\dfrac{1}{\cos x}(1 + \sin x)} =$
$\dfrac{\cos^3 x}{1 + \sin x}$

55. $\dfrac{\cos x}{1 - \sin x} \cdot \dfrac{1 + \sin x}{1 + \sin x} = \dfrac{\cos x(1 + \sin x)}{1 - \sin^2 x}$
$= \dfrac{\cos x(1 + \sin x)}{\cos^2 x} = \dfrac{1 + \sin x}{\cos x} = \dfrac{1}{\cos x} + \dfrac{\sin x}{\cos x}$
$= \sec x + \tan x$

57. $\dfrac{\cos x \cot x}{\cot x - \cos x} \cdot \dfrac{\sin x}{\sin x} = \dfrac{\cos x \cos x}{\cos x - \cos x \sin x}$
$= \dfrac{\cos^2 x}{\cos x(1 - \sin x)} = \dfrac{\cos x}{1 - \sin x} \cdot \dfrac{1 + \sin x}{1 + \sin x}$
$= \dfrac{\cos x(1 + \sin x)}{1 - \sin^2 x} = \dfrac{\cos x + \sin x \cos x}{\cos^2 x} \cdot \dfrac{\dfrac{1}{\sin x}}{\dfrac{1}{\sin x}}$
$= \dfrac{\dfrac{\cos x}{\sin x} + \cos x}{\cos x\left(\dfrac{\cos x}{\sin x}\right)} = \dfrac{\cot x + \cos x}{\cos x \cot x}$

59. $\cot x = \dfrac{1}{\tan x}$, so $\log_{10}(\cot x) = \log_{10}\left(\dfrac{1}{\tan x}\right)$
$= -\log_{10}(\tan x)$

61. $\csc x + \cot x = (\csc x + \cot x) \cdot \dfrac{(\csc x - \cot x)}{(\csc x - \cot x)}$
$= \dfrac{\csc^2 x - \cot^2 x}{\csc x - \cot x} = \dfrac{1}{\csc x - \cot x}$;
so, $\log_{10}(\csc x + \cot x) = \log_{10}\left(\dfrac{1}{\csc x - \cot x}\right)$
$= -\log_{10}(\csc x - \cot x)$

63. $-\tan x \tan y(\cot x - \cot y)$
$= -\tan y(\tan x \cot x) + \tan x(\tan y \cot y)$
$= -\tan y + \tan x = \tan x - \tan y$

65. $\dfrac{\cos x - \sin y}{\cos y - \sin x} \cdot \dfrac{\cos x + \sin y}{\cos x + \sin y}$
$= \dfrac{\cos^2 x - \sin^2 y}{(\cos y - \sin x)(\cos x + \sin y)}$
$= \dfrac{(1 - \sin^2 x) - (1 - \cos^2 y)}{(\cos y - \sin x)(\cos x + \sin y)}$
$= \dfrac{\cos^2 y - \sin^2 x}{(\cos y - \sin x)(\cos x + \sin y)}$
$= \dfrac{(\cos y - \sin x)(\cos y + \sin x)}{(\cos y - \sin x)(\cos x + \sin y)} = \dfrac{\cos y + \sin x}{\cos x + \sin y}$

Section 8.2, page 525

1. $\dfrac{\sqrt{6} - \sqrt{2}}{4}$ **3.** $2 - \sqrt{3}$ **5.** $2 - \sqrt{3}$

7. $-2 - \sqrt{3}$ **9.** $-2 - \sqrt{3}$ **11.** $\dfrac{\sqrt{6} + \sqrt{2}}{4}$

13. $\cos x$ **15.** $-\sin x$ **17.** $-1/\cos x$

19. $-\sin 2$ **21.** $\cos x$ **23.** $-2 \sin x \sin y$

25. $\dfrac{4 + \sqrt{2}}{6}$ **27.** $\dfrac{2\sqrt{6} - \sqrt{3}}{10}$ **29.** $-.393$

31. $.993$ **33.** -2.34

35. $\dfrac{f(x + h) - f(x)}{h} = \dfrac{\cos(x + h) - \cos x}{h}$
$= \dfrac{\cos x \cos h - \sin x \sin h - \cos x}{h}$
$= \dfrac{\cos x \cos h - \cos x}{h} - \dfrac{\sin x \sin h}{h}$
$= \cos x\left(\dfrac{\cos h - 1}{h}\right) - \sin x\left(\dfrac{\sin h}{h}\right)$

37. $\sin(x + y) = -44/125$; $\tan(x + y) = 44/117$; $x + y$ is in the third quadrant.

39. $\cos(x + y) = -56/65$; $\tan(x + y) = 33/56$; $x + y$ is in the third quadrant.

41. $\sin(u + v + w) = \sin u \cos v \cos w$
$+ \cos u \sin v \cos w$
$+ \cos u \cos v \sin w - \sin u \sin v \sin w$

43. Since $y = \pi/2 - x$, $\sin y = \sin(\pi/2 - x) = \cos x$. Hence, $\sin^2 x + \sin^2 y = \sin^2 x + \cos^2 x = 1$

45. $\sin(x - \pi) = \sin x \cos \pi - \cos x \sin \pi$
$= (\sin x)(-1) - (\cos x)(0) = -\sin x$

47. $\cos(\pi - x) = \cos \pi \cos x + \sin \pi \sin x$
$= (-1)\cos x + (0)\sin x = -\cos x$

49. $\sin(x + \pi) = \sin x \cos \pi + \cos x \sin \pi$
$= (\sin x)(-1) + (\cos x)(0) = -\sin x$

51. By Exercises 49 and 50, $\tan(x+\pi) = \dfrac{\sin(x+\pi)}{\cos(x+\pi)}$
$= \dfrac{-\sin x}{-\cos x} = \tan x$

53. $\frac{1}{2}[\cos(x-y) - \cos(x+y)] = \frac{1}{2}[\cos x \cos y + \sin x \sin y - (\cos x \cos y - \sin x \sin y)]$
$= \frac{1}{2}(2 \sin x \sin y) = \sin x \sin y$

55. $\cos(x+y)\cos(x-y)$
$= (\cos x \cos y - \sin x \sin y)(\cos x \cos y + \sin x \sin y)$
$= (\cos x \cos y)^2 - (\sin x \sin y)^2$
$= \cos^2 x \cos^2 y - \sin^2 x \sin^2 y$

57. $\dfrac{\cos(x-y)}{\sin x \cos y} = \dfrac{\cos x \cos y + \sin x \sin y}{\sin x \cos y}$
$= \dfrac{\cos x}{\sin x} + \dfrac{\sin y}{\cos y} = \cot x + \tan y$

59. Not an identity

61. $\dfrac{\sin(x+y)}{\sin(x-y)}$
$= \dfrac{\sin x \cos y + \cos x \sin y}{\sin x \cos y - \cos x \sin y} \cdot \dfrac{1/\cos x \cos y}{1/\cos x \cos y}$
$= \dfrac{\dfrac{\sin x}{\cos x} + \dfrac{\sin y}{\cos y}}{\dfrac{\sin x}{\cos x} - \dfrac{\sin y}{\cos y}} = \dfrac{\tan x + \tan y}{\tan x - \tan y}$

63. Not an identity

65. Not an identity

Excursion 8.2.A, page 530

In Answers 1–11, all angles in radians.

1. .64 **3.** 2.47 **5.** $\pi/2$

7. 1.37 or 1.77 **9.** $\pi/4$ or $3\pi/4$ **11.** 1.39 or 1.75

Section 8.3, page 537

1. $\dfrac{\sqrt{2+\sqrt{2}}}{2}$ **3.** $\dfrac{\sqrt{2+\sqrt{2}}}{2}$

5. $2 - \sqrt{3}$ **7.** $\dfrac{\sqrt{2+\sqrt{3}}}{2}$

9. $\dfrac{\sqrt{2-\sqrt{2}}}{2}$ **11.** $-\sqrt{2} + 1$

13. $\frac{1}{2}\sin 10x - \frac{1}{2}\sin 2x$ **15.** $\frac{1}{2}\cos 6x + \frac{1}{2}\cos 2x$

17. $\frac{1}{2}\cos 20x - \frac{1}{2}\cos 14x$ **19.** $2 \sin 4x \cos x$

21. $2 \sin 2x \cos 7x$

23. $\sin 2x = \dfrac{120}{169}, \cos 2x = \dfrac{119}{169}, \tan 2x = \dfrac{120}{119}$

25. $\sin 2x = \dfrac{24}{25}, \cos 2x = -\dfrac{7}{25}, \tan 2x = -\dfrac{24}{7}$

27. $\sin 2x = \dfrac{24}{25}, \cos 2x = \dfrac{7}{25}, \tan 2x = \dfrac{24}{7}$

29. $\sin 2x = \dfrac{\sqrt{15}}{8}, \cos 2x = \dfrac{7}{8}, \tan 2x = \dfrac{\sqrt{15}}{7}$

31. $\sin\dfrac{x}{2} = .5477, \cos\dfrac{x}{2} = .8367, \tan\dfrac{x}{2} = .6547$

33. $\sin\dfrac{x}{2} = \dfrac{1}{\sqrt{10}}, \cos\dfrac{x}{2} = \dfrac{-3}{\sqrt{10}}, \tan\dfrac{x}{2} = \dfrac{-1}{3}$

35. $\sin\dfrac{x}{2} = \sqrt{\dfrac{\sqrt{5}+2}{2\sqrt{5}}}, \cos\dfrac{x}{2} = -\sqrt{\dfrac{\sqrt{5}-2}{2\sqrt{5}}},$
$\tan\dfrac{x}{2} = -\sqrt{9+4\sqrt{5}} = -\sqrt{5} - 2$

37. $\sin 2x = .96$ **39.** $\cos 2x = .28$

41. $\sin\dfrac{x}{2} = .3162$

43. $\cos 3x = 4\cos^3 x - 3\cos x$

45. $\cos x$ **47.** $\sin 4y$ **49.** 1

51. $\sin 16x = \sin[2(8x)] = 2 \sin 8x \cos 8x$

53. $\cos^4 x - \sin^4 x = (\cos^2 x - \sin^2 x)(\cos^2 x + \sin^2 x) = \cos^2 x - \sin^2 x = \cos 2x$

55. Not an identity

57. $\dfrac{1+\cos 2x}{\sin 2x} = \dfrac{1}{\dfrac{\sin 2x}{1+\cos 2x}} = \dfrac{1}{\tan\left(\dfrac{2x}{2}\right)} = \cot x$

59. $\sin 3x = \sin(2x + x) = \sin 2x \cos x + \cos 2x \sin x$
$= (2 \sin x \cos x)\cos x + (1 - 2\sin^2 x)\sin x$
$= 2 \sin x \cos^2 x + \sin x - 2\sin^3 x$
$= 2 \sin x(1 - \sin^2 x) + \sin x - 2\sin^3 x$
$= \sin x(2 - 2\sin^2 x + 1 - 2\sin^2 x)$
$= \sin x(3 - 4\sin^2 x)$

61. Not an identity

63. $\csc^2\left(\dfrac{x}{2}\right) = \dfrac{1}{\sin^2\left(\dfrac{x}{2}\right)} = \dfrac{1}{\dfrac{1-\cos x}{2}} = \dfrac{2}{1-\cos x}$

65. $\dfrac{\sin x - \sin 3x}{\cos x + \cos 3x} = \dfrac{-2\cos 2x \sin x}{2\cos 2x \cos x} = \dfrac{-\sin x}{\cos x}$
$= -\tan x$

67. $\dfrac{\sin 4x + \sin 6x}{\cos 4x - \cos 6x} = \dfrac{2 \sin 5x \cos x}{2 \sin 5x \sin x} = \dfrac{\cos x}{\sin x} = \cot x$

69. $\dfrac{\sin x + \sin y}{\cos x - \cos y} = \dfrac{2\sin\left(\frac{x+y}{2}\right)\cos\left(\frac{x-y}{2}\right)}{-2\sin\left(\frac{x+y}{2}\right)\sin\left(\frac{x-y}{2}\right)}$

$= \dfrac{-\cos\left(\frac{x-y}{2}\right)}{\sin\left(\frac{x-y}{2}\right)} = -\cot\left(\frac{x-y}{2}\right)$

71. (a) It suffices to prove the equivalent identity $(1 - \cos x)(1 + \cos x) = (\sin x)(\sin x)$. We have $(1 - \cos x)(1 + \cos x) = 1 - \cos^2 x = \sin^2 x = (\sin x)(\sin x)$.

(b) By the half-angle identity proved in the text and part (a), $\tan\dfrac{x}{2} = \dfrac{1 - \cos x}{\sin x} = \dfrac{\sin x}{1 + \cos x}$

Section 8.4, page 546

In Answers 1–63, $k = 0, \pm 1, \pm 2, \pm 3, \ldots$.

1. $x = .5275$ or $1.6868 + k\pi$

3. $x = .4959$ or 1.2538 or 1.5708 or 1.8877 or 2.6457 or $4.7124 + 2k\pi$

5. $x = .1671$ or 1.8256 or 2.8867 or $4.5453 + 2k\pi$

7. $x = 1.2161$ or $5.0671 + 2k\pi$

9. $x = 2.4620$ or $3.8212 + 2k\pi$

11. $x = .5166$ or $5.6766 + 2k\pi$

13. (a) The graph of $f(x) = \sin x$ on the interval from 0 to 2π shows that $\sin x = 1$ only when $x = \pi/2$. Since $\sin x$ has period 2π, all other solutions are obtained by adding or subtracting integer multiples of 2π from $\pi/2$, that is,

$\frac{\pi}{2} + 2\pi = \frac{5\pi}{2}, \frac{\pi}{2} + 2(2\pi) = \frac{9\pi}{2},$

$\frac{\pi}{2} + 3(2\pi) = \frac{13\pi}{2}$, etc., and $\frac{\pi}{2} - 2\pi = \frac{-3\pi}{2},$

$\frac{\pi}{2} - 2(2\pi) = \frac{-7\pi}{2}, \frac{\pi}{2} - 3(2\pi) = \frac{-11\pi}{2}$, etc.

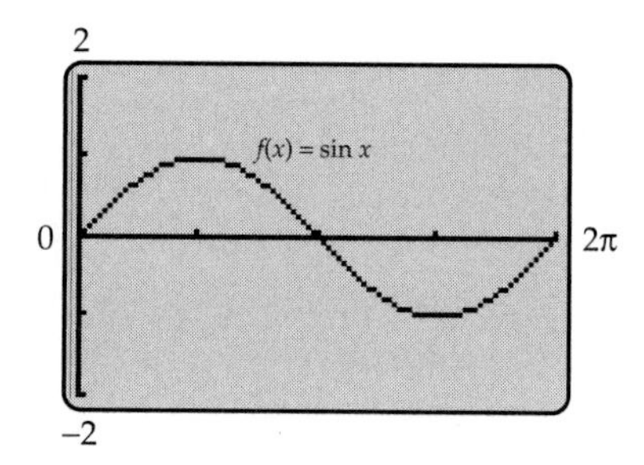

(b) Similarly, the graph shows that $\sin x = -1$ only when $x = 3\pi/2$, so that all solutions are obtained by adding or subtracting integer multiples of 2π from $3\pi/2$:

$\frac{3\pi}{2} + 2\pi = \frac{7\pi}{2}, \frac{3\pi}{2} + 2(2\pi) = \frac{11\pi}{2},$

$\frac{3\pi}{2} + 3(2\pi) = \frac{15\pi}{2}$, etc., and $\frac{3\pi}{2} - 2\pi = \frac{-\pi}{2},$

$\frac{3\pi}{2} - 2(2\pi) = \frac{-5\pi}{2}, \frac{3\pi}{2} - 3(2\pi) = \frac{-9\pi}{2}$, etc.

15. $x = \frac{\pi}{3}$ or $\frac{2\pi}{3} + 2k\pi$

17. $x = -\frac{\pi}{3} + k\pi$

19. $x = \pm\frac{5\pi}{6} + 2k\pi$

21. $x = -\frac{\pi}{6}$ or $\frac{7\pi}{6} + 2k\pi$

23. $x = .1193$ or 3.0223

25. $x = 1.3734$ or 4.5150

27. $\theta = 82.83°, 262.83°$

29. $\theta = 114.83°, 245.17°$

31. $\theta = 210°, 270°, 330°$

33. $\theta = 60°, 120°, 240°, 300°$

35. $\theta = 120°, 240°$

37. 65.38°

39. 30°

41. 27.57°

43. 14.18°

45. $x = -.4836$ or $3.6252 + 2k\pi$

47. $x = \pm 2.1700 + 2k\pi$

49. $x = -.2327 + k\pi$

51. $x = .4101 + k\pi$

53. $x = \pm 1.9577 + 2k\pi$

55. $x = -\frac{\pi}{6}$ or $\frac{2\pi}{3} + k\pi$

57. $x = \pm\frac{\pi}{2} + 4k\pi$

59. $x = -\frac{\pi}{9} + \frac{k\pi}{3}$

61. $x = \pm .7381 + \frac{2k\pi}{3}$

63. $x = 2.2143 + 2k\pi$

65. $x = 3.4814, 5.9433$

67. $x = \frac{3\pi}{4}, \frac{7\pi}{4}, 2.1588, 5.3004$

69. $x = \frac{\pi}{4}, \frac{\pi}{2}, \frac{5\pi}{4}, \frac{3\pi}{2}$

71. $x = \frac{\pi}{6}, \frac{\pi}{2}, \frac{5\pi}{6}, \frac{3\pi}{2}$

73. $x = .8481, 1.7682, 2.2935, 4.9098$

75. $x = .8213, 2.3203$

77. $x = .3649, 1.2059, 3.5065, 4.3475$

79. $x = 1.0591, 2.8679, 4.2007, 6.0095$

81. No solution

83. $x = \frac{\pi}{2}, \frac{7\pi}{6}, \frac{3\pi}{2}, \frac{11\pi}{6}$

85. $x = \frac{\pi}{6}, \frac{5\pi}{6}$

87. $x = \frac{\pi}{2}, \frac{7\pi}{6}, \frac{11\pi}{6}$

89. $x = 0, \frac{\pi}{3}, \frac{5\pi}{3}$

91. (a) March 1 (day 60); October 9 (day 282)
(b) June 20 (day 171)

93. .5475 or 1.0233

95. 13.25° or 76.75°

97. Sin $x = k$ and cos $x = k$ have no solutions when $k > 1$ or $k < -1$.

99. The solutions $x = 0, \pi$ are missed due to dividing by $\sin x$.

Section 8.5, page 554

1. $\frac{\pi}{2}$ **3.** $-\frac{\pi}{4}$ **5.** 0 **7.** $\frac{\pi}{6}$ **9.** $-\frac{\pi}{4}$

11. $-\frac{\pi}{3}$ **13.** $\frac{2\pi}{3}$ **15.** .3576 **17.** -1.2728

19. .7168 **21.** $-.8584$ **23.** 2.2168

25. $\cos u = 1/2$; $\tan u = -\sqrt{3}$

27. $\frac{\pi}{2}$ **29.** $\frac{5\pi}{6}$ **31.** $-\frac{\pi}{3}$ **33.** $\frac{\pi}{3}$

35. $\frac{\pi}{6}$ **37.** $\frac{4}{5}$ **39.** $\frac{4}{5}$ **41.** $\frac{5}{12}$

43. $\cos(\sin^{-1} v) = \sqrt{1 - v^2} \quad (-1 \leq v \leq 1)$

45. $\tan(\sin^{-1} v) = \dfrac{v}{\sqrt{1 - v^2}} \quad (-1 < v < 1)$

47.

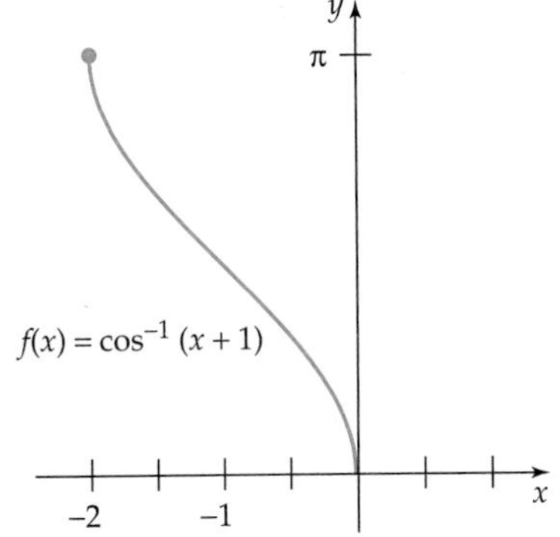

49.

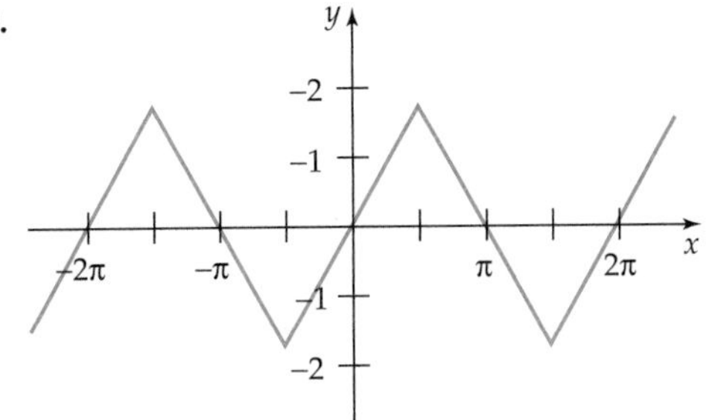

51. (a) $t = \dfrac{1}{2\pi f} \sin^{-1}\left(\dfrac{V}{V_{max}}\right) + \dfrac{k}{f}$ or

$t = \dfrac{1}{2\pi f}\left[\pi - \sin^{-1}\left(\dfrac{V}{V_{max}}\right)\right] + \dfrac{k}{f}$

$(k = 0, \pm 1, \pm 2, \ldots)$

(b) $t = .0005822$

53. (a) $\theta = \tan^{-1}\left(\dfrac{25}{x}\right) - \tan^{-1}\left(\dfrac{10}{x}\right)$ or

$\theta = \cos^{-1}\left(\dfrac{(x^2 + 10^2) + (x^2 + 25^2) - 15^2}{2\sqrt{x^2 + 10^2}\sqrt{x^2 + 25^2}}\right)$

(b) 15.8 ft

55. No horizontal line intersects the graph of $g(x) = \sec x$ more than once when $0 \leq x \leq \pi$ (see Figure 6–62). Hence, the restricted secant function has an inverse function, as explained in Section 3.7.

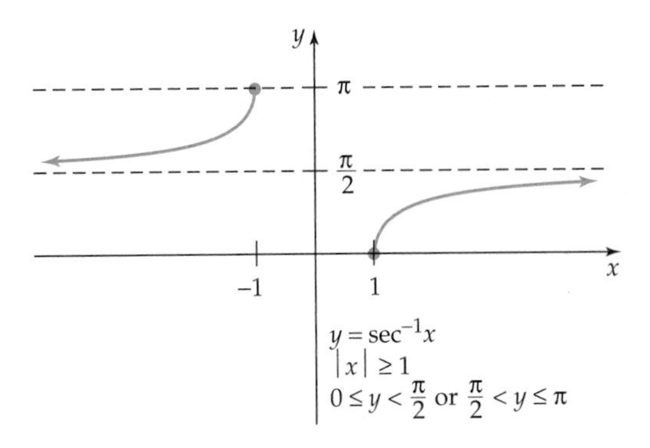

57. No horizontal line intersects the graph of $f(x) = \cot x$ more than once when $0 < x < \pi$ (see Figure 6–63). Hence the restricted cotangent function has an inverse function, as explained in Section 3.7.

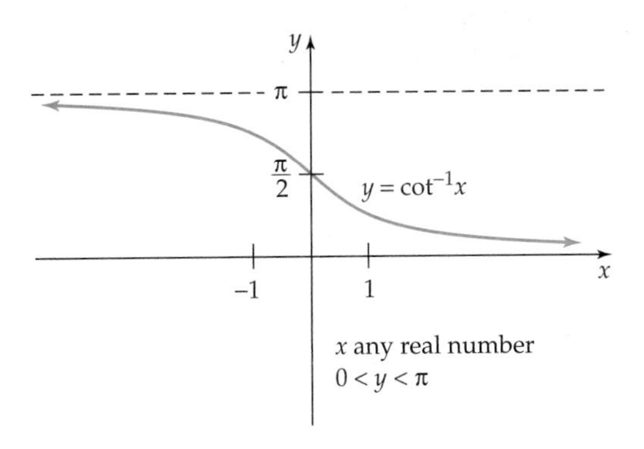

59. Let $u = \sin^{-1}(-x)$. Then $-\pi/2 \leq u \leq \pi/2$ and $\sin u = -x$; hence, $x = -\sin u = \sin(-u)$. Therefore, $\sin^{-1} x = -u$ (since $-\pi/2 \leq -u \leq \pi/2$) so that $\sin^{-1}(-x) = u = -\sin^{-1} x$.

61. Let $u = \cos^{-1}(-x)$. Then $0 \leq u \leq \pi$ and $\cos u = -x$. Since $0 \leq u \leq \pi$, we have $0 \leq \pi - u \leq \pi$. Since $\cos(\pi - u) = -\cos u = x$, $\cos^{-1} x = \pi - u = \pi - \cos^{-1}(-x)$. Therefore, $\cos^{-1}(-x) = \pi - \cos^{-1} x$.

63. If $0 < x < \pi$, then $-\frac{\pi}{2} < \frac{\pi}{2} - x < \frac{\pi}{2}$. Since $\tan\left(\frac{\pi}{2} - x\right) = \cot x$, $\tan^{-1}(\cot x) = \frac{\pi}{2} - x$.

65. Let $u = \sin^{-1} x$. Then $-\pi/2 \le u \le \pi/2$ and $\sin u = x$. Then $\cos u = \pm\sqrt{1 - \sin^2 u} = \pm\sqrt{1 - x^2}$. Since $-\pi/2 \le u \le \pi/2$, $\cos u$ is positive, so $\cos u = \sqrt{1 - x^2}$. Therefore, $\tan u = \dfrac{\sin u}{\cos u} = \dfrac{x}{\sqrt{1 - x^2}}$. Hence, $u = \tan^{-1}\left(\dfrac{x}{\sqrt{1 - x^2}}\right)$, so that $\sin^{-1} x = u = \tan^{-1}\left(\dfrac{x}{\sqrt{1 - x^2}}\right)$.

67. Let $u = \tan^{-1} x$. Then $\tan u = x$ and $\dfrac{\pi}{2} - \tan^{-1} x = \dfrac{\pi}{2} - u$. Let $\tan^{-1} v = \dfrac{\pi}{2} - u$. Then $v = \tan\left(\dfrac{\pi}{2} - u\right) = \cot u = \dfrac{1}{\tan u} = \dfrac{1}{x}$. Hence, $\tan^{-1}\left(\dfrac{1}{x}\right) = \tan^{-1} v = \dfrac{\pi}{2} - u = \dfrac{\pi}{2} - \tan^{-1} x$, so that $\tan^{-1} x + \tan^{-1}\left(\dfrac{1}{x}\right) = \dfrac{\pi}{2}$.

69. No; the graph of the left side function differs from the graph of the right side function.

Chapter 8 Review, page 558

1. $\dfrac{1}{3} + \cot t$ **3.** $\sin^4 x$

5. $\sin^4 t - \cos^4 t = (\sin^2 t - \cos^2 t)(\sin^2 t + \cos^2 t) = [\sin^2 t - (1 - \sin^2 t)](1) = 2\sin^2 t - 1$

7. $\dfrac{\sin t}{1 - \cos t} = \dfrac{\sin t}{1 - \cos t} \cdot \dfrac{(1 + \cos t)}{(1 + \cos t)} = \dfrac{\sin t(1 + \cos t)}{1 - \cos^2 t} = \dfrac{\sin t(1 + \cos t)}{\sin^2 t} = \dfrac{1 + \cos t}{\sin t}$

9. Not an identity

11. $(\sin x + \cos x)^2 - \sin 2x$
$= \sin^2 x + 2\sin x\cos x + \cos^2 x - 2\sin x\cos x$
$= \sin^2 x + \cos^2 x = 1$

13. $\dfrac{\tan x - \sin x}{2\tan x} \cdot \dfrac{\cos x}{\cos x} = \dfrac{\sin x - \sin x\cos x}{2\sin x}$
$\dfrac{\sin x(1 - \cos x)}{2\sin x} = \dfrac{1 - \cos x}{2} = \sin^2\left(\dfrac{x}{2}\right)$

15. $\cos(x + y)\cos(x - y)$
$= [\cos x\cos y - \sin x\sin y][\cos x\cos y + \sin x\sin y]$
$= \cos^2 x\cos^2 y - \sin^2 x\sin^2 y$
$= \cos^2 x(1 - \sin^2 y) - (1 - \cos^2 x)\sin^2 y$
$= \cos^2 x - \cos^2 x\sin^2 y - \sin^2 y + \cos^2 x\sin^2 y$
$= \cos^2 x - \sin^2 y$

17. $\dfrac{\sec x + 1}{\tan x} = \dfrac{(\sec x + 1)(\sec x - 1)}{\tan x(\sec x - 1)}$
$= \dfrac{\sec^2 x - 1}{\tan x(\sec x - 1)} = \dfrac{\tan^2 x}{\tan x(\sec x - 1)} = \dfrac{\tan x}{\sec x - 1}$

19. $\dfrac{1 + \tan^2 x}{\tan^2 x} = \dfrac{1}{\tan^2 x} + 1 = \cot^2 x + 1 = \csc^2 x$

21. $\tan^2 x - \sec^2 x = -(\sec^2 x - \tan^2 x) = -1$
$= -(\csc^2 x - \cot^2 x) = \cot^2 x - \csc^2 x$

23. $\dfrac{120}{169}$ **25.** $-\dfrac{56}{65}$ **27.** $\dfrac{3\sqrt{5} + 1}{8}$

29. Yes. $\sin 2x = 2\sin x\cos x = 2(0)\cos x = 0$

31. $\cos\left(\dfrac{\pi}{12}\right) = \cos\left[\dfrac{1}{2}\left(\dfrac{\pi}{6}\right)\right]$
$= \sqrt{\dfrac{1 + \cos\frac{\pi}{6}}{2}} = \sqrt{\dfrac{1 + \sqrt{3}/2}{2}}$
$= \sqrt{\dfrac{2 + \sqrt{3}}{4}} = \dfrac{\sqrt{2 + \sqrt{3}}}{2}$; $\cos\left(\dfrac{\pi}{12}\right) =$
$\cos\left(\dfrac{\pi}{4} - \dfrac{\pi}{6}\right) = \cos\dfrac{\pi}{4}\cos\dfrac{\pi}{6} + \sin\dfrac{\pi}{4}\sin\dfrac{\pi}{6}$
$= \dfrac{\sqrt{2}}{2}\left(\dfrac{\sqrt{3}}{2}\right) + \dfrac{\sqrt{2}}{2}\left(\dfrac{1}{2}\right) = \dfrac{\sqrt{6} + \sqrt{2}}{4}$. So,
$\dfrac{\sqrt{2 + \sqrt{3}}}{2} = \dfrac{\sqrt{6} + \sqrt{2}}{4}$ or $\sqrt{2 + \sqrt{3}} = \dfrac{\sqrt{2} + \sqrt{6}}{2}$

33. $\dfrac{\sqrt{6} + \sqrt{2}}{4}$ **35.** (a)

37. .96 **39.** $\pi/4$

In Answers 41–59, $k = 0, \pm 1, \pm 2, \pm 3, \ldots$.

41. $x = k\pi$

43. $x = .8419$ or 2.2997 or 4.1784 or $5.2463 + 2k\pi$

45. $x = \dfrac{\pi}{6}$ or $\dfrac{5\pi}{6} + 2k\pi$ **47.** $x = -\dfrac{\pi}{4} + k\pi$

49. $x = .7754$ or $2.3662 + 2k\pi$

51. $x = 1.4940 + k\pi$ **53.** $x = \dfrac{\pi}{6}$ or $\dfrac{5\pi}{6} + 2k\pi$

55. $x = -\dfrac{\pi}{6}$ or $\dfrac{7\pi}{6} + 2k\pi$

57. $x = \pm\dfrac{\pi}{3} + k\pi$

59. $x = .8959$ or $2.2457 + 2k\pi$

61. $\theta = 225.50°, 314.50°$ **63.** $9.06°$ or $80.94°$

65. $\dfrac{\pi}{4}$ **67.** $\dfrac{\pi}{3}$ **69.** $\dfrac{5\pi}{6}$ **71.** .75 **73.** $\dfrac{\pi}{3}$

75.

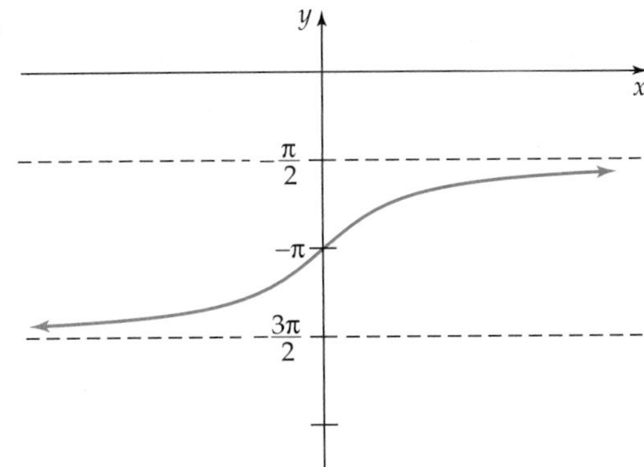

77. $\sqrt{15}/4$

Chapter 9

Section 9.1, page 569

1.–7.

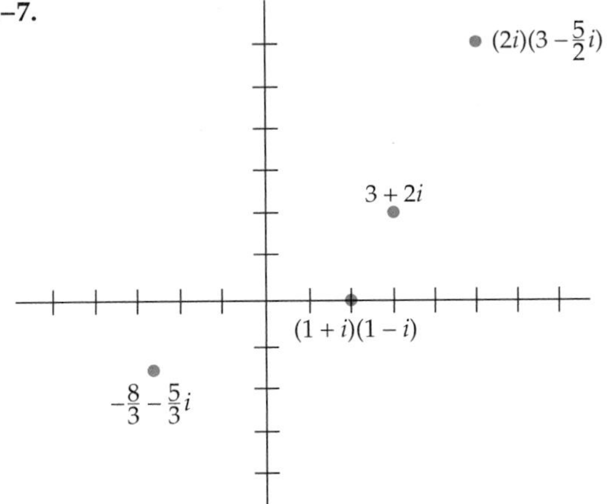

9. 13
11. $\sqrt{3}$
13. 12
15. Many correct answers, including $z = 1$, $w = i$

17.

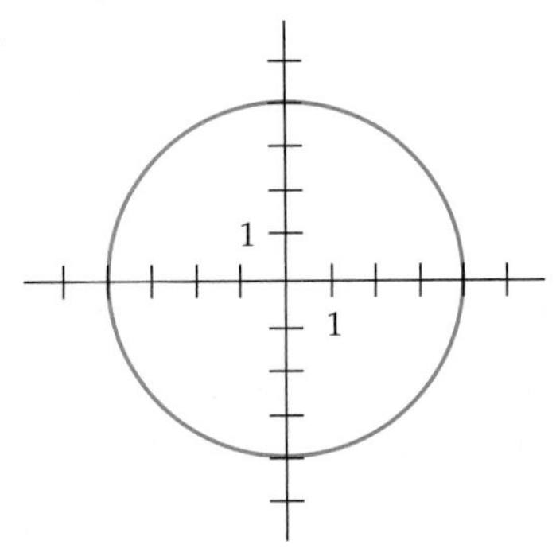

19.

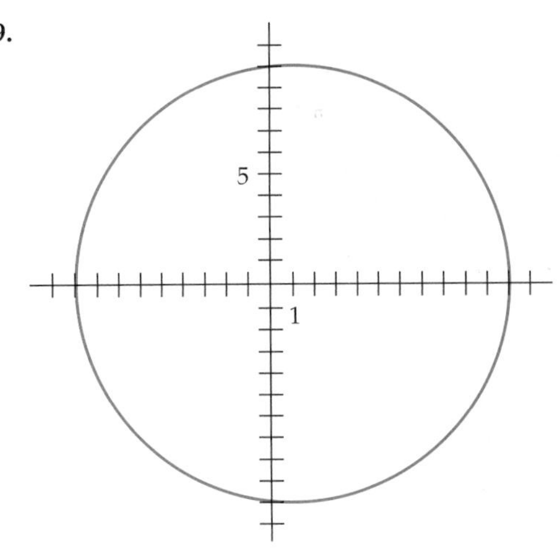

21.

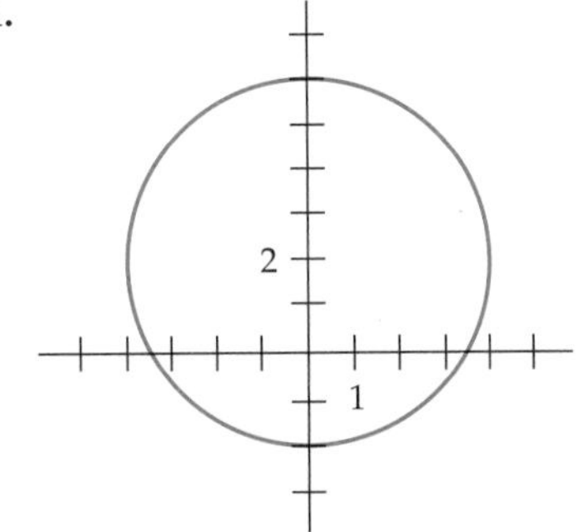

23.

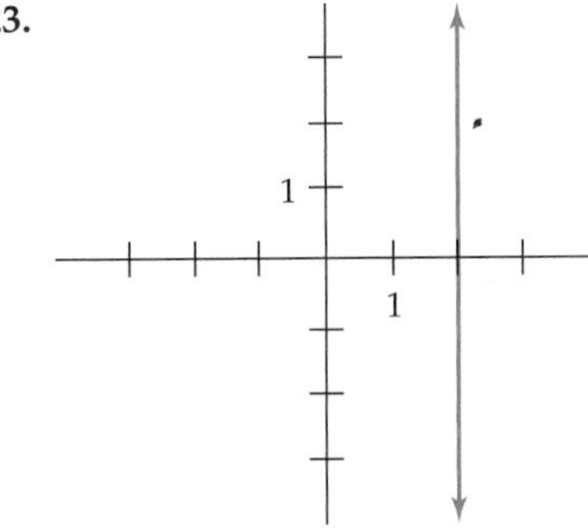

25. $5(\cos .9273 + i \sin .9273)$
27. $13(\cos 5.1072 + i \sin 5.1072)$
29. $\sqrt{5}(\cos 1.1071 + i \sin 1.1071)$
31. $\sqrt{18.5}(\cos 2.1910 + i \sin 2.1910)$
33. $2\left(\cos \frac{2\pi}{3} + i \sin \frac{2\pi}{3}\right) = -3 + 3\sqrt{3}i$
35. $42\left(\cos \frac{7\pi}{6} + i \sin \frac{7\pi}{6}\right) = -21\sqrt{3} - 21i$

37. $\frac{3}{2}\left(\cos\frac{\pi}{4} + i\sin\frac{\pi}{4}\right) = \left(\frac{3\sqrt{2}}{4}\right) + \left(\frac{3\sqrt{2}}{4}\right)i$

39. $2\sqrt{2}\left(\cos\frac{7\pi}{12} + i\sin\frac{7\pi}{12}\right)$

41. $\cos\frac{\pi}{2} + i\sin\frac{\pi}{2}$

43. $12\left(\cos\frac{2\pi}{3} + i\sin\frac{2\pi}{3}\right)$

45. $2\sqrt{2}\left(\cos\frac{19\pi}{12} + i\sin\frac{19\pi}{12}\right)$

47. The polar form of i is $1(\cos 90° + i\sin 90°)$. Hence, by the Polar Multiplication Rule

$$zi = r\cdot 1(\cos(\theta + 90°) + i\sin(\theta + 90°)).$$

You can think of z as lying on a circle with center at the origin and radius r. Then zi lies on the same circle (since it too is r units from the origin), but 90° farther around the circle (in a counterclockwise direction).

49. **(a)** $\frac{b}{a}$ **(b)** $\frac{d}{c}$ **(c)** $y - b = \left(\frac{d}{c}\right)(x - a)$

(d) $y - d = \frac{b}{a}(x - c)$

(f) $(a + c, b + d)$ lies on L since $(b + d) - b = \frac{d}{c}[(a + c) - a]$ and $(a + c, b + d)$ lies on M since $(b + d) - d = \frac{b}{a}[(a + c) - c]$.

51. **(a)** $r_2(\cos\theta_2 + i\sin\theta_2)(\cos\theta_2 - i\sin\theta_2) = r_2(\cos^2\theta_2 + \sin^2\theta_2) = r_2$

(b) $r_1(\cos\theta_1 + i\sin\theta_1)(\cos\theta_2 - i\sin\theta_2) = r_1[(\cos\theta_1\cos\theta_2 + \sin\theta_1\sin\theta_2) + i(\sin\theta_1\cos\theta_2 - \cos\theta_1\sin\theta_2)] = r_1[\cos(\theta_1 - \theta_2) + i\sin(\theta_1 - \theta_2)]$

Section 9.2, page 577

1. i **3.** $\frac{-243\sqrt{3}}{2} - \frac{243}{2}i$ **5.** -64

7. $\frac{1}{2} - \frac{\sqrt{3}}{2}i$ **9.** i **11.** $1, -1, i, -i$

13. $4\left(\cos\frac{\pi}{15} + i\sin\frac{\pi}{15}\right), 4\left(\cos\frac{11\pi}{15} + i\sin\frac{11\pi}{15}\right), 4\left(\cos\frac{7\pi}{5} + i\sin\frac{7\pi}{5}\right)$

15. $3\left(\cos\frac{\pi}{48} + i\sin\frac{\pi}{48}\right), 3\left(\cos\frac{25\pi}{48} + i\sin\frac{25\pi}{48}\right), 3\left(\cos\frac{49\pi}{48} + i\sin\frac{49\pi}{48}\right), 3\left(\cos\frac{73\pi}{48} + i\sin\frac{73\pi}{48}\right)$

17. $\left(\cos\frac{\pi}{5} + i\sin\frac{\pi}{5}\right), \left(\cos\frac{3\pi}{5} + i\sin\frac{3\pi}{5}\right), (\cos\pi + i\sin\pi), \left(\cos\frac{7\pi}{5} + i\sin\frac{7\pi}{5}\right), \left(\cos\frac{9\pi}{5} + i\sin\frac{9\pi}{5}\right)$

19. $\left(\cos\frac{\pi}{10} + i\sin\frac{\pi}{10}\right), \left(\cos\frac{\pi}{2} + i\sin\frac{\pi}{2}\right), \left(\cos\frac{9\pi}{10} + i\sin\frac{9\pi}{10}\right), \left(\cos\frac{13\pi}{10} + i\sin\frac{13\pi}{10}\right), \left(\cos\frac{17\pi}{10} + i\sin\frac{17\pi}{10}\right)$

21. $\sqrt[4]{2}\left(\cos\frac{\pi}{8} + i\sin\frac{\pi}{8}\right), \sqrt[4]{2}\left(\cos\frac{9\pi}{8} + i\sin\frac{9\pi}{8}\right)$

23. $x = \frac{\sqrt{3}}{2} + \frac{1}{2}i$ or $\frac{\sqrt{3}}{2} - \frac{1}{2}i$ or $-\frac{\sqrt{3}}{2} + \frac{1}{2}i$ or $-\frac{\sqrt{3}}{2} - \frac{1}{2}i$ or i or $-i$

25. $\frac{\sqrt{3}}{2} + \frac{1}{2}i$ or $-\frac{\sqrt{3}}{2} + \frac{1}{2}i$ or $-i$

27. $x = 3i$ or $\frac{3\sqrt{3}}{2} - \frac{3}{2}i$ or $-\frac{3\sqrt{3}}{2} - \frac{3}{2}i$

29. $\sqrt[4]{2}\left(\frac{\sqrt{3}}{2} + \frac{1}{2}i\right)$ or $\sqrt[4]{2}\left(-\frac{1}{2} + \frac{\sqrt{3}}{2}i\right)$ or $\sqrt[4]{2}\left(-\frac{\sqrt{3}}{2} - \frac{1}{2}i\right)$ or $\sqrt[4]{2}\left(\frac{1}{2} - \frac{\sqrt{3}}{2}i\right)$

31. $1, .6235 \pm .7818i, -.2225 \pm .9749i, -.9010 \pm .4339i$

33. $\pm 1, \pm i, .7071 \pm .7071i, -.7071 \pm .7071i$

35. $1, .7660 \pm .6428i, .1736 \pm .9848i, -.5 \pm .8660i, -.9397 \pm .3420i$

37. $x^6 - 1 = (x - 1)(x^5 + x^4 + x^3 + x^2 + x + 1)$, so the solutions of $x^5 + x^4 + x^3 + x^2 + x + 1 = 0$ are the sixth roots of unity other than 1; namely, $-1, \frac{1}{2} + \frac{\sqrt{3}}{2}i, \frac{1}{2} - \frac{\sqrt{3}}{2}i, -\frac{1}{2} + \frac{\sqrt{3}}{2}i, -\frac{1}{2} - \frac{\sqrt{3}}{2}i.$

39. 12

41. For each i, u_i is an nth root of unity, so $(u_i)^n = 1$. Hence $(vu_i)^n = (v^n)(u_i)^n = v^n\cdot 1 = r(\cos\theta + i\sin\theta)$ and vu_i is a solution of the equation. If $vu_i = vu_j$, then multiplying both sides by $1/v$ shows that $u_i = u_j$. In other words, if u_i is not equal to u_j, then $vu_i \neq vu_j$. Thus, the solutions $vu_1, \ldots, vu_n$ are all distinct.

Section 9.3, page 590

1. $3\sqrt{5}$ **3.** $\sqrt{34}$ **5.** $\langle 6, 6\rangle$

7. $\langle -6, 10\rangle$ **9.** $\langle 13/5, -2/5\rangle$

11. $\mathbf{u}+\mathbf{v}=\langle 4, 5\rangle$; $\mathbf{u}-\mathbf{v}=\langle -8, 3\rangle$; $3\mathbf{u}-2\mathbf{v}=\langle -18, 10\rangle$

13. $\mathbf{u}+\mathbf{v}=\langle 3+4\sqrt{2}, 1+3\sqrt{2}\rangle$; $\mathbf{u}-\mathbf{v}=\langle 3-4\sqrt{2}, -1+3\sqrt{2}\rangle$; $3\mathbf{u}-2\mathbf{v}=\langle 9-8\sqrt{2}, -2+9\sqrt{2}\rangle$

15. $\mathbf{u}+\mathbf{v}=\langle -23/4, 13\rangle$; $\mathbf{u}-\mathbf{v}=\langle -9/4, 7\rangle$; $3\mathbf{u}-2\mathbf{v}=\langle -17/2, 24\rangle$

17. $\mathbf{u}+\mathbf{v}=14\mathbf{i}-4\mathbf{j}$; $\mathbf{u}-\mathbf{v}=2\mathbf{i}+4\mathbf{j}$; $3\mathbf{u}-2\mathbf{v}=12\mathbf{i}+8\mathbf{j}$

19. $\mathbf{u}+\mathbf{v}=-\frac{5}{4}\mathbf{i}-\frac{3}{2}\mathbf{j}$; $\mathbf{u}-\mathbf{v}=-\frac{11}{4}\mathbf{i}-\frac{3}{2}\mathbf{j}$; $3\mathbf{u}-2\mathbf{v}=-\frac{15}{2}\mathbf{i}-\frac{9}{2}\mathbf{j}$

21. $-7\mathbf{i}$ **23.** $-2\mathbf{i}+\frac{1}{2}\mathbf{j}$ **25.** $6\mathbf{i}-\frac{13}{4}\mathbf{j}$

27. $\mathbf{v}=\langle 4, 0\rangle$ **29.** $\mathbf{v}=\langle -5\sqrt{2}, -5\sqrt{2}\rangle$

31. $\mathbf{v}=\langle 4.5963, 3.8567\rangle$

33. $\mathbf{v}=\langle -0.1710, -0.4698\rangle$

35. $\|\mathbf{v}\|=4\sqrt{2}, \theta=45°$ **37.** $\|\mathbf{v}\|=8, \theta=180°$

39. $\|\mathbf{v}\|=6, \theta=90°$

41. $\|\mathbf{v}\|=2\sqrt{17}, \theta=104.04°$

43. $\left\langle \frac{4}{\sqrt{41}}, \frac{-5}{\sqrt{41}}\right\rangle$ **45.** $\frac{1}{\sqrt{5}}\mathbf{i}+\frac{2}{\sqrt{5}}\mathbf{j}$

47. Direction: 46.1°; magnitude: 108.2 lb

49. Direction: 213.4°; magnitude: 17.4 kg

51. Resultant force $=\langle -8, -2\rangle$; $\mathbf{v}=\langle 8, 2\rangle$

53. $\mathbf{v}+\mathbf{0}=\langle c, d\rangle+\langle 0, 0\rangle=\langle c, d\rangle=\mathbf{v}$, and $\mathbf{0}+\mathbf{v}=\langle 0, 0\rangle+\langle c, d\rangle=\langle c, d\rangle=\mathbf{v}$

55. $r(\mathbf{u}+\mathbf{v})=r(\langle a, b\rangle+\langle c, d\rangle)=r\langle a+c, b+d\rangle=\langle r(a+c), r(b+d)\rangle=\langle ra+rc, rb+rd\rangle$, and $r\mathbf{u}+r\mathbf{v}=r\langle a, b\rangle+r\langle c, d\rangle=\langle ra, rb\rangle+\langle rc, rd\rangle=\langle ra+rc, rb+rd\rangle$

57. $(rs)\mathbf{v}=(rs)\langle c, d\rangle=\langle rsc, rsd\rangle$; $r(s\mathbf{v})=r(s\langle c, d\rangle)=r\langle sc, sd\rangle=\langle rsc, rsd\rangle$; and $s(r\mathbf{v})=s\langle rc, rd\rangle=\langle src, srd\rangle=\langle rsc, rsd\rangle$

59. 48.58 lb

61. Parallel to plane: 32.1 lb; perpendicular to plane: 38.3 lb

63. 66.4°

65. Ground speed: 253.2 mph; course: 69.1°

67. Ground speed: 304.1 mph; course: 309.5°

69. Air speed: 424.3 mph; direction: 62.4°

71. 69.08°

73. 170.32 lb on **u**; 341.77 lb on **v**

75. (a) $\mathbf{v}=\langle x_2-x_1, y_2-y_1\rangle$; $k\mathbf{v}=\langle kx_2-kx_1, ky_2-ky_1\rangle$
(b) $\|\mathbf{v}\|=\sqrt{(x_2-x_1)^2+(y_2-y_1)^2}$; $\|k\mathbf{v}\|=\sqrt{(kx_2-kx_1)^2+(ky_2-ky_1)^2}$
(c) $\|k\mathbf{v}\|=\sqrt{k^2(x_2-x_1)^2+k^2(y_2-y_1)^2}=\sqrt{k^2}\sqrt{(x_2-x_1)^2+(y_2-y_1)^2}=|k|\|\mathbf{v}\|$
(d) $\tan\theta=\frac{y_2-y_1}{x_2-x_1}$; $\tan\beta=\frac{ky_2-ky_1}{kx_2-kx_1}=\frac{k(y_2-y_1)}{k(x_2-x_1)}=\frac{y_2-y_1}{x_2-x_1}$. Since the values of the tangent function are repeated only at intervals of π, it follows that θ and β differ by a multiple of π, and so represent either the same direction or opposite directions.
(e) If $k>0$, the signs of the components of $k\mathbf{v}$ are the same as those of **v**, so **v** and $k\mathbf{v}$ lie in the same quadrant. Therefore **v** and $k\mathbf{v}$ must have the same direction rather than opposite directions. If $k<0$, the signs of $k\mathbf{v}$ are opposite those of **v**, so **v** and $k\mathbf{v}$ do not lie in the same quadrant. They cannot have the same direction, so must have opposite directions.

77. (a) Since $\mathbf{u}-\mathbf{v}=\langle a-c, b-d\rangle$, the box on page 581 shows that its magnitude is $\sqrt{(a-c)^2+(b-d)^2}$. On the other hand, the distance formula shows that the length (magnitude) of **w** is $\sqrt{(a-c)^2+(b-d)^2}$. Hence, $\|\mathbf{u}-\mathbf{v}\|=\|\mathbf{w}\|$.
(b) $\mathbf{u}-\mathbf{v}$ lies on the straight line through (0, 0) and $(a-c, b-d)$ which has slope $\frac{(b-d)-0}{(a-c)-0}=\frac{b-d}{a-c}$. Similarly, **w** lies on the line through (a, b) and (c, d), which also has slope $\frac{b-d}{a-c}$. So, $\mathbf{u}-\mathbf{v}$ and **w** are parallel. Verify that they actually have the same direction by considering the relative positions of (a, b), (c, d), and $(a-c, b-d)$. For instance, if $\mathbf{u}-\mathbf{v}$ points upward to the right, then (a, b) lies to the right and above (c, d). Hence $c<a$ and $d<b$, so that $a-c>0$ and $b-d>0$, which means that the endpoint of **w** lies in the first quadrant, that is, **w** points upward to the right.

Section 9.4, page 601

1. $\mathbf{u}\cdot\mathbf{v}=-7$, $\mathbf{u}\cdot\mathbf{u}=25$, $\mathbf{v}\cdot\mathbf{v}=29$

3. $\mathbf{u}\cdot\mathbf{v}=6$, $\mathbf{u}\cdot\mathbf{u}=5$, $\mathbf{v}\cdot\mathbf{v}=9$

5. $\mathbf{u}\cdot\mathbf{v}=12$, $\mathbf{u}\cdot\mathbf{u}=13$, $\mathbf{v}\cdot\mathbf{v}=13$

7. 6 **9.** 20 **11.** -28 **13.** 1.75065 radians

15. 2.1588 radians **17.** $\pi/2$ radians

19. Orthogonal **21.** Parallel **23.** Neither

25. $k = 2$ **27.** $k = \sqrt{2}$

29. $\text{proj}_{\mathbf{u}}\mathbf{v} = \langle 12/17, -20/17\rangle$; $\text{proj}_{\mathbf{v}}\mathbf{u} = \langle 6/5, 2/5\rangle$

31. $\text{proj}_{\mathbf{u}}\mathbf{v} = \langle 0, 0\rangle$; $\text{proj}_{\mathbf{v}}\mathbf{u} = \langle 0, 0\rangle$

33. $\text{comp}_{\mathbf{v}}\mathbf{u} = 22/\sqrt{13}$ **35.** $\text{comp}_{\mathbf{v}}\mathbf{u} = 3/\sqrt{10}$

37. $\mathbf{u}\cdot(\mathbf{v} + \mathbf{w}) = \langle a, b\rangle\cdot(\langle c, d\rangle + \langle r, s\rangle)$
$= \langle a, b\rangle\cdot\langle c + r, d + s\rangle = a(c + r) + b(d + s)$
$= ac + ar + bd + bs$
$\mathbf{u}\cdot\mathbf{v} + \mathbf{u}\cdot\mathbf{w} = \langle a, b\rangle\cdot\langle c, d\rangle + \langle a, b\rangle\cdot\langle r, s\rangle$
$= (ac + bd) + (ar + bs) = ac + ar + bd + bs$

39. $\mathbf{0}\cdot\mathbf{u} = \langle 0, 0\rangle\cdot\langle a, b\rangle = 0a + 0b = 0$

41. If $\theta = 0$ or π, then **u** and **v** are parallel, so $\mathbf{v} = k\mathbf{u}$ for some real number k. We know that $\|\mathbf{v}\| = |k|\|\mathbf{u}\|$ (Excercise 75, Section 9.3). If $\theta = 0$, then $\cos\theta = 1$ and $k > 0$. Since $k > 0$, $|k| = k$ and so $\|\mathbf{v}\| = k\|\mathbf{u}\|$. Therefore, $\mathbf{u}\cdot\mathbf{v} = \mathbf{u}\cdot k\mathbf{u} = k\mathbf{u}\cdot\mathbf{u} = k\|\mathbf{u}\|^2 = \|\mathbf{u}\|(k\|\mathbf{u}\|) = \|\mathbf{u}\|\|\mathbf{v}\| = \|\mathbf{u}\|\|\mathbf{v}\|\cos\theta$. On the other hand, if $\theta = \pi$, then $\cos\theta = -1$ and $k < 0$. Since $k < 0$, $|k| = -k$ and so $\|\mathbf{v}\| = -k\|\mathbf{u}\|$. Then $\mathbf{u}\cdot\mathbf{v} = \mathbf{u}\cdot k\mathbf{u}, = k\,\mathbf{u}\cdot\mathbf{u} = k\|\mathbf{u}\|^2 = \|\mathbf{u}\|(-k\|\mathbf{u}\|) = -\|\mathbf{u}\|\|\mathbf{v}\| = \|\mathbf{u}\|\|\mathbf{v}\|\cos\theta$. In both cases we have shown $\mathbf{u}\cdot\mathbf{v} = \|\mathbf{u}\|\|\mathbf{v}\|\cos\theta$.

43. If $A = (1, 2)$, $B = (3, 4)$, and $C = (5, 2)$, then the vector $\overrightarrow{AB} = \langle 2, 2\rangle$, $\overrightarrow{AC} = \langle 4, 0\rangle$, and $\overrightarrow{BC} = \langle 2, -2\rangle$. Since $\overrightarrow{AB}\cdot\overrightarrow{BC} = 0$, $\overrightarrow{AB}$ and $\overrightarrow{BC}$ are perpendicular, so the angle at vertex B is a right angle.

45. Many possible answers: One is $\mathbf{u} = \langle 1, 0\rangle$, $\mathbf{v} = \langle 1, 1\rangle$, and $\mathbf{w} = \langle 1, -1\rangle$.

47. 300 lb ($= 600\sin 30°$)

49. 13 **51.** 24

53. The force in the direction of the lawnmower's motion is $30\cos 60° = 15$ lb. Thus, the work done is $15(75) = 1125$ ft-lb.

55. 1368 ft-lb

Chapter 9 Review, page 604

1. $\sqrt{10} + \sqrt{20}$

3. The graph is a circle of radius 2 centered at the origin.

5. $2\left(\cos\frac{\pi}{3} + i\sin\frac{\pi}{3}\right)$ **7.** $4\sqrt{2} + 4\sqrt{2}i$

9. $2\sqrt{3} + 2i$ **11.** $\frac{81}{2} - \frac{81\sqrt{3}}{2}i$

13. $1, \cos\frac{\pi}{3} + i\sin\frac{\pi}{3}, \cos\frac{2\pi}{3} + i\sin\frac{2\pi}{3}, -1,$
$\cos\frac{4\pi}{3} + i\sin\frac{4\pi}{3}, \cos\frac{5\pi}{3} + i\sin\frac{5\pi}{3}$

15. $\cos\frac{\pi}{8} + i\sin\frac{\pi}{8}, \cos\frac{5\pi}{8} + i\sin\frac{5\pi}{8},$
$\cos\frac{9\pi}{8} + i\sin\frac{9\pi}{8}, \cos\frac{13\pi}{8} + i\sin\frac{13\pi}{8}$

17. $\langle 11, -1\rangle$ **19.** $2\sqrt{29}$ **21.** $-11\mathbf{i} + 8\mathbf{j}$

23. $\sqrt{10}$ **25.** $\left\langle\frac{5\sqrt{2}}{2}, \frac{5\sqrt{2}}{2}\right\rangle$

27. $-\frac{1}{\sqrt{5}}\mathbf{i} + \frac{2}{\sqrt{5}}\mathbf{j}$

29. Ground speed: 321.87 mph; course: 126.18°

31. -26 **33.** 3 **35.** .70 radians

37. $\text{proj}_{\mathbf{v}}\mathbf{u} = \mathbf{v} = 2\mathbf{i} + \mathbf{j}$

39. $(\mathbf{u} + \mathbf{v})\cdot(\mathbf{u} - \mathbf{v}) = \mathbf{u}\cdot\mathbf{u} - \mathbf{u}\cdot\mathbf{v} + \mathbf{v}\cdot\mathbf{u} - \mathbf{v}\cdot\mathbf{v} = \mathbf{u}\cdot\mathbf{u} - \mathbf{v}\cdot\mathbf{v} = \|\mathbf{u}\|^2 - \|\mathbf{v}\|^2 = 0$ since **u** and **v** have the same magnitude.

41. 1750 lb

Chapter 10

Section 10.1, page 620

1. $-5 \le x \le 6$ and $-2 \le y \le 2$

3. $-3 \le x \le 4$ and $-2 \le y \le 3$

5. $0 \le x \le 14$ and $-15 \le y \le 0$

7. $-2 \le x \le 20$ and $-11 \le y \le 11$

9. $-12 \le x \le 12$ and $-12 \le y \le 12$

11. $-2 \le x \le 20$ and $-20 \le y \le 4$

13. $-25 \le x \le 22$ and $-25 \le y \le 26$

15. $y = 2x + 7$ **17.** $y = 2x + 5$ **19.** $y = \ln x$

21. $x^2 + y^2 = 9$ **23.** $16x^2 + 9y^2 = 144$

25. Both give a straight line segment between $P = (-4, 7)$ and $Q = (2, -5)$. The parametric equations in (a) move from P to Q, and the parametric equations in (b) move from Q to P.

27. Solving $x = a + (c - a)t$ for t gives $t = \frac{x - a}{c - a}$ and substituting in $y = b + (d - b)t$ then gives $y = b + (d - b)\frac{x - a}{c - a}$, or $y = b + \frac{d - b}{c - a}(x - a)$. This is a linear equation and therefore gives a straight line. You can check by substitution that (a, b) and (c, d) lie on this straight line; in fact, these points correspond to $t = 0$ and $t = 1$, respectively.

29. $x = -6 + 18t, y = 12 - 22t \quad (0 \le t \le 1)$

31. $x = 18 - 34t, y = 4 + 10t \quad (0 \le t \le 1)$

33. $x = 9 + 5\cos t, y = 12 + 5\sin t \quad (0 \le t \le 2\pi)$

35. $x = 2\cos t + 2, y = 2\sin t + 3 \quad (0 \le t \le 2\pi)$

37. Local minimum at $(-6, 2)$

39. Local maximum at $(4, 5)$

41. (a)

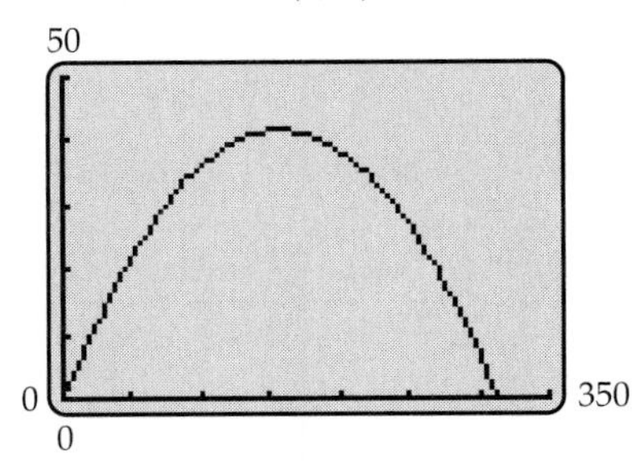

$x = (110 \cos 28°)t$
$y = (110 \sin 28°)t - 16t^2$
$0 \leq t \leq 3.5$

(b) About 3.2 sec

(c) 41.67 ft

43. (a)

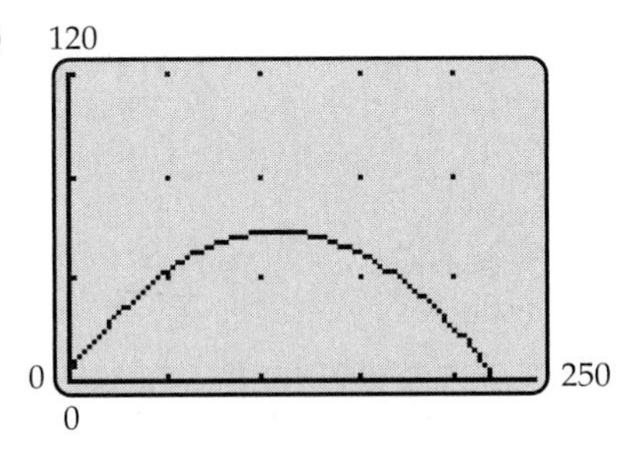

$x = (88 \cos 48°)t$
$y = (88 \sin 48°)t - 16t^2 + 4$
$0 \leq t \leq 4.5$

(b) Yes

45. $v = 80\sqrt[4]{3} \approx 105.29$ ft/sec

47. (a)

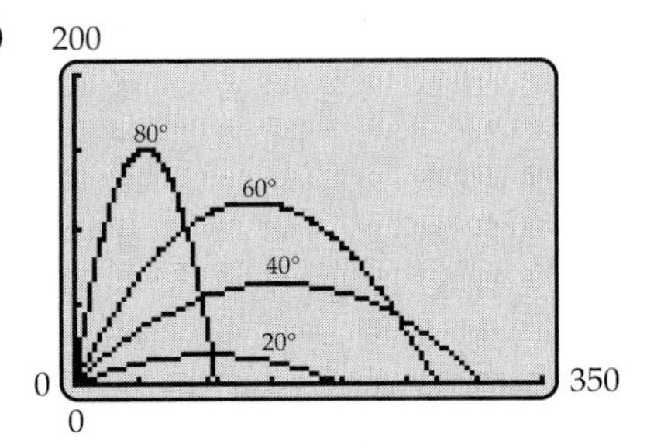

(b) 40°

(c) 200

45°
40°

0 350
0

An angle of 45° seems to result in the longest distance.

49. (b) $\cos(t - \pi/2) = \cos t \cos \pi/2 + \sin t \sin \pi/2 = (\cos t)(0) + (\sin t)(1) = \sin t$. Therefore, $3t - 3\cos(t - \pi/2) = 3t - 3 \sin t = 3(t - \sin t)$. $\text{Sin}(t - \pi/2) = \sin t \cos \pi/2 - \cos t \sin \pi/2 = (\sin t)(0) - (\cos t)(1) = -\cos t$. Therefore, $3 + 3\sin(t - \pi/2) = 3 - 3\cos t = 3(1 - \cos t)$.

51. (b) $\cos(t - 3\pi/2) = \cos t \cos 3\pi/2 + \sin t \sin 3\pi/2 = (\cos t)(0) + (\sin t)(-1) = -\sin t$. Therefore, $3t + 3\cos(t - 3\pi/2) = 3t - 3\sin t = 3(t - \sin t)$. $\text{Sin}(t - 3\pi/2) = \sin t \cos 3\pi/2 - \cos t \sin 3\pi/2 = (\sin t)(0) - (\cos t)(-1) = \cos t$. Therefore, $3 - 3\sin(t - 3\pi/2) = 3 - 3\cos t = 3(1 - \cos t)$.

53. (a)

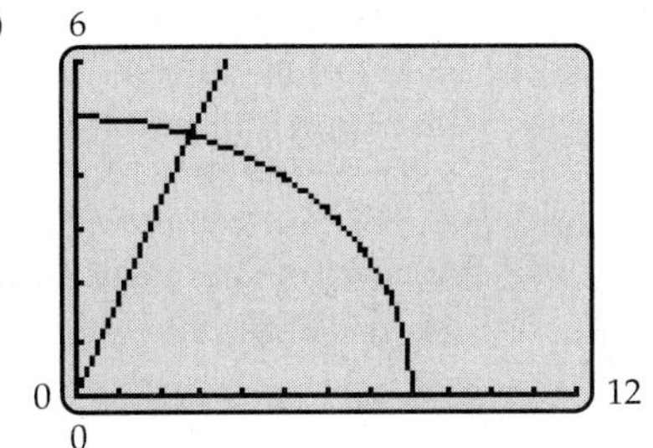

The particles do not collide.

(b) $t \approx 1.1$

(c)

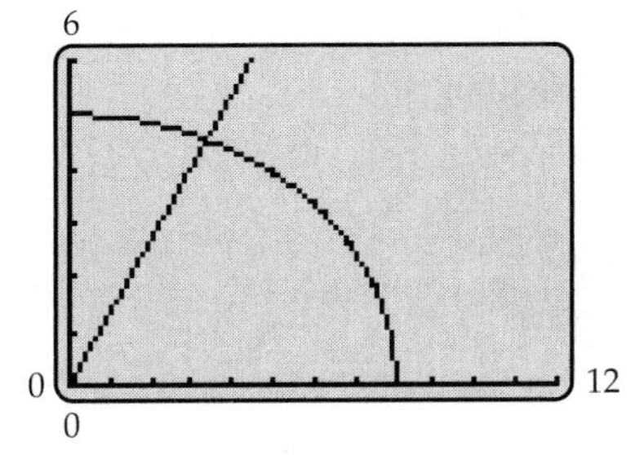

The particles do not collide; they are closest when $t \approx 1.13$.

(d)

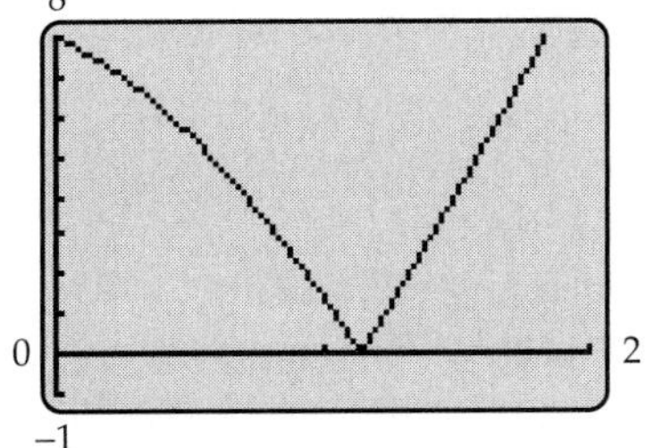

d is smallest when $t \approx 1.1322$.

55. (a) As shown in the diagram below the center Q of the small circle is always at distance $a - b$ from the origin O. Suppose t is the angle that OQ makes with the x-axis. Then the coordinates of Q are $x = (a - b)\cos t$, $y = (a - b)\sin t$. Examining the smaller circle in detail, we see that the change in x-coordinate from Q to P is $b \cos u$, where u is the angle that PQ makes with the positive x-axis. Likewise, the change in y-coordinate from Q to P is $-b \sin u$. Therefore, the coordinates of P are

$$x = (a - b)\cos t + b \cos u$$
$$y = (a - b)\sin t - b \sin u.$$

The angles t and u are related by the fact that the inner circle must roll without "slipping." This means the arc length that P has moved around the inner circle must equal the arc length that the inner circle has moved along the circumference of the larger circle. In other words, the arc length from P to W must equal the arc length from S to V. Since the length of a circular arc is the radius times the angle, this means $bu = (a - b)t$, or $u = (a - b)t/b$. Substituting this for u in the above equations will give the desired parametric equations.

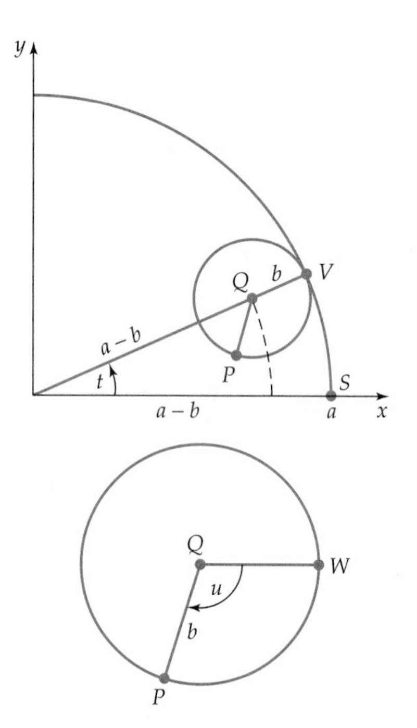

(b)

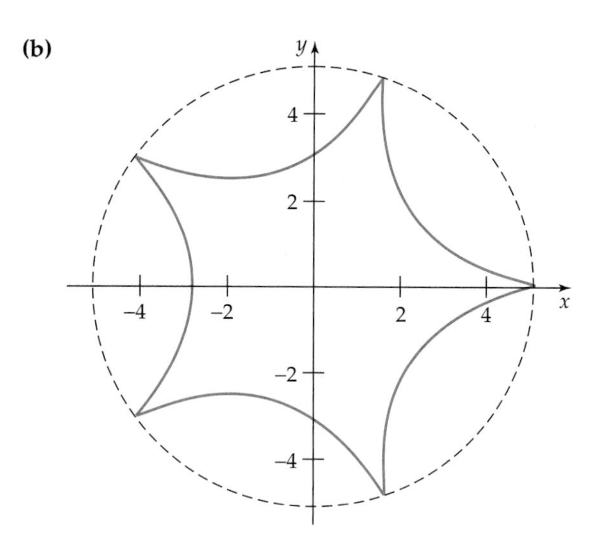

(c)

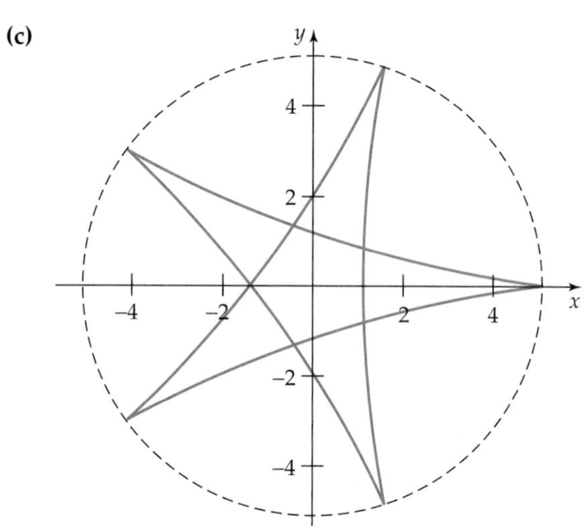

Section 10.2, page 636

1. $\dfrac{x^2}{49} + \dfrac{y^2}{4} = 1$ **3.** $\dfrac{x^2}{36} + \dfrac{y^2}{16} = 1$

5. $\dfrac{x^2}{9} - \dfrac{y^2}{36} = 1$ **7.** $\dfrac{x^2}{4} - y^2 = 1$

9. $y = 3x^2$ **11.** $y^2 = 20x$

13. $\dfrac{x^2}{9} + \dfrac{y^2}{49} = 1$ **15.** $6x = y^2$

17. $2x^2 - y^2 = 8$ **19.** $x^2 + 6y^2 = 18$

21. Ellipse

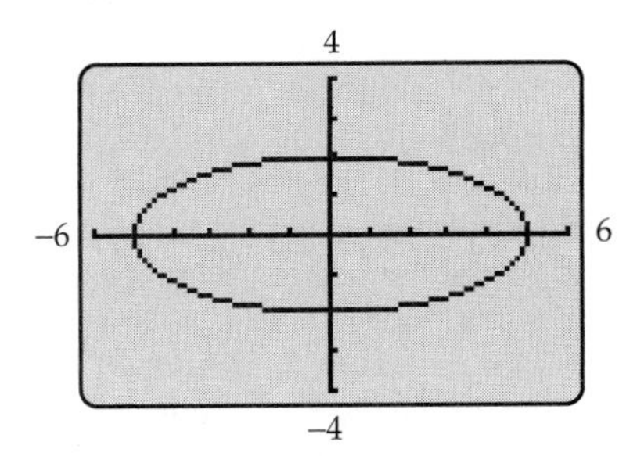

23. Ellipse

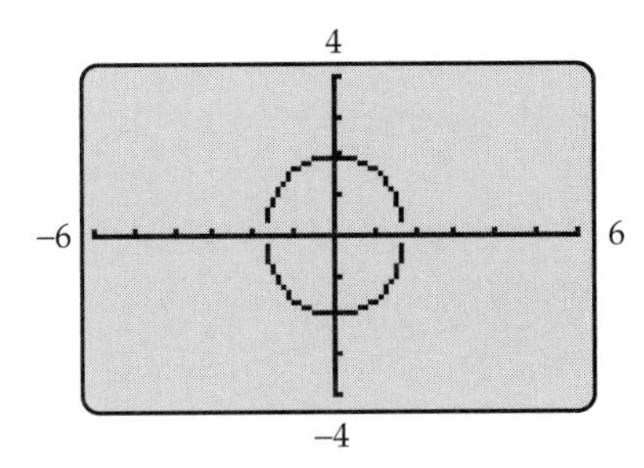

25. Hyperbola

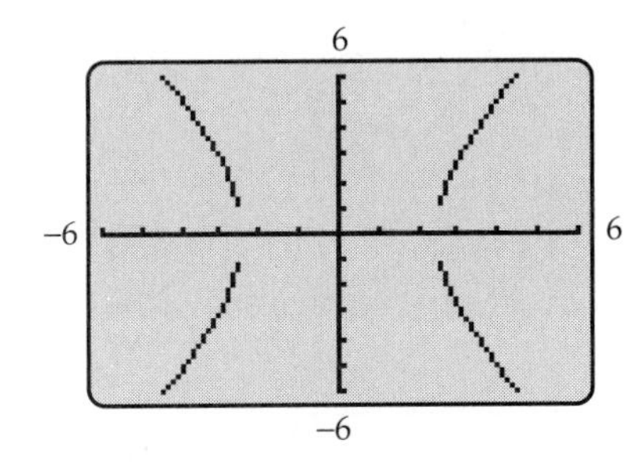

Because of limited resolution, this calculator-generated graph does not show that the top and bottom halves of the graph are connected.

27. Hyperbola

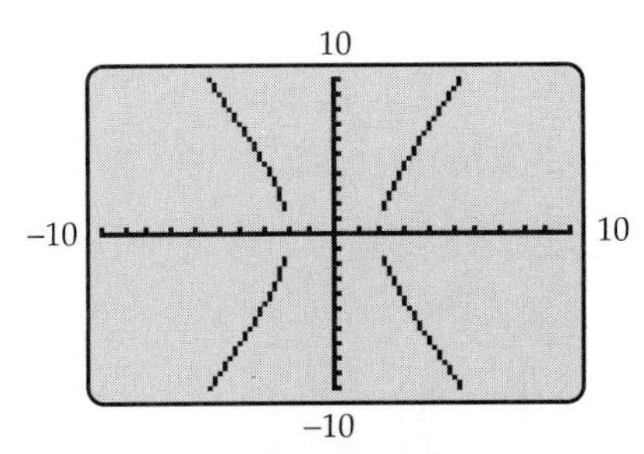

Because of limited resolution, this calculator-generated graph does not show that the top and bottom halves of the graph are connected.

29. Parabola

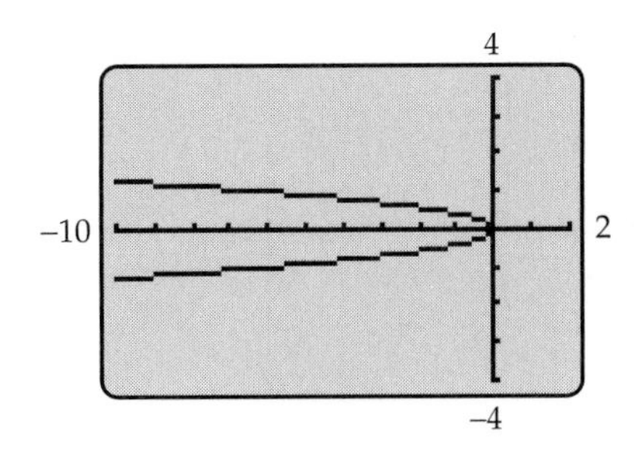

31. Hyperbola

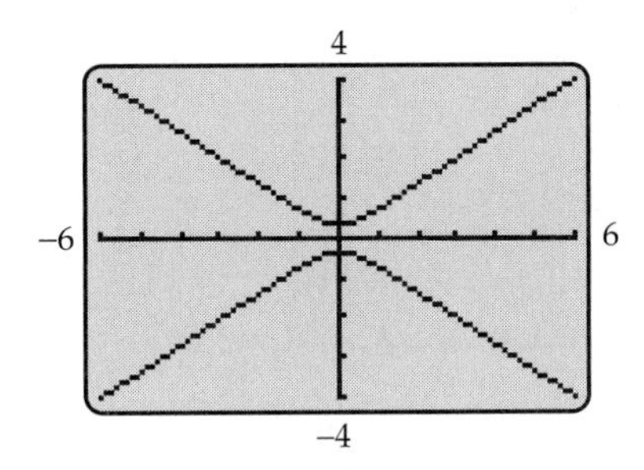

33. $x = \sqrt{10}\cos t, y = 6\sin t \quad (0 \le t \le 2\pi)$

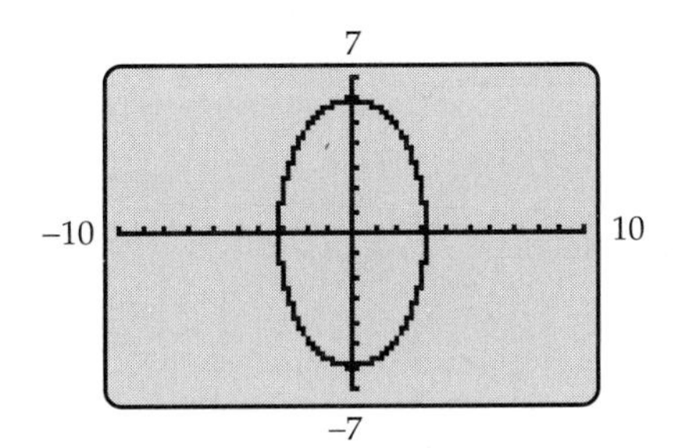

35. $x = \frac{1}{2}\cos t, y = \frac{1}{2}\sin t \quad (0 \le t \le 2\pi)$

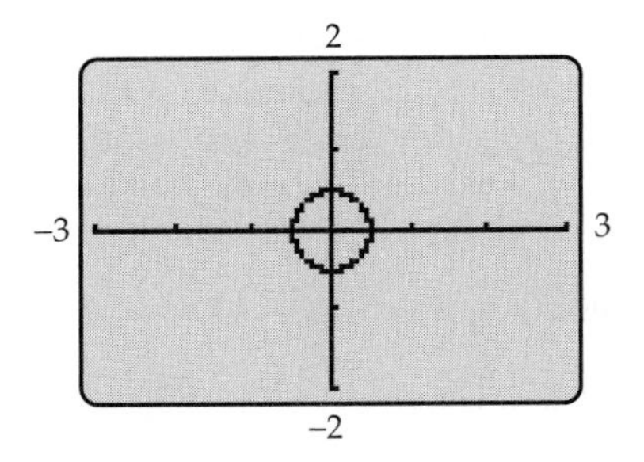

37. $x = \sqrt{10}/\cos t,\ y = 6\tan t \quad (0 \le t \le 2\pi)$

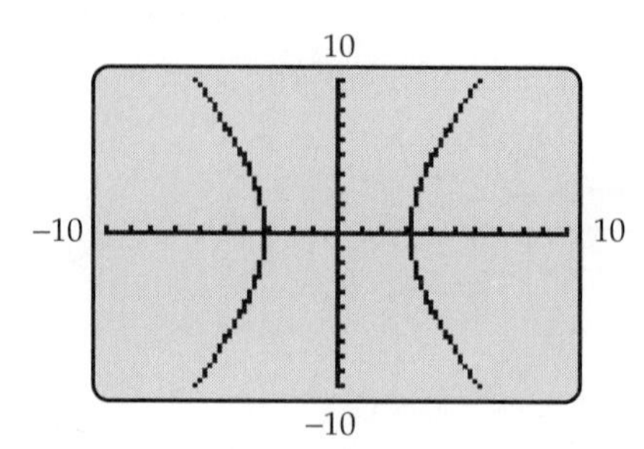

39. $x = 1/\cos t,\ y = \frac{1}{2}\tan t \quad (0 \le t \le 2\pi)$

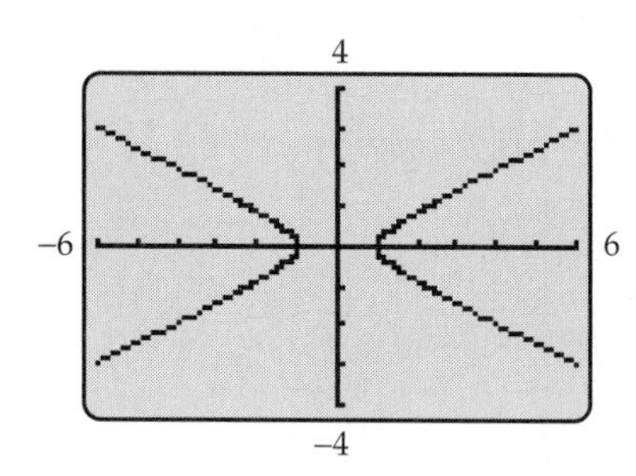

41. $x = t^2/4,\ y = t$ (t any real number)

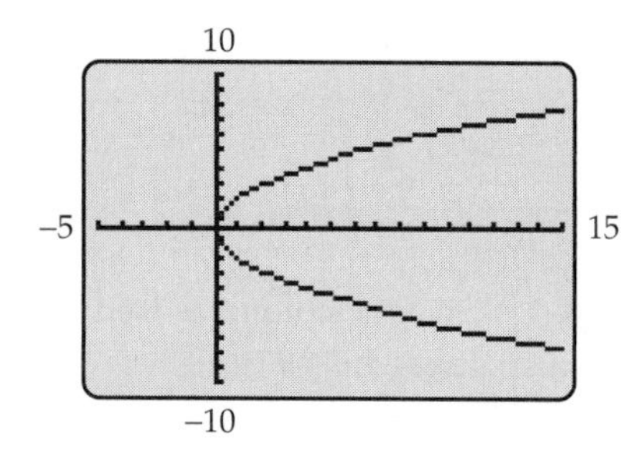

43. 8π **45.** $2\sqrt{3}\pi$ **47.** $7\pi/\sqrt{3}$

49. Focus: $(0, 1/12)$; directrix: $y = -1/12$

51. Focus: $(0, 1)$; directrix: $y = -1$

53. If $a = b$, then $\dfrac{x^2}{a^2} + \dfrac{y^2}{a^2} = 1$. Multiplying both sides by a^2 gives $x^2 + y^2 = a^2$, the equation of a circle of radius a with center at the origin.

55. The two branches of the hyperbola are very "flat" when b is large. With very large b and a small viewing window, the hyperbola may look like two horizontal lines, but it isn't because its asymptotes $y = \pm\dfrac{2}{b}x$ are not horizontal (their slopes, $\pm 2/b$, are close to, but not equal to, 0 when b is large).

57. $x = t^2/4p = y^2/4p$; hence, $y^2 = 4px$ and the graph is a parabola.

59. Approximately 226,335 mi and 251,401 mi

61. Let P denote the punch bowl and Q the table. In the longest possible trip starting at point X, the sum of the distance from X to Q and the distance from X to P must be 100 (since the distance from Q to P is 50). Thus, the fence should be an ellipse with foci P and Q and $r = 100$, as described on page 625 (with $c = 25$). Verify that the length of its major axis is 100 ft and the length of its minor axis is approximately 86.6 ft.

Section 10.3, page 648

1. $\dfrac{(x-2)^2}{4} + \dfrac{(y-3)^2}{16} = 1$

3. $\dfrac{4(x-7)^2}{25} + \dfrac{(y+4)^2}{36} = 1$

5. $\dfrac{(y-3)^2}{4} - \dfrac{(x+2)^2}{6} = 1$

7. $\dfrac{(x-4)^2}{9} - \dfrac{(y-2)^2}{16} = 1$

9. $y = 13(x-1)^2$ **11.** $x - 2 = 3(y-1)^2$

13. $\dfrac{(x-3)^2}{36} + \dfrac{(y+2)^2}{16} = 1$

15. $x + 3 = 4(y+2)^2$

17. Ellipse **19.** Parabola **21.** Hyperbola

23. $x = 2\cos t + 1,\ y = 3\sin t + 5 \quad (0 \le t \le 2\pi)$

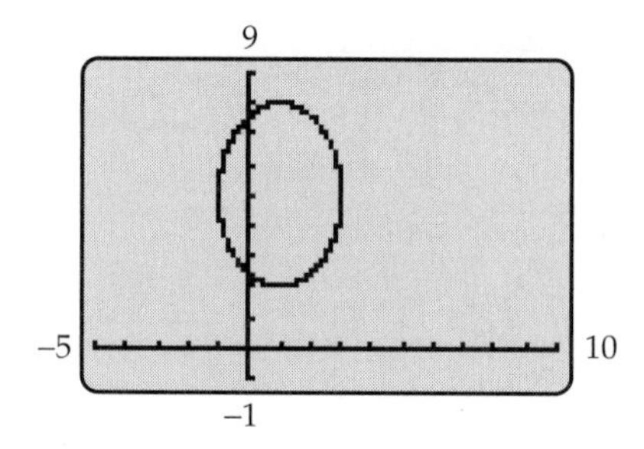

25. $x = 4\cos t - 1,\ y = \sqrt{8}\sin t + 4 \quad (0 \le t \le 2\pi)$

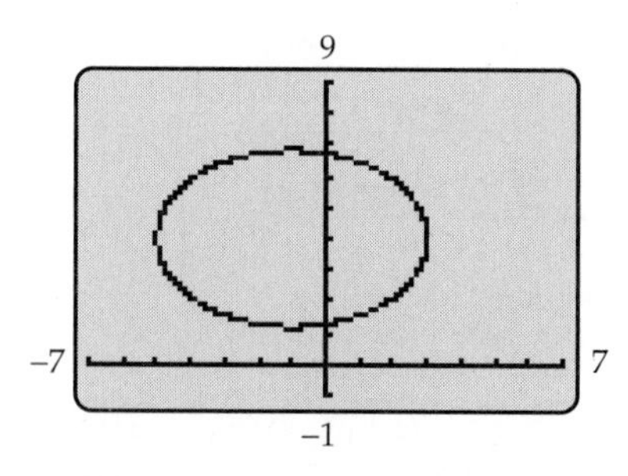

27. $x = t, y = 4(t - 1)^2 + 2$ (any real number t)

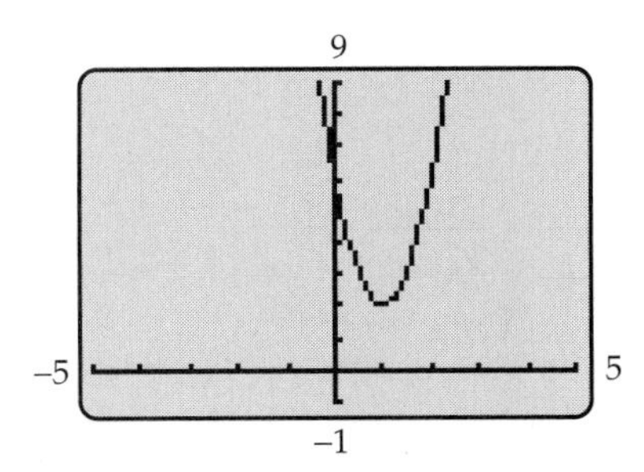

29. $x = 2(t - 2)^2, y = 1$ (any real number t)

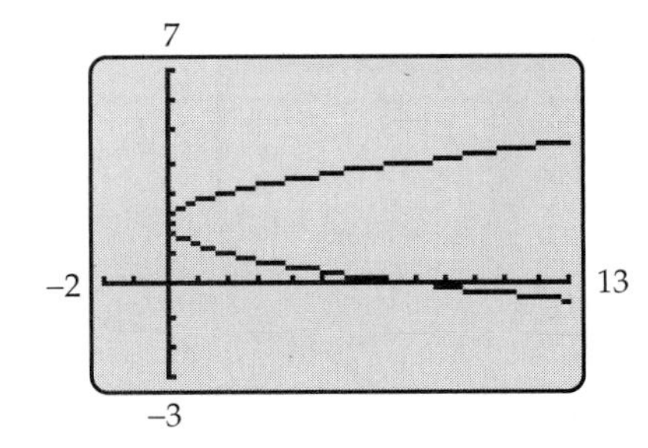

31. $x = 4\tan t - 1, y = 5/\cos t - 3$ $(0 \le t \le 2\pi)$

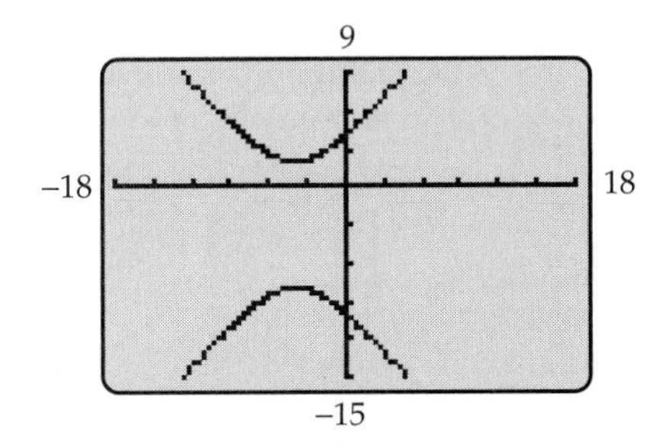

33. $x = 1/\cos t - 3, y = 2\tan t + 2$ $(0 \le t \le 2\pi)$

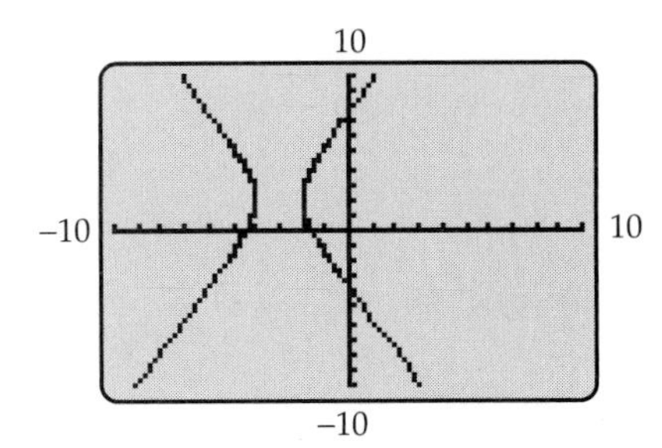

35. Ellipse; $-6 \le x \le 3$ and $-2 \le y \le 4$

37. Hyperbola; $-7 \le x \le 13$ and $-3 \le y \le 9$

39. Parabola; $-1 \le x \le 8$ and $-3 \le y \le 3$

41. Ellipse; $-1.5 \le x \le 1.5$ and $-1 \le y \le 1$

43. Hyperbola; $-15 \le x \le 15$ and $-10 \le y \le 10$

45. Parabola; $-19 \le x \le 2$ and $-1 \le y \le 13$

47. Hyperbola; $-15 \le x \le 15$ and $-15 \le y \le 15$

49. Ellipse; $-6 \le x \le 6$ and $-4 \le y \le 4$

51. Parabola; $-9 \le x \le 4$ and $-2 \le y \le 10$

53. $\dfrac{(x+5)^2}{49} + \dfrac{(y-3)^2}{16} = 1$ or $\dfrac{(x+5)^2}{16} + \dfrac{(y-3)^2}{49} = 1$

55. The asymptotes of $\dfrac{x^2}{a^2} - \dfrac{y^2}{a^2} = 1$ are $y = \pm\dfrac{a}{a}x$ or $y = \pm x$, with slopes $+1$ and -1. Since $(+1)(-1) = -1$, these lines are perpendicular.

57. $b = 0$

59. $(9, -\frac{1}{2} \pm \frac{1}{2}\sqrt{34})$

61. $\dfrac{x^2}{1{,}210{,}000} - \dfrac{y^2}{5{,}759{,}600} = 1$ (measurement in feet). The exact location cannot be determined from the given information.

Excursion 10.3.A, page 654

1. $\left(\dfrac{5\sqrt{2}}{2}, -\dfrac{\sqrt{2}}{2}\right)$

3. $\left(\dfrac{\sqrt{3}}{2}, -\dfrac{1}{2}\right)$

5. $\dfrac{u^2}{2} - \dfrac{v^2}{2} = 1$

7. $\dfrac{u^2}{4} + v^2 = 1$

9. $\theta \approx 53.13°; x = \dfrac{3}{5}u - \dfrac{4}{5}v; y = \dfrac{4}{5}u + \dfrac{3}{5}v$

11. $\theta \approx 36.87°; x = \dfrac{4}{5}u - \dfrac{3}{5}v; y = \dfrac{3}{5}u + \dfrac{4}{5}v$

13. (a) $(A\cos^2\theta + B\cos\theta\sin\theta + C\sin^2\theta)u^2 + (B\cos^2\theta - 2A\cos\theta\sin\theta + 2C\cos\theta\sin\theta - B\sin^2\theta)uv + (C\cos^2\theta - B\cos\theta\sin\theta + A\sin^2\theta)v^2 + (D\cos\theta + E\sin\theta)u + (E\cos\theta - D\sin\theta)v + F = 0$

(b) $B' = B\cos^2\theta - 2A\cos\theta\sin\theta + 2C\cos\theta\sin\theta - B\sin^2\theta = 2(C - A)\sin\theta\cos\theta + B(\cos^2\theta - \sin^2\theta)$ since B' is the coefficient of uv

(c) Since $\sin 2\theta = 2\sin\theta\cos\theta$ and $\cos 2\theta = \cos^2\theta - \sin^2\theta$, $B' = (C - A)\sin 2\theta + B\cos 2\theta$.

15. (a) From Exercise 13 (a) we have $(B')^2 - 4A'C' = (B\cos^2\theta - 2A\cos\theta\sin\theta + 2C\cos\theta\sin\theta - B\sin^2\theta)^2 - 4(A\cos^2\theta + B\cos\theta\sin\theta + C\sin^2\theta)(C\cos^2\theta - B\cos\theta\sin\theta + A\sin^2\theta) = [B(\cos^2\theta - \sin^2\theta) + 2(C - A)\cos\theta\sin\theta]^2 - 4(A\cos^2\theta + B\cos\theta\sin\theta + C\sin^2\theta)(C\cos^2\theta - B\cos\theta\sin\theta + A\sin^2\theta) = B^2(\cos^2\theta - \sin^2\theta)^2 + 4(C - A)^2\cos^2\theta\sin^2\theta + 4B(C - A)(\cos^2\theta - \sin^2\theta)\cos\theta\sin\theta - [4AC(\cos^4\theta + \sin^4\theta) + 4(A^2 + C^2 - B^2)\cos^2\theta\sin^2\theta - 4AB(\cos^3\theta\sin\theta - \cos\theta\sin^3\theta) + 4BC(\cos^3\theta\sin\theta - \cos\theta\sin^3\theta)] = B^2(\cos^4\theta - 2\cos^2\theta\sin^2\theta + \sin^4\theta + 4\cos^2\theta\sin^2\theta) - 4AC(2\cos^2\theta\sin^2\theta + \cos^4\theta + \sin^4\theta)$ (everything

else cancels) =
$(B^2 - 4AC)(\cos^4\theta + 2\cos^2\theta\sin^2\theta + \sin^4\theta) =$
$(B^2 - 4AC)(\cos^2\theta + \sin^2\theta)^2 = B^2 - 4AC$

(b) If $B^2 - 4AC < 0$, then also $(B')^2 - 4A'C' < 0$. Since $B' = 0$, $-4A'C' < 0$ and so $A'C' > 0$. By Exercise 14, the graph is an ellipse. The other two cases are proved in the same way.

Section 10.4, page 663

1. $P = (2, \pi/4)$, $Q = (3, 2\pi/3)$, $R = (5, \pi)$, $S = (7, 7\pi/6)$, $T = (4, 3\pi/2)$, $U = (6, -\pi/3)$ or $(6, 5\pi/3)$, $V = (7, 0)$

3. $(5, 2\pi), (5, -2\pi), (-5, 3\pi), (-5, -\pi)$, and others

5. $(1, 5\pi/6), (1, -7\pi/6), (-1, 11\pi/6), (-1, -13\pi/6)$, and others

7. $(3/2, 3\sqrt{3}/2)$

9. $(\sqrt{3}/2, -1/2)$

11. $(6, -\pi/6)$

13. $(2\sqrt{5}, 1.1071)$

15. $(\sqrt{31.25}, 2.6779)$

17.

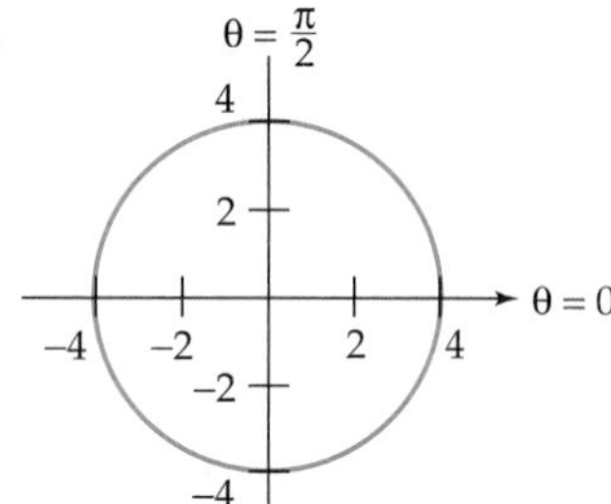

19.

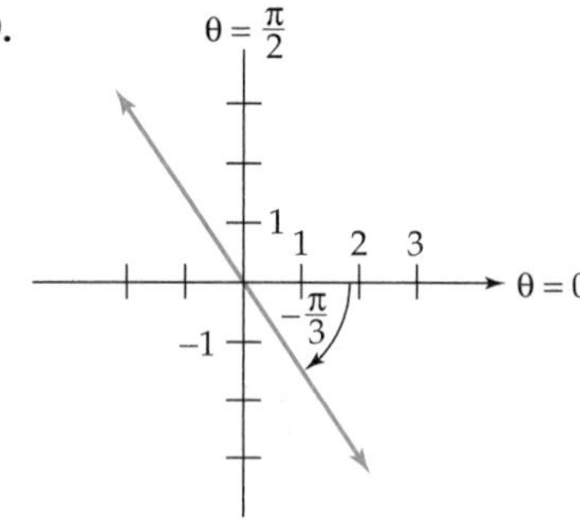

21.

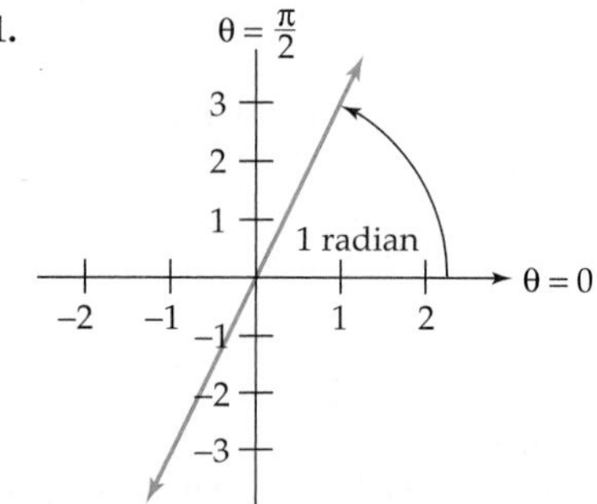

23.

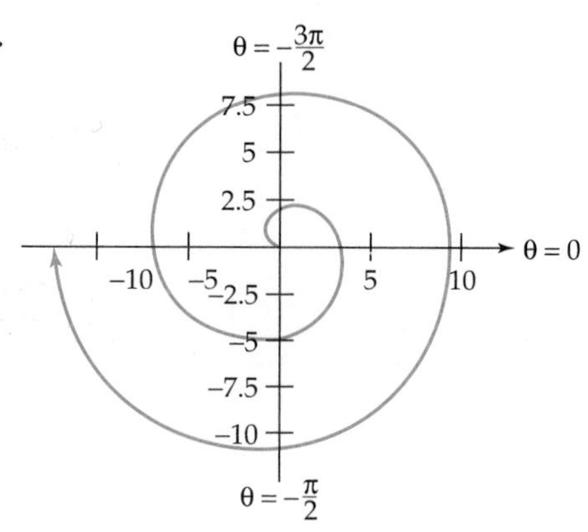

25.

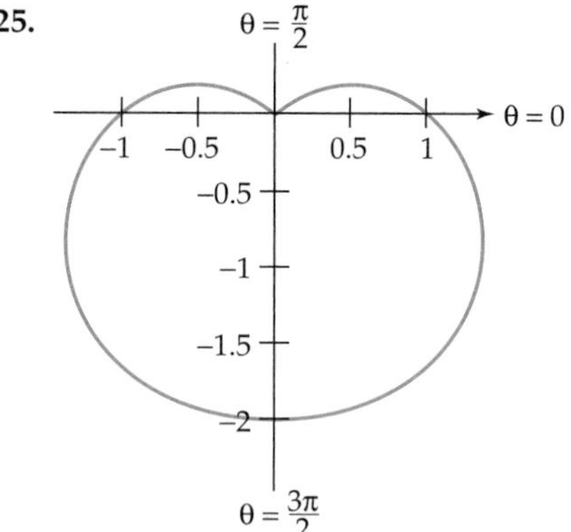

27.

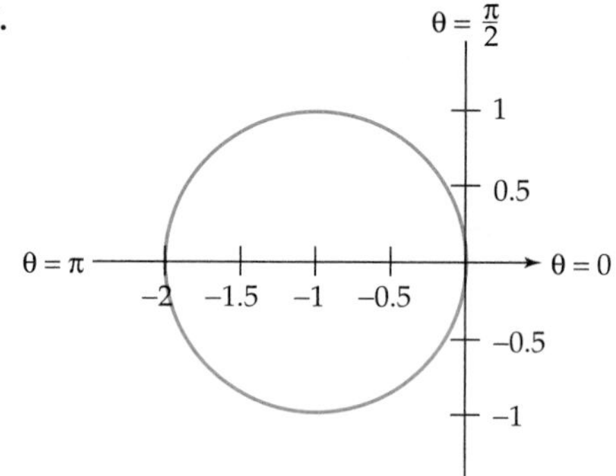

29.

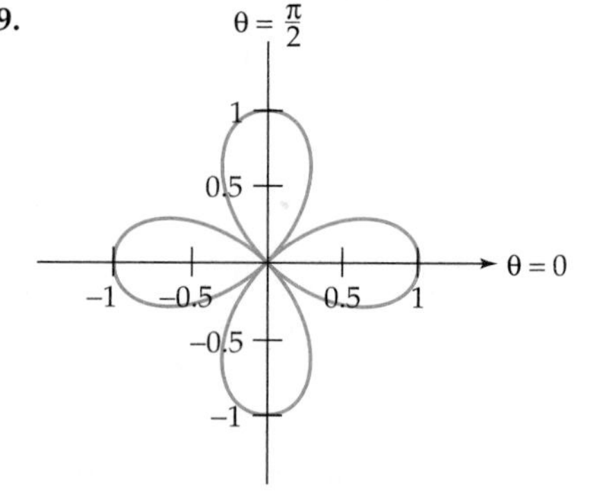

31.

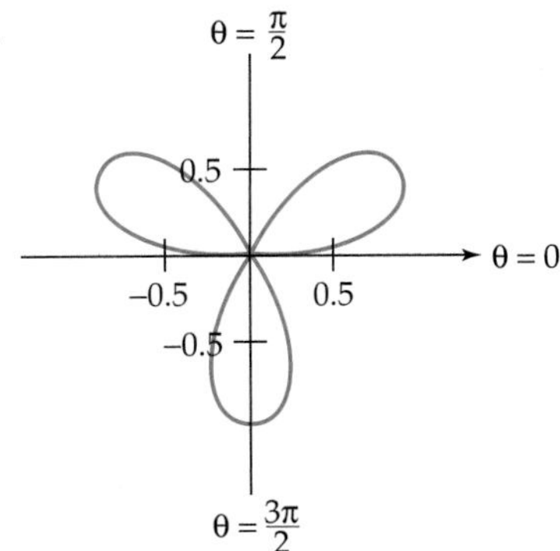

33.

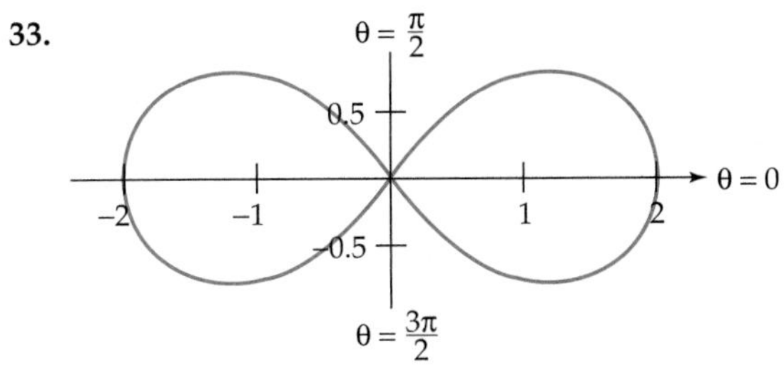

35.

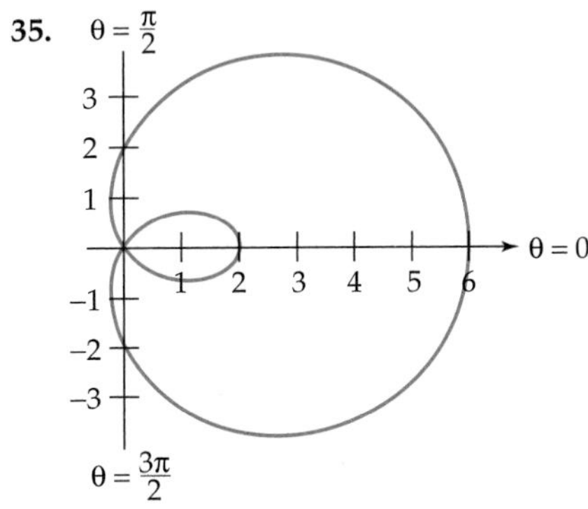

37.

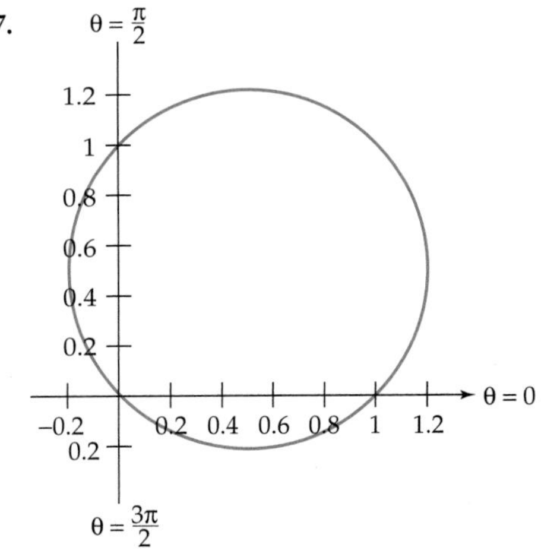

39.

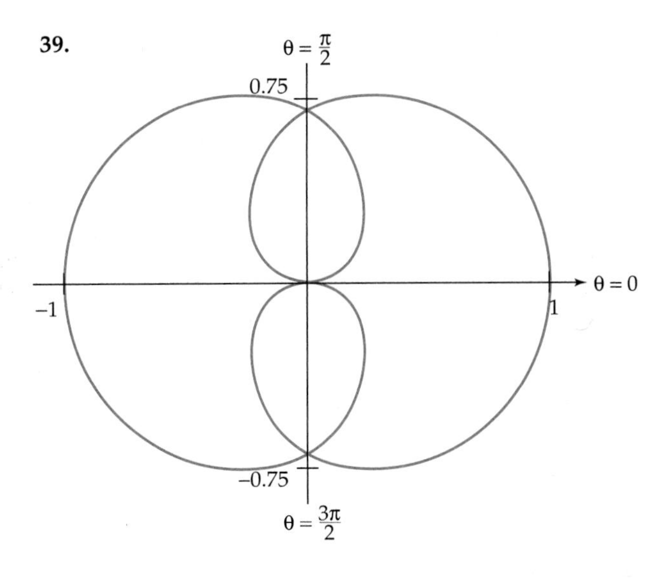

41.

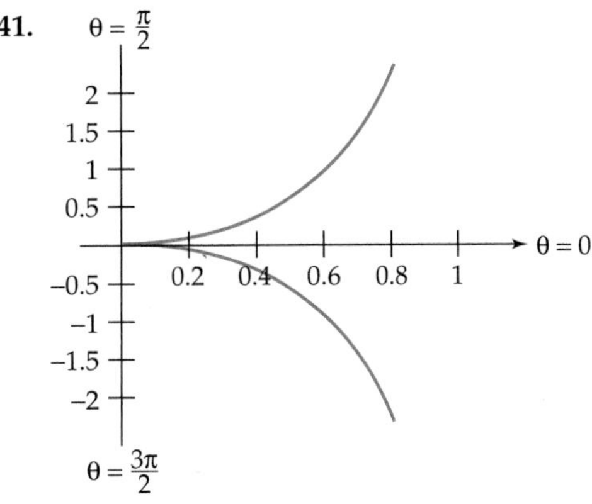

43.

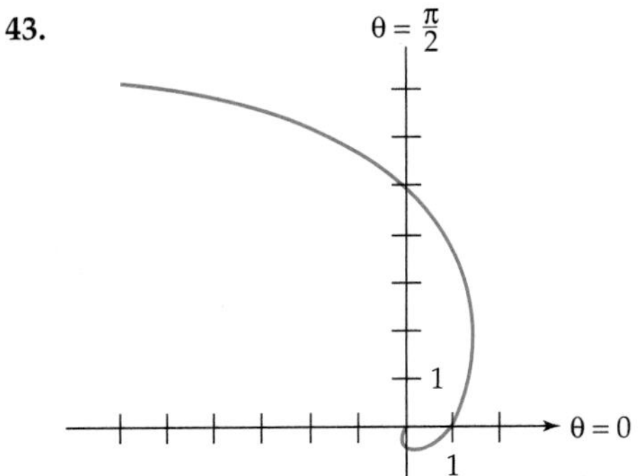

45.

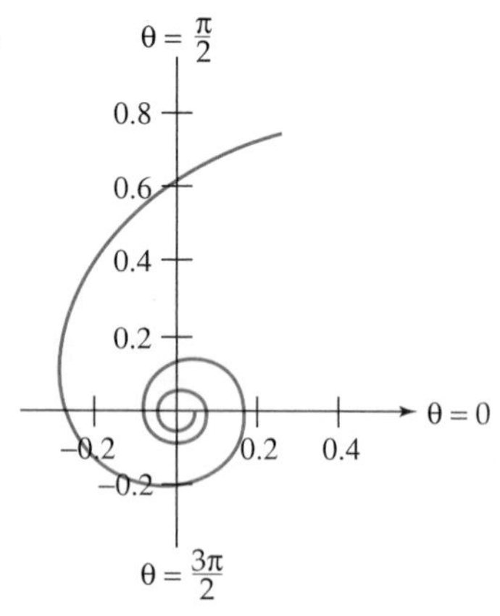

47. (a)

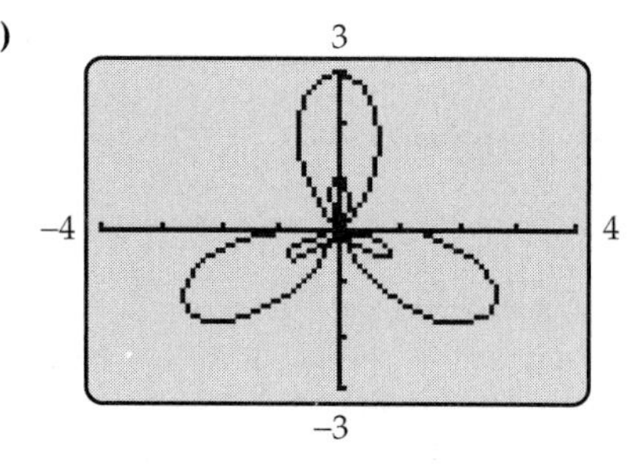

(b)

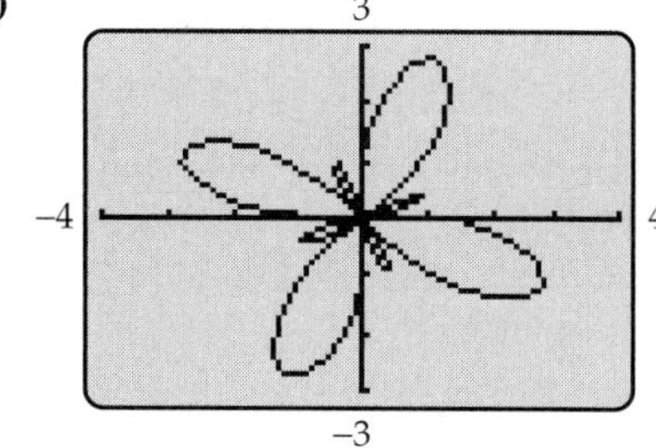

49. $r = a\sin\theta + b\cos\theta \Rightarrow r^2 = ar\sin\theta + br\cos\theta \Rightarrow$ $x^2 + y^2 = ay + bx \Rightarrow x^2 - bx + y^2 - ay = 0 \Rightarrow$ $(x^2 - bx + b^2/4) + (y^2 - ay + a^2/4) = (a^2 + b^2)/4 \Rightarrow$ $(x - b/2)^2 + (y - a/2)^2 = (a^2 + b^2)/4$, a circle with center $(b/2, a/2)$ and radius $\sqrt{a^2 + b^2}/2$.

51. Using the Law of Cosines in the following diagram, $d^2 = r^2 + s^2 - 2rs\cos(\theta - \beta)$, so $d = \sqrt{r^2 + s^2 - 2rs\cos(\theta - \beta)}$.

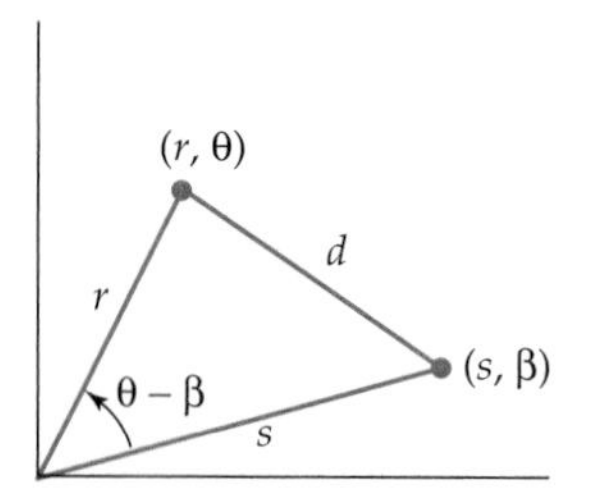

Section 10.5, page 673

1. (d) **3.** (c) **5.** (a)

7. Hyperbola, $e = 4/3$ **9.** Parabola, $e = 1$

11. Ellipse, $e = 2/3$

13. .1 **15.** $\sqrt{5}$ **17.** 5/4

19. (b) $\sqrt{15}/4, \sqrt{10}/4, \sqrt{2}/4$

(c) The smaller the eccentricity, the closer the shape is to circular.

In the graphs for Exercises 21–31, the x- and y-axes with scales are given for convenience, but coordinates of points are in polar *coordinates.*

21.

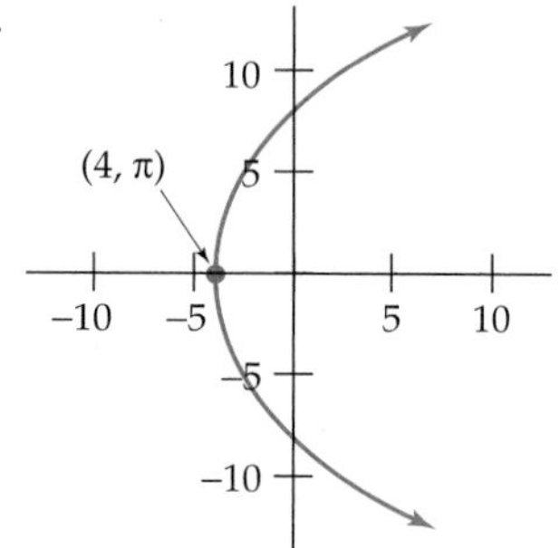

23.

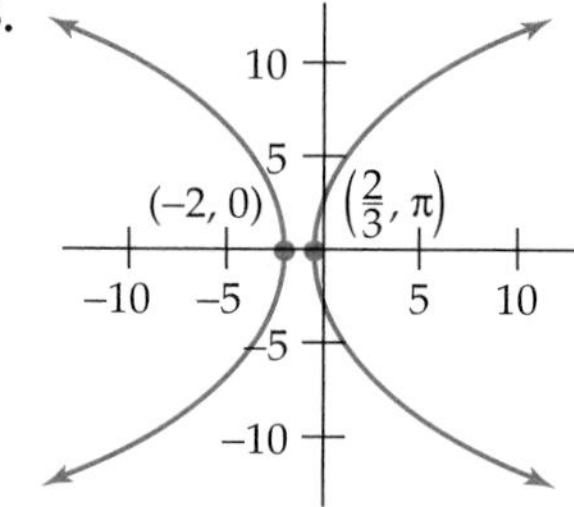

25.

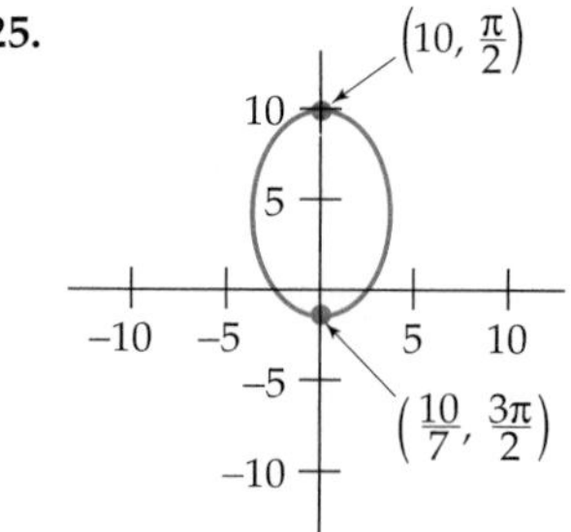

27.

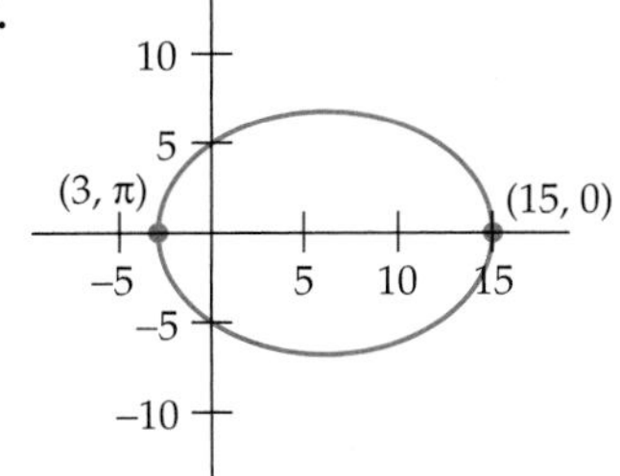

29.

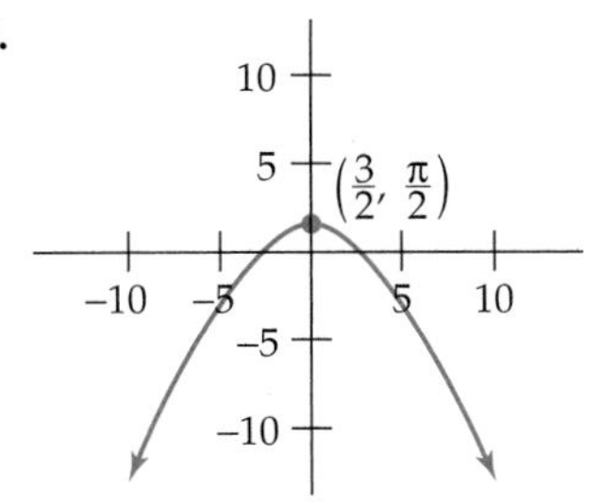

31.

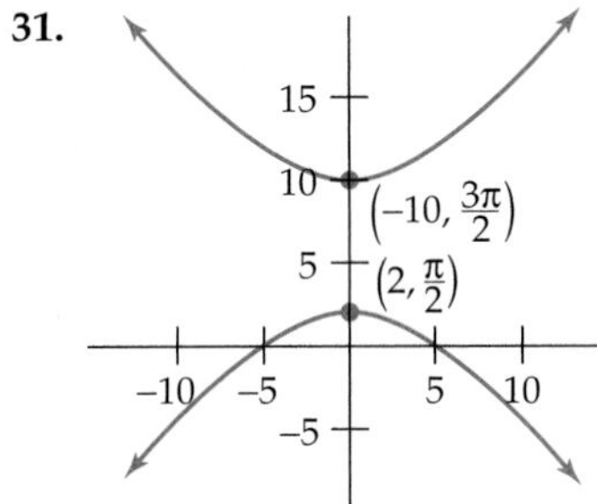

33. $r = \dfrac{6}{1 - \cos\theta}$

35. $r = \dfrac{16}{5 + 3\sin\theta}$

37. $r = \dfrac{3}{1 + 2\cos\theta}$

39. $r = \dfrac{8}{1 - 4\cos\theta}$

41. $r = \dfrac{3}{1 - \sin\theta}$

43. $r = \dfrac{2}{2 + \cos\theta}$

45. $r = \dfrac{2}{1 - 2\cos\theta}$

47. Since $0 < e < 1, 0 < 1 - e^2 < 1$ as well. The formulas for a^2 and b^2 show that $a^2 = b^2/(1 - e^2)$, so $a^2 > b^2$. Since a and b are both positive, $a > b$.

49. $r = \dfrac{3 \cdot 10^7}{1 - \cos\theta}$

Chapter 10 Review, page 676

1. $-15 \le x \le 15$ and $-10 \le y \le 10$

3. $-15 \le x \le 8$ and $-6 \le y \le 10$

5. $x = 3 - 2y$ or $y = -\dfrac{1}{2}x + \dfrac{3}{2}$

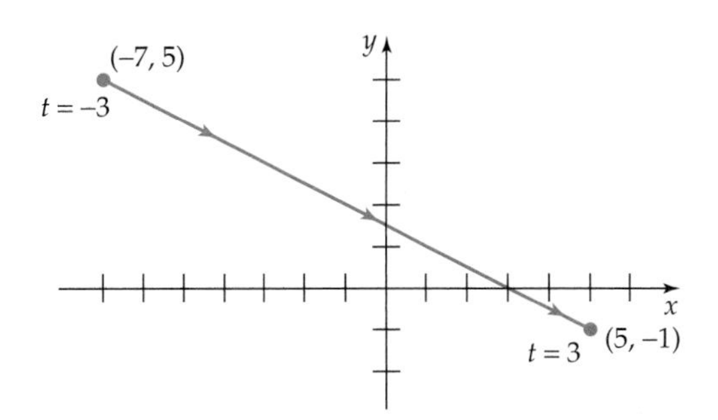

7. $y = -2x^2 + 2$

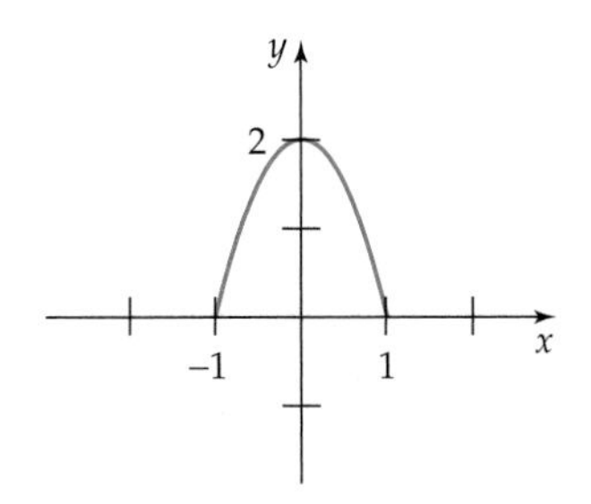

Point moves from $(1, 0)$ to $(-1, 0)$ as t goes from 0 to π. Then point retraces its path, moving from $(-1, 0)$ to $(1, 0)$ as t goes from π to 2π.

9. (b) and (c)

11. Ellipse, foci: $(0, 2)$, $(0, -2)$, vertices: $(0, 2\sqrt{5})$, $(0, -2\sqrt{5})$

13. Ellipse, foci: $(1, 6)$, $(1, 0)$, vertices: $(1, 7)$, $(1, -1)$

15. Focus: $(0, 5/14)$, directrix: $y = -5/14$

17.

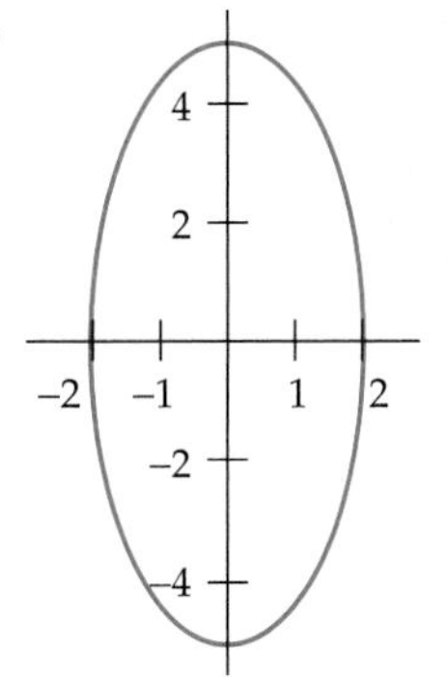

19.

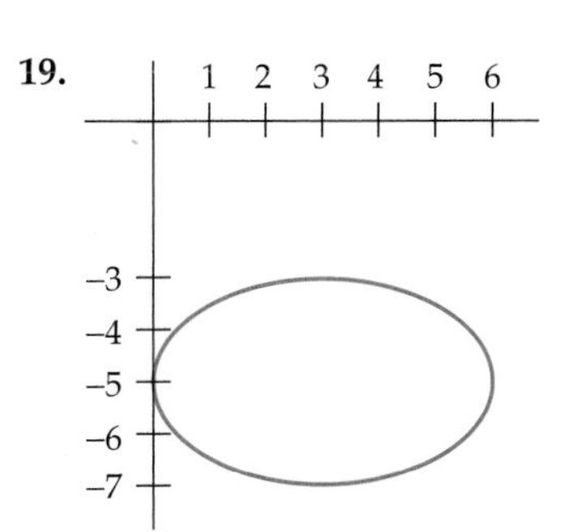

21. Asymptotes:

$y + 4 = \pm\frac{5}{2}(x - 1)$

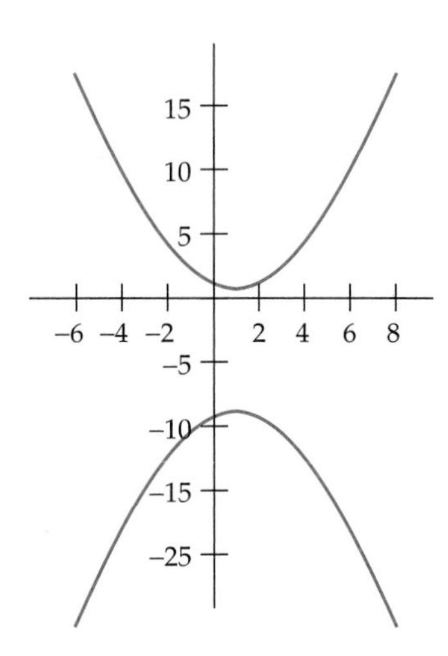

23.

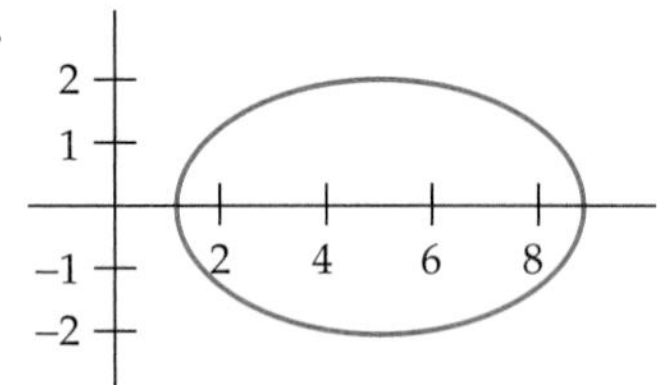

25.

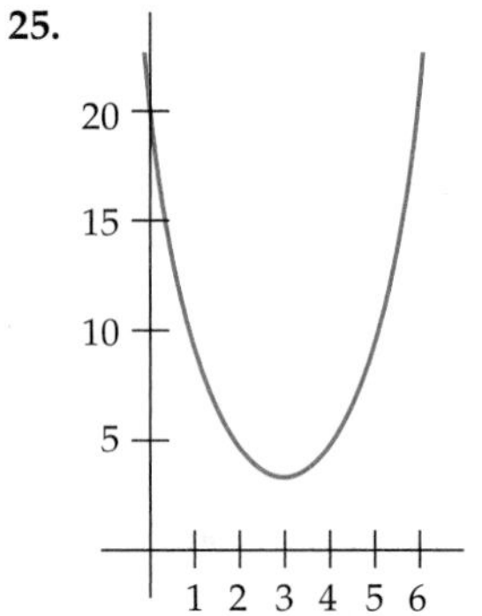

27.

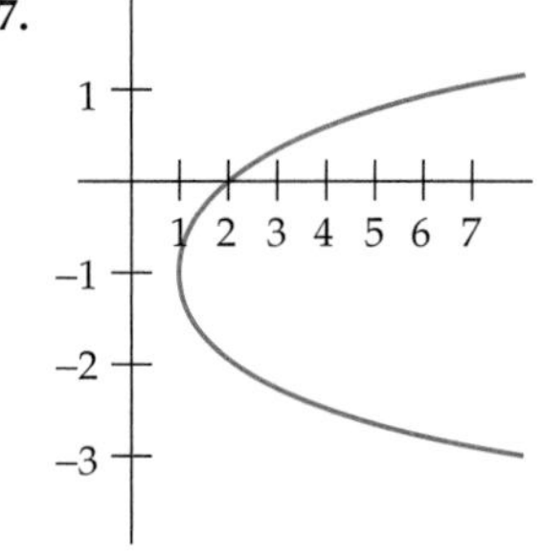

29. Center: $(4, -6)$

31. $\frac{(x-3)^2}{4} + \frac{(y-1)^2}{2} = 1$

33. $\frac{y^2}{4} - \frac{(x-3)^2}{16} = 1$

35. $\left(y + \frac{1}{2}\right)^2 = -\frac{1}{2}\left(x - \frac{3}{2}\right)$

37. Ellipse

39. Hyperbola

41. $-6 \le x \le 6$ and $-4 \le y \le 4$

43. $-9 \le x \le 9$ and $-6 \le y \le 6$

45. $-15 \le x \le 10$ and $-10 \le y \le 20$

47. $x = \frac{1}{2}u - \frac{\sqrt{3}}{2}v$

$y = \frac{\sqrt{3}}{2}u + \frac{1}{2}v$

49. 45°

51.

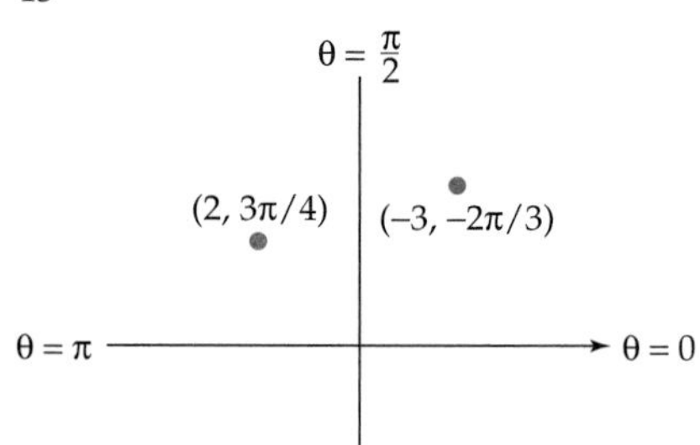

53.

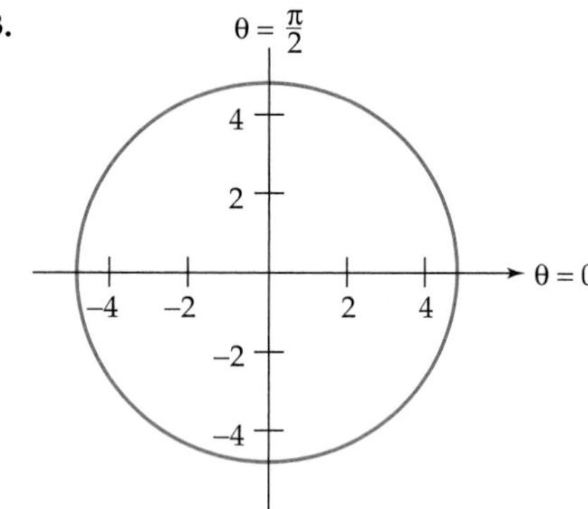

55.

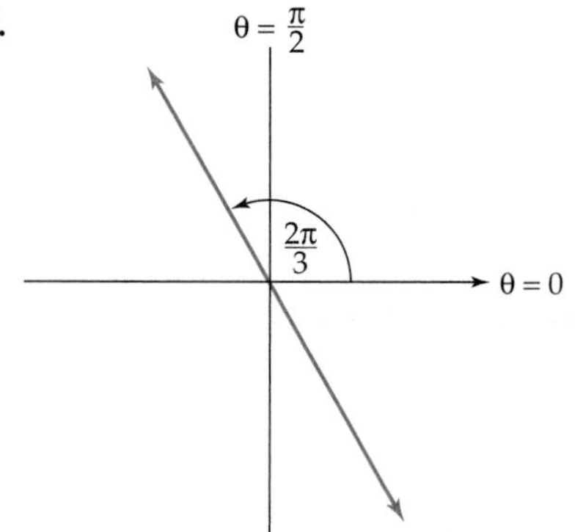

57.

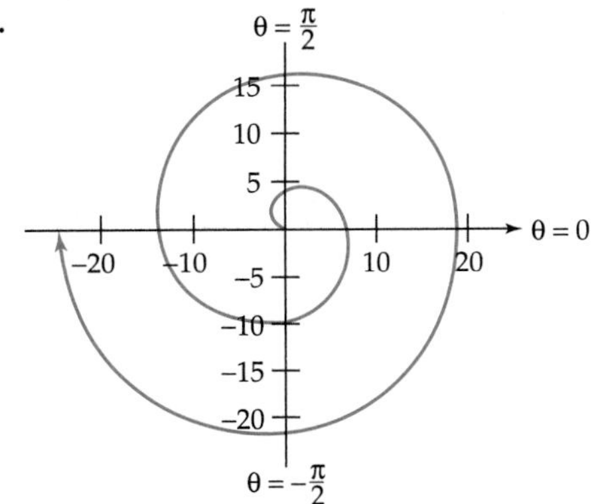

59.

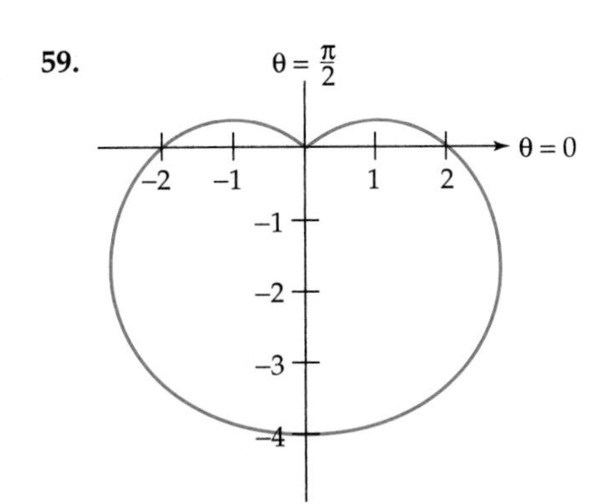

61.

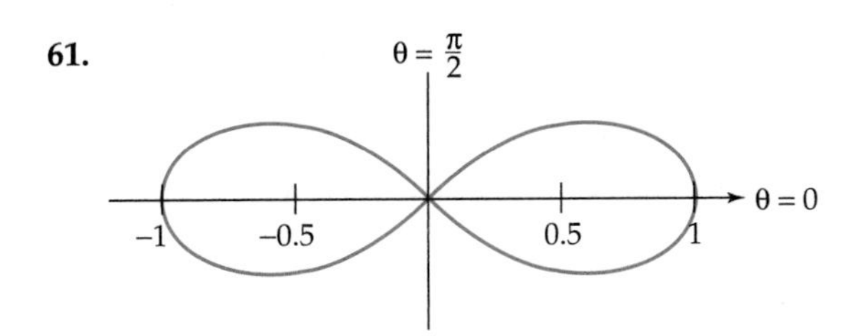

63. $\left(-\frac{3}{2}, -\frac{3\sqrt{3}}{2}\right)$ **65.** Eccentricity $= \sqrt{\frac{2}{3}} \approx 0.8165$

67. Ellipse

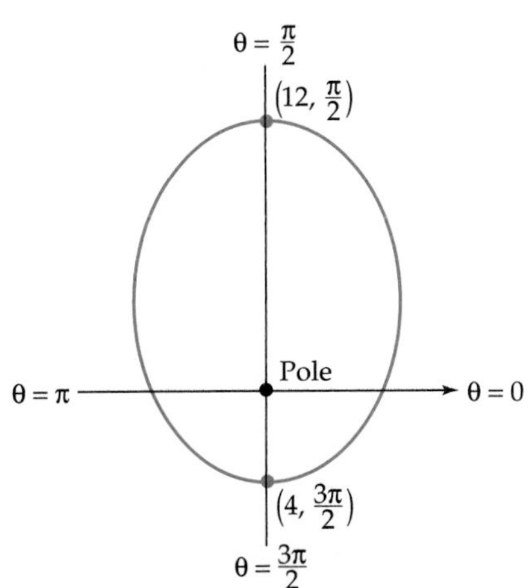

69. Hyperbola

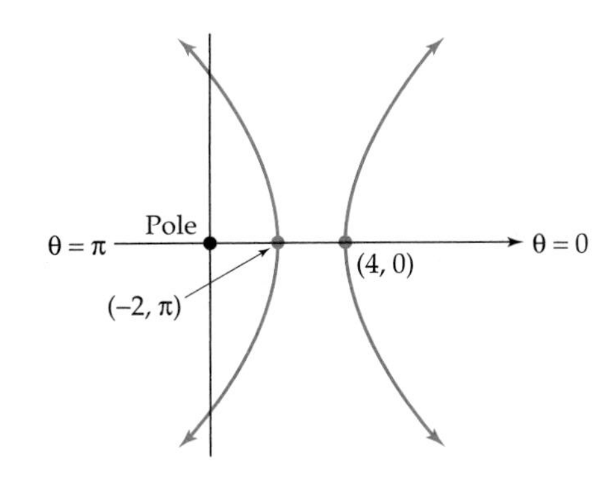

71. $r = \dfrac{24}{5 + \cos\theta}$ **73.** $r = \dfrac{2}{1 + \cos\theta}$

Chapter 11

Section 11.1, page 690

1. Yes **3.** Yes **5.** No

7. $x = \frac{11}{5}, y = -\frac{7}{5}$

9. $x = \frac{2}{7}, y = -\frac{11}{7}$

11. $r = \frac{5}{2}, s = -\frac{5}{2}$

13. $x = \frac{3c}{2}, y = \frac{-c + 2d}{2}$

15. $x = 28, y = 22$ **17.** $x = 2, y = -1$

19. Inconsistent

21. $x = b, y = \frac{3b - 4}{2}$, where b is any real number

23. $x = b, y = \frac{3b - 2}{4}$, where b is any real number

25. Inconsistent

27. $x = -6, y = 2$ **29.** $x = \frac{66}{5}, y = \frac{18}{5}$

31. $x = \frac{7}{11}, y = -7$ **33.** $x = 1, y = \frac{1}{2}$

35. $x = -\frac{3}{7}, y = \frac{42}{13}$ **37.** $x = .185, y = -.624$

39. $x = \frac{rd - sb}{ad - bc}, y = \frac{as - cr}{ad - bc}$

41. $x = \frac{3c + 8}{15}, y = \frac{6c - 4}{15}$ is the only solution.

43. $c = -3, d = \frac{1}{2}$

45. (a) Electric: $y = 960x + 2000$; solar: $y = 114x + 14{,}000$
(b) Electric: \$6800; solar: \$14,570
(c) Costs same in fourteenth year; electric; solar

47. (a) $y = 7.50x + 5000$
(b) $y = 8.20x$
(c)

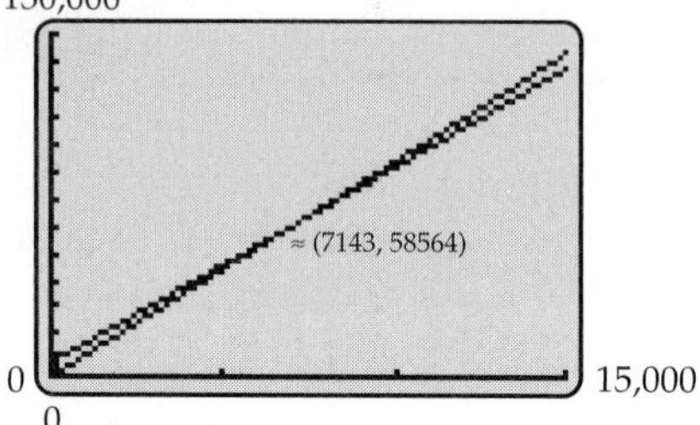

Costs equal at approximately 7143 cases.
(d) The company should buy from the supplier any number of cases less than 7143 and produce their own beyond that quantity.

49. 140 adults, 60 children

51. \$14,450 at 9% and \$6450 at 11%

53. 3/4 lb cashews and $2\frac{1}{4}$ lb peanuts

55. Boat speed 18 mph; current speed 2 mph

57. 12 liters of the 18%; 18 liters of the 8%

59. 24 g of 50% alloy; 16 g of 75% alloy

61. (a) $x = -\frac{1}{2}z_1 - \frac{2}{3}z_2 + 62{,}000$
(b) $y = -\frac{1}{2}z_1 - z_2 + 76{,}000$

Section 11.2, page 705

1. $\begin{pmatrix} 2 & -3 & 4 & 1 \\ 1 & 2 & -6 & 0 \\ 3 & -7 & 4 & -3 \end{pmatrix}$

3. $\begin{pmatrix} 1 & -\frac{1}{2} & \frac{7}{4} & 0 \\ 2 & -\frac{3}{2} & 5 & 0 \\ 0 & -2 & \frac{1}{3} & 0 \end{pmatrix}$

5. $\begin{aligned} 2x - 3y &= 1 \\ 4x + 7y &= 2 \end{aligned}$

7. $\begin{aligned} x \qquad + z \qquad &= 1 \\ x - y + 4z - 2w &= 3 \\ 4x + 2y + 5z \qquad &= 2 \end{aligned}$

9. $x = 3/2, y = 5, z = -2, w = 0$

11. $x = 2 - t, y = -3 - 2t, z = 4, w = t$, where t is any real number

13. $x = \frac{3}{2}, y = \frac{3}{2}, z = -\frac{3}{2}$

15. $z = t, y = -1 + \frac{1}{3}t, x = 2 - \frac{4}{3}t$, where t is any real number

17. $x = -14, y = -6, z = 2$

19. $x = 100, y = 50, z = 50$

21. $z = t, y = \frac{1}{2} - 2t, x = t$, where t is any real number

23. $x = 1, y = 2$

25. No solutions

27. $z = t, y = t - 1, x = -t + 2$, for any real number t

29. $x = 0, y = 0, z = 0$

31. $x = -1, y = 1, z = -3, w = -2$

33. $x = \frac{7}{31}, y = \frac{6}{31}, z = \frac{1}{31}, w = \frac{29}{31}$

35. $x = 1/2, y = 1/3, z = -1/4$

37. $A = -1, B = 2$

39. $A = -3/25, B = 3/25, C = 7/5$

41. $A = 2, B = 3, C = -1$

43. $x = -s - t, y = s + 2t, z = s, w = t$, where s and t are any real numbers

45. $x = t, y = 0, z = t, w = 2t$, where t is any real number

47. 10 quarters; 28 dimes; 14 nickels

49. \$3000 from her friend; \$6000 from the bank; \$1000 from the insurance company

51. \$15,000 in the mutual fund; \$30,000 in bonds; \$25,000 in food franchise

53. Three possible solutions:
18 bedroom, 13 living room, 0 whole house
16 bedroom, 8 living room, 2 whole house
14 bedroom, 3 living room, 4 whole house

55. 8 hr for Tom; 24 hr for Dick; 12 hr for Harry

57. 2000 chairs; 1600 chests; 2500 tables

59. 20 model A; 15 model B; 10 model C

Section 11.3, page 717

1. AB defined, 2×4; BA not defined

3. AB defined, 3×3; BA defined, 2×2

5. AB defined, 3×2; BA not defined

7. $\begin{pmatrix} 3 & 0 & 11 \\ 2 & 8 & 10 \end{pmatrix}$

9. $\begin{pmatrix} 1 & -3 \\ 2 & -1 \\ 5 & 6 \end{pmatrix}$

11. $\begin{pmatrix} 1 & -1 & 1 & 2 \\ 4 & 3 & 3 & 2 \\ -1 & -1 & -3 & 2 \\ 5 & 3 & 2 & 5 \end{pmatrix}$

13. $AB = \begin{pmatrix} 17 & -3 \\ 33 & -19 \end{pmatrix}$; $BA = \begin{pmatrix} -4 & 9 \\ 24 & 2 \end{pmatrix}$

15. $AB = \begin{pmatrix} 8 & 24 & -8 \\ 2 & -2 & 6 \\ -3 & -21 & 15 \end{pmatrix}$; $BA = \begin{pmatrix} 19 & 9 & 8 \\ -10 & 2 & 0 \\ 0 & 0 & 0 \end{pmatrix}$

17. $\begin{pmatrix} -2 & 1 \\ 3/2 & -1/2 \end{pmatrix}$ **19.** No inverse

21. No inverse **23.** $\begin{pmatrix} -3 & 2 & -4 \\ -1 & 1 & -1 \\ 8 & -5 & 10 \end{pmatrix}$

25. $x = -1, y = 0, z = -3$

27. $x = -8, y = 16, z = 5$

29. $x = -.5, y = -2.1, z = 6.7, w = 2.8$

31. $x = 10.5, y = 5, z = -13, v = 32, w = 2.5$

33. $x = -1149/161, y = 426/161, z = -1124/161, w = 579/161$

35. $x = 0, y = 0, z = 0, v = 0, w = 0$

37. Inconsistent system; no solutions

39. $x = -2 - 3w, y = 5/4 - (1/4)z$, where z and w are any real numbers.

41. $a = \frac{35}{8}, b = \frac{13}{4}, c = \frac{-117}{8}$

43. $a = 2, b = -3, c = 1$

45. $a = 3, b = -1, c = 2$

47. $a = 1, b = -4, c = 1$

49. 15,000 of A; 18,000 of B; 54,000 of C

Section 11.4, page 724

1. $x = 3, y = 9$ or $x = -1, y = 1$

3. $x = \frac{-1 + \sqrt{73}}{6}, y = \frac{37 - \sqrt{73}}{18}$ or $x = \frac{-1 - \sqrt{73}}{6}, y = \frac{37 + \sqrt{73}}{18}$

5. $x = 7, y = 3$ or $x = 3, y = 7$

7. $x = 0, y = -2$ or $x = 6, y = 1$

9. $x = 2, y = 0$ or $x = 4, y = 2$

11. $x = 4, y = 3$ or $x = -4, y = 3$ or $x = \sqrt{21}, y = -2$ or $x = -\sqrt{21}, y = -2$

13. $x = -1.6237, y = -8.1891$ or $x = 1.3163, y = 1.0826$ or $x = 2.8073, y = 2.4814$

15. $x = -1.9493, y = .4412$ or $x = .3634, y = .9578$ or $x = 1.4184, y = .5986$

17. $x = -.9519, y = -.8145$ **19.** No solutions

21. $x = \frac{13 - \sqrt{105}}{8}, y = \frac{-3 - \sqrt{105}}{8}$ or $x = \frac{13 + \sqrt{105}}{8}, y = \frac{-3 + \sqrt{105}}{8}$

23. $x = -4.8093, y = 19.3201$ or $x = -3.1434, y = 7.7374$ or $x = 2.1407, y = 7.7230$ or $x = 2.8120, y = 11.7195$

25. $x = -3.8371, y = -2.2596$ or $x = -.9324, y = -7.7796$

27. $x = -1.4873, y = .0480$ or $x = -.0480, y = 1.4873$ or $x = .0480, y = -1.4873$ or $x = 1.4873, y = -.0480$

29. There is no solution when the wire is 70 ft long because the graphs of $y = 70 - 2x$ and $y^2 = 4\pi(100 - x^2/8)$ do not intersect.

31. Two possible boxes: one is 2 by 2 by 4 m and the other is approximately 3.123 by 3.123 by 1.640 m.

33. -4 and -12 **35.** 1.6 and 2.6

37. 15 and -12 **39.** 12 ft by 17 ft

41. 8×15 inches **43.** $y = 6x - 9$

Chapter 11 Review, page 726

1. $x = -5, y = -7$ **3.** $x = 0, y = -2$

5. $x = -35, y = 140, z = 22$ **7.** $x = 2, y = 4, z = 6$

9. $x = -t + 1, y = -2t, z = t$ for any real number t

11. 37 and -19 **13.** (c) **15.** 100

17. $\begin{pmatrix} 1 & -2 & 3 & 4 \\ 2 & 1 & -4 & 3 \\ -3 & 4 & -1 & -2 \end{pmatrix}$

19. $\begin{pmatrix} 2 & -1 & -2 & 2 \\ 1 & 3 & -2 & 1 \\ -1 & 4 & 2 & -3 \end{pmatrix}$ **21.** $\frac{3}{x + 2} + \frac{1}{x - 3}$

23. $\begin{pmatrix} -2 & 3 \\ -4 & -1 \end{pmatrix}$ **25.** Not defined

27. $\begin{pmatrix} -9 & 7 \\ -4 & 3 \end{pmatrix}$ **29.** $\begin{pmatrix} 1 & 2 & -2 \\ -1 & 3 & 0 \\ 0 & -2 & 1 \end{pmatrix}$

31. $x = -1/85, y = -14/85, z = -21/34, w = 46/85$

33. $y = 5x^2 - 2x + 1$

35. $x = 3, y = 9$ or $x = -1, y = 1$

37. $x = 1 - \sqrt{7}, y = 1 + \sqrt{7}$ or $x = 1 + \sqrt{7}, y = 1 - \sqrt{7}$

39. $x = -1.692, y = 3.136$ or $x = 1.812, y = 2.717$

41. 30 lb corn, 15 lb soybeans, 40 lb by-products

Chapter 12

Section 12.1, page 740

1. 8, 10, 12, 14, 16 **3.** $1, \frac{1}{8}, \frac{1}{27}, \frac{1}{64}, \frac{1}{125}$

5. $-\sqrt{3}, 2, -\sqrt{5}, \sqrt{6}, -\sqrt{7}$

7. 3.9, 4.01, 3.999, 4.0001, 3.99999

9. 2, 7, 8, 13, 14 **11.** $\sum_{i=1}^{11} i$

13. $\sum_{i=7}^{13} \frac{1}{2^i}$ (other answers possible)

15. 45 **17.** 224 **19.** 15,015

21. $a_n = (-1)^n$ **23.** $a_n = \frac{n}{n+1}$ **25.** $a_n = 5n - 3$

27. 4, 11, 25, 53, 109 **29.** 1, −2, 3, 2, 3

31. $2, 3, 3, \frac{9}{2}, \frac{27}{4}$ **33.** 3, 1, 4, 1, 5

35. Third − 10; sixth − 2 **37.** Third 5; sixth 0

39. $\sum_{n=1}^{6} \frac{1}{2n+1}$ (other answers possible)

41. $\sum_{n=8}^{12} (-1)^n\left(\frac{n-7}{n}\right)$ (other answers possible)

43. 2.613035 **45.** $\frac{1051314}{31415} \approx 33.465$

47. 1.5759958

49. **(a)** 1993: \$0.34; 1995: \$0.45; 1998: \$0.615
(b) \$3.38

51. **(a)** 1993: 17 billion; 2000: 21.2 billion
(b) 185.4 billion

53. **(b)** 59, 61, 67, 71

55. 4, 9, 25, 49, 121 **57.** 3, 7, 13, 19, 23

59. **(a)** 1, 1, 2, 3, 5, 8, 13, 21, 34, 55
(b) 1, 2, 4, 7, 12, 20, 33, 54, 88, 143
(c) nth partial sum $= a_{n+2} = 1$

61. $n = 1$: $5(1)^2 + 4(-1)^1 = 1 = 1^2$;
$n = 2$: $5(1)^2 + 4(-1)^2 = 9 = 3^2$;
$n = 3$: $5(2)^2 + 4(-1)^3 = 16 = 4^2$;
$n = 4$: $5(3)^2 + 4(-1)^4 = 49 = 7^2$; etc.

63. We have $a_1 = a_3 - a_2$; $a_2 = a_4 - a_3$; $a_3 = a_5 - a_4$; $\ldots a_{k-1} = a_{k+1} - a_k$; $a_k = a_{k+2} - a_{k+1}$. If these equations are listed vertically, the sum of the left-side terms is $\sum_{n=1}^{k} a_k$. On the right side, one term in each line is the same as a term in the next line, except for sign. So the sum of the right-side terms is $a_{k+2} - a_2$. Since $a_2 = 1$, we conclude that $\sum_{n=1}^{k} a_n = a_{k+2} - 1$.

Section 12.2, page 746

1. 13; $a_n = 2n + 3$ **3.** 5; $a_n = n/4 + 15/4$

5. 8; $a_n = -n/2 + 21/2$ **7.** 87

9. −21/4 **11.** 30

13. $a_n - a_{n-1} = (3 - 2n) - [3 - 2(n-1)] = -2$; arithmetic with $d = -2$

15. $a_n - a_{n-1} = \frac{5+3n}{2} - \frac{5+3(n-1)}{2} = \frac{3}{2}$; arithmetic with $d = \frac{3}{2}$

17. $a_n - a_{n-1} = (c + 2n) - [c + 2(n-1)] = 2$; arithmetic with $d = 2$

19. $a_5 = 14$; $a_n = 2n + 4$ **21.** $a_5 = 25$; $a_n = 7n - 10$

23. $a_5 = 0$; $a_n = -15/2 + 3n/2$ **25.** 710

27. $156\frac{2}{3}$ **29.** 2550 **31.** 20,100

33. \$77,500 in tenth year; \$437,500 over ten years

35. 428 **37.** 23.25, 22.5, 21.75, 21, 20.25, 19.5, 18.75

Section 12.3, page 750

1. Arithmetic **3.** Geometric **5.** Arithmetic

7. Geometric **9.** $a_6 = 160$; $a_n = 2^{n-1} \cdot 5$

11. $a_6 = \frac{1}{256}$; $a_n = \frac{1}{4^{n-2}}$

13. $a_6 = -\frac{5}{16}$; $a_n = \frac{(-1)^{n-1} \cdot 5}{2^{n-2}}$

15. 315/32 **17.** 381

19. $\frac{a_n}{a_{n-1}} = \frac{\left(-\frac{1}{2}\right)^n}{\left(-\frac{1}{2}\right)^{n-1}} = -\frac{1}{2}$; geometric with $r = -\frac{1}{2}$

21. $\frac{a_n}{a_{n-1}} = \frac{5^{n+2}}{5^{(n-1)+2}} = 5$; geometric with $r = 5$

23. $a_5 = 1$; $a_n = \frac{(-1)^{n-1}64}{4^{n-2}} = \frac{(-1)^{n-1}}{4^{n-5}}$

25. $a_5 = \frac{1}{16}$; $a_n = \frac{1}{4^{n-3}}$

27. $a_5 = -\frac{8}{25}$; $a_n = -\frac{2^{n-2}}{5^{n-3}}$ **29.** 254

31. $-\frac{4921}{19{,}683}$ **33.** $\frac{665}{8}$

35. **(a)** Since for all n, the ratio r is
$$\frac{a_{n+1}}{a_n} = \frac{1.71(1.91^{n+1})}{1.71(1.91^n)} = \frac{1.71(1.91^n)(1.91)}{1.71(1.91^n)} = 1.91,$$
the sequence is geometric.
(b) \$217.47

37. 23.75 ft

39. $\sum_{n=1}^{31} 2^{n-1} = \frac{1-2^{31}}{1-2} = (2^{31} - 1)$ cents $=$ \$21,474,836.47

41. \$1898.44

43. $\log a_n - \log a_{n-1} = \log \frac{a_n}{a_{n-1}} = \log r$

45. The sequence is $\{2^{n-1}\}$ and $r = 2$. So for any k, the kth term is 2^{k-1}, and the sum of the preceding terms is the $(k-1)$th partial sum of the sequence, $\sum_{n=1}^{k-1} 2^{n-1} = \frac{1-2^{k-1}}{1-2} = 2^{k-1} - 1$.

47. 37 payments

Excursion 12.3.A, page 756

1. 1 **3.** $.06/.94 = 3/47$ **5.** $\frac{500}{.6} = 833\frac{1}{3}$

7. $4 + 2\sqrt{2}$ **9.** $2/9$ **11.** $597/110$

13. $10{,}702/4995$ **15.** $174{,}067/99{,}900$

17. (a) $a_1 = 2(1.5)^1 = 3$. For each $n \geq 2$, $a_{n-1} = 2(1.5)^{n-1}$ and $a_n = 2(1.5)^n$, so that the ratio $\frac{a_n}{a_{n-1}}$ is $\frac{2(1.5)^n}{2(1.5)^{n-1}} = 1.5$. Therefore, this is a geometric series.
(b) $S_k = -6 + 6(1.5)^k$ and $f(x) = -6 + 6(1.5)^x$
(c) The graph shows that as x gets large, $f(x)$ gets huge, so there is no horizontal asymptote. Hence, the series does not converge.

Section 12.4, page 763

1. 720 **3.** 220 **5.** 0 **7.** 64 **9.** 3,921,225

11. $x^5 + 5x^4y + 10x^3y^2 + 10x^2y^3 + 5xy^4 + y^5$

13. $a^5 - 5a^4b + 10a^3b^2 - 10a^2b^3 + 5ab^4 - b^5$

15. $32x^5 + 80x^4y^2 + 80x^3y^4 + 40x^2y^6 + 10xy^8 + y^{10}$

17. $x^3 + 6x^2\sqrt{x} + 15x^2 + 20x\sqrt{x} + 15x + 6\sqrt{x} + 1$

19. $1 - 10c + 45c^2 - 120c^3 + 210c^4 - 252c^5 + 210c^6 - 120c^7 + 45c^8 - 10c^9 + c^{10}$

21. $x^{-12} + 4x^{-8} + 6x^{-4} + 4 + x^4$ **23.** 56

25. $-8i$ **27.** $10x^3y^2$ **29.** $35c^3d^4$

31. $\frac{35}{8}u^{-5}$ **33.** 4032 **35.** 160

37. (a) $\binom{9}{1} = \frac{9!}{1!8!} = 9;\ \binom{9}{8} = \frac{9!}{8!1!} = 9$
(b) $\binom{n}{1} = \binom{n}{n-1} = \frac{n!}{1!(n-1)!} = \frac{n(n-1)!}{(n-1)!} = n$

39. $2^n = (1+1)^n = 1^n + \binom{n}{1}1^{n-1}\cdot 1 + \binom{n}{2}1^{n-2}\cdot 1^2 + \binom{n}{3}1^{n-3}\cdot 1^3 + \cdots + \binom{n}{n-1}1^1\cdot 1^{n-1} + 1^n = \binom{n}{0} + \binom{n}{1} + \binom{n}{2} + \binom{n}{3} + \cdots + \binom{n}{n-1} + \binom{n}{n}$

41. $\cos^4\theta + 4i\cos^3\theta\sin\theta - 6\cos^2\theta\sin^2\theta - 4i\cos\theta\sin^3\theta + \sin^4\theta$

43. (a) $f(x+h) - f(x) = (x+h)^5 - x^5 = \left[x^5 + \binom{5}{1}x^4h + \binom{5}{2}x^3h^2 + \binom{5}{3}x^2h^3 + \binom{5}{4}xh^4 + h^5\right] - x^5 = \binom{5}{1}x^4h + \binom{5}{2}x^3h^2 + \binom{5}{3}x^2h^3 + \binom{5}{4}xh^4 + h^5$
(b) $\frac{f(x+h) - f(x)}{h} = \binom{5}{1}x^4 + \binom{5}{2}x^3h + \binom{5}{3}x^2h^2 + \binom{5}{4}xh^3 + h^4$
(c) When h is *very* close to 0, so are the last four terms in part (b), so $\frac{f(x+h) - f(x)}{h} \approx \binom{5}{1}x^4 = 5x^4$.

45. $\frac{f(x+h) - f(x)}{h} = \frac{(x+h)^{12} - x^{12}}{h} = \binom{12}{1}x^{11} + \binom{12}{2}x^{10}h + \binom{12}{3}x^9h^2 + \binom{12}{4}x^8h^3 + \cdots + \binom{12}{10}x^2h^9 + \binom{12}{11}xh^{10} + h^{11} \approx \binom{12}{1}x^{11} = 12x^{11}$, when h is very close to 0.

47. (a) $(n-r)! = (n-r)(n-r-1)(n-r-2)(n-r-3)\cdots 2\cdot 1 = (n-r)[n-(r+1)](n-(r+1)-1)\cdots 2\cdot 1 = (n-r)[n-(r+1)]!$
(b) Since $(n+1) - (r+1) = n - r$, $[(n+1) - (r+1)]! = (n-r)!$
(c) $\binom{n}{r+1} + \binom{n}{r} = \frac{n!}{(r+1)![n-(r+1)]!} + \frac{n!}{r!(n-r)!} = \frac{n!(n-r) + n!(r+1)}{(r+1)!(n-r)!} = \frac{n!(n+1)}{(r+1)!(n-r)!} = \frac{(n+1)!}{(r+1)![(n+1)-(r+1)]!} = \binom{n+1}{r+1}$
(d) For example, rows 2 and 3 of Pascal's triangle are

1 2 1

1 (3) 3 1

that is,

$\binom{2}{0}$ $\binom{2}{1}$ $\binom{2}{2}$

$\binom{3}{0}$ $\binom{3}{1}$ $\binom{3}{2}$ $\binom{3}{3}$

The circled 3 is the sum of the two closest entries in the row above: 1 + 2. But this just says that $\binom{3}{1} = \binom{2}{0} + \binom{2}{1}$, which is part (c) with $n = 2$ and $r = 0$. Similarly, in the general case, verify that the two closest entries in the row above $\binom{n+1}{r+1}$ are $\binom{n}{r}$ and $\binom{n}{r+1}$ and use part (c).

Section 12.5, page 772

1. *Step 1:* For $n = 1$ the statement is $1 = 2^1 - 1$, which is true. *Step 2:* Assume that the statement is true for $n = k$: that is, $1 + 2 + 2^2 + 2^3 + \cdots + 2^{k-1} = 2^k - 1$. Add 2^k to both sides, and rearrange terms:
$$1 + 2 + 2^2 + 2^3 + \cdots + 2^{k-1} + 2^k = 2^k - 1 + 2^k$$
$$1 + 2 + 2^2 + 2^3 + \cdots + 2^{k-1} + 2^{(k+1)-1} = 2(2^k) - 1$$
$$1 + 2 + 2^2 + 2^3 + \cdots + 2^{k-1} + 2^{(k+1)-1} = 2^{k+1} - 1$$
But this last line says that the statement is true for $n = k + 1$. Therefore, by the Principle of Mathematical Induction the statement is true for every positive integer n.

   Note: **Hereafter, in these answers, step 1 will be omitted if it is trivial (as in Exercise 1), and only the essential parts of step 2 will be given.**

3. Assume that the statement is true for $n = k$: $1 + 3 + 5 + \cdots + (2k - 1) = k^2$. Add $2(k + 1) - 1$ to both sides: $1 + 3 + 5 + \cdots + (2k - 1) + [2(k + 1) - 1] = k^2 + 2(k + 1) - 1 = k^2 + 2k + 1 = (k + 1)^2$. The first and last parts of this equation say that the statement is true for $n = k + 1$.

5. Assume that the statement is true for $n = k$:
$$1^2 + 2^2 + 3^2 + \cdots + k^2 = \frac{k(k+1)(2k+1)}{6}$$
Add $(k + 1)^2$ to both sides:
$$1^2 + 2^2 + 3^2 + \cdots + k^2 + (k+1)^2$$
$$= \frac{k(k+1)(2k+1)}{6} + (k+1)^2$$
$$= \frac{k(k+1)(2k+1) + 6(k+1)^2}{6}$$
$$= \frac{(k+1)[k(2k+1) + 6(k+1)]}{6}$$
$$= \frac{(k+1)(2k^2 + 7k + 6)}{6}$$
$$= \frac{(k+1)(k+2)(2k+3)}{6}$$
$$= \frac{(k+1)[(k+1)+1][2(k+1)+1]}{6}$$
The first and last parts of this equation say that the statement is true for $n = k + 1$.

7. Assume that the statement is true for $n = k$:
$$\frac{1}{1\cdot 2} + \frac{1}{2\cdot 3} + \cdots + \frac{1}{k(k+1)} = \frac{k}{k+1}.$$
Adding $\frac{1}{(k+1)[(k+1)+1]} = \frac{1}{(k+1)(k+2)}$ to both sides yields:
$$\frac{1}{1\cdot 2} + \frac{1}{2\cdot 3} + \cdots + \frac{1}{k(k+1)} + \frac{1}{(k+1)(k+2)}$$
$$= \frac{k}{k+1} + \frac{1}{(k+1)(k+2)}$$
$$= \frac{k(k+2)+1}{(k+1)(k+2)} = \frac{k^2+2k+1}{(k+1)(k+2)}$$
$$= \frac{(k+1)^2}{(k+1)(k+2)} = \frac{k+1}{k+2} = \frac{k+1}{(k+1)+1}$$
The first and last parts of this equation show that the statement is true for $n = k + 1$.

9. Assume the statement is true for $n = k$: $k + 2 > k$. Adding 1 to both sides, we have: $k + 2 + 1 > k + 1$, or equivalently, $(k + 1) + 2 > (k + 1)$. Therefore, the statement is true for $n = k + 1$.

11. Assume the statement is true for $n = k$: $3^k \geq 3k$. Multiplying both sides by 3 yields: $3\cdot 3^k \geq 3\cdot 3k$, or equivalently, $3^{k+1} \geq 3\cdot 3k$. Now since $k \geq 1$, we know that $3k \geq 3$ and hence that $2\cdot 3k \geq 3$. Therefore, $2\cdot 3k + 3k \geq 3 + 3k$, or equivalently, $3\cdot 3k \geq 3k + 3$. Combining this last inequality with the fact that $3^{k+1} \geq 3\cdot 3k$, we see that $3^{k+1} \geq 3k + 3$, or equivalently, $3^{k+1} \geq 3(k + 1)$. Therefore, the statement is true for $n = k + 1$.

13. Assume the statement is true for $n = k$: $3k > k + 1$. Adding 3 to both sides yields: $3k + 3 > k + 1 + 3$, or equivalently, $3(k + 1) > (k + 1) + 3$. Since $(k + 1) + 3$ is certainly greater than $(k + 1) + 1$, we conclude that $3(k + 1) > (k + 1) + 1$. Therefore, the statement is true for $n = k + 1$.

15. Assume the statement is true for $n = k$; then 3 is a factor of $2^{2k+1} + 1$; that is, $2^{2k+1} + 1 = 3M$ for some integer M. Thus, $2^{2k+1} = 3M - 1$. Now $2^{2(k+1)+1} = 2^{2k+2+1} = 2^{2+2k+1} = 2^2\cdot 2^{2k+1} = 4(3M - 1) = 12M - 4 = 3(4M) - 3 - 1 = 3(4M - 1) - 1$. From the first and last terms of this equation we see that $2^{2(k+1)+1} + 1 = 3(4M - 1)$. Hence, 3 is a factor of $2^{2(k+1)+1} + 1$. Therefore, the statement is true for $n = k + 1$.

17. Assume the statement is true for $n = k$: 64 is a factor of $3^{2k+2} - 8k - 9$. Then $3^{2k+2} - 8k - 9 = 64N$ for some integer N so that $3^{2k+2} = 8k + 9 + 64N$. Now $3^{2(k+1)+2} = 3^{2k+2+2} = 3^{2+(2k+2)} = 3^2 \cdot 3^{2k+2} = 9(8k + 9 + 64N)$. Consequently,

$$\begin{aligned}3^{2(k+1)+2} - 8(k+1) - 9 &= 3^{2(k+1)+2} - 8k - 8 - 9\\ &= 3^{2(k+1)+2} - 8k - 17\\ &= [9(8k + 9 + 64N)] - 8k - 17\\ &= 72k + 81 + 9\cdot 64N - 8k - 17\\ &= 64k + 64 + 9\cdot 64N = 64(k + 1 + 9N).\end{aligned}$$

From the first and last parts of this equation we see that 64 is a factor of $3^{2(k+1)+2} - 8(k+1) - 9$. Therefore, the statement is true for $n = k + 1$.

19. Assuming that the statement is true for $n = k$: $c + (c + d) + (c + 2d) + \cdots + [c + (k-1)d] = \dfrac{k[2c + (k-1)d]}{2}$. Adding $c + kd$ to both sides, we have

$$\begin{aligned}c + (c + d) + (c + 2d) + \cdots &+ [c + (k-1)d] + (c + kd)\\ &= \frac{k[2c + (k-1)d]}{2} + c + kd\\ &= \frac{k[2c + (k-1)d] + 2(c + kd)}{2}\\ &= \frac{2ck + k(k-1)d + 2c + 2kd}{2}\\ &= \frac{2ck + 2c + kd(k-1) + 2kd}{2}\\ &= \frac{(k+1)2c + kd(k - 1 + 2)}{2}\\ &= \frac{(k+1)2c + kd(k+1)}{2} = \frac{(k+1)(2c + kd)}{2}\\ &= \frac{(k+1)(2c + [(k+1) - 1]d)}{2}\end{aligned}$$

Therefore, the statement is true for $n = k + 1$.

21. **(a)** $x^2 - y^2 = (x - y)(x + y)$; $x^3 - y^3 = (x - y)(x^2 + xy + y^2)$; $x^4 - y^4 = (x - y)(x^3 + x^2y + xy^2 + y^3)$

(b) *Conjecture:* $x^n - y^n = (x - y)(x^{n-1} + x^{n-2}y + x^{n-3}y^2 + \cdots + x^2y^{n-3} + xy^{n-2} + y^{n-1})$. *Proof:* The statement is true for $n = 2, 3, 4$, by part (a). Assume that the statement is true for $n = k$:

$$x^k - y^k = (x - y)(x^{k-1} + x^{k-2}y + \cdots + xy^{k-2} + y^{k-1}).$$

Now use the fact that $-yx^k + yx^k = 0$ to write $x^{k+1} - y^{k+1}$ as follows:

$$\begin{aligned}x^{k+1} - y^{k+1} &= x^{k+1} - yx^k + yx^k - y^{k+1}\\ &= (x^{k+1} - yx^k) + (yx^k - y^{k+1})\\ &= (x - y)x^k + y(x^k - y^k)\\ &= (x - y)x^k + y(x - y)(x^{k-1} + x^{k-2}y\\ &\qquad + x^{k-3}y^2 + \cdots + xy^{k-2} + y^{k-1})\\ &= (x - y)x^k + (x - y)(x^{k-1}y +\\ &\qquad x^{k-2}y^2 + x^{k-3}y^3 + \cdots + xy^{k-1} + y^k)\\ &= (x - y)[x^k + x^{k-1}y + x^{k-2}y^2 +\\ &\qquad x^{k-3}y^3 + \cdots + xy^{k-1} + y^k]\end{aligned}$$

The fist and last parts of this equation show that the conjecture is true for $n = k + 1$. Therefore, by mathematical induction, the conjecture is true for every integer $n \geq 2$.

23. False; counterexample: $n = 9$

25. True: *Proof:* Since $(1 + 1)^2 > 1^2 + 1$, the statement is true for $n = 1$. Assume the statement is true for $n = k$: $(k + 1)^2 > k^2 + 1$. Then $[(k + 1) + 1]^2 = (k + 1)^2 + 2(k + 1) + 1 > k^2 + 1 + 2(k + 1) + 1 = k^2 + 2k + 2 + 2 > k^2 + 2k + 2 = k^2 + 2k + 1 + 1 = (k + 1)^2 + 1$. The first and last terms of this inequality say that the statement is true for $n = k + 1$. Therefore, by induction the statement is true for every positive integer n.

27. False; counterexample: $n = 2$

29. Since $2\cdot 5 - 4 > 5$, the statement is true for $n = 5$. Assume the statement is true for $n = k$ (with $k \geq 5$): $2k - 4 > k$. Adding 2 to both sides shows that $2k - 4 + 2 > k + 2$, or equivalently, $2(k + 1) - 4 > k + 2$. Since $k + 2 > k + 1$, we see that $2(k + 1) - 4 > k + 1$. So the statement is true for $n = k + 1$. Therefore, by the Extended Principle of Mathematical Induction, the statement is true for all $n \geq 5$.

31. Since $2^2 > 2$, the statement is true for $n = 2$. Assume that $k \geq 2$ and that the statement is true for $n = k$: $k^2 > k$. Then $(k + 1)^2 = k^2 + 2k + 1 > k^2 + 1 > k + 1$. The first and last terms of this inequality show that the statement is true for $n = k + 1$. Therefore, by induction, the statement is true for all $n \geq 2$.

33. Since $3^4 = 81$ and $2^4 + 10\cdot 4 = 16 + 40 = 56$, we see that $3^4 > 2^4 + 10\cdot 4$. So the statement is true for $n = 4$. Assume that $k \geq 4$ and that the statement is true for $n = k$: $3^k > 2^k + 10k$. Multiplying both sides by 3 yields: $3\cdot 3^k > 3(2^k + 10k)$, or equivalently, $3^{k+1} > 3\cdot 2^k + 30k$. But

$$3\cdot 2^k + 30k > 2\cdot 2^k + 30k = 2^{k+1} + 30k.$$

Therefore, $3^{k+1} > 2^{k+1} + 30k$. Now we shall show that $30k > 10(k + 1)$. Since $k \geq 4$, we have

$20k \geq 20 \cdot 4$, so that $20k > 80 > 10$. Adding $10k$ to both sides of $20k > 10$ yields: $30k > 10k + 10$, or equivalently, $30k > 10(k + 1)$. Consequently,

$$3^{k+1} > 2^{k+1} + 30k > 2^{k+1} + 10(k + 1).$$

The first and last terms of this inequality show that the statement is true for $n = k + 1$. Therefore, the statement is true for all $n \geq 4$ by induction.

35. **(a)** 3 (that is, $2^2 - 1$) for $n = 2$; 7 (that is, $2^3 - 1$) for $n = 3$; 15 (that is, $2^4 - 1$) for $n = 4$.

(b) *Conjecture:* The smallest possible number of moves for n rings is $2^n - 1$. *Proof:* This conjecture is easily seen to be true for $n = 1$ or $n = 2$. Assume it is true for $n = k$ and that we have $k + 1$ rings to move. In order to move the *bottom* ring from the first peg to another peg (say, the second one), it is first necessary to move the top k rings off the first peg *and* leave the second peg vacant at the end (the second peg will have to be used *during* this moving process). If this is to be done according to the rules, we will end up with the top k rings on the third peg in the *same* order they were on the first peg. According to the induction assumption, the least possible number of moves needed to do this is $2^k - 1$. It now takes one move to transfer the bottom ring [the $(k + 1)$st] from the first to the second peg. Finally, the top k rings now on the third peg must be moved to the second peg. Once again by the induction hypothesis, the least number of moves for doing this is $2^k - 1$. Therefore, the smallest total number of moves needed to transfer all $k + 1$ rings from the first to the second peg is $(2^k - 1) + 1 + (2^k - 1) = (2^k + 2^k) - 1 = 2 \cdot 2^k - 1 = 2^{k+1} - 1$. Hence, the conjecture is true for $n = k + 1$. Therefore, by induction it is true for all positive integers n.

37. *De Moivre's Theorem:* For any complex number $z = r(\cos\theta + i\sin\theta)$ and any positive integer n, $z^n = r^n[\cos(n\theta) + i\sin(n\theta)]$. *Proof:* The theorem is obviously true when $n = 1$. Assume that the theorem is true for $n = k$, that is, $z^k = r^k[\cos(k\theta) + i\sin(k\theta)]$. Then

$$z^{k+1} = z \cdot z^k = [r(\cos\theta + i\sin\theta)](r^k[\cos(k\theta) + i\sin(k\theta)]).$$

According to the multiplication rule for complex numbers in polar form (multiply the moduli and add the arguments) we have:

$$\begin{aligned} z^{k+1} &= r \cdot r^k[\cos(\theta + k\theta) + i\sin(\theta + k\theta)] \\ &= r^{k+1}\{\cos[(k+1)\theta] + i\sin[(k+1)\theta]\}. \end{aligned}$$

This statement says the theorem is true for $n = k + 1$. Therefore, by induction, the theorem is true for every positive integer n.

Chapter 12 Review, page 775

1. $-3, -1, 1, 3$ **3.** $1, \frac{1}{4}, \frac{1}{9}, \frac{1}{16}$ **5.** -368

7. 129 **9.** $a_n = 9 - 6n$

11. $a_n = 6n - 11$ **13.** $a_n = 2 \cdot 3^{n-1}$

15. $a_n = \dfrac{3}{2^{n-8}}$ **17.** -55

19. $\dfrac{121}{4}$ **21.** 8.75, 13.5, 18.25

23. Second method is better. **25.** 2

27. $(1.02)^{51} = (1 + .02)^{51} = 1^{51} + \binom{51}{1}1^{50}(.02) + \binom{51}{2}1^{49}(.02)^2 + \text{other positive terms} = 2.53 + \text{other positive terms} > 2.5$

29. 455 **31.** $n + 1$ **33.** 1140 **35.** 80

37. True for $n = 1$. If the statement is true for $n = k$, then $1 + 5 + \cdots + 5^{k-1} = \dfrac{5^k - 1}{4}$ so that $1 + 5 + \cdots + 5^{k-1} + 5^k = \dfrac{5^k - 1}{4} + 5^k = \dfrac{5^k - 1 + 4 \cdot 5^k}{4} = \dfrac{5 \cdot 5^k - 1}{4} = \dfrac{5^{k+1} - 1}{4}$. Hence, the statement is true for $n = k + 1$ and therefore true for all n by induction.

39. Since the statement is obviously true for $x = 0$, assume $x \neq 0$. Then the statement is true for $n = 1$. If the statement is true for $n = k$, then $|x^k| < 1$. Then $|x^k| \cdot |x| < |x|$. Thus, $|x^{k+1}| = |x^k| \cdot |x| < |x| < 1$. Hence, the statement is true for $n = k + 1$ and therefore true for all n by induction.

41. True for $n = 1$. If the statement is true for $n = k$, then $1 + 4 + \cdots + 4^{k-1} = \frac{1}{3}(4^k - 1)$. Hence, $1 + 4 + \cdots + 4^{k-1} + 4^k = \frac{1}{3}(4^k - 1) + 4^k = \frac{1}{3}(4^k - 1) + \dfrac{3 \cdot 4^k}{3} = \dfrac{4^k - 1 + 3 \cdot 4^k}{3} = \dfrac{4 \cdot 4^k - 1}{3} = \frac{1}{3}(4^{k+1} - 1)$. Hence, the statement is true for $n = k + 1$ and therefore for all n by induction.

43. If $n = 1$, then $9^n - 8n - 1 = 0$. Since $0 = 0 \cdot 8$, the statement is true for $n = 1$. If the statement is true for $n = k$, then $9^k - 8k - 1 = 8D$, so that $9^k - 1 = 8k + 8D = 8(k + D)$. Consequently, $9^{k+1} - 8(k + 1) - 1 = 9^{k+1} - 8k - 8 - 1 = 9^{k+1} - 9 - 8k = 9(9^k - 1) - 8k = 9[8(k + D)] - 8k = 8[9(k + D) - k]$. Thus, 8 is a factor of $9^{k+1} - 8(k+1) - 1$ and the statement is true for $n = k + 1$. Therefore it is true for all n by induction.

Algebra Review

Section 1.A, page 782

1. 36 3. 73 5. -5
7. $-125/64$ 9. $1/3$ 11. -112
13. $81/16$ 15. $-211/216$ 17. $129/8$
19. x^{10} 21. $.03y^9$ 23. $24x^7$
25. $9x^4y^2$ 27. $-21a^6$ 29. $384w^6$
31. ab^3 33. $8x^{-1}y^3$ 35. a^8x^{-3}
37. $3xy$ 39. 2^{12} 41. 2^{-12}
43. x^7 45. ce^9 47. $b^2c^2d^6$
49. $a^{12}b^8$ 51. $1/(c^{10}d^6)$ 53. $1/(108x)$
55. a^7c/b^6 57. c^3d^6 59. $a + \frac{1}{a}$
61. Negative 63. Negative 65. Negative
67. 3^s 69. $a^{6t}b^{4t}$ 71. b^{rs+st}/c^{2rt}
73. Many possible examples, including $3^2 + 4^2 = 9 + 16 = 25$, but $(3 + 4)^2 = 7^2 = 49$
75. Many possible examples, including $3^2 \cdot 2^3 = 9 \cdot 8 = 72$; but $(3 \cdot 2)^{2+3} = 6^5 = 7776$
77. Many possible examples, including $2^6/2^3 = 64/8 = 8$, but $2^{6/3} = 2^2 = 4$
79. False for all nonzero a; for instance, $(-3)^2 = (-3)(-3) = 9$, but $-3^2 = -9$

Section 1.B, page 786

1. $8x$ 3. $-2a^2b$
5. $-x^3 + 4x^2 + 2x - 3$ 7. $5u^3 + u - 4$
9. $4z - 12z^2w + 6z^3w^2 - zw^3 + 8$
11. $-3x^3 + 15x + 8$ 13. $-5xy - x$ 15. $15y^3 - 5y$
17. $12a^2x^2 - 6a^3xy + 6a^2xy$
19. $12z^4 + 30z^3$ 21. $12a^2b - 18ab^2 + 6a^3b^2$
23. $x^2 - x - 2$
25. $2x^2 + 2x - 12$ 27. $y^2 + 7y + 12$
29. $-6x^2 + x + 35$ 31. $3y^3 - 9y^2 + 4y - 12$
33. $x^2 - 16$ 35. $16a^2 - 25b^2$
37. $y^2 - 22y + 121$ 39. $25x^2 - 10bx + b^2$
41. $16x^6 - 8x^3y^4 + y^8$ 43. $9x^4 - 12x^2y^4 + 4y^8$
45. $2y^3 + 9y^2 + 7y - 3$
47. $-15w^3 + 2w^2 + 9w - 18$
49. $24x^3 - 4x^2 - 4x$ 51. $x^3 - 6x^2 + 11x - 6$
53. $-3x^3 - 5x^2y + 26xy^2 - 8y^3$
55. 3 57. -6 59. 6 61. 1 63. 5
65. $x - 25$ 67. $9 + 6\sqrt{y} + y$
69. $\sqrt{3}x^2 + 4x + \sqrt{3}$ 71. $3ax^2 + (3b + 2a)x + 2b$
73. $abx^2 + (a^2 + b^2)x + ab$
75. $x^3 - (a + b + c)x^2 + (ab + ac + bc)x - abc$
77. 3^{4+r+t} 79. $x^{m+n} + 2x^n - 3x^m - 6$
81. $2x^{4n} - 5x^{3n} + 8x^{2n} - 18x^n - 5$
83. Example: if $y = 4$, then $3(4 + 2) \neq (3 \cdot 4) + 2$; correct statement: $3(y + 2) = 3y + 6$
85. Example: if $x = 2$, $y = 3$, then $(2 + 3)^2 \neq 2 + 3^2$; correct statement: $(x + y)^2 = x^2 + 2xy + y^2$
87. Example: if $x = 2$, $y = 3$, then $(7 \cdot 2)(7 \cdot 3) \neq 7 \cdot 2 \cdot 3$; correct statement: $(7x)(7y) = 49xy$
89. Example: if $y = 2$, then $2 + 2 + 2 \neq 2^3$; correct statement: $y + y + y = 3y$
91. Example: if $x = 4$, then $(4 - 3)(4 - 2) \neq 4^2 - 5 \cdot 4 - 6$; correct statement: $(x - 3)(x - 2) = x^2 - 5x + 6$
93. If x is the chosen number, then adding 1 and squaring the result gives $(x + 1)^2$. Subtracting 1 from the original number x and squaring the result gives $(x - 1)^2$. Subtracting the second of these squares from the first yields: $(x + 1)^2 - (x - 1)^2 = (x^2 + 2x + 1) - (x^2 - 2x + 1) = 4x$. Dividing by the original number x now gives $\frac{4x}{x} = 4$. So the answer is always 4, no matter what number x is chosen.
95. Many correct answers

Section 1.C, page 791

1. $(x + 2)(x - 2)$ 3. $(3y + 5)(3y - 5)$
5. $(9x + 2)^2$ 7. $(\sqrt{5} + x)(\sqrt{5} - x)$
9. $(7 + 2z)^2$ 11. $(x^2 + y^2)(x + y)(x - y)$
13. $(x + 3)(x - 2)$ 15. $(z + 3)(z + 1)$
17. $(y + 9)(y - 4)$ 19. $(x - 3)^2$
21. $(x + 5)(x + 2)$ 23. $(x + 9)(x + 2)$
25. $(3x + 1)(x + 1)$ 27. $(2z + 3)(z + 4)$
29. $9x(x - 8)$ 31. $2(x - 1)(5x + 1)$
33. $(4u - 3)(2u + 3)$ 35. $(2x + 5y)^2$
37. $(x - 5)(x^2 + 5x + 25)$
39. $(x + 2)^3$ 41. $(2 + x)(4 - 2x + x^2)$
43. $(-x + 5)^3$ 45. $(x + 1)(x^2 - x + 1)$
47. $(2x - y)(4x^2 + 2xy + y^2)$
49. $(x^3 + 2^3)(x^3 - 2^3) = (x + 2)(x^2 - 2x + 4)(x - 2)(x^2 + 2x + 4)$
51. $(y^2 + 5)(y^2 + 2)$ 53. $(9 + y^2)(3 + y)(3 - y)$

55. $(z + 1)(z^2 - z + 1)(z - 1)(z^2 + z + 1)$

57. $(x^2 + 3y)(x^2 - y)$ 59. $(x + z)(x - y)$

61. $(a + 2b)(a^2 - b)$

63. $(x^2 - 8)(x + 4) = (x + \sqrt{8})(x - \sqrt{8})(x + 4)$

65. If $x^2 + 1 = (x + c)(x + d) = x^2 + (c + d)x + cd$, then $c + d = 0$ and $cd = 1$. But $c + d = 0$ implies that $c = -d$ and hence that $1 = cd = (-d)d = -d^2$, or equivalently, that $d^2 = -1$. Since there is no real number with this property, $x^2 + 1$ cannot possibly factor in this way.

Section 1.D, page 795

1. $\frac{9}{7}$ 3. $\frac{195}{8}$ 5. $\frac{x - 2}{x + 1}$

7. $\frac{a + b}{a^2 + ab + b^2}$ 9. $\frac{1}{x}$ 11. $\frac{29}{35}$

13. $\frac{121}{42}$ 15. $\frac{ce + 3cd}{de}$ 17. $\frac{b^2 - c^2}{bc}$

19. $\frac{-1}{x(x + 1)}$ 21. $\frac{x + 3}{(x + 4)^2}$ 23. $\frac{2x - 4}{x(3x - 4)}$

25. $\frac{x^2 - xy + y^2 + x + y}{x^3 + y^3}$

27. $\frac{-6x^5 - 38x^4 - 84x^3 - 71x^2 - 14x + 1}{4x(x + 1)^3(x + 2)^3}$

29. 2 31. $2/(3c)$ 33. $3y/x^2$ 35. $\frac{12x}{x - 3}$

37. $\frac{5y^2}{3(y + 5)}$ 39. $\frac{u + 1}{u}$ 41. $\frac{(u + v)(4u - 3v)}{(2u - v)(2u - 3v)}$

43. $\frac{35}{24}$ 45. $\frac{u^2}{vw}$ 47. $\frac{x + 3}{2x}$

49. $\frac{x^2y^2}{(x + y)(x + 2y)}$ 51. $\frac{cd(c + d)}{c - d}$

53. $\frac{y - x}{xy}$ 55. $\frac{-3y + 3}{y}$

57. $\frac{-1}{x(x + h)}$ 59. $\frac{xy}{x + y}$

61. Example: if $a = 1$, $b = 2$, then $\frac{1}{1} + \frac{1}{2} \neq \frac{1}{1 + 2}$; correct statement: $\frac{1}{a} + \frac{1}{b} = \frac{b + a}{ab}$

63. Example: if $a = 4$, $b = 9$, then $\left(\frac{1}{\sqrt{4} + \sqrt{9}}\right)^2 \neq \frac{1}{4 + 9}$; correct statement: $\left(\frac{1}{\sqrt{a} + \sqrt{b}}\right)^2 = \frac{1}{a + 2\sqrt{ab} + b}$

65. Example: if $u = 1$, $v = 2$, then $\frac{1}{2} + \frac{2}{1} \neq 1$; correct statement: $\frac{u}{v} + \frac{v}{u} = \frac{u^2 + v^2}{vu}$

67. Example: if $x = 4$, $y = 9$, then $(\sqrt{4} + \sqrt{9}) \cdot \frac{1}{\sqrt{4} + \sqrt{9}} \neq 4 + 9$; correct statement: $(\sqrt{x} + \sqrt{y}) \cdot \frac{1}{\sqrt{x} + \sqrt{y}} = 1$

Index of Applications

Biology and Life Science

Business and Manufacturing

Business and Manufacturing
(continued)

Chemistry and Physics

Chemistry and Physics
(continued)

Construction

Consumer Affairs

Consumer Affairs
(continued)

Financial

Geometry

INDEX